WITHDRAWN
UTSA LIBRARIES

RENEWALS 458-4574
DATE DUE

EMISSION TOMOGRAPHY
The Fundamentals of PET and SPECT

EMISSION TOMOGRAPHY
The Fundamentals of PET and SPECT

Miles N. Wernick, Ph.D.
*Illinois Institute of Technology
and Predictek, LLC*

John N. Aarsvold, Ph.D.
*Veterans Affairs Medical Center—Atlanta
and Emory University*

ELSEVIER
ACADEMIC
PRESS

Amsterdam • Boston • Heidelberg • London • New York • Oxford
Paris • San Diego • San Francisco • Singapore • Sydney • Tokyo

Cover Photo Credit: Image courtesy of MIICRO, Inc.

Elsevier Academic Press
525 B Street, Suite 1900, San Diego, California 92101-4495, USA
84 Theobald's Road, London WC1X 8RR, UK

This book is printed on acid-free paper.

Copyright 2004, Elsevier Inc.

All rights reserved.
No part of this publication may be reproduced or transmitted in any form or by any means, electronic or mechanical, including photocopy, recording, or any information storage and retrieval system, without permission in writing from the publisher.

Permissions may be sought directly from Elsevier's Science & Technology Rights Department in Oxford, UK: phone: (+44) 1865 843830, fax: (+44) 1865 853333, e-mail: permissions@elsevier.com.uk. You may also complete your request on-line via the Elsevier Science homepage (http://elsevier.com), by selecting "Customer Support" and then "Obtaining Permissions."

Library of Congress Cataloging-in-Publication Data

Emission tomography: the fundamentals of PET and SPECT/[edited by] Miles Wernick, John Aarsvold.
    p. ; cm.
   Includes bibliographical references and index.
   ISBN 0-12-744482-3 (alk. paper)
   1. Tomography, Emission. I. Wernick, Miles. II. Aarsvold, John.
   [DNLM: 1. Tomography, Emission-Computed. 2. Tomography, Emission-Computed, Single-Photon. WN 206 E53 2004]
RC78.7.T62E455 2004
616.07′575–dc22

                                         2003052370

British Library Cataloguing in Publication Data
A catalogue record for this book is available from the British Library

ISBN: 0-12-744482-3

For all information on all Academic Press publications
visit our website at www.books.elsevier.com

PRINTED IN CHINA
04 05 06 07 08 9 8 7 6 5 4 3 2 1

Contents

Contributors xiii
Foreword xv
Preface xvii
Acknowledgements xix

1. Imaging Science Bringing the Invisible to Light
ROBERT N. BECK

I. Preamble 1
II. Introduction 1
III. Imaging Science 3
IV. Fundamental and Generic Issues of Imaging Science 5
V. Methodology and Epistemology 8
VI. A View of the Future 8

2. Introduction to Emission Tomography
MILES N. WERNICK AND JOHN N. AARSVOLD

I. What is Emission Tomography? 11
   A. The Tracer Principle 11
   B. Tomography 12
II. The Making of an Emission Tomography Image 13
   A. Single-Photon Emission Computed Tomography 15
   B. Positron Emission Tomography 17
   C. Image Reconstruction 18
   D. Image Analysis 19
III. Types of Data Acquisition: Static, Dynamic, Gated, and List Mode 20
IV. Cross-Sectional Images 21
V. Radiopharmaceuticals and Their Applications 21
VI. Developments in Emission Tomography 22

3. Evolution of Clinical Emission Tomography
A. BERTRAND BRILL AND ROBERT N. BECK

I. Introduction 25
II. The Beginnings of Nuclear Medicine 25
   A. Developments Before 1945 25
   B. The Next 25 Years (1945–1970) 26
III. Early Imaging Devices 26
   A. Early Scanning Imagers 26
   B. Early Devices Based on Gamma Cameras 28
   C. Early Dedicated Positron-Emission Imagers 30
IV. Evolution of Emission Tomography and Initial Applications 31
   A. Anger Camera Developments 31
   B. Radiopharmaceutical Targeting Agents 31
   C. Early Theoretical Developments 33
V. Clinical Applications 38
   A. Brain Imaging 39
   B. Thyroid Imaging and Therapy 40
   C. Parathyroid Disease 41
   D. Cardiac Imaging 42
   E. Lung Imaging 43
   F. Kidney Imaging 44
   G. Bone Imaging 45
   H. Reticuloendothelial System Imaging 45
   I. Pancreas Imaging 46
   J. Tumor Imaging 46
   K. Molecular Biology Applications 48
VI. Summary 49

4. Basic Physics of Radioisotope Imaging
CRAIG LEVIN

I. Where Do the Nuclear Emissions Used in Imaging Come From? 53
   A. Nuclear Constituents 53

B. Nuclear Forces and Binding Energy   54
   C. Nuclear Energy Levels   54
   D. Nuclear De-excitation   54
   E. Nuclear Stability   55
   F. Nuclear Transmutation   56
   G. Nuclear Decay Probability   56
II. Relevant Modes of Nuclear Decay for Medical Radionuclide Imaging   57
   A. Isomeric Transitions: Gamma-Ray Emission   57
   B. Isobaric Transitions: Beta (β) Emission   57
III. Production of Radionuclides for Imaging   60
   A. The Nuclear Reactor   60
   B. The Nuclear Generator   62
   C. The Cyclotron   63
IV. Interactions of Nuclear Emissions in Matter   64
   A. Interactions of Electrons and Positrons in Matter   65
   B. Interactions of High-Energy Photons in Matter   70
V. Exploiting Radiation Interactions in Matter for Emission Imaging   75
   A. Statistics of Ionization Production   76
   B. Detector and Position-Sensitive Detector Basics   76
VI. Physical Factors That Determine the Fundamental Spatial Resolution Limit in Nuclear Emission Imaging   83
   A. Overview of Physical Factors That Affect Emission Imaging Performance   83
   B. What Is the Fundamental Spatial Resolution Limit in Radioisotope Imaging?   83

5. Radiopharmaceuticals for Imaging the Brain

JOGESHWAR MUKHERJEE, BINGZHI SHI, T. K. NARAYANAN, B. T. CHRISTIAN, AND YANG ZHI-YING

I. Introduction   89
II. Biochemical Processes in the Brain   90
III. New Radiopharmaceutical Development   91
   A. Steps in Radiopharmaceutical Development   91
   B. Criteria for Pharmaceuticals Used for *In Vivo* Receptor Imaging   91
IV. Neuroscience Studies   92
   A. Dopaminergic System   92
   B. Cholinergic System   93
   C. Serotonergic System   94
   D. Other Systems   95
V. Applications of Imaging Studies: Dopamine System   96
   A. Distribution of Receptors, Transporters, and Enzymes   96
   B. Distribution and Pharmacokinetics of Radiolabeled Drugs   96
   C. Receptor Occupancy of Therapeutic Drugs   96
   D. Integrity of the Presynapse   96
   E. Alterations in the Levels of Endogenous Ligand Dopamine   96
   F. Study of Substance Abuse Drugs   97
   G. Interactions with Other Neurotransmitter Systems   97
   H. Posttreatment Monitoring of the System   97
   I. Etiological Studies of the System   97
   J. Neurotoxicity of Drugs   98
VI. Oncology Studies   98
VII. Genomic Studies   98
   A. Imaging Protein Products   99
   B. Use of Antisense RNA to Image mRNA   99
VIII. Summary   99

6. Basics of Imaging Theory and Statistics

CHIEN-MIN KAO, PATRICK LA RIVIÈRE, AND XIAOCHUAN PAN

I. Introduction   103
II. Linear Systems   104
   A. Linear Shift-Invariant Systems   104
   B. Point Spread Functions   104
   C. Fourier Transform and Spectrum   105
   D. System Transfer Functions   106
III. Discrete Sampling   107
   A. Sampling Theorem and Aliasing   108
   B. Discrete Fourier Transform   110
   C. Relationship among FT, DFT, and Sampling   111
   D. Interpolation   111
IV. Noise and Signal   114
   A. Random Variables   114
   B. Stochastic Processes   115
   C. Noise Models   115
   D. Power Spectrum   116
V. Filtering   117
   A. Pseudoinverse   117
   B. Tikhonov Regularization   117
   C. Wiener Filtering   118
VI. Smoothing   119
   A. Smoothing Filters   119
   B. Median Filters   120
   C. Smoothing Splines   121
VII. Estimation   121
   A. Bayesian Parameter Estimation   122
   B. Maximum Likelihood Estimation   122
   C. Bias and Variance   123
VIII. Objective Assessment of Image Quality   123
   A. ROC Analysis   124
   B. Ideal Observer: Likelihood Ratio   124
   C. Noise Equivalent Quanta and Detective Quantum Efficiency   125
   D. Model Observer: (Channelized) Hotelling Observer   125

7. Single-Photon Emission Computed Tomography

GENGSHENG LAWRENCE ZENG, JAMES R. GALT, MILES N. WERNICK, ROBERT A. MINTZER, AND JOHN N. AARSVOLD

I. Planar Single-Photon Emission Imaging 127
   A. Thyroid Studies 128
   B. Ventilation/Perfusion Studies 128
   C. Whole-Body Bone Studies 128
   D. Other Nuclear Medicine Studies 129
II. Conventional Gamma Cameras 130
   A. The Anger Camera 130
   B. Principles of Collimation 131
   C. Collimator Types 132
   D. Scintillation Detection 133
   E. Event Positioning 135
   F. Photon Interactions 135
   G. Camera Performance 135
III. Tomography 136
   A. Tomographic Imaging 136
   B. Transmission Imaging 137
   C. Emission Imaging 138
   D. Transmission Tomography 138
   E. Emission Tomography 139
IV. Single-Photon Emission Computed Tomography Systems 140
   A. System Configurations 140
   B. Gantry Motions 140
   C. Transmission-Source Tomography 141
   D. SPECT/CT 142
   E. System Performance 144
V. Tomographic Single-Photon Emission Imaging 145
   A. Bone SPECT 145
   B. Brain SPECT 146
   C. Myocardial Perfusion SPECT 146
VI. Other Detectors and Systems 147
   A. Multiple-Pinhole Coded Apertures 147
   B. Multisegment Slant-Hole Collimators 149
   C. Rotating-Slat Collimators 149
   D. Compton Cameras 149
   E. Segmented-Scintillator Detectors 150
   F. Solid-State Detectors 150
VII. Summary 150

8. Collimator Design for Nuclear Medicine

DONALD L. GUNTER

I. Basic Principles of Collimator Design 153
II. Description of the Imaging System and Collimator Geometry 154
III. Description of Collimator Imaging Properties 156
IV. Septal Penetration 160
V. Optimal Design of Parallel-Hole Collimators 161
VI. Secondary Constraints 164
   A. Collimator Weight 164
   B. Septal Thickness 165
   C. Visibility of Collimator Hole-Pattern 166
VII. Summary 168

9. Annular Single-Crystal SPECT Systems

SEBASTIAN GENNA, JINSONG OUYANG, AND WEISHI XIA

I. Overview: Annular Single-Photon Emission Computed Tomography Systems 169
II. Principles and Design of CeraSPECT 170
III. Annular SensOgrade Collimators 171
IV. Modification of Light Optics in a Scintillation Camera 172
V. NeurOtome, A Bridge between Single-Photon Emission Computed Tomography and Positron Emission Tomography 173
VI. MammOspect, an Annular Breast Single-Photon Emission Computed Tomography Camera 175
VII. Small Animal Single-Photon Emission Computed Tomography Using an Annular Crystal 177
VIII. Discussion 178

10. PET Systems

THOMAS LEWELLEN AND JOEL KARP

I. Basic Positron Emission Tomography Principles 179
II. Detector Designs 182
III. Tomography System Geometry 184
IV. Positron Emission Tomography Scintillators 186
V. Positron Emission Tomography System Electronics 187
VI. Attenuation Correction 188
VII. Scatter Correction 190
VIII. Noise Equivalent Count Rate 191
IX. Future Trends 192

11. PET/CT Systems

CHARLES C. WATSON, DAVID W. TOWNSEND, AND BERNARD BENDRIEM

I. Introduction 195
II. Motivation 197
III. Initial Development 197
IV. Design 198
   A. General Considerations 198
   B. Detectors 199
   C. Gantry 200
   D. Patient Handling System 200
   E. Software and Computers 202
V. Protocols 202
VI. Image Registration and Fusion 206

- VII. Attenuation Correction 206
- VIII. Dosimetry 209
  - A. CT Effective Dose 209
  - B. Effective Dose in PET 210
  - C. Total Effective Dose in PET/CT 210
- IX. The Future 210

12. Small Animal PET Systems

SIMON R. CHERRY AND ARION F. CHATZIIOANNOU

- I. Introduction 213
  - A. Why Use PET in Laboratory Animal Research? 213
  - B. Laboratory Animal Models of Human Disease 214
  - C. Opportunities for PET in Small Animal Imaging 214
- II. Challenges in Small Animal PET 215
  - A. Spatial Resolution 215
  - B. Sensitivity 215
  - C. Injected Dose and Injected Mass 215
  - D. Specific Activity of Tracers 216
  - E. Measurement of the Input Function 216
  - F. Anesthesia 216
  - G. Other Issues 217
- III. Early Development of Animal PET Scanners 217
  - A. Large Animal PET Scanners 217
  - B. Hammersmith RAT-PET 218
- IV. New Generation Small Animal PET Scanners 218
  - A. Sherbrooke Animal PET 218
  - B. microPET 219
  - C. HIDAC Animal PET 219
  - D. Other Small Animal PET Systems 220
- V. Applications of Small Animal PET 221
  - A. Measurement of Glucose Metabolism in the Rat Brain and Heart 221
  - B. Studies of the Dopaminergic System in Rat Brain 222
  - C. Animal PET in Oncology 222
  - D. Imaging of Gene Expression by PET 223
- VI. Future Opportunities and Challenges 223
  - A. Small Animal PET Research Challenges 223
  - B. Resolution Limits of Small Animal PET 224
  - C. Multimodality Small Animal Imaging 225
  - D. Use of PET for High-Throughput Phenotyping and Drug Screening 225
- VII. Summary 225

13. Scintillators

FRANK WILKINSON III

- I. Introduction 229
- II. Gamma-Ray Interactions in Scintillation Crystals 229
  - A. The Photoelectric Process 230
  - B. The Compton Effect 231
  - C. Pair Production 233
- III. The Characteristics and Physical Properties of Scintillators 233
  - A. Light Output 234
  - B. Scintillator Energy Resolution 239
  - C. Material Density 240
  - D. Optical Properties 241
  - E. Mechanical Properties and Intrinsic Background 241
- IV. Scintillation Detectors: Design and Fabrication 242
  - A. Detector Design 242
  - B. Detector Components 244
  - C. Detector Fabrication 245
- V. Measurements with Scintillators 246
  - A. Measurement Systems 246
- VI. Summary and Comments 253

14. Photodetectors

BERND J. PICHLER AND SIBYLLE I. ZIEGLER

- I. Introduction 255
  - A. Light Emission of Common Scintillators 255
  - B. Important Performance Factors of Light Detectors 255
- II. Photomultiplier Tubes 256
  - A. The Photocathode of a PMT 256
  - B. Quantum Efficiency of PMTs 257
  - C. Electron Multiplication, Gain, and Dynode Structures 257
  - D. Electronic Properties of PMTs 257
  - E. Effects of Magnetic Fields 259
  - F. Position-Sensitive Photomultiplier Tubes 259
  - G. Photomultiplier Tubes: Future Trends 259
- III. Semiconductor Diode Detectors 260
- IV. PIN Diodes 261
- V. Avalanche Photodiodes 262
  - A. Structure and Properties of APDs 262
  - B. Theory of Avalanche Multiplication 262
  - C. Noise Behavior of APDs 263
  - D. Guard Structures 264
- VI. Comparison of PMT and APD Properties 264
- VII. Drift Diodes 265
- VIII. Direct Detection of Gamma Rays: CdTe and CdZnTe Detectors 266

15. CdTe and CdZnTe Semiconductor Detectors for Nuclear Medicine Imaging

DOUGLAS J. WAGENAAR

- I. Introduction 270
  - A. Semiconductors as Radiation Detectors 270

B. Crystal Growth and Contacts 270
C. Advantages of CdTe and CZT 272
D. Relevant Societies and Scientific Exchange 273
II. Energy Spectrum Performance 274
A. Energy Resolution 275
B. The Low-Energy Tail 275
C. Neighboring Pixel Effects 275
D. Depth-of-Interaction Effects 276
E. The Small Pixel Effect 277
III. Imaging Performance 277
A. Electronics and Signal Processing 278
B. Energy Windowing 279
C. Uniformity and Other Corrections 279
D. Cooling 279
E. Timing Resolution 280
F. Spatial Resolution 280
G. Collimation 280
IV. Nuclear Medicine Applications 281
A. Prototypes and Products 282
B. Cardiac Imaging 283
C. Breast Imaging 283
D. Handheld and Surgical Imagers 283
E. Dual-Modality Single-Photon Emission Computed Tomography and X-ray Computed Tomography 284
F. Small Animal Imaging 284
G. Compton Camera 284
V. Conclusion 284

16. Application-Specific Small Field-of-View Nuclear Emission Imagers in Medicine
CRAIG LEVIN

I. Overview of Application-Specific Small Field-of-View Imagers 293
A. Motivation for the Small Camera Concept 293
B. General Design Principles 294
II. Scintillation Detector Designs of Small Field-of-View Imagers 300
A. Scintillation Crystal Design 300
B. Collimation for Scintillation Crystal Designs 306
C. Photodetector Design 306
D. Electronic Readout of Position-Sensitive Photodetectors 310
E. Electronic Processing and Data Acquisition for Imaging 311
F. Event Positioning Schemes and Image Formation 312
III. Semiconductor Detector Designs of Small Field-of-View Imagers 315
A. Semiconductor versus Scintillation Crystals 315
B. Semiconductor Imaging Array Configurations 316
IV. Review of Current Designs and Applications for Small Field-of-View Imagers 317
A. Small-FOV Gamma-Ray Imagers in Medicine 317
B. Small-FOV Coincidence Imagers in Medicine 325
C. Small-FOV Beta Imagers in Medicine 328

17. Intraoperative Probes and Imaging Probes
EDWARD J. HOFFMAN, MARTIN P. TORNAI, MARTIN JANECEK, BRADLEY E. PATT, AND JAN S. IWANCZYK

I. Introduction 335
II. Early Intraoperative Probes 336
A. Geiger-Müller Counters 336
B. Scintillation Detectors 337
C. Solid-State Detectors 338
D. Detector Configurations 339
E. Summary of Intraoperative Probe Instrumentation 341
III. Clinical Applications 341
A. Radioimmunoguided Surgery 341
B. Location of Sentinel Nodes 342
C. Summary of Clinical Applications 343
IV. The Future: Imaging Probes? 344
A. Early Intraoperative Imaging Probes 344
B. Beta Imaging Probes 345
C. Gamma Imaging Probes 351
D. Imaging Probe Summary 352
V. Discussion 353
VI. Conclusion 353

18. Noble Gas Detectors
ALEXANDER BOLOZDYNYA

I. Why Noble Gas Detectors are Interesting for Medical Gamma-Ray Imaging 359
A. Brief History of Gas Detectors 359
B. Intrinsic Energy Resolution 360
C. Position Resolution 361
D. Technical Features 361
II. Basic Processes of Energy Dissipation and Generation of Light Signals 362
A. Ionization and Scintillation 362
B. Drift of Charge Carriers 363
C. Gas Gain and Electroluminescence 364
D. Electron Emission from Condensed Phases 364
III. Earlier Developments of Gas Detectors for Medical Applications 366
A. Ionization Chambers 366
B. Analog Imaging 366

C. Digital Imaging with Multiwire Proportional Drift Chambers   366
IV. Luminescence Detectors   367
   A. Electroluminescence Drift Chambers   367
   B. Multilayer Electroluminescence Chamber   371
   C. Liquid Xenon Detectors   372
V. Technical Features of Luminescence Detectors   373
   A. Gas Purification   373
   B. Photosensors   374
   C. UV Wavelength Shifting   374
   D. Construction   374
VI. Applications for Single-Photon Emission Computed Tomography   375
   A. Small Gamma Camera   375
   B. Cylindrical Gamma Camera   376
   C. Compton Camera   378
VII. Concluding Remarks   378

19. Compton Cameras for Nuclear Medical Imaging
W. L. ROGERS, N. H. CLINTHORNE, AND A. BOLOZDYNYA

I. Introduction   383
   A. Method and Motivation   383
   B. Brief History   383
II. Factors Governing System Performance   385
   A. Geometric Effects   385
   B. Statistical and Electronic Effects   386
   C. Physics Effects   386
   D. Combined Effects   389
III. Analytical Prediction of System Performance   390
   A. Noise Propagation and Lower Bound   390
   B. Observer Performance   393
   C. Predicted Efficiency Gains for Various Geometries   394
IV. Image Reconstruction for Compton Cameras   397
   A. Background   397
   B. The Forward Problem   398
   C. The Inverse Problem   399
   D. Overview of Compton Reconstruction Methods   399
   E. Direct or "Analytic" Solutions   400
   F. Statistically Motivated Solutions   402
   G. Regularization   403
V. Hardware and Experimental Results   403
   A. Silicon Pad Detectors   403
   B. C-Sprint   404
   C. Scintillation Drift Chambers as Compton Cameras   407
VI. Future Prospects for Compton Imaging   411
   A. Compton Probes   411
   B. Combined PET-SPECT Imaging   412
   C. Very High Resolution Animal PET   412
   D. Coincidence SPECT   413
   E. Imaging of High Energy Radiotracers   414
VII. Discussion and Summary   415

20. Analytic Image Reconstruction Methods
PAUL E. KINAHAN, MICHEL DEFRISE, AND ROLF CLACKDOYLE

I. Introduction   421
II. Data Acquisition   422
   A. Two-Dimensional Imaging   423
   B. Fully Three-Dimensional Imaging   424
   C. The Relationship between the Radon and X-Ray Transforms   425
III. The Central Section Theorem   426
   A. The Two-Dimensional Central Section Theorem   427
   B. The Three-Dimensional Central Section Theorem for X-Ray Projections   427
   C. Other Versions of the Central Section Theorem   427
IV. Two-Dimensional Image Reconstruction   429
   A. Backprojection   429
   B. Reconstruction by Backprojection Filtering   430
   C. Reconstruction by Filtered Backprojection   431
   D. Reconstruction by Direct Fourier Methods   432
   E. Other Data Acquisition Formats   433
   F. Regularization   434
V. Three-Dimensional Image Reconstruction from X-Ray Projections   435
   A. Spatial Variance and the Three-Dimensional Reprojection Algorithm   435
   B. Three-Dimensional Backprojection   435
   C. Three-Dimensional Reconstruction by Backprojection Filtering   436
   D. Three-Dimensional Reconstruction by Filtered Backprojection   437
   E. Other Three-Dimensional Reconstruction Methods   439
VI. Summary   441

21. Iterative Image Reconstruction
DAVID S. LALUSH AND MILES N. WERNICK

I. Introduction   443
II. Tomography as a Linear Inverse Problem   444
   A. Linear Model of the Imaging Process   444
   B. Statistical Model of Event Counts   445
   C. Spatiotemporal (4D) Imaging Model   446
III. Components of an Iterative Reconstruction Method   446
IV. Image Reconstruction Criteria   446
   A. Constraint Satisfaction   446
   B. Maximum-Likelihood Criterion   447
   C. Least-Squares and Weighted-Least-Squares Criteria   447

- D. Shortcoming of Maximum-Likelihood, Least-Squares, and Weighted-Least-Sqaures Methods 448
- E. Bayesian Methods 448
- F. Criteria for Reconstruction of Image Sequences: 4D Reconstruction 451
V. Iterative Reconstruction Algorithms 453
- A. General Structure of Iterative Algorithms 453
- B. Constraint-Satisfaction Algorithms 453
- C. The Maximum-Likelihood Expectation-Maximization Algorithm 455
- D. Least-Squares and Weighted-Least Squares Algorithms 457
- E. Maximum A Posteriori Reconstruction Algorithms 459
- F. Subset-Based Reconstruction Algorithms 461
- G. Iterative Filtered Backprojection Algorithms 462
VI. Evaluation of Image Quality 463
- A. Bias and Variance 464
- B. Effective Spatial Resolution 464
- C. Numerical Observers 464
VII. Summary 465
VIII. Appendices 465
- A. Modeling the Projection of a Pixel 465
- B. Projections onto Convex Sets 467
- C. Details of the Maximum-Likelihood Expectation-Maximization Algorithm 467

22. Attenuation, Scatter, and Spatial Resolution Compensation in SPECT

MICHAEL A. KING, STEPHEN J. GLICK, P. HENDRICK PRETORIUS, R. GLENN WELLS, HOWARD C. GIFFORD, MANOJ V. NARAYANAN, AND TROY FARNCOMBE

I. Review of the Sources of Degradation and Their Impact in SPECT Reconstruction 473
- A. Ideal Imaging 473
- B. Sources of Image Degradation 474
- C. Impact of Degradations 476
II. Nonuniform Attenuation Compensation 477
- A. Estimation of Patient-Specific Attenuation Maps 477
- B. Compensation Methods for Correction of Nonuniform Attenuation 480
- C. Impact of Nonuniform Attenuation Compensation on Image Quality 483
III. Scatter Compensation 484
- A. Scatter Estimation Methods 485
- B. Reconstruction-Based Scatter Compensation Methods 487
- C. Impact of Scatter Compensation on Image Quality 489
IV. Spatial Resolution Compensation 490
- A. Restoration Filtering 490
- B. Modeling Spatial Resolution in Iterative Reconstruction 491
- C. Impact of Resolution Compensation on Image Quality 492
V. Conclusion 494

23. Kinetic Modeling in Positron Emission Tomography

EVAN D. MORRIS, CHRISTOPHER J. ENDRES, KATHLEEN C. SCHMIDT, BRADLEY T. CHRISTIAN, RAYMOND F. MUZIC JR., AND RONALD E. FISHER

I. Introduction 499
- A. What's in a Compartment? 500
- B. Constructing a Compartmental Model 501
II. The One-Compartment Model: Blood Flow 502
- A. One-Tissue Compartmental Model 502
- B. One-Compartment Model in Terms of Flow 502
- C. Volume of Distribution/Partition Coefficient 503
- D. Blood Flow 504
- E. Dispersion, Delay Corrections 505
- F. Noninvasive Methods 505
- G. Tissue Heterogeneity 505
III. Positron Emission Tomography Measurement of Regional Cerebral Glucose Use 506
- A. The Basic Compartmental Model 506
- B. Protocol for Measurement of Regional Cerebral Glucose Use 507
- C. Estimation of Rate Constants 508
- D. The Lumped Constant 508
- E. Tissue Heterogeneity 508
IV. Receptor-Ligand Models 512
- A. Three-Compartmental Model 512
- B. Modeling Saturability 514
- C. Parameter Identifiability: Binding Potential or Other Compound Parameters 514
- D. Neurotransmitter Changes—Nonconstant Coefficients 517
- E. Neurotransmitter Levels 519
V. Model Simplifications 519
- A. Reference Region Methods 519
- B. Compartmental Model Simplifications 519
- C. Logan Analysis 520
- D. Limitations and Biases of the Reference Region Methods 522
- E. Practical Use 522
VI. Limitations to Absolute Quantification 522
- A. Spatial Resolution Effects 523
- B. Correcting for the Spatial Resolution Effect 524
VII. Functional Imaging of Neurochemistry—Future Uses 527

A. Can Neurotransmitter Activation Imaging Really Work?   527
   B. Activation of the Dopamine System by a Task Can Be Imaged   529
   C. Clinical Uses of Neurotransmitter Activation Studies   529
   D. Caution in the Interpretation of Neurotransmitter Activation Studies   530
   E. PET Imaging and Kinetic Modeling Issues in Gene Therapy   531
VIII. A Generalized Implementation of the Model Equations   532
   A. State Equation   532
   B. Output Equation   534
   C. Implementing an Arbitrary Model   535

24. Computer Analysis of Nuclear Cardiology Procedures

ERNEST V. GARCIA, TRACY L. FABER, C. DAVID COOKE, AND RUSSELL D. FOLKS

I. Introduction   541
II. Advances in Single-Photon Emission Computed Tomography Instrumentation   541
III. Advances in Computer Methods   541
   A. Automatic Oblique Reorientation and Reslicing   542
   B. Stress-Rest Studies   543
   C. Automated Perfusion Quantification   543
   D. Image Display   543
   E. Automatic Quantification of Global and Regional Function   546
   F. Artificial Intelligence Techniques Applied to SPECT   546
   G. Software Registration of Multimodality Imagery: Image Fusion   548
IV. Conclusion   549

25. Simulation Techniques and Phantoms

MICHAEL LJUNGBERG

I. Introduction   551
   A. Why Simulation? Limitations and Benefits   551
   B. Probability Distribution Functions   552
   C. The Random Number Generator   552
II. Sampling Techniques   552
   A. The Distribution Function Method   552
   B. The Rejection Method   552
III. Mathematical Phantoms   552
   A. Analytical Phantoms   553
   B. Voxel-Based Phantoms   553
   C. Other Types of Phantoms   554
IV. Photon and Electron Simulation   554
   A. Cross-sectional Data for Photons   554
   B. Path-Length Simulation   554
   C. Sampling Interaction Types   554
   D. Photoabsorption   555
   E. The Compton Process   555
   F. Bound Electrons   555
   G. Coherent Scattering   555
   H. Electron Simulation   555
V. Detector Simulation   556
   A. Simulation of Energy Resolution   556
   B. Simulation of Temporal Resolution   557
   C. Backscattering of Photons behind the NaI(Tl) Crystal   557
   D. Collimator Penetration and Scattering   557
VI. Variance Reduction Methods   557
   A. The Idea behind the Weight   557
   B. Forced Detection (Angular and Spatial)   558
VII. Examples of Monte Carlo Programs for Photon and Electrons   559
   A. The SIMIND Program   559
   B. The SimSET Program   559
   C. The EGS4 Package   559
   D. The MCNP4 Program   560
VIII. Examples of Monte Carlo Applications in Nuclear Medicine Imaging   560
   A. General Detector Parameters and Energy Spectrum Analysis   560
   B. Evaluation of Scatter and Attenuation Correction Methods   560
   C. Collimator Simulation   560
   D. Transmission SPECT Simulation   561
   E. MC Calculations and SPECT in Dose Planning of Radionuclide Therapy   561
IX. Conclusion   561

Contributors

John N. Aarsvold, Nuclear Medicine Service, Atlanta VA Medical Center, Decatur, GA.

Robert N. Beck, Department of Radiology, The University of Chicago, Chicago, IL.

Bernard Bendriem, CPS Innovation, Knoxville, TN.

A. Bolozdynya, Constellation Technology Corporation, Largo, FL.

A. Bertrand Brill, Vanderbilt University, Nashville, TN.

Arion F. Chatziioannou, Crump Institute for Biological Imaging, UCLA School of Medicine, Los Angeles, CA.

Simon R. Cherry, Biomedical Engineering, UC Davis, Davis, CA.

B. T. Christian, PET/Nuc Med, Kettering Medical Center, Wright State University, Dayton, OH.

Rolf Clackdoyle, Medical Imaging Research Lab, University of Utah, Salt Lake City, UT.

N. H. Clinthorne, Division of Nuclear Medicine, Department of Internal Medicine, University of Michigan Medical Center, Ann Arbor, MI.

C. David Cooke, Emory University School of Medicine, Atlanta, GA.

Michel Defrise, University Hospital AZ-VUB, Division of Nuclear Medicine, National Fund for Scientific Research, Brussels, Belgium.

Christopher J. Endres, Department of Radiology, Johns Hopkins University, Baltimore, MD.

Tracy L. Faber, Emory University School of Medicine, Atlanta, GA.

Troy Farncombe, McMaster University, Hamilton, ON, Canada.

Ronald E. Fisher, Division of Neuroscience, Baylor College of Medicine, Houston, TX.

Russell D. Folks, Emory University School of Medicine, Atlanta, GA.

James R. Galt, Department of Radiology, Emory University, Atlanta, GA.

Ernest V. Garcia, Department of Radiology, Emory University Hospital, Atlanta, GA.

Sebastian Genna, Digital Scintigraphics, Inc., Waltham, MA.

Howard C. Gifford, Department of Nuclear Medicine, University of Massachusetts Medical Center, Worcester, MA.

Stephen J. Glick, Department of Nuclear Medicine, University of Massachusetts Medical Center, Worcester, MA.

Donald L. Gunter, Department of Physics and Astronomy, Vanderbilt University, Nashville, TN.

Edward J. Hoffman, Crump Institute for Biological Imaging, UCLA School of Medicine, Los Angeles, CA.

Jan S. Iwanczyk, Photon Imaging, Northridge, CA.

Martin Janecek, Division of Nuclear Medicine, Department of Pharmacology, UCLA School of Medicine, Los Angeles, CA.

Chien-Min Kao, Department of Radiology, University of Chicago, Chicago, IL.

Joel Karp, University of Pennsylvania, Philadelphia, PA.

Paul E. Kinahan, University of Washington, Seattle, WA.

Michael E. King, Department of Nuclear Medicine, University of Massachusetts Medical Center, Worcester, MA.

Patrick La Rivière, Department of Radiology, University of Chicago, Chicago, IL.

David S. Lalush, Department of Biomedical Engineering, North Carolina State University, Raleigh, NC.

Craig Levin, Nuclear Medicine Division and Molecular Imaging Program at Stanford, Stanford University School of Medicine, Stanford, CA.

Thomas Lewellen, Division of Nuclear Medicine, University of Washington Medical Center, Seattle, WA.

Michael Ljungberg, Department of Radiation Physics, The Jubileum Institute, Lund University, Lund, Sweden.

Robert A. Mintzer, Department of Radiology, Emory University, Atlanta, GA.

Evan D. Morris, Department of Radiology and Biomedical Engineering, Indiana University School of Medicine, Indianapolis, IN.

Jogeshwar Mukherjee, Department of Nuclear Medicine, Kettering Medical Center, Wright State University, Dayton, OH.

Raymond F. Muzic, Jr., Nuclear Medicine, Radiology & Biomedical Engineering, University Hospitals of Cleveland, Cleveland, OH.

Manoj V. Narayanan, Department of Nuclear Medicine, University of Massachusetts Medical Center, Worcester, MA.

T. K. Narayanan, Department of Nuclear Medicine, Kettering Medical Center, Wright State University, Dayton, OH.

Jinsong Ouyang, Digital Scintigraphics, Inc., Waltham, MA.

Xiaochuan Pan, Department of Radiology, University of Chicago, Chicago, IL.

Bradley E. Patt, Photon Imaging, Northridge, CA.

Bernd J. Pichler, Nuklearmedizinische Klinik, Klinikum rechts der Isar, Technische Universitat Munchen, Munchen, Germany.

P. Hendrik Pretorius, Department of Nuclear Medicine, University of Massachusetts Medical Center, Worcester, MA.

W.L. Rogers, Division of Nuclear Medicine, University of Michigan, Ann Arbor, MI.

Kathleen C. Schmidt, Laboratory of Cerebral Metabolism, National Institute of Mental Health, Bethesda, MD.

Bingzhi Shi, Department of Nuclear Medicine, Kettering Medical Center, Wright State University, Dayton, OH.

Martin P. Tornai, Department of Radiology, Duke University Medical Center, Durham, NC.

David W. Townsend, Department of Medicine and Radiology, University of Tennessee, Knoxville, TN.

Douglas J. Wagenaar, Nuclear Medicine Group, Siemens Medical Solutions USA, Inc., Hoffman Estates, IL.

Charles C. Watson, CPS Innovation, Knoxville, TN.

R. Glenn Wells, Department of Nuclear Medicine, University of Massachusetts Medical Center, Worcester, MA.

Miles N. Wernick, Departments of Electrical and Computer Engineering and Biomedical Engineering, Illinois Institute of Technology, and Predictek, LLC, Chicago, IL.

Frank Wilkinson III, Alpha Spectra, Inc., Grand Junction, CO.

Weisha Xia, Digital Scintigraphics, Inc., Waltham, Ma.

Yang Zhi-Ying, Department of Psychiatry, University of Chicago, Chicago, IL.

Gensheng Lawrence Zeng, Radiology Research, University of Utah, Salt Lake City, UT.

Sibylle I. Ziegler, Nuklearmedizinische Klinik, Klinikum rechts der Isar, Technische Universitat Munchen, Munchen, Germany.

Foreword

As explained in the Preface, this book was inspired by a symposium held to honor my mentor, friend, and colleague, Robert N. Beck, upon his retirement from the University of Chicago, Department of Radiology, after a lifetime of outstanding contributions to the field of emission imaging.

For many of us who have had the good fortune to work with Beck, his vision and passion for nuclear medicine and imaging science have inspired our own lifelong commitment to research in these areas and have made our own careers in science more enjoyable and meaningful. For those less familiar with him, let me recall briefly some of his accomplishments.

Beck began his career in imaging science in the 1950s, during the era of the Atoms for Peace program, an effort by the U.S. government to promote peaceful uses of atomic energy. As a young man with broad interests in the arts and sciences, he sought to develop constructive uses of this new technology for biomedical applications. While studying mathematics and physics at the University of Chicago, he joined a research team at the Argonne Cancer Research Hospital (ACRH) in 1954. During the following decades, he made many key contributions to radionuclide imaging.

In the early 1960s, Beck was the first to develop a mathematical theory to determine the optimum gamma-ray energy for specific imaging applications. His theory led to the first clinical application of ^{99m}Tc for brain tumor detection. ^{99m}Tc quickly became the most commonly used radioisotope in nuclear medicine, which is still true today. Also in the early 1960s, Beck developed the first comprehensive theory of optimum collimator design, which is the basis for many modern-day commercial collimators. In the mid-1960s, he was the first to adapt Fourier methods and to employ the modulation transfer function (MTF) to characterize the spatial resolution of radionuclide imaging systems. In the early 1970s, he was the first to develop a theory of optimum weights for multichannel spectral data in imaging.

Beginning in the mid-1970s, as Director of Radiological Sciences and the Franklin McLean Memorial Research Institute (FMI, formerly ACRH), Beck was responsible for the development of the FMI PET Center for Brain Research (with Malcolm D. Cooper), the Frank Center for Image Analysis (with Malcolm D. Cooper and me), the Maurice Goldblatt Center for Magnetic Resonance Imaging (with David N. Levin), and the Center for Imaging Science, a joint venture between the University of Chicago and Argonne National Laboratory (with Albert V. Crewe). Beck also helped to establish the Graduate Program in Medical Physics at the University of Chicago, in which he mentored students and lectured on the physical principles of radionuclide imaging.

In 1991, Beck received the Computerworld Smithsonian Nominee Award "in recognition of visionary use of information technology in the field of imaging science" based on his conceptualization of this emerging new academic discipline. For his pioneering work in nuclear medicine, he has received the Society of Nuclear Medicine's Paul Aebersold Award and the IEEE Medical Imaging Scientist Award.

His current interests are focused on the development of quantitative imaging methods for brain research, advanced software for the presentation of visual and verbal materials to accelerate learning, and the impact of multimedia technology on our culture.

For those closest to him, Beck's kind heart has brought a special atmosphere of cooperation and partnership to every activity in which he participated. To us, he is not only a pioneer of nuclear medicine research, but also a friend whom we will treasure always.

In closing, let me take this opportunity to congratulate Miles Wernick and John Aarsvold for their vision and persistence in bringing together an outstanding book on emission tomography, which I expect will be a valuable resource for this and future generations of scientists who will continue the traditions begun by Beck and his contemporaries.

Chin-Tu Chen

Preface

The term *emission tomography* (ET) encompasses two leading medical imaging techniques: positron emission tomography (PET) and single-photon emission computed tomography (SPECT). PET and SPECT are used to study biological function in humans and animals and to detect abnormalities that are characteristic of disease. PET and SPECT, which measure the distribution of radiotracer materials administered to the subject, are currently the leading examples of molecular imaging in medicine. PET and SPECT have a wide array of uses, including the detection of cancer, mapping of the brain, and diagnosis of heart disease.

The aim of this book is to provide a comprehensive presentation of the physics and engineering concepts behind these important imaging technologies. The book is intended for graduate students in physics and engineering, researchers, clinical medical physicists, and professional physicists and engineers in industry.

INSPIRATION FOR THIS BOOK

Before discussing the organization and suggested use of the book, let us first say a few words about how the book came about.

This book was inspired by a symposium held at the University of Chicago in 1999 to honor Robert N. Beck, a pioneer in emission imaging, for a lifetime of scientific achievement and academic mentorship. The editors of this book, as well as several of the contributing authors, were co-workers in the Franklin McLean Institute (FMI) at the University of Chicago, a research organization led by Beck for many years.

The symposium, which was entitled *Future Directions in Nuclear Medicine Physics and Engineering*, included several days of invited presentations by leading researchers in emission imaging, many of whom also contributed to the writing of this book. The aim of the symposium was to review where emission imaging has been, including its early history (see Chapter 3) and to share ideas about its future directions on the eve of the twenty-first century.

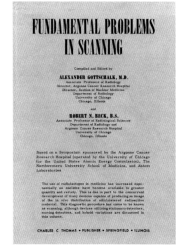

Both the symposium and this book echo back to similar efforts in which Beck participated in 1965, in particular a book (co-edited with Alexander Gottschalk) based on a similar symposium, also held at the University of Chicago. The volume, entitled *Fundamental Problems in Scanning* (see figure), included four papers by Beck and many other important papers by the early pioneers in the field. It is still cited today and provides a wonderful window into the early developments of emission imaging as talented researchers of the day worked out the basic ideas that underpin today's technology.

The 1999 *Future Directions* symposium was a splendid experience, with researchers on the cutting edge of the field sharing their ideas in a relaxed setting. Building on the symposium's success, we undertook to compile a comprehensive overview of emission tomography technology, including both basic tutorial information and timely research directions. With this in mind, we recruited some 50 authors, each an expert in his or her specialization, to put together a picture of modern emission tomography. This book is the result of that effort.

WHERE TO START

In each chapter we have endeavored to include a basic introduction to the topic, as well as more advanced methods and state-of-the-art concepts. To gain a basic appreciation of emission tomography, the reader is encouraged to begin by reading Chapter 2, which provides an introduction to the subject, followed by Chapters 7 and 10, which further explain SPECT and PET, respectively. The remaining chapters cover specific aspects of ET and may be approached in various ways, depending on the interests of the reader.

ORGANIZATION OF THE BOOK

The book is organized as follows.

Part I, Introduction provides an overview of the topic, as well as scientific and mathematical foundations that are helpful background material for appreciating the later chapters. These sections provide useful material for coursework dealing with ET. Chapter 1 puts ET in the context of the larger field of imaging science and puts imaging in the context of human progress toward a better understanding of the physical world. Chapter 2 serves as the introduction to the book, explaining the basic idea of ET and its uses. Chapter 3 describes the historical development of ET and its clinical applications. Chapter 4 describes the basic physics that underpins ET. Chapter 5 describes the development and application of radiopharmaceuticals, which are used to produce ET images. Chapter 6 introduces the mathematical preliminaries that are relevant to ET.

Part II, Single-Photon Emission Computed Tomography (SPECT) covers topics specific to SPECT. Chapter 7 explains gamma cameras, SPECT systems, and their basic uses. Chapter 8 describes the design and function of collimators, which are an important component of gamma cameras. Chapter 9 is an overview of annular SPECT systems.

Part III, Positron Emission Tomography (PET) discusses three aspects of PET imaging systems. Chapter 10 introduces the basic ideas of PET, including the design and function of PET systems. Chapter 11 describes hybrid systems that can perform both X-ray computed tomography and PET. Chapter 12 describes PET systems for imaging of small animals.

Part IV, Gamma-Ray Detectors and Their Components covers the hardware used to detect emitted gamma rays, which are the basis of ET imaging. Chapter 13 discusses scintillators, which turn gamma rays into light; Chapter 14 describes detectors that sense the light created by the scintillator. Chapter 15 explains semiconductor detectors, an emerging alternative to conventional detectors.

Part V, Other Emission Imaging Technologies covers a number of important and promising developments in emission imaging that differ from conventional PET and SPECT. Chapter 16 describes small portable gamma cameras. Chapter 17 describes counting and imaging probes, which are used, for example, to obtain count measurements and images during surgery. Chapter 18 describes noble gas detectors. Chapter 19 explains Compton cameras, which are based on an alternative approach to emission imaging.

Whereas Parts II–V deal mainly with hardware systems and components, Parts VI–VII cover the mathematical and software methods used to create, analyze, and evaluate ET images.

Part VI, Image Reconstruction explains the techniques and principles of image reconstruction, which is the process by which ET images are computed from acquired gamma-ray data. This subject is divided into three parts. Chapter 20 describes analytic image reconstruction methods, which include the standard two-dimensional filtered backprojection (FBP) method, as well as more advanced techniques. Chapter 21 describes iterative methods, mostly based on statistical estimation, which are becoming widely used alternatives to FBP. Chapter 22 explains methods of correcting for complicating factors in the SPECT image reconstruction process, namely attenuation and scatter.

Part VII, Image Analysis and Performance Evaluation covers techniques for computerized analysis of ET images and data, and methods of simulating ET data and images for purposes of evaluation and design of imaging systems and algorithms. Chapter 23 describes kinetic modeling, which uses mathematical techniques to explain the behavior of tracer compounds in the body and which is capable of summarizing important information about the body's physiology. Chapter 24 explains how cardiac SPECT images are processed by computer to provide physicians with valuable displays and analyses of the images. Chapter 25 explains how imaging systems and algorithms can be tested and evaluated through the use of computer simulations.

We hope this book provides the reader with a solid foundation for making his or her own contributions to this important field.

Miles N. Wernick
John N. Aarsvold

Acknowledgments

We are very grateful to the U.S. Department of Energy for underwriting the publication of this book and, especially, Dean Cole, who immediately saw the value of the project and whose patience and help were key to making the book a reality.

Of course, our deepest thanks go to our families, whose support helped us to see the project through to the end. M.N.W. thanks especially his wife, Hesna; his daughters, Kyra and Marisa; and his mother, Katherine. J.N.A. thanks especially his mother, Dorothy. In memory of our fathers, Sherman (M.N.W.) and Milton (J.N.A.), we gratefully acknowledge their important roles in our lives.

The idea for the book was inspired by a symposium held at the University of Chicago to mark the retirement of Robert N. Beck after 50 years of dedication to the field of emission imaging. We express our appreciation to Beck for bringing together and leading a splendid and talented group of people at the University of Chicago, with whom we thoroughly enjoyed working during our time there. Beck has contributed many valuable insights to our own thoughts about imaging science and, of course, was a major contributor to the field of emission tomography, as explained in the Foreword.

We next thank the following people and organizations for making the Beck symposium possible. First, we acknowledge the hard work of the organizing committee, which, in addition to us, consisted of Ariadne Beck, Chin-Tu Chen, Ruthie Cornelius, Melvin L. Griem, Donald L. Gunter, Chien-Min Kao, Kenneth L. (Kip) Matthews II, Patrick LaRivière, Robert A. Mintzer, Caesar E. Ordonez, Xiaochuan Pan, Bill O'Brien-Penney, Cindy Peters, Benjamin M. W. Tsui, and Nicholas J. Yasillo.

We are extremely grateful to all the financial sponsors of the symposium, which included the National Science Foundation (NSF), the Whitaker Foundation, ADAC Laboratories, BICRON Corporation, CTI PET Systems, GE Medical Systems, Hamamatsu Corporation, Hitachi Medical Corporation of America, Picker International, Siemens Medical Systems, SMV America, Toshiba America Medical Systems, and the Department of Radiology at the University of Chicago.

We thank all the invited speakers who gave generously of their time to make detailed and insightful presentations on their vision for the future of nuclear medicine imaging in the twenty-first century. In addition to M.N.W., and J.N.A., who served as symposium moderator, the speakers included: H. Bradford Barber, Harrison H. Barrett, Robert N. Beck, A. Bertrand Brill, Thomas F. Budinger, Simon R. Cherry, Malcolm D. Cooper, Ernest V. Garcia, Donald L. Gunter, Edward J. Hoffman, Ronald H. Huesman, Joel S. Karp, Paul E. Kinahan, Michael A. King, Ronald J. Jaszczak, Thomas K. Lewellen, Jogeshwar Mukherjee, W. Leslie Rogers, David W. Townsend, and Benjamin M. W. Tsui.

In addition to the dozens of authors and co-authors who contributed selflessly to the writing of the book, we also thank the following people who provided helpful thoughts and suggestions about its content: Mark A. Anastasio, Stephen E. Derenzo, Daniel Gagnon, Guido Germano, Hank F. Kung, Bradley E. Patt, Janet R. Saffer, Dennis E. Persyk, Charles W. Stearns, Hiroshi Watabe, Robert E. Zimmerman, and George I. Zubal.

Miles N. Wernick
John N. Aarsvold

CHAPTER 1

Imaging Science: Bringing the Invisible to Light

ROBERT N. BECK

Department of Radiology, The University of Chicago, Chicago, Illinois

I. Preamble
II. Introduction
III. Imaging Science
IV. Fundamental and Generic Issues of Imaging Science
V. Methodology and Epistemology
VI. A View of the Future

All men by nature desire to know. An indication of this is the delight we take in our senses; for even apart from their usefulness they are loved for themselves; and above all others the sense of sight.—The reason is that this, most of all the senses, makes us know and brings to light many differences between things.

Aristotle (1941, 689)

I. PREAMBLE

This book concerns emission tomography (ET), a form of medical imaging that uses radioactive materials. The purpose of this chapter is to place the highly specialized field of emission tomography within the broader context of imaging science, which has to do with natural (i.e., unaided) vision and its extension into invisible realms by means of a great variety of imaging systems. The reason is that the fundamental issues of ET are virtually identical to those that underlie other, more mature imaging modalities. Recognition of this fact has enabled researchers in ET to make use of principles, concepts, strategies, and methods developed in other areas of imaging and, thereby, accelerate progress in this field. We must expect this process to continue.

And as the quotation from Aristotle suggests, it is also of value to recognize the place of imaging science within the much broader context of epistemology—the study of the origin, nature, methods, and limits of knowledge. The reason is that much of what we have learned about ourselves and the world around us, especially during the twentieth century, has been gained through visual means, utilizing newly developed imaging methods. The ultimate limits of knowledge that can be gained by such means are unclear but appear to be bounded primarily by human imagination and ingenuity.

Finally, to foster an appreciation of the uniqueness of our times and of the opportunities for further development of biomedical imaging methods, it is of value to discuss, very briefly, both imaging science and epistemology from evolutionary and historical perspectives.

II. INTRODUCTION

The history of imaging science is, indeed, brief. In fact, we might say that imaging science does not yet exist, in the sense that it is not yet recognized as an academic discipline in most universities, with a well-defined curriculum leading to a degree. (A notable exception is the Rochester Institute of Technology, which was the first institution to offer a PhD degree in Imaging Science.) On the other hand, the survival value of visual knowledge is recognized as both obvious and very ancient. For example, it is generally accepted that mollusks had very sophisticated visual systems more than 450 million years ago (Strickberger, 1995), and it is difficult to imagine the evolution of humans without vision.

In contrast, the history of epistemology and the emphasis on vision/sight as a major source of knowledge date back to antiquity. Although some may disagree with Aristotle's claim that sight is loved above all other senses, most would agree that the desire to know is universal and, moreover, that sight/vision provides an important means for knowing. An understanding of the strengths and limitations of this mode of knowing requires, in the first instance, an understanding of the phenomena involved in natural vision.

Questions regarding the nature of light and sight were asked for several millennia and answered in similar, although incorrect, ways. For example, with minor variations on the concepts and terms employed, influential scholars such as Zoroaster, Plato, Euclid, Ptolemy, Augustine and others (Polyak, 1957; Duke-Elder, 1958; Lindberg, 1976; Park, 1997) believed that the phenomenon of sight results from the emanation of a substance from the eyes, which traverses and comingles with the intervening medium—in effect, touching objects that are seen and, in some cases, causing them harm. Although Kepler (1604/2000) finally gave a more acceptable, ray theory of light and vision in 1604, belief in the evil eye still persists in some cultures.

A more complete historical review of concepts of the nature of light would include not only the emanatists cited here, but also the notions of Newton (corpuscular theory/color), Huygens (longitudinal wave theory), Maxwell (electromagnetic wave theory), Planck (early quantum theory of radiation), Einstein (quanta/photons), and Feynman (quantum-electrodynamics).

For the purposes of this discussion, Maxwell's (1873) theory of electromagnetism provides a very useful model for understanding the nature of light and its propagation through space and transparent material media, as well as phenomena such as reflection, refraction, and scattering. In addition, we now have at least a basic understanding of the structure and functions of the eye–brain system and of natural vision. In particular, we now know that natural vision is due to the response of the eye to a very narrow portion of the electromagnetic (EM) spectrum, called *visible light*, with wavelengths, approximately, from 400 to 750 nm, lying between ultraviolet and infrared.

Moreover, we are aware of certain imperfections of the visual system, which give rise to a variety of visual misperceptions and illusions, some of which are commonly employed by artists and magicians. As a consequence, although many of us may not know precisely how an artist creates the impression of a three-dimensional object on a two-dimensional surface or how a magician is able to deceive us with his visual tricks, nevertheless, we remain confident that such experiences can be explained in rational terms based on current theories of light and vision. In fact, in most circumstances of daily life we do not question what it means to see something; rather, we take our normal visual experiences largely for granted.

Despite the acknowledged value of sight/vision, in most academic circles, language is regarded as the basis for knowledge. It is important to recognize that language can be extended, elaborated and embellished endlessly without the use of instruments of any kind and used to explain the meaning and significance of what we see. Even so, certain aspects of our visual experiences are of such complexity as to defy detailed verbal description, despite the fact that language is believed to have existed throughout all human cultures for more than 40,000 years (Holden, 1998) and has been expanded continuously through the creation of new words as well as analogies and metaphors.

In contrast to language, any extension of natural vision requires the development and use of some form of instrument, or imaging system, that performs the function of mapping invisible object properties into visible images, as indicated in Figure 1. Apart from simple magnifiers and spectacles, which were being sold by street vendors in Amsterdam during the 1500s, the first significant extension of natural vision came with the development of the optical telescope and the optical microscope in the 1600s. In the hands of Galileo (1610/1989), Hooke (1665/1987), and many others, these instruments altered dramatically humans' understanding of themselves and the world around them, revealing material objects and certain properties that are associated with visible light, but which are quite invisible to natural vision—objects too small to be seen or large but very distant. These instruments gave rise to major advances in the physical and biological sciences, which are grounded in the observation and measurement of object properties.

As a result, confidence in the methods of science and the state of understanding of the physical world reached a high point by the late 1800s, when Albert Michelson (in Loevinger, 1995) is said to have made the following statement in a lecture given at The University of Chicago in 1894, just 1 year before Röntgen (1934) discovered X rays.

> While it is never safe to affirm that the future of Physical Science has no marvels in store even more astonishing than those of the past, it seems probable that most of the grand underlying principles have been firmly established and that further advances are to be sought chiefly in the rigorous application of these principles to all the phenomena which come under our notice.—The future truths of Physical Science are to be looked for in the sixth place of decimals.

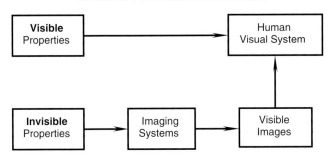

FIGURE 1 Visible properties of objects may be viewed directly. Imaging systems may be used to map certain invisible object properties into visible images.

Röntgen discovered X rays the following year, 1895. This event is generally associated with an X-ray image he made of his wife's hand. The significance of this image is that it demonstrated for the very first time that it is possible to use *invisible* radiation (X rays) to make a *visible* image (recorded on film and viewed with visible light) of an *invisible* object (bones inside the intact hand). This was the very first nonoptical imaging method,[1] and Röntgen was awarded the first Nobel Prize in physics for his achievement.

Röntgen's discovery created enormous excitement among scientists and helped to set off an avalanche of fundamental advances in physics (e.g., the discovery of radioactivity and the development of theories regarding photoelectric effect, atomic structure, special and general relativity, quantum mechanics, and nuclear and particle physics), as well as inventions that led to the remarkable array of imaging methods that we see today. Optical imaging systems map object properties that are associated with visible light into visible images, whereas nonoptical imaging systems map object properties associated with invisible radiation into visible images (see Figure 2). For use in biomedical research and clinical applications, the list of current nonoptical imaging systems includes, among many others, positron emission tomography (PET), single photon emission computed tomography (SPECT), X-ray computed tomography (CT), magnetic resonance imaging (MRI), functional magnetic resonance imaging (fMRI), ultrasound, electron microscopy, and atomic-force microscopy.

Thus, within the brief span of 100 years, imaging methods have been developed that make use of the electromagnetic spectrum (from radio waves to gamma rays) and acoustic spectrum (from vibrations to ultrasound), as well as various particle beams (electrons, protons, neutrons, etc.) and scanning styli. Most important, they extend the range of vision into the realm of object properties that are not associated directly with visible light—its emission, transmission, reflection, or scattering—and are, therefore, totally inaccessible to natural, unaided vision. Nevertheless, many such object properties are of vital interest to physical and biological scientists, engineers, and physicians.

For example, in the field of nuclear medicine, images of the distribution of an administered radiotracer may reveal invisible properties of the human body, such as the local values of blood volume, blood flow, perfusion, tissue metabolism, oxygen utilization, and receptor binding, all of which may be of scientific and diagnostic value. In particular, radionuclide imaging provides a powerful method for investigating the effects of drugs on such parameters. Exploitation of this method promises to provide new insights into the physiological processes involved in drug effects and to reduce dramatically the cost and time required for testing of putative new drugs.

III. IMAGING SCIENCE

Every object, including those of biomedical interest (e.g., molecules, cells, organs, and intact organisms), may be described, in principle, by its physical–chemical–isotopic composition at each point in space and time. Associated with its composition are certain properties that are detectable (i.e., they give rise to signals in suitable detectors), some of which can be localized in space and time and represent local values of static or dynamic, and structural or functional, object properties and the processes they undergo (Beck, 1993c).

As indicated, the fundamental concept underlying all imaging systems and procedures is that of mapping; that is, any object property that can be detected and localized in space and time may be mapped into image space, where it may be viewed with visible light. In the case of radionuclide imaging procedures, the object to be imaged is the spatiotemporal distribution *in vivo* of radioactive material that has been administered to the patient and which may be described by the local concentration of the material averaged over the observation period. Tomography permits us to map invisible views of the body—cross-sectional slices through its interior—into visible representations, which can be displayed by computer.

Recent advances in digital computer technology enable us to deal with any image as an array of numbers that represents the local values of the object property that has been imaged and to make quantitative measurements of object properties from their images. More important for the purposes of this chapter, the same principles, concepts, strategies, and computer-based methods may be

---

[1] We use the term *optical imaging* to refer to methods based on visible light.

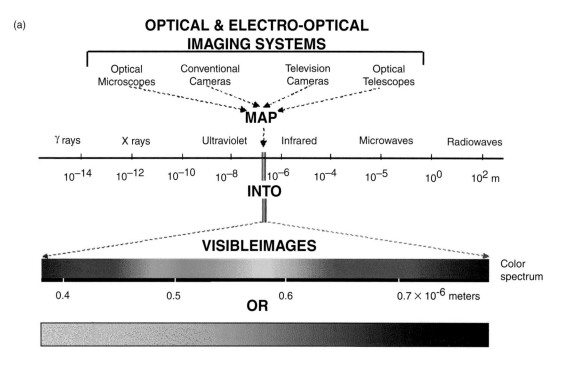

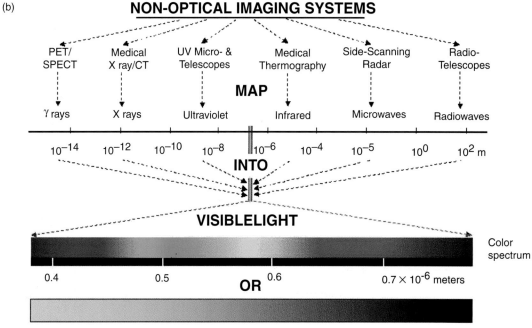

FIGURE 2 (a) Optical and electrooptical and (b) nonoptical imaging systems map visible and invisible radiations, respectively, from an object into a visible image that can be displayed in color or gray scale.

employed to deal with the generic issues of imaging science—image-data acquisition, image recovery, recording, distribution, display, observation, analysis, interpretation, evaluation, and optimization—whether the image is acquired using a microscope, a telescope, or a biomedical imaging system.

These mappings, or images, are in several ways, always imperfect representations of the object or the detected object property. In particular, they are:

- Incomplete representations because any given imaging system may detect and map only a limited number of object properties.

- Inaccurate representations because blurring, nonlinear distortion, spatial deformation, and artifacts are always present in images.
- Irreproducible because random fluctuation, or noise, always accompanies the signal associated with the detected object property.

As a consequence, all measurements of object properties from their images are correspondingly imperfect. These imperfections are addressed within the field of imaging science, the goals of which include:

- Increasing the number of properties that can be imaged, registered spatially, and superimposed as desired, so that our representations of objects are more nearly complete and useful.
- Increasing the accuracy of measured values of object properties by improving spatial resolution (to reduce spatial averaging or blurring), by increasing the sensitivity of the imaging system (to reduce the observation time and temporal averaging or motion blurring), by improving temporal resolution (to reduce, for example, the nonlinear saturation effects when photon counters are used), by improving energy resolution (to reduce the effects of scattered radiation), and by reducing distortion, deformation, artifacts, and so on.
- Increasing the reproducibility of measured values of object properties by reducing noise, for example, by increasing the sensitivity of the detector system; by use of improved radiotracers to increase the local signal amplitude and contrast, as well as to reduce toxicity and the absorbed dose to the patient; by use of advanced image-reconstruction techniques; or, as a last resort, by increasing the observation time.

The achievement of these goals, which involves balanced trade-offs among multiple competing measures of performance, will require a comprehensive understanding of all of the steps, or generic issues, that are involved in the imaging process because all of these steps may affect the accuracy and reproducibility of measurements made during finite periods of observation. In particular, such an understanding would enable us, in principle, to explore alternative strategies and to optimize parameter values associated with each of these steps, based on some appropriate (goal-related) criterion for the evaluation of image quality. As yet, no imaging system used in any field of science or medicine has been optimized fully in this sense.

A student or researcher who wishes to improve some particular imaging modality may benefit from an awareness of what has been done in several other, more advanced fields of imaging in which virtually identical and useful principles, concepts, strategies, and methods have been explored and developed. The matrix in Figure 3 may help to identify topics for further research. To assist in such pursuits, we summarize next the generic issues of imaging science and some of the relevant topics and key words associated with each.

IV. FUNDAMENTAL AND GENERIC ISSUES OF IMAGING SCIENCE

Image formation, in the most general sense of the term, as it is used here, involves not only the concept of the image as a mapping of some object property into image space, but also the classification of objects in terms of their detectable physical, chemical, and isotopic properties; alternative strategies for performing the mapping; the mathematical concepts and assumptions used for modeling these strategies; and the classification of imperfections inherent in all such mappings (Beck, 1993b).

Relevant topics include:

Mapping in real space: by geometrical means (e.g., ray tracing), the solutions of forward and inverse problems, point measurements (scanning), and projections.

Linear and nonlinear systems, and shift-invariance.

Images of point and line elements, sensitivity, spread functions, and convolution integrals and the convolution theorem.

Mapping in frequency space: diffraction imaging; continuous and discrete Abel, Fourier (FFT), and Hankel transforms; spatial-frequency response; cascaded stages; conventional and generalized transfer functions; and wavelets.

Image imperfections that limit the accuracy and reproducibility of measurements of object properties: blurring, distortion, interference, artifacts, and noise.

Development of the comprehensive understanding of image formation that is needed to optimize mappings requires attention to all of the steps involved, as well as their interdependencies. Virtually all digital imaging procedures involve the following steps.

Image-data acquisition requires consideration of the physical principles that govern the detection and spatiotemporal localization of a particular object property, as well as the particular strategy to be used in image formation. These considerations govern the design of the front-end hardware of the imaging device (e.g., the optical components of a microscope or telescope, the magnetic field of an MRI device, and the detectors used in SPECT and PET) which, in turn, determine the sensitivity, resolution, and other quality characteristics of the imaging system.

Relevant topics include:

Radiation sources and their properties (particles and waves): thermal (black-body), laser, nuclear, synchrotron, acoustic, seismic, and so on.

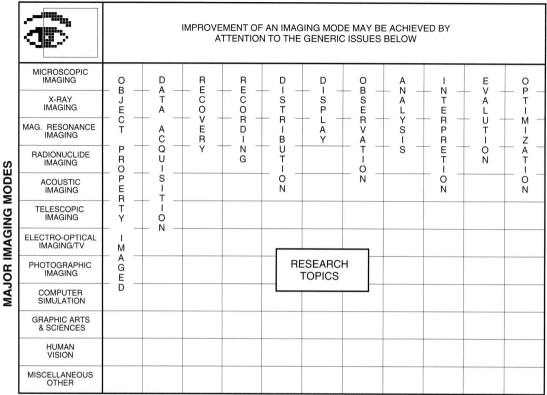

FIGURE 3 Each generic issue provides research opportunities for each imaging modality.

Wave equations, propagation of radiation, and the effects of turbulent media: absorption, scattering, refraction, reflection, diffraction, coherence, and interference.

Radiometry and photometry, collimators, coded apertures, lenses, optical elements, and adaptive optics.

Radiation detection: film, semiconductors, charge-coupled device (CCD) arrays, scintillation detectors, transducers, antennas, and phased arrays.

Efficiency, linearity, and dynamic range of radiation detectors.

Sampling theory: Whittaker-Shannon theorem and Nyquist limit, ideal versus integral sampling, convex projection, and aliasing.

Poisson processes and shot noise, spectral density, correlation function, Wiener-Khintchine theorem, and signal-noise ratio.

Spatial, temporal, and energy resolution.

Trade-offs between sensitivity and resolution, and noise and contrast.

Windowing and multichannel spectroscopic approaches.

Use of mathematical models and computer simulation of stochastic and nonstochastic processes for optimization of parameters of the image-data acquisition (detector) system for a particular class of objects.

Image recovery—reconstruction and processing—involves the mathematical concepts and algorithms for producing an image from the acquired data (e.g., image reconstruction from projections) and for processing the image to improve its quality or usefulness (e.g., to sharpen edges and to reduce distortion, interference, artifacts, and image noise).

Relevant topics include:

Reconstruction via analytic, statistical, and probabilistic algorithms: inverse problems, the Radon transform, limited-angle reconstruction/ill-conditioned problems, singular-value decomposition, wavelets, regularization, filtered back-projection, Fourier methods, expectation-maximization/maximum-likelihood (EM/ML) algorithms, Bayesian approaches; use of *a priori* knowledge and heuristic assumptions, Gibbs priors, maximum entropy, and closure phase.

Corrections for attenuation, scatter and dead time.

Noise smoothing: running mean and median filters, and matched filters.

Edge sharpening: Wiener and Metz filters, unsharp masking, contrast enhancement, and motion deblurring.

Multimodality image registration and fusion.

Weighted aggregation of images formed using primary and scattered radiation.

Nonlinear operations: background subtraction and histogram equalization.

Computer simulation of algorithms.

Image recording and distribution involve the mathematical algorithms for image compression (i.e., reduction of redundant information in images that are to be stored, networked, or transmitted), image indexing schemes and search strategies for rapid retrieval, and network architectures and strategies for controlling access to and sharing of stored images, software, and facilities.

Relevant topics include:

Data compression: orthogonal transforms of Walsh, Hadamard, and Karhunen-Loève; Huffman and Q-coding; wavelet analysis; and fractals.

Storage media: film, magnetic tape/disks, optical disks, CD/ROM, and DVD/R.

Entropy and information theory, error correction, and packet switching.

Electrical, optical, and microwave links and interfaces.

Networks and protocols.

Image formats and standards, and documentation.

Picture archiving and communications systems (PACS).

Neural networks and associative memories.

Image display/visualization involves strategies for display of images of real objects and dynamic processes or of hypothetical objects and processes obtained from computer simulations based on mathematical models, as well as the graphical display of multiparameter data sets representing complex phenomena, (e.g., multimodality image fusion).

Relevant topics include:

Physical modeling; surface and volume rendering; depth coding; shape, texture, shading, and motion/animation; perspective; and parallax.

3D display via stereoscopy and holography (static and dynamic).

Virtual reality.

Dynamic range of display: black-white, pseudocolor, and color.

Conventional CRT, HDTV, touch-screen, and liquid crystal displays.

Hard copy: impact, ink jet, and laser printers; lithography; xerography; and facsimile.

Graphics and imaging workstations.

Interactive display of multimodality images, registered and fused.

Image observation/human vision involves the response characteristics of the eye–brain system and measures of the ability of the observer to perform visual tasks, as well as strategies for the development of interactive, analytic, diagnostic, and adaptive software to facilitate observer performance, based on knowledge of human vision.

Relevant topics include:

Spatiotemporal response of the human visual system, photopic and scotopic vision, and spectral sensitivity.

Visual-attentional and perceptual biases; and contrast, color, and depth perception.

Visual cognition and memory, fading of stabilized retinal images, visual masking, and visual illusions.

Image interpretation and understanding.

Mathematical modeling of the human visual system.

Observer performance/receiver operating characteristics (ROC) analysis, which may also be used as a criterion for image evaluation.

Image analysis involves strategies for the extraction of qualitative and quantitative information about an object from its image(s) and the automation of these strategies to assist or to replace the observer.

Relevant topics include:

Signal-detection theory and statistical-decision theory.

The ideal observer.

Extraction of information: edge detection; perimeter, distance, area, and volume measures; segmentation; region growing; shape from shading, texture, and motion; morphologic analysis; pattern recognition; correlation; feature/factor extraction; multivariate analysis and discriminant functions; moments; derived parametric/functional images; object recognition; image understanding; expert systems and artificial intelligence schemes; neural nets; computer vision; automated image analysis; and robotics.

Image interpretation generally involves a verbal description of the meaning and the significance of an image or set of images. To be most useful, this involves knowledge of the method used in the production of the image(s) and may make use of quantitative measures of image features as well as subjective impressions.

Relevant topics include:

Motivation, goals, and methods for producing, analyzing and displaying the image(s).

Prior training and experience with the interpretation of similar images, including an awareness of the consequences of misinterpretation.

Strategies for incorporating information from other sources.

Image evaluation/optimization involves measures of image quality that can be used as criteria for the (goal-dependent) optimization of parameters associated with each of the previous steps. In general, the goals and the criteria used for the evaluation of image quality are different in science, medicine, education, graphic arts, advertising, and the news and entertainment media. In medicine, diagnostic accuracy is generally the most important criterion.

Relevant topics include:

Image quality measures based on: signal-to-noise ratio, correspondence between object and image (e.g., correlation; least squares), likelihood ratio, Hotelling trace, information content, complexity, accuracy and reproducibility of measures of object properties; and risk-benefit and cost-benefit ratios.

Optimization of parameters: method of steepest decent and simulated annealing.

Observer performance/ROC analysis.

V. METHODOLOGY AND EPISTEMOLOGY

Currently, much of the valuable diagnostic information derived from imaging procedures is qualitative in nature and conclusions are frequently expressed in ordinary language. The linguistic aspects of the stated goals of imaging procedures, and of the meaning and significance of the resultant images, are largely beyond the scope of this chapter, except to note that verbal communications using ordinary language (Yngve, 1996) frequently include subtle metaphorical, emotional, and rhetorical components that may or may not reflect what the speaker or writer believes to be true. The most notable exception is communication in the relatively unemotional, nonrhetorical, stable, and self-consistent terms of mathematics, the language of science. Perhaps for this reason, some have regarded mathematical modeling as the only reliable means for knowing Truth about material objects in the real, physical world. However, in addressing this point, Einstein (1921) has offered a cautionary observation with which most scientists would agree:

> So far as the laws of mathematics refer to reality, they are not certain. And so far as they are certain, they do not refer to reality.

Nevertheless, imaging methods that provide quantitative information are needed for purposes of scientific biomedical research, and recognition of this has motivated the use of mathematical models and digital computer simulations to reduce dramatically the number of experiments that need to be performed. Such approaches are used increasingly in imaging research:
- To simulate certain properties of objects, both visible and invisible.
- To model certain steps in the imaging process so as to optimize parameters associated with these steps, based on some goal-specific criterion of image quality that is expressed in mathematical terms.
- To display the numerical results as simulated images.

It is important to recognize that all mathematical models of object properties and steps in the imaging process involve the use of simplifying assumptions and idealizations and, therefore, are imperfect representations of real material objects and imaging systems. As a consequence, the same must be said of the images produced by computer simulation. In short, both images and words (including the terms employed in all mathematical models) provide only imperfect, although complementary and frequently useful, representations of material objects and their images.

VI. A VIEW OF THE FUTURE

Advances in imaging methods have resulted in an exponential rate of discovery of new knowledge of all material objects—from atoms and molecules to genes and cells, to the human body and brain, to Earth, the solar system and the distant galaxies—and are now accelerating the emergence of imaging science as an academic discipline on a par with linguistics and computer science.

The further development of imaging science, and of imaging systems optimized to achieve specific goals of the sciences and medicine, will undoubtedly accelerate the production of new knowledge in these fields. However, the dissemination of knowledge to students and the general public is, perhaps, of even greater, long-term importance. Images that are carefully selected and effectively presented can increase learning rate and retention time substantially, not only in the sciences and medicine, but also in general education.

In the future, computer terminals linked to archival storage media via high-speed networks will facilitate the sharing of images for purposes of research, education, and clinical practice. This process has begun already with the use of the World Wide Web as a global communications medium. Moreover, we must expect the increased availability of such facilities to foster the establishment of multidisciplinary teams to develop more advanced computer-aided instruction (CAI) paradigms—for example, those incorporating both visual and verbal materials and using software that is not only interactive and analytic, but also diagnostic and adaptive to the needs of individual students, thus simulating some important qualities of a private tutor.

The very power of educational tools such as these—which could be used effectively for general education by the scrupulous and the unscrupulous alike—requires that they be used thoughtfully and responsibly if the impact on our culture is to be constructive. We are all too familiar with the use of emotionally charged images for manipulating public opinion and for persuading rather than informing. When coupled with advanced software of the very sort needed for educational purposes, which would tailor the presentation of visual and verbal materials in a way that is conditioned by the responses of the individual student, the power to manipulate would be increased many-fold. As a consequence, issues concerning the ethical use of these tools need to be addressed and the public made aware of the potential for their abuse.

Education in the future must be based on recognition of the fact that both images and words may contain important information and provide means for knowing and for communicating what we know to others. However, even when they are not used rhetorically, each provides only incomplete and otherwise imperfect information, and each may therefore give false or misleading impressions. In short, both images and words may be regarded as limited, although complementary, means for education. In the future, formal education must prepare us to deal with both in an objective, insightful, critical, and ethical manner. How this can be accomplished is unquestionably one of the most important and challenging issues for future consideration (Beck, 1993a, 1994). Imaging science will not have emerged as a fully mature and separate new academic discipline until the conceptual, organizational, educational, cultural, and ethical issues it raises have been addressed.

Acknowledgments

The symposium that inspired this book was designed, in part, to update an earlier symposium entitled Fundamental Problems in Scanning (Gottschalk and Beck, 1968) held in Chicago in 1965 and, in part, as a celebration of my first 50 years at The University of Chicago. A huge measure of thanks must go to John Aarsvold and the organizing committee, who worked tirelessly to make this a highly successful international event with outstanding presentations and a very special poster session. I was especially honored to have the organizers recognize my student status by including enlarged pages of my first theoretical paper (Beck, 1961) as a poster, along with excellent contributions by other students and trainees. In short, the outpouring of recognition I received was truly overwhelming. In particular, I was deeply touched by the ways in which my work has been meaningful and useful to the presenters, and I thank them most sincerely for their kind remarks.

An appropriate acknowledgement of essential contributions to my life-long pursuit of learning, education, and research would require many pages. Let it suffice to say that, on entering The University of Chicago in 1948, I discovered an ideal environment in which to pursue my eclectic interests in the arts and sciences in an entirely self-directed manner. In 1954, I had the great good fortune to begin many years of stimulating, challenging, and rewarding research at the Argonne Cancer Research Hospital (ACRH), established by the U.S. Atomic Energy Commission to conduct research in the diagnosis and treatment of cancer, using radioactive materials and radiation beams. Dr. Leon O. Jacobson, the Founding Director of the ACRH, created an exciting and supportive environment in which I could pursue and develop my idiosyncratic interests in applied mathematics and physics. I began as a very junior member of a multidisciplinary team of talented and generous collaborators, focused on the development of imaging systems for use in biomedical research and clinical practice in nuclear medicine. In the early years, these individuals included Donald Charleston, Paul Eidelberg, Alexander Gottschalk, Paul Harper, Paul Hoffer, Lawrence Lanzl, Katherine Lathrop, Lester Skaggs, Nicholas Yasillo, Lawrence Zimmer, and others too numerous to mention. In 1976, I was given responsibility for directing this research facility, which had been renamed the Franklin McLean Memorial Research Institute (FMI). In more recent years, I have had the opportunity to work in collaborative and supportive roles with a uniquely gifted group including John Aarsvold, Francis Atkins, Chin-Tu Chen, Malcolm Cooper, Donald Gunter, Chien-Min Kao, Patrick LaReviere, David Levin, Kenneth Matthews, John Metz, Robert Mintzer, Jogeshwar Mukherjee, Caesar Ordonez, Xiaochuan Pan, Benjamin Tsui, Miles Wernick, Chung-Wu Wu, Nicholas Yasillo, and many, many others, both within the university and beyond. Each of these individuals provided unique insights and expertise that made it possible for me to do whatever I have done. In addition, much of this would have been impossible without the departmental and divisional support of Robert Moseley, Robert Uretz, Donald King, and the multidivisional and institutional support of John Wilson, Stuart Rice, Walter Massey, Albert Crewe, Hanna Gray, Fred Stafford, Alan Schriesheim, Jack Kahn, Barbara Stafford and a host of others who contributed to the founding of the UC/ANL Center for Imaging Science and to the implementation of its various programs and projects. I am deeply grateful to all of these individuals for enriching my life, both professionally and personally.

Special thanks for support of my research and development activities must go to the federal funding agencies (U.S. AEC/ERDA/DOE, NIH), the state of Illinois, the Keck, Maurice Goldblatt, and Brain Research Foundations, Kiwanis International, the Clinton Frank family, as well as Siemens and IBM Corporations.

Finally, Miles Wernick and John Aarsvold deserve special recognition and sincere thanks for their willingness to edit this volume.

References

Aristotle. (1941). Metaphysica. *In* "The Basic Works of Aristotle" (R. McKeon, ed. W. D. Ross, trans.) Random House, New York.

Beck R. N. (1961). A theoretical evaluation of brain scanning systems. *J. Nucl. Med.* **2**: 314–324.

Beck R. N. (1993a). Issues of imaging science for future consideration. *Proc. Natl. Acad. Sci.* **90**: 9803–9807.

Beck R. N. (1993b). Overview of imaging science. *Proc. Natl. Acad. Sci.* **90**: 9746–9750.

Beck R. N. (1993c). Tying science and technology together in medical imaging. *In* "AAPM Monograph No. 22" (W. R. Hendee and J. H. Trueblood, eds.), pp. 643–665. Medical Physics Publishing. Madison, WI.

Beck R. N. (1994). The future of imaging science. *In* "Advances in Visual Semiotics: The Semiotic Web" (Thomas Sebeok and Jean Umiker-Sebeok, eds), pp. 609–642. Walter de Gruyter, Berlin.

Duke-Elder, S. (1958). "System of Ophthalmology, Vol. I, The Eye in Evolution." C. V. Mosby, St. Louis.

Einstein, A. (1921). Geometry and experience. An address to the Prussian Academy of Sciences, Berlin.

Galileo (1610/1989). "Sidereus Nuncius or Siderial Messenger" (A. Van Helden, trans.). University of Chicago Press, Chicago.

Gottschalk, A., and Beck, R. N. (eds.) (1968). "Fundamental Problems in Scanning." Charles C. Thomas, Springfield, IL.

Holden, C. (1998). No last word on language origins. *Science* **282**: 1455.

Hooke, R. (1665/1987). "Micrographia" (History of Microscopy Series). Science Heritage Ltd. Lincolnwood, IL.

Kepler, J. (1604/2000). "Ad Vitellionem Paralipomena, quibus Astronomiae pars Optica Traditur." (William H. Donahue, trans). Green Lion Press, Santa Fe, NM.

Lindberg, D. C. (1976). "Theories of Vision from Al-Kindi to Kepler." University of Chicago Press, Chicago.

Loevinger, L. (1995). The paradox of knowledge. *The Skeptical Inquirer*, Sept/Oct. **19**: 18–21.

Maxwell, J. C. (1873). "Treatise on Electricity and Magnetism 2 Volumes," Clarendon, Oxford (reprinted, Dover, 1965).

Park, D. (1997). "The Fire within the Eye." Princeton University Press, Princeton, NJ.

Polyak, S. L. (1957). "The Vertebrate Visual System." University of Chicago Press, Chicago.

Roentgen, W. C. (1934). On a new kind of ray, in Otto Glasser, "Wilhelm Conrad Roentgen and the Early History of the Roentgen Rays" (Otto Glasser, ed.). Charles C. Thomas, Springfield, IL.

Strickberger, M. W. (1995). "Evolution," 2nd ed. Jones & Bartlett, Sudbury, MA.

Yngve, V. H. (1996). "From Grammar to Science: New Foundations for General Linguistics." John Benjamins, Philadelphia.

CHAPTER 2

Introduction to Emission Tomography

MILES N. WERNICK* and JOHN N. AARSVOLD[†]

*Departments of Electrical and Computer Engineering and Biomedical Engineering,
Illinois Institute of Technology, and Predictek, LLC, Chicago, Illinois
[†]Atlanta VA Medical Center, Decatur, Georgia

I. What Is Emission Tomography?
II. The Making of an Emission Tomography Image
III. Types of Data Acquisition: Static, Dynamic, Gated, and List Mode
IV. Cross-Sectional Images
V. Radiopharmaceuticals and Their Applications
VI. Developments in Emission Tomography

I. WHAT IS EMISSION TOMOGRAPHY?

Emission tomography (ET)[1] is a branch of medical imaging that encompasses two main techniques—positron emission tomography (PET) and single-photon emission computed tomography (SPECT)[2]—which use radioactive materials to image properties of the body's physiology. For example, ET images can represent the spatial distribution of properties such as glucose metabolism, blood flow, and receptor concentrations. Thus, ET can be used to detect tumors, locate areas of the heart affected by coronary artery disease, and identify brain regions influenced by drugs.

ET is categorized as a *functional imaging* approach to distinguish it from methods such as X-ray computed tomography (CT) that principally depict the body's architectural structure (anatomy). PET and CT images of the same patient, shown in Figure 11–7, illustrate the complementary nature of CT's anatomical depiction of the body and PET's functional representation.

As the term *emission tomography* suggests, this form of imaging is a marriage of two basic principles: imaging through the use of gamma-ray *emission* (called the *tracer principle*) and volumetric imaging of the body's interior (called *tomography*).[3] We introduce these fundamental concepts in the following sections.

A. The Tracer Principle

ET is founded on an important insight, known as the tracer principle, which was developed in the early 1900s by

---

[1] *Emission tomography* is also known as *emission computed tomography* (ECT) and is a subset of the field known generally as *nuclear medicine*. Today, most nuclear medicine imaging studies are tomographic; hence, the terms emission tomography and nuclear medicine are often used interchangeably. However, nuclear medicine also encompasses planar emission imaging and therapeutic uses of radioactive compounds.

[2] In Europe, SPECT is often referred to as *single-photon emission tomography* (SPET).

[3] In this book, we use the term *emission tomography* to refer specifically to methods based on gamma-ray emission, but the term can also be used to describe any tomographic method that is based on radiation emitted from within an object. Two alternative categories of tomographic imaging methods are those that use *reflection* of radiation from within the object (as in ultrasound imaging) and *transmission* of radiation through the object (as in X-ray CT).

George de Hevesy, who received the Nobel Prize in Chemistry in 1943. The tracer principle is based on the fact that radioactive compounds participate in an organism's physiological processes in the same way as nonradioactive materials. Because radioactive materials can be detected by way of their emission of gamma rays, these materials can be used to track the flow and distribution of important substances in the body.

In de Hevesy's early work, he placed bean plants in a solution containing lead salts, partly composed of a radioactive isotope of lead. He found that, after the plants absorbed the solution, the distribution of lead within each portion of the plant could be deduced from radioactivity readings taken with a Geiger counter. De Hevesy used the local concentration of radioactive lead atoms in the plant as a representative measure of the local concentration of stable (nonradioactive) lead. He found that this approach to measuring chemical concentrations was more sensitive and accurate than conventional chemical analysis techniques. In his experiments, he discovered that the lead concentration remained greater in the roots than in the leaves, shedding light on the physiology of bean plants.

The use of radioactive materials as a representative marker, or tracer, of natural nonradioactive substances is the foundation of present-day PET and SPECT. However, since its initial discovery, the tracer principle has been greatly enhanced by the development of artificially radioactive compounds. Imaging agents, called *radiopharmaceuticals* or *radiotracers*, can be designed to act as markers for a great variety of substances that participate in the body's natural processes; thus, ET can yield valuable diagnostic information.

As illustrated by de Hevesy's bean plant experiments, the tracer principle offers two powerful benefits as a basis for imaging biological processes in humans. First, because one can readily detect even minute quantities of radioactive material, the tracer principle can be used to measure molecular concentrations with tremendous sensitivity. Second, tracer measurements are noninvasive—the concentration of tracer is deduced from counts of gamma rays (high-energy photons) emitted from within the body as the result of radioactive decay of the administered tracer.

To illustrate how the tracer principle is used today, let us introduce an important radiopharmaceutical, ^{18}F-fluorodeoxyglucose (^{18}F-FDG). ^{18}F-FDG, which is a tracer for measuring glucose metabolism, consists of two components (1) FDG (an analog of glucose) and (2) a fluorine-18 label (^{18}F) that permits us to detect the tracer by counting the gamma-ray emissions it produces. ^{18}F-FDG enters cells in the same way as glucose, but is metabolized by the cell to create a new compound (a metabolite) that remains trapped within the cell. Therefore, the concentration of the radioactive metabolite grows with time in proportion to the cell's glucose metabolic rate. In this way, injection of ^{18}F-FDG

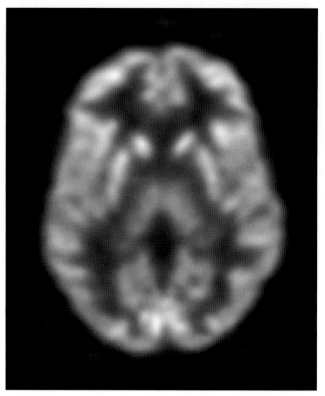

FIGURE 1 Transaxial slice through a PET image of the brain, showing regional glucose metabolism related to the level of neuronal activation. The brightness at each point in the image is proportional to the concentration of the radiotracer. (Image courtesy of MIICRO, Inc.)

into the body allows us to form images depicting local levels of glucose metabolism.

^{18}F-FDG is a valuable tool for brain imaging (Fig. 1) because glucose metabolism is related to the regional level of neuronal activation. In recent years, ^{18}F-FDG has also become an extremely important agent for cancer imaging because malignant tumor cells can exhibit greatly increased glucose metabolism in comparison with normal cells. Thus, malignant tumors are highlighted in ET images as bright regions against the relatively dark background of surrounding normal tissues (Fig. 2).

Let us now turn to the second important principle behind ET—tomography.

B. Tomography

Unlike visible light, gamma rays can pass through the body with relative ease. Thus, if our eyes were sensitive to gamma rays, the body would appear to us to be translucent but dark, whereas the radiotracer would appear to be a glowing translucent substance. Upon injection of the radiotracer, we would see the material quickly moving throughout the bloodstream and then concentrating in some tissues, making these tissues appear brighter than others.

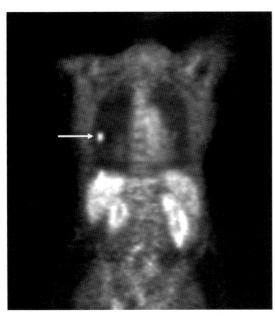

FIGURE 2 Portion of a PET image, obtained using ^{18}F-FDG, showing metastatic lesion as a bright spot in the lung (indicated by arrow). (Adapted from images courtesy of the Ahmanson Biological Imaging Center, University of California, Los Angeles.)

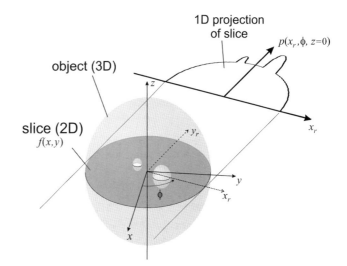

FIGURE 3 One-dimensional parallel projection of a two-dimensional slice through an object.

In ET, there are two principal means of visualizing this radiotracer distribution: projection imaging and tomography. Roughly speaking, a projection is an image as seen by an observer who is located outside the body and looking *through* it. In contrast, a tomographic image is a picture of a cross-sectional slice of the body (the word *tomography* combines the Greek words for "slice" and "drawing"). Such an image cannot be physically realized except by making a planar cut through the body and then looking at the exposed surface. The magic of tomography is that it allows us to visualize the inside of the body from this unique and valuable perspective without the need for surgery.

In ET, the data recorded by the imaging hardware are in the form of projections. Tomographic images cannot be directly observed but rather are calculated by computer from the projection measurements in a process called *image reconstruction*, which we return to later in this chapter and then in greater depth in Kinahan *et al.* (Chapter 20 in this volume) and Lalush and Wernick (Chapter 21 in this volume).

To better understand projection images and tomographic images, and the relationship between them, let us begin by introducing some notation. For simplicity, let us consider only the portion of the body lying within a particular slice plane. We identify points within this plane using two frames of reference (see Fig. 3). The first is the stationary frame of reference of the patient, which is described by coordinates (x, y). The second is a rotating frame of reference, described by coordinates (x_r, y_r) and the rotation angle ϕ. This second frame of reference is that of our hypothetical observer, who is permitted to move about the patient in order to view the body's interior from various perspectives. Finally, let us define $f(x, y)$ as the distribution of radiotracer within the slice of the body defined by the xy plane.

Let us now introduce an important type of projection, called the *parallel projection*, which is defined as the line integral of $f(x, y)$ along the y_r axis when viewed at angle ϕ, that is, $p(x_r, \phi) = \int f(x, y) dy_r$ (see Fig. 3). At a given angle ϕ, the projection is the integral of the radiotracer distribution along lines of sight parallel to the y_r axis.

In an ET study, measurements of the projections are obtained from many perspectives about the body (i.e., for many values of ϕ). The aim of tomography is to infer the radiotracer distribution $f(x, y)$ from these projection data. In all modern ET imaging systems, projections are acquired for many different slice planes simultaneously (i.e., many values of the z coordinate in Fig. 3), thus allowing a three-dimensional (3D) radiotracer distribution $f(x, y, z)$ to be constructed. In fact, most imaging systems collect projection data for lines of sight not confined to these parallel planes. To make the best use of this more general class of data, *fully 3D* image-reconstruction methods have been developed that treat the body as a single 3D object rather than a stack of 2D slices.

II. THE MAKING OF AN EMISSION TOMOGRAPHY IMAGE

The most common application of ET is the diagnosis of disease in humans. The main steps of a clinical ET imaging study are outlined in Figure 4.

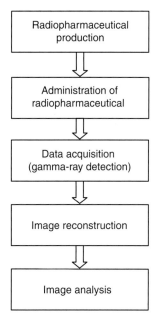

FIGURE 4 Key steps in an emission tomography study.

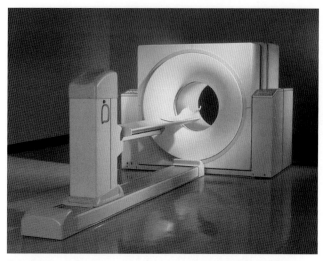

FIGURE 5 PET scanner. The patient is placed inside the aperture of the scanner and imaged by detector hardware contained within the surrounding housing. (Photo courtesy of CPS Innovations.)

Radiopharmaceutical production. The first step in an ET study is the production of a radiopharmaceutical that is appropriate for imaging the disease in question. Details of radiopharmaceuticals and their production are explained in Levin (Chapter 4 in this volume) and Mukherjee *et al.* (Chapter 5 in this volume).

Administration of radiopharmaceutical. Next, the radiopharmaceutical is introduced into the body, usually by injection but sometimes by inhalation. The quality of the image increases with the amount of administered radiopharmaceutical; however, radiation exposure of internal organs imposes a safety limit on the dose.

Data acquisition. The time between the administration of the radiopharmaceutical and the beginning of data acquisition depends on the purpose of the imaging study and the nature of the tracer. In some studies, data acquisition begins immediately; in others, it can begin hours or days after administration of the radiotracer. Data acquisition takes place while the patient lies still on a bed. The radioactive isotope with which the radiopharmaceutical has been labeled emits gamma rays as a product of radioactive decay. These emissions emanate in all directions from within the patient's body at a rate proportional to the local concentration of radiopharmaceutical. For example, many more gamma rays are emitted from a tumor being imaged with ^{18}F-FDG than from surrounding tissues if the tumor exhibits elevated glucose metabolism. As the gamma rays emanate from the patient, they are detected and recorded by imaging hardware, such as the PET system shown in Figure 5, which surrounds or revolves around the patient bed. Positional and directional information about each gamma ray are measured, and the results are tallied into a histogram of discrete position-direction bins. The resulting histogram bins contain measurements of the projections.

Image reconstruction. In the image-reconstruction step, the projection data acquired by the imaging system are used to estimate the desired tomographic images. The conventional method of image reconstruction is a mathematical technique called *filtered backprojection* (FBP). Analytic approaches (such as FBP), which compute the image directly, are explained in detail in Kinahan *et al.* (Chapter 20 in this volume). In recent years iterative techniques have emerged, such as the expectation-maximization (EM) algorithm and ordered-subsets EM (OS-EM) algorithm. These methods, which are based predominantly on statistical estimation procedures such as maximum-likelihood (ML) and maximum *a posteriori* (MAP) estimation, are described in Lalush and Wernick (Chapter 21 in this volume).

Image analysis. Image analysis in medicine traditionally consisted entirely of visual inspection and evaluation of images by physicians, but this has changed in recent years. Computerized analysis of medical images now plays a significant role in modern medicine, providing the physician with supplementary information that can help guide the decision-making process. For example, computers can be used to estimate parameters of the dynamics of a radiopharmaceutical distribution, as explained in Morris *et al.* (Chapter 23 in this volume), and can provide analyses and advanced visualization of cardiac function based on ET images, as described in Garcia (Chapter 24 in this volume).

In addition to covering these key steps in the imaging process, this book will briefly introduce the important issues of image quality (discussed in Kao *et al.*, Chapter 6, and Kinahan *et al.*, Chapter 20 in this volume) and the simulation of imaging data (Ljungberg, Chapter 25 in this volume), both of which are critical to the development and assessment of hardware components, system designs, and reconstruction algorithms.

The following sections provide a brief overview of the two main ET techniques: PET and SPECT. These sections are intended as an introduction to the detailed treatments provided in Zeng (Chapter 7) and Lewellyn and Karp (Chapter 10 in this volume).

A. Single-Photon Emission Computed Tomography

PET and SPECT are distinguished mainly by the type of radioisotope incorporated in the tracer. SPECT studies use radiopharmaceuticals labeled with a single-photon emitter, a radioisotope that emits one gamma-ray photon with each radioactive decay event. PET requires the labeling isotope to be a positron emitter. Upon decay, a positron-emitting isotope ejects a positron (anti-electron) from its nucleus. When the positron encounters an electron in the surrounding medium, the two particles mutually annihilate, resulting in the release of two photons. This distinction between the numbers of emitted photons dictates the type of hardware required to detect and localize each event. We begin with a description of the imaging approach used by SPECT.

Recall that a SPECT study, like any ET study, begins with the administration of a radiopharmaceutical, usually by injection. Next, the imaging hardware begins detecting and recording gamma rays emitted by the radiopharmaceutical as a product of radioactive decay. Because of safety limitations on the amount of radiopharmaceutical that can be administered to the patient, the rate of gamma-ray emissions is relatively low (typically, ~10^4 counts/s/ml of a tissue of interest in a clinical SPECT study). Therefore an ET study requires a relatively long period of time for data collection. For example, a SPECT cardiac study takes 15–20 min, whereas a modern X-ray CT scan can be completed in seconds.

1. Collimators

To form an image of the radiopharmaceutical distribution, we must be able to create a correspondence between points in the object and points in the image. In an ordinary photographic camera, this is accomplished by a lens, which diverts light rays by refraction to form an image on the film or detector array. Gamma rays are too energetic to be imaged with a conventional lens; instead, the image is formed by a component called a *collimator*, which is a thick

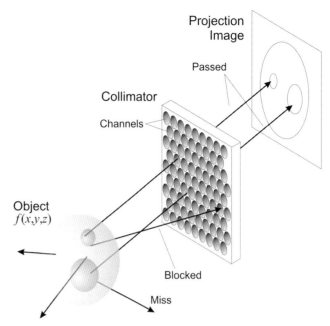

FIGURE 6 A collimator forms an image of the object by passing only rays traveling in a particular direction.

sheet of a heavy material (such as lead), perforated like a honeycomb by long thin channels (Fig. 6).

Rather than diverting the rays like a lens, the collimator forms an image by selecting only the rays traveling in (or nearly in) a specific direction, namely the direction in which the channels are oriented. Gamma rays traveling in other directions are either blocked by the channel walls or miss the collimator entirely (see Fig. 6). The collimator illustrated in Figure 6 is a *parallel-hole collimator*, in which all the channels are arranged parallel to one another. When the normal to the face of the collimator is oriented in the ϕ direction, the image formed is composed of the projections defined earlier. For a number of reasons that we discuss shortly, the projections formed by a collimator are not exactly line integrals of the object distribution; however, the general idea of SPECT imaging can be understood in terms of the simple projection. To learn more about collimators and their characteristics, the reader is referred to Gunter (Chapter 8 in this volume).

2. Detection of Gamma Rays

So far we have shown how the collimator selects the gamma rays that form an image; however, we have not yet discussed how these selected gamma rays are detected. The conventional *gamma camera*, sometimes called the *Anger camera* (after its inventor, Hal Anger, 1958), incorporates a system of detection hardware on the opposite side of the collimator from the object, as shown in Figure 7.

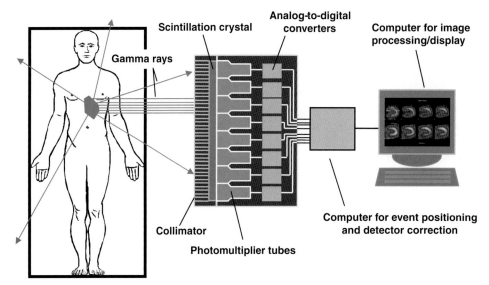

FIGURE 7 Schematic diagram of a conventional gamma camera used in SPECT. The collimator forms an image of the patient on the scintillation crystal, which converts gamma rays into light. The light is detected by photomultiplier tubes, the outputs of which are digitized and used to compute the spatial coordinates of each gamma-ray event (with respect to the camera face). A computer is used to process, store, and display the images.

A gamma ray that successfully traverses the collimator strikes the first element of the detection apparatus, a piece of crystal material called a *scintillator*. The scintillator uses energy from the high-energy gamma ray to produce many optical-wavelength photons. These photons are detected by a collection of photomultiplier tubes (PMTs), devices based on the photoelectric effect. From a single photoelectron, a PMT can produce a cascade of electrons, which yields a measurable electrical current. This current is sensed by accompanying electronics, which register the occurrence of an event. Relative readings from PMTs near the point of impact of the gamma ray are used to compute two-dimensional spatial coordinates of the gamma-ray event relative to the face of the camera. The events are tallied in a histogram based on their position. At the end of the imaging period, after many events have been tallied, the resulting histogram represents a projection image of the object. The measured projection image is noticeably noisy, meaning that it is corrupted by random fluctuations that reduce image fidelity. (We explain the principal source of this noise shortly.)

The projection images needed for image reconstruction can be obtained by repeatedly imaging the patient from many points of view by positioning the gamma camera at many orientations about the patient. Current SPECT systems accelerate this process and capture a greater number of emitted gamma rays by positioning two or three cameras at once about the patient so as to measure multiple projection images simultaneously. The specifics of SPECT system designs, including multiple-camera systems, are explored in Zeng (Chapter 7 in this volume).

3. Factors Influencing SPECT Image Reconstruction

As we have hinted, SPECT projection images are not precisely described by the idealized parallel projections we introduced earlier. Therefore, they cannot be properly reconstructed if the data are assumed to be simple line integrals of the object. In particular, three principal factors must be incorporated in an accurate description of the SPECT imaging process:

Attenuation. Gamma rays can be absorbed within the object or scattered outside the field of view and thus lost to the imaging process. Gamma rays emitted from deep within the object have a greater chance of being absorbed, so the effect is depth-dependent. If not accounted for, attenuation causes the reconstructed image to appear dark in the interior of the object.

Scatter. The term *scattering* refers to a category of interactions that can take place between gamma rays and the matter composing the patient's body. Scatter interactions cause a deflection of the radiation, which can cause a misleading indication of each gamma ray's initial propagation direction. This is manifested as a nonlinear blurring effect in the image.

Depth-dependent blur. If the channels of a collimator were infinitely long and thin, each would pass rays traveling only in a single direction, thus approximating the line-integral model of projection imaging. However, a real collimator channel has a finite spatial extent and sees a cone-like region of the object emanating from the channel. This is manifested as a blurring of the image, which is most severe for portions of the object that are farthest from the collimator.

These factors and methods for accounting for them are explained in King *et al.* (Chapter 22 in this volume).

4. Noise in Emission Tomography Data

In addition to the factors just mentioned, all ET data are corrupted by noise, which is usually the factor that most limits the image quality obtainable by ET. To understand noise in ET, we must look more closely at the detection process.

Gamma rays are photons and thus are governed by the laws of quantum physics and the randomness that this entails. Therefore, to be precise, we should think of the projection image at the face of the gamma camera as a probability density function governing the photon detection process. Specifically, the probability that any given event will take place within a given region of the image (e.g., a pixel) is proportional to the average rate of gamma-ray arrivals within that region.[4] Thus, on average, more events will be recorded in bright areas of the projection image than in dark areas during any given time interval.

However, the actual number of events recorded within a region of the image during any fixed time interval is a random number. Therefore, if we were to image the same object again and again, we would each time obtain a slightly different result. This variation, which is manifested as a speckled appearance within each projection image, is called *noise*. The noise in ET images is called *photon noise* because it arises from randomness inherent in the photon-counting process. Photon noise is also called *Poisson noise*, because the number of events recorded in any fixed interval of time obeys the well-known Poisson probability distribution, which is commonly associated with counting processes.[5]

B. Positron Emission Tomography

Like SPECT, PET aims to measure the (x_r, ϕ) (sinogram) coordinates of gamma rays emitted from the body in preparation for the subsequent image-reconstruction step. PET differs from SPECT principally in that it uses positron-emitting radioisotopes instead of single-photon emitters. As explained in Lewellyn and Karp (Chapter 10 in this volume), the decay of each nucleus of a positron-emitting isotope leads to the emission of two gamma rays (Chapter 10, Figure 2) that travel in nearly opposite directions from one another. This feature of PET allows the gamma-ray coordinates to be measured without a physical collimator by a scheme called *electronic collimation*, which works as follows.

---

[4] This assumes that the camera has uniform spatial sensitivity.

[5] If one were instead to specify in advance the total number of gamma rays to be detected and stop counting once that number was reached, then the numbers of detected events in the pixels of the projection image would obey a multinomial distribution, and not a Poisson distribution.

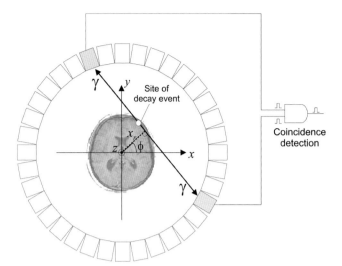

FIGURE 8 Schematic diagram of electronic collimation (coincidence detection) in PET. Two gamma rays emitted as a result of positron annihilation are sensed at two detectors at roughly the same instant. Thus, it can be inferred that a decay event occurred along the line segment connecting the participating detectors. This line segment is described by coordinates (x_r, ϕ).

The patient is surrounded by a collection of gamma-ray detectors, which are connected to circuitry that senses the timing of the gamma-ray detections. The (x_r, ϕ) coordinates of each decay event are inferred by using the fact that positron emitters yield pairs of gamma rays and that the gamma rays constituting each pair travel in nearly opposite directions. When two gamma rays are detected roughly simultaneously, it is inferred that these two gamma rays must have resulted from the same decay event. Thus, it is assumed that the originating decay event must have taken place somewhere along the line segment connecting the two participating detectors. This line segment defines the (x_r, ϕ) coordinates, as shown in Figure 8.

Although PET generally produces better images than SPECT, SPECT remains the more widely available imaging method. SPECT radiopharmaceuticals are easier and less costly to produce, are based on longer-lived isotopes and use a wider variety of isotopes and tracer compounds. SPECT systems are also less expensive than PET systems. However, the number of PET systems in use has been growing rapidly in recent years, as has the number of local production facilities providing PET radiopharmaceuticals. PET is by far the preferred technique for research applications.

1. Factors Influencing PET Image Reconstruction

PET projection data are usually better represented by the simple parallel projection model than are SPECT data, but PET also has several physical factors that complicate the imaging model. These are explained in Lewellyn and Karp (Chapter 10 in this volume), and are illustrated graphically in Figures 3 and 4 of that chapter. As in SPECT, PET data

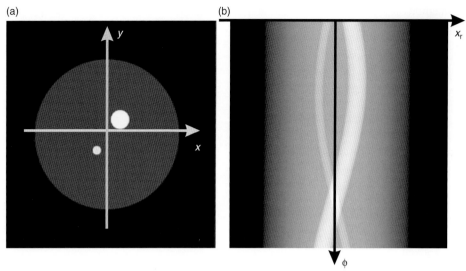

FIGURE 9 (a) Image of object slice $f(x, y)$ and (b) sinogram $p(x_r, \phi)$ of this slice.

are affected by attenuation, scatter, and several mechanisms of blurring. These factors are generally easier to correct in PET than in SPECT.

C. Image Reconstruction

After the gamma-ray data have been acquired, the next step is to compute, or *reconstruct*, images from these data. We introduce image reconstruction by briefly describing the classical approach to the problem, in which 2D slice images are computed one by one to form a 3D picture of the body.

1. Method of Filtered Backprojection

Conventional approaches to image reconstruction from PET and SPECT data are based on the method of FBP. FBP is a mathematical technique based on an idealized model of PET and SPECT data that ignores many significant features of real data. Specifically, FBP assumes that the number of detected gamma-ray events traveling along a particular direction approximates an integral of the radiotracer distribution along that line, that is, the parallel projection $p(x_r, \phi)$ defined earlier. In spite of its approximate nature, FBP has enjoyed widespread use and great longevity, largely because of its computational simplicity.

The reader should bear in mind that in introducing FBP here, we neglect all the important effects that FBP fails to consider, such as noise, attenuation, scatter, and blur. Suboptimal, but reasonable, results can be obtained in practice using FBP, even if attenuation, scatter, and blur are not accounted for. (Indeed, clinical images are often made in this manner.) However, noise must always be accounted for in some way, and this is usually achieved in FBP by smoothing the projections prior to reconstruction or by smoothing the image afterward.

To illustrate the workings of the FBP algorithm, let us suppose we wish to image the 2D slice of the simple object shown in Figure 3. The figure illustrates the parallel projection of this slice at angle ϕ. To form a complete data set, PET and SPECT systems measure projections from many points of view about the object (i.e., projections for many values of angle ϕ). Our particular slice, which is shown as an image in Figure 9a, yields a set of projections $p(x_r, \phi)$ that can itself be viewed as a 2D image (Fig. 9b) with dimensions and x_r and ϕ. This image is often referred to as a *sinogram* because a point source traces out a sinusoidal path through this diagram, as explained in Kinahan *et al.* (Chapter 20 in this volume).

To turn the sinogram into an image, the FBP algorithm uses a procedure called *backprojection*, which consists of smearing each projection back into the object region along the direction ϕ in which it was measured. Figure 10a shows the process of idealized forward projection, and Figure 10b shows the process of backprojection for a single angle ϕ. By backprojecting the projection for every angle into the image region and adding together the results of all these smearing operations, one can obtain a reasonably good representation of the original object, but the result will be somewhat blurry. The mathematical analysis given in Kinahan *et al.* (Chapter 20 in this volume) tells us why. It turns out that backprojection is nearly the correct way to produce the image from the sinogram; however, backprojection alone produces an image that is blurred with a point-spread (blurring) function $1/r$, where $r = \sqrt{x^2 + y^2}$. Thus, the FBP algorithm involves the application of a sharpening operation to the projections that exactly cancels the blurring effect. This sharpening operation has the form of a 1D linear ramp filter in the Fourier domain ($|v_{xr}|$, where v_{xr} is the spatial frequency variable corresponding to x_r), as explained in Kinahan *et al.* (Chapter 20 in this volume).

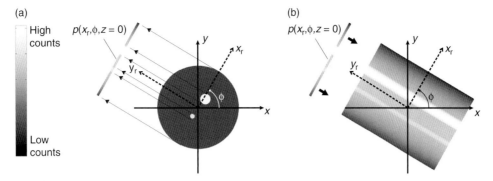

FIGURE 10 (a) Idealized forward projection of image slice for a particular angle and (b) backprojection for the same angle. Bright areas of the image and projection denote high count rates, as indicated by the color bar at far left. Actual PET and SPECT data are not well described by this simplified projection model because of attenuation, blur, scatter, and noise; however, the model does capture the essential process of forming an image from projection data. With proper corrections, the filtered backprojection method produces reasonably good images in spite of its adherence to an idealized model of the imaging process.

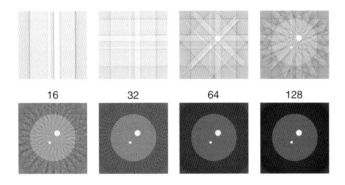

FIGURE 11 Images reconstructed by filtered backprojection from different numbers of equally spaced projections. The number of views included in each reconstruction is shown above its image. The original object is shown in Figure 9.

Figure 11 shows the backprojection of noise-free projections (of the object in Fig. 9) that have been sharpened with a ramp filter. As the number of projections (number of viewing angles) increases, the fidelity of the reconstructed image improves. As shown in Kinahan *et al.* (Chapter 20 in this volume), an image slice can be exactly reconstructed by FBP, provided that the data indeed follow the parallel-projection imaging model and that sufficiently many projections are acquired.

In the presence of noise, the FBP algorithm must be slightly modified by the introduction of a smoothing step. This smoothing can be implemented either as a 1D filter applied to the projections prior to backprojection, or as a 2D filter applied to the image after backprojection. When applied to the projections, the smoothing filter is usually combined with the ramp filter, because the filters are linear. Several well-known filter types are introduced in Kao *et al.* (Chapter 6 in this volume).

2. Analytic vs. Iterative Image Reconstruction

The method of FBP is an example of a category of image reconstruction approaches referred to in this book as *analytic methods* (Kinahan *et al.*, Chapter 20 in this volume) to distinguish them from *iterative methods* (Lalush and Wernick, Chapter 21 in this volume). Analytic methods typically neglect noise and complicating physical factors in an effort to obtain frameworks that yield explicit inversion formulas for the reconstruction problem. Analytic methods usually produce solutions that are relatively practical to compute and provide insight about data-acquisition issues such as sampling.

Iterative methods, which are primarily used to implement statistical estimation approaches, allow for explicit inclusion of realistic factors in the reconstruction process. These factors include noise (due to randomness in the numbers of detected gamma-ray counts) and many physical features of the imaging process such as detector response characteristics, scattering, and attenuation of gamma-ray emissions.

D. Image Analysis

In many applications, clinicians make diagnostic and prognostic conclusions about their patients by visually interpreting PET or SPECT images produced by the reconstruction step. In these cases, the computer is used to acquire the data and reconstruct the images, but the rest of the image analysis is done by the physician.

However, in recent years, the medical-imaging field has seen a rapid expansion of the computer's role in image interpretation. In its most modest role, the computer is used to produce various kinds of displays and renderings of images that make them easier to interpret visually. In its most ambitious role, the computer can be used to produce diagnostic decisions, such as the probability that a given image implies disease; however, this capability has been the slowest to be adopted in clinical practice.

An important role of the computer in image analysis is in deriving from the images various numerical parameters that may be difficult to discern by simple visual inspection. Some of these parameters are *global,* serving to summarize important information about the entire imaging study (e.g., cardiac ejection fraction and other measures of cardiac function). Other kinds of parameters are *local,* meaning they are calculated for every location (or region) in the image. These measurements can be used to produce parametric images, which depict variations in biological function throughout the imaged area of the body.

For example, the field of functional neuroimaging or brain mapping aims to identify regions of the brain that experience elevated or suppressed activity in response to external stimuli (e.g., drugs), normal brain activity (e.g., cognitive tasks), or disease. Figure 12 is a parametric image showing brain regions affected by a therapeutic drug. The colored regions, which are computed from FDG PET images through a series of image-processing and statistical calculations, are overlaid on magnetic resonance images (MRI) to provide anatomical context.

Another important example of parametric analysis of ET images is kinetic modeling (Morris *et al.*, Chapter 23 in this volume). In this approach, sets of coupled differential equations are used to describe the time evolution of a radiotracer distribution and its interaction with the body's physiology. Estimation techniques are used to determine the rate constants in these equations as well as other relevant parameters. Kinetic modeling can be used to calibrate ET images in units of blood flow, glucose metabolism, receptor concentration, or other parameters specific to a given study.

III. TYPES OF DATA ACQUISITION: STATIC, DYNAMIC, GATED, AND LIST MODE

ET studies can be categorized by the manner in which time variations of the radiotracer distribution are treated in the data-acquisition process. In a *static* study, which is analogous to a long-exposure photograph, the images represent a time average of the radiotracer distribution over the entire period of data acquisition. A static study is suitable when physiologically induced time variations in the radiotracer distribution are very slow, when observation of these variations is not a goal of the study, or when the gamma-ray count rate is so low that a long acquisition is required to achieve satisfactory image quality.

There are two principal alternatives to a static imaging study, dynamic studies and gated studies, both of which produce a sequence of images that can be viewed as a video. A *dynamic* study, which is directly analogous to an ordinary movie or video, aims to capture the radiotracer distribution as a function of time. Dynamic studies are used in visual assessments of organ function and are required for most kinds of kinetic parameter estimation.

In a *gated* study the data acquisition is synchronized to the rhythm of the patient's cardiac or breathing cycles. The most important application of gated imaging is in cardiac studies, in which the result is a short movie (usually 8–16 frames) that depicts the motion taking place during one cardiac cycle of the patient. This single heartbeat is actually a composite representation formed by averaging together the imaging data collected during a very large number of cardiac cycles. The purpose of this compositing operation is to combat the effect of noise. Each cardiac cycle is much too short (~1 s) to acquire sufficient gamma-ray counts to produce a useful image, but the composite image has acceptable quality because it is constructed from all the counts acquired during the study.

In gated cardiac imaging, an electrocardiogram (ECG) signal is used to control the synchronization of the data acquisition, as shown in Figure 13. The gamma-ray counts acquired just after the onset of each cardiac cycle are combined to form a composite of the first image frame. Similarly, each successive frame is formed by combining all the counts collected at corresponding points in the cardiac cycle.

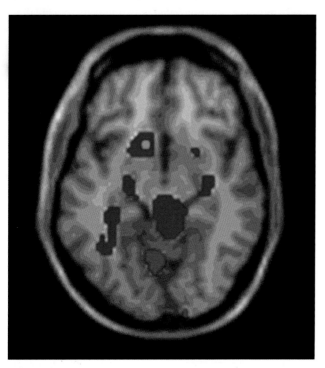

FIGURE 12 Parametric image derived from PET scans showing brain regions affected by an antidepressant drug. Blue regions show diminished glucose metabolism; red regions show elevated metabolism. (Courtesy of Predictek, LLC, and MIICRO, Inc.)

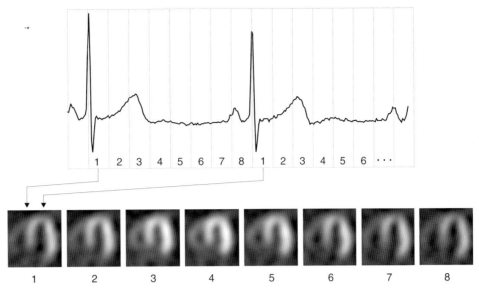

FIGURE 13 Synchronization of data acquisition to electrocardiogram signal in gated imaging. A gated image sequence, which represents one cardiac cycle, is actually a composite representation of thousands of cardiac cycles. The cardiac cycle is divided into N time intervals (frames) to obtain a video sequence of N images. The data corresponding to frame 1, for example, consist of all gamma rays collected during the time intervals labeled 1 in the figure.

In regards to time information, the most detailed mode of data collection is called list-mode acquisition. In *list mode*, the time of detection of every gamma-ray event is recorded in addition to its spatial coordinates and other parameters such as photon energy. Thus, data collected in list mode can be reformatted to obtain static, dynamic, or gated images by appropriate regrouping of the data.

IV. CROSS-SECTIONAL IMAGES

Static ET studies aim to measure the spatial distribution of the radiotracer, which can be described by a function $f(x, y, z)$, where, for the moment, let us define the (x, y, z) coordinates as follows with respect to a person who is standing. Let the x coordinate correspond to the left-right axis, the y coordinate to the forward-backward (anterior-posterior) axis, and the z coordinate to the up-down (superior-inferior) axis. Gated and dynamic studies add the extra dimension of time, thus producing 4D images, which can be denoted by $f(x, y, z, t)$. However, in practice, ET images are often viewed in the form of sets of 2D slices, owing in part to the difficulty of visualizing 3D and 4D data.

Slices coinciding with the (x, y, z) coordinate system have special names in medical imaging: *axial* (or *transaxial*) slices lie parallel to the xy plane, *sagittal* slices lie parallel to the yz plane, and *coronal* slices lie parallel to the xz plane (see Chapter 11, Figure 7). The use of these terms is relatively standard throughout medical imaging; however, the orientation in which the slice images are displayed varies across areas of medical specialization (e.g., axial slices can be viewed as if looking from the feet toward the head, or vice versa).

It should be noted that particular specialties also use their own particular coordinate systems. For example, cardiac image slices are usually oriented with reference to the long axis of the left ventricle, not the long axis of the body; and the field of functional brain mapping uses its own standardized coordinate system defined by brain landmarks.

V. RADIOPHARMACEUTICALS AND THEIR APPLICATIONS

ET imaging of a particular organ or disease begins with development of a radiopharmaceutical that will interact with the body in a way that produces an informative image. A radiopharmaceutical consists of two parts: the tracer compound (e.g., FDG) that interacts with the body and the radioactive label (e.g., ^{18}F) that allows us to image it.

Labeling isotopes fall into two broad categories: *positron emitters*, used mainly in PET, and *single-photon emitters*, used in SPECT. The most commonly used positron emitters are ^{18}F, ^{82}Rb, ^{11}C, ^{15}O, and ^{13}N. Some commonly used single-photon emitters are ^{99m}Tc, ^{201}Tl, ^{123}I, ^{131}I, ^{111}In, and ^{67}Ga. The radioactive half-life of each of

TABLE 1 Values of the Half-Life of Some Commonly Used Radioisotopes

Isotope	Half-Life ($t_{1/2}$)
^{11}C	20.4 min
^{13}N	9.96 min
^{15}O	124 s
^{18}F	110 min
^{67}Ga	78.3 h
^{82}Rb	1.25 min
^{99m}Tc	6.02 h
^{111}In	2.83 days
^{123}I	13.2 h
^{131}I	8.02 days
^{201}Tl	73.1 h

TABLE 2 Clinical PET Radiopharmaceuticals and Their Applications[a]

PET Radiopharmaceutical	Application
^{18}F FDG	Characterization, diagnosis, staging, and restaging of many forms of cancer
	Solitary pulmonary nodule assessment
	Epilepsy (refractory seizures)
	Myocardial perfusion or viability assessment
^{82}Rb RbCl	Myocardial perfusion or viability assessment

[a] Many other PET radiopharmaceuticals exist and are routinely used in research, but are not yet used widely in clinical practice. The number of radiopharmaceuticals in use in the United States is limited by eligibility for reimbursement under health insurance programs.

TABLE 3 Clinical Applications of SPECT and Radiopharmaceuticals Used in These Applications[a]

SPECT Radiopharmaceutical	Application
^{67}Ga citrate	Infection or lymphoma detection
^{111}In capromab pentetide	Prostate cancer detection
^{201}Tl TlCl	Myocardial perfusion or viability assessment
^{99m}Tc TlCl	Brain lymphoma detection
^{99m}Tc sestamibi or tetrofosmin	Myocardial perfusion or viability assessment
^{99m}Tc MDP	Metastases or fracture detection
^{99m}Tc HMPAO	Brain perfusion assessment
^{99m}Tc RBCs	Liver hemangioma detection
^{99m}Tc sulfur colloid	Liver/spleen assessment, lymphoscintigraphy
^{99m}Tc sestamibi or tetrotosmin	Parathyroid localization

[a] HMPAO, hexamethylpropylene amine oxime; MDP, methylene diphosphonate; RBCs, red blood cells.

TABLE 4 Clinical Applications of Single-Photon Planar (Projection) Imaging and Radiopharmaceuticals Used in These Applications[a]

SPECT Radiopharmaceutical	Application
^{123}I or ^{131}I MIBG	Neuroendocrine tumor detection
^{123}I NaI	Thyroid function assessment
^{123}I NaI	Thyroid function assessment
^{99m}Tc RBCs	Gastrointestinal bleed detection
^{99m}Tc sulfur colloid	Sentinel lymph node localization for melanoma and breast cancer
^{99m}Tc DTPA, MAG3	Renal function or obstruction assessment
^{99m}Tc MAA	Lung ventilation assessment
^{99m}Tc DTPA	Lung perfusion assessment
^{99m}Tc RBCs, MUGA	Left ventricular function characterization
^{133}Xe xenon gas	Lung ventilation assessment

[a] MIBG, meta-iodobenzylguanidine; DTPA, diethylenetriamine pentaacetate; MAG3, mercaptoacetyltriglycine; MAA, macroaggregated albumen; MUGA, multiple gated acquisition.

these commonly used isotopes is given in Table 1 (the *half-life* is the average time in which the nuclei of one-half of a given population of atoms will undergo radioactive decay).

The most common clinical applications of PET by far are in oncology, with neurology and cardiology accounting for most other clinical PET studies. SPECT is used routinely in a wider variety of applications. Tables 2–4 contain highly abbreviated lists of some of the more commonly used radiopharmaceuticals and their applications. Further discussion of the clinical uses of PET and SPECT can be found in Brill and Beck (Chapter 3 in this volume), and a discussion of the development of radiopharmaceuticals for brain imaging is given in Mukherjee *et al.* (Chapter 5 in this volume).

VI. DEVELOPMENTS IN EMISSION TOMOGRAPHY

Like any technology, ET is constantly evolving, with improvements taking place principally in four broad areas: radiotracers, imaging systems and hardware components, image reconstruction and analysis, and applications and protocols. These four areas are inextricably intertwined, and each depends on the others for its success. In ET, the body is imaged only indirectly by way of the measured distribution of radiotracer; therefore, it is critical that the radiotracer be designed to maximize the information its distribution conveys about the body. Imaging hardware components must be designed so as to make optimal use of the gamma rays emitted by the radiotracer, by maximizing sensitivity, count rate, spatial resolution, and other related factors. Similarly important is overall imaging system design, which involves improvements to existing imaging configurations, but also novel approaches, such as Compton imaging (Rogers *et al.*, Chapter 19 in this volume), which are not yet in clinical use. Image reconstruction seeks to make optimal use of the data acquired by the imaging instruments, whereas image analysis aims to extract useful information from the reconstructed images. Steering the entire process is the intended application, be it in the laboratory or the clinic. New ways of using ET images and new ways of structuring imaging studies are still being developed, despite the relative maturity of this technology. The rest of this book is intended to help researchers and practitioners to understand where ET began, how it works, and where the field is headed.

Acknowledgments

The authors thank Jovan G. Brankov and James R. Galt for providing material used in the figures and Robert Mintzer for many helpful comments.

References

Anger, H. O. (1958). Scintillation camera. *Rev. Sci. Instrum.* **29**, 27–33.

CHAPTER 3

Evolution of Clinical Emission Tomography

A. BERTRAND BRILL* and ROBERT N. BECK[†]
*Radiology Department, Vanderbilt University, Nashville, Tennessee
[†]Department of Radiology, The University of Chicago, Chicago, Illinois

I. Introduction
II. The Beginnings of Nuclear Medicine
III. Early Imaging Devices
IV. Evolution of Emission Tomography and Initial Applications
V. Clinical Applications
VI. Summary

I. INTRODUCTION

The practice of medicine relies on the ability to detect disease-related changes in anatomy, biochemistry, and physiology. The development of tomography—both emission and transmission—has provided physicians with an unprecedented tool for detecting such changes. We have moved from the days when a surgeon would go in and have a look to a new era in which physicians have many noninvasive diagnostic and image-guided therapeutic strategies at their command.

Imaging technology benefits patient care through its role in diagnosis, therapy, and drug development and testing. As a diagnostic tool, imaging fosters preventive medicine by helping to detect disease in its early stages when therapy is most effective. By providing accurate spatial localization of abnormal processes, tomography plays a critical role in planning surgery and radiotherapy. Nuclear medicine also contributes directly to therapy through the use of receptor-targeted densely ionizing particles that selectively kill cancer cells. Because of the high specificity of radioactive-tracer targeting, nuclear medicine is the primary modality for producing noninvasive functional images of normal and disease processes in the body.

In this chapter, we trace the beginnings of nuclear medicine imaging and recall some of the history, experience, and theoretical developments that guided the development of the technology. Then we describe many of the major clinical applications of emission tomography and discuss the history of how these uses of the technology evolved.

II. THE BEGINNINGS OF NUCLEAR MEDICINE

A. Developments Before 1945

The first application of the tracer principle was made in 1911 by Hevesy (1948). The earliest nuclear medicine investigations used naturally occurring radioactive tracers, which made it possible to amass increasingly detailed knowledge about the biochemical pathways common to all biological systems. With the development of radioactive-isotope production systems in the 1930s, a wider range of tracers became available and more applications emerged. The potential usefulness of radioactive isotopes for radiation therapy was recognized early on, and therapy was one of the earliest applications. The bridge to modern nuclear medicine and its emphasis on imaging awaited the development of imaging devices, which first appeared in the late 1940s.

The thyroid was the first organ studied by nuclear medicine techniques. It had been known since the early 1900s that thyroid hormone has high iodine content. A short-lived iodine isotope (~30-min half-life ^{128}I) was produced by Fermi (1934) in Italy using a Po/Be neutron source, and thyroid studies followed shortly thereafter. Within 2 years of the discovery of artificial radioactivity, the search was on for medical applications. A longer-lived iodine isotope was needed for clinical studies, and this requirement was satisfied in 1937 when Seaborg used the Berkeley cyclotron to search for, and find, the 8-day half-life ^{131}I, which is still in use today in many important clinical and research applications (Livingood, 1938). Hamilton (1938; Hamilton and Soley, 1939) working at Berkeley, made many of the earliest physiological studies with the newly available radioactive tracers produced on the Berkeley cyclotron. Soon thereafter, the MIT cyclotron was developed in response to strong medical initiatives.

Where indicated, thyroid therapy almost immediately followed the initial diagnostic and physiological observations. The Harvard/MIT/MGH group (Hertz and Roberts, 1942) and the Berkeley group (Hamilton and Lawrence, 1942) published the earliest accounts of the use of ^{131}I in the treatment of hyperthyroid patients. Small hand-held Geiger-Müller (GM) detectors, moved manually over the neck, produced the first crude two-dimensional numerical maps of thyroid function based on counts from ^{131}I taken up by the thyroid gland. The displays were crude, and some investigators color-coded the images to create an analog visual image that was more easily interpreted (Kakehi et al., 1962). Dot tappers, early plotting devices that showed the locations of detected events, were used routinely to record the output of clinical scanners (Andrews et al., 1962) (Fig. 1).

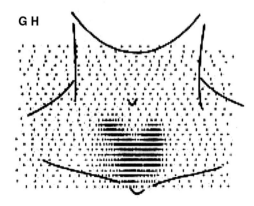

FIGURE 1 Well differentiated thyroid cancer in the right lower lobe of the thyroid. This image was recorded by a dot tapper, an early method of recording nuclear medicine images (Andrews, 1962).

B. The Next 25 Years (1945–1970)

Progress in nuclear medicine slowed during World War II, when the production of radioactive materials was temporarily halted. Following the war, there was a resurgence in the field, partly due to new technology resulting from the Manhattan Project, which had developed the atomic bomb. In particular, cyclotron and reactor sources became available that delivered large amounts of radioactive isotopes at low cost, improved methods of radiolabeling were developed, and new instrumentation was produced for the detection and measurement of radiation.

In the United States, the Atomic Energy Commission (AEC) provided strong financial and political support for the development of peacetime applications of nuclear technology, including a strong emphasis on medical diagnosis and therapy. Major programs were sponsored at the national laboratories in collaboration with industrial partners, as well as university clinical and research institutions.

Initial biomedical work focused on normal physiology, including nutritional requirements of trace elements (e.g., radioactive iron, calcium, zinc, and magnesium) and vitamins (cobalt-labeled vitamin B_{12}) in normal and diseased individuals. Body composition was measured using common elements (e.g., sodium and potassium) and labeled constituents (e.g., albumin for plasma volume, labeled red blood cells, and various fluid space markers) (Stannard, 1988a, 1988b). Methods for labeling different compounds with radioactive carbon isotopes (^{11}C and ^{14}C) were actively investigated, and many ^{14}C-labeled compounds were developed and used in animal and human research. Methods for labeling proteins with ^{131}I led the way. Many of the spin-offs of the technology were accomplished by the AEC-sponsored national laboratories, several of which had on-site research hospitals.

The development of nuclear medicine imaging awaited the availability of tracers and detector hardware that would meet all the needed requirements. The physical half-life of the radioactive tracer needed to be long enough to allow the tracer to be labeled, shipped, and received in time for clinical use. The energy of the gamma-ray emissions needed to be in the range of the imaging device, and the decay scheme had to be such that the dose to the patient was in a range believed to be safe. Sufficient photon flux was needed to produce an adequate image in an acceptably short time.

III. EARLY IMAGING DEVICES

A. Early Scanning Imagers

In the mid- to late 1940s, NaI(Tl) scintillators, counting circuits, and scintillation detectors emerged. When coupled to motor-drive systems, clinical scanning devices were developed. The first planar gamma-ray scanning device, developed by Cassen et al. (1949), used a calcium tungstate scintillator. This was followed within 2 years by a commercial NaI (Tl)-based scanner that was used clinically (Reed-Curtis sold a system based on the Cassen design) and, by 1956, positron-annihilation imaging scanners (Aronow and Brownell, 1956) were developed and used for human studies.

The initial gamma scanning devices used single detectors, whereas the positron scanners used opposed pairs of small detectors with and without electronic coincidence. Various designs of field-limiting apertures (collimators) were explored to achieve different compromises between spatial resolution and sensitivity, and these concepts are discussed in Section IV C 3. The development of focused collimators (Newell et al., 1952) made it possible to increase the sensitivity at the focal point of the gamma-ray imaging detector, whereas the positron scanners had a relatively uniform response with depth. In both cases, scanning required moving the detector sequentially over the area to be imaged (which took many minutes) and it was not possible to correct for patient movement during the procedure. When the detector was too heavy or too cumbersome to move, the patient was moved instead.

The most commonly used gamma-emitting tracer was reactor-produced ^{131}I, which also emits a number of energetic beta rays. To keep radiation doses acceptably low, only small amounts can be administered for diagnostic studies, especially when using long-lived beta- and gamma-emitting tracers. To efficiently stop and collect the energy from ^{131}I requires large thick detectors, and 2-in-thick 3-, 5- and 8-in-diameter NaI(Tl) crystals came into standard use in later years. The collimators used with the larger-area detectors produced hourglass-shaped response profiles with best resolution at the focal point of the system. The smaller-diameter collimated detectors had response profiles shaped like elongated ellipses, and these provided the more uniform depth response needed for cross-sectional tomographic imaging, which was then emerging.

The efficiency of positron and gamma-ray imaging systems depends basically on the amount of scintillator and the geometry used in the imaging system. Stopping the 511-keV annihilation photons requires thicker crystals than are needed for low-energy gamma rays. The number, kind, and volume of the detectors, their geometric aarrangement, and the criteria for accepting a valid event influence the relative sensititivity of the two methods, and these factors are discussed in the early theoretical section of this chapter. To overcome sensitivity limitations, Anger designed and built the first multiprobe positron-imaging detector. This whole-body imaging device consisted of two opposed banks of 100 NaI crystal arrayed above and below the subject, through which the patient was moved to collect data for whole-body imaging and tracer-metabolism studies. Using coincidence electronics, the device produced whole-body emission (and transmission) images of annihilation photons; although, in principle, the device could have been used with banks of focused collimators for imaging single gamma-ray emissions, as well.

A great deal of work was done to optimize collimators for scanning and camera-based imaging systems. Camera systems typically employ parallel-hole collimators, in which hole size, septal thickness and collimator length are designed based on a trade-off between the desired spatial resolution (at appropriate distances) and sensitivity for the energy range to be used. Similar decisions are needed for the design of slant-hole, converging, multi- and single-pinhole collimators (Myhill and Hine, 1967).

Positron-emitting radionuclides are mostly cyclotron-produced, and hence early developments with short-lived emitters were limited to places with on-site cyclotrons. Longer-lived positron emitters and generator-produced positron emitters were used at more remote sites. The scanner mode of operation was further limited to imaging static or very slow dynamic processes. By the end of the 1950s, it was appreciated that mapping the distribution of positron-emitting tracers could have a significant role for detecting and localizing brain tumors in patients. The advantage over single-photon imaging was due to the better tumor-to-background ratio associated with the higher tumor avidity of the positron-emitting radiopharmaceuticals that were then available (Aronow, 1962). Much of the early work used tracers with several-day half-lives, mainly for brain-tumor imaging. Initial studies produced high-contrast diagnostic images using As-74 (17.5-day) and Cu-64 (12.8-h) tracers, and enthusiasm for their development and use was high.

Pulmonary physiology studies used stationary pairs of opposed counting probes to map regional transients of respired gases. The kinetics of ^{15}O, ^{13}N, $C^{15}O_2$ and $^{11}CO_2$ gases were measured in the upper, middle and lower thirds of the human lung in the field of view of the paired detector sets (West and Dollery, 1960). The lack of a large-field-of-view positron-imaging device precluded the further development of pulmonary imaging for many years.

The first computer-controlled dual-detector scanner, which produced longitudinal and transverse section images of gamma rays, was developed by Kuhl and Edwards at the University of Pennsylvania (Kuhl, 1964). In stepwise developments, they improved system sensitivity and resolution by increasing the number of detectors and by using a more efficient scanning geometry (Kuhl et al., 1977). The device produced high-quality transverse-section images of the brain that distinguished tumor masses from the wedge-shaped lesions characteristic of stroke. The device became a brain-only imaging system, but none of the companies at that time were willing to produce and market a device for what was perceived to be a limited market.

Commercial manufacturers took advantage of research innovations and produced a number of multidetector devices in which arrays of focused detectors were moved above and below patients to produce maps of organ uptake, aswell as whole-body distribution images. Some of the devices had multiple scanning detectors arrayed at different angles, each of which moved in different patterns about the patient. The major aim of these developments was to increase system sensitivity and resolution to augment image quality.

B. Early Devices Based on Gamma Cameras

Camera-type imaging devices have come to be the main instruments used in modern nuclear medicine. The Anger camera (Anger, 1957), which was invented in 1957, became commercially available in the early 1960s; by the late 1960s, it had become the instrument of choice for static and dynamic gamma-ray imaging tasks. A number of other devices were developed as extensions of the Anger camera; still others were designed as competing strategies.

Anger himself demonstrated many extensions and applications of the device. He used pinhole collimation to increase resolution in small regions. He created a surface-projection image by rotating the patient in front of the camera, allowing the viewer to distinguish superficial lesions from deep ones by their differential rates of rotation. He developed a scanning focused-collimator system, which produced longitudinal tomographic images of several planes simultaneously. This device was well-suited for imaging high-energy gamma rays and continues to be used in whole-body imaging of I-131 and Ga-67 in patients.

In 1965, Harper *et al.* (1968) used a moving camera and coupled film recording of an oscilloscope output to create the first whole-body gamma camera image of a patient (see Fig. 2). Shortly thereafter, Anger (1968) developed the tomoscanner (see Fig. 3), employing a similar approach, in which a mechanical method was used to image planes at different depths within the patient (unlike modern tomographic methods, which use image reconstruction). A schematic diagram of the workings of the tomoscanner is shown in Figure 4, in which the scanner is shown focused on plane C within the patient. In operation, the tomoscanner, moved laterally across the patient; thus, the images of objects within the patient (except at the focal plane) moved across the camera face, with a given rate and direction of motion for each plane within the patient. Both optical and computer-based recording/display systems were developed by Anger (1968), providing simultaneous, in-focus images of multiple planes. Motion across the camera face, consistent with radioactivity at a given depth, could be calculated and this information used to produce simultaneous, in-focus images of activity at different depths. What is not generally

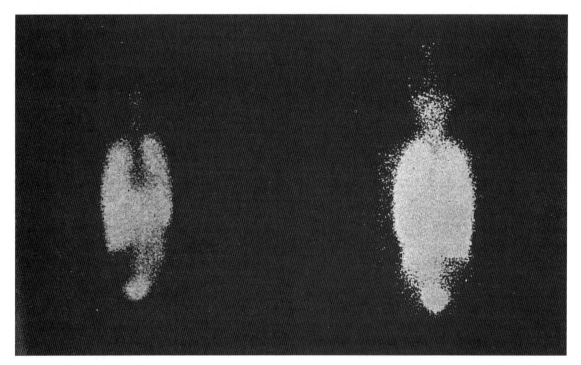

FIGURE 2 Whole body scans of a six-year-old child done about 1 hour after the injection of 2.5 millicuries of technetium 99m macro aggregated albumin. Each scintiphotograph represents a single 2–3 minute scan pass. *Left:* Anterior scan. Note that not only are the lungs visualized, but in addition, liver and spleen; and thyroid, stomach, and bladder are shown. This indicates that the macro aggregates have been broken down into smaller particles, and that some free pertechnetate is also present. *Right:* Posterior scan with the same technique. This scintiphotograph is purposely overexposed to permit a better demarcation of the soft tissue of the arm, base of the skull, and sagittal sinus due to the pertechnetate. (Harper *et al*, 1968)

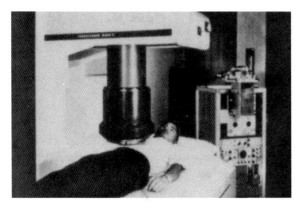

FIGURE 3 Tomoscanner, developed by Anger, which used mechanical scanning motions to produce slice images (Anger, 1974).

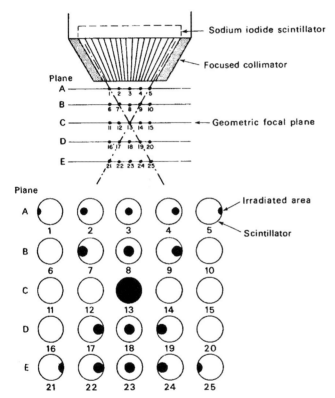

FIGURE 4 Scintillator areas irradiated through focused multichannel collimator by point source at 25 different locations. Entire scintillator is irradiated from point 13 at geometric focus. At other points, limited areas of the scintillator are irradiated as shown (Anger, 1968).

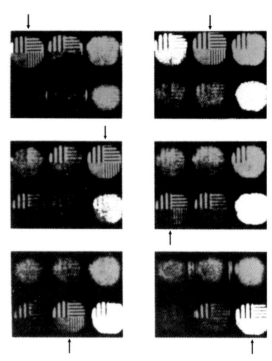

FIGURE 5 Images of a test object obtained by tomoscanner. Each image shows six planes of the object. For each image, the tomoscanner was tuned to best visualize the plane indicated by the arrow (Anger, 1974).

appreciated is that every one of these images contains *all* the detected gamma rays, from activity in all planes; hence the low contrast of longitudinal tomography. In-focus images are simply based on different assumptions about the motions of detected gamma rays from different depths. Although objects at other depths also appeared in the resulting image, these objects were motion-blurred and, therefore, less conspicuous. Figure 5 shows six images obtained in this way, each depicting six planes through a test object. Each image shows that the plane of interest (indicated by an arrow) is the sharpest.

Thereafter, a series of innovations were made with rotating gamma cameras (Jaszczak *et al.*, 1976; Keyes *et al.*, 1977). Other workers developed and used special collimators to produce real-time tomographic images. Anger cameras with seven-pinhole (7PH), single (Vogel *et al.*, 1978) and dual, orthogonal 7PH collimators (Bizais *et al.*, 1983) were able to produce dynamic longitudinal tomography images. Near real-time longitudinal tomography was obtained using rotating slant-hole collimators in front of Anger cameras to sample temporal transients rapidly (5- 30-s intervals/rotational cycle) (Muellehner, 1971), but distracting artifacts were present in the images, due to limitations in the reconstruction methods available at the time. Longitudinal tomography provides high temporal resolution at the penalty of poor axial resolution. Multiple, small, modular, pinhole-collimated NaI(Tl) gamma cameras provide a cost-effective means of surrounding the body with detectors used to collect dynamic brain and heart single-photon emission computed tomography (SPECT) images from patients (Klein *et al.*, 1996), and these methods are now being extended by the same investigators to small-animal studies using multiple high-resolution semiconductor arrays (Barber *et al.*, 1997).

A number of other attempts were made to develop gamma-ray imaging cameras. The first attempt used an image intensifier 6 years before the Anger camera was invented (Mortimer and Anger, 1952). Later attempts were made by Ter-Pogossian *et al.* (1966), who adapted a commercial CsI x-ray image-intensifier system for use in nuclear imaging. Carlson (1970) followed this development by embedding NaI crystals in a lead collimator placed in front of an image-intensifier tube (the Quantiscope, produced by Harshaw Chemical Co.; Hine and Sorenson, 1974, 48). In this device, a pulse-height gating circuit was included to reject scattered radiation. A series of upgrades of the autofluoroscope, with an array of discrete NaI detectors pioneered by Bender and Blau resulted in a commercial system (Baird Atomic), that permitted high-data-rate studies particularly well-suited to heart imaging (Grenier *et al.*, 1974). The system was also used for high-energy gamma-ray imaging (511 keV) due to the 2-in-long NaI crystals it contained. Despite all these efforts, no camera-type imaging device has yet surpassed or replaced the Anger camera for clinical nuclear medicine imaging.

C. Early Dedicated Positron-Emission Imagers

The evolution of positron-imaging detectors followed a different path from the development of single-photon detectors. By 1959, Anger had added a positron-imaging capability to the gamma camera (Anger and Rosenthal, 1959), and this device became commercially available in the late 1960s. The device had a focal-plane detector operated in coincidence with an opposed uncollimated gamma camera. The device produced high-quality planar projection images and, although the field of view was small, the device was used to image organ, regional, and whole-body distributions. The latter were obtained by creating overlapping snapshots of adjacent body regions. This approach was used to image ^{52}Fe distributed in the bone marrow of patients with hematological diseases (Anger and Vah Dyke, 1964) (Fig. 6).

Stationary positron-emitter imaging systems using other approaches were pioneered by Robertson (Bozzo *et al.*, 1968; Robertson and Bozzo, 1964), Brownell (Burnham and Brownell, 1972), Ter-Pogossian *et al.* (1975), and others. The Robertson hemispherical probe array anticipated subsequent detector geometries, but computer hardware and software technologies were not sufficiently developed at that time to make it a useful device (Robertson *et al.*, 1973). With subsequent modifications, however, the device was used for brain imaging in Montreal, Canada (Thompson *et al.*, 1979). By the mid-1970s, these new classes of devices replaced the Anger-type positron camera and were used in a growing number of major research centers. By the early 1980s, commercial production resulted in a wider dissemination of clinically useful positron emission tomography (PET) imaging devices.

Dedicated medical cyclotrons first appeared in medical institutions in the 1960s when the Hammersmith Hospital (London, England) and Washington University (St. Louis,

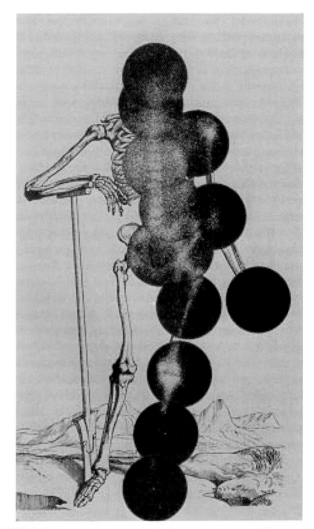

FIGURE 6 Erythropoietic marrow distribution in a patient with hemolytic anemia showing expansion of red marrow into long bone sites (VanDyke *et al.*, 1965).

MO) established their facilities. Since then, positron imaging applications have expanded rapidly, and now major teaching hospitals in most countries have facilities to image gamma- and positron-emitting radioactive pharmaceuticals.

In recent years, a series of instrumentation improvements have been introduced based on the use of new scintillators that extend the imaging possibilities (Melcher, 2000). The introduction of coded bismuth germanate (BGO) modules (Casey and Nutt, 1986) made it possible to obtain images with improved sensitivity and spatial resolution. Fast computers at low cost have made it feasible to use different coding schemes to further decrease device costs while increasing the field of view and improving spatial resolution. Smaller detector elements permit increased sampling; better scintillators provide greater light output. Composite scintillators and avalanche photodiodes (mounted at both ends of the crystal) make it possible to discriminate depth of interaction (Shao *et al.*, 2002). The use of multielement coded arrays provides high sensitivity at lower readout cost than completely discrete detectors. Much remains to be gained from innovations in front-end instrumentation.

IV. EVOLUTION OF EMISSION TOMOGRAPHY AND INITIAL APPLICATIONS

Why did 3D imaging develop in nuclear medicine before it did in X-ray based approaches? First, nuclear medicine procedures have significantly lower data rates than X-ray procedures. The slower data rates could be captured by the data acquisition and processing systems that were then available (circa 1960). Medical applications attracted the attention of bright scientists moving from other fields at a time when computing technology was rapidly developing. The University of Pennsylvania (U. Penn.) had played a key role in the development of computers used in World War II. Kuhl, a U. Penn. student, was interested in cross-sectional imaging and, during his training, developed an analog means of producing cross-sectional X-ray images. He and Edwards, an electrical engineer, went on to develop a special-purpose data collection and display system that they used to produce the first nuclear medicine transverse- and longitudinal-section tomographic images (Kuhl, 1964).

A. Anger Camera Developments

The Anger camera continues to evolve. The newest systems are digital throughout the chain of detection and processing hardware. This makes it possible for the first time to use the same Anger camera for either coincidence or single-photon imaging. Previous attempts to combine these functions resulted in limited success because the optimal electronic settings were quite different for coincidence imaging (fast timing) and single-photon imaging (long integration to get good energy resolution). Use of digital signal processing allows the faithful recording of very high event rates (2–3 Mevents/s) with minimal spatial and energy degradations (Wong, 1998). This should be useful in increasing the count-rate capability of the next generation of PET/SPECT cameras. A number of innovative geometries have been developed, and some of the new SPECT systems have combined computed tomography (CT) capabilities (Patton et al., 2000). PET/CT systems have also been developed, as described in Watson et al. (Chapter 11 in this volume).

B. Radiopharmaceutical Targeting Agents

Given the availability of gamma-ray imaging systems, nuclear medicine requires radiopharmaceuticals that accomplish the desired clinical tasks. Adequate photon flux

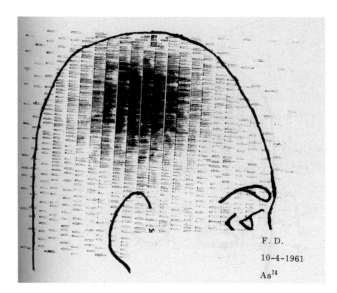

FIGURE 7 Brain tumor imaged with ^{74}As (Aronow, 1962).

is needed to produce an acceptable image, and this requires the administration of sufficient radioactivity while keeping the absorbed patient dose low enough for diagnostic acceptability. The earliest brain scans that produced adequate images of brain tumors were positron-emitter scanning devices. The tracers used were more avid for tumors than those used for gamma-ray scanning (Fig. 7). The 3-in scanner brain images of ^{203}Hg chlormerodrin were photon poor and very difficult to interpret, as were the contemporary images obtained with an ^{131}I label. The low image quality was due to the low count rate from relatively high-energy tracers and the poor tumor uptake. In 1957, the development of the 2.8 day ^{99}Mo/^{99m}Tc generator by Richards supplied the vital partner needed for the gamma camera to emerge as the premier clinical imaging tool (Brookhaven National Laboratory, 1960). The use of radionuclide generators made it possible to have ready access to tracers to label important radioactive compounds at sites remote from production devices. A single ^{99m}Tc generator furnished enough radioactivity for over a week or more, and recipes for producing a wide range of clinically important compounds developed rapidly. The 140-keV energy of the ^{99m}Tc gamma ray was optimally imaged by the thin NaI(Tl) scintillation crystals initially used in the Anger camera. Furthermore, the short half-life and low-energy ^{99m}Tc photon imparts a low dose to the patient, which permits the administration of relatively large amounts of the nuclide. The use of 6 hour ^{99m}Tc made it possible to acquire higher-photon-density images than ever before, and high-quality static and dynamic images could be obtained in reasonable times from multiple projections. The widespread clinical use of ^{99m}Tc awaited only the development of easily labeled kits that could be used in routine hospital practice. Early brain scans produced with the Argonne Cancer Research Hospital

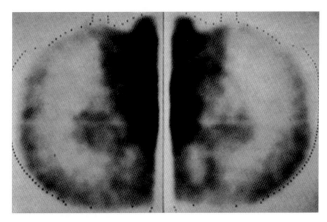

FIGURE 8 The first ^{99m}Tc scans, done with the ACRH-BS, were done in September of 1963. Examples appear in the reprint dated January, 1964, entitled "Optimization of a Scanning Method Using ^{99m}Tc" (Beck *et al.*, 1967).

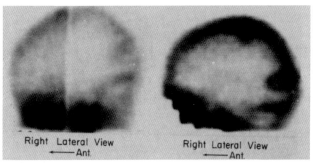

FIGURE 9 Images obtained with an ACRH scanner on the left, and a Ge(Li) scanner on the right. A more accurate delineation of the occipital doughnut-shaped lesion is obtained with the Semiconductor scanner, although a longer imaging time was needed. Film exposures were not optimal (Hoffer *et al.*, 1971).

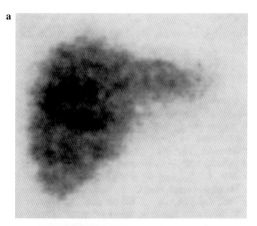

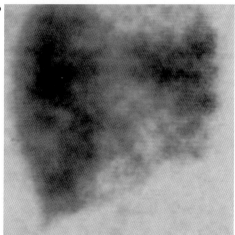

FIGURE 10 (a) Liver scan made using Tc_2S_7 in sulfur colloid in normal subject (Harper *et al.*, 1965). (b) Liver scan made using Tc_2S_7 in sulfur colloid in a patient with metastases from a breast tumor (Harper *et al.*, 1965).

(ACRH) scanner with ^{99m}Tc-pertechnetate are shown in Figure 8. Planar brain images obtained with the NaI ACRH scanner and those obtained in the same patient with a high-resolution germanium semiconductor device are illustrated in Figure 9. ^{99m}Tc-sulfur colloid was also developed at ACRH as a liver-scanning agent (Fig. 10a, normal; Fig 10b, abnormal). Liver-scanning agents were the first of a long series of successful ^{99m}Tc-based radiopharmaceuticals that became available (Harper, Lathrop, *et al.*, 1964). To this day, the vast majority of clinical nuclear medicine imaging procedures are based on the Anger camera and ^{99m}Tc. With the availability of ^{99m}Tc- labeled compounds, nuclear medicine experienced a rapid increase in the kinds and quality of procedures that could be performed in routine hospital practice. The need for a longer-lived parent for use in remote areas led to the development of the ^{113}Sn/^{113m}In generator. The long half-life of the parent was attractive, and many labeling kits were developed for use with the ^{113m}In daughter. Unfortunately, the 390-keV energy was not well suited to the Anger camera, and the tracer is no longer in common use.

The same pattern of requirements and developments apply to positron imaging. The earliest studies used planar imaging scanners and relatively long-lived radiotracers. The advantages of positron emitters are (1) ease of accurate attenuation correction (due to the almost 180° path between the two emerging annihilation photons and (2) the biochemical validity of compounds labeled with the short-lived positron emitters ^{11}C, ^{13}N, and ^{15}O. These tracers, when labeled to naturally occurring substances, can substitute in the chemical compound, making it possible to image the biodistribution kinetics of the labeled material. Short-lived radioactive fluorine (^{18}F) is also a useful tracer when labeled to a number of important compounds by replacing stable fluorine in the compounds. In other cases, it substitutes for a hydroxyl group (as in ^{18}F-2-deoxy-D-glucose, FDG, discussed later in this chapter).

The use of the radioactive gases, ^{13}N, ^{11}C, ^{15}O, ^{11}CO, $C^{15}O$, $^{11}CO_2$, or $C^{15}O_2$ requires an on-site cyclotron. The radioactive gases are easy to produce, and because these are natural substances no special licensing is required. The 110-min half-life of ^{18}F is long enough for ^{18}F to be shipped as the fluoride or as the labeled compound. There are a few generators one can use in the absence of a cyclotron, namely the ^{82}Sr/^{82}Rb generator, which has been used for cardiac studies, and the ^{68}Ge/^{68}Ga system, which is used as a transmission source for attenuation correction but has not been widely used as a labeling agent. The main positron emitter that has achieved widespread use is FDG. The major advantage of PET imaging derives from the biochemical elegance of the tracers (e.g., the kinetics of ^{11}C-labeled agents cannot be distinguished from stable ^{12}C compounds found in nature). The entire field of brain research has experienced a tremendous boost based on extensive studies with FDG and labeled receptor-targeting agents.

The major benefit of nuclear medicine lies in the selectivity and sensitivity of the method: Many labeled compounds are available that have high specific activity, they go where they are targeted, and they stay intact during the imaging period. Given a compound that is excellent on all scores, fair instrumentation suffices; but, when the tracer is marginal, a very excellent imaging device is needed. Where the tracer is not effective, the best imaging instrument is of little value.

C. Early Theoretical Developments

The generic issues of radionuclide imaging for applications in clinical nuclear medicine are highlighted in Figure 3 of Chapter 1. However, much of the early work of theorists in quantitative radionuclide imaging focused more sharply on only a few of these fundamental issues.

1. The object property to be mapped or imaged (i.e., how best to describe *in vivo* the distribution of radioactivity administered to the patient).
2. Image-data acquisition (i.e., how best to parameterize and model mathematically the response of a radiation-detector system to this distribution).
3. The recorded image (i.e., how best to describe the process and result of the mapping operation performed by the radiation detector to produce an image).
4. Image quality (i.e., how best to evaluate image quality in goal-specific terms so as to provide a quantitative basis for the optimal design of collimated detectors and other components for specific applications).

Next, with minor modifications of certain symbols used to represent the relevant parameters, we summarize some of these issues and early theoretical developments and indicate how they are based on concepts adapted from work in other fields of science and engineering. This summary is focused primarily on planar (attenuated projection) imaging from a single view; thus, we do not discuss the very important

developments of emission and transmission tomography made by David E. Kuhl a decade before the invention of X-ray CT by Hounsfield and Cormack (see Kuhl and Edwards, 1963, 1964, 1968a, 1968b; Kuhl *et al.*, 1966, 1972). These and other fundamental issues of radionuclide imaging (i.e., image reconstruction and processing; image compression, storage, retrieval, and distribution; image display and visualization; image analysis and interpretation; and image evaluation and optimization) are discussed in (Gottschalk and Beck, 1968) and are extensively updated in other chapters of this book. Early methods of quantitative image analysis and color display appear in Charleston *et al.* (1964, 1969). A summary of early radionuclide imaging systems appears in Beck (1975).

1. Object to Be Imaged

A fairly complete description of the object is provided by the dynamic 3D distribution of administered radioactivity within the patient, which redistributes and decays over time t. Such a distribution can be described by $A(r_O, t; E_O)$ (in curies per cubic centimeters) or by ρ the expected number of gamma rays with energy E_O (or positrons) emitted per second per cubic centimeter at each point r_O within the region to be imaged.

$$\rho(r_O \; t; \; E_O) \equiv \rho(x_O, y_O, z_O, t; E_O) \quad (1)$$

This general description can be simplified considerably and still yield useful results. For example, in many cases we may assume that the distribution of activity is static and that the total observation time, T, is much shorter than the half-life, $T_{1/2}$, of the radioactive material. Furthermore, in the simplest mathematical model, the object is assumed to be a uniform volume distribution (i.e., p = constant) but for the presence of a small right-circular cylindrical tumor region with radius R_T and extending from depth $z_O = d_1$ to d_2. Within the tumor, the concentration of radioactivity is assumed also to be uniform and given by U, the uptake ratio, times the concentration in normal tissue.

It is important to note that a further simplification is possible, based on observations made by Mayneord (1950), Brownell (1959), and Beck (1961, 1964a); namely, as a uniform sheet distribution of radioactive material in the *xy* plane is moved (in air) along the *z* axis of a focused collimator system or a coincidence system, the expected count rate is constant. This is valid provided only that the sheet is large enough to cover the entire region of view of the collimator and that a negligible number of gamma rays (or annihilation photons) penetrate the collimator and surrounding shielding.

A uniform source of thickness H (in centimeters) along the z axis may be regarded as a stack of uniform sheets. In that case, the observations just cited enable us to define an equivalent sheet distribution, $O_N(x_O, y_O)$, when the collimated detector views normal tissue. Integrating over all sheets and taking into account photon attenuation in the intervening tissue, which has an attenuation coefficient µ (per centimeter), we obtain:

$$O_N(x_O, y_O) = \int_0^H \rho e^{-\mu z_O} dz_O = \rho \left[\frac{1-e^{-\mu H}}{\mu}\right] \quad (2)$$

Similarly, when the detector views the embedded tumor, which is assumed to be at least as large as the collimator field of view (i.e., $R_T > R_{DETECTOR}$), the equivalent uniform sheet distribution is given by:

$$O_T(x_O, y_O) = \rho\left[\left(\frac{1-e^{-\mu H}}{\mu}\right) + (U-1)e^{-\mu d_1}\left(\frac{1-e^{-\mu|d_2-d_1|}}{\mu}\right)\right] \quad (3)$$

If the object is a positron emitter, H (in centimeters) is the total distance traveled through tissue by the oppositely directed annihilation photons. Therefore the equivalent sheet distribution for a coincidence system viewing normal tissue is given by:

$$O_{NC}(x_O, y_O) = \rho H e^{-\mu H} \quad (4)$$

and for tumor tissue by:

$$O_{TC}(x_O, y_O) = \rho[H + (U-1)|d_2 - d_1|]e^{-\mu H} \quad (5)$$

2. Detector Systems

Early theoretical work focused on measures of detector sensitivity as well as on spatial, temporal, and energy resolution, in response to both primary and scattered photons from point, line, plane, and volume distributions of radioactivity. For a scanning system employing a focused collimator, the detector sensitivity to an equivalent sheet distribution of activity is designated S_D (in counts per second per photons emitted per second per squared centimeter) and can be described by:

$$S_D = G\eta\psi(1 + f_P + f_S) \quad (6)$$

where G (in square centimeters) is the geometrical efficiency of the collimator, η is the photopeak crystal efficiency (the product of the interaction ratio and the photofraction: see Beck, 1968b), ψ is the window efficiency (the fraction of photopeak pulses within the window of the pulse height analyzer), and f_P and f_S are the penetration and scatter fractions, respectively.

For a focused collimator, Beck (1961) showed that G is given by:

$$G = \frac{N_H \pi R^4}{2(8F^2 + R^2)\left(\frac{R}{r_H} + 2\right)^2} \quad (7)$$

and for a coincidence system by:

$$G_C = \frac{\pi R^4}{2(8F^2 + R^2)} \quad (8)$$

3. The Numerical Image

For the simple object just described, the expected count rates over normal and tumor tissues are given by $C_N = O_N S_D$ and $C_T = O_T S_D$ respectively (both in counts per second). And if T (in seconds) is the total observation time used to scan an area A_O (in square centimeters), the observation time per unit area is simply τ [s/cm^2] = T/A_O. But for edge effects, the expected number of counts detected over a tumor with area $A_T = \pi R_T^2$ is given by:

$$N_T = \tau A_T C_T = \tau \pi R_T^2 O_T S_D \quad (9)$$

and over an area of normal tissue of the same size by

$$N_N = \tau A_T C_N = \tau \pi R_T^2 O_N S_D \quad (10)$$

4. Image Quality and Tumor Detectability

Observed values of N_T and N_N are Poisson distributed, with standard deviations given by $(N_T)^{1/2}$ and $(N_N)^{1/2}$, respectively (see Evans 1955). We may regard the expected difference, $(N_T - N_N)$, as a measure of the signal to be detected, and the standard deviation of observed differences, $(N_T + N_N)^{1/2}$ as a measure of the statistical noise that accompanies this signal. In that case, the signal-to-noise ratio, SNR is given by:

$$\text{SNR} = \frac{(N_T - N_N)}{(N_T + N_N)^{1/2}} = \frac{(C_T - C_N)\tau^{1/2}}{(C_T + C_N)^{1/2}} \quad (11)$$

If q is the number of standard deviations required to ensure the desired reliability of tumor detection, the quantity:

$$Q \equiv \frac{(\text{SNR})^2}{\tau} = \frac{(C_T - C_N)^2}{(C_T + C_N)} \geq \frac{4q^2}{\tau} \quad (12)$$

may be interpreted as the count rate at which tumor detectability can be achieved. This quantity may be used as a figure of merit for comparing and evaluating detector systems, as well as for optimizing various parameters of system design, such as the trade-off between detector sensitivity and spatial resolution. This criterion is similar to that introduced by Rose (1948) for the detection of features in noise-limited television images. In general, if the count rates in Eq. (12) are such that the inequality holds, then the imaging procedure is feasible; otherwise, it is not (unless, of course, the observation time per unit area, τ, is increased or the required statistical reliability q is decreased).

It should be recognized that maximum SNR in the image is not the only intuitively appealing basis for the evaluation of image quality. Alternatively, and with some justification, we might have chosen maximum correspondence between object and image, maximum information in the image, or maximum likelihood that an observer will detect the presence of a tumor. All these criteria have been used in similar contexts. (As examples, see Beck, 1964a; Beck and Harper, 1968; Beck et al., 1971, 1972; Beck, Zimmer, Charleston, Shipley, et al., 1973; Beck, Zimmer, Charleston, Harper, et al., 1973; Kullback, 1959; Linfoot, 1961; Gregg, 1965; Metz, 1995.)

5. Other Early Advances

a. Collimator Design

The first scanner in the United States, built by Cassen and colleagues (Cassen and Curtis, 1951; Cassen et al., 1950, 1951), made use of a small cadmium tungstate scintillation detector with a cylindrical single-bore collimator. As a result, the detector sensitivity was low, requiring a large dose of administered radioactivity and a long examination time, even for thyroid scanning with fairly large doses of I-131 (364 keV).

When substantially larger NaI(Tl) crystals became available in the early 1950s, Newell et al. (1952) proposed the use of a multichannel focused collimator to increase the sensitivity for the detection of midline brain tumors with I-131. They developed a mathematical model to characterize the sensitivity and spatial resolution of focused collimators and introduced an estimate of the effective path length through the septa.

Making use of Newell's estimate of path length, Beck (1961) formulated an equation to approximate the septal penetration fraction, f_P (the number of primary gamma rays that penetrate the septa divided by the number that pass through the collimator holes properly):

$$f_P \approx \left[\frac{27 D_C^4 \left(1 + \frac{R^2}{8F^2}\right)}{64\pi^2 r_H^4 N_H \left(1 + \frac{3\sqrt{3} D_C^2 R^2}{64 r_H^2 F^2}\right)} - 1 \right] e^{-\left[\frac{2 r_H F \lambda}{R}\left(1 - \frac{8 N_H \pi r_H^2}{3\sqrt{3} D_C^2}\right)\right]} \quad (13)$$

which is expressed in terms of the collimator parameters and the attenuation coefficient of collimator material.

The goal of the collimator design procedure, based on Eq. (13), is to find the number of holes in hexagonal array, N_H, and the hole radius, r_H, that yield an acceptable value of f_P (say 0.1) for specified values of the crystal diameter, D_C, desired focal length, F, radius of view at the focal distance, R, and attenuation coefficient, λ, in collimator material for a specified gamma-ray energy, E_O.

This transcendental equation cannot be solved explicitly for r_H; however, f_P can be plotted as a function of r_H to find the hole radius that yields the acceptable value of f_P. This hole radius is then used to calculate the geometrical efficiency of the collimator, G, in response to a uniform sheet distribution of activity, given by Eq. (7).

The procedure is repeated for each value of N_H corresponding to a hexagonal array; namely, $N_H = (3n^2 + 1)/4$, for $n = 1, 3, 5, 7, \ldots$. The optimum collimator is characterized by the values of N_H and r_H that yield the largest value of G.

This procedure was used to design optimal focused collimators for gamma rays in the range 200–511 keV with the goal of determining the optimal energy for the detection of a midline brain lesion, based on the statistical criterion given in Eq. (12). (The existence of an optimal gamma-ray energy is obvious. If the energy is too low, a large fraction of the gamma rays will be absorbed in the intervening tissue and will not emerge to be detected; if the energy is too high, a large fraction of the gamma rays will penetrate the scintillation crystal without interacting.)

Specifically, Beck (1961) compared detectability of midline brain lesions using an opposed pair of scintillation detectors with optimal focused collimators and an identical pair of scintillation detectors operated in coincidence for 511 keV annihilation photons. The second pair of detectors made use of tapered single-hole collimators having the same radius of view for coincidence detection as the focused collimators at the focal distance. In that case, the geometrical efficiency in response to a uniform sheet distribution is given by Eq. (8).

Two unexpected conclusions were drawn from this study.
1. For a pair of NaI(Tl) scintillation crystals (5 cm in diameter × 5 cm long), a focused collimator system can be designed for counting 511 keV photons in the singles mode that is superior to a coincidence system by factors of 2 to 10 (depending on spatial resolution) for the detection of midline brain lesions. (Note that this conclusion is probably still valid for a single pair of detectors. However, modern PET systems make use of a large array of scintillation detectors surrounding the patient. In that case, the sensitivity of the system is proportional to the number of pairs of detectors operated in coincidence and is, therefore, quite high compared with the singles mode, in which sensitivity is simply proportional to the number of detectors employed. Similarly, modern Anger-type cameras employ large-area scintillation crystals that provide a large increase in sensitivity for the detection of single gamma rays in the SPECT mode.)
2. Because detectability was seen to increase monotonically with decreasing gamma-ray energy down to 200 keV (the low-energy limit of this study), it appeared likely that the advantages of low-energy gamma radiation had not yet been fully exploited.

This study was immediately extended down to 88 keV (the absorption edge of Pb), indicating that the optimum gamma-ray energy for the detection of midline brain lesions is approximately 100 keV. This conclusion was not taken very seriously because the effects of scattered radiation were not taken into account (i.e., f_S was assumed to be zero), and it was well understood that the deleterious effects of scatter increase with decreasing gamma-ray energy. Nevertheless, this study stimulated enough excitement to lead directly to the first brain scan with ^{99m}Tc (140 keV) in 1961 (using a commercial scanner) and to the design and construction of the ACRH brain scanner, completed in 1963. This system produced the first spectacularly successful brain scans with ^{99m}Tc in 1963, as reported by Harper and colleagues (Harper, Beck, et al., 1964; Harper et al., 1965, 1966). Meanwhile, Beck (1964a) made use of the Klein-Nishina equation for Compton-scattered photons to estimate the effects of scattered radiation on image contrast and figure of merit, with the result that the optimum gamma-ray energy for the detection of midline brain lesions appeared to be approximately 150 keV. Beck et al. (1967) later reported design considerations for the ACRH brain scanner and produced a more comprehensive study of the effects of scattered radiation (Beck et al. 1969). The latter study was extended by Atkins and colleagues (Atkins and Beck, 1975; Atkins et al., 1977) for an Anger-type scintillation camera.

b. Spatial Resolution of Scintillation Detectors

The developments just summarized take into account only the sensitivity of a detector to an equivalent uniform sheet of activity, with little consideration of spatial resolution (beyond the use of the collimator radius of view, R, as a design parameter and the requirement that $R \leq R_T$ at the tumor depth). To compute the image of an arbitrarily shaped object, a more general and detailed description of the spatial resolution is needed.

For a scanning system with the detector axis at (x_D, y_D), the response to a point source of unit intensity can be described by the point-source response function, $p_D(r_D; r_O)$ (in number of counts at r_D/s divided by the number of gamma rays emitted at r_O/s):

$$p_D(r_D; r_O) = p_G(r_D; r_O) + p_P(r_D; r_O) + p_S(r_D; r_O) \quad (14)$$

where the terms with subscripts G, P, and S represent the geometrical, penetration, and scatter responses, respectively; and p_G, p_P, and p_S are the probabilities that a primary gamma ray emitted at r_O will be detected and recorded after passing through the intervening medium and a collimator hole (G), after penetrating the collimator septa (P), or after being scattered (S) within the intervening medium, where r_O is at a depth $z_M = z_O - z_C$ within the scattering medium, and z_C is the clearance between the patient and the collimator face.

The total probability is a measure of detector sensitivity and is given by:

$$\iint p_D(r_D; r_O) dr_D = S_D(z_M) = G\eta\psi e^{-\mu_M} \quad (15)$$
$$[1 + f_P + f_S(z_M)]$$

where the integration extends over the entire image plane. This equation provides a formal definition of the penetration fraction and the scatter fraction; namely:

$$f_P \equiv \frac{\iint p_P(r_D; r_O) dr_D}{\iint p_G(r_D; r_O) dr_D} \quad (16)$$

and

$$f_S \equiv \frac{\iint p_S(r_D; r_O) dr_D}{\iint p_G(r_D; r_O) dr_D} \quad (17)$$

Moreover, the shape of the point-source response function, called the point-spread function, provides a measure of spatial resolution and is given by:

$$\hat{p}_D(r_D; r_O) = \frac{p_D(r_D; r_O)}{\iint p_D(r_D; r_O) dr_D} \quad (18)$$

This function is normalized in the sense that:

$$\iint \hat{p}_D(r_D; r_O) dr_D = 1 \quad (19)$$

Equation (14) provides a fairly complete description of the detector response in the sense that we can use this quantity to compute the expected image, $n(r_D)$, of an arbitrary object.

First, however, consider a planar object $O(r_O) = O(x_O, y_O) = \rho(x_O, y_O; z_O) dz_O$ embedded in the z_O plane. If T (in seconds) is the total time spent in scanning the total area A_S (in square centimeters), which contains the object, the expected image is given by:

$$n(r_D; z_O) = \frac{T}{A_S} \iint_{A_S} p_D(r_D; r_O) O(r_O) dr_O \quad (20)$$

Having assumed that the detector is linear, the expected image, $n(r_I)$, of a volume distribution of activity is given by the superposition or sum of images of all object planes:

$$n(r_I) = n(r_D) = \int n(r_D; z_O) dz_O \quad (21)$$

where it is assumed that there is no magnification between the detector and image planes.

If, in addition to being linear, the shape of the point-spread function is the same over the entire object plane (as it is for a scanning system), the detector is said to be shift-invariant. In that case, Eq. (20) can be written as a convolution integral:

$$n(r_D) = \frac{T}{A_S} \int \left[\iint_{A_S} p_D(r_D - r_O) O(r_O) dr_O \right] dz_O \quad (22)$$

This fact suggests an alternative way of describing the imaging process—namely, in terms of the spatial frequency spectra of the object and image and the spatial frequency response of the imaging system. To do this, we make use of the Convolution Theorem, described next. (For a more elegant and complete discussion of the spatial frequency domain and the use of Fourier methods in imaging, see

Chapter 6. Here, we touch only on some of the early applications of Fourier concepts and methods.)

c. The Spatial Frequency Domain

Every audiophile is familiar with the concept of the temporal frequency response of an audio reproduction system. The notion of the spatial frequency response of an imaging system dates at least as far back as Abbe's (1873) theory of the microscope. However, this notion did not enjoy widespread use until the 1950s, following its reintroduction by Duffieux (1946). During the 1960s and 1970s, the concept was used increasingly to describe the response of optical systems (with both coherent and incoherent light), as well as the response of lenses, film, and radiographic and television systems. The concept was first introduced into radionuclide imaging by Beck and colleauges (Beck, 1964a, 1964b; Beck, Zimmer, Charleston, Harper, *et al.*, 1973), who adapted it from papers by Hopkins (1956) on optical systems and by Perrin (1960) on methods for appraising photographic systems.

Briefly, both the object and the image can be described in terms of their components at spatial frequencies, ν (in cycles per centimeter), and the imaging system can be described in terms of the efficiency with which it is able to transfer object modulation at each spatial frequency to the image, that is, the modulation transfer function, $\text{MTF}(\nu)$, defined by:

$$\text{MTF}(\nu) \equiv \frac{\text{Image modulation }(\nu)}{\text{Object modulation }(\nu)} \quad (23)$$

where, in the terms of electrical engineering, modulation is defined by the ratio of the AC and DC components of the sinusoid at each spatial frequency, ν.

A more formal approach makes use of the Convolution Theorem to recast Eq. (22) in the spatial frequency domain. Thus, if the spatial frequency spectra of the z_O plane of the object and its image are given by:

$$\tilde{O}(\nu; z_O) \equiv \text{FT}[O(x_O, y_O; z_O)]$$
$$\equiv \iint O(x_O, y_O; z_O) e^{-j2\pi(\nu_x x_O + \nu_y y_O)} dx_O\, dy_O \quad (24)$$

and

$$\tilde{N}(\nu; z_O) \equiv \text{FT}[n(x_D, y_D; z_O)]$$
$$\equiv \iint n(x_D, y_D; z_O) e^{-j2\pi(\nu_x x_D + \nu_y y_D)} dx_D\, dy_D \quad (25)$$

respectively, and the spatial frequency response of the detector system is given by:

$$\tilde{D}(\nu; z_O) \equiv \text{FT}[p_D(r_D; z_O)]$$
$$\equiv \iint p_D(x_D, y_D; z_O) e^{-j2\pi(\nu_x x_D + \nu_y y_D)} dx_D\, dy_D \quad (26)$$

then the Convolution Theorem states that:

$$\tilde{N}(\nu; z_O) = \frac{T}{A_S} \tilde{D}(\nu; z_O)\, \tilde{O}(\nu; z_O) \quad (27)$$

Thus, the convolution integral in Eq. (22) has been replaced by the (simpler) product of two functions that represent the object spectrum and the spatial frequency response of the detector. The expected numerical image of the z_O plane of the object is then given by the inverse Fourier transform of its spectrum:

$$n(r_D; z_O) = \text{FT}^{-1}\{\tilde{N}(\nu; z_O)\}$$
$$\equiv \iint \tilde{N}(\nu_x, \nu_y; z_O) e^{j2\pi(\nu_x x_D + \nu_y y_D)} d\nu_x\, d\nu_y \quad (28)$$

Finally, the numerical image of a volume distribution is found by applying Eq. (21).

The detector transfer function has provided the basis for a number of more definitive studies of collimator design for both scanners and scintillation cameras, taking into account geometrical, penetration, and scatter effects, by Beck and colleagues (Beck, 1968a, 1968c; Beck and Gunter, 1985), Tsui *et al.* (1978, 1988), and Gunter and colleagues (Gunter and Beck, 1988; Gunter, Bartlett, *et al.*, 1990; Gunter, Hoffmann, *et al.*, 1990).

d. Effects of Septal Penetration and Scatter on the Detector Transfer Function

A number of insights emerge from our simple formulation. For example, making use of Eqs. (14)–(19) and (26), it is easy to show that the detector transfer function (which is frequently called the optical transfer function, even in nonoptical contexts) can be expressed in the form:

$$\tilde{D}(\nu; z_O) = \frac{\tilde{D}_G(\nu; z_O) + f_P \tilde{D}_P(\nu; z_O) + f_S \tilde{D}_S(\nu; z_O)}{1 + f_P + f_S} \quad (29)$$

and the modulation transfer function of the detector in the form:

$$\text{MTF}(\nu; z_O) = |\tilde{D}(\nu; z_O)| \quad (30)$$

The point-source response functions for septal penetration and scattered radiation tend to be quite broad when compared to the geometrical component and, as a consequence, the spatial frequency response functions for penetration and scatter decrease rapidly for increasing values of spatial frequency, ν, and Eq. (29) approaches:

$$\tilde{D}(\nu; z_O) = \frac{\tilde{D}_G(\nu; z_O)}{1 + f_P + f_S} \quad (31)$$

Thus, the principal effect of septal penetration and scatter is reduced image contrast. In fact, in virtually all the early work in radionuclide imaging, penetration and scatter were regarded as equivalent to background radiation, which is normally reduced by use of a single-channel pulse-height analyzer (PHA). In that case, it was shown (Beck *et al.*, 1969) that the optimum baseline setting of the PHA minimizes the quantity $(1 + f_S)/\psi$. (This is the factor by which the radiation dosage or the scanning time must be increased to obtain the SNR of an ideal system, with $f_S = 0$ and $\psi = 1$.)

On the other hand, for small values of ν (i.e., in the low-frequency domain of large structures) the penetration and scatter components transfer significant modulation or contrast from the object to the image. This has implications for how we might use the scatter component of detector response in optimum-weighted multichannel imaging, described next.

e. Optimum-Weighted Multichannel Imaging

For the most common case of noise-limited images having structures near the threshold of detectability, the maximum SNR may be an appropriate criterion for use in optimization procedures. Following the formulation by Beck *et al.* (1969, 1972; Beck, Zimmer, Charleston, Shipley, *et al.*, 1973) we let $w_j S_j$ and $w_j N_j$ designate the weighted signal and noise associated with the jth channel, respectively. If the signals for all channels add linearly, so that:

$$S = \sum_j w_j S_j \quad (32)$$

and noise adds in quadrature, so that

$$N = \sqrt{\sum w_j^2 N_j^2} \quad (33)$$

then the total signal-to-noise ratio, SNR, is given by:

$$\text{SNR} = \frac{\sum w_j S_j}{\sqrt{\sum w_j^2 N_j^2}} \quad (34)$$

Setting $d(\text{SNR})/dw_j = 0$ and solving for the optimum weighting factors, we obtain:

$$w_j(\text{optimum}) = \frac{S_j}{N_j^2} \quad (35)$$

To determine the optimum weighting factors we must state explicitly how the signal and noise are defined. For example, we may make use of the formulation in Eq. (11) and define the signal S_j as the expected difference in counts accumulated in the jth channel over the tumor and normal regions, and define the noise, N_j, accompanying this signal as the standard deviation of this difference. In that case, the optimum weighting factors are given by:

$$w_j(\text{optimum}) = \frac{S_j}{N_j^2} = \frac{(N_{Tj} - N_{Nj})}{(N_{Tj} + N_{Nj})} \quad (36)$$

This quantity is commonly used as a measure of image contrast; thus, the optimum weighting factor for the jth channel is simply proportional to the image contrast produced by the photons detected in this channel. For large structures, this contrast may be quite high even for photons that have been scattered through large angles. Summing over all channels results in an aggregated image contrast that is lower than the contrast associated with unscattered photons in the photopeak alone; however, the aggregated SNR is increased over that obtained with photopeak pulses alone.

Using optimum weights in Eq. (34) yields the maximum value of the SNR:

$$(\text{SNR})_{\text{max}} = \sqrt{\sum_j (S_j / N_j)^2} \quad (37)$$

In the special case in which the object structure is a sinusoid, ν (in cycles per centimeter), with maximum and minimum values in the image given by N_{Tj} and N_{Nj}, respectively, the optimum weighting factors are simply proportional to the detector transfer function at this spatial frequency.

It should be noted that the optimum weighting factors are independent of the number of recorded events making up the image and that all weighting factors are positive when $N_{Tj} > N_{Nj}$. However, this is not the case when certain (equally appealing) criteria for image quality are used in place of maximum SNR. For example, had we chosen as a criterion maximum correspondence between the object and image, based on a least-squares fit, the situation would be more complex. For very small numbers of recorded events making up the image, all weighting factors are positive and near the value 1. As the event density increases, the optimum weighting factors for the scattered radiation channels decrease and approach 0. As the event density continues to increase toward infinity (i.e., for a noiseless image), these weighting factors become negative while the weighting factors for unscattered radiation remain positive. This results in an aggregated bipolar response function, which tends to sharpen the image, compensating for deterministic factors such as collimator blurring and thereby increasing the correspondence between object and image.

Clearly, the optimum-weighted multichannel mode shows the greatest improvement in image quality when implemented in an imaging system with detectors having excellent energy resolution—much better than that of early imaging systems. By the 1980s, energy resolution had been improved sufficiently to encourage Siemens Gammasonics to implement a fairly versatile version of multichannel weighted imaging in their scintillation cameras. In the future, we hope that the superior energy resolution of semiconductor detectors may be used to advantage in such applications; however, detector sensitivity will probably remain a problem (Beck *et al.* 1971).

V. CLINICAL APPLICATIONS

The development of new imaging technology has revolutionized the diagnostic and therapeutic processes. The understanding of biochemical pathways and physiological processes was the first focus of nuclear medicine research. Electrolyte metabolism and body-space measurements were obtained by blood assays initially because body-counting and -imaging instrumentation was yet to be developed. Thyroid function and treatment of thyroid disease were the first important clinical applications, beginning before World War II. The first clinical therapy attempt was to improve the treatment of leukemia using ^{32}P in 1938. Brain function and cancer research have been the most intensively studied areas, and these continue to occupy a great deal of clinical and research effort. In this section we consider these in greatest depth, along with a brief introduction to the other major clinical areas in which nuclear medicine has played a role.

A. Brain Imaging

Emission tomography has made a tremendous impact on brain research (see Wagner *et al.*, 1995, 485–594). The brain is composed of many small, uniquely important and specialized functioning structures, nerve cell bodies, and fiber tracts, which are best characterized by what they do. The labeling of neuroreceptors and neurotransmitters has provided a sensitive and specific tool for probing brain function in health and disease. (The details of this subject are given in Lalush and Wernick [Chapter 21 in this volume]). This has led to enormous advances in our knowledge of normal and abnormal brain function and of neurochemical pathways, having importance for the diagnosis and treatment of individual patients and for the development of new treatment strategies. Drug development, for example, is guided by the qualitative and quantitative effect of drugs on the outcome of the treatment of neurological disorders, cancer in different organs, and cardiovascular and renal diseases. Nuclear medicine also can provide early evidence of response to therapy. Early surrogate end points can make it possible to adjust therapy (amount and kind) to find one that works better in a given patient. Such information can also improve the drug development and clinical evaluation cycle.

The earliest nuclear medicine brain research used end-window Geiger-Müller (GM) detectors to measure tracer uptake in brain tumors. Probes were first placed on the surface of the skull and then on the surface of the brain at operation, followed by the development and use of needle-shaped GM detectors inserted into the brain at operation to probe uptake at depth. Because many of the tracers were beta-gamma emitters it was important to get the detector as close to the lesion as possible. Nonspecific tracers were used, mainly ^{131}I-human serum albumen. To obtain registered sequential measurements, a bathing cap with inscribed circles was used by some to guide detector repositioning. The temporal behavior of the tracer was plotted and the data from the suspect tumor region were used to differentiate malignant from benign lesions (Planiol, 1961).

As in many of the nuclear imaging paradigms, different tracer-kinetic patterns were studied to characterize normal and abnormal regional processes.

Measurements of global blood flow were first accomplished with stable and later with radioactive gases (Kety and Schmidt, 1948). Regional blood flow measurements were originally probe-based, but later Lassen and others extended the work using radioactive xenon and specialized imaging devices (Lassen *et al.*, 1978; Stokely *et al.*, 1980). Brain blood flow measured with various labeled gases, ligands, and microspheres remains of keen interest for diagnosis and correlation with functional brain activity.

With the advent of ^{99m}Tc, the imaging opportunities changed dramatically. The administration of 10 mCi of ^{99m}Tc yielded approximately 50 times as many counts as 300 μCi of ^{203}Hg in a given study, and this made a tremendous difference in image quality. Imaging was first performed with scanning instruments that required that the operator preselect image display factors. More skill and *a priori* knowledge were needed to produce images that were as good as the Anger camera images. As cameras replaced scanners in the late 1960s and early 1970s, the complexity of image capture greatly decreased, the quality of routinely produced images improved, and clinical nuclear medicine rapidly expanded up to the time when X-ray CT became a well-established widely disseminated alternative.

Planar brain imaging studies included a series of first-pass images, followed by static images from multiple projections when equilibrium had been achieved. Typically flow studies were done from the front and right-left differences in image features, including time of appearance, transit, and clearance times, were noted. As computers were coupled to gamma cameras, transient analysis was commonly employed, and functional images were used to compress the large series of images into a single view (Croft, 1981). The temporal change in count rate through each (x, y) (pixel) location in successive frames revealed localized kinetic features. Three- and four-dimensional filters were used to smooth the noisy data; half-values, times to peak, peak amplitudes, and clearance parameters were extracted; and a series of functional images was formed (trixel images). These parametric images were useful in visualizing, localizing, and characterizing temporal and spatial asymmetries. Parameters of the trixel curves were presented as intensity or color-coded (x, y) images, which were useful quantitative images that condensed a large amount of quantitative data in an easily interpreted format. Color provided a semiquantitative view of functional images.

Initial brain emission tomography studies used ^{99m}Tc-labeled tracers that delineated brain blood flow. An abnormal flow pattern was early evidence of a subdural hematoma and a useful guide to the surgeon as to where to drill a bur hole in the skull over the affected area in order to reduce brain compression and thus avoid serious and potentially lethal complications. Stroke is best treated early; therefore early detection is critical. The flame shape of a typical cortical stroke is best seen on transverse-section tomographic images. Abnormal flow patterns are also useful in identifying unsuspected vascular shunts and arteriovenous malformations, which may require urgent therapy.

The first commercial gamma cameras were small (approximately 28 cm in diameter), making them more suitable for brain imaging than for the imaging of larger organs. Anatomy also works in favor of tomographic imaging of the brain: The presence of the patient's neck makes it possible to rotate the camera around the head without hitting the shoulders. Because spatial resolution falls off rapidly with distance, it was important that the distance between the collimator and the patient's head be kept small. The challenge in the 1980s was to compete with the high anatomic resolution of transverse-section X-ray CT images and magnetic resonance (MR) images that were in wide and routine use by that time. Although the best images were produced by specialized, brain-only systems, such as those developed by Kuhl and Edwards (1968b), commercial manufacturers were not convinced to produce and market dedicated nuclear medicine brain-imaging systems. Important prognostic information regarding a particular patient's potential for recovery from a stroke can be obtained by SPECT studies using ^{123}I-iodoamphetamine, ^{99m}Tc-sestamibi, or FDG brain images. The ability of the brain to trap iodoamphetamine in the early days following a stroke provides good evidence that tissue function is returning and functional recovery likely. Early SPECT images of ^{99m}Tc-sestamibi uptake are also more sensitive for detecting stroke than CT or 2D nuclear images, but even today such images are not commonly used. Another practical reason that 3D SPECT is not more widely used for stroke or other emergency applications is that CT and MRI tend to be more readily available during off hours.

The surgical treatment of intractable epilepsy requires precise knowledge of the anatomical site that triggers the focal discharge. The high rate of glucose use (FDG uptake) in the periictal state has proven to be the best noninvasive means of identifying the location. Surface electrodes at surgery add tissue-level localization information as final guides for the surgical resection. PET and SPECT imaging of FDG, along with ^{99m}Tc-sestamibi imaging, have proven effective, and surgical excision of functionally active regions has been clearly shown to yield dramatic patient benefits. The appropriate radiopharmaceuticals and imaging equipment must be available on standby, and it is challenging to capture the transient phenomena with rotational gamma imaging.

There are many research questions that are being probed with ^{11}C- and ^{18}F- labeled congeners of neurobiologically active molecules. Radiolabeled drugs used in psychopharmacology are being used to determine their binding sites in normal and diseased subjects. Emission tomography is absolutely necessary for the quantitative analysis of data from these studies. Some of the drugs that contain fluorine in their natural form are ideally studied with ^{18}F. Other molecules can be labeled with ^{18}F without significant change in their biological distribution, partly because of the small fluoride ionic radius. The occupancy of targeted receptors can be determined. Differential rates and pool sizes can be assessed and related to the degree of benefit that patients derive from particular drugs. Conceivably, the classification of disorders that remain ill-defined, such as schizophrenia, can be established operationally, based on which drugs are beneficial, and such a classification could be more useful than symptom-descriptive terms.

The treatment of Alzheimer's disease is a growing area of concern because effective drugs are not yet established and the disease is becoming more prevalent as people live longer. The observation of altered glucose metabolism patterns in cross-sectional images can be an early sign of this disease. If it becomes possible to treat patients before amyloid plaques, tangles, and choline esterase level changes are advanced, one can expect improved benefit from therapy. A large number of important neuroreceptor studies have been conducted in patients with dementia (see Wagner *et al.*, 1995, 485–594).

PET is rarely used on an emergency basis because of the complexity of producing short-lived PET tracers and imaging around the clock. However, there are two possible exceptions: generator produced PET agents (Robinson, 1985) (^{82}Rb or ^{68}Ga generators are commercially available) and FDG imaging using material synthesized at the end of the day (^{18}F has a relatively long half-life of 110 min).

B. Thyroid Imaging and Therapy

The thyroid was the first organ imaged and the first for which effective therapy was developed. The high concentration of iodine in the thyroid was known for more than 100 years; thus it was natural that radioactive isotopes of iodine were the first to be used in patients. Studies were done to determine the rate of iodine uptake in thyroid; shortly thereafter, iodine's high concentration in thyroid was used to treat overactive or malignant glands. Uptake was determined by using GM counters placed on the neck. The development of NaI(Tl) scintillators, spectrometers, and collimated detectors led to the widespread use of thyroid scanning. The goal was to determine the size, shape, and temporal distribution of the tracer in the gland. Low uptake or absence of tracer concentration in nodules suggested a potentially malignant growth. The presence of increased uptake was evidence of hyperfunctioning thyroid tissue. Abnormal distribution of focal uptake outside the thyroid was taken as evidence of ectopic normal thyroid tissue (sublingual thyroid) or of malignant tissue metastatic from the thyroid. The initial images were produced with dot tappers from mechanical devices, and these were used commonly though the 1960s.

The calibration of global thyroid uptake measurements was investigated in detail by Marshall Brucer at the Oak Ridge Institute of Nuclear Science (ORINS). Neck phantoms containing known amounts of mock iodine were distributed to hospitals and physics labs around the world. The reported results varied very widely and revealed the need for much greater attention to the proper use of equipment and the need for standardization of methods. (This was at a time when even the definition of the millicurie, a unit of radioactivity, differed among major institutions.) However, by using known phantoms, protocols evolved that produced reliable results so that measurements could be compared among laboratories.

A problem in therapy of overactive thyroids (which is still not entirely solved) was to determine how much to administer a given patient based on thyroid uptake, size of the gland, and its inherent radiosensitivity. Similar problems in tumor therapy also remain.

Nodule detection and radioiodine quantitation remain problems. Emission tomography does not yet have sufficient resolution to accurately quantitate the activity in small (< 5–10 mm) nodules. The problem is that underlying and overlapping activity in the gland obscures local variations in the uptake in the nodules. With large functioning nodules, as in Plummer's disease, ^{99m}Tc scans show clear uptake in the nodule (Fig. 11). However, for detection of small nodules (as small as 2 mm) ultrasound is the best of the current imaging modalities and in fact has taken over many of the anatomical tasks previously conducted for intrathyroidal nodule detection. For further characterization, fine-needle aspiration is then used to characterize the lesion.

In addition to radioactive iodine, ^{99m}Tc-pertechnetate is often used to image the thyroid. Its uptake is increased in proportion to iodine uptake in most circumstances, it is readily available, and it is more effectively imaged than ^{131}I. It is not useful in planning for therapy because of its unpredictable uptake in malignant lesions. The measurement of stable iodine contained in the gland was made popular by Hoffer and colleagues at the ACRH in the 1970s using X-ray fluorescence systems (Jonckheer and Deconinck, 1983; Hoffer et al., 1968). Some systems used X-ray tubes as exciting sources, whereas most used large amounts of ^{241}Am. Problems in calibration made it difficult to compare thyroid iodine content between large and small glands. This has now been overcome by the use of ultrasound-correlated studies (Reiners Chr 1998; Mardirossian et al., 1996). Quantitation of iodine content in nodules has been reported, but the error bounds are large (Patton et al., 1976). One of the early roles of nuclear medicine was to assist the surgeon in locating tumor-involved nodes in the operative field. The ORINS (subsequently known as ORAU) group developed a small, gold-collimated intraoperative probe (IOP). Several days prior to surgery, the patient was given ^{131}I. The probe was then used by the surgeon to ensure that he had removed all of the tumor that was accessible in the operative field.

Surgical probes are enjoying a renewed level of interest, mainly for similar applications with thyroid, breast, and melanomas. In these cases, tumor and lymph-node dissemination pathways are revealed, and surgical-management plans are developed using this information. Three-dimensional imaging of anatomy prior to surgery can be fused with properly registered IOP images and could be useful in surgery in the near future for a number of clinical applications.

C. Parathyroid Disease

The four small parathyroid glands, usually closely approximated to the posterior surface of the thyroid glands when diseased, may become overactive, and they may also become malignant. The biochemical diagnosis is based on abnormal calcium and parathyroid hormone levels in blood, and anatomical information obtained using MRI and CT is not often very helpful when the tumors are small. Radiolabeled tracers have been used with varying success for many years. ^{57}Co B$_{12}$ was used for a short time but, more recently, the late uptake of ^{99m}Tc-sestamibibi has proven more useful. Differences in the time of appearance of activity in the thyroid and parathyroid help to distinguish abnormal uptake in the closely approximated parathyroid glands. Not infrequently, several years after the primary tumor has ben removed, the symptoms and signs recur. The prior surgery makes reoperation more difficult and, furthermore, the hyperfunctioning gland may be in different locations due to the diverse embryological sites from which the gland originates. In some cases, emission tomograms can be useful to the surgeon in planning an operative approach. Figure 12 presents a planar ^{99m}Tc-sestamibi scan in which a functioning parathyroid tumor is located in the heart region. The changing time distribution over the 20–30 min following injection has precluded standard SPECT imaging, but dSPECT could be a useful approach (Celler, 2000).

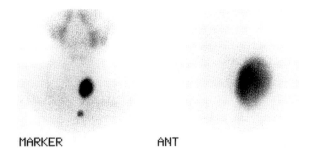

FIGURE 11 Functioning thyroid nodule in the left lobe of the thyroid in a patient with Plummer's disease (Department of Radiology, Vanderbilt University).

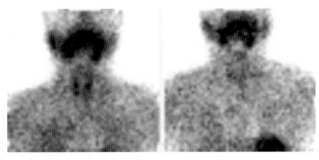

FIGURE 12 ^{99m}Tc-sestamibi parathryroid scan showing increased uptake in normal regions in the thyroid and salivary glands, plus increased uptake in parathyroid adenoma in the chest in the heart region (upper mediastinum) (Department of Radiology, Vanderbilt University).

D. Cardiac Imaging

The heart has been the subject of much investigation, and tomography has now become an important part of routine patient care. The very first nuclear cardiology study measured the transit time through the heart-lung circuit. Blumgart and Weiss in the late 1920s administered radon in solution into an arm vein and recorded its transit time across the cardiopulmonary circuit in normal subjects and patients with congestive heart failure.

In later years, the known avidity of the heart for cations—K, Rb, Cs, and Tl—has led to many investigations of the pattern of normal and abnormal uptake of ^{42}K, ^{43}K, ^{82}Rb, ^{131}Cs, and ^{201}Tl. Similarly, the avidity of the heart for radioactive glucose and fatty acids as metabolic substrates has also been studied extensively. The use of flow tracers has also received much attention, including several ^{99m}Tc-based tracers (e.g., pertechnetate, sestamibi, and tetrafosmin).

Early studies of coronary artery perfusion used selective catheterization of the right and left coronaries with tracers of distinguishable energy. In this way, the distribution of flow through the individual arteries could be viewed in a single image in which each of the injections could be coded in a different color and anomalies in their biodistribution and mixtures of pathways could be noted.

The relative merits of different myocardial ischemia protocols depend on the radionuclidic and pharmacokinetic properties of the tracers, their cost, and their availability. For clinical evaluation of myocardial ischemia, ^{201}Tl is most commonly favored.

All procedures employ differences in the distribution at rest and exercise or the response to pharmacological strategies simulating the effect of exercise on coronary perfusion. Exercise brings out the compliance of the normal vessels, which increases blood flow to keep up with the increased cardiac metabolic needs of the perfused muscle regions in the heart. Abnormal vessels do not dilate normally, and the resulting distribution differences are key to the diagnosis of myocardial ischemia. Because the heart depends normally on fatty acid metabolism and secondarily on glucose for its energy source, FDG and fatty acid analogs have been the subject of much research and clinical interest.

FDG use is a well-established indicator of recoverable myocardium in damaged heart regions. ^{11}C-labeled palmitate and acetate have also been used to map the distribution of aerobic cardiac energy use. Beta-methyl-substituted fatty acid derivatives have also been used to assess uptake by the heart. The beta-methyl group blocks the oxidative breakdown of the fatty acid congeners, which makes it possible to image its integrated uptake, as in the case of FDG accumulation (Kropp et al., 1999).

Because two simultaneously administered tracers cannot be distinguished using PET imaging systems, other options have been explored. One is to use Anger cameras with thick crystals and high-energy collimators. One energy window can be used to image a 511-keV positron-emitting tracer (^{18}F- or ^{11}C-labeled compounds) while a second tracer, such as ^{99m}Tc-sestamibi or ^{123}I-labeled beta-methyl fatty acid congener, can be imaged in a second energy window. Alternatively, one can administer the same or a different short-lived PET tracer repeatedly to the patient in successive states. The main limitation in the second approach is that the patient may not be in the same physiological state or anatomical position in the two study sessions.

An interesting approach explored in the 1970s used an in vivo ^{81}Rb-^{81m}Kr generator (H. S. Winchell, pers. comm.). The strategy was to inject ^{81}Rb (446 keV, 4.6 h), which was preferentially extracted by the healthy well-perfused tissue. The short-lived ^{81m}Kr (190 keV, 13 s) daughter grew into the Rb-occupied regions and then was washed out in proportion to regional blood flow. Changes in the Rb/Kr ratio were a measure of regional myocardial wall uptake and perfusion. The poor ability of the Anger camera to image the 446-keV photons, scatter from the Rb into the Kr window, and the limited availability of the agent caused the method to fail. However, new detectors and on-site cyclotrons could make it a feasible option once again.

Other interesting tracers used in cardiac imaging included ^{43}K and ^{42}K, but their energies were too high for the Anger camera. Also, the uptake of generator-produced ^{52m}Mn, the daughter of ^{52}Fe, was used to map the distribution of myocardial mitochondrial processes, but the availability of ^{52}Fe generators was never achieved commercially.

Parametric imaging condenses information and has been important for the display of the results of dynamic studies, especially in nuclear cardiology. Color-coded images have been used to advantage in several circumstances. The first was in the evaluation of shunts and valvular regurgitation (Patton and Cutler, 1971). The temporal kinetics of flow between the different cardiac chambers provides information on timing and extent of shunting. To flag a region of interest and determine the activity in individual heart chambers is difficult as they partly overlap in different planar projections. A bolus of tracer is injected, and sequential images of tracer moving through the cardiac chambers are stored. When replayed, the initial appearance in the right heart was coded red, the intermediate pulmonary phase green, and the last (ventricular) phase blue. The summed color mixtures revealed places where chambers overlapped, whereas pure colors revealed separated regions of interest in different chambers. Uptake in the myocardium was obtained in a similar fashion using the time-coded distribution of radiolabeled particles injected directly into the coronary circulation (Ashburn et al., 1971). The use of color has been disputed by many radiologists, which may be related to the fact that many of them often claimed to be color-blind.

Two of the current applications of color coding of nuclear cardiology images are for delineating wall motion (cine display) and ischemia mapping bulls-eye display). Wall motion is an index of the forcefulness with which blood is ejected by the heart and is best displayed by color-coded intensity in gated movies of the cardiac cycle. Displays of myocardial perfusion gradients compare polar images of tracer uptake at rest and following exercise in the bulls-eye presentation. These functional maps condense the activity on successive tomographic planes into a single view that summarizes much of the information. Looking at the individual slices in highlighted bulls-eye regions provides additional information and can resolve doubts concerning the validity of the apparent lesions, which could be due to artifacts induced by patient motion or artifacts caused by activity in adjacent organs (liver), attenuation by the arms and breasts, or obesity.

Dead myocardial tissue has been imaged with ^{111}In-antimyosin. The myosin fibrils are ordinarily not in contact with blood proteins, but when the heart cell membranes are damaged, antimyosin can react with and image the dead tissue.

Emission tomography of the heart has made it possible to accurately identify and localize defects in myocardial perfusion and wall motion abnormalities. The effects of exercise and medical therapy can be assessed; recovery and response to medications can quantitated using 3D information. The competition with real-time 3D ultrasound, as well as MRI, is keen, and the diffusion of technology and skill in its use will dictate their future pattern of use in this rapidly developing area.

E. Lung Imaging

The lung is responsible for the control of gas exchange and storage, activation, and release of metabolically active substances from its large vascular bed. Thus, studies of these processes continue to be important in health and disease.

The first clinical lung studies were directed at the diagnosis and localization of pulmonary emboli. The lung is a filter that protects the brain from small clots that occur at a low rate in normal individuals. Thus, the use of small radioactive aggregates that plug up a small fraction of the capillaries in perfused segments of the lung have been useful means of identifying normally perfused regions. The absence of radioactivity in lung regions is taken as presumptive evidence of impaired perfusion. The presence of wedge-shaped lesions is a frequent finding in small emboli in the lung periphery, and these are best appreciated on transverse-section images. Large central lesions can be validated radiographically, whereas small peripheral lesions are not easily verified and the fraction that are false positives is not known. The ventilation in the affected lung segments is not affected in the early hours following an embolus. Thus, the finding of intact ventilation in a poorly perfused segment is supportive of an embolic event.

Ventilation defects are identified based on low content of Xe-133 in impaired lung regions. Scatter from surrounding well-ventilated regions obscures defects and poor scatter rejection in the low-energy regions (^{133}Xe, $E = 77$ keV) makes it difficult to detect even large (2 to 4-cm) poorly ventilated regions in the lung. Blur distortion from lung and heart motion further lessen the sensitivity of detecting small intrapulmonary (and intracardiac) lesions. For imaging perfusion defects due to pulmonary emboli, one compares the decreased uptake of ^{99m}Tc MAA, a macroaggregate that plugs up capillaries that normally receive blood, with normal uptake of radioactive xenon gas in the well-ventilated lung segments compared to the embolized poorly perfused segments. Multiple views of the lung are imaged and disparities are looked for on the regionally matched image pairs. Such an image sequence is shown in Figure 13 in a patient with a cardiac lesion who had an otherwise normal lung scan.

SPECT and PET imaging of the lungs are still relatively underdeveloped. The problems are optimal tracer availability and difficulty in stop-motion lung imaging. Dual-isotope imaging would be the ideal means of imaging blood-flow and gas-exchange abnormalities simultaneously. The ideal combination would be to have a higher-energy gas-flow tracer that is washed out with lung clearance imaged simultaneously with a ^{99m}Tc-based macroaggregate that is excluded from inflow into embolized regions. The gas-flow tracers suggested for use in this combination have been ^{127}Xe (203 keV) and ^{81m}Kr (190 keV). Production problems and cost issues have retarded the development of these studies. The fact that PET systems can image only one tracer at a time has limited its contribution.

The first PET coincidence system used effectively for lung imaging was developed by the Massachusetts Institute of Technology (MIT) group (Burnham et al., 1970). The system, which comprised two opposed large-area detectors,

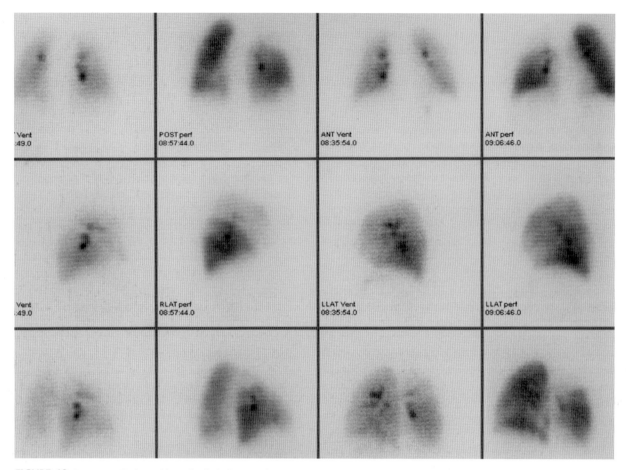

FIGURE 13 Lung scan: Patient with mediastinal disease, with a normal lung scan (Radiology Department, Vanderbilt University).

collected and displayed 3D images of radioactive gas distribution in the lungs (Chesler et al., 1975). The limited field of view of the subsequently developed ring-geometry PET systems did not contain the full lung field, and little work on lung imaging has been pursued in more recent years. Motion distortion is a major problem degrading resolution when imaging the heart and lungs, the interaction of their time-varying cycles affecting imaging of both organs. Lung gating has been possible in animals and in normal subjects who are able to cooperate better than sick people. Flow monitors and expansile measurements of chest wall motion have been useful along with electrocardiogram (ECG) gating to correlate radioactive images in the lung and heart at selected points in their respective cycles. Such studies require more time and patient cooperation than can ordinarily be expected, a problem that deserves further attention now that large-area PET/SPECT cameras are available.

A novel approach to embolus detection used ^{11}CO, which tightly couples to hemoglobin when mixed in a syringe with the patient's blood. When injected intravenously, the ^{11}CO-labeled red blood cells are distributed to the different perfused lung regions. When the tracer reaches the blocked segments, it remains there, whereas the rest of the ^{11}CO-labeled blood is cleared from normally perfused segments (West, 1966). Except for the problems of coordinating cyclotron time with patient imaging, this should be a useful procedure.

F. Kidney Imaging

Kidney imaging with radioactive tracers is directed to looking for abnormalities in the renal location, intra- and perirenal masses, and functional indicators of glomerular and tubular function. The anatomic studies employ tracers that localize in the kidney and stay there. ^{203}Hg- and ^{197}Hg-chlormerodrin were the early agents used, but ^{99m}Tc-DMSA replaced these because of its favorable imaging and dosimetry properties. These procedures were done with planar imaging, but are best done using 3D SPECT systems for increased anatomic detail.

Renal blood flow can be studied with any number of tracers that stay in the blood stream or are extracted from the blood in passage through the organ. Clearance of tracers by the cortical glomeruli is studied most commonly with ^{99m}Tc DTPA. Tubular function was originally studied with ^{131}I-labeled hippuran, but this has been replaced by ^{99m}Tc-

MAG3, a better imaging agent that also imparts a lower absorbed patient dose.

Dynamic images of tracers are typically made using rapidly collected sequential planar images of the heart, kidneys, and bladder, assuming all can be included in the camera field of view. The heart information (probe or camera-based) gives the input function, and the system transit times are derived from the kidney and bladder kinetics. The use of new slip-ring rotating camera detectors now makes it possible to accomplish complete rotations of one or more detectors on the gantry in less than 1 min. Thus cortical, medullary, renal pelvis, and bladder distributions can be separated and quantitated. It has also been shown that mathematical analysis of the changing distributions from more slowly sampled data can also be reconstructed accurately (Celler *et al.*, 2000).

Quantitation of renal physiology studies has been done using curve and compartmental analysis. The study of hypertension induced by kidney disease, the radiographic study of narrowing of the renal arteries, and the biochemical analysis of blood levels of rennin and angiotensin are still being used clinically. Changes in renal function induced by captopril in renovascular disease are used to bring out masked abnormalities.

Kidney imaging has also been done to provide an internal landmark when the primary question lies elsewhere. Thus, imaging of hyperfunctioning adrenal lesions with radioiodine-labeled iodocholesterol has often been used with simultaneous imaging of the kidney to provide a good anatomic landmark for lesion localization. These studies can be done with planar or emission tomography systems.

G. Bone Imaging

^{47}Ca-, ^{85}Sr-, and ^{18}F-fluoride imaging were used in early studies of bone uptake to detect primary and secondary disease. The high gamma-ray energy of these tracers was unfavorable for the gamma camera and, when ^{99m}Tc-labeled phosphonate derivatives became available, they quickly took over (Subramanian *et al.*, 1977). Figure 14 illustrates the uneven and abnormal distribution seen in bone in a patient with hypertrophic pulmonary osteoarthropathy. Inflammatory lesions, tumors, and fracture repair all have a similar appearance on static images, but the temporal evolution of changes can often help resolve these differences. Labeled white blood cells, FDG, and antigranulocyte antibodies have been used to target inflammatory lesions in blood vessel walls, soft tissue, as well as in bone. An example of abnormal uptake of ^{111}In-labeled white blood cells in a vascular graft is illustrated in Figure 15. SPECT and PET studies are important well-established nuclear medicine procedures. Because of the high bone-to-soft-tissue contrast, the visualization of rotating projections of bone scans often provides sufficient clues based on the

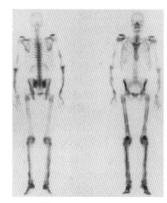

FIGURE 14 Images of ^{99m}Tc $^-$MDP in a patient with a lung disease manifest by abnormal distribution especially in the distal long bones (Radiology Department, Vanderbilt University).

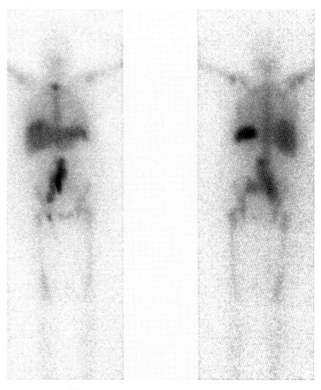

FIGURE 15 ^{111}In-white blood cell (WBC) image of an infected graft (Radiology Department, Vanderbilt University).

differential rates of rotation of structures at varying depths to be useful, even without tomographic image reconstruction.

H. Reticuloendothelial System Imaging

Radioactive particulates are cleared by phagocytic cells in the liver, spleen, bone marrow, and lymph nodes and by macrophages in the circulation and extravascular spaces. Typically, these studies are done using a series of planar images that track the movement of the tracer following systemic or local instillation.

Although there is no clear separation based on size, there is preferential clearance of very small particulates (average 100 nm) by phagocytic elements in the bone marrow. Intermediate sizes are best cleared by the lymph nodes, and these are used for sentinel node imaging. Liver and spleen uptake is the preferred path for 500-nm to 5-μm particles. Imaging of hepatocyte function in the liver is based on tracers that target the parenchymal cells. Agents such as rose bengal, or bromsulphalein labeled with ^{131}I were used for many years. More recently, tracers that are cleared by the liver and exit through the gall bladder are used for imaging both pathways, along with kinetics to distinguish normal from abnormal transit times and response to gall-bladder contraction stimuli. SPECT imaging is useful in revealing anatomic detail not otherwise available from planar imaging.

I. Pancreas Imaging

The pancreas continues to be a difficult organ to image. The use of protein synthetic precursors started with ^{75}Se-labeled methionine, but the procedure had low sensitivity and specificity. Work with ^{11}C-labeled amino acids had some success, but the pancreas remains a difficult organ in which to differentiate malignant from inflammatory disease (Yamamoto et al., 1990). The differential uptake and washout of D- versus L-amino acids showed some promise, but the inefficiency of the labeling and separatory process never allowed it to become a practical reality (Buonocore and Hubner, 1979).

J. Tumor Imaging

Tumors that have received most attention are brain, breast, prostate, thyroid, and lung (see Wagner et al., 1995, 1041–1136). The high false-positive rate of X-ray mammography has stimulated nuclear medicine to develop new devices and tracers to improve breast cancer detection and characterization. So far, ^{99m}Tc-sestamibi and ^{18}F-FDG have been used most prominently for breast-imaging tasks.

Antibody targeting has been widely explored since the 1970s. The main problem with intravenous-injected antibodies is the large dilution factor and the poor transport of large molecules through the capillaries, with limited tumor-cell access. Injection into the skin surrounding the lesion provides good access of the injected tracer to the pathways of dissemination, and tumor-involved nodes can be well delineated in this way. This is useful for tumor staging including sentinel-node imaging.

Prostate tumor imaging started with unsuccessful attempts to image the gland following injection of radioactive zinc (^{65}Zn), a normal constituent. More recently, imaging has been done with ^{111}In-ProstaScint, a commercially available antibody that targets a specific membrane protein on the prostate cell wall. Because of poor access characteristic of large molecules and resultant low prostate uptake, these studies have not lived up to their initial promise.

Lung cancer imaging has been achieved with FDG. Other agents that have been used include several radiolabeled antibodies, along with ^{11}C-labeled methionine (Kubota et al., 1989) and ^{57}Co-bleomycin (Slosman et al., 1989). Currently, the most common nonspecific agent used for clinical imaging of cancer is FDG (imaged in positron mode), although imaging of the 511-keV gamma rays with collimated detectors is also used clinically when PET cameras are not available. Many examples of FDG images are presented in Lalush and Wernick (Chapter 21 in this volume).

There are several ways in which nuclear medicine plays a role in oncology, including staging (delineating involved tumor sites), tumor characterization (kind and amount of uptake), prognosis (based on changes in tumor uptake during/after different forms of therapy), and delivery of tumor therapy per se from injected heavily ionizing drugs. Radioactive tracers can target specific metabolic pathways and cytoplasmic and nuclear receptors with high selectivity. The high specificity of targeting provides unique information that makes it possible to identify the features of particular tumor cell types. Because we know that cancer is not a single process and that each tumor has very specific properties, this high selectivity is very important for understanding the behavior of tumors in clinical and research applications.

Much industrial and academic research has been devoted to the development of tracers that target tumor-specific receptors. Typically, these use antibodies or peptides that are distributed in kits as ligands to which one adds the desired label prior to their administration for diagnostic or therapeutic purposes. New combinatorial chemistry methods make it possible to develop ligands that target a specific receptor. From a diagnostic standpoint, labeling such molecules with short-lived photon emitting tracers is a very powerful tool for diagnosis.

Tumor-seekers labeled with beta emitters (positron or negatrons) that emit densely ionizing particles when injected in large amounts are useful in delivering tumor therapy. The advantage of positron emitters in this application is that one can do accurate dosimetry on the therapy dose by imaging the 511-keV annihilation rays, while therapy is delivered by energy absorbed in stopping the positron in the tumor region.

Imaging plays a key role in diagnosis and therapy in several ways. First, the findings from diagnostic imaging are a guide to therapy, answering questions such as: Where is the tumor (and its metastases)? Is the tumor well perfused? Is the tumor uniformly or asymmetrically active? Do chemotherapy agents localize in the tumor, and if so to what extent? Given that radiation therapy is to be administered, can radiolabeled tumor seekers be used to deliver therapy?

Given that therapy has been administered, what metabolic changes in tumor occur following chemotherapy? Following surgery, was it all removed? Following radiation therapy, is the tissue remaining in the tumor bed necrotic or is it metabolically active residual tumor?

When nuclear medicine provides the primary therapy, imaging plays a major role in deciding on the dose to be given to a particular patient (known as patient-specific dosimetry). This involves the choice of the optimum tumor-specific agent based on kinetics of the distribution derived from multiple images of the radioactive-drug distribution, on the location of the tumor sites, and their relation to radiosensitive organs. In these cases, emission tomography can provide the accurate quantitative information needed to compute dose distributions and plan therapy. In general, whole-body 2D scans are done with PET or SPECT systems to delineate suspect regions of interest. Thereafter, 3D images provide the anatomic information in greater detail in tumor-involved regions.

Blood flow and tumor metabolism are two processes that nuclear medicine delineates. Angiogenesis is an important aspect of tumor biology, and there is keen interest in mapping gradients in flow to and within tumors at baseline and following different forms of drug therapy including anti-angiogenesis drugs. Tumors grow by the proliferation of new blood vessels at their surface. Blood flow to the tumor is important because tumors depend on blood flow for their high growth rate. Further, distribution of flow within the tumor is important because hypoxic regions require higher doses than well-perfused regions to achieve a good response to radiation therapy. Tumors tend to have hypoxic or necrotic centers when tumor-growth requirements exceed the tumor blood supply. Drugs, such as labeled misonidazoles, localize in the hypoxic tissue and have been used in such circumstances, but the rapid urinary clearance and poor perfusion of the central tumor region has limited their clinical role ("Molecular (functional) imaging," 2003).

Tumor blood flow can be measured using first-pass uptake of tracers such as radioactive microspheres injected into the arterial tumor inflow, which get stuck in the first capillary bed to which they are delivered. Because their distribution changes very slowly (by degradation and release of the label), they can then be imaged with PET/SPECT systems. Flow can also be delineated by chemical microspheres, agents for which the tumor bed has a particular affinity (Kung, 1990). A kinetic study of the distribution of many labeled compounds can also be used to assess tumor blood flow. Where there is an in-house cyclotron, the use of natural substances such as ^{15}O-water and $C^{15}O_2$ can provide flow information without having to get approval from the Food and Drug Administration (FDA) for in-house produced drugs.

Tracers that specifically target tumors are of greatest interest because of their investigative role in tumor biology along with diagnosis and therapy. Initial work was done using tracers that delineated normal organ pathways. The nonspecific localization of areas of diminished normal organ function directed attention to suspicious abnormalities that could be diagnosed by other tests, including image-guided open or closed biopsies. Modern radiochemistry now provides more specific radiopharmaceuticals that target particular tumors. Radiolabeled antibodies directed to tumor antigens have not been as successful as had initially been hoped due to both nonspecific and low uptake of these large molecules. Receptor-targeted agents, such as somatostatin analogs, play an important role in localizing and in treating neuroendocrine tumors (Krenning et al., 1993). Agents directed against the carcinoembryonic antigen (anti-CEA) have been useful in delineating some tumors (colorectal and medullary thyroid), as well as bone marrow and inflammatory sites (based on labeling CEAs on the surface of circulating granulocytes) (Goldenberg et al., 1989). In contrast to the use of tumor-specific tracers, FDG has proved to be a useful means of detecting a wide range of tumors of different types in primary and metastatic sites (Delbeke et al., 1998). The tracer is picked up by tumors that are metabolically active, based on the prominent glycolytic pathway in tumors that favors glucose over the oxygen-dependent fatty acid pathway. Thus, FDG is useful in targeting all but a few tumors and is able to help distinguish metabolically active tumors from residual fibrotic tissue remaining after otherwise successful therapy. Cost-effectiveness issues are being resolved in favor of its use in conjunction with other anatomical 3D imaging methods (Jerusalem et al., 2003). Tumor imaging is important for directing further therapy (surgery, external beam or internal emitter, or no treatment). The tracer studies are important in each situation. For surgery, knowledge that there are distant metastases based on SPECT images, and confirmed by CT or MRI, changes the decisions on the type and extent of further treatment.

Tumor-directed imaging studies help lead the surgeon going into an operation to the region(s) of interest. Once in surgery, the surgeon can use a hand-held IOP to determine the location of tumor by its increased radioactive content and use this evidence to create an appropriate operative margin. Alternatively, given a lesion that is inoperable due, for example, to its adherence to vital structures, radioactive colloids can be injected into the tumor region defined by SPECT and guided by CT images without the need for surgery. This has been used with benefit in advanced pancreatic cancers (Order et al., 1994).

In determining whether a melanoma has spread, imaging the lymphatic distribution is a routine procedure. The area adjacent to the tumor is injected with radioactive particulates. The distribution of the observed pathways of dissemination reveals the site of the first node in the chain. The surgeon removes the lymph node, which, if positive, leads to a revised surgical plan with adjuvant treatment. With new imaging probes being developed for small animal imaging, it is possible that high-resolution images of the intranodal distribution of the tracer in excised enlarged nodes could assist the pathologist in microscopic evaluation and play an added role in sentinel node imaging in melanoma and breast cancer characterization.

The nuclear medicine physician treating cancer patients needs to be able to work with the other members of the oncology team to determine the dosage that is needed to destroy the tumor and the amount that can be administered with safety based on the patient's disease status and other cytotoxic treatments received or planned. The amount of activity that should be administered depends on tumor size and uptake per gram, and these require knowledge of the anatomical size of the tumor (CT/MRI defined) and the functional (PET/SPECT defined) uptake kinetics and the distribution of the therapeutic radionuclide within the tumor. Further, the dosage to the bone marrow and other dose-limiting organs needs to be assessed. These determinations require accurate quantitative measurements of the amount of radioactive drug in the tumor and normal organs as a function of time. From this, one determines radionuclide residence times in the different organs. Dose contributions to different organs and the tumor are obtained from calculations using dose-conversion factors based on anthropomorphic reference phantoms; more recently, it has become possible to do so using patient-specific anatomic data (Kolbert *et al.*, 1997; Bolch *et al.*, 1999). This requires registration of the radionuclide images with the particular patient's anatomical maps (3D voxel-based image correlations) to determine absorbed radiation dose to the tumor and normal organ. These are all tasks that can be accomplished and the further development of 3D treatment-planning software is the subject of much ongoing investigative effort.

The choice of optimum therapeutic radionuclide depends on radionuclide properties, on the intratumoral distribution, and on the time course of its retention in the tumor and normal organs. Accurate biodistribution data are obtained from SPECT/PET images registered with anatomical data. The logistics of acquiring multiple images are a problem in that sick patients cannot lie still for the multiple SPECT and whole-body images that ideally would be obtained for the most accurate analyses. Higher-sensitivity lower-resolution systems are needed for multiple SPECT-based dosimetry studies to become a practical reality. The use of electronic collimation minimizes the mechanical collimator losses that decrease the sensitivity of SPECT imaging. The possibility of obtaining higher-sensitivity (high-resolution) images with the Compton coincidence imaging of high-energy gamma rays remains a viable option that is the subject of continuing exploration (Rogers, Chapter 19 in this volume).

The radiotherapist has an increasing need for the 3D nuclear images beyond tumor localization. Modern radiotherapy is attempting to use knowledge of blood flow and functional activity distribution within the tumor. With the advent of the new multileaf collimators and intensity-modulated conformal treatment-planning systems, it is possible to shape the therapy field to focus on the active parts of the tumor and to deliver lower doses to less-active parts of the tumor (when they are adjacent to highly radiosensitive normal structures). Thus, in treating prostate cancer with external beam radiotherapy, radiotherapists try to deliver low doses to the inactive portion of the prostate adjacent to the rectum. The combination of CT with SPECT gantries provides direct fusion of images, which assists in the treatment-planning process.

The chemotherapist also has emerging needs for nuclear medicine imaging data. It is recognized that multi-drug-resistant genes exist that decrease tumor retention of chemotherapy drugs by pumping them out of the tumor (which means that larger amounts of the drugs are needed to treat that tumor or that a search should be made for alternative drugs that are less affected). ^{99m}Tc-sestamibi uptake and release kinetics from the tumor provide information on the P-glycoprotein transporter pathway responsible for this drug-rejection mechanism in individual patients (Luker *et al.*, 1997). The use of radioactive platinum or ^{13}N-labeled Cisplatin could be used for such applications, as could the actual drug being used if it could be labeled and analyzed for drug-delivery decisions for individual patients. Patient-specific treatment planning is important for chemotherapy, as well as for other forms of radioactive and drug therapy (Evans *et al.*, 1998).

K. Molecular Biology Applications

Modern molecular biology uses the detailed evolving knowledge of the gene and genetic control processes to trace the biochemical steps underlying gene action. The great interest and support for basic and applied research in molecular biology has stimulated rapid advances in high-resolution imaging technology designed to delineate sites and modes of gene expression in small animals, especially mice. Mice are especially useful because there are many mouse models of disease and many site-specific molecular markers and probes that are commercially available.

Great interest lies in the use of molecular methods to locate and characterize tumor cells with the intent to change these processes by gene manipulations. Because one now has complete information on the large number of genes in the mammalian genome, knowledge of what they code for and where the processes are located are critical issues. Radioactive and fluorescent reporter molecules are used to locate sites to which the genes are targeted and to identify the expression products of these genes (Gambir *et al.*, 1999). Great interest lies in methods of optimizing the delivery of genes to desired targets, and various methods and vehicles are under active investigation. Specialized 3D imaging systems have been developed for PET imaging of mice for use in these studies (Cherry *et al.*, 1997) and state-of-the-art devices are discussed elsewhere in this volume. Submillimeter-pixel, high-resolution, planar, and SPECT systems using multipinhole-collimated solid-state devices are under development (Barber *et al.*, 1997). Compton scatter and attenuation are not the major problem for mouse imaging that they are in human studies. If scatter is not a serious problem and depth information is not needed for imaging gene-delivery systems in mice, the use of a large planar Anger-type cameras, collimated with small pinholes, can provide the spatial and temporal resolution needed for many purposes. For brain research in rodents, the requirements for high spatial resolution are especially stringent, and 3D imaging of PET tracers in rat brains is at the limit of what appears feasible with presently available microPET imaging systems. Higher-resolution functional images can be obtained from *in vitro* studies obtained at sacrifice and correlated with the last *in vivo* images obtained. Autoradiography of thin sections can be registered with and compared to the optical images of specimens stained with immunofluorescent markers of receptors of interest, as well as standard histochemical agents.

With the new understanding of genes and tumor biology, new drug development methods are proceeding at a rapidly increasing rate. New combinatorial chemistry methods add a new dimension to drug discovery using randomly associated oligomers selected for their receptor-binding properties, instead of specifically synthesized intact fragments of DNA that normal enzymes degrade. Automated screenings of randomly generated molecules, which adhere most strongly to a receptor bound to a separatory column, are selected based on their binding constants. Polymerase chain reaction (PCR) amplification makes it possible to select the most strongly reacting molecules and make large numbers of them for use in the desired targeting tests. The method is being applied to the development of tumor-diagnostic and -therapeutic agents, as well as to the development of agents that target drug-resistant bacteria.

A new understanding of cancer biology and treatment options is advancing rapidly using new *in vivo* and *in vitro* imaging technologies. The rapidly growing understanding of genes and their role in biological control and regulation is leading to important new approaches to diagnosis and therapy, and nuclear medicine is playing an important role in these developments. As the human genome map is completed, much attention is being focused on the development of drugs to perform very specific functions, and imaging studies will play an important role in the validation and characterization of their actions (Phelps, 2000).

VI. SUMMARY

Emission tomography has made it possible to increase the visibility of low-contrast structures and provides quantitative information on the structure, uptake, turnover, and metabolic properties of biological systems, along with the pharmacokinetic behavior of radiolabeled drugs. Emission tomography has made it possible to move seamlessly from destructive 3D studies in animals to noninvasive nuclear imaging procedures in humans that are enhancing knowledge in many areas of medicine (Phelps, 2000). Excellent recent reviews of many of the clinical issues can be found, with extensive illustrations, in contemporary nuclear medicine textbooks (Sandler *et al.*, 2003).

References

Abbe, E. (1873). Schultzes. *Archiv. Mikroskopische Anat.* **9:** 413.

Andrews, G. A., Sitterson, B. W., and Ross, D. A. (1962). The use of thyroid scanning in thyroid cancer. *In* "Progress in Medical Radioisotope Scanning" (R. M. Knisely, G. A. Andrews, and C. C. Harris, eds.), pp. 312–343. U.S. Atomic Energy Commission, Washington, D.C.

Anger, H. O. (1957). "A New Instrument for Mapping Gamma-Ray Emitters". Biology and Medicine Quarterly Report. UCRL 3653.

Anger, H. O. (1972). Multiplane tomographic gamma–ray scanner. *In* "Medical Radioisotope Scintigraphy," Vol. 1, pp. 203–216. International Atomic Energy Agency, Vienna.

Anger, H. O. (1968). Tomographic gamma-ray scanner with simultaneous read-out of several planes. *In* "Fundamental Problems of Scanning" (A. Gottschalk and R. N. Beck, eds.), p. 195–211. Charles C. Thomas, Springfield, IL.

Anger, H. O. (1974). Tomography and other depth-discrimination techniques. *In* "Instrumentation in Nuclear Medicine, 2nd ed." (Hine, G. J., and Sorenson, J. A., eds.), Vol. 2, pp. 61–100. Academic Press, 1974.

Anger, H. O., and Rosenthal, D. J. (1959). Scintillation camera and positron camera. *In* "Medical Radioisotope Scanning," pp. 59–75. International Atomic Energy Agency, Vienna.

Anger, H. O., and Van Dyke, D. C. (1964). Human bone marrow distribution shown in vivo by iron-52 and the positron scintillation camera. *Science* **144:** 1587.

Aronow, S. (1962). Positron brain scanning. *In* "Progress in Medical Radioisotope Scanning" (R. M. Knisely, G. A. Andrews, and C. C. Harris, eds.), pp. 371–387. U.S. Atomic Energy Commission, Washington, D.C.

Aronow, S., Brownell, G. L. (1956). An apparatus for brain tumor localization using positron emitting radioactive isotopes. *IRE Conv. Rec.* **9:** 8–16.

Ashburn, W. L., Braunwald, E., Simon, A. L., *et al.* (1971). Myocardial perfusion imaging with radiolabeled particles injected directly into the coronary circulation in patients with coronary artery disease. *Circulation* **44:** 851–865.

Atkins, F., and Beck, R. (1975). Effects of scatter subtraction on image contrast. *J. Nucl. Med.* **16**(1): 102–104.

Atkins, F., Hoffer, P., Palmer, D., and Beck, R. N. (1977). Dependence of optimum baseline setting on scatter fraction and detector response function. *In* "Medical Radionuclide Imaging," Vol. 1, pp. 101–118. International Atomic Energy Agency, Vienna.

Barber, H. B., Marks, D. G., Apotovsky, B. A., Augustine, F. L., Barrett, H. H., Butler, E. L., Dereniak, F. P., Doty, F. P., Eskin, J. D., Hamilton, W. J., Matherson, K. J., Venzon, J. E., Woolfenden, J. M., and Young, E. T. (1997). Semiconductor pixel detectors for gamma-ray imaging in nuclear medicine. *Nucl. Inst. Meth. A* **395**: 421–428.

Beck, R. N. (1961). A theoretical evaluation of brain scanning systems. *J. Nucl. Med.* **2**: 314–324.

Beck, R. N. (1964a). A theory of radioisotope scanning systems. *In* "Medical Radioisotope Scanning," Vol. 1, pp. 35–56. International Atomic Energy Agency, Vienna.

Beck, R. N. (1964b). Collimators for radioisotope scanning systems. *In* "Medical Radioisotope Scanning," Vol. 1, pp. 211–232. International Atomic Energy Agency, Vienna.

Beck, R. N. (1968a). Collimation of gamma rays. *In* "Proceedings of the Symposium on Fundamental Problems in Scanning," pp. 71–92. Charles C Thomas, Springfield, IL.

Beck, R. N. (1968b). A note on crystal efficiency. *In* "Proceedings of the Symposium on Fundamental Problems in Scanning," pp. 229–230. Charles C Thomas, Springfield, IL.

Beck, R. N. (1968c). The scanning system as a whole—general considerations. *In* "Proceedings of the Symposium on Fundamental Problems in Scanning," pp. 17–39. Charles C Thomas, Springfield, IL.

Beck, R. N. (1975). Instrumentation and information portrayal. *In* "Clinical Scintillation Imaging" (L. M. Freeman and P. M. Johnson eds.), 2nd ed. pp. 115–146. Grune & Stratton, New York.

Beck, R. N., Charleston, D. B., Eidelberg, P. E., and Harper, P. V. (1967). The Argonne Cancer Research Hospital's brain scanning system. *J. Nucl. Med.* **8**: 1–14.

Beck, R. N., and Gunter, D. L. (1985). Collimator design using ray-tracing techniques, *IEEE Trans. Nucl. Sci.* **32**: 865–869.

Beck, R. N., and Harper, P. V. (1968). Criteria for comparing radioisotope imaging systems. *In* "Proceedings of the Symposium on Fundamental Problems in Scanning," pp. 348–384. Charles C Thomas, Springfield, IL.

Beck, R. N., Schuh, M. W., Cohen, T. D., and Lembares, N. (1969). Effects of scattered radiation on scintillation detector response. *In* "Medical Radioisotope Scintigraphy," Vol. 1, pp. 595–616. International Atomic Energy Agency, Vienna.

Beck, R. N., Zimmer, L. T., Charleston, D. B., Harper, P. V., and Hoffer, P. B. (1973). Advances in fundamental aspects of imaging systems and techniques. *In* "Medical Radioisotope Scintigraphy," pp. 3–45. International Atomic Energy Agency, Vienna.

Beck, R. N., Zimmer, L. T., Charleston, D. B., and Hoffer, P. B. (1972). Aspects of imaging and counting in nuclear medicine using scintillation and semiconductor detectors. *IEEE Trans. Nucl. Sci.* **19**(3): 173–178.

Beck, R. N., Zimmer, L. T., Charleston, D. B., Hoffer, P. B., and Lembares, N. (1971). The theoretical advantages of eliminating scatter in imaging systems. *In* "Semiconductor Detectors in the Future of Nuclear Medicine" (Hoffer, P. B., Beck, R. N., Gottschalk, A., eds.), pp. 92–113.

Beck, R. N., Zimmer, L. T., Charleston, D. B., Shipley, W. W., and Brunsden, B. S. (1973). A theory of optimum utilization of all detected radiation. *In* "CONF-730321," pp. 87–106. U.S. Atomic Energy Commission, Washington, D.C.

Bizais, Y., Rowe, R. W., Zubal, I. G., Bennet, G. W., and Brill, A. B. (1984). Coded aperture tomography revisited. *In* "Information Processing in Medical Imaging." (F. Deconinck ed.), pp. 63–93. Matinus Nijohoff, Boston.

Bolch, W. E., Bouchet, J. S., Robertson, J. S., Wessels, B. W., Siegel, J. A., Howell, R. W., Erdi, A. K., Aydogan, B., Costes, S., and Watson, E. W. (1999). MIRD pamphlet no. 17: The dosimetry of nonurniform activity distributions-radionuclide S-values at the voxel level. *J. Nucl. Med.* **40**: 11S–36S.

Bozzo, S., Robertson, J. S., and Milazzo, J. P. (1968). A data processing method for a multidetector positron scanner. *In* "Fundamental Problems in Scanning". (A Gottschalk and RN Beck, eds.), pp. 212–225. CC Thomas, Springfield, IL.

Brookhaven National Laboratory. (1960). Isotope Sales Catalog.

Brownell, G. L. (1959). Theory of isotope scanning. *In* "Medical Radioisotope Scanning," pp. 1–12. International Atomic Energy Agency, Vienna.

Buonocore, E., and Hubner, K. (1979). Positron-emission computed tomography of the pancreas: A preliminary study. *Radiology* **133**: 195–201.

Burnham, C. A., and Brownell, G. L. (1972). A multi-crystal positron camera. *IEEE, Trans. Nucl. Sci.* **19**: 201–205.

Burnham, C. A., Aronow, S., and Brownell, G. L. (1970). A hybrid positron scanner. *Phys. Med. Biol.* **15**(3): 517–528.

Casey, M. E., and Nutt, R. (1986). A multicrystal two dimensional BGO detector system for positron emission tomography. *IEEE Trans. Nucl. Sci.* **33**: 460–463.

Cassen, B. (1968). Problems of quantum utilization. *In* Gottschalk, A., and Beck, R. N. (Editors): "Fundamental Problems in Scanning" (A. Gottschalk and R. N. Beck, ed.), pp. 50–63. Charles C. Thomas, Springfield, IL.

Cassen, B., and Curtis, L. (1951). "The In Vivo Delineation of Thyroid Glands with an Automatically Scanning Recorder." UCLA report no. 130. University of California, Los Angeles.

Cassen, B., Curtis, L., and Reed, C. W. (1949). "A Sensitive Directional Gamma Ray Detector". UCLA report 49, University of California, Los Angeles.

Cassen, B., Curtis, L., and Reed, C. (1950). A sensitive directional gamma-ray detector. *Nucleonics* **6**: 78–80.

Cassen, B., Curtis, L., Reed, C., and Libby, R. (1951). Instrumentation for I-131 use in medical studies. *Nucleonics* **9**: 46–50.

Celler, A., Farncombe, T., Bever, C., Noll, D., Maeght, J., Harrop, R., and Lyster, D. (2000). Performance of the dynamic single photon emission computed tomography (dSPECT) method for decreasing or increasing activity changes. *Phys. Med. Biol.* **45**(12): 3525–3543.

Charleston, D. B., Beck, R. N., Eidelberg, P. E., and Schuh, M. W. (1964). Techniques which aid in quantitative interpretation of scan data. *In* "Medical Radioisotope Scanning," Vol. 1, pp. 509–525. International Atomic Energy Agency, Vienna.

Charleston, D. B., Beck, R. N., Wood, J. C., and Yasillo, N. J. (1969). A versatile instrument for photoscan analysis which produces color display from black and white photoscans. *In* "Radioactive Isotopes in the Localization of Tumours," pp. 56–57. William Heinemann Medical Books, London.

Cherry, S. R., Shao, Y., and Silverman, R. W. (1997). MicroPET: A high resolution scanner for imaging small animals. *IEEE Trans. Nucl. Sci.* **44**: 1109–1113.

Chesler, D. A., Hales, C., Hnatowich, D. J., and Hoop, B. (1975). Three-dimensional reconstruction of lung perfusion image with positron detection. *J. Nucl. Med.* **16**(1): 80–82.

Croft, B. Y. (1981). Functional imaging. *In* "Functional Mapping of Organ Systems and Other Computer topics." P. D. Esser (ed.). Society of Nuclear Medicine.

Delbeke, D., Patton, J. A., Martin, W. H., and Sandler, M. P. (1998). Positron imaging in oncology: present and future. *In* "Nuclear Medicine Annual" (L. M. Freeman, ed.), pp. 1–49. Lippincott-Raven, Philadelphia.

Duffieux, P. M. (1946). "L'Intégrale de Fourier et ses Applications à L'Optique." Faculté des Sciences, Besançon, France.

Evans, R. D. (1955). "The Atomic Nucleus." McGraw-Hill, New York.

Evans, W. E., Relling, M. V., and Rodman, J. H. (1998). Conventional compared with individualized chemotherapy of childhood acute lymphoblastic leukemia. *N. Eng. J. Med.* **338**: 499–505.

Fermi, E. (1934). Radioactivity induced by neutron bombardment (latter). *Nature* **133**: 757.

Gambir, S. S., Barrio, J. R., and Phelps, M. E. (1999). Imaging adenoviral-directed reporter gene expression in living animals with positron emission tomography. *Proc. Natl. Acad. Sci. U.S.A.* **96**: 2333–2338.

Goldenberg, D. M., Goldenberg, H., Sharkey, R. M., Lee, R. H., Higginbotham-Ford, E., Horowitz, J. A., Hall, T. C., Pinskey, C. M., and Hansen, J. H. (1989). Imaging of colorectal carcinoma with radiolabeled antibodies. *Semin. Nucl. Med.* **4**: 262–281.

Gottschalk, A., and Beck, R. N. (eds.) (1968). "Fundamental Problems in Scanning." Charles C. Thomas, Springfield, IL.

Gregg, E. (1965). Information capacity of scintiscans. *J. Nucl. Med.* **6**: 441.

Grenier, R. P., Bender, M. A., and Jones, R. H. (1974). A computerized multi-crystal scintillation gamma camera. *In* "Instrumentation in Nuclear Medicine" (G. J. Hine and J. A. Sorenson (eds.), Vol 2, pp. 101–134. Academic Press, New York.

Gunter, D. L., Bartlett, A., Yu, X., and Beck, R. N. (1990). Optimal design of parallel hole collimators. *J. Nucl. Med.* **31**: 728.

Gunter, D. L., and Beck, R. N. (1988). Generalized collimator transfer function, *Radiology* **169**: 324.

Gunter, D. L., Hoffmann, K. R., and Beck, R. N. (1990). Three-dimensional imaging utilizing energy discrimination I. *IEEE Trans. Nucl. Sci.* **37**: 1300–1307.

Hamilton, J. G. (1938). The rates of absorption of the radioactive isotopes of sodium, potassium, chlorine, bromine, and iodine in normal human subjects. *Am. J. Physiol.* **124**: 667–678.

Hamilton, J. G., and Lawrence, J. H. (1942). Recent clinical developments in the therapeutic application of radio-phosphorus and radio-iodine (abstract). *J. Clin. Invest.* **21**: 624.

Hamilton, J. G., and Soley, M. H. (1939). Studies in iodine metabolism by the use of a new radioactive isotope of iodine. *Am. J. Physiol.* **127**: 557–572.

Harper, P. V., Beck, R. N., Charleston, D. B., and Lathrop, K. A. (1964). Optimization of a scanning method using ^{99m}Tc. *Nucleonics*, **22**(1): 50–54.

Harper, P. V., Fink, R. A., Charleston, D. B., Beck, R. N., Lathrop, K. A., and Evans, J. P. (1966). Rapid brain scanning with technetium-99m. *Acta Radiol.* **5**: 819–831.

Harper, P. V., Gottschalk, A., Charleston, D. B., and Yaillo, N. (1968). Area scanning with the Anger camera. *In* "Fundamental Problems in Scanning" (A. Gottschalk, R. N. Beck., and C. C. Thomas, eds.), pp. 140–147. Charles C. Thomas, Springfield, IL.

Harper, P. V., Lathrop, K. A., Andros, G., McCardle, R., Goodman, A., Beck, R. N., and Covell, J. (1965). Technetium-99m as a clinical tracer material. *In* "Radioaktive Isotope in Klinik und Forschung," Vol. 6, pp. 136–145.

Harper, P. V., Lathrop, K., and Richards, P. (1964). ^{99m}Tc as a Radiocolloid. *J. Nucl. Med.* **5**: 382.

Hertz, S., and Roberts, A. (1942). Application of radioactive iodine in therapy of Grave's Disease (abstract). *J. Clin. Invest.* **21**: 624.

Hevesy, G. (1948). "Radioactive Indictors: Their Application in Biochemistry, Animal Physiology and Pathology". Interscience Publishers, New York.

Hine, G. J., and Sorenson, J. A. (eds.). (1974). "Instrumentation in Nuclear Medicine," Vol. 2. Academic Press, New York.

Hoffer, P. B., and Beck, R. N. (1971). Effect of minimizing scatter using Ge(Li) detectors on phantom models and patients. *In* "The Role of Semiconductor Detectors in the Future of Nuclear Medicine" (Hoffer, P. B., Beck, R. N., Gottschalk, A., eds.), pp. 131–143. The Society of Nuclear Medicine, New York.

Hoffer, P. B., Jones, W. B., Crawford, R. B., Beck, R., and Gottschalk, A. (1968). Fluorescent thyroid scanning: A new method of imaging the thyroid. *Radiology* **90**: 342–344.

Hopkins, H. H. (1956). The frequency response of optical systems. *Proc. Phys. Soc.* (London) **B69**: 452.

Jaszczak, R. J., Huard, D., Murphy, P., and Burdine, J. (1976). Radionuclide emission tomography with a scintillation camera. *J. Nucl. Med.* **17**: 551.

Jonckheer, M. H., and Deconinck, F. (eds.) (1983). "X-ray Fluorescent Scanning of the Thyroid." Martinus Nijhoff, Boston.

Jerusalem, G., Hustinx , Beguin, Y., and Fillet, G. (2003). PET scan imaging in oncology. *Eur. J. Cancer* **39**(11): 525–534.

Kakehi, H., Arimizu, M., and Uchiyama, G. (1962). Scan recording in color. *In* "Progress in Medical Radioisotope Scanning" (R. M. Knisely, G. A. Andrews, and C. C. Harris, eds.), pp. 111–131. U.S. Atomic Energy Commission, Washington, D.C.

Kety, S. S., Schmidt, C. F. (1948). The nitrous oxide method for the quantitative determination of cerebral blood flow in man: Theory, procedure, and normal values. *J. Clin. Invest.* **24**: 476–483.

Keyes, J. W., Oleandea, N., Heetderks WJm Leonard, P. F., and Rogers, W. L. (1977). The humongotron: A scintiallation camera transaxial tomography. *J. Nucl. Med.* **18**: 381–387.

Klein, W. P., Barrett, H. H., Pang, I. W., Patton, D. D., Rogulski, M. M., and Sain, J. D. (1996). FASTSPECT: Electrical and mechanical design of a high-resolution dynamic SPECT imager. *1995 Conf. Rec. IEEE Nucl. Sci. Symp. Med. Imaging Conf.* **2**: 931–933.

Kolbert, K. S., Sgouros, G., Scott, A. M., Bronstein, J. E., Malane, R. A., Zhang, J., Kalaigian, H., McNamara, S., Schwartz, and Larsen S. M. (1997). Implementation and evaluation of patient-specific three-dimension internal dosimetry. *J. Nucl. Med.* **38**: 301–308.

Krenning, E. P., Kwekkeboom, D. J., and Bakker, W. H. (1993). Somatostatin receptor scintigraphy with 111-In-D-ThA-D-Phe and 123-I-Tyrl-octreotide: The Rotterdam experience with more than 1000 patients. *Eur. J. Nucl. Med.* **20**: 716–731.

Kropp, J., Eisenhut, M., Ambrose, K. R., Knapp, F. F. Jr, and Franke, W. G. (1999). Pharmacokinetics andmetabolism of the methyl-branched fatty acid (BMIPP) in animals and humans. *J. Nucl. Med.* **40**: 1484–1491.

Kubota, K., Matsuzawa, T., Fujiwara, T., Ito, M., Watanaki, S., and Ido, T. (1989). Comparison of C-11 methionine and F-18 fluorodeoxyglucose for the differential diagnosis of lung tumor. *J. Nucl. Med.* **30**: 788–789.

Kuhl, D. E. (1964). A clinical radioisotope scanner for cylindrical and section scanning. *In* "Medical Radioisotope Scanning," Vol. 1, pp. 273–289. International Atomic Energy Agency, Vienna.

Kuhl, D. E., and Edwards, R. Q. (1963). Image separation radioisotope scanning. *Radiology* **80**: 653–662.

Kuhl, D. E., and Edwards, R. Q. (1964). Cylindrical and section radioisotope scanning of the liver and brain. *Radiology* **83**: 926–936.

Kuhl D. E., and Edwards R. Q. (1968a). Digital techniques for on-site scan data processing. *In* "Fundamental Problems in Scanning" (A. Gottschalk and R. N. Beck, eds.), Charles C. Thomas, Springfield, IL.

Kuhl D. E., and Edwards R. Q. (1968b). Reorganizing data from transverse section scans of the brain using digital processing. *Radiology* **91**: 926–935.

Kuhl D. E., Edwards R. Q., Ricci A. R., *et al.* (1972). Quantitative section scanning using orthogonal tangent correction. *J. Nucl. Med.* **13**: 447–448.

Kuhl D. E., Hale J., and Eaton W. L. (1966). Transmission scanning: A useful adjunct to conventional emission scanning for accurately keying isotope deposition to radiographic anatomy. *Radiology* **87**: 278–284.

Kuhl D. E., Hoffman E. J., Phelps M. E., *et al.* (1977). Design and application of mark IV scanning system for radionuclide computer tomography of the brain. *In* "Medical Radionuclide Imaging," pp. 309–320.

Kullback, S. (1959). "Information Theory and Statistics." John Wiley, New York.

Kung H. F. (1990). New technetium 99m labeled brain perfusion imaging agents. *Semin. Nucl. Med.* **20**(2): 150–158.

Lassen N. A., Ingvar D. H., and Skinhoj E. (1978). Brain function and blood flow. *Sci. Am.* **239**: 62–71.

Linfoot, E. H. (1961). Equivalent quantum efficiency and information content of photographic images. *J. Phot. Sci.* **9**: 188.

Livingood J. J., (1938). Seaborg GT Radioactive iodine isotopes. *Phys. Rev.* **53**: 775.

Luker G. D., Fracasso P. M., Dobkin J., and Pinwica-Worms D. (1997). Modulation of the multidrug resistance P-glycoprotein: Detection with technetium-99m-sestamibi in vivo. *J. Nucl. Med.* **38**: 369–372.

Mardirossian G., Matsushita T., Lei K., Clune T., Luo D., Karellas A., Botz E., and Brill A. B. (1996). Calibration of X-ray fluorescence thyroid imaging system. *Phys. Med.* **12**(2): 83–92.

Mayneord, W. V. (1950). Some applications of nuclear physics to medicine. *Br. J. Radiol.* (Suppl:) 2, 168.

Melcher C. L. (2000). Scintillation crystals for PET. *J. Nucl. Med.* **41**: 1051–1055.

Metz C. E. (1995). Evaluation of radiologic imaging systems by ROC analysis. *Med. Imag. Inform. Sci.* **12**: 113–121.

Metz, C. E., Tsui, B. M. W., and Beck, R. N. (1974). Theoretical prediction of the geometric transfer function for focused collimators. *J. Nucl. Med.* **15**(12): 1078–1083.

Molecular (functional) imaging for radiotherapy applications. (2003). *Int. J. Radiat. Oncl. Biol. Phys.* **55**(2): 319–325.

Muehllehner G. (1971). A tomographic scintillation camera. *Phys. Med. Biol.* **16**: 87–96.

Myhill, J., and Hine, G. J. (1967). Multihole collimators for scanning. *In* "Instrumentation and Nuclear Medicine" (G. J. Hine, ed.), Vol. 1, pp. 429–460. Academic Press, New York.

Newell R. R., Saunders W., and Miller E. (1952). Multichannel collimators for gamma-ray scanning with scintillation counters. *Nucleonics* **10**(7): 36–40.

Order S. E., Siegel J. A., and Lustig R. A. (1994). Infusional brachytherapy in the treatemt of non resectable pancreatic cancer: A new radiation modality. *Antibody, Immunoconjugates, Radiopharmaceuticals* **7**: 11–17.

Patton D. D., and Cutler J. E. (1971). Gamma camera dynamic flow studies on a single film by color coding of time. *J. SPIE* **9**: 140–143.

Patton J. A., Delbeke D., and Sandler M. (2000). Image fusion using an integrated dual head coincidence camera with X-ray tube based attenuation maps. *J. Nucl. Med.* **41**: 1364–1368.

Patton J. A., Hollifield J. W., Brill A. B., Lee G. S., and Patton D. D. (1976). Differentiation between malignant and benign thyroid nodules by fluorescent scanning. *J. Nucl. Med.* **17**: 17–21.

Perrin, F. H. (1960). Methods of appraising photographic systems. *J. Soc. Motion Picture Television Eng.* **69**: 151–156, 239–249.

Phelps M. E. (2000). PET: The merging of biology and medicine into molecular imaging. *J. Nucl. Med.* **41**: 661–681.

Planiol T. (1961). Radio-isotopes et affections du systeme nerveux central: Diagnostic et bases biologiques. "Journee de la Federation Mondial de Neurologie," pp. 1–104. Ed. Libraire de L'Academie de Medecinee, Paris.

Reiners, C., Hänscheid, H., Laßmann, M., Tilmann, M., Kreißl, M., Rendl, J., Bier, D. (1998). Fluorescence analysis of thyroidal iodine content with an improved measuring system. *Exp. Clin. Endocrinol.* **106**(Suppl. 3): 31–33.

Robertson J. S., and Bozzo S. R. (1964). Positron scanner for brain tumors. *In* "Proceedings of the 6th IBM Medical Symposium," pp. 631–645.

Robertson J. S., Marr R. B., Rosenblum M., Radeka V., and Yamamoto Y. L. (1973). 32-crystal positron transvers section detector. *In* "Tomographic Imaging in Medicine" (G. S. Freedman, ed.), pp. 142–153. Society of Nuclear Medicine, New York.

Robinson G. D., Jr. (1985). Generator systems for positron emitters. *In* "Positron Emission Tomography" (M. Reivich and M. Alavi, eds.), pp. 81–102. Alan R. Liss, New York.

Rose, A. (1948). The sensitivity performance of the human eye on an absolute scale. *J. Opt. Soc. Am.* **38**(2): 196.

Sandler, M. P., Coleman, R. E., Patton, J. A., Wackers, F. J. T., and Gottschalk, A. (eds.) (2003). "Diagnostic Nuclear Medicine." 4th ed. Lippincott, Williams and Wilkins, Philadelphia.

Shao Y., Meadors K., Silverman R. W., Farrell R., Cirignano L., Grazioso R., Shah K. S. and Cherry S. R. (2002). Dual APD array readout of LSO crystals: Optimization of crystal surface treatment *IEEE Trans. Nucl. Sci.*

Slosman D., Polla B., Townsend D., *et al.* (1985). ^{57}Co-labelled bleomycin scintigraphy for the detection of lung cancer: a prospective study. *Eur. J. Respir. Dis.* **67**(5): 319–325.

Stannard, J. N. (1988a). The early days of manmade radioisotopes. *In* "Radioactivity and Human Health: A History (R. W. Baalman, Jr., ed.), pp. 279–294. U.S. Department of Energy, Washington, D.C.

Stannard, J. N. (1988b). New dimension number one—the fission products, 1939–1950. *In* "Radioactivity and Human Health: A History (R. W. Baalman, Jr., ed.), pp. 295–330. U.S. Department of Energy, Washington, D.C.

Stokely E. M., Sveinsdottir E., Lassen N. A., and Rommer P. (1980). A single-photon dynamic computer-assisted tomograph for imaging brain function in multiple cross-sections. *J. Comp. Assist. Tomogr.* **4**: 230–240.

Subramanian G., McAfee J. G., Blair R. J., and Thomas F. D. (1977). Radiopharmaceuticals for bone and bone marrow imaging: A review. *In* "Medical Radionuclide Imaging," pp. 83–104. International Atomic Energy Agency, Vienna.

Ter-Pogossian M. M., Niklas W. F., Ball J., and Eichling J. O. (1966). An image tube scintillation camera for use with radioisotopes emitting low energy photons. *Radiology* **86**: 463–469.

Ter-Pogossian M. M., Phelps M. E., Hoffman E. J., and Mullani N. (1975). A positron-emission transaxial tomograph for nuclear imaging (PETT). *Radiology* **114**: 89–98.

Thompson C. J., Yamamoto Y. L., and Meyer E. (1979). Positome II: A high efficiency positron imaging device for dynamic brain studies. *IEEE, Trans. Nucl. Sci.* **26**: 583–589.

Tsui, B. M. W., Beck, R. N., Metz, C. E., Atkins, F. B., and Starr, S. J. (1978). A comparison of optimum collimator resolution based on a theory of detection and observer performance. *Phys. Med. Biol.* **23**: 654–676.

Tsui, B. M. W. Gunter, D. L., and Beck, R. N. (1988). The physics of collimator design. *In* "Diagnostic Nuclear Medicine" A. Gottschalk, P. B. Hoffer, and E. J. Potchen, (eds.), pp. 42–54. Williams and Wilkins, Baltimore.

Van Dyke, D., and Anger, H. O. (1965). Patterns of marrow hypertrophy and atrophy in man. *J. Nucl. Med.* **6**: 109.

Vogel R. A., Kirch D., LeFree M., and Steele P. (1978). A new method of multiplanar emission tomography using a seven pinhole collimator and an Anger scintillation camera. *J. Nucl. Med.* **19**: 648–654.

Wagner, H. N., Szabo, Z., and Buchanan, J. W. (eds.) (1995). "Principles of Nuclear Medicine." 2nd ed. Saunders, Philadelphia.

West J. B. (1966). Distribution of pulmonary blood flow and ventilation measured with radioactive gases. *Scand. J. Resp. Dis.* (Suppl) **2**: 9.

West J. B., and Dollery C. T. (1960). Distribution of blood flow and ventilation perfusion ratio in the lung, measured with readioactive CO_2. *J. Appl. Physiol.* **15**: 405–410.

Wong. (1998). A high count rate position decoding and energy measuring method for nuclear cameras using Anger logic detectors. *IEEE Trans. Nucl. Sci.* **45**: 1122–1132.

Yamamoto K., Shibata T., Saji H., Kubo S., Aoki E., Fujita T., Yonekura Y., Konishi J., and Yokoyama A. (1990). Human pancreas scintigraphy using iodine-123-labeled HIPDM and SPECT. *J. Nucl. Med.* **31**: 1015–1019.

CHAPTER 4

Basic Physics of Radionuclide Imaging

CRAIG LEVIN

Molecular Imaging Program, Stanford University School of Medicine, Stanford, California

I. Where Do the Nuclear Emissions Used in Radionuclide Imaging Come From?
II. Relevant Modes of Nuclear Decay for Medical Radionuclide Imaging
III. Production of Radionuclides for Imaging
IV. Interactions of Nuclear Emissions in Matter
V. Exploiting Radiation Interactions in Matter for Emission Imaging
VI. Physical Factors That Determine the Fundamental Spatial Resolution Limit in Nuclear Imaging

Radioisotope imaging is the *in vivo* imaging of physiological function using radiation emanating from radionuclides in tracer quantity inside the subject. Physics is involved in almost every step, from the production and decay of the radionuclide to the formation and analysis of the resulting images. In this chapter we discuss some basic physics topics that are relevant to imaging of nuclear emissions. More in-depth discussions can be found in the references given at the end of this chapter, especially in Rollo (1977) and Sorenson and Phelps (1987). Because later chapters deal with the topics of radiation detectors (Part IV) and imaging systems (Parts II, III, and V) used in nuclear emission imaging, we do not cover the topics of radiation detector and imaging system instrumentation physics in full detail in this chapter.

I. WHERE DO THE NUCLEAR EMISSIONS USED IN IMAGING COME FROM?

A. Nuclear Constituents

The atomic nucleus consists of neutrons and protons, also known as *nucleons*. The proton has one unit of fundamental electronic charge, whereas the neutron is electrically neutral. The size and shape of the nucleus depends on the number of neutrons (N) and protons (Z), their particular energy states, and the angular momentum of the nucleus. A typical nuclear diameter is on the order of 10^{-12}–10^{-13} cm and the density of nuclear matter is roughly 10^{14} g/cm^3. The total number of nucleons in a nucleus (A) is known as the *mass number*. A *nuclide* has a particular nuclear composition with mass number A, atomic number Z and *neutron number N*. The notation we use to identify a particular nuclear composition of an atomic element E with atomic number $A = Z + N$ is $^{A}_{Z}E$, $_{Z}E^{A}$, or just ^{A}E. Atomic species with identical chemical properties (same Z) but distinct masses (different A) are called *isotopes*. Nuclides with the same A are called *isobars*; those with the same N are called *isotones*. Because chemical reactions involve primarily the outermost orbital electrons of the atom, in general, labeling a compound with a radioactive isotope will not change its chemical behavior. Likewise, the chemical state of an atom does not affect its nuclear radioactive characteristics. These are key concepts to the development of radiotracers for nuclear medicine that are discussed in the next chapter.

B. Nuclear Forces and Binding Energy

Three out of the four fundamental interactions in nature play important roles within the nucleus. The *electromagnetic* interaction is responsible for repulsive Coulomb forces among the protons within a nucleus (because the neutron is electrically neutral, it does not interact electrically). Because a typical distance between protons in a nucleus is $\sim 10^{-11}$ cm, these electrostatic repulsion forces are immense. The *strong* nuclear interaction is an attractive force occurring between nucleons within the nucleus irrespective of their electronic charge and is responsible for holding together the nucleus in the face of the large electrostatic repulsion between protons. This interaction's name comes from the fact that it is typically over a factor of 100 times stronger than the electromagnetic interaction between protons within the nucleus. The *weak* nuclear force is responsible for certain types of radioactive decay of which spontaneous nuclear β decay is an example. The weak interaction is typically a factor of 10^{-3}–10^{-4} weaker than the electromagnetic interaction. In current theories of elementary particles, the weak and electromagnetic forces are considered different manifestations of the same fundamental electroweak interaction. The fourth fundamental force of nature, gravity, has negligible effects within the nucleus because the gravitational forces between the relatively light nucleons ($\sim 10^{-27}$ kg) are roughly a factor of 10^{-36} weaker than the typical electromagnetic forces between them.

The *binding energy* is the energy (in the form of work) required to overcome the forces holding an atomic species together and separate it into its individual, unbound constituents: neutrons, protons, and orbital electrons. Basically, if you add up the masses of these individual components, the sum is greater than the mass of the combined nucleus, and the difference is the binding energy. Conversely, the binding energy is the energy that would be released if the separated constituents were assembled into a nucleus. The energy required just to strip the orbital electrons from the atom is negligible compared to that portion of the binding energy required to separate the nucleons, called the *nuclear binding energy*. From Einstein's mass–energy equivalence equation, $E = mc^2$, we can therefore write the binding energy $B_{tot}(A,Z)$ as:

$$B_{tot}(A, Z) = [ZM_H + NM_n - M(A, Z)]c^2 \quad (1)$$

where M_H and M_n are the masses of the hydrogen atom and neutron, respectively, and we have written the expression in terms of atomic rather than nuclear mass for convenience.

C. Nuclear Energy Levels

The motions of nucleons under the influence of the fundamental forces within the nucleus are quite complex. One of the most successful models describing the energy states of nucleons within a nucleus is the *shell model*. In this model, nucleons move in orbits about one another in analogy to Bohr's atomic shell model, which describes the motion of orbital electrons about the nucleus. The types of motions allowed are *quantized*, that is, described by discrete nuclear quantum number parameters. The result is a unique set of allowed discrete energy states, called *energy levels*, for each of the nucleons in a nucleus. The most stable arrangement of nucleons in which the entire system has the lowest overall energy is called the ground state. Excited states of the nucleus, in which the nucleons occupy elevated energy states, exist typically for extremely short times before transforming or *decaying* to a different lower energy state. Those excited states of an atomic species E that are maintained for a relatively long period of time (≥ 1 ns) before decay are called *metastable* states and denoted by ^{Am}E. If one nuclide is a metastable state of another, they are termed *isomers*. An example of a metastable state is ^{99m}Tc, which is currently the most commonly used gamma-ray emitter in nuclear medicine imaging. Changes in nucleon configurations are nuclear transitions between the discrete nuclear energy states or levels in analogy to rearrangements of orbital electrons in the atomic shell structure.

Figure 1 shows a simplified version of the nuclear *energy level diagram* relevant to the generation and decay of ^{99m}Tc. A typical level diagram shows the ground and excited energy states of the nucleus (depicted as horizontal lines with corresponding radiation energies) as well as the allowed transitions between higher and lower energy levels (vertical arrows with relative transition probabilities listed).

D. Nuclear De-excitation

A fundamental property in nature is that systems tend to evolve from higher to lower energy states. Nuclear transitions from higher to lower states of the same nuclide are electromagnetic in nature and result in the emission of electromagnetic radiation. The electric and magnetic fields are created for a very short time as one or more nucleons in the nucleus rearrange themselves during a transition from an initial to a final state. The energy available from this reconfiguration can be released in two ways. The first is in the form of *gamma-ray* photons. Gamma rays are the most commonly used nuclear emissions in radioisotope imaging. The second transition process by which an excited nucleus can relax is known as *internal conversion*. In this process, the nuclear energy is transferred directly to an atomic electron (usually inner shell). These transitions are somewhat analogous to that which occurs in inner shell

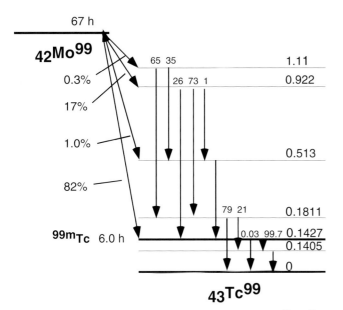

FIGURE 1 Relevant nuclear energy level diagram of the ^{99}Mo–^{99}Tc system for the generation and decay of ^{99m}Tc. Level energies shown are in MeV. The 6.0-hour half-life decay of ^{99m}Tc results in a two-photon cascade of 2.2 and 140.5 keV gamma rays. The 2.2 keV photon is absorbed before reaching the detector in most imaging applications. (Adapted from Lederer et al., 1967.)

atomic transitions in high-Z materials. When excited electrons in an inner shell of a high-Z atom transition from higher to lower energy levels, electromagnetic emission of a *characteristic* X-ray or ejection of an *Auger electron* occurs to carry away the available excitation energy. (Note that Auger electrons, however, are more likely released from outer atomic shells.) Thus, the competition between gamma-ray and internal-conversion electron emission in nuclear transitions is somewhat analogous to the alternate emission of an X-ray or Auger electron as a result of atomic transitions involving inner-shell electrons of a high-Z atom. Internal-conversion electrons are rapidly absorbed in tissue and are thus not directly useful in radioisotope imaging.

When energy is released in a nuclear transition from a higher to lower energy state, from say a gamma-ray emission, the binding energy of the nucleus increases and the mass decreases. Thus, some of the available nuclear mass is converted into energy. The energy released is called the *transition energy*. Usually, the larger the energy difference between initial and final states, the greater the transition rate. Differences in nuclear *spin* and *parity* of the two states involved also play a crucial role in determining the transition rate, in analogy to atomic transitions. For transitions to occur from lower to higher energy states, external energy must be furnished, as is the case in *nuclear reactions*. This topic is discussed in a subsequent section when we discuss radioisotope production.

E. Nuclear Stability

There is a tendency toward instability especially in systems comprising a large number of identical particles confined in a small volume. An unstable nucleus emits photons and/or particles in order to transform itself into a more stable nucleus. This is what is meant by *radioactive decay*, and nuclides undergoing such transformations are called *radionuclides*. Studies of radioactivity are the basis for our understanding of the atomic nucleus. The process of nuclide formation favors nuclides with higher binding energy. Larger binding energy and lower mass mean more-stable nuclei. However, the more nucleons, the greater the total binding energy. Thus, a more-appropriate measure of stability is the average *binding energy per nucleon*, B_{tot}/A. Higher B_{tot}/A values indicate greater stability. For nuclides with mass number $A > 20$, B_{tot}/A lies between 8 and 9 MeV per nucleon, with a maximum at $A \approx 60$ and a slow decrease for higher values of A.

There are certain nucleon configurations that result in more-stable nuclei, and there are a few key factors contributing to this stability. Figure 2 shows schematic plots of neutron number N versus atomic number Z for experimentally determined stable (higher binding energy, lower mass) odd-A and even-A nuclei. The different line thickness for odd- and even-A plots represents the fact that there are more stable isotopes and isotones for even-A nuclei. For light nuclides, the average *line of stability* clusters around $N \approx Z$ (equal number of protons and neutrons); for heavier ones, it deviates from this ($N \approx 1.5Z$) because of the increasing contribution of the Coulomb repulsive force toward instability for higher Z. An excess of neutrons and

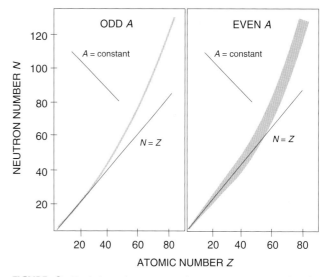

FIGURE 2 Depiction of neutron number versus proton number for stable odd-A (left) and even-A (right) nuclides. (Adapted from Meyerhof, 1967.)

the accompanying strong nuclear force is required in heavy nuclei to overcome the long-range Coulomb repulsive forces between a large number of protons.

For odd A, usually only one stable isobar exists. For even A, usually only even N–even Z nuclides exist. Even–even nuclides occur most frequently. Of the 281 known stable nuclides, 165 of them are even N–even Z, 53 are even N–odd Z, 57 are odd N–even Z, and only 6 are odd N–odd Z. Particularly high stability (higher binding energy) and high abundance with respect to neighboring species are seen for nuclides with N or Z equal to 2, 8, 20, 28, 50, 82, and 126, called the *magic numbers*. In the shell model it is supposed that the magic numbers reflect effects in nuclei very similar to the closing of electronic shells in atoms. The additional stability associated with even nucleon number (even N or Z or both) reflects the tendency of nuclear stability through the *pairing up* of identical nucleons (with opposite spins), similar to electrons in atomic orbitals. The further a nuclide is from the line of stability, the more likely it is to be unstable. All nuclides heavier than ^{209}Bi are unstable. Unstable nuclides lying below the line of stability are said to be *neutron deficient* or *proton rich*; those above the line are *neutron rich* or *proton deficient*. Unstable nuclides generally undergo transformations into those species lying closer to the line of stability. In the radioactive decay of nuclides, the initial unstable nucleus is called the *parent*; the final, more stable nucleus is termed the *daughter*.

F. Nuclear Transmutation

There exists a semiempirical mass formula that can be used to determine the mass of a nucleus. This formula is derived from a *liquid drop model* of the nucleus. This model explains nuclear-binding-energy increases and decreases in terms of factors such as the size and surface area of the nucleus, the Coulomb repulsion of the protons in a nucleus, the symmetry term favoring equal number of neutrons and protons, the pairing effect of nucleons, and the shell term or proximity of N or Z to magic numbers. The shape of the resulting semiempirical mass formula is parabolic as a function of Z. That is, if you plot atomic mass versus Z for isobaric nuclides, it follows a parabola, as depicted in Figure 3. The lower the mass of the isobar along this empirical curve, the more stable that nuclide, and nuclear transformations take the nuclides from higher to lower mass positions on this mass curve. These transitions occur through nuclear *beta (β) decay*, of which *electron (β^-)*, *positron (β^+)*, and *electron capture* (EC) decay are examples. In electron or positron decay, the charged particle is spontaneously created in and ejected directly from the nucleus. In electron capture, an electron is captured from an inner atomic shell. Because these decays result in changing the atomic number Z by 1 (see Figure 3), different chemical elements result, and a nuclear *transmutation* has occurred. Examples of decay by $\beta^\pm$ emission or electron capture are given in Section IIB of this chapter.

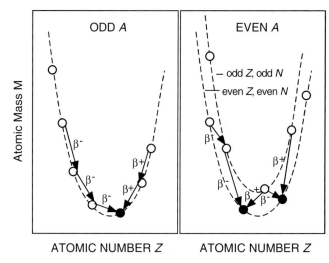

FIGURE 3 Depiction of isobaric mass parabola for odd-A (left) and even-A (right) nuclei. Open and filled circles represent unstable and stable nuclides, respectively. Transformations (represented by arrows) from lower to higher Z occur through β^- decay. Those from higher to lower Z are by β^+ or electron-capture decay (only β^+ is shown). (Adapted from Meyerhof, 1967.)

The isobars with masses on the left of the parabola minimum are neutron rich (lie above the curve of stability) and are candidates for β^- decay. Those with masses to the right of minimum are proton rich (lie below the stability curve) and will undergo positron decay or electron capture. The isobar with the lowest mass on the parabolic curve is the most stable. It turns out that the minimum of this curve as a function of A follows the exact shape of the empirical line of stability. The deviation of the stability line from $N = Z$ or $Z = A/2$ is caused by the competition between the Coulomb energy, which favors $Z < A/2$ (more neutrons than protons) and the N–Z asymmetry energy which favors $N = Z$ or $Z = A/2$.

For odd-A isobars, the pairing effect is negligible, and there is only a single mass parabola, and transitions occur from higher to lower positions along this curve. For even-A isobars, two closely lying mass parabolas are generated due to the different binding energies (and, hence, masses) of isobars with even-N and even-Z compared to those with odd-N and odd-Z, due to the significantly different pairing effects. In this case, transitions occur between isobars along both mass parabolas, as depicted in Figure 3.

G. Nuclear Decay Probability

If N_0 identical parent nuclei were present initially, the number N most probably surviving after a time t is given by:

$$N = N_0 e^{-\lambda t} \quad (2)$$

where λ is the *decay constant*. This exponential decay law is characterized by the fact that a constant fraction of radioactive atoms disappears in a given constant time interval. The corresponding probability that the nucleus has decayed is proportional to the number of *disintegrations* that have occurred after a time t, $N_0(1 - e^{-\lambda t})$, which is also the probability of populating states (metastable, excited, or ground) of the daughter nucleus. If a single nuclide can decay by more than one independent process, the probabilities of the individual decay modes are additive, and the λ in Eq. (2) would be interpreted as the sum of the decay constants of the individual decay modes.

The *half-life* $t_{1/2}$ of a radioactive decay is the time interval in which the original number of nuclei is reduced to one-half, which is determined by setting $N = N_0/2$ in Eq. (2) and solving for t:

$$t_{1/2} = 0.693/\lambda \quad (3)$$

The mean life τ is the average survival time of a radioactive nucleus:

$$\tau = \frac{\int_0^\infty t P(t) dt}{\int_0^\infty P(t) dt} = \frac{1}{\lambda} \quad (4)$$

where we have substituted Eq. (2) as the survival probability $P(t)$ in a time interval dt. The *activity* $A(t)$ is the rate of generation of radioactive decay particles and also follows the same exponential decay form:

$$A(t) = -\frac{dN}{dt} = \lambda N_0 e^{-\lambda t} = \left(\frac{dN_0}{dt}\right) e^{-\lambda t} \quad (5)$$

The basic unit of activity is the curie (Ci). 1 Ci = 3.7×10^{10} disintegrations per second (dis/s). The SI unit of activity is the bequerel (Bq). 1 Bq = 1 dis/s = 2.7×10^{-11} Ci.

II. RELEVANT MODES OF NUCLEAR DECAY FOR MEDICAL RADIONUCLIDE IMAGING

A. Isomeric Transitions: Gamma-Ray Emission

If a radioactive parent nucleus decays into an isomeric (metastable) rather than excited state of the daughter, the decay of the daughter is an *isomeric transition*. This isomeric decay can result in the emission of a gamma ray (γ-ray) or ejection of an internal conversion electron. These transitions do not involve a change in the atomic number Z, and so the parent and daughter are chemically identical. If the energies of the metastable and ground states are E_i and E_f, respectively, the gamma ray is emitted with the full transition energy $\Delta E = E_i - E_f$. The internal conversion electron is ejected with a kinetic energy:

$$T_e = E_i - E_f - E_b \quad (6)$$

where E_b is the binding energy of the electron in the atomic shell from which it has been ejected and the recoil energy of the atom has been neglected. Internal conversion should be viewed as an alternative and independent process by which a nucleus can release excitation energy besides gamma-ray emission. The likelihood of this process to occur instead of gamma emission increases with higher Z and lower ΔE. Using quantum mechanics terminology, internal conversion is also more likely to occur for inner atomic shell electrons (K, L, M, N) due to the more significant overlap of the inner atomic shell and nucleon *wave functions*. The two K-shell electrons' wave functions have the largest overlap with the nucleus. It should be noted that because the atom is left in an excited state of energy E_b, internal conversion is always accompanied by a secondary process by which the atomic excitation energy is released through X-ray or Auger electron emission (with net energy release E_b). This secondary process is discussed again in a subsequent section. For an in-depth discussion of the classification of gamma-ray emissions in terms of nuclear *spin* and *parity* states, angular momentum carried off by the *multipole radiation moments*, and the associated *selection rules* for transitions, see Meyerhof (1967) and Evans (1972).

Gamma-ray emission from typical short-lived nuclear deexcitation is of little use for imaging. Because of their relatively long *lifetimes*, metastable radionuclides are important to nuclear emission imaging. If the *half-life* of the decay is anywhere from a few minutes to a few months, the isomeric gamma-ray emissions can be used for *in vivo* imaging. Figure 1 highlights the isomeric decay scheme of ^{99m}Tc that leads to the well-known 140-keV photon emission that is favorable for use with a gamma-ray camera. This emission has a reasonable half-life (6 h), delivers a relatively low radiation dose, and the isotope can be used to label a wide variety of imaging agents. The internal conversion electrons, however, cannot be imaged *in vivo* with radiation detectors positioned outside the body due to the rapid absorption of electrons in tissue. We discuss the absorption of charged particles in matter later in this chapter and in Levin *et al.* (Chapter 16 in this book).

B. Isobaric Transitions: Beta (β) Emission

Beta (β) decay is a process by which a neutron n in a nucleus is transformed into a proton p, or vice versa, involving the spontaneous emission or absorption of a β particle (electron, e^- or positron, e^+) and electron neutrino (ν_e, or anti-neutrino, $\bar{\nu}_e$) within a nucleus. There are four processes that are all considered forms of β decay from parent (P) into daughter (D) nuclei:

P(A,Z) → D(A,Z + 1) $n \to p + e^- + \bar{\nu}_e$ (electron emission)

P(A,Z) → D(A,Z − 1) $p \to n + e^+ + \nu_e$ (positron emission)

P(A,Z) → D(A,Z − 1) $e^- + p \to n + \nu_e$ (electron capture)

P(A,Z) → D(A,Z − 1) $\bar{\nu}_e + p \to n + e^+$ (inverse beta decay)

Because the atomic number Z changes by 1 in beta decay, parent and daughter nuclei correspond to different chemical elements and a transmutation of elements has occurred. Note that in β decay, because the mass number A does not change, the parent and daughter are always isobars.

In principle, decay by electron emission can originate from within the nucleus or for a free, unbound neutron. However, because the mass of the neutron (939 MeV/c^2) is slightly larger than that of the proton (938 MeV/c^2), decay by positron emission occurs only from within a nucleus, which supplies the necessary energy. Energetically, ignoring the recoil energy of the daughter nucleus, for isobars with atomic mass $M(A,Z) > M(A,Z + 1)$, electron decay takes place from Z to Z+1 (see Figure 3). Pure β⁻ emitters include ^{3}H, ^{14}C, and ^{32}P.

For those nuclei with $M(A,Z) > M(A,Z − 1) + 2m_0$ (m_0 = electron rest mass), positron decay can take place from Z to Z − 1 (see Figure 3). Note that this $2m_0$ term comes about because we have written the inequality in terms of atomic rather than nuclear masses. After ejection from the nucleus, the positron loses its kinetic energy in collisions with atoms of the surrounding matter and comes to rest. In tissue, this occurs typically within a few millimeters from its point of emission and within a nanosecond. Near or at the point at which the positron stops, it combines with an atomic electron of the surrounding material and, as is typical for the union of particle and antiparticle, *annihilation* occurs. The annihilation probability is greatest for very slow positrons. In this annihilation reaction, the mass energy of the electron–positron system at or near rest ($2m_0c^2$ = 1022 keV) is converted into electromagnetic energy (photons) called *annihilation radiation*. If the positron annihilates with a free electron, conservation of linear momentum requires that at least two photons be emitted. It is roughly a factor of 1000 more likely for two rather than three photons to be emitted. For the two-photon process, each has an energy equal to 511 keV and is oppositely directed (180° apart) from the other. If the electron is bound in an atom, annihilation with the production a single photon can occur because the atom can take up the necessary momentum, but this is an extremely rare process. Example positron emitters are ^{11}C, ^{13}N, ^{15}O, and ^{18}F.

EC is analogous to positron decay except an atomic electron is captured by a proton within a nucleus resulting in the emission of a neutrino. The most probable capture is from the atomic K shell (two electrons) because a K-shell electron has the greatest probability of being inside the nucleus. In quantum mechanics terms, the K-shell electron wavefunction has largest overlap with that of the nucleons, similar in concept to the preference of the K-shell electrons for internal conversion of a nucleus. If $M(A,Z)c^2 > M(A,Z − 1)c^2 + E_b$, ($E_b$ = binding energy of the missing electron in the daughter nucleus) electron-capture decay can take place from Z to Z − 1. For the much less probable valence electron capture, $E_b \approx 0$. Electron capture from an inner atomic shell is followed by a secondary process in which the available energy associated with filling the resulting orbital vacancy is carried off by the emission of X-rays or Auger electrons from the daughter atom. Example electron capture radionuclides are ^{57}Co, ^{67}Ga, ^{111}In, ^{123}I, ^{125}I, and ^{201}Tl. In some cases, EC competes with positron emission to decrease the atomic number by 1 and produce a more stable configuration. It turns out that EC is more frequent for heavier elements because it is more likely for their inner shell electrons to be closer to the nucleus at any given time.

Inverse beta decay occurs because neutrinos have a very small, but finite, weak interaction probability with nuclei (typically ~10^{-14}–10^{-19} smaller than for strong interaction *cross sections*). Unlike the other forms of beta decay, this reaction requires an external beam of neutrinos and thus is not relevant to medical nuclear emission imaging. In theory, this reaction is essentially the inverse of ordinary neutron beta decay because the creation of an electron is identical to the disappearance of a positron. Such a process can occur energetically if $M(A,Z) > M(A,Z − 1)$.

Figure 1 depicts the *decay scheme* of the parent ^{99}Mo into the daughter ^{99m}Tc that occurs through β⁻ emission. The line representing the parent in a decay scheme is above and to the left of the daughter nucleus. The decay proceeds from left to right for β⁻ decay because the atomic number Z increases by 1. For β⁺ emission, electron capture, and inverse beta decay, the decay proceeds from right to left because Z decreases by 1. Figure 4 shows the β⁺ decay schemes of ^{11}C, ^{13}N, ^{15}O, and ^{18}F. The vertical distance between the lines represents the total amount of energy released (the transition energy) for the decay process and is also known as the *Q value*.

β decay can transform the parent into an excited, metastable, or ground state of the daughter. For the ^{99}Mo–^{99}Tc pair (Figure 1), both the metastable and excited states are formed from β⁻ decay. If an excited state is formed, the daughter nucleus promptly decays to a more stable nuclear arrangement by the emission of a γ-ray. This sequential decay process is called a β–γ *cascade*. The γ transition may be to another lower excited state and additional γ-rays may be emitted before the ground state is reached. This is known as a gamma-ray *cascade*. The sum of all the resulting gamma-ray, beta, and neutrino energies emitted is equal to the full transition energy between parent

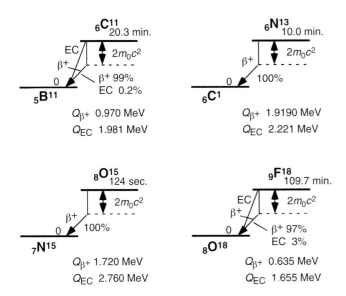

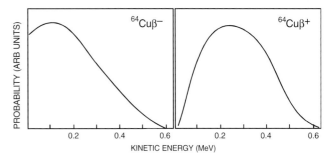

FIGURE 5 Depiction of kinetic energy spectra for β^- and β^+ emission of ^{64}Cu. Adapted from Meyerhof (1967).

FIGURE 4 Decay schemes of common β^+ emitters. Q_{β^+} and Q_{EC} represent the total energy available to positron and electron capture decays, respectively. The maximum available energy for β^+ decay is the full transition energy from parent to daughter minus twice the electron rest mass (1.022 MeV). (Adapted from Lederer et al., 1967.)

and daughter nuclei. Other example radionuclides that involve β–γ cascades are ^{131}I, ^{133}Xe, and ^{137}Cs.

Note that the energy released in electron or positron decay is randomly shared between the β particle and the neutrino. Maximum β emission energy is equal to the full transition energy. This occurs when the neutrino is ejected with negligible kinetic energy, which is highly unlikely. For *allowed transitions*, where zero orbital angular momentum is carried off by the β–ν system, the theoretical distribution of energies or *energy spectrum* for betas is of the form (Wu and Moskowski, 1966):

$$N(E)dE = gF(Z,E)pE(E_{max} - E)^2 dE \qquad (7)$$

where, $N(E)$ = number of decays at energy E, E = total β energy in units of mc^2, E_{max} = maximum (end-point) energy of the β particle in units of mc^2, p = momentum of β in units of mc, g is a coupling constant, $F(Z,E)$ = Fermi function, and Z = atomic number of β decay daughter. The Fermi function takes into account the Coulomb interaction between the β and the daughter nucleus. A nonrelativistic approximation for $F(Z,E)$, valid for allowed transitions of lighter elements (Wu 1966; Daniel 1968) is:

$$F_{allowed}(Z,E) = 2\pi\eta/(1 - e^{-2\pi\eta}) \qquad (8)$$

with $\eta = \pm Z\alpha E/p$ for $\beta^{-/+}$ decay, and $\alpha = 1/137$ is the fine structure constant. The typical short-lived radionuclides used in nuclear medicine undergo allowed transitions.

Because of the opposite charge of the electron and positron, the shapes of their energy spectra differ significantly. This is illustrated best for a nucleus that undergoes both β^- and β^+ decay modes. Figure 5 depicts the electron kinetic energy spectra for β^- and β^+ decay of ^{64}Cu. The *end-point energy* or maximum β emission energy for the two modes are similar (0.58 MeV for β^-, and 0.65 MeV for β^+), but the lower energies are enhanced for the β^- and higher energies are favored for the β^+. These effects are caused by nuclear Coulomb effects; the negatively charged electron is actually somewhat held back by the electric field of the nucleus, limiting the range of velocities; the positively charged positrons are repelled by the same electric field that facilitates acceleration of the particles leaving the nucleus. For an in-depth discussion of the classification of beta-ray emissions in terms of nuclear *spin* and *parity* states, angular momentum carried off by the β–ν pair, and the associated *selection rules* for transitions, see Meyerhof (1967) and Evans (1972).

Because electron decay and internal conversion both result in the ejection of an electron from the nucleus, initial investigations confused the two processes. Chadwick (1914) distinguished the two by demonstrating that the former has a continuous energy distribution for a given nuclide (Figure 5) and the latter is monoenergetic, as can be seen from Eq. (6). Gamma rays are also emitted with discrete energy values that are characteristic of the excitation energies of the radionuclide.

The neutrino is a particle with immeasurable mass and charge that was first postulated to exist by Pauli (1934) to explain the continuous energy distribution of electrons seen in electron decay without violating the laws of energy, momentum, and angular momentum conservation. The neutrino interacts only through the weak force, and so all the β decays are due to small perturbations of weak (as opposed to strong) or electromagnetic interactions within the nucleus. In elementary particle theory, nucleons comprise fundamental particles called *quarks* and these weak interaction decays are considered to occur between individual quarks. Whether a neutrino or antineutrino is emitted in a decay is determined by another fundamental particle law know as *lepton number conservation* in which electrons and neutrino are assigned a lepton number of +1 and their antiparticles are assigned a value of –1. The chargeless and essentially massless neutrinos and antineutrinos can be distinguished experimentally because the former have their intrinsic spins oppositely directed to the direction of travel, whereas the latter have their spins in the direction of travel.

III. PRODUCTION OF RADIONUCLIDES FOR IMAGING

Radionuclide imaging requires appropriate radioactive emissions. Naturally occurring radionuclides are not useful for tracers in biomedical diagnostic imaging because they are typically slowly decaying, long-lived constituents of heavy elements. The radionuclides used in nuclear medicine require some sort of artificial step in production that usually involves the bombardment of stable nuclei with subnuclear particles. The resulting *nuclear reactions* convert stable nuclei to higher-atomic-mass radioactive species that can be used directly to label compounds for imaging or for further generation of a desired imaging isotope. Nuclear reactions involve a bombarding particle a, target nucleus X, and the production of two products, a light reaction product b and heavy reaction product Y. Such a reaction is written as $a + X \rightarrow b + Y$, or in shorter notation $X(a,b)Y$. In this section we describe some basic methods and principles for radionuclide production for nuclear emission imaging. We define a produced radioactive sample that does not contain stable isotopes of the desired radionuclide to be *carrier free*. The ratio of sample radionuclide activity to the total sample mass is the *specific activity*. Obviously those samples with high specific activity and that are carrier free are most desirable for tracer imaging. In the next subsections we discuss the three most common methods for medical radioisotope production.

A. The Nuclear Reactor

1. Nuclear Fission in a Reactor

A nuclear reaction $X(a,b)Y$ is called fission if b and Y have comparable masses. Some nuclei fission spontaneously. Typically fission occurs only when sufficient energy is given to a nucleus by bombardment with subnuclear particles, such as neutrons, protons, or gamma rays, or by *capture* of a slow neutron. The common fissionable material used in a nuclear reactor is uranium isotopically enriched in ^{235}U content. ^{235}U undergoes spontaneous nuclear fission ($t_{1/2} \sim 10^9$ years) into two lighter nuclei or *fission fragments*. In the process, two or three prompt fission neutrons are also released. The total energy released per nuclear fission is approximately 200 MeV, which goes mostly into the kinetic energy of the fission fragments and the neutrons emitted. These fission products are of use in radionuclide imaging. The radioactive fission fragments can be used in tracers or the neutrons released can be used for further radionuclide production through *neutron activation*. The fission fragments are not stable because in the fission process both fragments keep the same N/Z ratio as the original nucleus, which lies close to the stability line. The fragments are very neutron rich and prompt neutron emission is probable. Also, the neutron-rich (lying above the line of stability) fragments will eventually undergo β^- and gamma decay to take them closer to the stability line.

The spontaneous fission of ^{235}U is not a useful source of energy or neutrons. The fission neutrons emitted, however, can stimulate additional fission events when they bombard ^{235}U and ^{238}U. The most important reaction typically for a reactor is:

$$_{92}U^{235} + \text{Thermal } n \rightarrow {}_{92}U^{236*} \rightarrow Y_1 + Y_2 \rightarrow Y_1' + Y_2' + \text{Neutrons} + \text{Energy} \quad (9)$$

where the compound nucleus ^{236}U* created is highly unstable and promptly decays in a manner independent of its mode of formation into the two fission products Y_1 and Y_2. The fission products have the same N/Z ratio as ^{236}U and undergo prompt neutron emission and β^- and γ-ray decay to a more stable configuration for the final fission products Y_1' and Y_2'. All the elements near the center of the periodic table can appear as fission fragments, but fragments with mass numbers near 90 and 140 are more probable. For ^{235}U, example fission fragments are $_{42}$Mo^{101} and $_{50}$Sn^{133} with an average of 2.5 prompt neutrons ejected. The total energy release (~ 200 MeV) for the final fission products includes the energy carried off by neutrons, the energy release by βs, γs, and νs, and thermal energy, and is fairly independent of the pair of fission fragments.

2. Nuclear Reactor Design

The goal of a reactor is to provide a controlled fission *chain reaction* involving the sequential absorption of fission neutrons ejected in the reaction of Eq. (9). The reactor core contains a critical amount of fissionable ^{235}U and ^{238}U in fuel cells. In order to use fission neutrons efficiently to initiate further fission events, they must be slow or *thermal neutrons*. A moderator material, typically containing deuterium and graphite, surrounds the fuel cells and functions to slow down the prompt fission neutrons ejected from the cell. Rather than having one large fuel cell, in order to control the amount of sequential fission occurring, parallel control rods either shield or expose several small, parallel, rod-shaped fuel cells from one another. These control rods contain materials that have a high neutron capture cross section. For control of the reaction it is desired that each fission event stimulate no more than one additional fission event. The fuel cells and control rods are correctly positioned so as to establish the critical conditions for a controlled chain reaction. Much of the energy release in a reactor is ultimately dissipated in the form of thermal energy. Reactor *meltdown* occurs when the positioning of these components is incorrect, causing multiple fission events to be initiated for each fission event. For safe containment and to control the temperature, the reactor core sits inside a high pressure vessel with coolant circulating throughout the volume. The entire vessel is surrounded by neutron and gamma-ray radiation shielding. The vessel also has pneumatic lines for insertion of target samples for production of radionuclides.

3. Reactor-Produced Radionuclides

a. Reactor-Generated Neutron Activation

Reactor-generated neutrons can also be used for production of radionuclides through *neutron activation*. Target nuclei sitting in a reactor core may capture neutrons and produce useful radioactive product nuclei through (n,p) and (n,γ) *reactions*. In the (n,p) reaction, the target nucleus captures a neutron and promptly releases a proton, transforming the target to a different chemical element. The isobaric product is carrier free and of high specific activity. In the (n,γ) reaction, the target nucleus captures a neutron and is converted into an excited state of the product nucleus, which on deexcitation promptly emits a gamma ray. For an (n,γ) reaction the target and product are the same chemical element (isotopes). Most reactor-produced radionuclides are of the (n,γ) variety. However, the products so produced are not carrier-free because they are the same chemical element as the target material and are of lower specific activity than radionuclides produced by fission products or fast neutron bombardment. Because neutrons are added to the target nuclides, the products lie above the line of stability and will tend to decay by β^- emission. Example important reactor-produced (n,γ) reactions are $^{50}\text{Cr}(n,\gamma)^{51}\text{Cr}$, $^{98}\text{Mo}(n,\gamma)^{99}\text{Mo}$, $^{124}\text{Xe}(n,\gamma)^{125}\text{Xe}\rightarrow{}^{125}\text{I}$, and $^{130}\text{Te}(n,\gamma)^{131}\text{Te}\rightarrow{}^{131}\text{I}$. ^{99}Mo is of special interest because it is the parent precursor for the ^{99m}Tc nuclear generator. The ^{99}Mo–^{99m}Tc generator is discussed in the next section.

The activity produced when a sample is irradiated by a beam of particles depends on the intensity of the beam (Number of particles/Area/Time), the number of target nuclei in the sample that is seen by the beam, and the probability of interaction of the bombarding particle with a target nucleus. Probabilities for interactions in nuclear reactions are expressed in terms of the cross section. The cross section has units of area (barns; 1 barn = 10^{-24} cm^2) and represents an effective area presented by a target nucleus in the path of the bombarding particle beam. It is a fictitious area that is not necessarily related to the cross-sectional area of the struck nucleus. A nucleus may present a large area for one type of reaction and a small area for another. Because of their importance in radionuclide production, cross sections for thermal neutron capture have been extensively measured. The interaction cross section depends on the type and energy of the incoming particles and the type of interaction occurring. For example, the probability of neutron absorption or neutron capture cross section is highest for thermal neutron velocities. Neutron fluxes ranging from 10^{10} to 10^{14} neutrons/cm^2/s (depending on the position within the reactor) are available. Even with intense neutron fluxes, only a small percentage ($\leq 0.0001\%$) of the target nuclei are activated. Thus, activation by (n,γ) tends to produce low specific activity samples because of the presence of significant nonactivated target material.

Using classic arguments, we can arrive at a simple expression for the number of activations produced per area per time. Assume a uniform beam of cross-sectional area A consisting of I neutrons/cm^2/s bombards a very thin target sample (no more than one target nucleus is struck by the same particle) with thickness Δx and number density n atoms/cm^3, with an interaction cross section σ. We pretend that with each target nucleus there is an associated cross-sectional area σ seen by the incoming beam such that if the center of a bombarding particle strikes inside σ there is a hit and a reaction is produced, and if the center misses σ no reaction is produced. Then the number of target activations per second per square centimeter, N, is equal to the number of particles removed from the beam per square centimeter of beam area per second, ΔI, due to interactions with target nuclei (assume isotropic distribution). The probability that any one incoming particle has a hit is equal to N/I and is also equal to the projected total cross section of all target nuclei ($nA\Delta x\sigma$) lying within the area of the beam divided by the total beam area A:

$$N/I = -\Delta I/I = n\sigma\Delta x \quad \text{or} \quad N = n\sigma\Delta x I \qquad (10)$$

In addition to this constant rate of production of the sample, the product will also be continually decaying as well, so Eq. (10) also includes an exponential decay term. The product activity starts at zero and gradually increases with irradiation time t until it reaches equilibrium, when the production rate equals the decay rate. This saturation rate is just equal to the final expression in Eq. (10). So the activation rate N as a function of time t is expressed as:

$$N(t) = n\sigma\Delta x I(1 - e^{-\lambda t}) \qquad (11)$$

where λ is the decay constant of the product radioactive nucleus. Dividing Eq. (11) by the mass of the sample gives an expression for the specific activity as a function of irradiation time.

b. Reactor-Generated Fission Fragments

The fission products formed in a reactor are rich in neutrons and decay by sequential β⁻ emission. This series of transmutations produces high-specific-activity radioactive offspring that can also be used in nuclear medicine. A common example of a useful fission product is the ^{99}Mo parent precursor for the ^{99m}Tc nuclear generator; ^{99}Mo can also be obtained through (n,γ) activation, but this produces a lower specific activity. A problem with fission-produced ^{99}Mo is that it may be contaminated with other high-specific-activity fission products. The ^{99}Mo–^{99m}Tc generator is discussed in the next section.

B. The Nuclear Generator

A radionuclide generator is a system that repeatedly produces and replenishes a sample of short-lived daughter nuclei from the decay of a long-lived parent. The apparatus allows the clean separation and extraction of the daughter from the parent. The expressions for the decay rate of the parent dN_p/dt and the accumulation and decay of the daughter dN_d/dt radionuclides are:

$$\begin{aligned} dN_p/dt &= -\lambda_p N_p \\ dN_d/dt &= \lambda_p N_p - \lambda_d N_d \end{aligned} \quad (12)$$

where λ_p and λ_d are the decay constants of the parent and daughter samples, respectively. Solving the first differential equation for N_p, substituting the result into the second, solving for N_d, and using the relation $A(t) = \lambda N(t)$ for the activity—see Eqs. (2) and (5)—we arrive at an expression for the activity of the daughter sample A_d as a function of time:

$$A_d(t) = A_d(0)e^{-\lambda_d t} + A_p(0)\frac{\lambda_d}{\lambda_d - \lambda_p}(e^{-\lambda_p t} - e^{-\lambda_d t}) \quad (13)$$

where $A_d(0)$ and $A_p(0)$ are the daughter and parent sample activities at time $t = 0$, respectively. If the daughter sample did not exist initially, $A_d(0) = 0$ and only the second term remains in Eq. (13). With $A_p(t) = A_p(0)e^{-\lambda_p t}$ and rearranging terms, we obtain:

$$A_d(t) = A_p(t)\frac{\lambda_d}{\lambda_d - \lambda_p}(1 - e^{-(\lambda_d - \lambda_p)t}) \quad (14)$$

In a radionuclide generator, the parent half-life is several times longer than that of the daughter. From Eq. (13) or (14) we see that the activity of the daughter sample first increases and, as the parent is decaying, eventually exceeds that of the parent. As time goes on, the daughter activity reaches a maximum and then starts to decrease and follow the decay curve of the parent. This equilibrium point at which the ratio of the decay rate of the daughter to that of the parent is constant is called *transient equilibrium*. At transient equilibrium, using Eq. (14), $\lambda_d \gg \lambda_p$ and large t, we obtain $A_d/A_p = \lambda_d/(\lambda_d - \lambda_p)$ = constant. Note that if only

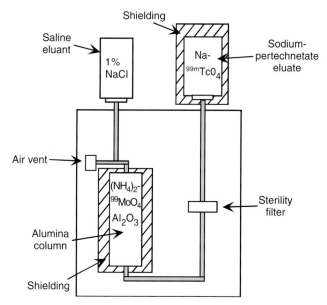

FIGURE 6 Schematic drawing of a ^{99}Mo–^{99m}Tc generator. (Adapted from Rollo, 1977.)

a fraction of the parent atoms decay to the daughter, the daughter activity is given by Eq. (13) or (14) multiplied by that fraction. This is the case for the ^{99m}Tc–^{99}Mo system in which this fraction or *branching ratio* is roughly 92%.

Because of the widespread clinical use of the isotope ^{99m}Tc, the most important nuclear generator to nuclear medicine currently is the ^{99}Mo–^{99m}Tc system. A cross section through a ^{99}Mo–^{99m}Tc generator is schematically depicted in Figure 6. The generator is based on the concept of an ion-exchange column. The ^{99}Mo is absorbed onto an alumina (Al_2O_3) column in monovalent binding sites in the form of $(NH_4)_2{}^{99}MoO_4$ (ammonium molybdate). The ^{99}Mo in this compound exists in the form of the molybdate ion $^{99}MoO_4^{2-}$ in an ionic state of –2 units of charge. When ^{99}Mo in the molybdate ion decays, it is transformed into ^{99m}Tc in the form of the pertechnetate ion $^{99m}TcO_4^-$ in an ionic state of –1. Because the $^{99}MoO_4^{2-}$ ion has two ionic charges it binds firmly to two binding sites on the alumina column, whereas the $^{99m}TcO_4^-$ ion with only one ionic charge weakly binds to only one site. Elution with isotonic saline solution (*eluant*) removes the weakly bound $^{99m}TcO_4^-$, leaving the $^{99}MoO_4^{2-}$ on the alumina column. The chemical form of the final solution (*eluate*) is $Na^{99m}TcO_4$ (sodium pertechnetate). This process of pulling saline solution through the alumina column into an evacuated vial for the extraction of the ^{99m}Tc is referred to as *milking* the generator. Roughly 80% of the available ^{99m}Tc activity is extracted in one elution. After elution, the ^{99m}Tc activity in the generator again grows with time. The time t_m at which maximum ^{99m}Tc activity is available is determined by setting $dA_d/dt = 0$ in Eq. (14), using $\lambda = \ln 2/t_{1/2}$, and solving for t:

$$t_m = 1.44 \frac{t_{1/2p}\, t_{1/2d}}{t_{1/2p} - t_{1/2d}} \ln\left(\frac{t_{1/2p}}{t_{1/2d}}\right) \qquad (15)$$

Using $t_{1/2}$ = 6 and 67 hours, respectively, for ^{99m}Tc and ^{99}Mo in Eq. (15), we arrive at roughly 24 hours for maximum available activity. Note that the maximum available ^{99m}Tc activity after elution as a function of time follows the decay curve of the original ^{99}Mo activity. Typical use of the generator involves repeated ^{99m}Tc buildup and elution phases until the ^{99}Mo decays away (~1 week).

One main problem with the ^{99}Mo–^{99m}Tc generator is that there is a possibility that some ^{99}Mo is removed as well (^{99}Mo breakthrough) during elution due to imperfections in the production of the alumina column. This radionuclide impurity gives the patient unnecessary radiation exposure and will degrade the imaging performance due to a misrepresentation of the tracer distribution. The amount of this radionuclide impurity in the Na^{99m}TcO$_4$ sample can be assayed for by surrounding the vial with 5-mm-thick lead shielding and measuring its activity in a well counter. ^{99}Mo has high-energy gamma-ray emissions that will not be effectively stopped in the lead and will register activity, whereas the 140-keV ^{99m}Tc emissions cannot penetrate the lead. The Nuclear Regulatory Commission (NRC) limits for ^{99}Mo breakthrough are 1 µCi ^{99}Mo/1 mCi ^{99m}Tc. The level of other radionuclide contaminants depends on the source of the ^{99}Mo. ^{235}U-fission-produced ^{99}Mo has the highest specific activity. However, the ^{99}Mo may be contaminated by other high-specific-activity fission products. ^{235}U-fission neutron-activated ^{99}Mo has lower specific activity but in general has lower amounts of radionuclide contaminants. Another concern is breakthrough of the Al^{3+} ion from the alumina column. The presence of this aluminum ion can interfere with the labeling process and cause clumping of red blood cells. Chemical test kits are available that test for the presence of aluminum ion in the final solution. Current regulatory limits are 10 µg Al^{3+}/ml Na^{99m}TcO$_4$.

C. The Cyclotron

A powerful method of radionuclide production is with the use of charged particle accelerators that bombard target samples with protons, deuterons (^{2}H nuclei), α particles (^{4}He nuclei), and other positively charged particles to produce useful nuclear reactions. However, unlike electrically neutral neutrons that are neither attracted or repelled by atomic nuclei during neutron activation of a target, charged particles have to overcome the repulsive nuclear Coulomb forces, or *Coulomb barrier*, before a nuclear reaction can occur. Thus, a relatively high energy particle beam of > 10 MeV is required. The most popular charged particle accelerator used in medicine is the cyclotron due to its relatively high compactness. Linear accelerators for medical isotope production are available as well but will not be discussed.

A schematic diagram of a cyclotron is given in Figure 7. The cyclotron comprises two hollow semicircular metal electrodes, referred to as Ds, with a narrow gap between them. The Ds are placed between the poles of a large DC electromagnet and are themselves connected to a high-frequency (~5- to 10-MHz) and high-voltage (~200-kV AC) electromagnetic oscillator that produces a time-varying opposite electric polarity on the two Ds. A positive ion source near the center of the Ds is used to generate positively charged particles for acceleration in the

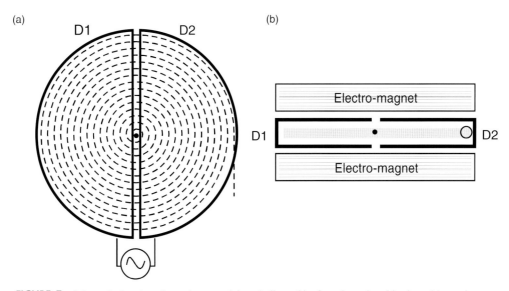

FIGURE 7 Schematic drawing of a cyclotron and the spiraling orbit of accelerated positive ions, (a) top view and (b) side view. Ions are injected into the cyclotron near the center of two D-shaped electrode shells. The Ds sit between poles of two electromagnets. The entire apparatus is evacuated.

cyclotron. For protons, a typical ion source consists of hydrogen gas irradiated by a beam of electrons produced from a tungsten filament. For deuterons or alpha particles, deuterium or helium gas, respectively, is used. During operation, the positive ions are created in bursts or pulses and injected into the center of the gap. Care must be taken to inject the ions at the exact center and in the same horizontal plane of the Ds. Once injected, the positive ions are immediately attracted toward the negative polarity D. As soon as the ions move, the magnetic $q\mathbf{v} \times \mathbf{B}$ force acts causing the ion group to follow a circular trajectory. As the curving ions approach the inside of the attracting D, because there is no electric field within the D the charge continues to move in the arc of a circle inside the D at constant speed until it again reaches the gap between the Ds. The AC frequency is tuned perfectly so that just as the particles emerge from the D, the Ds change polarity and the electric field is in the opposite direction and at its maximum amplitude. This causes the charged particles to accelerate toward the other D, which in the presence of the constant magnetic field causes the ions to move in a larger diameter circular orbit. The ions enter the second D with a larger velocity that remains constant until the particles curve around the other side and are again accelerated by the oscillating electric field. Each time the ions cross the gap, they gain energy qV where q is the ion charge and V is the voltage across the gap and the circular path diameter increases. The result is an outwardly spiraling accelerating path of the positive ions.

When the cyclotron is tuned perfectly, the increasing speed of the particles generated within the gap exactly compensates for the longer path traveled within a given D, and the ions continually arrive at the gap exactly in phase with the AC voltage. These conditions can be understood by considering the relevant force equations. A particle of charge q and mass m moving at velocity $\mathbf{v}$ within the cyclotron under influence of a magnetic field $\mathbf{B}$ experiences a force of magnitude qvB at right angles to the magnetic field and direction of its motion which is in the same plane as the Ds. This forces the particle into a circular orbit of radius r such that the centrifugal force equals the magnetic deflection force:

$$mv^2/r = qvB \qquad (16)$$

The distance traveled while inside a D is $\pi r = \pi mv/qB$ and the time it takes is $\pi r/v = \pi m/qB$, independent of velocity as long as the charge to mass ratio of the accelerating particles and the magnetic field are constant. However, we know from special relativity that once particles are accelerated to relativistic velocities, the mass will change as a function of velocity, the time to traverse a D will change, the particle acceleration will no longer be in step with the AC voltage polarity switch, and the acceleration process is no longer efficient. For this reason, the cyclotron in the form described here cannot be used to accelerate particles to ultra-high energies or to accelerate low mass particles, such as the electron (relatively little energy is required to take the electron to relativistic velocities).

When the radius of the ion orbit approaches that of the Ds, the particles can be directed onto a target placed either internally or externally by being deflected into a long and straight beam tube through a channel in one of the Ds. The energy available to charge particles in the cyclotron when the orbit radius reaches that of the Ds depends on the radius R of the Ds. From Eq. (16), the nonrelativistic velocity of a particle circulating at this radius is given by $v = qBR/m$. So the available kinetic energy E is:

$$E = q^2 B^2 R^2 / 2m \qquad (17)$$

For a typical biomedical cyclotron magnetic field strength of 1.5 tesla, and a D radius of 40 cm, common achievable deuteron kinetic energies are approximately 10 MeV and those of alpha particles and protons are 15 MeV.

The positively charged accelerated particles hitting a target will add positive charge to the target nuclei and change their atomic number. The resulting cyclotron-activated product will tend to lie below the line of stability (see Figure 2) and decay by β^+ emission or electron capture. This is of interest for production of radioisotopes for positron emission tomography (PET). Four of the most commonly used PET isotopes ^{11}C, ^{13}N, ^{15}O, and ^{18}F may be produced in a cyclotron through (p,n), (d,n), or (α,n) reactions: ^{11}B$(p,n)^{11}$C, ^{12}C$(d,n)^{13}$N, ^{14}N$(d,n)^{15}$O, and ^{20}Ne$(d,\alpha)^{18}$F (see Figure 4 for decay schemes). The products are usually carrier-free because the atomic number of the target nuclei changes, so high specific activities are available. The expression for activation rate for cyclotron-produced reactions is similar to that given in Eq. (11), except the charged particle activation cross sections and cyclotron beam intensities are in general lower than those available from reactor neutron activation and so cyclotrons produce less specific activity per irradiation time than reactors. The resulting target then undergoes further radiochemistry processing steps to produce the radiopharmaceutical.

IV. INTERACTIONS OF NUCLEAR EMISSIONS IN MATTER

We have seen that unstable nuclides that occur naturally or are artificially produced release energy in the form of photons or other particles. These emissions from a radioactive nucleus are too high in energy and too small in size to see by the naked eye. We therefore observe their existence and use their energy only through the effects they produce in matter. These effects are caused by the various forces and interactions the nuclear emissions experience when they are confronted with the atoms and molecules of the material they traverse. Such processes are the basis of current radionuclide radiation detection and imaging devices and determine the efficiency and sensitivity of a detector. These reactions also may actually interfere with a measurement by disturbing the physical state of the radiation, causing, for example, deflection, absorption, or loss of radiation energy before detection is accomplished. Depending on the type of radiation, its energy, and the material in which it traverses, interactions may occur with the atom or nucleus as a whole or constituent particles such as electrons, neutrons, protons, and quarks. For radiation passing through matter at the relatively low energies of interest in biomedical radioisotope imaging, the most common interactions are electromagnetic in nature, involving the atom and nucleus as a whole as well as atomic electrons. In this section, we focus our discussion on the interactions of beta and gamma rays, the most commonly used forms of radiation in biomedical nuclear emission imaging. For the design and development of detectors in radioisotope imaging and for understanding how radiation interactions in the subject can effect imaging, it is important to understand how these emissions deposit their energy in matter.

A. Interactions of Electrons and Positrons in Matter

Betas encountered in radioisotope imaging either are directly emitted from positron or electron emitting nuclei or are ejected from atoms as a result of X-ray or γ-ray interactions. For the energies of interest in diagnostic biomedical radioisotope imaging, charged particles such as the electron or positron mainly lose energy and slow down through electromagnetic (Coulomb force) collisions with atoms and molecules. These interactions are characterized by both a loss of energy and deflection of the particle from its incident direction. These effects are primarily a result of multiple inelastic collisions with the atomic electrons of the material and elastic scattering from nuclei. Because these electromagnetic forces between the charged beta particles and the atoms in the material traversed are always present, the energy loss is a continuous process. Because the beta particle mass is equal to that of the orbital electrons with which it is interacting, large deviations in its path can occur through these collisions and a large fraction of its energy can be lost in a single encounter. So, a beta particle may not always slow down in a constant and continuous manner. Because the impulse and energy transfer for incident particles of equal mass are the same independent of charge, the energy loss, multiple scattering, and shape of the primary tracks are essentially identical for electrons and positrons of the same initial energy. Positrons, however, annihilate with atomic electrons near the end of their trajectory, which leads to the emission of annihilation photons that will deposit energy far from the positron end point. For the beta energies of interest (10–1000 keV), the deflection of the electrons is due almost entirely to the elastic collisions with the atomic nuclei, while the energy loss (except that due to an effect known as Bremsstrahlung, which is practically negligible) results from the interaction with the atomic electrons. The result is a random zigzag, tortuous path for a beta particle as it slows down. Because the diagnostic energy range of interest is well below typical binding energies per nuclei, nuclear reactions in general will not occur as a result of beta interactions.

1. Radiative Energy Loss Effects

Rare energy loss effects that can occur in the common diagnostic emission imaging energy range of interest (≤ 511 keV) are the emission of Cerenkov radiation and Bremsstrahlung. *Cerenkov radiation* is a result of charged particles traveling faster than the speed of light in a medium, c/n, where c is the speed of light in a vacuum and n is the medium index of refraction. This phenomenon is an electromagnetic shock wave that results in the emission of a continuous frequency spectrum of polarized light that propagates at a fixed angle with respect to the direction of the particle.

Bremsstrahlung (German for "breaking radiation") is the emission of a continuous spectrum of photon radiation energies due to the deceleration and deflection of the charge particle in the presence of the strong electric field of the nucleus. The loss of energy appears as a photon with energy equal to the difference between initial and final energies of the deflected beta particle. The Bremsstrahlung cross section σ_b, which gives the radiation emission probability, depends on the strength of the nuclear Coulomb force, which in turn depends on the atomic number and how close the encounter is. The Bremsstrahlung cross section varies as:

$$\sigma_b \sim Z^2 r_e^2 f(E) = Z^2 \left(\frac{e^2}{m_e c^2}\right)^2 f(E) \quad (18)$$

where Z is the nucleus atomic number, and $r_e \equiv e^2/m_e c^2$ is the classical electron radius = 2.82×10^{-13} cm. The function $f(E)$ is a strongly increasing function of initial energy E that becomes appreciable only for $E > m_e c^2$, the electron rest mass energy. The resulting average energy loss per length traversed in a medium due to the emission of Bremsstrahlung radiation varies as:

$$\left(-\frac{dE}{dx}\right)_{rad} \sim NE\sigma_b \quad (19)$$

where E is the incoming beta energy and N is the number density of atoms, $N = \rho N_a/A$ (ρ is density, N_a is Avogadro's number, and A is atomic mass). Due to the $1/m^2$ and Z^2 dependence (Eq. 18), radiative energy loss essentially only occurs for electrons and positrons and for high Z materials. It does not play a role in energy loss until the beta energy is high enough for the deceleration to be appreciable. At very high beta energies (> 20 $m_e c^2$ for lead and > 200 $m_e c^2$ for water) the energy loss is dominated by Bremsstrahlung. Figure 8a (later in the chapter) depicts the approximate radiation loss component of the beta energy loss per unit length as it traverses water and lead as a function of energy. In both media, the radiation loss is *negligible* for diagnostic energy ranges of interest (< 511 keV) for which the energy loss is almost entirely due to inelastic atomic collisions.

2. Inelastic Collision Energy Loss: Ionization and Excitation

The inelastic collisions with atomic electrons are mostly responsible for the energy loss of charged particles in matter. Energy is transferred from the particle to the atom causing ionization or excitation. Excitation is more probable than ionization, even at high energies. The ratio of probabilities of ionization to excitation and the average energy transferred to an ionized electron are nearly independent of the primary particle energy. Ionization and excitation can involve both inner and outer atomic shells of the atoms in the absorbing medium, but most of these inelastic collisions involve outer orbital electrons. If an inner-shell electron is ejected, an accompanying X-ray or Auger electron will also be emitted as a result of filling the inner-shell vacancy, as discussed in Section ID. Resulting ionization electrons typically have a mean kinetic energy of only a few electron volts. The maximum energy that can be transferred to a free atomic electron by a primary beta (electron or positron) is simply one-half the kinetic energy of the primary. Although the average energy transferred per inelastic collision is very small in comparison to the incoming particle's total kinetic energy, the typical number of collisions is large and a substantial cumulative energy loss is observed even in relatively thin layers of material. A charged particle continuously interacts with atomic electrons, and so the velocity continues to decrease (although not necessarily in a continuous manner) until the particle is completely stopped.

These atomic collisions are categorized as *hard* if the energy transferred is sufficient to produce an ionization, corresponding to a relatively close distance of approach with the atom. The collisions are *soft* if only an excitation results, corresponding to a not-so-close encounter and smaller energy losses than for ionization events. The energy transferred to an atom in excitation is dissipated in atomic and molecular vibrations and in the emission of low-energy radiation such as visible, ultraviolet, and infrared photons. In general, ionization electrons ejected will be rapidly absorbed. If the primary encounter is so hard that the energy transferred to the recoiling ionization electron is sufficient for it to produce its own secondary ionization, this recoiling electron is termed a *delta* or *knock-on* electron.

Because the nuclear masses are much greater than those of incoming charged particles such as electrons or positrons, collisions with atomic nuclei are usually elastic in nature and the particle does not lose energy and is only deflected. Such collisions are termed *elastic scattering*. Elastic collisions with atomic nuclei occur less frequently than the inelastic interactions with atomic electrons.

The average energy lost per distance traversed by a beta particle dE/dx due to inelastic processes (ionization and excitation) is estimated by the Bethe-Bloch formula for betas (Leo, 1987; Knoll, 1989):

$$\frac{dE}{dx} = 4\pi r^2_0 \frac{mc^2}{\beta^2} NZ(A+B) \qquad (20)$$

with

$$A = \ln\left(\frac{\beta\gamma\sqrt{\gamma-1}mc^2}{I}\right)$$

and

$$B = \frac{1}{2\gamma^2}\left(\frac{(\gamma-1)^3}{8} + 1 - (2\gamma^2 + 2\gamma - 1)\ln 2\right)$$

where r_0 is the classical electron radius = 2.82×10^{-13} cm, $\gamma = (1-\beta)^{-1/2}$, $\beta = v/c$, N is the number density of atoms for the medium (water-equivalent tissue, for example), Z is the atomic number, and I is the mean excitation potential of the medium in electron volts (which for $Z_{\text{eff}} < 12$ is approximately $I = 12Z + 7$; for water, assuming $Z_{\text{eff}} \approx 7.22$, $I \approx 94$ eV; Leo, 1987). This average energy loss includes both soft and hard collisions and excitation processes including emission of Cerenkov radiation (Leo, 1987). This expression essentially holds for both electrons and positrons. However, for incoming electrons the collisions are between identical particles, so the calculation of dE/dx must take into account their indistinguishability, which is unnecessary for incoming positrons. This leads to a different expression for the factor B for electrons and positrons. Fortunately, this difference is only slight for the energies of interest and we ignore it here.

The form of dE/dx for beta particles traversing lead as a function of their energy is shown in Figure 8a. There are a few features to note from Eq. (20) and Figure 8a. Because the terms A and B vary slowly with particle energy, the $1/\beta^2$ term dominates, especially for lower energies. This particular velocity dependence is due to the fact that the slower the incoming beta, the more time it spends in the vicinity of the atomic electron it interacts with and the relatively larger the momentum imparted to that electron through the Coulomb force. This transferred energy goes as the square of the momentum imparted. Near 500 keV, where the kinetic energy is of the order of the rest mass

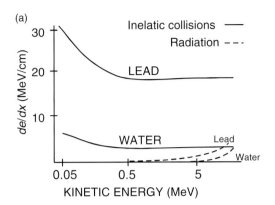
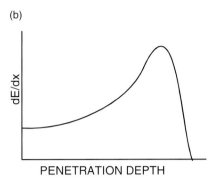

FIGURE 8 (a) Shape of inelastic collision energy loss per length dE/dx for betas as a function of kinetic energy. Also shown is the slight contribution due to radiative emission loss that becomes important only at very high energies. (Adapted from Leo, 1987; Knoll, 1989; Heitler, 1960; Siegbahn, 1955.) (b) Typical shape of Bragg curve for betas depicts the variation of dE/dx as a function of electron penetration depth in matter. The particle is more ionizing toward the end of its trajectory.

energy, the beta velocity approaches the constant speed of light and is relativistic enough for the factors A and B to have an effect, and the curve begins to level off. dE/dx reaches a minimum near 1 MeV ($v \sim 0.9$ c) where the beta is *minimum-ionizing*, the tracks are straighter, and the particle energy is absorbed less quickly through inelastic collisions with atomic electrons. Because the velocity asymptotically approaches a constant, so do the momentum imparted and the energy transferred to an atomic electron, and the energy loss per collision is approximately constant.

As the energy increases beyond the minimum, the $1/\beta^2$ factor approaches a constant value 1 and dE/dx begins to very slowly rise due to the logarithmic dependence of the factor A. The slight increase in inelastic collisional dE/dx after its minimum at higher energies for fast electrons is mostly due to the production of Cerenkov radiation. In dense media, however, this energy loss mechanism is partially compensated by a polarization effect for fast electrons, whereby some of the electronic charge is effectively shielded by polarization of the medium. This explains the almost insignificant increase of dE/dx in the high energy region for lead seen in Figure 8a.

Because dE/dx rapidly increases as the beta slows down, most of the energy deposited and the highest ionization density created along a beta track will be near its end. A depiction of the energy deposited per unit length for a charged particle as a function of penetration depth demonstrates this feature and is known as a *Bragg curve* (Figure 8b). Assuming the beta loses energy in many small, equal decrements per inelastic collision, this nonlinear energy deposition means that the maximum extent or *range* of the trajectory varies with energy in a nonlinear fashion. Because the spectrum of particle energies in beta decay is peaked toward lower energies, a significant fraction of beta trajectories will end in less than 0.5 mm. Because of the dependence of dE/dx on the product NZ of the ionization medium, high-atomic-number, high-density materials will have the greatest stopping power. High-energy betas lose energy at a lower rate (see Figure 8a) and follow a straighter path than do lower-energy betas.

When dE/dx is expressed in terms of *mass thickness* (mg/cm^2) by dividing by the density ρ (mg/cm^3), it is found to vary little over a wide range of materials. This is due to the fact that the number of electrons per gram is nearly the same for all materials. Equation 20 applies to pure elements. For mixtures of n atoms an approximate expression for dE/dx can be found by averaging dE/dx over each element in the compound weighted by the fraction of electrons in each element. This is known as *Bragg's rule*:

$$\frac{1}{\rho}\frac{dE}{dx} = \frac{f_1}{\rho_1}\left(\frac{dE}{dx}\right)_1 + \frac{f_2}{\rho_2}\left(\frac{dE}{dx}\right)_2 + \ldots \frac{f_n}{\rho_n}\left(\frac{dE}{dx}\right)_n \quad (21)$$

where f_i is the fraction by weight of element i in the compound. This relation is another justification for expressing energy loss as mass stopping power $(1/\rho)(dE/dx)$, rather than linear stopping power dE/dx.

3. Hard Collisions: Knock-on (δ) Electron Production

The probability per unit length of emitting a delta ray of energy E_δ due to a beta particle with incident kinetic energy E (with $\beta = v/c$) traversing a medium of electron density N is given by (Ritson, 1961):

$$P(E_\delta) = \frac{2\pi r_e^2 m_e N}{\beta^2}\frac{1}{E_\delta^2} \quad (22)$$

where r_e is again the classical electron radius. The number of δ-rays, N_δ, created per unit length with energies greater than ε in an average track is obtained by integrating Eq. (22) over that energy range:

$$N_\delta(E_\delta > \varepsilon) = \frac{2\pi r_e^2 m_e N}{\beta^2} \frac{1}{\varepsilon} \quad (23)$$

4. Multiple Coulomb Elastic Scattering from Nuclei: Moliere's Theory

Because of the large mass of the nucleus compared to that of betas, the energy lost by betas in Coulomb collisions with nuclei is negligible, and the collisions are essentially elastic. However, because of the larger target mass, transverse elastic scattering of betas is appreciable in the Coulomb field of the nucleus. The electron and positron are also susceptible to *multiple scattering* or repeated elastic Coulomb scatters by nuclei, although with a smaller probability than for multiple inelastic collisions with atomic electrons. If the absorber is so extremely thin that the probability of more than one Coulomb scatter is small (as was the case with the well-known Rutherford experiment with alpha particles and a gold foil), then the angular distribution follows the Rutherford formula:

$$\frac{d\sigma}{d\Omega} = Z^2 r_e^2 \frac{mc/\beta p}{4\sin^4(\theta/2)} \quad (24)$$

where Z is the atomic number of the nucleus, $\beta = v/c$ and $p = mv$ for the incoming beta particle, and θ is the scattering angle. Thus, most collisions cause only small angle deflections.

The Moliere theory of multiple Coulomb scattering is appropriate for most beta energies encountered in medical radioisotope imaging, where the number of elastic collisions with the nucleus of the absorber is greater than roughly 20 and includes the probability of large angle scatters and back scatters. Because of its small mass, a beta particle is susceptible to large angle deflections by scattering from nuclei. In fact, betas may be turned around and backscattered out of an absorber. This effect is particularly strong for low-energy electrons and increases with the Z of the material and angle of incidence. For independent multiple elastic scatters off of nuclei in the absorbing medium in which the number of collisions $\Omega_0 > 20$, a standard description is given by Bethe's (1953) treatment of Moliere's theory of multiple Coulomb scattering, which describes the scattering of fast-charged particles in a screened Coulomb field. In Bethe's work, the probability that an electron of momentum p and velocity v is scattered into the angle θ and angular interval $d\theta$ after traversing a thickness t in a material of atomic number Z and number density of atoms N is given by:

$$f(\theta)\theta d\theta = \lambda d\lambda \int_0^\infty y\,dy\,J_0(\lambda y)\exp\left[\frac{1}{4}y^2\left(-b + \ln\frac{1}{4}y^2\right)\right] \quad (25)$$

where y is a dummy variable, J_0 is the zeroth-order Bessel function, $\lambda = \theta/\chi_c$, b is defined by

$$e^b = \frac{\chi_c^2}{1.167\chi_a^2} \quad (26)$$

where χ_c characterizes the minimum single scattering angle possible,

$$\chi_c^2 = \frac{4\pi N t e^4 Z(Z+1)}{(pv)^2} \quad (27)$$

and χ_a parameterizes the screening angle, given in Moliere's approximation as

$$\chi_a^2 = \chi_0^2(1.13 + 3.76\alpha^2) \quad (28)$$

where χ_0 represents the critical scatter angle below which differences from the Rutherford scattering law (with the characteristic $1/\theta^4$ dependence) become apparent due to nuclear effects; χ_0 is given by:

$$\chi_0 = \frac{\lambda'}{(0.885 a_0 Z^{-1/3})}, \quad (29)$$

$\alpha = Ze^2/hv$, $\lambda' = h/p$, the electron DeBroglie wavelength, h is Planck's constant, a_0 is the Bohr radius, J_0 is the zeroth-order Bessel function.

Bethe assumed $\chi_0 \ll \chi_c$ in the derivation of $f(\theta)$, which holds for relatively thick t, but fails for $y \sim \chi_c/\chi_0 \sim e^{b/2}$. The quantity $e^b \sim (\chi_c/\chi_a)^2$ is roughly the number of collisions Ω_0 that occur in the thickness t of material. To determine whether Moliere's theory of multiple scatter holds for a particular ionization medium, we check to see whether e^b is greater than 20. Moliere considered his model to be valid for $\Omega_0 > 20$ and when the parameter B (defined later) > 4.5. For example, in water-equivalent tissue at room temperature, an 80-keV beta that travels a maximum distance of 0.14 mm from its point of emission has $\Omega_0 \approx 30$ and $B \approx 5$; so we expect Moliere's theory to apply. However, a 50-keV beta with a maximum range of 0.06 mm in water has $\Omega_0 \approx 20$, which is near the region where Moliere's theory breaks down. So we expect there to be a discrepancy between theory and measurement below 50 keV in water.

Moliere evaluated $f(\theta)$ for all angles by a change of variable $\vartheta = \theta/(\chi_c B^{1/2})$. B is a constant defined by: $b = B - \ln B$. With these definitions, $f(\theta)$ can be expanded in a power series in B^{-1}:

$$f(\theta)\theta d\theta = \vartheta d\vartheta[f^{(0)}(\vartheta) + B^{-1}f^{(1)}(\vartheta) + B^{-2}f^{(2)}(\vartheta) + \ldots] \quad (30)$$

where

$$f^{(n)}(\vartheta) = \frac{1}{n!}\int_0^\infty u\,du\,J_0(\vartheta u)\exp\left(-\frac{u^2}{4}\right)\left[\frac{u^2}{4}\ln\left(\frac{u^2}{4}\right)\right]^n \quad (31)$$

where $u = B^{1/2}y$. In the limit of large angles, the distribution function tends toward the Rutherford single scattering law: $f_R(\theta)\theta\,d\theta = (2/B)\,d\vartheta/\vartheta^3$. The ratio of Moliere to Rutherford scattering probabilities is $R = f/f_R = 1/2\,\vartheta^4(f^{(1)} + B^{-1}f^{(2)} + \cdots)$, which gives asymptotic expressions for f^1 and f^2. Together with $f^{(0)}$ obtained from $f^{(n)}$ we have:

$$f^{(0)}(\vartheta) = 2e^{-\vartheta}$$
$$f^{(1)}(\vartheta) = \frac{2(1-5\vartheta^{-2})^{-4/5}}{\vartheta^4} \qquad (32)$$
$$f^{(2)}(\vartheta) = \frac{16(\ln\vartheta + \ln 0.4)}{\vartheta^6(1-9\vartheta^{-2}-24\vartheta^{-4})}$$

For $\vartheta > 4$ (large θ) these expressions will hold. For $\vartheta < 4$ (small θ) we may use the values in Table II of Bethe (1953) to determine $f^{(1)}$ and $f^{(2)}$. For most calculations, it is not necessary to go past $f^{(2)}$.

It is difficult to use approximations for $f(\theta)$ that will hold for both small and large values of θ in multiple scatter calculations because it is not a simple function for the entire angular range of interest. In Monte Carlo simulation calculations of electron trajectories, we may use rejection techniques for generating $f(\theta)$ (Levin and Hoffman, 1999). Figure 9 shows the form of the distribution of scattering angles for multiple scattering of betas in water as estimated by incorporating Eqs. (25)–(32) into a Monte Carlo simulation (Levin and Hoffman, 1999). A vast majority of the collisions result in a small angular deflection of the particle. For small-angle scatters (< 5°), this distribution is approximately Gaussian in form. For larger scattering angles, the deviation from a Gaussian is significant due to the possibility of single large-angle deflections. For small scattering angles, the multiple-scatter cross section is largest. In any given layer of material, the net elastic Coulomb scattering is mostly the result of a large number of small deviations, each independent of the others.

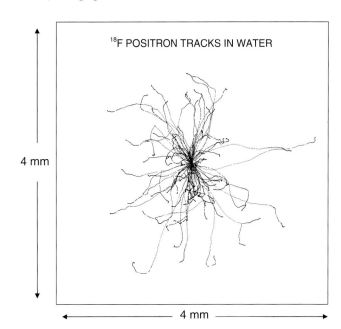

FIGURE 10 Projections onto the plane of 100 Monte Carlo calculated positron trajectories from a point source in water. (Adapted from Levin and Hoffman, 1999.)

Figure 10 shows projections onto a plane of 100 positron trajectories from a ^{18}F point positron source in water as calculated by a Monte Carlo simulation calculation (Levin and Hoffman, 1999) that incorporates the effects described in Eqs. (7)–(8) (beta decay energy spectrum), (20)–(23) (inelastic collisions: ionization, excitation, and delta electron production), and (25)–(32) (multiple Coulomb elastic scattering). The multiple scatter angle is inversely related to the energy. Thus, small-angle multiple scatter is dominant along the beginning of a track because at higher energies the scattering involves primarily small-angle deflections. Toward the end of the track where the energy is lowest, the scatter angle is largest. dE/dx also rapidly increases as the electron slows down. Thus, toward the end of the trajectory, scattering at large angles becomes more frequent and the electron's path begins to bend around and show more curvature or backscatter. This curving around for lower energies is most prominent for high Z materials. Note the signs of δ electron production along some of the tracks plotted in Figure 10. From conservation of energy and momentum, δ electrons are ejected at large angles from the primary track.

5. Range of Beta Trajectories

The *range* of a beta trajectory is defined as the maximum distance it penetrates before it loses all of its energy. In the case of beta decay where there is a continuous spectrum of possible energies, the range refers to the maximum distance penetrated for all possible beta decay energies. Note that the total path length traversed by the beta is considerably

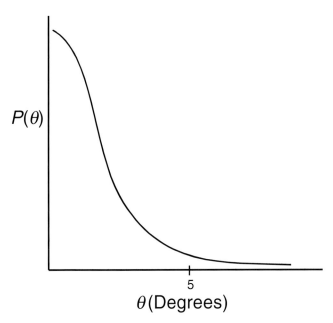

FIGURE 9 Form of the angular distribution for multiple scattering of betas in water.

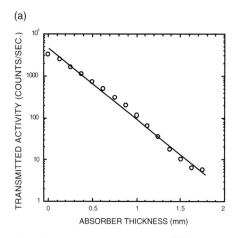
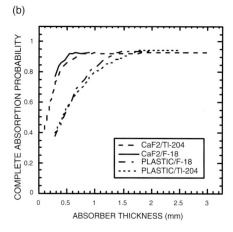

FIGURE 11 (a) Measured transmitted ^{204}Tl beta activity as a function of plastic absorber thickness. (b) Calculated probability for complete absorption of all beta tracks in CaF$_2$(Eu) and plastic beta detectors as a function of material thickness for ^{204}Tl and ^{18}F. (Adapted from Levin *et al.*, 1996.)

greater than the range as defined here. If the energy loss were a continuous process, this range would have the same well-defined value for particles of the same mass and initial energy for a given material traversed. In reality, the energy loss is not continuous but stochastic or statistical in nature and two identical beta particles with the same energy will not undergo the same number and type of collisions in a given medium. Because of multiple elastic scattering by nuclei, the range of betas is in general very different from the calculated path length obtained from an integration of the *dE/dx* formula. Energy loss by betas also has large fluctuations or *energy straggling* due to the large energy transfer collisions allowed. It is even possible for a few single collisions to absorb the major part of the electron's energy. This results in great variation in range or *range straggling* for betas. However, it is a fact that beta ranges expressed in grams per square centimeter are nearly identical for different materials.

Because of their continuous spectrum of energies, the absorption of allowed beta decay electrons and positrons as a function of absorber depth is well approximated by an exponential:

$$I = I_0 e^{-\mu x} \qquad (33)$$

where I is the beta source intensity after a thickness x of the absorbing medium is traversed, I_0 is the initial intensity, and μ is the beta absorption coefficient, which is a function of the material and beta decay end-point energy. Figure 11a shows results from transmission–absorption measurements for a ^{204}Tl beta source (E_{max} = 765 keV) using plastic as an absorber medium. Thin plastic sheets were successively stacked and placed in between the source and a beta detector. The log of the transmitted activity as a function of absorber thickness follows a straight line, indicating that Eq. (33) is a good approximation for describing the absorption properties of beta source emissions. For this measurement, we see that 1.7-mm of plastic stops nearly all ^{204}Tl betas. Figure 11b shows the results of a Monte Carlo calculation of the probability for complete absorption of all beta emissions from ^{204}Tl and ^{18}F in both plastic (ρ = 1.19 g/cm^3, $Z_{eff} \approx$ 6.6; close to absorption parameters of water) and CaF$_2$(Eu) (ρ = 3.17 g/cm^3, $Z_{eff} \approx$ 16.9), which are common beta scintillation detection materials, as a function of absorber thickness. Because the two beta sources have similar end-point energies, nearly all beta emissions from these sources are completely contained in approximately 1.8 mm of plastic and 0.7 mm of CaF$_2$(Eu). Note the probability for complete absorption of all tracks in Figure 11b never reaches 1 due to effects such as electron scatter out of the absorber and Bremsstrahlung. Figure 12 shows results from Monte Carlo simulations of ^{18}F beta point source interaction trajectories in CaF$_2$(Eu) from which the exponential form of absorption as a function of depth (Eq. 33) can be visualized. The corresponding positron energy spectrum is shown in the inset of Figure 12.

B. Interactions of High-Energy Photons in Matter

High-energy photons encountered in biomedical radioisotope imaging can be gamma rays resulting from nuclear deexcitation, annihilation photons resulting from positron annihilation, characteristic X-rays resulting from inner atomic shell transitions of atoms ionized by gamma- or beta-ray interactions, or Bremsstrahlung X-rays resulting from the deceleration of betas in the Coulomb field of the nucleus. Depending on the nuclear energy levels involved, the typical gamma-ray energies are larger than 100 keV. Annihilation photons carry away the mass-energy of the positron-electron system at rest, which means each has 511 keV. Characteristic X-rays typically have discrete energies under 100 keV, depending on the binding energies of the atomic shells involved. Bremsstrahlung X-rays

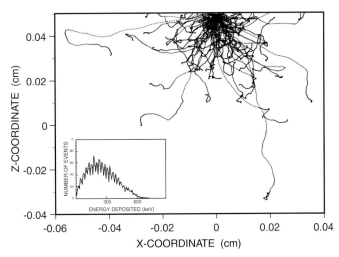

FIGURE 12 Results from Monte Carlo simulations of positron interaction trajectories in $CaF_2(Eu)$. All interactions are shown for 100 positron events from an ^{18}F point source placed on top of the scintillator. Inset: Calculated spectrum of energies deposited in the crystal from 2500 ^{18}F positron events (fit to data is superimposed). (Adapted from Levin et al., 1996.)

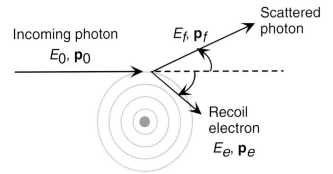

FIGURE 13 Depiction of the two-body Compton scatter process in the plane that probably involves outer-shell electrons.

exhibit a continuous spectrum of energies up to a few hundred keV due to the nature of electro-magnetic collisions between betas and the nucleus. For the beta energies and materials of interest in nuclear medicine, Bremsstrahlung is not an important source of high-energy photons, as discussed in the previous section.

Because photons do not carry electric charge they interact differently with the molecules, atoms, electrons, and nuclei and with a lower probability than do beta particles. However, because photons carry the electromagnetic field, they will interact with electric charge. For the photon energies of interest in nuclear medicine, the interactions are mainly with individual atomic electrons. Unlike the nearly continuous Coulomb interactions that betas undergo as they traverse matter, high-energy photons lose their energy to the individual atomic electrons discretely. Once energy is transferred to an individual atomic electron, that charged particle traverses matter as described in the previous section. The two main interactions with individual atomic electrons high energy-photons experience for the energies of interest are the Compton effect and the photoelectric effect.

1. Compton Effect

If the incident photon energy and momentum are, respectively, much larger than the binding energy and momentum of a struck electron, then for purposes of calculations, the electron can be considered approximately unbound (free) or loosely bound and at rest. From energy and momentum conservation considerations, because the photon is massless and the electron is not, a photon cannot be completely absorbed by and transfer all of its energy to a loosely bound electron at rest. Therefore, for the interaction of a high-energy photon with such an electron to take place, the photon must scatter and sustain a loss in energy. Such a scatter of a high-energy photon off a free or loosely bound atomic electron is termed *Compton scatter*. Figure 13 depicts the initial and final velocity vectors of the photon and electron in the plane the scatter takes place. For the photon energy range of interest in medicine, Compton scatter essentially involves only outer-atomic-shell electrons, as depicted, because they are more loosely bound. From basic conservation of energy and momentum equations with the parameters shown in Figure 13, and using $E_0 = p_0 c$ and $E_f = p_f c$ (photons), we find an expression for the final photon energy in terms of the initial photon energy, the electron rest mass, and the photon scatter angle:

$$E_f = \frac{E_0}{1 + \frac{E_0}{m_e c^2}(1 - \cos\theta)} \quad (34)$$

Figure 14a plots the final photon energy as a function of scatter angle for 140- and 511-keV incident photon energies. As expected, the final photon energy falls off smoothly with increasing angle corresponding to a more energetic recoiling electron. In terms of wave parameters, because photon energy is directly related to its wavelength, the photon undergoes a wavelength shift as a result of Compton scatter.

High-energy photon detectors are sensitive only to the ionization produced by the recoil electrons. The interactions of recoil electrons in matter have been described in the previous section. For loosely bound electrons, the recoil electron energy is given by the difference between the initial and final photon energies, $E_0 - E_f$. For the gamma-ray energies of interest in nuclear medicine, the recoil electron is absorbed within a few millimeters or less, depending on the absorbing medium and electron energy. The recoil electron energy T_{max} will be maximum when the scattered photon energy is lowest, which is when the photon is backscattered ($\theta = 180°$ in Eq. 34):

$$T_{max} = E_0 - E_f(\theta = 180) = \frac{2E_0^2}{2E_0 + m_e c^2} \quad (35)$$

This equation defines what is known as the Compton *edge* in gamma-ray spectroscopy. For example, for 140- and 511-keV incident photon energies, the theoretical maximum recoil electron energies for the Compton edge are 50 and 340 keV, respectively. In reality, because of detector energy resolution blurring effects, the measured maximum Compton electron energy will not be sharply defined. For an appreciable range of angles θ near $\theta = 180°$, $\cos(\theta)$ is near -1 and so the final photon energy E_f in Eq. (34) remains close to the value $E_0/[1 + (2E_0/m_e c^2)]$ and the recoil energy distribution forms a peak near T_{max} in Eq. (35). Note that regardless of scatter angle, the Compton-scattered photon energy never vanishes. A photon cannot transfer all of its energy to an unbound electron in a scatter event.

The Compton scatter photon angular distribution is given by the Klein-Nishina formula for the differential scattering cross section for a photon with an unbound electron (Knoll, 1989):

$$\frac{d\sigma_c}{d\Omega} = r_0^2 \frac{1 + \cos^2\theta}{2} \frac{1}{[1 + \alpha(1 - \cos\theta)]^2}$$
$$\left\{ 1 + \frac{\alpha^2(1 - \cos\theta)^2}{(1 + \cos^2\theta)[1 + \alpha(1 - \cos\theta)]} \right\} \quad (36)$$

where θ is the polar angle the photon scatters into, r_0 is the classical electron radius, and $\alpha = E_0/m_e c^2$. Because Compton scatter involves free electrons, all materials absorb essentially the same amount of radiation per electron through this process. Because most materials have nearly the same number of electrons per gram, the Compton absorption probability per gram is nearly the same for all materials. Thus, the shapes of the curves in Figure 14 are independent of Z. The total probability of Compton scatter per atom depends on the number of scatter electron targets available, which increases linearly with Z, the atomic number of the medium. Because we have assumed that all the electrons are essentially free and independent of one another, to obtain the differential cross section per atom we simply multiply Eq. (36) by Z. To obtain the absorption coefficient, τ_c per centimeter, we multiply by N, the atomic number density. That is, the absorption coefficient is proportional to the total number of electrons NZ per unit volume. The first factor in Eq. (36) is essentially the formula for classical *Thomson scatter*, where the electron does not recoil. The next factors take into account the electron recoil and the number of photons scattered into unit solid angle $d\Omega$ at θ is reduced from the classical case by these factors. Figure 14b shows the Compton scatter angle distribution. Small-angle scatter involving low energy and momentum transfer to the atomic electron is most likely, especially for higher incident photon energy.

For lower incident photon energies, the distribution has a trend toward that of the classical Thompson scattering distribution with the characteristic $1 + \cos^2\theta$ dependence. The classical Thompson scatter phenomenon describes the emission of radiation by a charged particle that has been accelerated by the absorption of incoming electromagnetic radiation. Classically, when electromagnetic radiation encounters a loosely bound electron, the induced acceleration causes the electron to reradiate some of the electromagnetic energy. Because the electron's acceleration will be greatest for a direct hit, emission in the backward direction is favored. For grazing hits, much of the incident electromagnetic energy will be preserved (less will be transferred to the electron) and concentrated in the forward direction. Thus, in the classical case, the flux of radiation is higher in the forward and backward directions, as can be seen in Figure 14b. The Compton effect is the quantum mechanics extension of this classical phenomenon. Thus, for lower incident photon energies, forward and backward

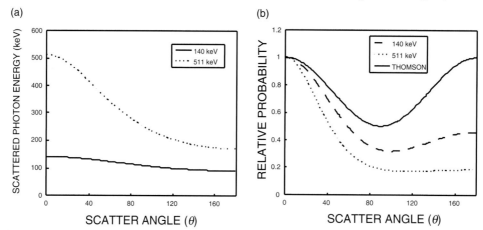

FIGURE 14 (a) Compton scatter photon energy as a function of scatter angle for 140- and 511-keV incident photon energy. (b) Compton scatter probability as a function of scatter angle for 140- and 511-keV incident photon energy and for very low photon energy where the distribution follows the classical Thomson distribution.

Compton scattering is more likely than scattering at 90°, as can be seen by the fall and rise of the scatter probability with increasing angles for the low energy (140-keV) curve. Integration of Eq. (36) over $d\Omega$ gives an expression for the total cross section for a Compton scatter to occur with an individual electron (Heitler, 1960):

$$\sigma_c = 2\pi r_e^2 \left\{ \frac{1+\alpha}{\alpha^2} \left[\frac{2(1+\alpha)}{1+2\alpha} - \frac{1}{\alpha} \ln(1+2\alpha) \right] + \frac{1}{2\alpha} \ln(1+2\alpha) - \frac{1+3\alpha}{(1+2\alpha)^2} \right\} \quad (37)$$

which starts off roughly at 6 barns (1 barn = 10^{-24} cm^2) at 10 keV and strongly decreases with higher energy. Values for the Compton scatter cross section are larger for lower energies. This function for lead and water is depicted in Figure 15. For photon energies greater than 100 keV, > 99% of all interactions in water are due to Compton scatter. The situation is drastically different in lead. For 140 keV, photoelectric absorption is roughly 10 times more likely than Compton scatter, and, at 511 keV, the two processes are roughly equally likely.

a. Coherent and Incoherent Scatter

If the struck atomic electron is instead strongly bound to and not removed from the atom, Eq. (34) still holds, but the photon momentum and energy is transferred to the entire atom which recoils with the bound electron. Hence the m_e in Eq. (34) should be replaced by the entire mass of the atom because the photon scatters off the entire atom. In this case, the scattering is essentially elastic and there is no significant change in photon energy (or wavelength). This type of scattering is termed *Rayleigh scattering*. The scatter probability increases with higher atomic number Z of the scatter medium because a larger fraction of atomic electrons are considered strongly bound. The angular distribution of Rayleigh scattering is unlike that expressed in Eq. (36) because the radiation scattered from all strongly bound atomic electrons interferes coherently. In other words, all the electrons in the atom participate in a coherent manner and the scattering angle is very sharply peaked around $\theta = 0°$. As such, Rayleigh scattering is considered *coherent scattering*. In coherent scattering, the energy of the scattered radiation is essentially the same as the incident energy, the electron is neither excited nor ionized, and the phase of the scattered waves from different electrons are well correlated and only the direction of the photon is changed. To determine the total coherent scatter intensity from the atom, the individual scatter amplitudes from each bound electron are added first and then squared. Compton scatter, on the other hand, is considered *incoherent scattering*. In incoherent scattering, the scattered photon energy is less than the incident value and the difference is transferred to the recoil electron, which either jumps to an excited atomic level or is ejected from the atom. In this case there is no phase relation between the radiation scattered off the individual atomic electrons and the total intensity is obtained by summing the individual electron scatter intensities. The total differential scattering cross section for a photon from an atomic electron is then the sum of the contributions from coherent and incoherent parts. Because coherent scatter is not an effective mechanism for transferring photon energy to matter, it is of little importance in emission imaging.

2. Photoelectric Effect

A gamma ray can transfer all its energy to a bound electron because the atom can absorb some of the recoil momentum, allowing both energy and momentum to be conserved. This process is known as the *photoelectric effect*, the resulting ejected electron is known as a *photoelectron*, and the incident photon disappears. The process is depicted in Figure 16. Because a free electron cannot completely absorb a photon, this effect involves only bound electrons and interactions of this kind mostly take place with electrons in either the inner K shell (as depicted), three L subshells, five M subshells, or seven N subshells. Because a third body, the nucleus, is necessary for conserving momentum, the probability of photoelectric absorption increases rapidly with the strength of the electron binding. That is, the more tightly bound electrons are important for photoelectric absorption. At energies between the binding energies of the K (two-electron) and L (eight-electron)

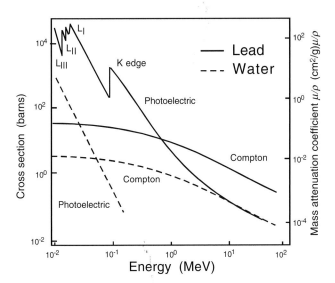

FIGURE 15 Form of Compton scatter and photoelectric absorption cross sections (left axis) and corresponding mass absorption coefficients (right axis) in lead and water as a function of incident photon energy. For gamma-ray energies near the binding energies of K and L shells (88 and 14 keV), the maximum photoelectric absorption occurs when the photon has just enough energy to eject the respective bound electron. (Adapted from Leo, 1987; Knoll, 1989; Heitler, 1960; Siegbahn, 1955.)

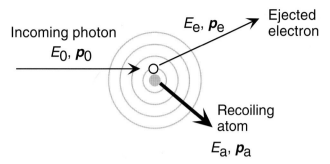

FIGURE 16 Depiction of the photoelectric effect in the plane, which more likely involves an inner-shell electron. Atomic recoil is usually neglected.

shells, the absorption due to outer shells is negligible. For photon energies greater than the binding energy of the K shell, the photoabsorption more likely occurs with a K-shell electron because it is the closest match in energy. In fact, for photon energies above the K shell roughly 80% of all photoelectric absorption involves K-shell electrons (Johns and Cunningham, 1978). For these energies, because the resulting photoelectron is ejected from and no longer bound to the atom, all energies can be absorbed (continuous absorption spectrum). Below the K-shell energy, most of the photoabsorption is due to photons interacting with L and M electrons.

From energy and momentum considerations, the recoil kinetic energy E_a of the atom is of the order $(m_e/M_a)E_e$, where M_a is the atomic mass. Because typically $m_e/M_a \sim 10^{-4}$, we can neglect E_a, and then conservation of energy implies for the kinetic energy of the electron, E_e:

$$E_e = E_0 - E_B \qquad (38)$$

where E_B is the binding energy of the atomic electron shell and E_0 is the incident photon energy. The maximum absorption occurs when the photon has just enough energy to eject a bound electron. For incident photon energies greater than the K-shell electron binding energy and less than the rest mass energy of the electron, the cross section for photoelectric emission per atom, σ_p, is given by (Heitler, 1960):

$$\sigma_p = \frac{32}{3}\sqrt{2}\pi r_e^2 \alpha^4 Z^5 \left(\frac{m_e c^2}{E_0}\right)^{7/2} \qquad (39)$$

where r_e is again the classical electron radius and α the fine structure constant. For E_0 near the K-, L-, or M-shell binding energy, quantum effects require Eq. (39) to be multiplied by a correction factor that is on the order of and proportional to $(1/Z^2)(m_e c^2/E_0)^{1/2}$, which essentially amplifies the cross section near $E_0 = E_B$ creating a sharp discontinuity called the *absorption edge*. For $E_0 \geq m_e c^2$, relativistic effects modify Eq. (39) with a factor that is proportional to $(m_e c^2/E_0)^{3/2}$. For the K-, L-, M-shell absorption coefficient, τ_K, τ_L, τ_M, we multiply the cross section by the atomic number density N. Thus, for relevant diagnostic photon energies, the photoabsorption probability varies with atomic number Z somewhere between the third and fifth power, depending on energy, and, away from the absorption edges, the photoabsorption probability varies as $E_0^{-3.5}$. The shape of the photoelectric absorption probability in lead and water as a function of incident photon energy was shown in Figure 15. For diagnostic photon energies, the photoelectric effect is more likely at low energies and Compton scatter is more likely at higher energies. The photoelectric effect is over a factor of 1000 more likely in lead than water for a given energy (Figure 15), even though the ratio of atomic numbers (82 and ~7.4) is only slightly over a factor of 10. The ejected electron interacts with matter as described in the Section IVA of this chapter. For photon energies of interest, the ejected photoelectron is absorbed within a few millimeters or less, depending on the absorption material. Because the photoelectric effect creates a vacancy in an inner orbital electron shell, as a result of filling that vacancy, characteristic X-rays or Auger electrons will be emitted.

3. Attenuation of High-Energy Photons

We have seen that gamma rays and annihilation photons are many times more penetrating in matter than are charged particles such as electrons, due to the much smaller cross section of the main interaction processes (Compton scatter and photoelectric absorption for diagnostic photon energies of interest) compared to that of inelastic and elastic beta Coulomb collisions. Unlike charged particles, a beam of photons is not degraded in energy as it passes through matter but, rather, is attenuated in intensity because Compton scatter and the photoelectric effect remove the photon from the beam entirely through either scattering or absorption. Those photons that pass straight through a material are those that have not interacted at all and they therefore retain their full energy. The total number of photons, however, is reduced by the number that have interacted. As for the case of betas, the attenuation of a narrow photon beam in matter follows an exponential form with the penetration depth, except with weaker absorption coefficients:

$$I = I_0 e^{-\mu x} \qquad (40)$$

where I_0 is the nonattenuated incident beam intensity, x the absorber thickness, and μ the attenuation coefficient. In contrast to betas, photons do not have a definite maximum range. The quantity I/I_0 can also be viewed as the fraction of photons remaining after the beam has traversed a depth x into the absorber. The μ value is directly related to the total interaction cross section (see Figure 15), which depends on the density and effective atomic number of the absorbing material and the energy of the incoming photons. The total probability of interaction is the sum of the individual Compton scatter ($\sigma_{Compton}$) and photoelectric absorption (τ_{Photo}) cross-section contributions:

$$\sigma_T = Z\sigma_{\text{Compton}} + \tau_{\text{Photo}} \qquad (41)$$

where, to obtain the total Compton scatter contribution, we have multiplied the interaction cross section by the atomic number Z, representing the number of electrons per atom that may undergo a scatter interaction. Multiplying σ_T by the atomic density N gives us the interaction probability per unit length, or the total linear absorption coefficient, $\mu = N\sigma_T = \sigma_T(N_A\rho/A)$, where N_A is Avogadro's number, ρ the density of the matter, and A the molecular weight. The inverse of μ can be thought of as the mean free path or average distance traveled for the photon between interactions. It turns out that μ increases linearly with ρ. By dividing μ by ρ, these effects are removed. The result is know as the *mass attenuation coefficient* and has units of square centimeters per gram. Tables of μ or μ/ρ for various materials can be found in Siegbahn (1955). For mixtures of elements, the total absorption coefficient follows a form analogous to Eq. (21):

$$\frac{\mu}{\rho} = f_1 \frac{\mu_1}{\rho_1} + f_2 \frac{\mu_2}{\rho_2} + \ldots \qquad (42)$$

where f_i is the weight fraction of each element making up the mixture.

Figure 17 depicts Monte Carlo–simulated photon interactions in a pixellated array of CsI(Tl) ($Z_{\text{eff}} = 54$, $\rho = 4.51$ g/cm^3) crystals from 140-keV photons emanating from a point source located above the array. The exponential behavior of the photon absorption is qualitatively seen in the figure from the higher density of interactions near the top of the array. The corresponding spectrum of absorbed energies is shown at the right of Figure 17. From this spectrum we see the total absorption energy peak at 140 keV, the Cs or I K-shell X-ray escape peak near 105 keV, and a combination of Compton scatter and X-ray interactions at low energies (≤ 50 keV). A visual comparison with the Monte Carlo–calculated data for beta interactions shown in Figure 12 demonstrates the discrete versus continuous nature of photon and beta interactions, respectively.

V. EXPLOITING RADIATION INTERACTIONS IN MATTER FOR EMISSION IMAGING

With a basic understanding of how relevant nuclear emissions interact in matter we next turn to how those interactions can be detected and localized in various position-sensitive detectors used in biomedical radioisotope imaging. Parts II–VIII of this book describe in detail imaging detector designs and methodology that have been developed for clinical and research applications. So as not to duplicate this information, in this section we only describe some of the basic preliminary concepts involved in event positioning. A more comprehensive review of available detector technologies can be found in Leo (1987) and Knoll (1989).

All current biomedical radioisotope detection methods use the ionization or excitation created in matter as a result of the partial or full absorption of a nuclear particle's energy. The total amount of energy deposited and recorded in the detector represents the energy of the incoming particle. For charged-particle emission detection, the energy deposition is a direct result of the interaction of the particle's electromagnetic field with atomic electrons of the detection medium and the resultant ionization and excitation are manifested as a track of separated charge or excited electron states centered about the particle's trajectory. For gamma rays, either Compton scatter or photoelectric absorption must first occur, whereby part or all of the energy is transferred to an atomic electron in the absorbing material, which in turn produces the ionization or

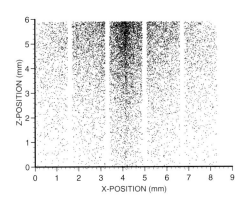
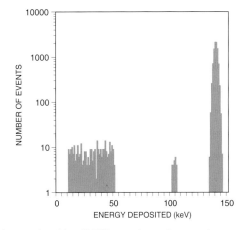

FIGURE 17 Side view of Monte Carlo–simulated interactions produced in a CsI(Tl) crystal array from a point source of 140-keV photons 1 cm above the top of the array. Interactions from 12,000 events are shown. The corresponding calculated spectrum of absorbed interaction energies in the array is shown at the right.

excitation track. In gas-filled detectors (see Bolozdynya et al., Chapter 18 in this book), the deposited energy results in the formation of electron-ion pairs and emission of light photons. For semiconductors (see Wagenaar, Chapter 15 in this volume), the energy deposited is manifested as electron-hole pairs, where an electron from a lattice site is elevated from the valence to the conduction band of the semiconductor leaving a hole behind. We refer to electron-ion or electron-hole pair production as *charged pair creation*. For scintillators (see Wilkinson, Chapter 13 in this volume) the absorbed energy produces either electron-hole pairs if the scintillation material is an inorganic crystal or excitation of molecular valence electrons if it is an organic compound. The ionization and excitation produced with the absorption of a particle in the different types of detectors will create the necessary energy, timing, and positioning information for imaging.

A. Statistics of Ionization Production

For gas-filled and semiconductor detectors the total ionization produced is that which is measured and represents the energy of the incoming particle. An important quantity for a detection medium is the w-value, which is defined as the average energy required to create an electron-ion or electron-hole pair in gases or semiconductors, respectively. In a given ionization medium the w-value is nearly identical for beta or gamma radiation over a large range of energies. This constancy allows a good proportionality between incoming particle energy and amount of ionization produced. Table 1 lists some w-values for common gases and semiconductors along with the ionization potential or band gap. Because much of the absorbed energy is in the form of excitation or other processes that do not result in the liberation of electrons from the parent atomic sites, the w-value for a material is always higher than the binding energy, which is the ionization potential in gases or band gap in semiconductors.

The limit in precision of the energy measurement, or *energy resolution*, depends on the statistical fluctuation of the number N of charged pairs produced. However, if the detector medium completely absorbs the particle's energy, then the standard deviation of N, σ_N is smaller than that predicted by Poisson statistics ($\sigma_N = \sqrt{N}$) because the creation of each charged pair is correlated with all other charged pairs created. This correlation is a result of the fact that the energy deposited at each interaction and for every event is constrained by the fixed value of the total particle energy absorbed. Thus, in reality $\sigma_N = \sqrt{FN}$, where $F < 1$ is called the *Fano factor* and depends on absorber characteristics. Because of the lower w-values (and therefore larger N) and smaller Fano factors, σ_N/N will be lower and the energy resolution limit higher for semiconductor compared to gaseous detectors.

The factors limiting the energy resolution in scintillation detectors are considerably different. The electronic signals produced by scintillation detectors are a convolution of five factors: (1) the absorption of the particle's energy and excitation/ionization of the scintillator material, (2) light yield from a scintillator due to the deexcitation of molecular energy levels, (3) the collection of that light by the photo-multiplier tube (PMT), (4) the absorption of light photons in and emission of photoelectrons from the photocathode of the PMT, and (5) the collection and amplification of the photoelectron signal in the dynode stages of the PMT. The result is that the fluctuation in the number of electrons measured at the PMT anode is close to that expected from Poisson statistics.

B. Detector and Position-Sensitive Detector Basics

Details of biomedical emission imaging detector designs are given in Parts II–V. Here we briefly describe some basics of signal formation and positioning in gaseous, semiconductor, and scintillation detectors used in medical nuclear emission imaging. In nuclear medicine, it is necessary to image the distribution of photon (gamma ray or annihilation photon) and sometimes beta emissions from the body. For photons, this is achieved by (1) proper *collimation*, which gives a preferred direction to the photons entering the camera for a high spatial correlation between those photons detected and those emitted from the body and (2) proper electronic segmentation of the particular radiation detector used. In the case of gamma-ray imaging, a physical collimator is used to absorb those photons not entering the camera in a particular direction (see Zeng, Chapter 8 in this book). In the case of a coincident annihilation photon imaging system (e.g. a PET camera), the directionality is provided electronically from the coin-cident detector interactions of the two oppositely directed annihilation photons (see Lewellen, Watson, and Cherry, Chapters 10–12 in this book). In the case of direct beta imaging (see Rogers et al., Chapter 19 in this book), no collimation is necessary. The short range of betas in tissue and detector materials provides a natural collimation mechanism.

In this section, we describe the position-sensitive radiation detector portion of an emission imaging system. An emission *camera* senses the two-dimensional coordinates of the particles of interest emitted from the body through the

TABLE 1 w-Values and Ionization Potential or Band Gap Values for Some Common Gases and Semiconductors

Detector Material	w-Value (eV)	Ionization Potential or Band Gap (eV)
Argon gas	26.3	15.8
Xe gas	21.9	12.1
Si	3.62	1.12
CdTe	4.43	1.47

interactions occurring in the detector. By accumulating and positioning many such events, an image of the projected radioisotope distribution may be formed.

1. Ionization Detection in Gases

a. Ionization Chambers

The traversal of an ionizing particle through a volume of gas creates a track of electron and positive-ion pairs. The quantity of ionization produced can be measured by applying an appropriate electric field across the medium with the use of anode and cathode electrodes configured with the cathode at a negative potential with respect to the anode. This is known as an *ion chamber*. The electric field must be strong enough to overcome the mutual attraction between electrons and ions or else they will recombine. The electrons will drift toward the anode and the positive ions toward the cathode. These motions will induce charge on the respective electrodes and will cause the anode potential to be changed slightly. The voltage induced on the anode by the electron's motion is proportional to the fraction of voltage through which the electron drops before collection, and the same is true for the ion. Because the total voltage drop traversed for the particles is equal to that applied across the system, in principle the total charge induced on the anode is independent of the original position of the ion pair. However, because the positive ion is much more massive than the electron, it moves much slower and the voltage signal induced has a much smaller slope than that for the electron. Thus, there will be a systematic variation in shape of the signal induced with the location and orientation of the ionization track. It is also important that the chamber be free from impurities. The presence of impurities adds to the variation of the signal with drift distance due to trapping of drifting charge. The anode is usually maintained at a virtual ground by means of a very large resistor value R_L connected to a voltage source that allows the induced charge to leak off the chamber capacitance C at a rate controlled by the time constant $R_L C$.

If all the electrons and ions are completely collected, the final voltage signal will be proportional to the total ionization produced and hence the particle energy, independent of the distribution of ionization events in the chamber. In reality, however, because of the typically long ion drift times (~1 ms), it would be impractical to wait for complete ion signal collection, and so usually the detector signals are derived form the electron induced component only, which is on the order of 100 to 1000 times faster. This is accomplished by using pulse-shaping circuits with an appropriately short time constant. However, now the proportionality between the ionization produced and induced voltage is lost because the voltage through which the electrons fall prior to collection varies with position. The design of electron signal–only ion chambers aims to maintain a constant voltage drop for the majority of electron tracks produced from ionizing radiation. Figure 18 shows a ^{207}Bi energy spectrum measured in a small parallel-plate grid ionization chamber (Levin *et al.*, 1993) filled with 60 atm (1.4 g/cm^3) of xenon gas at room temperature. Due to the high energy resolution, the system easily resolves the daughter ^{207}Pb K-shell internal conversion electron lines (at 570 and 976 keV). The system also resolves the full energy peaks (at 570 and 1063 keV) consisting of either the absorbed K-shell electron summed with the detected ^{207}Pb K-shell X-rays or Auger electrons or, with lower probability, the absorbed gamma rays.

One of the most common ion chamber configurations is a cylindrical cathode with a thin anode wire at the axis. Because the voltage at radius r in the chamber, which is essentially a cylindrical capacitor, is $\sim \log(1/r)$, most of the voltage drop for drifting electrons is confined to very near the anode wire independent of where the ionization is created. Because of the relatively low gain in gas ionization chambers, amplifier noise can make a significant contribution to the overall noise and low-noise, charge-sensitive preamplifiers are needed for the charge–voltage conversion and impedance matching to read out the anode signal.

b. Proportional Chambers

A disadvantage of the ion chamber is the relatively small output signal. A high detector gain is desired so that the

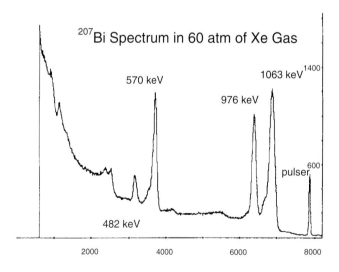

FIGURE 18 ^{207}Bi pulse height spectrum in a 42-cm^3 xenon gas ionization chamber pressurized to 60 atm (1.4 g/cm^3) at room temperature (Levin *et al.*, 1993). The ^{207}Bi source was deposited directly on the cathode within the chamber. The drift electric field was 3.0 kV/cm. The noise subtracted energy resolution of the 976 keV conversion electron peak is 17% full width at half maximum (FWHM). The noise contribution was measured with an electronic pulser.

amplifier noise is not determining the energy resolution and threshold. This can be accomplished by increasing the drift electric field strength. If the drift electric field is increased substantially, the drifting electrons can acquire enough kinetic energy to cause further ionization as a result of molecular collisions. Again, if the electric field is strong enough, this accelerating ionization charge can produce further ionization and so on, with the result of an avalanche of electron and positive ion pairs that can consist of 10^4–10^6 times the original ionization charge. If the electric field is not too high, the charge induced on the electrodes is still proportional to the initial ionization charge, which is proportional to the incoming energy deposited. The resulting configuration is called a *proportional chamber*. At too high an electric field, the amount of ionization charge created can actually modify the electric field and lower the gain for each event and the proportionality between input energy and output charge is not preserved. This also can happen if the incoming radiation flux is very high and the ion drift velocity is relatively low to the point that the electric field is altered. These problems of ion accumulation are referred to as *space charge buildup*. Because it is desired that the charge multiplication factor is fairly independent of the origin of ionization creation, this suggests the need for a high electric field gradient.

To attain high fields, usually a thin wire anode and coaxial cylindrical cathode are used as depicted in Figure 19a and most of the multiplication takes place in the intense field region near the anode wire. This implies that of the total electrons created, most undergo a relatively small voltage drop as they drift toward the anode. Because the anode is at a high positive electric potential with respect to the cathode, the positive ions, on the other hand, undergo a much larger voltage drop as they drift toward the cathode. Because the signal induced on the anode from a drifting charge carrier depends on the voltage drop encountered, most of the signal induced on the anode is due to the motion of positive ions away from the anode rather than from electron motion toward the anode and a negative signal is induced on the anode. In addition to ionization produced in the avalanche process, excited gas atoms and molecules that are formed deexcite, giving rise to photons with high enough energy to ionize materials inside the chamber. This problem can be solved by adding a polyatomic gas, such as methane, to act as a quencher that absorbs the unwanted photons and dissipates their energy into other processes besides ionization.

c. Position-Sensitive (Imaging) Gas Detectors

We have discussed basic configurations used to collect ionization produced in a gas in order to measure an incoming particle's energy. For imaging, the location of the ionization created must also be precisely defined. The basic technique to achieve imaging in gaseous detectors is through *segmentation* of the anode and cathodes. The spatial resolution is thus determined by the particular electrode configurations. The gas-filled detector designs that are being used in biomedical emission imaging (Jeavons et al., 1999; Visvikis et al., 1997) are essentially position-sensitive proportional chambers that are a combination of basic elements of a multiwire proportional chamber (MWPC) and drift chamber (Leo, 1987; Knoll, 1989).

A basic configuration of a MWPC is shown in Figure 19b. Each wire acts as an independent proportional chamber. If the anode wire spacing is a few millimeters and the anode to cathode spacing is just under a centimeter, the electric field lines are essentially constant and perpendicular to the cathode surface, except for the region very close to the anode wires where the field takes on the characteristic $1/r$ dependence of a proportional chamber. Electrons and ions created in the constant field region drift along the field lines toward the nearest anode wire and cathode plane, respectively. As with the basic proportional chamber, most of the induced anode signal is due to the positive ion motion away from the anode. The wire closest to the ionization event records the largest (negative) signal. This basic configuration allows ionization positioning information in one dimension. For information about the second coordinate, a second parallel plane of anode wires that are oriented perpendicular to the first is integrated into the same chamber. The cathode planes may also be formed as sets of segmented strips in alternating orthogonal orientations, as for the anode (Leo, 1987; Knoll, 1989). An alternate to the planar design of a MWPC shown in Figure 19b may be formed by configuring a set of parallel closely packed cylindrical proportional cells that each look similar to that depicted in Figure 19a, except instead of the cathode surface being a solid shell it is composed of wires (Leo, 1987; Knoll, 1989). Adjacent cells then share cathode wires.

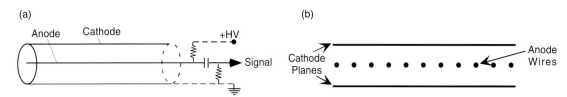

FIGURE 19 (a) Schematic drawing of a basic gas proportional ionization cell. (b) Basic multiwire proportional chamber (MWPC) configuration. In the multianode position-sensitive MWPC, typically a negative voltage is applied to the cathode.

There are three basic readout methods for obtaining positioning information in MWPCs. They all exploit the fact that the avalanche at the anode is highly localized along the length of the wires. The *center of gravity* method uses cathode strip signals. Because the induced signals are the largest on the closest strip to the avalanche, the avalanche point in any one direction is obtained by calculating the center of gravity of the induced signals. The *charge division* method uses the fact that the anode wire is resistive and the signal measured on either end of the wire depends on how far from the respective end the avalanche appears. The *delay line* technique measures the time difference between the arrival of the signals at the ends of a delay line that is coupled to the cathode or anode for a measure of the avalanche location along the respective electrode segment. Spatial and temporal resolution in these systems are determined by the anode wire spacing. Spatial resolution is on the order of one-half the wire spacing. The temporal resolution is related to the time it takes for electrons to drift one-half the wire spacing, which for 2-mm spacing and typical drift velocities of 5 cm/μs is about 20 ns.

The efficiency and sensitivity of MWPCs depend on factors such as the activity strength, dE/dx in the gas, the gas density, the presence of impurities, the anode cathode spacing, the applied high voltage, the gating and electronic threshold for noise rejection, and other factors that affect the total charge created and collected. Typically, flux rates no higher than 10^3–10^4 counts per second per mm of wire can be measured. The MWPCs may be stacked together at various orientations to provide *x-y* position information for a variety of beta and high energy photon ionization imagers.

In a drift chamber (Leo, 1987; Knoll, 1989), spatial information is obtained by measuring the electron drift time in an ionization event. If a trigger is available to determine a time reference point for the event and the electron drift velocity is known, the distance from the anode readout wire to the ionization origin may be precisely determined. If the electric field is constant, the drift velocity is constant and there is a linear relationship between drift time and distance to the anode. A drift cell comprises a cathode at high voltage, the drift region, and an anode, similar in principle to a basic proportional counter. However, for a constant electric field over a relatively large region, a series of field-shaping wires surround the drift region. For the trigger, typically detection of internal scintillation light created from the fill gas itself or an added scintillator is implemented. A typical drift distance is several centimeters. Cells are typically stacked next to or on top of one another to cover a large area (Leo, 1987; Knoll, 1989).

For radioisotope imaging there are many possible configurations of MWPCs and drift detectors and associated readout methods. Hybrid detectors referred to as *gas proportional scintillation counters* combine properties of the proportional counter and scintillation detector. More information on various MWPC designs can be found in Leo (1987), Knoll (1989), and references therein. Some gaseous and hybrid detector designs currently used in emission tomography and beta autoradiography are described in Jeavons *et al.* (1999), Visvikis *et al.* (1997), and Bolozdynya *et al.* (Chapter 18 in this book).

2. Electron-Hole Pair Detection in Semiconductors

The basic operating principle of semiconductor detectors is analogous to that in gas ionization chambers except that a semiconductor crystal instead of a gas fills the space between the anode and cathode. When ionizing radiation is absorbed in the crystal, electron-hole pairs are created rather than electron-ion pairs. The object is again to collect both positive and negative charge carriers on electrodes through the application of an electric field. A semiconductor crystal has a quantized electron energy *band structure* that describes the allowed electron energy states in the crystal lattice (see Figure 20). In a pure crystal, the electron can only lie in the many close-lying electron energy levels of the valence or conduction band and not in the well-defined energy gap E_g between the two energy bands. The interaction of a nuclear particle raises electrons from the valence band, where the electrons are bound to a particular

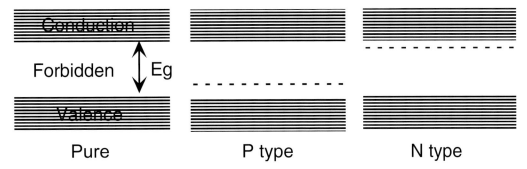

FIGURE 20 Depiction of electron energy band structure for a pure semiconductor crystal and for that with acceptor (P-type) and donor (N-type) impurities. The dotted lines represent impurity energy levels in the forbidden gap of the pure crystal.

lattice site, across the energy gap to the conduction band where the electron is free to move about the crystal lattice. This action leaves a positive *hole* or absence of a valence electron. Under an appropriate electric field the conduction electron will move toward the anode and the positive hole will essentially move toward the cathode through an effective field-induced hopping of valence electrons between neighboring atoms. As for gases, a small energy gap will yield a large number of electron-hole pairs from a particle interaction. Typically only a planar geometry is used for any given single semiconductor device. As is the case for gas detectors, the goal is the complete collection of the electrons and holes created from an ionizing event so that the amplitude of the electrode signals induced is proportional to the number of electrons and holes created.

Not all semiconductors make good detectors. Because of the higher atomic density in semiconductors, trapping of charge carriers at defects or impurities is a bigger problem for most semiconductors than for gases. The extent of these traps determines the lifetimes of the charge carriers. The mobilities of the electron and hole should be high and the lifetimes long for fast signal rise time and efficient charge collection. The conductivity produces a significant leakage current across the cell that could swamp any ionization-produced charge carrier signal. So the purity of semiconductor detector materials must be extraordinarily high (< 1 part in 10^8 impurity level). The main advantage of a semiconductor crystal as an ionization medium is that the w-value is roughly 10 times smaller (see Table 1) and, thus, 10 times more ionization is created than for gases for a given particle energy, which in principle can lead to better energy resolution. Also, semiconductors typically have a higher effective Z and density and, thus, a higher stopping power than for gases. The most common semiconductor crystals that are found in current nuclear medicine cameras are Si and CdZnTe. Properties of some of these materials and a more detailed discussion of semiconductor detector configurations and principles of operation are given in Wagenaar and Levin (Chapters 15 and 16 in this book), and can also be found in Leo (1987) and Knoll (1989).

In reality a semiconductor always contains impurity centers in the crystal lattice and electrons can occupy energy states of an impurity site that fall within the forbidden band of the pure crystal (see Figure 20a). Impurities having energy levels that are initially vacant just above the top of the valence band are termed *acceptors* and are referred to as *P type*. These atoms have an outer electronic structure with one electron less than the host material. A vacant impurity level may be easily filled by thermal excitation of a valence electron from a nearby atom, leaving a hole in that atom. If the number of such holes exceeds that available from the intrinsic thermal excitation of valence electrons, the conductivity will increase. Those impurities that introduce energy levels that are initially filled just below the conduction band are *donor* or *N type* impurities and are produced by impurities with an outer electronic structure with one more electron than the host material. In that case, an electron occupying such an impurity level may be more easily thermally excited into the conduction band compared to a valence electron, and the conductivity would therefore increase. Because hopping of neighboring host atom valence electrons is energetically unfavorable, the impurity holes left behind are not free to move about and the only contribution to conduction by holes is through intrinsic thermal excitation of valence electrons.

Usually semiconductors contain both acceptor and donor impurities and their effects on conductivity will cancel out. If a material tends to have more acceptors than donors, or vice versa, it is possible to compensate for the increase of conductivity by the addition of an appropriate amount of an impurity so that the material becomes closer to the less conductive intrinsic material. For low leakage current, the conductivity should be low and so the material should be well compensated. Figure 21 shows ^{99m}Tc energy spectra taken in a 3.3-mm^2 × 500-μm-thick HgI$_2$ pixel at room temperature and cooled (Patt *et al.*, 1996). The cooled spectrum has better energy resolution due to lower dark current. The low energy tail on the 140-keV peak is due to incomplete hole collection and incomplete energy deposition in the small detector. Note this high resolution detector resolves some of the *K*-shell *X*-ray lines (in Hg, I, and Pb) and X-ray escape peaks associated with ^{99m}Tc transitions and subsequent detection in HgI$_2$.

a. Position-Sensitive (Imaging) Semiconductor Detectors

As was the case for gaseous detectors, position sensitivity with semiconductor detectors is obtained from the localization of the charge created from a particle interaction. Position sensitivity in semiconductors can be established by segmenting or pixellating the anode or cathode electrodes. For radioisotope imaging the segmentation is typically sets of square pixels or parallel rectangular strips. The electron-hole pairs created travel along the established electric field lines and induce a signal on the closest electrode(s). In the square-pixel anode readout, the position is determined by that pixel location. In a cross-strip readout, the anode and cathodes are two orthogonal strips of electrodes, and the *x-y* position is determined by the two individual orthogonal strips that register the largest signal. Unlike in gases, it is difficult to configure many electrode planes within the same monolithic crystal, and usually only one plane of cathode and anode pixels is configured in any given device. Because the typical gain of semiconductor devices is relatively low, low-noise charge-sensitive preamplifier electrons are required for charge-to-voltage conversion and impedance matching for each semiconductor electronic channel. Various configurations of planar electrode patterns used in medical

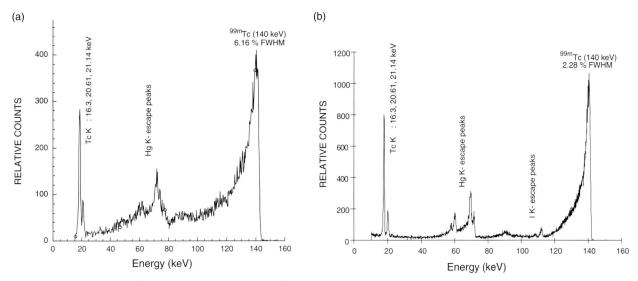

FIGURE 21 ^{99m}Tc spectra measured in a 3-mm^2 HgI$_2$ detector (a) at room temperature and (b) cooled to 0°C. (From Patt et al., 1996.)

imaging are described in more detail in Wagenaar and Levin (Chapters 15 and 16 in this book).

3. Light Detection in Scintillators

Scintillation detectors are currently the most commonly used technology in nuclear medicine. Most often inorganic scintillators are used because they have highest stopping power for gamma rays and annihilation photons, are relatively fast and bright, have emission spectra that are well matched to the absorption spectra of PMTs, and are relatively inexpensive to manufacture. For direct beta imaging, organic plastic scintillators can be used. PMTs are fast and generate gains $> 10^5$ with very little noise. The principles of PMT operation are described in detail in (Chapters 13 and 14 in this book). Alternative semiconductor photodetector designs such as the photodiode (PD), avalanche photodiode (APD), and drift photodiode (DPD) are discussed in Pichler and Levin (Chapters 14 and 16 in this book). Because of the high charge amplification ($\geq 10^5$ at ~1 kV bias), and low capacitance associated with the PMT, front-end amplifier noise is not important in limiting the signal-to-noise ratio (SNR) and energy threshold. A basic current-to-voltage converter and line driver used to match impedance for input into the next amplification stage are required for a PMT. However, because of the relatively low gain and high noise contribution of a semiconductor photodetector, amplifier noise can make a significant contribution. Thus, for each semiconductor photodetector channel used, special low-noise, charge-sensitive preamplifiers are needed for the charge–voltage conversion and impedance matching (see Chapters 14 and 16 in this book). Figure 22 shows a collimated ^{57}Co source spectrum measured in a NaI(Tl)-PMT scintillation detector. The 136.5-keV ^{57}Co line is

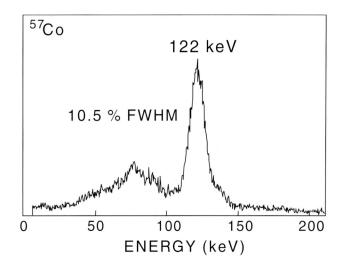

FIGURE 22 ^{57}Co (122-keV) gamma-ray spectrum in a 6 × 6 × 0.6 cm^3 NaI(Tl) crystal coupled to a PMT.

the cause of the bump on the high energy portion of the 122-keV peak. The hump centered at around 75 keV is due to the absorption of Pb K-shell X-rays (~72 keV) and the backscattered photon peak at 90 keV (140 – 50 = 90 keV; see Eq. 34).

a. Mechanism of Light Production in Inorganic Scintillation Crystals

For completeness we give a brief description of the inorganic crystal scintillation mechanism. More detail can be found in Wilkinson (Chapter 13 in this book) and in Leo (1987), Knoll (1989), and Birks (1964). As for semiconductor crystals (see Figure 20), inorganic scintillation crystals have an electronic band structure that dictates the allowed electron energy states in the lattice. A particle interaction in a pure inorganic crystal results in the liberation of electrons from the valence band into the conduction band. After some time, the conduction electrons will recombine with holes with the resulting excitation energy E_g converted into light. However, this process is inefficient and the wavelength of light corresponding to E_g is below the visible range, which is not useful for a standard PMT. The luminescent efficiency is improved and the wavelength of emission increased into the visible range by the addition of a small amount of a suitable activator impurity, which effectively produces electron energy levels in the previously forbidden gap between the valence and conduction bands. The excitation energy in the host material is then transferred to the energy levels of the impurity centers in the crystal lattice, which deexcite with the emission of longer wavelength photons that overlap with the visible range. Properties of various scintillators are given in Wilkinson (Chapter 13 in this book).

b. Position-Sensitive (Imaging) Scintillation Detectors

The most common camera currently found in radioisotope imaging is based on position-sensitive scintillation detectors. Position-sensitive scintillation detectors localize the light flash created within the crystal. This may be achieved by segmenting the crystal, the photodetector, or both. Figure 23 depicts three basic configurations of two-dimensional position-sensing scintillation detectors known as *gamma-ray cameras*. In this figure we also show a gamma-ray collimator (see Gunter, Chapter 8 in this book), which absorbs photons not entering the camera in a preferred direction. The position-sensitive scintillation detector used in coincident annihilation photon imaging does not require such a physical collimator, but the light-pulse positioning concepts are similar.

The position-sensitive scintillation light detector consists of a scintillation crystal(s) and a position-sensitive photodetector. The scintillation crystal may be a single-crystal slab or a set of small discrete or pseudodiscrete crystals. For the single crystal design (Fig. 23a) the scintillation light diffuses and reflects throughout the crystal and the light pulse is positioned by an appropriate weighted mean of the individual photodetector signals over their respective positions.

This weighted-mean position calculation is also used with the multiplexed discrete crystal design (Fig. 23b), which comprises an array of optically isolated crystals coupled through a light diffuser to photodetectors. In this crystal design, the light created is confined to one crystal and collected onto a smaller area on the photodetector plane compared to that shown in Figure 23a. With a weighted-mean calculation, the detector intrinsic spatial resolution in this design will be finer than the photodetector anode readout pitch and on the order of the crystal pixel size. Because more crystals can be decoded than there are photodetectors, this is an efficient technique for reducing the number of electronic processing channels required. A variation on this theme is used in the PET block detector (Casey and Nutt, 1986), where pseudo-discrete crystals (partially optically coupled) are used and the light is diffused within the scintillation crystal itself and the PMT entrance window rather than in a separate light diffuser. In the completely discrete system (Fig. 23c), the imaging array is formed by many small, optically isolated crystals, each coupled one to one with a separate photodetector channel. All light created in one crystal is focused onto one photodetector. As for the multiplexed discrete crystal design, the detector spatial resolution is determined by the crystal width. The position-sensitive photodetector array (depicted at the bottom of each configuration in Figure 23) might be an array of individual PMTs, position sensitive PMT(s) with segmented electrodes,

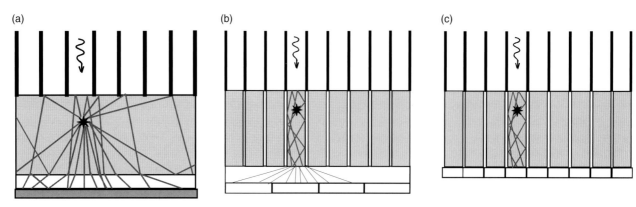

FIGURE 23 Possible coupling schemes of crystals to photodetector arrays in scintillation cameras. (a) Light sharing over the position-sensitive photodetector is natural in single-crystal slab design, but a light diffuser may be needed for optimal light spread. (b) Light multiplexing is necessary when there are fewer readout channels than individual crystals. A light diffuser might be required in this situation as well. (c) No light multiplexing is necessary when the crystals are coupled one to one with the photodetector pixels. Light created in a crystal is focused onto one photodetector element. Photodetector shown at bottom of each drawing may be either an array of individual PMTs, a multianode PMT, or a semiconductor photodetector array. Only fraction of the detector field of view (FOV) and a parallel-hole collimator design are shown.

or a semiconductor photodetector array. Because of its high light yield, appropriate emission spectrum, and reasonable cost, NaI(Tl) is the crystal of choice for commercial systems. Because of its relatively low cost compared to arrays, a single continuous crystal is typically used in clinical systems. Because of its high gain, low noise, and reasonable cost the PMT is currently the photodetector of choice in nuclear medicine.

VI. PHYSICAL FACTORS THAT DETERMINE THE FUNDAMENTAL SPATIAL RESOLUTION LIMIT IN NUCLEAR EMISSION IMAGING

A. Overview of Physical Factors That Affect Emission Imaging Performance

There are many physical factors that effect radioisotope imaging performance. Many of these topics are discussed in detail in later chapters of this book. In planar gamma ray imaging and single-photon emission computed tomography (SPECT) (see Zeng, Gunter, Genna, and Levin, Chapters 7, 8, 9, and 16 in this book), the sensitivity and typically spatial resolution are limited by the properties of the physical collimator used. In certain cases, the spatial resolution is strongly affected by the intrinsic spatial resolution and object scatter as well. For a source deep within tissue, photon attenuation can significantly reduce available counts and cause nonuni-form image artifacts. King (Chapter 22 in this book) discusses approaches to attenuation correction in SPECT. Image contrast between different activity structures is affected by the incorrectly positioned Compton scatter photons. The largest positioning errors originate from scatter that occurs from within the imaging subject. Because Compton scatter photons are less energetic than the initial photon, Compton scatter effects can be reduced by accepting only those events that have total energy within a small range (window) around the photopeak energy. To achieve this scatter reduction while still maintaining high sensitivity, the energy resolution of the system should be high; hence, the scintillation crystal should be bright and the electronic noise low. King (Chapter 22 in this book) discusses other scatter rejection methods for SPECT. Having high crystal photofraction will contribute to improved sensitivity by ensuring that a high fraction of detected events fall in the photopeak.

In coincident annihilation photon imaging systems such as PET (see Lewellen, Watson, and Cherry, Chapters 10–12 in this book), because no physical collimator is used, the system sensitivity is determined by the system intrinsic and geometric efficiencies for 511-keV annihilation photons. For good efficiency the detector material should have high Z and density for good photon-stopping power and cover as large an area as possible around the imaging subject. As is the case for single-photon imaging, annihilation photon attenuation within the body can significantly reduce available counts and produce nonuniform image artifacts. This problem is actually worse for coincident annihilation photon imaging because there are two detected photons per event, and hence there is a higher probability for an interaction to occur. Attenuation correction techniques for PET are discussed in Lewellen (Chapter 10 in this book). The spatial resolution is determined by the fluctuations of positron range and annihilation photon noncollinearity, and intrinsic detector resolution. Image contrast is affected by Compton scatter, random or accidental coincidences, and detector dead time. Compton scatter positioning errors are more prominent in annihilation photon imaging due to the broader scatter distribution and reduced (for 2D acquisition) or absence of (in 3D acquisition) physical collimation (see Lewellen, Watson, and Cherry, Chapters 10–12 in this book). Compton scatter can be somewhat reduced by energy windowing, but again, to maintain high sensitivity (fewer rejected events), the energy resolution and photofrac-tion should be high so that a narrow photopeak window will contain a high fraction of the incoming events. Other scatter rejection techniques for PET are discussed in Lewellen, Watson, and Cherry (Chapters 10–12 in this book).

Random coincidences in annihilation photon imaging are background events in which two different nuclei each contribute one detected annihilation photon within the coincidence resolving time of the system. The effect of random coincidences on image contrast can be reduced by having excellent coincident time resolution and by reducing the detected activity. High coincident resolving time requires a fast, bright scintillation crystal or detector and low electronic noise. Standard techniques for random-coincidence reduction are discussed in Lewellen, Watson, and Cherry (Chapters 10–12 in this book). Detector dead time due to pulse pile-up degrades energy and spatial resolutions, resulting in reduced PET image contrast at high count rates. Random-coincidence and dead-time effects can be reduced if fast detectors are implemented. Standard dead-time correction for PET systems is discussed in Lewellen, Watson, and Cherry (Chapters 10–12 in this book).

B. What Is the Fundamental Spatial Resolution Limit in Radioisotope Imaging?

The importance of rodent (mouse and rat) models for human disease has spawned an interest in high spatial resolution medical imaging. To be able to resolve fine structures, high spatial sampling of the signal from the object is required. There has especially been a recent interest in the development of high spatial resolution radioisotope imaging systems for small animals (see Cherry and Levin, Chapters 12 and 16 in this book and the references therein) due to the inherent functional biological information that may be obtained. These developments motivate the question: What is the fundamental spatial resolution that can be achieved with radioisotope imaging? It is well known that other radiology imaging modalities such as X-ray computed tomography (CT) or magnetic resonance imaging (MRI) can achieve submillimeter spatial resolution. This allows tiny structures of the animal subject to be resolved with those systems. Could this submillimeter spatial resolution be possible for nuclear medicine techniques such as SPECT and PET? If so, would the sensitivity per resolution element be adequate enough to fully use this spatial resolution potential? These physical questions are addressed in this section.

1. Spatial Resolution Potential for Single-Photon Emission Imaging Systems

In most gamma-ray imaging systems, the achievable spatial resolution is limited by the collimator properties. The well-known collimator equations for spatial resolution and sensitivity in parallel-hole, converging, diverging, and pinhole collimators are presented in (Chapter 8, Gunter in this book). The collimator spatial resolution is quite variable. For the parallel-hole design, the spatial resolution depends on the hole size, height, linear attenuation coefficient at the photon energy of interest, and distance of the radioisotope source from the collimator. For the pinhole design, in addition to similar parameters, the resolution also depends on the apex angle of the pinhole cone. The total system spatial resolution is a convolution of the collimator and intrinsic detector resolutions. We assume here that these contributions are approximately Gaussian in shape and we ignore scatter effects on the full width at half maximum (FWHM) spatial resolution.

Figure 24 plots the system point source spatial resolution for different source–collimator distances as a function of intrinsic detector spatial resolution for parallel-hole and pinhole collimators at 140 keV. In these calculations, we have assumed a hole diameter of 1.2 mm, a collimator height of 4.0 cm, $\mu = 27 cm^{-1}$ for lead at 140 keV, a septal thickness of 0.2 mm for the parallel-hole design, and for the pinhole an apex angle of 45° and source locations along its axis. The results will change for different values of these parameters, but the assumed parameters promote high spatial resolution performance and will serve as a basis to understand spatial resolution limitations. For the pinhole calculations we have included the magnification factor x/h, where x is the source–aperture distance and h the height of the collimator. The magnifying property of pinhole collimators can provide an improvement in overall system spatial resolution in the image plane when the source–aperture distance is less than the collimator height because the resulting image magnification factor reduces the effective contribution of the intrinsic detector resolution.

Higher intrinsic detector resolutions can be achieved by segmenting the detector into tiny pixels. In this way, the intrinsic spatial resolution is selected to be on the order of the detector pixel size (see Chapter 16 in this book). For fine sampling of the radioiso-tope distribution, these pixels must be closely spaced with little dead area in between pixels.

Figure 24 shows the well-known feature that in single-photon imaging the spatial resolution improves with decreasing source–collimator distance and hole diameter. For source–collimator distances less than 1 cm (i.e., shallow structures), the spatial resolution achievable is essen-tially determined by the hole diameter (1.2 mm for Figure 24) and the intrinsic detector resolution. To achieve submillimeter spatial resolution with collimated gamma-ray imaging, we could decrease the hole diameter d into the submillimeter range, be extremely close to the object, and use a finely pixellated detector design. However, because the geometric efficiency is on the order of d^2 for both the parallel- and

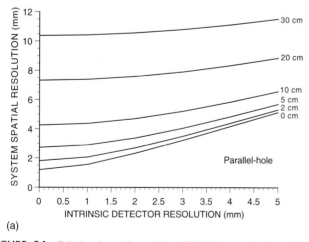
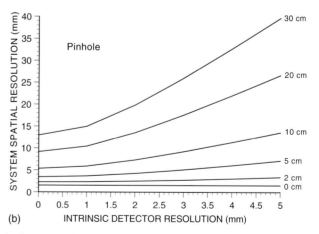

FIGURE 24 Calculated spatial resolution (FWHM) as a function of intrinsic detector spatial resolution for various source–collimator distances for (a) parallel-hole and (b) pinhole collimators at 140 keV. We have assumed a collimator hole diameter of 1.2 mm, height of 4.0 cm, and $\mu = 27 cm^{-1}$ for lead at 140 keV; and for the pinhole an apex angle of 45° and the magnification factor have been included.

pinhole collimators (see Chapter 8 in this book, for efficiency equations), the system sensitivity would also significantly drop. For example, with our parameters ($d = 1.2$ mm) and $x = 2$ cm, the system spatial resolutions at 140 keV are 2.1 and 2.3 mm with corresponding geometric efficiencies of 5×10^{-5} and 2×10^{-4}, respectively, for the parallel- and pinhole collimators using 1-mm intrinsic spatial resolution. This gives sensitivities of 2.0 and 7.3 cps/µCi, respectively, for the parallel- and pinhole designs for a 140-keV point source in air and a single planar detector array. With one-half the hole diameter (600 µm), the system spatial resolutions at 2.0 cm would be 1.3 and 1.4 mm, respectively, for the parallel- and pinhole designs, but the collimator efficiency values would drop to 1×10^{-5} and 6×10^{-5}, respectively. This yields low sensitivity values of 0.4 and 2.2 cps/µCi, respectively, for the two designs for 140-keV photons.

For planar imaging, because the standard deviation per image pixel is $\sqrt{N}$, where N is the number of events in that pixel, the image pixel SNR is $\sqrt{N}$. If the system resolution drops by a factor of two, the number of events must increase by a factor of four to maintain the same average SNR per image pixel. In tomographic (SPECT) imaging, the pixel values are highly correlated. It turns out that if the resolution cell size decreases by two, the required number of events for constant uncertainty must to increase by roughly a factor of eight (Budinger et al., 1978). So, the significantly reduced sensi-tivity values for the 600-µm-hole collimators may not allow these systems to attain their resolution potential of approxi-mately 1 mm due to poor SNR per image pixel.

The sensitivity in a parallel-hole collimator system does not change significantly with source-to-collimator distance. However, the pinhole sensitivity drops off as $1/x^2$, where x is source–aperture distance. Thus, moving the detector closer to the source allows us to regain sensitivity in a pinhole design. Another basic solution to improving the overall system sensitivity while maintaining high resolution is to increase the number of pinhole or parallel-hole detector arrays and surround the object so that more area surrounding the radioisotope distribution is covered. Such a small-diameter ring system consisting of multiple, finely spaced, pinhole-collimated, pixellated detector arrays is under development (Pang et al., 1994).

Note that the resolution values plotted in Figure 24 are for 140-keV photons. Significantly improved spatial resolution may also be achieved by using lower energy sources (e.g., ^{125}I) with a submillimeter pinhole collimator. However, due to the high degree of attenuation for such low-energy photons, such a system would be limited to imaging very shallow structures within tissue.

2. Spatial Resolution Potential for Coincident Annihilation Photon Imaging Systems

The main advantages of annihilation photon imaging systems are the improved sensitivity and spatial resolution potential because no physical collimator is required. The direction of annihilation photon emission is defined electronically by the line connecting the two interactions on opposing detectors. However, both oppositely directed photons must be detected, so the extent of the sensitivity improvement depends on the geometric and intrinsic efficiency of the detector gantry. Although there is no collimator resolution component in annihilation photon imaging, there are two additional effects that contribute to spatial resolution blurring: positron range and annihilation photon noncollinearity.

In radioisotope emission imaging, we are after the exact location of the emitter nucleus. Unlike in single-photon emission imaging, photons from positron annihilation do not originate from the same location as the particle that was emitted. There is a fluctuation in penetration or range of the energetic positron in tissue due to the varying positron emission energy and direction and attenuating properties of the tissue (see Figures 5 and 10). More energetic particles penetrate deeper. There is a fluctuation in the end point of the positron's track and therefore in the origin of the annihilation photons. Thus, the precision in localizing a positron emitter is compromised by the positron's range. For best spatial resolution, the positron emitter chosen should have the smallest maximum decay energy. Attempts to measure (Derenzo, 1979; Phelps et al., 1975) and calculate (Levin and Hoffman, 1999; Palmer and Brownell, 1992) the effect of positron range on resolution have been made. A histogram of calculated positron annihilation endpoint coordinates along any given direction is cusplike, as shown in Figure 25 for ^{18}F positrons. These calculated data result from Monte Carlo simulations of electron trajecto-ries from a point source that includes effects of the positron energy spectrum, multiple Coulomb elastic scattering from the nucleus, hard elastic collisions with electrons, and energy loss due to ionization and excitation (see Section IVA). The cusplike shape is due to the fact that many of the positrons are emitted with low (< 50 keV) energy and thus will be absorbed very close to the location of emission. This cusplike structure can be approximated by a sum of two decaying exponential functions, one with a fast decay constant (short ranges) and the other with a slow decay constant (large ranges) (see Derenzo et al., 1979 and Levin et al., 1999):

$$F(x) = Ce^{-k_1 x} + (1-C)e^{-k_2 x} \quad x \geq 0 \quad (42)$$

The best fit parameters to this equation for the simulated cusplike positron range functions for ^{18}F, ^{11}C, ^{13}N, and ^{15}O are given in Table 2. Because ^{18}F positron emission has a relatively low maximum and mean kinetic energy of 635 and 250 keV, respectively, the calculated maximum range is only ~ 2 mm with a FWHM in one dimension on the order of 0.1 mm. For positron emitters with higher maximum positron emission energy, the blurring due to positron range

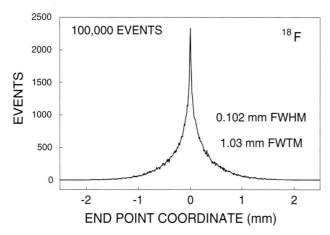

FIGURE 25 A histogram of x-coordinates of positron annihilation endpoints from a Monte Carlo simulation of ^{18}F positron trajectories in water for a point source (see Figure 10). FWHM, full width at half maximum; FWTM, full width at tenth maximum. (Adapted from Levin and Hoffman, 1999.)

TABLE 2 Best-Fit Parameters of Equation (42) to the Simulated 1D Cusplike Positron Range Distributions in Water for Four Common Positron Emitters.[a]

	^{18}F	^{11}C	^{13}N	^{15}O
C	0.52	0.49	0.43	0.38
k_1 (mm^{-1})	37.9	23.8	20.2	18.1
k_2 (mm^{-1})	3.10	1.81	1.42	0.90

[a] See Fig. 25; $x \geq 0$ and normalized for y-intercept of 1.

TABLE 3 Calculated Range-Blurring Contributions for Common Positron Emitters[a]

Isotope	Average Positron Kinetic Energy (keV)	Maximum Positron Kinetic Energy (keV)	FWHM (mm)	FWTM (mm)
^{18}F	250	635	0.10	1.03
^{11}C	390	970	0.19	1.86
^{13}N	490	1190	0.28	2.53
^{15}O	740	1720	0.50	4.14

[a] From [16].

will be greater. Table 3 gives the FWHM and full width at tenth maximum (FWTM) of Monte Carlo simulation–calculated positron range cusp functions for four of the most commonly used positron emitters (Levin and Hoffman, 1999). It should be noted that for a cusplike function, FWHM and FWTM do not have a well-defined meaning. Note the positron range blurring contribution depends only on the isotope and material traversed and is independent of system parameters.

Positioning algorithms in annihilation photon coincidence imaging assume that annihilation photons are emitted in completely opposite directions (180° apart). In reality if the positron–electron system is not at rest at annihilation, the two annihilation photons cannot be completely collinear as a result of momentum conservation. It turns out from detailed experiments that the positron annihilation photon angular deviation from collinearity is small and the angular distribution is approximately Gaussian-shaped with 0.25° or 0.0044 radian FWHM for a point source (Levin and Hoffman, 1999). The extent of resolution blurring from annihilation photon noncollinearity depends on the system detector spacing (system diameter in the case of dedicated-ring PET) because the further the detection points are from the annihilation origin, the larger the positioning error introduced by the assumption of photon collinearity. Using the formula for arc length, we multiply the system radius by the angular deviation to determine the linear deviation introduced. The linear deviation (FWHM) is roughly 0.0022 and D, where D is the system diameter in millimeters, independent of the isotope used. For a 20- and 80-cm detector separation, for example, the deviation at FWHM introduced is roughly 0.4 and 1.8 mm, respectively. Thus, for optimal spatial resolution, the system detector separation, (or diameter) should be small. Note that as the detector separation is decreased, the sensitivity significantly increases in coincidence photon imaging.

The last important contribution to the annihilation photon system spatial resolution is due to the intrinsic detector resolution, which is also an important factor in close proximity single-photon imaging. For annihilation photon coincidence imaging, however, a measurement of point-source response (i.e., *point spread function*) is triangular in shape with FWHM on the order of one-half the detector element size, assuming source-to-detector distances much greater than the detector element size. Thus, for example, the intrinsic detector resolution contributes on the order of 2 mm when the crystal pixel width is 4 mm. For better spatial resolution, finer detector elements should be used. The system spatial resolution will always be greater than the detector half width contribution due to positron range and noncollinearity contributions. Other blurring effects in annihilation photon imaging such as off-axis detector penetration, detector Compton scatter, and inadequate sampling of the signal for image reconstruction can be compensated for with proper detector design and are covered in Lewellen and Cherry, Chapters 10 and 12 in this book.

Thus, the fundamental spatial resolution in annihilation photon coincidence imaging is determined by positron range, photon noncollinearity and detector pixel size. The total system resolution is a mathematical convolution of the three distributions. Figure 26 plots the distributions (blurring functions) for these three physical factors as well as the total system spatial resolution function for an ^{18}F point source in a 20-cm-diameter animal system and an 80-cm-diameter

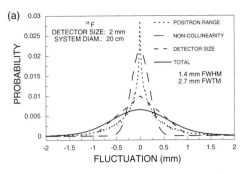
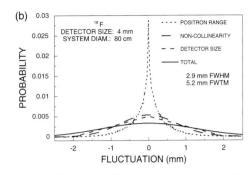

FIGURE 26 Spatial resolution blurring functions for positron range, photon noncollinearity, and detector size using an ^{18}F point source for (a) 20-cm- and (b) 80-cm-diameter systems. (Adapted from Levin and Hoffman, 1999.)

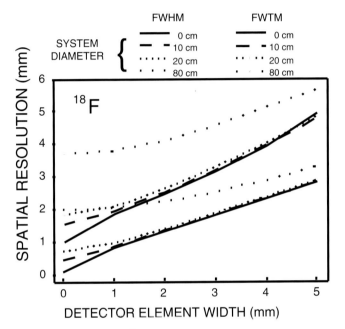

FIGURE 27 Calculated ^{18}F PET system spatial resolution (FWHM and FWTM) as a function of detector element width for various system diameters. For 1-mm detectors, 20-cm detector separation, the spatial resolution potential is ~ 750 mm FWHM for coincident annihilation photon imaging.

human system. Figure 27 plots the ^{18}F system resolution (FWHM and FWTM) as a function of detector element size for various system diameters. Figure 27 may be useful for the design of high spatial resolution ^{18}F coincidence imaging systems. Resolution limit curves for other common PET isotopes are given in Levin and Hoffman (1999).

Figure 26 shows that for small system diameters, the system spatial resolution for ^{18}F is limited by the detector element width and for large diameters by the noncollinearity effect. From Figure 27, we see that it is only possible to push the spatial resolution of PET into the micron range by using a low-energy position emitter such as ^{18}F and ≤1-mm detector pixels. For a 10-cm-diameter PET system with 1-mm-wide detectors, the resolution limit would be roughly 750 μm, which is certainly comparable to or better than most CT or MRI system spatial resolutions. Further improvements in spatial resolution for ^{18}F imaging would be possible by using highly pixellated, submicron semiconductor detector arrays with adequate stopping power for 511-keV photons. The SNR varies as the inverse cube root of the detector sampling distance (Sorenson and Phelps, 1987). So, by decreasing the intrinsic detector resolution by a factor of two, the sensitivity drops by roughly a factor of eight. The sensitivity of a PET system consisting of a ring of detectors for a point source at its center is roughly $2\pi RH/4\pi R^2 \sim 1/R$, where R is the system radius and H is the axial extent of the ring. Thus, submillimeter spatial resolution may be realized with adequate image SNR by moving the detectors as close as possible to the imaged object.

References

Bethe, H. (1953). Molierc's theory of multiple scattering. *Phys. Rev.* **89**, 1256–1266.
Birks, J. B. (1964). Theory and Practice of Scintillation Counting. Pergamon, Oxford.
Budinger, T. F., Greenberg, W. L., Derenzo, S. E., et al. (1978). Quantitative potentials of dynamic emission computed tomography. *J. Nucl. Med.* **19**, 309.
Casey, M. E., and Nutt, R. (1986). A Multicrystal Two Dimensional BGO Detector System for Positron Emission Tomography. *IEEE Trans. Nucl. Sci.* **33**(1), 460–463.
Chadwick, J. (1914). The intensity distribution in magnetic spectrum of b-rays of radium B & C. *Verhandl. Deut. Physik. Ges.* **16**, 383.
Daniel, H. (1968). Shapes of beta-ray spectra. *Rev. Med. Phys.* **40**, 659–672.
Derenzo, S. E. (1979). Precision measurement of annihilation point spread distributions for medically important positron emitters. In 'Positron Annihilation'. (R. R. Hasiguti, and K. Fujiwara, eds.), pp. 819–823, Japan Institute of Metals, Sendai, Japan.
Evans, R. D. (1972). "The Atomic Nucleus." McGraw-Hill, New York.
Heitler, W. (1960). "The Quantum Theory of Radiation." 3rd ed. Oxford University Press.
Jeavons, A. P., Chandler, R. A., and Dettmar, C. A. R. (1999). A 3D HIDAC-PET Camera with Sub-millimetre Resolution for Imaging Small Animals. *IEEE Trans. Nucl. Sci.* **46**(3), 468–473.
Johns, H. E., and Cunningham, J. R. (1978). 3rd Ed. The Physics of Radiology. C Thomas, Springfield.

Knoll, G. F. (1989). "Radiation Detection and Measurement," 2nd ed. Wiley, New York.

Lederer, C. M., Hollander, J. M., and Perlman, I. (1967). "Table of Isotopes," 6th ed. Wiley, New York.

Leo, W. R. (1987). "Techniques for Nuclear and Particle Physics." Springer-Verlag. Berlin.

Levin, C. S., Germani, J., Markey, J. (1993). Charge collection and energy resolution studies in compressed xenon gas near its critical point. *Nucl. Instr. Meth. Phys. Res.* A. **332,** 206–214.

Levin, C. S., and Hoffman, E. J. (1999). Calculation of positron range and its effect on the fundamental limit of positron emission tomography system spatial resolution. *Phys. Med. Biol.* **44,** 781–799.

Levin, C. S., MacDonald, L. R., Tornai, M. P., and Hoffman. E. J. (1996). Optimizing light collection from thin scintillators used in a beta-ray camera for surgical use. *IEEE Tran. Nucl. Sci.* **43,** 2053–2060.

Meyerhof, W. E. (1967). "Elements of Nuclear Physics." McGraw-Hill, New York.

Palmer, M. R., and Brownell, G. L. (1992). Annihilation density distribution calculations for medically important positron emitters. *IEEE Trans. Med. Imaging* **11,** 373–378.

Pang, I., Barrett, H. H., Chen, J-C *et al.* (1994). Physical evaluation of a four-dimensional brain imager. [Abstract]. *J. Nucl. Med.* **35**(5), 28P.

Patt, B. E., Iwanczyk, J. S., Wang, Y. J., *et al.* (1996). Mercuric Iodide Photodetector Arrays for Gamma-Ray Imaging. *Nucl. Inst. & Meth.* A **380,** 295–300.

Pauli, W. (1934). "Rapports du 7^e Conseil de Physique Solvay, Brussels, 1933." Gauthier-Villars, Paris.

Phelps, M. E., Hoffman E. J., and Huang, S. (1975). Effects of positron range on spatial resolution. *J. Nucl. Med.* **16,** 649–652.

Ritson, D. M. (1961). Techniques of High Energy Physics. Interscience, New York.

Roll, F. David (ed.) (1977). "Nuclear Medicine Physics, Instrumentation and Agents." CV Mosby, Saint Louis.

Siegban, K. (ed.) (1955). Beta and Gamma Ray Spectroscopy. Interscience, New York.

Sorenson, J. A., and Phelps, M. E. (1987). "Physics in Nuclear Medicine," 2nd ed. W. B. Saunders, Philadelphia.

Visvikis, D., Ott, R. J., Wells, K., and Flower, M. A. *et al.* (1997). Performance characterisation of large area BaF_2-TMAE detectors for use in a whole body clinical PET camera. *Nucl. Instr. Meth. Phys. Res.* A **392,** 414–420.

Wu, C. S., and Moskowski, S. A. (1966). "Beta Decay." Interscience, New York.

CHAPTER 5

Radiopharmaceuticals for Imaging the Brain

JOGESHWAR MUKHERJEE,* BINGZHI SHI,* T. K. NARAYANAN,* B. T. CHRISTIAN,* and YANG ZHI-YING[†]
*Department of Nuclear Medicine, Kettering Medical Center, Wright State University, Dayton, Ohio
[†]Department of Psychiatry, University of Chicago, Chicago, Chicago, Illinois

I. Introduction
II. Biochemical Processes in the Brain
III. New Radiopharmaceutical Development
IV. Neuroscience Studies
V. Applications of Imaging the Dopamine System
VI. Oncology Studies
VII. Genomic Studies
VIII. Summary

I. INTRODUCTION

Functional imaging of the brain has taken on new dimensions with the advent of sophisticated radiopharmaceuticals that can be used in noninvasive imaging studies using positron emission tomography (PET) and single-photon emission computed tomography (SPECT). These new technologies have essentially unlimited potential because they are tools to study an array of physiological processes in human health and disease. Development of specifically targeted radiopharmaceutical enables us to study the exquisite biochemistry of the brain due to the very high sensitivity of these imaging methods. The ability to study biomolecules at a nanomolar to picomolar concentration *in vivo* allows a systematic study of various physiological processes. These advancements in imaging technology have further enhanced our capability to study the various processes in small brain regions.

The goal of this chapter is to provide a general background on radiopharmaceuticals and review briefly some of the most common radiopharmaceuticals that are used clinically and in clinical research for studies related to the brain. Emphasis is placed on radiopharmaceutical development related to neurotransmitter receptor systems, an area in which a substantial effort has been placed and major advancements in the field have occurred. However, functions of the brain indicate that we have only begun to scratch the surface of various biochemical processes that are taking place. New protein targets continue to be identified and their function and pathophysiological implications are yet to be clearly defined. The dopamine neurotransmitter receptor system has been most widely studied using these imaging methods; applications of the dopamine radiopharmaceuticals are discussed. With the completion of mapping of the human genome, efforts are now underway to take imaging a step further into genomics. This presents unique challenges for radiopharmaceutical development; an assessment of the issues is presented. Major advances have been made in several areas of SPECT radiopharmaceuticals but are not covered in this chapter due to space limitations; the reader is referred to reviews and other book chapters that have covered various aspects of SPECT radiopharmaceuticals (Saha *et al.*, 1994; Heinz *et al.*, 2000).

II. BIOCHEMICAL PROCESSES IN THE BRAIN

Metabolic studies using ^{18}F-2-fluoro-2-deoxyglucose (^{18}F-FDG), various amino acids such as ^{11}C-methionine, and perfusion studies have now been carried out for a number of years. These studies have been summarized and the reader is referred to several excellent reviews and book chapters (Phelps *et al.*, 1986). Neurochemical processes in the brain are the primary targets for many new radiopharmaceuticals. As we might imagine, there are numerous neurochemical pathways in the brain. Each pathway is immense with a number of proteins as potential targets that may be involved in health and disease. Typically, neurons communicate using neurotransmitters and receptors. This requires the presence of a neurotransmitter, its production pathways, its storage facilities, release mechanisms, retrieval of unused or excess neurotransmitter, receptors for relaying the messages, amplifiers of the message, second-messenger systems, and message termination processes that include various feedback loops.

Summarized in Figure 1 are the various biochemical processes that have been the targets of radiopharmaceutical development:

1. Enzymes in the cytoplasm either produce a neurotransmitter (such as DOPA decarboxylase) or metabolize a radiotracer (such as hexokinase). These measurements allow us to assess the extent of enzyme activity.
2. Neurotransmitters synthesized in the cytoplasm are transported into vesicles through the vesicular transporter for storage. Radiotracers can be specifically targeted to bind to the vesicular transporter in order to assess the anomalies in vesicular storage.
3. Radiolabeled neurotransmitter precursors can be used to assess the storage capability of neurons.
4. Membrane transporters located on the presynaptic membrane are involved in the transport of unused or excess neurotransmitter released in the synaptic junction. Radiotracers targeted specifically to these sites allow the measurement of the integrity of the presynaptic membrane as well as assess up- or downregulation of the membrane transporters.
5. Postsynaptic receptors can be targeted by radiotracers in order to assess distribution, up- or downregulation, and competition with neurotransmitters.
6. Postsynaptic membranes also contain receptor ionophores and ion channels that can be direct targets for radiotracers.
7. Second-messenger systems that relay information received by the receptors have also been the targets for radiotracer development.
8. Enzyme systems located in the mitochondria, such as monoamine oxidases that are responsible for degrading the monoamine neurotransmitters, have also been targets for radiotracer development.

It is important to note that processes in the brain are dynamic, which adds another dimension to the development and use of radiopharmaceuticals. Some mechanisms by which radiotracers are capable of providing useful biochemical information are as follows:

1. Inert diffusible: R* <=====> R* (not retained)
   The radiotracer (R*) enters and exits the tissues without being retained by any specific biochemical process. Examples include ^{15}O-water which is used to measure blood flow.
2. Metabolite trapping: R* ----Enzyme----> B* (intermediate trapped)
   The radiotracer (R*) enters the tissue and is converted into a labeled metabolite (B*) by the specific process under study. Due to slow clearance of the metabolite out of the cell, retention of the radiolabeled metabolite occurs. Examples include ^{18}F-fluorodeoxyglucose (^{18}F-FDG) which is metabolized by the enzyme hexokinase to ^{18}F-FDG-6-phosphate. The ^{18}F-FDG-6-phosphate is retained in the cell and serves as a measure of glucose consumption by the cell.
3. Metabolite storage: R* -----Enzyme-----> B* -----> B* (stored in vesicles)
   The radiotracer (R*) enters the tissue and is converted into a labeled metabolite (B*) by the specific process under study. Retention of the radiolabeled metabolite (B*) occurs due to a specific storage mechanism. Examples include ^{18}F-fluorodihydroxyphenylalanine (^{18}F-FDOPA) which is metabolized by the enzyme DOPA decarboxylase to give ^{18}F-fluorodopamine, which is a neurotransmitter analog of dopamine. The ^{18}F-fluorodopamine is therefore stored in dopamine containing vesicles in the dopaminergic neurons.
4. Suicide trapping: R* ------Enzyme------> R*-E (trapped)

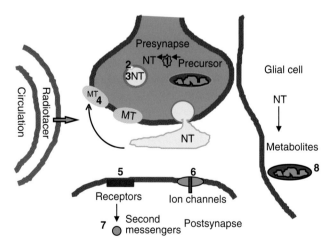

FIGURE 1 Schematic showing sites of action of radiopharmaceuticals in a synaptic junction.

The radiotracer (R*), which is designed to be a good substrate for a specific enzyme, enters the tissue and binds the enzyme. Due to the nature of the radiotracer design, suicide inhibition of the enzyme occurs and the radiotracer is thus trapped covalently in the enzyme (R*-E). Examples include the use of monoamine oxidase A and B inhibitors, ^{11}C-clorgyline and ^{11}C-deprenyl, which are propargylamine analogs of phenethylamines and bind to the enzymes and covalently modify them.

5. Protein inhibition: R* + Protein ----> R*P (retained)----> R* + Protein

The radiotracer (R*) is designed to be a high-affinity and selective antagonist for the protein. Upon entry into the tissues the radiotracers binds to the protein and, because of the slow dissociation rate constant of the radiotracer–protein complex, retention of the radiotracer occurs. Examples include the numerous postsynaptic receptor, membrane transporter, and vesicular transporter radiotracers.

III. NEW RADIOPHARMACEUTICAL DEVELOPMENT

A. Steps in Radiopharmaceutical Development

The development of diagnostic radiopharmaceuticals comprises of a number of steps. It generally starts out with the identification of a biological hypothesis and specific biochemical target(s) associated with this hypothesis. This is an essential element of the development process because knowledge of the biochemical properties as well as the molecular nature of the target greatly assists in the identification of the most suitable radiopharmaceutical. Subsequent to the identification of a given target, the process of radiopharmaceutical design begins. Most of the times lead compounds are available from the pharmaceutical industry, which may have drugs for therapeutic purposes that have an affinity for the target or related target molecules (for example, for receptor subtypes). This knowledge can then be used to further refine the structure by a limited structure–activity relationship (SAR) study and can be further assisted by molecular modeling methods.

Once potential compounds have been identified, the synthesis of the various compounds can be carried out. These compounds are tested *in vitro* in order to evaluate their affinity and selectivity for binding to the target sites. This *in vitro* pharmacology assessment assists in screening out compounds that may be suboptimal because of their low affinity or low selectivity for the target sites. Once a lead compound has been identified, methods to radiolabel the chosen compound are developed. Depending on the kind of radionuclide that is planned, radiosynthesis steps are developed. Critical in designing radiosynthesis schemes is the emphasis on obtaining high radiochemical yields in a short duration of time. The developer has to factor in the time required for chromatographic purification of the radiopharmaceutical as well. After the completion of the radiosynthesis of the radiopharmaceutical, *in vitro* and *in vivo* evaluations of the new agent are carried out. *In vitro* studies may include autoradiographic evaluations in brain slices in order to study distribution of the radiopharmaceutical. Studies can also be carried out on the subcellular distribution of the radiopharmaceutical as well as kinetic studies on association and dissociation of the radiopharmaceutical.

In vivo studies are carried out to investigate *in vivo* stability, blood–brain barrier (BBB) permeability, selective localization, and retention in the regions of interest. Pharmacological studies such as competition with other unlabeled drugs interacting with the same target sites may also be carried out in order to demonstrate binding of the radiopharmaceutical to the desired target region. If the radiopharmaceutical shows promise in the rodent studies, the selected agents are taken to nonhuman primate PET studies. Studies using PET can allow the very detailed *in vivo* evaluation of the distribution and kinetics of the radiopharmaceutical in the various regions of the entire brain. Pharmacological studies can be carried out to further understand the subtleties of neurochemistry. After covering appropriate regulatory issues, the radiopharmaceutical can be taken into initial human studies.

B. Criteria for Pharmaceuticals Used for *In Vivo* Receptor Imaging

In order to develop an agent for purposes of *in vivo* imaging, there are several criteria that need to be met. These are (1) high affinity (<10 nM) for the receptors, which allows for adequate retention of the radiopharmaceutical in the tissues during *in vivo* experiments; (2) high selectivity for the receptors, which allows for the specific study of a given target site; (3) reversibility of binding *in vivo*, which allows competition experiments *in vivo* in order to demonstrate the selective binding of the agent to target sites as well as allowing the agent to come to a pseudoequilibrium; (4) permeability and establishment of steady state over the BBB *in vivo*, which are essential in order to demonstrate that the BBB is not an impediment in the uptake of the radiopharmaceutical; (5) low *in vivo* accumulation of labeled or active ligand metabolites, which is important in order to avoid complications in the analysis of data resulting from the metabolite as well as the radiopharmaceutical; (6) low nonspecific binding *in vivo*, which enables the acquisition of data sets that provide a good signal-to-noise ratio and can be critical due to the short duration of the imaging period; and (7) saturable binding *in vivo*, which is an essential feature in order to demonstrate that the radiopharmaceutical is binding to specific target sites found in small concentrations.

IV. NEUROSCIENCE STUDIES

A. Dopaminergic System

The dopamine system comprises of presynaptic and postsynaptic processes for which a number of radiopharmaceuticals are currently available. The presynapse consists of steps to produce the neurotransmitter dopamine (involving enzymes such as tyrosine hydroxylase and DOPA decarboxylase), dopamine storage vesicles, vesicular transporter, and presynaptic membrane transporter. Radiopharmaceutical studies have been carried out with ^{18}F-fluoroDOPA, which provides a suitable marker for studying DOPA decarboxylase activity as well as storage of the resulting neurotransmitter analog, ^{18}F-fluorodopamine in the presynaptic vesicles (Garnett *et al.*, 1983). Design processes have resulted in ^{18}F-fluoro-*m*-tyrosine as an analog of ^{18}F-fluoroDOPA in order to reduce metabolites and thus lower nonspecific binding (DeJesus *et al.*, 1997). Vesicular transporter located on the storage vesicles has been studied by imaging agents such as ^{11}C-tetrabenazine (Kilbourn, 1997). Currently there are no radiotracers for tyrosine hydroxylase, an enzyme that is the rate-limiting step in the production of dopamine. Imaging of tyrosine hydroxylase may provide vital information on the ability and extent of production of the precursor of dopamine, L-DOPA. Development of imaging agents for the dopamine membrane transporter were initiated with the realization that cocaine was able to selectively bind to these sites *in vivo*. Since then a number of agents, based primarily on the tropane skeleton of cocaine have been prepared for studying the dopamine transporter located on the presynaptic cell membrane in the synapse (reviewed in Carroll *et al.*, 1995). Applications of the various presynaptic radiotracers are continually being expanded in the study of health and disease (reviewed in Verhoeff, 1999).

Postsynaptically, there are at least five dopamine receptors that have been identified and characterized to various degrees. The dopamine D-2 receptor has been studied to a large extent by imaging methods with a number of available imaging agents. The interest in this receptor was largely triggered by the fact that neuroleptics interacted with this receptor system and therefore imaging of this receptor would have diagnostic as well as therapeutic applications. Radiotracer design strategies have addressed subtle issues such as efficient signal-to-noise ratio, reversibility of binding, competition with dopamine, binding to brain regions containing small concentrations of the receptors, and radiolabeling with various radionuclides. Thus, currently there are a number of radiopharmaceuticals for the D-2 receptor based on the butyrophenones and substituted benzamides (Figure 2). Due to their enhanced selectivity, the substituted benzamides are more commonly used than other agents. One as yet unresolved feature of substituted benzamides is their affinity for the less abundant D-3 receptors. The substituted benzamide, ^{11}C-raclopride, and ^{123}I-IBZM are commonly used in PET and SPECT studies of D-2 receptors found in the caudate and putamen (Verhoeff, 1999). There is now an interest in the study of D-2 receptors in

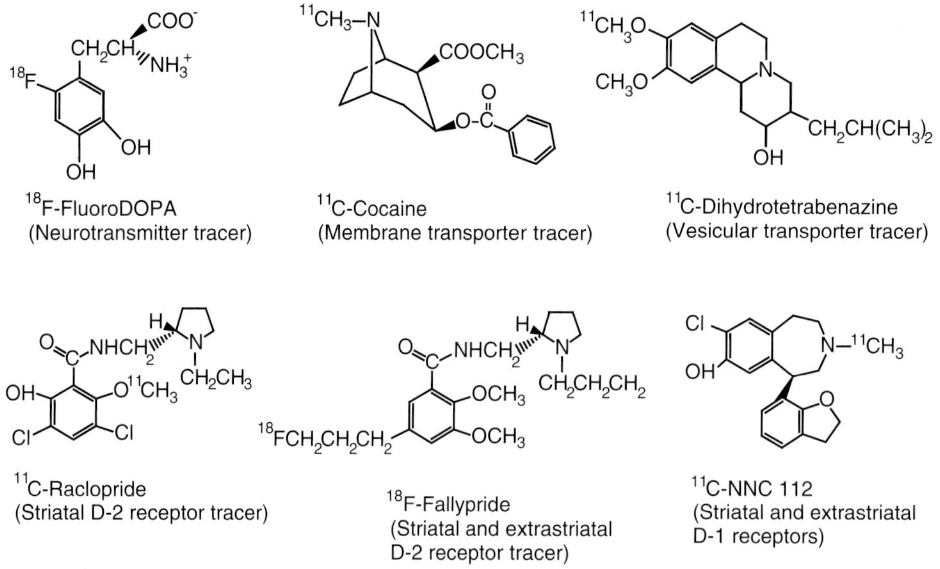

FIGURE 2 Selected dopamine neurotransmitter receptor system radiopharmaceuticals.

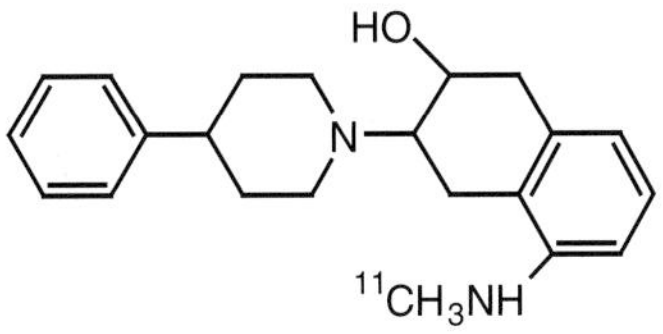

^{11}C-MABV
(Vesicular acetylcholine transporter tracer)

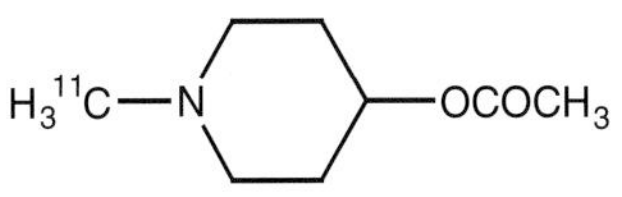

^{11}C-Methylpiperidyl acetate
(Acetylcholinesterase activity tracer)

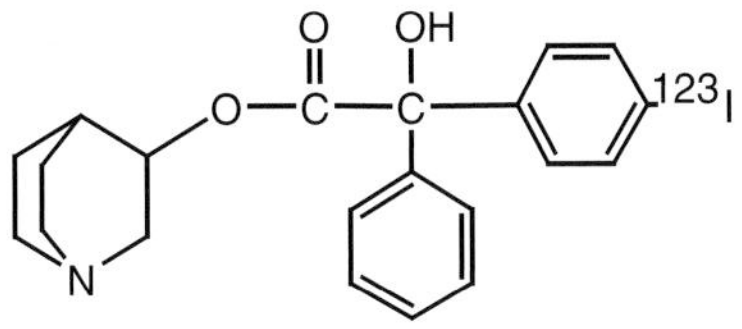

^{123}I-IQNB
(Muscarinic receptor tracer)

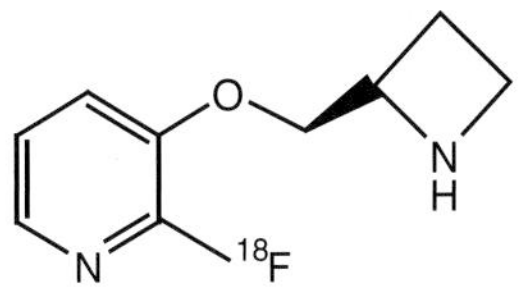

^{18}F-A85380
(nACh receptor tracer)

FIGURE 3 Selected cholinergic neurotransmitter receptor system radiopharmaceuticals.

areas of the brain containing smaller concentrations of the receptors. The high-affinity selective agent ^{18}F-fallypride is being used for the study of the D-2 receptors in striatal and extrastriatal regions (Mukherjee *et al.*, 2002). The next widely studied dopamine receptor is the D-1 subtype for which the substituted benzazepines have been developed as imaging agents. Improvements in the design process have addressed issues of signal-to-noise ratios, labeling with various radionuclides, binding to brain regions containing small concentrations of the receptors, and competition with dopamine. Most studies have been carried out with ^{11}C-SCH 23390 and, more recently, ^{11}C-NNC-112 has been used for the study of striatal and extrastriatal D-1 receptors (Verhoeff, 1999). A fluorine-18 labeled D-1 receptor radiotracer, F-FPr SCH38548, has also been reported (Yang *et al.*, 1996). Although there are good radiopharmaceuticals for the D-1 receptors and the receptors are present in high concentrations, imaging studies on this receptor have been fewer compared to the D-2 receptor system.

The D-3, D-4, and D-5 receptors are less abundant and there is an ongoing effort to develop selective radiopharmaceuticals for these subtypes. Preliminary reports are emerging on the development of radiopharmaceuticals for these receptors. Significant challenges for imaging *in vivo* are selective localization and retention of the radiopharmaceuticals due to the small concentrations of these receptors.

B. Cholinergic System

A number of efforts have been made in order to study the cholinergic system. Of the many neurological areas in which this system is implicated, Alzheimer's disease has been at the forefront. Presynaptically, a significant amount of effort has gone into the development of radiotracers for the vesicular transporter of acetyl choline (VAChT). The vesamicol class of compounds have been developed as good radiotracers (Figure 3). These compounds have been radiolabeled with iodine-123 for use in SPECT imaging as well as carbon-11 and fluorine-18 for use in PET imaging. Efforts are also underway to develop imaging tools for the enzyme acetylcholinesterase, which is involved in degrading acetyl choline back to choline; however, no studies have been reported in attempts to study choline acetyltransferase.

Two classes of postsynaptic receptors exist in the cholinergic system—the muscarinic (mAChR) and the nicotinic (nAChR) receptors. In the case of the muscarinic system, several receptor subtypes, M-1, M-2, M-3, M-4, and M-5, have been identified. Abnormalities in the muscarinic receptors have been reported in Alzheimer's disease, Parkinson's disease, and Huntington's disease. The quinuclidinyl series of compounds, such as iodine-123-labeled IQNB, have been reported to have high affinity for the muscarinic receptors; however, these compounds have low selectivity (Gibson, 1992). More recently, fluorinated analogs of the quinuclidinyl backbone have been synthesized and are being evaluated as *in vivo* imaging agents (Knapp *et al.*, 1997).

Interest in the nicotinic system has gained in the last few years with the identification of epibatidine as an agonist for the nAChRs (Badio and Daly, 1994). A number of agonists, such as ^{18}F-A85380, for nAChRs are undergoing evaluation for their *in vivo* properties (Sihver *et al.*, 2000; Kimes *et al.*, 2003). These radiopharmaceuticals are being used in pharmacological studies of nicotine effects in the brain. Development of nAChR antagonists such as ^{18}F-nifrolidine is currently being pursued for use in PET studies (Chattopadhyay et al., 2003).

C. Serotonergic System

The serotonin system has attracted a significant amount of attention due to its perceived role in a number of neuropsychiatric illnesses. Presynaptically, the serotonin transporter has been an area of significant study for the development of radiopharmaceuticals. A number of antidepressants are known to target the serotonin transporter. Although developmentally a fair amount of effort has been placed on the serotonin transporter, very few successful radiopharmaceuticals have been developed. Efforts were made to develop radiolabeled fluoxetine, paroxetine, and citalopram as radiotracers for the serotonin transporter. However, in vivo specific binding of these agents was not clearly demonstrated, due to the high lipophilicity and prolonged retention of these radiotracers in vivo (reviewed in Fletcher et al., 1995). Until recently, ^{11}C-Mc-5562 was the most selective compound for the serotonin transporter suitable for human studies (Szabo et al., 1999). Included in Figure 4 is a new class of substituted phenylthiobenzylamines (^{11}C-DASB) that are being developed as serotonin transporter agents and show promise as selective agents (Houle et al., 2000).

Although, some effort has gone into the use of ^{11}C-methyltryptophan for the study of serotonin synthesis, applications have only been limited due to questions regarding its ability to measure serotonin synthesis (Diksic et al., 2000). No effort has gone into the development of radiopharmaceuticals for the serotonin vesicular transporter. A number of postsynaptic serotonin receptors have been identified (Peroutka, 1994). The serotonin 5HT-2A receptor has attracted a lot of attention due to its role in neuropsychiatric illnesses. The butyrophenones, such as spiperone, have been used although they bind to other receptor systems, such as the dopamine D-2 receptor, as well. Two commonly used radiopharmaceuticals for the 5HT-2A receptor are ^{18}F-setoperone (Yatham et. al., 2000) and ^{18}F-altanserin (Meltzer et al., 1999). ^{18}F-altanserin is the most commonly used agent for the 5HT-2A receptors. However, there is still room for improving the signal-to-noise ratio and other characteristics of the 5HT-2A radiotracers. Other receptors such as the 5HT-1A has attracted much attention recently with the development of selective antagonists such as WAY-100635 and their potential role in anxiety as well depression. Imaging studies have been carried out with ^{11}C-WAY-100635, and other analogs of this compound have also been developed in order to improve imaging characteristics, reduce metabolites, and incorporate the longer half-life fluorine-18. Work continues on the development of improved in vivo characteristics of the WAY-100635 (Pike et al., 2000). There are a number of other serotonin receptors, such as 5HT-1B, 5HT-1C, 5HT-1D, 5HT-3, 5HT-4, 5HT-6, and 5HT-7, that have been thought to be of interest from a therapeutic point of view. However, little effort has been directed toward these targets.

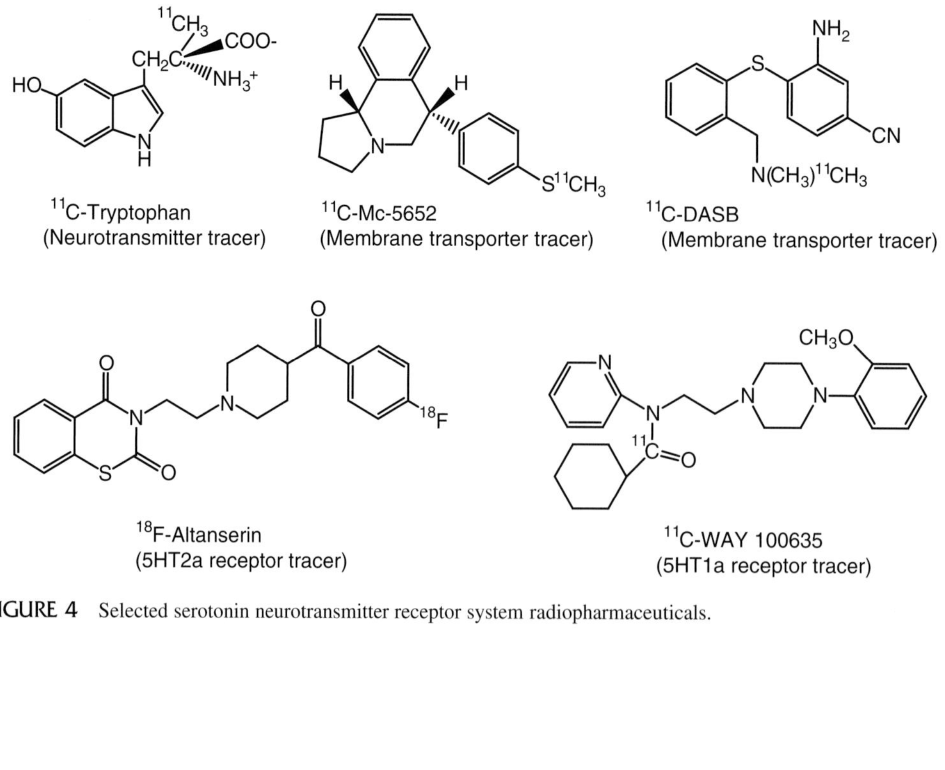

^{11}C-Tryptophan
(Neurotransmitter tracer)

^{11}C-Mc-5652
(Membrane transporter tracer)

^{11}C-DASB
(Membrane transporter tracer)

^{18}F-Altanserin
(5HT2a receptor tracer)

^{11}C-WAY 100635
(5HT1a receptor tracer)

FIGURE 4 Selected serotonin neurotransmitter receptor system radiopharmaceuticals.

D. Other Systems

Opiod system: Opiate has been known to be involved in pain and mood alteration. Three postsynaptic opiate receptors, mu, kappa, and delta have been pursued by imaging studies using agents such as ^{11}C-diprenorphine, ^{11}C-carfentanil, and ^{11}C-methylnaltrindole, shown in Figure 5 (reviewed in Mayberg and Frost, 1990; Frost, 1993; Smith *et al.*, 1999; Zubieta *et al.*, 1999).

GABAergic system: The benzodiazepine site of the GABA complex has attracted the most interest (reviewed in Sadzot and Frost, 1990). A number of studies have been carried out with ^{11}C-flumazenil and ^{11}C-PK 11195, which are known to bind selectively to central and peripheral benzodiazepine sites, respectively. There is a significant amount of interest in the study of the benzodiazepine sites in epilepsy due to the hypothesis that there may be a decrease in the inhibitory pathways that involve the GABA complex (Lamusuo *et al.*, 2000).

Glutamate system: The glutamate system is an important excitatory amino acid system that has attracted attention for radiopharmaceutical development but with little success. The *N*-methyl-D-aspartate (NMDA) ion-channel complex has been implicated in long-term potentiation with significance to learning and memory. The complex has a calcium ion channel at which drugs such as phenylcyclohexyl piperidine (PCP), thienylcyclohexyl piperidine (TCP), and dizoclipine (MK-801) are known to bind. The ionotropic glutamate receptors have been further refined into kainate, quisqualate, AMPA, and NMDA subclasses. Efforts have been made to develop imaging agents based on these compounds, but with little success (Ouyang *et al.*, 1996). In recent years, several other metabotropic glutamate receptors have also been identified. Although the glutamate system is known to be vital in various brain functions as well as excitotoxicity, there are currently no radiopharmaceuticals to study this important system.

Adrenergic system: Both the central and peripheral adrenergic systems have attracted attention due to their roles in various physiological functions. Several proteins in this system present as potential targets (presynaptic vesicular transporter, membrane transporter, and several postsynaptic receptors divided into two major subclasses of alpha- and beta-receptors). Efforts are being made to develop good *in vivo* imaging agents of both classes. Preliminary studies have been carried out with the prazosin analog, ^{11}C-GB67 for the alpha-1 receptors (Law *et al.*, 2000). Pulmonary beta-adrenoreceptors have been studied in humans using ^{11}C-CGP-12177 (Qing *et al.*, 1996).

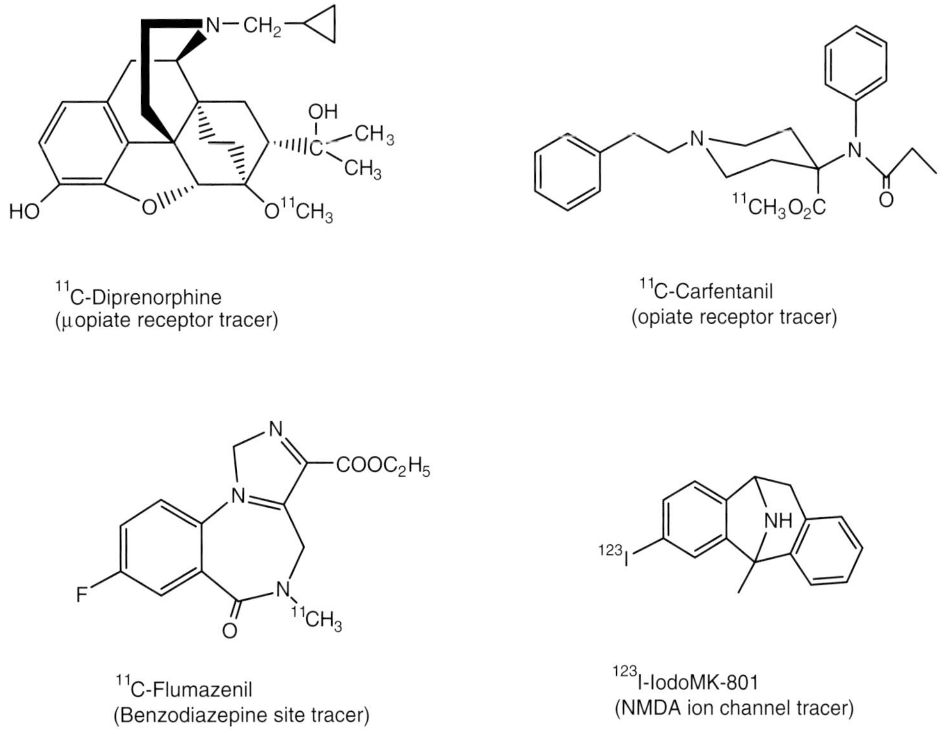

^{11}C-Diprenorphine
(μ opiate receptor tracer)

^{11}C-Carfentanil
(opiate receptor tracer)

^{11}C-Flumazenil
(Benzodiazepine site tracer)

^{123}I-IodoMK-801
(NMDA ion channel tracer)

FIGURE 5 Selected radiopharmaceuticals for the opiate, GABAergic and glutamate neurotransmitter receptor systems.

Other systems: There are several other systems that are being explored for new radiopharmaceuticals, such as the histaminergic system, consisting of histidine decarboxylase, vesicular transporter, membrane transporter, and postsynaptic receptors. Second-messenger systems are being investigated and some effort has gone into the development of radiotracers based on phorbol esters and forskolin. Very little effort has gone into the development of radiopharmaceuticals for neuroactive peptides.

V. APPLICATIONS OF IMAGING STUDIES: DOPAMINE SYSTEM

A significant amount of effort has been placed on the dopaminergic system which has resulted in a number of applications. Excellent reviews dealing with the dopaminergic system have appeared in the literature (Verhoeff, 1999; Volkow *et al.*, 1996). A summary of these applications is discussed in this section.

A. Distribution of Receptors, Transporters, and Enzymes

Selective radiopharmaceuticals for a particular protein can be used for mapping the distribution in the entire brain. These methods have been applied to various receptor and neurotransmitter systems. *In vitro* mapping of the distribution of the various receptors is carried out with radioligands labeled with various radioisotopes such as tritium, carbon-14, iodine-125, and others. Distribution of the receptors *in vitro* can be studied more cleanly because nonspecific binding can be removed by fine-tuning washing protocols. Also, when using relatively nonselective agents, undesired binding can be blocked out by using available antagonists. Thus, the remaining selective binding of the radioligand can be quantitated in the various brain regions. In the case of *in vivo* mapping, the constraints are somewhat more stringent. *In vivo* kinetics of the radiopharmaceutical, distribution of the radiopharmaceutical in small brain regions, and complications from nonspecific binding can all contribute to difficulties in obtaining good distribution maps of the proteins. Design of good radiopharmaceuticals that address the various concerns can be critical for successful use. Brain mapping with the dopamine D-2 receptor radiopharmaceutical ^{18}F-fallypride and the D-1 receptor radiopharmaceutical ^{11}C-NNC-112, as well as mapping dopamine using ^{18}F-fluoroDOPA, have been successful. Shown in Figure 6 is the distribution of D-2/D-3 receptors in the human brain using ^{18}F-fallypride. The high affinity and selective nature of this radiopharmaceutical allow the exquisite delineation of these receptors in the entire brain (Mukherjee *et al.*, 2002).

B. Distribution and Pharmacokinetics of Radiolabeled Drugs

One of the advantages of radiolabeling a drug molecule is the ability to carry out the whole-body distribution of the new drugs. This information leads to the distribution of the drugs in the various organs and pharmacokinetics of the drugs. Issues of toxicity as well as dosimetry can be rapidly evaluated in the whole body quite rapidly.

C. Receptor Occupancy of Therapeutic Drugs

Using radiopharmaceuticals for a given receptor or protein that is a target for therapeutic drugs, competition experiments can be carried out with the various drugs. The studies can evaluate the adequate BBB permeability of the therapeutic drug, drug–response curve with respect to a given receptor, and pharmacokinetics of the drug at the receptor site. These findings can eventually provide an appropriate dosing regimen for therapeutic drugs which will provide maximal benefit with minimal side effects (Mukherjee *et al.*, 2001).

D. Integrity of the Presynapse

Two particular target sites are currently being used to investigate the integrity of the presynapse. Dopamine that has been synthesized from L-DOPA in the presynapse is stored in vesicles in the presynapse. Thus, ^{18}F-fluoroDOPA is a good marker for studying the process of storing dopamine in the vesicles. This radiotracer therefore has found widespread application in the study of Parkinson's disease, which involves a degeneration of the dopamine presynaptic terminals. More recently, the presynaptic membrane–bound dopamine transporter has been a target for the study of substance abuse drugs as well as to measure the integrity of the presynapse. Various tropane compounds are currently being evaluated as clinically suitable imaging agents for the dopamine transporter.

E. Alterations in the Levels of Endogenous Ligand Dopamine

An important application of dopamine D-2 receptor radiopharmaceuticals is the study of competition with endogenous dopamine, released pharmacologically or behaviorally. Dopamine is known to have a high affinity for the D-2 receptor and thus is able to compete with radiopharmaceuticals that bind to these receptors. By virtue of this competition, efforts have been made to assess the amount of synaptic dopamine that may be present (released) during various pharmacological or behavioral challenges. Several radiopharmaceuticals, such as ^{11}C-raclopride, ^{18}F-fallypride, ^{18}F-FCP, ^{18}F-desmethoxyfallypride and ^{123}I-IBZM have been used to study release of dopamine.

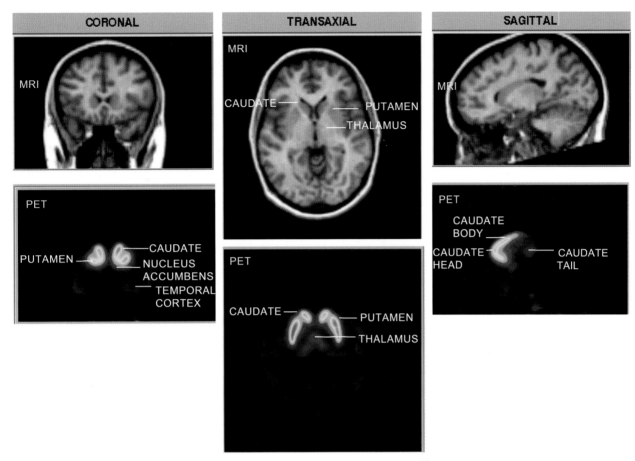

FIGURE 6 MRI and corresponding PET image slices of the human brain showing binding of ^{18}F-fallypride to dopamine D-2/D-3 receptors.

F. Study of Substance Abuse Drugs

Substance abuse drugs, such as cocaine, amphetamine, and methamphetamine, are known to interact with dopaminergic nerve terminals. It is this interaction (the release of dopamine by either the blockade of the transporter or by rupture of the vesicles) that results in the increased dopamine concentration in the synapse and accounts for the drug high. Drugs such as cocaine and methylphenidate have been labeled with carbon-11 and studies have indicated a good correspondence in the drug high and *in vivo* binding of ^{11}C-cocaine. Efforts are now underway to develop potential therapies for cocaine abuse, and imaging studies have played a critical role in this development.

G. Interactions with Other Neurotransmitter Systems

Because the brain is a complex organ, a number of neurotransmitter receptor systems are interconnected and nuclear imaging offers the possibility of studying these interactions *in vivo*. The first interaction to be studied was between dopamine and serotonin receptor systems. Efforts have also been made to study the dopamine system with the GABAergic system and the NMDA calcium ion channel.

H. Posttreatment Monitoring of the System

Therapies that involve improvements or modifications of a neurotransmitter–receptor system can be evaluated by the use of imaging radiopharmaceuticals. This has been applied in the case of dopaminergic system in which fetal implants have been grafted on to brains deficient in the ability to produce enough dopamine. By using ^{18}F-fluoroDOPA or ^{18}F-fluoro-m-tyrosine, the increase in dopamine-producing cells has been monitored.

I. Etiological Studies of the System

The dopaminergic system, including dopamine, the receptors, and the transporters, has been implicated in a number of neurological and psychiatric disorders. A number of radiopharmaceuticals have been targeted to the various sites in order to study the nature of anomalies associated with various disease states.

J. Neurotoxicity of Drugs

A number of substance abuse drugs have been known to cause long-term damage to brain cells. By using selective radiopharmaceuticals, specific processes in the brain can be studied in order to evaluate neurotoxicity. Methamphetamine, for example, has been known to be quite neurotoxic to dopamine and serotonin nerve terminals. Efforts have been made to study the dopamine transporter in methamphetamine abusers.

VI. ONCOLOGY STUDIES

Diagnostic radiopharmaceuticals have played an important role in the study of various brain tumors. Perfusion studies using ^{15}O-water and metabolic studies using ^{18}F-FDG have been used extensively. Clinically, ^{18}F-FDG is used routinely for the study of brain tumors. Amino acids such as ^{11}C-methionine are also applied in conjunction with ^{18}F-FDG. Methionine is transported and incorporated into protein synthesis quite rapidly and therefore serves as a marker for tumors. Attempts have also been made to use ^{11}C-choline, a lipid marker in cases of brain tumor (Figure 7). Although brain uptake of ^{11}C-choline is low due to poor BBB permeability, uptake in the tumors seems to be quite high. Radiopharmaceuticals that assess hypoxia, sigma receptors, estrogen receptors, and androgen receptors are also being evaluated for oncology studies (reviewed in Varagnolo *et al.*, 2000). Other radiotracers for the study of DNA duplication processes, such as ^{18}F-fluorothymidine and ^{18}F-FMAU, are also being evaluated as potential *in vivo* radiopharmaceuticals (Gambhir *et al.*, 1999).

VII. GENOMIC STUDIES

With the advances made in the identification and cloning of a number of genes and progress in a plethora of molecular biology techniques, the potential for developing diagnostic and therapeutic tools at the genetic level is increasingly becoming a reality. The pathway from a gene to a protein is a complex one. Within the nucleus of a cell, a gene is expressed selectively in certain cells due to factors that initiate RNA polymerase–assisted transcription of the DNA to form pre-mRNA. Pre-mRNA undergoes a process of methylation at the 5′-end (referred to as capping) and is polyadenylated at the 3′-end (referred to as polyadenylation), both of which are believed to enhance the stability of the pre-mRNA and assist in the transport across the nuclear membrane. Prior to transport, the pre-mRNA is spliced to remove the introns, which results in the final mRNA, which leaves the nucleus. Clearly, for evaluating gene expression for a given gene, there are a set of targets within the nucleus that ought to be evaluated. In particular, probes for the factors that trigger gene expression will be very valuable. However, there are several issues to contend with in developing imaging agents for events with the nucleus, the main ones being (1) target size because of the extremely small concentration of genes (one per cell), (2) design of appropriate probes (possibly probes that may form a triplex with the DNA sequence) and (3) transport into the nuclear membrane.

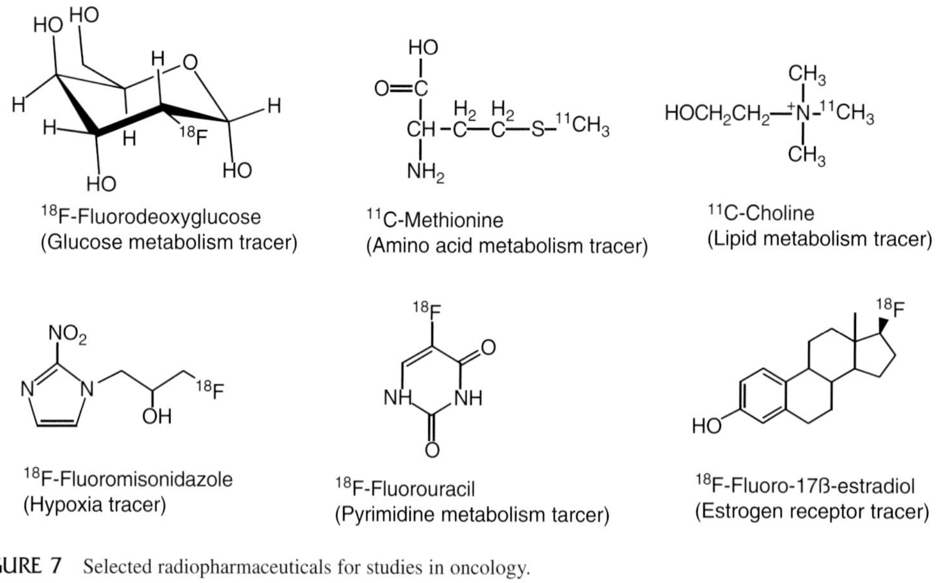

FIGURE 7 Selected radiopharmaceuticals for studies in oncology.

The mRNA from the nucleus binds to the ribosome, which, using tRNA, translates the nucleotide sequence into a peptide sequence. Levels of mRNA are significantly higher than the gene itself and may therefore offer as potential target sites. Although not measuring the gene transcription directly (due to the number of steps involved from the gene to mRNA levels in the cytoplasm), the assessment of mRNA levels in the cytoplasm is possibly a few steps closer to measuring gene expression. Imaging mRNA levels *in vivo* is still a formidable task. Although specific gene sequences of mRNA are available, the development of selective probes that will maintain their integrity *in vivo* and transport to their target sites while displaying appropriate pharmacokinetics in other regions is going to be a challenge. The following approaches are being used in studying gene expression.

A. Imaging Protein Products

Efforts have been underway to image protein products as a gauge of the level of gene expression (reviewed by Gambhir *et al.*, 1999). A gene system using such an approach is thymidine kinase, which has relevance to the treatment of herpes simplex virus infection. Acyclovir is phosphorylated by thymidine kinase, and acyclovir triphosphate leads to chain termination when it is incorporated into DNA. Several different substrates have been used to evaluate expression of this protein (Gambhir *et al.*, 1999). Although these approaches are a good measure of evaluating the presence of mRNA resulting from gene expression, the target sites are translated protein products and thus only indirectly reflect mRNA levels and gene expression profile. This approach has not been applied to brain studies yet.

B. Use of Antisense RNA to Image mRNA

Inhibition of transcription and translation events of genes by synthetic fragments of DNA and RNA has opened up the field of antisense nucleotides for diagnosis and therapy (Stephenson and Zamecnik, 1978; Murray, 1992). There are now numerous examples of the *in vitro* use of oligonucleotides that have been used for *in situ* hybridization experiments using fluorescent, chemiluminescent, and radiolabeled probes. The ability to evaluate levels of mRNA provides an important index of gene expression and these findings can be correlated to the levels of the protein. Imaging of mRNA levels is likely to be a significant challenge. The high-resolution imaging studies possible with PET using minute amounts of radiotracers makes the development of a positron-emitting oligonucleotide possible. There is emerging enthusiasm for the potential of antisense mRNA–based oligodeoxynucleotide imaging agents, although reports have raised a number of issues that need to be overcome in order to accomplish this task successfully (Dewanjee *et al.*, 1994; Hnatowich, 1999; Piwnica-Worms, 1994).

A significant amount of attention is being afforded to identifying oligonucleotides called aptamers, that will bind with high affinity to specific targets (Jayasena, 1999). This approach uses a systematic evolution of ligands by exponential enrichment (also referred as SELEX) and develops a series of aptamers that bind tightly to the three-dimensional structure of the target site (Brody *et al.*, 1999; Drolet *et al.*, 1999). This approach will identify rapidly, using combinatorial approaches, high-affinity oligomers for mRNA. Although it has been assumed that an oligomer 17 nucleotides long should find a unique target within the 3×10^9 base pairs of the human genome (Hélène, 1994; Ho and Parkinson, 1997), it remains to be seen whether the SELEX approach can identify the fingerprint regions within a given mRNA using smaller oligomers. Development of such smaller oligomers may be essential for the development of brain-imaging agents.

VIII. SUMMARY

Radiopharmaceutical development has progressed and evolved substantially since the 1980s. This chapter has covered only a small fraction of the radiopharmaceuticals that are being continually investigated for various uses and has focused on the neuroscience and, more briefly, oncology studies. Due to the substantially large number of targets in the brain, a number of efforts have gone into developing radiopharmaceuticals for neurotransmitter–receptor systems. Within this class, some systems such as the dopamine, has been more widely studied, whereas other equally important systems such as glutamate have not advanced to any significant degree. Clearly, the number of targets in the brain are numerous and emphasis on radiopharmaceutical development has to be prioritized according to relevance to health and disease. In oncology, metabolic studies with ^{18}F-FDG have taken a front seat, but there is an increasing effort to understand the molecular basis of cancer. Although efforts have been made in the past to develop receptor-based targeting of cancer, current efforts are now being placed increasingly on genomic studies. In cardiology, studies have primarily emphasized either metabolic (using glucose or fatty acids) or perfusion studies. Neurotransmitter studies of the heart are clearly wanting if we are to understand the relationship of the heart to the mind.

References

Badio, B., and Daly, J. W. (1994). Epibatidine, a potent analgetic and nicotinic agonist. *Mol. Pharmacol.* **45:** 563–569.

Brody, E. N., Willis, M. C., Smith, J. D., Jayasena, S., Zichi, D., and Gold, L. (1999). The use of aptamers in large arrays for molecular diagnostics. *Mol. Diagnosis* **4:** 381–388.

Carroll, F. I., Scheffel, U., Dannals, R. F., Boja, J. W., and Kuhar, M. J. (1995). Development of imaging agents for the dopamine transporter. *Med. Res. Rev.* **15**: 419–444.

Chattopadhyay, S., Bagnera, R., Leslie, F. M., Xue, B., Christian, B., Shi, B., Narayanan, T. K., Potkin, S. G., and Mukherjee, J. (2003). ^{18}F-Nifrolidine: PET imaging agent for $\alpha 4\beta 2$ nicotinic receptors. *J. Nucl. Med.* **44**: 606 suppl.

DeJesus, O. T., Endres, C. J., Shelton, S. E., Nickles, R. J., and Holden, J. E. (1997). Evaluation of fluorinated m-tyrosine analogs as PET imaging agents of dopamine nerve terminals: Comparison with 6-fluorodopa. *J. Nucl. Med.* **38**: 630–636.

Dewanjee, M. K., Ghafouripour, A. K., Kapadvanjwala, M., Dewanjee, S., Serafini, A. N., Lopez, D. M., and Sfakianakis, G. N. (1994). Noninvasive imaging of c-myc oncogene messenger RNA with indium-111-antisense probes in a mammary tumor-bearing mouse model. *J. Nucl. Med.* **35**: 1054–1063.

Diksic, M., Tohyama, Y., and Takada, A. (2000). Brain net unidirectional uptake of alpha-[C-14]methyl-L-tryptophan (alpha-MTrp) and its correlation with regional serotonin synthesis, tryptophan incorporation into proteins, and permeability surface area products of tryptophan and alpha-MTrp. *Neurochem. Res.* **25**: 1537–1546.

Drolet, D. W., Jenison, R. D., Smith, D. E., Pratt, D., and Hicke, B. J. (1999). A high throughput platform for systematic evolution of ligands by exponential enrichment (SELEX(TM)). *Comb. Chem. High Throughput Screening* **2**: 271–278.

Fletcher, A., Pike, V. W., and Cliffe, I. A. (1995). Visualization and characterization of 5-HT receptors and transporters in vivo and in man. *Sem. Neurosci.* **7**: 421–431.

Fowler, J. S. (1990). Enzyme activity:Monoamine oxidase. In "Quantitative Imaging: Neuroreceptors, Neurotransmitters and Enzymes" (J. J. Frost and H. N. Wagner, Jr., eds.), pp. 179–192, Raven Press, New York.

Frost, J. J. (1993). Receptor imaging by PET and SPECT-Focus on the opiate receptor. *J. Receptor Res.* **13**: 39–53.

Gambhir, S. S., Barrio, J. R., Herschman, H. R., and Phelps, M. E. (1999). Assays for noninvasive imaging of reporter gene expression. *Nucl. Med. Biol.* **26**: 481–490.

Garnett, E. S., Firnau, G., and Nahmias, C. (1983). Dopamine visualization in the basal ganglia of living man. *Nature* **305**: 137–138.

Gibson, R. E., Moody, T., Schneidau, T. A., Jagoda, E. M., and Reba, R. C. (1992). The in vitro dissociation kinetics of (R,R)-[125I]4IQNB is reflected in the in vivo washout of the radioligand from rat brain. *Life Sci.* **50**: 629–637.

Heinz, A., Jones, D. W., Raedler, T., Coppola, R., Knable, M. B., and Weinberger, D. R. (2000). Neuropharmacological studies with SPECT in Neuropsychiatric disorders. *Nucl. Med. Biol.* **27**: 677–682.

Hélène, C. (1994). Control of oncogene expression by antisense nucleic acids. *Eur. J. Cancer* **30A**: 1721–1726.

Hnatowich, D. J. (1999). Antisense and nuclear medicine. *J. Nucl. Med.* **40**: 693–703.

Ho, P. T., and Parkinson, D. R. (1997). Antisense oligonucleotides as therapeutics for malignant diseases. *Sem. Oncol.* **24**: 187–202.

Houle, S., Ginovart, N., Hussey, D., Meyer, J. H., and Wilson, A. A. (2000). Imaging the serotonin transporter with positron emission tomography: Initial human studies with [C-11]DAPP and [C-11]DASB. *Eur. J. Nucl. Med.* **27**: 1719–1722.

Jayasena, S. D. (1999). Aptamers: An emerging class of molecules that rival antibodies in diagnostics. *Clin. Chem.* **45**: 1628–1650.

Kilbourn, M. R. (1997). In vivo radiotracers for vesicular neurotransmitter transporters. *Nucl. Med. Biol.* **24**: 615–619.

Kimes, A. S., Horti, A. G., London, E. D., Cheffer, S. I., Contoreggi, C., Ernst, M., Friello, P., Koren, A. O., Kurian, V., Matochik, J. A., Pavlova, O., Vaupel, D. B., and Mukhin, A. G. (2003). 2-[18F]F-A-85380: PET imaging of brain nicotinic acetylcholine receptors and whole body distribution in humans. *FASEB J.* **17**: 1331–1333.

Knapp, F. F., McPherson, D. W., Luo, H., and Zeeburg, B. (1997). Radiolabeled ligands for imaging the muscarinic-cholinergic receptors of the heart and brain. *Anticancer Res.* **17**: 1559–1572.

Lamusuo, S., Pitkanen, A., Jutila, L., Ylinen, A., Partanen, K., Kalviainen, R., Ruottinen, H. M., Oikonen, V., Nagren, K., Lehikoinen, P., Vapalahti, M., Vainio, P., and Rinne, J. O. (2000). [C-11]Flumazenil binding in the medial temporal lobe in patients with temporal lobe epilepsy: Correlation with hippocampal MR volumetry, T2 relaxometry, and neuropathology. *Neurology* **54**: 2252–2260.

Law, M. P., Osman, S., Pike, V. W., Davenport, R. J., Cunninggham, V. J., Rimoldi, O., Rhodes, C. G., Giardina, D., and Camici, P. G. (2000). Evaluation of [C-11]GB67, a novel radioligand for imaging myocardial alpha1-adrenoceptors with positron emission tomography. *Eur. J. Nucl. Med.* **27**: 7–17.

Mayberg, H. S., and Frost, J. J. (1990). Opiate receptors. In "Quantitative Imaging: Neuroreceptors, Neurotransmitters and Enzymes" (J. J. Frost and H. N. Wagner, Jr., eds.), pp. 81–95, Raven Press, New York.

Meltzer, C. C., Price, J. C., Mathis, C. A., Greer, P. J., Cantwell, M. N., Houck, P. R., Mulsant, B. H., Ben-Eliezer, D., Lopresti, B., DeKosky, S. T., and Reynolds, C. F. (1999). PET imaging of serotonin type 2A receptors in late-life neuropsychiatric disorders. *Amer. J. Psychiatry* **156**: 1871–1878.

Mukherjee, J., Christian, B. T., Dunigan, K., Shi, B., Narayanan, T. K., Satter, M., and Mantil, J. (2002). Brain Imaging of ^{18}F-fallypride in normal volunteers: Blood analysis, kinetics, distribution, test-retest studies and preliminary assessment of sensitivity to aging effects on dopamine D-2/D-3 receptors. *Synapse* **46**: 170–188.

Mukherjee, J., Christian, B.T., Narayanan, T.K., Shi, B., and Mantil, J. (2001). Evaluation of dopamine D-2 receptor occupancy in vivo by clozapine, risperidone, and haloperidol in rodents and non-human primates using ^{18}F-fallypride. *Neuropsychopharmacology* **25**: 476–488.

Murray, J. M., and Nishikura, K. (1991). Application of antisense nucleic acids in oncology. In "Antisense Nucleic Acids and Proteins" (J. N. M. Mol and A. R. van der Krol, eds.), pp. 95–112. Marcel Dekker, New York.

Ouyang, X. H., Mukherjee, J., and Yang, Z. Y. (1996). Syntheses, radiosyntheses, and biological evaluation of thienylcyclohexyl piperidine derivatives as potential radiotracers for the N-methyl-D-aspartate receptor-linked calcium ionophore. *Nucl. Med. Biol.* **23**: 315–324.

Peroutka, S. J. (1994). Molecular biology of serotonin (5-HT) receptors. *Synapse* **18**: 241–260.

Phelps, M. E., Mazziotta, J. C., and Schelbert, H. R. (1986). "Positron Emission Tomography and Autoradiography." Raven Press, New York.

Pike, V. W., Halldin, C., Wikstrom, H., Marchais, S., McCarron, J. A., Sandell, J., Nowicki, B., Swahn, C-G., Osman, S., Hume, S. P., Constantinou, M., Andree, B., and Farde, L. (2000). Radioligands for the study of brain 5-HT1a receptors in vivo: Development of some new analogues of WAY. *Nucl. Med. Biol.* **27**: 449–455.

Piwinca-Worms, D. (1994). Making sense out of antisense: Challenges of imaging gene translation with radiolabeled oligonucleotides. *J. Nucl. Med.* **35**: 1064–1066.

Qing, F., Rhodes, C. G., Hayes, M. J., Krausz, T., Fountain, S. W., Jones, T., Hughes, and J, M. B. (1996). In vivo quantification of human pulmonary beta-adrenoreceptor density using PET: Comparison with in vitro radioligand binding. *J. Nucl. Med.* **37**: 1275–1281.

Sadzot, B., and Frost, J. J. (1990). Benzodiazepine receptors. In "Quantitative Imaging: Neuroreceptors, Neurotransmitters and Enzymes" (J. J. Frost and H. N. Wagner, Jr., eds.), pp. 109–127, Raven Press, New York.

Saha, B., MacIntyre, W. J., Brunken, R. C., Go, R. T., Raja, S., Wong, C. O., and Chen, E. Q. (1996). Present assessment of myocardial viability by nuclear imaging. *Sem. Nucl. Med.* **26**: 315–335.

Saha, B., MacIntyre, W. J., and Go, R. T. (1994). Radiopharmaceuticals for brain imaging. *Sem. Nucl. Med.* **24**: 324–349.

Sihver, W., Nordberg, A., Langstrom, B., Mukhin, A. G., Koren, A. O., Kimes, A. S., and London, E. D. (2000). Development of ligands for in vivo imaging of cerebral nicotinic receptors. *Behav. Brain Res.* **113:** 143–157.

Smith, J. S., Zubieta, J. K., Price, J. C., Flesher, J. E., Madar, I., Lever, J. R., Kinter, C. M., Dannals, T. F., and Frost, J. J. (1999). Quantification of delta-opiod receptors in human brain with N1'-([C-11]methyl)naltrindole and positron emission tomography. *J. Cereb. Blood Flow Metab.* **19:** 956–966.

Stephenson, M. L., and Zamecnik, P. C. (1978). Inhibition of Rous sarcoma viral RNA translation by a specific oligodeoxyribonucleotide. *Proc. Natl. Acad. Sci., U.S.A.* **75:** 285–288.

Szabo, Z., Scheffel, U., Mathews, W. B., Ravert, H. T., Szabo, K., Kraut, M., Palmon, S., Ricuarte, G. A., and Dannals, R. F. (1999). Kinetic analysis of [11C]McN5652: a serotonin transporter radioligand. *J. Cereb. Blood Flow Metab.* **19:** 967–981.

Varagnolo, L., Stokkel, M. P. M., Mazzi, U., and Pauwels, E. K. J. (2000). 18F-Labeled radiopharmaceuticals for PET in oncology, excluding FDG. *Nucl. Med. Biol.* **27:** 103–112.

Verhoeff, N. P. L. G. (1999). Radiotracer imaging of dopaminergic transmission in neuropsychiatric disorders. *Psychopharmacology,* **147:** 217–249.

Volkow, N. D., Fowler, J. S., Gatley, S. J., Logan, J., Wang, G. J., Ding, Y. S., and Dewey, S. (1996). PET evaluation of the dopamine system of the human brain. *J. Nucl. Med.* **37:** 1242–1256.

Yang, Z. Y., Perry, B. D., and Mukherjee, J. (1996). Fluorinated Benzazepines 1. Syntheses, radiosyntheses, and biological evaluation of a series of substantial benzazepines as selective and high affinity radiotracers for positron emission tomographic studies of dopamine D-1 receptors. *Nucl. Med. Biol.* **23:** 793–805.

Yatham, L. N., Steiner, M., Liddle, P. F., Shiah, I. S., Lam, R. W., Zis, A. P., and Coote, M. (2000). A PET study of brain 5-HT2 receptors and their correlation with platelet 5-HT2 receptors in healthy humans. *Psychopharmacology* **151:** 424–427.

Zubieta, J. K., Dannals, R. F., and Frost, J. J. (1999). Gender and age influences on human brain mu-opiod receptor binding measured by PET. *Amer. J. Psychiatry* **156:** 842–848.

C H A P T E R 6

Basics of Imaging Theory and Statistics

CHIEN-MIN KAO, PATRICK LA RIVIÈRE, and XIAOCHUAN PAN
Department of Radiology, University of Chicago, Chicago, Illinois

I. Introduction
II. Linear Systems
III. Discrete Sampling
IV. Noise and Signal
V. Filtering
VI. Smoothing
VII. Estimation
VIII. Objective Assessment of Image Quality

I. INTRODUCTION

Generally speaking, imaging involves the creation of pictorial representations of objects or processes. Nuclear medicine imaging, for example, seeks to produce representations of the distribution of radioactive tracers within the body by detecting the photons emitted due to decay of the tracer. *Planar imaging* involves the creation of two-dimensional (2D) projection images of the three-dimensional (3D) distribution of tracers, whereas *tomographic imaging* involves the use of numerous such projection images to reconstruct a 3D representation of tracer distribution.

In practice, one never measures data that flawlessly represent the underlying distribution or process; measured data are always corrupted by deterministic degradations such as blurring and sampling as well as random degradations such as noise. Figure 1 provides examples of such degradations. Figure 1a depicts an ideal image of a mathematical phantom.

Figure 1b depicts the same image degraded by blur, Figure 1c depicts degradation by noise, and Figure 1d depicts degradation by both blur and noise. Clearly, image blur leads to loss of image contrast and definition, whereas image noise results in intensity fluctuations, making small low-contrast structures difficult to resolve.

Sampling effects (Figure 2) are associated with the practical need to work with discrete images consisting of samples of underlying continuous-domain images. Figure 2 displays discrete images, generated by four different sampling grids, of a spatial pattern that comprises alternating black and white sectors. The center of the spatial pattern contains increasingly finer (i.e., higher-resolution) structures. Therefore, near the image center the structures become too small for a given sampling scheme to capture the variations faithfully, and this failure results in *aliasing errors* in the sampled images. Such aliasing errors can be observed in all the images shown in Figure 2. As one would expect, the errors are observed to be more pronounced when coarser sampling is used.

Imaging theory is concerned with the mathematical description of the imaging process, accounting for these deterministic and random degradations, so that their effects on image quality can be characterized and, if possible, corrected for. This chapter provides an overview of the basic imaging theory, including an introduction to deterministic linear systems; statistical characterizations of random processes; and fundamental theories of interpolation, filtering, estimation, and detection. In addition to blur, noise, and sampling, nonlinear degradations can also occur in the

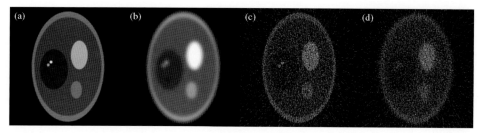

FIGURE 1 Effects of blur and noise on image quality. Images shown are (a) an ideal image of a mathematical phantom, (b) the same image degraded by blur, (c) the same image degraded by noise, and (d) the same image degraded by both blur and noise. As shown, image blur leads to loss of image contrast and definitions. Image noise results in intensity fluctuations, also making small low-contrast structures difficult to resolve.

FIGURE 2 Illustration of sampling effects. From left to right, the images shown are the same spatial pattern sampled by 512×512, 256×256, 128×128, and 64×64 rectangular grids. Aliasing errors can be observed at the centers of these images. These errors become more pronounced as the grid size decreases.

imaging process. However, we restrict our discussions to linear systems because they describe the main characteristics of most imaging systems and form the basis of many image-processing techniques. In our discussions, we consider only 2D images and systems; however, the underlying principles can be easily extended to other dimensions. Moreover, we consider only the cases in which the inputs and outputs of the linear systems and operators are defined on the same space. As noted, in emission tomography one is also concerned with reconstructing images from projections, and these objects are not defined on the same space. This reconstruction problem is described later in the book (Chapters 20–22). Finally, the material presented is limited to principles and techniques that are relevant to nuclear-medicine imaging. For in-depth discussions of the general theory of image processing, readers can consult an excellent textbook, Jain (1989).

II. LINEAR SYSTEMS

A. Linear Shift-Invariant Systems

In this chapter, we use the term *system* to refer to both physical imaging systems and to abstract mathematical image-processing operations. A system $\mathcal{H}\{\cdot\}$ is called *linear* if, for any images $f_1(x,y)$ and $f_2(x,y)$, $x,y \in \mathbb{R}$; the following equality is true for any $\alpha_1, \alpha_2 \in \mathbb{C}$:

$$\mathcal{H}\{\alpha_1 f_1(x,y) + \alpha_2 f_2(x,y)\} = \alpha_1 g_1(x,y) + \alpha_2 g_2(x,y) \quad (1)$$

where $g_1(x,y) = \mathcal{H}\{f_1(x,y)\}$ and $g_2(x,y) = \mathcal{H}\{f_2(x,y)\}$. Linear systems thus obey the principle of *superposition*—a weighted sum of two inputs produces a weighted sum of the two respective outputs.

In addition, the system is called *shift-invariant*, or *space-invariant*, if the equality

$$\mathcal{H}\{f(x-x', y-y')\} = g(x-x', y-y') \quad (2)$$

holds for any $x', y' \in \mathbb{R}$. This means that shifting the input simply produces a shifted version of the output, without any other change. A system that is not shift-invariant is called a *shift-variant*, or *space-variant*, system.

Linear shift-invariant (LSI) systems play a fundamental role in imaging theory. These systems have favorable mathematical properties that make analysis of their properties tractable. As a result, LSI systems have been thoroughly studied, and numerous theories and methods exist to describe them. To a good approximation, many physical imaging systems are LSI. Even for non-LSI systems, approximations of these systems by LSI systems can often provide important insights into their properties.

B. Point Spread Functions

The deterministic blurring properties of an LSI system $\mathcal{H}\{\cdot\}$ can be characterized by a single function—the *point spread function* (PSF), denoted by $h(x,y)$. This is the system response to a point source placed at the origin of the image; that is,

$$h(x,y) = \mathcal{H}\{\delta(x,y)\} \quad (3)$$

and knowledge of it is sufficient to allow determination of the system response to arbitrary inputs.

The output of an LSI system is simply given by an operation called the *convolution* of its input and its PSF. The *convolution* of two functions $f_1(x,y)$ and $f_2(x,y)$, denoted as $f_1(x,y) * f_2(x,y)$, is defined as

$$f_1(x,y) * f_2(x,y) = \iint_{-\infty}^{\infty} f_1(x',y') f_2(x-x', y-y') dx' dy' \quad (4)$$

It follows directly from this definition that convolution is commutative:

$$f_1(x,y) * f_2(x,y) = f_2(x,y) * f_1(x,y) \quad (5)$$

Using Eqs. (1)–(4), the following demonstrates that the response of an LSI system is described by a convolution:

$$\begin{aligned} g(x,y) = \mathcal{H}\{f(x,y)\} &= \iint_{-\infty}^{\infty} f(x',y') \mathcal{H}\{\delta(x-x', y-y')\} dx' dy' \\ &= \iint_{-\infty}^{\infty} f(x',y') h(x-x', y-y') dx' dy' \\ &= f(x,y) * h(x,y) \end{aligned} \quad (6)$$

Equation (6) states that the output of an LSI system is given by a weighted superposition of its shifted PSFs, and that, at each shifted position, the weight is equal to the amplitude of the input at that location. On the other hand, a linear shift-variant system does not possess this simple property because in general the system response to a point source will depend on the source location. Therefore, one should express the response of the system to a point at (x',y') as $\mathcal{H}\{\delta(x-x', y-y')\} = h(x,y;x',y')$, which is sometimes referred to as the *local PSF* of the linear imaging system. In the shift-variant case,

$$g(x,y) = \mathcal{H}\{f(x,y)\} = \iint_{-\infty}^{\infty} f(x',y') h(x,y;x',y') dx' dy' \quad (7)$$

C. Fourier Transform and Spectrum

A powerful mathematical tool for investigating and analyzing linear systems is the Fourier transform (FT). The FT of a function $f(x,y) \in L^2$, real- or complex-valued, is defined by

$$F(v_x, v_y) = \mathcal{F}\{f(x,y)\} = \iint_{-\infty}^{\infty} f(x,y) e^{-j2\pi(v_x x + v_y y)} dx\, dy \quad (8)$$

where $j = \sqrt{-1}$, and $v_x, v_y \in \mathbb{R}$ are the spatial frequencies associated with the coordinates x and y, respectively. The inverse Fourier transform (IFT) is given by

$$f(x,y) = \mathcal{F}^{-1}\{F(v_x, v_y)\} = \iint_{-\infty}^{\infty} F(v_x, v_y) e^{j2\pi(v_x x + v_y y)} dv_x dv_y \quad (9)$$

This equation states that one can decompose a 2D image $f(x,y)$ into complex exponential (sinusoidal) functions $\exp\{j2\pi(v_x x + v_y y)\}$, with the Fourier transform $F(v_x,v_y)$ specifying the amplitude of the components. Therefore, $F(v_x,v_y)$ is also known as the *frequency spectrum*, or simply the *spectrum*, of $f(x,y)$. Because of the one-to-one correspondence between $f(x,y)$ and $F(v_x,v_y)$ it is customary to speak of an image and its spectrum interchangeably.

The fundamental importance of the FT is that one can describe the characteristics of an LSI system entirely by the system's effect on the spectrum of the input, independently at each spatial frequency. The following properties of the Fourier transform can be easily derived from its definition.

(P1) Linearity:

$$\begin{aligned} \mathcal{F}\{\alpha_1 f_1(x,y) + \alpha_2 f_2(x,y)\} &= \alpha_1 F_1(v_x, v_y) \\ &+ \alpha_2 F_2(v_x, v_y), \quad \forall \alpha_1, \alpha_2 \in \mathbb{C} \end{aligned} \quad (10)$$

(P2) Shifting property:

$$\mathcal{F}\{f(x-x', y-y')\} = F(v_x, v_y) e^{-j2\pi(v_x x' + v_y y')} \quad (11)$$

$$\mathcal{F}\{f(x,y) e^{j2\pi(v'_x x + v'_y y)}\} = F(v_x - v'_x, v_y - v'_y) \quad (12)$$

(P3) Scaling property: for $a,b \neq 0$,

$$\mathcal{F}\{f(ax, by)\} = \frac{1}{|ab|} F\left(\frac{v_x}{a}, \frac{v_y}{b}\right) \quad (13)$$

(P4) Differentiation:

$$\mathcal{F}\left\{\frac{\partial f(x,y)}{\partial x}\right\} = j2\pi v_x F(v_x, v_y) \quad \mathcal{F}\left\{\frac{\partial f(x,y)}{\partial y}\right\} = j2\pi v_y F(v_x, v_y) \quad (14)$$

$$\mathcal{F}\{xf(x,y)\} = \frac{j}{2\pi} \frac{\partial F(v_x, v_y)}{\partial v_x} \quad \mathcal{F}\{yf(x,y)\} = \frac{j}{2\pi} \frac{\partial F(v_x, v_y)}{\partial v_y} \quad (15)$$

(P5) Integration:

$$\begin{aligned} \mathcal{F}\left\{\int_{-\infty}^{x} f(x',y) \mathcal{D}x'\right\} &= \frac{F(v_x, v_y)}{j2\pi v_x} \\ \mathcal{F}\left\{\int_{-\infty}^{y} f(x,y') \mathcal{D}y'\right\} &= \frac{F(v_x, v_y)}{j2\pi v_y} \end{aligned} \quad (16)$$

$$\begin{aligned} \mathcal{F}\left\{\frac{f(x,y)}{x}\right\} &= -j2\pi \int_{-\infty}^{v_x} F(v'_x, v_y) dv'_x \\ \mathcal{F}\left\{\frac{f(x,y)}{y}\right\} &= -j2\pi \int_{-\infty}^{v_y} F(v_x, v'_y) dv'_y \end{aligned} \quad (17)$$

(P6) Convolution theorem:

$$\mathcal{F}\{f(x,y) * g(x,y)\} = F(v_x, v_y) G(v_x, v_y) \quad (18)$$

$$\mathcal{F}\{f(x,y) g(x,y)\} = F(v_x, v_y) * G(v_x, v_y) \quad (19)$$

(P7) Parseval's theorem:

$$\iint_{-\infty}^{\infty} f(x,y) g^*(x,y) dx\, dy = \iint_{-\infty}^{\infty} F(v_x, v_y) G^*(v_x, v_y) dv_x dv_y \quad (20)$$

$$\iint_{-\infty}^{\infty} |f(x,y)|^2 \, dx \, dy = \iint_{-\infty}^{\infty} |F(v_x,v_y)|^2 \, dv_x \, dv_y \quad (21)$$

From these properties, we can make the following observations. First, we note that $f(ax,by)$ with $a,b > 1$ is a contraction of $f(x,y)$ by scaling factors a and b in the x and y directions. According to the scaling property, such a contraction of $f(x,y)$ leads to a dilation of its spectrum $F(v_x,v_y)$ by the scaling factors $1/a$ and $1/b$ in the v_x and v_y directions. Similarly, a dilation of $f(x,y)$ will lead to a contraction of $F(v_x,v_y)$. Therefore, the scaling property suggests that, in general, a wider spatial function will have a narrower spectrum, and vice versa. Similarly, properties P4 and P5 indicate that the differentiation/integration operation emphasizes/suppresses high-frequency components of a function. Finally, according to the Parseval's theorem $|F(v_x,v_y)|^2$ is the energy distribution of $f(x,y)$ in the frequency space. Therefore, $|F(v_x,v_y)|^2$ is sometimes known as the *energy spectrum* of $f(x,y)$.

Because physically realizable functions must have finite energy, they all belong to the L^2 class and have well-defined Fourier transforms. Furthermore, some useful idealizations of physical functions that are not proper L^2 functions, but that are instrumental in analysis such as the Dirac delta function $\delta(x,y)$, can also be treated as limiting cases in the Fourier theory to obtain well-defined generalized Fourier transforms. Several Fourier transform pairs useful in imaging analysis are given here:

$$\delta(x,y) \Leftrightarrow 1 \quad (22)$$

$$\frac{1}{2\pi}\exp\left\{-\frac{1}{2}(x^2+y^2)\right\} \Leftrightarrow \exp\{-2\pi(v_x^2+v_y^2)\} \quad (23)$$

$$\frac{1}{4}\exp\{-|x|-|y|\} \Leftrightarrow (1+4\pi^2 v_x^2)^{-1}(1+4\pi^2 v_y^2)^{-1} \quad (24)$$

$$\text{rect}(x)\text{rect}(y) \Leftrightarrow \frac{\sin(\pi v_x)}{\pi v_x}\frac{\sin(\pi v_y)}{\pi v_y} \quad (25)$$

$$\text{rect}\left(\sqrt{x^2+y^2}\right) \Leftrightarrow \frac{\pi}{2}\frac{J_1\left(\pi\sqrt{v_x^2+v_y^2}\right)}{\pi\sqrt{v_x^2+v_y^2}} \quad (26)$$

$$\text{comb}(x,y) \Leftrightarrow \text{comb}(v_x,v_y) \quad (27)$$

where

$$\text{rect}(x) = \begin{cases} 1 & |x| \leq 1/2 \\ 0 & \text{otherwise,} \end{cases} \quad (28)$$

$J_1(x)$ is the Bessel function of the first kind and first order, and

$$\text{comb}(x,y) = \sum_{m=-\infty}^{\infty}\sum_{n=-\infty}^{\infty}\delta(x-m,y-n) \quad (29)$$

is the 2D impulse lattice. Note that by making the substitutions $x \Leftrightarrow v_x$ and $y \Leftrightarrow v_y$ in Eqs. (22)–(27), the results are also valid Fourier transform pairs. For example, applying these substitutions to Eq. (22), we obtain the Fourier transform pair $1 \Leftrightarrow \delta(v_x,v_y)$.

Properties *(P1)–(P6)* are useful for generating new Fourier transforms from known ones. For example, the application of the shifting property given by Eq. (11) to Eq. (22) yields

$$\mathcal{F}\{\delta(x-x',y-y')\} = \mathcal{F}\{\delta(x,y)\}e^{-j2\pi(v_x x'+v_y y')} = e^{-j2\pi(v_x x'+v_y y')} \quad (30)$$

Similarly, by applying Eq. (12) and using the identity $\cos(x) = (e^{jx} + e^{-jx})/2$, we obtain the modulation theorem given by

$$\mathcal{F}\{f(x,y)\cos(2\pi v_x' x)\} = \frac{F(v_x-v_x',v_y)+F(v_x+v_x',v_y)}{2} \quad (31)$$

Finally, using Eqs. (13) and (27) we can immediately show that, for $x_s > \emptyset, y_s > \emptyset$,

$$\mathcal{F}\left\{\text{comb}\left(\frac{x}{x_s},\frac{y}{y_s}\right)\right\} = \frac{1}{v_x^{(s)}v_y^{(s)}}\text{comb}\left(\frac{v_x}{v_x^{(s)}},\frac{v_y}{v_y^{(s)}}\right) \quad (32)$$

where $v_x^{(s)} = x_s^{-1}$ and $v_y^{(s)} = y_s^{-1}$. By use of the equality

$$\delta(\alpha x,\beta y) = \frac{1}{|\alpha\beta|}\delta(x,y), \quad \forall \alpha,\beta \neq 0 \quad (33)$$

and the definition of the comb function given by Eq. (29), this equation then becomes

$$\mathcal{F}\left\{\sum_{m=-\infty}^{\infty}\sum_{n=-\infty}^{\infty}\delta(x-mx_s,y-ny_s)\right\}$$
$$= v_x^{(s)}v_y^{(s)}\sum_{m=-\infty}^{\infty}\sum_{n=-\infty}^{\infty}\delta(v_x-mv_x^{(s)},v_y-nv_y^{(s)}) \quad (34)$$

Fourier theory provides the mathematical foundation for the analysis of linear systems and is of paramount importance. Our treatment of this subject is necessarily brief due to space limitations. Readers are encouraged to consult the classic monograph by Bracewell (1965) for a more thorough discussion of this important subject.

D. System Transfer Functions

Using the convolution theorem of the Fourier transform, the imaging model of an LSI system given by Eq. (6) becomes

$$G(v_x,v_y) = H(v_x,v_y)F(v_x,v_y) \quad (35)$$

where $F(v_x,v_y)$, $G(v_x,v_y)$, and $H(v_x,v_y)$ are the Fourier transforms of $f(x,y)$, $g(x,y)$, and $h(x,y)$, respectively. Therefore, the deterministic input–output characteristics of an LSI system in the frequency space are uniquely determined by its *system transfer function* $H(v_x,v_y)$ with a simple multiplicative relationship. In the literature, $H(v_x,v_y)$ is also sometimes referred to as the *optical transfer function* and denoted by OTF(v_x,v_y). The magnitude of the OTF is called the *modulation transfer function* (MTF), and MTF(v_x,v_y) = |OTF(v_x,v_y)| = |H(v_x,v_y)|. Without loss of generality, a PSF $h(x,y)$ is often normalized so that MTF(0,0) = 1.

As a result of Eq. (35), it is often more convenient to study or define an LSI system by examining or specifying its transfer function $H(v_x,v_y)$ rather than its PSF $h(x,y)$. For example, it is straightforward to see that the transfer function $H(v_x,v_y)$ = rect($v/2B$), where $v = \sqrt{v_x^2 + v_y^2}$ and $B > 0$, defines a lowpass filter because such a system will pass spectral components of the input below B unaltered, but completely removes all higher-frequency components. The output of the system is therefore a limited-resolution version of its input. Figure 3 demonstrates this operation in the spatial and the frequency domains with $B = 0.15$ (pixel size)$^{-1}$. The bottom row of Figure 3 clearly shows that the transfer function retains only low-frequency components of the input spectrum in the output. In the spatial domain, examination of the input and output images in Figure 3 reveal that this lowpass filtering results in loss of structural details and intensity contrast. Although it is evident from $H(v_x,v_y)$, the lowpass filtering interpretation is not readily seen by examining the PSF $h(x,y)$ shown in Figure 3. Figure 4 similarly demonstrates the operation of highpass filtering, by which intensity transitions (edges) in an image are emphasized.

Working in frequency space can make imaging theorems more understandable and can make image-processing techniques easier to define or develop. In particular, it provides a rigorous connection between a continuous-domain image and its discrete samples. This important subject is discussed in the following section.

III. DISCRETE SAMPLING

In our discussions so far, we have considered only continuous-domain functions. In medical imaging, however,

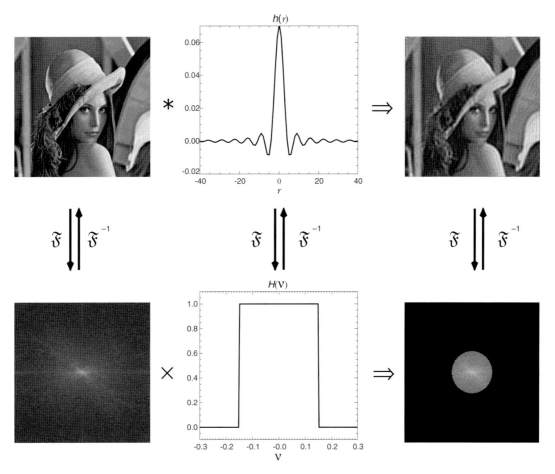

FIGURE 3 Ideal lowpass filtering of an image. Figures in the top and bottom rows demonstrate the filtering operation in the spatial and frequency domains, respectively. The left column shows the input image and the magnitude of its Fourier transform. The middle column are the circular symmetric PSF and its transfer function, where $v = \sqrt{v_x^2 + v_y^2}$ and $r = \sqrt{x^2 + y^2}$. Images in the right column are the output image and the magnitude of its Fourier transform.

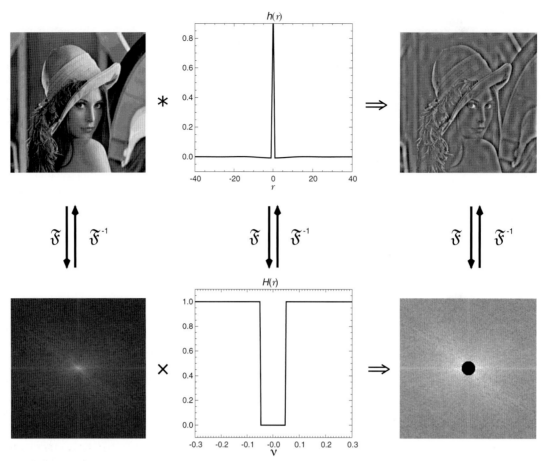

FIGURE 4 Ideal highpass filtering of an image. Figures in the top and bottom rows demonstrate the filtering operation in the spatial and frequency domains, respectively. The left column shows the input image and the magnitude of its Fourier transform. The middle column shows the circular symmetric PSF and its transfer function, where $v = \sqrt{v_x^2 + v_y^2}$ and $r = \sqrt{x^2 + y^2}$. Images in the right column are the output image and the magnitude of its Fourier transform.

data are often available only in discrete form. We must thus consider the issue of faithfully representing a continuous-domain image by its discrete samples. In particular, we have demonstrated in Figure 2 that aliasing artifacts can occur when the sampling distance is too large to capture faithfully the spatial variations of the underlying continuous-domain image. In this section, we discuss the fundamental theorem of sampling and introduce mathematical tools for manipulating discrete images.

A. Sampling Theorem and Aliasing

The conditions under which a continuous-domain function can be faithfully represented by its samples and the procedure for reconstructing the function from its samples are governed by the Whittaker-Shannon sampling theorem, stated as follows.

Whittaker-Shannon Sampling Theorem *Let $f(x,y)$ be a band-limited function so that $F(v_x, v_y) = 0$ unless $|v_x| < B_x$ and $|v_y| < B_y$, where $B_x > 0$ and $B_y > 0$ are the bandwidths of $f(x,y)$ in the v_x and v_y coordinates, respectively. The function $f(x,y)$ can then be completely represented by its regular samples $f_{mn} = f(mx_s, ny_s)$, $m,n \in \mathbb{Z}$, provided that the sampling distances are sufficiently small to satisfy the sampling conditions $x_s \leq (2B_x)^{-1}$ and $y_s \leq (2B_y)^{-1}$. Furthermore, in this case one has*

$$f(x,y) = \sum_{m=-\infty}^{\infty} \sum_{n=-\infty}^{\infty} f_{mn} \frac{\sin(\pi(x/x_s - m))}{\pi(x/x_s - m)} \frac{\sin(\pi(y/y_s - n))}{\pi(y/y_s - n)} \quad (36)$$

To gain a better understanding of this theorem, the following proof is particularly insightful. We begin our proof by constructing the sampled function of $f(x,y)$ from the discrete samples f_{mn}, $m,n \in \mathbb{Z}$, by

$$f_s(x,y) = \sum_{m=-\infty}^{\infty} \sum_{n=-\infty}^{\infty} f_{mn} \delta(x - mx_s)\delta(y - ny_s) \quad (37)$$

$$= f(x,y) \cdot \sum_{m=-\infty}^{\infty} \sum_{n=-\infty}^{\infty} \delta(x - mx_s)\delta(y - ny_s) \quad (38)$$

Application of Eqs. (19) and (34) to Eq. (38) then yields that

$$F_s(v_x,v_y) = F(v_x,v_y) * \left\{ v_x^{(s)}v_y^{(s)} \sum_{m=-\infty}^{\infty}\sum_{n=-\infty}^{\infty} \delta(v_x - mv_x^{(s)}, v_y - nv_y^{(s)}) \right\} \quad (39)$$

$$= v_x^{(s)}v_y^{(s)} \sum_{m=-\infty}^{\infty}\sum_{n=-\infty}^{\infty} F(v_x - mv_x^{(s)}, v_y - nv_y^{(s)}) \quad (40)$$

Therefore, as demonstrated in Figure 5, sampling of the function $f(x,y)$ at spacings x_s and y_s will result in periodic repetitions of its spectrum $F(v_x,v_y)$ in frequency space at spacings $v_x^{(s)}$ and $v_y^{(s)}$. When the conditions $v_x^{(s)} \geq 2B_x$ and $v_y^{(s)} \geq 2B_y$ are met, that is, when $x_s \leq (2B_x)^{-1}$ and $y_s \leq (2B_y)^{-1}$, these replications are well separated (shown in Figure 5a). In this case, one can easily recover the spectrum $F(v_x,v_y)$ from $F_s(v_x,v_y)$ simply by multiplying it with the lowpass filter given by

$$H_L(v_x,v_y) = \frac{1}{v_x^{(s)}v_y^{(s)}} \operatorname{rect}\left(\frac{v_x}{v_x^{(s)}}\right) \operatorname{rect}\left(\frac{v_y}{v_y^{(s)}}\right)$$

Application of Eqs. (13), (18), (25), and (37) to this equation then yields

$$f(x,y) = \mathcal{F}^{-1}\{F_s(v_x,v_y)H_L(v_x,v_y)\}$$

$$= f_s(x,y) * \left\{ \frac{\sin(\pi v_x^{(s)}x)}{\pi v_x^{(s)}x} \frac{\sin(\pi v_y^{(s)}y)}{\pi v_y^{(s)}y} \right\}$$

$$= \sum_{m=-\infty}^{\infty}\sum_{n=-\infty}^{\infty} f_{mn} \frac{\sin(\pi(x/x_s - m))}{\pi(x/x_s - m)} \frac{\sin(\pi(y/y_s - n))}{\pi(y/y_s - n)}$$

and the theorem is proved.

On the other hand, Figure 5b depicts the situation when $v_x^{(s)} < 2B_x$ and $v_y^{(s)} < 2B_y$. The sampling condition is violated in this case, and the spectral replications begin to overlap with one another. As a result, high-frequency components of $F(v_x,v_y)$ above $0.5v_x^{(s)}$ and $0.5v_y^{(s)}$ are folded over into lower-frequency components of $v_x^{(s)} - v_x$ and $v_y^{(s)} - v_y$ below $0.5v_x^{(s)}$ and $0.5v_y^{(s)}$ and assume their identities (see Figure 6a). The situation is more clearly illustrated by Figure 6b which demonstrates that samples of a high-frequency sinusoidal function will appear to belong to a lower-frequency one when the sampling rate is too low. This phenomenon, known as *aliasing*, occurs when exact recovery of $f(x,y)$ from its samples $f(mx_s,ny_s)$, $m,n \in \mathbb{Z}$, cannot be achieved. The errors caused by this spectral overlapping are known as *aliasing errors* or *foldover errors*. For Eq. (36) to hold, the minimum necessary sampling frequencies are $v_x^{(N)} = 2B_x$ and $v_y^{(N)} = 2B_y$, which are known as the *Nyquist frequencies*.

Finally, it is noted that in general the Whittaker-Shannon sampling theorem requires an infinite number of samples for exact recovery of the original continuous-domain images. In practice, however, one always works with a finite number of samples acquired in a limited sampling region. In this case, one typically assumes that the underlying continuous-domain image is zero outside the sampling region; the infinite summation in Eq. (36) therefore becomes a finite one.

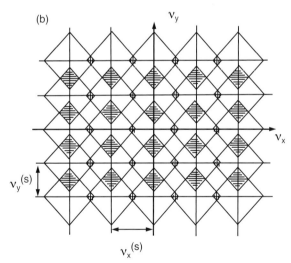

FIGURE 5 Illustrations of the spectrum $F_s(v_x,v_y)$ of the sampled function $f_s(x,y) = \Sigma_{mn} f_{mn} \delta(x - mx_s, y - ny_s)$, where $f_{mn} = f(mx_s,ny_s)$. The diamond shapes in these drawings indicate the support of the original spectrum $F(v_x,v_y)$, assuming to have the bandwidths B_x and B_y in the v_x and v_y coordinates, respectively. As shown, sampling creates replications of the spectrum at the lattice with the spacings $v_x^{(s)} = x_s^{-1}$ and $v_y^{(s)} = y_s^{-1}$ in the v_x and v_y coordinates. When the sampling distances are sufficiently small so that the sampling conditions $v_x^{(s)} \geq 2B_x$ and $v_y^{(s)} \geq 2B_y$ are satisfied, the spectral replications are well separated; see (a). In this case, it is possible to recover $F(x,y)$ exactly from $F_s(v_x,v_y)$ by lowpass filtering. On the other hand, when the sampling distances violate the sampling condition, the spectral replications begin to overlap with each other, indicated by hashed regions in (b). Exact recovery of $F(v_x,v_y)$ from $F_s(v_x,v_y)$ is no longer possible, and aliasing errors arise.

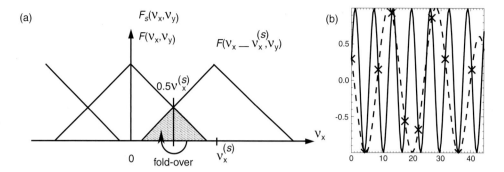

FIGURE 6 Illustration of aliasing errors. Panel (a) demonstrates that frequency components above $0.5v_x^{(s)}$ are folded over into lower-frequency components below $0.5v_x^{(s)}$ and assume their identities when the bandwidth B_x is larger than the $0.5v_x^{(s)}$. Panel (b) shows that when a sinusoidal function (solid line) is not sampled at a sufficiently small spacing (samples marked by ×), the resulting samples appear to belong to a lower-frequency sinusoidal function (dashed line).

B. Discrete Fourier Transform

When working with a sampled image, it is useful to make use of a discrete counterpart to the Fourier transform known as the discrete Fourier transform (DFT). In the DFT, a finite 2D array

$$f_{mn}, \quad m = 0, \cdots, M-1, \quad n = 0, \cdots, N-1 \tag{41}$$

is considered to constitute one period of a periodic 2D array. (Without loss of generality, we assume that both M and N are even numbers in the following discussion.) The DFT of this periodic array is defined as

$$F_{mn} = \frac{1}{\sqrt{MN}} \sum_{m'=0}^{M-1} \sum_{n'=0}^{N-1} f_{m'n'} e^{-j2\pi mm'/M} e^{-j2\pi nn'/N} \tag{42}$$

It is not difficult to see that F_{mn} is also a periodic 2D array with the same period as f_{mn}. Therefore, it suffices to compute F_{mn} for one period and by convention the period given by $0 \le m \le M - 1$ and $0 \le n \le N - 1$ is typically used. The inverse DFT is given by

$$f_{mn} = \frac{1}{\sqrt{MN}} \sum_{m'=0}^{M-1} \sum_{n'=0}^{N-1} F_{m'n'} e^{j2\pi mm'/M} e^{j2\pi nn'/N} \tag{43}$$

As in the continuous case, this equation states that one can decompose a 2D finite (or periodic) image array f_{mn} into sinusoidal components $\exp\{j2\pi mm'/M\} \exp\{j2\pi nn'/N\}$ and the discrete Fourier transform of f_{mn}, denoted by F_{mn}, gives the amplitude in this decomposition. Therefore, one can also speak of a finite (or periodic) image array and its spectrum interchangeably.[1]

---

[1] In the literature, the DFT and its inverse of a discrete image f_{mn}, $m = 0, ..., M - 1$, $n = 0, ..., N - 1$, are also often defined using an alternate normalization:

$$F_{mn} = \sum_{m'=0}^{M-1} \sum_{n'=0}^{N-1} f_{m'n'} e^{-j2\pi mm'/M} e^{-j2\pi nn'/N} \tag{44}$$

$$f_{mn} = \frac{1}{MN} \sum_{m'=0}^{M-1} \sum_{n'=0}^{N-1} F_{m'n'} e^{j2\pi mm'/M} e^{j2\pi nn'/N} \tag{45}$$

It is noted that, because of the periodic structure, larger values of the subscript indices m and n on F_{mn} do not indicate higher frequency components. In fact, because $\exp\{j2\pi mm'/M\} = \exp\{j2\pi(m - M)m'/M\}$, the DFT component F_{mn} with $M/2 < m < M$ is actually equal to $F_{m'n}$ with $m' = m - M$, which is negative and smaller than m in magnitude. The same consideration also applies to the n index. Therefore, we conclude that F_{mn} with $m = M/2$ and $n = N/2$ is the highest frequency component in the period given by $0 \le m \le M - 1$ and $0 \le n \le N - 1$. It is interesting to note that this periodic structure of F_{mn} is quite similar to that of $F_s(v_x,v_y)$, with the discrete indexes $m = M/2$ and $n = N/2$ in F_{mn} playing the role of $v_x = v_x^{(s)}/2$ and $v_y = v_y^{(s)}/2$ in $F_s(v_x,v_y)$. The relationship between F_{mn} and $F_s(v_x,v_y)$ is made clear in the next section.

Many properties of the Fourier transform also hold for the discrete Fourier transform. For example, one can show that

$$\sum_{m=0}^{M-1} \sum_{n=0}^{N-1} f_{mn} g_{mn}^* = \sum_{m=0}^{M-1} \sum_{n=0}^{N-1} F_{mn} G_{mn}^* \tag{46}$$

and in particular,

$$\sum_{m=0}^{M-1} \sum_{n=0}^{N-1} |f_{mn}|^2 = \sum_{m=0}^{M-1} \sum_{n=0}^{N-1} |F_{mn}|^2 \tag{47}$$

Therefore, $|F_{mn}|^2$ can be interpreted as the energy spectrum of f_{mn}. In addition, the convolution theorem becomes

$$g_{mn} = h_{mn} * f_{mn} \Leftrightarrow G_{mn} = H_{mn} F_{mn} \tag{48}$$

where G_{mn}, H_{mn}, and F_{mn} are the DFTs of g_{mn}, h_{mn}, and h_{mn}, respectively, and $*$ denotes the circular convolution defined by

$$h_{mn} * f_{mn} = \sum_{i=0}^{M-1} \sum_{j=0}^{N-1} h_{ij} f_{(m-i)(n-j)} \tag{49}$$

where, as previously mentioned, all arrays are implicitly assumed to be periodic. As in the continuous case, h_{mn} and H_{mn} are, respectively, the point spread function and the system transfer function of a discrete, linear, periodically shift-invariant system.

Finally, we note that DFTs for image arrays of special sizes can be computed by use of fast Fourier transform (FFT) algorithms (Press et al., 1992). These algorithms are computationally efficient, making frequency-space-based image operations practical.

C. Relationship among FT, DFT, and Sampling

Obviously, the Fourier transform of a continuous-domain image $f(x,y)$ and the discrete Fourier transform of the discrete image $f_{mn} = f(mx_s, ny_s)$, $m = 0, \ldots, M - 1$, $n = 0, \ldots, N - 1$, are not unrelated. As already mentioned, we have assumed that $f(x,y)$ has a finite spatial support so that $f(x,y) = 0$ outside of the sampling region defined by $0 \leq x < Mx_s$ and $0 \leq y < Ny_s$. Therefore, the sampled function

$$f_s(x,y) = \sum_{m=-\infty}^{\infty} \sum_{n=-\infty}^{\infty} f_{mn}\delta(x - mx_s)\delta(y - ny_s)$$
$$= \sum_{m=0}^{M-1} \sum_{n=0}^{N-1} f_{mn}\delta(x - mx_s)\delta(y - ny_s) \quad (50)$$

now has the Fourier transform given by

$$F_s(v_x, v_y) = \sum_{m'=0}^{M-1} \sum_{n'=0}^{N-1} f_{m'n'} e^{-j2\pi v_x m' x_s} e^{-j2\pi v_y n' y_s} \quad (51)$$

Comparing with Eq. (42), we conclude that

$$F_{mn} = \frac{1}{\sqrt{MN}} F_s\left(\frac{m}{M} v_x^{(s)}, \frac{n}{N} v_y^{(s)}\right) = \frac{v_x^{(s)} v_y^{(s)}}{\sqrt{MN}} \sum_{k=-\infty}^{\infty} \sum_{l=-\infty}^{\infty} F\left(\left(\frac{m}{M} - k\right)v_x^{(s)}, \left(\frac{n}{N} - l\right)v_y^{(s)}\right) \quad (52)$$

where the second equality follows from Eq. (40). Therefore, within a constant scale, the DFT coefficients F_{mn}, $m,n \in \mathbb{Z}$, are discrete samples of $F_s(v_x, v_y)$ with the sampling distances $v_x^{(s)}/M = (Mx_s)^{-1}$ and $v_y^{(s)}/N = (Ny_s)^{-1}$ in the v_x and v_y coordinates, respectively.

Now consider the case when aliasing errors are negligible.[2] Under this condition, we must have $F(v_x, v_y) \approx 0$ for $|v_x| \geq 0.5 v_x^{(s)}$ and $|v_y| \geq 0.5 v_y^{(s)}$ (see Figure 5a). Hence,

$$\sum_{k=-\infty}^{\infty} \sum_{l=-\infty}^{\infty} F(v_x - k v_x^{(s)}, v_y - l v_y^{(s)}) = F(v_x - k_0 v_x^{(s)}, v_y - l_0 v_y^{(s)}) \quad (53)$$

where k_0 and l_0 are integers such that $-0.5 v_x^{(s)} < v_x - k_0 v_x^{(s)} \leq 0.5 v_x^{(s)}$ and $-0.5 v_y^{(s)} < v_y - k_0 v_y^{(s)} \leq 0.5 v_y^{(s)}$. Combination of these two equations then yields

---

[2] Strictly speaking, the finite spatial-support assumption made here has required $f(x,y)$ to be non-bandlimited. However, one can still assume $f(x,y)$ to be essentially bandlimited so that, for some B_x and B_y, $F(v_x, v_y) \approx 0$ unless $|v_x| < B_x$ and $|v_y| < B_y$. For such $f(x,y)$, it is therefore possible to find a sufficiently small sampling distances to generate samples containing negligible aliasing errors. Many images of practical interest have finite spatial support and are essentially bandlimited.

$$F_{mn} = \frac{v_x^{(s)} v_y^{(s)}}{\sqrt{MN}} F\left(\frac{m}{M} v_x^{(s)}, \frac{n}{M} v_y^{(s)}\right) \quad (54)$$

for the period given by $-M/2 < m \leq M/2$ and $-N/2 < n \leq N/2$. Therefore, when aliasing errors in f_{mn} are negligible, within a constant scale, F_{mn} are samples of $F(v_x, v_y)$ in the region defined by $-0.5 v_x^{(s)} < v_x \leq 0.5 v_x^{(s)}$ and $-0.5 v_y^{(s)} < v_y \leq 0.5 v_y^{(s)}$. As previously mentioned, the period given by $m = 0, \ldots, M - 1$ and $n = 0, \ldots, N - 1$ is more often considered in the literature. In this period, one can obtain from Eq. (54) that

$$F_{mn} = \frac{v_x^{(s)} v_y^{(s)}}{\sqrt{MN}} F\left(\frac{c(m;M)}{M} v_x^{(s)}, \frac{c(n;N)}{N} v_y^{(s)}\right) \quad (55)$$

where

$$c(a; A) = \begin{cases} a & 0 \leq a \leq A/2 \\ a - A & A/2 < a < A \end{cases} \quad (56)$$

Consequently, when aliasing errors are negligible one can rewrite the continuous imaging model given by Eq. (35) to obtain the discrete model:

$$G_{mn} = H_{mn} F_{mn}, m = 0, \ldots, M-1, n = 0, \ldots, N-1 \quad (57)$$

where F_{mn} and G_{mn} are the DFTs of the discrete images $f_{mn} = f(mx_s, ny_s)$ and $g_{mn} = g(mx_s, ny_s)$, respectively, and

$$H_{mn} = H\left(\frac{c(m;M)}{M} v_x^{(s)}, \frac{c(n;N)}{N} v_y^{(s)}\right) \quad (58)$$

Therefore, the continuous LSI system can be digitally implemented by working on the discrete samples f_{mn} and g_{mn}.

D. Interpolation

Interpolation is the inverse process of sampling: It considers the reconstruction of a continuous-domain function $f(x,y)$ from its samples $f_{mn} = f(mx_s, ny_s)$. In digital image processing, this operation is frequently needed in order to change the size and orientation of an image and to zoom in on a given region of an image. It is also a necessary intermediate step in many image processing methods, including 3D rendering and alignment of images. It is also used in image reconstruction as explained in Kinahan et al., Chapter 20. In this section, we discuss several basic techniques for image interpolation (Lehmann, 1999; Thèvenaz, 2000)

1. Convolution Interpolation

A classical interpolation formula has the convolutional form given by

$$\hat{f}(x,y) = \sum_{m=-\infty}^{\infty} \sum_{n=-\infty}^{\infty} f_{mn} I_2\left(\frac{x}{x_s} - m, \frac{y}{y_s} - n\right) \quad (59)$$

where $\hat{f}(x,y)$ is an estimate of $f(x,y)$, and $I_2(x,y)$ is the 2D interpolator. Note that, by definition, one must have $\hat{f}(mx_s, ny_s) = f_{mn}$. This condition leads to the requirement that

$$I_2(x,y) = \begin{cases} 1 & \text{for } (x,y) = (Z,0) \\ 0 & \text{for } (x,y) \in z^2, (x,y) \neq (0,0) \end{cases} \quad (60)$$

Other interpolation techniques include polynomial interpolations (Press et al., 1992) and model-based techniques (Vaseghi, 2000). In medical image processing, these methods are not as widely used as the convolutional form; therefore, they are not discussed in this chapter.

Quite often, the 2D interpolator $I_2(x,y)$ is obtained from a one-dimensional (1D) interpolator $I_1(x)$ by

$$I_2(x,y) = I_1(x)I_1(y) \quad (61)$$

Several important interpolators having this separable form are discussed next.

a. Ideal Bandlimited Interpolation

According to the sampling theorem, exact recovery of a bandlimited image from its samples can be achieved by use of the *sinc interpolator* (see Eq. [36])

$$I_1(x) = \text{sinc}(x) \equiv \frac{\sin(\pi x)}{\pi x} \quad (62)$$

provided that the sampling condition are satisfied. Unfortunately, this interpolator has an infinite length and decays slowly. Therefore, a long summation is needed in order to obtain good approximation of the infinite summation in Eq. (59), making this interpolation computationally unfavorable. In addition, real-world images are not bandlimited and using this interpolator generates aliasing errors that often appear as oscillatory artifacts. As demonstrated by Figure 7b, such aliasing errors are highly unfavorable for image perception. For these reasons, this ideal interpolator for bandlimited images is seldom used in practice. However, for special cases this interpolation can be achieved by DFT zero-padding. This implementation, to be discussed later, is computationally efficient and has been often used in practice.

b. Nearest-Neighbor Interpolation

In this method, to obtain $\hat{f}(x,y)$ one simply chooses the sample that is closest to (x,y) in location as its value. When there are multiple closest samples, the choice is arbitrary. For example, one can define $\hat{f}(x,y) = f_{mn}$ when $(m - 0.5)x_s \leq x < (m + 0.5)x_s$ and $(n - 0.5)y_s \leq y < (n + 0.5)y_s$. The corresponding interpolator of this simple assignment rule employs

$$I_1(x) = \begin{cases} 1 & -\tfrac{1}{2} \leq x < \tfrac{1}{2} \\ 0 & \text{otherwise} \end{cases} \quad (63)$$

This interpolation is the easiest to compute. However, it generates images containing *stair-step artifacts* (see Figure 7c).

c. Linear Interpolation

In linear interpolation, $\hat{f}(x,y)$ is obtained by weighted interpolation of its four nearest samples $f_{m,n}$, $f_{m,n+1}$, $f_{m+1,n}$ and $f_{m+1,n+1}$, where m and n are integers satisfying $mx_s \leq x < (m + 1)x_s$, $ny_s \leq y < (n + 1)y_s$, by use of

$$I_1(x) = \wedge(x) = \begin{cases} 1-|x| & 0 \leq |x| \leq 1 \\ 0 & \text{otherwise} \end{cases} \quad (64)$$

It is not difficult to show that

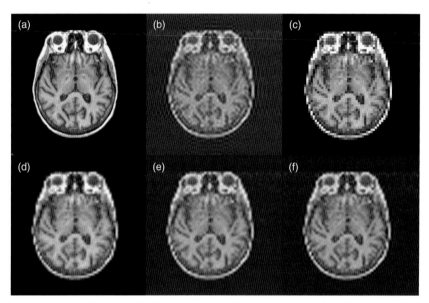

FIGURE 7 Comparison of interpolation methods. The original image is a 256 × 256 MRI brain image shown in (a). This image is subsampled to generate a 64 × 64 image, which is then upsampled by five interpolation methods to reproduce the original image size, including (b) sinc interpolation implemented by DFT zero-padding, (c) nearest-neighbor interpolation, (d) linear interpolation, (e) cubic convolution interpolation, and (f) cubic B-spline interpolation.

$$\hat{f}(x,y) = (1-w_x)(1-w_y)f_{m,n} + (1-w_x)w_y f_{m,n+1} \\ + w_x(1-w_y)f_{m+1,n} + w_x w_y f_{m+1,n+1} \quad (65)$$

where $w_x = x/x_s - m$ and $w_y = y/y_s - n$. This interpolation is computationally more demanding than the nearest-neighbor method, but it is still quite easy to compute and can reduce the stair-step artifacts observed in the nearest interpolation to generate smooth images (see Figure 7d). For these reasons, linear interpolation has been widely used in practice. A shortcoming of this method is that the resulting images have discontinuous first-order derivatives at the sampled points. Such discontinuity can cause significant difficulties when, for example, the interpolated images are used in 3D rendering.

d. Cubic Convolution Interpolation

This method employs a finite-length interpolator given by

$$I_1(x) = \begin{cases} (a+2)|x|^3 - (a+3)|x|^2 + 1 & 0 \le |x| \le 1 \\ a(|x|^3 - 5|x|^2 + 8|x| - 4) & 1 \le |x| \le 2 \\ 0 & 2 \le |x| \le \infty \end{cases} \quad (66)$$

where a is a free parameter that can be chosen to satisfy various interpolation criteria. A typical choice is $a = -3/4$ because the resulting image will have continuous second-order derivatives at the sampled points. Clearly, this method uses 16 adjacent samples for computation of $\hat{f}(x,y)$ and is computationally much more demanding than linear interpolation. However, as demonstrated by Figure 7e it can generate smoother images than linear interpolation.

e. Cubic B-Spline Interpolation

The methods described so far all employ interpolators that satisfy Eq. (60). In contrast, the cubic B-spline interpolation is defined by

$$\hat{f}(x,y) = \sum_{m=-\infty}^{\infty} t_{mn} h_4(x/x_s - m) h_4(y/y_s - n) \quad (67)$$

where

$$h_4(x) = \begin{cases} (4 - 6|x|^2 + 3|x|^3)/6 & 0 \le |x| \le 1 \\ (2 - |x|)^3 / 6 & 1 \le |x| \le 2 \\ 0 & 2 \le |x| \le \infty \end{cases} \quad (68)$$

is the cubic B-spline.[3] Because $h_4(x)h_4(y)$ does not satisfy Eq. (60), t_{mn} does not equal f_{mn} in general and must be obtained by solving the linear system of equations generated by the conditions $\hat{f}(mx_s, ny_s) = f_{mn}, \forall m,n$. B-spline interpolation also generates images that have continuous second-order derivatives and the interpolated images are

---

[3] The B-splines (basic splines) are recursively defined as $h_N(x) = h_0 * h_{N-1}(x)$, for $N = 1,2,3,\ldots$, where $h_0(x) = \text{rect}(x)$. For $N \to \infty$, $h_\infty(x)$ converges to a Gaussian function.

found favorable in many medical image applications (see Figure 7f). Finally, it can be shown that Eq. (67) can also be put into the classical form with an interpolator of infinite length given by (Lehmann, 1999; Chui, 1992):

$$I_1(x) = h_4(x) * \sum_{m=-\infty}^{\infty} \sqrt{3}(\sqrt{3} - 2)^{|m|} \delta(x+m) \quad (69)$$

2. DFT Zero-Padding

DFT zero-padding is another popular approach for solving a special interpolation problem: *up-sampling* (Vaseghi, 2000). In up-sampling, rather than calculating $f(x, y)$ at every location (x,y) one is interested in calculating from $f_{mn} = f(mx_s, ny_s)$, $m = 0, \ldots, M - 1, n = 0, \ldots, N - 1$, a new image

$$\tilde{f}_{kl} = f(k\tilde{x}_x, l\tilde{y}_s), k = 0, \cdots, \alpha M - 1, l = 0, \cdots, \alpha N - 1 \text{ with } \tilde{x}_s \\ = \frac{x_s}{\alpha}, \tilde{y}_s = \frac{y_s}{\alpha} \quad (70)$$

where α is a positive integer. To see that DFT zero-padding is a valid approach for accomplishing this task, we assume that aliasing errors are negligible in f_{mn}, and, without loss of generality, that M and N are even numbers. Therefore, we must have

$$F(v_x, v_y) = 0 \text{ for } |v_x| \ge 0.5 v_x^{(s)} \text{ or } |v_y| \ge 0.5 v_y^{(s)} \quad (71)$$

Because $\tilde{f}_{kl}$ employs smaller sampling distances than f_{mn}, aliasing errors are also negligible in $\tilde{f}_{kl}$ by assumption. Therefore, according to Eq. (54) we have

$$\tilde{F}_{kl} = \frac{\tilde{v}_x^{(s)} \tilde{v}_y^{(s)}}{\sqrt{\alpha^2 MN}} F\left(\frac{k}{\alpha M} \tilde{v}_x^{(s)}, \frac{l}{\alpha N} \tilde{v}_y^{(s)}\right) \\ = \frac{\alpha v_x^{(s)} v_y^{(s)}}{\sqrt{MN}} F\left(\frac{k}{M} v_x^{(s)}, \frac{l}{N} v_y^{(s)}\right) \quad (72)$$

for the period of $\tilde{f}_{kl}$ defined by $-\alpha M/2 < k \le \alpha M/2$ and $-\alpha N/2 < l \le \alpha N/2$. In this period, from Eqs. (54), (71), and (72) we then obtain

$$\tilde{F}_{kl} = \begin{cases} \alpha F_{kl} & -M/2 < k \le M/2 \text{ and } -N/2 < l \le N/2 \\ 0 & \text{otherwise} \end{cases} \quad (73)$$

In other words, within a constant scale one can zero-pad F_{mn} at high frequencies to increase its size to $\alpha M \times \alpha N$. This zero-padded DFT is then inverse transformed to generate the up-sampled image. Our analysis shows that the up-sampled image is exact if the sampling condition is met by f_{mn}. When considering the period $0 \le k < \alpha M$ and $0 \le l < \alpha N$, Eq. (73) becomes

$$\tilde{F}_{kl} = \begin{cases} \alpha F_{kl} & 0 \le k \le M/2, \ 0 \le l \le N/2 \\ \alpha F_{k'l} & \alpha M - M/2 < k < \alpha M, \ 0 \le l \le N/2 \\ \alpha F_{kl'} & 0 \le k \le M/2, \alpha N - N/2 < l < \alpha N \\ \alpha F_{k'l'} & \alpha M - M/2 < k < \alpha M, \alpha N - N/2 < l < \alpha N \\ 0 & \text{otherwise} \end{cases} \quad (74)$$

where $k' = k - (\alpha - 1)M$ and $l' = l - (\alpha - 1)N$.

3. Approximators and Nonuniform Sampling

The interpolation methods discussed so far all reproduce the original samples f_{mn} as the interpolated values. When dealing with noisy samples, this exact reproduction may not be desirable. Instead, one may wish to perform approximate interpolation to reduce the influence of noise on the interpolation results. Approximate interpolation can still be obtained by using Eq. (59), but with kernels that do not satisfy Eq. (60). One way to obtain such kernels, called the *approximators*, is to apply smoothing windows to the ideal sinc interpolator. Alternatively, B-splines are also useful for approximate interpolation. This topic is revisited in Section VI.

Finally, it worth noting that with the advances in approximation and wavelet theories, general sampling theorems and interpolation techniques concerned with nonbandlimited functions and nonregular samples have also been developed. Interested readers are referred to Unser (2000) for an excellent review of this advanced subject.

IV. NOISE AND SIGNAL

Up to this point, we have been concerned with deterministic image-quality degradations: the blur caused by the system PSF and aliasing caused by sampling. Most images also suffer random degradations of some kind collectively referred to as noise. Noise can be defined as random, unwanted signal that interferes with the processing or measurements of the desired signal. In most applications, one typically assumes the noise is *additive* so that, for an LSI system, one can extend Eq. (6) to obtain

$$g(x,y) = h(x,y) * f(x,y) + n(x,y) \quad (75)$$

where $n(x)$ denotes the additive noise. In frequency space, Eq. (75) becomes

$$G(v_x, v_y) = H(v_x, v_y)F(v_x, v_y) + N(v_x, v_y) \quad (76)$$

To achieve optimal handling of noisy data, a rigorous mathematical characterization of the statistical nature of the noise is necessary. In the following sections, we present the basic definitions and concepts that are essential for dealing with random quantities in medical imaging.

A. Random Variables

Probability theory lays the foundation for a formal description of noise; some of its basic concepts and definitions are summarized here in order to introduce terminology that is needed in the following discussions. For an in-depth treatment of this subject, readers can consult the classic textbook by Papoulis (1991).

Intuitively, a *random variable* (RV) is a variable whose value assignment is probabilistic. A *realization*, on the other hand, refers to the actual value taken by an RV at a particular instant. In the following discussions, when the distinction becomes necessary we denote RVs by italic letters (such as u) and the (nonrandom) values an RV takes by script letters (such as $\mathfrak{u}$).

We begin our discussion with real RVs. Given a real RV u, its *cumulative distribution function* (cdf), $P_u(\mathfrak{u})$, is defined as

$$P_u(\mathfrak{u}) = \text{Probability}\{u \leq \mathfrak{u}\} \quad (77)$$

The derivative $p_u(\mathfrak{u}) = dP_u(\mathfrak{u})/d\mathfrak{u}$ is called the *probability density function* (pdf) of the RV u, and $p_u(\mathfrak{u})d\mathfrak{u}$ equals the probability that the value of u occurs in $[\mathfrak{u}, \mathfrak{u} + d\mathfrak{u}]$. The *expectation* (or the *mean*) of an RV u is defined as

$$E\{u\} = \int_{-\infty}^{\infty} \mathfrak{u} p_u(\mathfrak{u}) d\mathfrak{u} \equiv \bar{u} \quad (78)$$

Given a function $f(u): \mathbb{R} \to \mathbb{R}$, the expression $v = f(u)$ denotes a new real RV having the cdf $P_v(\mathfrak{v}) = \text{Probability}\{f(u) \leq \mathfrak{v}\}$ and the pdf $p_v(\mathfrak{v}) = dP_v(\mathfrak{v})/d\mathfrak{v}$. It can then be shown that

$$E\{v\} = \int_{-\infty}^{\infty} \mathfrak{v} p_v(\mathfrak{v}) d\mathfrak{v} = \int_{-\infty}^{\infty} f(\mathfrak{u}) p_u(\mathfrak{u}) d\mathfrak{u} \quad (79)$$

The *variance* of u, denoted by σ_u^2, can then be defined as $\sigma_u^2 = E\{|u - \bar{u}|^2\}$. The quantity $\sigma_u = \sqrt{\sigma_u^2}$ is called the *standard deviation* of u.

Next, we consider the statistical properties of two real RVs. The concepts and definitions discussed here can be easily extended to situations concerning with more than two real RVs. Given two real RVs u and v, their joint cdf is

$$P_{uv}(\mathfrak{u}, \mathfrak{v}) = \text{Probability}\{u \leq \mathfrak{u} \text{ and } v \leq \mathfrak{v}\} \quad (80)$$

and their joint pdf is $p_{uv}(\mathfrak{u},\mathfrak{v}) = \partial^2 P_{uv}(\mathfrak{u},\mathfrak{v})/\partial\mathfrak{u}\partial\mathfrak{v}$. Given a function $f(u,v): \mathbb{R}^2 \to \mathbb{R}$, $w = f(u,v)$ is a new real RV with the cdf $P_w(\mathfrak{w}) = \text{Probability}\{f(u,v) \leq \mathfrak{w}\}$. Again, it can be shown that

$$E\{w\} = \int_{-\infty}^{\infty} \mathfrak{w}\, p_w(\mathfrak{w}) d\mathfrak{w} = \iint_{-\infty}^{\infty} f(\mathfrak{u},\mathfrak{v}) p_{uv}(\mathfrak{u},\mathfrak{v}) d\mathfrak{u} d\mathfrak{v}, \quad (81)$$

where $p_w(\mathfrak{w}) = dP_w(\mathfrak{w})/d\mathfrak{w}$ is the pdf of $\mathfrak{w}$. Two RVs u and v are called *independent* if $p_{uv}(\mathfrak{u},\mathfrak{v}) = p_u(\mathfrak{u})p_v(\mathfrak{v})$, and *uncorrelated* if $E\{uv^*\} = E\{u\}E\{v^*\}$. The *correlation* and *covariance* of u and v are defined as $R_{uv} = E\{uv^*\}$ and $C_{uv} = E\{(u - \bar{u})(v - \bar{v})^*\} = R_{uv} - \bar{u}\bar{v}^*$, respectively. Here, complex conjugates are employed so that the definitions also apply to complex RVs to be introduced later.

One may also consider the statistical properties of an RV u subject to the occurrence of another RV v having a particular realization $\mathfrak{v}$. Following this definition, the *conditional pdf* of u given v can be expressed as

$$p_{u|v}(\mathfrak{u}|\mathfrak{v}) = \frac{p_{uv}(\mathfrak{u},\mathfrak{v})}{p_v(\mathfrak{v})} \quad (82)$$

Conditional statistics of u given v, such as conditional mean and conditional variance, are then defined as before in terms of this conditional pdf. From Eq. (82), one can immediately see that two RVs u and v are independent if $p_{u|v}(u|v) = p_u(u)$ so that, as one would naturally expect, the distribution of u remains the same irrespective of which value v takes. It also follows from Eq. (82) that $p_{uv}(u,v) = p_{u|v}(u|v)p_v(v) = p_{v|u}(v|u)p_u(u)$; hence,

$$p_{u|v}(u|v) = \frac{p_{v|u}(v|u)p_u(u)}{p_v(v)} \quad (83)$$

This equation is known as *Bayes's theorem*; this theorem has been widely used for developing statistical signal and image processing techniques and is the basis for many iterative image reconstruction methods.

The extension of these concepts and definitions to complex RVs is straightforward. A complex RV is the sum $z = u + jv$, where $j = \sqrt{-1}$, and u and v are real RVs. The statistics of a complex RV z are then defined as before for a real RV u, but with the joint cdf $p_z(\delta) = p_{uv}(u,v)$ in place of the cdf $p_u(u)$. Similarly, for a collection of complex RVs $z_i = u_i + jv_i$, $i = 1, \ldots, m$, their joint statistics are to be defined in terms of the joint cdf $p_{u_1v_1u_2v_2\ldots u_mv_m}(u_1,v_1,u_2,v_2,\ldots,u_m,v_m)$.

B. Stochastic Processes

A *stochastic process* $u(\alpha)$ is a collection of RVs indexed by a deterministic variable α; the collection of all realizations of a stochastic process is known as the *ensemble*. In the following discussions, the indexing variable α is either a 2D spatial coordinate, $\alpha = (x,y)^T$, or a 2D frequency coordinate, $\alpha = (v_x, v_y)^T$.

The statistical properties of a real stochastic process $u(\alpha)$ are completely determined in terms of its ensemble joint cdf given by

$$P_{u(\alpha)}(u_1,\ldots,u_m; \alpha_1,\ldots,\alpha_m) = \text{Probability}\{u(\alpha_1) \leq u_1,\ldots,u(\alpha_m) \leq u_m\} \quad (84)$$

for $m = 1, 2, \ldots, \infty$. The statistical properties for a complex stochastic processes can be similarly specified. However, this specification is rarely useful in practice. Instead, one often considers only its first- and second-order statistics, including the mean

$$\bar{u}(\alpha) = E\{u(\alpha)\} \quad (85)$$

the *autocorrelation*

$$R_u(\alpha_1,\alpha_2) = R_{u(\alpha_1)u(\alpha_2)} = E\{u(\alpha_1)u^*(\alpha_2)\} \quad (86)$$

and the *autocovariance*

$$C_u(\alpha_1,\alpha_2) = C_{u(\alpha_1)u(\alpha_2)} = R_u(\alpha_1,\alpha_2) - \bar{u}(\alpha_1)\bar{u}^*(\alpha_2) \quad (87)$$

It is not difficult to see that $R_u(\alpha,\alpha)$ and $\sigma_u^2(\alpha) \equiv C_u(\alpha,\alpha)$ are real and nonnegative quantities. Given two processes $u(\alpha)$ and $v(\alpha)$, one can similarly define their *cross-correlation* as

$$R_{uv}(\alpha_1,\alpha_2) = R_{u(\alpha_1)v(\alpha_2)} = E\{u(\alpha_1)v^*(\alpha_2)\} \quad (88)$$

and their *cross-covariance* as

$$C_{uv}(\alpha_1,\alpha_2) = C_{u(\alpha_1)v(\alpha_2)} = R_{uv}(\alpha_1,\alpha_2) - \bar{u}(\alpha_1)\bar{v}^*(\alpha_2) \quad (89)$$

C. Noise Models

The key noise model for emission tomography is Poisson noise, which describes the behavior of the photon-counting process. Gaussian noise has also been widely considered in reconstruction technique due to its mathematical tractability. In some situations, Gaussian noise can also provide a good approximation to Poisson noise.

1. Poisson Noise

It is known that the detected photon count z follows a Poisson distribution with parameter λ given by

$$\text{Probability}\{z = k\} = \text{Poisson}\{k;\lambda\} = \lambda^k e^{-\lambda}/k!, \quad k = 0,1,2,\ldots \quad (90)$$

It is straightforward to verify that $\bar{z} = \lambda$ and $\sigma_z^2 = \lambda$. Therefore, in Eq. (75) the stochastic process $g(x,y)$ has the pdf

$$\text{Probability}\{g(x,y) = k\} = \text{Poisson}\{k;\bar{g}(x,y)\}, \quad k = 0, 1, 2, \ldots \quad (91)$$

where $\bar{g}(x,y) = h(x,y) * f(x,y)$, and the resulting zero-mean process $n(x,y) = g(x,y) - \bar{g}(x,y)$ is said to be a *Poisson noise*. It is easy to see that $\bar{n}(x,y) = 0$ and $\sigma_n^2(x,y) = \bar{g}(x,y)$. In medical imaging, it is also often true that $n(\alpha_1)$ and $n(\alpha_2)$ are uncorrelated for any $\alpha_1 \neq \alpha_2$, where $\alpha_i = (x_i,y_i)^t$. The resulting process is *white noise*, with the autocorrelation $R_n(\alpha_1,\alpha_2) = \bar{g}(x,y)\delta(x_1 - x_2)\delta(y_1 - y_2)$.

These results indicate that Poisson noise is signal dependent. In particular, we have

$$\bar{g}^2(x,y)/\sigma_n^2(x,y) = \bar{g}(x,y) \quad (92)$$

that is, the local signal-to-noise ratio (SNR) of an image corrupted by Poisson noise is equal to the mean photon count at the local position. Therefore, as demonstrated by Figure 8, for a Poisson-noise-corrupted image, its quality deteriorates rapidly as the photon count decreases. Therefore, nuclear medicine image quality is directly related to the acquisition time that is available for collecting the gamma-ray photons.

2. Gaussian Noise

A noise process $n(\alpha)$ is called *Gaussian* (or *normal*) if $n(\alpha_1), \ldots, n(\alpha_j)$ are jointly Gaussian (normal) distributed for any $\alpha_1, \ldots, \alpha_j$, $j = 1, 2, \ldots, \infty$. Specifically, defining a random vector $\mathbf{n} = (n(\alpha_1), \ldots, n(\alpha_j))^T$ with the mean $\bar{\mathbf{n}} = (\bar{n}(\alpha_1), \ldots, \bar{n}(\alpha_j))^T$, the joint pdf of a Gaussian process $n(\alpha)$ is given by

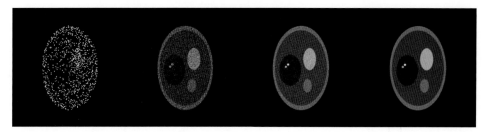

FIGURE 8 Example noisy images of a mathematical phantom, corrupted by Poisson noise. From left to right, the images contain 10^3, 10^5, 10^6, and 10^{11} total photon counts.

$$p_n(\mathfrak{n}_1,\ldots,\mathfrak{n}_j;\boldsymbol{\alpha}_1,\ldots,\boldsymbol{\alpha}_j) = \frac{1}{\sqrt{(2\pi)^j|\Sigma|}}\exp\left\{-\frac{1}{2}(\mathfrak{n}-\overline{\mathfrak{n}})\Sigma^{-1}(\mathfrak{n}-\overline{\mathfrak{n}})'\right\} \quad (93)$$

for $j = 1, 2, \ldots, \infty$, where $\boldsymbol{n} = (n_1, \ldots, n_j)^T$ and

$$\Sigma = \begin{bmatrix} C_n(\boldsymbol{\alpha}_1,\boldsymbol{\alpha}_1) & C_n(\boldsymbol{\alpha}_1,\boldsymbol{\alpha}_2) & \cdots & C_n(\boldsymbol{\alpha}_1,\boldsymbol{\alpha}_j) \\ C_n(\boldsymbol{\alpha}_2,\boldsymbol{\alpha}_1) & C_n(\boldsymbol{\alpha}_2,\boldsymbol{\alpha}_2) & \cdots & C_n(\boldsymbol{\alpha}_2,\boldsymbol{\alpha}_j) \\ \vdots & \vdots & \ddots & \vdots \\ C_n(\boldsymbol{\alpha}_j,\boldsymbol{\alpha}_1) & C_n(\boldsymbol{\alpha}_j,\boldsymbol{\alpha}_2) & \cdots & C_n(\boldsymbol{\alpha}_j,\boldsymbol{\alpha}_j) \end{bmatrix} = E\{\mathbf{nn}^T\} - \overline{\mathbf{n}}\overline{\mathbf{n}}^T \quad (94)$$

is the *autocovariance matrix* of $\mathbf{n}$. Therefore, the statistics of Gaussian noise are completely determined in terms of its mean $\overline{n}(\boldsymbol{\alpha})$ and autocovariance $C_n(\boldsymbol{\alpha}_i,\boldsymbol{\alpha}_j)$. In particular, it follows from Eq. (93) that, for a zero-mean Gaussian noise with $\overline{n}(\boldsymbol{\alpha}) = 0$, $\forall \boldsymbol{\alpha}$, its first-order density function is the normal density

$$p_n(\mathfrak{n};\boldsymbol{\alpha}) = \frac{1}{\sqrt{2\pi\sigma_\alpha^2}}\exp\left\{-\frac{\mathfrak{n}^2}{2\sigma_\alpha^2}\right\} \quad (95)$$

where $\sigma_\alpha^2 = C_n(\boldsymbol{\alpha},\boldsymbol{\alpha})$, and its second-order density is the jointly normal density

$$p_n(\mathfrak{n}_1,\mathfrak{n}_2;\boldsymbol{\alpha}_1,\boldsymbol{\alpha}_2) = \frac{1}{2\pi\sigma_{\alpha_1}\sigma_{\alpha_2}\sqrt{1-r_{12}^2}}\exp\left\{-\frac{1}{2(1-r_{12}^2)}\left[\frac{\mathfrak{n}_1^2}{\sigma_{\alpha_1}^2} - \frac{2r_{12}\mathfrak{n}_1\mathfrak{n}_2}{\sigma_{\alpha_1}\sigma_{\alpha_2}} + \frac{\mathfrak{n}_2^2}{\sigma_{\alpha_2}^2}\right]\right\} \quad (96)$$

where $\sigma_{\alpha_i}^2 = C_n(\boldsymbol{\alpha}_i,\boldsymbol{\alpha}_i)$, $i = 1, 2$, and $r_{12} = C_n(\boldsymbol{\alpha}_1,\boldsymbol{\alpha}_2)/\sigma_{\alpha_1}\sigma_{\alpha_2}$.

A Gaussian noise process is called *Gaussian white noise* if it is also white, that is, if $C_n(\boldsymbol{\alpha}_i,\boldsymbol{\alpha}_j) = 0$ for any $\boldsymbol{\alpha}_i \neq \boldsymbol{\alpha}_j$.

D. Power Spectrum

A stochastic process $u(\boldsymbol{\alpha})$ is called *strict-sense stationary* (SSS) if its statistical properties are invariant to a shift of the origin, that is, when the two sets of random variables $\{u(\boldsymbol{\alpha}_1), \ldots, u(\boldsymbol{\alpha}_j)\}$ and $\{u(\boldsymbol{\alpha}_1 + \boldsymbol{\beta}), \ldots, u(\boldsymbol{\alpha}_j + \boldsymbol{\beta})\}$ have the same statistics for any $\boldsymbol{\beta} \in \mathbb{R}^2$ and $j = 1, 2, \ldots, \infty$. On the other hand, it is called *wide-sense stationary* (WSS) if its mean is constant and its autocorrelation depends only on $\boldsymbol{\tau} = \boldsymbol{\alpha}_1 - \boldsymbol{\alpha}_2$, that is, if one can write

$$\overline{u}(\boldsymbol{\alpha}) = \overline{u} \quad \text{and} \quad R_u(\boldsymbol{\alpha}+\boldsymbol{\tau},\boldsymbol{\alpha}) = R_u(\boldsymbol{\tau}) \quad (97)$$

It follows directly from the definition that $R_u(-\boldsymbol{\tau}) = R_u^*(\boldsymbol{\tau})$ if $u(\boldsymbol{\alpha})$ is WSS. It is obvious that SSS implies WSS, but the converse is generally not true. However, because the statistics of Gaussian noise are completely determined by the mean and autocovariance, a WSS Gaussian noise is automatically SSS.

Note that by definition a WSS process should look similarly in mean and variance at all locations. Therefore, generally speaking, a WSS process does not have finite energy and its Fourier transform does not exist. However, a WSS process can have finite power and the concept of *power spectrum* can be defined. The *power spectrum* (or the *spectral density*) of a WSS process $u(x,y)$, denoted by $S_u(v_x,v_y)$, is the Fourier transform of its autocorrelation $R_u(\tau_x,\tau_y)$:

$$S_u(v_x,v_y) = \int\!\!\int_{-\infty}^{\infty} R_u(\tau_x,\tau_y)e^{-j2\pi(v_x\tau_x+v_y\tau_y)}d\tau_x d\tau_y \quad (98)$$

provided that the transform exists. It can be shown that $S_u(v_x,v_y)$ is real and nonnegative, that is, that $S_u(v_x,v_y) \geq 0$, $\forall v_x,v_y$, for any complex process $u(x,y)$. Conversely, for any nonnegative function $S(v_x,v_y) \in L^2$ one can always find a WSS process $u(x,y)$ such that $S_u(v_x,v_y) = S(v_x,v_y)$. Furthermore, it can be shown that (Papoulis, 1991)

$$S_u(v_x,v_y) = \lim_{T\to\infty}\frac{1}{4T^2}E\left\{|U_T(v_x,v_y)|^2\right\} \quad (99)$$

where $U_T(v_x,v_y) = \int_{-T}^{T}\int_{-T}^{T} u(x,y)e^{-j2\pi(v_xx+v_yy)}dxdy$. This equality justifies the interpretation of $S(v_x,v_y)$ as being the power spectrum of $u(x,y)$.

Similarly, consider two stochastic processes $u(x,y)$ and $\upsilon(x,y)$. If the cross-correlation depends only on $\boldsymbol{\tau} = \boldsymbol{\alpha}_1 - \boldsymbol{\alpha}_2$, that is, $R_{u\upsilon}(\boldsymbol{\tau}) \equiv R_{u\upsilon}(\boldsymbol{\alpha}+\boldsymbol{\tau},\boldsymbol{\alpha})$, then their *cross-power spectrum*, denoted by $S_{u\upsilon}(v_x,v_y)$, is defined as the Fourier transform of their $R_{u\upsilon}(\boldsymbol{\tau})$. In general, the function $S_{u\upsilon}(v_x,v_y)$ is complex even when both $u(x,y)$ and $\upsilon(x,y)$ are real. Because $R_{u\upsilon}(\boldsymbol{\tau}) = R_{\upsilon u}^*(-\boldsymbol{\tau})$, it follows that $S_{u\upsilon}(v_x,v_y) = S_{\upsilon u}^*(v_x,v_y)$ for any $u(x,y)$ and $\upsilon(x,y)$.

With these definitions, for the linear system given by

$$v(x,y) = h(x,y) * u(x,y) + w(x,y) \quad (100)$$

where $u(x,y)$ and $w(x,y)$ are uncorrelated zero-mean WSS stochastic processes and $h(x,y)$ is a deterministic function, one can show that

$$R_{uv}(x,y) = h'(x,y) * R_u(x,y) \quad (101)$$

$$R_v(x,y) = h(x,y) * h'(x,y) * R_u(x,y) + R_w(x,y) \quad (102)$$

where $h'(x,y) = h*(-x,-y)$. Consequently,

$$S_{uv}(v_x,v_y) = H^*(v_x,v_y) S_u(v_x,v_y) \quad (103)$$

$$S_v(v_x,v_y) = |H(v_x,v_y)|^2 S_u(v_x,v_y) + S_w(v_x,v_y) \quad (104)$$

1. Power Spectra of White Noise

By the definitions of white noise and WSS, a zero-mean WSS white noise process $n(x,y)$ must have autocorrelation $R_n(\tau_x,\tau_y) = q\,\delta(\tau_x,\tau_y)$, $q > 0$. Therefore the power spectrum of a WSS white noise process is constant, given by $S_n(v_x,v_y) = q$. Notice that, strictly speaking, a Poisson noise is not WSS unless it has a constant mean at all spatial locations. In spite of this, the WSS white noise model has been employed in nuclear medicine image processing.

V. FILTERING

The diagnostic accuracy that can be achieved from medical images depends critically on the image quality. Therefore, it is always desirable to suppress the degrading effects of blur and noise in medical images, that is, to *restore* the images. This amounts to the task of estimating $f(x,y)$ from the acquired data $g(x,y)$ by solving Eq. (75). Unfortunately, it is widely known that inversion of Eq. (75) is *ill-posed* in the sense that the presence of a small amount of noise $n(x,y)$ in the acquired data can result in a substantial departure of the estimate $\hat{f}(x,y)$ from the true solution $f(x,y)$. Techniques to control this deteriorating effect of noise on the inverse solution are known as *regularization* techniques, and we discuss a few of these in this section. Figure 9 compares some example results.

A. Pseudoinverse

The pseudoinverse of Eq. (75), denoted by $f^\dagger(x,y)$, is the minimum-norm solution defined by

$$f^\dagger(x,y) = \min_{\arg f(x,y)} |g(x,y) - h(x,y) * f(x,y)|^2 \quad (105)$$

subject to $|f(x,y)|^2$ is minimum

In other words, the pseudoinverse is the minimum-energy solution that best agrees with the acquired data.

According to Parseval's theorem, in frequency space this equation becomes

$$F^\dagger(v_x,v_y) = \min_{\arg F(v_x,v_y)} |G(v_x,v_y) - H(v_x,v_y)F(v_x,v_y)|^2 \quad (106)$$

subject to $|F(v_x,v_y)|^2$ is minimum

where $F^\dagger(v_x,v_y) = \mathcal{F}\{f^\dagger(x,y)\}$. It is not difficult to obtain from Eq. (106) that

$$F^\dagger(v_x,v_y) = \begin{cases} H^{-1}(v_x,v_y)G(v_x,v_y) & H(v_x,v_y) \neq 0 \\ 0 & H(v_x,v_y) = 0 \end{cases} \quad (107)$$

To illustrate the ill-posed nature of the inverse problem, we can consider the noise characteristics of the pseudoinverse solution. By substituting Eq. (76) into Eq. (107), one obtains

$$F^\dagger(v_x,v_y) =$$
$$\begin{cases} F(v_x,v_y) + H^{-1}(v_x,v_y)N(v_x,v_y) & H(v_x,v_y) \neq 0 \\ 0 & H(v_x,v_y) = 0 \end{cases} \quad (108)$$

Therefore, $F^\dagger(v_x,v_y)$ is a noisy estimate of the true solution $F(v_x,v_y)$, bandlimited to the bandwidth of $H(v_x,v_y)$. Note that, even with a small $N(v_x,v_y)$, the noise term $H^{-1}(v_x,v_y)N(v_x,v_y)$ can have extremely large magnitude when $H(v_x,v_y) \approx 0$, making pseudoinverses notoriously sensitive to noise. Because $H(v_x,v_y) \approx 0$ typically occurs at high frequencies, pseudoinverses are typically corrupted by large-amplitude high-frequency fluctuations.

B. Tikhonov Regularization

To reduce the deleterious effects of noise on the inverse solution, Tikhonov (in Galsko, 1984) proposed to estimate the solution of Eq. (75) by

$$\hat{f}_{\text{Tik}}(x,y) = \min_{\arg f(x,y)} \left\{ |g(x,y) - h(x,y) * f(x,y)|^2 + \gamma |f(x,y)|^2 \right\} \quad (109)$$

or, equivalently,

$$\hat{F}_{\text{Tik}}(v_x,v_y) = \min_{\arg F(v_x,v_y)} \left\{ |G(v_x,v_y) - H(v_x,v_y)F(v_x,v_y)|^2 + \gamma |F(v_x,v_y)|^2 \right\} \quad (110)$$

where $\gamma > 0$ is a *regularization parameter* and $\hat{F}_{\text{Tik}}(v_x,v_y) = \mathcal{F}\{\hat{f}_{\text{Tik}}(x,y)\}$. Applying the calculus of variations to Eq. (110), one obtains

$$\hat{F}_{\text{Tik}}(v_x,v_y) = \frac{H^*(v_x,v_y)}{|H(v_x,v_y)|^2 + \gamma} G(v_x,v_y) \quad (111)$$

It is easy to see that

$$\begin{cases} \hat{F}_{\text{Tik}}(v_x,v_y) \approx H^{-1}(v_x,v_y)G(v_x,v_y) = F^\dagger(v_x,v_y) & \text{if } |H(v_x,v_y)|^2 \gg \gamma \\ |\hat{F}_{\text{Tik}}(v_x,v_y)| \approx |\gamma^{-1}H^*(v_x,v_y)G(v_x,v_y)| \ll |F^\dagger(v_x,v_y)| & \text{if } |H(v_x,v_y)|^2 \ll \gamma \end{cases} \quad (112)$$

Therefore, with an appropriate small, but finite regularization parameter γ, one can greatly reduce the amplitude of noise that is often observed in the pseudoinverse solution at frequencies where $H(v_x,v_y) \approx 0$.

More generally, one can consider the regularized inverse solution defined by

$$\hat{f}_{\text{Tik}}(x,y) = \min_{\arg f(x,y)} \left\{ |g(x,y) - h(x,y) * f(x,y)|^2 + \gamma |\mathcal{D}\{f(x,y)\}|^2 \right\} \quad (113)$$

where $\mathcal{D}\{\cdot\}$ is an operator so that the term $|\mathcal{D}\{f(x,y)\}|^2$ gives a measure of the roughness in $f(x,y)$. For example, the Laplacian operator $\partial^2/\partial x^2 + \partial^2/\partial y^2$ is a frequently used operator, yielding

$$\hat{f}_{\text{Tik}}(x,y) = \min_{\arg f(x,y)} \left\{ |g(x,y) - h(x,y) * f(x,y)|^2 + \gamma \left| \frac{\partial^2 f(x,y)}{\partial x^2} + \frac{\partial^2 f(x,y)}{\partial y^2} \right|^2 \right\} \quad (114)$$

or, equivalently,

$$\hat{F}_{\text{Tik}}(v_x,v_y) = \min_{\arg F(v_x,v_y)} \left\{ |G(v_x,v_y) - H(v_x,v_y)F(v_x,v_y)|^2 + 16\pi^4 \gamma v^4 |F(v_x,v_y)|^2 \right\} \quad (115)$$

where $v^2 = v_x^2 + v_y^2$. This equation has the solution

$$\hat{F}_{\text{Tik}}(v_x,v_y) = \frac{H^*(v_x,v_y)}{|H(v_x,v_y)|^2 + \beta v^4} G(v_x,v_y) \quad (116)$$

where $\beta = 16\pi^2 \gamma > 0$. The term βv^4 therefore favorably decreases the degree of noise regularization at low frequencies, where the signal is expected to dominate the noise, but increases the degree of noise regularization at high frequencies, where the noise is expected to dominate the signal.

C. Wiener Filtering

So far, we have considered deterministic inverse techniques that do not explicitly make use of the statistical nature of the noise. Wiener filtering, on the other hand, is a widely used statistical technique for generating regularized inverse solutions. In its derivation, one considers the input $f(x,y)$ to be stochastic as well. Generally, one assumes that both $f(x,y)$ and $n(x,y)$ are zero-mean WSS processes and seeks to find a solution of the form $\hat{f}_{\text{Wiener}}(x,y) \equiv w(x,y) * g(x,y)$, or equivalently $\hat{F}_{\text{Wiener}}(v_x,v_y) \equiv W(v_x,v_y)G(v_x,v_y)$, so that the mean-square error of the estimate,

$$\epsilon^2 = E\left\{ |\hat{f}(x,y) - f(x,y)|^2 \right\} \quad (117)$$

is minimized. In other words, in Wiener filtering one seeks to obtain the linear minimum mean-square error solution (LMMSE) of Eq. (75).

To solve for the *Wiener filter* $w(x,y)$, one can apply the calculus of variations to Eq. (117) to obtain the *Wiener-Hopf* equation:

$$\begin{aligned} R_{fg}(x,y) &= \int\int_{-\infty}^{\infty} w(x',y')R_g(x-x',y-y')dx'dy' \\ &= w(x,y) * R_g(x,y) \end{aligned} \quad (118)$$

In frequency space, this equation becomes $S_{fg}(v_x,v_y) = W(v_x,v_y)S_g(v_x,v_y)$; hence,

$$W(v_x,v_y) = \frac{S_{fg}(v_x,v_y)}{S_g(v_x,v_y)} \quad (119)$$

When the input signal $f(x,y)$ and noise $n(x,y)$ are uncorrelated, by making use of Eqs. (103) and (104) the Wiener filter becomes

$$\begin{aligned} W(v_x,v_y) &= \frac{H^*(v_x,v_y)S_f(v_x,v_y)}{|H(v_x,v_y)|^2 S_f(v_x,v_y) + S_n(v_x,v_y)} \\ &= \frac{H^*(v_x,v_y)}{|H(v_x,v_y)|^2 + \Phi(v_x,v_y)} \end{aligned} \quad (120)$$

where $\Phi(v_x,v_y) = S_n(v_x,v_y)/S_f(v_x,v_y)$ is the noise-to-signal power ratio at (v_x,v_y).

By comparing this result with the Tikhonov filters given in Eqs. (111) and (116), we see that the Wiener filter can be considered as a generalization of the Tikhonov regularization with the regularization parameter α being replaced with a more general frequency-dependent regularization term $\Phi(v_x,v_y)$. Hence, from Eq. (112) we have

$$\begin{cases} \hat{F}_{\text{Wiener}}(v_x,v_y) \approx F^\dagger(v_x,v_y) & \text{if } |H(v_x,v_y)|^2 \gg \Phi(v_x,v_y) \\ |\hat{F}_{\text{Wiener}}(v_x,v_y)| \ll |F^\dagger(v_x,v_y)| & \text{if } |H(v_x,v_y)|^2 \ll \Phi(v_x,v_y) \end{cases} \quad (121)$$

Note that the condition $|H(v_x,v_y)|^2 \gg \Phi(v_x,v_y)$ is equivalent to $|H(v_x,v_y)|^2 S_f(v_x,v_y) \gg S_n(v_x,v_y)$. Therefore, the Wiener filter attempts to exactly recover $f(x,y)$ only at frequencies at which the signal power in the acquired data $g(x,y)$ [equal to $|H(v_x,v_y)|^2 S_f(v_x,v_y)$] is much greater than the noise power in $g(x,y)$ [given by $S_n(v_x,v_y)$]. On the other hand, it applies smoothing to reduce image noise at frequencies when noise power dominates the signal power. In typical situations, one has $S_f(v_x,v_y) \geq S_n(v_x,v_y)$ at low frequencies. As $|v_x|$ or $|v_y|$ increases, one often observes that $S_f(v_x,v_y) \to 0$ rapidly while $S_n(v_x,v_y)$ remains relatively constant. Consequently, as frequency increases one has $\Phi^{-1}(v_x,v_y) = S_f(v_x,v_y)/S_n(v_x,v_y) \to 0$ and hence $W(v_x,v_y) \approx$

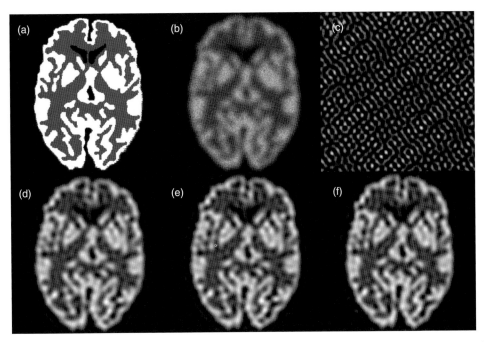

FIGURE 9 Comparison of various inverse filters. The original 256 × 256 computer-generated brain image (the Hoffman phantom; Hoffman et al., 1990) is shown in (a), and a blurred (Hanning filter cut off at 0.1 (pixel size)$^{-1}$, see Table 1) and noisy version (Poisson noise with 10^6 total counts) of the image is shown in (b). The other images are restored from the degraded image by use of (c) the pseudoinverse, (d) the Tikhonov regularization given by Eq. (111) with $\gamma = 10^{-3}$, (e) the Tikhonov regularization given by Eq. (116) with $\gamma = 1$, and (f) the Wiener filter. Clearly, the pseudoinverse solution is extremely sensitive to noise and is of little practical use. In Wiener filtering, we assume a white noise and the known original image and blurring model is used to estimate the noise-to-signal power ratio.

$H^*(v_x,v_y)\Phi^{-1}(v_x,v_y) \to 0$ as well. Note that when using the Wiener filter, the signal power spectrum $S_f(v_x,v_y)$ must be either known *a priori* or be estimated from the data $g(x,y)$.

One limitation of all the approaches considered is that they either assume that the underlying statistics are stationary or not known at all. Real images typically have nonstationary characteristics (e.g., in general, an image containing Poisson noise is not stationary); in these cases a better approach would be able to adapt its operations to the local characteristics of the image. For example, it is often desirable to preserve edges and boundaries in the restored images while reducing noisy fluctuations in flat regions. In theory, this could be achieved by using a roughness operator $\mathcal{D}\{\cdot\}$ in Eq. (113) that will not penalize discontinuities associated with edges and boundaries, by using Bayesian techniques with similarly defined *a priori* density function for the solution, or by the use of wavelet transforms. These subjects, however, are beyond the scope of this chapter.

VI. SMOOTHING

A special case of filtering is *smoothing*, in which we ignore blur or assume it is negligible (i.e., assume $h(x,y) = \delta(x,y)$ or $H(v_x,v_y) = 1$) and focus solely on reducing the impact of noise. The resulting imaging equation therefore becomes

$$g(x,y) = f(x,y) + n(x,y) \qquad (122)$$

As already mentioned in the discussion of Wiener filtering, noise is often observed to dominate signal at high frequencies. As a result, noise-reduction filters are typically lowpass filters that preferentially suppress image noise present at high frequencies.

A. Smoothing Filters

Although smoothing filters can be obtained by using the previously derived filters with $H(v_x,v_y) = 1$, in practice it is more common to apply an *ad hoc* smoothing filter $s(x,y)$ so that

$$\hat{f}(x,y) = s(x,y) * g(x,y) \qquad (123)$$

In medical imaging applications, several frequency-space *window* functions have been widely considered for achieving image smoothing. These windows include the rectangular window, the generalized Hanning window, the Butterworth window, the Shepp-Logan window, and the Gaussian window; their transfer functions are defined in

Table 1. Clearly, these window functions are all lowpass filters, and they reduce image noise at the expense of image resolution. Window functions that employ a smaller cutoff frequency v_c reject more high-frequency components of the original image to produce smoother images. Therefore, the cutoff frequency needs to be subjectively selected to achieve good noise reduction without substantially reducing image resolution. Example results obtained by use of these filters are shown in Figure 10.

In addition to these window functions, image smoothing is also often achieved by the use of the convolution form of Eq. (123) directly. A commonly used smoothing kernel is

$$s(x,y) = \text{rect}\left(\frac{x}{W}\right)\text{rect}\left(\frac{y}{W}\right) \quad (124)$$

where W is a suitably chosen window size. Therefore, pixel intensity in the smoothed image is simply the local average of the intensities of its neighboring pixels within the specified window size (see Figure 10g for an example result). Obviously, a large window size can produce stronger smoothing, and the optimal window size to use again depends on the trade-off between noise reduction and image resolution. Intuitively, two pixels that are farther away are expected to have a lesser degree of intensity correlation. Therefore, local weighted averages in which the amplitude of the smoothing kernel decreases with $|x|$ and $|y|$ in some fashion have also been widely considered.

TABLE 1 Several Commonly Used One-Dimensional Lowpass Filters[a]

Filter	Transfer Function $L(v)$
Rectangular	$\text{rect}\left(\dfrac{v}{2v_c}\right)$
Generalized Hanning	$\left(\gamma+(1-\gamma)\cos\left(\dfrac{\pi v}{v_c}\right)\right)\text{rect}\left(\dfrac{v}{2v_c}\right),\ 0\le\gamma\le 1$
Butterworth	$\left(1+\left(\dfrac{v}{v_c}\right)^{2n}\right)^{-1},\ n>0$
Shepp-Logan	$\text{sinc}\left(\dfrac{v}{v_c}\right)\text{rect}\left(\dfrac{v}{2v_c}\right)$
Gaussian	$\exp\left\{-\dfrac{v^2}{2v_c^2}\right\}$

[a] $v_c > 0$ denotes the *cutoff frequency* of the filter. As the frequency increases, transfer functions of the Butterworth and Gaussian filters roll off to zero continuously while the others becomes identically zero at v_c. The additional parameter n in the Butterworth filter is known as its *order*. The generalized Hanning filters obtained with $\gamma = 0.5$ and $\gamma = 0.54$ are also known as the Hanning and Hamming filters, respectively. Two-dimensional filters can be obtained by either $S(v_x,v_y) = L(\sqrt{v_x^2+v_y^2})$ or $S(v_x,v_y) = L(v_x)L(v_y)$.

B. Median Filters

The smoothing filters are all stationary; they obviously cannot adapt to local image statistics to achieve the best noise reduction while retaining image resolution. A median filter, on the other hand, is a useful nonstationary smoothing filter for processing discrete data. Assume that measurements $g_{mn} = g(mx_s, ny_s)$, $m,n \in \mathbb{Z}$, are available. The median filter produces estimates of $f_{mn} = f(mx_s, ny_s)$ by

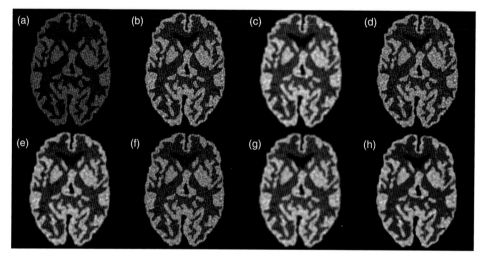

FIGURE 10 Comparison of various smoothing methods. Image (a) shows a noisy brain image that contains Poisson noise having 10^6 total counts (the original noise-free image is shown in Fig. 9a). The image is smoothed by (b) a rectangular window, (c) a Hanning window, (d) a sixth-order Butterworth window, (e) a Shepp-Logan window, and (f) a Gaussian window. All of these windows employ the symmetrical form $S(v_x,v_y) = L(\sqrt{v_x^2+v_y^2})$, where $L(v)$ is defined in Table 1, with a cut-off frequency of 0.25 (pixel size)$^{-1}$. Images shown in (g) and (h), on the other hand, are obtained by local average and by a median filter, both employing a 5×5-pixel window size.

$$\hat{f}_{mn} = \text{median}\{g_{m'n'} : m - K \leq m' \leq m + K, n - K \leq n' \leq n + K\} \quad (125)$$

where $K > 0$ is the window size of the median filter. The median of a set of numbers is a nonlinear statistic of the set, with the useful property that it is insensitive to the presence of a few samples with unusually large values (the *outliers*) in the set. An important property of the median filter is that its preserves edges or stepwise discontinuities in the image; thus, it is useful for removing salt-and-pepper image noise (impulsive noise) without blurring the images. An example result obtained with median filtering is shown in Figure 10h. The performance of median filters may be improved by employing an adaptive threshold so that a sample is replaced by the median only if the difference between the sample and the median is above the threshold.

C. Smoothing Splines

Another smoothing technique worth noting involves the use of cubic splines. In this technique, one aims to fit a curve to observed noisy samples g_i, $0 \leq i \leq N - 1$, of a 1D function $f(x)$. In 2D, this fitting is first applied to individual rows of the image array to the obtain the desired new sampling spacing. The procedure is then repeated to the newly formed image columns. As with the Tikhonov regularization previously discussed, this curve-fitting problem is framed as a trade-off between achieving agreement with the measured data and achieving a smooth fit curve. In one popular formulation, assuming a finite number N of measurements, the fit curve $\hat{f}(x)$ is obtained as the minimizer of the penalized weighted least squares (PWLS) objective function

$$\Phi(f(x), \mathbf{g}) = \sum_{i=0}^{N-1} w_i [g_i - f(x_i)]^2 + \gamma \int_{-\infty}^{\infty} [f''(x)]^2 \, dx \quad (126)$$

where $\mathbf{g} = (g_0, \ldots, g_{N-1})^t$, x_i is the sampling position of g_i, w_i are weights reflecting the relative certainty of each measurement, $f''(x)$ denotes the second-order derivative of $f(x)$, and $\gamma > 0$ is a smoothing parameter that controls the relative influence of the goodness-of-fit and smoothing terms in this equation (Green and Silverman, 1994). Note that the sampling points (x_i, y_i) need not be uniformly spaced—only distinct. The weights w_i can be chosen *a priori* or estimated from the data. It can be shown that the curve minimizing this objective is a natural cubic spline (NCS)—a piecewise cubic polynomial that is continuous up to and including the second derivative at the knots between pieces. The details of determining this NCS are given in Green and Silverman (1994).

The choice of the smoothing parameter γ profoundly influences the appearance of the fit curve $\hat{f}(x)$ because γ determines the relative influence of the two terms in the penalized-likelihood expression, the first rewarding goodness-of-fit to the data, the second rewarding smoothness. A small value of γ leads to a ragged curve, whereas a large value of γ leads to a smooth curve. In most applications, a value between these two extremes is desirable, and, although this can be found through trial and error for most datasets, a more principled and automatic approach would clearly be preferred.

One such automatic approach to choosing the smoothing parameter is based on the principle of cross validation (CV), which has been discussed in the context of image processing by Galatsanos and Katsaggelos (1992). The approach is grounded in the assumption that the choice of γ should yield a fit curve $\hat{f}(x)$ that accurately predicts the outcomes of further observations. Remarkably, the predictive accuracy of the fit curve can be quantified solely on the basis of the fit values and the measured data (Wabba, 1990). This CV score can be expressed as

$$CV(\gamma) = \frac{1}{N} \sum_{i=0}^{N-1} \left(\frac{g_i - \hat{f}(x_i)}{1 - A_{ii}(\gamma)} \right)^2 \quad (127)$$

where A_{ii} are the diagonals of the hat matrix $A(\gamma)$, which links the values of the estimate at the measured points to the values of the observations at those points: $\hat{\mathbf{f}} = A(\gamma) \mathbf{g}$, where $\hat{\mathbf{f}}$ is an $N \times 1$ vector with elements $\hat{f}_i = \hat{f}(x_i)$. The value of γ minimizing this curve is considered optimal and can generally be found fairly quickly using a golden section search minimization approach. Efficient algorithms exist for carrying out this process (Hutchinson and de Hoog, 1985).

A final approach to smoothing, which we do not discuss in great detail, is similar in spirit to the approach just presented, but with a more explicitly statistical point of view. Specifically, the approach involves replacing the weighted least squares goodness-of-fit term with a statistical likelihood function appropriate for the noise statistics being modeled (Green and Silverman, 1994). This produces a penalized likelihood objective function, which can again be shown to be minimized by a natural cubic spline. Currently, the approach can accommodate the class of exponential distributions, which includes the Poisson distribution, that is of great interest in medical imaging. From this point of view, Eq. (126) emerges as a special case when the measurement noise is normally distributed. Further details, including an application to emission tomography, can be found in La Rivière and Pan (2000).

VII. ESTIMATION

In the problem of parameter estimation, one is concerned with the task of finding the "best" guess of the underlying parameters that give rise to particular observations. In solving this problem, the parameters of interest can be treated as either random or nonrandom. The parameters to be estimated can be of any type and do not necessarily have to be defined in the same functional space as the observations. The only requirement is that there exists a well-defined mathematical relationship between the parameters and the observations. In fact, the filtering problem considered previously can also be considered to be an estimation problem in which the parameters to be estimated are the image pixel intensities f_{mn} and the observations are the degraded data g_{mn}. Similarly, the tomographic image reconstruction problem (which is not discussed in this chapter) is also an estimation problem in which the observations are the degraded sinograms and the parameters to be estimated are the pixel intensities of the source images.

In the following, we discuss some general formulations for solving the estimation problem, along with an introduction to the concept of bias and variance for quantitative characterization of a given estimation.

A. Bayesian Parameter Estimation

In the Bayesian formulation, the parameters to be estimated are treated as random variables. An observation $\mathbf{y}$ in the observation space $\mathcal{Y} \subseteq \mathbb{R}^M$ is statistically related to $\mathbf{x}$ in the parameter space $\chi \subseteq \mathbb{R}^N$ through a *known* family of pdfs given by $\{p_{y|x}(\mathfrak{y}|\mathfrak{x}): \mathfrak{x} \in \chi\}$. Given this model, the goal of parameter estimation is to find a function $\hat{\mathbf{x}}$ such that $\hat{\mathbf{x}}(\mathfrak{y})$ is the best guess of the true value $\mathfrak{x}$ that gives rise to that particular observation $\mathbf{y} = \mathfrak{y}$. This task is often achieved by the construction of a cost function $C: \chi \times \chi \to \mathbb{R}^+$ such that $C(\mathfrak{x}_1, \mathfrak{x}_2)$, $\mathfrak{x}_1, \mathfrak{x}_2 \in \chi$, is the cost associated with generating the estimated value $\mathfrak{x}_1$ when the truth is $\mathfrak{x}_2$. One can then define a Bayes risk for an estimate $\hat{\mathbf{x}}$ as

$$r(\hat{\mathbf{x}}) = \int_{\mathcal{Y}} \int_{\chi} C(\hat{\mathbf{x}}(\mathfrak{y}), \mathfrak{x}) p_{xy}(\mathfrak{x}, \mathfrak{y}) \, d\mathfrak{x} \, d\mathfrak{y} \quad (128)$$

where $p_{xy}(\mathfrak{x}, \mathfrak{y})$ is the joint pdf of $\mathbf{x}$ and $\mathbf{y}$.

The Bayes estimate of $\mathbf{x}$ is the one that minimizes the Bayes risk $r(\hat{\mathbf{x}})$. Because $p_{xy}(\mathfrak{x}, \mathfrak{y}) = p_{x|y}(\mathfrak{x}|\mathfrak{y}) p_y(\mathfrak{y})$, Eq. (128) can also be written

$$r(\hat{\mathbf{x}}) = \int_{\mathcal{Y}} \left\{ \int_{\chi} C(\hat{\mathbf{x}}(\mathfrak{y}), \mathfrak{x}) p_{x|y}(\mathfrak{x}|\mathfrak{y}) \, d\mathfrak{x} \right\} p_y(\mathfrak{y}) \, d\mathfrak{y} \quad (129)$$

Therefore, minimization of $r(\hat{\mathbf{x}})$ can be achieved by minimizing the *posterior cost*,

$$r(\hat{\mathbf{x}}|\mathfrak{y}) = \int_{\chi} C(\hat{\mathbf{x}}(\mathfrak{y}), \mathfrak{x}) p_{x|y}(\mathfrak{x}|\mathfrak{y}) \, d\mathfrak{y} \quad (130)$$

independently at each element $\mathfrak{y}$ in the observation space $\mathcal{Y}$.

The use of a different cost function in the Bayesian estimation yields different estimates. Two popular cost functions are considered next.

1. Minimum Mean-Square Error (MMSE) Estimation

When $\chi = \mathbb{R}^N$, one commonly used cost function is $C(\mathfrak{x}_1, \mathfrak{x}_2) = |\mathfrak{x}_1 - \mathfrak{x}_2|^2$, $\mathfrak{x}_1, \mathfrak{x}_2 \in \mathbb{R}^N$, to yield the minimum mean-square error (MMSE) estimate given by

$$\hat{\mathbf{x}}_{\mathrm{MMSE}}(\mathfrak{y}) = \int \mathfrak{x} p_{x|y}(\mathfrak{x}|\mathfrak{y}) \, d\mathfrak{x} = E\{\mathbf{x}|\mathfrak{y}\} \quad (131)$$

Hence, the MMSE estimate is the conditional mean of $\mathbf{x}$ given the observation $\mathbf{y} = \mathfrak{y}$. As already mentioned, the Wiener filtering generates the LMMSE estimate of the inverse solution to the stochastic imaging model given by Eq. (75).

2. Maximum A Posteriori Probability (MAP) Estimation

Another often used estimate, particularly in emission image reconstruction, is the maximum *a posteriori* probability (MAP) estimate. This estimate is obtained by considering a uniform cost function given by

$$C(\mathfrak{x}_1, \mathfrak{x}_2) = \begin{cases} 0 & \text{if } |\mathfrak{x}_1 - \mathfrak{x}_2| \leq \Delta \\ 1 & \text{otherwise} \end{cases} \quad (132)$$

where Δ is small positive number. It follows that the posterior cost is approximately equal to $1 - p_{x|y}(\hat{\mathbf{x}}(\mathfrak{y})|\mathfrak{y})\Delta$ provided that $p_{x|y}(\mathfrak{x}|\mathfrak{y})$ is smooth with respect to the scale Δ. Thus, the Bayesian estimate in this case becomes the MAP estimate defined by

$$\hat{\mathbf{x}}_{\mathrm{MAP}}(\mathfrak{y}) = \arg \max_{\mathfrak{x} \in \chi} p_{x|y}(\mathfrak{x}|\mathfrak{y}) \quad (133)$$

Using the Bayes's rule $p_{x|y}(\mathfrak{x}|\mathfrak{y}) = p_{y|x}(\mathfrak{y}|\mathfrak{x}) p_x(\mathfrak{x}) / p_y(\mathfrak{y})$ and the fact that logarithm is a monotonically increasing function, Eq. (133) can also be written

$$\hat{\mathbf{x}}_{\mathrm{MAP}}(\mathfrak{y}) = \arg \max_{\mathfrak{x} \in \chi} \{\log p_{y|x}(\mathfrak{y}|\mathfrak{x}) + \log p_x(\mathfrak{x})\} \quad (134)$$

A necessary condition for the maximization is that the MAP estimate must solve the *MAP equation*:

$$\frac{\partial}{\partial \mathfrak{x}} \log p_{y|x}(\mathfrak{y}|\mathfrak{x}) \bigg|_{\mathfrak{x} = \hat{\mathbf{x}}_{\mathrm{MAP}}(\mathfrak{y})} = -\frac{\partial}{\partial \mathfrak{x}} \log p_x(\mathfrak{x}) \bigg|_{\mathfrak{x} = \hat{\mathbf{x}}_{\mathrm{MAP}}(\mathfrak{y})} \quad (135)$$

Evidently, the MAP estimate depends critically on the *a priori* probability $p_x(\mathfrak{x})$. With the use of an appropriately defined $p_x(\mathfrak{x})$, MAP methods are particularly useful in situations when the information provided by observations alone is inadequate to generate a good estimation of the underlying parameters, such as when the observations are extremely noisy or incomplete. Such situations are often encountered in emission tomography and MAP methods are widely used for reconstruction of emission images. In general, the MAP equation is highly nonlinear, especially when nontrivial *a priori* probabilities are used and analytical solutions are not feasible. Therefore, the MAP solution is typically obtained by use of iterative algorithms, which generate a sequence of estimates that converges to the true solution.

B. Maximum Likelihood Estimation

In many situations, we may have nonrandom but unknown parameters or we may not have sufficient information to assign an *a priori* probability to the unknown parameter. In these cases, the *maximum-likelihood (ML) estimate* given by

$$\hat{\mathbf{x}}_{ML}(\mathfrak{y}) = \arg\max_{\mathfrak{x} \in \mathfrak{X}} p_{y|x}(\mathfrak{y}|\mathfrak{x}) = \arg\max_{\mathfrak{x} \in \mathfrak{X}} \log p_{y|x}(\mathfrak{y}|\mathfrak{x}) \quad (136)$$

where $p_{y|x}(\mathfrak{y}|\mathfrak{x})$ is known as the *likelihood function* in the literature, has been found useful. A necessary condition of the ML estimate is that it must satisfy the *likelihood equation*:

$$\frac{\partial}{\partial \mathfrak{x}} \log p_{y|x}(\mathfrak{y}|\mathfrak{x}) \bigg|_{\mathfrak{x} = \hat{\mathbf{x}}_{ML}(\mathfrak{y})} = 0 \quad (137)$$

A comparison between Eqs. (134) and (136) indicates that the ML estimate corresponds to the limiting case of a MAP estimate in which the *a priori* probability $p_x(\mathfrak{x})$ is constant. As with MAP solutions, the ML solution is typically obtained by using iterative methods.

C. Bias and Variance

Bias and variance are two statistics that are commonly used for quantifying the performance characteristics of given estimates. Suppose the actual value of x is known and equal to $\mathfrak{x}$. The quantity

$$E(\hat{x}|\mathfrak{x}) = \int \hat{x}(\mathfrak{y}) p_y(\mathfrak{y}|\mathfrak{x}) d\mathfrak{y} \quad (138)$$

is therefore the expected value of the estimate $\hat{x}$. The bias and variance of the estimate $\hat{x}$ are defined as

$$\text{Bias}(\hat{x}|\mathfrak{x}) = E\{\hat{x}|\mathfrak{x}\} - \mathfrak{x}$$
$$\text{Var}(\hat{x}|\mathfrak{x}) = E\{(\hat{x} - E\{\hat{x}|\mathfrak{x}\})^2\} \quad (139)$$

Therefore, the bias measures the *accuracy*, that is, the difference between the estimate and the true on average, of an estimate, whereas the variance measures its *precision*, that is, the spread of the estimate. Clearly, a good estimate should have simultaneously small bias and variance. Unfortunately, it is typically true that the bias and variance of an estimate cannot be reduced at the same time. Rather, one can only reduce the bias at the expense of increasing the variance, and vice versa. Therefore, when evaluating the performance of an estimate one needs to consider both bias and variance at the same time.

Two important concepts in characterizing estimates are unbiasedness and efficiency. An estimate $\hat{x}$ is called *unbiased* if bias$\{\hat{x}|\mathfrak{x}\} = 0$. Unbiased estimators are desirable because they will, on average, generate accurate results. However, an unbiased estimator having a large variance will generate results with large fluctuations; therefore, such an estimator does not necessarily generate an accurate result from a given observation. In view of this, it is favorable to have an unbiased estimator having the minimum possible variance. An important theorem in the estimation theory states that, provided that $\partial \log p_{y|x}(\mathfrak{y}|\mathfrak{x})/\partial \mathfrak{x}$ and $\partial^2 \log p_{y|x}(\mathfrak{y}|\mathfrak{x})/\partial \mathfrak{x}^2$ exist and are absolutely integrable, if x is an unbiased estimate of x, then

$$\text{var}(\hat{x}|\mathfrak{x}) \geq I^{-1}(x|\mathfrak{x}) \quad (140)$$

where $I(x)$ is defined by

$$I(x|\mathfrak{x}) = E\left\{\left(\frac{\partial \log p_{y|x}(\mathfrak{y}|\mathfrak{x})}{\partial \mathfrak{x}}\right)^2\right\} = -E\left\{\frac{\partial^2 \log p_{y|x}(\mathfrak{y}|\mathfrak{x})}{\partial \mathfrak{x}^2}\right\} \quad (141)$$

In the literature, $I(x|\mathfrak{x})$ is known as the the *Fisher's information* and Eq. (140) is called the *Cramer-Rao (CR) inequality*.

An unbiased estimate that satisfies the CR bound with an equality is called an *efficient* estimate. One reason that motivates the use of ML estimate in the case of nonrandom parameters is the observation that an efficient estimate, if it exists, must be an ML estimate. The converse, however, is generally not true. In addition, under general conditions it can also be shown that an ML estimate is (1) *consistent* (i.e., it converges in probability to the true parameter), (2) *asymptotically unbiased*, and (3) *asymptotically efficient*. Readers can consult Poor (1994) for more detailed information on this subject.

VIII. OBJECTIVE ASSESSMENT OF IMAGE QUALITY

The assessment of image quality is an important and often misunderstood area of medical imaging research. Although representative images and resolution or noise property measurements are frequently presented to illustrate the strengths of some new system or image-processing algorithm, these measures do not provide complete and objective characterization of image quality. It is generally agreed that the objective assessment of image quality must be task-based; quality is measured on the basis of how well an imaging system allows observers to perform a specified task (International Commission on Radiation Units and Measurements [IC], 1996; Barrett, 1990).

Imaging tasks can, in general, be divided into two principal categories: estimation tasks and classification tasks. Estimation tasks involve the use of an image or sequence of images to determine the value of one or more numerical parameters of interest, such as the average activity level in a brain region of interest (ROI) or a patient's cardiac ejection fraction. Classification involves determining to which of two or more categories an image belongs. The simplest example of a classification task is the detection of a specified signal, in which the aim is simply to determine whether the image belongs to the class of images containing the signal of interest or to the class of images not containing the signal. This section focuses exclusively on the detection task to give a somewhat detailed exposition of one specific approach to objective assessment of image quality. Discussions of additional classification and estimation tasks can be found in IC (1996) and Barrett (1990).

The most common question in assessing medical images is how well the images allow a human observer to perform the specified task, such as detection of a lung nodule. The most meaningful and reproducible measure of such performance is obtained by use of *receiver operating characteristic* (ROC) analysis, in which the trade-off between false-positive and true-positive classifications is plotted for an observer using an increasingly permissive detection threshold.

A. ROC Analysis

Imagine that an observer is presented with a stack of images and asked to determine, for each image, which of two hypotheses is true. The first hypothesis, denoted H_0 and called the negative or null hypotheses, corresponds to a statement such as "a lesion does not exist at a particular location on the image." The second hypothesis, denoted H_1 and called the positive hypothesis, corresponds to the converse statement "a lesion does exist at a particular location on the image." For each image, the observer calculates, either explicitly or implicitly, a scalar quantity $z \in \mathbb{R}$ that represents his or her degree of agreement with hypothesis H_1. The observer then compares this quantity to some threshold γ and decides in favor of hypothesis H_1 if $z > \gamma$ and in favor of hypothesis H_0 if $z \leq \gamma$.

In general, the observer will classify some of the images correctly and some of them incorrectly. For a particular value of the threshold γ, classification performance can be summarized by the following two quantities:

$$\begin{cases} FPF = \int_{Z_1} p(z|H_0) \, dz \\ TPF = \int_{Z_1} p(z|H_1) \, dz \end{cases} \quad (142)$$

where Z_1 is the set $\{z > \gamma\}$. Clearly, *FPF* and *TPF* are the probabilities of a false positive and of a true positive, respectively. There is, of course, a correlation involved between these two quantities. Using a more permissive threshold will, in general, admit more true positives but will also admit more false positives. A plot of all possible pairs of *TPF* and *FPF* obtained as the threshold γ is varied is called an ROC curve and is commonly used for comparing imaging systems and assessing observer performance. Figure 11 shows three example ROC curves. Because at the same *FPF* it is desirable for a decision to generate a higher *TPF*, curve 1 in Figure 11 is often considered to exhibit the best performance among the three curves shown.[4] For the

---

[4] More generally, the performance of a curve should be evaluated by considering the overall risk, defined as $R = C_{00}P_0\Pr(H_0|H_0) + C_{10}P_0\Pr(H_1|H_0) + C_{11}P_1\Pr(H_1|H_1) + C_{01}P_1\Pr(H_0|H_1)$ where $\Pr(H_i|H_j)$ and C_{ij} are, respectively, the probability of and the risk in making a H_i decision while the truth is H_j, and P_i is the probability of occurrence (the prevalence) of H_i, $i,j = 0, 1$. The prevalences and the risks can drastically change the performance interpretation of a ROC curve.

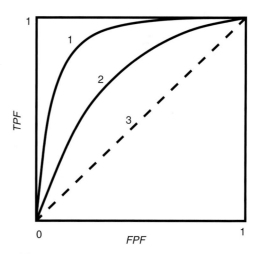

FIGURE 11   Example ROC curves.

same reason, the performance of a curve is also customarily characterized by the area A_z under it in the *FPF-TPF* plot, with a larger A_z ($A_z \leq 1$) suggesting a better performance.

B. Ideal Observer: Likelihood Ratio

Because ROC experiments are time-consuming, they are impractical for use in system design and optimization, in which one often wishes to study the dependence of image quality on numerous system parameters. In this situation, it is desirable to consider the use of mathematical observers implemented on computers that attempt to reproduce the performance of human observers in some way. The simplest, although in some way most sophisticated, model observer is the ideal or Bayesian observer (IC, 1996; Barrett *et al.*, 1993).

Like the human observers previously discussed, all model observers can be thought of as performing a sequence of operations on the measured image to obtain a scalar decision variable z. The observer then compares this decision variable to a predetermined threshold in order to reach a decision about which category an image belongs to. The ideal observer is assumed to know the conditional probability densities of the data for the two categories and uses the logarithm of the ratio of these densities—the log likelihood ratio—as the decision variable z. Specifically, if we adopt a discrete notation for the measured image, representing it through lexicographical ordering of pixel values by an $N \times 1$ vector **g**, the ideal observer's decision variable is given by

$$z(\mathbf{g}) = \log \left[\frac{p(\mathbf{g}|1)}{p(\mathbf{g}|2)} \right] \quad (143)$$

where $p(\mathbf{g}|i)$ is the likelihood function for the *i*th category; that is, it is conditional probability density of obtaining the observed image **g** for the *i*th category. The ideal observer thus has complete information about the odds of a given category producing a given image and can use this information to achieve the best achievable classification performance. The ideal observer performance thus represents an upper bound on the performance of any observer.

In the case of an LSI system applied to the detection of a known signal in a known background (the signal known exactly/background known exactly [SKE/BKE] task), the ideal observer performance can be computed in closed form and related to basic image resolution and noise measures. This is most simply expressed in the Fourier domain and through the adoption of a continuous notationz (IC, 1996). Defining the square of the ideal observer signal-to-noise ratio (SNR) to the square of the difference of the means of the decision variable under each hypothesis (signal present or signal absent) divided by the average of the decision variable variances, one can show that

$$\text{SNR}_I^2 = \iint_{-\infty}^{\infty} |F(v_x, v_y)|^2 \left\{ \frac{K^2 \text{MTF}^2(v_x, v_y)}{S_n(v_x, v_y)} \right\} dv_x dv_y \quad (144)$$

where K is the system's large-scale transfer function, MTF (v_x, v_y) is the system modulation transfer function, $S_n(v_x, v_y)$ is the system noise power spectrum, and $F(v_x, v_y)$ is the spectrum of the signal of interest.

C. Noise Equivalent Quanta and Detective Quantum Efficiency

Equation (144) has a pleasing structure in that it represents a frequency–space integration over the product of two factors, one of which, $(|F(v_x, v_y)|^2)$, represents the signal or task of interest and the other of which (the terms in curly braces) represents inherent properties of the imaging system (Barrett *et al.*, 1995). If the signal and power spectrum are expressed not in absolute but in relative units (such as fractional change in activity level), we obtain

$$\text{SNR}_I^2 = \iint_{-\infty}^{\infty} |[F(v_x, v_y)]_{\text{rel}}|^2 \left\{ \frac{\text{MTF}^2(v_x, v_y)}{[S_n(v_x, v_y)]_{\text{rel}}} \right\} dv_x dv_y$$

(the factor of K^2 is canceled out by the change of unit factors), and the set of terms in curly braces in referred to as the *noise equivalent quanta* (NEQ) of the system (IC, 1996; Barrett *et al.*, 1995). The name arises because the factor has units of quanta per unit area and can be thought of as the quantum value of the image at that particular spatial frequency; regardless of how many quanta actually contributed to the image at frequency (v_x, v_y), the image has noise levels equivalent to that of one formed with NEQ(v_x, v_y) quanta. The ratio of NEQ(v_x, v_y) to the actual number of quanta contributing $(\bar{Q})$ measures how efficiently the system uses incident quanta and is known as the *detective quantum efficiency* (DQE) (IC, 1996):

$$\text{DQE}(v_x, v_y) = \text{NEQ}(v_x, v_y) / \bar{Q} \quad (145)$$

Bear in mind that this definition of NEQ emerges from consideration of an LSI system applied to the SKE/BKE task. For nonlinear but linearizable systems, such as screen-film radiography operating under the small signal assumption, a very similar expression can be derived that incorporates relevant linearizing constants. For more general tasks and systems, the mathematical definition given here no longer applies. However, expressions for ideal and other model observers' SNRs can often be factored into components depending only on the task and components depending only on the system properties, and in these cases the system property factors can often be usefully considered as a generalized NEQ (Barrett *et al.*, 1995).

D. Model Observer: (Channelized) Hotelling Observer

The ideal observer may provide an upper bound on the performance of a human observer, but it is not, in general, a particularly good predictor of actual human performance. Moreover, calculation of the ideal-observer performance is not generally tractable, except in the special case presented in the previous section, when the ideal observer's operations on the data happen to be linear.

Given the computational and predictive shortcomings of the ideal observer, a reasonable alternative strategy is to constrain attention to model observers that perform only linear operations on the data. Using only sample statistics, the optimal linear classifier is given by the Fisher discriminant (Barrett *et al.*, 1993). However, the Fisher discriminant involves inversion of a covariance matrix that has dimensions $N \times N$ and thus requires more than N^2 realizations in the sample population for the inverse of the sample estimate to exist. When dealing with images, where N equals the total number of image pixels, this is an unreasonable demand. (Barrett *et al.*, 1993).

A more tractable approach is to make use of simulated images and consider a model observer—the Hotelling observer (Barrett *et al.*, 1993)—that works with ensemble rather than sample statistics. The observer works with a decision variable given by

$$z_{\text{Hot}}(\mathbf{g}) = (\bar{\mathbf{g}}_2 - \bar{\mathbf{g}}_1) S_2^{-1} \mathbf{g}$$

where $\bar{\mathbf{g}}_k$ is the class mean for the kth class and S_2 is the intraclass scatter matrix, given by

$$S_2 = \sum_{k=1}^{2} P_k \langle (\mathbf{g} - \bar{\mathbf{g}}_k)(\mathbf{g} - \bar{\mathbf{g}}_k) \rangle_k$$

where P_k, $k = 1, 2$, is the probability of occurrence for the kth class, and $\langle \cdot \rangle_k$ represent an ensemble average over all objects in class k and all noise realizations.

Although the Hotelling observer is usually mathematically tractable and better tracks human observer performance than the ideal observer, it still fails to match human performance when confronted with postfiltering of images that introduces noise correlations (Barrett *et al.*, 1993). One modification that provides better performance in this context is to introduce a channel model of vision into the Hotelling observer model (Barrett *et al.*, 1993). In this case, the Hotelling discriminant is applied not to the images themselves but rather to images that have been integrated over a relatively small number of spatial frequency channels, a well-established feature of the human visual system. This channelized Hotelling observer has been found to correlate well with most known psychophysical results (Barrett *et al.*, 1993).

References

Barrett, H. H. (1990). Objective assessment of image quality: effects of quantum noise and object variability. *J. Opt. Soc. Am. A* **7**: 1266–1278.

Barrett, H. H., Denny, J. L., Wagner, R. F., and Myers, K. J. (1995). Objective assessment of image quality II: Fisher information, Fourier crosstalk, and figures of merit for task performance. *J. Opt. Soc. Am. A* **12**: 834–852.

Barrett, H. H., Yao, J., Rolland, J. P., and Myers, K. J. (1993). Model observers for assessment of image quality. *Proc. Natl. Acad. Sci. U.S.A.* **90**: 9758–9765.

Bracewell, R. N. (1965). "The Fourier Transform and Its Applications." McGraw-Hill, New York.

Chui, C. K. (1992). "An Introduction to Wavelets." Academic Press, San Diego.

Galatsanos, N. P., and Katsaggelos, A. K. (1992). Method for choosing the regularization parameter and estimating the noise variance in image restoration and their relation. *IEEE Trans. Imag. Proc.* **1**: 322–336.

Galsko, V. B. (1984). "Inverse Problems of Mathematical Physics." American Institute of Physics, New York.

Green, P. J., and Silverman B. W. (1994). "Nonparametric Regression and Generalized Linear Models. Chapman Hall, London.

Hoffman, E. J., Cutler, P. D., Digby, W. M., and Mazziotta, J. C. (1990). 3-D phantom to simulate cerebral blood flow and metabolic images for PET. *IEEE Trans. Nucl. Sci.* **37**: 616–620.

Hutchinson, M. F., and de Hoog, F. R. (1985). Smoothing noisy data with spline functions. *Numer. Math.* **47**: 99–106.

International Commission on Radiation Units and Measurements (IC). (1996). "Medical Imaging: The Assessment of Image Quality." Bethesda, MD.

Jain, A. K. (1989). "Fundamentals of Digital Image Processing." Prentice-Hall, Englewood Cliffs, NJ.

La Rivière, P. J., and Pan, X. (2000). Nonparametric regression sinogram smoothing using a roughness-penalized Poisson likelihood objective function. *IEEE Trans. Med. Imaging* **19**: 773–786.

Lehmann, T. M. (1999). Survey: Interpolation methods in medical imaging processing. *IEEE Trans. Med. Imaging* **18**: 1049–1075.

Papoulis, A. (1991). "Probability, Random Variables, and Stochastic Processes," 3rd ed. McGraw-Hill, New York.

Poor, H. V. (1994). "An Introduction to Signal Detection and Estimation," 2nd ed. Springer-Verlag, New York.

Press, W. H., Teukolsky, S. A., Vetterling, W. T., and Flannery, B. P. (1992). "Numerical Recipe in C," 2nd ed. Cambridge University Press, New York.

Thèvenaz, P. (2000). "Interpolation revisited [medical images applications]." *IEEE Trans. Med. Imaging* **19**: 739–758.

Unser, M. (2000). Sampling—50 years after Shannon. *Proc. IEEE* **88**: 569–587.

Vaseghi, S. V. (2000). "Advanced Digital Signal Processing and Noise Reduction," 2nd ed. John Wiley & Sons, Chichester, U.K.

Wahba, G. (1990). "Spline Models for Observational Data." SIAM Press, Philadelphia.

CHAPTER 7

Single-Photon Emission Computed Tomography

GENGSHENG LAWRENCE ZENG,* JAMES R. GALT,[†] MILES N. WERNICK,[‡]
ROBERT A. MINTZER,[§] and JOHN N. AARSVOLD[¶]

Radiology Research, University of Utah, Salt Lake City, Utah
[†] *Department of Radiology, Emory University, Atlanta, Georgia*
[‡] *Departments of Electrical and Computer Engineering and Biomedical Engineering, Illinois Institute of Technology, and Predictek, LLC, Chicago, Illinois*
[§] *Department of Radiology, Emory University, Atlanta, Georgia*
[¶] *Nuclear Medicine Service, Atlanta VA Medical Center, Department of Radiology, Emory University, Atlanta, Georgia*

I. Planar Single-Photon Emission Imaging
II. Conventional Gamma Cameras
III. Tomography
IV. SPECT Systems
V. Tomographic Single-photon Emission Imaging
VI. Other Detectors and Systems
VII. Summary

Single-photon emission computed tomography (SPECT) is a diagnostic imaging technique in which tomographs of a radionuclide distribution are generated from gamma photons detected at numerous positions about the distribution. SPECT, as routinely performed in nuclear medicine clinics, uses for photon detection/data acquisition an imaging system composed of one or more rotating gamma cameras (Fig. 1). In a process known as *image reconstruction*, tomographs are computed from the data using software that inverts a mathematical model of the data acquisition process.

In this chapter, we focus on data acquisition in SPECT, leaving the details of image reconstruction and analysis for later chapters (Kinahan *et al.*, Chapter 20; Lalush and Wernick, Chapter 21; King *et al.*, Chapter 22; and Garcia, Chapter 24 in this volume). The reader, however, should bear in mind that the processes of data acquisition and image reconstruction are inextricably intertwined and should not be considered in isolation or in the absence of consideration of other important factors in SPECT, such as radioisotopes and their emission properties (e.g., energy and half-life), pharmaceuticals and their biodistribution properties (e.g., uptake and washout rates), and the living body and its physical properties (e.g., absorption and scattering properties) and biological characteristics (e.g., the motions of life—the beating heart, flowing blood, and expanding and contracting lungs).

The reader is encouraged to be constantly mindful that clinical SPECT, practiced daily for 3 decades, is the result of clever identification and exploitation of science, mathematics, and engineering principles that result in the generation of clinically useful images of human physiology. Through careful attention to the many factors that influence the processes and outcomes of SPECT, one can appreciate the evolution of gamma cameras and SPECT systems and gain insight that leads to novel investigative research.

I. PLANAR SINGLE-PHOTON EMISSION IMAGING

There are two common forms of single-photon emission imaging: planar and tomographic. A planar image depicts a single view (a projection) of a radiotracer distribution in a patient; a tomographic image is a slice or volume image of the radiotracer distribution computed from multiple images acquired from multiple camera positions. Both imaging methods are used routinely in nuclear medicine clinics, and both use a gamma camera to collect the data. Planar single-photon imaging requires a gamma camera and a means

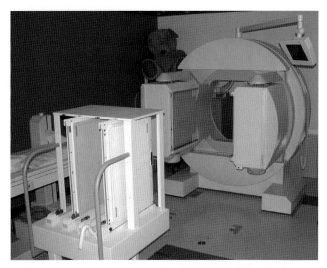

FIGURE 1 Dual-head SPECT system: two gamma cameras, gantry, imaging table, collimators, and collimator cart. Planar imaging is performed with the two gamma cameras in fixed positions. The two gamma cameras are separated by 180°, positioned at 0° and 180°. The system is variable angle, implying the two heads can be positioned at any two angular positions. The imaging table is set left of gantry center so that the collimator cart in the foreground can be positioned for the exchange of collimator sets. SPECT data acquisition involves rotation of the two cameras about a patient lying on the imaging table.

of displaying the acquired images; tomographic imaging requires a camera, a method of display, a gantry for rotating the camera about the patient, and a means of performing image reconstruction.

To illustrate basic uses of gamma cameras and provide examples of planar images, we begin with a brief discussion of three of the most commonly performed planar-imaging studies: thyroid studies, ventilation/perfusion (V/P or V/Q) studies, and whole-body bone studies.

A. Thyroid Studies

Most thyroid studies are performed using ^{123}I-NaI (sodium iodide) administered orally in capsule form (7.4–14.8 MBq; 200–400 µCi). Because time is required for the radiotracer to progress from the stomach through blood to the thyroid, the radiotracer is allowed 24 hours to accumulate in the thyroid (and clear from most of the rest of the body) before imaging commences.

Study imaging using a gamma camera with a pinhole aperture generally takes 20–60 min and involves acquisition of three planar images taken from three different views: anterior, left anterior oblique (LAO), and right anterior oblique (RAO). Normal thyroid anatomy and physiology yield images showing homogeneous uptake throughout the thyroid, whereas abnormal thyroid anatomy or physiology yields heterogeneous uptake.

A planar thyroid study is a relatively simple nuclear medicine study, yet even it requires significant attention to detail.

For example, (1) a thyroid's uptake of iodine may be affected by the patient's consumption of foods, vitamins, medications, or radiological contrast agents; (2) positioning of the pinhole aperture must be such that the entire thyroid is within the camera's field of view (FOV), with appropriate magnification; and (3) high-quality diagnostic images of thyroids require that patients be positioned in very uncomfortable neck-extended positions. Factors such as these, which are not always immediately appreciated, must be addressed for successful performance of many emission studies.

B. Ventilation/Perfusion Studies

The V/P lung study, which is somewhat more technically complicated than the thyroid study, involves two successive studies—a ventilation study and a perfusion study—yielding two sets of images. Interpretation of the images includes nuclear medicine physician comparison of the two sets of images, specifically the comparison of the uptakes in the 12 segments of each lung in each study. The purpose of the V/P study is detection of pulmonary emboli (PEs). Mismatches of segment uptakes between ventilation and perfusion studies suggest the presence of PEs (Fig. 2).

In the ventilation study, the patient breathes an aerosol of ^{99m}Tc-DTPA (diethylenetriaminepentaacetic acid; 37 MBq; 1 mCi) for several minutes, after which the distribution of the DTPA retained in the patient's lungs is imaged. Image acquisition is generally count-based, with each image acquisition ending when 200,000 gamma photons have been acquired (~5 min). Immediately thereafter, a perfusion study begins. In the perfusion study, the patient is administered ^{99m}Tc-MAA (macroaggregated albumin; 80–180 MBq; 2–5 mCi) via intravenous (IV) injection and the lungs are again imaged (750,000 total counts/image; ~5 min). This perfusion study can be conducted immediately after the ventilation study, in part because the MAA uptake is significantly greater than the retained DTPA dose and therefore contributes most of photons to the perfusion images.

Typical ventilation and perfusion studies include anterior, posterior, LAO, RAO, left lateral (LL), and right lateral (RL) views. V/P studies are often performed using systems with two gamma cameras so that total data acquisition time can be reduced through simultaneous acquisition of some of the views. Total study time is 30–60 min. Gamma cameras with collimators (discussed later), rather than pinhole apertures, are used in most V/P studies.

C. Whole-Body Bone Studies

Whole-body bone studies start with IV injection of ^{99m}Tc-MDP (methylene diphosphonate; 740–1110 MBq; 20–30 mCi) or a similar compound, with imaging commencing 2–5 hours after injection. Uniform skeletal uptake usually indicates a normal study. Focal uptake (localized uptake that

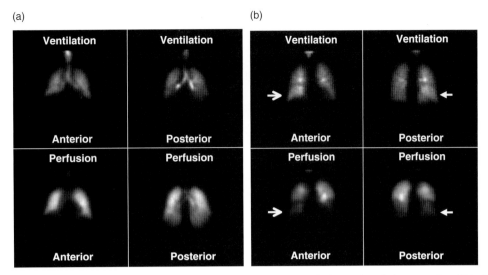

FIGURE 2 V/P lung studies. (a) Anterior and posterior views of ventilation and perfusion scans in which the ventilation and perfusion uptakes are matched. The matched nature of the uptakes suggests the probability is low that the patient has one or more PEs. (b) Anterior and posterior views of ventilation and perfusion scans in which ventilation and perfusion uptakes are not matched. The mismatched nature of the uptakes suggests the probability is high that the patient has one or more PEs (the arrows on the images indicate lobes with mismatched uptakes).

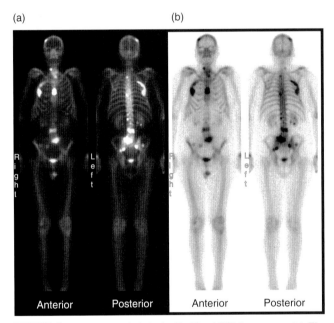

FIGURE 3 Abnormal whole-body Tc-99m MDP bone scan. (a) The white-on-black and (b) black-on-white images represent identical data. The two pairs of images have been displayed with different color tables and different gray scales. Each of the white regions in (a) and black regions in (b) represent abnormally high uptake and, in this case, indicate locations at which the patient has bone metastases.

is different in magnitude from uptake in neighboring bone) may represent an abnormality. If the focal uptake is greater than uptake in neighboring bone, this may indicate arthritis, a fracture, or a metastasis (Fig. 3). Focal uptake that is less than uptake in neighboring bone may indicate a necrotic tumor, a lytic lesion, or a sequela of radiation therapy.

No commercially available gamma camera is large enough to acquire an image of an entire average-size adult without moving the camera or the patient. Thus, whole-body bone scans are performed by moving the camera along the long axis of a patient or, equivalently, by moving the patient longitudinally past the camera. To perform a bone scan using the system in Figure 1, the cameras are positioned at 90° and 270°, and the patient is positioned supine on the table with feet toward the gantry. Then the table and patient are moved into the acquisition start position, which places the patient's head in the FOV of the cameras. During the study, the patient and table move so that the patient is imaged from head to foot, in that order. Note that acquisition of whole-body images that represent accurately the relative values of radiotracer uptake in the head, the chest, the abdomen, and the feet requires accurate encoding and matching of acquisition and motion parameters. Most whole-body bone scans are performed with systems comprising two gamma cameras so that anterior and posterior views can be acquired simultaneously.

D. Other Nuclear Medicine Studies

In these introductory examples, we have mentioned just a small fraction of the isotopes, radiopharmaceuticals, and imaging studies that relate to planar single-photon emission imaging. Some additional isotopes, radiopharmaceuticals, and studies are listed in Wernick and Aarsvold (Chapter 2, Tables 1 and 4, in this volume). More extensive discussions of clinical nuclear medicine studies can be found in Henkin *et al.* (1996), Mettler and Guiberteau (1991), Sandler *et al.* (2003), and Taylor *et al.* (2000).

We note that planar imaging is not limited to static imaging of radiotracers. The assessment of function is sometimes more easily accomplished through evaluation and comparison of image sequences rather than through assessment of a single image. Such is the case if certain radiotracers are used to assess organs such as the kidneys, the heart, or the brain. An example is renal imaging for assessment of kidney function using ^{99m}Tc-DTPA or ^{99m}Tc-MAG3 (mercaptoacetyl-triglycine). In a study of this kind, one observes how the body processes the radiotracer; that is, how the tracer is extracted from the blood by the kidney and excreted from the kidney through the ureters to the bladder in urine (Taylor *et al.*, 2000).

As planar imaging is not limited to static imaging, so also nuclear medicine is not limited to imaging. In some situations, numerical parameters only are evaluated instead of images. Although this book focuses on tomographic imaging, it is important to recognize that simple planar imaging, or even parameter estimation without the display of any images at all, may be more appropriate for a given task than is generation of images.

Having introduced some of the imaging studies that can be performed using a standard gamma camera, we now discuss the workings of the gamma camera, the principal component of a SPECT system.

II. CONVENTIONAL GAMMA CAMERAS

A. The Anger Camera

In 1958, Hal Anger, a scientist at the University of California at Berkeley, developed an instrument for imaging gamma emissions (Anger, 1958, 1964, 1967, 1974; Anger and Davis, 1964). Although many innovations have been made since 1958, today's clinical gamma cameras are often called *Anger cameras* because they share many of the essential features of Anger's early designs. The basic operation of an Anger camera is as follows (see Fig. 4).

First, an aperture or collimator mechanically selects for possible detection gamma photons traveling in specific directions. An aperture or collimator does this by absorbing gamma photons traveling in directions other than those specified by the aperture or collimator (Figs. 5 and 6). (The principle of image formation by collimation is introduced in Wernick and Aarsvold, Chapter 2 in this volume, and is further discussed later in this chapter. The reader may also refer to Gunter, Chapter 8 in this volume, for a detailed discussion of collimators.) The selected gamma photons then encounter a scintillation detector. Some pass through the detector without interacting with it. Those that do interact with the detector generate electronic signals used to estimate the location (the spatial coordinates in the image

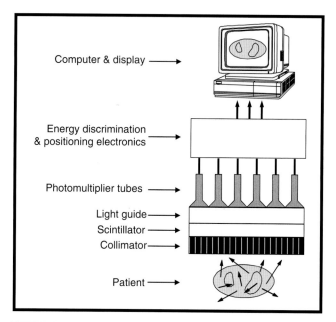

FIGURE 4 Fundamental components of a conventional gamma camera. Most gamma cameras have a collimator, a scintillation crystal, a light guide, an array of photomultiplier tubes, radiation shielding, energy discrimination and positioning electronics, and a computer and display for acquisition, processing, and display of data and images.

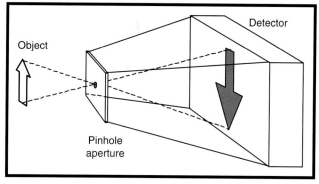

FIGURE 5 Imaging using a single-pinhole aperture. Pinhole imaging with gamma photons is fundamentally the same as pinhole imaging with light photons. The primary differences are the size of the pinhole (generally larger for gamma photon imaging because gamma photons are at a premium in most emission imaging settings) and the material (lead alloy or tungsten for gamma photon imaging).

plane) of the photon-detector interaction and to estimate the energy deposited by the photon.

Detected gamma photons that exhibit energy that is less than the known energy of the primary emission of the radioisotope being imaged are usually rejected. Diminished energy is an indication that a photon has undergone a scattering interaction inside the patient or collimator or detector and, thus, has been deflected from its original path. Photons with diminished energy have limited information about

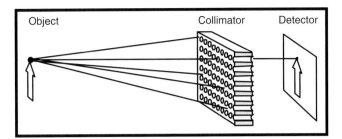

FIGURE 6 Imaging using a parallel-hole collimator. An object emits gamma photons in all directions. An image is formed primarily from those photons that have paths of travel that are parallel to the bores of the collimator—such paths are also paths normal to the detector surface. Most photons that reach the collimator will not be traveling the required paths and thus will be absorbed by the collimator.

FIGURE 7 Scintillation-detector components. Displayed are components of a scintillation detector of a gamma camera, including scintillation crystal, light guide with masking, circular-face PMTs, magnetic shielding of PMTs, and signal-processing electronics.

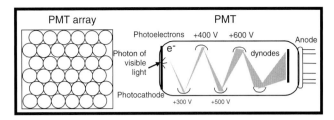

FIGURE 8 Schematics of PMT packing and PMT internal structures. PMT arrays are generally packed such that every nonedge PMT has six neighbors. Each PMT has a photocathode, 6–10 dynodes, and an anode. During operation of a PMT, voltage differences are maintained among its photocathode, dynode stages, and anode.

their original point of origin, and thus their inclusion in an image reduces the quality of the image unless further processing is performed.

The detection process itself consists of two steps. In the first step, a gamma photon that survives collimation interacts with and deposits energy in the scintillation crystal. This energy is converted to multiple visible light photons. The light photons propagate through the crystal and light guide to an array of photomultiplier tubes (PMTs) (Fig. 7). The PMTs detect the light photons; PMTs are sensitive high-voltage devices that produce measurable electrical current from as little as a single photon of light (Fig. 8). Each PMT outputs an electrical current proportional to the number of light photons detected. We note radiation shielding and light shielding are required for the operation of a scintillation detector.

The light output from the scintillator is generally spatially broad and registered by several PMTs. Dedicated electronics and software are used to infer the likely point of impact of the gamma photon based on the output of each PMT in the array. Historically, this was a simple centroid calculation performed entirely with hardware; it is now done more precisely using statistical estimation techniques implemented using combinations of hardware and software and measured calibration data.

In keeping with digital imaging technology, the image formed by a standard gamma camera image is represented on a grid of pixels. The value assigned to each pixel is the number of gamma photons that have been detected within the pixel's spatial extent. Thus, the image produced by a gamma camera is actually a histogram of the spatial locations of all counts detected. Kao *et al.* (Chapter 6, Fig. 8, in this volume) illustrates the appearance of planar gamma-ray images that have different numbers of recorded photons. As the number of detected unscattered gamma photons increases, image noise decreases; thus, it is important to detect as many unscattered gamma photons as possible.

We next discuss the details of the various components of gamma cameras.

B. Principles of Collimation

In an ordinary photographic camera, an image is formed by a lens, which diverts light rays by refraction. A lens is an efficient imaging instrument; every ray that enters the lens from an object is diverted to its desired location in the image. Unfortunately, gamma rays are too energetic to be significantly refracted; therefore cruder methods must be used for image formation.

A pinhole aperture for gamma-ray imaging works on the same principle as the simple box camera one might use to view a solar eclipse. Known originally as a *camera obscura*, this method of image formation dates back to the 1500s. As shown in Figure 5, all rays coming from the object are mapped into their respective locations in the image. The main disadvantage of this approach is that only those rays passing through the tiny opening are used. Other rays are lost; thus, the sensitivity of this method is very low. If one enlarges a pinhole, sensitivity increases but spatial resolution degrades. In addition, pinhole apertures produce significant image distortion.

A more common device for image formation with gamma photons is the collimator. As shown in Figure 6, a parallel-hole collimator is an array of parallel bores of uniform size, surrounded by septa. A parallel-hole collimator works by passing all photons traveling in (or nearly in) a particular direction. Although use of a collimator yields much greater sensitivity than use of a pinhole aperture, it also imposes a significant limitation on the sensitivity of a gamma camera because it depends on the selective blocking (absorption) of photons.

Ideally, gamma photons not traveling along the direction of the bores (i.e., not traveling normal to the face of the scintillation crystal) will be absorbed. In practice, however, because the bore diameter is not infinitesimally small, there is a small range of incidence angles that will be accepted by the collimator. The larger the bore diameter, the larger will be the range of accepted angles and the poorer will be the resulting spatial resolution of the gamma camera.

So far, we have described collimators and pinholes as if they were capable of blocking all undesired photons. In reality, however, some photons will pass unhindered through the material of the collimator or pinhole and others will scatter within the collimator material and be absorbed by the scintillator. In both instances, undesired photons may be included in the final image, leading to degraded image quality.

In spite of the aforementioned shortcomings of collimators, no practical alternative has yet been developed for clinical gamma-photon imaging and collimators remain a critical component of most gamma cameras.

C. Collimator Types

Collimator septa are composed of highly absorbing material, that is, material that has a high atomic number Z and a high density ρ. Alloys of lead ($Z = 82$, $\rho = 11.34$ g/cm^3) are the most common septal materials. Alloys of tungsten ($Z = 74$, $\rho = 19.4$ g/cm^3) and gold ($Z = 79$, $\rho = 19.3$ g/cm^3) are also used, but not widely, because tungsten is difficult to work and gold is relatively expensive.

Collimators with various hole shapes—circular, square, triangular, or hexagonal—have been developed. Hexagonal holes are the most common because they are generally the most efficient (Fig. 9). Although most collimators have parallel holes, some collimators have bore patterns that converge or diverge, as illustrated in Figure 10. Converging collimators magnify an image on a camera face and thus can yield finer resolution and/or higher sensitivity images than those resulting from use of parallel-hole collimators, particularly when a converging collimator is used for imaging objects that are small relative to a detector face. Figure 10b shows a small object imaged with a large-FOV detector and a parallel-hole collimator. In this case, a large portion of the crystal area is unused. On the other hand, in Figure 10a, in

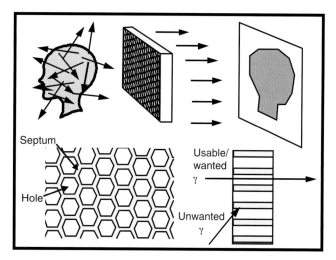

FIGURE 9 Hexagonal parallel-hole collimator. A parallel-hole collimator is a mechanical selector of gamma photons. It passes what become usable photons—photons that enter the crystal from known directions. The collimator absorbs most other photons—those that interact with septa. The resolution and sensitivity of collimators are determined by septal thickness, septal material, hole size, and bore length. Not all photons that pass through the collimator are wanted photons. Some photons passed by a collimator have desired directions of travel but do not have desired energies.

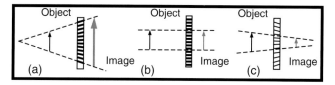

FIGURE 10 Three collimator geometries. The geometries shown are: (a) converging, (b) parallel, and (c) diverging.

which a converging collimator is used, a large-FOV crystal detector is used more efficiently. Due to the increased efficiency resulting from the relationship of the object size and the converging collimator, more gamma photons are acquired using the converging collimator than are acquired using the parallel-hole collimator. The improvement in sensitivity can be used to produce images with more counts or images with higher resolution, depending on the collimator design (Jaszczak, Chang, and Murphy, 1979; Tsui et al., 1986; Hawman and Hsieh, 1986; Nohara et al., 1987; Gulberg et al., 1992; Metz et al., 1980; Tsui and Gulberg, 1990; Sorenson and Phelps, 1987).

If one uses a converging collimator to image a large object, the image data will be truncated (portions of the object will be outside of the field of view) and the image will be significantly distorted (object magnification is dependent on distance from the collimator). Care should be taken when using converging collimators for planar imaging because one needs to mentally compensate for these effects when viewing the images. When converging collimators are used for SPECT, the distortions can be removed during the image reconstruction process, but there is some computational cost in doing so. In clinical use, converging collimators are used primarily for brain SPECT because the brain is relatively small compare to the FOVs of most cameras and because the shape and location of the head allow for close positioning of the camera to all sides of the head.

Four types of parallel-hole collimators find routine use in nuclear medicine clinics: LEHR (low-energy, high-resolution), LEAP (low-energy, all-purpose), MEAP (medium-energy, all-purpose), and HEAP (high-energy, all-purpose). Each designation is an indication of the energy range, and thus the radioisotopes, for which it is designed. Each designation is also an indication of the collimator's trade-off between resolution and sensitivity, which are affected by septal material, bore diameter, bore length, and septal thickness. LE collimators are designed for radioisotopes such as ^{57}Co (122 keV), ^{123}I (159 keV), ^{99m}Tc (140 keV), and ^{201}Tl (69 to 81 keV); ME for radioisotopes such as ^{67}Ga (93 keV, 184 keV, and 296 keV) and ^{111}In (172 keV and 247 keV); and HE for radioisotopes such as ^{131}I (284 keV and 364 keV). All gamma-camera manufacturers offer collimators labeled with these designations, and similar labels indicate similar uses and similar characteristics. Parameters of similarly labeled collimators, however, need not be identical nor are the imaging characteristics of the collimators necessarily the same. Variations in sensitivity and resolution are often evident when like-labeled collimators are compared. The septa of some collimators are fabricated as rows of accurately bent lead-alloy foils; the septa of others are cast from molten lead alloy. This difference alone leads to different imaging characteristics among like-labeled collimators.

Imaging with an inappropriate collimator can cause serious degradation of image quality. Figure 11 shows two ^{67}Ga scans. One was acquired using LEAP collimators, inappropriate collimators for ^{67}Ga imaging; the other was acquired using MEAP collimators, appropriate collimators for ^{67}Ga imaging. The LEAP images have very limited resolution because the LEAP collimators do not have sufficiently thick septa; many of the 184- and 296-keV emissions of ^{67}Ga that are used for image formation pass through the LEAP septa unhindered, independent of their directions of travel. The result is detection by the crystal of excessive unwanted photons—photons having appropriate energies but indeterminable directions of travel.

Collimator design and performance are discussed in greater detail in Gunter (Chapter 8 in this volume).

D. Scintillation Detection

Photon detection in a gamma camera is performed using a scintillation detector, which comprises a scintillation crystal, a light guide, an array of PMTs, and positioning electronics and software (Fig. 7).

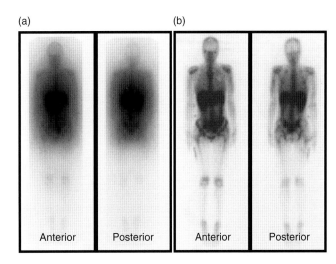

FIGURE 11 Incorrect and correct collimators. (a) Images of a Ga-67 citrate distribution acquired with a LEAP collimator—an incorrect collimator for the imaging of this radioisotope. (b) Images of the same Ga-67 citrate distribution acquired with a MEAP collimator—a correct collimator for the imaging of this radioisotope. Gamma photons of Ga-67 penetrate the septa of the LEAP collimator and the mechanical selection process fails.

A scintillation crystal absorbs gamma photons by one or more collision processes and converts some of their energy into visible light and ultraviolet (UV) photons. This is done through a process know as scintillation. In a sense, the crystal acts as a wavelength shifter. It generates long wavelength radiation (visible and UV) from short wavelength radiation (gamma photons). Because the light output of a crystal depends on the energy of a gamma photon that interacts with it, a gamma camera can be used for energy-selective counting. The thicker the crystal, the more gamma photons it can stop (and thus detect), and the more efficient a detector based on such a crystal will be. However, there is a trade-off: the thicker the crystal, the coarser the intrinsic resolution of a detector with such a crystal is. Despite these issues, it is desirable for a scintillation crystal to have high atomic number and high density so as to increase the probability of gamma photons interacting within the crystal and to increase the production and output of detectable light photons.

In nuclear medicine, the most commonly used scintillation crystal is NaI(Tl) (thallium-doped sodium iodide). NaI(Tl) is a relatively efficient scintillator, producing one visible light photon per approximately 30 eV of gamma-photon energy absorbed. Ideally, a scintillator should be highly absorbing for gamma photons, while being relatively transparent to visible and UV light (otherwise, the crystal will be self-defeating, absorbing the very light it is creating). NaI(Tl) performs well on both counts, producing the highest signal per amount of crystal-absorbed radiation of all scintillators commonly used in nuclear medicine. Most clinical gamma cameras incorporate a single large-area (e.g., 40 × 50 cm) NaI(Tl) crystal. In most cases, the thickness is 3/8 inch, which is sufficient to stop most of the 140-keV photons of ^{99m}Tc. A few systems, designed for use in the imaging of photons with energies up to 511 keV, have 1/2-, 5/8-, or 3/4-inch-thick crystals. A few other systems each have a 1-inch-thick large-area NaI(Tl) crystal with a scored surface—a surface with a grid cut 1/4 to 1/2 inch deep. The thickness of such a crystal ensures significant stopping power; the scoring is a means of controlling optical properties and thus of improving resolution. (For additional information on scintillators, see Wilkinson, Chapter 13 in this volume.)

Once a gamma photon passes through the collimator and is absorbed by the crystal, visible light photons are emitted and propagate throughout the crystal. The image of *an event* (a gamma-photon scintillation-crystal interaction) at the exit face of the crystal is not a tiny bright dot as one might expect but rather is a broadly spatially distributed region of light, with the brightest part of the region coinciding roughly with the location at which the gamma photon interacted with the crystal.

The light output of the scintillator is transferred to an array of PMTs through a plate of optically transparent material called a *light guide*. The thickness, and sometimes the shape and masking, of the light guide are chosen so that the scintillation light from one gamma-photon event is distributed among PMTs in a manner that enables the accurate determination of the location of the interaction for all locations in the crystal.

In a typical gamma camera, there are 37–93 PMTs in a hexagonal array. PMTs are fabricated with glass faceplates, which can be circular, square, or hexagonal. Circular PMTs are shown in Figure 7. A PMT is a vacuum-tube light detector that is an evacuated glass envelope containing an anode, a cathode, and several intermediate electrodes called dynodes (Fig. 8). The cathode (or *photocathode*) is a material coated on the inner surface of the PMT faceplate that photoelectrically absorbs light photons impinging on it. It usually absorbs 20–25% of impinging light photons—a significant percentage. Each such absorption results in the emission, from the photocathode, of one photoelectron. The electrons are called *photoelectrons*, but they are no different than any other electrons (the prefix *photo* simply refers to the manner in which they were liberated). The dynodes are coated with a material that emits secondary electrons when struck by another electron. The first dynode has a higher voltage than the photocathode, and each subsequent dynode has a higher (less negative) voltage than the preceding one. Thus, electrons are propagated through the tube in a stepwise fashion from one dynode to another. At each step, the number of electrons is multiplied.

There may be 8, 10, or more dynodes in a PMT. The strength of the output signal (an electric current) of a PMT is proportional on average to the number of visible light photons that impinge on the PMT's photocathode. As a result, its output is a greatly amplified version of the signal initiated at the photocathode when triggered by light photons from a gamma-photon event in the scintillator.

The sum of all of a gamma camera's PMT outputs that result from a single photon–crystal interaction is proportional to the total energy deposited by the photon. This sum signal is analyzed and only photons within a set energy window (a range of photon energies) or windows are included in the image. All other detected photons—those with energies outside the specified energy range or ranges—are rejected and thus not included in the image. Figure 12 is a typical ^{99m}Tc energy spectrum; a 20% photopeak energy window is shown in light gray about the 140-keV peak representing the primary emission of ^{99m}Tc. Prior to the start of a clinical imaging study, the energy window boundaries are specified. Specification generally requires two parameters: a window median value and a window range specified as a percentage thereof. For example, a window may be specified as a 140 keV, 20% window. This means that the range of accepted energies is 126–154 keV and the full window width is 28 keV, which is 20% of the median or central energy of 140 keV. Imaging of radioisotopes with two primary emissions, such as ^{111}In, usually involves specification of acceptance windows about the energies of both emissions.

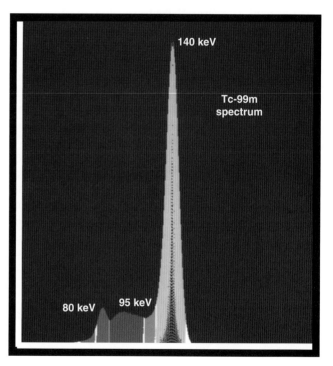

FIGURE 12 Tc-99m spectrum. For most emission imaging, one is interested in the primary emissions of an isotope. Imaging of Tc-99m is usually done using a 20% window about the 140-keV photopeak. Here, the photopeak window is shown in light gray.

E. Event Positioning

The interaction of a gamma photon with the scintillator produces a flash of visible and UV light that strikes several of a scintillation detector's PMTs. Early gamma cameras employed passive circuits to calculate weighted sums of the PMT output signals, producing analog signals x, y, and Z, representing the x and y position of the centroid of the event's scintillation light distribution and the total energy Z of the event. Pulse-height analyzer circuitry then allowed acceptance of only detected events with Z pulses within the photopeak energy range of interest. Only these selected signals were used in images displayed on persistence-phosphor cathode-ray tubes (CRTs) or in the acquisition of integrated images on photographic film exposed by a flying-spot CRT.

Spatial linearity and uniformity of sensitivity (known as flood uniformity) in gamma cameras can be optimized by clever light-guide design and by the tuning of PMT gains, but such purely analog camera designs are limited. Digitization of the x, y, and Z signals not only allows for image histogramming, processing, and storage in computer memory, it also allows for the incorporation of calibration data and software algorithms in linearity and uniformity correction schemes. Methods incorporating such information were developed for commercial systems early on and have been improved with each generation of camera to present cameras. Spatial linearity can be improved by using data collected with gamma source beams at known locations, with corrections stored in lookup tables. System energy resolution can be improved by correcting for variations in total signal response with position. Intrinsic flood uniformity (i.e., detector uniformity in the absence of collimation) and uniformity of detector sensitivity with each collimator can be improved using measured high-count flood data. Modern systems compensate for some of the more significant sources of nonideal characteristics in Anger cameras, such as nonuniformity of PMT photocathode response, variations in optical properties of the scintillator crystal surfaces, and variations in optical properties due to optical design limitations. Compensations for other sources of error continue to be sought.

Current gamma camera systems represent the state of the art in Anger camera design. In recent designs, improvements in integrated circuit technology, electronic component stability, PMT stability and gain stabilization circuitry, and individual PMT signal digitization technology have been combined with significant computer processing speed. The improved technologies and implementations allow gamma cameras to acquire image data at very high event rates while incorporating correction algorithms based on calibration data and sophisticated modeling. Manufacturers probably have nearly achieved the maximum performance possible using Anger's design based on a lead-alloy collimator, a NaI(Tl) scintillation crystal, and an array of PMTs.

F. Photon Interactions

To fully understand the imaging process, it is important to understand the interactions that gamma photons can have within a patient's body and within a gamma camera. Gamma photons emitted from radioisotopes within a patient may escape the patient with no change of energy or direction, or they may interact with the patient's body one or more times prior to their escapes, or they may be fully absorbed by the body. Three forms of interactions affect image formation: photoelectric absorption, Compton scattering, and coherent scattering. Photoelectric absorption, which is sometimes accompanied by emission of low-energy fluorescent X-rays that do not typically escape the patient, is the complete absorption of a photon by an atom. Photoelectric absorption in a gamma camera's collimator and crystal are common interactions of gamma photons that escape an object and interact with a gamma camera during imaging. The absorption of a photon by the collimator means it goes undetected; the absorption of a photon by the scintillation crystal means it is detected.

Compton scattering of photons by electrons changes the directions and the energies of the photons. Coherent scattering, on the other hand, results in a typically lesser change in the direction of propagation and insignificant change in energy. At the gamma-emission energies of radionuclides used in diagnostic nuclear medicine, coherent scatter in tissue accounts for less than a few percent of all scattered photons. It is not easy to characterize the path of a scattered photon and impossible to determine precisely its initial propagation direction. Most scattering events reduce the energy of the photon, a fact that can be used to exclude the undesired contributions of such photons to an image. On the other hand, the energy resolution of scintillation detectors is not sufficient to discriminate all scattered photons from unscattered photons. As a result, Compton scattered photons, as well as coherently scattered photons, are often included in images and cause losses in image contrast when they are. (Gamma photons, traveling in specific directions, pass through a gamma camera's collimator. The camera's scintillation detector absorbs or detects many, if not most, of these. However, not all detected photons are included in images. Some detected photons are rejected because their energies are not within specified energy ranges.)

G. Camera Performance

To conclude our discussion of the gamma camera, we discuss some of the ways that their performance is measured and characterized. Performance measures of gamma cameras are outlined in the National Electrical Manufacturers Association ([NEMA], 2001) standards document NU-1-2001, *Performance Measurements of Scintillation Cameras*. The measures include intrinsic (detector) spatial resolution, spatial linearity, energy resolution, uniformity (integral and differential), multiple-window spatial registration, and count-rate performance. They also include camera (aperture/collimator and detector) spatial resolution with and without scatter, planar and volume sensitivities, uniformity (integral and differential), count-rate performance with scatter, and detector shielding.

The intrinsic spatial resolution of a gamma camera—normally specified in terms of the full width at half-maximum (FWHM) or full width at tenth-maximum (FWTM)—refers to the resolution limitation of the detector only (i.e., the resolution of the position-sensitive detector, assuming perfect collimation). The camera (collimator and position-sensitive detector) spatial resolution is the combined effect of the collimator resolution and the intrinsic resolution. The intrinsic resolution is the result of several factors, including, (1) the noise in the Poisson process of scintillation photon production, (2) the finite energy resolution of the scintillation crystal due to variations in crystal composition and variations in yields of scintillation photons as a function of energy, (3) the finite energy resolution of PMTs due primarily to fluctuations in the photoelectron yield at the first dynode, (4) the variation in mean detector response due primarily to variation in depths of photon-crystal interactions and variations in surface optical properties, and (5) the multiple scatterings of some gamma photons within the crystal result in full depositions of energies but average depositions not located at the site of initial interaction.

The Poisson noise in scintillation photon production is the primary degrading factor of the detector intrinsic resolution. As a consequence, lower energy photons lead to poorer intrinsic resolution because fewer visible light photons are generated, which in turn implies a lower signal-to-noise ratio. In general, the intrinsic resolution of a detector in a gamma camera is approximately 3–5 mm and depends on the thickness of the crystal, the number of PMTs, and the position-localization algorithms used.

Collimator resolution is defined by the design of the collimator. If other factors are constant, longer bore length implies finer resolution and less sensitivity. If other factors are constant, larger bore diameter implies higher sensitivity and poorer resolution. Collimator resolution is a function of the distance from the collimator face; thus, specification of collimator resolution includes the specification of the location at which the stated resolution was measured. Collimator resolution is usually specified at 10 cm from the face of the collimator and ranges from a few millimeters to a few centimeters, depending on the collimator design.

Intrinsic and camera spatial resolutions are parameters important to characterization of image quality and quantitative accuracy. Most modern gamma cameras have similar values for these parameters, and most of the values are close to optimal for present collimator/scintillation-crystal/photodetector technology.

The energy resolution of a detector is determined by the degree of fluctuation of PMT outputs resulting from variations in the signal generation process that starts with a photon-crystal interaction. The range of energy resolutions of NaI(Tl)/PMT-array scintillation detectors in modern gamma cameras is 9–11%. Energy resolution is an important parameter that affects image quality because finer energy resolution leads to better rejection of scattered photons.

Gamma cameras, until recently, could only process the signals from one interaction at a time. This restriction is due to finite scintillation rise and decay times and corresponding requirements on the detector electronics. Because this is the case, interactions that result in an overlap of scintillation and signal processing cannot be distinguished and must be rejected. This effect produces dead time in a camera and limits its count rate. In a typical camera, the dead time is on the order of microseconds. It is desirable for dead time to be as short as possible, so as to maximize the achievable count rate and thus reduce the severity of Poisson noise in the image.

In recent years, gamma camera manufacturers have developed electronics that segment electronically a camera's FOV. A photon-crystal interaction takes place; a small number of PMTs receive most of the generated light photons; the detector identifies these PMTs and uses them to estimate an event location; and the detector releases the remaining PMTs for detection of other events while the signals from the first event are processed. This advance in technology has had some impact on single-photon emission imaging, but was implemented mostly so that gamma camera systems could be used to perform positron emission tomography (PET)—with data acquisition achieved by coincidence detection and gantry rotation.

III. TOMOGRAPHY

A. Tomographic Imaging

So far, we have discussed gamma cameras in the context of planar emission imaging only. In some studies, a planar image is the ultimate goal; in others, planar images are acquired for the purpose of computing tomographic images by image reconstruction. Wernick and Aarsvold (Chapter 2 in this volume) explain the basic ideas of tomography; here, we describe further the details of tomography as they relate to SPECT.

Tomography in SPECT is a process of creating two-dimensional (2D) slice images or three-dimensional (3D) volume images from 2D planar (projection) images obtained by one or more gamma cameras. Tomographic images differ from planar images in that each pixel (picture element) or voxel (volume element) in a tomographic image represents a measurable parameter at only one point in space, whereas a pixel in a planar image represents the result of integration of the parameter over all locations along a roughly line-shaped volume through an object. In SPECT, two types of images may be acquired: emission images (which show the radiotracer distribution within the body) and transmission images (which show the attenuation-coefficient distribution of the body). Although transmission tomography is the primary purpose of X-ray computed tomography (CT), it is typically an auxiliary step in SPECT. In SPECT, transmission tomography can be used to aid in the accuracy of the reconstruction of emission tomographs by providing information that can lead to more accurate models of the process of acquisition of emission data and thus to more accurate inversion software.

A chest X-ray is an example of planar transmission imaging. Figure 2 shows a lung perfusion scan, which is an example of planar emission imaging. An abdomen CT study is an example of transmission tomography (TT). A bone SPECT study is an example of emission tomography (ET) (see Figures 32 and 33). Planar transmission and emission images are projections of the imaged object. In them, the gray-scale value at each location conveys a mixture of information from all points along a path through the object's interior. The purpose of image reconstruction is to undo this mixture so as to determine the object's properties at every individual point within the body.

In ET, an image represents the distribution of radiotracer within the patient. In TT, an image represents the attenuation coefficients of tissues within the patient. In SPECT, the principal goal is to produce ET images—images of a radiotracer within a body; the acquisition of TT images, when done, is done for the purpose of correcting ET images for the effect of gamma-ray attenuation within the patient. ET images that are not attenuation corrected are not quantitative, only qualitative estimates/images of radiotracer distributions. Transmission imaging is performed using photon sources of known strengths (X-ray or sealed radioactivity) at known positions outside the patient. Emission imaging is performed using photon sources of unknown strengths (radioisotopes) at unknown positions inside the patient.

B. Transmission Imaging

As explained in Wernick and Aarsvold (Chapter 2 in this volume), the value of each pixel in a planar (projection) image can be described approximately as a line integral, representing a sum of the object distribution along each line through the body. In transmission imaging, the property being measured is the attenuation coefficient.

In the following discussion, as in Wernick and Aarsvold (Chapter 2 in this volume) we use two coordinate systems (defined in Fig. 13), which refer to two frames of reference. The patient's frame of reference, specified by spatial coordinates $\mathbf{x} = (x,y,z)$, is assumed to be stationary.

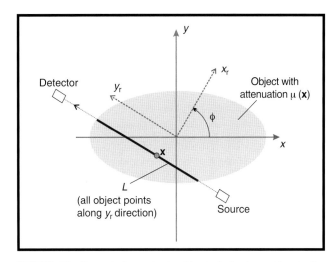

FIGURE 13 Transmission projection. This is the fundamental projection for planar and tomographic transmission imaging.

The gamma camera's frame of reference, defined by coordinates (x_r, y_r, z), rotates about the z axis (which is the long axis of the patient) with rotation angle ϕ. At an angular position ϕ of the camera, we consider only radiation traveling in the y_r direction toward a single point specified by coordinates (x_r, z).

To illustrate transmission imaging, we suppose that a collimated beam of radiation of intensity I_0 travels in the y_r direction, as shown in Figure 13. In this case, the measured radiation I that emerges from the patient is related to I_0 by Beer's law and has the form:

$$I(x_r, z) = I_0 e^{-\int_L \mu(x) dy_r} \quad (1)$$

where $\mu(\mathbf{x})$ is the attenuation-coefficient distribution of the body; L is the line segment shown in Figure 13, which extends from the source to the detector; and dy_r is integration in the y_r direction along L.

Because $\mu(\mathbf{x})$ is truly the quantity of interest, it is common to define a modified transmission measurement $T(x_r,z)$ as the negative of the argument of the function in the exponential in Eq. (1):

$$T(x_r, z) = \int_L \mu(\mathbf{x}) dy_r = -\ln \frac{I(x_r,z)}{I_0} \quad (2)$$

Eq. (2) is obtained by simple manipulation of Eq. 1.

Eq. (1) represents the transmission measurment at the single point (x_r,z). Eq. (2) represents the modified transmission measurement at (x_r,z). The integral in Eq. (2) is the Radon integral of $\mu(\mathbf{x})$ along the single line L measured at the single point (x_r,z). The integral in Eq. (2) is the fundamental integral of planar and tomographic transmission imaging.

The collection of line integrals of $\mu(\mathbf{x})$, [one at each image location (x_r,z)] in the y_r direction corresponds to one planar image (projection) of the patient. The collection of all such integrals in all y_r directions for a fixed z are the data [the 2D Radon transform of $\mu(\mathbf{x})$ at level z] needed to perform tomgraphic reconstruction of the transaxial slice of the distribution of attenuation coefficients at level z in the body. The collection of all such integrals in all y_r directions for all z are the data needed to perform a 3D tomographic reconstruction of the distribution of attenuation coefficients in the body.

C. Emission Imaging

Now we consider a single emission measurement for a spatial distribution $f(\mathbf{x})$ of radiotracer. Ideally, assuming an idealized (blur-free) parallel-hole collimator and neglecting effects of noise, an emission measurement $E(x_r,z)$ along the line L in the y_r direction has the form:

$$E(x_r,z) = \int_L f(\mathbf{x}) e^{-\int_{L'(\mathbf{x})} \mu(\mathbf{x}) dy_r'} dy_r \qquad (3)$$

where $\mu(\mathbf{x})$ is the distribution of attenuation coefficients and dy_r' is integration along the line segment $L'(\mathbf{x})$ of line L from $\mathbf{x}$ to the detector in the y_r direction. Eq. (3) is one line integral of the radiotracer distribution $f(\mathbf{x})$. The source is weighted by an exponential term that represents the attenuation of the source by the body. As shown in Figure 14, the integration of the attenuation term extends only along the line segment $L'(\mathbf{x})$, which is the segment from location $\mathbf{x}$ to the detector, the path taken by gamma photons that reach the detector having been emitted from $\mathbf{x}$. The integral in Eq. (3) is the attenuated Radon integral of $f(\mathbf{x})$ and $\mu(\mathbf{x})$ along the single line L measured at the single point (x_r,z). The integral in Eq. (3) is the fundamental integral in planar and tomographic single-photon emission imaging.

The collection of line integrals, Eq. (3) of $f(\mathbf{x})$ and $\mu(\mathbf{x})$ [one at each image location (x_r,z)] in the y_r direction corresponds to one planar image (projection) of the patient. The collection of all such integrals in all y_r directions for a fixed z are the data [the 2D attenuated Radon transform of $f(\mathbf{x})$ and $\mu(\mathbf{x})$ at level z] needed to perform tomographic reconstruction of the transaxial slice of the distribution of radiotracer at level z in the body. The collection of all such integrals in all y_r directions for all z are the data needed to perform a 3D tomographic reconstruction of the distribution of radiotracer in the body.

In most clinical SPECT performed over the past several decades, $\mu(\mathbf{x})$ has been assumed to be a constant function, usually with value zero. If $\mu(\mathbf{x})$ is assumed to be zero, Eq. (3) reduces to:

$$E(x_r,z) = \int_L f(\mathbf{x}) dy_r \qquad (4)$$

We note that Eq. (4) has the same form as Eq. (2) except that the integrand in Eq. (2) is $\mu(\mathbf{x})$ whereas the integrand in Eq. (4) is $f(\mathbf{x})$. That is, the first is a Radon integral of attenuation coefficients and the second is a Radon integral of radiotracer activity. The integral in Eq. (4) is the Radon integral of $f(\mathbf{x})$ along the single line L measured at the single point (x_r,z).

Eq. (4) is a crude approximation of the process of emission measurement, yet the equation has been and is used as the basis of data acquisition models routinely and is successfully applied in reconstruction in clinical SPECT. The tomographic images produced, although not quantitatively accurate, are sufficiently qualitatively accurate that physicians make accurate diagnoses from them. Eq. (4) has been used rather than Eq. (3) because there have not been convenient approaches to inversion of models based on the more complex expression of Eq. (3).

Use of nonzero constant $\mu(\mathbf{x})$ provides some quantitative information for some emission tomographs, particularly of volumes of the body that have attenuation coefficients that are relatively uniform, such as the brain. But, in general, quantitative accuracy in SPECT requires that $\mu(\mathbf{x})$ be fully determined and incorporated into the inversion process. In the last several years, commercial technology has been developed to pursue such clinically.

D. Transmission Tomography

Knowledge of $\mu(\mathbf{x})$, the distribution of attenuation coefficients in the body, is necessary for accurate quantitative SPECT. Thus, TT is a part of quantitative SPECT. One approach to measuring $\mu(\mathbf{x})$ is acquisition of a standard CT image acquired on a separate machine at a separate time. This requires that the CT be registered to the emission data. Another approach is acquisition of TT data at the same time as acquisition of ET data. Versions of this approach are in place and being further developed. The TT data might be CT data or they might be data acquired with a sealed source of a radioisotope. In either case, the goal is an estimate of $\mu(\mathbf{x})$ that can be included in the process of reconstruction of the emission data. By using transmission

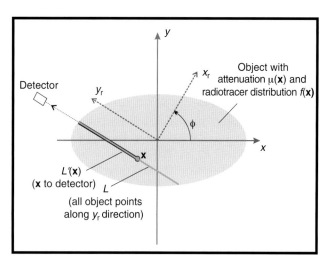

FIGURE 14 Emission projection. This is the fundamental projection for planar and tomographic emission imaging.

imaging to measure $\mu(\mathbf{x})$ one can more accurately reconstruct the radiotracer distribution $f(\mathbf{x})$ by using an imaging model such as that in Eq. (3). This use of $\mu(\mathbf{x})$ is called *attenuation correction* and is further explained in King *et al.* (Chapter 22 in this volume).

High-quality TT has a second use in SPECT protocols. In the form of CT, TT can be a high-resolution representation of the anatomy of a body. If CT and SPECT studies are registered or if CT is performed in tandem with SPECT by SPECT/CT systems, which acquire both types of images, then anatomical and physiological images can be displayed simultaneously (see Fig. 33 later in this chapter). The simultaneous display of SPECT and CT provides the nuclear medicine physician a means to more accurately locate abnormal tracer uptake and to more easily interpret a difficult SPECT study.

E. Emission Tomography

In most clinical applications, ET image reconstruction continues to be treated as a 2D problem, in which images of the body are computed slice by slice (Fig. 15). Most of the fundamentals of ET are similar to those of TT, but ET is more complicated because $\mu(\mathbf{x})$ is needed to reconstruct the radiotracer distribution accurately. Figures 16 and 17 indicate that TT and ET both require sampling of data at numerous detector positions. In most cases, data are acquired at 120 or 128 angular positions about an object. This is true for TT and ET as performed in SPECT protocols.

Figure 18 shows reconstructions of a radiotracer distribution $f(\mathbf{x})$ of a phantom obtained using various assumptions about $\mu(\mathbf{x})$. The images in the left column were reconstructed without attenuation correction, that is, assuming $\mu(\mathbf{x}) = 0$, whereas the images in the right column are properly attenuation corrected using TT data. The principal difference in the images in this example is the diminished

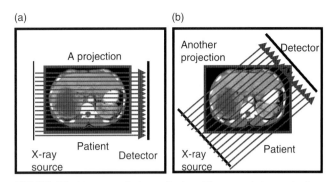

FIGURE 16 Two transmission projections. The transmission projections are at positions (a) 0° and (b) 45°. See also Figure 15.

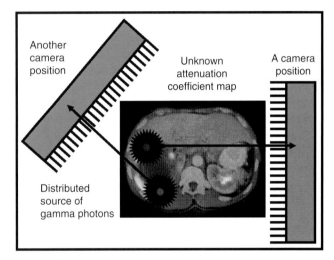

FIGURE 17 Two emission projections. The emission projections are at positions 0° and 135°. The patient slice represents two parameters, the tissue densities and the radiopharmaceutical concentrations (visible here as the white pixels on the edges of the enlarged left kidney and in the center of the right kidney). See also Figure 15.

activity in the object's interior when attenuation correction is not applied. This is typical of emission images that are not attenuation corrected.

There have been many developments in attenuation correction, and the benefits of proper attenuation correction have been demonstrated (see King *et al.*, Chapter 22 in this volume). In addition, it has recently been demonstrated that the attenuated Radon transform can be inverted analytically (Novikov, 2000; Natterer, 2001; Kunyansky, 2001; Noo and Wagner, 2001; Sidky and Pan, 2002). In spite of these developments, most clinical studies continue to be performed without attenuation correction. This is beginning to change due to the development and proliferation of various attenuation-correction technologies (transmission-source hardware and reconstruction algorithms). However, attenuation correction of emission studies will be clinically routine only when it is demonstrated that it results in sufficient added value to patient management.

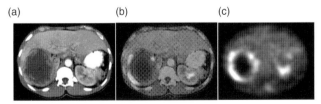

FIGURE 15 Attenuation coefficients and radiotracer concentrations. Displayed are transaxial slices of (a) a patient's attenuation coefficients, (c) radioactivity distribution, and (b) radioactivity distribution registered and superimposed on the attenuation coefficients. The goal of ET is the accurate representation of radiotracer concentration. Quantitative ET requires an accurate estimation of the attenuation coefficients and the incorporation of this information into the process of ET reconstruction. This slice is at the level of the kidneys. The right kidney (left side of the images) has a large tumor that is displacing the liver through its enlargement. The left kidney (right side of the images) is normal.

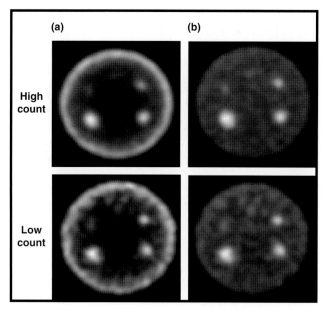

FIGURE 18 Phantom reconstructions (a) without and (b) with attenuation correction. All four images are reconstruction of a cylindrical phantom with five spheres of various diameters. The spheres were filled with a higher concentration of radiotracer than was the background. The top images are reconstruction of relatively high-count data; the bottom images of relatively low-count data. The bottom images are significantly noisier than the top images. The images in (a) have low counts in the center and excessive counts on the object outer contour.

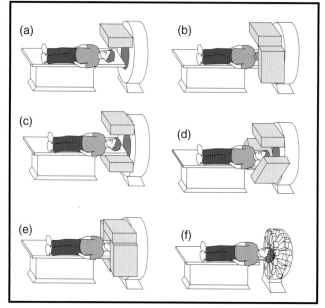

FIGURE 19 Several SPECT system configurations. Shown schematically are systems with, (a) a single head, (b) two orthogonal heads, (c) two opposed heads; (d) three heads, (e) four heads; and (f) multiple small-FOV scintillation detectors.

IV. SINGLE-PHOTON EMISSION COMPUTED TOMOGRAPHY SYSTEMS

The goal in SPECT data acquisition is the acquisition of projection data sufficient for tomographic image reconstruction. To accomplish this, a gamma camera is rotated about the object and data are acquired for multiple slices simultaneously. In standard protocols, during a camera rotation, 64 or 128 measurements (representing 64 or 128 detector elements) are acquired for 64 or 128 transaxial slices of the object at each angle. That is, for each angular stop in the acquisition protocol, a 64 × 64 or 128 × 128 image or projection is acquired.

Early SPECT systems required that the patient sit in a special chair that rotated in front of a stationary gamma camera (Muehllehner and Wetzel, 1971; Budinger and Gulberg, 1974). However, almost all SPECT systems since then have been such that the camera orbits the stationary patient (Jaszczak et al. 1977; Keyes et al., 1977).

A. System Configurations

SPECT systems differ in the number of gamma cameras that are incorporated (Jaszczak, Chang, Stein, et al., 1979; Lim et al., 1980, 1985; Chang et al., 1992; Rowe et al., 1993; Milster et al., 1990). A single-camera system and several multicamera systems are illustrated in Figure 19. In principle, the greater the number of cameras, the greater the sensitivity of the system because gamma photons not traveling normal to the face of any given camera are simply not observed. When multiple cameras are simultaneously capturing data, the proportion of the gamma photons that are used is increased.

At this writing, almost all SPECT systems sold commercially have two cameras, with single-camera and three-camera systems being the only exceptions. In the past, there have been commercial systems with four and six heads, but the added cost of the additional heads is not justifiable in most clinical settings.

B. Gantry Motions

There are two commonly used forms of data acquisition, which differ in the way the gantry is moved: continuous acquisition and step-and-shoot acquisition. In a continuous acquisition, data are acquired as the gantry continuously rotates the camera about the patient. Data are binned into discrete equivalent angular ranges. In a step-and-shoot acquisition, the gantry stops at various positions, waiting at each position while data are acquired and then moving on to the next position.

C. Transmission-Source Tomography

Most clinical SPECT studies are performed without corrections for finite-size collimator bores, for photon scatter, or for attenuation. Methods to correct for these effects are not easily developed or implemented without significant computing power—power that only became readily available on clinical systems starting in the 1990s. Details of the issues in the development of scatter and attenuation correction and iterative image reconstruction are discussed in King *et al.*, (Chapter 22) and Lalush and Wernick (Chapter 21 in this volume), respectively. Here we briefly discuss some hardware technologies that enable attenuation correction data to be performed routinely.

As we have discussed earlier, there are two unknown functions in SPECT imaging, the radiotracer distribution $f(\mathbf{x})$ and the attenuation map $\mu(\mathbf{x})$. Transmission-source attenuation-correction technology has been developed for the purpose of measuring $\mu(\mathbf{x})$, enabling quantitative calculation of $f(\mathbf{x})$.

Attenuation correction is important in SPECT, particularly when quantitative accuracy is desired. Thus, transmission-scan capabilities have been developed and are now available on most SPECT systems. Some transmission-source technologies require sequential emission- and transmission-data acquisition; others allow the simultaneous acquisition of emission and transmission data. If the transmission and emission projections are acquired simultaneously, the transmission source must emit photons with photon energies different from the energies of the primary emissions of the radioisotope or radioisotopes being imaged. Otherwise, emission and transmission data cannot be distinguished.

There are several configurations for transmission scanning (Celler *et al.*, 1997, 1998; Tung *et al.*, 1992; Tan *et al.*, 1993; Bailey *et al.*, 1987; Manglos, 1992; Manglos *et al.*, 1987; Frey *et al.*, 1992; Beekman *et al.*, 1998; Gulberg *et al.*, 1998; Kamphius and Beekman, 1998; Zeng *et al.*, 2001). When a parallel-hole collimator is used, a sheet source, multiple line sources, or a scanning line source can be positioned on the opposite side of the patient to irradiate the patient and thus to obtain transmission measurements (Figs. 20 and 21).

In a multiple-line-source system (Figs. 20b and 21a), to reduce the cost of replacing entire sets of the sources at once (which degrade due to radioactive decay), one can replace the sources one at a time. In this case, the line sources will have unequal strength at any given time, and appropriate normalizations are used to account for these variations. To make best use of these sources, the ones with greatest strength are usually positioned at the center of the camera's FOV because the thickest path through a patient's body usually passes through the center of the camera's FOV.

In a scanning line-source system, data acquisition is synchronized with the scanning motion of the source, through the use of a scanning virtual window on the camera, as shown in Figures 20c and 21b.

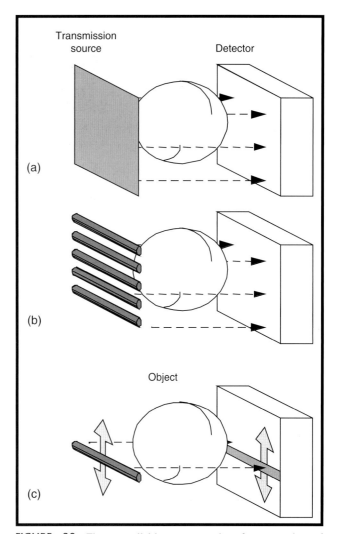

FIGURE 20 Three parallel-beam approaches for generation of transmission data on a gamma camera with a parallel-hole collimator. The approaches use (a) a sheet source, (b) multiple line sources, and (c) a scanning line source with a scanning electronic window. The transmission sources are on the opposite side of the object from the camera with collimator. The transmission photons detected are those that pass through the collimator bores.

When using a fan-beam or offset fan-beam collimator, an external line source or a scanning point source can be used for transmission measurements, as shown in Figure 22. Truncation problems can and usually do occur in a fan-beam transmission geometry because the FOV of the transmission images is usually too small to capture the entire patient. However, a 360°-rotation, offset-fan-beam geometry can be implemented so that an object's transmission sinogram is sampled sufficiently to minimize problems caused by truncation. This can also be done when using a cone-beam or an offset cone-beam collimator. In the cone-beam case, a fixed external point source is used for transmission measurements (Fig. 23).

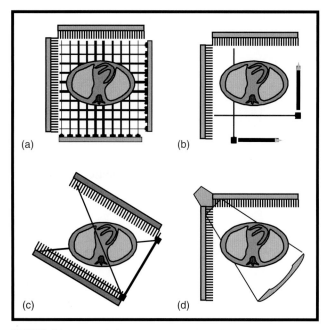

FIGURE 21 Transmission-source geometries. Displayed are schematics for four forms of transmission-source technologies. They are (a) multiple line sources (e.g., Gd-153) to scintillation detectors with parallel-hole collimators, (b) scanning line sources (e.g., Gd-153) to scintillation detectors with parallel-hole collimators, (c) scanning point sources collimated to form fan-beam transmission (e.g., Ba-133) to scintillation detectors with parallel-hole collimators, and (d) fan-beam CT-to-CT detector.

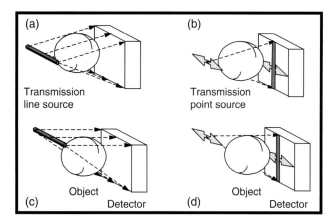

FIGURE 22 Four fan-beam collimator approaches to generation of transmission data. The approaches use (a) a fan-beam collimator with a fixed line source, (b) a fan-beam collimator with a scanning point source and scanning electronic window, (c) an offset fan-beam collimator with a fixed line source, and (d) an offset fan beam collimator with a scanning point source and scanning electronic window.

An offset-fan-beam transmission geometry can be implemented when using a parallel-hole collimator. The setup is similar to that shown in Figure 21c, except that the collimator is a parallel-hole collimator. Such a setup has been implemented with a medium-energy scanning point source. Figure 24 is a photograph of a system with this technology.

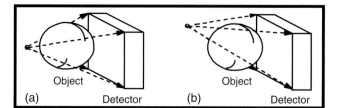

FIGURE 23 Two approaches to generation of transmission data using fixed medium-energy point sources and parallel-hole collimators. The approaches use (a) a cone-beam geometry and (b) an offset cone-beam geometry. These strategies are practical only when a low-energy collimator that allows significant penetration of the medium-energy transmission photons is used because acquisition times are excessive if higher-energy collimators are used.

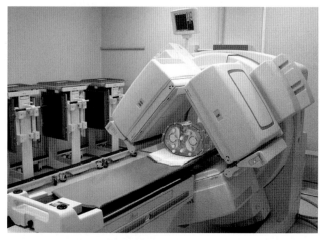

FIGURE 24 Triple-head SPECT/PET system with transmission-source attenuation-correction technology. The units on the sides of heads I and II are this system's transmission-source technology. Each unit houses a scanning point source collimated to produce fan-beam transmission data during operation. On the left are three sets of collimators on collimator carts. On the patient table is a fillable anthropomorphic chest phantom. The electronics of this system allow it to be used for both single-photon detection (for SPECT) and coincidence detection (for PET).

Figures 25 and 26 are examples of TT and ET data, respectively, for a cardiac SPECT study. More detailed presentations of ET studies are provided later.

D. SPECT/CT

When using SPECT, a common disease signature is the presence of an abnormal focus or focal deficit of radiotracer (i.e., an abnormally bright or dark region on an image). Because SPECT images have relatively poor spatial resolution, physicians often compare SPECT images against X-ray CT images to better identify the precise location of the abnormality in relation to the surrounding anatomical structures. To aid in this process, imaging systems have been developed that can perform both SPECT

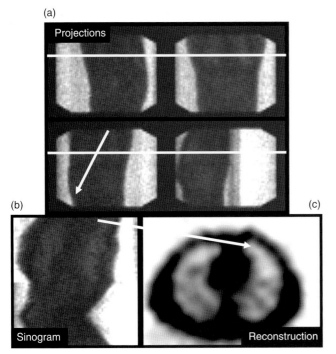

FIGURE 25 Transmission-source: TT projections, sinogram, and reconstruction. (a) Four TT projections of the torso obtained using Gd-153 line sources. (b) The sinogram of the transmission data for the slice, indicated by the white line through the sample projections. (c) Reconstruction of the displayed sinogram. These data are the basis for attenuation correction of related emission data.

and CT using the same gantry (Hasegawa *et al.*, 1993; Lang *et al.*, 1992; Blankespoor *et al.*, 1996). These imagers, called SPECT/CT systems, provide a means to obtain well-aligned SPECT and CT images that facilitate anatomical localization; in addition, the CT images can be used to perform attenuation correction of the SPECT images (Figure 21d).

Some open questions remain in the development of SPECT/CT systems, including the following: (1) Should the CT system be mounted for rapid rotation, or should it be mounted on the SPECT gantry so that SPECT and CT data are acquired over similar time frames? (2) Should SPECT and CT data acquisition be simultaneous or sequential? (3) Should the CT be a state-of-the-art system or is a moderate-resolution, limited-cost system sufficient? (4) How should the acquired, reconstructed, and registered data be presented to reviewing physicians? and (5) What is necessary to implement attenuation correction of the SPECT data and reconstructions using the CT data and reconstructions?

A SPECT/CT system is commercially available (Fig. 27), and studies to assess the added value of multimodal and attenuation-corrected images are underway; therefore, the open issues will come increasingly into focus in the coming years. An example study, a red blood cell (RBC) liver study, is shown in Figure 28. More discussion on dual-modality imaging (in particular, the marriage of PET and CT) is provided in Watson *et al.* (Chapter 11 in this volume).

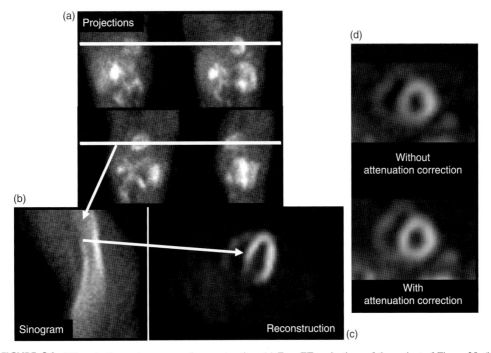

FIGURE 26 ET projections, sinogram, and reconstruction. (a) Four ET projections of the patient of Figure 25. (b) The sinogram of the transmission data for the slice, indicated by the white line through the sample projections. (c) Reconstruction of the displayed sinogram. (d, top) Reconstruction of the ET data without transmission-source attenuation correction. (d, bottom) Reconstruction of the ET data with transmission-source attenuation correction using the data of Figure 25.

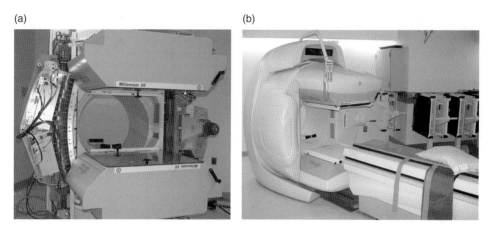

FIGURE 27 Dual-head SPECT/CT system. In both photographs, one gamma camera is at the top and the second is at the bottom, the CT X-ray source is on the right, and the CT detectors are on the left. Collimators on carts can be seen on the right side of the right image. (a) Fiberglass covers have been removed from the gantry so that the CT source and detectors are visible. (b) The gantry covers are in place, and the system is ready to acquire clinical studies.

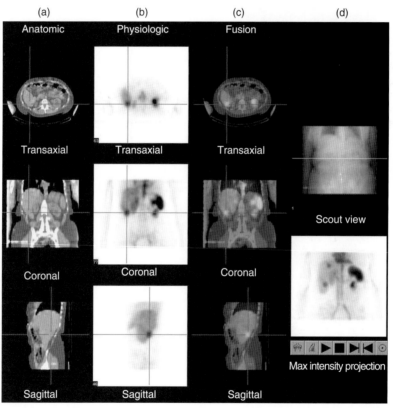

FIGURE 28 SPECT/CT RBC liver study. (Column a) CT reconstruction: transaxial, coronal, and sagital slices. (Column b) SPECT reconstruction: transaxial, coronal, and sagital slices. (Column c) Registered SPECT and CT reconstructions: transaxial, coronal, and sagital slices. (Column d) Transmission scout scan and emission maximum-intensity projection.

E. System Performance

An important part of clinical SPECT is the evaluation of system designs, and the routine maintenance and calibration of clinical scanners. SPECT system acceptance testing and quality assurance requires first that performance measurements of each gamma camera of a SPECT system be performed as discussed in the NEMA NU-1-2001 standard

(NEMA, 2001). Assurance of high-quality SPECT images requires that performance measures of the SPECT system also be performed. These are, like the gamma camera performance measures, outlined in NEMA NU-1-2001 standards document. The measurements recommended include those to determine reconstructed spatial resolution with and without scatter, reconstructed image uniformity, and assessment of accuracy of the center of rotation (COR).

Scattered photons degrade image quality, and quantitative SPECT, particularly of radiotracers emitting photons of multiple energies, requires that corrections for scatter be applied. Techniques for accomplishing scatter correction are described in King *et al.*, (Chapter 22 in this volume).

In addition to camera and SPECT quality assurance measurements, it is necessary to perform TT quality assurance measurements on many modern SPECT systems, because these systems now have integrated hardware for the acquisition and processing of transmission data for attenuation correction.

V. TOMOGRAPHIC SINGLE-PHOTON EMISSION IMAGING

Having introduced the basic technology of SPECT, let us now provide some examples of the most common uses of SPECT in clinical practice. It is worthwhile to compare these studies, and the images they produce, to the planar-imaging examples presented earlier in this chapter.

A. Bone SPECT

A bone SPECT study, using ^{99m}Tc-MDP, is an imaging study performed in addition to a planar whole-body bone study when the reviewing physician seeks to better assess and locate possible abnormalities detected in the planar study. Patient preparation for a bone SPECT study is identical to that for a planar bone study (see Section I.C.). When bone SPECT is conducted immediately subsequent to a planar bone study, no additional radiotracer is administered. SPECT data acquisition parameters, for a study performed on a dual-head SPECT system with the two heads separated by 180° using a step-and-shoot acquisition, are 60 or 64 steps (120 or 128 total views), ~3°/step, and 20 s/step (total acquisition time is approximately 20 min/bed position). Figure 29 is a summary of the data (projections and sinograms) for an example normal study. Figure 30 shows sample reconstructions (transaxial slices, sagittal slices, and coronal slices) of the normal bone scan. An abnormal bone scan exhibits foci having uptakes that are above or below normal. The scan shown in Figure 30 has relatively uniform uptake throughout the skeletal system. The data

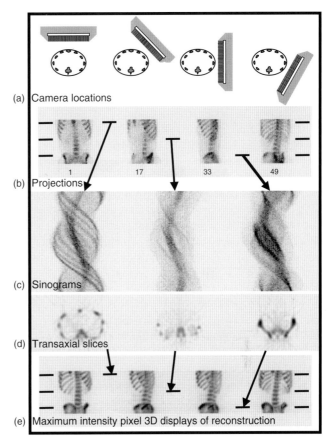

FIGURE 29 Bone SPECT data and reconstructions. Bone SPECT protocols generally include acquisition of projections at 128 angular positions over 360°. Shown are 4 of 128 projections (row b) corresponding to the 4 indicated camera positions (row a). Shown also are three sinograms (row c) representing the data acquired for the three transaxial slices indicated by the black lines (row b). The three displayed transaxial slices (row d) were reconstructed from the three displayed sinograms (row c). A 3D reconstruction was obtained by reconstruction and stacking of 128 transaxial slices; four maximum-intensity projections (row e) are representative of the 3D reconstruction. The three sinograms are displayed with the horizontal axis as detector location and the vertical axis as angular location. See also Figure 30.

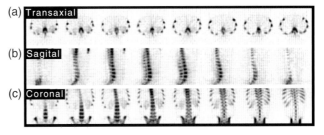

FIGURE 30 Bone SPECT reconstruction. (a) Transaxial slices, (b) sagital slices, and (c) coronal slices of the reconstructed images of a bone SPECT study. The images are of a normal bone SPECT study. See also Figure 29.

for the presented reconstructions were acquired at one bed position.

B. Brain SPECT

^{99m}Tc-HMPAO (hexamethylpropylene amine oxime; also known as exametazime) and ^{99m}Tc-ECD (ethyl cysteinate dimer) are two radiopharmaceuticals that cross the blood–brain barrier (BBB) and distribute in the brain in proportion to blood flow. These two agents are the most common agents used in brain perfusion SPECT studies. The most common reasons for performing a brain perfusion SPECT study are clinical indications of cerebrovascular disease, dementia, or seizure. Brain perfusion studies are abnormal if there are regions of hypoperfusion or hyperperfusion. Each of the stated indications, if detected by a perfusion SPECT study, will present as a region of ischemia i.e., hypoperfusion. Because an ECD or HMPAO study provides only information about perfusion, interpretation of the study requires that the data be considered in the context of the clinical indications that led to the request of the study. In a brain SPECT study, either ECD or HMPAO (1110 MBq; 30 mCi) is injected IV with imaging commencing 1 hour after administration. Such a study typically uses two heads (gamma cameras) separated by 180°, continuous acquisition with data assigned to 120 or 128 total views, ~3°/step, and a total acquisition time of approximately 20 minutes. Figures 31 and 32 summarize the data (projections and sinograms) and reconstructions (transaxial slices, sagittal slices, coronal slices, and volumes) of an abnormal ECD brain SPECT study.

C. Myocardial Perfusion SPECT

Myocardial perfusion rest/stress SPECT studies are the most common of all SPECT studies. A common version of such studies is a dual-isotope version that involves the characterization of a patient's heart with the patient at rest using ^{201}Tl-TlCl and under stress using ^{99m}Tc-sestamibi or tetrofosmin. Interpretation of the studies involves comparison of the reconstructions of the two data sets.

The protocol is as follows. With the patient at rest, ^{201}Tl-TlCl (148 MBq, 4 mCi) is injected IV. After 15 minutes, imaging commences. Data acquisition parameters, for a rest study performed on a dual-head SPECT system with the two heads separated by 90° using a step-and-shoot acquisition, are 30 or 32 steps (60 or 64 total views; total angular range 180° from LPO through RAO), ~3°/step, and 30 s/stop (total acquisition time approximately 15 minutes). Upon completion of the resting study, the patient's heart is placed under stress conditions. This is accomplished either by vigorous walking on a treadmill for approximately 10 minutes or by the injection of a pharmacological cardiac stress agent such as dipyridamole or dobutamine. With the patient under

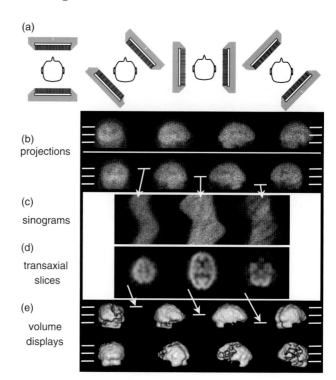

FIGURE 31 Brain SPECT data and reconstructions. Brain SPECT protocols generally include acquisition of projections at 128 angular positions over 360°. Shown are 8 of 128 projections (row b) corresponding to the eight indicated camera positions (row a). Shown also are three sinograms (row c) representing the data for the three transaxial slices indicated by white lines (row b). The three displayed transaxial slices (row d) were reconstructed from the three displayed sinograms (row c). Approximately 40 transaxial slices were reconstructed and stacked to form 3D surface representations (row e). The three sinograms are displayed with the horizontal axis as detector location and the vertical axis as angular location.

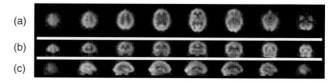

FIGURE 32 Brain SPECT reconstructions. (a) Transaxial slices, (b) coronal slices, and (c) sagital slices of reconstructions in a brain SPECT study. The images are of an abnormal brain SPECT study. There are subtle asymmetries of uptake that are more easily seen in a surface rendered 3D display.

stress, ^{99m}Tc-sestamibi or tetrofosmin (925 MBq; 25 mCi) is injected IV. Both of these agents go rapidly to cardiac cells where they remain for several hours. Both have slow biological washout rates from myocardium. As this is the case, images acquired within a couple hours of injection represent myocardial perfusion at the time of stress, not at the time of imaging. Imaging usually commences 30 minutes post injection. Data-acquisition parameters for a stress study performed on a dual-head SPECT system with the two heads separated by 90° are the same as those for a rest study, except that 20 s/stop (total acquisition time approximately 11 minutes) are used instead of 30 s/stop. The imaging time is reduced because the photon flux of the ^{99m}Tc dose is measurably greater than that of the ^{201}Tl dose.

The 180° anterior orbit of cardiac imaging is different from that of most SPECT studies, which have 360° orbits. The 180° orbit is used because of the heart's position near the chest wall and the fact that $\mu(\mathbf{x})$ is usually assumed to be identically zero. Attenuation effects in anterior projections are much less than those of posterior projections. Resolution is also better in anterior projections than posterior. The use of 180° anterior SPECT for cardiac imaging produces better contrast between defects (abnormal tissue) and normal tissue with fewer artifacts than does 360° SPECT, if attenuation correction is not performed.

There is no difficulty in imaging the ^{99m}Tc, even though both ^{201}Tl and ^{99m}Tc are in the myocardium at the same time. This is because the emissions of ^{99m}Tc that are of interest have sufficiently higher energy than the emissions of ^{201}Tl. By using energy discrimination in the electronics, one can prevent contamination of the ^{99m}Tc data by ^{201}Tl photons; however, there would be downscatter contamination difficulties if ^{201}Tl were imaged after an injection of ^{99m}Tc. Thus, the rest and stress studies cannot be performed in the reverse order if these two radioisotopes are used as indicated.

The reconstructions produced in cardiac studies are usually reformatted to the coordinate system of the left ventricle of the heart, rather than that of the body, to facilitate comparison of the rest and stress images. Figure 33 shows examples of projections, sinograms, and reconstructed cardiac images. Figure 34 is an illustration of the cardiac coordinate system and display orientations.

As explained in Garcia *et al.* (Chapter 24 in this volume), the characterization and interpretation of cardiac studies involve the use of various forms of cardiac-image processing, analysis, and display software that generate semi-quantitative comparisons and indices of and from the data.

Often included in myocardial rest/stress studies is an electrocardiogram-gated acquisition of the stress study. The data then require slightly different processing, but the reconstructed images are interpreted in the same manner as those in a nongated study. Gated data provide information that can be used to assess wall motion and wall thickening and to estimate left-ventricular ejection fraction (LVEF).

In addition to the studies we have mentioned, other common studies are ones used to detect and assess abnormalities that might be cancer. Many of these studies use radioisotopes of higher energy than those used in the studies we have described here. Thus, these studies require the use of collimators that are appropriate for higher-energy photons.

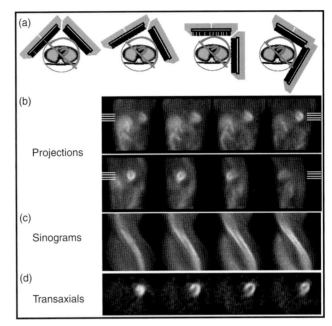

FIGURE 33 Cardiac SPECT data and reconstructions. The cardiac SPECT data displayed were acquired with a protocol that included the acquisition of projections at 64 angular positions over 180° using two gamma cameras separated by 90°. The range is indicated by the semicircle displayed in the camera position schematics (row a). Shown are 8 of 64 projections (row b) corresponding to the eight indicated camera positions (row a). Shown also are four sinograms (row c) representing the data of the four transaxial slices, indicated by white line segments (row b). The four transaxial slices (row d) were reconstructed from the four sinograms (row c). The cardiac volume is contained in approximately 20 transaxial slices in a standard cardiac study. In nuclear medicine, the left side of the body is displayed on the right side of an image and the right side of the body is displayed on the left side of an image—note the left and right ventricles in the camera position diagrams. The four sinograms are displayed with the horizontal axis as detector location and the vertical axis as angular location.

VI. OTHER DETECTORS AND SYSTEMS

Thus far, we have emphasized the technologies associated with clinical imaging hardware that are commercially available at this writing. Next, we briefly summarize some technologies that are being investigated as alternatives or adjuncts to current devices and components, and which in most cases have been demonstrated at least in the form of prototype designs. The aim of most of these new approaches is primarily to improve sensitivity and secondarily to improve spatial resolution.

A. Multiple-Pinhole Coded Apertures

Figure 4 shows a pinhole aperture for a gamma camera. Pinhole apertures are used to image small organs such as thyroids because a gamma camera with a pinhole aperture is an efficient imager if the object being imaged is close to the pinhole.

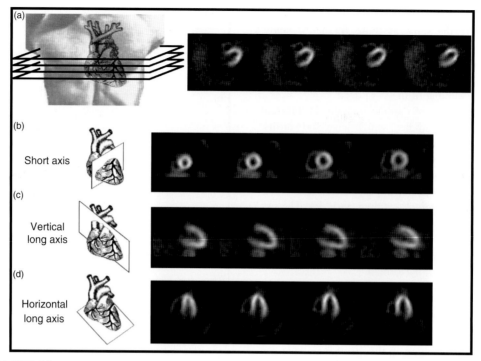

FIGURE 34 Standard oblique orientation for display of cardiac studies. Shown first are transaxial slices (row a) obtained using standard 2D slice-by-slice reconstruction. The left side of the body is to the right in each slice, with the anterior side at the top. Slices are displayed from inferior to superior, left to right. Short-axis slices (row b) are displayed from apex to base, from left to right. The anterior wall of the left ventricle is toward the top, the inferior wall is toward the bottom, and the septal wall is toward the left. Vertical long-axis slices (row c) are displayed from medial to lateral, left to right. The base of the left ventricle is toward the left side the image and the apex is toward the right. Horizontal long-axis slices (row d) are displayed from inferior to anterior, from left to right. The base of the left ventricle is toward the bottom of the image and its apex is toward the top. The right ventricle appears on the left side of the image. (Figure courtesy of Tracy L. Faber, Ph.D., Emory University School of Medicine Atlanta, GA.)

The use of multiple-pinhole apertures has been proposed as a means of increasing sensitivity over that obtained by use of a single-pinhole aperture (Simpson and Barrett, 1980; Barrett *et al.*, 1974; Brown, 1974; Chang *et al.*, 1974; Rogers *et al.*, 1980). Multiple-pinhole apertures have also been proposed as a means of performing multiple angular sampling simultaneously, an approach that can be used to acquire dynamic tomographic data rapidly.

As shown in Figure 35, the distribution of the multiple holes is usually coded. This means that the hole pattern satisfies certain mathematical rules; for example, the autocorrelation function of the distribution of holes should approximate a delta function. The choice of the aperture pattern determines the system response function and detector spatial resolution. The increased sensitivity of multiple pinholes over a single pinhole is proportional to the number of pinholes. The challenge in the use of a multiple-pinhole coded-aperture system for planar or tomographic imaging is that the recorded images or projection data are composed of multiple overlapped images or projections and the creation of interpretable images requires that the overlap be decoded. It is generally not a trivial task, and it is sometimes impossible to decode a coded

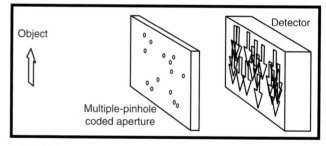

FIGURE 35 Imaging using a multiple-pinhole coded aperture. This system acquires several copies of the object. Some copies are isolated from all others; some overlap. The data are coded; thus, generation of a planar image requires data processing (i.e., data decoding) as well as data acquisition.

image or projection. Nevertheless, operational decoding schemes for some codes have been developed, and the development of both planar and tomographic systems based on multiple-pinhole coded apertures proceeds.

Systems based on multiple-pinhole technology generally have detector geometries that are different from the geometries of conventional gamma cameras and SPECT systems.

Thus, although multiple-pinhole systems have a higher sensitivity than single-pinhole systems for equivalent detectors, sensitivity is not, in most cases, the only motivation to use multiple pinholes in novel single-photon emission imaging systems.

B. Multisegment Slant-Hole Collimators

Multisegment slant-hole collimators provide a means of designing a SPECT system with higher sensitivity than that of conventional SPECT systems (Chang *et al.*, 1982; Dale and Bone, 1990; Wessell *et al.*, 1998; Clack *et al.*, 1995; Bal *et al.*, 1999). Figure 36 shows the geometric configuration of slant-hole collimators with one, two, and four segments. When a two-segment slant-hole collimator is used, one detector can be used to image simultaneously an object at two different crystal locations. This results in increased system sensitivity over a one-segment collimator. Similarly, three- and four-segment slant-hole collimators increase system sensitivity further. The trade-off for additional sensitivity, however, is that one can only use a multisegment slant-hole collimator to image small objects. Imaging large

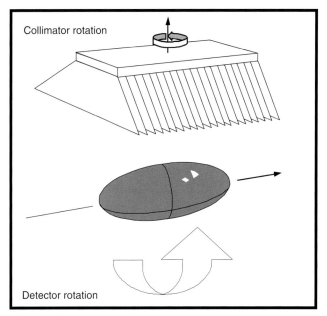

FIGURE 37 Rotating-slat collimator. A slat-collimator/scintillation-detector system acquires one-dimensional data sets. SPECT requires rotation of the camera at each angular position and conventional rotation of the camera about the object being imaged.

objects results in truncation and overlapping of segments. Data acquisition for SPECT with slant-hole collimators requires two motions: collimator rotation and detector rotation. For a four-segment collimator, one needs to rotate the collimator $360°/4 = 90°$ for each detector position. The number of detector positions to which one needs to rotate depends on the slant angle of the collimator holes.

C. Rotating-Slat Collimators

In 1975, Keyes proposed a rotating-slat collimator that could be mounted on a gamma camera. This collimator does not consist of holes. Instead, it is made of parallel plates, as shown in Figure 40. One advantage of a slat collimator is that it has greater geometric efficiency than a hole collimator (Webb *et al.*, 1992, 1993; Lodge *et al.*, 1995, 1996). A slat collimator, however, only measures 1D profiles and cannot be used to directly produce a 2D image. Even planar images cannot be produced using this technology unless image reconstruction of the 1D profiles is performed. As with multisegment collimator systems, data acquisition for SPECT with a slat-collimator system requires two motions: collimator rotation about its central axis and detector rotation about the transaxial axis of the imaged object.

D. Compton Cameras

Rogers *et al.*, (Chapter 19 in this volume) is dedicated to Compton cameras, but we briefly introduce them here.

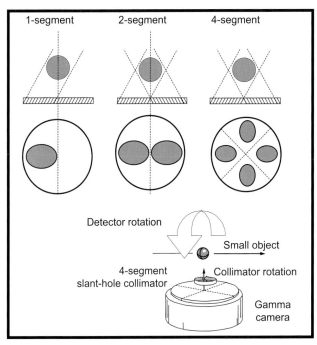

FIGURE 36 Multisegment slant-hole collimator. Imaging strategies using this technology focus on system sensitivity. These strategies are most useful when small objects are to be imaged with medium- to large-area detectors. Sensitivity of a multisegment system is higher than that of a single-segment system when objects that are small relative to detector area are imaged. This is because a multisegment collimator makes it possible to use available detector area more effectively. When multisegment slant-hole collimators are used for SPECT, two rotations are used: rotation of the collimator at each angular position and conventional rotation about the object being imaged.

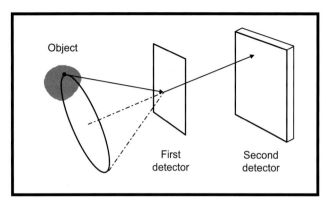

FIGURE 38 Compton camera. A Compton camera comprises a scattering detector and an absorption detector. Image formation requires interaction locations and energy deposited in each detector. These data determine for each detected photon a conic surface from which the detected photon could have been emitted.

Figure 38 shows a schematic representation of a Compton camera. The motivation for the Compton camera is to avoid the loss of photons inherent to conventional collimation, in which only about 1 out of every 10,000 photons emitted from the patient is actually detected. Compton cameras have two gamma-ray detectors operating in coincidence to acquire data for single-photon emission imaging. The energy deposited by Compton scattering in the first detector combined with spatial information from, and the energy deposited in, the second detector are used to determine directional information and the total energy deposited. Such information provides a means of estimating the point of origin of each detected photon. The key research areas in Compton camera development are optimization of detector hardware and development of appropriate image-reconstruction algorithms.

E. Segmented-Scintillator Detectors

In a conventional Anger camera, a single large-area FOV NaI(Tl) crystal and an array of PMTs are used to detect incident gamma photons and measure their energies and positions of interaction. An alternative to the single large-area crystal is an array of small, optically isolated scintillation crystals. Such detectors are physically segmented, in contrast to conventional detectors in which pixels are defined virtually in the electronics and software. Light output of the scintillating material in such an array can be detected and the pixel location determined by various means, including arrays of PMTs, position-sensitive PMTs, and pixellated solid-state photodiode arrays (Fiorini et al., 2000; Patt et al., 1998). Such detectors are similar to block detectors and pixellated-scintillator PMT-array detectors used in some PET scanners. One desirable property of a camera based on photodiodes is that it can be significantly thinner than one based on PMTs. A system with an array of CsI(Tl) crystals and photodiodes is now commercially available. It has a 20 cm × 20 cm FOV and is used primarily for cardiac imaging. The reader is referred to Pichler and Ziegler (Chapter 14 in this volume) for more information on photodetectors including photodiodes and Wilkinson (Chapter 13 in this volume) for information on scintillators.

F. Solid-State Detectors

Solid-state detectors detect gamma photons directly, rather than indirectly through the use of a scintillator material. The charge produced by interaction of a gamma photon in the detector material of a solid-state detector is collected directly from the material (Mauderli et al., 1979; Urie et al., 1979; Butler et al., 1998; Matherson et al., 1998; Singh and Doria, 1984). The appeal of solid-state detectors lies in their significantly superior energy resolution.

Solid-state detectors based on Ge and Si require cryogenic cooling and have the drawback of being composed from low-Z materials; therefore, their ability to stop gamma photons is limited. A CdZnTe (CZT) crystal is a room-temperature semiconductor, which is being investigated extensively by nuclear medicine researchers.

The application of semiconductor detectors to gamma imaging may make it possible to design systems that are more rugged, reliable, and portable than conventional systems. A CdZnTe semiconductor gamma-ray detector has better energy resolution (as good as 4% at 140 keV) than a Na(Tl)/PMT camera (10% at 140 keV). For semiconductor detector devices to find wide use, their cost must become competitive with that of scintillation detectors. Readers should note the possibility of replacing NaI(Tl)/PMT detectors with semiconductor detectors has been investigated throughout the last 20 years, yet NaI(Tl)/PMT detectors remain the technology of choice for clinical single-photon emission imaging. For a thorough discussion on semiconductor detectors, see Wagenaar (Chapter 15 in this volume).

VII. SUMMARY

In this chapter, we have endeavored to explain the fundamentals of imaging systems, cameras and their components, and common clinical applications.

For years, gamma cameras of a standard design, first proposed by Anger, have proven to be reliable, useful, and cost-effective. It is expected that they will serve the nuclear medicine community as the fundamental device for single-photon emission imaging for many years to come. Gamma-camera and SPECT-system emission technologies are relatively mature, and many standard clinical SPECT protocols exist for the management of patients with various diseases and ailments. Yet the advancement of emission detection and imaging technology continues with active efforts in vital areas of research. Of particular interest is development and proliferation of quantitative SPECT techniques, which exploit advances being made in transmission imaging, image reconstruction, and multimodal capabilities.

Acknowledgments

We thank Sean Webb and Desireé N. Jangha for assistance in the preparation of this chapter.

References

Anger, H. O. (1958). Scintillation camera. *Rev. Sci. Instrum.* **29**: 27.
Anger, H. O. (1964). Scintillation camera with multichannel collimators. *J. Nucl. Med.* **5**(6): 515–531.
Anger, H. O. (1967). Radioisotope cameras. *In* "Instrumentation in Nuclear Medicine" (G. J. Hine, ed.), Vol. 1, pp. 485–552. Academic Press, New York.
Anger, H. O. (1974). Sensitivity and resolution of the scintillation camera. *In* "Fundamental Problems in Scanning" (A. Gottschalk and R. N. Beck, eds.), pp. 117–144. Thomas, Springfield, IL.
Anger, H. O., and Davis D. H. (1964). Gamma-ray detection efficiency and image resolution in sodium iodide. *Rev. Sci. Instrum.* **35**: 693.
Bailey, D. L., Hutton B. F., and Walker P. J. (1987). Improved SPECT using simultaneous emission and transmission tomography. *J. Nucl. Med.* **28**(5): 844–851.
Bal, G., Clackdoyle, R., and Zeng, G. L. (1999). Evaluation of different three-dimensional image reconstruction methods used in rotating slant-hole (RSH) SPECT. *J. Nucl. Med.* **40**(5): 284P.
Barrett, H. H., Stoner, W. W., Wilson, D. T., and DeMeester, G. D. (1974). Coded apertures derived from the Fresnel zoneplate. *Opt. Eng.* **13**(6): 539–549.
Beekman, F. J., Kamphuis, C., Hutton, B. F., and van Rijk, P. P. (1998). Half-fanbeam collimators combined with scanning point sources for simultaneous emission-transmission imaging. *J. Nucl. Med.* **39**(11): 1996–2003.
Blankespoor, S. C., Xu, X., Kaiki, K., Brown, J. K., Tang, H. R., Cann, C. E., and Hasegawa, B. H. (1996). Attenuation correction of SPECT using X-ray CT on an emission-transmission CT system: Myocardial perfusion assessment. *IEEE Trans. Nucl. Sci.* **43**(4): 2263–2274.
Brown, C. M., (1974). Multiplex imaging with multiple-pinhole cameras. *J. Appl. Phys.* **45**: 4.
Budinger, T. F., and Gullberg, G. T. (1974). Three-dimensional reconstruction in nuclear medicine emission imaging. *IEEE Trans. Nucl. Sci.* **21**(3): 2–20.
Butler, J. F., Lingren, C. L., Friesenhahn, S. J., Doty, F. P., Ashburn, W. L., Conwell, R. L., Augustine, F. L., Apotovsky, B., Pi, B., Collins, T., Zhao, S., and Isaacson, C. (1998). CdZnTe solid-state gamma camera. *IEEE Trans. Nucl. Sci.* **45**(3): 354–358.
Celler, A., Sitek, A., and Harrop, R. (1997). Reconstruction of multiple line source attenuation maps. *IEEE Trans. Nucl. Sci.* **44**(4): 1503–1508.
Celler, A., Sitek, A., Stoub, E., Hawman, P., Harrop, R., and Lyster, D. (1998). Multiple line source array for SPECT transmission scans: simulation, phantom and patient studies. *J. Nucl. Med.* **39**(12): 2183–2189.
Chang, L. T., Kaplan, S. N., Macdonald, B., Perez-Mendez, V., and Shiraishi, L. (1974). A method of tomographic imaging using a multiple pinhole-coded aperture. *J. Nucl. Med.* **15**(11): 1063–1065.
Chang, W., Huang, G., Tian, Z., Liu, Y., Kari, B., and Madsen, M. T. (1992). Initial characterization of a prototype multi-crystal cylindrical SPECT system. *IEEE Trans. Nucl. Sci.* **39**(4): 1084–1087.
Chang, W., Lin, S. L., and Henkin, R. E. (1982). A new collimator for cardiac tomography: The quadrant slant-hole collimator. *J. Nucl. Med.* **23**(9): 830–835.
Clack, R., Christian, P. E., Defrise, M., and Welch, A. E. (1995). Image reconstruction for a novel SPECT system with rotating slant-hole collimators 1994. *Conf. Rec. IEEE Nucl. Sci. Symp. Med. Imaging Conf.* 1948–1952.
Dale, S., and Bone, D. (1990). Tomography using a rotating slant-hole collimator and a large number of projections. *J. Nucl. Med.* **31**(10): 1675–1681.
Fiorini, C., Longoni, A., Perotti, F., Labanti, C., Rossi, E., Lechner, P., and Struder, L. (2000). First prototype of a gamma-camera based on a single CsI:(Tl) scintillator coupled to a silicon drift detector array. *IEEE Trans. Nucl. Sci.* **47**(6): 1928–1932.
Frey, E. C., Tsui, B. M., and Perry, J. R. (1992). Simultaneous acquisition of emission and transmission data for improved thallium-201 cardiac SPECT imaging using a technetium-99m transmission source. *J. Nucl. Med.* **33**(12): 2238–2245.
Gullberg, G. T., Morgan, H. T., Zeng, G. L., Christian, P. E., Di Bella, V. R., Tung, C.-H., Maniawski, P. J., Hsieh, Y.-L., and Datz, F. L. (1998). The design and performance of a simultaneous transmission and emission tomography system. *IEEE Trans. Nucl. Sci.* **45**(3): 1676–1698.
Gullberg, G. T., Zeng, G. L., Christian, P. E., Datz, F. L., Tung, C.-H., and Morgan, H. T. (1992). Review of convergent beam tomography in single photon emission computed tomography. *Phys. Med. Biol.* **37**(3): 507–534.
Hasegawa, B. H., Lang, T. F., Brown, E. L., Gingold, E. L., Reilly, S. M., Blankespoor, S. C., Liew, S. C., Tsui, B. M. W., and Ramanathan, C. (1993). Object-specific attenuation correction for SPECT with correlated dual-energy X-ray CT. *IEEE Trans. Nucl. Sci.* **40**(5): 1242–1252.
Hawman, E. G., and Hsieh, J. (1986). An astigmatic collimator for high sensitivity SPECT of the brain. *J. Nucl. Med.* **27**(6): 930.
Henkin, R. E., Boles, M. A., Dillehay, G. L., Halama, J. R., Karesh, S. M., Wagner, R. H., and Zimmer, A. M. (eds.) (1996). "Nuclear Medicine," Vols. 1–2. Mosby, St. Louis, MO.
Jaszczak, R. J., Chang, L. T., and Murphy, P. H. (1979). Single photon emission computed tomography using multi-slice fan beam collimators. *IEEE Trans. Nucl. Sci.* **26**(1): 610–619.
Jaszczak, R. J., Chang, L. T., Stein, N. A., and Moore, F. E. (1979). Whole-body single-photon emission computed tomography using dual, large-field-of-view scintillation cameras. *Phys. Med. Biol.* **24**(6): 1123–1143.
Jaszczak, R. J., Murphy, P. H., Huard, D., and Burdine, J. A. (1977). Radionuclide emission computed tomography of the head with ^{99m}Tc and a scintillation camera. *J. Nucl. Med.* **18**(4): 373–380.
Kamphuis, C., and Beekman, F. J. (1998). The use of offset cone-beam collimators in a dual head system for combined emission transmission brain SPECT: A feasibility study. *IEEE Trans. Nucl. Sci.* **45**(3): 1250–1254.
Keyes, J. W., Jr, Orlandea, N., Heetderks, W. J., Leonard, P. F., and Rogers, W. L. (1977). The humongotron—a scintillation-camera transaxial tomograph. *J. Nucl. Med.* **18**(4): 381–387.
Keyes, W. I. (1975). Correspondence: The fan-beam gamma camera. *Phys. Med. Biol.* **20**(3): 489–491.
Kunyansky, L. A. (2001). A new SPECT reconstruction algorithm based on the Novikov explicit inversion formula. *Inverse Problems* **17**(2): 293–306.
Lang, T. F., Hasegawa, B. H., Liew, S. C., Brown, J. K., Blankespoor, S. C., Reilly, S. M., Gingold, E. L., and Cann, C. E. (1992). Description of a prototype emission-transmission computed tomography imaging system. *J. Nucl. Med.* **33**(10): 1881–1887.
Lim, C. B., Chang, L. T., and Jaszczak, R. J. (1980). Performance analysis of three camera configurations for single photon emission computed tomography. *IEEE Trans. Nucl. Sci.* **27**(1): 559–568.
Lim, C. B., Gottschalk, S., Walker, R., Schreiner, R., Valentino, F., Pinkstaff, C., Janzo, J., Covic, J., Perusek, A., Anderson, J., Kim, K. I., Shand, D., Coolman, K., King, S., and Styblo, D. (1985). Triangular

SPECT system for 3-D total organ volume imaging: design concept and preliminary imaging results. *IEEE Trans. Nucl. Sci.* **32**(1): 741–747.

Lodge, M. A., Binnie, D. M., Flower, M. A., and Webb, S. (1995). Experimental evaluation of a prototype rotating slat collimator for planar gamma-camera imaging. *Phys. Med. Biol.* **40**(3): 427–448.

Lodge, M. A., Webb, S., Flower, M. A., and Binnie, D. M. (1996). A prototype rotating slat collimator for single photon emission computed tomography. *IEEE Trans. Med. Imaging* **15**(4): 500–511.

Manglos, S. H. (1992). Truncation artifact suppression in cone-beam radionuclide transmission CT using maximum likelihood techniques: Evaluation with human subjects. *Phys. Med. Biol.* **37**(3): 549–562.

Manglos, S. H., Jaszczak, R. J., Floyd, C. E., Hahn, L. J., Greer, K. L., and Coleman, R. E. (1987). Nonisotropic attenuation in SPECT: Phantom tests of quantitative effects and compensation techniques. *J. Nucl. Med.* **28**(10): 1584–1591.

Matherson, K. J., Barber, H. B., Barrett, H. H., Eskin, J. D., Dereniak, E. L., Woolfenden, J. M., Young, E. T., and Augustine, F. L. (1998). Progress in the development of large-area modular 64×64 CdZnTe imaging arrays for nuclear medicine. *IEEE Trans. Nucl. Sci.* **45**(3): 354–358.

Mauderli, W., Luthmann, R. W., Fitzgerald, L. T., Urie, M. M., Williams, C. M., Tosswill, C. H., and Entine, G. (1979). A computerized rotating laminar radionuclide camera. *J. Nucl. Med.* **20**(4): 341–344.

Mettler, F. A., and Guiberteau, M. J. (1991). "Essentials of Nuclear Medicine Imaging". 3rd ed. W. B. Saunders, Philadelphia.

Metz, C. E., Atkins, F. B., and Beck, R. N. (1980). The geometric transfer function component for scintillation camera collimators with straight parallel holes. *Phys. Med. Biol.* **25**(6): 1059–1070.

Milster, T. D., Aarsvold, J. N., Barrett, H. H., Landesman, A. L., Mar, L. S., Patton, D. D., Roney, T. J., Rowe, R. K., and Seacat, R. H. 3rd, (1990). A full-field modular gamma camera. *J. Nucl. Med.* **31**(5): 632–639.

Muehllehner, G., and Wetzel, R. A. (1971). Section imaging by computer calculation. *J. Nucl. Med.* **12**(2): 76–84.

National Electrical Manufacturers Association. (2001). "Performance Measurements of Scintillation Cameras." NEMA Standards Publication no. NU-1-2001. NEMA, Washington, D.C.

Natterer, F. (2001). Inversion of the attenuated Radon transform. *Inverse Problems* **17**(1): 113–119.

Nohara, N., Murayama, H., and Tanaka, E. (1987). Single photon emission computed tomography with increased sampling density at central region of field of view. *IEEE Trans. Nucl. Sci.* **34**(1): 359–363.

Noo, F., and Wagner J.-M. (2001). Image reconstruction in 2D SPECT with 180° acquisition. *Inverse Problems* **17**(5): 1357–1371.

Novikov, R. G. (2000). An inversion formula for the attenuated x-ray transformation. Preprint CNRS, UMR 6629, Departement de Mathematiques, Universite de Nantes.

Patt, B. E., Iwanczyk, J. S., Rossington Tull, C., Wang, N. W., Tornai, M. P., and Hoffman, E. J. (1998). High resolution CsI(Tl)/Si-PIN detector development for breast imaging. *IEEE Trans. Nucl. Sci.* **45**(4): 2126–2131.

Rogers, W. L., Koral, K. F., Mayans, R., Leonard, P. F., Thrall, J. H., Brady, T. J., and Keyes, J. W., Jr. (1980). Coded-aperture imaging of the heart. *J. Nucl. Med.* **21**(4): 371–378.

Rowe, R. K., Aarsvold, J. N., Barrett, H. H., Chen, J. C., Klein, W. P., Moore, B. A., Pang, I. W., Patton, D. D., and White, T. A. (1993). A stationary hemispherical SPECT imager for three-dimensional brain imaging. *J. Nucl. Med.* **34**(3): 474–480.

Sandler, M. P., Coleman, R. E., Patton, J. A., Wackers, F. J. Th., and Gottschalk, A., (eds.) (2003)."Diagnostic Nuclear Medicine." 4th ed. Lippincott Williams & Wilkins, Philadelphia.

Sidky, E., and Pan, X. (2002). Variable sinogram and redundant information in SPECT with non-uniform attenuation. *Inverse Problems* **18**(6): 1483–1479.

Singh, M., and Doria, D. (1984). Germanium-scintillation camera coincidence detection studies for imaging single-photon emitters. *IEEE Trans. Nucl. Sci.* **31**(1): 594–598.

Simpson, R. G., and Barrett, H. H. (1980). Coded-aperture imaging. In "Imaging in Diagnostic Medicine" (S. Nudelman, ed.), pp. 217–311. Plenum, New York.

Sorenson, J. A., and Phelps, M. E. (1987). "Physics in Nuclear Medicine." 2nd ed. W. B. Saunders, Philadelphia.

Tan, P., Bailey, D. L., Meikle, S. R., Eberl, S., Fulton, R. R., and Hutton, B. F. (1993). A scanning line source for simultaneous emission and transmission measurements in SPECT. *J. Nucl. Med.* **34**(10): 1752–1760.

Taylor, A., Schuster, D. M., and Alazraki, N. (2000). "A Clinician's Guide to Nuclear Medicine." Society of Nuclear Medicine, Reston, VA.

Tsui, B. M. W., and Gullberg, G. T. (1990). The geometric transfer function for cone and fan beam collimators. *Phys. Med. Biol.* **35**(1): 81–93.

Tsui, B. M., Gullberg, G. T., Edgerton, E. R., Gilland, D. R., Perry, J. R., and McCartney, W. H. (1986). Design and clinical utility of a fan beam collimator for SPECT imaging of the head. *J. Nucl. Med.* **27**(6): 810–819.

Tung, C.-H., Gullberg, G. T., Christian, P. E., Datz, F. L., and Morgan, H. T. (1992). Nonuniform attenuation correction using simultaneous transmission and emission converging tomography. *IEEE Trans. Nucl. Sci.* **39**(4): 1134–1143.

Urie, M. M., Mauderli, W., Fitzgerald, L. T., and Williams, C. M. (1979). Cadmium telluride gamma camera. *IEEE Trans. Nucl. Sci.* **26**(1): 552–558.

Webb, S., Binnie, D. M., Flower, M. A., and Ott, R. J. (1992). Monte Carlo modelling of the performance of a rotating slit-collimator for improved planar gamma-camera imaging. *Phys. Med. Biol.* **37**(5): 1095–1108.

Webb, S., Flower, M. A., and Ott, R. J. (1993). Geometric efficiency of a rotating slit-collimator for improved planar gamma-camera imaging. *Phys. Med. Biol.* **38**(5): 627–638.

Wessell, D. E., Tsui, B. M. W., Frey, E. C., and Karimi, S. S. (1998). Rotating quadrant slant-hole SPECT scintimammography: An initial investigation 1997. *Conf. Rec. IEEE Nucl. Sci. Symp. Med. Imaging Conf.* **2**: 1145–1149.

Zeng, G. L., Gullberg, G. T., Christian, P. E., Gagnon, D., and Tung, C.-H. (2001). Asymmetric cone-beam transmission tomography. *IEEE Trans. Nucl. Sci.* **48**(1): 117–124.

CHAPTER

8

Collimator Design for Nuclear Medicine

DONALD L. GUNTER

Department of Physics and Astronomy, Vanderbilt University, Nashville, Tennessee

I. Basic Principles of Collimator Design
II. Description of the Imaging System and Collimator Geometry
III. Description of Collimator Imaging Properties
IV. Septal Penetration
V. Optimal Design of Parallel-Hole Collimators
VI. Secondary Constraints
VII. Summary

The work reported in this chapter was begun in the 1980s while the author was working for Professor Robert N. Beck at The University of Chicago. New technologies are changing the problems of collimator design, but the fundamental principles are the same as those discovered by Beck in the 1960s. Together with his colleagues at the Franklin McLean Memorial Institute, Beck oversaw the development of collimator design from simply drilling holes in lead to a precise science. Those of us raised with computer technology stand in awe of the results produced without a single computer simulation, when each new experiment involved casting a new collimator and each Fourier transform meant fitting data and consulting integral tables. Collimator design may be more sophisticated today, but the innovations and insights of Beck are the basis of all our current progress.

This chapter is intended for graduate students who are encountering collimators for the first time. (A more mathematical account of this material can be found in Gunter, 1996.) The most important ideas in collimator design concern the relationship between the collimator geometry and the resulting imaging properties. Once this relationship is understood, we can devise strategies for the optimal design of collimators. Only in the 1990s have such strategies had a significant impact on collimator construction. The optimization strategy explained in Section V is the outgrowth of 40 years of research by many workers in the field. However, not all the problems are solved. For many applications, the optimal collimator design is precluded by secondary constraints (as described in Section VI). Thus, despite the fact that collimator design is now a mature science, there remain numerous problems for the enterprising graduate student.

I. BASIC PRINCIPLES OF COLLIMATOR DESIGN

In nuclear medicine, the collimator is a crucial component of the gamma-ray camera. A collimator resembles a lead (Pb) honeycomb that is placed between the patient and the gamma-ray detector. Unlike optical photons, gamma rays cannot be refracted and focused. Consequently, the images in a gamma-ray camera are formed by selective absorption. Radiation emitted by radiopharmaceuticals within the patient must pass through the collimator to reach the detector. Gamma rays that hit the collimator walls (called septa) are generally absorbed. As a result, only gamma rays traveling in the direction of the hole axes appear in the images. Although this selective absorption is necessary for the creation of the images, it is also very inefficient because most of the gamma rays are absorbed in the collimator. For the collimators used in nuclear medicine, only about one in ten thousand of the emitted gamma rays pass through the collimator. Thus, the design of the collimator has a significant effect on the overall performance of the gamma camera. Not only does the collimator determine the resolution of the camera, but it is also the major limitation on the number of counts detected.

Collimator design requires detailed knowledge of the relation between the collimator geometry and the resulting imaging properties. In the nuclear medicine clinic, collimators are characterized by their imaging properties, resolution and sensitivity. On the other hand, the collimator manufacturer must know the geometric parameters such as collimator thickness, septal thickness, and hole diameter. As we expect, simple geometry dictates most of the imaging properties, so the relation appears straightforward. Unfortunately, collimators are plagued by numerous secondary problems that are only indirectly related to the geometry and, consequently, collimator design has remained a vexing problem since the 1960s. Among these complicating problems are the penetration of radiation through the collimator septa, the visibility of collimator hole pattern in the images, weight constraints imposed by the camera gantry, the minimal septal thickness imposed by the limitations of fabrication, and the effects of scattering within the collimator. If a collimator is properly designed, these secondary effects are not noticeable. However, naive designs based solely on geometry without careful attention to the secondary problems may produce either images that are significantly degraded or collimators that are impractical in the clinic.

II. DESCRIPTION OF THE IMAGING SYSTEM AND COLLIMATOR GEOMETRY

Before attempting to design a collimator, we must know four fundamental properties of the overall imaging system and the gamma-ray camera. First, the energy E of the incident radiation must be known because the attenuation within the collimator septa is determined by the energy. Second, the typical distance F between the source (within the patient) and the front face of the collimator is important in collimator design (see Figure 1). In most clinical applications, the distance F is in the range of 10 to 30 cm. However, for imaging small animals, F may be 2 to 5 cm, and in astronomy $F = \infty$ is reasonable. Third, the distance B, which describes the gap between the back of the collimator and the imaging plane within the camera, is very important. In general, small values of B improve the collimator resolution, so that camera designers have attempted to minimize this

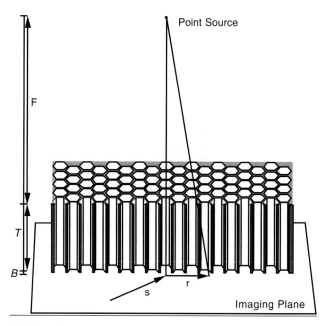

FIGURE 1 Basic collimator geometry and notation are shown. The distance B is the gap between the imaging plane and the back of the collimator, T is the thickness of the collimator, and F is the distance between a typical source and the front of the collimator. The vector **s** denotes a point in the imaging plane directly beneath the source. The vector **r** denotes a displacement in the imaging plane from the source position.

gap. However, the effects of collimator hole pattern in images can be reduced by the careful selection of B. As a result, for special applications in which collimator hole pattern is a problem, larger values of B may be desirable. Finally, the intrinsic resolution of the gamma camera must be known. Most gamma cameras generate images by Anger logic. The intrinsic resolution of such cameras can be characterized by Gaussian blurring that is parameterized by the full width at half maximum (FWHM$_{cam}$). Typically, FWHM$_{cam}$ of commercial gamma cameras ranges between 2 and 5 mm. (Recently, digital gamma cameras that generate pixelated images have been made with solid-state detectors. At first glance, the pixel size of these cameras determines the intrinsic resolution. However, due to the interplay between the pixel lattice and the collimator hole array, the imaging properties of these pixelated cameras can be very complicated and require careful analysis.) In general, these four parameters (E, F, B, and FWHM$_{cam}$) describe the overall imaging system and cannot be adjusted by the collimator designer.

The designer can control the collimator material and geometry. Collimator materials must satisfy two important properties. First, the material must have a high attenuation coefficient μ at the energy E so that incident gamma rays are absorbed in the septa. Generally, this implies high-density materials because the attenuation coefficient is proportional to the density. Second, the interactions of the gamma radiation within the collimator material should produce few secondary photons (which can contaminate the images). Thus, photoelectric interactions (that produce electrons and low-energy secondary photons via fluorescence) are preferable to either Compton or coherent scattering (that produce scattered gamma rays). For gamma-ray energies in the range 80–511 keV (which are typical in nuclear medicine), the fraction of photoelectric interactions increases with atomic number Z. Thus, the best collimator materials have high density and high Z. For most practical applications at low energies ($E > 230$ keV), Pb is an excellent material for collimators. Because Pb is ductile and has a relatively low melting point, collimators can be fabricated by either folding foil or casting. For high-energy radiation ($E \geq 230$ keV) other materials such as tungsten, gold, and tantalum have been used for collimation, but the cost and fabrication problems have generally limited their commercial use. Tungsten is frequently discussed for collimation of high-energy radiation, but the hard, brittle metal has a high melting point that prohibits casting and renders construction very difficult and expensive. Advances in fabrication methods have overcome some of these problems. However, these techniques remain relatively expensive and are not yet generally available. As a result, the collimator designer usually has little choice: the collimator material is almost always lead.

Collimator geometry is the domain in which the designer can practice ingenuity. The geometry can be adjusted either globally or locally. Global geometry refers to the alignment and arrangement of the holes throughout the collimator, whereas local geometry refers to the specific dimensions (e.g., hole diameter, septal thickness, and collimator thickness) required for construction. Generally, the global geometry of collimators is classified based on the alignment of the hole axes. The two basic design classifications are parallel-hole and converging-hole designs. Most clinical collimators have parallel holes aligned normal to the camera face in a regular lattice array. Among various collimator manufacturers, the hole shape and lattice array vary greatly. Hexagonal holes in a hexagonal array are the most popular, but triangular, square, and circular holes in various different arrays are also common. In general, the imaging properties and design of parallel-hole collimators are well understood, although serious problems still remain for collimators intended for high-energy radiation. The most challenging problems are associated with converging-hole collimators. The advantage of a converging-hole collimator is that, within a limited field of view, the detection efficiency (and, therefore, the number of detected gamma rays) increases. The disadvantage is that the gamma camera response is no longer translationally invariant. As a result, the planar images are distorted and SPECT reconstructions are much more complicated. Because the design of converging-hole collimators is extremely difficult, these collimators will not be discussed further in this chapter.

Parallel-hole collimators are specified by the lattice structure of the hole pattern and three parameters. Figure 2 shows the cross sections of three typical hole patterns used in collimator design: (a) hexagonal holes in a hexagonal array, (b) square holes in a square array, and (c) triangular holes in a point-to-point array. Because the hole pattern is periodic, the geometry need only be described inside a single lattice cell with lattice vectors l_1 and l_2. The hole pattern can then be tessellated over the entire collimator face by translations of integer numbers of lattice vectors. Once the hole pattern is selected, three dimensions are sufficient to determine the collimator geometry: (1) the collimator thickness T, (2) the hole separation HOLSEP, and (3) the hole size S. The collimator thickness T is unambiguous, but the hole separation HOLSEP and hole size S require further explanation. The hole separation HOLSEP is defined as the distance between the centers of adjacent lattice cells. Care is required for triangular holes because holes of opposite orientation appear in each lattice cell. The hole separation must be measured between the centers of similarly oriented holes as shown in Figure 2c. The definition of hole size also requires care and consistent notation. Generally, the hole cross sections are equilateral polygons, so that the length of a hole side S uniquely determines the hole size. For hexagonal holes, on the other hand, the distance face to face (F2F) across the hole (see Figure 2a) is more useful for estimations of collimator performance. (Obviously, the F2F distance is interchangeable with S because F2F = $\sqrt{3}\,S$.) The average distance across the hole plays a crucial role in the resolution of the collimator. For hexagonal holes, F2F provides a better estimate of the distance across the hole than does S. As a consequence of this observation, a hybrid parameter, called the hole diameter D and defined by the relation

$$\text{Area of the hole} = \frac{\pi}{4} D^2 \qquad (1)$$

(for all hole shapes) is very useful in collimator design. This parameter D describes a mean geometric diameter associated with the different hole shapes. Table 1 shows the relation between S and D for various holes. For the remainder of this chapter, D is used rather than S. The septal thickness (SPT) is determined by HOLSEP and D (as shown in Table 1). The three dimensions (T, D, HOLSEP) completely describe the collimator. The space of possible collimator designs is restricted by the positivity of these physical lengths, so that

$$0 < T, \quad 0 < D, \quad 0 < \text{HOLSEP}, \quad 0 < \text{SPT} \qquad (2)$$

The problem of designing a parallel-hole collimator is that of selecting a particular point in this three-dimensional (3D) parameter space. The criterion for this selection must be based primarily on the imaging properties of the collimator—namely, the collimator resolution and sensitivity. Therefore,

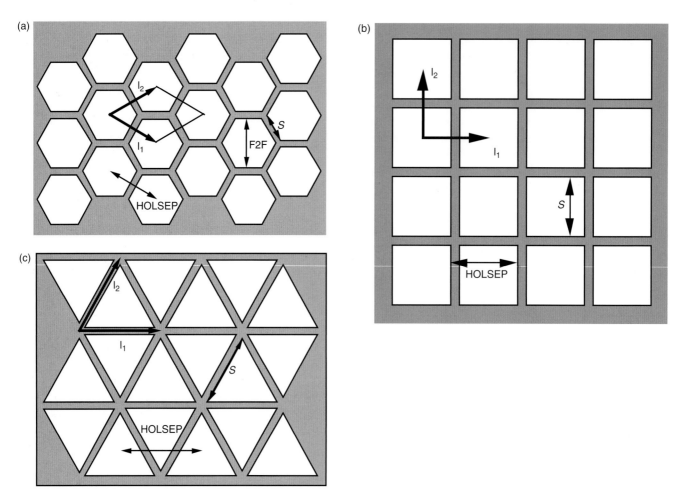

FIGURE 2 Three basic collimator hole patterns are shown. In each case, the distance between adjacent hole centers are shown and indicated as HOLSEP. (a) Hexagonal holes are shown in a hexagonal array. The hole diameter is commonly given in terms of the face-to-face distance (F2F) across the hole or the length of a hole side S. (b) Square holes are shown in a square array. (c) Triangular holes are shown in a tip-to-tip array.

we must know the relation between the collimator dimensions (T, D, HOLSEP) and the imaging properties.

III. DESCRIPTION OF COLLIMATOR IMAGING PROPERTIES

For most workers in nuclear medicine, collimator performance is characterized by two quantities: sensitivity and resolution. These two quantities are adequate if the collimator (1) has parallel holes, (2) does not permit significant penetration, and (3) has hole separation less than the intrinsic resolution of the camera. In general, however, the description of collimator performance requires the detailed measurement of a complicated function called the point source response function (PSRF).

The PSRF describes the image produced by a point source as a function of the source position and the location in the imaging plane. The source position is described by a three-dimensional vector; however, because of the special role of the imaging plane, the source position is usually decomposed into a two-dimensional vector $\mathbf{s}$ in the imaging plane and the distance z in front of the imaging plane (see Figure 1). The location within the imaging plane is denoted by a vector $\mathbf{x}$. The point $\mathbf{x} = \mathbf{s}$ corresponds to the position directly beneath the source where we would expect the image from a parallel-hole collimator to be maximum. Consequently, the vector $\mathbf{r} = \mathbf{x} - \mathbf{s}$ is more convenient than $\mathbf{x}$ for specifying positions in the imaging plane. The point $\mathbf{r} = 0$ corresponds approxi-

TABLE 1

Parameter\Hole Shape	Triangular Holes	Square Holes	Hexagonal Holes
Diameter (D)	$D = \dfrac{\sqrt[4]{3}}{\sqrt{\pi}} S = 0.7425\, S$	$D = \dfrac{2}{\sqrt{\pi}} S = 1.128\, S$	$D = \dfrac{\sqrt[4]{108}}{\sqrt{\pi}} S = 1.819\, S$
Area of lattice cell	$\dfrac{\sqrt{3}}{2}\, \text{HOLSEP}^2$	HOLSEP^2	$\dfrac{\sqrt{3}}{2}\, \text{HOLSEP}^2$
Septal thickness (SPT)	$\left(\dfrac{1}{\sqrt{3}} \text{HOLSEP} - \dfrac{\sqrt{\pi}}{\sqrt[4]{27}} D \right)$	$\left(\text{HOLSEP} - \dfrac{\sqrt{\pi}}{2} D \right)$	$\left(\text{HOLSEP} - \dfrac{\sqrt{\pi}}{\sqrt[4]{12}} D \right)$
Sensitivity ($\$$)	$\dfrac{4}{\sqrt{3}} \dfrac{\pi D^4}{64\, T^2\, \text{HOLSEP}^2}$	$\dfrac{\pi D^4}{64\, T^2\, \text{HOLSEP}^2}$	$\dfrac{2}{\sqrt{3}} \dfrac{\pi D^4}{64\, T^2\, \text{HOLSEP}^2}$

mately to the central maximum of the PSRF. The PSRF is defined in terms of these vectors by the relation

$$\text{PSRF}(\mathbf{r}, \mathbf{s}, z) = \frac{\left[\begin{array}{c}\text{Number of counts/cm}^2 \text{ detected at point}\\ \mathbf{x} = (\mathbf{s} + \mathbf{r}) \text{ in the imaging plane}\\ \text{from a point source located at } (\mathbf{s}, z)\end{array}\right]}{[\text{Number of counts emitted from}\atop \text{the point source located at } (\mathbf{s}, z)]} \quad (3)$$

where the PSRF has units cm^{-2} (area^{-1}). The PSRF completely describes the imaging process. Unfortunately, a comparison of the PSRFs of two collimators is very difficult because the PSRF is a function on a five-dimensional space (two 2D vectors, $\mathbf{s}$ and $\mathbf{r}$, and a distance parameter, z). For parallel-hole collimators, the PSRF is periodic with respect to the source position because translations from one hole to the next do not affect the imaging properties of the system. Thus, we can write

$$\text{PSRF}(\mathbf{r}, \mathbf{s} + m\mathbf{l}_1 + n\mathbf{l}_2, z) = \text{PSRF}(\mathbf{r}, \mathbf{s}, z) \quad (4)$$

where m and n are integers. Without much loss of information, therefore, we can average over a lattice cell of the collimator and define

$$\psi(\mathbf{r}, z) = \int_0^1 dt_1 \int_0^1 dt_2\, \text{PSRF}(\mathbf{r}, t_1\mathbf{l}_1 + t_2\mathbf{l}_2, z) \quad (5)$$

The source position $\mathbf{s}$ is averaged out, thereby leaving $\psi(\mathbf{r}, z)$ as a function on a 3D space. For purposes of comparison, the distance of the point source in front of the collimator F is generally set equal to a typical value for clinical applications (10 or 15 cm is common in nuclear medicine); so that, $z = F+T+B$ (see Figure 1) is constant and the PSRF reduces to a function of the two-dimensional vector $\mathbf{r}$. The PSRF is anisotropic with respect to $\mathbf{r}$ because the hole shape and lattice structure break the rotational symmetry, so that further dimensional reduction is not possible in a rigorous analysis. However, the PSRF depends on the magnitude $|\mathbf{r}|$ much more than on the direction of $\mathbf{r}$. Consequently, the PSRF is frequently reported as a simple one-dimensional function, that is, $f(r) = \psi(r\mathbf{e}_1, z)$. Although this function f is not sufficient for a complete description of the imaging properties of the collimator, it is indicative of collimator performance. Unfortunately, even this simplified function is too complicated for comparisons between collimators. Most clinical users want just one or two simple parameters that will characterize the collimator, not a complicated function. The sensitivity and resolution provide such a summary of collimator performance based on the most important properties of the PSRF.

Collimator sensitivity is defined as the fraction of emitted photons that pass through the collimator and reach the imaging plane. In general, the sensitivity depends on the position of the source. The collimator sensitivity is commonly denoted by the symbol $\$$ and defined by the ratio

$$\$(\mathbf{s}, z) = \frac{\left[\begin{array}{c}\text{Number of counts detected from a point}\\ \text{source located at } (\mathbf{s}, z)\end{array}\right]}{[\text{Number of counts emitted from}\atop \text{the point source located at } (\mathbf{s}, z)]} \quad (6)$$

$$= \iint d^2\mathbf{r}\, \text{PSRF}(\mathbf{r}, \mathbf{s}, z)$$

(This dollar sign $\$$ notation for sensitivity was first introduced by Professor Robert Beck.) According to this definition, the collimator sensitivity $\$$ is a fraction that must lie between zero and one-half, $(0.0 \leq \$ \leq 0.5)$. The upper limit (0.5) is obtained because one-half of the emitted photons will travel away from the collimator and cannot possibly be detected in the imaging plane. In practice, the collimator sensitivity is much less than one-half; a typical value is 3×10^{-4}. This definition of sensitivity is just the efficiency of photons passing through the collimator. Indeed, the sensitivity $ is called the (geometric) efficiency of the collimator by some authors. This fraction, however, is not convenient in the nuclear medicine clinic, where doses are administered in millicuries (mCi) and counted for minutes. Consequently, another unit of sensitivity is often used in the clinic (and quoted in the commercial sales literature). This sensitivity is denoted by the symbol ¢ and is defined as the number of counts passing through the collimator per minute per microcurie (µCi) of source (often denoted as cpm/µCi). Because 1 curie = 3.7×10^{10} decays/s, the conversion between $ and ¢ is given by the relation ¢ = (2.22×10^6) $. Thus, the sensitivity of a typical collimator, $ = 2×10^{-4}, is equivalent to ¢ = 444 cpm/µCi. For a clinical dose of 10 mCi, we would, therefore, expect 4.4 million counts per minute. This count rate is much higher than observed in the clinic because only a fraction of the clinical dose is deposited within the field of view of the collimator and because the patient tissue absorbs part of the emissions. A significant warning is required concerning comparison of $ and ¢ for 511 keV annihilation radiation. For each positron annihilation two photons are emitted. The collimator sensitivity $ is unaffected by this double emission because both the number of photons emitted and the number detected are doubled, so that the ratio in Eq. (5) is unchanged. However, the clinical sensitivity ¢ doubles because the number of emissions per decay is doubled (*N.B.*: µCi refers to the number of decays, not the number of emissions). Extreme care must be taken in comparisons quoted in the literature because some researchers interchange the definitions without noting this factor of 2. For the purposes of collimator design, $ is preferable because the imaging properties of the collimator are unrelated to the number of photons emitted. On the other hand, the clinical utility of ¢ cannot be denied.

For general collimators, the sensitivity depends on the source position (**s**, z). The definition of sensitivity can be further simplified for parallel-hole collimators. Because of the periodic lattice structure of the collimator holes, the average sensitivity can be defined by

$$\langle \$(z) \rangle = \int_0^1 dt_1 \int_0^1 dt_2 \, \$\left(t_1 \mathbf{l}_1 + t_1 \mathbf{l}_2, z\right) \quad (7)$$

For parallel-hole collimators, Eq. (7) reveals that the average sensitivity depends only on the distance of the source from the imaging plane z. Actually, the average sensitivity of parallel-hole collimators is virtually independent of source position. This observation is extremely important and should be understood. When a photon enters the collimator face, the collimator effectively asks the question, "What is the direction of the photon momentum with respect to the hole axis?" If the photon is traveling parallel to the hole axis, it will probably pass through unimpeded. On the other hand, if the photon momentum makes a large angle with respect to the hole axis, it will probably hit a septal wall and be absorbed. Thus, photons are filtered by collimators based on the direction of the momentum. For a parallel-hole collimator, the photon momentum is compared with the same direction (the hole axis) at every point on the collimator face. Assuming that no scattering occurs within the patient, the direction of photon momentum is determined at the time of emission, has an isotropic distribution, and is independent of source position. Therefore, the probability of a gamma ray passing through the collimator does not depend on the source position. The only major limitation on this observation arises if the source lies near the edge of the field-of-view of the camera.

For parallel-hole collimators with perfect septal absorption (i.e., no septal penetration) and convex hole shapes, the PSRF can be evaluated analytically in terms of the collimator geometry. All the subsequent collimator properties can then be extracted from the PSRF. This mathematical exercise is beyond the scope of this chapter (see Gunter 1996 for complete mathematical derivations). The general formula for parallel-hole collimator sensitivity can derived in this manner and is very useful:

$$\langle \$ \rangle = \frac{\sum_{i \,=\, \text{collimator hole within a single lattice cell}} [\text{Area of hole } (i)]^2}{4\pi T^2 [\text{Area of a lattice cell}]} \quad (8)$$

For hexagonal and square holes, each lattice cell contains only one hole; so that only one term arises in the numerator of Eq. (8). However, for triangular holes, two holes of opposite orientation are contained within each lattice cell (see Figure 2c), so that two terms are required in the numerator. In general, Eq. (8) yields

$$\langle \$ \rangle = C_{\text{hole shape}} \frac{\pi D^4}{64 \, T^2 [\text{HOLSEP}]^2} \quad (9)$$

where

$$C_{\text{square}} = 1.0, \; C_{\text{hex}} = \frac{2}{\sqrt{3}} \approx 1.155, \; C_{\text{triangle}} = \frac{4}{\sqrt{3}} \approx 2.309 \quad (10)$$

We should note that if all the collimator dimensions (T, D, HOLSEP) are multiplied by a scaling factor, the expression for average collimator sensitivity <$> is unchanged. Thus, rescaling a collimator has no effect on the sensitivity. If we set typical values of $T = 12\,D$ and HOLSEP = $1.05\,D$, then $ = 3.6×10^{-4} for hexagonal-hole collimators.

The resolution of a parallel-hole collimator requires much more careful analysis than the sensitivity. Resolution is generally reported as the smallest separation at which two point sources can be distinguished. This minimal separation is significantly dependent on the distance of the sources from the imaging plane. Thus, two distances are required for specification of the collimator resolution—the minimum resolving distance and the distance of the source in front of the collimator F. A detailed analysis of the resolution requires all the information contained in the PSRF. In practice, such detailed analysis is not feasible. As a result, we generally assume that, for fixed z, $\psi(\mathbf{r}, z)$ is a Gaussian function of $\mathbf{r}$, that is,

$$\text{PSRF}(\mathbf{r}, \mathbf{s}, z) = \psi(\mathbf{r}, z) \propto \exp\left[-\frac{4\ln 2 |\mathbf{r}|^2}{\text{FWHM}_{\text{col}}^2}\right] \quad (11)$$

Thus, the PSRF can be characterized by a single parameter, FWHM_{col}, for a point source located at a distance z in front of the imaging plane (or F in front of the collimator face). Two sources closer than FWHM_{col} at distance F from the collimator face cannot be distinguished. The resolution, therefore, is reported as "the FWHM_{col} at distance F in front of the collimator face."

Unfortunately, an obvious conceptual error is glossed over in the Gaussian approximation. In averaging over the lattice cell in the definition of $\psi(\mathbf{r}, z)$ in Eq. (5), we assume that the lattice structure of the hole pattern is not observed. However, the lattice structure of the collimator holes is often observable, especially for point sources. A particularly bothersome problem arises when a single point source is located near the collimator face and directly over a septum. The resulting image will display the shadow of the septum appearing between two local maxima corresponding to the adjacent holes. This bifurcation of a point source is a major problem. If a single point source can produce two apparently disjoint images, how can we distinguish two point sources at distances greater than FWHM_{col}? Are we seeing two sources or one source divided by the collimator hole pattern? Generally, in collimator design the concept of resolution is isolated from the issue of hole pattern visibility and treated separately. In reality, however, the two issues must be treated together in the context of the full PSRF.

Once the assumption of a stationary Gaussian PSRF is made, the FWHM_{col} can be estimated very easily. Figure 3 illustrates a simple, back-of-the-envelope, calculation of the FWHM_{col}. The geometric resolution of the collimator (denoted by R_g) is defined as the radius $|\mathbf{r}|$ such that $|\mathbf{r}| > R_g$ implies that no ray can pass through the collimator without penetrating a septum. A simple argument based on similar triangles leads to the relation

$$R_g(F) = \frac{D}{T}z = \frac{D}{T}(T+B+F) \approx \frac{D}{T}F \quad (12)$$

The geometric resolution R_g once again illustrates the scaling property of collimators. The only collimator dimensions that appear in R_g are the lengths D and T. For most collimator designs ($B \ll T \ll F$), the geometric resolution is determined by the ratio (D/T) and the distance F. Thus, a rescaling of the collimator parameters has only a minor effect on the resolution. For parallel-hole collimators, R_g is a good estimate of the FWHM_{col} because (for all hole shapes)

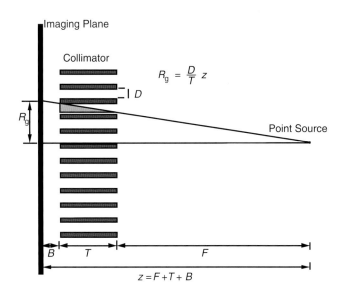

FIGURE 3 A geometric estimation of collimator resolution is shown. The crucial ray in this diagram enters the front of the collimator on one side of a hole and exits on the other side of the hole. Thus, the ray travels diagonally across the hole. Any ray with more inclination will hit a septum and be absorbed. Thus, all gamma rays emitted by the point source and passing through the collimator must lie within the radius R_g in the imaging plane. Consequently, R_g is a good estimation of the collimator resolution. Similar triangles give the desired algebraic relation.

$$R_g \cdot \text{FWHM}_{\text{col}} < 2R_g \quad (13)$$

so that

$$\text{FWHM}_{\text{col}} = K_{\text{hole shape}}\frac{D}{T}z = K_{\text{hole shape}}\frac{D}{T}(T+B+F) \quad (14)$$

where $K_{\text{hole shape}}$ is a constant depending only on the hole shape and satisfies

$$1 \leq K_{\text{hole shape}} < 2 \quad (15)$$

The most important result of this calculation is that FWHM_{col} increases linearly with z, which explains why it is crucial that any description of resolution must include both FWHM_{col} and F.

The most famous observation concerning parallel-hole collimators arises from the combination of Eqs. (9) and (14):

$$\langle \$ \rangle = \frac{C_{\text{hole shape}}}{K_{\text{hole shape}}^2}\left(\frac{\pi D^2}{64[\text{HOLSEP}]^2}\right)\left[\frac{\text{FWHM}_{\text{col}}}{z}\right]^2 \quad (16)$$

From this relation, we conclude that $\langle \$ \rangle$ can be increased in two ways. First, the term in parentheses ( ) can be increased by reduction of the septal thickness. However, reductions in septal thickness are severely limited by the problems of septal penetration and collimator fabrication. Second, the term in the square brackets [ ] in Eq. (16) can be increased by increasing the FWHM_{col} at fixed distance z. Thus, we can increase the collimator sensitivity by degrading the resolution. Conversely, the resolution can be improved (i.e., $FWHM_{col}$ decreased) only if the sensitivity is decreased. This trade-off between sensitivity and resolution is justly famous and is generally written as the proportionality

$$\langle \$ \rangle \propto FWHM_{col}^2 \qquad (17)$$

This result can be understood intuitively because the sensitivity can only be increased if the collimator accepts gamma rays that enter at larger angles with respect to the hole axis. Thus, the trade-off is a general result for all parallel-hole collimators and is independent of detailed considerations of collimator dimensions or hole shape.

IV. SEPTAL PENETRATION

All the results in the last section are based on the assumption that no radiation penetrates through the septal walls of the collimator. On the other hand, collimator sensitivity is maximized by the thinnest possible septa. Thus, the collimator designer naturally wants to build collimators with thin septa. Unfortunately, thin septa can lead to septal penetration, and penetration is extremely detrimental to diagnostic performance. Although the penetrating radiation increases the sensitivity of collimator, the additional counts do not come in the sharp central peak of the PSRF [or $\psi(\vec{r}, z)$]. Instead, penetration contributes a long tail to the PSRF, so that an overall background is added to the images. This penetration background reduces contrast and, therefore, masks subtle signals that might otherwise be visible.

We might expect that penetration would increase the FWHM, thereby decreasing the resolution of the collimator. But such an effect is seldom observed unless the collimator exhibits truly catastrophic penetration. Most of the penetration appears as an increase in counts at radii $|\mathbf{r}|$ larger than the FWHM, so that neither the central peak nor the half-width are significantly affected. For this reason, a collimator can have apparently good sensitivity and resolution, but, nonetheless, produce unacceptable images due to septal penetration. Consequently, another parameter is required as an indicator of penetration. The full width at tenth maximum (FWTM) is often quoted as a measurement of penetration. In the absence of penetration, we expect that

$$R_g < FWHM < FWTM < 2R_g \qquad (18)$$

Thus, if the FWTM is greater than twice the FWHM, we have a clear indication of penetration. Although the measured FWTM is a good indicator of septal penetration, the collimator designer must note two drawbacks in its application. First, there are secondary problems associated with penetration that are not measured by the FWTM. In particular, septal penetration is generally anisotropic in the images. The amount of penetration is determined by the distance

FIGURE 4 An image of radiation from a point source of I-131 (365 keV) is shown for a standard high-resolution collimator (designed for 140 keV). A star-pattern of preferential penetration is found in the directions for which the septa are thinnest.

traveled through the absorbing septal material. This distance is affected by the angle at which the radiation hits the septal wall. Rays striking the septal walls at small angles are more likely to be absorbed, whereas those hitting perpendicular to the wall are most likely to penetrate. As a result, penetration can appear as a starlike pattern (see Figure 4) with outward spokes perpendicular to the septal faces. Hexagonal- and square-hole collimators are most susceptible to these artifacts. Obviously, penetration is worse for high-energy radiation for which the attenuation coefficients are smaller. The worst case of penetration ever observed by the author was a commercial collimator supposedly designed for 511 keV annihilation radiation. A whole-body scan using this collimator revealed a large source in the bladder. Unfortunately, the collimator was completely inadequate for 511 keV and a penetration artifact could be seen extending from the bladder to the umbilicus (at least 10 cm). The second problem associated with the FWTM is that its estimation from the collimator geometry is very difficult. During the early years of nuclear medicine, heuristic criteria for acceptable penetration were devised. These criteria were often inaccurate.

During the 1980s, Professor Beck's research group at The University of Chicago began an extensive study of collimator penetration using computer simulations. A ray-tracing program called RAYTRC was written (by the author) that produced the full PSRF (including the penetration effects) of any parallel-hole collimator design. For each collimator design, over 500,000 rays were traced through the collimator. The attenuation of each ray was determined by the distance traveled through the septal material. The PSRF of the collimator was then compiled from the results of the individual rays. Over a 5-year period, tens of thousands of collimators were evaluated using this program. Much of that data analysis was based on the rescaling property of collimators.

For ray-tracing programs, rescaling of the collimator is a particularly useful operation. The attenuation of each ray is determined by the distance traveled in the septal material, which is rescaled like all the other dimensions. As a result, the ray-tracing operation need be performed only once and the PSRFs for all the rescaled collimators are determined simultaneously. As noted earlier, in the absence of penetration, the rescaling of the collimator dimensions has no effect on sensitivity and little effect on resolution. However, rescaling the collimator does have a dramatic effect on septal penetration. Above a certain critical size, septal penetration is negligible and the sensitivity and resolution are constant. But as the septal thickness is reduced by the rescaling, eventually the septa become too thin and penetration becomes significant. Below this critical size, the FWTM increases dramatically, followed by similar increases in the collimator sensitivity $ and resolution $FWHM_{col}$. Because the penetration arises from the exponential attenuation in the septal walls, the transition from acceptable to unacceptable penetration is extremely sharp. Indeed, the transition is so sharp that the analogy of flipping a light switch seems appropriate. As a result, knowledge of this penetration threshold is crucial for the design of collimators. Conceptually, this penetration threshold is a 2D surface in the 3D space (T, D, HOLSEP) of collimator designs. Designs above the penetration threshold are acceptable, while those below it are not. (The penetration threshold is crucial in the discussion of optimal collimators in Section V.) Based on the results of the ray-tracing analysis, an empirical formula for the penetration threshold was determined by a chi-squared fit. The outcome of this fitting procedure was The University of Chicago Penetration Criterion, which can be stated as the inequality

$$P_{\text{hole shape}} \leq \mu T \left[\frac{(\text{Volume of absorbing material in the collimator})}{(\text{Volume of the collimator})} \right]$$
$$= \mu T \left[1 - \frac{(\text{Area of the holes in a lattice cell})}{(\text{Area of a lattice cell})} \right] \quad (19)$$

where μ is the attenuation coefficient of the septal material (and, therefore, dependent on the energy of the incident radiation). The empirical parameter $P_{\text{hole shape}}$ is a dimensionless constant dependent only on the hole shape and lattice arrangement. Values of $P_{\text{hole shape}}$ for various hole shapes are given in Table 2. Collimators that satisfy Inequality (19) have little penetration, whereas those that fail the criterion

TABLE 2 Values of Penetration Parameter for Various Hole Shapes

Hole Pattern	Penetration Parameter $P_{\text{hole shape}}$
Hexagonal holes/hexagonal array	11.44 ± 0.51
Square holes/square array	12.57 ± 0.53
Triangular holes/point-to-point array	12.20 ± 0.35
Circular holes/hexagonal array	12.85 ± 1.05

show unacceptable penetration. Collimators that satisfy Inequality (19) as an equality are considered on the penetration threshold. Because The University of Chicago Penetration Criterion is based on empirical data fits, some care is required in its application. The criterion provides an excellent estimation of the penetration threshold for normal designs in which $5D < T < 25D$; however, no collimator should be constructed without the determination of the full PSRF by ray-tracing (or Monte Carlo) analysis.

One immediate conclusion can be drawn from the Inequality (19), the collimator thickness T must always be greater than the parameter $\tau_{\text{hole shape}}$:

$$\frac{P_{\text{hole shape}}}{\mu} \equiv \tau_{\text{hole shape}} < T \quad (20)$$

Because the attenuation coefficient μ decreases as the energy of the radiation increases, the parameter $\tau_{\text{hole shape}}$ and, therefore, collimator thickness must increase for higher energies. For Pb collimators, τ_{hex} (140 KeV) = 0.429 cm and τ_{hex} (511 KeV) = 6.9 cm. These values reveal the major problem in designing collimators for 511 keV; namely, the collimators must be very thick. If the collimator is thick and the resolution held fixed, the hole diameter must be large. Because the hole diameter is large, the hole separation must be large (HOLSEP > D). If the hole separation is large, the hole pattern becomes visible. Thus, hole visibility is a major collimator design problem for 511 keV radiation. Furthermore, as we see in Section VI.1, such collimators generally weigh more than permitted by standard camera gantries. As a result, almost all commercial collimators designed for 511 keV radiation violate the UC Penetration Criterion.

V. OPTIMAL DESIGN OF PARALLEL-HOLE COLLIMATORS

In nuclear medicine, the concept of optimal design must ultimately be related to the observer performance of the diagnostic task. For collimator design, the major factor affecting observer performance is the trade-off between sensitivity and resolution. For some tasks we need high sensitivity and can accept low resolution. For other tasks, we need high resolution and must sacrifice sensitivity. Most gamma-ray cameras come with an assortment of collimators with a range of sensitivities and resolutions. Which collimator is appropriate for each clinical task is beyond the scope of this chapter. The optimization problem faced by the collimator designer is more restricted. One assumes that the user specifies the required collimator resolution and then designs the collimator with the maximum sensitivity. Obviously, we find a different design for each resolution, so that we discover a family of collimators varying from low resolution (high sensitivity) to high resolution (low sensitivity). Which collimator in this family is appropriate for the specific imaging task must be discovered by clinical testing. However, for a collimator designer the problem of optimal design is that of finding the family of collimators with maximum sensitivity for specific resolution.

Mathematically, a parallel-hole collimator is specified by three parameters (T, D, HOLSEP) that can be viewed as a 3D space. For the optimization problem, we as designers fix the resolution (FWHM$_{col}$ at distance F), which restricts the collimators to a 2D subspace. Next, we observe that the optimal collimator should be immune to septal penetration, which means that the optimal design should satisfy the UC Penetration Criterion. More specifically, the optimal collimator should be on the penetration threshold. If the collimator is above the penetration threshold, then the collimator septa are too thick and more photons than necessary are absorbed in the septa. If the collimator is below the penetration threshold, penetration will degrade the images. The penetration threshold is another 2D surface in the 3D space of collimator designs. The optimal design must be on the intersection of the constant resolution surface and the penetration threshold. The intersection of these 2D surfaces is a 1D curve. The optimal design is the point on this curve with the maximum sensitivity.

The geometric picture of the optimization problem described in the last paragraph is made more explicit by simple algebra. The expression for the FWHM$_{col}$ in Eq. (14) can be rewritten as

$$D = \frac{\text{FWHM}_{col}}{K_{\text{hole shape}}} \frac{T}{(T+B+F)} \quad (21)$$

which expresses the hole diameter D in terms of the collimator thickness T. The penetration threshold, given by the equality in Inequality (19), can be rewritten as

$$(\text{Area of a lattice cell}) = \frac{T}{\left(T - \tau_{\text{hole shape}}\right)} \times \quad (22)$$

(Area of the holes in a lattice cell)

[The HOLSEP can be recovered from the (Area of a lattice cell) by simple geometry, as in Table 1.] Finally, substitution of the last two equations for D and the (Area of a lattice cell) into Eq. (8) for the collimator sensitivity yields

$$\langle \$ \rangle = \frac{1}{16} \frac{\text{FWHM}_{col}^2}{K_{\text{hole shape}}^2} \frac{(T - \tau_{\text{hole shape}})}{T(T+B+F)^2} \quad (23)$$

which expresses the sensitivity in terms of the collimator thickness T. Simple calculus reveals that the collimator sensitivity is maximized ($d\langle \$ \rangle/dT = 0$) if the collimator thickness is given by

$$T_{opt} = \frac{3}{4}\tau_{\text{hole shape}} + \sqrt{\frac{1}{2}\tau_{\text{hole shape}}(F+B) + \frac{9}{16}\tau_{\text{hole shape}}^2} \quad (24)$$

Once this optimal thickness is known, the hole diameter D and hole separation HOLSEP are determined by the substitution of T_{opt} for T in Eqs. (21) and (22), respectively. Thus, Eq. (24) provides the family of optimal collimators for any desired resolution FWHM$_{col}$ at distance F. The major surprise in this optimization is that all the optimal designs (for imaging with radiation of a particular energy and for source distance F) have the same thickness, independent of the desired FWHM$_{col}$. The FWHM$_{col}$ affects the hole diameter D and the hole separation HOLSEP, but not the collimator thickness. The collimator thickness is affected by the collimator material and the energy of the radiation (via the attenuation coefficient implicit in $\tau_{\text{hole shape}}$), the typical source distance F, and the gap between the collimator and imaging plane B, but not the FWHM$_{col}$!

The dependence of T_{opt} on F and not FWHM$_{col}$ is, at first glance, somewhat perplexing. After all, a collimator with resolution FWHM$_{col}$ at distance F should be approximately equivalent to a collimator with resolution 2(FWHM$_{col}$) at distance $2F$. Nonetheless, there is a crucial difference. The source distance F buffers the effect of collimator thickness T on the resolution. The actual distance between the imaging plane and the source is $(T + B + F)$ rather than simply F. The factor $(T + B + F)$ appears in Eq. (14) for the resolution, and consequently the distances T and F both cause loss of resolution. If F is significantly greater than T, small changes in the collimator thickness will not affect resolution. If, on the other hand, T is much greater than F, the collimator thickness T will be the predominant factor degrading the resolution. Therefore, a small T is desirable for resolution. On the other hand, we can increase the sensitivity of a collimator by making the collimator very thick, with thin septa The competition between these two contradictory tendencies produces the optimal design. As a result, the selection of the appropriate source distance F for the particular imaging application is very important in optimal design. An extreme example arises in collimators designed for X-ray astronomy ($F = \infty$). In that case, $T_{opt} = \infty$, so that the optimal design is a long tube (a single-hole collimator). More realistic examples of the optimal design strategy are shown in Tables 3a–c.

Tables 3a–c demonstrate the results of the optimal design strategy for hexagonal-hole collimators. In Table 3a the colli-

TABLE 3A Optimal Pb Collimators with Hexagonal Holes Designed for Resolution FWHM$_{col}$ = 0.80 at Distance F = 10.0 cm and B = 0.75 cm[a]

Optimal (Pb) collimator design for hexagonal holes

B = 0.75 cm	For the 140-keV design:
F = 10.00 cm	F = 15.0 cm
FWHM$_{col}$ = 0.80 cm	FWHM$_{col}$ = 1.117 cm

Energy (keV)	Thickness (cm)	F2F (cm)	HOLSEP (cm)	SPT (cm)	D (cm)	$
110.0	1.306	0.0825	0.0910	0.0085	0.087	2.26E-04
140.0	1.875	0.1132	0.1289	0.0157	0.119	1.93E-04
170.0	2.538	0.1455	0.1710	0.0255	0.153	1.64E-04
200.0	3.336	0.1804	0.2190	0.0386	0.189	1.37E-04
230.0	4.043	0.2082	0.2589	0.0507	0.219	1.18E-04
260.0	4.901	0.2386	0.3042	0.0657	0.251	1.00E-04
300.0	6.319	0.2820	0.3721	0.0900	0.296	7.89E-05
380.0	8.878	0.3446	0.4756	0.1310	0.362	5.45E-05
511.0	13.173	0.4195	0.6081	0.1886	0.441	3.33E-05

[a]Increased penetration at higher energies requires that the collimators become thicker with larger hole diameters and thicker septa. As a result, the sensitivity drops significantly as energy increases. N.B.: The collimator design for E = 110 keV requires thinner septa (0.0085 cm) than can currently be produced commercially. Also, the collimator design for 140 keV yields resolution FWHM$_{col}$ = 1.117 cm at F = 15.0 cm.

TABLE 3B Optimal Pb Collimators with Hexagonal Holes Designed for Resolution FWHM$_{col}$ = 1.0 cm at Distance F = 10.0 cm and B = 0.75 cm[a]

Optimal (Pb) collimator design for hexagonal holes

B = 0.75 cm	For the 140-keV design:
F = 10.00 cm	F – 15.0 cm
FWHM$_{col}$ = 1.000 cm	FWHM$_{col}$ = 1.396 cm

Energy (keV)	Thickness (cm)	F2F (cm)	HOLSEP (cm)	SPT (cm)	D (cm)	$
110.0	1.306	0.1031	0.1138	0.0106	0.108	3.53E-04
140.0	1.875	0.1414	0.1611	0.0196	0.149	3.02E-04
170.0	2.538	0.1819	0.2138	0.0319	0.191	2.56E-04
200.0	3.336	0.2255	0.2738	0.0482	0.237	2.14E-04
230.0	4.043	0.2603	0.3237	0.0634	0.273	1.85E-04
260.0	4.901	0.2982	0.3803	0.0821	0.313	1.57E-04
300.0	6.319	0.3525	0.4651	0.1125	0.370	1.23E-04
380.0	8.878	0.4307	0.5945	0.1637	0.452	8.52E-05
511.0	13.173	0.5244	0.7601	0.2358	0.551	5.20E-05

[a]These collimators provide less resolution (and more sensitivity) than those in Table 3a; nonetheless, the thickness of these collimators are the same as those in Table 3a. However, the hole diameter and septal thickness are both larger than the corresponding parameter in Table 3a.

TABLE 3C Optimal Pb Collimators with Hexagonal Holes Designed for Resolution FWHM$_{col}$ = 1.117 cm at Distance F = 15.0 cm and B = 0.75 cm[a]

Optimal (Pb) collimator design for hexagonal holes

B = 0.75 cm	For the 140-keV design:
F = 15.00 cm	F = 10.0 cm
FWHM$_{col}$ = 1.117 cm	FWHM$_{col}$ = 0.806 cm

Energy (keV)	Thickness (cm)	F2F (cm)	HOLSEP (cm)	SPT (cm)	D (cm)	$
110.0	1.539	0.0947	0.1027	0.0081	0.099	2.21E-04
140.0	2.189	0.1298	0.1448	0.0150	0.136	1.95E-04
170.0	2.934	0.1671	0.1915	0.0244	0.175	1.70E-04
200.0	3.819	0.2076	0.2448	0.0372	0.218	1.46E-04
230.0	4.593	0.2402	0.2894	0.0492	0.252	1.30E-04
260.0	5.522	0.2761	0.3404	0.0642	0.290	1.13E-04
300.0	7.039	0.3286	0.4179	0.0893	0.345	9.28E-05
380.0	9.736	0.4064	0.5397	0.1334	0.427	6.81E-05
511.0	14.188	0.5041	0.7036	0.1995	0.529	4.47E-05

[a]The design for 140 keV in this table yields the same resolution at F = 15.0 as the design in Table 3a, but it has greater sensitivity. However, for F = 10.0 cm the design in Table 3c yields worse resolution, FWHM$_{col}$ = 0.806 cm, than the corresponding collimator in Table 3a.

mators are designed for resolution FWHM$_{col}$ = 0.8 cm at F = 10.0 cm and B = 0.75 cm. Each row in the table gives the collimator design parameters for a specific energy of radiation (ranging from 110 keV to 511 keV). Table 3b shows the optimal designs for collimators with less resolution; namely, FWHM$_{col}$ = 1.0 cm at F = 10.0 cm and B = 0.75 cm. As previously noted, the collimator thickness depends on the energy of the radiation, but not FWHM$_{col}$. As a result, the collimator thicknesses (column 2) arc the same in Tables 3a and 3b. Note that the collimator for 511 keV is over 13-cm thick with a hole diameter greater than 0.5 cm and a septal thickness of nearly 2 cm. (As we see in the next section, such a collimator is plagued with problems of weight and hole-pattern visibility.) As expected, the sensitivities are higher for the lower-resolution collimators in Table 3b than for the higher-resolution collimators in Table 3a. We also expect that the resolution will degrade linearly with distance from the collimator face, so that a collimator with FWHM at F = 10 cm will yield 1.5 (FWHM) and F = 15 cm. Because of the collimator thickness, however, this relation is only approximate. In the top portions of Tables 3a and 3b, the resolution of the 140-keV collimator is shown for F = 15 cm. In each case, the resolution at F = 15 cm is better (i.e., smaller) than expected because the actual distance between the source and imaging plane is (F + T + B) rather than F. This example illustrates a general rule. If we design a collimator for FWHM$_0$ at F_0, then for $F > F_0$ the resultant

FWHM is less than the expected value (F/F_0)FWHM$_0$. On the other hand, for $F_0 > F$ the resultant FWHM is greater than the expected value (F/F_0)FWHM$_0$. This rule is further illustrated in Table 3c. Table 3c shows optimal designs for resolution FWHM$_{col}$ = 1.117 cm at F = 15.0 cm and B = 0.75 cm, which is the resolution we find for the 140-keV collimator in Table 3a at F = 15 cm. The collimators of Table 3c are thicker than those in Tables 3a and 3b, because the F is larger. As a result, the hole diameters are larger and the septa thinner; and, consequently, the sensitivities are higher in Table 3c than Table 3a. We might think that the collimators in Table 3c are better than those in Table 3a; after all, the resolutions (for the 140-keV design) at F = 15 are the same and Table 3c gives higher sensitivity. But the problem is that for Table 3c the resolution at F = 10 cm is worse (FWHM = 0.806 cm) than Table 3a (FWHM = 0.8 cm). We pay for the increased sensitivity with a loss of resolution for objects closer to the camera than F. For this reason, we must select F carefully. If F is too small, we lose sensitivity; if F is too large we lose resolution near the camera. Consequently, the source distance in front of the collimator F must be chosen as representative of the imaging system. For small animal studies, F may be small (2–5 cm); whereas, for SPECT studies F should be near the radius of rotation.

VI. SECONDARY CONSTRAINTS

The optimization derived in the previous section incorporates our knowledge of the geometric imaging properties of parallel-hole collimators and the limitations imposed by septal penetration. If septal penetration were the only problem affecting the collimator performance, then the design problem would be solved. Unfortunately, other difficulties arise that impose secondary constraints on the collimator. These secondary constraints often exclude the optimal designs and, in some cases, preclude any design. Three such secondary constraints are imposed by: (1) limits on the collimator weight, (2) the limits on the minimal septal thickness, and (3) the visibility of the collimator hole pattern. Full discussion of these constraints is beyond the scope of this chapter. But, anyone using or designing collimators should understand the basic imaging problems associated with these constraints.

A. Collimator Weight

Collimator weight is conceptually the simplest problem associated with collimator design. Nonetheless, weight is a major problem for collimators designed for high-energy (especially 511-keV) radiation. For 140-keV radiation, a typical collimator weighs 25–40 lb, which permits the technician to mount the collimator manually. If the collimator weighs 100 lb, for medium-energy collimators, special racks permit the technician to slide the collimator onto the camera. However, if the collimator weight exceeds 400 lb, few camera gantries will support the weight. Unfortunately, for 511-keV radiation, the collimator design can easily exceed the weight capacity of the gantry. Heavy collimators are required for the prevention of septal penetration by the high-energy radiation. A lower limit on the mass of the collimator can be calculated directly from the UC Penetration Criterion, Eq. (19) and (20):

$$\rho \tau_{\text{hole shape}} \leq \frac{[\text{Mass of the collimator}]}{[\text{Area of collimator face}]} \quad (25)$$

where ρ is the density of the collimator material. The equality holds in Inequality (25) if the collimator is on the penetration threshold (which is true of all the optimal collimators designed in the last section). For the two most commonly considered materials, we find

$$\rho_{\text{Pb}} \tau_{\text{Hex}}(511 \text{ keV, Pb}) = 78.4 \text{gm/cm}^2$$

and

$$\rho_W \tau_{\text{Hex}}(511 \text{ keV, W}) = 91.2 \text{gm/cm}^2$$

(*N.B.*: Pb collimators can be made with less mass than W collimators. The weight constraint is one of the few design considerations for which Pb is better than W.) For a square 50 cm × 50 cm camera face, the minimum collimator masses are given by

[Minimum collimator mass (511 keV, Pb)] = 196 kg = 431 lb

and

[Minimum collimator mass (511 keV, W)] = 228 kg = 502 lb

(The same calculation for 140 keV yields a minimal collimator weight for Pb of 26 lb.) Because of the weight constraint, none of the commercially available collimators for 511-keV radiation satisfies the UC Penetration Criterion. We can understand the reluctance of the manufacturers to redesign the camera gantry to support a 500-lb collimator that is seldom used in current clinical practice. Nonetheless, the collimators currently marketed for 511 keV should be considered examples of deceptive advertising. The task of the collimator designer, however, is not to indict overzealous marketing departments but to overcome the technical problem. Is there a way to design a collimator for 511-keV radiation that will weigh less than 300 lb and still be immune to septal penetration? The answer appears to be a qualified yes.

Inequality (25) is a direct outgrowth of the UC Penetration Criterion and cannot be avoided. The only way this inequality can be satisfied is by making sacrifices elsewhere. The required sacrifice is a reduction of the area of the collimator face. If we reduce the area of the collimator face, then the mass is reduced. But how can we reduce the area of the collimator face? Camera manufacturers have struggled for years growing larger NaI crystals to increase the sensitive area in the imaging plane of the camera. Reduction of this sensitive area for the occasional application with 511-keV radiation is not acceptable. However, one need not use the entire camera crystal for imaging with 511-keV radiation. If only half the area of the camera face is used for imaging, then the collimation for this half will weigh only 200 lb. The problem, of course, is that the remaining half of the camera face must be shielded from the 511-keV radiation. The key observation is that shielding requires less mass than collimation. We can shield the remaining half of the camera face with 100 lb of lead, so that the total mass of the collimator and shielding is only 300 lb. The obvious question is: Why does shielding require less Pb than collimation? Shielding stops all radiation, but collimation must selectively admit radiation aligned with the hole axes. Photons that are slightly misaligned from the hole axes can penetrate the septa and pose a problem for collimators. These slightly misaligned photons must be stopped with as little penetration as observed in the fully shielded areas. This selective absorption requires more Pb in the septa than is required by simple shielding. Thus, collimators are always heavier than shielding; and, consequently, collimator weight can be reduced by shielding part of the imaging area of the gamma camera.

B. Septal Thickness

Septal thickness provides a second constraint on collimator construction. Generally, collimators are fabricated by either casting or folding Pb foil. The casting process generally requires septa thicker than 0.3–0.5 mm (depending on the casting process and the company fabricating the collimator); whereas, folded-foil collimators can have septal thickness as small as 0.15 mm. In casting, the problem is removing the collimator from the mold without destroying the septa. If the Pb septa are too thin, they lack structural strength and easily tear out as the cast is removed. Usually, the quality of cast collimators is better than folded-foil collimators because the folded-foil collimators are more prone to construction defects. Indeed, these construction defects can cause noticeable artifacts; however, the defects can be eliminated with sufficient quality control. For obvious reasons, the collimator designer prefers folded-foil construction that permits thinner septa. Unfortunately, the optimal designs for low energy ($E < 140$ keV) require thinner septa than are available by either process. Under these conditions, the collimator designer must abandon the optimal design and construct a design with thicker septa.

A naive solution begins with the optimal design of the previous section and simply increases the septa to the minimal thickness σ that is technologically feasible (e.g., $\sigma = 0.15$ mm for folded-foil collimators). This strategy gives the same resolution as the optimal design with a significant loss in sensitivity. A more rigorous approach requires a complete reanalysis of the optimization. In our previous analysis, the optimal design was on the penetration threshold because penetration was the limiting factor in the collimator performance. For very low-energy radiation, the limitation is septal thickness not penetration. Consequently, septal thickness replaces penetration as the constraint on the optimization, so that the collimator lies above (not on) the penetration threshold. The optimization proceeds in the same manner as Section V. We select the desired resolution [$FWHM_{col}$ at F] and, as before, conclude from Eq. (14) that

$$D = \frac{FWHM_{col}}{K_{hole\ shape}} \frac{T}{(T+B+F)} \quad (26)$$

Next, we assert that the septal thickness takes the minimal value σ. For the hexagonal holes we conclude that (see Table 1 for the septal thickness SPT in terms of D and HOLSEP)

$$HOLSEP = \frac{\sqrt{\pi}}{\sqrt[4]{12}} D + \sigma = \frac{\sqrt{\pi}}{\sqrt[4]{12}} \frac{FWHM_{col}}{K_{hole\ shape}} \frac{T}{(T+B+F)} + \sigma \quad (27)$$

so that the sensitivity is given by

$$\langle \$ \rangle = \frac{C_{hex}\ \pi\ W^4 T^2}{64(T+B+F)^2 \left[(W+\sigma)T + \sigma(B+F)\right]^2} \quad (28)$$

where

$$W = \frac{\sqrt{\pi}}{\sqrt[4]{12}} \frac{FWHM_{col}}{K_{hex}} \quad (29)$$

Once again, the sensitivity depends only on the collimator thickness T and is maximized ($d\langle\$\rangle/dT = 0$) by

$$T_{opt} = (B+F) \left[\frac{\sigma}{W+\sigma}\right]^{1/2} \quad (30)$$

so that

$$HOLSEP_{opt} = (W+\sigma)^{1/2} \sigma^{1/2} \quad (31)$$

and, for hexagonal holes,

$$F2F_{opt} = \sigma^{1/2} \left[(W+\sigma)^{1/2} - \sigma^{1/2}\right] \quad (32)$$

Equations (30)–(32) provide an optimal hexagonal-hole collimator design if the septal thickness as determined by the optimization in Section V is smaller than σ. For example, in Table 3a, the collimator design for 110 keV requires a septal thickness of 0.0085 cm, which is less than minimal septal thickness $\sigma = 0.015$ cm permitted by folded-foil collimators. Consequently, the analysis here can be used for the optimal design in the case $\sigma = 0.015$ cm, with resolution $FWHM_{col} = 0.8$ cm at $F = 10$ cm and $B = 0.75$ cm. The resulting collimator is given by $T = 1.4938$ cm, HOLSEP = 0.108 cm, and F2F = 0.093 cm; so that the sensitivity is $ = 1.98E-4. This sensitivity is only 87% of that achieved by the optimal design of Section V, but is marginally better than the naive strategy of increasing the septal thickness to σ (86%). A 13% loss of sensitivity may seem a small cost for maintaining minimal septal thickness, but state-of-the-art collimator design is concerned with 2–4% improvements in sensitivity. By this standard, the insistence on a minimal septal thickness has a staggering effect on collimator design.

One of the biggest controversies in collimator design since the 1980s pits the advocates of cast collimators against those of folded-foil collimators. Advocates of cast collimators argue correctly that quality control is better for cast colli mators and that folded-foil collimators are prone to visual artifacts caused by fabrication defects. On the other hand, the advocates of folded-foil collimators can boast of thinner septal thickness that permits better design. For the collimator designer, the crucial difference between the two techniques is the minimal septal thickness σ. Most clinical collimators are designed for the 140 keV emissions of Tc99m. In Tables 3a–c, all the optimal collimators at 140 keV require septal thicknesses between 0.015 cm and 0.02 cm. These collimator designs can be constructed with folded foil, but not by casting, for which σ = 0.03 cm. The strategy of the last paragraph can be used for σ = 0.03 cm with resolution $FWHM_{col}$ = 0.8 cm at F = 10 cm and B = 0.75 cm. The resulting design is T = 2.092 cm, HOLSEP = 0.1541 cm, and F2F = 0.1242 cm; which yields $ = 1.57E-4. The optimal design in Table 3a gives $ = 1.93E-4, which means that the best cast collimator design gives only 81.5% of the sensitivity of the best folded-foil. In a field in which 2–4% changes in sensitivity are considered crucial, we cannot discard 19% and expect to remain competitive. From the view of collimator design, therefore, folded-foil collimators are the current winner. However, technology is continually changing. If casting technology can reduce the mininum septal thickness σ to 0.015 cm, casting will regain its dominance in collimator construction.

C. Visibility of Collimator Hole-Pattern

Hole visibility is one of the most difficult problems associated with collimator design. Even the most casual visual examination of a collimator reveals the hole pattern. If we remove the thin aluminum covering that protects its surface, a collimator looks like a honeycomb made of lead. We might expect to see this hole pattern in the images. However, because of the limited resolution of the gamma camera and the diffuse nature of most clinical sources, the hole pattern is seldom visible. As a result, the practicing radiologist may become complacent and assume the hole pattern has little or no effect on the images. This is a major mistake. Under certain conditions the hole pattern can drastically affect the images with serious diagnostic consequences. Prevention of these hole-pattern artifacts is a major challenge for the collimator designer.

Clinical images formed with 140-keV radiation (Tc99m) seldom display artifacts of the collimator hole pattern. For this fortunate situation, the collimator designer can take no credit; the hole pattern is invisible because the typical camera resolution ($FWHM_{cam}$ = 4 mm) is insufficient for observation of the collimator holes. A careful examination of the imaging equations (Gunter, 1996) reveals that the hole pattern will be invisible provided that

$$\text{HOLSEP} < \frac{FWHM_{cam} \, FWHM_{obj}}{\sqrt{FWHM_{cam}^2 + FWHM_{obj}^2}} \quad (33)$$

where $FWHM_{obj}$ is the size of the object being imaged. This inequality is called the *Classical Hole-Pattern Criterion* because no details of the hole shape or configuration are involved; only HOLSEP appears. In clinical imaging the typical objects (organs or tumors) are much larger than the camera resolution ($FWHM_{cam} \ll FWHM_{obj}$), so the classical hole-pattern criterion becomes HOLSEP < $FWHM_{cam}$. According to Table 3a, the optimal collimator for 140 keV has HOLSEP = 1.289 mm which is much less than the typical camera resolution of 4 mm. Consequently, the hole pattern is invisible in most clinical images. This satisfactory result fails if the object is smaller than HOLSEP in any dimension. Thus, images of narrow capillary tubes filled with Tc99m will display significant hole-pattern artifacts (see Figure 5). Because narrow veins are seldom imaged in nuclear medicine, this poses few clinical problems. However, imaging with high-energy radiation does pose a major problem. According to Table 3a, HOLSEP increases significantly as we design for radiation of higher energies. In general, HOLSEP will exceed $FWHM_{cam}$ (4 mm) for

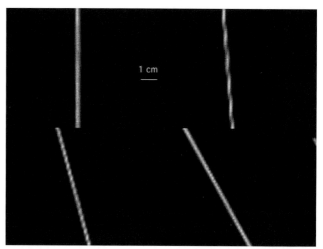

FIGURE 5 Hole-pattern artifacts are demonstrated by the "braiding" structure in these images of capillary tubes (I.D. = 1 mm) filled with Tc-99m placed on the face of a LEAP collimator.

energies greater than 300 keV; so the optimal collimators for these high energies will all exhibit hole-pattern artifacts. Even if we abandon the optimal strategy of Section V, the UC Penetration Criterion requires thick septa that, in turn, require the violation of the Classical Hole-Pattern Criterion. Thus, the avoidance of hole-pattern artifacts in collimators for energies greater than 300 keV is one of the most challenging problems in nuclear medicine.

The most common method of handling the hole-pattern artifacts is simple denial—we pretend that the artifacts do not exist. In the case of SPECT imaging, this strategy apparently works because the artifacts are smeared out in the reconstruction process (although the author knows of no rigorous study proving this claim). For planar imaging, we assume that the clinical radiologist is capable of distinguishing between the hole-pattern artifacts (which are presumably periodic) and real biological structures (which are seldom periodic). The problem with this strategy is that the hole-pattern artifacts can have very long wavelengths (much larger than HOLSEP) that can be mistaken for biological structure and affect the diagnostic results. Perhaps the best method of removing hole-pattern artifacts is by averaging out the hole pattern with collimator motion (with respect to the imaging plane) during the acquisition. This idea was first suggested by Wilks et al. (1969) and is called "collywobbling." Various experimental applications of this idea have been successfully implemented over the years—both linear translations and rotations of the collimator have eliminated the hole-pattern artifacts. Currently, however, no commercial gamma camera offers a collywobbling option. The mechanical engineering required for collywobbling is generally considered too expensive for the limited number of applications requiring high-energy radiation. So once again, the collimator designer must attempt to overcome the problem with clever tricks.

For hole-pattern artifacts, the clever trick is penumbral masking. Although the detailed mathematics is beyond this chapter, the basic idea can be understood with an analogy from astronomy. During a solar eclipse, the Moon obscures the disk of the Sun. At some places on Earth (the umbral region), the Moon completely obscures the Sun and a total eclipse results. However, at most places on Earth (the penumbral region), the solar disk is only partially covered and a partial eclipse results. The size of the umbral region on Earth's surface is determined by the distance between Earth and the Moon. If the Moon is close to Earth, the umbral region is large and the total eclipse is visible across a wide path on Earth's surface. If, on the other hand, the Moon is far from Earth, the disk of the Moon is too small to obscure the entire solar disk and an annular eclipse results. In the case of an annular eclipse, almost all observers on Earth see a reduction in light, but no observer experiences total darkness. The analogous situation in collimator design occurs when the septa block radiation from the imaging plane. If the septa completely block radiation from regions in the imaging plane that are separated by more than the camera resolution, the collimator hole-pattern becomes visible. However, by expanding the gap B between the imaging plane and the back of the collimator, we can arrange that the radiation from adjacent holes overlap in the imaging plane and produce a penumbral region behind the septa rather than a completely dark umbral region. As a result, the shadow effects in the imaging plane are reduced and the hole pattern becomes less visible. The crucial idea is that the collimator designer must now control gap B between the collimator and the imaging plane. The overall collimator design problem becomes four dimensional; that is, the designer must select four parameters (T, D, HOLSEP, B). Whereas the optimization in Section V assumed that B was fixed by the camera manufacturer, now the gap B is a parameter that is selected by the collimator designer for minimum hole-pattern visibility.

The selection of the gap B that minimizes the hole-pattern visibility requires a detailed mathematical analysis of the hole-pattern harmonics discussed in the chapter (Gunter, 1996). Although such an analysis is beyond the scope of this chapter, we can summarize the results (Beck and Redtung, 1988; Formiconi et al., 2000). If the gap vanishes ($B = 0$), the hole pattern is extremely visible. As the gap B increases, the hole-pattern visibility gradually decreases to a minimum and then begins to increase again. Conceptually, the hole pattern is minimized if the penumbra from adjacent holes mesh exactly. As B is further increased, the penumbra will overlap and produce ringing that can be observable as hole pattern. Further study reveals that the hole pattern appears and then disappears periodically as B increases. The optimal value of B corresponds to the first minimum in the hole pattern visibility (i.e., the smallest value of B exhibiting minimal hole pattern). An analytic expression can be derived for this optimal B that is dependent on the hole shape and lattice structure. For square holes in a square array (the simplest case), the optimal gap B_{opt} is given by

$$B_{opt} = T \frac{\left(2\,\text{HOLSEP} - \sqrt{\pi} D\right)}{\sqrt{\pi} D} \quad (34)$$

Similiar expressions can be derived for different hole shapes. The optimal design strategy of Section V must be completely revised because B_{opt} is now a function of the other collimator dimensions. This revision is not trivial. If we consider the optimal collimator for 511 keV in Table 3a, the optimal gap is $B_{opt} \approx 6$ cm. Such a large gap has a serious effect on collimator resolution. To compensate for this large gap and retain the same resolution at distance F, we must significantly reduce the ratio (D/T). Consequently, for the same resolution we must accept serious losses in collimator sensitivity. Because the gap B_{opt} is so large for 511 keV, no one has ever seriously implemented this strategy (except in computer simulations). Most commercial manufacturers choose to ignore the problem of hole-pattern visibility and hope that no one notices the artifacts.

VII. SUMMARY

This survey of the current problems and techniques in collimator design was intended as an introduction for graduate students working in nuclear medicine. Despite the mature state of collimator technology, many problems remain. In particular, all the problems of converging-hole collimation have been ignored in this chapter. Furthermore, new developments in material science and fabrication are revolutionizing the possible geometries available to the collimator designer. As in any field, collimator design will remain interesting as long as the researchers emulate Professor Robert Beck and avoid the conventional solutions.

References

Beck, R. N., and Redtung, L. D. (1988). Generalized collimator transfer function. *Radiology* **169:** 324.

Formiconi, A. R., Di Martino, F., Volterrani, D., and Passeri, A. (2002). Study of high-energy multihole collimators. *IEEE Trans. Nucl. Sci.* **49:** 25–30.

Gunter, D. L. (1996). Collimator characteristics and design. *In* "Nuclear Medicine" (R. E. Henkin *et al.*, eds.), pp. 96–124. Mosby-Year Book, St. Louis.

Wilks, R. J., Mallard, J. R., and Taylor, C. G. (1969). Instrumental and technical notes: The collywobbler—a moving collimator image-processing device for stationary detectors in radioisotope scanning. *Brit. J. Radiol.* **42:** 705–709.

CHAPTER 9

Annular Single-Crystal Emission Tomography Systems

SEBASTIAN GENNA, JINSONG OUYANG, and WEISHI XIA
Digital Scintigraphics, Inc., Waltham, Massachusetts

I. OVERVIEW: ANNULAR SINGLE-PHOTON EMISSION COMPUTED TOMOGRAPHY SYSTEMS

Instead of rotating scintillation cameras, an annular camera system uses a stationary annular crystal within which a concentric ring collimator is rotated. The first such system was CeraSPECT, a brain-imaging camera developed by Genna and Smith in 1988 and 1995 (Genna and Smith, 1988, 1995; Smith and Genna, 1988). Brain imaging was a particularly convenient first application of an annular camera, primarily because it was possible to encircle a patient's head with an annular single-crystal using conventional technology. The main advantage of such a system over a rotating-camera single-photon emission computed tomography (SPECT) is its ability to transaxially sample gamma rays more efficiently over a 2π crystal geometry. Other performance advantages of an annular camera include patient friendliness due its stationary design, system stability, and the elimination of rotation ring artifacts arising from crystal nonuniformity. CeraSPECT typically uses a relatively thin (10-mm) crystal in order to restrict light spread for high-resolution imaging and a triad of parallel-hole collimators to sample gamma-ray projections over the crystal's surface. More recently, new collimator designs incorporate continuously varying converging focuses to provide still higher sensitivity and resolutions by magnifying gamma-ray projections and sampling gamma rays more efficiently (El Fakrhri, *et al.*, 2004).

An annular camera may also be designed as a hybrid bridge between SPECT and PET, although the requirements for each are often contradictory. Good low-energy SPECT typically requires thin crystals, whereas high-energy coincidence detection demands thick ones. Thick crystals are usually antithetical to the narrow light spread needed for good intrinsic resolution and the low parallax requirements of positron emission tomography (PET). Parallax is further degraded as the diameter of the crystal annulus is decreased. In contrast, the sensitivity of annular SPECT increases as the crystal diameter is made smaller by approximately $1/r$. The degradation of light spread by thick crystals also reduces the count rate because of pulse pile-up, an important consideration for PET imaging. In a hybrid annular brain camera, therefore, (1) the crystal must be thick for high sensitivity, (2) the diameter must be small for good SPECT resolution, (3) the light spread must be narrow for good intrinsic resolution and low pulse pile-up, and (4) a means must exist for determining the depth of interaction (DOI) at which gamma rays are converted into light scintillations in order to reduce the high parallax in PET resulting from small-diameter, thick crystal detectors. All these require radical optical design modifications to control the spreading of light from its scintillation points of origin to the camera's photosensors.

The making of annular scintillation cameras to image other body organs, such as the breast, although conceptually attractive, was not pursued until recently (Genna *et al.*, 1997) due to the fact that conventional gamma camera optics, constrained by crystal-to-glass window light-coupling requirements, precluded position analyses near the edges of the camera. A fundamental requirement for tomographic breast imaging is that as much breast tissue as possible should be included in the camera's field of view. Typically, a scintillation camera cannot image within 30–50 mm of its edge, so a camera cannot encircle a breast and image close to its chest wall. It is possible, however, to modify the light optics of a single-crystal scintillation annular camera, in combination with axially slanting collimation, to enable it to image to within 5 mm of the chest.

More recently, as a result of increased interest in small animal imaging, innovative methods of using an annular crystal in combination with rotating pinhole collimators have been developed (Jones *et al.*, 2003). For this application, projected images are magnified through a multiplicity of pinholes to achieve high intrinsic resolution in combination with high sensitivity.

II. PRINCIPLES AND DESIGN OF CERASPECT

Except for geometry, the methods used to construct the first annular camera detector followed those used to make conventional planar cameras. CeraSPECT's annular crystal, measuring 310 mm inside diameter, 130 mm long, and 10 mm thick, is cut from a single crystal. It is then coupled to an outer ring of glass using the same light-coupling silicon gel that is used in planar cameras, and the crystal is hermetically sealed in a toroidal envelope comprising the outer glass window ring joined to an inner annular aluminum shell. As illustrated schematically in Figure 1, photomultipliers (PMs) are coupled with optical grease to the glass window to complete the optical link between scintillations in the crystal and the camera's photosensors, as is conventional practice. A collimator system that rotates internally to the annular crystal is typically structured with three identical parallel-hole collimator segments oriented in 120° offsets.

The internal optical characteristics of the CeraSPECT annular single-crystal cameras are similar to those of planar systems. One difference is that in an annular camera light emissions from a scintillation event traveling in its angular direction reach the photosensors through shorter path lengths because of crystal curvature. Also, because its circular optical pathway is continuous, there are no edge distortions in the angular direction. Both of these conditions narrow the light spreads from scintillation events and result in continuous angular PM response functions' having improved position analysis characteristics. In an annular camera having the dimensions of CeraSPECT, angular optical continuity is paramount, lest projected data gaps in the stationary crystal overly compromise reconstruction fidelity.

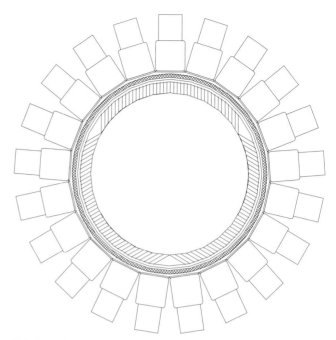

FIGURE 1 CeraSPECT annular brain camera. The three-segment parallel-hole rotating collimator system samples gamma rays uniformly across the camera's field of view.

The three-element parallel-hole collimator system shown in Figure 1 does not make full and efficient use of the 2π crystal detector. Each collimator has to have a field of view at least 20 mm larger than a human head to accommodate all patients. This results in sizable gaps at each of its three collimated junctions. These gaps reduce the amount of crystal surface that is exposed to projected gamma rays. Collimator channels pointing toward outer regions of the brain also make less efficient use of valuable crystal surface area than do centrally disposed holes because their axes subtend larger angles with the normal to the crystal surface. Note also that the information content in projected data collected from peripheral regions of the head is less than that from data projected through central brain regions, which sample a larger number of cells. Nevertheless, as is also commonplace with most rotating cameras, all regions are sampled equally. For equivalent-overall resolution per unit volume of brain matter, the camera of Figure 1 provides sensitivities that are comparable to those of triple-headed fan-beam rotating cameras, primarily because the closer proximity of an annular detector to the brain compensates for the focused sensitivity of planar rotating cameras. Still higher performance is obtainable by modifying the collimator designs in annular cameras, as discussed in Section III.

With respect to image artifacts and system stability, an annular SPECT camera system employing a stationary crystal with an independently rotating collimator enjoys certain other advantages relative to rotating cameras. System artifacts caused by crystal nonuniformities are virtually nonexistent in a stationary crystal camera (Smith and Genna, 1988), thus greatly reducing the need for quality assurance. In a rotating gamma camera, crystal nonlinearities and nonuniformities rotate with the detector, resulting in bull's-eyes, rings, and other artifacts. Some sources of nonuniformities are magnetic-field variations as a function of camera position and, often more troublesome, temporal changes in linearity calibrations that arise primarily as a result of temperature-sensitive PM gain drifts. For rotating cameras, this often requires daily quality control, whereas for annular cameras weekly PM gain matching is usually sufficient.

III. ANNULAR SENSOGRADE COLLIMATORS

In order to make better use of the 2π geometry of an annular camera; its collimator should sample central regions of its field of view preferentially. In a first design (Genna and Smith, 1995), a Multifield collimator comprising six discrete parallel-hole collimator segments, with transaxial widths that varied from approximately 50 to 220 mm, replaced the three equal-width collimators of Figure 1. This provided a collimator that sampled projections through a diameter of the brain six times more efficiently than peripherally projected rays. Kijewski et al. (1997) subsequently showed by simulations that a ring collimator with sampling distributions of that magnitude would have a variance of less than 60% of the variance of a three-field collimator when measuring small objects within a 200-mm cylindrical phantom. In practice, although the magnitudes of the sensitivity distributions of the Multifield collimator were equivalent those of Kijewski et al., the image discontinuities at the junctions of its discrete fields caused unacceptable ring artifacts that were uncorrectable and the increased number of junctions made inefficient use of the crystal's surface.

By way of illustration, Figure 2 shows schematically a class of SensOgrade collimators with continuously graded sensitivities. The collimators are single-field collimators having hole geometries that wrap around its ring symmetrically about a diameter. The upper channel points downward along the vertical diameter, at $\theta = 0$, and converges from the outer perimeter of the collimator to a distance, $\rho(0)$, which is inside the field of view but larger than the radius of the collimator. Depending on the design of the collimator, the distance of convergence, $\rho(\theta)$, may be held constant or made to functionally increase with θ. In either case, the parameters must be chosen so as to point the channels both continuously and increasingly toward the periphery of the field of view, starting with the vertically pointing channel along $\theta = 0$ and progressing smoothly with increasing θ so as to finally project tangentially to the edge of the fields, as θ approaches π, so

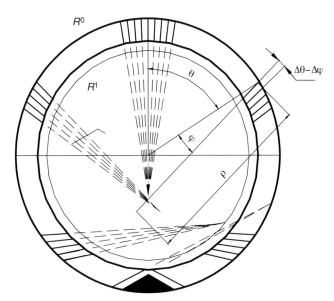

FIGURE 2 The angular direction of the collimator holes of the SensOgrade collimator, $\theta - \phi$, varies from zero to the maximum value needed to tangentially encompass the field of view. The functional variation of the angles $\theta - \phi$ over the circumference of the collimator may be set to sample the interior of the field of view more heavily than its perimeter.

that only one junction of the crystal is allocated to peripheral regions of the field of view. In a first class of collimators, referred to as A collimators, the distance of convergence $\rho(\theta)$ is held constant, whereas for B class collimators $\rho(\theta)$ increases with the θ position of the converging holes. In the latter, the angle of convergence, $\Delta\theta - \Delta\phi$, of the collimator holes decreases as a function of its θ position.

Figure 3a illustrates the sensitivity performance of a type A SensOgrade collimator versus the performance of the three-element parallel-hole collimator in Figure 1. Each of the three curves in this figure is shown relative to the uniform sensitivity of the parallel-hole collimator system in air. Curve (A) shows the relative projected sensitivity of the type A collimator in air. Curve (B) gives the ratio of the attenuated to nonattenuated projections of 140-keV gamma rays when using the parallel-hole collimator and a uniform 180-mm-diameter water phantom. For comparison, the effect of attenuation on the type A SensOgrade collimator is shown as curve (C). Clearly, its projected sensitivity is almost constant for all but the edges of the field of view demonstrating its ability to compensate for attenuation.

Figure 3b shows relative sensitivities for a type B collimator, for which the transaxial distribution of sensitivity is focused still more densely in central regions of its field of view. The back-projected sensitivity profile is everywhere higher than the uniform profile of CeraSPECT's parallel-hole collimator system. Also shown in Figure 3b are the relative signal-to-noise ratios in a reconstructed 200-mm water-filled phantom of uniform activity as a function of the radial displacement from the center of the phantom. The

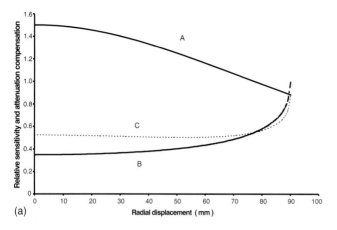

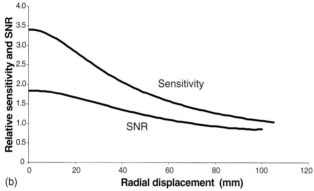

FIGURE 3 (a) Projected sensitivity profiles of a type A SensOgrade collimator compared to that of the parallel-hole collimator configuration of CeraSPECT, for which the total sensitivity in air is set equal to 1. Whereas the sensitivity of the parallel-hole collimator is uniform, the relative sensitivity of the multiply-converging type A collimator in air (curve [A]) closely compensates for the effect of attenuating 140-keV gamma rays (curve [C]) in a 180-mm-diameter water phantom. Curve (B) illustrates the effect of attenuation on the projected sensitivity of a parallel-hole collimator camera system. (b) Transaxial reconstructed sensitivity and signal-to-noise profiles of a type B SensOgrade collimator, shown relative to the sensitivity and signal-to-noise (SNR) from the parallel-hole collimator configuration of CeraSPECT. The more highly converging holes of this collimator show dramatic increases in central regions of the phantom without degrading peripheral sensitivity.

relatively high sensitivity ratio in the center of the phantom (approximately 3.5 times that of the three-element parallel-hole collimator) results in a signal-to-noise ratio improvement of approximately 1.8. This collimator is particularly useful for imaging centrally disposed regions of the brain.

Recent phantom and primate studies at the Massachusetts General Hospital have demonstrated the efficacy of SensOgrade collimation (El Fakrhri *et al.*, 2004). Images of a cylindrical phantom containing nine spherical sources of Tc-99m are shown in Figure 4. The first image (Fig. 4a) was produced with a standard CeraSPECT parallel-hole collimator, whereas the second image (Fig. 4b) was taken with a SensOgrade collimator, as in Figure 2. Sphere diameter ranged from 1.3 to 2.2 cm and with source-to-background

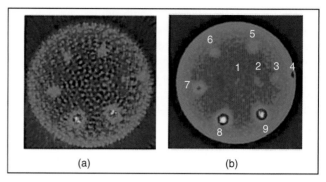

FIGURE 4 Shows a comparison of phantom images taken with (a) parallel-hole collimator to (b) a SensOgrade collimator. The phantom (b) contained nine sources ranging from 1.3 to 2.2 cm diameter.

activity ratios of 10:1. The SNR for estimating sphere activity concentration improved by 27% to 89% for spheres located more than 5 cm from the phantom center. Images acquired with the standard collimator were too noisy in more central regions to make comparisons, whereas the images taken with SensOgrade were discernable.

IV. MODIFICATION OF LIGHT OPTICS IN A SCINTILLATION CAMERA

In 1997, Genna (1997) proposed coupling an annular crystal to its PMs by means of a liquid interface. In place of the rigid crystal-to-glass interfacing of Figure 1, the crystal and its surrounding PMs are mounted independently within a toroidal structure that is filled with a hydrophobic light-transmitting oil. This structure provides a simple means for geometrically configuring the optical components of the camera optimally in a more secure and stable environment for the crystal and PM. The hydrophobic fluid seals and protects the crystal from oxidation. It also ensures reliable optical coupling, unaffected by differential temperature expansions of the crystal and glass and precludes the drying and separating of the optical bond between the glass and its PM, which may occur in conventional cameras. Also notable is the fact that the temperatures of the PMs are kept uniform by immersing them in an oil bath. As a result, temperature-related gain drifts, which are a major cause of system instability, are virtually eliminated. Liquid interfacing also permits the use of unorthodox light collection and transmission geometries, including unconventional PM orientations and light reflectors interposed between the crystal and its PM.

The light-transfer characteristics of a scintillation camera substantially affect its performance. In general, it is advantageous to design a system so that the half-width of the light transfer function (LTF), which represents the light collected by a PM as a function of the displacement of a scintillation event from the axis of the PM, approximates a PM width.

It is also preferable that this function falls sharply with distance so light is essentially confined to three PM widths. Typically, the LTF is best in thin-crystal cameras that are approximately 10 mm thick and deteriorates rapidly with increasing thickness. A 30-mm-thick crystal camera, for example, typically yields a light spread that is more than 60% wider than that of a 10-mm-thick crystal with more slowly descending tails. A first task of good optical design of thick-crystal cameras is to narrow its light spread. A second task is to tailor this light spread so that that the transferred light can be also used to determine the DOI of a gamma ray within the crystal.

Figure 5 illustrates three methods of modifying light transfer from scintillation events by altering crystal reflection and refraction within the crystal and at its boundaries. The dispersion of light by back-reflection from the gamma-ray entrance face of a crystal may be reduced through the use of retro-reflectors. Figure 5a illustrates retro-reflectors shaped as right-angle pyramidal surfaces cut into the face of a planar crystal. These pyramids, when coated or backed-up with light-reflecting material, cause light to be refocused back toward its origin. The light-narrowing effect of such retro-reflectors were first simulated by Strobel et al. (1997) and demonstrated by Genna et al. (1997). Figure 5b shows a still more recent method of narrowing light spread by introducing scattering structures within the crystal. Shown is a family of cylindrical columns extending through both faces of the crystal. These holes may be filled with air or other material of index or refraction different from that of the crystal, such that light is scattered by reflection and refraction at their boundaries. It can be shown that the cross section for the scatter of light from the columns is proportional to $\sin\theta$, where θ is the angle that a light ray makes with the axes of the columns. Thus, light traveling perpendicular to the axial directions has the highest probability for scatter, and light traveling parallel to them has the lowest, leading, on multiple successive scatters, to a narrowing of the LTF. A 30-mm-thick crystal, for example, having right-circular cylinder scattering columns occupying approximately 15% of the crystal volume in combination with retro-reflectors yields a LTF having widths that are comparable to those of a conventional 10-mm-thick crystal. This combination also results in LTFs that are independent of depth within the cylinder.

Figure 5c shows a family of scattering columns that are tapered. The tapered columns present cross sections that vary linearly with depth within the crystal. Thus, the probability of scattering a light ray decreases linearly with depth, causing the width of the camera's LTF to increase linearly. This feature can be used to tailor the LTF of a camera system to be depth-dependent, thus providing a means of analyzing a scintillation event for its DOI, a useful parameter for reducing parallax in thick crystals.

V. NEUROTOME, A BRIDGE BETWEEN SINGLE-PHOTON EMISSION COMPUTED TOMOGRAPHY AND POSITRON EMISSION TOMOGRAPHY

Figure 6 shows a partial design for NeurOtome, an annular camera proposed for thick-crystal applications with DOI capability. Its crystal makes use of the composite retro-reflectors and tapered scattering columns of Section IV to provide good LTF characteristics in combination with a means for determining DOI from light spread. The design specifies an inner diameter of 360 mm, an axial length of 240 mm, and a thickness of 30 mm. The schematic illustration provides for 35 columns of 40-mm-wide PMs, with each column consisting of six adjacent 40-mm square PMs having end windows separated by 10 mm from the outer surface of the crystal. The tapered scattering columns are cut through the crystal volume and are preferentially

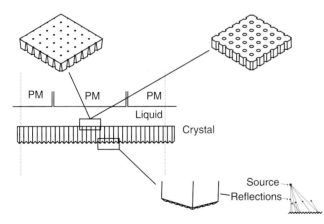

FIGURE 5 Three methods of narrowing the light spread transmitted by scintillation events to a camera's photomultipliers. (a) Pyramidal retro-reflectors direct light rays back toward their origins while light scattering columns interrupt and scatter light preferentially toward the crystal surfaces. (b) Circular columns yield light spreads that are independent of the DOI of gamma rays within the crystal. (c) Cone scattering structures make the light spread depth dependent so as to enable DOI analysis.

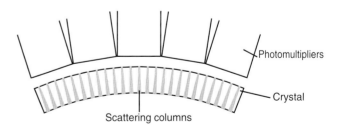

FIGURE 6 Partial design of a segment of a hybrid NeurOtome camera. Light spreads produced by the tapered scattering columns, in combination with pyramidal retro-reflectors, are an effective means of reducing light spread while making its width dependent on the depth that scintillations are produced by gamma rays. When used for SPECT, type A and B SensOgrade collimators provide high sensitivity and resolution.

filled with air, which, because of its low index of refraction, provides efficient scattering at column-to-air interfaces. Alternatively, the holes may be filled with silicon gel containing a suspension of high index particles to enhance secondary scattering within the columns.

Figure 7 shows Monte Carlo simulation determinations of the LTF of the 40-mm PM in the camera in Figure 6, as a function of the lateral displacements of scintillation events from the PM axis, at three different depths within the crystal. (DOI = 0 means that the event is midway through the crystal.) Also shown is a simulation study for a 10-mm-thick crystal as conventionally employed (without pyramidal retroreflectors or scattering columns). Both assume the same crystal-to-PM geometry and interfacing light transmission and reflection coefficients. However, the back-reflection of the conventional dry assembled systems is effectively higher because of the air layer between the crystal and reflective coating. Evident are the improved light-spread characteristics throughout the 30-mm crystal and the increased light collection of its PM relative to that of the thinner 10-mm crystal. Although not illustrated here, it should also be noted that when the scattering columns are right-circular cylinders light spread in the 30-mm-thick crystal is also substantially narrowed and becomes independent of depth. Clearly, the light spread within a crystal can be reduced. It can be made uniform with internally disposed right circular cylinder scattering columns or functionally dependent on DOI by tapering the columns.

It was suggested by Gagnon (1996) that depth dependence may be a useful tool for quantitatively estimating DOI as a function of light spread. He proposed that second- or higher-order moments of the PM charges collected during a scintillation event could be used to characterize light spreads. The third-order moment:

$$M_3^x = \left(\frac{\sum_i q_i |x_i - \bar{x}|^3}{\sum_i q_i} \right)^{1/3}$$

for example, is sensitive to light spread. Figure 8 illustrates the Monte Carlo simulations of third-order moments as a function of the DOI in the 30-mm-thick crystal assembly referred to in the discussion of Figure 6. The analyses are for light received by a 3 × 3 cluster of 40-mm square tubes, where $i = 0$ and $j = 0$ identify the central tube in the cluster and the crystal scintillation occurs within the xy domain of the center tube. Shown are the x moments of the cluster about points along two parallel lines, one on the axis of the cluster and another near to its edge. Results indicate that x and y moments can determine the DOI to within an approxi-

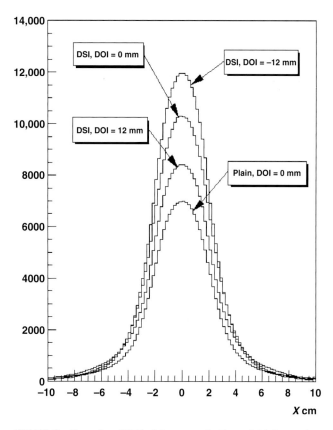

FIGURE 7 Examples of LTF of the camera in Figure 5. Light transfers to a 40-mm PM are shown as a function of the displacement of scintillation events from the axis of the PM at three different depths within the crystal. Curves labeled "DOI" refer to a camera fitted with a crystal having tapered-hole circular scattering columns, as shown in Figure 4a, whereas "Plain" refers to a conventionally uniform crystal camera. The spread of the LTF is narrower near the gamma-ray entrance face of the crystal, where the cone widths are broader. Also shown is the LTF from a 10-mm crystal, prepared as in conventional scintillation cameras.

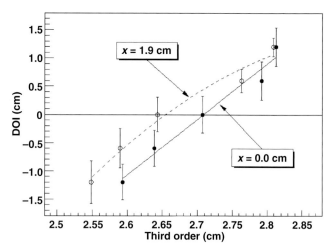

FIGURE 8 Monte Carlo simulations of the third-order moment of the light received by a 3 × 3 cluster of 40-mm square PMs of the camera in Figure 5 as a function of their DOI. The scintillation events are assumed to lie within the xy domain of the central PM of the three-tube cluster. Shown with arrows are two xy positions for which the curves are plotted. One scintillation (x = 0.0 cm) is assumed to fall along the central axis of the cluster and the other (x = 1.9 cm) along a parallel line near the edge of the central PM.

mately 3-mm standard deviation, thus reducing parallax to less than that of a 10-mm crystal. In these analyses, it was assumed that the light yield per 511-keV gamma-ray was not statistically degraded by pulse shortening (Wear et al., 1998) or pulse correction (Wong and Li, 1998; Wong et al., 1998) methods.

At high count rate, the third-order moment, used to characterize the DOI, is corrupted by the overlapping of light from scintillation events in the vicinity of the cluster. Figure 8, for example, shows the results of a simulated pile-up experiment under the camera conditions of Figure 6. In this study, it is assumed that scintillations from a primary event are emitted on the axis of the central PM midway into the crystal (DOI = 0), for which the third-order x moment $M_1^j = 2.7$. Simultaneously, a second scintillation of equal or lower energy is emitted at a point laterally separated from the primary by Δx. Figure 9 plots the moment, as a function of the separation, Δx, between the first and second event, the latter being assigned energy depositions of 100, 50, or 10% of the primary event. The moment rises rapidly as the second source is moved toward the edge of the central PM ($\Delta x = 0.5$ tube widths) and does not return to the uncorrupted primary emission value until the secondary emission is several tube widths away ($\Delta x = 1$ to 3.5 tube widths). The DOI determination is thus strongly dependent on the strength, proximity, and timing of pile-up events. Secondary events, having low fractions of the energy of the primary event (e.g., 10%) can make the determination of DOI meaningless if the events occur within the edge-ring of a cluster, for example. It should be noted, however, that because both x and y moments can be correlated separately with DOI, the probability of the y moment being corrupted by an x-displaced source, and contrariwise, is low. Thus, both moments should be used to analyze DOI. The count-rate range within which the moment analysis of DOI is applicable strongly depends on pile-up conditions and correction procedures.

The optical overlapping of light from scintillation events registered by PMs may be reduced by either reducing their temporal or spatial widths. The first approach employs analog means to reduce the width of PM pulses, such as in the work of Wear et al. (1998), in which the pulse is clipped by a delay line. This typically yields approximately 40% of the energy of an unclipped pulse. On the other hand, pile-up can be reduced by reducing the light spread of scintillation by increasing columnar scattering. Thus, for example, if the density of the columns in a large-area scintillation camera is increased so that its light spread matches a PM size of 28 mm instead of 40 mm, pile-up will be reduced by a factor of 2. In contrast to pulse clipping, however, the light yield and energy resolution in this case is preserved.

Pulse pile-up may also be reduced by dynamic methods, such as in the method of Wong and colleagues (Wong and Li, 1998; Wong et al., 1998), which recover the pulse shape of a primary event by removing remnant signals from prior and subsequent events. These methods carry statistical burdens, which increase with the count rate. Thus, in principle, pulse pile-up increases the uncertainty of a DOI determination but not its functional dependence on a cluster moment. Thus, although in the absence of pile-up DOI may be determined within a full width at half maximum (FWHM) of under 10 mm, its uncertainty increases with count rate. The amount of this increase and range of usefulness of the method are thus dependent on the methods of spatial and temporal treatment of pile-up in addition to the geometry of the camera.

When used for SPECT, NeurOtome may be fitted with type A and B SensOgrade collimators (see Section III). The former usefully compensates for attenuation in general purpose whole-brain applications such as cerebral blood-flow imaging. Type B collimators (Fig. 3b), on the other hand, are most suitable for focusing more intently on specific regions of the brain, such as for Parkinson's disease, attention deficit and hyperactivity disorders (ADDH), and psychoneurological disorders. In addition to providing higher sensitivity and lower variance, the converging transaxial projections reduce the effective transaxial intrinsic resolutions of the crystal to near-millimeter levels by virtue of their image magnifications.

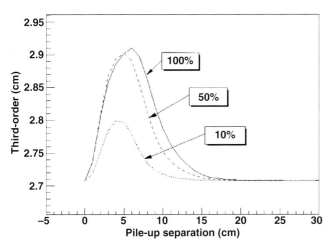

FIGURE 9 Effect of pulse pile-up on the third-order moment that is used to define light spread (in accordance with Figs. 5–7). A primary event, occurring midway into the 30-mm crystal in Figure 5 and along the central axis of a 3 × 3 cluster of 40-mm square PM, is assumed to trigger cluster analysis. Simultaneously, a secondary event of equal magnitude is assumed to occur at a displacement Δx from this center. The integration of light received from these two events shows that the LTF rises rapidly with Δx until the fraction of light reaching the cluster is ineffectual.

VI. MAMMOSPECT, AN ANNULAR BREAST SINGLE-PHOTON EMISSION COMPUTED TOMOGRAPHY CAMERA

MammOspect is camera designed for annular SPECT imaging of a pendant breast. The camera annulus encircles

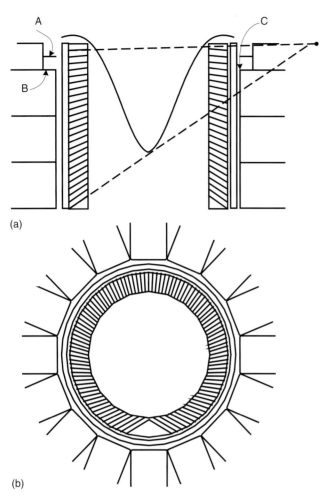

FIGURE 10 (a) Transaxial and (b) axial views of a MammOspect camera. Crystal surface modifications, nonuniform photomultiplier configuration, and reflecting baffles are used to image close to the camera's edge. The SensOgrade collimator design is configured to have continuously varying focused elements, as in a type A collimator. Axial focusing will further increase the sensitivity of the camera and reduce its effective intrinsic resolution.

the breast in close proximity to its chest wall, as illustrated in Figure 10. Its principal features are its axial detector optics, designed to image close to the edge of the camera, and its doubly convergent SensOgrade collimator, also contributing to edge imaging and to higher sensitivity and resolution.

The camera is a liquid-interfaced annular camera, which has a 10-mm-thick crystal with an inner retro-reflecting surface, prepared as in Section IV. This reflector narrows the spread of the transmission of light through the crystal. Also narrowing the propagation of light as it transmits through the outer surface of the annulus are 60° pyramidal structures (not shown in Fig. 10). These transmitting pyramids function as multiple mini-lenses, each of which focuses the light forward, much as a conventional converging lens does. The four surfaces of the pyramids also enhance light transmission by secondarily reflecting rays that are not refracted by one surface of a pyramid to other surfaces. Scattering columns may also be used to enhance transmission, but they are not currently employed.

Also contributing to edge performance of the camera are the dimensions and configurations of the PMs and the reflecting baffles interposed between them and the crystal. Of particular importance are the first two ring baffles (A and B in Fig. 10a) and the dimensions and placements of the PMs. The PMs in the leading ring of tubes, measuring 28×28 mm, are smaller than the 50-mm square PMs populating the outer three rings. The small tube ring is 17 mm from the outer surface of the crystal annulus and the 50-mm square PM is positioned 10 mm from it. In the first of these 50-mm tubes, a 6-mm-wide side window of the photocathodes, labeled C, is exposed to light, in addition to its end window. It is estimated that the efficiency of the conversion of light to photoelectrons by this side window is approximately 50% less than by its end window. Although its lower efficiency degrades energy resolution, light collection by this end window contributes to edge imaging by extending the field of its PMs. In combination, these optical modifications extend the field of view toward the crystal edge by shifting the LTF of the PMs so that the LTF of the first 28-mm tube intersects that of the adjacent 50-mm tube at approximately 10 mm from the edge of the crystal. Such shifting moves the higher-slope portions of the LTF toward the edge of the camera, thereby also improving its resolution in that region. Figure 11 shows representative axial intrinsic crystal resolutions achieved in a test camera. A useful intrinsic resolution distance of approximately 5-mm FWHM, obtained at a distance $z = 6$ mm from the edge of the crystal, decreases rapidly to an average of approximately 3.5 mm FWHM in the central field of view.

Internally concentric to the detector is a collimation system that is designed to focus transaxially, as in collimator type A in Figure 2. This collimator also focuses axially along a ring interior to the body so as to provide still higher sensitivity and to peer further toward the chest wall while being shielded from direct heart activity. The combination of transaxial and axial focusing, which increases both sensitivity and intrinsic resolution, is an extension of the methods of cone (Jaczyzak et al., 1986) and astigmatic (Hawman and Hsie, 1987) collimation. For a breast, shown cradled in a 150-mm holder, the camera's central axis sensitivity is double that obtainable by using three parallel-hole collimator segments of equal hole sizes, similar to those used in CeraSPECT (Fig. 1), for example. The higher sensitivities in central regions of the field are designed to compensate for both attenuation and reduced resolution. As with brain SPECT using a SensOgrade collimator, the FWHM of the transaxial intrinsic resolution is reduced to small values because of the magnification of its sharply

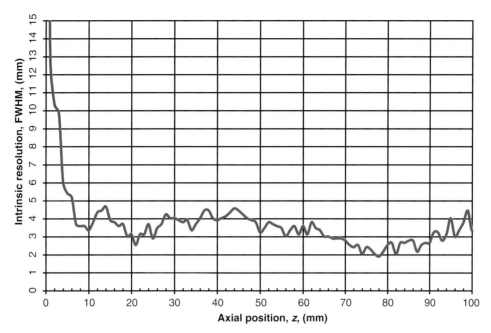

FIGURE 11 Plot of the intrinsic resolution of a test camera (without magnification) as a function of the displacement of a Tc-99m from the edge ($z = 0$) of the camera's crystal.

focusing collimator. Axial intrinsic resolution is also reduced by magnification, but to a lesser extent. The average reconstructed resolution for this system using an ultra-high-resolution collimator is expected to vary from approximately 3 mm FWHM at the surface of the 150-mm field to approximately 4 mm at its center.

VII. SMALL ANIMAL SINGLE-PHOTON EMISSION COMPUTED TOMOGRAPHY USING AN ANNULAR CRYSTAL

An annular camera may also be used for small-animal imaging, while retaining the advantages of a stationary camera outfitted with a rotating collimator. One such system, designed for rodent imaging, has been developed at the National Institutes of Health (D. W. Jones et al., 2003). It employs tungsten pinholes to projects eight magnified images on to the crystal of a CeraSPECT camera system. Figure 12 illustrates schematically a section of the camera. Each of the projected images is magnified by approximately 4.5, thereby reducing the FWHM of its projected intrinsic resolutions to submillimeter levels. The imaging field of view (FOV) is 2.54 cm, sufficient to encompass a 30-g mouse. Initial images reconstructed using a cone-beam OSEM algorithm exhibit spatial resolution of 1.7 mm FWHM and point-source sensitivity of 13.8 cps/microCi. A complete tomographic data set of non-overlapping projected images may be collected in 10 s during a collimator rotation of 45 degrees. The system is thus capable of dynamic imaging using list-mode acquisition with continuous collimator rotation through multiple cycles.

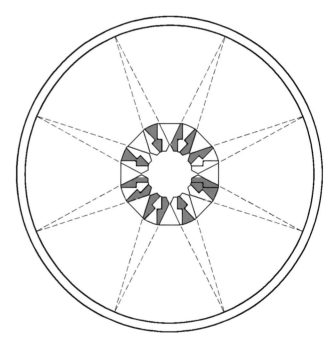

FIGURE 12 Section of a pinhole geometry in a small-animal annular camera. Each of eight equally spaced pinholes projects images simultaneously on to the annular crystal of a CeraSPECT camera with a magnification of 4.5. The FWHM of the intrinsic resolutions of eight equally spaced pinholes are thereby reduced to submillimeter levels. The system offers high-speed collection of complete sets of tomographic projection data with system resolution of 1.7 mm FWHM and a point source sensitivity of 13.8 cps/microCi. (From Jones et al., 2003.)

VIII. DISCUSSION

Annular cameras are an alternative to rotating cameras. When used for SPECT, in combination with SensOgrade collimators, they clearly can provide higher sensitivities and signal-to-noise ratios. Annular single-crystal cameras are also applicable to PET and to hybrid imaging. To bridge the gap between PET and SPECT, a camera should have an annular crystal that is thick enough for adequate PET sensitivity and small enough for good SPECT resolution. It also should be able to correct for parallax using DOI analyses and have narrow light spreads to limit pulse pile up. Sensitivity, resolution, and count rate are performance parameters that can be modified by optical design to meet specific performance requirements. Annular single crystals are most suited for small dedicated cameras, such as for brain and breast imaging. When applied to breast imaging, the optics has to be further modified to image close to the chest walls of the detector. SensOgrade collimators are uniquely fit for annular SPECT cameras by virtue of the continuity of its crystal's 2π geometry. Together they provide higher resolutions and sensitivities for brain and breast imaging. The annular camera geometry also can be used for efficient high-resolution small animal SPECT by using a rotating multiple-pinhole collimator.

References

El Fakrhri, G., Kijewski, M. F., Ouyang, J., Genna, S., Zimmerman, R. E., Xia, W., Moore, S. C., and Fishman, A. J. (2004). A novel collimator for high-sensitivity brain SPECT. *J. Nucl. Med.* (In press.)

Gagnon, D. (1996). Depth-of-interaction normalization of signals. U.S. Patent 5,576,546.

Genna, S. (1997). Liquid Interface Scintillation Camera. U.S. Patent 5,652,249.

Genna, S., Gayshan, V., and Smith, A. P. (1997). Scintillation camera for proximate SPECT imaging of a pendant breast. *J. Nucl. Med.* **41**: 32P.

Genna, S., and Smith, A. P. (1988a). The development of ASPECT, an annular single-crystal brain camera for high efficiency SPECT. *IEEE Trans. Nucl. Sci.* **34**: 654–658.

Genna, S., and Smith, A. P. (1988b). Acquisition and calibration principles for ASPECT—ASPET camera using digital position analysis. *IEEE Trans. Nucl. Sci.* **35**: 654–658.

Genna, S., and Smith, A. P. (1995). Annular single crystal SPECT. *In* "Principles of Nuclear Medicine" (H. N. Wagner, Jr., and Z. Szabo, eds.), 2nd ed., pp. 328–332. W. B. Saunders, Philadelphia.

Hawman, E. G., and Hsie, J. (1987). Astimatic collimator. *J. Nucl. Med.* **28**: 1308–1314.

Jaczyzak, R. J., Floyd, C. E. Manglos, S. M., Greer, K. L., and Coleman, R. E. (1987). Cone beam collimator for single photon emission computed tomography: Analysis, simulation, and image reconstruction using filtered back projection, *Med. Phys.* **13**: 484–489.

Jones, D. W., Goertzen, A. L., Riboldi, J., Seidel, J., and M. V., and Green, M. V. (2003). A novel mouse SPECT scanner using an annular scintillation crystal. *Mol Imaging Biol* **5**: 109.

Kijewski, M. F., Müller, S. P., and Moore, S. C. (1997). Nonuniform collimator sensitivity: Improved precision for quantitative SPECT. *J. Nucl. Med.* **38**: 151–156.

Smith, A. P., and Genna, S. (1988). Acquisition and calibration principles for ASPECT—ASPECT camera using digital position analysis. *IEEE Trans. Nucl. Sci.* **35**: 740–743.

Strobel, J., Clinthorne, N. H., and Rogers, W. L. (1997). A thick crystal gamma camera design for PET and SPECT using retrereflectors. *J. Nucl. Med.* **41**: 32P.

Wear, J. L., Karp, J. S., Freifelder, R., Mankoff, D. A., and Muehellehner, G. (1998). A model of the high count rate performance of NaI(Tl)-based PET detectors. *IEEE Trans. Nucl. Sci.* **45**: 1231–1237.

Wong, W. H., and Li, H. (1998). A high count rate position decoding and energy measuring method for nuclear cameras using anger logic detectors. *IEEE Trans Nucl Sci.* **45**: 838–842.

Wong, W. H., Li, H., and Uribe, J. (1998). A scintillation detector signal processing technique with active pileup prevention for extending scintillation count rates. *IEEE Trans. Nucl. Sci.* **45**: 1122–1127.

CHAPTER 10

PET Systems

THOMAS LEWELLEN* and JOEL KARP†

Department of Radiology, University of Washington Medical Center, Seattle, Washington
†Department of Radiology, University of Pennsylvania, Philadelphia, Pennsylvania

I. Basic Positron Emission Tomography Principles
II. Detector Designs
III. Tomography System Geometry
IV. Positron Emission Tomography Scintillators
V. Positron Emission Tomography System Electronics
VI. Attenuation Correction
VII. Scatter Correction
VIII. Noise Equivalent Count Rate
IX. Future Trends

I. BASIC POSITRON EMISSION TOMOGRAPHY PRINCIPLES

The goal of positron emission tomography (PET) is to generate images of the distribution of positron emitters *in vivo*. PET systems rely on the detection of annihilation gamma rays that follow positron decay. The gamma rays are detected in coincidence by detectors that surround the patient. Figure 1 depicts a detector ring system with a diagram to illustrate the event rates from two detectors. Note that only a small number of the events processed by each detector are in coincidence. The rate of events processed by each detector is often referred to as the single event rate for that detector, whereas the coincidence event rate includes true events, scattered events, and random events. We discuss in more detail issues of scanner geometry and their effect on count rate in following sections.

The spatial resolution of PET imaging is limited by the fundamental nature of positron annihilation (Phelps *et al.*, 1975; Muehllehner *et al.*, 1976; Hoffman and Phelps, 1978; Derenso, 1979). As positrons travel through human tissue, they give up their kinetic energy principally by Coulomb interactions with electrons. Because the rest mass of the positron is the same as that of the electron, the positrons may undergo large deviations in direction with each Coulomb interaction and they follow a tortuous path through the tissue as they give up their kinetic energy (see Fig. 2). When the positrons reach thermal energies, they interact with electrons by the formation of a hydrogen-like orbiting pair called positronium. Positronium is unstable and eventually decays, via annihilation, into a pair of anti-parallel 511-keV photons (emitted at 180° relative to one another). Note that a small percentage, less than 2%, annihilate without forming positronium. Although the radial distribution of annihilation events is sharply peaked at the origin (site of positron creation), a calculation of the radius that includes 75% of all annihilation events gives a realistic comparison of the impact of the maximum positron energy on the spatial resolution of PET imaging (Hoffman and Phelps, 1976). Table 1 lists the major emitters used in PET imaging, along with positron energy and range in water.

In addition to the positron range, the variation in the momentum of the positron also leads to a limitation of the spatial resolution of PET imaging. One would normally expect the annihilation gamma rays to be anti-parallel. However, the variation in momentum of the positron results in an

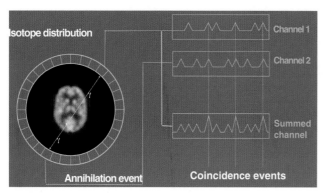

FIGURE 1 Block diagram of a basic PET scanner with illustration of events in coincidence.

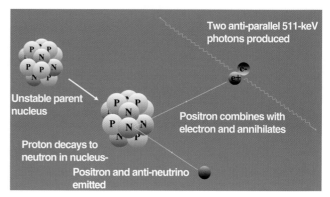

FIGURE 2 Physics of positron decay and annihilation, which results in two 511-keV gamma rays.

TABLE 1 Half-life and positron range of isotopes commonly used in PET[a]

Isotope	Half-Life (min)	Maximum Positron energy (MeV)	Positron Range in Water (mm)
^{11}C	20.3	0.96	2.1
^{13}N	9.97	1.19	—
^{15}O	2.03	1.7	—
^{18}F	109.8	0.64	1.2
^{68}Ga	67.8	1.89	5.4
^{82}Rb	1.26	3.15	12.4

[a]The range is defined as the radius that includes 75% of all annihilation events.

TABLE 2 Resolution Loss Due to Noncolinearity versus Tomograph Detector Ring Diameter[a]

Ring Diameter, D (cm)	Resolution Loss, Δx (mm)
60	1.3
80	1.7
100	2.2

[a]Calculated using the relationship $\Delta x = 0.5\, D \tan 0.25°$.

TABLE 3 Loss of System Spatial Resolution Due to the Combination of Positron Range and Gamma-Ray Noncolinearity for a 80-cm Ring Diameter and as a Function of Coincidence Detector Pair Resolution

Detector Pair Resolution (mm)	^{11}C (mm)	^{68}Ga (mm)	^{82}Rb (mm)
0	2.3	3.1	3.3
2	3.2	4.1	7.4
6	6.6	7.4	9.2

angular uncertainty in the direction of the 511-keV photons that is approximately 4 mrad (0.23°) (Beneditti *et al.*, 1950). This is referred to as noncolinearity. Table 2 shows the resolution loss due to noncolinearity as a function of detector ring diameter.

A third significant factor limiting PET image resolution is the intrinsic spatial resolution of the detector (Hoffman *et al.*, 1982). The resolution of a single detector is often quantified by the full width at half-maximum (FWHM) of the position spectrum obtained for a collimated point source placed before the detector at a fixed distance from it. The coincidence detector-pair resolution is normally specified as the FWHM of the point-spread function (PSF) obtained from the convolution of the two individual detector PSFs (Muehllehner *et al.*, 1986). For a detector composed of small discrete crystals, all interactions are assumed to occur at the center of individual crystals for the purpose of backprojection and image reconstruction. As a result, the PSF for such detectors is similar to a step function with a total width equal to the size of a crystal. The coincident PSF is, therefore, a triangular function whose base width is again equal to a crystal size. Thus, the FWHM of the coincident detector PSF is one-half the crystal size.

Together, the positron range, noncolinearity, and detector PSF limit the resolution of PET tomographs. Table 3 shows the impact of positron range on the detector-pair resolution for a 80-cm ring diameter (Nohara *et al.*, 1985). The resolution broadening effects due to positron range and noncolinearity for three isotopes (three different maximum positron energies) are tabulated for an infinitively small point source between two detectors for three coincidence detector-pair resolutions (0, 2, and 6 mm).

A final factor affecting PET image resolution is referred to as the parallax error, which results from the uncertainty of the depth of interaction (DOI) of the gamma rays in the crystal. Gamma rays travel some (unknown) distance in the crystal (or adjacent crystas) before being completely absorbed (see Fig. 3). As a result, if the gamma ray enters the crystal at an oblique angle, the location of the interaction will not be the same as the point of entry into the crystal;

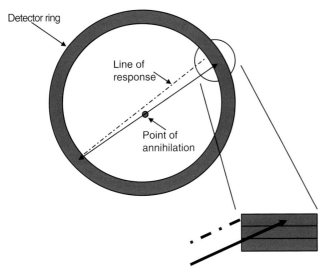

FIGURE 3 Parallax error. The gamma ray (solid line) interacts in a crystal after penetrating one or more adjacent crystals in the detector ring. Without depth-of-interaction information, the detection electronics will incorrectly assign the line of response (the dotted line) based on the front of the interaction crystal.

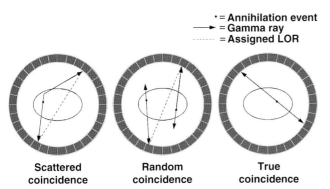

FIGURE 4 The three types of coincidence events measured in a PET scanner.

TABLE 4 Options to Increase Detected Coincidence Events in a PET Study

Increase the patient dose
Use more efficient scintillators/detectors
Use more of the energy spectrum
Increase solid angle

the crystal of the interaction may not even be the same as the one first entered. Thus, unless the DOI within a crystal can be accurately determined, an incorrect line of response (LOR) will be assigned to this interaction because the LOR is normally assigned to a position at the front of the crystal of interaction. The parallax effect worsens as the source position moves radially away from the center of the scanner because a larger fraction of the gamma rays enter the crystals at oblique angles.

Figure 4 illustrates three kinds of coincidence events that the tomograph accepts: (1) scattered events in which one or both gamma rays scatter within the patient; (2) random coincidences, in which two separate decays result in the detection of only one gamma ray from each one and the two events are close enough in time to be in coincidence; and (3) true coincidences, in which gamma rays are detected from a single decay that have not scattered in the patient. Our goal in PET imaging is to measure and reconstruct the distribution of true coincidences while minimizing the scattered and random coincidences and correcting for the bias (but not necessarily the noise) associated with scattered and random coincidences. Note that both true and scattered events are referred to as prompt events because they come from the decay of a single nucleus and, thus, the gamma rays are detected almost simultaneously. Random, or accidental, coincidences occur if two separate decays occur close enough in time to look like a single decay to the system electronics. The prompt rate (trues + scatters) is related linearly to the activity in the patient. However, the randoms rate increases as the square of the activity in the patient and becomes more dominant at higher activity levels.

Increasing the number of true coincidences leads to less noise in the image and allows one to reconstruct the data with high spatial resolution of the distribution of decay events, given the physical limitations already discussed. Specific design goals are generally determined by the application (e.g., small animal imaging, human neurological scanning, or human whole-body imaging). But for any design goal, if the variance in the data is too high, then the reconstruction algorithm will have to employ a filter to reduce the image variance at the expense of spatial resolution (Phelps *et al.*, 1982). The four major options for improving counts per unit dose in a study are listed in Table 4.

Before looking at some of the consequences for detector and electronics design when we increase sensitivity, let us consider the basic options in Table 4. Many centers with dedicated PET scanners are already using the maximum dose limits allowed in their institutions (between 10 and 20 mCi for ^{18}F-FDG, fluoro deoxyglucose); increasing the patient dose is not a practical approach because higher doses would result in higher radiation exposures to the patient. This would only be advantageous for 2D systems because 3D volume-imaging PET systems (without interplane septa) generally reach the peak count-rate performance below 20 mCi due to random coincidence or because the detector and electronics become count-rate limited at high activity levels. In fact, 3D scanners consisting of large, continuous NaI(Tl) detectors use less than 10 mCi, due to count-rate limitations.

Many research facilities are searching for better scintillators, but materials with higher stopping power than bismuth germanate (BGO) have proved to be elusive. The main advantages of new scintillators (discussed in more detailed later) such as lutetium oxyorthosilicate (LSO), in comparison to BGO, is the combination of high light output and fast decay. BGO-based systems typically use detectors 30 mm deep, which provide approximately 90% detection efficiency for 511-keV gamma rays (therefore, 82% coincidence efficiency). Although 100% detection efficiency would improve sensitivity even further, no systems have yet been built with basic detection efficiency greater than that provided by 30-mm-deep BGO crystals.

One approach to gain more sensitivity is to use more of the energy spectrum—to accept events with energy less than the photopeak. This approach has been studied in dual-headed gamma cameras when they are operated as coincidence detectors because the sensitivity of the NaI(Tl) detectors is relatively low compared to those in dedicated PET scanners (Nelleman et al., 1995). Two energy windows are set—one around the photopeak and one around a region of the Compton scatter spectrum. The lower window captures those events that Compton-scatter in the detector and then escape, producing events with energies in the Compton region of the spectrum. However, the inclusion of the Compton region also increases the amount of patient scatter, which also produces events with energies below the 511-keV photopeak. Although this technique can lead to an increase in sensitivity of close to a factor of 2, the increased scatter is problematic and, overall, the technique has not been shown to be clinically useful.

The fourth method to increase the solid angle is the most useful approach for dedicated PET systems. Techniques being developed include the removal of all axial collimation (3D volume imaging), reducing ring diameters, and extending the axial dimension of the detector array. For example, a typical whole-body ring system can realize an increase in the true sensitivity by up to a factor of 5 (for brain imaging) by removing the axial collimation. The penalty is a large increase in the acceptance of scattered photons (scatter fractions increasing by factors of 3 to 4) and increased singles rates from activity outside the field of view (FOV) of the system. We discuss some of the options for scatter correction and other issues in machine performance later in this chapter. First, we review the basics of detector designs used in PET systems.

II. DETECTOR DESIGNS

Early PET systems (Phelps et al., 1978; Ter-Pogossian et al., 1978; Brooks et al., 1980) used one photomultiplier tube (PMT) per scintillation crystal. There are several major limitations to this approach. For high spatial resolution, one needs to use small crystals and existing PMTs are too large to allow a full ring of very small crystals; also, it becomes difficult or impossible to make multiring systems. For a large number of crystals, this type of design also leads to a large number of electronic channels and PMTs, thus increasing the cost of the system.

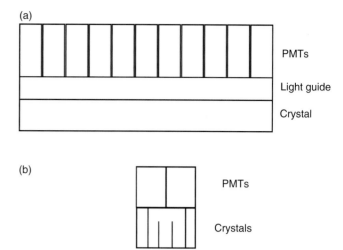

FIGURE 5 The major crystal-PMT decoding geometry options currently in use. (a) The continuous Anger-logic approach uses an array of PMTs to decode a large continuous crystal or an array of crystals. (b) The block detector uses four PMTs to decode an array of crystals with various combinations of reflectors and surface treatments between the crystals.

Alternative detector designs that shared crystals between PMTs began appearing in the 1980s. Examples of these designs can be found in the review article Muehllehner et al. (1986). Figure 5 shows two major design approaches currently in use. The first is based on the original Anger-logic approach (Karp et al., 1990) with large area detectors viewed by an array of PMTs; the other uses smaller discrete crystals arranged in a block viewed by a limited number of smaller PMTs (Casey and Nutt, 1986).

The use of Anger-logic NaI(Tl) detectors for PET imaging (Muehllehner et al., 1976) requires a thicker scintillation crystal than that normally used for single-photon imaging. The Philips Medical Systems C-PET scanner (Adam et al., 2001) is a modern dedicated PET scanner that uses NaI(Tl) detectors. All manufacturers offer dual- (or triple-) headed NaI(Tl) cameras that can be used for either PET imaging (electronics set for coincidence detection) or single-photon emission computed tomography (SPECT) imaging (no coincidence and with collimators on detectors). Although a 3/8-in-thick detector stops over 90% of the gamma rays at 140 keV, it stops only ~10% at 511 keV. This, in turn, leads to only ~1% coincidence detection efficiency. For PET imaging, a 1-in-thick detector is normally used to achieve higher sensitivity; approximately 56% of the gamma rays at 511 keV are stopped, lending to 31% coincidence detection efficiency. Although a thicker detector leads to higher detection efficiency, the intrinsic spatial resolution degrades as the detector is made thicker. The loss of spatial resolution for thick detectors is mainly due to Compton scattering within the scintillator, which spreads the area over which the gamma-ray energy is deposited. Spatial resolution also depends on the design of the light guide that serves to spread the scintillation light to an appropriate number of

PMTs. It is desirable to restrict the spread of light to a small number of PMTs, because this will reduce the chance of event pile-up at higher count rates.

In addition to using a thicker scintillation crystal, modifications are required to the detector and the electronics used in large NaI(Tl) systems in order to allow higher count rates. The need for a very high count-rate capability in a PET scanner is not immediately evident. It is mainly due to the fact that a relatively low fraction of single-photon events are in coincidence and available for image generation. A high percentage of detected gamma rays are not in coincidence with another gamma ray but contribute to the singles count rate seen by the system. Many techniques used to achieve a high count-rate capability have been developed. These techniques include pulse shortening (Amsel *et al.*, 1969), local centroid positioning (Karp *et al.*, 1986), and the use of multiple trigger channels (Mankoff *et al.*, 1990). Pulse shortening is required because the scintillation decay time of NaI(Tl) is relatively long. The characteristic decay constant is 240 ns, which means that it takes about 900 ns to collect all the light. For 140 keV, a long integration time is used and an intrinsic energy resolution of 11% and spatial resolution of ~3.5 mm are achieved. At 511 keV, more than three times as much light is emitted in the scintillator, and one can afford to clip the electronic pulse corresponding to the light emission to approximately 200 ns and still achieve approximately 11% energy resolution. Even though the light from the first event continues to be emitted after 200 ns, the electronics can analyze another event after 200 ns, thereby reducing the dead time in the detector and increasing the count-rate capability.

The event-positioning electronics in most modern systems are digitally controlled at almost all levels. Once the PMT signals are digitized, the integration period is digitally controlled, as well as the number of PMTs included in the position calculation. Traditionally, the position of an event in an Anger camera (with continuous detector, as in Fig. 5a) is determined by calculating the centroid of the emitted light:

$$X = \sum_i (x_i S_i) / \sum_i (S_i)$$

where x_i is the weight associated with the *i*th PMT and S_i is the integrated PMT signal. A similar calculation is made to determine the orthogonal direction Y to achieve a 2-D position. In an analog circuit, the summation occurs over all PMTs, but with digital control one can specify the number of PMTs in the immediate vicinity of the scintillation event. With C-PET, for example, seven PMTs (in a 2D hexagonal pattern) are used to calculate the position and energy of each event. The local cluster of PMTs is determined by searching for the PMT with the highest signal. The local centroid calculation allows multiple events to be processed in the detector simultaneously, as long as they are spatially separated.

In addition to reducing pulse pile-up from the combination of pulse shortening and the local centroid calculation, the electronic dead time is reduced by dividing the detector into multiple regions, each of which is connected to an independent set of electronics. For a coincidence event to be recorded, it is necessary for two trigger channels, from approximately opposite sides of the detector ring, to fire within a short period of time. Typically the coincidence-resolving time, referred to as 2τ, is 8 ns for NaI(Tl)-based PET systems. Multiple trigger channels in each detector allow simultaneous coincident events to be processed.

The large continuous crystal approach is generally not possible for a range of scintillators currently in use for PET scanner construction. In fact, a majority of the current PET scanners being built use discrete crystals or small blocks of scintillators rather than the large continuous crystal designs developed for NaI(Tl) systems. These scintillators that have higher stopping power (e.g., BGO, LSO, and GSO) can not currently be grown into large crystals and, with the exception of LSO, do not produce sufficient light output to achieve good spatial localization as a continuous detector.

For the small block arrays (Fig. 5b), a limited number (four) of small PMTs are used. The interface between the crystals is used to shape the light-spread function (LSF) to allow the decoding of crystal positions. The LSF can be controlled with different coupling compounds at the interface and surface finishing or by using different lengths of reflector between the crystals. For example, in the simplified block depicted in Figure 5b the outer crystals have reflectors running the full length of the block. Thus, any light produced in those crystals is detected in a single PMT. The remaining crystals have reflectors going only partway and coupling compound the rest of the way. Light from these crystals will be shared by the PMTs. By changing the lengths of the reflectors, the amount of light shared between the PMTs can be controlled. Figure 6 shows a 2D crystal map from a detector block using a 6 × 6 array of BGO crystals. Four PMTs are placed over the crystals in a rectangular pattern:

$$\begin{bmatrix} A & | & B \\ \hline C & | & D \end{bmatrix}$$

To form the ratios, the two PMTs across the top of the block are summed for the x value ($A + B$) and the two PMTs along the left side are summed for the y value ($A + C$). For both ratios, the sum of the signals from all four PMTs are summed to obtain the denominator:

$$x = \frac{A+B}{A+B+C+D}; \quad y = \frac{A+C}{A+B+C+D}$$

Due to variances in PMT performance, normal production techniques, and the methods to shape the LSF, the light isocontour plots are not symmetric around the block. Search routines are used to find the minima between the peaks (the thick white lines in Fig. 6) and any events within a given region are assigned to the crystal position associated with

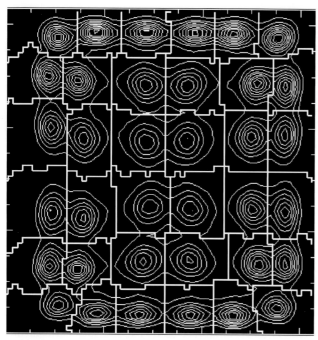

FIGURE 6 A 2D block map of a 6 × 6 BGO crystal array viewed by four PMTs. The relative light output is indicated by the isocount curves. The heavy white lines indicate which regions wil be assigned to each crystal position.

reveals that there is some overlap between adjacent crystals (the ratio does not go to zero between the peak centers).

Another example of the block design is the HR+ block (Adam *et al.*, 1997), which uses an 8 × 8 array of 4.5 × 4.1 mm^2 crystals in a block with four 19-mm PMTs performing signal readout. A modification of the block design is the quadrant-sharing block design (Wong *et al.*, 1995), which reduces the number of PMTs needed for a given number of crystals and has been used with arrays of smaller crystals than those used, for example, in the HR+ design. For example, one quadrant-sharing design uses blocks that make up a 7 × 7 array of 2.8 × 2.8 mm^2 crystals. The 19-mm PMTs now straddle four block quadrants so that each block is still viewed by four PMTs but each PMT is also shared by four blocks. The decoding ratio (number of crystals/PMT) of the standard block is 16:1, whereas that of the quadrant-sharing block is 49:1. However, because the PMTs are shared by four blocks, an interaction in any given block of the quadrant-sharing design will deaden the adjacent (eight) blocks. In comparison, an interaction in a block of the standard block design will deaden only one block area. Thus, the dead time of the quadrant-sharing block design using 2.8 × 2.8 mm^2 crystals will be more than a factor of 2 higher than that of the standard HR+ block design. The LSO panel detectors developed by CPS Innovations also use a quadrant-sharing block design, using 4 × 4 mm^2 crystals.

A different approach is used for the GSO detector modules used in the Phillips Allegro scanner. Although the detector uses discrete crystals (4 × 6 × 20 mm^3), the crystals are not grouped into small blocks, and the light guides for each of the 28 detector modules are coupled to allow light sharing between adjacent modules. Rather than using four PMTs to decode each event, the Allegro design uses seven 39-mm diameter PMTs in a hexagonal arrangement. However, unlike the block detector, the group of seven PMTs does not correspond to a specific block of crystals because as it is chosen electronically from among a large array of PMTs coupled to the detector.

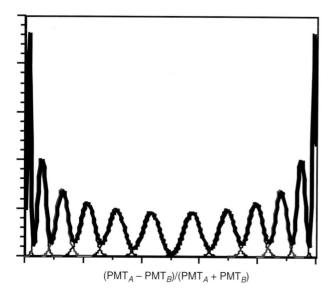

$(PMT_A - PMT_B)/(PMT_A + PMT_B)$

FIGURE 7 Plot of the PMT ratio (two PMTs) for a linear array of 12 LSO crystals. The *y* axis is the relative number of events detected.

that region. The graph in Figure 7 shows crystal identification (the ratio of the difference in the PMT outputs over the sum of the PMT outputs) versus source position for an experimental detector block using LSO crystals. This block allows the decoding of 12 crystals with two PMTs. As can be seen in the plot, the proper shaping of the LSF allows relatively easy decoding of the crystals. Careful inspection also

III. TOMOGRAPHY SYSTEM GEOMETRY

As we have discussed, current detector module designs use either large continuous NaI(Tl) crystals or arrays of discrete crystals. The NaI(Tl) systems offer a lower price due to the lower cost of the scintillator and the use of fewer electronics channels, whereas the discrete crystal machines offer higher sensitivity due to the higher stopping power of the scintillators used (e.g., BGO, LSO, or GSO) and much higher count-rate performance (e.g., lower dead time and less pulse pileup). In either case, the detectors can be configured as full rings that completely surround the patient or as partial rings with rotational motion to obtain the

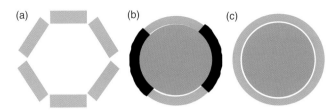

FIGURE 8 The three basic geometries found in modern dedicated PET systems. (a) Array of large detectors, either continuous detectors (flat or curve-plate NaI(Tl)) or plates of discrete crystals. (b) Partial ring of detector blocks that rotates. (c) Full ring of discrete crystals configured as small blocks or larger detector modules.

needed angular sampling (Fig. 8). Current designs with NaI(Tl) use either six large 25-mm-thick curve plates (full ring) or SPECT systems with two or three detector heads with 18- to 25-mm-thick detectors (partial ring). Current discrete crystal designs use BGO (25–30 mm thick), LSO (typically 25 mm thick), or GSO (typically 20 mm thick) arranged in full or partial rings about the patient.

Full-ring NaI(Tl) systems are currently being built with an axial extent of 25 cm (Adam *et al.*, 2001). Because the NaI(Tl) system use continuous detectors, the axial sampling is determined by the intrinsic spatial resolution of the system and the choice in how the data are collected. These systems typically sort the data into 128 axial image planes (2 mm thick) for brain studies, and 64 axial planes (4 mm thick) for whole-body studies. These full-ring NaI(Tl) systems are operated exclusively in positron volume imaging (PVI) mode (also known as 3D mode) to achieve better sensitivity.

Full-ring discrete crystal systems currently all use BGO, LSO, or GSO. As described in the preceding section, some common block sizes include 6 × 6 arrays of 4 × 8 × 30 mm³ BGO crystals (GE Advance) or 6.25 × 6.25 × 30 mm³ BGO crystals (GE Discovery ST) and 8 × 8 arrays of 6.75 × 6.75 × 20 mm³ BGO crystals (Siemens/CTI EXACT) or 4.5 × 4.8 × 30 mm³ BGO crystals (EXACT HR+). In addition, the ECAT ACCEL uses an 8 × 8 array of 6.75 × 6.75 × 25 mm³ LSO crystals. The Phillips Allegro uses 4 × 6 × 20 mm³ GSO crystals assembled in 28 modules that are optically coupled to one another and viewed by a hexagonal array of PMTs. All the dedicated stand-alone (i.e., not combined with computed tomography, CT, scanners) BGO and LSO systems have ring diameters of 80–90 cm and are equipped with removable axial collimators, allowing the systems to operate in either 2D or 3D mode. The GSO systems do not include axial collimators and only operate in 3D mode, as do the NaI(Tl) scanners. In addition, many of the PET/CT scanners now being offered (discussed later in this book) also do not include axial collimators, thereby allowing only 3D acquisition.

Discrete BGO and LSO systems are also available in partial-ring designs, the Siemens/CTI ECAT ART (BGO) and EMERGE (LSO). Because block detectors generally use

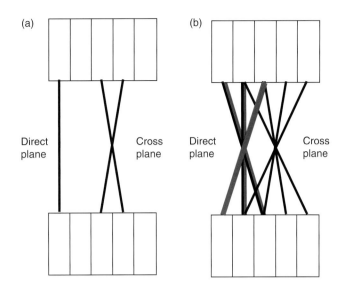

FIGURE 9 Two methods of binning 2D data in dedicated PET scanners with axial collimation. (a) High-resolution (HR) mode uses ring differences of 0 for the direct planes and ring differences of 1 for the cross planes. (b) To gain more sensitivity, data are most often binned in high-sensitivity (HS) mode. In this case the direct planes use ring differences of 0 and 2 and the cross planes use ring differences of 1 and 3. In both cases, the number of sinograms in 2n-1, where n is the number of detector rings.

smaller PMTs than the continuous crystal designs, they are more expensive (due to the increased number of PMTs and electronic channels). In an effort to reduce the cost of these systems, partial-ring designs were developed. These machines essentially take a portion of the blocks used in a full-ring system and mount them on a rotating plate. Current designs use blocks with 8 × 8 arrays of 6.75 × 6.75 × 20 mm³ (BGO) or 6.75 × 6.75 × 25 mm³ (LSO) crystals. A typical arrangement uses 66 blocks arranged with three blocks in the axial direction providing 24 partial rings of crystals and an axial extent of 16.2 cm. The ring diameters are 82.5 cm. To compensate for the loss of sensitivity from only using partial rings of detectors, these systems operate only in PVI mode.

When a system operates in 2D mode (axial collimation), the data normally are binned directly into stacked 2D sinograms. Figure 9 illustrates the two most common methods of binning the data when axial collimation is used. One scheme, termed in the GE Advance the high-resolution mode, uses ring differences of 0 for the direct planes and ring differences of 1 for the cross planes (Lewellen *et al.*, 1996). To gain more sensitivity, data are often binned in what is often termed high-sensitivity mode. In this case the direct planes use ring differences of 0 and 2 and the cross planes use ring differences of 1 and 3. In both cases, the number of sinograms is $2n - 1$, where n is the number of detector rings. Other schemes that combine the detector data into sinograms achieve a different trade-off between sensitivity and resolution.

When the axial collimators are removed, the system collects full 3D data sets and operates as a PVI system. As

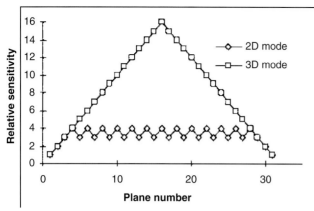

FIGURE 10 Comparison between 2D (axial collimation in place) and 3D (axial collimation removed) sensitivity.

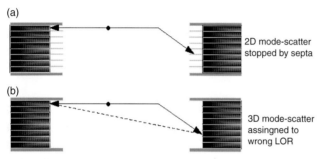

FIGURE 11 Illustration of the difference in axial acceptance of scattered events between 2D and 3D modes of operation of a PET system. (a) 2D mode, scatter stopped by septa. (b) 3D mode, scatter assigned to wrong LOR.

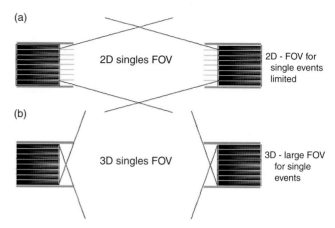

FIGURE 12 Illustration of the increase in the singles axial FOV in 3D mode (b) compared to the limited FOV in 2D mode (a) in a PET system. The increased singles FOV results in a higher randoms rate for the same amount of activity.

we have already noted, PVI offers an increase in sensitivity compared to 2D acquisitions (Fig. 10). However, there are some limitations to operating a system in 3D mode. Because the increased sensitivity is not uniformly distributed axially, the sensitivity advantage of 3D falls off rapidly as one approaches the edges of the axial FOV. The removal of collimation also increases the number of scattered events detected by the detector array (Fig. 11). The increase in scattered radiation depends on the lower-level energy discriminator (LLED), which in turn depends on the energy resolution of the system. A higher LLED setting can be used with scintillators that have better energy resolution. Typically, the LLED is set at 300–375 keV for BGO systems, 350–400 keV for LSO systems, 420 keV for GSO systems, and 435 keV for NaI(Tl) systems. The removal of the collimation also allows more photons from outside the axial FOV to be detected (Fig. 12). This increase in the number of single events detected versus the number of coincidence events detected leads to higher dead time and randoms rates.

In all system designs, the ring diameter and design of end shields and collimators have an impact on the accepted scatter and the trues-to-singles ratio (the number of true coincident events processed compared to the number of single events the detector must process). Thus, the system designer has to balance the increase in trues with larger axial FOVs and smaller detector rings against the impact of the out-of-field activity on the system dead time and accidental coincidence rate. Figure 12 illustrates the significant increase in the singles rate, and therefore randoms rate as well, that result from activity outside the FOV in 3D systems (without septa) compared to 2D systems (with septa). Of course, the system performance also depends heavily on the design of the supporting electronics, in addition to the detector configuration. As discussed later, another important parameter in 3D is the speed of the scintillator because fast scintillators such as LSO and GSO allow the use of a smaller coincidence window (2τ), which in turn leads to lower random coincidence rates.

IV. POSITRON EMISSION TOMOGRAPHY SCINTILLATORS

Table 5 lists some of the properties of scintillators that have been used for PET imaging systems. Other than BaF_2, which was used in early time-of-flight systems (Wong et al., 1984), all these scintillators are currently in use in commercial PET systems. System designers want a scintillator that has a very high light output, has a high energy resolution, is very dense with a large photoelectric cross section, has a fast decay constant, is nonhydroscopic, and is easy to grow into large crystals. Unfortunately, no single scintillator has all these ideal properties. We have already mentioned that the high light output of NaI(Tl) makes it a good choice for large area detectors. The fact that it can be grown into large crys-

Table 5 Properties of Scintillators Used in PET

	NaI(Tl)	BaF$_2$	BGO	LSO	GSO
Effective atomic number (Z)	51	54	74	66	59
Linear attenuation coefficient (cm^{-1})	0.34	0.44	0.92	0.87	0.62
Index of refraction	1.85	–	2.15	1.82	1.85
Light yield (% NaI(Tl))	100	5	15	75	30
Peak wavelength (nm)	410	220	480	420	430
Decay constant (ns)	230	0.8	300	40	65
Fragile	Yes	Slight	No	No	No
Hydroscopic	Yes	No	No	No	No

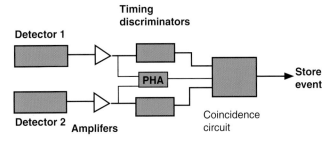

FIGURE 13 Basic electronic components for a PET system. The amplifiers integrate the detector signals. The pulse-height analyzer (PHA) selects the energy range of events to be accepted. The timing discriminators generate the timing pulse for the coincidence circuit.

tals at modest cost is another primary reason for its use in imaging systems. BGO has been popular due to its high stopping power and density. However, its low light output precludes its use for large area detectors with large PMTs. GSO and LSO both have good stopping power and fast decay, which lead to higher count-rate capability. However, due to their very high melting point (~2000°C) these crystals generally cost significantly more than either NaI(Tl) or BGO.

Currently, the high-end PET scanners being offered are all based on BGO, LSO, and GSO. Efforts are ongoing at many laboratories to develop better scintillators and reduce the cost of growing the more interesting candidates. A list of current candidates is beyond the scope of this chapter, but there are many exciting materials under development that may replace the current scintillators in future scanners (see van Eijk, 2002).

The intrinsic performance of these scintillators relates directly to overall scanner performance. For example, the energy resolution of a scintillator is dependent on its light output and the intrinsic scintillator resolution, which is a function of the variability in the light yield (e.g., how much light gets out of the scintillator as a function of where the scintillation occurs in the crystal). A high light yield from a scintillator normally improves the energy resolution for the detector because it reduces the Poisson noise of the PMT signal. Good energy resolution allows the use of a narrow photopeak energy window (high LLED) to maximize true coincidences and minimize scatter (and randoms) in the image, thereby improving contrast. High scintillator light yield also yields a high crystal decoding ratio, which permits the discrimination of smaller crystals with fewer, larger PMTs. The high density of a crystal (high linear attenuation coefficient) increases the detector sensitivity to coincident events. A high linear attenuation coefficient for a scintillator will also help reduce the parallax error in the scanner due to the reduced variability of the DOI of incident gamma rays in the detector. A short signal decay time helps reduce the width of the coincidence timing window (2τ), thereby reducing the rate of random coincidences within the scanner. Very short signal decay time and very good coincidence timing resolution (≤ 600 ps) are the main requirements for a scintillator to be used in a time-of-flight scanner.

V. POSITRON EMISSION TOMOGRAPHY SYSTEM ELECTRONICS

Figure 13 shows the basic components of the electronics system for a tomograph. The signals from the PMTs are integrated. In a BGO block system, which typically has many PMTs and readout channels, the major contribution to dead time is from these integrating amplifiers. The amplified signals are routed to a pulse-height analyzer (PHA) to select the energy range for the events to be processed. The signals also go to timing discriminators to generate timing signals for the coincidence system. Normally, the timing discriminators include lower-energy discriminators to reject events that have too low an energy to be passed by the PHA. Thus, the amplifiers see the full detector count rate, which is generally much higher than the output of the timing discriminators. For PET systems with large area detectors, each detector element (the crystal) subtends a large solid angle and has a high singles rate. As a result, the maximum count-rate performance of large area detector systems is usually limited by the pulse pile-up in the detectors, even when high count-rate techniques are employed (see previous discussion regarding pulse shortening, local centroid calculation, and multiple triggers). In a block system with discrete crystals, each detector element (the block that sends signals to an amplifier) subtends a relatively small solid angle. In these systems with many readout channels, the count-rate performance is usually limited by the random event rates rather than detector dead time.

The actual electronics in a full tomograph are more complicated than indicated in Figure 13. There are modules to decode the event position, to correct for the variance in photopeak position in various detector elements, to collect data for dead-time correction, and to sort the data and store it on disk. Major aspects of the system design are the details of the implementation of the coincidence electronics and the method to perform randoms correction. There are two standard approaches for measuring the randoms. Figure 14 illustrates the delayed-window approach. The number of events is plotted versus the time difference between detection of events by two detectors. The peak represents the prompt events *plus* random events. Our goal is to remove the random events from the prompt window. A delayed window can be defined that only samples random events. Data from the randoms window can either be saved or scaled and subtracted from the prompt window data during acquisition. This latter option (real-time subtraction) is often used by commercial PET systems. The disadvantage of real-time subtraction is that the randoms data for any given LOR may have a high variance. Saving the randoms data and using appropriate variance-reduction techniques can lead to a better quality data set. Another approach to reducing the variance is to calculate the randoms from the single event rates in the detectors. Because the singles rates are always much higher than the coincidence rates, the data have a low variance. The randoms can then be calculated using the relation:

$$R = 2\tau S_1 S_2$$

where τ is the time resolution of the system (and 2τ is the coincidence time window), and S_1 and S_2 are the singles rates of two detectors defining a given LOR. The major complications of this approach are (1) proper accounting of all sources of dead time so that the correct randoms rate is calculated for subtraction from the prompt data and (2) accounting for any variance in τ between different detectors. Whatever approach is used, the system designer usually separates the electronics into a series of parallel channels to reduce dead time in the coincidence electronics.

Once the data have passed the coincidence system, they are usually histogrammed into sinograms in high-speed memory. Such an arrangement easily allows dynamic studies to be acquired. Once the data have been saved to disk, the designer again has several options. The current system designs all use standard UNIX or Windows workstations to operate the scanner and function as image-display stations. When the data are reconstructed, two approaches are currently used. One approach uses a central reconstruction engine— a central computer that supports reconstruction requests from all workstations. Such an architecture typically uses a multi-CPU array of processors. The other approach is to let each workstation reconstruct its own images, with or without additional computational power added to the standard workstation. The data processing/reconstruction software must take into account many corrections, including dead time, randoms (if not subtracted during acquisition), detector normalization (correction for detector efficiency and geometric factors), scatter, and attenuation.

VI. ATTENUATION CORRECTION

A particular advantage of PET imaging is that attenuation correction factors do not depend on the location of a source along a given LOR. Consider Figure 15. For an event to be registered, both gamma rays from the decay must be detected. Thus, the probability that a gamma ray will reach detector 1 is:

$$p_1 = e^{-\int_0^{x'} \mu(x)dx}$$

and the probability that a gamma ray will reach to detector 2 is:

$$p_2 = e^{-\int_{x'}^{a} \mu(x)dx}$$

The total probability that the pair of gamma rays is detected is then:

$$p_d = p_1 p_2 = e^{-\int_0^{a} \mu(x)dx}$$

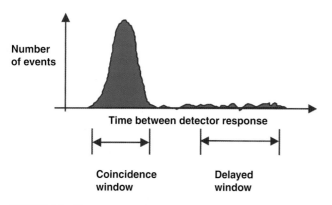

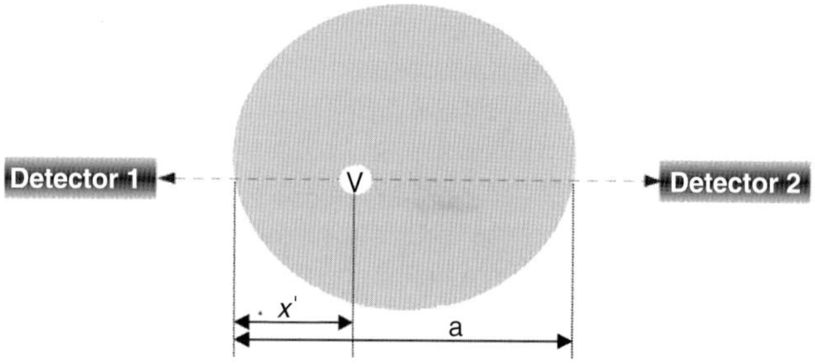

FIGURE 14 Time spectrum of the coincidence events in a PET scanner. The peak is termed the coincidence window and represents prompts + randoms. The delayed window samples the randoms distribution.

FIGURE 15 Attenuation in PET. The attenuation correction does not depend on where along the LOR the positron decay occurs.

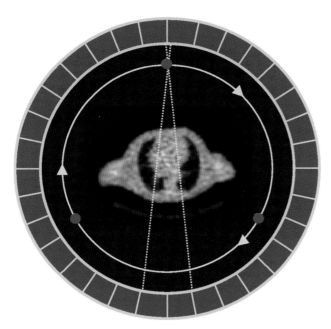

FIGURE 16 A configuration of transmission sources in a dedicated PET scanner. A rotating rod or point source (depending on whether the system is in 2D or 3D mode of operation) is rotated around the patient. Systems with one, two, or three rod/point sources have been developed commercially.

Thus, the probability of detection (of not being attenuated) is independent of the position X' of the decay site along the LOR. If a source is placed outside of the patient (Figure 16), then the attenuation measured along any LOR with the source will be the proper correction for the emission data (the transmission scan). The correction is simply measured by taking the natural logarithm of the ratio between the transmission scan and a blank scan (a scan with the external source without any objects in the scanner's FOV).

There are several options for acquiring the transmission data. Most of the dedicated ring scanners use a coincidence-based attenuation scan (Carroll et al., 1983; Daube-Witherspoon et al., 1988). The data are acquired in coincidence mode as the external source rotates around the scanner FOV. In this mode of operation, the data are normally restricted (or gated) to events where the LOR between the detector pair also goes through the current position of the external source. In that way, gamma rays that are scattered within the FOV are rejected (they do not pass through the position of the external source at the time of decay). If there is activity within the object during the transmission scan (postinjection transmission scanning), the use of the gated-source approach also rejects most of the events arising from the isotope within the patient. Only small corrections are needed from an emission scan to measure the attenuation correction factors accurately. Systems using this approach have been built using one, two, or three rotating rod sources, usually containing several millicuries of ^{68}Ge. One disadvantage of this approach is that it puts a high-activity source very close to detectors. The near detector (the detector closest to the source) will experience a large dead time, which limits the count rate that can be achieved in the coincidence data acquisition. The use of two or three rods in modern machines is a partial solution to the problem (distributing the activity so the count rate due to any single rod is lower). However, transmission scan times are still long (typically 10–15 min) if the data are used directly. This scan time corresponds to a single bed position, which typically covers an axial extent equal to the axial FOV. Therefore, to acquire a whole-body scan (approximately 70 cm), typically seven bed positions are acquired.

To shorten the overall scan time, several approaches have been developed. One method to reduce the transmission scan time is to ignore the near detector and collect the data in singles mode (disable the coincidence system). Such an approach allows the use of higher activity sources (5–20 mCi) to acquire high-quality data in < 1 min per FOV. The designer of such a system can also use a long-lived source such as ^{137}Cs (30 years) because the source does not need to be a positron emitter—an advantage in terms of cost of operation of the scanner (Karp et al., 1995). The disadvantage of this is that the attenuation data are contaminated by scatter and correction for events from the isotope already in the patient; however, a correction can be made to avoid bias errors in the attenuation correction. Another correction must be made to account for the fact that the measured attenuation coefficients from ^{137}Cs at 662 keV are approximately 10% lower than those for 511 keV, the energy of the emission gamma rays from a positron emitter (Smith et al., 1997). These systems use segmentation to identify different tissue types in the attenuation images and then reassign pixels to the proper values for those tissues (Bilger et al., 2001). The segmented data can then be used for calculating the attenuation correction factors. This approach has also been applied to coincidence-mode data (Xu et al., 1994; Weng and Bettinadi, 1993) with the result that transmission scan times of 1–3 min can be realized with either singles- or coincidence-mode transmission scan schemes.

Another approach is to use a small reference detector in conjunction with a collimated coincidence source that is rotated around the patient (Watson et al., 2001). Using a fast scintillator for the reference detector (e.g., LSO), fast count rates can be obtained with minimal problems with emission contamination from the patient. This technique has been used in some research systems developed by CPS Innovations.

Still another approach to the measurement of transmission data is to combine a CT scanner with a PET system. The resulting data are taken at an energy that is very different from 511 keV, so the data must be scaled to use for attenuation correction of the emission data. More information on this approach is included in Watson *et al.* (Chapter 11 in this volume).

VII. SCATTER CORRECTION

Scatter correction techniques include integral transform, function fitting, energy-based subtraction, and analytic calculation (Bergstrom *et al.*, 1983; Grootoonk *et al.*, 1991; Bailey and Meikle, 1994; Ollinger, 1995). The accuracy of each approach is dependent on, among other things, the mode of operation (e.g., 2D vs. 3D), the details of the scanner design and scintillator used, and the size of the object being scanned (e.g., brain vs. whole body). The integral transform method uses a simple function such as an exponential model for the scatter response to a point source within the object. The parameters for the model (usually amplitude and rate of fall-off) are normally derived from a series of phantom studies. To partially address the issue that the scatter response is not stationary and that the parameters are normally derived from simple cylindrical phantoms, the parameters are often modified by the measured attenuation length for a given LOR. Integral transform techniques are common for scanners operating in 2D mode, in which modest errors in scatter estimation (e.g., 10–20%) do not have a large impact because the overall scatter fraction (scatter/total events) is also modest (10–20%). However, in 3D mode, errors in scatter estimation have a much larger impact because the scatter fraction is significantly larger (35–60%).

The function-fitting method fits a function to the data acquired from outside the object. Here, the assumption is that all the events outside the object after randoms correction are due to scatter. In this approach, a modified Gaussian or parabolic function is often used to describe the scatter profile in sinogram space, and that function is fitted to the data using the events outside the object. As in the integral transform approach, the shape of the function is normally determined from a series of simple phantom studies. Although this method is simple, it has the advantage that it takes into account activity outside the FOV, unlike the integral transform method. The fitting method, however, does not take into account local variations of the activity distribution or attenuation; thus, it provides only a smooth estimate of the scatter distribution.

Energy-based approaches can take many forms. The most common has been to acquire data with two energy windows, one over the photopeak and one over the Compton scatter region (Grootoonk *et al.*, 1991; Harrison *et al.*, 1991). The assumption is that a scaled subtraction of the Compton window events from the photopeak window events for each LOR will provide the proper scatter correction. Part of the attraction of this approach in 3D is that the energy information, in principle, should reflect the impact of scatters from activity both outside and inside the FOV. However, in dedicated PET systems, the Compton region of the energy spectrum is dominated by multiple-scattered and large-angle single-scattered gamma rays. The spatial distribution of these gamma rays is different from the small-angle single scatters that dominate the scattered events accepted into the photopeak window. Thus, the parameters derived for the dual-energy window approach are object-dependent and can lead to inaccuracies, particularly in body imaging. A variant is to place a small window just on the high side of the energy peak (Adam *et al.*, 2000). This approach provides a better estimate of the correct spatial distribution, but it is very sensitive to small changes in PMT gains and, as a result, difficult to implement. A third energy-based approach is to acquire data in several energy windows and use a model of the unscattered events and various types of scatter to deconvolve the energy spectrum and extract the unscattered events. The disadvantages of this approach are that it requires a great deal of data to be acquired and the resulting arrays are sparse, making the accurate extraction of the unscattered events very difficult.

For 3D PET, another approach that has been tried is to acquire short 2D scans in which the scatter correction is reasonable and then use the differences between the 3D and 2D scans to estimate the 3D scatter correction. However, this approach has not been pursued for most scanners and, instead, a scatter correction based on analytic calculations of the single scatter events is gaining acceptance. Using the Klein-Nishina formula, the algorithms calculate the single scatter events starting with the attenuation data and an initial reconstruction of the emission data (Ollinger, 1995; Watson *et al.*, 1996). Because the original data included scatter, the algorithms are normally iterative—using the initial scatter calculation to produce a scatter sinogram for comparison to the original data. Based on that comparison, the estimated true distribution is updated and a new scatter calculation made (Fig. 17). Implementations with or without explicit estimates of multiple scatters have been used. To account for the impact of activity from outside the FOV, two approaches have been used. One is to calculate the scatter correction and then scale it to force the mean value of events outside of the object (in sinogram space) to be zero. The other approach is to include short scans of the volumes on either side of the volume being reconstructed. Because the algorithms generally undersample the data fairly coarsely, a small amount of data from the overscanned areas is required. Some variations of this algorithm work in image space rather than sinogram space. This analytic scatter approach (often referred to as model-based corrections) is still an area of development, but it has already been implemented on many dedicated PET systems.

VIII. NOISE EQUIVALENT COUNT RATE

We have already looked at some aspects of comparing 2D to 3D imaging modes in PET scanners. 3D mode greatly improves the sensitivity for any scanner design, but it comes at the cost of nonuniform axial sensitivity, increased scatter fraction, and increased impact of activity outside the FOV. One measurement that has been used to give some insight into the quality of the primary data is the noise equivalent count (NEC) rate (Strother *et al.*, 1990). Although this metric does not directly relate to final image quality, it does demonstrate the impact of dead time, randoms rates, and scatter on the overall ability of the scanner to measure the true events count rate. Generally, the NEC is defined as:

$$\text{NEC} = \frac{T^2}{T + S + KR}$$

where T is the trues count rate, S the scatter count rate, R the randoms count rate, and K is a factor that depends on the method used for randoms correction. In general, the real-time subtraction of randoms leads to the largest variances thus $K = 2$ is often used for real-time subtraction, whereas $K = 1$ is used when the randoms have first been smoothed before subtraction. Although a varying K is often used to distinguish among the different methods of randoms subtraction, this approach is normally not taken with scatter subtraction because it is more difficult to generalize the impact of the scatter correction accuracy on NEC.

Figure 18 shows the count-rate curves, the NEC for both the NEMA NU2 1994 (Karp *et al.*, 1991) and 2001 standards (Daube-Witherspoon *et al.*, 2002). The 1994 standard uses a 20-cm diameter by 19-cm long cylindrical phantom, which is representative of brain imaging. The 2001 standard uses a 20-cm diameter by 70-cm long phantom, which is more appropriate for whole-body imaging. This measurement is used so that the impact of out-of-field activity

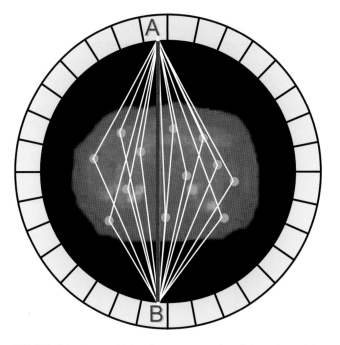

FIGURE 17 The model-based scatter correction defines the activity distribution from the emission image and the scattering medium distribution from the transmission image. The scatter points are distributed randomly in the volume, and the contributions of scatter are calculated using the Klein-Nishina formula for each selected LOR (AB).

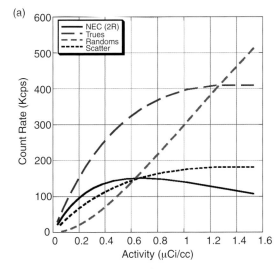
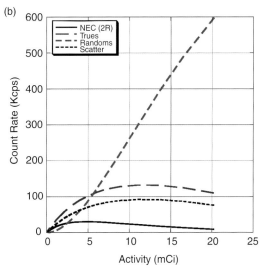

FIGURE 18 An example of noise equivalent count (NEC) rate for a dedicated PET scanner. (a) Data taken according to the NEMA 1994 protocol (20-cm-diameter × 19-cm-long cylinder of activity). (b) Data taken with the 2001 protocol (line source in a 20-cm-diameter × 70-cm-long cylinder). The data are from a BGO scanner with a lower energy threshold of 375 keV.

is included, as occurs naturally with whole-body patient studies. The NEC curves also illustrates the impact of randoms as the activity is increased. Families of such curves can provide some insight into the impact of different activity distributions on a given scanner's data collection (e.g., different body sizes and activity levels).

IX. FUTURE TRENDS

As we noted at the beginning of this chapter, system designers are always looking at ways to improve spatial resolution, and these generally require improvements in system sensitivity in addition to higher resolution detector arrays. One technology that aids in this quest is DOI detector systems. A DOI system is designed to give information about the depth in the crystal at which an interaction takes place. The system designer can the exploit that information to correct for parallax errors, allowing a reduction in the system ring diameter. Reducing the ring diameter allows either a potential reduction in cost (fewer scintillator and readout channels) or an extension of the axial FOV. In either case, there will be an increase in sensitivity due to the larger solid angle subtended by the detector array. Figure 19 depicts one approach being developed (Saoudi *et al.*, 1999). By using scintillators with different decay times, it is possible to select between different detector elements using pulse-shape discrimination (PSD). Systems with up to four different layers have been proposed. Systems with two layers (e.g., LSO with different time constants, or LSO/GSO and LSO/NaI(Tl) combinations) have been built (Schmand *et al.*, 1999). A second approach is illustrated in Figure 20. In this case, the detector surface is modified to force a different efficiency of light collection with depth, effectively causing the photopeak position to change with depth. Although some early development was done on this approach, it has not been adopted for any full systems designs thus far.

A third approach, that of exploiting how light is shared in the detector block, is illustrated in Figure 21. Three main designs have been developed. The first uses two layers of crystals that are offset, allowing the decoding of the crystal layer by which photocathodes in a multicathode PMT receive light (Vaquero *et al.*, 1998). The second uses two light collectors at either end of the crystals (usually an array of photodiodes coupled one on one to the crystals and a large-area PMT coupled to many crystals). The ratio of the light collected at each end is translated into the DOI location (Moses and Derenzo, 1994). The third uses surface treatments to change the sharing of light between two or more crystals as a function of the depth (Miyaoka *et al.*, 1998).

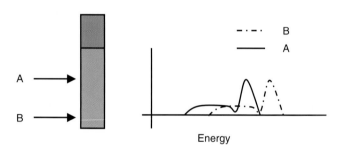

FIGURE 20 Depth of interaction by using the change of light collection with depth in long narrow crystals. The photopeak shift is related to the depth, as demonstrated by events incident at two depths, A and B.

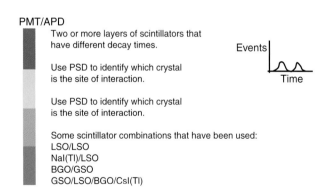

FIGURE 19 Depth of interaction by pulse shape analysis. Stacks of two or more layers of crystals with different timing properties allow the determination of which crystal is the source of the signal, based on analysis of the signal decay time, of pulse-shape discrimination (PSD). Some scintillator combinations that have been used are YLSO/LSO, NaI(Tl)/LSO, and BGO/GSO. The signals have been collected with PMTs, and avalanche photodiodes (APD).

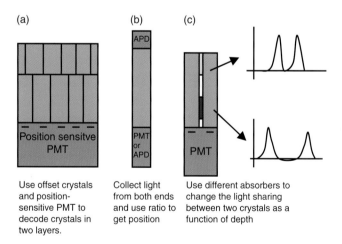

FIGURE 21 Depth of interaction by light sharing. The distribution of light is used to locate the depth of interaction with the array. (a) Using offset crystals and mutlianode PMTs to determine which crystal was the source. (b) Using PMT and APD at opposite ends of a crystal to measure the light ratio at both ends of the crystal to locate the interaction point. (c) Using different absorbers to modify the sharing of light between crystals to determine the interaction point by looking at the ratio of light between PMTs.

In this approach, the ratio value normally used for determining which crystal is the source of the interaction is also used to determine the DOI (the peaks, such as illustrated in Fig. 6, shift within the crystal zones as a function of the DOI). All three of these light-sharing approaches are being developed for possible use in imaging systems.

Another area of design that is active is the development of multimodality scanners. Currently, systems that combine a dedicated PET system with a CT scanner are in production. These systems generally use one detector array for PET imaging and a second, positioned opposite from an X-ray tube, for the CT scanning. Other laboratories are looking into the possibility of developing detector stacks that will allow a single detector array to function for several different modalities (X-ray, SPECT, and PET).

The designs for PET systems seem to be separating into three main groups of applications: (1) small animal imaging, (2) neurological imaging, and (3) whole-body imaging. This is largely due to the widely varying design criteria these different applications entail. Whole-body imaging is the application in which the majority of commercial system development is taking place, although there are also recent interest and development in small animal systems. The small animal systems can take advantage of new, costly scintillators because they generally need small volumes of detector material compared to a human whole-body scanner. To some extent, specialized neurological scanners can more easily take advantage of newer, more costly scintillators (e.g., LSO and GSO) (Weinhard *et al.*, 2002; Karp *et al.*, to appear) due to the reduction in scintillator volume compared to a whole-body system. Perhaps more important, a dedicated neurological scanner can reduce the ring diameter and select an axial FOV that provides significantly more sensitivity than can be obtained with a whole-body system. However, the cost of LSO and GSO has been reduced to the point that these materials are also being used in commercial whole-body scanner systems. The majority of the commercial development is likely to be in whole-body systems because the demand for body oncology imaging is high. The challenge is to increase the system's sensitivity to improve image quality (and decrease scan time) without increasing the cost of the system.

References

Adam, L. E., Karp, J. S., Daube-Witherspoon, M. E., and Smith, R. J. (2001). Performance of a whole-body PET scanner using curve-plate NaI(Tl) detectors. *J. Nucl. Med.* **48**: 1821–1830.

Adam, L. E., Karp, J. S., and Freifelder, R. (2000). Energy based scatter correction for 3D PET scanners using NaI(Tl) detectors. *IEEE Trans. Med. Imag.* **19**: 513–521.

Adam, L. E., Zaers, J., Ostertag, H., Trojan, H., Bellemann, M. E., and Brix, G. (1997). Performance evaluation of the whole-body PET scanner ECAT EXACT HR+ following the IEC standard. *IEEE Trans. Nucl. Sci.* **44**: 1172–1179.

Amsel, G., Brosshard, R., and Zajde, C. (1969). Shortening of detector signals in passive filters for pile-up reduction. *Nucl. Instr. Meth.* **71**: 1–12.

Bailey, D., and Meikle, S. (1994). A convolution-subtraction scatter correction method for 3D PET. *Phys. Med. Biol.* **39**: 411–424.

Beneditti, S. D., *et al.* (1950). On the angular distribution of two-photon annihilation radiation. *Phys. Rev.* **77**: 205–212.

Bergstrom, M., Eriksson, L., Bohm, C., Blomqvist, G., and Litton, J. (1983). Correction for scattered radiation in a ring detector positron camera by integral transformation of the projections. *J. Comput. Assist. Tomogr.* **7**: 42–50.

Bilger, K., Adam, L. E., and Karp, J. S. (2001). Segmented attenuation correction using Cs-137 single photon transmission. *2001 Conf. Rec. IEEE Nucl. Sci. Symp. Med. Imaging Conf.*

Brooks, R. A., Sank, U. J., DiChiro, G., Friauf, W. S., and Leighton, S. B. (1980). Design of a high resolution positron emission tomograph: The neuro-PET. *J. Comput. Assist. Tomogr.* **4**: 5–13.

Carroll, L., Kertz, P., and Orcut, G. (1983). The orbiting rod source: Improving performance in PET transmission correction scans. *In* "Emission Computed Tomography—Current Trends" (p. Esser, ed.), pp. ?. Society of Nuclear Medicine, New York.

Casey, M., and Nutt, R. (1986). A multicrystal two dimensional BGO detector system for positron emission tomography. *IEEE Trans. Nucl. Sci.* **33**: 460–463.

Daube-Witherspoon, M., Carson, R., and Green, M. (1988). Postinjection transmission attenuation measurments for PET. *IEEE Trans. Nucl. Sci.* **35**: 757–761.

Daube-Witherspoon, M. E., Karp, J. S., Casey, M. E., DiFilippo, F. P., Hines, H., Muehllehner, G., Simcic, V., Stearns, C. W., Vernon, P., Adam, L. E., Kohlmyer, S., and Sossi, V. (2002). PET performance measurements using the NU 2-2001 standard. *J. Nucl. Med.* **43**: 1398–1409.

Derenzo, S. E. (1979). Precision measurement of annihilation point spread distributions for medically important positron emitters. *In* "Proceedings of the 5th International Conference on Positron Annihilation" pp. 819–824.

Grootoonk, S., Spinks, T. J., Michel, C., and Jones, T. (1991). Correction for scatter using dual energy window technique in a tomograph operated without septa. *1991 Conf. Rec. IEEE Nucl. Sci. Symp. Med. Imaging Conf.* **3**: 1569–1573.

Harrison, R. L., Haynor, D. R., and Lewellen, T. K. (1991). Dual energy window scatter corrections for positron emission tomography. *1991 Conf. Rec. IEEE Nucl. Sci. Symp. Med. Imaging Conf.* **3**: 1700–1704.

Hoffman, E. J., Huang, S., Plummer, D., and Phelps, M. E. (1982). Quantitation in positron emission computed tomography: 6 effect of nonuniform resolution. *J. Comput. Assist. Tomogr.* **6**(5), 987–999.

Hoffman, E. J., and Phelps, M. E. (1976). An analysis of some of the physical aspects of positron axial tomography. *Comput. Biol. Med.* **6**: 345–360.

Hoffman, E. J., and Phelps, M. E. (1978). Resolution limits for positron imaging devices. *J. Nucl. Med.* **18**: 491–492.

Karp, J. S., Daube-Witherspoon, M. E., Hoffman, E. J., *et al.* (1991). Performance standards in positron emission tomograpy. *J. Nucl. Med.* **32**: 2342–2350.

Karp, J., Muehllehner, G., Beerbohm, D., and Mankoff, D. (1986). Event localization in a continuous scintillation detector using digital processing. *IEEE Trans. Nucl. Sci.* **33**: 550–555.

Karp, J. S., Muehllehner, G., Mankoff, D. A., Ordonez, C. E., Ollinger, J. M., Daube-Witherspoon, M. E., Haigh, A. T., and Beerbohm, D. J. (1990). Continuous-Slice PENN-pet: A positron tomograph with volume imaging capability. *J. Nucl. Med.* **31**: 617–627.

Karp, J. S., Muehllehner, G., Qu, H., and Yan, X.-H. (1995). Singles transmission in volume imaging PET with a Cs-137 source. *Phys. Med. Biol.* **40**: 929–944.

Karp, J. S., Surti, S., Freifelder, R., Daube-Witherspoon, M. E., Cardi, C., Adam, L. E., and Muehllehner, G. (to appear). Performance of a brain PET camera based on Anger-logic GSO detectors. *J. Nucl. Med.*

Lewellen, T. K., Kohlmyer, S. G., Miyaoka, R. S., Kaplan, M. S., Stearns, C. W., and Schubert, S. F. (1996). Investigation of the performance of the General Electric ADVANCE positron emission tomograph in 3D mode. *IEEE Trans. Nucl. Sci.* **43:** 2199–2206.

Mankoff, D. A., Muehllehner, G., and Miles, G. E. (1990). A local coincidence triggering system for PET tomographs composed of large-area position-sensitive detectors. *IEEE Trans. Nucl. Sci.* **37:** 730–736.

Miyaoka, R.S., Lewellen, T.K., Yu, H., and McDaniel, D.L. (1998). Design of a depth of interaction (DOI) PET detector module. *IEEE Trans. Nucl. Sci.* **45:** 1069–1073.

Moses, W. W., and Derenzo, S. E. (1994). Design studies for a PET detector module using a PIN Photodiode to measure depth of interaction. *IEEE Trans. Nucl. Sci.* **41:** 1441–1445.

Muehllehner, G., Buchin, M. P., and Dudek, J. H. (1976). Performance parameters of a positron imaging camera. *IEEE Trans. Nucl. Sci.* **23:** 528–537.

Muehllehner, G., and Karp, J. (1986). Positron emission tomography—technical considerations. *Sem. Nucl. Med.* 35–50.

Nelleman, P., Hines, H., Braymer, W., Muehllehner, G., and Geagan, M. (1995). Performance characteristics of a dual head SPECT scanner with PET capability. *1995 Conf. Rec. IEEE Nucl. Sci. Symp. Med. Imaging Conf.* 1751–1755.

Nohara, N., Tanaka, E., Tomitani, T., Yamamoto, M., and Murayama, H. (1985). Analytical study of performance of high resolution positron emission computed tomographs for animal study. *IEEE Trans. Nucl. Sci.* **32:** 818–821.

Ollinger, J. M. (1995). Model-based scatter correction for fully 3D PET. *Phys. Med. Biol.* **41:** 153–176.

Phelps, M. E., Hoffman, E. J., Huang, S. C., and Ter-Pogossian, M. M. (1975). Effect of positron range on spatial resolution. *J. Nucl. Med.* **16:** 649–652.

Phelps, M. E., Hoffman, E. J., Huang, S. C., *et al.* (1978). A new computerized tomographic imaging system for positron-emitting radiopharmaceuticals. *J. Nucl. Med.* **19:** 635–647.

Phelps, M. E., Huang, S. C., Hoffman, E. J., Plummer, D., and Carson, R. (1982). An analysis of signal amplification using small detectors in positron emission tomography. *J. Comput. Assist. Tomogr.* **6:** 551–565.

Rogers, J. G., and Gumplinger, P. (1999). A pixelated 3D Anger camera with light-loss compensation. *IEEE Trans. Nucl. Sci.* **46:** 973–978.

Saoudi, A., Pepin, C. M., Dion, F., Bentourkia, M., Lecomte, R., Andreaco, M., Casy, M., Nutt, R., and Dautet, H. (1999). Investigation of depth-of-interaction by pulse shape discrimination in multicrystal detectors read out by acvalanec photodiodes. *IEEE Trans. Nucl. Sci.* **46**(3): 462–467.

Schmand, M., Weinhard, K., Casey, M. E., Eriksson, L., *et al.* (1999). Performance evaluation of a new LSO High Resolution Research Tomograph—HRRT. *1999 Conf. Rec. IEEE Nucl. Sci. Symp. Med. Imaging Conf.* M4–2.

Smith, R., Karp, J. S., Muehllehner, G., Gualtieri, E., and Benard, F. (1997). Singles transmission scans performed post-injection for quantitative whole-body PET imaging. *IEEE Trans. Nucl. Sci.* **44:** 1329–1335.

Strother, S. C., Casey, M. E., and Hoffman, E. J. (1990). Measuring PET scanner sensitivity: Relating countrates to image signal-to-noise ratios using noise equivalent counts. *IEEE Trans. Nucl. Sci.* **37:** 783–788.

Ter-Pogossian, M., Mullani, N. A., Hood, J. T., Higgins, C. S., and Ficke, D. C. (1978). Design considerations for a positron emission tomograph (PETT-V) for imaging of brain. *J. Comput. Assist. Tomogr.* **2:** 539–544.

van Eijk, C. W. E. (2002). Inorganic scintillators in medical imaging. *Phys. Med. Biol.* **47:** R85–R106.

Vaquero, J. J., Seidel, J., Siegel, S., and Green, M. V. (1998). A depth-encoding PET detector module with improved spatial sampling. *1998 Conf. Rec. IEEE Nucl. Sci. Symp. Med. Imaging Conf.* M6–29.

Watson, C. C., Eriksson, L., Casey, M. E., *et al.* (2001). Design and performance of collimated coincidence point sources for simultaneous transmission measurements in 3-D PET. *IEEE Trans. Nucl. Sci.* **48:** 673–679.

Watson, C. C., Newport, D., and Casey, M. E. (1996). A single scatter simulation technique for scatter correction in 3D PET. *In* "Three-Dimensional Image Reconstruction in Radiology and Nuclear Medicine" (P. Grangeat and J.-L. Amans, eds.), pp. 255–268. Kluwer Academic, Dordrecht.

Weinhard, K., Schmand, M., Casey, M. E., *et al.* (2002). The ECAT HRRT: Performance and first clinical application of the new high resolution research tomography. *IEEE Trans. Nucl. Sci.* **49:** 104–110.

Weng, S., and Bettinadi, B. (1993). An automatic segmentation method for fast imaging in PET. *Nucl. Sci. Techniques* **4**(2): 114–119.

Wong, W. H., Mullani, N. A., *et al.* (1984). Characteristics of small barium fluoride (BaF2) scintillator for high intrinsic resolution time-of-flight positron emission tomography. *IEEE Trans. Nucl. Sci.* **31:** 381–386.

Wong, W., Uribe, J., Hicks, K., and Hu, G. (1995). An analog decoding BGO block detector using circular photomultipliers. *IEEE Trans. Nucl. Sci.* **42:** 1095–1101.

Xu, M., Luk, W. K., *et al.* (1994). Local threshold for segmented attenuation correction of PET imaging of the thorax. *IEEE Trans. Nucl. Sci.* **41:** 1532–1537.

CHAPTER 11

PET/CT Systems

CHARLES C. WATSON,* DAVID W. TOWNSEND,† and BERNARD BENDRIEM*
CPS Innovations, Knoxville, Tennessee
†*Department of Medicine and Radiology, University of Tennessee, Knoxville, Tennessee*

I. Introduction
II. Motivation
III. Initial Development
IV. Design
V. Protocols
VI. Image Registration and Fusion
VII. Attenuation Correction
VIII. Dosimetry
IX. The Future

I. INTRODUCTION

A PET/CT system combines the functionality of positron emission tomography (PET) and X-ray computed tomography (CT) into a device with a single patient table and a unified operating system so that a patient may be scanned in both modalities during one session without moving from the table. This may seem like a fairly modest advancement of medical imaging technology compared to individual PET and CT scans, but in fact it is a new imaging paradigm that appears to be revolutionizing the use of PET in nuclear medicine and that will probably make a significant impact on CT imaging practice in radiology as well. (*Time* magazine in fact declared PET/CT to be the medical invention of the year in 2000; Jaroff, 2000). The reason for this is the speed, convenience, and accuracy with which such a combined system can produce a fused image having co-registered functional and anatomical components, permitting the physician to evaluate both types of information together in one view of the body. With the fastest scanners currently available (multislice CT combined with lutetium oxyorthosilicate [LSO] PET detectors), it is possible to obtain whole-body (neck to pelvis) CT and ^{18}F-FDG emission surveys that are intrinsically spatially registered to within less than 1 mm in a single procedure lasting 15 min or less. In fact, the use of CT-based attenuation correction of the emission data makes this scan significantly more rapid than a comparable stand-alone PET scan.

From the point of view of the nuclear medicine physician, the main benefit of PET/CT is the ability to accurately localize regions of high tracer uptake—for instance, to locate the primary tumor corresponding to a poorly differentiated metastasis in the head and neck. Some examples are shown in Figure 1. For the radiologist, the benefit is the enhanced sensitivity to malignancy that comes from the metabolic information in cases where there may yet be little morphological change, as well as an improved ability to distinguish benign neoplasm. For the health care administrator, the potential benefit is the improved cost-effectiveness of the single versus dual procedures. The patient will realize the benefit of all of these and in addition only have to endure a single procedure on one visit to the hospital or clinic.

At this time commercial PET/CT systems that provide both diagnostic-quality PET and CT have been in clinical use for about 3 years. Although the enhanced utility of these two-component fused images is intuitively obvious to nearly

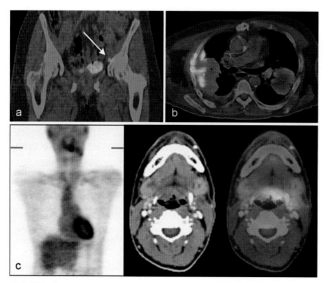

FIGURE 1 Clinical images illustrating the power of PET/CT. (a) A coronal section through the pelvic region of a 50-year-old female with a history of adenocarcinoma of the fallopian tube is shown. The PET/CT image localizes a focus of abnormal FDG uptake in the region of the left obturator lymph node (arrow). Accurate localization would have been difficult from the PET scan alone. The lymph node was removed during laparotomy and pathology confirmed malignancy. (b) A 68-year-old male evaluated for a post-transplant lymphatic disorder displayed heterogeneous uptake of FDG in large bilateral lung masses. A 6-cm left-lung mass showed a ring of high FDG uptake around the periphery of the lesion suggesting a necrotic center, but no involvement of the chest wall. The large right lung lesion, however, showed heterogeneous FDG uptake throughout with uptake extending into the right chest wall. (c) Carcinoma of the head and neck with unknown primary—one of the most important applications of PET/CT. A 63-year-old male with a large left neck mass. Biopsy indicated poorly differentiated squamous cell carcinoma (SCCA), but panendoscopy found no mucosal abnormalities. Left: Coronal section of PET image. Center: Transverse CT slice through suspect region. Right: Fused PET/CT image demonstrating the primary lesion as SCCA of the tongue. The CT alone was negative for disease, and the PET alone was not adequately anatomically localized. [(a) and (b) from the University of Pittsburgh prototype PET/CT, courtesy University of Pittsburgh. (c) from a Siemens/CPS biograph, courtesy of Dr. Todd Blodgett, University of Pittsburgh Medical Center.]

everyone who views them, definitive studies of their efficacy and cost effectiveness relative to individual PET and CT procedures are not yet in hand. Indeed, the optimal design of PET/CT protocols is currently the subject of considerable research and development and will remain so for some time to come. Initial reports in the literature have primarily been retrospective oncology studies from the nuclear medicine perspective comparing PET/CT to PET only or independent PET and CT (Charron et al., 2000; Kleutz et al., 2000; Meltzer, Martinelli, et al., 2001; Meltzer, Snyderman, et al., 2001). These and subsequent reports (Bar-Shalom et al., 2002; Dizendorf, Ciernik, et al., 2002; Freudenberg, Antoch, et al., 2002; Hany et al., 2002; Yeung et al., 2002) generally describe a significant impact on diagnosis, staging, and treatment of cancer, especially in the head and neck, and in the abdomen and pelvis, regions in which spatial localization of malignant focal uptake and its distinction from nonspecific uptake are frequently very difficult. Consequences for patient management in the case of lung nodules also seem very positive (Keidar et al., 2002; Steinert et al., 2002; Osman et al., 2002), although significant potential problems with motion and breathing artifacts are recognized (Beyer et al., 2000; Blodgett et al., 2002; Chin et al., 2002; Goerres et al., 2002). Overall these preliminary results suggest that PET/CT procedures potentially improve diagnostic accuracy and tumor staging by 48–60%, significantly change clinical management decisions in 12–27% of all cases compared to independent PET and CT, and lead to modification of radiation treatment plans in up to 63% of patients examined. Thus it already seems clear that PET/CT will have a major impact on health care in oncology.

Potential drawbacks to PET/CT systems include the somewhat higher initial capital investment required compared to a dedicated PET machine and the larger footprint, requiring more (possibly expensive) floor space for installation. It is also likely that additional patient staging areas, and other infrastructure, must be provided to accommodate the higher throughput these machines can attain. Concern has been raised that use of the CT component of the PET/CT may be inefficient due to the shorter time required for the CT scan (2–4 min) compared to the matching PET acquisition (15–30 min), and schemes have been proposed for performing supplementary CT-only scanning to help defray costs. However, unless the two components can be adequately shielded from one another, safety constraints prohibit simultaneous use on different patients, so any CT-only use would reduce the time available for PET/CT scans. This issue will become less important as PET-tomograph manufactures continue to improve their speed, potentially reducing the PET whole-body scan time to 5 min or less (Nahmias et al., 2002).

These hybrid machines have raised new issues concerning the training and licensing of the technologists who operate them, falling as they do at the boundary between traditional nuclear medicine and radiology. At some sites participation by both a nuclear medicine technologist and a CT technologist is required for operation of the PET/CT, but having a single technologist trained in both aspects of the operation of the scanner would clearly be a better solution. Standards for the training and licensing of PET/CT technologists are currently being developed. A consensus statement on this question has already been issued by a joint conference of the Society of Nuclear Medicine Technologist Section and the American Society of Radiologic Technologists.

Insurance reimbursement for PET/CT scans is a point of some discussion in the medical community as well. Specific reimbursement codes for PET/CT procedures have not yet been approved, and in most cases they are billed simply as PET procedures. The CT portion of the exam cannot be billed separately if it follows a preceding standalone CT exam too closely in time. This situation will probably be resolved by recognizing PET/CT scanning as a new procedure with added benefit.

Our goal in this chapter is to inform the reader of the current state of the art of PET/CT imaging systems, as well as the challenges involved in the ongoing technological development of this new medical imaging modality.

II. MOTIVATION

The addition of CT to PET has three potential technical advantages that have been realized to varying degrees in different designs. It provides (1) a fast, low-noise attenuation correction of PET emission data, (2) high-resolution, accurately co-registered anatomical reference information to help locate tumor-specific radiopharmaceutical uptake and distinguish it from nonspecific uptake, and (3) supplementary diagnostic information, for example, accurate tumor size or the detection of lesions not visible in PET due to lack of tracer uptake. From a different point of view, the addition of PET to CT provides, in effect, a metabolic contrast agent that helps distinguish malignant from benign tissue and aids in the early detection of tumors not yet visible with CT only. True synergistic diagnostic applications of the metabolic and morphologic information, such as, perhaps, the correction of standardized uptake values (SUV) for partial volume effects, remain to be developed, although the benefit of incorporating the anatomical information from CT in PET reconstruction has already been demonstrated (Comtat *et al.*, 1998).

Even before the advent of dual modality scanners, it was common practice in nuclear medicine to read PET images side by side with the most recent CT or MRI images of the patient. Although the utility of the anatomical information was clearly recognized, particularly in oncology (Wahl *et al.*, 1993; Eubank *et al.*, 1998), the accuracy of the localization obtained with such separately acquired images suffered from the fact that the scans were generally performed with the patient in different positions and different physiological states. Algorithms and software have been developed in an attempt to improve the co-registration accuracy of such image sets (Pietrzyk *et al.*, 1990; Woods *et al.*, 1993; Tai *et al.*, 1997). Considerable success has been reported with these techniques in brain imaging, but they are more problematic in other parts of the body because PET images frequently exhibit few anatomical landmarks and the appropriate transformation is generally not rigid-body (Mehta *et al.*, 2002). No fully automated software solutions for the registration of independently acquired CT and PET images are in widespread use for routine clinical whole-body oncology. From this perspective, the value of a hardware solution for accurate co-registration seems clear.

III. INITIAL DEVELOPMENT

The concept of combining emission tomography and X-ray CT in a single device appears to have arisen first in nuclear medicine. It may be seen as a natural extension of the use of an external transmission source measurement to provide a correction for the attenuation of the emission radiation. Conventionally, such transmission scans are accomplished through the use of a gamma-ray or positron-emitting nuclide external to the patient, producing radiation similar in energy to the emission radiation. X-ray CT provides closely related attenuation information that may be used for correction of the emission data if appropriately adjusted for the difference in photon energy. The primary advantage of using an X-ray tube and detector system rather than a radioactive source is the much higher photon flux achievable. The tube of a modern clinical scanner (40 kW or higher) may have an instantaneous output roughly equivalent to 10^{15} Bq of radioactivity, whereas standard radioactive sources for transmission measurements are typically limited to less than 10^9 Bq, due both to radiation shielding concerns, and the count rate capabilities of conventional detector technology employed in PET and single-photon emission computed tomography (SPECT) scanners. The statistical data-quality advantage of the CT is reduced by the higher attenuation of X-rays in the body compared to annihilation radiation (the linear attenuation coefficient is approximately a factor of 2 higher for X-rays), but the photon flux reaching the detectors is still at least 10^4 times higher for the X-ray measurement. The main disadvantage of the low-energy X-ray CT source for a high-energy PET gamma-ray attenuation measurement is the variation in interaction cross sections with photon energy in biological materials. However, it appears that this problem can be adequately addressed in most cases, as we discuss later.

Lang and Hasegawa (Lang *et al.*, 1991, 1992) pioneered the development of a prototype SPECT/CT system at the University of California, San Francisco, employing a low-power X-ray device and a special-purpose solid-state detector capable of discriminating both the gamma-ray and X-ray radiation simultaneously in pulse-counting mode. An algorithm for processing dual-energy X-ray data to correct ^{99m}Tc gamma-ray emission data at 140 keV for attenuation was also proposed (Hasegawa *et al.*, 1993). Using a quite different design strategy, Blankespoor *et al.* (1996) built a higher performance prototype SPECT/CT system from off-the-shelf components. This comprised a commercial CT scanner (the GE 9800 Quick™) integrated with a standard single-head SPECT camera (the GE 600 XR/T™) so as to use a single patient table for scanning in both machines. The first (and so far, only) SPECT/CT system marketed commercially was the Hawkeye™ camera introduced by GE Medical Systems in 2000 (Patton *et al.*, 2000; Bocher et al., 2000). This was actually a hybrid system also capable of coincidence imaging. It consisted of a conventional dual-headed NaI gamma camera with coincidence detection capability (GE's Millenium VG™), to which was attached a low-power CT tube and detector. The CT was employed for attenuation correction for both SPECT and PET and for anatomical reference in fused images, but was not of diagnostic quality. This device has been used with some success for clinical ^{18}F-FDG coincidence imaging (Delbeke et al., 2001; Israel et al., 2002), although the Centers for Medicare and Medicaid Services currently restricts reimbursement of PET procedures performed on such hybrid gamma cameras compared to dedicated PET systems (Centers for Medicare & Medicaid Services, 2002).

The first work on a unified PET/CT system comprising both dedicated PET and clinical diagnostic-quality CT was that of Townsend and colleagues (Townsend et al., 1998; Beyer et al., 2000). This prototype system, shown in Figure 2, was based on the Siemens/CPS ECAT ART™ PET scanner (a partial-ring rotating tomograph with bismuth germanate (BGO) block detectors and no septa; Bailey et al., 1997), combined with a Siemens Somatom AR.SP™ CT scanner (a third-generation single-slice spiral CT operated at 30 rpm). Patient scanning on this device began in 1998 at the University of Pittsburgh Medical Center and continued through 2001, with over 300 patients imaged. Similar to the SPECT/CT systems, the CT measurements were employed to perform an attenuation correction for the 511-keV annihilation radiation (Kinahan et al., 1998). The unique goal and achievement of this device, however, was to obtain full clinical diagnostic quality for both the PET and CT components of the scan and present these to the reading physicians in a timely manner as an intrinsically fused image. The reading of fused PET and CT images became a clinical routine for the first time and quickly demonstrated the potential of this new modality (Charron et al., 2000; Kleutz et al., 2000). This success led manufacturers to focus their products toward higher performance for both the CT and PET, as opposed to viewing the CT in a more limited role, such as simply an attenuation correction mechanism for PET. Most new commercial offerings follow this strategy. The emphasis on CT image quality has brought the attention of radiologists to the device. Consequently, interest in the modality has spread rapidly from nuclear medicine to the larger radiology community. The first commercial dedicated PET/CT was the biograph™ introduced by Siemens/CPS in November 2000, soon followed by the Discovery LS™ scanner from GE Medical Systems, the biograph LSO™ from Siemens/CPS, and the Gemini™ from Philips Medical Systems.

IV. DESIGN

A. General Considerations

The design of the first generation of clinical PET/CT scanners has been influenced to a large extent by their primary intended PET application, namely ^{18}F-FDG whole-body oncology surveys. Oncology-related procedures currently account for over 90% of all clinical PET studies. Appropriate characteristics of the PET component include a large aperture, high sensitivity, and good spatial resolution. A large patient port (70 cm in diameter is standard for CT) is especially important for radiation treatment planning (RTP) applications because these patients frequently must be precisely positioned using bulky appliances such as special head and arm holders. A co-scan range of 100 cm or greater, covered by both the PET and CT without repositioning the patient, is essential.

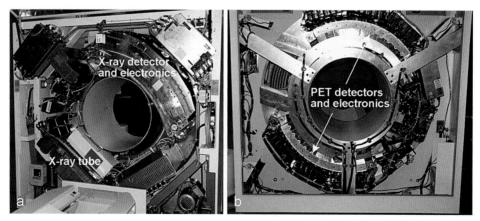

FIGURE 2 The University of Pittsburgh prototype PET/CT. (a) Front, showing the standard CT components. (b) The PET detectors and electronics mounted on the rear of the CT rotational plate.

The count rate performance of the PET component is preferably optimized for injection of 350–550 MBq of ^{18}F-FDG and an uptake period of 45–90 min, reflecting current clinical practice, although good capabilities at even lower activity levels are desirable, especially considering recent research on the benefits of dual-time-point ^{18}F-FDG PET scans for malignant tumor detection (Matthies et al., 2002). The duration of the PET scan is perhaps more critical for PET/CT than for standalone PET due to the need to limit patient motion between the CT and PET scans, and to make the best use of the CT. Dynamic scanning capability (repeated short time frames) is desirable for more sophisticated model-based quantitation (Saleem et al., 2000), although not yet widely used clinically. Gated acquisitions have historically not played a role in PET oncology applications (however, see Nehmeh et al., 2002), but this may change with PET/CT due to the interest in respiratory gating for reduction of motion artifacts (discussed later).

A very important characteristic of the CT scanner is that it should make efficient use of the X-ray dose to the patient because this is in addition to the dose received from the emission activity, because the patient could have already had a diagnostic CT prior to being referred for a PET/CT, and because it may develop that optimal clinical protocols require multiple CT scans over at least part of the body. The duration of the CT scan is less critical because total scan time is dominated by the PET acquisition, not the CT. Nevertheless, the speed and perceived image quality of the new multislice CT designs are appealing to many users, and radiologists especially feel that high-end CT performance is important for PET/CT.

B. Detectors

On conventional clinical scanners, the detectors and readout electronics used for positron annihilation radiation detection and processing are quite different in their design and function from those used for X-ray transmission measurements. PET data are acquired in pulse-counting mode at much lower rates than for CT, typically using multicrystal block scintillators with high stopping power, such as BGO, coupled to photomultiplier tubes (PMTs). Each detected photon must be examined to determine its energy, spatial position, and whether it is in time coincidence with another detected photon. This requires significant processing time per coincidence event. Typical maximum coincidence reporting rates are about 3 Mcps for clinical whole-body scanners. CT machines, on the other hand, employ ceramic detectors read out by photodiodes operated in current mode at high flux rates. Because the source position is known and no energy discrimination is performed, the electronic processing is quite different from that for PET and the data rate is much higher. The current from each detector may be sampled 1000 times per rotation. With a rotation period of 0.5 s and several thousand detector channels in a multislice system, the total sampling rate may exceed 10 Mcps. Each sample may represent the current produced by several thousand individual photons absorbed by the detector. Although detection systems capable of processing both the emission and transmission radiation have been developed (Lang et al., 1992), the difficulty of achieving high performance for either PET or CT with such dual-purpose detectors has so far prevented them from being adopted commercially. Nevertheless, research is ongoing to combine these detector technologies, and it is likely that a practical solution will eventually be found. For the foreseeable future, however, clinical PET/CT scanners will continue to rely on separate PET and CT detector systems.

The performance of the PET component depends to a large extent on the design of its detector system. This technology is still rapidly evolving. Manufacturers are working to include more detector material in their scanners to improve sensitivity, expanding the detectors axially. Costs are kept down by decoding more crystals per PMT, hence using fewer PMTs. BGO is being replaced by newer scintillator materials such as LSO and gadolinium oxyorthosilicate (GSO), that have higher light output, better energy resolution, and shorter scintillation decay times. The higher light output enables the use of fewer PMTs, and together with the shorter decay time improves the timing resolution of detected photons, thus enabling a smaller coincidence window to be used for better rejection of random events. The better energy resolution improves the detector's ability to reject scattered events. The shorter decay time means less dead time for the system and hence greater throughput capabilities. Many of these advances are already being implemented in PET/CT systems, and we can expect others to be available in succeeding generations of scanners. NaI, although still used in some standalone PET tomographs, has poorer performance than either BGO or the newer scintillators and is currently not employed for dedicated PET/CT.

PET detector shielding design is another important issue. Although phantom-based standardized tests of standalone PET scanner performance (noise equivalent count rate versus activity; National Electrical Manufacturers Association, 2001) show superior results in the clinical imaging range for full three-dimensional (3D) acquisitions (no septa in the field of view) compared to 2D acquisitions (one septum per crystal ring), actual clinical results are more equivocal, particularly for large patients. Further, as the patient port diameter and the axial extent of the detectors increase, the acceptance of radiation from outside of the imaging field of view (FOV) may also increase, potentially degrading performance. Shielding solutions intermediate between full 2D septa and 3D (no septa) are being investigated to determine whether performance can be enhanced in a clinically practical way, but no definitive answer is yet established. It seems likely that the optimal design may depend both on the scanner's geometry and on the object to be imaged.

C. Gantry

Most clinical CTs are currently based on a partial-ring detector fan-beam geometry rotating together with an X-ray tube; this configuration is known as a third-generation design. The rotation is continuous for spiral acquisition (Kalender *et al.*, 1990). The technology for continuous detector rotation has also been developed independently for emission tomography and is employed on one commercial dedicated PET tomograph (Bailey *et al.*, 1997). Consequently, an obvious approach to building a combined scanner is to mount the PET and X-ray CT hardware components in a single gantry, using only one rotational bearing and slip ring (the slip ring carries power and communications between the moving and stationary parts). This would potentially reduce manufacturing costs, make the machine as compact as possible, and bring the CT and PET imaging fields of view close together. In fact, this was the design strategy employed for the University of Pittsburgh prototype, in which a partial ring of PET detectors and electronics were installed on the rear of the CT mounting plate in a modified CT gantry (Figure 2).

Extending this approach to higher performance CT and PET, however, presents many engineering challenges. Clinical CTs are capable of rotating at 120 rpm or higher, whereas PET detectors, and their photomultiplier tubes in particular, are currently not designed to withstand this type of acceleration (about 6.5 g), so independent rotation mechanisms may be required for the two components. On the other hand, for highest sensitivity, a PET scanner requires a full ring of detectors, which do not then necessarily need to be rotated. Another factor to be considered is that CT detectors currently have an axial extent of up to 32 mm, whereas PET detectors cover from 150 to 500 mm axially. Given these differences in the technology, and considering the large number of auxiliary components such as power supplies, cooling systems, electronics, shielding and the X-ray tube itself, full integration of the CT and PET components is difficult to accomplish without introducing significant axial separation in their fields of view. Even in the University of Pittsburgh prototype design, the axial separation between the CT and PET fields of view remained around 60 cm.

The fundamental engineering problem of PET/CT, however, is not necessarily how to combine the PET and CT into a single gantry but how to obtain the two scans with an accurately known relation between the patient's position in the two fields of view, independent of the patient's weight and the region of the body scanned. An alternative to an integrated scanner therefore is to use two essentially separate gantries placed adjacent to one another and to solve the problem of transporting the patient between them (Blankespoor *et al.*, 1996). This is the approach that has been taken by manufacturers for the first generation of commercial PET/CTs. It reduces the amount of engineering development needed for the data acquisition systems, but presents greater challenges for the design of the patient handling system (PHS). This dual-gantry strategy also has the advantages of being more flexible with respect to incorporating new PET and CT technologies, both of which are still rapidly evolving, and allowing the components to be more easily serviced. It does not permit significant savings on manufacturing costs, however, and results in fairly large systems. It is likely that the need to reduce size and cost will eventually push the designs back toward the more integrated solutions that were originally conceived for SPECT/CT and PET/CT.

The designs of the various current commercial products are similar in many ways. A schematic of the PET/CT developed by CPS Innovations is shown in Figure 3. The CT and PET components are based on units also available for standalone use. The CT is a single- or dual-slice Siemens Somatom Emotion™ spiral scanner and the PET is a full-ring scanner based on either the ECAT EXACT HR+™ with BGO block detectors or the ECAT Accel™ with LSO detectors. The PET gantries are modified to increase their tunnel diameter from 58 to 70 cm to match the CT and to eliminate the redundant radioactive transmission sources. They have no inter-ring septa. The length of the patient tunnel is 110 cm, and the separation between the CT and PET fields of view is 80 cm. The horizontal travel of the patient bed is sufficient to provide a metal-free co-scan range of 145 cm in both the CT and PET. The total length of the system (gantry plus PHS) is approximately 540 cm. Recommended room dimensions for installation are roughly 500 cm × 730 cm.

Note that the CT is placed in front of the PET toward the patient table. One reason for this is related to the acquisition protocol. The CT scan is generally performed before the PET, and the CT image data are then used to automatically position the patient for the PET acquisition. Thus, the patient only needs to be positioned once manually in the CT, which is more convenient if the CT is toward the front. A further advantage of having the PET to the rear is that if the patient is placed on the table in the head-first position, relatively little time during the scan is spent with the head in the tunnel. Service access to the machine is obtained by sliding the two gantries apart.

D. Patient Handling System

The PHS design may seem like a secondary issue, as is usually the case in nuclear medicine imaging, but in fact it is the key mechanical development for the dual-gantry configuration. The PET/CT presents a unique problem for the PHS design due to the separation of the two fields of view. Normally, for a medical imaging table, the pallet extends from the front of the table over a fixed cantilever point. Thus, its downward deflection increases as it is extended, depending on patient weight, as indicated in Figure 4. This is not a critical problem for a single field of view close to the end of the table, but it can be very significant when two axially

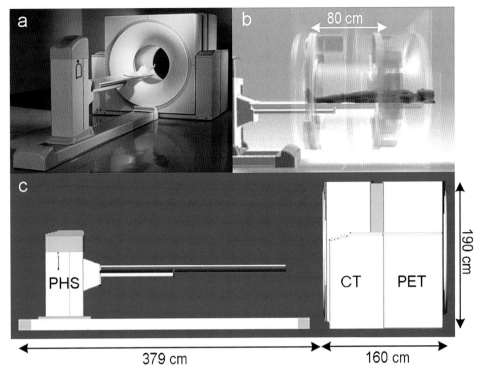

FIGURE 3 Commercial PET/CT. (a) PET/CT manufactured by CPS Innovations. (b) Patient positioned within the tunnel for a PET acquisition. (c) Dimensions.

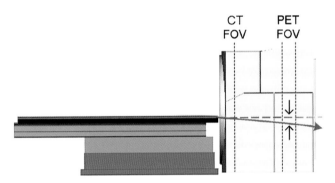

FIGURE 4 A conventional PHS would permit weight-dependent deflection of the pallet as it extends, causing varying intrinsic offset between the CT and PET images.

separated fields of view are involved, resulting in object-dependent variations in image offsets. It is necessary to accomplish this patient transport over a distance of 150 cm or more with less than 1 mm of variation. Several solutions to this problem have been implemented. The CPS PHS (shown in Figure 3) has a pallet attached at one end to a pedestal that moves along floor mounted rails. This avoids variable deflection of the pallet, but requires a long base. The PHS solution offered by GE Medical Systems uses a standard CT table that is repositioned on a floor-mounted track between the CT and PET scans to move the cantilever point to the edges of the fields of view, for the purpose of obtaining similar deflection across each. This may be somewhat more compact, but potentially still allows some relative deflection of the pallet across the PET field of view. The Philips system employs a support track that extends through the tunnel, along which the pallet moves. We can expect to see further evolution of patient transport solutions, not only for PET/CT applications, but possibly also for PET/MRI or trimodality imaging.

It is of interest that PET and CT are conventionally acquired with different modes of patient motion during scanning. Continuous bed motion for spiral scanning has become the most common acquisition mode for CT. PET, however, is still performed clinically with a step-and-shoot procedure. Continuous bed motion in PET has several advantages, including eliminating the dead time during bed steps and improving the axial uniformity of the data (Dahlbom *et al.*, 2001). Implementation requires some modification of the PET data-acquisition systems, but the advent of PET/CT has spurred renewed interest in the concept (Townsend *et al.*, 2002), and it is likely that continuous motion will be employed for both modalities in the near future. Due to the higher spatial resolution and sampling rates of CT, however, the specifications for the accuracy of the table motion will still be governed by the requirements of CT. Typically, spatial positioning repeatability of ±0.25 mm and a velocity accuracy of ±2% from 1 to 150 mm/s are needed.

E. Software and Computers

Software is a key factor in the PET/CT integration. It is essential that the user interface provide a flexible, seamless connection between the CT and PET. This is difficult to achieve when combining two operating systems originally developed independently for the two modalities, and the first-generation systems lack full flexibility. Nevertheless, considerable integration of the operator's workflow has been achieved within the scope of a basic whole-body protocol encompassing acquisition, reconstruction, fusion, and visualization. This includes CT-based attenuation correction applied to PET, the ability to reconstruct PET data in parallel with PET acquisitions, a common database shared by the two modalities so that all data for a patient appear under a single study, full compliance with the Digital Imaging and Communications in Medicine (DICOM) standard for both types of data, and integrated PHS control. Manufacturers are working to develop operating systems designed specifically for PET/CT, and we can expect their capabilities to increase rapidly. One significant advance in usability compared to standalone PET that is already available is the use of the CT topogram to simultaneously plan both the CT and PET acquisitions so that they are compatible. The topogram is a low-dose, very rapid (< 10 s) plane-projection X-ray performed initially on the patient in order to plan the position of subsequent spiral CT acquisitions. The spiral range can be positioned and sized graphically on this image to exactly encompass an integral number of PET bed positions with no wasted dose, as illustrated in Figure 5. Following the spiral CT, the matching PET intervals can be acquired with automatic bed positioning.

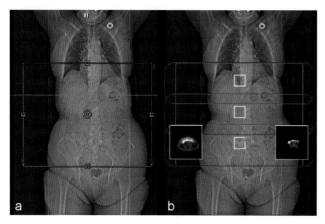

FIGURE 5 CT topogram–based PET acquisition planning. (a) A spiral CT range is graphically defined on the topogram image by dragging and stretching the rectangular box. For PET/CT acquisitions, the axial length of the spiral is constrained to include an integral number of PET bed positions (in this case 3), including their standard overlap, as indicated in (b). Following the CT scan, the PHS automatically positions the patient for each of these PET acquisitions, and the progress of the acquisition and reconstruction is indicated on the screen (first PET bed position is completed).

Currently the PET and CT employ separate computers for acquisition and reconstruction of their data. The acquisition systems are likely to remain separate for some time due to the differences in the nature of the configuration and readout of the electronics, as previously described. On the other hand, there will probably be some unification of the reconstruction engines. However, high-end CTs tend to use specialized hardware to perform very rapid filtered back-projection reconstruction, whereas PET in recent years has moved to iterative reconstruction algorithms due to their superior noise-rejection properties. Differences in the computer platforms may persist to the extent that optimized numerical algorithms require specialized architecture.

V. PROTOCOLS

Data acquisition and processing protocol design can be complex for PET/CT, particularly when it involves multiple CT spirals or PET bed positions, as is required for whole-body oncology applications. So far only the simplest protocols have been implemented for clinical use. An example is illustrated in Figure 6 (timing is for an LSO PET/dual-slice CT system). A patient is injected with a radiopharmaceutical as for a standard PET scan and placed on the scanning table following a typical uptake period of 45–90 min. The injected dose does not affect the CT measurement (Beyer et al., 2000). The procedure begins with the topogram for acquisition planning purposes (1; 10 s). The CT spiral acquisition is defined, usually with a slice width of approximately 5 mm to better match the PET resolution for attenuation correction purposes, and a moderate dose of 80–130 mAs per 360° rotation. Lower doses (e.g., 40 mAs) may be employed if diagnostic-quality CT images are not required. The spiral is performed (2; 60 s) and CT image reconstruction is begun. Simultaneously with this reconstruction, the PET acquisition is begun. For a scanner with a 16-cm axial FOV and a standard 25% overlap, six bed positions at 2 min per bed would cover 76 cm axially (3; 12 min). As the PET data are being acquired, the CT reconstruction completes and the PET attenuation correction factors (ACFs) are computed (4). Once the ACFs are available, PET reconstruction begins on earlier beds as later beds are still being acquired. Reconstruction of the final bed position and assembly of the data into a whole-body volume are finished about 60 s after the acquisition is completed (5). The images may then be transferred to a visualization workstation for display (6), archived offline or sent to an online picture archiving and communication system. Counting the overhead for bed motion and scan initiation, the total time for the acquisition and reconstruction would be approximately 15 min for a total scan length of 76 cm. Additional time may be needed for patient registration, initial positioning, and

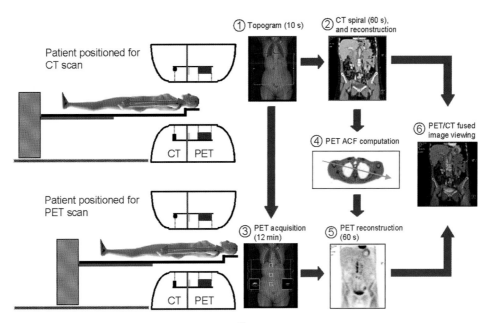

FIGURE 6 A standard PET/CT protocol. (1) Following ^{18}F-FDG injection and uptake, the patient is positioned for the CT scan and a topogram performed (10 s). (2) Acquisition planning is performed on the topogram, the CT spiral is a acquired and reconstruction begins (60 s). (3) While CT reconstruction completes, the patient is automatically positioned for the PET scans, and acquisition begins (12 min for six bed positions). (4) CT reconstruction completes and the PET attenuation correction factors (ACF) are computed. (5) Each PET bed position is reconstructed as its data become available. Assembly of the whole-body image is completed 60 s following the last acquisition. (6) Fused images are reviewed.

acquisition planning. The results of a clinical protocol similar to the one described here are shown in Figure 7.

An important protocol design constraint is that it is not feasible to acquire the CT and PET data simultaneously, particularly when the dual-gantry configuration is used, because the patient is not likely to be properly positioned in both fields of view at the same time. Further, the scatter of X-rays into the PET detectors creates a large background signal, leading to high random event rates and pulse pile-up if the X-ray tube is turned on during the emission acquisition (Beyer et al., 2000; Eriksson et al., 2002). It is not practical to completely shield the PET detectors from this background, thus the CT and PET acquisitions must be performed sequentially.

New technology, including faster multislice CTs and, especially, faster PET scintillator detector materials such as LSO and GSO, will appreciably reduce the data acquisition times. Clinical imaging at the rate of 5 cm axially per minute of total scan time is already possible with the fastest systems on the market, as outlined in the example. Image reconstruction times are not negligible, but with continuing improvements to algorithms and computer performance, seem to be keeping pace with the acquisition times, especially when acquisition and reconstruction can be performed in parallel. As the scan time decreases relative to the patient preparation time, which is dominated by the 45 to 90-min uptake period (for FDG), patient logistics will become more critical. Multiple staging areas with parallel preparation will be required.

Going beyond the simple clinical protocol described here, there are many options to be considered. If a patient has recently received a diagnostic-quality CT, then perhaps only a low-dose CT for attenuation correction purposes is required. If not, then higher-dose CT with special techniques may be required, but must still be usable for PET attenuation correction. It is conceivable that diagnostic quality would be desirable for certain sections, but only quality sufficient for attenuation correction needed for other regions. It may not be necessary that the CT and PET always cover the same axial range. It could be desirable, for instance, to perform the PET survey (without attenuation correction) before the CT in order to identify focal uptake and other regions of interest which could then be scanned with the CT for diagnosis and attenuation correction purposes. Interleaving the CT and PET acquisitions would have the advantage of reducing the potential for patient motion artifacts, but would incur some additional dead time moving the patient between the fields of view. It would also reduce the efficiency of the data processing and reconstruction because the CT images must be reconstructed and ACFs computed before the PET data can be reconstructed.

Another challenge for PET/CT protocols is respiratory gating. CT imaging of the thorax is conventionally performed with breath hold in order to avoid artifacts and resolution loss due to the motion of the lungs and diaphragm. It is of course not feasible for patients to hold their breath

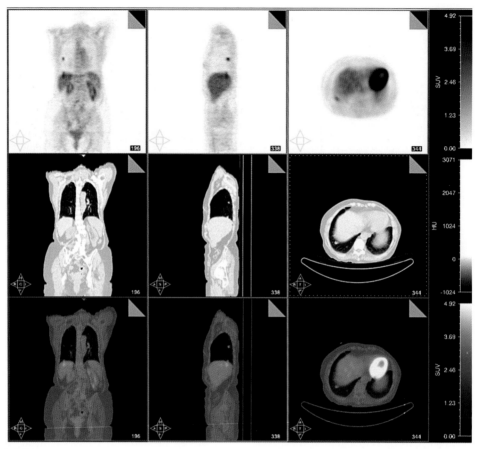

FIGURE 7 Visualization of the results of a clinical PET/CT protocol. PET, CT, and fused images are shown in coronal, sagittal, and transverse slices. The scan was performed with a CTI/CPS Reveal RT™ (LSO PET/dual-slice CT). CT parameters: 130 kVp, 140 mAs, 2 × 5 mm slice width. PET parameters: 629-MBq injection of ^{18}F-FDG, 1-h uptake, six bed positions at 2 min/bed (12 min/76 cm total), CT-based attenuation correction. Clinical summary: 76-year-old female with history of colon cancer was referred for evaluation for malignancy or possible metastatic disease. Foci of increased uptake were identified in the right lung and mediastinum, probably consistent with metastatic disease. The lower lesion in the right lung appears somewhat elongated axially in the PET image, with uncertain location, probably as a consequence of diaphragm motion during breathing. The CT image shows it to be in the base of the lung, above the diaphragm. The fused image shows the breathing-induced misregistration of this lower lesion between the PET and CT, with the centroid of the FDG uptake appearing to be approximately 10 mm below the CT position, in the liver. The co-registration of the upper lung lesion, on the other hand, is very accurate. (Courtesy of the Ahmanson Biological Imaging Center, University of California, Los Angeles.)

throughout a multiminute PET scan. Breath holding during the CT-only part of the scan can exacerbate differences in the shape of the thorax between the CT and PET scans, leading to problems with CT-based attenuation correction and image registration (Townsend *et al.*, 1998; Chin *et al.*, 2002). A compromise is often made by having the patient engage in shallow breathing through both the CT and PET scans in order to minimize artifacts due to the mismatch in the PET ACFs produced from the CT images. However, this procedure is suboptimal for CT image quality, and artifacts can persist in both the CT and PET scans, particularly in the lower lung and anterior chest wall, as shown in Figure 8. Such artifacts are reduced with faster, multislice CT scanning (Blodgett *et al.*, 2002), but "freezing" the lung motion with faster CT acquisition may enhance attenuation correction mismatch and PET image registration problems in a way similar to breath holding. CT and PET acquisitions gated on respiration could potentially reduce these artifacts.

Although the present focus of PET/CT is on whole-body oncology, potentially significant applications in cardiology are also being investigated. One possibility is the combination of ^{18}F-FDG PET viability with CT cardiac perfusion to evaluate heart attack victims as candidates for revascularization (Stearns, 2002). CT cardiac perfusion imaging is under development by manufacturers of fast multislice scanners. By monitoring the dispersion of a contrast agent through the heart's vasculature, a map of blood flow in the myocardium can be obtained and ischemic regions identified. If an ^{18}F-FDG PET scan indicates that this region is still metabolically active, then the patient may be a good candidate for bypass

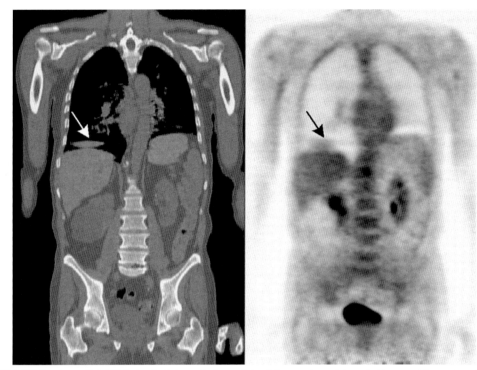

FIGURE 8 An artificial high-density lobe at the base of the right lung (arrow) is probably caused by motion of the diaphragm during the CT scan (left). This results in an overestimate of the emitter activity in the attenuation corrected emission image (right). (Images courtesy of the University of Pittsburgh.)

surgery. This protocol is well known in nuclear medicine, where the PET perfusion agent is usually cyclotron-produced ^{13}N-ammonia with a half-life of 10 min or generator-produced but costly ^{82}Rb with a half-life of 76 s. Two PET scans, one with the perfusion agent and one with ^{18}F-FDG are required, making the procedure lengthy and somewhat complex. Further, these perfusion isotopes are not readily available outside research institutions. Thus their replacement by CT-based perfusion measurement could make this type of evaluation faster, less costly, and more widely available.

One of the most exciting new applications of PET/CT is to RTP and follow-up. Because radiation therapy kills both benign and malignant cells in the treatment volume, it is obviously beneficial to have the delivered dose tailored to the cancerous growth as precisely as possible. Conventionally, treatment planning is performed on CT images that show the anatomical boundaries of a tumor but not the distribution of metabolic activity within it. The benefit of having information on ^{18}F-FDG uptake in targeted lung tumors has been recognized (Kiffer et al., 1998; Erdi et al., 2002). Some tumors may in fact have heterogeneous activity such as a necrotic center. By employing a PET image showing metabolic activity precisely superimposed on the CT image, it is possible to define a metabolic treatment planning volume, and possibly even a desirable inhomogeneous dose distribution over this volume that may be feasible to deliver with intensity modulated radiation therapy technology.

Potentially, this could maximize the effectiveness of the delivered dose, significantly reduce collateral damage to sensitive neighboring organs, and otherwise improve treatment planning (Dizendorf, Ciernik, et al., 2002). PET/CT imaging may also prove valuable in follow-up evaluation of treatment results because morphological and metabolic response can be different (Erdi et al., 2000). These RTP applications are still in their early stages of development, but many workers feel they could be some of the most important uses of PET/CT in the future.

A set of standard PET/CT clinical protocols, analogous to those in radiology and nuclear medicine, will probably emerge over the course of the next few years. Yet there remain many possible applications to be explored, including the use of radiotracers other than ^{18}F-FDG. Early work has shown a benefit for ^{124}I PET/CT imaging in the staging of thyroid cancer, for instance (Freudenberg, Mueller, et al., 2002). Unlike FDG, newer tracers may be so specific for certain tumor types that there is very little nonspecific physiological uptake to provide anatomical clues, making the CT even more essential for localization. It is important for manufacturers to provide adequate flexibility of their systems so as to not unduly constrain these developments. The implementation of software and hardware enhancements capable of supporting the full potential of PET/CT protocols is a very challenging engineering development that will undoubtedly continue to evolve rapidly.

VI. IMAGE REGISTRATION AND FUSION

Accurate registration of the CT and PET images is critical not only for viewing the fused images, but also for the calculation and application of CT-based ACFs. With the dual-gantry designs, the two gantries and PHS are installed separately and must be aligned with one another. This can be difficult to do precisely (to within 1 mm or less) with pieces of equipment that weigh a ton or more each. Further, mechanical alignment does not guarantee alignment of the image spaces. Thus it is necessary to calibrate the offset of the CT and PET images. This is done by means of an activated calibration phantom, scanned in both CT and PET, whose position and orientation in the image volumes can be accurately determined. The relative spatial offsets and rotations, if necessary, can be extracted from these data and are then applied during the reconstruction process, typically to the PET data, to bring it into alignment with the CT images. If the PHS does not move perfectly parallel to itself, transformation data as a function of axial position may be required.

Another consideration is that the CT and PET scanners usually have different FOV diameters, different spatial sampling and different conventions for image reconstruction. CT images are usually reconstructed into 512×512 matrices with a pixel size less than 1 mm, whereas PET transaxial images are typically reconstructed into 128×128 matrices with a pixel size on the order of 5 mm. Axially, PET image spacing is constrained by the detector's crystal pitch, which may not be commensurate with allowed CT reconstruction intervals. This means that in order to represent the two image volumes on a common grid, interpolation of one or both is necessary. This may be done either as a separate processing step, with storage of the resampled images, dynamically by the fused image display software, or perhaps using a combination of both.

The display of the PET and CT data as a co-registered, fused image is the most important application of the PET/CT. The CT image data are stored as 12-bit integers and displayed as CT numbers in the range of -1024 to 3071, usually as a grayscale image. PET image data are stored as 16-bit integers, but with a floating point scale factor used to scale all values in the image volume to represent a physical quantity such as the specific activity of the emitter (Bq/cc) or perhaps SUV. PET/CT vendors support both the CT and PET image Information Object Definitions from the DICOM standard for storage and communication of these data. Several algorithms are available for combining the two images on a single video display, but perhaps the most common is the alpha-blending technique that averages the images pixel by pixel. Consider a 24-bit depth display with 8 bits each for red, green, and blue (RGB). Given two images, each with its own values of RGB, and a mixing ratio or transparency factor α between 0 and 1, the blended image would have a red value (R) of:

$$R_\alpha = \alpha R_{PET} + (1-\alpha) R_{CT},$$

and similar equations for green and blue. The value of α is adjustable by the viewer. Thus the blended image would appear as a CT image with more or less of the PET image contributing ($\alpha = 0$, CT only; $\alpha = 1$, PET only), as shown in the series of images in Figure 9. Of course, this would not work well if both the CT and PET images were grayscale, so the PET image is typically rendered in red or blue, leaving the CT gray. Prior to blending, the PET and CT data may be independently windowed to restrict their dynamic range in order to improve contrast or focus on specific tissue types or lesions. Standard window definitions are commonly used in CT. It should be mentioned that the fused images are only used for display purposes and not stored or archived, because for a fixed value of α they do not contain the complete information content of the images—the individual CT and PET images cannot be recovered from the fused image. This fused display is a very powerful visualization tool for understanding the joint information content of the two images, as illustrated in Figure 7.

VII. ATTENUATION CORRECTION

One of the most important aspects of PET/CT is the use of the CT images to correct the PET emission data for effects of attenuation. CT-based attenuation correction is

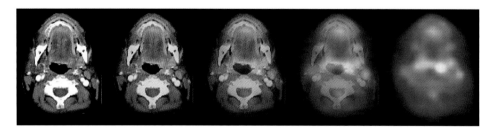

FIGURE 9 An example of alpha-blending image fusion. Left to right: $\alpha = 0.$, 0.25, 0.5, 0.75, and 1.0. (Images courtesy of Dr. Jeffery Yap, University of Pittsburgh.)

not only much faster than a standard PET transmission measurement (by a factor of 10 to 20), it is capable of producing high-resolution estimates of the ACFs that are nearly free of statistical uncertainty compared to measurements based on radioactive sources. Noise in the transmission data is an important contribution to image variance in conventional PET (Dahlbom and Hoffman, 1987); thus its reduction could lead to significant improvement in image quality. The one drawback to CT-based attenuation correction is the potential for bias it may introduce due to the need to transform the linear attenuation coefficients of the lower-energy, polychromatic X-ray beam to those appropriate for the 511-keV annihilation photons. (A potential for bias also exists in standard PET transmission measurements performed postinjection, due to the emission background in the transmission data.)

In conventional PET, the ACFs are estimated by taking a ratio of blank scan to transmission scan data with very little processing needed. This is not appropriate with X-ray transmission due to differences in energy, sampling, and data corrections. Instead, the ACFs are computed from the reconstructed CT images. A CT image is a map of the effective linear attenuation coefficient, μ, of the X-ray beam at each point in space, represented as CT numbers in Hounsfield units (HU):

$$HU = 1000\,(\mu/\mu_{H2O} - 1),$$
$$\mu = \mu_{H2O}(HU/1000 + 1).$$

The normalization to the μ of water keeps the HU scale independent of changes in the X-ray energy spectrum (due to different tube voltages, for instance), for mixtures of water and air, and thus nearly constant for specific soft tissues. Once the CT image has been transformed to a map of $\mu(E_\gamma)$, where $E_\gamma = 511$ keV, it can be forward projected along any desired PET line of response (LOR) and the appropriate ACF can be computed:

$$ACF = \exp\left(\int_{LOR} \mu(E_\gamma)dx\right)$$

However, the transformation of the CT μ to $\mu(E_\gamma)$ is not completely straightforward. A fundamental difference between CT and PET transmission is that the X-ray tube is not a monoenergetic source but produces an emission spectrum that is a combination of a continuous bremsstrahlung distribution and characteristic lines, as illustrated in Figure 10. The spectrum depends on a number of factors including the tube potential, target material and design, and filtering. For a tube potential of 130 kVp, the emitted spectrum may extend from 40 to 130 keV, with an effective energy, E_x, of approximately 80 keV. The effective energy may be understood as the energy of a monochromatic beam that would give the same observed attenuation as the energy integrated polychromatic beam (McCullough, 1975). Due to the strong energy

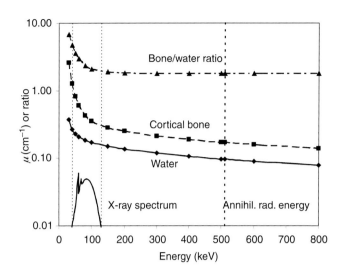

FIGURE 10 Linear attenuation coefficients of photons in water and cortical bone versus energy. The higher photoelectric absorption of bone increases the bone/water ratio below 150 keV. A representation of a 130-kVp CT X-ray spectrum is shown between 40 and 130 keV (arbitrary units). Because the total attenuation cross section is rapidly varying over this interval, the attenuation of the X-ray beam is sensitive to changes in the shape of this spectrum.

dependence of photoelectric absorption, this may be somewhat material dependent. A related problem is the well-known beam-hardening effect whereby the effective energy increases as the beam penetrates the object due to the preferential absorption of the lower-energy radiation. This leads to an apparent reduction in the μ values along lines of high attenuation. CT manufacturers generally provide corrections for beam-hardening valid in soft tissue (and include the skull for brain imaging), but the effects are not entirely eliminated.

Other effects leading to CT image artifacts include respiratory motion, scattered radiation, and the truncation of objects that extend beyond the transverse field of view of the scanner. The last two effects are more common when the patient is positioned with the arms down in the imaging FOV. This is more common practice with PET/CT than with CT only, due to the greater time required for the PET. On the other hand, the shorter scan times compared to standard PET make imaging with the arms down less common for PET/CT than standalone PET. Scattered radiation leads to an apparent reduction in the narrow-beam attenuation similar to the beam-hardening effect, which is especially pronounced in LORs passing through both arms. Likewise, truncation of the arms in the CT image will cause the total attenuation along such LORs to be underestimated. Truncation is more of a problem for CT than for PET because the CT transverse FOV is typically 50 cm in diameter, whereas for PET it is usually closer to 60 cm. Because typically only a small fraction of all projections are truncated, however, the effect on the image μ values tends to be minimal, except in the vicinity of the arms themselves. A simple procedure for correcting CT truncation effects for the purpose of PET ACF computation by extrapolating the data at the edges of the sinogram has been proposed by Carney et al. (2001). These effects, although usually not severe, could make the CT images difficult to interpret in terms of μ values at a single energy and lead to artifacts that propagate through to the estimated ACFs.

The emission background from the injected radiopharmaceutical appears to have a negligible effect on the CT data (Beyer et al., 2000), due to the very low flux of emission radiation relative to the X-ray component when the tube is on. Thus the CT may be performed postinjection, without requiring the patient to remain immobile on the scanner's table during the uptake period.

Assuming that the linear attenuation coefficients are known at some effective X-ray energy, E_x, they must be transformed to E_γ. At X-ray energies, 10–150 keV, photon attenuation in the body is dominated by inelastic (Compton) scattering and photoelectric absorption. At 511 keV, only Compton scattering is appreciable for materials usually found in the body. The total linear attenuation coefficients for water and bone are plotted against energy in Figure 10. In a given material, the attenuation coefficients may be expressed as:

$$\mu(E_\gamma) = \rho_e \sigma_c(E_\gamma),$$
$$\mu(E_x) = \rho_e[\sigma_c(E_x) + \tau(E_x, Z_{eff})].$$

Here, ρ_e is the electron density, σ_c and τ are the Compton and photoelectric cross section per electron, respectively, and Z_{eff} is an effective atomic number for the material. This is a shorthand way of representing τ, which is really a sum over the partial cross sections of the atomic constituents (Jackson and Hawkes, 1981). Neglecting atomic incoherent scattering functions, σ_c is independent of Z, but τ depends strongly on Z_{eff}, approximately as $\tau \sim Z_{eff}^{3.6}$ (Rutherford et al., 1976). It can be seen that in general a measurement of $\mu(E_x)$ is not sufficient to uniquely determine $\mu(E_\gamma)$ because ρ_e and Z_{eff} may change independently. Thus ρ_e might decrease and Z_{eff} increase in such a way that two materials would have the same attenuation at E_x, but different values of $\mu(E_\gamma)$. Fortunately, in the body constraints on tissue composition exist that largely remove this degeneracy (although there are some significant exceptions). In particular, it appears that soft tissue can be well represented in terms of its attenuation properties as a mixture of water and air. This is true because even at E_x, the photoelectric cross section is still a small fraction of the total interaction and also because other soft-tissue materials such as fat have Z_{eff} very similar to water. Thus only the density of the tissue is varying significantly. Bone, on the other hand, does not follow this trend because it contains substantial amounts of calcium and phosphorous and thus has higher Z_{eff}.

A mixture of two unique component materials has a linear attenuation coefficient that is a simple linear function of the concentration (volume fraction), c_1, of one of the components:

$$\mu_{mix} = c_1 \mu_1 + (1 - c_1) \mu_2.$$

This is true for any energy. Consequently, for mixtures $\mu(E_\gamma)$ can be expressed as a linear function of $\mu(E_x)$, in terms of the μ of the endpoint components:

$$\mu_y - \mu_{2y} = (\mu_x - \mu_{2x}) \frac{(\mu_{1y} - \mu_{2y})}{(\mu_{1x} - \mu_{2x})},$$

where $\mu_{2\gamma} = \mu_2(E_\gamma)$, and so on, and 1 and 2 refer to the endpoint components. This assumes, of course, that the endpoint materials are known. In the body it appears that for $\mu < \mu_{H_2O}$, tissues (e.g., lung) are equivalent to mixtures of water plus air, whereas for $\mu > \mu_{H_2O}$ tissues (e.g., spongiosa bone) are equivalent to mixtures of water plus cortical bone. Therefore, for a given $\mu(E_x)$, only one possibility exists for the material's composition. Thus, to a good approximation the relation between $\mu(E_x)$ and $\mu(E_\gamma)$ appears to be piecewise linear and uniquely defined for biological tissues. This relation is shown in Figure 11 for an effective energy of $E_x = 80$ keV. These coefficients have been computed from measured cross-section data (Hubbell and Seltzer, 1997) and based on standard compositions of biological materials (International Commission on Radiation Units and Measurements, 1989). A similar piecewise linear relation has been employed to scale CT attenuation values to 140 keV for use in SPECT (Blankespoor et al., 1996). There is not universal agreement on this relationship in the literature,

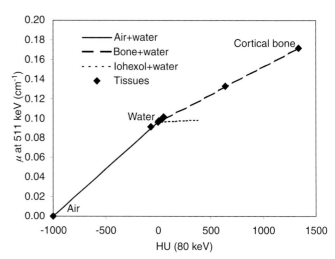

FIGURE 11 Plot of X-ray linear attenuation coefficients in HU computed at an effective energy of 80 keV versus μ at 511 keV. The linear trends corresponding to mixtures of air and water and of bone plus water are shown, together with several values for human tissues and tissue equivalents. Also shown is the trend for a dilute aqueous solution of iohexol, an iodinated CT contrast agent.

however, and some authors have used a quadratic form (Tai et al., 1997; Burger et al., 2001), whereas others have defined the break between soft tissue and bonelike material at 100 or 300 HU rather than 0 HU (Watanabe, 1999; Kinahan et al., 1998). In fact, theoretical values for blood and many soft-tissue components appear to fall along the air–water trend line, but at a slightly higher density than water (i.e., with HU in the range 0–60). Thus a model based on air–blood and blood–bone mixtures might prove somewhat more accurate.

When the X-ray to γ-ray transformation is expressed in terms of HU, the soft tissue region (HU < 0) is not sensitive to the X-ray tube's potential due to the normalization to the μ of water; however, the bonelike tissue trend will have a slope that varies with the kVp setting of the CT scanner, which must be accounted for.

This approach to transforming the CT image to ACFs can break down when high-Z materials that cannot be represented as combinations of water, bone, and air are introduce into the body. Examples include metallic implants, dental fillings, and, most important, CT contrast media. The attenuation of metallic objects can in fact be too high to be measured by a standard medical CT scanner, effectively saturating the CT image. Such artifacts are usually not corrected, and the user must be aware of their characteristic signature (Cohade et al., 2002).

CT contrast agents contain sufficient barium or iodine to drastically boost their photoelectric absorption of X-rays compared to water, with little effect on density or γ-ray attenuation. An example is shown in Figure 11 for a dilute aqueous solution (0–21 mgI/ml) of a common iodinated agent, iohexol ($C_{19}H_{26}I_3N_3O_9$), which is administered both orally and intravascularly. Note that as the concentration of I increases, $\mu(E_x)$ increases by up to 40% while $\mu(E_\gamma)$ increases by only 2% due to the varying density of the mixture. If this material were present in the body, CT image values in the range of 0–400 HU could not be unambiguously interpreted in terms of $\mu(E_\gamma)$ unless it were known whether they corresponded to contrast media or bonelike tissue. Experience so far suggests that this is generally not a problem for intravenously given contrast that tends to be rapidly dispersed in the vasculature of an organ and hence is averaged out somewhat in the estimation of the ACFs. Artifacts have been observed, however, associated with large blood vessels during bolus passage (Antoch et al., 2002).

Orally administered contrast medium, on the other hand, may collect and even concentrate in the bowel. These macroscopic regions of enhanced attenuation may result in ACF artifacts if transformed as bonelike material (i.e., using values of $\mu(E_\gamma)$ that are too high). This could result in overestimation of activity in abdominal lesions, although initial clinical studies have so far failed to find significant problems related to this effect (Dizendorf, Treyer, et al., 2002). Dual energy tomography could potentially provide a way to distinguish contrast agent from bone, but is currently not judged practical for routine clinical use (Kalender, 2000). This problem is the subject of current research, and solutions based on image segmentation have been proposed (Carney et al., 2002).

A complicating aspect of contrast agent use is the fact that the distribution of the contrast medium in the body may change between the CT and PET scans. This is especially true for intravascular contrast. However, this is not a particular problem if the agent displaces and is replaced by fluids of similar attenuation at 511 keV, such as blood. As we have seen, a dilute solution of a contrast agent may differ by only about 2% from water in its attenuation of 511-keV photons. Of course, the problems with contrast agents could be avoided if the contrast scan were performed after the PET scan rather than before, but this would mean doing two CT scans, one with contrast and one without, thus increasing patient dose.

VIII. DOSIMETRY

Patient radiation dose is of particular concern for PET/CT because it combines two procedures that may each deliver significant dose to the patient and because it is not yet clear to what extent this combined procedure may reduce the need for other, standalone, procedures and their associated doses. The radiological dose to patients from a PET/CT diagnostic procedure consists of two components: the X-ray dose from the CT and the internal dose from the injected radiopharmaceutical. The total dose is the sum of these two components. We consider them first individually. This discussion refers to adult patients.

A. CT Effective Dose

The dose delivered to a patient during a CT exam varies widely, depending on the machine characteristics, technical factors used for the examination, the patient size, and the type of procedure (organs scanned). Typical effective doses per diagnostic procedure are 1–30 mSv (International Commission on Radiological Protection [ICRP], 2001). One of the primary uses of the PET/CT is for whole-body cancer surveys and staging. Whole-body scans are not one of the typical scans discussed in the literature. To estimate an effective dose for a whole-body procedure, we add the doses expected for separate scans of the head (0.9 mSv), chest (6.4 mSv), abdomen (6.8 mSv), and pelvis (3.9 mSv) for a total effective dose of 18 mSv (Kalender, 2000). These numbers are approximate and only reflect rough estimates of typical scans. It should also be recognized that these doses are appropriate for diagnostic-quality procedures. Frequently with the PET/CT the intention is not to employ the CT images for diagnosis but only for anatomical reference and attenuation correction of the PET. In this case, the effective dose may be reduced by a factor of 2 or more (e.g., 9 mSv). Dose reduction in CT scanning is an active area of development among the major vendors, with emphasis on improved collimation, more efficient use of detectors on multislice machines (Toth *et al.*, 2000), and the dynamic modulation of tube current with changes in the attenuation through the patient's body—less dose being employed when attenuation is lower (Kalender *et al.*, 1999).

B. Effective Dose in PET

If there is no transmission source in the gantry (other than the CT), the effective dose to the patient during the PET part of the procedure does not depend directly on the PET scanner but only on the injected radiopharmaceutical. However, advances in PET tomograph hardware and software technology, including detectors, electronics, shielding (elimination of septa), data corrections, and reconstruction algorithms continue to improve the use of the injected activity, potentially permitting lower doses to be administered. Essentially all whole-body oncology surveys use ^{18}F-FDG, and we discuss here only this radiopharmaceutical. Effective doses for ^{18}F-FDG and other radiopharmaceuticals can be found in ICRP publication 80 (ICRP, 1998). This publication reports the average effective dose for ^{18}F-FDG to be 0.019 mSv per MBq of injected activity. However, a study based on calibrated ^{18}F-FDG PET scans in combination with magnetic resonance imaging (MRI) reported a mean effective dose of 0.029 ± 0.009 mSv/MBq (Deloar *et al.*, 1998), and we use this higher number in our estimate. A comprehensive report on ^{18}F-FDG absorbed dose estimates using the Medical Internal Radiation Dose model alternative may be found in Hays *et al.* (2002). A very typical injected activity of ^{18}F-FDG is 370 MBq, thus implying an average effective dose of approximately 11 mSv. It should be appreciated, however, that the dose distribution in the body from radiopharmaceutical injection is far from uniform, and certain organs may receive much higher doses than others. The bladder wall is typically the critical organ in this regard, with an estimated absorbed dose of 0.31 mGy/MBq (Deloar *et al.*, 1998). The additional dose due to the CT scan is of special concern for these organs.

C. Total Effective Dose in PET/CT

The total PET/CT effective dose is the sum of the doses from the PET and CT procedures. Thus, for a whole-body (head through pelvis) ^{18}F-FDG PET/CT scan with diagnostic-quality CT, a typical total effective dose is expected to be on the order 30 mSv. If diagnostic-quality CT is not required, the total dose could be reduced to 20 mSv. If further the injected radiopharmaceutical is reduced to a minimum (perhaps 185 MBq) and/or if the CT scan range is reduced to encompass less than the whole body, the effective dose could possibly be as low as 10–15 mSv. To put this in perspective, the average natural background exposure for individuals is on the order of 3 mSv per year.

IX. THE FUTURE

PET/CT technology will continue to incorporate the highest clinical performance in both CT and PET and evolve in parallel with these modalities for some time. High capital costs will be recovered by the increased patient throughput and medical efficacy, and effective cost per image will decrease. We can expect current multislice CT machines to develop into even faster, true 3D volume imaging scanners as cone-beam reconstruction is perfected, techniques for dealing with scatter are developed, and other issues are resolved. On the PET side, the resolution, sensitivity, and speed of scanners will continue to improve. Indeed, 5-min whole-body scans are already within the realm of possibility (Nahmias *et al.*, 2002). Tighter integration of the PET and CT at the software and mechanical level is likely to occur within 2 to 5 years. Integration of the detection and electronic systems for commercial machines is probably farther in the future. We can also easily foresee the development of clinical PET/MRI (Townsend and Cherry, 2001), and of microPET/CT for animal scanning.

For the near future, PET/CT will continue to be employed primarily for cancer staging and treatment follow-up procedures. In fact, it seems likely that PET/CT will displace standalone PET as the main tool for these applications. We can also expect PET/CT use in RTP to grow very rapidly and perhaps become the standard. We have touched on several potential applications of PET/CT and there are certainly many others yet to be conceived. PET/CT is the first full-performance dual-technology medical imaging modality. As such, it opens up a whole new dimension of possibilities for technological development in medical imaging.

Acknowledgments

We thank Jonathan Carney and Jeffery Yap of the University of Tennessee and W. Curtis Howe of CPS Innovations for valuable contributions to this chapter.

Somatom AR.SP, *biograph* and *biograph* LSO are trademarks of Siemens Medical Systems. 9800 Quick, 600 XR/T, Millenium VG, Hawkeye and Discovery LS are trademarks of GE Medical Systems. Gemini is a trademark of Philips Medical Systems. ECAT ART, ECAT HR$^+$, ECAT Accel, and Reveal RT are trademarks of CTI.

References

Antoch, G., Freudenberg, L. S., Debatin, J. F., Mueller, S. P., Beyer, T., Bockisch, A., and Stattaus, J. (2002). A radiologist's perspective on dual-modality PET/CT: Optimized CT scanning protocols and their effect on PET quality. *J. Nucl. Med.* **43:** 307P.

Bailey, D. L., Young, H., Bloomfield, P. M., Meikle, S. R., Glass D., Meyers, M. J., Spinks, T. J., Watson, C. C., Luk, P., Peters, A. M., and Jones, T. (1997). ECAT ART—a continuously rotating PET camera: Performance characteristics, initial clinical studies, and installation considerations in a nuclear medicine department. *Eur. J. Nucl. Med.* **24**: 6–15.

Bar-Shalom, R., Keidar, Z., Guralnik, L., Yefremov, N., Sachs, J., and Israel, O. (2002). Added value of fused PET/CT imaging with FDG in diagnostic imaging and management of cancer patients. *J. Nucl. Med.* **43**: 32P.

Beyer, T., Townsend, D. W., Brun, T., Kinahan, P. E., Charron, M., Roddy, R., Jerin, J., Young, J., Byars, L., and Nutt, R. (2000). A combined PET/CT scanner for clinical oncology. *J. Nucl. Med.* **41**(8): 1369–1379.

Blankespoor, S. C., Wu, X., Kalki, K., Brown, J. K., Tang, H. R., Cann, C. E., and Hasegawa, B. H. (1996). Attenuation correction of SPECT using x-ray CT on an emission-transmission CT system: Myocardial perfusion assessment. *IEEE Trans. Nucl. Sci.* **43**(4): 2263–2274.

Blodgett, T., Beyer, T., Antoch, G., Mueller, S., Freudenberg, L., and Akhurst, T. (2002). The effect of respiratory motion on PET/CT image quality. *J. Nucl. Med.* **43**: 58P.

Bocher, M., Balan, A., Krausz, Y., Shrem, Y., Lonn, A., Wilk, M., and Chisin, R. (2000). Gamma-camera mounted anatomical x-ray tomography: Technology, system characteristics and first images. *Eur. J. Nucl. Med.* **27**(6): 619–627.

Burger, C., Schoenes, S., Goerres, G., Buck, A., and Von Schulthess, G. K. (2001). Transformation of CT-values into 511 keV attenuation coefficients for usage in PET attenuation correction. *J. Nucl. Med.* **42**: 57P.

Carney, J., Beyer, T., Brasse, D., Yap, J. T., and Townsend, D. W. (2002). Clinical PET/CT scanning using oral CT contrast agents. *J. Nucl. Med.* **43**: 57P.

Carney, J., Townsend, D. W., Kinahan, P. E., Beyer, T., Kachelriess, M., Kalender, W. A., De Man, B., and Nuyts, J. (2001). CT-based attenuation correction: The effects of imaging with the arms in the field of view. *J. Nucl. Med.* **42**: 56P.

Centers for Medicare & Medicaid Services. (2002). Positron emission tomography (PET) scans. National Coverage Determination, no. 6, October 1, 2002. Centers for Medicare & Medicaid Services, Baltimore, MD.

Charron, M., Beyer, T., Bohnen, N. N., Kinahan, P. E., Dachille, M., Jerin, J., Nutt, R., Meltzer, C. C., Villemagne, V., and Townsend, D. W. (2000). Image analysis in patients with cancer studied with a combined PET and CT scanner. *Clin. Nucl. Med.* **25**(11): 905–910.

Chin, B. B., Nakamoto, Y., Kraitchman, D. L., Clark, P., and Wahl, R. L. (2002). Quantitative differences in ^{18}FDG uptake due to respiratory motion in PET CT: Attenuation correction using CT in end inspiration and end expiration versus ^{68}Ge correction. *J. Nucl. Med.* **43**: 58P.

Cohade, C., Osman, M., Marshall, L., and Wahl, R. L. (2002). Metallic object artifacts on PET-CT: Clinical and phantom studies. *J. Nucl. Med.* **43**: 308P.

Comtat, C., Kinahan, P. E., Fessler, J. A., Beyer, T., Townsend, D. W., Defrise, M., and Michel, C. (2002). Clinically feasible reconstruction of 3D whole-body PET/CT data using blurred anatomical labels. *Phys. Med. Biol.* **47**(1): 1–20.

Dahlbom, M., and Hoffman, E. J. (1987). Problems in signal-to-noise ratio for attenuation correction in high resolution PET. *IEEE Trans. Nucl. Sci.* **34**: 288–293.

Dahlbom, M., Reed, J., and Young, J. (2001). Implementation of true continuous bed motion in 2-D and 3-D whole-body PET scanning. *IEEE Trans. Nucl. Sci.* **48**: 1465–1469.

Delbeke, D., Martin, W. H., Patton, J. A., and Sandler, M. P. (2001). Value of iterative reconstruction, attenuation correction, and image fusion in the interpretation of FDG PET images with an integrated dual-head coincidence camera and x-ray-based attenuation maps. *Radiology* **218**: 163–171.

Deloar, H. M., Fujiwara, T., Shidahara, M., Nakamura, T., Watabe, H., Narita, Y., Itoh, M., Miyake, M., and Watanuki, S. (1998). Estimation of absorbed dose for 2-[F-18]fluoro-2-deoxy-d-glucose using whole-body positron emission tomography and magnetic resonance imaging. *Eur. J. Nucl. Med.* **25**: 565–574.

Dizendorf, E., Ciernik, I. F., Baumert, B., von Schulthess, G. K., Luetolf, U. M., and Steinert, H. C. (2002). Impact of integrated PETCT scanning on external beam radiation treatment planning. *J. Nucl. Med.* **43**: 33P.

Dizendorf, E. V., Treyer, V., Von Schulthess, G. K., and Hany, T. F. (2002). Application of oral contrast media in coregistered positron emission tomography-CT. *Am. J. Roentgenol.* **179**: 477–481.

Erdi Y. E., Macapinlac H., Rosenzweig K. E., Humm J. L., Larson S. M., Erdi A. K., and Yorke E. D. (2000). Use of PET to monitor the response of lung cancer to radiation treatment. *Eur. J. Nucl. Med.* **27**: 861–866.

Erdi, Y. E., Rosenzweig, K., Erdi, A. K., Macapinlac, H. A., Hu, Y. C., Braban, L. E., Humm, J. L., Squire, O. D., Chui, C. S., Larson, S. M., and Yorke, E. D. (2002). Radiotherapy treatment planning for patients with non-small cell lung cancer using positron emission tomography (PET). *Radiother. Oncol.* **62**: 51–60.

Eriksson, L., Watson, C. C., Eriksson, M., Casey, M. E., Loope, M., Schmand, M., Bendriem, B., and Nutt, R. (2002). On the effect of x-rays on the PET detectors in a PET/CT system. *2002 IEEE Nucl. Sci. Symp. Conf. Rec.* paper M11–40.

Eubank, W. B., Mankoff, D. A., Schmiedl, U. P., Winter, T. C., III, Fisher, E. R., Olshen, A. B., Graham, M. M., and Eary, J. F. (1998). Imaging of oncologic patients: Benefit of combined CT and FDG PET in the diagnosis of malignancy. *Am. J. Roentgenol.* **171**: 1103–1110.

Freudenberg, L. S., Antoch, G., Mueller, S. P., Stattaus, J., Eberhardt, W., Debatin, J., and Bockisch, A. (2002). Preliminary results of whole body FDG-PET/CT in lymphoma. *J. Nucl. Med.* **43**: 30P.

Freudenberg, L. S., Mueller, S., Antoch, G., Beyer, T., Knust, J., Goerges, R., Jentzen, W. J., Brandau, W., Debatin, J., and Bockisch, A. (2002). Value of ^{124}I PET/CT in staging of patients with differentiated thyroid cancer. *J. Nucl. Med.* **43**: 280P.

Goerres, G. W., Kamel, E., Seifert, B., Burger, C., Buck, A., Hany, T. F., and von Schulthess, G. K. (2002). Accuracy of image coregistration of pulmonary lesions in patients with non-small cell lung cancer using an integrated PET/CT system. *J. Nucl. Med.* **43**: 1469–1475.

Hany, T. F., Steinert, H. C., Goerres, G. W., Buck, A., and von Schulthess, G. K. (2002). PET diagnostic accuracy: Improvement with in-line PET-CT system, initial results. *Radiology* **225**: 575–581.

Hasegawa, B. H., Lang, T. F., Brown, J. K., Gingold, E. L., Reilly, S. M., Blankespoor, S., Liew, S. C., Tsui, B. M. W., and Ramanathan, C. (1993). Object-specific attenuation correction of SPECT with correlated dual-energy x-ray CT. *IEEE Trans. Nucl. Sci.* **40**(4): 1242–1252.

Hays, M. T., Watson, E. E., Thomas, S. R., and Stabin, M. (2002). MIRD dose estimate report no. 19: Radiation absorbed dose estimates from 18F-FDG. *J. Nucl. Med.* **43**: 210–214.

Hubbell, J. H., and Seltzer, S. M. (1997). Tables of X-ray mass attenuation coefficients and mass energy-absorption coefficients (version 1.03), [Online]. Available: http://physics.nist.gov/xaamdi [2002, August 16]. National Institute of Standards and Technology, Gaithersburg, MD.

International Commission on Radiation Units and Measurements. (1989). "Tissue Substitutes in Radiation Dosimetry and Measurement." Report no. 44. Nuclear Technology Publishing, Ashford, UK.

International Commission on Radiological Protection. (1998). "Radiation Dose to Patients from Radiopharmaceuticals." Publication 80. Elsevier Science, Tarrytown, New York.

International Commission on Radiological Protection. (2001). "Managing Patient Dose in Computed Tomography." Publication 87. Elsevier Science, Tarrytown, New York.

Israel, O., Mor, M., Gaitini, D., Keidar, Z., Guralnik, L., Engel, A., Frenkel, A., Bar-Shalom, R., and Kuten, A. (2002). Combined functional and structural evaluation of cancer patients with a hybrid camera-based PET/CT system using ^{18}F-FDG. *J. Nucl. Med.* **43**: 1129–1136.

Jackson, D. F., and Hawkes, D. J. (1981). X-ray attenuation coefficients of elements and mixtures. *Phys. Rep.* **70**: 169–233.

Jaroff, L. (2000). A winning combination. *Time* **156**(23): 72–74.

Kalender, W. A. (2000). "Computed Tomography." Publicis MCD Verlag, Munich.

Kalender, W. A., Seissler, W., Klotz, E., and Vock, P. (1990). Spiral volumetric CT with singe-breath-hold technique, continuous transport, and continuous scanner rotation. *Radiology* **176**: 181–183.

Kalender, W. A., Wolf, H., Suess, C., Gies, M., Greess, H., and Bautz, W. A. (1999). Dose reduction in CT by on-line tube current control: Principles and validation on phantoms and cadavers. *Eur. Radiol.* **2**: 323–328.

Keidar, Z., Bar-Shalom, R., Guralnik, L., Yefremov, N., Kagana, O., Gaitini, D., and Israel, O. (2002). Hybrid imaging using PET/CT with ^{18}F-FDG in suspected recurrence of lung cancer: Diagnostic value and impact on patient management. *J. Nucl. Med.* **43**: 32P.

Kiffer, J. D., Berlangieri, S. U., Scott, A. M., Quong, G., Feigen, M., Schumer, W., Clarke, C.P., Knight, S. R., and Daniel, F. J. (1998). The contribution of 18F-fluoro-2-deoxy-glucose positron emission tomographic imaging to radiotherapy planning in lung cancer. *Lung Cancer* **19**: 167–177.

Kinahan, P. E., Townsend, D. W., Beyer, T., and Sashin, D. (1998). Attenuation correction for a combined 3D PET/CT scanner. *Med. Phys.* **25**: 2046–2053.

Kluetz, P. G., Meltzer, C. C., Villemagne, V. L., Kinahan, P. E., Chander, S., Martinelli, M. A., and Townsend, D. W. (2000). Combined PET/CT imaging in oncology: Impact on patient management. *Clin. Positron Imag.* **3**: 223–230.

Lang, T. F., Hasegawa, B. H., Liew, S. C., Brown, J. K., Blankespoor, S., Reilly, S. M., Gingold, E. L., and Cann, C. E. (1991). A prototype emission-transmission imaging system. *1991 IEEE Nucl. Sci. Symp. Conf. Rec.* **3**: 1902–1906.

Lang, T. F., Hasegawa, B. H., Liew, S. C., Brown, J. K., Blankespoor, S., Reilly, S. M., Gingold, E. L., and Cann, C. E. (1992). Description of a prototype emission-transmission computed tomography imaging system. *J. Nucl. Med.* **33**: 1881–1887.

Matthies, A., Hickeson, M., Cuchiara, A., and Alavi, A. (2002). Dual time point 18F-FDG PET for the evaluation of pulmonary nodules. *J. Nucl. Med.* **43**: 871–875.

Mehta, L., Echt, E. A., Nelson, A. D., and Devlin, A. L. (2002). Limitations of CT/PET fusion using software based algorithms. *J. Nucl. Med.* **43**: 305P.

Meltzer, C. C., Martinelli, M. A., Beyer, T., Kinahan, P. E., Charron, M., McCook, B., and Townsend, D. W. (2001). Whole-body FDG PET imaging in the abdomen: value of combined PET/CT. *J. Nucl. Med.* **42**: 35P.

Meltzer, C. C., Snyderman, C. H., Fukui, M. B., Bascom, D. A., Chander, S., Johnson, J. T., Myers, E. N., Martinelli, M. A., Kinahan, P. E., and Townsend, D. W. (2001). Combined FDG PET/CT imaging in head and neck cancer: Impact on patient management. *J. Nucl. Med.* **42**: 36P.

McCullough, E. C. (1975). Photon attenuation in computed tomography. *Med. Phys.* **2**: 307–320.

Nahmias, C., Nutt, R., Hichwa, R. D., Czernin, J., Melcher, C., Schmand, M., Andreaco, M., Eriksson, L., Casey, M., Moyers, C., Michel, C., Bruckbauer, T., Conti, M., Bendriem, B., and Hamill, J. (2002). PET tomograph designed for five minute routine whole body studies. *J. Nucl. Med.* **43**: 11P.

National Electrical Manufacturers Association. (2001). "Performance Measurements of Positron Emission Tomographs." NEMA Standards Publication NU 2–2001. NEMA, Rosslyn, Virginia.

Nehmeh, S. A., Erdi, Y. E., Ling, C. C., Rosenzweig, K. E., Schoder, H., Larson, S. M., Macapinlac, H. A., Squire, O. D., and Humm, J. L. (2002). Effect of respiratory gating on quantifying PET images of lung cancer. *J. Nucl. Med.* **43**: 876–881.

Osman, M. M., Cohade, C., Leal, J., and Wahl, R. L. (2002). Direct comparison of FDG-PET and PET-CT imaging in staging and re-staging patients with lung cancer. *J. Nucl. Med.* **43**: 151P.

Patton, J. A., Delbeke, D., and Sandler, M. P. (2000). Image fusion using an integrated, dual-head coincidence camera with X-ray tube-based attenuation maps. *J. Nucl. Med.* **41**: 1364–1368.

Pietrzyk, U., Herholz, K., and Heiss, W. D. (1990). Three-dimensional alignment of functional and morphological tomograms. *J. Comput. Assist. Tomogr.* **14**: 51–59.

Rutherford, R. A., Pullman, B. R., and Isherwood, I. (1976). Measurement of effective atomic number and electron density using an EMI scanner. *Neuroradiology* **11**: 15–21.

Saleem, A., Yap, J., Osman, S., Brady, F., Suttle, B., Lucas, S. V., Jones, T., Price, P. M., and Aboagye, E. O. (2000). Modulation of fluorouracil tissue pharmacokinetics by eniluracil: In-vivo imaging of drug action. *Lancet* **355**: 2125–2131.

Stearns, C. W. (2002). PET/CT scanner design. Paper presented at the SNM Mid-winter Educational Symposium, Scottsdale, Arizona, Feb.

Steinert, H. C., Hany, T. F., Kamel, E., Lardinois, D., Weder, W., and von Schulthess, G. K. (2002). Impact of integrated PETCT scanning on pre-operative staging of lung cancer. *J. Nucl. Med.* **43**: 151P.

Tai, Y.-C., Lin, K. P., Hoh, C. K., Huang, S. C. H., and Hoffman, E. J. (1997). Utilization of 3-D elastic transformation in the registration of chest x-ray CT and whole body PET. *IEEE Trans. Nucl. Sci.* **44**: 1606–1612.

Toth, T. L., Bromberg, N. B., Pan, T. S., Rabe, J., Woloschek, S. J., Li, J., and Seidenschnur, G. E. (2000). A dose reduction x-ray beam positioning system for high-speed multislice CT scanners. *Med. Phys.* **27**: 2659–2668.

Townsend, D. W., Beyer, T., Kinahan, P. E., Brun, T., Roddy, R., Nutt, R., and Byars, L. G. (1998). The SMART scanner: A combined PET/CT tomograph for clinical oncology. *1998 IEEE Nucl. Sci. Symp. Conf. Rec.*, paper M5–1.

Townsend, D. W., and Cherry, S. R. (2001). Combining anatomy with function: The path to true image fusion. *Eur. Radiol.* **11**: 1968–1974.

Townsend, D. W., Newport, D. F., Brasse, D., Carney J. P., Yap, J. T., Reynolds, C., Reed, J., Bao, J., Luk, P., and Beyer, T. (2002). Whole-body combined PET/CT scanning with continuous bed movement. *J. Nucl. Med.* **43**: 11P.

Wahl, R. L., Quint, L. E., Cieslak, R. D., Aisen, A. M., Koeppe, R. A., and Meyer, C. R. (1993). "Anatometabolic" tumor imaging: Fusion of FDG PET with CT or MRI to localize foci of increased activity. *J. Nucl. Med.* **34**: 1190–1197.

Watanabe, Y. (1999). Derivation of linear attenuation coefficients from CT numbers for low-energy photons. *Phys. Med. Biol.* **44**: 2201–2211.

Woods, R. P., Mazziotta, J. C., and Cherry, S. R. (1993). MRI-PET registration with an automated algorithm. *J. Comput. Assist. Tomogr.* **17**: 536–546.

Yeung, H. W., Schoder, H., and Larson, S. M. (2002). Utility of PET/CT for assessing equivocal PET lesions in oncology—initial experience. *J. Nucl. Med.* **43**: 32P.

CHAPTER 12

Small Animal PET Systems

SIMON R. CHERRY* and ARION F. CHATZIIOANNOU[†]

Department of Biomedical Engineering, University of California, Davis, Davis, California
[†] *Crump Institute for Molecular Imaging and Department of Molecular and Medical Pharmacology, UCLA School of Medicine, Los Angeles, California*

I. Introduction
II. Challenges in Small Animal PET
III. Early Development of Animal PET Scanners
IV. New Generation Small Animal PET Scanners
V. Applications of Small Animal PET
VI. Future Opportunities and Challenges
VII. Summary

I. INTRODUCTION

Many of the traditional medical imaging technologies, including positron emission tomography (PET), are being adapted for use in small laboratory animal imaging. This chapter discusses the motivation for developing dedicated small animal PET scanners and the opportunities for bringing together high-resolution PET imaging systems with the techniques of modern molecular biology, functional genomics, and drug development. Important factors in the design of dedicated small animal PET scanners are identified and the geometry and performance of several first-generation systems are reviewed. Newer small animal PET systems that are under development also are described. Last, we illustrate some of the early applications of PET for studying small laboratory animals and identify the challenges of, opportunities for, and ultimate limitations in applying PET to small animal imaging.

A. Why Use PET in Laboratory Animal Research?

PET can be viewed as an *in vivo* counterpart to autoradiography, tissue dissection, and other techniques that involve imaging or counting excised tissue samples taken from animals into which a radioactively labeled tracer has been introduced prior to sacrifice. The advantage of a noninvasive imaging technique such as PET is that the entire time course of the biodistribution of a radiolabeled tracer can be determined in a single living animal. Furthermore, that animal can be studied again at a later time, permitting longitudinal, within-subject study designs to follow disease models and interventions over periods of days, weeks, and even months. Because the same animal is used at every time point, each animal serves as its own control and variability due to interanimal differences is effectively removed. Therefore, a single animal studied multiple times by PET may in some instances provide the same data that would have required tens of animals using traditional invasive techniques that require sacrifice of the animal. This clearly is in keeping with the desire to reduce the number of laboratory animals used in experiments, but equally important, it has the potential to dramatically reduce the cost of experiments and to speed up the availability of results. It may also improve the quality of the data (because of the within-subject design), although this has yet to be unequivocally demonstrated.

A large number of positron-labeled compounds have been synthesized (Fowler and Wolf, 1986) thus enabling a wide range of biological processes to be measured quantitatively, non-invasively and repeatedly using PET (Phelps, 2000). Combined with the very high sensitivity of radiotracer methods, this flexibility to interrogate living biological systems at the level of specific enzymes, proteins, receptors, and genes makes PET extremely attractive for studies in laboratory animals.

A final important advantage of using medical imaging techniques such as PET in small animal models of disease is that imaging provides a bridge between the animal model and human studies. A valid concern in the use of animal models relates to how well that model predicts what will happen in the human. Techniques such as PET provide the opportunity to perform exactly the same experiments in mouse and human, facilitating direct comparison and appropriate interpretation of the animal model data.

At this time, the range of applications for PET in small laboratory animals are still quite limited due largely to high initial equipment cost, limited availability of animal PET scanners, limited spatial resolution, and lack of access to PET tracers. However, awareness of this technology, and interest in using it, is growing rapidly throughout the biological sciences and in the pharmaceutical and biotechnology industries.

B. Laboratory Animal Models of Human Disease

The most widespread application of PET in animal models prior to the development of small animal PET systems was studying brain function in nonhuman primates. This was driven in large part by the expense of these animals, which precludes large numbers of terminal studies. Furthermore, their large brain size makes them amenable to imaging studies using human PET scanners. Some specialized PET systems have been developed specifically for nonhuman primate imaging (Cutler *et al.*, 1992; Watanabe *et al.*, 1992, 1997), but for many studies, current-generation clinical scanners provide adequate spatial resolution. PET studies of the nonhuman primate central nervous system will continue to play an important role in the investigation of new pharmaceuticals and in the development of new PET tracers; the focus of this chapter is on the use of PET for imaging smaller laboratory animals. The animals of most interest are the mouse and the rat because of the tools available for genetic manipulation in these species.

The mouse has become the animal of choice for creating human disease models and for trying to understand mammalian biology. It is estimated that some 25 million mice were used worldwide for research in 2000, accounting for more than 90% of all mammals used in research (Malakoff, 2000). The relatively low cost of maintaining mouse colonies and the rapid rate of reproduction (up to 250 descendents from a pair of mice per year) are some of the economic driving forces that favor the mouse. In addition, mice are physiologically and genetically similar to humans. Most human genes have a related mouse gene, allowing mice to be used to mimic many human diseases. The technology now exists to knock-out or disable genes and to insert or knock-in new ones, essentially at will. Mice become fertile within 2 months of birth, allowing transgenic animals with these knock-in or knock-out genes to be created in a period of just a few months. Significantly, the mouse genome is the second mammalian genome to be fully sequenced following completion of the human genome and a draft of the mouse genome has already been completed.

One issue related to research in the mouse is worth emphasizing because it bears directly on the use of expensive imaging technologies such as PET. It is a common misconception that studies in mice are inexpensive—studies employing large numbers of genetically modified mice can quickly become very expensive. Many of these mice require special care and specialized facilities that can push the cost per mouse up to hundreds or even thousands of dollars per animal. Studies requiring large numbers of animals can easily run into the tens of thousands of dollars when the cost of buying, breeding, housing, and monitoring animals are added up. Any technique, such as PET, that can dramatically reduce the number of animals required for a particular study has the potential to result in considerable cost savings.

Rat models also are of importance in several fields of biology and are still favored as experimental animals, particualrly in neuroscience. This is partly for historical reasons (many useful experimental models were established in the rat prior to the advent of mouse genomics) and partly for practical reasons related to the ease of surgical manipulation and anatomical and developmental studies in the larger rat brain (roughly 3.3 grams compared to the 0.45 gram mouse brain). Thus, the rat brain will probably remain an important experimental system in the foreseeable future and is another appropriate target for small animal PET systems.

C. Opportunities for PET in Small Animal Imaging

The challenge for PET in light of the prior discussion is clear. A compact, affordable PET scanner with sufficient performance (spatial and temporal resolution, and sensitivity) to quantitatively measure biological processes in the mouse and in the rat would be a powerful tool in modern biology. The ability to study transgenic animals *in vivo*, the use of imaging as an aid in rapid phenotyping, the development of techniques with PET to measure endogenous and reporter gene expression *in vivo*, and the possibilities for *in vivo* screening of new pharmaceuticals open up almost unlimited possibilities for research on small animal models. But the challenge to PET is not a trivial one because the resolution and sensitivity requirements are stringent, the cost and complexity of PET need to be addressed, and access to PET tracers is an obvious limiting factor. However, the potential of PET to play an important role in reaping the benefits of the genome project by providing a tool to assist in developing, validating, and monitoring new genetic and molecular therapies, from the mouse to the human, makes these challenges worthy of a concerted research effort.

II. CHALLENGES IN SMALL ANIMAL PET

There are a number of issues that must be carefully considered in animal PET studies. Some of these have much in common with designing and optimizing PET scanners for human imaging; others are problems specific to small animal imaging. The underlying challenge, as always, is to obtain as many counts as possible and to localize these counts as accurately as possible. Accurate localization of counts depends primarily on the spatial resolution of the detectors and the ability to remove or correct events such as scatter, accidental coincidences, and pile-up that are incorrectly positioned. Maximizing the detected counts requires injection of the maximum radioactivity possible based on mass and specific activity considerations, and using an imaging system with high-efficiency detectors and large solid-angle coverage. The system must also be able to run at high counting rates so that no counts are lost to dead time and have a narrow timing window to minimize accidental coincidences. Some of these challenges are discussed next.

A. Spatial Resolution

The first challenge to PET imaging technology clearly comes from the vast difference in physical size between the subject for which clinical PET systems have been developed, the human (weight ~70 kg), and the laboratory rat (weight ~ 300 g). This represents more than a 200-fold decrease in volume. Laboratory mice, at ~ 30 g, account for another order of magnitude decrease in volume. Therefore, to achieve similar image quality and to address the same biological questions in mice that can currently be studied in humans, PET systems must be developed with similar improvements in spatial resolution. This suggests a reconstructed spatial resolution <1 mm in all directions (<1 μl in volume) as opposed to the ~10 mm (~1 ml in volume) reconstructed image resolution typical in human whole body studies. This stringent requirement calls for new approaches in both detector materials and design.

B. Sensitivity

The absolute detection sensitivity of the imaging instrument (the fraction of radioactive decays that result in a detected event) must be at least as good, and preferably much better than, the typical clinical PET scanner. The number of detected counts per resolution element directly determines the signal-to-noise level of the reconstructed images. If the sensitivity criterion is not satisfied, statistical noise in the reconstructed images will require spatial smoothing which will degrade the much sought after spatial resolution. Whole-body human PET scanners detect on the order of 0.3–0.6% of the coincident annihilation photons in two-dimensional (2D) mode and 2–4% in 3D acquisition mode (Bailey et al., 1991; Chatziioannou et al., 1999). Based on the same argument of scale presented for spatial resolution and to preserve the number of counts per resolution element, the sensitivity for mouse imaging would need to improve by a factor of 1000 relative to human imaging, which is clearly not possible. Even with perfectly efficient detectors and complete solid angle coverage around the animal, the best we can hope to achieve is about a 200-fold increase in 2D mode and a 30-fold increase in 3D mode.

Although this relative reduction in sensitivity from mouse to human can partly be dealt with by injecting larger amounts of radioactivity, there are some fundamental issues that limit how far the dose can be raised, as discussed in Section IIC. Another approach to compensate for the sensitivity problem is to use more sophisticated reconstruction algorithms that make better use of the available counts. Iterative statistical algorithms (Lalush and Wernick, Chapter 21 in this volume) that accurately model the physics of the scanner and the statistics of the raw data will probably play an important role in very-high-resolution PET studies because they can produce improvements in either resolution or signal-to-noise relative to analytic reconstruction algorithms. This is discussed further in Section VIA.

C. Injected Dose and Injected Mass

One might be tempted to think that because laboratory animals are not subject to the same radiation exposure rules and procedures applicable to humans, the injected dose per gram of tissue could be adjusted upward, thereby increasing the detected counts per resolution element and overcoming some of the sensitivity challenges outlined in Section IIB. However, the whole idea of a tracer kinetic experiment is that the mass levels are sufficiently low so as not to perturb the biological system under study. For a given specific activity (see Section IID) of the radiotracer, the injected mass is linearly proportional to the injected activity. There are many circumstances in which the tracer mass will limit the amount of radioactivity that can be injected into a mouse in the range of 1 to 100 µCi (Hume et al., 1998). The amount of mass that can be injected without violating tracer principles must be carefully determined on a case by case basis.

There are certain situations in which relatively large amounts of radioactivity can be injected into an animal, such as tracers of endogenous compounds that are naturally present at fairly high concentrations (e.g., analogs of glucose to measure glucose metabolism). In these cases, dead time and count rate performance of the animal PET scanner may become the limiting factor. The random coincidences are proportional to the square of the injected activity and increase rapidly as the injected dose is increased. This is problematic in small animal PET studies because in many circumstances the entire animal (and therefore the entire injected dose) is within the field of view and cannot be shielded if the whole body is being imaged.

The amount of radioactivity that can be injected may also be limited by the physical volume of the injectate. In a mouse, the blood volume is just 2.5 ml and the maximum volume that can be safely injected is approximately 0.25 ml. This may involve concentrating the injectate to reach an acceptable volume.

D. Specific Activity of Tracers

The specific activity of a radiotracer is related to the fraction of the molecules in the tracer solution which are radiolabeled at a given time and is expressed in units of Bq/g becquerels per gram (or curies per gram) (Ci/g) or, more commonly, in concentration units of Bq/mol becquerels per mole (or curies per mole) (Ci/mol). The specific activity declines over time as the radionuclide decays. Specific activity can be critical in animal PET studies that are mass limited because it directly determines the activity of the tracer that can be administered. The theoretical maximum specific activity for a radionuclide with a half-life (in hours) of $T_{1/2}$ is:

$$SA_{max}(Bq/mole) = 1.16 \times 10^{20} / T_{1/2} \qquad (10.1)$$

In practice, this specific activity is never reached due to the presence of unlabeled atoms in the air and in reaction vessels and, in some cases, the need to add mass levels of compounds to enable chemical reactions to proceed at a reasonable rate during the labeling procedure. For example, the maximum specific activity for ^{11}C-labeled compounds from Eq. 10.1 is 3.5×10^8 MBq/µmol, whereas in practice somewhere in the range of ~10^5 MBq/µmol is usually obtained (Hume et al., 1998). Increases in specific activity would allow more radioactivity to be injected into an animal and thus improve counting statistics and signal-to-noise in the images. To ultimately achieve adequate signal-to-noise in sub-millimeter resolution images from mass-limited tracers, it is likely that significant improvements in specific activity (of an order of magnitude or so) will need to be realized.

E. Measurement of the Input Function

In addition to achieving adequate spatial resolution and counting statistics, the performance of fully quantitative studies in small animals dictates the need for measurement of the blood input function and, in many cases, labeled metabolites in the blood. In the case of humans, blood samples may be an inconvenience, but are easily obtained. The rat has a blood volume of just ~30 ml, of which roughly 3 ml may be safely withdrawn over a period of an hour or so for external counting. Nevertheless, it has been possible to implant arterial lines into rats for blood withdrawal and to carry out longitudinal studies in these animals (Moore et al., 2000). A microsampling system for the rat that continuously draws small blood samples, separated by air bubbles, into plastic tubing and passes the samples through a plastic scintillation detector has also been developed (Lapointe et al., 1998). In the mouse, direct sampling becomes extremely difficult. The blood volume is just 1.3 ml, allowing only 0.13 ml to be safely removed and the vessels are tiny, requiring fairly sophisticated microsurgery for routine and repeated cannulation. Direct cardiac puncture and sampling has been successfully employed for a limited number of samples in the mouse (Green et al., 1998), but remains difficult to implement routinely.

It may be possible to estimate the input function in the mouse by rapid imaging of the left ventricular blood pool. This requires high-resolution, high-sensitivity, and rapid dynamic imaging along with appropriate corrections for spillover, partial volume effects, and heart motion. Similar techniques have been used successfully in the human and can probably be adapted for the mouse. Studies that require metabolite analysis are still problematic and may require additional modeling or assumptions such that only a small number of arterial or, preferably, venous blood samples will suffice. A practical approach to this difficulty in neuroreceptor studies is to use regions of the brain that have no specific binding as reference regions (Lammertsma and Hume, 1996). This allows receptor kinetics to be measured without requiring the input function.

F. Anesthesia

The animal must be kept still during the imaging study and therefore almost all animal PET studies involve anesthesia. This can be a major confounding variable, particularly in brain studies (Gjedde and Rasmussen, 1980; Saija et al., 1989). Further characterization of the effects of different anesthetics on biological systems and careful anesthetic selection will be required to minimize this difficulty.

In some cases, tracers that are irreversibly trapped can be used, thus enabling distribution and uptake of the tracer to occur while the animal is conscious, followed by scanning of the anesthetized animal after uptake is complete. Fluorodeoxyglucose (FDG) is a good example of a tracer that can be used in this manner. This approach also permits activation-stimulation-type studies in awake animals (Kornblum et al., 2000).

An alternative to anesthesia is to use complete restraint of the animal in a tube or body cast or to use paralyzing drugs that prevent motion but do not interfere with brain function. However, these methods cause enormous stress to the animal and also lead to highly unphysiological conditions that can confound experimental results just as much as anesthesia does (Lasbennes et al., 1986).

A number of groups are in the early stages of exploring the use of motion detection systems or mounting a ring of counterbalanced detectors directly onto a rat's head (Craig Woody, personal communication) to permit PET studies in freely moving animals.

G. Other Issues

There are two practical issues that confront PET as a technology when applied to small animal imaging, particularly if one envisions wide distribution of small animal PET scanners in core laboratories in the biological sciences and in the pharmaceutical industry. The first is related to the cost, both the cost of the initial equipment and the cost of performing studies. Small animal PET scanners are required to have far superior resolution and sensitivity characteristics, as already discussed, but realistically also need to be close to an order of magnitude less expensive than human PET scanners. Fortunately, there is some economy of scale in the sense that animal PET scanners have much smaller bore sizes. The second major practical issue relates to access to PET tracers. Most animal PET users will not want the cost and responsibility for setting up and operating a cyclotron. Therefore a network of PET isotope distribution centers and collaboration with existing PET centers and facilities will play an essential role in supporting a wider network of animal PET sites.

III. EARLY DEVELOPMENT OF ANIMAL PET SCANNERS

Clinical PET scanners have been used in the past for animal research (Agon et al., 1988; Fowler et al., 1989; Ingvar et al., 1991) but in order to meet the challenges mentioned in the previous section, several groups have developed PET scanners specifically for animal imaging. This section briefly reviews the design of some of the early animal PET systems.

A. Large Animal PET Scanners

The first animal PET tomographs were designed for imaging nonhuman primates and were constructed around 1990. One system (SHR-2000) was developed by Hamamatsu (Hamamatsu, Japan) and installed in their Central Research Laboratory (Watanabe et al., 1992); the other (ECAT-713) was developed by CTI PET Systems Inc. (Knoxville, TN) and installed at UCLA (Cutler et al., 1992). They offered a relatively large ring diameter and transaxial field of view, which was sufficient to accommodate adult vervet and rhesus monkeys and dogs. Both these systems were based on bismuth germanate (BGO) scintillator. The UCLA tomograph used an adaptation of the block detector designed for the ECAT HR scanner, configured in a 64-cm diameter detector ring. The detector elements were $3.5 \times 6.25 \times 30$ mm deep in a 6×8 matrix and mounted on two dual anode photomultiplier tubes (PMTs). The axial field of view (FOV) was 5.4 cm. The Hamamatsu tomograph used individual 1.7 mm $\times$ 10 mm $\times$ 17 mm deep BGO crystals mounted on a position sensitive photomultiplier tube (PSPMT). Four rows of 33 crystals were mounted to each PSPMT. Fifteen of these modules were arranged to form a 34.8-cm diameter scanner. The axial field of view was 4.6 cm. Both scanners were designed to be used in 2D mode and had interplane septa, although the UCLA system was later modified to enable the septa to be removed, allowing full 3D acquisition.

The reconstructed spatial resolution of the systems at the center of the field of view were reported as 3.8 mm transaxial and 4.2 mm axial for the UCLA system (Cutler et al., 1992), and 3.0 mm transaxial and 4.4 mm axial for the Hamamatsu system (Watanabe et al., 1992). This leads to volumetric resolutions, (transaxial resolution)^2 $\times$ (axial resolution), of 0.061 cc for the UCLA system and 0.040 cc for the Hamamatsu system. The reported sensitivities were 20.7 kcps/μCi/cc at a 300 keV lower energy discriminator measured with a 10-cm diameter cylinder for the Hamamatsu system and 58.1 kcps/μCi/cc at 350 keV lower energy discriminator with a 20-cm diameter cylinder for the UCLA system. These numbers cannot be directly compared due to the different phantoms used. A better measure of sensitivity for comparative purposes is the absolute sensitivity (Bailey et al., 1991), which was measured to be 0.36% in 2D mode on the UCLA scanner.

Hamamatsu have more recently developed a higher-resolution PET scanner, the SHR-7700, for nonhuman primate imaging (Watanabe et al., 1997), which is an extension of their previous design. It uses an 8×4 matrix of $2.8 \times 6.95 \times 30$ mm deep individual BGO crystals on a compact PSPMT. There are four rings, each consisting of 60 of these detector modules, leading to a ring diameter of 50.8 cm and an axial field of view of 11.4 cm. The system has retractable septa to permit both 2D and 3D imaging. The axial resolution (slice thickness) is 3.2 mm and the reconstructed transaxial resolution at the center of the FOV is 2.6 mm, yielding a volumetric resolution of 0.022 cc. The sensitivity measured with a 10-cm diameter cylinder is 84 kcps/μCi/ml in 2D mode and 845 kcps/μCi/ml in 3D mode. Thus both the resolution and sensitivity are greatly improved over the SHR-2000 system. Both Hamamatsu systems have a unique gantry that enables the detector ring to be tilted through 90°, enabling PET studies to be performed in conscious monkeys that have been trained to sit in a chair (e.g., Tsukada et al., 2000).

B. Hammersmith RAT-PET

The first PET system developed specifically for rodent imaging was the RATPET scanner (Bloomfield et al., 1995, 1997) developed by Hammersmith Hospital in collaboration with CTI PET Systems Inc. (Knoxville, TN). This system used an almost identical detector to the ECAT-713 already described, but with the BGO block segmented into an 8 × 7 matrix, rather then 8 × 6. This gives individual element dimensions measuring 3.5 mm (transaxially) × 5.95 (axially) × 30 mm (depth). Sixteen of these block detectors are arranged in a 11.5-cm diameter ring to form the scanner. Figure 1 shows a photograph of the system. The axial field of view is 50 mm and the system has no interplane septa, operating entirely in 3D mode. The measured spatial resolution in 3D at the center of the FOV is 2.4 mm transaxially by 4.6 mm axially, giving a volumetric resolution of 0.026 cc.

FIGURE 1 Picture of the RAT-PET scanner, the first PET system dedicated to rodent imaging. The ring diameter is 11.5 cm and the axial field of view is 50 mm. (Photograph courtesy of Peter Bloomfield, MRC Cyclotron Unit, Hammersmith Hospital, London, UK.)

The absolute sensitivity of the scanner is 4.3% for a lower energy threshold of 250 keV.

This pioneering system demonstrated for the first time the potential utility of a dedicated small animal PET scanner. Even with its relatively coarse spatial resolution, many useful neuroreceptor studies have been performed in the rat brain (see Section VB). For these types of dynamic studies using small injected doses, the relatively high sensitivity of this scanner is an important asset. Some of the challenges raised by this system relate to the large angular gaps where the detectors meet, which, if not treated correctly, can lead to artifacts in the reconstructed images. The resolution is also very nonuniform and degrades rapidly as you move away from the center of the field of view, due to depth of interaction effects. These effects are particularly pronounced in this scanner due to the large depth of the crystals (30 mm) and the small ring diameter (11.5 cm). At a radius of 4 cm, the volumetric resolution has degraded from 0.026 to 0.149 cc.

IV. NEW GENERATION SMALL ANIMAL PET SCANNERS

These initial animal PET systems demonstrated the potential value of small animal PET using existing detector technology based on BGO scintillator coupled to photomultiplier tubes. These detectors used light sharing to decode the position of the event, just as in most modern clinical PET scanners. It soon became apparent that due to the limited light output of BGO scintillator, particularly when cut into small crystals, substantial improvements in detector resolution would require new and different approaches. More recently, several different detector technologies have been proposed for small animal PET and a number of these have been developed into fully operational systems. New scintillators and photodetectors have been explored, and there are also approaches that do not employ any form of scintillation detector. This section reviews some of the systems that have been developed or are in the process of being developed for small animal PET imaging.

A. Sherbrooke Animal PET

The animal PET system developed at the Université de Sherbrooke was the first system to replace PMTs with solid state photon detectors (Lecomte et al., 1994). Each detector consists of an individual 3-mm (transaxial) × 5-mm (axial) × 20-mm (deep) BGO scintillator crystal coupled to its own avalanche photodiode (APD) photon detector. The primary advantages of the APD that make it attractive for PET applications are its high quantum efficiency and its compact size, which allows for a high packing fraction of small discrete detector elements for high resolution animal imaging. The drawbacks of APDs are that they require tight regulation of voltage and temperature to prevent gain changes and that a fast low-noise preamplifier is required on each APD output due to the lower gain (10^2–10^3 compared with 10^6 for PMTs). Chapter 14 (in this volume) provides more detail on APDs and their operation.

The Sherbrooke animal PET consists of two rings of the BGO detectors described, with 256 detectors per detector ring. The ring diameter is 31 cm. The system operates exclusively in 2D. The axial field of view is 10.5 mm and is covered by three image planes. The system employs a two position clam-shell wobble motion to improve spatial sampling (Lecomte *et al.*, 1994). At the center of the FOV, the reconstructed resolution is 2.1 mm full width at half maximum (FWHM) and the axial resolution (slice thickness) is 3.1 mm, producing a volumetric resolution of ~ 0.014 cc. The reported sensitivity is 2 kcps/μCi/ml for a 10.8-cm diameter uniform cylinder at 350 keV lower energy threshold. The absolute sensitivity is estimated to be 0.51% (Lecomte *et al.*, 1996).

B. microPET

The microPET scanner, originally developed at UCLA, was based on one-to-one coupling between individual 2-mm (transaxial) × 2-mm (axial) × 10-mm (deep) scintillator crystals and a multichannel PMT. The scintillator material used was the newly available lutetium oxyorthosilicate (LSO), which has a higher light output than BGO, similar stopping power, and a much shorter decay constant of ~ 40 ns (Melcher, 2000). In order to uncouple the physical size of the photodetector, with its bulky packaging, from the size of the scintillator crystal, optical fibers are used to transfer the scintillation light from the crystals to multichannel PMTs. The advantage of this design was that it provided freedom in the choice of scintillator crystal geometry and allowed for high intrinsic resolution detectors (Cherry *et al.*, 1996). The drawback was that the scintillation light signal suffers losses while traveling from the crystals to the PMT. Still, the use of the high light output LSO scintillator produced sufficient signal for a detector with a 2.4-ns timing resolution and an average energy resolution of 19% (Cherry *et al.*, 1997).

Each microPET detector consists of an 8 × 8 array of individual LSO scintillator crystals coupled to a 64-channel PMT. Thirty of these detector modules are arranged in a continuous ring with diameter of 17.2 cm. The imaging FOV is 11.2 cm in the transverse and 18.0 mm in the axial direction. The system has no interplane septa and operates exclusively in 3D mode. At the center of the FOV, the reconstructed resolution is 1.8 mm FWHM isotropically, producing a volumetric resolution ~0.006 cc, while the peak absolute sensitivity is 0.56% at a lower energy threshold of 250 keV (Chatziioannou *et al.*, 1999).

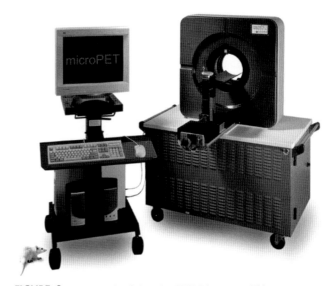

FIGURE 2 Photograph of the microPET P4 scanner. This system uses fiber-optically coupled LSO scintillator elements and a position-sensitive PMT. The ring diameter is 26 cm and the axial FOV is 78.9 mm. (Reprinted from Molecular Imaging and Biology, Vol. 4, Chatziioannou, PET Scanners Dedicated to Molecular Imaging of Small Animal Models, pages 47–63, 2002, with permission from Elsevier Science.)

Concorde Microsystems Inc. (Knoxville, TN) has recently developed similar microPET systems with four rings of detector modules, increasing the axial field of view to almost 8 cm. The P4 model, shown in Figure 2, has a 26 cm ring diameter to accommodate brain imaging in small nonhuman primates in addition to whole-body rodent imaging (Tai *et al.*, 2001). The R4 model, for small rodents only, has a 15-cm ring diameter. The larger axial field of view, in addition to allowing whole-body mouse positioning in a single bed position improves sensitivity at the center of the scanner to approximately 1.2 and 2.2% for the 26- and 15-cm diameter systems, respectively. These systems also incorporate a helical point source mechanism for acquiring transmission scans to aid in attenuation correction.

C. HIDAC Animal PET

Multiwire proportional chamber (MWPC) technology, developed initially in the 1970s for high-energy physics experiments and used in several experimental human PET systems in the 1970s and 1980s, has more recently found successful application in autoradiography (Englert *et al.*, 1995) and in small animal PET systems (Jeavons *et al.*, 1999). The HIDAC detector module developed by Oxford Positron Systems (Weston-on-the-Green, England) consists of a MWPC with the addition of conversion plates to convert the incoming annihilation photons into electrons. The conversion plates are made up of laminated layers of lead and insulating sheets that are drilled with small 0.4-mm holes on a 0.5-mm pitch. There is an electric field between the lead layers that creates a pinch electric field in the holes. Each conversion plate has an *x* or *y* cathode strip readout, with a plane of anode wires between each pair of converters. Incoming annihilation photons interact in the conversion plates and produce electrons that can escape into a hole, where they are focused and multiplied by the electric field. Further multiplication takes place in the avalanche region of the MWPC immediately surrounding the anode wires. The anode signal induces a signal on the *x* and *y* cathode strips, providing 2D localization of the signal. Although each of these modules has low detection efficiency, six of these 25-cm × 21-cm modules are stacked on top of one another to increase efficiency.

The HIDAC animal PET system (Jeavons *et al.*, 1999), shown in Figure 3, consists of two such detector heads that are rotated around the object in order to acquire complete projection data. The scanner has a variable detector separation between 10 and 20 cm and an axial FOV of 21 cm and operates exclusively in 3D mode. The reconstructed spatial resolution of the tomograph at the center of the FOV is ~0.95 mm FWHM in the transverse and ~1.2 mm FWHM in the axial direction using filtered backprojection. A newer version of the system, the quad-HIDAC, is available and has four detector heads, thus providing better solid-angle coverage and increased sensitivity.

The main advantages of these detectors are their high intrinsic spatial resolution and the depth of interaction information they provide. The relative inefficiency of the detectors is compensated for by large solid-angle coverage around the animal. The drawbacks of these detectors come from the conversion process that does not provide energy information, thus making it difficult to discriminate against scatter, and the large coincidence timing window of 60 ns, which ultimately limits count-rate performance. Both these factors will degrade the noise equivalent count (NEC) performance of the system.

D. Other Small Animal PET Systems

Several other scintillator- or PMT-based animal PET systems are currently in operation. An early system developed at the National Institutes of Health (NIH) used two NaI(Tl) plates coupled to position-sensitive PMTs to form an animal imaging system that broke the 2-mm resolution barrier for the first time (Seidel *et al.*, 1994). The same group has also built a complete PET tomograph called ATLAS that uses a dual-layer scintillator array, also known as a phoswich. The top layer of the array is made from 2- × 2- × 7.5-mm LGSO scintillator elements (a mixture of lutetium and gadolinium oxyorthosilicates) and the bottom layer is made of 2- × 2- × 7.5-mm elements of pure gadolinium oxyorthosilicate scintillator (Seidel *et al.*, 2000). By examining the decay time of the detector pulses, the layer in which the annihilation photons interact can be determined, providing one bit of depth of interaction information.

Two systems have been developed based on the scintillator yttrium aluminium perovskit (YAP). The TierPET system (Research Center Jülich) uses arrays of small 2- × 2- × 15-mm cross-section YAP crystals coupled to a position-sensitive photomultiplier tube. Four of these detectors are mounted on a gantry that allows variable separation to accommodate animals of different sizes. The reported reconstructed image spatial resolution of the TierPET tomograph was ~ 0.009 cc, with an absolute sensitivity peak of 0.32% at the center of the FOV for a 16-cm detector spacing (Weber *et al.*, 1996, 1999). A very similar tomograph, YAP-PET, based on the same scintillator material and PMT, with 30-mm long crystals was developed at the University of Ferrara (Del Guerra *et al.*, 1998). This system has a sensitivity as high as 1.7% for a detector-to-detector spacing of 15 cm when a very low energy threshold of 50 keV is applied.

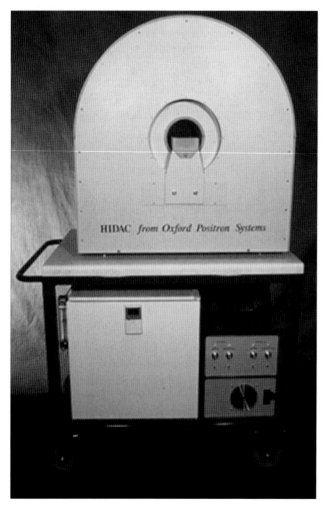

FIGURE 3 Photograph of the HIDAC system, an animal PET scanner that uses multiwire proportional chamber technology together with stacks of lead converters. These detectors have an intrinsic spatial resolution of <1 mm. (Reprinted from Molecular Imaging and Biology, Vol. 4, Chatziioannou, PET Scanners Dedicated to Molecular Imaging of Small Animal Models, pages 47–63, 2002, with permission from Elsevier Science.)

Massachussetts General Hospital has built a single-slice machine using LSO elements of dimensions 1 mm (transaxial) × 4.5 mm (axial) × 5 mm (deep) (Correia et al., 1999). This system produces a reconstructed resolution of 1.3 mm at the center of the FOV, but only has an absolute sensitivity of ~0.08% because of the single-ring design and the short length of the crystals. An extension to a multislice system is planned.

The development of APD arrays has opened up new opportunities for PET detectors. The coupling of individual LSO crystals to elements on an 8 × 2 APD array is being explored by the Max-Planck-Institute in Munich. A prototype system (MADPET) has been constructed with a reconstructed image resolution of 2.3 mm FWHM and a sensitivity for a central line source of 0.035% (Ziegler et al., 2001). The same group is currently developing two-layer modules that will provide depth of interaction information using a similar type of detector (Pichler et al., 1998).

An animal PET scanner that combines BaF_2 scintillator with a photosensitive gas detector was developed at the University of Brussels in Belgium (Bruyndonckx et al., 1996). The ultraviolet (UV) component of the scintillation light ionizes the gas in the detector, creating electrons that drift towards a two-stage amplification region, after which the location and timing of the events are determined. The FOV of the tomograph is 52 mm in the axial and 11 cm in the transaxial direction. At the center of the FOV, the reconstructed spatial resolution is 3 mm FWHM in the transverse direction and 3.5 mm in the axial direction, while the reported peak absolute system sensitivity is 3.5% (Bruyndonckx et al., 1997). This system, like the other gas-based detector system already described, has no energy resolution and the slow pulse drift region necessitates the use of a long coincidence timing window of 50 ns.

V. APPLICATIONS OF SMALL ANIMAL PET

There are now a number of reports documenting the use of dedicated small animal PET scanners for different applications. These include studies of the dopaminergic system in the rat brain, longitudinal studies of glucose metabolism in rat brain, investigation of the effect of photodynamic therapy in a mouse tumor model, and the development of methods for imaging reporter gene expression *in vivo*. A selection of applications are summarized next.

A. Measurement of Glucose Metabolism in the Rat Brain and Heart

Early attempts to use ^{18}F-fluorodeoxyglucose (FDG) and PET to measure glucose metabolism in the rat brain as an *in vivo* alternative to ^{14}C-2-deoxyglucose (2DG) autoradiographic studies (Sokoloff et al., 1977) were limited to whole-brain or coarse regional estimates of cerebral metabolic rate because of spatial resolution constraints (Brownell et al., 1991; Ingvar et al., 1991; Ogawa et al., 1996; Ouchi et al., 1996). A particularly confounding issue in these studies is high uptake of FDG in the Harderian glands, situated behind the orbits of the eye and close to the brain. The spillover of this activity into the brain due to partial volume effects make the quantification of uptake extremely difficult in lower-resolution PET scanners (Kuge et al., 1997). Furthermore, FDG is a very nonspecific tracer, making identification of the major structures in FDG PET rat brain images difficult unless the spatial resolution is ~ 2 mm or better.

In later generation animal PET systems, the resolution is sufficient that extracerebral activity in the Harderian glands and other structures is clearly separated from cerebral activity and structures such as the thalamus, striatum, and the overlying cortex are readily identified. A range of semiquantitative studies of brain plasticity and conscious brain activation have been performed (Kornblum et al., 2000) and a series of fully quantitative longitudinal studies in a traumatic brain injury model complete with direct comparison to quantitative 2DG autoradiography (Moore et al., 2000) have also been reported. A longitudinal series of images from a single rat showing quantitative cerebral metabolic rate for glucose utilization before and after traumatic brain injury is presented in Figure 4. It is apparent that small animal PET scanners can now be used to quantitatively monitor the metabolic rate for glucose in the major structures in the rat brain and that this will be a valuable tool in the study of brain development, brain plasticity, and brain injury.

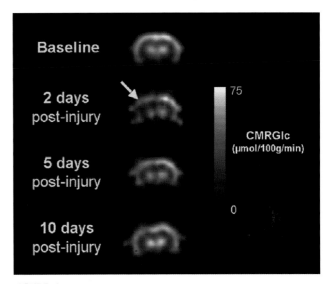

FIGURE 4 Coronal section through rat brain showing cerebral metabolic rate for glucose (CMRGlc) following fluid percussion traumatic brain injury. Note metabolic deficit at day 2 followed by gradual recovery at days 5 and 10. (Reprinted from Molecular Imaging and Biology, Vol. 4, Chatziioannou, PET Scanners Dedicated to Molecular Imaging of Small Animal Models, pages 47–63, 2002, with permission from Elsevier Science.)

Small animal PET scanners are also being employed to measure glucose metabolism and myocardial blood flow (with ^{13}N-ammonia) in rat heart models of ischemia and infarction. Initial studies comparing infarct extent (Morguet et al., 1998) and FDG uptake (Kudo et al., 1999) against in vitro measures are promising. The group at Sherbrooke University has also demonstrated the feasibility of acquiring gated PET images of the rat heart (Lapointe et al., 1999). Future studies will use PET to investigate the effects of protective interventions prior to ischemia or therapeutic interventions during or following ischemia.

B. Studies of the Dopaminergic System in Rat Brain

Some of the earliest small animal PET studies used a variety of tracers to probe the dopaminergic system in the rat brain. Tracers are available that reflect dopamine synthesis (e.g., ^{18}F-fluoro-meta-tyrosine), D2 receptor binding (e.g., ^{11}C-raclopride), and dopamine transporter concentration (e.g., ^{11}C-CFT). Because these tracers are only accumulated to a high degree in the striata and these structures are fairly large and well separated in the rat brain, successful studies can be performed even at moderate 3- to 4-mm spatial resolution (Hume et al., 1996). This pioneering work demonstrated, for example, the ability of PET to quantify D2 receptor binding in the rat brain (Hume et al., 1995), the survival and function of neural transplants (Torres et al., 1995; Fricker et al., 1997; Brownell et al., 1998), and the effects of drugs on D1 and D2 receptors (Tsukada et al., 1996; Unterwald et al., 1997). An example from one of these studies is shown in Figure 5. There has also been important work on issues related to quantification of such receptor studies and the pharmacological constraints associated with PET imaging in small animals (Hume et al., 1997, 1998). Further improvements in quantification (higher resolution leads to smaller partial volume effects and less spillover) and the ability to visualize smaller changes should now be possible with newer-generation animal PET scanners and similar studies now appear possible in the mouse (Chatziioannou et al., 1999), where the mass of the striatum is just 15 mg.

C. Animal PET in Oncology

Oncology applications offer perhaps some of the best opportunities for small animal PET. Tumor models result in high variability from one animal to the next, increasing the power of within-subject designs that become possible with PET. Furthermore, study of primary tumors is often straightforward because these can be situated away from major organs in the thigh, in the shoulder, or on the back of the animal. In this low background environment, quantification is easier and partial volume corrections can be estimated. The ability of PET to survey the entire animal also allows the spread of metastatic disease to be observed and monitored. A range of PET tracers are of interest in cancer models, including FDG, FLT (^{18}F-fluorothymidine, a marker of cell proliferation), and labeled antibodies and antibody fragments (Wu et al., 1999). The power of serial PET studies as a tool for assessing tumor response to therapy has been demonstrated in a study of photodynamic therapy (PDT) (Lapointe et al., 1999). A high-resolution animal PET scanner was used together with FDG to compare the relative efficacy of two PDT drugs in a murine tumor model (Figure 6).

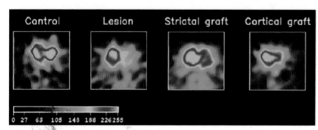

FIGURE 5 Coronal section through rat brain showing D2 receptor binding following injection of ^{11}C-raclopride. Four different rats are shown: (a) control rat; (b) lesioned rat (unilateral ibotenic acid lesion); (c) lesioned rat following graft of striatal cells; (d) lesioned rat following graft of cortical cells. Rats with grafts were scanned approximately 10 months postsurgery. Note loss of D2 binding signal on right side of image following lesion and partial recovery of the signal in the rat that received a striatal graft. (Reprinted from Molecular Imaging and Biology, Vol. 4, Chatziioannou, PET Scanners Dedicated to Molecular Imaging of Small Animal Models, pages 47–63, 2002, with permission from Elsevier Science.)

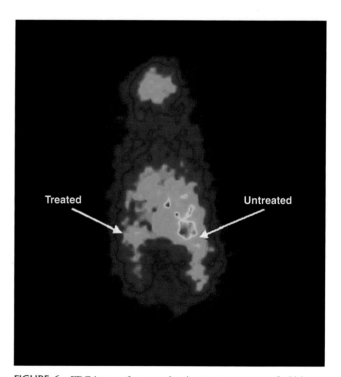

FIGURE 6 FDG image of a mouse bearing two tumors, one of which was treated by photodynamic therapy (PDT). This image was obtained 2 hours after PDT and FDG uptake is clearly reduced in the treated tumor. (Image courtesy of Dr. Roger Lecomte, Université de Sherbrooke, Canada.)

D. Imaging of Gene Expression by PET

An area attracting considerable interest is the merger of nuclear medicine with molecular biology to create methods to measure reporter gene expression *in vivo* with PET. In analogy to the way green fluorescent protein (GFP) is used as a standard reporter gene in molecular biology, a PET reporter gene is used that is able to trap (e.g., an enzyme that can convert a substrate to a trapped form) or bind (e.g., a receptor) a positron-labeled compound (Tjuvajev *et al.*, 1998; Gambhir *et al.*, 1999a, 1999b; Gambhir, Bauer, *et al.*, 2000). The reporter gene is driven by the same promoter (the promoter can be thought of as a switch that controls the level of expression of the gene) as the gene of interest, such that when the gene of interest is expressed, the reporter gene is also expressed. The retention of the positron-labeled probe by the protein product of the PET reporter gene has been shown to be proportional to the level of reporter gene expression, which in turn reflects the level of expression of the gene of interest (Gambhir, Bauer, *et al.*, 2000). In this way, the location, magnitude of expression and time course of expression levels of any gene that is introduced into a mouse can be monitored *in vivo*. This allows gene therapy protocols to be monitored *in vivo*, both in animal models and ultimately in humans, by PET. The same PET reporter gene approach can be used in transgenic mice, where every cell in the mouse carries the PET reporter gene, but the signal is only detected when the promoter driving the PET reporter gene is switched on. This now enables endogenous gene expression to be studied in mouse models. Figure 7 is an image of a transgenic mouse in which the PET reporter gene (in this case herpes simplex virus type 1 thymidine kinase, HSV1-tk) is driven by the albumin promoter. Albumin is only produced at high levels in the liver, and therefore the liver is the only place that the albumin promoter should be switched on. Figure 7 shows that ^{18}F-FHBG (9-[4-[^{18}F]fluoro-3(hydroxymethyl)butyl] guanosine), which is phosphorylated and trapped in cells that contain the HSV1-TK enzyme, is able to provide a PET image showing cells in which the albumin promoter is switched on.

There are widespread applications for these PET reporter gene methods (Gambhir, Herschman, *et al.*, 2000), for example, genetic tagging of tumor cells that can then be followed over time after injection into an animal, study of the efficiency of gene therapy vectors for delivering genes into experimental animals, interactions between cancer cells and the immune system, and study of gene expression patterns during development to see when certain genes are switched on or off. This opens up many powerful research opportunities that take advantage of the ability of PET to longitudinally measure gene expression in an entire mouse.

VI. FUTURE OPPORTUNITIES AND CHALLENGES

Small animal PET scanners have been successfully constructed and are now being used in a range of applications. Biologists are starting to become aware of PET and its potential to provide useful information in studies of animal models. Many pharmaceutical companies have set up imaging task forces and are considering animal PET as one of the technology options. The number of centers that have animal PET technology is growing rapidly, although at the time of writing (2002) the majority of these are operating within existing large research PET centers and feed off the existing infrastructure and expertise that is available there. Clearly, issues related to scanner performance, cost, access to PET tracers, and user friendliness will ultimately dictate the level and extent of PET's participation in the era of postgenomic biology. A number of specific research areas are worthy of further comment and are summarized next.

A. Small Animal PET Research Challenges

Many research groups are involved in further development of technology and methodology for small animal PET. On the detector front, the search for improved scintillators for PET applications continues. Multichannel and position-sensitive PMTs are improving in a number of ways, including larger active areas, reduced packaging size, and

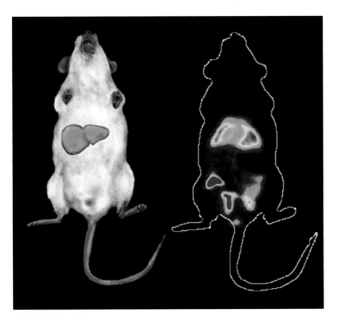

FIGURE 7 PET image of a transgenic mouse showing cells in which the albumin gene is switched on. The primary location for albumin production is in the liver. (Image courtesy of Dr. Sam Gambhir, Crump Institute for Molecular Imaging, UCLA School of Medicine, Los Angeles, CA.)

lower cost. Several groups are pursuing the use of very small LSO elements to reach submillimeter resolution with high sensitivity (Chatziioannou et al., 2001; Miyaoka et al., 2001). Some designs incorporate depth of interaction schemes and others employ novel decoding schemes to reduce the number of channels required to readout the matrix of scintillator elements. One design seeks to dramatically improve sensitivity by surrounding the entire animal with extremely efficient depth-encoding detectors (Huber and Moses, 1999). Solid-state detectors continue to advance, with multichannel APD arrays being actively investigated by several groups (Casey et al., 1998; Shao et al., 2000). The direct detection of gamma rays using relatively high-density semiconductor materials such as cadmium zinc telluride is another avenue that may be worthy of exploration. There are thus still plenty of opportunities to improve the performance of animal PET systems using these or other approaches.

This chapter has focused largely on instrumentation. As animal PET scanner technology starts to mature, other areas will be given their due attention. A particularly important area is the application of iterative statistical reconstruction algorithms that accurately model the physics of the scanner (e.g., depth of interaction effects, positron range and noncolinearity, and detector scatter), the geometry of the scanner (e.g., depth-dependent efficiency along lines of response), and the statistics of the acquired data. There is already strong evidence that these algorithms can be used to achieve better spatial resolution and improved signal-to-noise compared with analytic reconstruction methods (Brix et al., 1997; Johnson et al., 1997; Qi et al., 1998) while maintaining the quantitative nature of the data (Chatziioannou et al., 2000). An example of the application of a statistical reconstruction algorithm to data from a small animal PET scanner is shown in Figure 8. Given the sensitivity issues discussed in Section II, it is highly likely that these types of algorithms will be required to achieve submillimeter reconstructed images at acceptable signal-to-noise values in small animal PET studies, unless dramatic improvements in tracer specific activity and scanner sensitivity are forthcoming.

A third area of importance will be software for accurate quantification and robust, objective data analysis of small animal PET studies. Although many of the correction methods that are used in human PET scanners can be adapted for use in small animal PET, the relative magnitude of the corrections can be very different. In systems that use different detector technologies, new approaches to data correction may be required. Both hardware (headholders and other positioning devices) and software (image registration and image warping) tools that enable studies to be accurately registered with one another will be important. In brain studies of the mouse and rat, high-resolution atlases onto which the PET data can be mapped are likely to be useful.

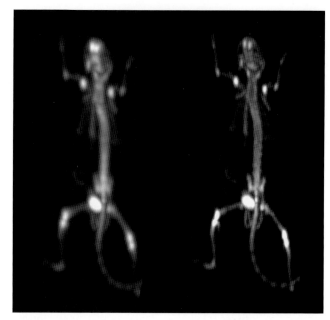

FIGURE 8 Whole-body bone scans (^{18}F-fluoride ion) of a mouse acquired using the UCLA microPET scanner. The image on the left was reconstructed using standard 3D filtered backprojection technique with a ramp filter. The image on the right was reconstructed from the same raw data using a MAP algorithm that accurately models the data collection statistics and the physics of the scanner (Qi et al., 1998). A substantial improvement in spatial resolution without noise amplification is clearly demonstrated. (Reprinted from Molecular Imaging and Biology, Vol. 4, Chatziioannou, PET Scanners Dedicated to Molecular Imaging of Small Animal Models, pages 47–63, 2002, with permission from Elsevier Science.)

B. Resolution Limits of Small Animal PET

For ^{18}F-labeled tracers, it is clear that the resolution limit for small animal PET studies lies below 1 mm. The blurring due to ^{18}F positron range is estimated to be between 0.1 and 0.2 mm FWHM, with a root mean-square range of between 0.3 and 0.4 mm (Derenzo, 1979; Levin and Hoffman, 1999), and the noncolinearity effect contributes a Gaussian blur of ~ 0.3 mm for a 15-cm diameter scanner. One of the existing animal PET scanners already has an intrinsic spatial resolution capability of under 1 mm (Jeavons et al., 1999), but has yet to demonstrate the ability to reconstruct in vivo images at this resolution. This is probably due to the requirement for very high counting statistics to routinely reconstruct images at reasonable signal-to-noise levels at submillimeter resolution. To routinely reach submillimeter resolution will probably require substantial increases in the sensitivity of small animal PET scanners, high specific-activity labeled compounds to keep injected mass levels low, and the use of a sophisticated reconstruction algorithms that properly model both the statistics of the data acquisition process and the physical factors that degrade spatial resolution in a real PET system. How far the resolution can ultimately be pushed is not currently known, but somewhere in the 0.5- to 0.75-mm range would seem a reasonable estimate for ^{18}F-labeled tracers and fractionally worse for the slightly more energetic positrons from ^{11}C-labeled tracers. This would provide ~3000 resolution elements in the mouse brain (compared with ~16,000 in the human brain at the 4–5 mm resolution typical of clinical PET scanners), which is adequate for many interesting applications.

C. Multimodality Small Animal Imaging

One of the strengths of radiotracer imaging can in some instances also turn out to be a limitation. PET images obtained using extremely specific molecular probes, which target only certain cell types or populations, can be very difficult to interpret on their own. Figure 7 is an example of such a study, where the outline of the mouse was used to provide some basic anatomical context for the image. In many cases, the anatomical correlates of areas of uptake can only be guessed at from the PET images and may require sacrifice and dissection of the animal for confirmation. A better approach would be to combine PET imaging with high-resolution anatomical imaging by X-ray computerized tomography (CT) or magnetic resonance imaging (MRI) to provide direct registration of function and anatomy. The most efficient and accurate way to do this, particularly in the whole-body mouse setting, is to have integrated imaging systems. A combined PET-CT scanner has already been developed for human imaging (Townsend *et al.*, 1998), and there has been considerable work on combining SPECT and CT (Lang *et al.*, 1992; Kalki *et al.*, 1997; Iwata *et al.*, 1999) for both animal and human applications. An integrated system for planar nuclear and X-ray imaging of mice has also been developed (M. Williams, University of Virginia, personal communication).

High-resolution dedicated microCT systems for *in vivo* small animal imaging have been successfully constructed and tested (Paulus *et al.*, 1999) and have a number of applications in their own right (Paulus *et al.*, 2000). We are in the process of integrating microPET and microCT technology to produce a combined system for *in vivo* mouse imaging (Goertzen *et al.*, 2002). There is also interest in combining PET with MRI (Raylman *et al.*, 1996; Shao, Cherry, Farahani, Slates, *et al.*, 1997) because of the excellent soft tissue contrast available with MRI and the tremendous flexibility of pulse sequence design to enhance and emphasize different tissue characteristics. We have built and evaluated a single-slice prototype MR-compatible PET scanner (Shao, Cherry, Farahani, Meadors, *et al.*, 1997; Slates *et al.*, 1999) and are currently constructing a larger-bore system for routine *in vivo* studies in mice and rats (Slates *et al.*, 1999).

The anatomical data do more than just provide landmarks for interpretation of the PET data. They can potentially be used to improve quantification of the PET signal by providing information for attenuation correction (Beyer *et al.*, 1995) and partial volume correction. It may also be possible to use the anatomical data as prior information in the PET image reconstruction (Leahy and Yan, 1991). It seems likely that animal PET scanners of the future will, at least, incorporate some means of acquiring an anatomical image or outline.

D. Use of PET for High-Throughput Phenotyping and Drug Screening

A challenge and opportunity for all small animal imaging technologies relates to the need for relatively high-throughput (10^2–10^3 mice per day) screening techniques for mouse phenotyping. Large numbers of mutant mice are being produced, both by random mutagenesis and by controlled gene activation or deletion in transgenic animals. A key question involves relating the known (or readily determined) genotype with phenotype. Imaging is one approach to answering this question, and CT has already been proposed for high-throughput anatomical screening (Paulus *et al.*, 1999). A combined microPET-CT scanner could be even more powerful by providing both anatomical and functional screens in a single setting, with the nature of the functional screen determined by the PET tracer used. The challenges include the need for extremely high reproducibility in the PET imaging (complicated by anesthesia and tracer administration issues); high signal-to-noise and resolution in short scan times; and sophisticated image-processing tools to handle vast quantities of data, transform them into a common coordinate system, and extract relevant parameters of interest in a highly automated fashion.

Small animal PET could also be used for *in vivo* screening of candidate drugs for lead compound selection. Following standard *in vitro* tests to reduce the number of candidate drugs for a particular target to a manageable number (say 10–20), PET imaging could be used to determine which of the candidate drugs are delivered and bind to their target *in vivo*. The concentration of the drug at the target can also be determined to assist in setting initial dose ranging for human trials. This approach could be very efficient if a well-characterized PET tracer already exists for the same target because displacement types of approaches can be used to get relative measures of drug-binding affinity across multiple drugs using a single PET tracer. This avoids radiolabeling each individual drug and permits *in vivo* screening of tens of compounds in a fairly straight-forward manner.

VII. SUMMARY

A number of small animal PET systems have been constructed and applied to a range of biological questions. New scanner designs are currently under development and, because the physical limitations of PET have yet to be reached and the resolution and sensitivity of small animal PET scanners are still limited by the detector technology and the scanner geometry, significant further improvements in performance can be anticipated. The role of PET in modern biology has yet to be defined, but, if some of the challenges identified in this chapter can be overcome, then it is likely to play a significant role in the study of animal models of disease.

References

Agon, P., Kaufman, J. M., Goethals, P., Van Haver, D., and Bogaert, M. G. (1988). Study with positron emission tomography of the osmotic opening of the dog blood-brain barrier for quinidine and morphine. *J. Pharm. Pharmacol.* **40:** 539–543.

Bailey, D. L., Jones, T., and Spinks, T. J. (1991). A method for measuring the absolute sensitivity of positron emission tomographic scanners. *Eur. J. Nucl. Med.* **18:** 374–379.

Beyer, T., Kinahan, P. E., Townsend, D. W., and Sashin, D. (1995). The use of X-ray CT for attenuation correction of PET data. *IEEE Trans. Nucl. Sci.* **42:** 1573–1577.

Bloomfield, P. M., Myers, R., Hume, S. P., Spinks, T. J., Lammertsma, A. A., and Jones, T. (1997). Three-dimensional performance of a small-diameter positron emission tomograph. *Phys. Med. Biol.* **42:** 389–400.

Bloomfield, P. M., Rajeswaran, S., Spinks, T. J., Hume, S. P., Myers, R., Ashworth, S., Clifford, K. M., Jones, W. F., Byars, L. G., Young, J., Andreaco, M., Williams, C. W., Lammertsma, A. A., and Jones, T. (1995). The design and physical characteristics of a small animal positron emission tomograph. *Phys. Med. Biol.* **40:** 1105–1126.

Brix, G., Doll, J., Bellemann, M. E., Trojan, H., Haberkorn, U., Schmidlin, P., and Ostertag, H. (1997). Use of scanner characteristics in iterative image reconstruction for high-resolution positron emission tomography studies of small animals. *Eur. J. Nucl. Med.* **24:** 779–786.

Brownell, A. L., Kano, M., McKinstry, R. C., Moskowitz, M. A., Rosen, B. R., and Brownell, G. L. (1991). PET and MR studies of experimental focal stroke. *J. Comput. Assist. Tomogr.* **15:** 376–380.

Brownell, A. L., Livni, E., Galpern, W., and Isacson, O. (1998). In vivo PET imaging in rat of dopamine terminals reveals functional neural transplants. *Ann. Neurol.* **43:** 387–390.

Bruyndonckx, P., Liu, X., Tavernier, S., and Zhang, S. (1996). Performance of a small animal PET scanner based on photosensitive wire chambers. *In* "1996 IEEE Nuclear Science Symposium and Medical Imaging Conference Record" (A. Del Guerra, ed.), Vol. 2, pp. 1335–1338. IEEE, Piscataway, NJ.

Bruyndonckx, P., Liu, X., Tavernier, S., and Zhang, S. (1997). Performance study of a 3D small animal PET scanner based on BaF^2 crystals and a photo sensitive wire chamber. *Nucl. Instrum. Methods Phys. Res., Sect. A* **392:** 407–413.

Casey, M. E., Dautet, H., Waechter, D., Lecomte, R., Eriksson, L., and Schmand, M. (1998). An LSO block detector for PET using an avalanche photodiode array. *In* "1998 IEEE Nuclear Science Symposium and Medical Imaging Conference Record" (A. Del Guerra, ed.), Vol. 2, pp. 1105–1108. IEEE, Piscataway, NJ.

Chatziioannou, A., Qi, J., Moore, A., Annala, A., Nguyen, K., Leahy, R. M., and Cherry, S. R. (2000). Comparison of 3D maximum a posteriori and filtered backprojection algorithms for high resolution animal imaging with microPET. *IEEE Trans. Med. Imag.* **19:** 507–512.

Chatziioannou, A. F., Cherry, S. R., Shao, Y., Silverman, R. W., Meadors, K., Farquhar, T. H., Pedarsani, M., and Phelps, M. E. (1999). Performance evaluation of microPET: A high-resolution lutetium oxyorthosilicate PET scanner for animal imaging. *J. Nucl. Med.* **40:** 1164–1175.

Chatziioannou, A., Tai, Y. C., Doshi, N., Cherry, S. R. (2001). Detector development for microPET II: A 1 μL resolution PET scanner for small animal imaging. *Phys. Med. Biol.* **46:** 2899–2910.

Cherry, S. R., Shao, Y., Silverman, R. W., Meadors, K., Siegel, S., Chatziioannou, A., Young, J. W., Jones, W., Moyers, J. C., Newport, D., Boutefnouchet, A., Farquhar, T. H., Andreaco, M., Paulus, M. J., Binkley, D. M., Nutt, R., and Phelps, M. E. (1997). MicroPET: A high resolution PET scanner for imaging small animals. *IEEE Trans. Nucl. Sci.* **44:** 1161–1166.

Cherry, S. R., Yiping, S., Siegel, S., Silverman, R. W., Mumcuoglu, E., Meadors, K., and Phelps, M. E. (1996). Optical fiber readout of scintillator arrays using a multi-channel PMT: A high resolution PET detector for animal imaging. *IEEE Trans. Nucl. Sci.* **43:** 1932–1937.

Correia, J. A., Burnham, C. A., Kaufman, D., and Fischman, A. J. (1999). Development of a small animal PET imaging device with resolution approaching 1 mm. *IEEE Trans. Nucl. Sci.* **46:** 631–635.

Cutler, P. D., Cherry, S. R., Hoffman, E. J., Digby, W. M., and Phelps, M. E. (1992). Design features and performance of a PET system for animal research. *J. Nucl. Med.* **33:** 595–604.

Del Guerra, A., Di Domenico, G., Scandola, M., and Zavattini, G. (1998). High spatial resolution small animal YAP-PET. *Nucl. Instrum. Methods Phys. Res., Sect. A* **409:** 508–510.

Derenzo, S. (1979). Precision measurement of annihilation point spread distributions for medically important positron emitters. *In* "Positron Annihilation" (R. R. Hasiguti and K. Fujiwara, eds.), 819–823. Japan Institute of Metals, Sendai, Japan.

Englert, D., Roessler, N., Jeavons, A., and Fairless, S. (1995). Microchannel array detector for quantitative electronic radioautography. *Cell. Mol. Biol.* **41:** 57–64.

Fowler, J. S., Volkow, N. D., Wolf, A. P., Dewey, S. L., Schlyer, D. J., Macgregor, R. R., Hitzemann, R., Logan, J., Bendriem, B., Gatley, S. J., *et al.* (1989). Mapping cocaine binding sites in human and baboon brain in vivo. *Synapse* **4:** 371–377.

Fowler, J. S., and Wolf, A. P. (1986). Positron emitter-labeled compounds: Priorities and problems. *In* "Positron Emission Tomography and Autoradiography" (M. Phelps, J. Mazziotta, and H. Schelbert, eds.), pp. 391–450. Raven Press, New York.

Fricker, R. A., Torres, E. M., Hume, S. P., Myers, R., Opacka-Juffrey, J., Ashworth, S., Brooks, D. J., and Dunnett, S. B. (1997). The effects of donor stage on the survival and function of embryonic striatal grafts in the adult rat brain. II. Correlation between positron emission tomography and reaching behaviour. *Neuroscience* **79:** 711–721.

Gambhir, S. S., Barrio, J. R., Herschman, H. R., and Phelps, M. E. (1999a). Assays for noninvasive imaging of reporter gene expression. *Nucl. Med. Biol.* **26:** 481–490.

Gambhir, S. S., Barrio, J. R., Herschman, H. R., and Phelps, M. E. (1999b). Imaging gene expression: Principles and assays. *J. Nucl. Cardiol.* **6:** 219–233.

Gambhir, S. S., Bauer, E., Black, M. E., Liang, Q., Kokoris, M. S., Barrio, J. R., Iyer, M., Namavari, M., Phelps, M. E., and Herschman, H. R. (2000). A mutant herpes simplex virus type 1 thymidine kinase reporter gene shows improved sensitivity for imaging reporter gene expression with positron emission tomography. *Proc. Natl. Acad. Sci. U.S.A.* **97:** 2785–2790.

Gambhir, S. S., Herschman, H. R., Cherry, S. R., Barrio, J. R., Satyamurthy, N., Toyokuni, T., Phelps, M. E., Larson, S. M., Balatoni, J., Finn, R., Sadelain, M., Tjuvajev, J., and Blasberg, R. (2000). Imaging transgene expression with radionuclide imaging technologies. *Neoplasia* **2:** 118–136.

Gjedde, A., and Rasmussen, M. (1980). Pentobarbital anesthesia reduces blood-brain glucose transfer in the rat. *J. Neurochem.* **35:** 1382–1387.

Goertzen, A. G., Meadors, A. K., Silverman, R. W., and Cherry, S. R. (2002). Simultaneous molecular and anatomical imaging of the mouse in vivo. *Phys. Med. Biol.* **47:** 4315–4328.

Green, L. A., Gambhir, S. S., Srinivasan, A., Banerjee, P. K., Hoh, C. K., Cherry, S. R., Sharfstein, S., Barrio, J. R., Herschman, H. R., and Phelps, M. E. (1998). Noninvasive methods for quantitating blood time-activity curves from mouse PET images obtained with fluorine-18-fluorodeoxyglucose. *J. Nucl. Med.* **39:** 729–734.

Huber, J. S., and Moses, W. W. (1999). Conceptual design of a high-sensitivity small animal PET camera with 4 pi coverage. *IEEE Trans. Nucl. Sci.* **46:** 498–502.

Hume, S. P., Brown, D. J., Ashworth, S., Hirani, E., Luthra, S. K., and Lammertsma, A. A. (1997). In vivo saturation kinetics of two dopamine transporter probes measured using a small animal positron emission tomography scanner. *J. Neurosci. Methods* **76:** 45–51.

Hume, S. P., Gunn, R. N., and Jones, T. (1998). Pharmacological constraints associated with positron emission tomographic scanning of small laboratory animals. *Eur. J. Nucl. Med.* **25:** 173–176.

Hume, S. P., Lammertsma, A. A., Myers, R., Rajeswaran, S., Bloomfield, P. M., Ashworth, S., Fricker, R. A., Torres, E. M., Watson, I., and Jones, T. (1996). The potential of high-resolution positron emission tomography to monitor striatal dopaminergic function in rat models of disease. *J. Neurosci. Methods* **67:** 103–112.

Hume, S. P., Opacka-Juffry, J., Myers, R., Ahier, R. G., Ashworth, S., Brooks, D. J., and Lammertsma, A. A. (1995). Effect of L-dopa and 6-hydroxydopamine lesioning on [11C]raclopride binding in rat striatum, quantified using PET. *Synapse* **21:** 45–53.

Ingvar, M., Eriksson, L., Rogers, G. A., Stone-Elander, S., and Widén, L. (1991). Rapid feasibility studies of tracers for positron emission tomography: High-resolution PET in small animals with kinetic analysis. *J. Cereb. Blood. Flow. Metab.* **11:** 926–931.

Iwata, K., Hasegawa, B. H., Heanue, J. A., Bennett, P. R., Shah, K. S., Boles, C. D., and Boser, B. E. (1999). CdZnTe detector for combined X-ray CT and SPECT. *Nucl. Instrum. Methods Phys. Res. A* **422:** 740–744.

Jeavons, A. P., Chandler, R. A., and Dettmar, C. A. R. (1999). A 3D HIDAC-PET camera with sub-millimetre resolution for imaging small animals. *IEEE Trans. Nucl. Sci.* **46:** 468–473.

Johnson, C. A., Seidel, J., Carson, R. E., Gandler, W. R., Sofer, A., Green, M. V., and Daube-Witherspoon, M. E. (1997). Evaluation of 3D reconstruction algorithms for a small animal PET camera. *IEEE Trans. Nucl. Sci.* **44:** 1303–1308.

Kalki, K., Blankespoor, S. C., Brown, J. K., Hasegawa, B. H., Dae, M. W., Chin, M., and Stillson, C. (1997). Myocardial perfusion imaging with a combined X-ray CT and SPECT system. *J. Nucl. Med.* **38:** 1535–1540.

Kornblum, H. I., Araujo, D. M., Annala, A. J., Tatsukawa, K. J., Phelps, M. E., and Cherry, S. R. (2000). In vivo imaging of neuronal activation and plasticity in the rat brain with microPET, a novel high-resolution positron emission tomograph. *Nature Biotechnology* **18:** 655–660.

Kudo, T., Annala, A. J., Cherry, S. R., Phelps, M. E., and Schelbert, H. R. (1999). Measurement of myocardial blood flow during occlusion/reperfusion in rats with dynamic microPET imaging. *J. Nucl. Med.* **40:** 6P.

Kuge, Y., Miyake, Y., Minematsu, K., Yamaguchi, T., and Hasegawa, Y. (1997). Effects of extracranial radioactivity on measurement of cerebral glucose metabolism by rat-PET with [18F]-2-fluoro-2-deoxy-D-glucose [letter; comment]. *J. Cereb. Blood. Flow. Metab.* **17:** 1261–1262.

Lammertsma, A. A., and Hume, S. P. (1996). Simplified reference tissue model for PET receptor studies. *Neuroimage* **4:** 153–158.

Lang, T. F., Hasegawa, B. H., Soo Chin, L., Keenan Brown, J., Blankespoor, S. C., Reilly, S. M., Gingold, E. L., and Cann, C. E. (1992). Description of a prototype emission-transmission computed tomography imaging system. *J. Nucl. Med.* **33:** 1881–1887.

Lapointe, D., Bentourkia, M., Cadorette, J., Rodrique, S., Ouellet, R., Benard, F., Van Lier, J. E., and Lecomte, R. (1999). High-resolution cardiac PET in rats. *J. Nucl. Med.* **40:** 185P.

Lapointe, D., Brasseur, N., Cadorette, J., La Madeleine, C., Rodrigue, S., van Lier, J. E., and Lecomte, R. (1999). High-resolution PET imaging for in vivo monitoring of tumor response after photodynamic therapy in mice. *J. Nucl. Med.* **40:** 876–882.

Lapointe, D., Cadorette, J., Rodrigue, S., Rouleau, D., and Lecomte, R. (1998). A microvolumetric blood counter/sampler for metabolic PET studies in small animals. *IEEE Trans. Nucl. Sci.* **45:** 2195–2199.

Lasbennes, F., Lestage, P., Bobillier, P., and Seylaz, J. (1986). Stress and local cerebral blood flow: Studies on restrained and unrestrained rats. *Exp. Brain Res.* **63:** 163–168.

Leahy, R., and Yan, X. (1991). Incorporation of anatomical MR data for improved functional imaging with PET. *In* "Information Processing in Medical Imaging" (D. Hawkes and A. Colchester, eds.), pp. 105–120. Wiley Liss, New York.

Lecomte, R., Cadorette, J., Richard, P., Rodrigue, S., and Rouleau, D. (1994). Design and engineering aspects of a high resolution positron tomograph for small animal imaging. *IEEE. Trans. Nucl. Sci.* **41:** 1446–1452.

Lecomte, R., Cadorette, J., Rodrigue, S., Lapointe, D., Rouleau, D., Bentourkia, M., Yao, R., and Msaki, P. (1996). Initial results from the Sherbrooke avalanche photodiode positron tomograph. *IEEE Trans. Nucl. Sci.* **43:** 1952–1957.

Levin, C. S., and Hoffman, E. J. (1999). Calculation of positron range and its effect on the fundamental limit of positron emission tomography system spatial resolution. *Phys. Med. Biol.* **44:** 781–799.

Malakoff, D. (2000). The rise of the mouse, biomedicine's model mammal. *Science* **288:** 248–253.

Melcher, C. L. (2000). Scintillation crystals for PET. *J. Nucl. Med.* **41:** 1051–1055.

Miyaoka, R. S., Kohlmyer, S. G., and Lewellen, T. K. (2001). Performance characteristics of micro crystal elements (MiCE) detectors. *IEEE Trans. Nucl. Sci.* **48:** 1403–1407.

Moore, A. H., Osteen, C. L., Chatziioannou, A. F., Hovda, D. A., and Cherry, S. R. (2000). Quantitative assessment of longitudinal metabolic changes in vivo following traumatic brain injury in the adult rat using FDG-microPET. *J. Cereb. Blood. Flow. Metab.* **20:** 1492–1501.

Morguet, A. J., Chatziioannou, A. F., Cherry, S. R., Phelps, M. E., and Schelbert, H. R. (1998). Evaluation of a newly developed small-animal PET scanner in experimental myocardial infarction. *J. Nucl. Med.* **39:** 9P.

Ogawa, M., Fukuyama, H., Ouchi, Y., Yamauchi, H., Matsuzaki, S., Kimura, J., and Tsukada, H. (1996). Uncoupling between cortical glucose metabolism and blood flow after ibotenate lesion of the rat basal forebrain: A PET study. *Neurosci. Lett.* **204:** 193–196.

Ouchi, Y., Fukuyama, H., Ogawa, M., Yamauchi, H., Kimura, J., Magata, Y., Yonekura, Y., and Konishi, J. (1996). Cholinergic projection from the basal forebrain and cerebral glucose metabolism in rats: A dynamic PET study [see comments]. *J. Cereb. Blood. Flow. Metab.* **16:** 34–41.

Paulus, M. J., Gleason, S. S., Kennel, S. J., Hunsicker, P. R., and Johnson, D. K. (2000). High resolution x-ray computed tomography: an emerging tool for small animal cancer research. *Neoplasia* **2:** 62–70.

Paulus, M. J., Sari-Sarraf, H., Gleason, S. S., Bobrek, M., Hicks, J. S., Johnson, D. K., Behel, J. K., Thompson, L. H., and Allen, W. C. (1999). A new X-ray computed tomography system for laboratory mouse imaging. *IEEE Trans. Nucl. Sci.* **46:** 558–564.

Phelps, M. E. (2000). PET: The merging of Biology and Imaging into Molecular Imaging. *J. Nucl. Med.* **41:** 661–681.

Pichler, B. J., Boning, G., Rafecas, M., Pimpl, W., Korenz, E., Schwaiger, M., and Ziegler, S. I. (1998). Feasibility study of a compact high resolution dual layer LSO-APD detector module for positron emission tomography. *In* "1998 IEEE Nuclear Science Symposium and Medical Imaging Conference Record" (A. Del Guerra, ed.), Vol. 2, pp. 1199–1203. IEEE, Piscataway, NJ.

Qi, J., Leahy, R., Cherry, S., Chatziioannou, A., and Farquhar, T. (1998). High-resolution 3D Bayesian image reconstruction using the microPET small-animal scanner. *Phys. Med. Biol.* **43:** 1001–1013.

Raylman, R. R., Hammer, B. E., and Christensen, N. L. (1996). Combined MRI-PET scanner: A Monte Carlo evaluation of the improvements in PET resolution due to the effects of a static homogeneous magnetic field. *IEEE Trans. Nucl. Sci.* **43:** 2406–2412.

Saija, A., Princi, P., De, P. R., and Costa, G. (1989). Modifications of the permeability of the blood-brain barrier and local cerebral metabolism in pentobarbital and ketamine-anaesthetized rats. *Neuropharmacology* **28:** 997–1002.

Seidel, J., Gandler, W. R., and Green, M. V. (1994). A very high resolution single-slice small animal PET scanner based on direct detection of coincidence line endpoints. *J. Nucl. Med.* **35:** 40P.

Seidel, J., Vaquero, J. J., Barbosa, F., Lee, I. J., Cuevas, C., and Green, M. V. (2000). Scintillator identification and performance characteristics of LSO and GSO PSPMT detector modules combined through common X and Y resistive dividers. *IEEE Trans. Nuc. Sci.* **47:** 1640–1645.

Shao, Y., Cherry, S. R., Farahani, K., Meadors, K., Siegel, S., Silverman, R. W., and Marsden, P. K. (1997). Simultaneous PET and MR imaging. *Phys. Med. Biol.* **42:** 1965–1970.

Shao, Y., Cherry, S. R., Farahani, K., Slates, R., Silverman, R. W., Meadors, K., Bowery, A., Siegel, S., Marsden, P. K., and Garlick, P. B. (1997). Development of a PET detector system compatible with MRI/NMR systems. *IEEE Trans. Nucl. Sci.* **44:** 1167–1171.

Shao, Y., Silverman, R. W., Farrell, R., Cirignano, L., Grazioso, R., Shah, K. S., Visser, G., Clajus, M., Tumer, T. O., and Cherry, S. R. (2000). Design studies of a high resolution PET detector using APD arrays. *IEEE Trans. Nucl. Sci.* **47:** 1051–1057.

Slates, R., Cherry, S., Boutefnouchet, A., Yiping, S., Dahlborn, M., and Farahani, K. (1998). Design of a small animal MR compatible PET scanner. *In* "1998 IEEE Nuclear Science Symposium and Medical Imaging Conference Record" (A. Del Guerra, ed.), Vol. 1, pp. 565–570. IEEE, Piscataway, NJ.

Slates, R. B., Farahani, K., Yiping, S., Marsden, P. K., Taylor, J., Summers, P. E., Williams, S., Beech, J., and Cherry, S. R. (1999). A study of artefacts in simultaneous PET and MR imaging using a prototype MR compatible PET scanner. *Phys. Med. Biol.* **44:** 2015–2027.

Sokoloff, L., Reivich, M., Kennedy, C., Des Rosiers, M. H., Patlak, C. S., Pettigrew, K. D., Sakurada, O., and Shinohara, M. (1977). The [14C]deoxyglucose method for the measurement of local cerebral glucose utilization: Theory, procedure and normal vlaues in the conscious and anesthetized albino rat. *J. Neurochem.* **28:** 897–916.

Tai, Y. C., Chatziioannou, A., Siegel, S., Young, J., Newport, D., Goble R. N., Nutt, R. E., and Cherry, S. R. (2001). Performance evaluation of the microPET P4: A PET system dedicated to animal imaging. *Phys. Med. Biol.* **46:** 1845–1862.

Tjuvajev, J. G., Avril, N., Oku, T., Sasajima, T., Miyagawa, T., Joshi, R., Safer, M., Beattie, B., DiResta, G., Daghighian, F., Augensen, F., Koutcher, J., Zweit, J., Humm, J., Larson, S. M., Finn, R., and Blasberg, R. (1998). Imaging herpes virus thymidine kinase gene transfer and expression by positron emission tomography. *Cancer Res.* **58:** 4333–4341.

Torres, E. M., Fricker, R. A., Hume, S. P., Myers, R., Opacka-Juffry, J., Ashworth, S., Brooks, D. J., and Dunnett, S. B. (1995). Assessment of striatal graft viability in the rat in vivo using a small diameter PET scanner. *Neuroreport* **6:** 2017–2021.

Townsend, D. W., Beyer, T., Kinahan, P. E., Brun, T., Roddy, R., Nutt, R., and Byars, L. G. (1998). The SMART scanner: a combined PET/CT tomograph for clinical oncology. *In* "1998 IEEE Nuclear Science Symposium and Medical Imaging Conference Record" (A. Del Guerra, ed.), Vol. 2, pp. 1170–1174. IEEE, Piscataway, NJ.

Tsukada, H., Harada, N., Nishiyama, S., Ohba, H., and Kakiuchi, T. (2000). Dose-response and duration effects of acute administrations of cocaine and GBR12909 on dopamine synthesis and transporter in the conscious monkey brain: PET studies combined with microdialysis. *Brain Research* **860:** 141–148.

Tsukada, H., Kreuter, J., Maggos, C. E., Unterwald, E. M., Kakiuchi, T., Nishiyama, S., Futatsubashi, M., and Kreek, M. J. (1996). Effects of binge pattern cocaine administration on dopamine D1 and D2 receptors in the rat brain: An in vivo study using positron emission tomography. *J. Neurosci.* **16:** 7670–7677.

Unterwald, E. M., Tsukada, H., Kakiuchi, T., Kosugi, T., Nishiyama, S., and Kreek, M. J. (1997). Use of positron emission tomography to measure the effects of nalmefene on D1 and D2 dopamine receptors in rat brain. *Brain Research* **775:** 183–188.

Watanabe, M., Okada, H., Shimizu, K., Omura, T., Yoshikawa, E., Kosugi, T., Mori, S., and Yamashita, T. (1997). A high resolution animal PET scanner using compact PS-PMT detectors. *IEEE Trans. Nucl. Sci.* **44:** 1277–1282.

Watanabe, M., Uchida, H., Okada, H., Shimizu, K., Satoh, N., Yoshikawa, E., Ohmura, T., Yamashita, T., and Tanaka, E. (1992). A high resolution PET for animal studies. *IEEE Trans. Med. Imag.* **11:** 577–580.

Weber, S., Herzog, H., Cremer, M., Engels, R., Hamacher, K., Kehren, F., Muehlensiepen, H., Ploux, L., Reinartz, R., Reinhart, P., Rongen, F., Sonnenberg, F., Coenen, H. H., and Halling, H. (1999). Evaluation of the TierPET system. *IEEE Trans. Nucl. Sci.* **46:** 1177–1183.

Weber, S., Terstegge, A., Engels, R., Herzog, H., Reinartz, R., Reinhart, P., Rongen, F., Muller-Gartner, H. W., and Halling, H. (1996). The KFA TierPET: Performance characteristics and measurements. *In* "1996 IEEE Nuclear Science Symposium and Medical Imaging Conference Record" (A. Del Guerra, ed.), Vol. 2, pp. 1335–1338. IEEE, Piscataway, NJ.

Wu, A. M., Yazaki, P. J., Tsai, S.-W., Anderson, A.-L. J., Williams, L. E., Shively, J. E., Wong, J. Y. C., Raubitschek, A. A., Nguyen, K., Toyokuni, T. T., Phelps, M. E., and Gambhir, S. S. (1999). MicroPET imaging of tumors in a murine model utilizing a Cu-64 radiolabeled genetically engineered anti-CEA antibody fragment. *J. Nucl. Med.* **40:** 80P.

Ziegler, S. I., Pichler, B. J., Boening, G., Rafecas, M., Pimpl, W., Lorenz, E., Schmitz, N., and Schwaiger, M. (2001). A prototype high-resolution animal positron tomograph with avalanche photodiode arrays and LSO crystals. *Eur. J. Nucl. Med.* **28:** 136–143.

CHAPTER 13

Scintillators

FRANK WILKINSON III

Alpha Spectra, Inc. Grand Junction, Colorado

I. Introduction
II. Gamma-Ray Interaction in Scintillation Crystals
III. The Characteristics and Physical Properties of Scintillators
IV. Scintillation Detectors: Design and Fabrication
V. Measurements with Scintillators
VI. Summary and Comments

"Ex luce lucellum."—*Robert Lowe*

I. INTRODUCTION

In 1948, Robert Hofstadter first described the use of the alkali halide crystal, NaI(Tl), as a scintillation detector. His first measurements with NaI(Tl) used 8 g of powder that contained crystals that were only 1–2 mm on a side (Hofstadter, 1948). In 1978, the technology had advanced to the point where Hofstadter and his collaborators (Crystal Ball Collaboration, 1980), along with a design team from The Harshaw Chemical Company, built the Crystal Ball detector that consisted of over 4 million g of NaI(Tl) material. The Crystal Ball was used at the positron-electron ring of the Stanford Linear Accelerator Center to investigate the building blocks that make up electrons and positrons.

Since the 1950s, scintillators of various materials have been used in diverse applications from investigating the smallest particles known to scientists to exploring the origins and the extent of the universe. Detectors have been used for oil, gas, and precious-mineral exploration; health physics; industrial gauges; and nuclear power plant applications. Government labs including the Department of Defense, the Department of Energy, and the Environmental Protection Agency use scintillation detectors. These are only a few of the many different and fascinating applications for scintillators in industry, medicine, government research, and academic research. One of the world's largest consumers of scintillation materials is the medical-imaging marketplace. New scintillation-detector configurations are continually being developed and tested in an effort to provide better diagnostic care for the patient.

The characteristics and physical properties of inorganic scintillators are presented in this chapter. The mechanism for the scintillation process is reviewed, and detector design issues are presented. Crystal growth issues and detector manufacturing methods are discussed, and a review of a typical scintillation counting system is presented. The chapter concludes with a brief discussion of the use of scintillators in medical imaging. Scintillators have been used in many different modes and applications; the discussion here is limited to their use in detecting X-rays and gamma-rays.

II. GAMMA-RAY INTERACTIONS IN SCINTILLATION CRYSTALS

The nuclear decay process of some isotopes can include the emission of photons called gamma-rays. Gamma-rays

TABLE 1 Radio Isotopes Useful in Medical Imaging Technology

Isotope	Half-Life ($t_{1/2}$)	Gamma Energies (keV)^a	Application and Notes
^{57}Co	271 days	122 (86), 136 (11)	Calibration and test positron emitter
^{137}Cs	30 years	662 keV of Ba-137^m (90)	Specifications, calibration, and testing
^{18}F	1.8 h	—	Clinical positron emitter
^{99m}Tc	6.0 h	140 (89)	Clinical gamma emitter
^{201}Tl	3.0 days	167 (11), 135 (3)	Clinical emitter
^{67}Ga	3.3 days	93 (37), 184 (20)	Clinical emitter
^{111}In	2.8 days	245 (94), 172 (90)	Clinical emitter
^{123}I	13.2 h	159 (83), 529 (14)	Clinical emitter
^{22}Na	2.6 years	1567 (100)	Calibration and test positron emitter
^{40}K	1.26×10^9 years	1461 (11)	Naturally occurring

are part of the electromagnetic spectrum. In a vacuum, these photons travel at the speed of light, 3×10^{10} cm/s. Gamma-ray energies are typically in the range of a few thousand electron volts to many million electron volts, whereas X-rays have energy less than 100 keV. Table 1 gives useful information about several radioisotopes relevant to medical imaging.

In order to understand the scintillation process and how scintillators detect gamma radiation, it is important to understand how gamma-rays interact with matter. Because the typical gamma-ray energy is greater than the average binding energy of an electron, gamma-rays have enough energy to ionize matter. There are three important mechanisms for interaction: the photoelectric effect, Compton scattering, and pair production. In each of these physical processes, the photon incident on the crystal gives up all or part of its energy to the electrons in the crystalline matrix.

A. The Photoelectric Process

In the photoelectric process, the incident photon gives up all of its energy to a bound electron. The incident photon is completely absorbed by a photoelectron. The photoelectron kinetic energy is given by:

$$E_{e^-} = h\nu - E_b \qquad (1)$$

where $h\nu$ represents the photon's energy and E_b is the binding energy of the bound electron before it leaves its atomic shell. In a metal, and other crystalline structures, this second term is the familiar work function of the material. The photoelectric process requires that the electron absorbing the energy be in close proximity to an atom. Effectively, it must be a bound electron in order to ensure that conservation of energy and momentum are satisfied during the interaction. Here, the recoil of the atom left behind conserves momentum, whereas Eq. (1) conserves energy. Figure 1 demonstrates this interaction schematically.

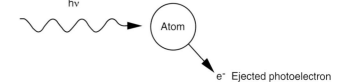

FIGURE 1 The photoelectric effect.

The likelihood that a photon is absorbed is strongly related to the atomic number of the nucleus. The probability of a reaction increases approximately as Z^4, where Z is the atomic number. It follows that a good scintillator will have a high atomic number.

After the electron is ejected from the atom, a vacancy exists in the atomic structure because the electron has absorbed the photon energy. The newly created atomic structure, or positively charged ion, will remain in an excited state for a short time. An electron from a higher energy level will fill the hole in the atomic structure in order to bring the atom to its lowest energy state. A photon is released when the electron moves from a higher energy state to the lower energy state. The released photon energy is exactly the energy difference between the initial and the final states of the electron; this photon is called a *characteristic X-ray*. The electron transition occurs in the crystalline lattice in approximately 1 ns.

The reabsorption of the X-ray energy by another atomic electron in an outer shell often occurs before the X-ray can leave the atom. This electron, called an *Auger electron*, has enough translational energy to leave the atom. Electrons from the outer shells will fill the two vacancies, causing the emission of additional characteristic X-rays and possibly Auger electrons.

The sharp rise in the mass attenuation coefficient at 32 keV (shown in Fig. 2) is equal to the binding energy of the K-shell electron in iodine. The structure in the curve

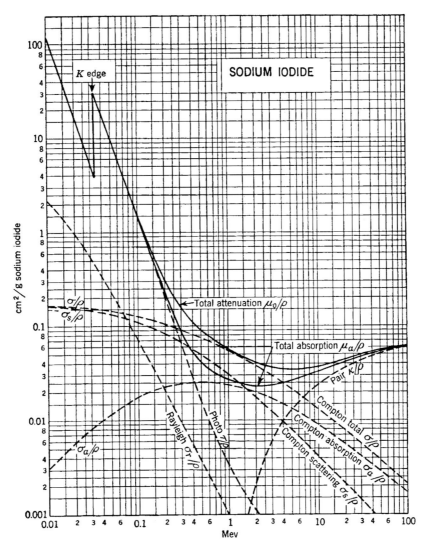

FIGURE 2 Mass attenuation coefficient curves for NaI(Tl). (From Evans, R.D. "The Atomic Nucleus." Krieger Publishing, Melbourne, FL, with permission.)

seen at 32 keV is an absorption edge. The characteristic X-ray emitted is usually reabsorbed by another atom near the initial event. When all the initial photon energy stays in the crystal, the total electron kinetic energy must be equal to the total energy of the incident photon. Ideally, the photoelectric events produced in the crystal are represented as a delta function in the gamma-ray spectrum (see Fig. 3).

B. The Compton Effect

In some cases, a photon enters the crystal lattice and scatters off an electron. The scattered photon and the electron will share the energy of the incident photon. Figure 4 shows a diagram of such a scattering event. The scattered photon energy, $h\nu'$, can be expressed in terms of the initial photon energy $h\nu$, the rest mass energy of the electron m_0c^2, and the photon scattering angle θ:

FIGURE 3 Ideal photopeak response.

$$h\nu' = \frac{h\nu}{1+(h\nu/m_0c^2)(1-\cos\theta)} \quad (2)$$

This expression is derived by conserving energy and momentum in a classical two-body collision and by using the relativistic relationships between momentum and total

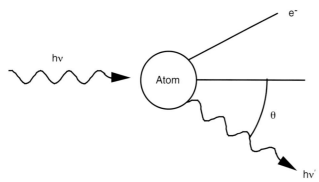

FIGURE 4 Scattering event diagram.

energy (Evans, 1955). In the case of the head-on collision, when $\theta \cong 180°$, the maximum amount of energy is transferred to the electron and the photon shows up in the lower end of the energy distribution (see Fig. 5). In the case of a negligible interaction, when $\theta \cong 0°$, a minimal amount of energy is transferred to an electron; these events represent the photons that are at the higher end of the Compton spectrum. The sharp edge at photon energies corresponding to $\theta \cong 0°$, when observed in a gamma-ray spectrum, is called the *Compton shoulder*. Figure 6 is a plot showing the scattering angle vs. the ratio of $h\nu$ to $h\nu'$ for several gamma-ray energies. The figure demonstrates that for very small angles ($\theta \cong 0°$) there is very little energy lost by the incident photon, whereas at larger scattering angles ($\theta \cong 180°$) the incident photon loses much more energy. Remember that the scattered photon while still in the crystal, is available for additional scattering events and possibly a final photoelectric event. When all the initial photon energy is deposited in the crystal, the sum total of all events related to the initial incident photon will show up in the measurement system as a full energy event and will be indistinguishable from photoelectric events.

The energy of the scattered electron is given by:

$$E_{e^-} = h\nu - h\nu' \qquad (3)$$

By substituting Eq. (3) into Eq. (2), we obtain:

$$E_{e^-} = \frac{(h\nu)(h\nu/m_0c^2)(1-\cos\theta)}{1+(h\nu/m_0c^2)(1-\cos\theta)} \qquad (4)$$

Equation (2) predicts the scattered photon energy; Eq. (4) predicts the kinetic energy of the electron for a given photon-scattering angle. The angular distribution of the scattered gamma-rays is obtained from the *Klein-Nishina formula*, which is written in terms of the differential scattering cross section:

$$\frac{d\sigma}{d\Omega} = Zr_0^2 \left[\frac{1}{1+\alpha(1-\cos\theta)}\right]^2 \left(\frac{1+\cos^2\theta}{2}\right) \left\{1+\frac{\alpha^2(1-\cos\theta)^2}{(1+\cos^2\theta)[1+\alpha(1-\cos\theta)]}\right\} \qquad (5)$$

where Z is the atomic number, α is the familiar term $h\nu/m_0c^2$, and r_0 is the classic electron radius. This expression gives us the probability that a photon will be scattered through a given angle in terms of the photon energy, the electron density, and the photon-scattering angle. Because the expression is derived as the probability for an interaction with an electron, the probability is given in unit area per number of photons. A polar plot of the scattering probability for several medical-imaging photon energies is shown in Figure 7. The greater the incident photon energy, the more likely that the photon will preferentially be scattered in the forward direction. Also, note that as the electron density increases in Eq. (5) with the Z of the material, the probability for a scattering interaction through a given angle increases linearly. Here, the amplitude factor $Z_0 r^2$ is set to unity. Scintillation detector materials and detector designs are influenced by the parameters discussed here.

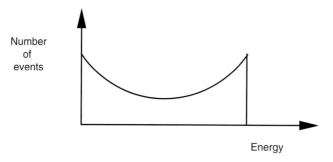

FIGURE 5 Ideal Compton scattering response function.

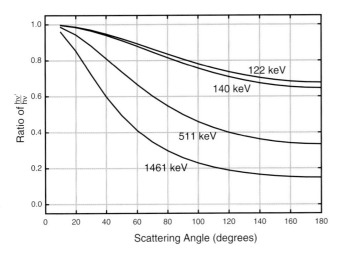

FIGURE 6 Ratio of scattered photon energy to incident photon energy (normalized) plotted against scattering angle.

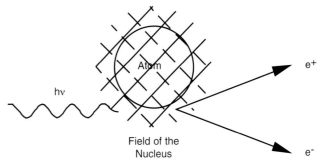

FIGURE 8 Pair production diagram.

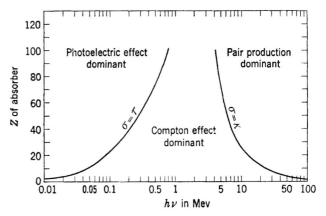

FIGURE 9 Three interaction mechanisms. (From Evans, R.D. "The Atomic Nucleus." Krieger Publishing, Melbourne, FL, with permission.)

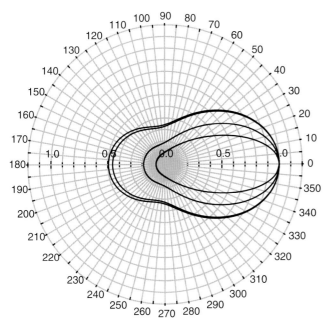

FIGURE 7 Klein-Nishina polar plot. The incident photon is approaching on axis from $\theta = 180°$.

C. Pair Production

In the case in which the incident photon energy exceeds two times the rest mass energy of the electron, pair production can occur; an electron and a positron appear in place of the incident photon. This process must occur in the Coulomb field of a nucleus in order to conserve momentum. Energy is conserved by:

$$E_{e^-} + E_{e^+} = h\nu - 2m_0c^2 \qquad (6)$$

This event is shown schematically in Figure 8. The angular distribution of the created electron-positron pair is predominantly forward for high-energy photons. The positron quickly recombines with any available electron, causing the creation of two 511 keV photons, which are emitted in coincidence and 180° apart. This process is called *positron annihilation*. This effect can cause some interesting effects on the observed gamma-ray spectrum. If one of the 511-keV photons escapes the crystal, then we observe additional peak in the gamma-ray spectrum at the full-energy photopeak minus 511 keV. We can observe another escape photo peak two times 511 keV. Having some knowledge of the radioisotopes present during a measurement reduces our confusion when interpreting a gamma-ray spectrum.

Whereas Figure 2 gives the mass attenuation coefficients for NaI(Tl), Figure 9 shows the incident photon energy with the Z of the material and plots the curve in which the interaction probability, or cross section, for each adjacent effect is equal. The figure demonstrates the regions in which the three interaction mechanisms dominate. Both figures demonstrate the strong dependence on the incident photon energy.

III. THE CHARACTERISTICS AND PHYSICAL PROPERTIES OF SCINTILLATORS

In this section, we first consider the properties that an ideal scintillator might possess. Knowledge of material properties is crucial in determining the appropriate material that will provide the needed performance for a given application.

The ideal scintillation detector material should convert all the incident photon energy into scintillation pulses. The photoconversion should respond linearly as a function of energy. Its light output should be maximized in order to optimize the detector's energy resolution. And the emission wavelength of the scintillator's light should match well with the response wavelength of commercially available photosensitive devices.

An ideal scintillation detector should have good stopping power; consequently a material that has high density and high-Z material would be beneficial. High count-rate applications and fast sampling rates require that the ideal crystal's light emission pulse have a fast rise time and fast decay time with no afterglow. This characteristic effectively removes signal pulse pile-up during electronic signal processing. At the same time, the material should be easy to grow in large volumes to meet the needs of any conceivable application. The ideal material should be nonhygroscopic, making the material easier to handle during manufacturing. The ideal material should be mechanically rugged so that it can withstand both mechanical and thermal shock. The index of refraction should be close to that of glass ($n \cong 1.5$) so that no light loss will occur when coupling the crystal to a photosensitive device. The crystal should have high optical light transmission and no self-absorption. The crystal should perform with a good signal-to-noise ratio and with no intrinsic radioactive background.

In reality, no scintillator behaves as the ideal scintillator described here. The selection of any scintillation material for a specific application entails a compromise in which some properties are optimized and other properties are diminished. NaI(Tl) is the scintillation material that comes closest to meeting most of these requirements; consequently, it is the most widely used.

A. Light Output

The intrinsic physical properties of the scintillating material determine its ability to produce light efficiently. Lempicki *et al.* (1999) discuss three significant quantities that determine a material's potential to have good light output: conversion efficiency, transfer efficiency, and luminescence efficiency. Each parameter influences the number of useful photons created per incident amount of photon energy deposited in the crystal. A detailed presentation of these parameters and how each affects the process of light emission can quickly become very complicated; a more detailed investigation is left to the reader (Rodnyi, 1997; Derenzo *et al.*, 1999).

For the purpose of this discussion, the amount of light that exits the crystal at the interface with the photosensitive device is called the *functional light output*. The functional light output is influenced by the detector design and by the assembly techniques used during the encapsulation of the crystal into the final detector assembly. These detector fabrication issues are discussed in a later section.

1. Luminescence in a Scintillator

Converting gamma-ray energy into useful light pulses can be described using a simple solid-state model that illustrates the band structure of an inorganic scintillator. Birks (1964) and Rodnyi (1997) give good presentations of this model of a scintillator. Figure 10 illustrates the band-gap structure in a scintillation crystal. The filled band represents the lower allowed energy levels that are usually occupied by electrons. The valence band is the highest filled band-gap and is made up of electrons that are effectively bound to the crystalline lattice sites. The conduction band contains mobile electrons after an ionizing event occurs in the scintillator. According to Schrödinger's equation, electrons can only exist in discrete energy levels; consequently, regions called forbidden gaps are essential to the model. In a pure crystal, no electrons would be found in these forbidden gaps.

The energy deposited in the lattice by a gamma-ray creates electron-hole pairs. For every electron that moves into a higher energy level in the conduction band a hole is created in the valence band. The number of electron-hole pairs created per unit of energy deposited by the photon is called the *conversion efficiency*.

When a single electron moves from the conduction band and returns to the valence band, the result is a visible light photon that has energy slightly less than that of the band-gap. A characteristic X-ray emission occurs when electrons move from the higher energy-filled bands to a hole in the lower core bands (not shown in Fig. 10). These X-rays have more energy than visible photon emissions. They are typically reabsorbed within the crystal before they can escape

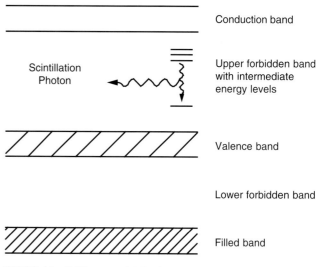

FIGURE 10 Solid-state model showing energy level bands in a scintillator.

and are available to be reabsorbed as part of the total integrated scintillation event in the crystal.

Impurities or lattice defects called *activators* are intentionally introduced into the crystalline lattice so that additional intermediate energy states are created in the forbidden band gap, becoming localized sites throughout the upper forbidden band gap. These defects, called *luminescence centers*, increase electron mobility by shifting the band-gap width enough to put the wavelength of the emitted photons into the visible region of the electromagnetic spectrum. The fractional amount of electron-hole energy that is transferred to the luminescence center is called the *transfer efficiency*. Ideally, transfer energy losses are zero; however, in reality some of the energy loss does occur when the holes or electrons migrate through the lattice or when electron-hole pairs recombine nonradiatively. The efficient transfer of electron energy in the luminescence center requires that holes in the valence band be readily available for recombination with electrons that are trapped in an activation center. If hole mobility is limited, the scintillation rise time will be slow. *Photoluminescence* occurs when photons with energies of less than 6 eV directly excite an activation center.

When a scintillator emits light after absorbing photons, this is called *luminescence*. The amount of thermal energy available in the material affects the luminescent efficiency or luminescent quantum efficiency. Consider the discrete levels present in the crystal lattice. The energy levels of the activation centers are influenced by the amount of thermal energy present in the crystalline lattice; as thermal energy is absorbed by the crystalline lattice, a shift from the equilibrium quantum states results. The result is that nonradiative transitions can occur by heat dissipation rather than by emitting radiation. This effect is called thermal quenching.

Each scintillation material has unique band gaps and activation centers with well-defined energy differences. The electronic transitions that occur between the energy states in the activation center dictate the wavelengths of the light emitted by the crystal. Accordingly, the model explains why we observe that each scintillator material has its own characteristic light emission curve. Good scintillators have emission band gaps that do not overlap with the optical absorption band gaps because an overlap causes the excessive self-absorption of light in the crystal.

Scintillators generate light over a wide range of wavelengths of the visible spectrum. As seen in Table 2, CsF emits in the ultraviolet region ($\lambda \cong 390$ nm), NaI(Tl) emits blue light ($\lambda \cong 415$ nm) at its maximum, and CsI(Tl) emits green light ($\lambda \cong 540$ nm) at its maximum.

The model presented here is specific to crystals that require an activator. These crystals are called *extrinsic scintillators*; examples are NaI(Tl), CsI(Tl), and LSO(Ce). Some crystal materials do not require a dopant to efficiently give off light; they are called *intrinsic scintillators*; examples are BGO, $CdWO_4$, and BaF_2. There are several clear advantages to some of the intrinsic scintillators, including typically higher radiation hardness, avoiding the

TABLE 2 Properties of the Common Scintillation Detector Crystal Materials[a]

Material[b]	Density (g/cm³)	Effective Atomic Number, Z_{eff}	Wavelength of Maximum Emission (nm)	Principal Decay Constant (μs)[c]	Pulse Rise Time (ps)	Index of Refraction at Emission Maximum, n	Hygroscopic
NaI(Tl)	3.67	51	415	0.23	—	1.85	Yes
$Bi_4Ge_3O_{12}$	7.13	76	505	0.30	30 ± 30	2.15	No
CsI(Na)	4.51	54	420	0.63	—	1.84	Slightly
CsI(Tl)[d]	4.51	54	540	0.68	9500, 41000	1.80	Slightly
Lu_2SiO_5(Ce)	7.40	65	420	0.04	30 ± 30, 350 ± 70	1.82	No
CaF_2(Eu)	3.19	17	435	0.9	40 ± 30	1.44	No
^{6}LiI(Eu)	3.49	54	470	1.4	—	1.96	Very
BaF_2[e]	4.89	53	310, 220	0.62, 0.0008	—	1.49	No
CsF	4.11	53	390	0.004	—	1.48	Very
$CdWO_4$	7.90	64	480	5.0	—	2.20	No
$GdSiO_5$(Ce)	6.71	59	430	0.06	—	1.85	No

[a]Data from Knoll (1989); *Harshaw Scintillation Phosphors* (1975); Melcher (2000); Derenzo et al. (2000).
[b]$Bi_4Ge_3O_{12}$, Lu_2SiO_5(Ce), and $GdSiO_5$(Ce) are often referred to as BGO, LSO, and GSO in the literature.
[c]Data are for room temperature.
[d]CsI(Tl) is better suited to be used with a photodiode because its emission spectrum does not match well with the response spectrum of a photomultiplier tube.
[e]BaF_2 has two dominant peaks in its emission spectrum. When two values appear in this row, the value for the faster scintillating component is given second.

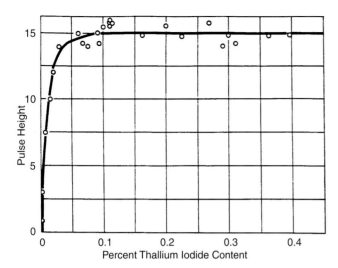

FIGURE 11 Crystal light output as a function of Tl activator concentration in NaI(Tl).

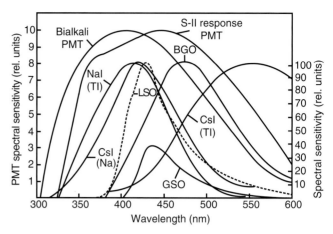

FIGURE 12 Light output intensity for various scintillators and PMT response versus wavelength.

problem of nonuniform distributions of the activator dopant, and good thermal stability in the scintillation characteristics.

The concentration and the distribution of the activator in an extrinsic scintillation crystal affects the light output and detector performance. The activator concentration is important because a threshold amount is necessary to enhance electron mobility and provide uniform light output in the bulk material. Figure 11 shows the activator concentration of thallium in NaI(Tl) as a function of crystal light output (Harshaw *et al.* 1952). The concentration of the activator throughout the scintillator should be as uniform as possible because small localized variations in the light output directly affect the energy resolution of the scintillator.

There are several mechanisms for light output in scintillators; the role of activation centers in an extrinsic scintillator is discussed here. The scintillation mechanism for intrinsic scintillators is described in Rodnyi (1997).

The energy difference observed in the activator band gap of an extrinsic scintillator is related to the wavelength of the photon emitted during the electron transition from the most excited state to the ground state. The maximum energy of the emitted photon can be calculated from the energy-wavelength relationship:

$$E_{max} = h\nu = \frac{hc}{\lambda_{max}} \qquad (8)$$

where λ_{max} is the maximum wavelength observed in the emission spectrum. Using data for NaI(Tl) taken from Figure 12, a quick calculation yields a maximum single-photon energy of 3.8 eV. As expected, the value is less than the band-gap value of 5.9 eV reported in the literature (Lempicki *et al.*, 1999). The difference between these values is the remaining energy difference between the valence and conduction bands.

Figure 12 shows emission curves that were determined experimentally for several common scintillation detector materials along with the response curves for two different photocathodes. The observed peak emission wavelength represents the most probable transition between the band gaps in the activation center.

It is known that band gaps are discrete quanta; however, we see that the measured emission spectra are smooth curves. The reason for this is that the band-gap structure is in fact more complicated than can be explained by the simple model presented here.

2. Scintillation Conversion Efficiency

We have seen that the electron-hole pairs are created in the crystal as the photon deposits energy into the crystal. Only a certain percentage of incident photon energy is actually converted into electron-hole pairs and, eventually, scintillation light. During the process, some of the energy is lost to X-ray emission, impurity quenching, concentration quenching, and (largely) to internal heat (phonons).

Lempicki *et al.* (1999) express light output L, the number of photons per million electron volts, as:

$$L = \frac{10^6}{2.3 E_g} \beta S Q \qquad (9)$$

where β is the conversion efficiency in number of electron-hole pairs created per million electron volts of incident ionizing radiation, S is the efficiency with which the electron-hole pairs transfer energy to the luminescence centers, and Q is the luminescence center quantum efficiency in its excited state. All these parameters have been examined here. The term, $\beta S Q$ is called the *total scintillation efficiency* and sometimes called η. This expression for light output shows that a good scintillator must have (1) efficient creation and trapping of electron-hole pairs, (2) enough electron-hole mobility to allow recombination, and (3) minimal thermal quenching of the excited states.

Because the light output is inversely related to the band gap, we expect the light output to increase continually as

TABLE 3 Additional Properties of the Common Scintillation Detector Crystal Materials[a]

Material[b]	Band-Gap Width (eV)	Total Light Yield (photons/MeV)	Absolute Scintillation Efficiency for Fast Electrons (%)	Output Relative to NaI(Tl) on a Bialkali PMT	Measured Energy Resolution at 662 keV	Calculated PMT Resolution from Photoelectron Yield	Calculated Limiting Resolution at 662 keV
NaI(Tl)	5.9	37,700	11.3	1.00	6.5	3.1	5.7 ± 0.2
$Bi_4Ge_3O_{12}$	5.0	8200	2.1	0.13	9.3	8.1	4.2 ± 0.4
CsI(Na)	6.4	38,500	11.4	1.11	7.4	3.3	6.6 ± 0.3
CsI(Tl)[c]	6.4	64,800	14.9[d]	0.49	7.3	4.3	5.9 ± 0.3
Lu_2SiO_5(Ce)	6	30,000	—	0.75	7.9	4.4	6.6 ± 0.4
CaF_2(Eu)	12.2	23,650	6.7	0.78	—	—	—
^{6}LiI(Eu)	6.1	11,000	2.8	0.23	—	—	—
BaF_2[e]	10.6/18.0	≈9950	4.5	0.13	7–8	6.2	4 ± 1
CsF	9.8	—	—	0.05	—	—	—
$CdWO_4$	—	15,300	3.8	0.18	6.8	5.2	4.4 ± 0.4
$GdSiO_5$(Ce)	—	10,000	—	0.25	7.8	6.2	2.7 ± 1.0

[a]Data from Knoll (1989); *Harshaw Scintillation Phosphors* (1975); Melcher (2000); Lempicki *et al.* (1999); Holl *et al.* (1988); Valentine *et al.* (1993); Dorenbos *et al.* (1975). PMT, photomultiplier tube.
[b]$Bi_4Ge_3O_{12}$, Lu_2SiO_5(Ce), and $GeSiO_5$(Ce) are often referred to as BGO, LSO, and GSO in the literature.
[c]CsI(Tl) is better suited to be used with a photodiode because its emission spectrum does not match well with the response spectrum of a PMT.
[d]Calculated based on total light yield (from Valentine *et al.*, 1993).
[e]BaF_2 has two dominant peaks in its emission spectrum. When two values appear in this row, the value for the faster scintillating component is given second.

the band gap decreases. In reality, we observe that as E_g decreases, Q also decreases because nonradiative transitions dominate the transfer of energy and the scintillator begins to self-absorb its light. Consequently, Eq. (9) is a good first-order approximation of the model, but other higher-order factors are required in this expression to correct for these physical characteristics.

The fractional amount of energy deposited into the crystal that eventually becomes scintillation light is the absolute scintillation efficiency (See Table 3). The measurements reported here were done using a photodiode because it has wide spectral response (Holl *et al.*, 1988; Sakai, 1987). Previously reported measurements were done using a photomultiplier tube (PMT) with an S-11 response curve (see Fig. 12). Notice that there is poor wavelength matching between the PMT response curve and CsI(Tl)'s light output curve.

Robbins (1980) shows that a minimum energy of $\xi_{min} = \beta E_{gap}$ ($\beta = 2.3$) is required to create a single electron-hole pair. Van Eijk (2000) reports $\beta \cong 2$–3 is possible, depending on the crystal material. For NaI, which has a band gap of 5.9 eV (Lempicki *et al.*, 1999), the energy per electron-hole pair is approximately 15.3 eV. By calculation, the maximum number of electron-hole pairs created by a 511-keV gamma-ray absorbed in the crystal is approximately 3.3×10^4. Using the absolute scintillation efficiency of NaI(Tl) given in Table 3 of 11.3%, for the same 511-keV gamma-ray we calculate approximately 1.9×10^4 photons with an average energy of 3 eV. This gives a ratio of 0.57 photons created for every electron-hole pair. This ratio is a measure of the ability of the scintillator to transfer energy to the activation centers, the transfer efficiency. The number calculated here agrees closely with the value of 0.59 derived by Lempicki *et al.* (1999).

3. Thermal Effects on Light Output

As previously discussed, the scintillator luminescent quantum efficiency, Q, is affected by thermal quenching in the excited states of the activation centers. Experimentally it is observed that the light output of a scintillator varies with temperature (Valentine *et al.*, 1993; Kobayashi *et al.*, 1989: Harshaw Radiation Detectors, 1984). Figure 13 shows the measured light output (normalized) of several scintillators as a function of temperature. For each material, notice the temperature at which the light output maximum occurs. Also, note the rate of change of the scintillator's light output around room temperature for each material shown. In some cases, it is possible for a crystal to have a temperature gradient over its volume. As this occurs, the light output throughout the volume of the crystal varies and the energy resolution of the detector will degrade.

4. The Scintillation Light Pulse

Each crystal material emits a well-defined light pulse during the scintillation process. Some scintillator materials have several components to the light emitted. A typical pulse shape is shown in Figure 14. The rise times and decay

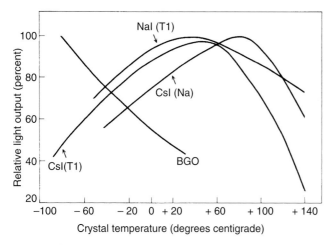

FIGURE 13 Measured light output of several crystal materials as a function of temperature.

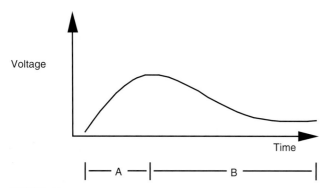

FIGURE 14 Typical scintillation pulse as seen at the PMT anode.

times for each material are well known. Table 2 gives values for the rise time and decay time of the common scintillation materials. Pulse rise times are expressed as the portion of the pulse between 10 and 90% of the total pulse rise time. Pulse decay times are expressed as the amount of time it takes to reach $1/e$ of the total duration of the decay.

The amplitude of the pulse shown in Figure 14 is dependent on several factors:
1. The amount of energy deposited in the crystal.
2. The factors influencing light output in the scintillator.
3. The factors influencing the functional light output of the scintillator.
4. The light collection efficiency of the photodevice coupled to the scintillator.
5. The signal processing capability of the system electronics.

It was stated earlier that in an ideal scintillator the light output would be in the form of a delta function (see Fig. 3). A counting system could then easily discriminate between two slightly different photon energies. Because the photopeak width varies with the parameters given here, obtaining an ideal spectrum is not possible.

Materials that have fast rise times and fast decay times are useful in applications in which high count-rates or timing is important (e.g., positron emission tomography, PET). Narrower pulse widths allow faster data acquisition to occur at higher rates. For example, during PET imaging less imaging time is required, providing some attractive benefits, including better image quality due to less patient motion, less patient discomfort, and more patient throughput.

The rise time of the light pulse follows the familiar exponential dependence:

$$I_A(t) = -I_0 e^{-\lambda_r t} \quad (10)$$

where λ_r is the characteristic time that it takes to occupy the activation centers for a given crystal material.

The decay time for all scintillators is greater than the rise time. The decay of the light pulse follows also follows an exponential dependence:

$$I_B(t) = -I_0 e^{-\lambda_d t} \quad (11)$$

where λ_d is the characteristic transition probability for the crystal material. The total waveform shape can be approximated by the sum of these two expressions.

Some scintillators have two or more components to the decay time. Ideally, decay time should be short in high-count-rate applications so that electronic pulse pile-up does not occur. Pulse pile-up occurs when two discrete pulses are superposed on top of one another in order to change the shape of the observed pulse. By taking the inverse of the total pulse duration, we obtain an idea of the count-rate limitation for each material. NaI(Tl) is count-rate-limited to approximately 700 kHz, whereas LSO is count-rate-limited at over 2000 kHz.

A closely related concern is scintillator afterglow. Afterglow is a postluminescence in the scintillation material following the removal of a radiation source. Scintillator afterglow results from high concentrations of unwanted defects and impurities in the crystalline lattice. These defects and impurities create luminescence centers in the scintillator, where electrons or holes can be trapped. The light pulse emitted from these anomalous centers typically has a longer decay time. The longer decay time is a problem because it contributes to DC baseline shift and pulse pile-up.

The light output associated with an allowed transition typically occurs within 10^{-8} s after excitation; this type of radiation is referred to as *fluorescence*. After observing the decay times given in Table 2, it is reasonable to inquire why the pulse decay times are reported in microseconds. The explanation is in how the quantum states are formed and what transitions are allowed. The transitions that occur through an impurity or defect site are associated with metastable transitions that are much longer decay times; this type of radiation is called *phosphorescence*.

The decay time of the scintillation pulse is effected by the ambient temperature. Measurements taken (Schweitzer and Ziehl, 1983) show that the decay time for NaI(Tl) decreases by a factor of 4 in the temperature range from

−25°C to +180°C. To further complicate matters, the activator concentration has an effect on the decay time. Eby and Jentscheke (1954) demonstrated the time dependence of the decay with the thallium concentration in NaI(Tl) to vary as much as 50% from the value of 230 ns given in Table 2.

B. Scintillator Energy Resolution

Scintillators are mainly used because of their ability to provide reasonable energy resolution and detection efficiency conveniently and at a reasonable cost. The following section on measurements discusses in more detail how applications are affected by pulse-height resolution (PHR). The previous discussion has shown that the most dominant factor in achieving good PHR is photon yield.

The energy resolution of the detector-PMT apparatus can be written as:

$$R^2 = R_t^2 + R_i^2 + R_n^2 + R_{PM}^2 \quad (12)$$

Where R_t is the scintillator transfer efficiency, R_i is the measure of the inhomogeneities in the scintillator, R_n is the nonproportional response of the scintillator, and R_{PM} is the photomultiplier tube resolution. As R decreases, the energy resolution improves because the system is then capable of separating gamma rays of different energies. The first three terms in the expression incorporate the contribution of the scintillator and the last includes the contribution of the photomultiplier tube. As each one of these three terms decreases, the energy resolution improves, as expected.

The transfer efficiency term includes the factors that affect how the light that is created in the crystal is converted to electrons in the PMT. Examples of these factors include light transmission in the crystal, nonuniform light collection due to geometric effects, nonuniform surface preparation of the scintillator, nonuniform reflector performance, improper light coupling between the crystal and the PMT, photocathode nonuniformities, poor crystal-PMT wavelength matching, and other PMT response nonuniformities.

The second term in Eq. (12), R_i, is related to the nonuniform distribution of luminescence centers in the crystal. For NaI(Tl), the Tl concentration can vary to contribute to localized variations in light output. Imperfections in the crystal lattice introduced during crystal growth, such as, flock, haze, bubbles, and other point defects, cause variations in the localized light output. Again, these factors degrade the PHR of the detector.

The third term in Eq. (12), R_n, includes the factors that cause the scintillator to respond nonlinearly as a function of energy deposited in the crystal. Numerous studies (Valentine et al., 1993; Kobayashi et al., 1989; Harshaw Radiation Detectors, 1984; Zerby et al., 1961; Narayan and Prescott, 1968; Meggitt, 1970; Dorenbos et al., 1975; Schweitzer and Ziehl, 1983; Eby and Jentschke, 1954; Prescott and Narayan, 1969; Valentine and Rooney, 1994; Fonte et al.,

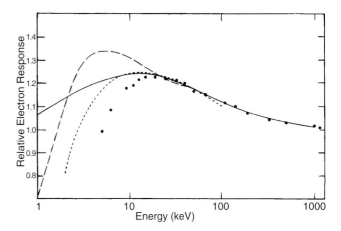

FIGURE 15 Electron response curves.

1991; Iredale, 1961; Murray, 1975) report experimental results that have been obtained for NaI(Tl) crystals. By impinging electrons of various energies onto the crystal the light output response function can be determined.

Prescott and Narayan (1969) present data showing that the measured electron response curve varies as a function of electron energy in NaI(Tl). The curves given in Figure 15 show that there is a nonlinear electron response over the energy range of 1 keV to 1 MeV. There is reasonably good flat response over the range from 10 to 1000 keV. Taking into account the various interaction mechanisms for a 662-keV photon in NaI(Tl), the electron response can vary as much as 50%. The nonlinear response to electrons in NaI(Tl) does degrade the intrinsic energy resolution of the scintillation detector. Dorenbos et al. (1975) review the response data from several scintillators including NaI(Tl), CsI(Tl), CsI(Na), BGO, CaF$_2$(Eu), CdWO$_4$, BaF$_2$, and LSO(Ce).

The electron response is a measure of mean light yield per unit of electron energy deposited in the crystal. The experimental curves given in Figure 15 show this ratio plotted against energy. This nonproportional or nonlinear response occurs as a result of the statistical process during the creation of secondary electrons. Recall that the incident photon energy with E_γ less than 1.0 MeV transfers energy to the electrons in the crystalline lattice by several possible combinations, including one or more Compton scattering events, a photoelectric event, and X-ray or Auger transitions. In this process a spectrum of electrons of varying energies is obtained. Because the experimental data show us that the total amount of energy absorbed for each incident photon varies, the light yield varies depending on the way in which the photon transfers its energy.

Another contributor to the nonproportional response in the scintillator occurs at the surface of the crystal. Some of the energy may be lost during multiple interaction events due to inefficient transfer or energy loss at the surface of the crystal (Meggitt, 1970).

The last term in Eq. (12), the PMT term, R_{PM}, includes statistical fluctuations due to the number of photoelectrons created at the photocathode. R_{PM} is related to the photon yield N by $1/\sqrt{N}$. It is obvious that as the photon yield increases the system resolution decreases, as expected.

The measured energy resolution for several scintillators is shown in Table 3. The limiting values for each scintillator were determined by Dorenbos *et al.* (1975) as follows:

$$R_{s\ limit} = (R^2_{s\ measured} + R^2_{PM\ calculated})^{1/2} \quad (13)$$

where $R_{PM\ calculated}$ is determined from experimentally determined photoelectron yields while correcting for variations in the PMT gain.

Of the three terms found in Eq. (12), R_t the transfer efficiency, contributes the greatest amount of spread to the photopeak width in NaI(Tl). The nonuniformities mentioned are all dominated by statistical fluctuations during the collection of the photoelectrons. As a result, the gamma-ray energy resolution is approximately inversely proportional to the square root of the photon energy. Statistical fluctuations alone indicate that the PHR for NaI at any energy is a straight line represented by:

$$\ln R_{s\ measured} \cong \ln K - 0.5 \ln E \quad (14)$$

where K ($\cong 0.14$) is a proportionality factor and E is given in units of mass times the square of the speed of light ($m_0 c^2$). Experimentally determined values for NaI(Tl) obtained by Beattie and Byrne (1972) were shown to be in good agreement with Eq. (14), except that the slope is not as steep as expected because other factors contribute to the widening of the photopeak. These data were gathered in 1972. Today the results follow the same relationship; however, the PHR for NaI(Tl) is much better than that reported by Beattie and Byrne.

C. Material Density

Scintillating materials attenuate gamma-rays by the following expression:

$$I(x) = I_0 e^{-\mu \rho x} \quad (15)$$

This expression is sometimes called the *pencil beam equation*. The initial intensity of a collimated monoenergetic photon beam is reduced by the exponential correction factor, which is dependent on the mass attenuation coefficient (μ), the density of the material (ρ), and the thickness of the material (x). The mass attenuation coefficient is a sum of the interaction mechanisms described earlier in this chapter. The stopping power of the material is greatly increased with small increases in the density or the thickness of the absorber. The effective atomic number, shown in Table 2, is good indicator of a scintillator's stopping power. For photons interacting at photoelectric energies, the effective atomic number is calculated by:

$$Z_{eff} = \left(\frac{\sum_i w_i A_i Z_i^4}{\sum_i w_i A_i} \right)^{1/4} \quad (16)$$

where w_i is the weighting factor, A_i is the atomic mass, Z_i is the atomic number, and the summations are over all of the molecular constituents.

The 1D expression given in Eq. (15) is useful for determining a first-order approximation of detector efficiency. Integrating over the all the potential photon path lengths and assuming equal weighting provides a more accurate result. The most accurate prediction involves a method of calculation called a *Monte Carlo simulation* (see Chapter 25 in this book).

Recall from the earlier discussion that the gamma-ray interaction probability is strongly dependent on both the electron density of the crystalline structure and the atomic number of the nucleus. The greater the number density of the electrons in the crystalline lattice, the more likely it is that the impinging photon will be attenuated by Compton scattering. The greater the atomic number, the greater the probability that a photon will be absorbed in the photoelectric process. The densities and the effective Z of several materials are given in Table 2.

The plot given in Figure 16 shows the attenuation curves for the 140 keV gamma ray of Tc-99m and the 511 keV gamma ray of Na-22 in NaI(Tl) and BGO. These results were obtained using the pencil beam expression. A SPECT (single-photon emission computed tomography) gamma camera's typical thickness is 9.5 mm. The plot shows that 90% of the 140-keV photons of Tc-99^m are attenuated in the NaI(Tl) crystal. A PET imaging device with a 19-mm-thick NaI(Tl) crystal attenuates 45% of the 511-keV gammas, whereas the same thickness of BGO attenuates 85% of the 511-keV photons.

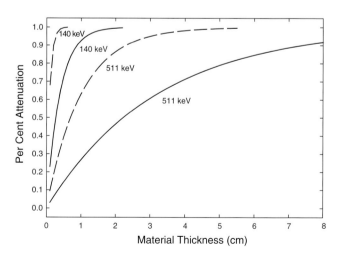

FIGURE 16 Attenuation curves for the 140 keV gamma ray of Tc-99m and the 511 keV gamma ray of Na-22 in NaI(Tl) (solid line) and BGO (dashed line).

A more detailed calculation takes into account the fact that the detector housing, or energy entrance window, reduces the number of photons available for detection according to the density and mass attenuation coefficient of the housing material. Typically, scintillator housings are built with thin-walled aluminum, so that the signal from the source is not reduced significantly. Aluminum is a low-Z material ($Z = 13$); however, for very low energy gamma-counting applications in the range of 5 to 30 keV, counting efficiency is increased substantially by using a thin piece of beryllium ($Z = 4$) as the energy entrance window.

D. Optical Properties

Several optical properties influence the performance of scintillators. The discussion here refers to the optical properties of light transmission, light absorption, and light reflection.

In order to optimize the performance of the scintillator, it is important that nearly all the light that is created in the scintillator is detected at the photocathode of the PMT. The light that is created in the scintillator must not be self-absorbed by the crystal; that is, the crystal must be transparent to its own light. As previously stated, good scintillators have emission band gaps that do not overlap with the optical absorption band gaps because an overlap causes excessive self-absorption of light in the crystal. Self-absorption can occur due to unwanted impurities quenching the light output of the scintillator. These impurities may be present in the original growth material or they may be introduced into the crystal during growth. Typically, elemental impurities are controlled to part-per-million concentrations.

Visual inspections for optical clarity often give a good indication of the quality of the crystal material. Materials that transmit in the visible regions of the spectrum and are water white in general perform quite well. Some scintillation materials such as LiI and ZnS must be used in thin sections because they have significant self-absorption.

Additional light losses can occur because the scintillation light reflects on the inside surface of the crystal. During the manufacturing process these crystal surfaces are prepared to minimize light loss at the surface and enhance uniform collection at the photocathode.

It is interesting to note that the optical properties of scintillators typically do not depend on the lattice orientation. Only in the case in which significant impurities have built up around domain boundaries does this become an issue. In particular, the image quality in gamma-camera plates is sometimes degraded when the light output across a grain boundary is affected.

The light available to enter the photocathode is influenced by the index of refraction of the material. The index of refraction (n) of air is close to unity, the index of glass is typically between 1.5 and 1.7, and the index of fused quartz is approximately 1.47. The index of refraction for most scintillation materials is between 1.44 and 2.20 (see Table 2). Remember that, if the index of refraction does not match well, then one component of the light is reflected and one component is refracted at the interface. This can either be beneficial or detrimental to the process of maximizing the light collected at the photocathode.

Total internal reflection of the light occurs in a material at the critical angle or Brewster's angle, which is given by:

$$\theta_c = \sin^{-1}(n_2/n_1) \quad (17)$$

where n_1 is the index of refraction of the material in which the light originates, and n_2 is the index for the material across the interface. That is, if the angle of incidence, as measured from the normal, is greater than θ_c, then all light incident at these angles remain in the crystal. NaI(Tl) has a critical angle of 32.7°, whereas BGO has a critical angle of 27.7°. Less light escapes a crystal with a higher index of refraction, unless some design feature is used to improve the light coupling to the PMT.

For angles of incidence less than the critical angle, it is important that a good diffuse reflector is used to return the light into the crystal. Reflectors that are used typically have coefficients of reflectivity greater than 0.95 at the wavelengths of concern. Some of the common materials used are Teflon, Al_2O_3, MgO, and high-reflectance papers. In the construction of scintillation detectors, the interface is actually crystal-air-reflector. Noting this is important. If the material adjacent the crystal has an index near that of the crystal material, then the light will more easily be coupled away from the crystal and some percentage of it lost due to the use of imperfect reflector materials.

At the optical interface to the PMT, ideally all incident light should pass directly through to the photocathode. Thin films of a silicon elastomer ($n \cong 1.43$) are used most often to glue the crystal to the PMT. These interface materials efficiently couple the light from the crystal to the PMT and they are selected because of good transparency at the appropriate wave length and good refractive-index match.

E. Mechanical Properties and Intrinsic Background

The mechanical properties of scintillation crystals have been characterized and reported (*Harshaw Radiation Detectors*, 1984; NASA, 1973, 1974, 1980; Ishii and Kobayashi, 1991). Some useful mechanical properties are given in Table 4.

Some crystal materials have unavoidable intrinsic background. A good example of this is the material LSO. The isotope of ^{176}Lu in LSO has a natural abundance of 2.6%, and has gamma-ray energies at 89, 202 and 307 keV. These photons contribute approximately 300 cts/s-cc of LSO material. In PET applications, the energy of interest is around 511 keV, so the background signal in LSO is not a problem.

TABLE 4 Useful Mechanical Properties of the Some of the Common Scintillation Detector Crystal Materials[1]

Material	Hardness (Mohs)	Cleavage Plane	Thermal Coefficient of Linear Expansion (%/°C × 10^{-6})	Melting Point (°C)	Radiation Hardness (rad)	Machinablitiy
NaI(Tl)	2	100	47.4	651	10^3	Carbide tools on standard lathes and mills
$Bi_4Ge_3O_{12}$	5	None	7	1050	10^{4-5}	Diamond tools at high speed
CsI(Tl)	2	None	54	621	10^3	Carbide tools on standard lathes and mills
CsI(Na)	2	None	54	621	—	Carbide tools on standard lathes and mills
CaF_2(Eu)	4	111	19.5	1418	—	Diamond tools at high speed
BaF_2	3	111	18.4	1354	10^{6-7}	Carbide tools on standard lathes and mills
LiI(Eu)	2	100	40	446	—	Carbide tools on standard lathes and mills
CsF	2	100	31.7	682	$<10^4$	Carbide tools on standard lathes and mills
$CdWO_4$	4–4.5	010	10.2	1325	10^3	Diamond tools at high speed
LSO	5.8	None	—	2070	10^{6-7}	Diamond tools at high speed

[1] Harshaw Radiation Detectors (1984); NASA 1973, 1974, and 1980; Kobayashi *et al.*, 1993.

Standard NaI(Tl) crystals, with less than 1 ppm potassium, exhibit some background due to a background contribution from ^{40}K. ^{40}K has a natural isotopic abundance of 0.01% and E_γ = 1461 keV with a branching ratio of 11% and E_{β^-} end-point energy of 1314 keV and a branching ratio of 89%. An activity calculation yields approximately 0.02 ^{40}K events/min/cc in NaI(Tl). When placed in a lead safe for several hours, a very large-volume crystal ($\cong$ 4000 cc) made from standard material exhibits a ^{40}K gamma spectrum. To reduce the amount of ^{40}K present in a crystal used in very low-background-counting applications, the crystal material can be grown twice. Assemblies that are carefully constructed with low-background materials can achieve count rates as low as 5 cps in the energy window of 200–3000 keV.

BGO can also have a background count rate as high as 7 events/s/cc if great care is not taken to ensure that the starting growth material is not contaminated with ^{206}Pb. Cosmic protons that transmute ^{206}Pb to ^{207}Bi, which has a $T_{1/2}$ = 38 years, will contribute gamma background at energies of E_γ = 570, 1060, 1630 (sum), and 2400 keV. If the starting material is obtained from lead-free ores, this contamination is not present.

IV. SCINTILLATION DETECTORS: DESIGN AND FABRICATION

The optimal design of any scintillation detector depends on a good understanding of physical properties and characteristics and their relation to scintillator performance. It is likely that several design iterations will occur on paper before a final design is found. The final detector design invariably is a compromise of the ideal design. This section discusses various design issues and then describes how a scintillation detector is manufactured.

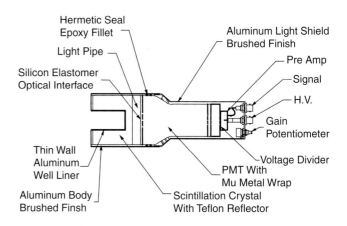

FIGURE 17 Scintillation detector diagram.

A. Detector Design

Figure 17 shows an integral-type well detector, which has the PMT mounted integrally to the crystal. Figure 17 shows all the other major detector components, including the reflector, silicon elastomer optical interface, hermetic seal, light shield, voltage divider, preamp, high-voltage input, and signal output. Well detectors are used in applications in which an increased solid angle is important to increase counting efficiency. Two common applications for well detectors are environmental sample counting and radioimmunoassay sample preparations.

The previous section explained in detail the importance of the light output of the scintillator. Because most scintillator applications involve spectroscopy, the most important design criterion is the energy resolution of the detector. Each experiment or application has its own set of design issues; the following provides a good starting point:

1. *At what energy is the gamma ray of interest? What is the anticipated signal strength (activity) of the sample to be measured?* The mass attenuation curves are a useful tool because we can determine an optimal thickness for the scintillator. To ensure that reasonable counting times are obtained, the geometry must be selected to achieve good counting efficiency. At the same time, too much material may cause unwanted background to be introduced into the spectrum and reduce the signal-to-noise ratio (SNR). If the gamma-ray energy is less than 100 keV, then the count rates and the total counting times will be improved if a low-Z material, such as a thin piece of beryllium, is used as the energy entrance window. At higher energies, gammas have a higher penetration probability, so there is much less signal lost to absorption in the crystal housing.

2. *Are there multiple gammas in the sample? If so, what PHR is required to separate the peaks?* These questions are important because the energy resolution must be good enough to separate the photopeaks at a reasonable confidence level. The design may require the selection of premium crystal material during the manufacturing process to ensure that adequate PHR is achieved. The best NaI(Tl) detectors that are produced will achieve slightly better than 6.5% at 662 keV. The selection of premium material will add cost to the detector. A good rule is to maintain a geometric aspect ratio of 1:1 to obtain the best energy resolution.

3. *Is there significant gamma background that will degrade the signal-to-noise ratio?* If significant background is present, the designer should consider shielding the detector or collimating the signal to improve the SNR. In some multiple-detector configurations, tungsten septa are placed between the individual elements to reduce cross-talk from Compton scattering. At the same time, the crystal geometry should be carefully selected. It should be optimized to the gamma energy of interest and to account for any directionality of the source.

4. *Where will the detector be used?* If the detector is used in laboratory conditions, the design requirements are much less stringent than if the detector is going to be incorporated into a space mission. Typically, detectors are manufactured to meet a certain level of shock and vibration specifications because the assembly must withstand handling during the shipping process. All designs must meet minimum mechanical and thermal-shock protection specifications, and they are warranted to meet these specifications. Other environmental conditions that affect the design are humidity and pressure.

5. *Is detector compactness important?* Some applications minimize the amount of material between multiple detector segments because the scientist is concerned about the loss of good signal. Introducing unwanted background through gaps in shield detectors can also be a concern. If compactness is a significant issue, the denser scintillation materials should be considered. The first PET imaging systems were built using NaI(Tl) crystals. Eventually, BGO and LSO became popular because they improved the overall performance due to their higher density. The higher density means more detector elements per solid angle subtended, better stopping power, and improved spatial resolution. An added benefit is that BGO and LSO are nonhygroscopic, which removes some significant engineering problems associated with providing a good hermetic seal.

6. *Should the PMT be demountable?* Some uses require that the crystal be demountable from the PMT assembly. This is useful when it is anticipated that damage can occur to either the crystal or the PMT; then only one component has to be repaired. This is especially true for detectors that are used in the field. When the assembly is designed with a demountable PMT, the PHR will degrade slightly due to the additional interface material and glass between the crystal and PMT photocathode.

7. *Will the application require multiple detectors?* At some point, the designer must consider the trade-offs between a large detector, multiple detectors, or detectors with multiple elements. The size of any detector is limited by the capabilities of the crystal-growth manufacturer. As expected, a larger crystal element increases the cost of the crystal. Some of the large crystal cost, which might have only one PMT, is offset by the cost savings associated with the cost of gain-matching the elements, the cost of the electronics to process the signal from multiple PMTs, or the cost of multiple detectors. Assembly costs also contribute to determining the best approach to minimize the cost of the detection system. Some detectors can require multiple, optically isolated elements to achieve the intended use.

8. *Is the scintillator pulse shape important to the application?* Some applications use the fast rise time as a technique for counting events. These applications, which are referred to as timing measurements, depend on fast electronics to obtain good data. Applications with high count rates can be concerned with the amount of afterglow present in a scintillator because afterglow can cause pulse pile-up. Afterglow lengthens the decay time due to impurities in the crystal material. At high count rates, the lengthened pulse shapes overlap and lead to pulse pile-up.

9. Finally, economics must be considered. Each design decision has a cost impact and a cost-benefit analysis should be done to determine the usefulness of all of the design options.

These design questions are typically reviewed by the manufacturer and customer before a project is initiated.

B. Detector Components

Each component that goes into a scintillation detector has numerous properties that can affect the overall performance of the detector. These component issues are continually being investigated to improve the quality of the product.

1. Crystal

Previous discussion has dealt with the many parameters important to obtaining the best performance crystal material can provide. To achieve the highest quality scintillation material, the starting material used to grow an ingot must be the highest purity available. Crystal growers along with material scientists are continually looking for ways to improve the quality of the scintillator by removing impurities from the starting material. The techniques used to grow the crystal ingots are also improved on with research and experience. Each material has quality issues that are affected by the growth process. Extrinsic scintillation materials must limit the activator or dopant nonuniformities. Gradients in the activator distribution cause nonuniform light output and the performance of the detector is degraded.

2. Reflector

A good reflector has high reflectivity at those wavelengths in which the scintillator emits light. It has the ability to form itself closely to the surface of the crystal. However, the reflector should not wet the surface because this will pipe light away from the crystal. Historically, MgO, Al_2O_3, and TiO_2 were used as reflector materials. Powders are difficult to dry, are messy, and do not have the best reflectance. Powders have been replaced with Teflon, which is hygrophobic, easy to form around the crystal, and not messy. Other paper reflectors are used in applications in which the Teflon is not stiff enough to hold its form.

3. Interface Materials

The best interface materials have the following properties: good light transmission at the wavelengths of interest, good index of refraction match to the crystal, good adhesive and cohesive strength, good function over a large temperature range, good electrical characteristics, a reasonable coefficient of thermal expansion, easy handling during assembly, and low cost. Both hard interfaces and soft interfaces are used, depending on the application. The hard interface materials are two-part epoxies, and the soft interface materials are two-part silicon elastomers. Applications in which the detector will experience a wide range of temperatures require that the interface, usually a silicon elastomer, be under constant pressure. A crystal grows and shrinks at greater rates than stainless steel because it has a much greater coefficient of thermal expansion. Partial interface separation, causing light loss, does not occur when the detector packaging keeps the interface under pressure.

4. Light Pipes

A glass or quartz window is sometimes placed between the crystal and PMT to improve PHR performance by reducing the nonuniform light response from the crystal or nonuniform response in the photocathode. In counting applications characterized by low count rate and low gamma-ray energy, a light pipe is sometimes used to shield a thin ($\approx$1/16-in-thick) crystal from the K-40 background that is present in some PMT glass. Light pipes can vary in thickness from 0.5 to 2 in thick. Quartz is more expensive to use than Pyrex or borosilicate glass, but it has better light transmission characteristics.

5. Crystal Housing

Crystal housings are usually made out of metals such as aluminum or stainless steel. Aluminum alloys are more machinable and are less expensive than stainless steel. Cylindrical crystal housings are usually spun or formed out of aluminum; however, stainless steel can also be spun. Stainless steel has the advantage that it is more rugged. In some low-background counting applications, it is used because it has less radioactive contaminants than aluminum. In ultralow-background counting applications, oxygen-free hydrogenated copper is used because of its extremely low background count rate.

6. Hermetic Seals and Light Seals

All hygroscopic crystal materials (see Table 2) require a reliable seal to keep the crystal from hydrating. Most detector seals use an epoxy filler between the PMT glass and the crystal housing. All-welded assemblies and special glass-to-metal seals are also very reliable hermetic seals in those applications in which the detector can experience severe mechanical shock and vibration.

As always, great care must be taken to minimize any extraneous background due to contaminated components. When the welded-seal technology was first introduced, it was determined that welding rods contaminated thorium introduced unwanted background into the spectrum. Today, tungsten inert gas welding is used.

Light seals also have an important role in a good detector assembly. If small amounts of ambient light leak to the photocathode, extraneous noise is introduced into the signal and damage may occur to the PMT.

7. Photomultiplier Selection

PMT characteristics that must be considered during the selection process include energy resolution (quantum efficiency), linearity, long- and short-term stability, electronic noise level, gain, photocathode uniformity, photocathode wavelength sensitivity, and PMT rise time. Manufacturers' handbooks and catalogs are very useful in making these selections.

Referring to Figure 17, observe that the PMT is wrapped with a special mu-metal foil. This foil significantly reduces the effect of external magnetic fields on the PMT's performance. If the PMT is stationary, motors, cathode ray tubes (CRTs) and other sources of magnetic fields can affect the PMT. If the PMT is in motion during counting, the Earth's weak magnetic field can affect the trajectory of the electrons from the photocathode to the first dynode.

The best PHR performance is achieved when the photocathode covers all the light exit face of the crystal. PMT photocathodes are usually circular in shape and fit well within a right cylinder.

8. Voltage Divider and Preamp

The PMT manufacturer's recommendations should be followed to obtain the best results. In some cases, the voltage-divider design should be modified for unusual count rates. With normal count rates, a lower-impedance ($\approx$ 5 MΩ) divider string is recommended. If a low count rate is expected and the power supply is current-limited, a high-impedance ($\approx$ 100 MΩ) divider string is recommended. The preamp is used to preserve the signal coming from the PMT anode and output the signal to a linear amplifier.

C. Detector Fabrication

Equally important as the design issues and component performance issues just discussed are the assembly practices and techniques used during the production of a detector. There are several techniques used to grow scintillation materials. The most commonly used growth techniques include Stockbarger-Bridgeman, Czochralski, horizontal Bridgman, radiofrequency (RF) induction, and *in situ* gradient freeze. Some growth techniques make use of growth from the melt; other techniques are from seeded growth; and others initiate nucleation spontaneously. All techniques involve loading salt into a crucible, melting the salt in an electric furnace, and then freezing the material at rates between 1 and 10 mm/h. After the charge is completely grown, it must be slowly ramped to room temperature to avoid thermal shock to the ingot. This process is called annealing the ingot.

Three important factors must be controlled to obtain the highest quality crystal ingots: (1) use the highest purity grade starting material; (2) use the best available growth equipment; and (3) the growth process must be continually monitored with computerized controls.

The assembly of the detector starts by sawing a crystal blank from the ingot. The crystal blank is then machined into the required shape with conventional milling or lathing machinery. Hard oxide crystals, such as BGO, CaF_2(Eu), and $CdWO_4$, require high-speed machinery with special cutters. After machining, the crystal is ready for its final surface preparation. Because some crystal materials are hygroscopic, they must be handled in humidity-controlled environments (refer to Table 2). Special dry rooms are used in the fabrication of NaI(Tl) detectors. A NaI(Tl) dry room operates at very low humidity levels, typically at dew points of -25 to $-50°C$. Operating under these conditions avoids the performance-degrading formation of hydration on the surface of the crystal.

After the hygroscopic crystal is machined, it is moved into the dry room to let the surface dehydrate. A hygroscopic crystal will have a thin layer of hydrate on its surface, and this opaque white powder can easily be removed with an anhydrous solvent and a mild abrasive cloth such as steel wool. Once the hydrate layer has been removed, an optically transparent surface is visible.

All water of hydration must be removed from the surface so that no light is lost as it reflects off the inner surface of the crystal. Any hydration at the surface will cause a loss of light because the hydrate is a poor reflector. Ideally, all the light should be internally reflected at these inside crystal surfaces and then eventually collected at the interface to the PMT because this gives the best energy resolution.

The cleaned surface of the crystal is then prepared by polishing or abrading the surface to optimize the light output of the crystal. This optimization process, called *compensating the crystal*, improves the PHR by making the light output response as uniform as possible along the length of the crystal. Technicians use a method called *mapping* to determine how uniform the response is for a given compensation. This technique is used frequently when the aspect ratio (length:width) is greater than 3:1. A uniform PHR to within 3% along the length of the crystal gives a good side on energy resolution. At the same time, end-on resolution provides a good indication of the material performance.

The face that is to be coupled to the PMT is polished. In general, all of the surface preparations are done manually due to the fragile mechanical characteristics of the crystal. During these manual operations, the technician must handle the crystal carefully so that the crystal will not fracture by mechanical or thermal shock. A hard oxide crystal may have all of its sides chemically etched, except for a polished face that is coupled to a PMT; sometimes a hard oxide may be polished all over.

At this point in the assembly process, the crystal is ready to be interfaced to a piece of glass or a PMT. The silicon elastomer material acts as an adhesive between the crystal and the glass surface. Elastomers have good mechanical properties because they allow some material movement in shear mode without cohesive or adhesive separation.

The crystal surface is then surrounded with a highly reflective material such as Teflon. The reflector must stay in good contact with the surface of the crystal to avoid light loss at the crystal-reflector boundary. Ideally, the best performance occurs if no light is lost at the surface of the crystal.

Finally the crystal is encapsulated in a metal container such as aluminum or stainless steel by creating a hermetic seal with an epoxy joint between the glass and the metal can. Figure 17 shows a typical detector assembly. Note, again, that it is important that the detector be constructed in such a way that no outside light interferes with the low light signals being generated in the crystal. Once the assembly is completed, it is tested to ensure that all the specifications have been met.

V. MEASUREMENTS WITH SCINTILLATORS

Previously in this chapter the various properties and characteristics of scintillation detectors (absorption of the incident photons, light output, thermal effects on light output, and scintillator energy resolution) have been discussed from a theoretical perspective. In this section, the discussion is related to the practical use of scintillation detectors. The reader can find additional useful information provided by Birks (1964), Knoll (1989), Hendee (1984), Sorenson and Phelps (1987), Hofstader (1975), and Heath et al. (1979).

A. Measurement Systems

1. Basic Spectroscopy Counting Systems

A simple gamma-ray counting system is shown in Figure 18. The block diagram includes all the essential components required to obtain a gamma-ray spectrum. The components that make up the signal-processing apparatus are discussed in this section.

a. Detector Assembly—Crystal/PMT/Voltage Divider

The detector assembly consists of the components shown in Figure 17 (see previous discussion). The PMT is required to convert the very low-level light created in the crystal to a signal that is conveniently processed. As the light enters the PMT, it is converted into an electron cloud at the photocathode. These electrons are directed to the first dynode of the PMT by an electric field that is produced by a voltage-potential difference between the photocathode and the first dynode of the PMT. By the appropriate distribution of the voltage between the dynodes in a PMT, (PMTs may have up to 10 dynodes), electrons are focused and multiplied such that a usable analog pulse is generated at the anode (see Fig. 14). The signal that is generated by the crystal/PMT/voltage divider at the anode is a negative-going pulse with amplitude around 50 mV. Typically, these assemblies have high impedance and low capacitance. Some portable low-count-rate assemblies have very high-impedance voltage dividers.

b. High-Voltage Input

High-voltage power supplies are designed to provide up to 2000 V DC with up to 2 mA if required. Most detection systems operate at positive high voltage. Good power supplies are designed to limit noise, limit drift, and hold these specifications over a wide temperature range. The high-impedance voltage divider used in low-current applications (low count rate) use a high-voltage power supply that can be operated with D-cell batteries. This is especially useful in counting systems that are hand-held field instruments.

c. Preamp

Preamps are low-noise electronic devices used to take the input and, without shaping the pulse input, preserve the pulse with the maximum SNR. In most cases, a counting system works quite well without a preamp. The use of a preamp circumvents problems that can occur when different length cables are introduced between the detector and linear amplifier. Changing cables affects the resistor-capacitor (RC) time constant, and this could mean that the linear-amplifier shaping time constants have to be adjusted. The RC differentiation that occurs in the preamp has a relatively long time constant to prevent pulse pile-up. There are three types of preamps: voltage-sensitive, charge-sensitive, and current-sensitive. Spectroscopy applications use a current-sensitive preamp in which the signal is taken from the anode of a PMT. The preamp is capable of taking the signal and driving several feet of cable so that the pulse can be accepted by a fast linear amplifier with minimum distortion. Preamps are designed to provide impedance matching between the detector assembly, cable, and linear amplifier.

d. Linear Amplifier

Signals from the preamp are amplified, filtered, and shaped in the linear amplifier. The linear amplifier takes a nominal 50 mV input and amplifies it to several volts, making the signal suitable for processing in a multichannel analyzer. Shaping and filtering the signal in the amplifier improves the SNR and the response time required for each pulse is reduced. Amplifiers are provided with coarse- and fine-gain adjustments, so that the signal output can be adjusted to a useful voltage. Shaping-time and integration-time adjustments help optimize the signal processing. The time-constant adjustments should be selected to match the pulse shape of the scintillator being used.

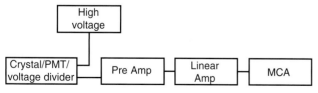

FIGURE 18 Gamma-ray counting system block diagram.

e. Multichannel Analyzer

A multichannel analyzer (MCA) is used to obtain the PHR of a scintillator. In an MCA, the analog signal output from the amplifier is digitized and displayed in the form of a histogram. This histogram represents an energy distribution for the events that occurred in the scintillator. MCAs have an adjustable low-level discriminator so that extraneous low voltage pulses will not cause significant counting dead time.

The counting system can be calibrated with an electronic pulser with a known voltage output or by using several radioactive sources to develop an energy-per-channel calibration factor. Because of the slight nonlinearity of the response of scintillation materials, the sources method should be used with caution.

Oscilloscopes are often used to analyze the waveform output form each component in a counting system to ensure that it is operating properly.

With this overview of the basic measurement system, the response characteristics of a basic counting system can now be discussed in more detail.

2. Measurement System Response

The properties of the scintillation crystal have the primary influence on the overall performance of the detection system. It is important that each performance characteristic is understood to ensure optimal system performance. The following discussion addresses the most important system response features.

a. Spectral Response Curve

Section III of this chapter explains that the amount of energy deposited in the crystal is equal to the incident photon energy and that the light output of the scintillator is nearly proportional to the energy deposited in the crystal. This is an essential aspect of detector technology. This feature gives us the opportunity to perform gamma ray spectroscopy. Now, gamma rays of different energies can be recorded and analyzed by energy and by activity.

A gamma-ray spectrum is shown in Figure 19. The spectrum represents a histogram of all the energies recorded by the counting system between 5 and 1000 keV. The histogram records the voltage pulse-height data for each pulse generated by the counting system. Each data point is stored in a bin or channel; each channel has a discrete voltage width. For scintillation counting applications, an MCA typically accepts voltage input from 0 to 10 V; the channel numbers (horizontal display) scale with the binary numbers 256, 512, or 1024. The vertical display scale can be adjusted from the log scale up to over 10^6.

The spectrum shown in Figure 19 is for ^{137}Cs. ^{137}Cs decays to ^{137}Ba by beta decay (β^-). ^{137m}Ba, with a half-life of 3 min, has a monoenergetic gamma-ray at 662 keV. Most studies state that the 662-keV gamma-ray is from ^{137}Cs, and

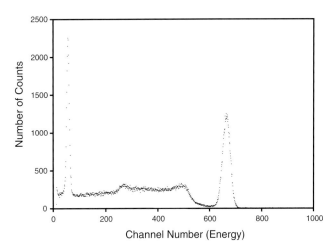

FIGURE 19 Pulse-height spectrum of 662-keV gamma ray of Cs-137.

we do the same here. The main features in the spectrum are the Ba X-rays around 30 keV (CH 35), the backscatter peak around 260 keV (CH 260), the Compton shoulder around 500 keV (CH 500), and the full-energy photopeak at 662 keV (CH664). Other interesting features include the tail of some electronic noise seen around 5 keV and the small number of counts seen in the valley around channel 600. The low-energy electronic noise, which is attributed to the PMT, sometimes limits the low-energy measurement capability of a counting system.

Table 5 shows the expected spectral response for various gamma-ray events in a scintillation detector. A gamma-ray spectrum has a number of predictable features that are summarized here. It is useful to know that the spectral features can be predicted because computer algorithms can be used to look for these response features. Monte Carlo codes are sometimes used to predict the behavior of a counting system.

b. Energy Resolution

Previous sections have discussed how light is produced in a scintillation crystal and the importance of efficient light collection. The ability of the crystal material to produce light and the ability of the detector assembly to collect the light and convert it to a usable signal are the two most important functions of the gamma-ray spectrometer. Each has a direct effect on the energy resolution of the counting system.

The energy resolution or PHR of a detector is defined as the ratio of the full width at half maximum (FWHM) of the photopeak to the mean of the photopeak. A typical PHR for NaI(Tl) is approximately 7.5% at 662 keV. Arithmetically, the PHR (in percent) is written as:

$$\text{PHR} = \frac{\text{FWHM(channels)}}{\text{Photopeak(channel number)}} \times 100\% \quad (18)$$

For example, examine Figure 19, which shows a gamma-ray spectrum for Cs-137 taken by a NaI(Tl) crystal detector

TABLE 5 Detector Spectral Responses

Interaction Event	Explanation and Observed Spectral Response
Photoelectric	Well-defined peak at E_γ
Compton scatter	A continuum of energies between $0 \leq h\nu' \leq h\nu'_{max}$ according to Eq. (2)
Photopeak/Compton valley	Region of the spectrum between the photopeak and the Compton shoulder that contains multiple Compton scattering events
Pair production (pp)	Peaks observed at 511 keV due to β^+ annihilation events that occur outside of the detector
Backscatter	Peak around 200 to 250 keV due to Compton scattering events that have occurred in materials surrounding the detector
X-ray peak	Peak associated with the emission of a characteristic X-ray from materials that surround the detector
Lead X-ray peak	Peak associated with the emission of a characteristic lead X-ray from lead material that may surround the detector, peak at 80–90 keV
Escape (I-single)	Peak associated with the creation of the K edge X-ray of iodine by photoelectric absorption interaction with I near the surface of the NaI(Tl), the 30 keV X-ray escapes from the crystal and a peak is observed at $E_\gamma - 30$ keV
Escape (pp-single)	Peak observed at $E_\gamma - 511$ keV; E_γ must be $> 2 \times 511$ keV
Escape (pp-single)	Peak observed at $E_\gamma - (2 \times 511$ keV); E_γ must be $> 2 \times 511$ keV
Sum	Peak associated with the coincident detection of two or more photons from simultaneous nuclear transitions
Background	Peaks associated with the presence of naturally occurring K, U and Th in the immediate area
Background	A continuum spectrum due to Compton scatter of naturally occurring background

assembly. This detector has a PHR of 6% at 662 keV. In other words, the photopeak at half maximum is 40 keV wide. The PHR is exceptionally good because it has high-grade NaI(Tl) crystal material and the light collection efficiency has been optimized by a favorable geometry. Most scintillation detectors are manufactured to a PHR specification quoted by the manufacturer.

Three major effects contribute to the width of the photopeak. First, the statistical nature of the photon production, collection, and multiplication of the electrons in the dynode string causes degradation in the photopeak width. Second, each scintillation pulse created has associated with it noise from dark current created at the photocathode. This electronic noise or electronic jitter degrades the width of the photopeak. Third, degradation in the photopeak width is caused by light collection. These variations in the pulse height are due to the nonuniform collection of the light created inside the crystal in spite of careful surface preparation of the crystal to minimize this problem.

Another minor contributor to the line-width spread is the previously discussed nonlinear response of the scintillation material as a function of energy deposited in the crystal (Zerby *et al.*, 1961). Temperature gradients in the bulk material and the temperature at which the measurement are performed also affect the PHR (see the previous discussion on scintillator energy resolution).

c. Peak-to-Valley Ratios

Another parameter that can be used to quantify the quality of a detector is the peak-to-valley ratio. This is simply the ratio of the number of photopeak events in the peak channel to the number of events recorded in a valley channel. For a large-volume detector (5000 cc), the peak-to-valley ratio for ^{60}Co, which has photon energies of 1170 keV and 1331 keV, can be approximately 7:1; whereas a smaller-volume detector (350 cc) can have a ^{60}Co peak-to-valley ratio of approximately 10:1. The detector used in obtaining the spectrum shown in Figure 19 has a ^{137}Cs peak-to-valley ratio of 70:1. The number of photopeak events measured is strongly dependent on the volume of the crystal, whereas the number counts recorded in the valley is strongly dependent on the PHR of the detector. The peak-to-valley ratio specification gives a direct measure of the ability of the detector to measure the separation of two peaks that have energies that are very close to one another.

d. Detector Counting Efficiency

Several factors determine the counting efficiency of a detection system. The factors include the energy of the incident photons, intrinsic efficiency of the scintillation material, geometry of the crystal, losses due to scatter or absorption in the source (e.g., a patient), and losses due to the scattering or absorption of the photons before they reach the crystal due to the finite thickness of the detector housing and other detector assembly components. As previously discussed, the intrinsic efficiency is sometimes related to the temperature at which the detector is operated.

The counting system electronics can also affect the counting efficiency. The electronic apparatus needs a finite time to record each pulse. If additional pulses arrive while the system is already processing a pulse, then they either pile on top of one another or the pulse is lost because the input gate is closed. In either case, a counting error occurs. Most MCAs are equipped with dead-time meters to record the amount of time that the system is busy processing signals so that a correction can easily be made.

In some counting applications, it is advantageous to reduce contributions due to unwanted background by counting only the events recorded in the photopeak. Photopeak or photofraction efficiency curves have been developed and are available in Knoll (1989), *Harshaw Scintillation Phosphors* (1975), and Birks (1964). This type of measurement is performed by setting a voltage window around the photopeak or region of interest. These measurements can be performed with a less-expensive electronic device called a *scaler*. A scaler simply records a count for each pulse that occurs between two voltage settings.

The geometric factor is addressed by using a crystal shape that improves the counting efficiency. Popular geometries include cylindrical end wells, cylindrical side wells, and stacking multiple detectors. Special rectilinear well geometries are also available. All well geometries cause degradation of the energy resolution; however, this is usually not a significant issue because the energy of the photon being measured is typically already known. For low-energy photons, counting efficiencies are improved by reducing the Z or the material thickness of the entrance window or the well liner. Thin sheets of either aluminum or beryllium are used often in planar counting applications.

e. Signal-to-Noise Ratio Issues

The SNR of a counting system is an important concern. Some applications have a limited signal available to be detected. This is true in medical-imaging applications, in which the dose to the patient is an important consideration. Good measurements require a good SNR. Each measurement requires an analysis in which an acceptable confidence level has been established. If the optimal signal strength or count rate is achieved, then the noise must be reduced to ensure an acceptable SNR. This section discusses some of these noise issues.

First, it must be understood that the radioactive decay process is random. The time between radioactive decay events cannot be predicted. Only the statistics of the next event can be stated. Because of the random nature of the radioactive decay process, the statistics of radiation measurement counting must be understood.

Sources of noise from the detector assembly include the scintillator, detector hardware, PMT, and the voltage divider assembly. Noise contributed by the scintillator can include any parameter that causes the PHR to degrade: intrinsic background due to radioactive impurities, Compton scattering from surrounding materials, and any background radiation present such as terrestrial or cosmic radiation.

The PHR directly affects the SNR in photofraction measurements because, as the energy window widens to include all the events in the photopeak, more background events from the Compton continuum are also included in the same counting window. The background from impurities can be limited by using cleaner starting materials during growth of the ingot. Background from surrounding materials can be limited by improving the shielding around the detector. In multiple-detector counting systems, Compton scatter or cross-talk between detector elements can be minimized with lead or tungsten septa.

The PMT also contributes to the noise in the counting system. One form of PMT noise is called dark current. When a PMT is operated in the dark, it still exhibits some amount of current at the anode. One contributor to dark current is thermionic noise, which is a result of the very low work functions of the photocathode and the dynode surfaces. Other sources of dark current include leakage current between the anode and other electrode surfaces, leakage current between the PMT leads at the surface of the glass envelope, and scintillation events in the PMT glass. Another source of PMT noise is improperly selected PMT glass that has a high concentration of potassium. The ^{40}K present increases the background noise in the detector.

PMT's can also add to the noise problem by degrading the PHR. This occurs with excessive photocathode nonuniformity, poor linearity, or poor stability because of long- or short-term drift. Improper voltage-divider design or assembly can also lead to electronic noise.

Many of the problems discussed here are component design issues and can be resolved by better component engineering. Some of the noise problems discussed here can be minimized with the appropriate counting system design, such as good shielding, gain stabilization techniques, and even a Compton suppression detector configuration; other sources of systematic noise can be reduced by spectral stripping techniques, such as background subtraction.

3. Detector Configurations in Measurement System Applications

Applications for scintillation detectors include nuclear medicine, dose calibration, health physics, whole-body counting, meteorology, oil and gas exploration, uranium exploration, precious mineral exploration, biology, archeology, food research, high-energy particle physics, environmental air and soil monitoring, water analysis, commercial density gauging, snow-pack analysis, nuclear power, space exploration, academic research, isotopic characterization, drug interdiction, nuclear weapons compliance verification, and portal monitors. This section discusses the common detector configurations and some special detector designs. The detector configurations discussed here are produced in many sizes and shapes. Some are designed to withstand extreme environmental conditions that require special ther-

FIGURE 20 Photograph of various scintillation detectors. (Photo courtesy of Alpha Spetra, Inc.)

mal and mechanical designs. The discussion demonstrates the range of capabilities of detector technology. An assortment of scintillation detectors is shown in Figure 20.

a. Simple Gross Counters

Simple gross-counting detectors are usually open-face or integral-type units that do not require good resolution. The integral type is constructed with a PMT, whereas the open-face type is supplied with an optical window. Gross-counting units are used in many field applications in which damage to the PMT or crystal can easily occur. With a demountable PMT, either the crystal or the PMT can easily be replaced without incurring the cost of replacing both components. Gross counting measurements are done by setting a lower-energy threshold (discriminator) and counting all the pulses that take place without recognizing the pulse height. Gross counters do not perform spectroscopy. Sometimes gross counters employ energy windows that are set up to measure a particular gamma-ray of interest.

b. Well Counters

Well counters are detectors that have end, side, or through holes bored into the crystal. For an example of a well detector, see Figure 17. Well detectors are designed to improve 4π counting efficiency. Wells can be machined in numerous shapes to accommodate the end-user's requirements. Well counters are used as dose calibrators, environmental sample measurements, and other applications in which good counting efficiency is required.

c. Thin Windows

Thin window assemblies are designed to detect low-energy photons between 5 and 100 keV. The detector crystal is usually a thin section of 1–2.5-mm in thickness. These thin sections are optimal for detecting photons at low energies. At the same time, the detection of Compton scattered photons from higher energies is minimized. Thin window detectors use low-Z entrance windows to enhance counting efficiency. Some NaI(Tl) thin windows use cleaved crystal material to eliminate the dead layer associated with the process of machining and polishing a crystal. A cleaved crystal is more sensitive to low-energy X-rays.

A very popular thin window configuration, a F.I.D.L.E.R. (field instrument for the detection of low-energy radiation) probe, uses a 5-in-diameter by 0.063-in-thick NaI(Tl) crystal with a 0.010-in-thick beryllium entrance window. This unit is very useful in measuring the very low energy (~15 keV) characteristic of *L*-shell X-rays.

d. Low-Background Counting Assemblies and Compton Suppression Counting

Some applications, such as measuring environmental air samples, activation analysis, tracing contaminants, and measuring low-count-rate branching ratios (Cecil and Wilkinson, 1984), require the capability of measuring activities that are less than the ambient background! Low background counting systems can be designed to measure either external or internal source samples. This discussion focuses on measuring internal samples that can be placed inside the detection assembly.

There are several methods of obtaining the best possible lower limit of detection (LLD). These counting techniques can easily improve the SNR by an order of magnitude. The first approach to measurements such as this is to reduce as much of the background from the detector assembly as possible by special construction techniques. The low background detector assembly is then placed in a lead-lined (sometimes up to 4 inches of lead) enclosure to reduce signal from ambient background. In addition, lead shields are sometimes copper-lined to reduce the contribution from gamma-ray interaction with the lead and causing X-ray background.

The second approach uses an annular detector that surrounds the main plug detector, usually a HPGe (hyper-pure germanium) solid-state-type detector. The plug detector is situated in an axial well of the annular detector and is shielded from background radiation by the annular detector and by surrounding lead shielding. The annular detector, which can be made from NaI(Tl) or BGO, acts as passive shield. A BGO annulus is more compact due to its higher density; this reduces the size and amount of shielding that is required.

There are three popular methods of using annular detectors: Compton suppression by the anticoincidence counting mode, the sum coincidence mode, and the pair spectrometer mode. The anticoincidence counting system rejects pulses that are coincident in both the main detector and the annular detector by electronically gating the signals in such a way that only the remaining pulses registered in the main detector are accepted. The disadvantage of this technique is that, in some counting experiments in which a nuclear transition has coincident events, a good signal is sometimes rejected. This technique has recently been used to measure the total electron energy resolution of a scintillation detector (Mengesha and Valentine, 1999).

The sum coincidence mode requires that an event occur in both the main detector and the annulus in coincidence in order to be accepted. This technique is effective because most of the Compton continuum spectrum is composed of single Compton scattering events followed by the escape of the scattered gamma-ray. Photopeak events consist of multiple scatters followed by a photoelectric absorption. In both instances, the events can be shared by the main detector and the surrounding annulus. The peak-to-Compton ratio can be improved if the counting system requires multiple scattering events to occur before an event is accepted. In practice, this is done by segmenting the detectors and gating the coincidence signal on all of the segments.

The pair spectrometer mode is only used in applications in which the gamma-ray energy is sufficient to cause pair production. This technique requires that three events be measured in coincidence: each annihilation photon created should be registered in opposite segments of the annular detector and the remaining energy of the incident photon, the second escape peak, should be counted in the main detector. This technique is very effective at eliminating cosmic-ray background; however, the detection efficiency in this technique is very poor.

e. Whole-Body Counting

Another low-background counting application is a whole-body counter. Workers that handle radioactive materials are frequently tested to assess the possibility of ingestion or inhalation of radioactive contamination. The detectors are placed in a room designed to provide a low-background counting environment. Several large-volume NaI(Tl) detectors of over 4000 cc each are positioned along the torso. The signals from all of the detectors are summed together. The data can also be stored so that additional analysis can provide information about the region of the body that may be exhibiting an abnormally high count rate. More recently, NaI(Tl) detectors have been used to screen workers while HPGe detectors are used to provide follow-up measurements.

f. Phoswich Measurements

Scintillators of different materials can be optically coupled and then interfaced to a PMT to form a phoswich, or phosphor sandwich as it was originally called (Mayhugh et al., 1978). This combination is useful because events that occur in each crystal material can be separated by pulse-shape discrimination (Birks, 1964; Wilkinson, 1952) because of the different decay times. NaI(Tl) and CsI(Na) are commonly used in this configuration. This is especially useful when a low-energy particle or photon can be absorbed fully in the first scintillator while the more penetrating higher-energy photons are detected in the second scintillator.

g. Ruggedized High Temperature

One of the more demanding applications for scintillation detectors is the geophysical measurement done in borehole logging. Some measurements are actually taken while the borehole tool is drilling through rock formations. Thermal shock, mechanical shock, and severe mechanical vibration to the detector and PMT are the challenges that the design engineer must face. Because the borehole environment is cylindrical, the detector geometry is cylindrical with a typical aspect ratio of approximately 4:1. Energy resolution is sacrificed in this application because light output is not optimized at these aspect ratios.

h. Timing Resolution

The moment at which the incident radiation is absorbed by the scintillator and the light pulse is created is well defined. The time between pulses can easily be resolved to several nanoseconds by using a scintillation material that has a fast decay time, a crystal that is properly sized, photomultiplier tubes that have fast rise times, and good pulse-counting electronics. This technique is useful in high-count-rate applications and coincident gamma counting in PET. This technique can be used to determine the annihilation-photon time of flight (TOF) with proper electronics.

i. Gain Stabilization Techniques

In measurement situations where temperature fluctuations are encountered, gain stabilization techniques called pulsers can be used. A pulser may be either an electronic device such as a light-emitting diode or a radioactive source (an ^{241}Am pulser) that is intentionally incorporated into the detector assembly. In either approach, the light pulse measured by the counting system is situated in the spectrum so that it does not interfere with the region of interest in the spectrum.

4. Medical Imaging Applications

The most extensive use of scintillation detectors is in nuclear medicine. Table 6 gives a list of the most commonly used scintillation materials in medical applications. The detectors have been used in numerous medical applications including isotope preparation, studies of biological samples, anatomical studies, whole-body counting, *in vivo* counting, and body-function studies. By far the greatest use is in imaging anatomical regions of the body.

Three types of devices are briefly discussed in this section: the Anger Camera or gamma camera, PET, and *in vivo* probes. The detectors used are of many different sizes and shapes and include some of the most difficult to fabricate. Since the 1950s, vast improvements have been made in detector technology and measurement techniques used in nuclear medicine. More recently, software

Table 6 Crystal materials commonly used in medical applications

Material	Applications	Advantages	Disadvantages
NaI(Tl)	Gamma cameras, SPECT, PET, probes, dose calibrators	Light output, cost	Fragile, hygroscopic
BGO	PET	Density	Light output
$CdWO_4$	CT	Density, low afterglow	Decay time
LSO	PET	Light output, density, decay time	Cost, intrinsic radio-active background

developments have significantly improved the performance of the measurement devices.

Each of these clinical measurement techniques requires that a radioactively labeled substance be injected into the patient's body. By carefully positioning the detector, it is used to measure the location and intensity of the tracer that is concentrated in the organ of interest. The ongoing challenge is to precisely image very small concentrations of radioactive tracer.

a. The Anger Camera—Single-Photon Emission Computed Tomography

The first scintillation camera or gamma camera proposed by Hal Anger (1957) consisted of a single-pinhole collimator, a 4-in-diameter by 1/4-in-thick NaI(Tl) crystal, seven 1.5-in PMTs, pulse-processing electronics, and an oscilloscope. The image was recorded with a Polaroid-Land camera. A later version of the basic Anger camera is shown in Figure 21. In his first paper, Anger reported the use of the counting apparatus in a clinical thyroid study; the immediate advantage seen over the existing scanning technology was the shorter imaging time.

Currently, gamma camera counting systems can have as many as three large field of view camera heads that have the capability of acquiring data while in motion. Currently available systems have the ability to take data over 360° and over multiple planes. Today SPECT systems are equipped with analog-to-digital convertors (ADCs) for signal processing. Computers are used to manipulate data during image reconstruction, store the data, and display the image. Typical spatial resolution is approximately 4 mm. Most studies are performed using ^{99m}Tc, which emits a 140-keV gamma ray.

Improvements in gamma camera technology are aimed at improving the spatial resolution, sensitivity, and uniformity of response as seen in the image. New hardware and software developments are continually being reviewed to improve the performance of gamma camera systems.

b. Positron Emission Tomography

The basis for PET is the coincident measurement of the 511-keV photons (created during positron annihilation) by a multi-element array of scintillation detectors. The first PET detector array built for tomographic imaging was the Brookhaven National Laboratory BNL-32 (Robertson *et al.*, 1972: Yamamoto, 1977). The BNL-32 crystal positron transverse section detector was built in 1972. Another early PET system was built with NaI(Tl) crystals in 1973 by Phelps *et al.* (1975). Soon after, the first BGO crystals were used because they are much more dense than NaI(Tl) crystals; this helped improve the system detection efficiency. More recently, LSO is being used due to its better light output and slightly higher density. With each generation of PET imaging device, the number of detector elements has increased. The current high-resolution research tomograph (HRRT) has approximately 120,000 pixelated elements (Schmand, 2000).

A schematic for a simple PET system is shown in Figure 22. A positron-emitting radioisotope is injected into the patient. The subsequent emission of the 511-keV photons is detected by two opposing detectors. The signals are processed and an image is reconstructed from the digitized data.

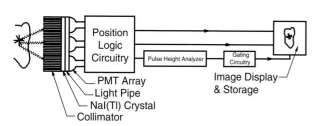

FIGURE 21 The basic Anger camera.

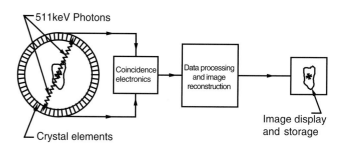

FIGURE 22 PET imaging system.

Recently a hybrid PET/SPECT instrument has been developed (Schmand, 2000), which incorporates NaI(Tl) to function in the SPECT modality and LSO to function in the PET modality.

c. In Vivo Measurements

In vivo counting systems are used to detect radioactivity that has been injected into the patient. Typically a single-element device, the *in vivo* probe measures a specific organ or anatomical region. Applications include a system for measuring thyroid uptake, cardiac studies, pulmonary studies, and surgical probes. Surgical probes are specifically designed to aid the physician in detecting cancerous tissue during surgery.

VI. SUMMARY AND COMMENTS

What will be the next improvement in scintillator technology? It seems that the initiative to develop new scintillators comes mainly from the medical imaging community and the high-energy physics community (Lecoq, 1999; Korzhik and Lecoq, 2000, 2001). We should recall the discussion regarding the ideal scintillator and how the technology has progressed (Heath *et al.*, 1979; Weber, 1999). The research community has developed some ingenious approaches in the search to find new scintillators (Derenzo *et al.*, 1990, 1994; Moses and Derenzo, 1990; Melcher and Schweitzer, 1992; Balcerzyk *et al.*, 2000). Unfortunately, at this time, the performance of any new potential materials cannot be predicted from theory.

Many useful materials have been developed since the 1950s; however, none performs well enough to completely replace NaI(Tl). There is a need for improvement in the crystal-growth techniques that are currently used. There is a need to improve the performance of light collection devices. There are numerous research projects in process with the continued hope of developing a much-improved scintillator material. A number of Ce-doped compound scintillators are currently being developed, and they show some promise due to improved speed and energy resolution. Examples include $LaBr_3(Ce^{3+})$, $LaCl_3(Ce^{3+})$, $K_2LaCl_3(Ce^{3+})$, and $RbGd_2Br_7(Ce^{3+})$. $LaBr_3(Ce^{3+})$ in very small pieces has demonstrated a PHR of approximately 3.3% at 662 keV, whereas both $K_2LaCl_3(Ce^{3+})$ and $RbGd_2Br_7(Ce^{3+})$ have demonstrated a PHR of approximately 4% at 662 keV (van Loef *et al.*, 2001). As with many other scintillators being developed, much work needs to be done with these materials to improve them until they become commercially available in large pieces and large quantities.

In this chapter, we have seen that scintillation detector technology uses many different skills and expertise: physics, chemistry, material science, thermodynamics, nuclear physics, quantum mechanics, solid-state physics, optics, mechanical engineering, electronic engineering, mathematics and statistics, and computing. Good detector design and fabrication require cooperation between experts in several scientific and engineering disciplines. For these reasons, the author has enjoyed working with scintillation detectors for over 25 years.

The word scintillation comes from the Latin verb *scintillare,* which means to sparkle or glitter. Consider this thought: The very small amount of light given off by a crystal has the ability to provide many useful tools, yet at the same time this light is lost to the human eye!

It is true that out of a little light there is much to be profited.

References

Anger, H.O. (1957). Scintillation camera. *Rev. Sci. Instrum.* **29**: 27–33.

Balcerzyk, M., Moszynski, M., Kapusta, M., Wolski, D., Pawelke, J., and Melcher, M., (2000). YSO, LSO, GSO and LGSO. A study of energy resolution and nonproportionality, *IEEE Trans. Nucl. Sci.* **47**: 1319–1324.

Beattie, R.J.D., and Byrne, J. (1972). A Monte Carlo program for evaluating the response of a scintillation counter to monoenergetic gamma rays. *Nucl. Instrum. Methods* **104**: 163–165.

Birks, J. B. (1964). "The Theory and Practice of Scintillation Counting," Pergamon Press, The MacMillan Company, New York.

Cecil, F.E., and Wilkinson, F.J. (1984). Measurement of the ground state gamma-ray branching ratio of the dt reaction at low energies. *Phy. Rev. Lett.* **53**: 767–770.

The Crystal Ball Collaboration. (1980). "A Brief Review of Recent Results from the Crystal Ball at SPEAR," SLC-PUB-2655, Nov. 20.

Derenzo, S.E., Klintenberg, M.K., and Weber, M.J. (1999). Quantum mechanical cluster calculations of critical scintillation processes. *Proceedings of the Fifth International Conference on Inorganic Scintillators and Their Applications.* p. 53–60.

Derenzo, S.E., Moses, W.W., Weber, M.J., and West, A.C. (1994). Scintillator and phosphor materials. *Mat. Res. Soc. Symp. Proc.* **348**: 39–49.

Derenzo, S.E., *et al.* (1990). Prospects for new inorganic scintillators. *IEEE Trans. Nucl. Sci.* **37**: 203–208.

Derenzo, S.E., *et al.* (2000). Measurements of the intrinsic rise times of common inorganic scintillators. *IEEE Trans. Nucl. Sci.* **47**: 860–864.

Dorenbos, J.T., de Haas, J.T.M., and van Eijk, C.W.E. (1975). Nonproportionality in the scintillation response and the energy resolution obtainable with scintillation crystals. *IEEE Trans. Nucl. Sci.* **42**: 2190–2202.

Eby, F., Jentschke, W., (1954). Fluorescent response of NaI(Tl) to nuclear radiations. *Phys. Rev.* **96**: 911.

Evans, R. D. (1955). "The Atomic Nucleus," McGraw-Hill, New York.

Fonte, R., Insolia, A., Lanzano, G., Pagano, A., Russo, G.V., Palama, G., de Jager, C.W., and de Vries, H. (1991). Response function of a BaF_2 detector to electrons. *Nucl. Instrum. Methods Phys. Res. A* **353**: 80–82.

Harshaw Scintillation Phosphors. (1975). Third Edition, (Company Catalog).

Harshaw, J.A., Stewart, E.C., and Hay, J.O. (1993). AEC Report NYO 1577, 1952. Preparation and Performance of Scintillation Crystals. *Research A* **333**: 304–311.

Harshaw Radiation Detectors. (1984). (Company Catalog).

Heath, R.L., Hofstadter, R., and Hughes, E.B. (1979). Inorganic scintillators, a review of techniques and applications. *Nucl. Instrum. Methods* **162**: 431–476.

Hendee, W.R. (1984). "Radioactive Isotopes in Biological Research," Krieger Publishing Company.

Hofstadter, R. (1948). Alkali halide scintillation counters. *Phys. Rev.* **74**: 100–101.

Hofstadter, R. (1975). Twenty-five years of scintillation counting. *IEEE Trans. Nucl. Sci.* **22**: 13–25.

Holl, I., Lorenz, E., and Mageras, G. (1988). A measurement of light yield of common inorganic scintillators. *IEEE Trans. Nucl. Sci.* **25**: 105–109.

Iredale, P. (1961). The effect of the non-proportional response of NaI(Tl) crystals to electrons upon the resolution for g-rays. *Nucl. Instrum. and Methods* **11**: 340–346.

Ishii, M. and Kobayashi, M. (1991). Single crystals for radiation detectors. *Prog. Crystal Growth and Characteristics* **23**: 245–311.

Knoll, G. F. (1989). "Radiation Detection and Measurement." John Wiley & Sons, New York.

Kobayashi, M., Carlson, P., and Berglund, S. (1989). Temperature dependence of CsI(Tl) scintillation yield for cosmic muons, 5 and 1.25 MeV γ-rays. *Nucl. Instrum. Methods Phys. Res. A* **281**: 192–196.

Kobayashi, M., Mitsuru, I., and Melcher, C. L. (1993). Radiation damage of a cerium-doped lutetium oxyorthosilicate single crystal. *Nucl. Instrum. Methods A* **335**: 509–512.

Korzhik, M., and Lecoq, P. (2001). Search of new scintillation materials for nuclear medicine application. *IEEE Trans. Nucl. Sci.* CD, article 6-1.

Korzhik, M., and Lecoq, P. (2000). Scintillator developments for high energy physics and medical imaging. 0-7803-5696-9/00. *IEEE Trans. Nucl. Sci.* **47**: N4, 1311.

Lecoq, P. (1999). How high energy physics is driving the development of new scintillators. *Proceedings of the Fifth International Conference on Inorganic Scintillators and Their Applications.* pp. 3–10

Lempicki, A., Wojtowicz, A.J., and Berman, E. (1993). Fundamental limits of scintillator performance. *Nucl. Instrum. Methods Phys. A* **333**: 304–311.

Mayhugh, M.R., Lucas, A.C., and Utts, B.K. (1978). Low background beta counting with CaF_2(Eu) in a Phoswich configuration. *IEEE Trans. Nucl. Sci.* **25**: 569–573.

Meggitt, G.C. (1970). The effect of the crystal surface on the derived electron scintillation response of NaI(Tl). *Nucl. Instrum. Methods* **83**: 313–316.

Melcher, C.L. (2000). Scintillation counters for PET. *J. Nucl. Med.* **41**: 1051–1055.

Melcher, C.L., Schweitzer, J.S. (1992). A promising new scintillator: cerium-doped lutetium oxyorthosilicate. *Nucl. Instrum. Methods Res. Phys. A* **314**: 212–214.

Mengesha, W., and Valentine, J.D. (1999). A technique for measuring scintillator electron energy resolution using a Compton coincidence technique. *Proceedings of the Fifth International Conference on Inorganic Scintillators and Their Applications.* pp. 173–178.

Moses, W.W. and Derenzo, S.E. (1990). Lead carbonate, a new fast heavy scintillator. *IEEE Trans. Nucl. Sci.* **37(2)**: 96–100.

Murray, R.B. (1975). Energy transfer in alkali halide scintillators by electron-hole diffusion and capture. *IEEE Trans. Nucl. Sci.* **22**: 54–57.

Narayan, G.H., and Prescott, J.R. (1968). The contribution of the NaI(Tl) crystal to the total linewidth of NaI(Tl) scintillation detectors. *IEEE Trans. Nucl. Sci.* **15**: 162–166.

Branch, Uber, J., Smyth, K., and Walch, C. (1980). NASA Interoffice Communication, Materials Control and Applications Branch. "MOR Bend Test results for NaI(Tl) and Additional Elastic Moduli Measurements."

Phelps, M.E., and Hoffman, E.J., Mullani, N.A., Ter-Pogossian, M.M. (1975). Application of annihilation coincidence detection to transaxial reconstruction tomography. *J. Nucl. Med.* **16**: 210–224.

Prescott, J.R. and Narayan, G.H. (1969). Electron response and intrinsic line-widths in NaI(Tl). *Nucl. Instrum. Methods* **75**: 51–55.

Robbins, D.J. (1980). On predicting the maximum efficiency of phosphor systems excited by ionizing radiation. *J. Electrochem. Soc.* **127**: 2694–2702.

Robertson, J.S., *et al.* (1972). In "Tomographic Imaging in Nuclear Medicine" (G.S. Freedman, ed.) pp. 142–153. The Society of Nuclear Medicine, Inc., New York.

Rodnyi, P. A. (1997). "Physical Processes in Inorganic Scintillators," CRC Press, Boca Raton, FL.

Sakai, E. (1987). Recent measurements on scintillator-photodetector systems. *IEEE Trans. Nucl. Sci.* **NS34**: 418–422.

Schmand, M. (2000). "Higher Resolution PET by Means of a New Scintillator LSO," Ph.D. thesis, Max-Planck-Institute for Neurological Research.

Schweitzer, J.S., and Ziehl, W. (1983). Temperature dependence of NaI(Tl) decay constant. *IEEE Trans. Nucl. Sci.* **NS30**: 380.

Snyder, R.S., and Clotfelter, W.N. (1974). NASA Technical Memorandum, NASA TM X-64898, Physical Property Measurements of Doped Cesium Iodide Crystals. George C. Marshall Space Flight Center, Alabama.

Sorenson, J.A., Phelps, M.E. (1987). "Physics in Nuclear Medicine," W.B. Saunders Company, New York.

Tidd, J.L., Dabbs, J.R., and Levine, N. (1973). NASA Technical Memorandum, NASA TM X-64741. Scintillator Handbook with Emphasis on Cesium Iodide. George C. Marshall Space Flight Center, Alabama.

Valentine, J.D., Wehe, D.K., Knoll, G.F., Moss, C.E. (1993). Temperature dependence of CsI(Tl) absolute scintillation yield. *IEEE Trans. Nucl. Sci.* **40**: 1267–1274.

Valentine, J.D., and Rooney, B.D. (1994). Design of a compton spectrometer experiment for studying scintillator non-linearity and intrinsic energy resolution. *Nucl. Instrum. Methods Phys. Res. A* **353**: 37–40.

van Eijk, C.W. (2000). Inorganic-scintillator development. *Nucl. Instrum. Methods A* **460**: 1–14.

van Loef, E.V.D., *et. al.* (2001). High-energy-resolution scintillator: Ce^{3+} activated $LaBr_3$. *Appl. Phys. Lett.* **79**: 1573–1575.

Weber, M.J. (1999). "Inorganic Scintillators Today and Tomorrow." IEEE Symposium Presentation, Lawrence Berkeley National Laboratory.

Wilkinson, D.H. (1952). The Phoswich-A multiple phosphor. *Rev. Sci. Instrum.* **23**: 414.

Yamamoto, Y.L. (1977). Dynamic position emission tomography for study of cerebral hemodynamics in a cross section of the head using positron-emitting ^{68}Ga-EDTA and ^{77}Kr. *J. Comput. Assis. Tomogr.* **1(1)**: 43–56.

Zerby, C.D., Meyer, A., and Murray, R.B. (1961). Intrinsic line broadening NaI(Tl) gamma-ray spectrometers. *Nucl. Instrum. Methods* **12**: 115–123.

C H A P T E R

14

Photodetectors

BERND J. PICHLER and SIBYLLE I. ZIEGLER

Nuklearmedizinische Klinik und Poliklinik Klinikum Rechts der Isar, Technische Universität München, Munich, Germany

I. Introduction
II. Photomultiplier Tubes
III. Semiconductor Diode Detectors
IV. PIN Diodes
V. Avalanche Photodiodes
VI. Comparison of PMT and APD Properties
VII. Drift Diodes
VIII. Direct Detection of Gamma Rays: CdTe and CdZnTe Detectors

I. INTRODUCTION

The detection of low levels of light is the key process in medical imaging techniques based on radiation detection with scintillators. For nearly 3 decades photomultiplier tubes (PMTs) have been used as photodetectors for single-photon emission computed tomography (SPECT) and positron emission tomography (PET) applications and are still the most commonly used light detector in this field. In recent years, new compact semiconductor devices have become available that offer new design options and are competitive to PMTs in terms of performance and cost. Especially dedicated high-resolution applications, such as small animal positron emission tomography, may benefit from these new detector technologies because compactness is a requirement. This chapter gives an introduction to the basic mechanism of light detection, the principles of PMTs, and modern compact light sensors such as avalanche photodiodes (APDs) and drift diodes. Finally, the potential of direct gamma-ray detection without a scintillator is discussed.

A. Light Emission of Common Scintillators

The light that is isotropically emitted in the scintillation crystal needs to be collected efficiently by the photodetector. For this purpose, a good reflector and an optimized optical coupling between scintillation crystal and light detector are essential. Depending on the crystal size and shape, only a small fraction of the emitted light reaches the photodetector, which can be as low as 60% for a 10-mm-thick crystal (Moszynski *et al.*, 1997). For common scintillators such as BGO, LSO, or NaI (Tl) the light yield is in the range of only ~ 9000 photons/MeV (BGO) to ~ 40000 photons/MeV (NaI(Tl)) (Dorenbos *et al.*, 1995). Because the light detector is at the beginning of the electronic chain, low noise and high internal gain are required to achieve a good signal-to-noise ratio (SNR).

B. Important Performance Factors of Light Detectors

Light detectors are designed to convert low light levels into an electric signal of reasonable amplitude to avoid deterioration of the signal by external noise or pick-up. To ensure a high-quality signal, a light detector must fulfill three main requirements:

- A high quantum efficiency to transfer as many of the photons as possible into charge carriers to ensure a high signal amplitude. The quantum efficiency ρ is the ratio of the number n_p of primarily produced electrons in a light detector and the number n_i of incident photons:

$$\rho = \frac{n_p}{n_i} \quad (1)$$

- A fast readout speed. This is essential for timing applications, such as PET, to obtain good time resolution.
- Good amplitude resolution that leads, in combination with a scintillator, to high energy resolution. The amplitude resolution is affected by the quantum efficiency, the internal gain, and the electronic noise of the light detector.

II. PHOTOMULTIPLIER TUBES

In emission tomography applications, PMTs are the most important light detectors. They are reliable and their high gain of about 10^6 provides a very good SNR. Nevertheless, their low quantum efficiency of approximately 25%, their sensitivity to even weak magnetic fields (Earth's magnetic field), and their size are drawbacks. Figure 1 shows the essential parts of a photomultiplier tube. The vacuum tube contains the light entrance window, the photocathode, the electron collection structure, the amplification stages (dynodes), and the anode. The light input window material is commonly borosilicate glass. It transmits light in the range from 300 nm up to near infrared. The light entrance window should match the index of refraction of the scintillator and should have a high transmission component for the wavelength of the incident light. The entrance window can be arranged on the top of the PMT (head-on type) or at the side of the PMT (side-on type). For imaging applications, the head-on type is the most commonly used. Their sizes range from 0.5-inch diameter up to 5-inch diameter and even larger for special applications.

A. The Photocathode of a PMT

The incident light generates photoelectrons on the photocathode by the photoelectric effect. There are two different types of photocathodes, the reflection mode cathode and the more commonly used transmission mode cathode. In the latter, the generated photoelectrons penetrate through the material to reach the first dynodes, whereas the reflection mode cathode requires a special dynode arrangement so that the reflected electrons can reach the first dynode. The spectral response of the PMT is determined by the material of the photocathode and the entrance window. Listed next are the most commonly used photocathode materials (Photonis, 1994).

- Bialkali materials have a spectral response from the ultraviolet (200 nm) up to the visible range (600 nm). Their maximum quantum efficiency is 25–30% at 370 nm; thus, bialkali photocathodes are widely used for scintillation light detection from NaI, BGO, and LSO crystals.
- Multialkali photocathodes have a flatter spectral response starting at 250 nm, going up to 850 nm, with a maximum quantum efficiency of 20% at 420 nm. The main advantage of multialkali cathodes is their sensitivity for longer wavelengths.
- Bialkali with an extended sensitivity in the green region is commonly used for CsI(Tl) scintillation crystal applications.
- GaAsP photocathodes activated with cesium reach a maximum quantum efficiency of approximately 50% at 550 nm. The spectral response starts at approximately 350 nm with a quantum efficiency of nearly 20% and goes up to 700 nm. GaAsP is used in hybrid photomultiplier tubes (HPMTs, see Section II G). HPMTs that are based on GaAsP are about 10 times more expensive than PMTs with bialkali or multialkali cathodes. The GaAsP cathode is a very recent development. Because of their high cost they will only be used for the detection of very weak light sources in special applications.
- InGaAs has properties comparable to GaAsP, but currently there has been little experience using this material.

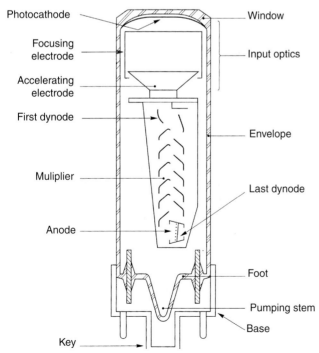

FIGURE 1 Basic structure of a photomultiplier tube. (From Photonis, 1994, *Photomultiplier Tubes: Principles and Applications*.)

B. Quantum Efficiency of PMTs

In general, the quantum efficiency of PMTs depends on the following factors:
- Light reflection at the glass window of the PMT.
- Light reflection at the photocathode (depends on the indices of refraction of the glass and of the cathode material).
- Photoelectric cross section of the photocathode material.
- Thickness of the photocathode.

The main factors affecting the quantum efficiency of PMTs are the material (photoabsorption efficiency of the cathode) and the thickness of the photocathode. If a cathode layer is too thin, not all photons will generate photoelectrons. On the other hand, if the cathode material is too thick, it is possible that the photoelectrons generated near the entrance window will recombine on their path through the cathode material.

The cathode radiant sensitivity also expresses the spectral response of a PMT. Whereas the quantum efficiency is a relative number, the unit of the radiant sensitivity is amperes per watt and describes the photoelectric current from the photocathode divided by the incident radiant power. Quantum efficiency is used more often in physics despite the fact that radiant sensitivity is an absolute value. Quantum efficiency, and therefore radiant sensitivity, depend on the wavelength λ of the incident light.

C. Electron Multiplication, Gain, and Dynode Structures

The photoelectrons (primary electrons) emitted by the photocathode are focused by the electron optical input system and accelerated by a high electric field on their trajectory to the first dynode. The dynode emits g_d secondary electrons when struck by an electron. The following dynodes, arranged in a cascade structure, further multiply the number of electrons. The electrons emitted by the last dynode are collected by the anode, yielding the signal current. In a PMT with N stages and n_p primary electrons, the total number n_{col} of electrons at the anode is therefore:

$$n_{col} = n_p \prod_{d=1}^{N} g_d \qquad (2)$$

The number of secondary electrons generated at a dynode depends on the energy of the primary electron (voltage between the dynodes) and the dynode material. A typical gain is ≥ 6 secondary electrons at the first dynode and approximately four electrons at the following dynode stages. A PMT with 10 dynode stages has a gain of 10^6–10^7 at a voltage between the cathode and anode of about 1200 V.

Dynodes are made of semiconductor or insulator materials, such as AgMg, CuBe, and NiAl with an MgO, BeO, or Al_2O_3 oxide (Photonis, 1994). The most commonly used

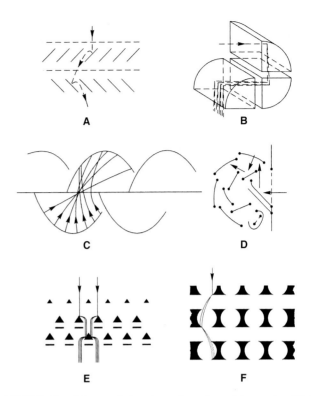

FIGURE 2 Scheme of dynode configurations: (a) venetian blind dynodes, (b) box dynode structure, (c) linear focusing dynodes, (d) circular cage dynodes, (e) mesh dynodes, and (f) foil dynodes. (From Photonis, 1994, *Photomultiplier Tubes: Principles and Applications*.)

material is a compound alloy of antimony-potassium-cesium (SbKCs). Figure 2 gives an overview of the arrangement of dynodes in a photomultiplier (PM) tube. The linear focusing type is the most frequently used.

D. Electronic Properties of PMTs

1. Electronic Noise

Because the emission of secondary electrons at the dynodes is a statistical process, their number can be assumed to vary according to Poisson statistics, although in reality PMTs have somewhat larger fluctuations. Therefore, the relative standard deviation of the number of secondary electrons produced for one incident electron at the first dynode is:

$$\frac{\sigma(g_1)}{g_1} = \frac{\sqrt{g_1}}{g_1} = \frac{1}{\sqrt{g_1}} \qquad (3)$$

Multiplication at all stages yields the amplitude of the output pulse at the anode. Assuming a Poisson process at each dynode, the SNR of the anode signal is mainly determined by the variation of the electron number at the first dynode, where the total number is smallest.

Although a PMT is a very low-noise device, thermal emission of electrons (thermionic noise) from the cathode or dynode material is the main reason for dark current in the PMT. This current is usually orders of magnitude smaller than the photocurrent. For example, from a PMT with a bialkali cathode approximately 30 single electrons per second are detected per square centimeter of cathode when the PMT is kept in the dark and under room-temperature conditions. This increases the dark current and therefore the noise.

2. Temperature Effects

The temperature mainly affects the thermionic noise, which means that in cooler conditions the noise is reduced. The quantum efficiency and the gain of a PMT show a weak dependence of temperature.

3. Time Response and Signal Shape

Three different time characteristics are important for PMT applications: the electron transit time, the rise time, and the fall time. The electron transit time describes the time between detection of an infinitely short (delta function) light pulse at the photocathode and the arrival of the output peak on the anode. The transit time spread is caused by time fluctuations of the primary electrons on their path from the photocathode to the first dynode and therefore depends on the size of the active area and the focusing structure of the PMT. This spread can range from 300 ps up to 5 ns. The signal rise time is defined as the time between 10% and 90% of the maximum anode signal amplitude. For PMTs, the rise time is approximately 0.5–3 ns for a delta input function. The signal fall time is defined as the time between 90% of the maximum of the anode pulse until the signal goes down to 10% of its maximum amplitude. In general, the time characteristics of PMTs are determined by the applied voltage and the dynode structure. Linear focusing and circular-cage dynode structures have a better time response than box or venetian-blind dynodes.

A PMT can be considered to be an ideal current source. Its fast time response can track the light output curve of a scintillation crystal. The output signal rise time t_{rise} of a PMT-scintillator detector is the combination of the rise time $t_{rise, PMT}$ of a multiplier tube and the scintillator rise time $t_{rise,scint}$:

$$t_{rise} = \sqrt{t_{rise,PMT}^2 + t_{rise,scint}^2} \quad (4)$$

Figure 3 shows the output signal of a LSO-PMT detector irradiated with a ^{68}Ge/^{68}Ga-source. The signal rise time of 6 ns is a combination of the rise time of the LSO scintillator and the PMT. The signal decay time, which is approximately 60 ns, follows the scintillation decay time of the LSO crystal (Melcher and Schweitzer, 1992).

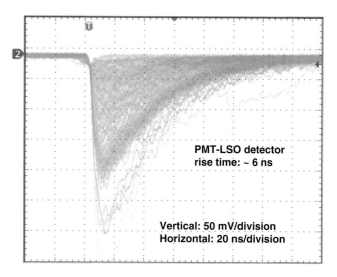

FIGURE 3 Output signal shape of a LSO-PMT detector.

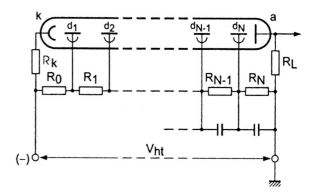

FIGURE 4 High-voltage PMT resistor network with the anode grounded and the cathode at negative high-voltage polarity. (From Photonis, 1994, *Photomultiplier Tubes: Principles and Applications*.)

4. Spatial Uniformity

The spatial uniformity of PMTs affects both the timing and the sensitivity. The PMT sensitivity and its time response vary with the position of the incident light on a photocathode. The sensitivity variation can be close to 10% for head-on-type PMTs and nearly 50% (less sensitivity on the edges) for side-on types. The variation of the time response depends on the size of the tube and the bias voltage; it can be as much as 4 ns from the center to the edges of an entrance window for a 2 × 2 inch tube. The time uniformity is much better for dedicated timing-PMTs with a spherical photocathode shape.

5. Voltage Divider and Polarity

The voltage divider is a resistor network that supplies the individual dynodes with the desired voltages. There are two different voltage divider circuits for PMT applications, for DC operation and for pulse operation. The voltage divider for pulse operation mode contains capacitors in parallel with the resistors of the last dynodes (see Figure 4).

These capacitors work as a charge storage and provide a homogeneous current flow (Photonis, 1994). The photocurrent of a PMT is large and can be in the milliampere range. Because the PMT gain depends strongly on the bias voltage, small deviations can cause large effects; thus the bias voltage has to be stabilized and the ripple and noise of the voltage has to be minimized. There are two ways to bias a PMT: connecting the anode to a positive high voltage and the cathode to ground, or connect the cathode with a negative polarity and the anode to ground. For most applications, the latter is used because no high-voltage decoupling of the cathode and the electronics is needed. The drawback is the higher chance of current leakage caused by discharges at the PMT glass.

E. Effects of Magnetic Fields

PMTs are very sensitive even to weak magnetic fields, such as earth's magnetic field. The most sensitive part of a PMT to magnetic fields is the electron collection structure in front of the first dynode. The photoelectrons are deflected from their trajectories, which results in decreased pulse height. The sensitivity to magnetic fields depends strongly on the type of the PMT, especially on the size of the entrance window, its shape, and the dynode types. Small round PMTs are less sensitive than large rectangular tubes, and meshed dynodes are less sensitive than other dynode types because of the short mean free path between two dynodes. To avoid this effect, a µ-metal shielding covering the tube and extending beyond the cathode is required (Photonis, 1994).

F. Position-Sensitive Photomultiplier Tubes

Several manufacturers have developed position-sensitive photomultiplier tubes (PSPMTs). These tubes provide spatial information about the detected light. They are based on the collection and multiplication of the charge cloud that is generated on the photocathode around the original position of the incident light. In medical imaging, especially in high-resolution applications (Cherry *et al.*, 1977; De Guerra *et al.*, 1998; Miyaoka *et al.*, 1998; Pani *et al.*, 1999; Seidel *et al.*, 2000; Weber *et al.*, 1999), these PMTs have proven very valuable. There are three main structural techniques: the multichannel dynode, the proximity mesh dynode, and the metal channel dynode (Pani *et al.*, 1997). The multichannel dynodes are usually combined with a multianode structure that minimizes the electronic cross talk of the anodes and provides well-defined position detection. The disadvantages are their large dead area, the limited number of anodes, and the limited size of their active area. PSPMTs with proximity mesh dynodes and crossed-wire anodes can be produced with an active surface of up to 13 cm and a total number of anodes up to 256. The metal channel dynodes are combined with a multianode or crossed-plate anode technique (Nagai *et al.*, 1999). They provide very low cross talk of only approximately 1% (Hamamatsu Photonics, 1999b) and a well-focused charge distribution. Their performance (rise time, gain, spectral response, and quantum efficiency) is comparable to conventional PMTs. The array sizes of current multianode PSPMTs based on metal channel dynodes are 6 × 6 or 8 × 8. The drawbacks of PSPMTs are their high gain variation within the array of approximately 1:3 (Hamamatsu Photonics, 1999b), their ratio of sensitive surface to total size of approximately 50–73% (Nagai *et al.*, 1999), and their relatively high cost.

G. Photomultiplier Tubes: Future Trends

Newer developments are aimed at improving the performance of PMTs and reducing their size. In this section, we discuss three new developments that may prove useful for medical imaging.

1. Hybrid Photomultiplier Tubes

A HPMT is a combination of a conventional PMT and photodiode (PIN or APD) (Mirzoyan *et al.*, 2000). The photoelectrons emitted by the photocathode are accelerated by a high voltage (8–15 kV) and focused onto the active surface of the diode, which is contained in a vacuum tube. As the photoelectrons exceed the required energy of $E_p \approx 2$ keV and penetrate into the photodiode, they are able to create electron–hole pairs. The photoelectrons are accelerated on their way from the cathode to the diode up to $E_A \approx 8$ keV. To produce one electron–hole pair in silicon, an energy of $E_{silicon} = 3.62$ eV is needed. This results in a gain of approximately

$$Gain_{HPMT} = \frac{E_A - E_P}{E_{silicon}} = \frac{8\text{ keV} - 2\text{ keV}}{3.6\text{ eV}} \approx 1600 \quad (5)$$

at 8000 V in a single amplification stage. This gain can further be increased by a factor of ~100 if an APD is used instead of a PIN diode. HPMTs provide good timing and excellent energy resolution because of their very low noise. This low noise can be explained by the high amplification at the first interaction of the photoelectrons (see Eq. 5).

Very compact HPMTs that have an effective area that is 8 mm in diameter and can be arranged in arrays have been developed (Suyama *et al.*, 1997, 1998). The small distance between photocathode and diode (~1 cm for Hamamatsu HPMTs and ~3 cm for Intevac HPMTs) ensures a compact design and allows their use in even relatively strong magnetic fields.

2. Channel Photomultiplier

A channel photomultiplier (CPM) (Perkin Elmer, Canada, formerly EG&G) is an ultrasensitive and low-noise photodetector with a unique design: the dynodes are replaced by a thin, bent semiconductor channel. The photoelectrons that are emitted by the photocathode are accelerated to the anode by a high voltage of approximately 2400 V, which results in a gain of approximately 5×10^7. The electron multiplication occurs in the channel. The dark noise in the channel is only 0.01–0.1 electrons/s. This results in a very low dark current of approximately 1/100 of the anode dark current in a typical PMT. The spectral response is in the range 115–850 nm and the time behavior is comparable to conventional PMTs. The length of the CPM is 75 mm, the diameter is 10.5 mm, and the entrance window diameter is 9 mm. The small size of this PMT may play a role in its future use in medical imaging.

3. Flat Panel Photomultiplier Tube

Another interesting development has been announced by Hamamatsu. The flat panel PMT (Hamamatsu Photonics, Japan) combines a multianode device with a very compact design. The total thickness of the tube is only 12.4 mm (compared to 20.1 mm of the PSPMT). This is possible due to the compact arrangement of the metal channel dynodes. The effective area is 49.7×49.7 mm^2 and comprises an 8×8 anode channel array. The width of the nonsensitive frame of the tube's top side is only 2 mm. The gain is about 10^6 and the spectral response is between 300 nm and 650 nm. The gain uniformity over the 64 channels is specified as 1:4 and is comparable to other multianode PMTs. The first results using this new PMT in imaging are expected in the near future.

III. SEMICONDUCTOR DIODE DETECTORS

In comparison to PMTs, semiconductor light detectors used for scintillator readout provide several advantages:
- They are very compact, allowing the production of miniaturized detectors.
- They are easy to produce as monolithic diode arrays for high-resolution applications.
- They are insensitive to magnetic fields.
- They are available with large active areas.
- They have a quantum efficiency of up to 90% in the visible range.

In semiconductors, light can be detected by generating electron–hole pairs in the substrate. The energy needed for this process exceeds the band gap energy of the semiconductor. In the case of silicon, the photons must deposit a minimum of 3.62 eV, even though the band gap is only 1.12 eV, because in this material excitation from conduction to valence band is an indirect process through phonon creation (lattice vibration). Semiconductor diode detectors are based on the properties of the junction between n-type and p-type silicon materials (Sze, 1981). The junction is generated within the material by doping characteristics. Charge diffusion results in a depletion layer at the junction, in which an electric field is generated by the electric potential difference. Within this field, electrons that are created near the junction are attracted to the p-type side, whereas holes move to the n-type region. Therefore, the concentration of charge in this zone is very low and the depletion region has a high resistivity compared to the n-type and p-type materials. Electron–hole pairs that are created in this region are moving out of the depletion layer and cause an electrical signal. Although this configuration can already function as a detector, the electrical field is not large enough to move the charge carriers effectively. Therefore, an external voltage is applied to increase the depletion layer.

Light in the visible and near-infrared range and low-energy gamma rays can be detected with semiconductor diode detectors. The absorption length, or penetration depth, of the light, which is the inverse of the absorption coefficient, strongly depends on the material and the wavelength of the incident light. For silicon, the absorption length is approximately 0.13 μm at a temperature of 300 K and light with a wavelength of 420 nm (Melchior, 1972; Sze, 1985). Two different cut-off wavelengths are important: the upper cut-off wavelength (1.1 μm in silicon) and the short wavelength cut-off. The upper cut-off wavelength is defined by the band gap energy and the short wavelength cut-off is caused by the very high absorption coefficient at low wavelengths. Photons are mostly absorbed close to the detector surface in this case, but not in or close to the pn-junction. Thus they cannot contribute to the signal. Light with a wavelength within these limits produces the sufficient photocurrent in the diode.

The quantum efficiency of a semiconductor diode depends on the following factors:
- Light reflection and absorption at the glass or epoxy protection layer of the diode.
- Light absorption in the passivation layer.
- Light loss due to reflection on the diode surface.
- Light loss due to recombination in the p-layers.
- Absorption coefficient of the material.

The charge collected in the semiconductor detector is too small without further amplification. Therefore, preamplifiers are used directly between the detector and the pulse analysis electronics. In general, there are two types of preamplifiers: the current-to-voltage converter and the charge-sensitive preamplifier. Both convert the charge Q_{in} from the detector into a voltage signal. The output signal U_{out} of a current-to-voltage converter is:

$$U_{out} = \frac{Q_{in}}{C_d}, \qquad (6)$$

therefore U_{out} depends on the detector's capacitance C_d.

Because the capacitance of a diode is inversely proportional to the depletion depth, which is a function of bias voltage, a preamplifier with a conversion factor independent of the detector capacitance is an advantage. For a charge-sensitive preamplifier, the output voltage is proportional to the total integrated charge in the pulse (see Figure 5). The output signal U_{out} of the charge-sensitive preamplifier is determined by the input charge and the feedback capacitor C_f. Based on this and the fact that integrated current-to-voltage amplifiers tend to oscillate with fast signals, most applications use charge-sensitive preamplifiers. The signal rise time of a photodiode-preamplifier detector is determined by the preamplifier (transconductance and capacitance of the preamplifier's input device) and the detector capacitance of the photodiode. Figure 6 shows the output signal of a LSO-APD detector read out by a fast charge-sensitive preamplifier with a rise time of 30 ns. In combination with the LSO scintillator, the resulting signal rise time is approximately 60 ns. Compared to the LSO-PMT detector, which is used without an amplifier, this is 10 times slower. The signal decay time in Figure 6 is determined by the feedback resistor and the feedback capacitor of the preamplifier (see Figure 5).

The noise of the detector–preamplifier system depends on the input capacitance (noise slope, which is expressed in electrons/pF). To increase the SNR, the capacitive load should be minimized by minimal detector capacitance and by using short connecting cables between the detector and amplifier.

IV. PIN DIODES

In a PIN diode, the p-side is connected to negative voltage relative to the n-side (reverse bias). This increases the potential difference across the junction and therefore the thickness of the depletion layer (i, intrinsic; see Figure 7). Therefore, the region within which radiation is detected is determined by the reverse bias. It can be made large for high quantum efficiency, but is limited by the maximum voltage before breakdown of the diode occurs.

Silicon PIN diodes have been produced since the 1980s with a high quantum efficiency of approximately 70–80%. Large PIN diodes are available in sizes from a few square millimeters to a few square centimeters. PIN diodes are cheap and reproducible. Their high quantum efficiency is an advantage compared to PMTs. Unfortunately, PIN diodes have no internal gain; their energy resolution is determined by electronic noise and not by photoelectron statistics, as is the case for PMTs. Therefore, their SNR is very low compared to PMTs. The structure of silicon PIN diodes allows the implementation of monolithic front-end transistors to improve their noise behavior (Fazzi et al., 2000). Nevertheless, to achieve a noise level of only a few hundred electrons, the shaping time of a PIN diode has to be above 1 μs. This limits the maximum

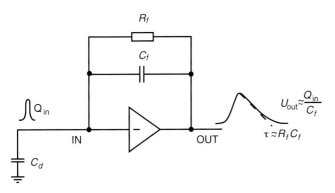

FIGURE 5 Basic circuit diagram of a charge-sensitive preamplifier. The charge of the detector capacitance C_d is converted into a voltage signal U_{out}.

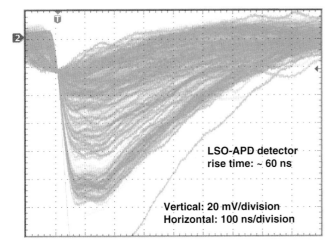

FIGURE 6 Output signal of a LSO-APD detector connected to a charge-sensitive preamplifier. The rise time is 10 times slower compared to the LSO-PMT detector. The signal decay time is determined by the feedback resistor and feedback capacitor of the amplifier.

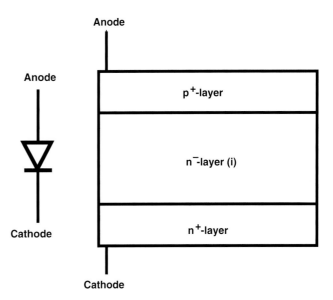

FIGURE 7 Scheme of a PIN diode (not to scale).

count rate and makes it impossible to use these devices for timing applications (Blanar et al., 1982). The leakage current strongly depends on the temperature and contributes significantly to the noise of PIN diodes. The shot noise current i_s^2 for a PIN diode is:

$$i_s^2 = 2qIB \quad (7)$$

where q is the electron charge, I is the current through the device, and B is the bandwidth.

The low bias voltage of up to 100 V and the possibility of producing PIN diodes with large areas or in compact arrays make this semiconductor device interesting for the identification of crystals within an array for depth-of-interaction measurements (Huber et al., 1997; Moses et al., 1995) and for applications with restricted space, such as breast imaging cameras (Gruber et al., 1998). Drawbacks of PIN diodes are their large noise and slow time response due to the shaping time constants.

V. AVALANCHE PHOTODIODES

Since the late 1980s, epitaxial wafers have become available that allow the production of high-quality diodes with internal gain—avalanche photodiodes (APDs). Their structure needs to be more complicated than that of PIN diodes to facilitate the avalanche process. Again, the compactness of APDs is an advantage in applications that require high spatial resolution, such as small animal scanners or dedicated human scanners with a space restriction.

Important properties of APDs are:
- High quantum efficiency
- High internal gain
- Low dark current
- Low bias voltage

A large difference between the bias and break-down voltage is important to ensure the stability of the diode and avoid breakdowns. Nevertheless, the diode current should be limited to a value that avoids the destruction of the diode.

A. Structure and Properties of APDs

Figure 8 shows the basic properties and structure of an APD. The bulk material (substrate) is low-resistivity n-doped silicon. This bulk material is the base for more epitaxially grown layers. The p^{++}-type layer on the top of the APD is the light-sensitive surface and the entrance window for the incident photons. The highly doped layer is necessary to achieve a low-resistivity anode contact. The incident light should be completely absorbed by the p-layer (conversion layer); thus the thickness of this layer must be adjusted to the wavelength of the light that should be detected. If the p-layer is too thin,

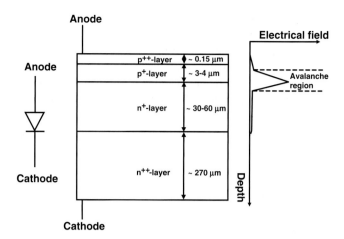

FIGURE 8 Structure of a blue-enhanced APD (not to scale). The layer thickness and arrangement are estimated for Hamamatsu APDs. In addition, the function of the electric field versus the depth is shown.

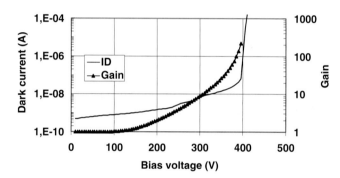

FIGURE 9 Gain and dark current of a 1.6×1.6 mm^2 Hamamatsu APD versus the high voltage.

the light penetrates into the adjacent pn-junction, where the avalanche multiplication process takes place. If light is absorbed in the high field region, it results in a gain loss because the generated charge does not have to transverse the complete avalanche region. On the other hand, a p-layer that is too thick can cause a loss of charge carriers due to recombination. The thickness of the APD determines its capacitance; thus to reduce the detector capacitance a low doped n-layer is inserted after the pn-junction. When the APD is biased, nearly the entire diode is depleted to reduce the capacitance. The diode dark current and the gain of the APD are a function of bias voltage (see Figure 9). Thus, small voltage drifts cause variations in the APD gain. The APD surface is covered with a SiO_2 or SiN_4 passivation layer and most of the APDs have an epoxy (Hamamatsu Photonics APDs) or glass window to protect the silicon against moisture and mechanical damage.

B. Theory of Avalanche Multiplication

Avalanche multiplication is based on a high electric field on the order of several volts per micrometer within the APD that accelerates the electrons. If the electron's energy is high enough, it is able to generate further electron–hole pairs by impact ionization. The probability of generating new electron–hole pairs within the path length dx is expressed by the ionization probabilities for electrons α and for holes β. The ionization probability depends on the electric field E:

$$\alpha(E), \beta(E) \propto \exp\left(-\frac{1}{E}\right) \quad (8)$$

For linear amplification and optimized noise, only one charge type—electron or holes—should be multiplied in the APD. Because $\alpha \gg \beta$ for silicon-based APDs, the electrons are the charge type that should be amplified. If $\alpha = \beta$, even small electric fields would cause a high current that destroys the diode. The ratio between the ionization probabilities of electrons and holes is called the ionization coefficient k:

$$k = \frac{\alpha}{\beta} \quad (9)$$

The gain M of an APD expresses the number of multiplied electron–hole pairs caused by one original charge pair. The multiplication process is started at position x within the depletion layer (length w). After McIntyre (1966), the APD gain can be calculated by the following integral:

$$M(x) = 1 + \int_0^x \alpha M(x') dx' + \int_x^w \beta M(x') dx' \quad (10)$$

Differentiation of Eq. (10) leads to a first-order linear differential equation. The solution of this equation is the gain of an APD as a function of the ionization probabilities:

$$M(x) = \frac{\exp\left[-\int_x^w (\alpha - \beta) dx'\right]}{1 - \int_0^w \alpha \exp\left[-\int_{x'}^w (\alpha - \beta) dx''\right] dx'} \quad (11)$$

C. Noise Behavior of APDs

The electronic noise of APDs depends on the following factors:

- Intrinsic APD parameters, such as ionization coefficients, that are expressed in the excess-noise factor.
- Bulk current I_b and surface current I_s of the APD.
- Parallel and series APD resistors.
- The internal APD gain.

The major noise factor of APDs is excess noise, which explains statistical fluctuations of the avalanche process in the depletion layer of the APD. The excess-noise factor $F(M)$ depends on the APD gain M and the ionization coefficient k:

$$F(M) = k \cdot M + (1-k)\left(2 - \frac{1}{M}\right) \quad (12)$$

According to Eq. (12), it follows that the excess-noise factor of APDs is at least $F(M) = 2$ because for each impact ionization only one electron–hole pair is generated.

The internal APD gain amplifies the bulk current I_b and the photo current I_p and therefore affects i_s^2, the current of the shot noise (Advanced Photonix, 1991):

$$i_s^2 = 2q\left[I_s + (I_b + I_p) M^2 F(M) B\right] \quad (13)$$

with the frequency bandwidth B and the electron charge q. Because the APD gain multiplies the photocurrent and the bulk current, the shot noise of APDs is higher than the shot noise of PIN diodes.

The surface current I_s does not undergo the avalanche multiplication process (see Eq. 13). The SNR, S/N, of an APD is (Webb, 1974):

$$\frac{S}{N} = \frac{I_p^2 M^2}{i_s^2 + i_t^2} = \frac{I_p^2 M^2}{2q\left[I_s + (I_b + I_p) M^2 F(M)\right] B + \frac{4kTB}{R_G}} \quad (14)$$

where $1/R_G = 1/R_{pn} + 1/R_l + 1/R_{in}$, R_{pn} = resistivity of the pn-junction, R_l = load resistance, R_{in} = input resistance of the preamplifier, $i_t^2 = \frac{4kTB}{R_G}$ = thermal noise, k = Boltzmann's constant, B = frequency band, and T = temperature in degrees K.

Figure 10 shows the scheme of an APD charge-sensitive preamplifier setup. The APD is reverse biased with a negative voltage on the anode and a grounded cathode. To decouple the preamplifier from the bias voltage potential, a high-voltage capacitor is needed. In general it would be possible to read out the cathode signal with the amplifier, but the first method is much more robust to electronic

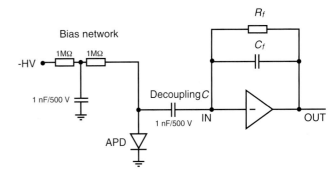

FIGURE 10 Reverse biasing of an APD. The figure shows the bias network, which limits the current and has a filter function. The charge-sensitive amplifier is decoupled from the APD by a high-voltage capacitor.

discharge—the amplifier is protected by the capacitor even in the case of a breakdown of the APD.

In combination with a charge-sensitive preamplifier and a Gaussian shaper, the electronic noise charge *ENC* can be calculated as (Martinez *et al.*, 2000; Radeka, 1988):

$$ENC^2 = \frac{e^2}{8q^2M^2}\left(2q[I_s + I_b M^2 F(M)]\tau + \frac{4kT}{R_f}\tau + \frac{4kTR_s C_T^2}{\tau}\right) \quad (15)$$

where $e = 2.71828$; q = electron charge; R_s = series noise resistor of the preamplifier; R_f = feedback resistor of the preamplifier; C_T = sum of the preamplifier input capacitance, the detector capacitance of the APD, and stray capacitors; and τ = shaping time of a Gaussian filter. This shows that increasing the gain of an APD and reducing the capacitance can drastically reduce the noise, while the contribution of surface currents is only minor.

D. Guard Structures

The APD structure affects the performance and the parameters (e.g., the required bias voltage, the dark current, the gain, the excess-noise factor, the signal time response, and the reliability of the diode). To produce blue-enhanced APDs, three structures are established.

1. *Guard ring structure.* On the junction of the n-layer and p-layer, very high electric fields occur. The edge zones are especially critical. To avoid an electrical breakdown and a resulting very high current that can destroy the diode, a guard ring structure (Sze, 1985) is implemented in the active area. The task of this guard ring structure is to remove the high electric fields and to drain the surface currents. Effective guard ring structures are usually not easy to implement.
2. *Beveled-edge structure.* An alternative to the guard ring structure is the beveled-edge structure—the sides of the diodes are beveled (Paulus *et al.*, 1995). Figure 11

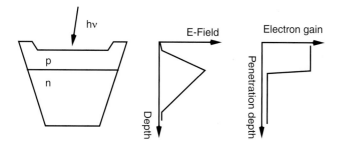

FIGURE 11 Basic structure of a beveled-edge APD, the corresponding electric field versus the layer depth, and electron gain versus the penetration depth of the photons.

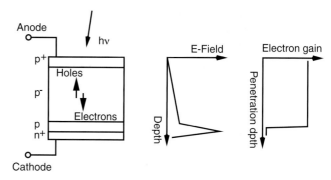

FIGURE 12 Basic structure of a reach-through APD, the corresponding electrical field versus the layer depth, and electron gain versus the penetration depth of the photons.

shows the beveled edge APD structure, a plot of the electric field, and the gain of the electrons versus the light penetration depth.

3. *Reach-through structure.* To avoid long transit times of the charge carriers and high bias voltages caused by the beveled-edge structure, the reach-through structure (McIntyre *et al.*, 1996) is used. Figure 12 shows the arrangement of the p- and n-layers, the vertical distribution of the electric field, and the electron amplification versus the penetration depth of the light. The pn-avalanche zone is arranged at the opposite side of the light entrance window, thus the electrons are accelerated in the p⁻-layer, the electrons and holes are separated, and the holes drift to the anode and mainly electrons will be multiplied in the avalanche region. Due to the thick depleted p⁻-layer, the light absorption is maximized and the diode capacitance is very low. Drawbacks are the high nuclear counter effect (photoelectric effect of gamma rays in the depletion layer) and the multiplication of the charge, which is thermally generated in the thick p⁻-layer.

APDs from Perkin Elmer, Canada, and APD arrays from Hamamatsu Photonics, Japan, are used in two small animal PET systems in combination with BGO or LSO scintillation crystals. Cherry and Chatziioannou (Chapter 12 in this volume) provide detailed information about these small animal PET scanners.

VI. COMPARISON OF PMT AND APD PROPERTIES

Both PMTs and APDs are photodetectors with internal gain and can be found in nuclear medical imaging. In Table 1, as an example, the parameters of the 13-mm-diameter Hamamatsu PMT R647-01 (Hamamatsu Photonics, 1998) are compared with the parameters of the 5-mm-diameter Hamamatsu silicon APD 1247.

TABLE 1 Parameters of the Hamamatsu PMT R647-01 compared to the Hamamatsu silicon APD 1247

Parameter	PMT	APD
Gain (typical)	1.4×10^6	~10^2
Bias voltage for this gain	1000 V	433 V
Quantum efficiency (max)	25% @ 370 nm	86% @ 420 nm
Quantum efficiency @ 400 nm	24%	66%
Spectral response	300–650 nm	350–1100 nm
Excess-noise factor	~1.2–1.5	~2–2.5
Breakdown voltage	—	441 V
Average anode current	0.1 mA	—
Dark current (max)	2 nA/132 mm^2	7.3 nA/19 mm^2
Rise time	2.5 ns	5 ns
Gain variations caused by magnetic fields	Up to 10% (Earth field)	No effect (10 T)
Detector capacitance	—	~111 pF

TABLE 2 Comparison of Position-Sensitive Detectors: Multianode PMT and an APD Array

Parameter	PMT	APD
Pixel arrangement	8×8 array	4×8 array
Pixel size	2×2 mm^2	1.6×1.6 mm^2
Dead space between pixels	0.3 mm	0.6 mm
Size ($l \times w \times h$)	$27.7 \times 27.7 \times 20.1$ mm^3	$19.5 \times 11.2 \times 1.5$ mm^2
Active surface/detector size	0.33	0.4
Gain (typical)	3×10^5	~200
Bias voltage for this gain	800 V	400 V
Gain variation	1:3 (max)	$\sigma < 15\%$
Quantum efficiency	20% @ 400 nm	66% @ 400 nm
Spectral response	300–650 nm	350–1100 nm
Detector capacitance	—	~10 pF
Breakdown voltage	—	412 V
Cross-talk	2%	<< 1%
Dark current (max)	0.2 nA	– 10 nA

A comparison of position-sensitive detectors, Hamamatsu Multianode PMT R5900-00-M64 (Hamamatsu Photonics, 1999a) and the 32-pixel Hamamatsu APD array (Pichler et al., to appear), is found in Table 2. Table 2 shows the superior performance of the compact APD array in most of the parameters fields. As discussed, the drawbacks of APDs are their low gain and slower signal rise time.

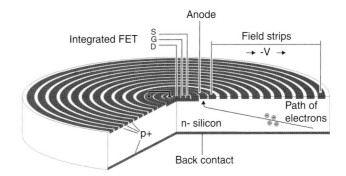

FIGURE 13 Structural view of a cylindrical silicon drift diode with integrated field effect transistor. (Courtesy of Dr. Peter Lechner, Max-Planck-Institute, Semiconductor Laboratory, Munich, Germany.)

VII. DRIFT DIODES

Silicon drift diodes have become available (Fiorini, Longoni, and Lechner, 2000; Fiorini, Longoni, and Perotti, 2000) that collect charge carriers in a way similar to gas drift detectors. Although they do not have an internal gain, their SNR is excellent because of very low noise characteristics. They consist of multiple highly doped concentric p$^+$-ring structures (cathodes), separated by oxide layers (Figure 13). The substrate is a highly resistant n$^-$-type material. An electrical field applied parallel to the detector surface transfers the free electrons to the n$^+$-collecting anode, which is in the center of the diode. The generated holes are absorbed by the p$^+$-ring structures (Fiorini, Longoni, and Perotti, 2000). An advantage of drift diodes compared to PIN diodes is their low output capacitance of approximately 0.1 pF, which is independent of the diode's size (Fiorini and Perotti, 1997). A n-channel junction field effect transistor (JFET) that is integrated in the detector silicon, close to the anode, minimizes the noise. In addition, the integrated JFET optimizes the capacitive matching between the drift diode and the front-end electronics. The noise of drift diodes is mainly dominated by the parallel white noise due to leakage currents. For shaping times larger than 1 μs, the root mean square (rms) noise is approximately a few hundred electrons at room temperature and can be reduced by a factor of two by cooling the detector to 0°C (Fiorini, Longoni, and Lechner, 2000). Silicon drift diodes can be produced with relatively large active area and very low noise. In combination with small CsI(Tl) crystals, an energy resolution of approximately 7.5% at 122 keV (full width at half maximum, FWHM) has been reached (Fiorini, Longoni, and Perotti, 2000). The application of drift diodes for the readout of large scintillation crystals with drift diode arrays is currently under evaluation, and it needs to be shown that similar performance can be reached within detector arrays (Fiorini, Longoni, and Perotti, 2000). Further studies will have to show the performance of this promising technology in a complete gamma camera with a larger field of view.

VIII. DIRECT DETECTION OF GAMMA RAYS: CDTE AND CDZNTE DETECTORS

The direct detection of gamma rays in a semiconductor detector without the need to first transfer the gamma energy into light would be most efficient. In silicon, the interaction probability of gamma rays is too low. Semiconductor materials with a large band gap (low thermionic noise) and high atomic number are needed for room temperature detectors. In cadmium telluride (CdTe) and cadmium zinc telluride (CdZnTe), the photoelectric absorption for typical gamma-ray energies is about 100 times higher than in silicon. CdZnTe has higher resistivity than CdTe and therefore lower noise. Approximately 7 mm of CdZnTe is needed to achieve the same detection efficiency as in a standard NaI(Tl) gamma camera. In a crystal with this thickness, the collection time for holes is much longer than for electrons. Developments in readout electronics have enabled the efficient collection of electrons only (Butler, 1997). The collection process affects the time response of CdZnTe; thus its application is currently limited to single-photon imaging. With the improvement of crystal production methods, pixellated CdZnTe detectors with sizes suitable for medical imaging can be manufactured (Butler *et al.*, 1998; Cirignano *et al.*, 1999). The energy resolution of approximately 5% at 140 keV is superior to the NaI(Tl)-PMT gamma camera (10%, FWHM).

References

Advanced Photonix. (1991). Noise characteristics of Advanced Photonix avalanche photodiodes. Advanced Photonix Application Note API/NOIS/1291/B.
Blanar, G., Dietl, H., Dobbins, J., Eigen, G., Lorenz, E., Pauss, F., Pimpl, W., Vogel, H., and Weissbach, P. (1982). Photodiode readout for scintillating crystals of BGO and NaI(Tl). *Nucl. Instr. Meth.* **203:** 213–221.
Butler, J. (1997). Novel detector design for single-carrier charge collection in semiconductor nuclear detectors. *Nucl. Instr. Meth.* **A396:** 396–399.
Butler, J., Lingren, C., Friesenhahn, S., Doty, F., Ashburn, W., Conwell, R., Augustine, F., Apotovsky, B., Pi, B., Collins, T., Zhao, S., and Isaacson, C. (1998). CdZnTe solid-state gamma camera. *IEEE Trans. Nucl. Sci.* **45:** 359–363.
Cherry, S., Shao, Y., Silverman, R., Meadors, K., Siegel, S., Chatziioannou, A., Young, J., Jones, W., Moyers, J., Newport, D., Boutefnouchet, A., Farquhar, T., Andreaco, M., Paulus, M., Binkley, D., Nutt, R., and Phelps, M. (1997). MicroPET: A high resolution PET scanner for imaging small animals. *IEEE Trans. Nucl. Sci.* **44:** 1161–1166.
Cirignano, L., Shah, K., Bennet, P., Klugerman, M., Dmitryev, Y., Squillante, M., Narita, T., Bloser, P., Grindlay, J., Hasegawa, B., and Iwata, K. (1999). Pixellated CdZnTe detector for emission/transmission computed tomography. *Nucl. Instr. Meth.* **A422:** 216–220.
Del Guerra, A., Di Domenico, G., Scandola, M., and Zavattini, G. (1998). YAP-PET: First results of a small animal positron emission tomograph based on YAP:Ce finger crystals. *IEEE Trans. Nucl. Sci.* **45:** 3105–3108.
Dorenbos, P., de Haas, J., and van Eijk, C. (1995). Non-Proportionality in the scintillation response and the energy resolution obtainable with scintillation crystals. *IEEE Trans. Nucl. Sci.* **42:** 2190–2202.
Fazzi, A., Pignatel, G., Dalla Betta, G., Boscardin, M., Varoli, V., and Verzellesi, G. (2000). Charge preamplifier for hole collecting PIN diode and integrated tetrode N-JFET. *IEEE Trans. Nucl. Sci.* **47:** 829–833.
Fiorini, C., Longoni, A., and Lechner, P. (2000). Single-side biasing of silicon drift detectors with homogeneous light-entrance window. *IEEE Trans. Nucl. Sci.* **47:** 1691–1695.
Fiorini, C., Longoni, A., and Perotti, F. (2000). New detectors for gamma-ray spectroscopy and imaging, based on scintillators coupled to silicon drift detectors. *Nucl. Instr. Meth.* **A454:** 241–246.
Fiorini, C., and Perotti, F. (1997). Scintillation detection using a silicon drift chamber with on-chip electronics. *Nucl. Instr. Meth.* **A401:** 104–112.
Gruber, G., Moses, W., Derenzo, S., Wang, N., Beuville, E., and Ho, M. (1998). A discrete scintillation camera module using silicon photodiode readout of CsI(Tl) crystals for breast cancer imaging. *IEEE Nucl. Instr. Meth.* **45:** 1063–1068.
Hamamatsu Photonics. (1998). Photomultiplier tubes and assemblies for scintillation counting and high energy physics. *Hamamatsu Photonics Catalog.* Japan.
Hamamatsu Photonics. (1999a). Multianode photomultiplier tube R5900-00-M64. *Hamamatsu Photonics Catalog.* Japan.
Hamamatsu Photonics. (1999b). Multianode photomultiplier tube R5900-00-M64. Hamamatsu Preliminary Data Sheet. Japan.
Huber, J., Moses, W., Derenzo, S., Ho, M., Andreaco, M., Paulus, M., and Nutt, R. (1997). Characterization of a 64 channel PET detector using photodiodes for crystal identification. *IEEE Trans. Nucl. Sci.* **44:** 1197–1201.
Martinez, M., Ostankov, A., Lorenz, E., Mirzoyan, R., and Schweizer, T. (2000). Constraints in using APDs in air Cherekov telescopes for γ-astronomy. *Nucl. Instr. Meth.* **A442:** 209–215.
McIntyre, R. (1966). Multiplication noise in uniform avalanche diodes. *IEEE/Trans. Electron. Devices* **ED-13:** 164–168.
McIntyre, R., Webb, P., and Dautet, H. (1996). A short-wavelength selective reach-through avalanche photodiode. *IEEE Trans. Nucl. Sci.* **43:** 1341–1346.
Melcher, C., and Schweitzer, J. (1992). Cerium-doped lutetium oxyorthosilicate: A fast, efficient new scintillator. *IEEE Trans. Nucl. Sci.* **39:** 502–505.
Melchior, H. (1972). Demodulation and photodetection techniques. In "Laser Handbook" (F. Arecchi and E. Schulz-Dubois, eds.), Vol. 1, pp. 725–835. North-Holland, Amsterdam.
Mirzoyan, R., Ferenc, D., and Lorenz, E. (2000). An evaluation of the new compact hybrid photodiodes R7110U-07/40 from Hamamatsu in high-speed light detection mode. *Nucl. Instr. Meth.* **A442:** 140–145.
Miyaoka, R., Lewellen, T., Yu, H., and McDaniel, D. (1998). Design of a depth of interaction (DOI) PET detector module. *IEEE Trans. Nucl. Sci.* **45:** 1069–1073.
Moses, W., Derenzo, S., Melcher, C., and Manante, R. (1995). A room temperature LSO/PIN photodiode PET detector module that measures depth of interaction. *IEEE Trans. Nucl. Sci.* **42:** 1085–1089.
Moszynski, M., Kapuska, M., Mayhugh, M., Wolski, D., and Flyckt, S. (1997). Absolute light output of scintillators. *IEEE Trans. Nucl. Sci.* **44:** 1052–1061.
Nagai, S., Watanabe, M., Shimoi, H., Liu, H., and Yoshizawa, Y. (1999). A new compact position-sensitive PMT for scintillation detectors. *IEEE Trans. Nucl. Sci.* **46:** 354–358.
Pani, R., Pergola, A., Pellegrini, R., Soluri, A., De Vincentis, G., Filippi, S., Di Domenico, G., Del Guerra, A., and Scopinaro, F. (1997). New generation position-sensitive PMT for nuclear medicine imaging. *Nucl. Instr. Meth.* **A392:** 319–323.

Pani, R., Soluri, A., Scafe, R., Pergola, A., Pellegrini, R., DeVincentis, G., Trotta, G., and Scopinaro, F. (1999). Multi-PSPMT scintillation camera. *IEEE Trans. Nucl. Sci.* **46**: 702–708.

Paulus, M., Rochelle, J., and Binkley, D. (1994). Comparison of the beveled-edge and reach-through APD structures for PET applications. IEEE Nuclear Science Symposium and Medical Imaging Conference 1994, pp. 1864–1868.

Photonis. (1994). Photomultiplier tubes: Principles and Applications. Photonis, Brive, France.

Pichler, B., Bernecker, F., Böning, G., Rafecas, M., Pimpl, W., Schwaiger, M., Lorenz, E., and Ziegler, S. (2001). A 4×8 APD array, consisting of two monolithic silicon wafers, coupled to a 32-channel LSO matrix for high-resolution PET. *IEEE Trans. Nucl. Sci.* **48**: 1391–1396.

Radeka, V. (1988). Low-noise techniques in detectors. *Ann. Rev. Nucl. Part. Sci.* **38**: 217–277.

Seidel, J., Vaquero, J., Barbosa, F., Lee, I., Cuevas, C., and Green, M. (2000). Scintillator identification and performance characteristics of LSO and GSO PSPMT detector modules combined through common X and Y resistive dividers. *IEEE Trans. Nucl. Sci.* **47**: 1640–1645.

Suyama, M., Hirano, K., Kawai, Y., Nagai, S., Kibune, A., Saito, T., Negi, Y., Asakura, N., Muramatsu, M., and Morita, T. (1998). A hybrid photodetector (HPD) with a III-V photocathode. *IEEE Trans. Nucl. Sci.* **45**: 572–575.

Suyama, M., Kawai, Y., Kimura, S., Asakura, N., Hirano, K., Hasegawa, Y., Saito, T., Morita, T., Muramatsu, M., and Yamamoto, K. (1997). A compact hybrid photodetector (HPD). *IEEE Trans. Nucl. Sci.* **44**: 985–989.

Sze, S. (1981). *In* "Physics of Semiconductor Devices," 2nd ed. John Wiley & Sons, New York.

Sze, S. (1985). *In* "Semiconductor Devices, Physics and Technology." John Wiley & Sons, New York.

Webb, P., McIntyre, R., and Conradi, J. (1974). Properties of avalanche photodiodes. *RCA Review* **35**: 234–278.

Weber, S., Herzog, H., Cremer, M., Engels, R., Hamacher, K., Kehren, F., Muehlensiepen, H., Ploux, L., Reinartz, R., Reinhart, P., Rongen, F., Sonnenberg, F., Coenen, H., and Halling, H. (1999). Evaluation of the TierPET system. *IEEE Trans. Nucl. Sci.* **46**: 1177–1183.

CHAPTER 15

CdTe and CdZnTe Semiconductor Detectors for Nuclear Medicine Imaging

DOUGLAS J. WAGENAAR
Siemens Medical Solutions, Nuclear Medicine Group, Hoffman Estates, Illinois

I. Introduction
II. Energy Spectrum Performance
III. Imaging Performance
IV. Nuclear Medicine Applications
V. Conclusion

The application of room-temperature semiconductor detectors to nuclear medicine imaging is presented here. Cadmium telluride (CdTe) and cadmium-zinc-telluride (CdZnTe, or CZT) are compound semiconductors that have been researched extensively in the fields of medical imaging, astronomy, and general gamma-ray spectroscopy. Much is understood about the growth, surface and contact treatment, and charge transport phenomena in this family of crystals. Improvements in the control of crystal-growing processes have produced steadily increasing spectroscopic crystal yields over the years. Advances in microelectronics, packaging, data-acquisition electronics, and computer power have combined to allow CdTe and CZT materials to be assembled into multiple-pixel modules. These modules can themselves be configured into convenient, flexible geometries for medical imaging and other applications.

The appeal of this family of detector materials for nuclear medicine is based on several factors. Most important is the elimination of scintillation photons and the need for photomultiplier tubes (PMTs). The PMTs in the conventional gamma camera design occupy considerable volume compared with semiconductors, and this volume must be shielded by lead. The complete semiconductor detector assembly therefore has considerable advantages in weight and maneuverability. Also, the compound semiconductor has extremely good energy resolution with modest or even no cooling. This allows for scatter rejection and multiple isotope imaging in nuclear medicine applications. Another attractive feature of the room-temperature semiconductors is the possibility of dialing the intrinsic spatial resolution due to the small cloud of electron–hole pairs created by the interaction of the gamma-ray or X-ray photon. This small size provides the detector designer the option of determining the location of the photon interaction to within a 0.5 mm × 0.5 mm × 0.5 mm cubic volume. This option comes with a price: the harnessing of thousands, even tens of thousands, of individual channels of electronics. Solutions have been developed and are making their way to imaging applications. In turn, the combinations of one or more of these features: size, energy resolution, intrinsic spatial resolution, and direct pixel readout, will give the designer the necessary tools to expand the current realm of nuclear medicine.

This chapter reviews the history of these detectors and the research behind the important aspects of their performance. Charge transport and signal generation schemes have been studied by many groups, and the current work describes those efforts that most directly impact medical imaging performance. Finally, the recent efforts of medical imaging equipment manufacturers to create prototypes from these materials are presented, along with comments on their potential future use in the emerging field of molecular imaging.

I. INTRODUCTION

A. Semiconductors as Radiation Detectors

Semiconductors have properties that make them desirable for high-performance detectors of X-rays and gamma rays. High energy resolution for nuclear spectroscopy is a significant advantage of semiconductors. The superior energy resolution results from the relatively large number of electron–hole pairs generated per kiloelectronvolt of photon energy deposited, coupled with a low Fano factor (Knoll, 1999), thereby producing low statistical variation in signal response to monoenergetic photon irradiation. The clouds of electrons and holes resulting from photon interactions in the nuclear medicine energy range (up to 511 keV) have been shown to be less than 250 microns after experiencing diffusion during drift in the electric field. This relatively small charge cloud size defines a spatial resolution limit on the order of 0.5 mm sampling elements. Semiconductors therefore combine the potential for both high energy and spatial resolution compared to the NaI(Tl) scintillator coupled to PMTs currently used in nuclear medicine imaging systems.

The history of semiconductor detectors involves extensive use of silicon- and germanium-based detectors in numerous spectroscopy applications. The drawback of using these devices is that they must be cooled to liquid-nitrogen temperatures to avoid excessive thermally generated electronic noise. Furthermore, the relatively low atomic numbers of Si ($Z = 14$) and Ge ($Z = 32$) make them unacceptably less efficient at the detection of nuclear medicine X-ray and gamma-ray photons than NaI(Tl).

The family of II-VI compound semiconductors is derived from the positions of the constituent elements within the periodic table. The II elements include zinc and cadmium, whereas the VI elements include selenium and tellurium. The II-VI compound semiconductors generally have a wide energy band-gap (compared with Si and Ge) and can operate at temperatures approaching room temperature without generating excessive electronic noise. Research reports on the family of cadmium telluride (CdTe) compound semiconductors from the II-VI family were disseminated beginning in the 1950s. The atomic numbers of Cd and Te are 48 and 52, respectively, and therefore the crystal has a definite advantage over silicon and germanium in the attenuation of X-rays and gamma rays in medical imaging. In addition, the density of the crystal is approximately 6.0 g/cc, compared with 3.67 g/cc for the conventional NaI(Tl) crystal, further enhancing the detection efficiency of this crystal family.

In semiconductor detectors, leakage current increases with temperature as more electrons are raised in energy to the conduction band by thermal excitation. *Resistivity* is the measure of the crystal's ability to avoid leakage current; the higher the resistivity the lower the leakage current. Leakage current contributes, in first order, to a degradation in energy resolution. It was found that the addition of a halogen as a dopant (i.e., the addition of chlorine to give CdTe:Cl) increased the resistivity and allowed room-temperature spectroscopic operation (Siffert et al., 1975). The substitution of the element Zn in lattice locations of Cd results in the compound semiconductor $Cd_{1-x}Zn_xTe$, where x is the fraction of Zn (typically less than 0.2). I refer to this ternary compound, also known as CdZnTe, as CZT in this chapter. Generally, the energy band-gap, and thus the resistivity of CZT, increases with increasing Zn content in the alloy (Yoon et al., 1999). Comprehensive reviews of CZT versus CdTe have been written by James et al. (1995) and Schlesinger et al. (2001).

The use of CdTe as a nuclear radiation detector was first emphasized at the First (in 1971) and Second (in 1976) International Conferences on CdTe held in Strasbourg, France (Zanio, 1978). Zanio reports that CdTe will not have the impact of silicon, but that research will reveal important science about defects, trapping energy levels, charge transport, and contacts in semiconductor crystals. This body of research has grown as predicted by Zanio since the 1970s.

B. Crystal Growth and Contacts

CdTe and CZT have been grown using a variety of methods. The traveling heater method (THM) and the vertical Bridgman method are long-standing growth protocols for the production of CdTe crystals. Both methods use sealed quartz containers operated at a pressure near 1 atmosphere. Verger et al. (1997) compares the performance of CdTe:Cl grown by THM and the vertical Bridgman method. Regarding the production of CZT, much has been written about the spectral performance advantages of the high-pressure Bridgman (HPB) method (Raiskin and Butler, 1988; Butler et al., 1992). The improved performance of HPB CZT usually is attributed to high resistivity and superior charge transport properties (quantified by a parameter known as mu-tau product). This method was promoted by Digirad Corporation (San Diego, CA) in the nuclear medicine industry and further refined by the eV Products Division of II-VI, Inc. (Saxonburg, PA). An update by eV Products (Szeles et al., 2001) describes improvements in conventional and HPB growing techniques that result in fewer crystal defects and larger yields of spectroscopic-grade crystal per boule. Eisen and Shor (1998) compare THM CdTe and HPB CZT in a review of detectors, mobility of charge carriers, leakage current, and energy resolution. Beginning in the late 1990s, the materials company Saint-Gobain (La Defense, France) began growing HPB CZT. For many years Acrorad (Okinawa, Japan) and Eurorad

(Strasbourg, France) have produced CdTe-family crystals by various growth methods and for many applications.

Franc *et al.* (1999) prepared CZT crystals in a vertical arrangement by gradual cooling of the melt—a method called the vertical gradient freeze method. The source material was high-purity (six "nines", or 6N) CZT polycrystals. Franc *et al.* also describe a Bridgman method of growing CdTe:Cl from Te-rich solution. CZT growth can be accomplished by the unseeded vertical Bridgman technique (Qifeng *et al.*, 2002) and chemical vapor deposition (Noda *et al.*, 2000). A new method of growing CZT, known as the modified vertical Bridgman has been reported by L. Li *et al.* (2001). Li *et al.* note high yield and good detector uniformity and energy resolution.

Weigel and Mueller-Vogt (1996) discuss indium (In)-doping of CZT grown by both the Bridgman process and THM. They state that undoped CdTe is p-type semiconductor, but addition of In creates n-type material. They define a narrow concentration range over which the CZT transforms from p-type to n-type. The horizontal Bridgman technique (Narita *et al.*, 2000) for growing CZT is used by Imarad Imaging Systems, Ltd. (Rehovot, Israel). The Imarad technique begins with lower-purity elemental constituents than most growth methods (i.e., fewer "nines", 99.99 … %) of purity are specified for their cadmium, zinc, and tellurium. Compensation for the impurities is accomplished through the doping of the compound semiconductor with In. This crystal has shown acceptable energy resolution performance. That is, although its resistivity is not as high as HPB material, the resultant photopeak energy resolution of approximately 5% at 140 keV is a significant improvement over current NaI-based gamma camera resolution of 9–10%. There appears to be a balance between compensation for impurities (Lee *et al.*, 1999) and charge trapping so that an adequate (although far from the theoretical minimum) energy resolution is achieved. The combination of improved energy resolution and high yield suggests that these crystals have realistic potential for general medical acceptance, and they are being used in many medical prototypes (see Section IV). On the other hand, the higher resistivity of eV Products' HPB material already has been used successfully in a variety of medical and industrial applications, achieving better energy resolution than the compensated crystals of Imarad. Only modest cooling measures are necessary for both crystal types.

The type of metal used for the contact is an important selection parameter. High-work-function metals such as gold or platinum will form ohmic (charge injecting) contacts on p-type semiconductors, whereas low-work-function metals such as indium will form ohmic contacts on n-type semiconductors. Schottky blocking contacts are formed if these material types are reversed (Schlesinger *et al.*, 2001; Rhoderick 1978). Gold and platinum are the most commonly used contact materials with CZT (Schlesinger *et al.*, 2001); they are applied through thermal evaporation, sputtering, or electroless deposition. Electroless Pt is found to have better properties than other materials for HPB CZT. Narita *et al.* (2000) have studied contacts on the Imarad CZT material. They state that Imarad uses indium contacts in the standard module in the year 2000 time frame. Narita *et al.* (p. 91) also conclude that "Imarad material is probably n-type CZT, since it forms an ohmic contact with a low work function metal such as indium, while it forms a blocking contact with a high work function metal such as gold." Narita *et al.* experimented with Au contacts on the n-type Imarad crystal (as blocking diode contacts), and these produced excellent spectra, excellent photopeak efficiency, and improved resistivity compared to the standard Imarad indium contacts. Nemirovsky *et al.* (2000) also used Imarad's n-type CZT with In ohmic contacts in a report on spectroscopic performance.

The performance of the detector is also influenced by the technique used to treat the surface. This includes both the preparation prior to the deposition of the contacts and the passivation of the surface left exposed between the contacts and on the edges of the semiconductor. Burger *et al.* (1999) discuss the prevention of the oxidation of the crystal surface prior to the deposition of ohmic contacts. Prettyman, Ameduri, *et al.* (2001) review surface preparation and show that the type of wet chemical processing significantly affects charge collection by altering the conductivity of the surface. Wet chemical processing involves polishing with mildly abrasive chemicals. For example Burger *et al.* (1997, 1999) prepared the CZT surface with 5% Br-MeOH followed by 2% Br–20% lactic acid in ethylene glycol prior to evaporated Au contact deposition. This technique is referred to as the BMLB method.

Surface passivation is the process by which the semiconductor surface is rendered inert and does not change semiconductor properties as a result of interaction with air or other materials in contact with the surface or edge of the crystal. Researchers have found that oxidation of the surface through ion implantation or boiling in hydrogen peroxide (James *et al.*, 1998) can reduce the surface leakage current. Mescher *et al.* (1999) demonstrate that a dry processing passivation technique involving oxygen plasma and sputtering of a silicon nitride layer reduces surface leakage current by as much as a factor of 20. Wright *et al.* (2000) conclude that the addition of ammonium fluoride to a solution of hydrogen peroxide produces a dielectric layer that also reduces the surface leakage current. Wright *et al.* recommend ammonium fluoride in hydrogen peroxide to replace hydrogen peroxide when passivating these crystal surfaces. Nemirovsky *et al.* (1997) studied contact size, metalization technique, surface effects, and passivation techniques in a review of this topic. Bolotnikov *et al.* (1999) examined surface conductivity and its effects on the drift of charge and signal induction in the contact circuit.

Aging of contacts in semiconductors is a concern, especially with the variety of contact types, bulk material preparation techniques, surface preparation techniques, and so on. Studies of contact aging necessarily must be performed after the completion of growth technique, contact and surface preparation, spectral performance, and charge transport research. Therefore, the field has progressed rapidly, but few studies on aging have been published. However, Dharmadasa *et al.* (1994) report aging effects on Sb contacts in n-type CdTe with observed intermixing of the Sb into the bulk material, thereby increasing the effective p-type doping in the n-type material. Wolf *et al.* (2001) observe the diffusion of silver atoms into the bulk of CdTe. Okamoto *et al.* (2000) note Cu diffusion and the degradation of photovoltaic performance over time.

It can be seen that there are numerous important decisions to be made regarding parameters when a CdTe-type detector is created. Scheiber and Giakos (2001) comment that material type (e.g., p-type, n-type, zinc presence, zinc content, material purity, etc.), contact material, surface preparation, and passivation all are key in defining the detector performance.

C. Advantages of CdTe and CZT

Beginning in the 1950s, electronic circuits based on vacuum tube technology began to be replaced by solid-state or transistor-based circuits. The NaI-PMT technology found in nearly all installed gamma cameras is one of the last remnants of the pretransistor vacuum tube era—the PMT, as sensitive as it is, is nevertheless a vacuum tube. The replacement of the PMT would allow the field of nuclear medicine into the solid-state world. There are currently two potential new technologies in nuclear medicine: (1) semiconductor detectors and (2) scintillators coupled to arrays of photosensors. In a semiconductor detector, the gamma-ray energy is converted into electron–hole pairs, which in turn generate electronic signals with high energy resolution. The scintillator detectors must go through an additional energy conversion step, namely the gamma-ray energy is converted into scintillation photons, which are then converted into electronic signals. The number of signal carriers produced in semiconductors is more than an order of magnitude higher for semiconductors (about 31,000 electron–hole pairs for a 140-keV ^{99m}Tc photon, compared with approximately 1000 photocathode electrons in NaI-PMT). Also, the scintillation photons have a Fano factor of 1, whereas the semiconductor detectors' Fano factor is six times lower. This results in a theoretical energy resolution of less than 0.5% for 140 keV, compared to approximately 5% for scintillators.

That the PMT is the last of the vacuum tubes is not a sufficient rationale for the development of two families of detector technology. Why do we pursue the development of solid-state technology? The large-field-of-view gamma camera technology has been perfected since the 1960s to the point that it is highly flexible—whole-body scanning, dynamic imaging of the kidneys, high-resolution imaging of the brain, and gated imaging of the heart can be done with this workhorse design. In the emerging world of molecular imaging, one in which quantitative evaluation of a small hot spot can determine the results of a virtual biopsy, single-photon NaI-PMT devices currently present spatial resolution, detection efficiency, and quantitative accuracy deficiencies when compared with positron emission tomography (PET). Solid-state cameras offer the opportunity to break the mold of all-purpose, large-field-of-view, parallel-hole collimated (low-sensitivity) single-photon imaging established by evolution of the dual-headed gamma camera and its clinical market since the 1980s. The small, lightweight nature of the solid-state camera is well suited to the new applications of molecular imaging with novel image formation (i.e., collimation) schemes, especially in the new imaging locales. Cardiologists are using nuclear medicine more than ever and are buying dedicated NaI-based cameras for their outpatient clinics and offices. They require smaller-footprint devices, and solid-state technology meets this need. Future applications for oncology and neurology specialists may require even smaller, dedicated devices because these specialists will need to image and quantify radiopharmaceutical uptake in tumors in known locations or specific regions of the brain. In addition, multiple isotope imaging and scatter rejection are enhanced features expected from this class of detectors. The potential of new molecular imaging markets is a primary rationale for solid-state cameras in nuclear medicine.

The CdTe crystal family is not the only semiconductor detector under investigation. It does have a significant lead in the development cycle, however, having begun in the 1950s. Other crystals under investigation are lead iodide, thallium bromide, bismuth iodide, and mercuric iodide. One can see that these semiconductor detector materials contain high-atomic-number elements and are potentially better radiation detectors for nuclear medicine. However, the problems and challenges in crystal growth and contact preparation that CdTe has are also present and in most cases worse in these heavier materials. The most progress has been made in the field of mercuric iodide (Richards *et al.*, 2001).

The CdTe family of crystals has been used in a variety of applications. This chapter deals specifically with the nuclear medicine spectroscopic application. Nonspectroscopic-grade CZT has been grown as a substrate for HgCdTe, a successful infrared imaging detector. The CdTe family has also generated considerable interest in the production of electrical current in photovoltaic solar cells (Levi *et al.*, 2000). High-quality spectroscopic-grade CdTe and CZT material is in various stages of development for use in X-ray and gamma-ray astronomy in balloons, orbiting satellites, and planetary explorers. Representative imaging systems are described by Ramsey *et al.* (2001), Prettyman, Feldman, *et al.* (2001), Parsons *et al.* (2001), Arques *et al.* (1999), and Rothschild *et al.* (2002). Prettyman, Feldman, *et al.* state the need a minimum of for 16 cm^3 of high performance material for an orbiting instrument. In astronomy applications, the distant pointlike sources of radiation enable coded aperture plates to be used instead of collimators for image formation. (Modified coded apertures to image isolated clinical hot spots are discussed later.) The compact charge cloud, the excellent energy resolution, and the ability to stop high-energy photons all combine to make the CdTe-family of detectors a popular choice in astronomy and space imaging applications. The International Atomic Energy Agency, the U.S. Department of Defense, and other agencies have studied spectroscopic-grade CdTe:Cl and CZT for use in the detection of special nuclear materials in the field of nuclear nonproliferation. Finally, CdTe crystals are finding new applications in the nanomaterial research field of quantum dots.

D. Relevant Societies and Scientific Exchange

The process of fabricating a detector begins with the crystal growth. The fundamental science of the growth and preparation of room-temperature semiconductor detectors is reported at biennial meetings of the Materials Research Society (MRS) and published either in the *Proceedings of the Materials Research Society* or the journal *Nuclear Instruments and Methods in Physics Research A* (see Table 1). Table 1 also includes the Symposium on Radiation Measurements and Applications, held every 4 years at the University of Michigan. Recent Michigan symposia have featured many presentations on the CdTe-family of detectors.

One can also find research reports presented at several other conferences of obvious relevance: International Symposia on II-VI Compounds, International Conferences on Crystal Growth, and Semiconductor and Compound Semiconductor Conferences. The Minerals, Metals, & Materials Society (TMS) holds annual U.S. Workshops on the Physics and Chemistry of II-VI Materials. The proceedings of these workshops appear in dedicated issues of the *Journal of Electronic Materials*. Also, some relevant and applicable information about the packaging of the crystals with the application-specific integrated circuit (ASIC) electronics can be found through the International Microelectronics and Packaging Society (IMAPS).

The dissemination of research results on the many topics related to room-temperature semiconductor detectors has grown steadily since the 1980s and currently occurs at annual and other regularly held symposia. CdTe and CZT detectors applied to medical imaging, nuclear physics experiments, nonproliferation enforcement, and astronomy imagers are annually featured at the Institute of Electrical and Electronics Engineers (IEEE) Nuclear Science Symposium/Medical Imaging Conference. The detector-related proceedings of this meeting are published in the *IEEE Transactions on Nuclear Science*.

The Society of Photo-optical Instrumentation Engineers (SPIE) also holds regular scientific conferences that include cutting-edge results in semiconductor detector research. The relevant SPIE conferences are entitled Hard X-Ray and Gamma-Ray Detector Physics, Physics of Semiconductor

TABLE 1 Conferences and Associated Publications Related to Room-Temperature Semiconductor Detectors

Conference Title	Location	Date	Publication	Volume
1st European Symposium on Semiconductor Detectors	Schloss Elmau, Germany	May 7–10, 1995	*Nucl. Instrum. Methods A*	377
International Workshop on Room Temp Semiconductor Detectors for Remote, Portable, and In Situ Radiation Measurement Systems	Jerusalem, Israel	July 26–29, 1998	*Nucl. Instrum. Methods A*	428 (1)
9th Symposium on Radiation Measurements and Applications	Ann Arbor, MI	May 11–14, 1998	*Nucl. Instrum. Methods A*	422 (1–3)
10th Symposium on Radiation Measurements and Applications	Ann Arbor, MI	May 21–23, 2002	*Nucl. Instrum. Methods A*	505 (1–2)
8th Semiconductor Room-Temperature Radiation Detectors.	San Francisco, CA	Apr. 12–16, 1993	*Mat. Res. Soc. Symp. Proc.*	302
9th International Workshop on Room Temperature Semiconductor X- and Gamma-Ray Detectors and Associated Electronics and Applications	Grenoble, France	Sept. 18–22, 1995	*Nucl. Instrum. Methods A*	380 (1–2)
10th Semiconductors for Room-Temperature Radiation Detector Applications II	Boston, MA	Nov. 30–Dec. 4, 1997	*Mat. Res. Soc. Symp. Proc.*	487
11th International Workshop on Room Temp Semiconductor X- and Gamma-Ray Detectors and Associated Electronics	Vienna, Austria	Oct. 11–15, 1999	*Nucl. Instrum. Methods A*	458 (1–2)
12th International Workshop on Room-Temperature Semiconductor X- and Gamma-Ray Detectors	San Diego, CA	Nov. 6–9, 2001	*IEEE Trans. Nucl. Sci.*	49
13th International Workshop on Room-Temperature Semiconductor X- and Gamma-Ray Detectors	Portland, OR	Oct. 20–24, 2003		

[a] These conferences contained sessions dedicated to CdTe and CZT crystal growth, preparation, charge transport parameters, and imaging performance.

TABLE 2 Society of Photooptical Instrumentation Engineers Meetings and Published Proceedings Containing Multiple Sessions Dedicated to CdTe and CZT Detectors

SPIE Conference Number Conference Title	Location	Date	Volume
Hard X-Ray and Gamma-Ray Detector Physics, Optics, and Applications	San Diego, CA	July 31–Aug. 1, 1997	3115
Hard X-Ray and Gamma-Ray Detector Physics and Applications	San Diego, CA	July 22–23, 1998	3446
1st Hard X-Ray, Gamma-Ray, and Neutron Detector Physics			3768
Penetrating Radiation Systems and Applications	Denver, CO	July 19–23, 1999	3769
2nd Hard X-Ray, Gamma-Ray, and Neutron Detector Physics II		July 31–Aug. 2, 2000	4141
Penetrating Radiation Systems and Applications II	San Diego, CA	Aug. 2–3, 2000	4142
3rd Hard X-Ray and Gamma-Ray Detector Physics III		July 30–Aug. 1, 2001	4507
Penetrating Radiation Systems and Applications III	San Diego, CA	Aug. 1–2, 2001	4508
4th X-Ray and Gamma-Ray Detectors and Applications IV		July 7–9, 2002	4784
Penetrating Radiation Systems and Applications IV	Seattle, WA	July 10–11, 2002	4786

[a]All volumes can be found in the publication series entitled *Proceedings of the SPIE*.

Detectors, X-Ray and Gamma-Ray Detectors and Applications, X-Ray and Gamma-Ray Instrumentation for Astronomy, and Penetrating Radiation Systems and Applications. Table 2 summarizes the SPIE meetings and associated proceedings containing CdTe- and CZT-related research reports. Because the detectors are developed into clinical prototypes, the disseminated research reports appear in the journals such as *Reviews of Scientific Instruments*, *Medical Physics*, and the *Journal of Nuclear Medicine*.

The physical properties of CdTe-type crystals meant for use as high-resistivity detector material have been studied extensively since the 1990s. Hundreds of CdTe-related research papers have been published on the topics of crystal growth, dopant concentrations, electron and hole mobility, contact electrode material, contact electrode geometry, signal generation, multichannel (pixel) electronics, detector performance, and clinical imaging applications. The Appendix at the end of this chapter lists the academic, governmental, and industrial groups that have publicly disclosed research progress in the CdTe family of radiation detectors. At least 1600 scientists have been listed as authors or co-authors on CdTe-related articles published since the early 1990s.

II. ENERGY SPECTRUM PERFORMANCE

Spectroscopic energy resolution is one of the intriguing features of CdTe detectors. Nuclear medicine scientists are lured by the SiLi-type peaks, with exquisite scatter rejection and high photofraction. One practical example is the issue of dual-isotope cardiac imaging in nuclear medicine with attenuation correction. This produces a multitude of problems, all of which could be alleviated with a high-photofraction spectroscopic peak with no low-energy tailing. The two radionuclides in this case are ^{99m}Tc and ^{201}Tl. ^{99m}Tc has a 140-keV photopeak, and ^{201}Tl has X-rays associated with its decay around 70 keV and 80 keV, along with higher-energy gamma rays of 135 and 165 keV. Furthermore, most attenuation correction schemes use ^{153}Gd, a radionuclide with a 100-keV gamma ray. Lead in the collimator and detector-shielding generates characteristic X-rays in the 75-keV and 85-keV range. These X-rays are created by photoelectric interactions of higher-energy photons such as those from the gadolinium source, the technetium, or the high-energy thallium gamma rays. Furthermore, the patient's tissues produce numerous backscatter photons of 90 keV from the technetium gamma rays.

Currently the NaI-PMT detector does a poor job of discriminating the jumble of photons entering the detector and having an energy less than 100 keV. A wide energy window is centered on the thallium peak of 75 keV, and this window includes photons from lead fluorescence as well as technetium backscatter, thallium 135 and 165 keV downscatter, and gadolinium gamma rays. Ideally, a CdTe-type detector would allow separate windows on the thallium X-rays (K_α and K_β), rejecting the K_α and K_β of Pb. The window around the 100-keV gadolinium peak would be very narrow, minimizing the acceptance of downscatter from the technetium photons in the patient's body. Imaging scientists would have better peak discrimination from downscatter and fluorescence and hence more information on which to make corrections based on models of interactions.

In this section, the energy spectrum of the CdTe and CZT detectors is discussed. The spectrum includes a low-energy tail that results from a combination of influences, and each of these is presented separately. The characterization of semiconductor performance should be standardized, as has been done with NaI. This effort is being spearheaded by Keyser and Fairstein (2001) through the IEEE (however, conformance to these new standards is not addressed in the following subsections).

A. Energy Resolution

As mentioned previously, high energy resolution is one of the main features of this class of detectors. The bandgap of these detectors is approximately 1.55 eV, and the energy required to create an electron–hole pair is generally three times this value, or approximately 4.5 eV. Therefore, a 140-keV photon results in 31,000 electron–hole pairs. Assuming we do not wish to use the holes for signal generation (due to their poor mobility), we have 31,000 electrons drifting toward the anode. The Poisson statistical variation on this number is 176, or 0.5% energy resolution for 140 keV. But the Fano factor is approximately 0.15, so the statistical variation on 31,000 is actually closer to $0.15 \times 176 = 26$, and the theoretical energy resolution is 0.09%. There are many factors that prevent us from realizing this resolution at the current state of technology, the main one being electron trapping during drift in the crystal. Akutagaw and Zanio (1969) extensively study the energy spectral response of CdTe crystals, examining the relationship between resolution and charge trapping and detrapping. Leakage current due to the finite resistivity and electronic noise in the amplifier circuits also contribute to the measured energy resolutions in the 2–5% range typically reported in the literature. Finally, Toney et al. (1998) derive an expression for the best obtainable energy resolution of a semiconductor detector, stating that an optimal band-gap for room-temperature operation is 2.0 eV.

B. The Low-Energy Tail

The excellent statistics concerning charge carrier numbers are counterbalanced by many effects working to prevent complete signal formation: (1) Imaging necessitates dividing the detector into resolution elements, and these elements (or pixels) have boundaries where signal collection is weaker than the pixel center; (2) electron and hole trapping creates a depth-of-interaction dependence on the generated signal; and (3) the finite size of the signal contact results in an induced signal that depends more on the relatively slow hole drift for signals generated by events closer to the anode. These three effects combine to create a tail of events on the low-energy side of the photopeak. Figure 1 demonstrates the magnitude of the tail observed for the 140-keV gamma ray of ^{99m}Tc. In experiments at Siemens Medical Solutions laboratories, our research group measured total-spectrum detection efficiency equivalent to that of a comparable detection thickness of NaI, but a *photofraction* of only 65% of that of NaI. This finding is generally in accordance with publications on this detector type. The data of Figure 1 were taken with the Imarad CZT module (described in the next section), a module version with

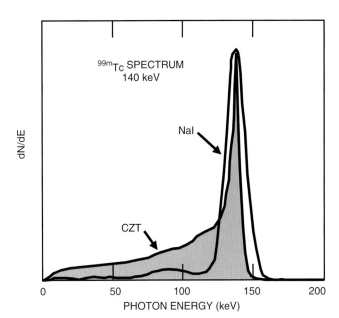

FIGURE 1 ^{99m}Tc intrinsic (noncollimated) spectra obtained with CZT and NaI(T1) detector systems. Note the low-energy tailing of the CZT, and the 110-keV escape peak in the CZT spectrum.

no design features intended to reduce the low-energy tail. Future versions of this module undoubtedly will attempt to reduce the tailing by employing some of the technology described in the following paragraphs.

C. Neighboring Pixel Effects

There are no boundaries separating the resolution elements within the Anger camera. Image matrices are defined by binning analog *xy* signals after events have undergone the Anger logic computation of position. Therefore, the Anger-type detector has a position-invariant[1] spectral response—a major advantage in large-field-of-view imaging. The continuously adjustable size of the Anger-camera pixel is set by the user's selection of matrix size, field of view, and zoom factor. Pixellated detectors are divided into separate volumes called *pixels*, and there are pixel–pixel boundaries. Experiments show that events directed into the boundaries contribute to the low-energy tail, whereas events directed into the center of the pixel contribute to the photopeak. Knowing the charge cloud has a small size (< 250 microns), the detector designer may be tempted to make small pixels of 1.0 mm or less. However, one should be aware that the smaller the pixel, the higher the fraction of boundary volume to center volume in the detector as a whole. Therefore the

---

[1]After corrections have been applied, only small differences in response remain between spectra acquired in the center of the PMT versus between two PMTs; large differences in response can be measured near the NaI crystal edge and this edge is routinely masked as outside the useful field of view.

low-energy tail increases as the pixel size is diminished due to the effects of the boundaries.

What exactly is happening at the boundary between pixels? At Siemens, we studied the boundary between two pixels with low-noise electronics in coincidence (Kalinosky et al., 2000). We found that signal sharing occurs in which a fraction of the photopeak signal is in one pixel and the complementary fraction occurs in the other pixel. Correlation plots have been created that clearly demonstrate this phenomenon (Kalinosky et al., 2000; Kalemci and Matteson, 2002). Sharing itself is due to three phenomena: (1) X-ray fluorescent scattering, which has a range of approximately 150 microns; (2) Compton scatter; and (3) charge cloud overlap, that is, the charge cloud straddling the boundary between pixels. Bolotnikov's group at the California Institute of Technology (Bolotnikov et al., 1999; Chen et al., 2002) has studied the nature of the electric fields and the surfaces between contacts. Their design includes a steering electrode between pixels to direct electrons toward the anode contacts and avoid losses of signal on the boundaries. Du et al. (1999) ran a Monte Carlo investigation into charge sharing in CZT gamma-ray spectrometers. Du et al. computed an optimum pixel size of 1 mm for 662-keV photons from ^{137}Cs (assuming a 1-cm drift length). This calculation includes the effects of electron range, drift distance, X-ray fluorescence, and diffusion of the charge cloud.

The most obvious step to reduce the boundary effects is to improve the data acquisition capability of the detector electronics so that the system responds to nearly simultaneous events in neighboring pixels. This requires very low-noise electronics because most of the boundary signal pairs will have one component less than 40 keV, a value that is typically the *low-energy threshold* for 140-keV operation. These circuits are currently under development. Another approach is to place the collimator walls in alignment with the boundaries, thereby covering them up. This prevents events from occurring in the boundary region. This approach is discussed in Section III as the registered collimator approach.

D. Depth-of-Interaction Effects

Nuclear medicine imaging is based on the detection of photons of known energy. For example, ^{99m}Tc has an energy of 140 keV and ^{131}I has an energy of 365 keV. The higher the energy, the longer the average penetration depth of the photon before it interacts—this parameter is known as the *mean free path* of the photon. In the standard configuration of CdTe-type detectors, the photon encounters first the cathode surface, then the detector itself, and last the anode array. This arrangement is designed to maximize the use of the drifting electrons to produce the signal and minimize the effect of the drifting holes. However, as the energy is increased, the mean free path increases, and the photons interact nearer to the anode array. In experiments at Siemens, we have documented increased tail at the energy of 140 keV, compared to 122 keV, presumably due to depth-of-interaction effects. Auricchio et al. (1998) also show signal degradation nearer the anode

The induced signals are created by the drift of the electrons and holes. The drift speed is determined by the electric field strength imposed from the external circuitry. The signals are also affected by the degree of electron and hole trapping. Shor et al. (2001) show that by tuning the electric field for a particular photon energy, one can achieve an optimum peak-to-tail ratio. This is because one achieves a balance between electron and hole trapping and the drift time. Shor et al. also show that by using a cathode signal as a trigger one can generate a correlation plot between signal amplitude and drift time. This correlation plot can be used to move tail events back into the photopeak in a process referred to as rise-time compensation. This process is a method of compensating for depth-of-interaction effects. This process improves the photofraction and lowers the tail, both of which are desirable in nuclear medicine imaging (and spectroscopy in general). However, the process adds noise and does not contribute to obtaining the optimal energy resolution. Rise-time event selection has been described by Lund et al. (1996). A biparametric variant of the rise-time technique (Mestais et al., 2001) results in greater detection efficiency and is in use in multichannel circuitry designed by the LETI group for use in a gamma camera prototype (see Section IV).

Knoll and McGregor (1993) describe the phenomenon of pseudopeaking in semiconductor detectors. The pseudopeak is formed by depth-dependent signal degradation due to trapping of holes and electrons, such that the width of the peak is dominated by the small differences in the trapping balance rather than by the statistical variations in the number of charge carriers being collected. This explains the order-of-magnitude difference between the theoretical and measured energy resolutions in these detectors. In their paper, Knoll and McGregor calculate pseudopeak spectra for a variety of extraction factors—the ratio of the average drift distance a charge travels before it is trapped to the maximum drift distance. Toney et al. (1998) review the current understanding of the spectral response of these detectors to monoenergetic photons based on mobility-lifetime products.

The elimination of the depth-of-interaction effect is accomplished by using the drift time information, usually with a cathode-based trigger signal. The spectrum can be improved by the removal of the influence of hole drift, such that the anode signal is only derived from the electrons. This is known as *single charge carrier processing* and is described in the following section. Electron trapping increases as drift distance grows and, although the signal amplitude can be corrected, the statistical information in the original charge cloud is lost. Therefore, the electron trapping represents a step away from the theoretical energy resolution limit that cannot be recovered (unless electron trapping itself is reduced). For the purposes of nuclear medicine imaging, depth-of-interaction correction not only accomplishes improvement in photofraction and reduction of the tail, but it provides the user knowledge of the depth of the photon interaction. The manner in which this localization information can be exploited remains to be determined. Depth-of-interaction information appears to be a potential advantage for the CdTe detectors relative to scintillators in imaging applications.

E. The Small Pixel Effect

A misconception about the generation of signals in ionization chambers and semiconductor detectors is that the signal is created when the charges arrive at the surface of the conductor. What actually happens is that the signal is induced during the drift of the charges—and the signal is generated when the induced charge on the surface of the electrode *changes* as a result of the different overall configuration of the drifting charges relative to the conductors surrounding it. One good way of seeing this is to imagine a 1-cm parallel-plate detector. If a photon interacts very close to the anode, then a cloud of electrons and holes is created. Initially, the electrons and holes contribute to a surface charge on the anode of 0 because the electrode senses no net difference. The electric field draws the clouds apart, generating a signal on the anode. However, because the holes have a poor mobility and are trapped quickly, they do not drift very far, and therefore the signal from their drift is small. The electrons, being near the anode initially, do not drift far either, and hence the signal near the anode is very poor. This phenomenon is seen in current CZT detectors. One way of avoiding this problem is to make the anodes very small and to design the detector such that only drifting electrons approaching the small anode generate a signal. This design exploits the concept known as the "small pixel effect."

The small pixel effect goes back to the 1930s, when the phenomenon was described by Shockley (1938) and Ramo (1939). In the 1950s, this effect was referred to as self-gridding in ionization chambers—the idea that if there are many electrodes, the induced charge on a single electrode is negligible for a distant charge cloud (i.e., a grid of electrodes shields itself). Wagenaar and Terwilliger (1995) report on this phenomenon as it relates to a multiple-cathode drift chamber. Barrett *et al.* (1995) describe the small pixel effect in semiconductors and conclude that large anodes will have hole contributions and small anodes will see signals due only to electrons. This approach is now known as the single charge carrier detector operation. Z. He (2001) reviews the Shockley-Ramo theorem as applied to semiconductor detectors and includes an energy conservation description of the charge drift and voltage supplies. Small strips of conductor also exploit the small pixel effect, and descriptions of strip detector designs are given by Macri *et al.* (1996) and Kurczynski *et al.* (1997).

In addition to the small pixels themselves, conductor surfaces can be added to the detector design to enhance the shield and amplify the small pixel effect. This is the Frisch grid approach. A voltage potential is usually applied to the grid such that the drifting charges are steered toward the anodes, thereby avoiding charge loss to the grid. The concept of the coplanar grid was introduced by Luke (1995; Luke and Eissler 1996). In this design, steering electrodes commingle on the anode plate surface with the anodes, shielding the anodes from the drifting charges until the charges are steered into the anode, generating the full signal only upon arrival near the anode. Z. He *et al.* (1997, 1998) have exploited the coplanar grid combined with a cathode trigger signal to obtain single carrier charge sensing. He has achieved excellent spectral response by correcting for depth of interaction and rejecting hole contributions, while also obtaining 3D volume detection through the recording of the depth of interaction. McGregor *et al.* (1998) describe a Frisch grid design in which each pixel has a separate grid located between the anode and the cathode. McGregor *et al.* (1999) also have designed an altered, wedge-shaped geometry in the pixel itself to further enhance the single charge carrier signal.

In summary, the small pixel effect can be used to remove the influence of the hole drift and create a single charge carrier detector. Tail signals due to hole contributions are thereby removed. Depth-of-interaction corrections can be achieved by using a cathode trigger to compute the drift time of the electrons. Boundaries between neighboring pixels also contribute to the spectral tail and can be reduced by covering the boundaries with collimator septa and by designing and implementing low-noise, two-pixel acquisition electronics.

III. IMAGING PERFORMANCE

Imaging with semiconductors requires the localization of the photon event. This can be accomplished through two means: (1) pixels at known locations, and (2) crossed strips of conductor on either side of the detector (the orthogonal strip approach). The strips (Hamel *et al.*, 1996) were intended to reduce the number of electronic channels in the mid-1990s. However, because of the boundary problems reported in the preceding section (placing a practical limit on the pixel size of approximately 1 mm), advances in packaging and interconnectivity, and relentless improvements in inexpensive microelectronic circuitry and computer power, the need to avoid multiple channels has faded. We mainly deal with pixellated imaging devices and the subsequent section on medical imaging prototypes. The general concept that has evolved in the design of imaging devices is based on making a module of many pixels and then assembling these modules into a geometry that is optimized for the imaging task at hand.

FIGURE 2 An Imarad CZT module. This module is the culmination of several earlier versions with steadily improving microelectronics and packaging designs. Many imaging companies and research institutions have acquired imaging experience with this general purpose, 256-pixel imaging platform.

An Imarad CZT module is shown in Figure 2. The research group at Siemens Medical Solutions USA, Inc. (Hoffman Estates, IL), has been collaborating with Imarad Imaging Systems, Ltd. (Rehovot, Israel), since 1996 on the development and testing of this module. Research groups at Elscint (Haifa, Israel) and General Electric Medical Systems (Waukesha, Wisconsin) have played parallel roles to Siemens in collaboration with Imarad. The module consists of 256 pixels arranged in a square array pattern of 16 × 16, each of 2.46 mm dimension. The thickness of the detector is 5.0 mm. The standard crystal size is 8 × 8 pixels, although larger crystals of 8 × 16 and 16 × 16 frequently are used in Imarad modules. Internal research at Philips Medical Systems has also characterized the Imarad modules (corroborating most of the published performance results). The Philips group have also concentrated on the higher-resistivity HPB material in a pixellated eV Products module (De Geronimo *et al.*, 2002) to build a high-spatial-resolution scanner that exploits a new collimation scheme (see Sections IIIG and IVA).

This section describes the steps necessary to perform imaging with this family of detectors. It is based primarily on our experience at Siemens, although reference is made to other systems and techniques when they apply. The detector performance of the Imarad module was studied by W. Li *et al.* (2001). The spectroscopic performance of Imarad material was reported by Schlesinger *et al.* (1999). Although an improvement compared to many existing and previous generations of semiconductors, Figure 1 shows that removal of the tail remains a major challenge for the clinical utility of these devices compared with existing NaI-PMT systems.

A. Electronics and Signal Processing

The concept of imaging with individual pixels in routine medical applications was unimaginable in the 1980s. Thousands of data channels simultaneously active were found only in high-budget national and international physics laboratories. The cost and size of microelectronic circuitry continuously declined throughout the 1990s, enabling the development of kilochannel imaging devices for widespread clinical use. During this time, Imarad began working with European Organization for Nuclear Research (CERN) engineers to design 64- and 128-channel self-triggered application-specific integrated circuits (ASICs). In 1999 Imarad began to collaborate with IDE AS (Hovik, Norway) on their concept of a 128-channel circuit.

The idea behind the circuitry is to record the pixel location and energy signal amplitude for each event above a preset threshold amplitude. The version of the Imarad module precludes the acquisition of signals from more than one pixel during the processing of any event, but there are designs on the drawing board for neighboring pixel circuitry to be developed. The anodes are connected to a routing board by conductive epoxy. This routing board is itself a marvel in packaging: it takes 256 signals, arranged in a square array, and routes them through multiple layers following routes in three dimensions until they arrive at the ASIC bonding pads on the opposite side of the routing board. This must be accomplished with minimal stray capacitance and signal cross-talk—problems that were encountered during the course of the module development. In the Imarad module, there are two 128-channel ASICs on the routing board, and gold wire bonding is used to connect the anode signals to the ASICs. The energy signal is digitized in circuitry located on boards external to the module itself, and the digitized energy and pixel location are buffered and interfaced to the acquisition computer. A feature of the IDE ASIC that is common to most CdTe architectures is the compensation for leakage current. Leakage current represents a low-frequency background that can be removed through dedicated circuitry.

Module architectures can be divided into multiplexed acquisition (periodic sequential readout) and self-triggered (event-driven readout). The IDE/Imarad module is a self-triggered architecture fabricated with a 1-micron CMOS process. The European Medipix2 project (Llopart *et al.*, 2002) is also a self-triggered, single-photon counting chip that uses a 0.25-micron CMOS fabrication technology, and it was designed for use with different semiconductor sensors at CERN. Medipix 2 is a 64 k (256 × 256) pixel readout chip with 55-micron-squared elements in single-photon counting mode. Instead of wire-bonding, it uses bump-bonding. The circuit also compensates for leakage current on a pixel-by-pixel basis. Jakobson and Nemirovsky (1997) also have designed a low-noise CMOS circuit for CdTe and CZT. Multiplexed architectures are employed by the Biomed consortium (Scheiber et al., 1999) and the University of Arizona (Kastis et al., 2000). In this technique the entire module's pixels are read out on a periodic basis (e.g., every few milliseconds). Cook et al. (2000) have designed a 1.2-micron CMOS, 64-channel circuit for astronomy applications. This ASIC uses In bump bonds for interconnection. eV Products, in collaboration with a group at Brookhaven National Laboratory (De Geronimo et al., 2002), has designed a circuit that can record individual events or a multiplexed raster readout. Very accurate feedback matching on the preamplifiers and a robust baseline restorer directly on the ASIC resulted in superior overall performance, achieving very low noise (noise-equivalent electron value of 35 e^- + 35 e^- per picoFarad compared to a specified 120 e^- + 20 e^- per picoFarad for the IDE ASIC). The low electronic noise and the higher-resistivity material (lower leakage current noise) results in an overall system noise reduction and better observed energy resolution. This combination is employed in Philips' Solstice prototype and Anzai's eZ-SCOPE product (see Section IV).

Irreversible deterioration of CdTe and CZT detectors occurs after heat treatment in the 150–200°C range (Chattopadhyay et al., 2000). Because of this temperature sensitivity, special care must be employed to use flip chip bonding and other preparations involving the crystal contacts. Riley and Bornstein (2001) describe gold-stud bump flip chips using special adhesives and underfills to avoid damaging temperatures in the crystal.

Most of the Anger camera circuitry in use was designed in the 1990s. Data acquisition systems are currently capable of listmode recording of events, that is, the recording of the time of each event in addition to the location and energy. Kastis et al. (2000) detail the listmode capabilities of the University of Arizona pixellated detector. More advanced modern acquisition systems can also record the rise time of the signal for depth-of-interaction correction (Mestais et al., 2001). The development of semiconductor electronics has allowed the field of nuclear medicine imaging to modernize its data acquisition capabilities of count rate, dead time, and listmode, which might otherwise not have been done.

B. Energy Windowing

Nuclear medicine imaging requires the selection of energy windows, upper and lower discriminator levels on the energy spectrum that define the detected photon energy range used to generate the image. In contrast to the Anger camera, the spectra from pixellated CdTe-type detectors have separate energy scales due to crystal differences and differences in electronic gain and offset between channels. Therefore, each pixel has its own energy spectrum and its own energy window. This is readily accomplished through calibration software that keeps track of the spectral response per pixel and sets the energy discriminator levels on each pixel individually.

C. Uniformity and Other Corrections

Uniformity correction is a major concern in the Anger camera due to the nature of the localization algorithm. The concept of linearity in the Anger camera results in a strong connection between the gain of the PMT and the uniformity of the full-field detector response to a uniform illumination. Therefore, linearity corrections and PMT gain drift must be constantly monitored and corrected in the Anger camera. In the pixellated detectors, the linearity problem is nonexistent. The localization of the event to a pixel is exact. However, between pixels there is a considerable variation in uniformity of response, primarily due to the imperfections in the crystal-growing process. The standard deviation of the histogram of photopeak counts per channel is typically 10% in CZT. Furthermore, pixels located on the edges and especially the corners of the square arrays generally are less responsive than internal pixels (Mardor et al., 2001). However, these nonuniformities do not tend to change with time, and therefore they are easily corrected through the use of uniformity correction maps. In our experience with the Imarad modules, nonuniformity artifacts due to edge pixels and other pixels of highly varying uniformity are not observed after uniformity correction has been applied.

Improvements in ASIC electronics and packaging since the late 1990s have dramatically reduced the number of dead pixels one encounters in a given module. The Imarad modules now typically have approximately 1 dead pixel per module (256 pixels in a module). This is either from crystal imperfections or electronic failure. Another form of failed pixel is referred to as a "screaming" pixel, one that constantly is outputting a meaningless signal and thereby interrupting the operation of the entire module. Screaming pixels must be turned off. Dead and screaming pixels are corrected by using a linear interpolation from the surrounding pixels.

D. Cooling

Obviously an advantage of room-temperature semiconductor detectors is that there is no need for cryogenic cooling to obtain acceptable spectroscopic performance. The University of Arizona detector cools to −15°C to reduce noise due to leakage current (Kastis et al., 2000). Cook et al. (1999, 2000) and Bolotnikov et al. (1999) note an 18% improvement in resolution by cooling to 5°C and a 28% improvement at −25°C for 60-keV photons. Although in general cooling produces less leakage current and therefore better energy resolution, it has been noted that the Imarad modules actually perform worse at low temperatures, indicating a high degree of electron trapping (Narita et al., 2000). It appears then that the selection of growth technique, impurity level, dopant, and concentration also impacts the performance as a function of temperature.

It has been found that Imarad's detectors must be cooled to a constant temperature above the dew point. This is because the performance of the crystals and the electronics is temperature-dependent, and it is more straightforward to maintain a constant temperature through the use of thermoelectric, water-flow, or air-flow coolers than it is to correct for temperature variations. Khusainov (1992, 2001) used a thermoelectric cooling system for a CdTe detector. In our Siemens prototype, we employ chilled water to maintain a constant temperature of approximately 12°C on the external surface of the detector housing. Within the detector itself, module designers attempt to draw the heat generated by the ASICs away from the crystal and toward the cooling system. Although the Imarad module offers an elegant solution for packaging detectors into modules, the proximity of the ASICs to the CZT crystal presents a challenge in thermal management. Other solutions to the thermal management task have been investigated by Bicron/LETI, NASA, and Philips Medical Systems. These generally involve removing heat-generating ASIC circuitry to perpendicular cooling fins through the use of low-noise connectors, allowing conventional air cooling to be used. These alternative packaging methods are displayed schematically in Figure 3.

E. Timing Resolution

In the late 1990s, it was thought that NaI Anger cameras and, by extension, any nuclear medicine imaging device must be capable of coincidence detection of positron annihilation photons of 511 keV. This led to studies of coincidence timing resolution in CdTe detectors. Shao et al. (2000) measured timing resolution with CdTe detectors of 2-mm thickness to be 14–24 ns and found the resolution to be dependent on the low-energy threshold—the lower the threshold, the worse the resolution. Okada et al. (2001) measured 9 ns and noted that the timing performance is best when the drift of the electrons is perpendicular to the entrance direction of the incident photons. This geometry is also better for making the detector deeper in the incident direction, a necessity for high-sensitivity detection of 511-keV photons. However, it is unlikely that CdTe or CZT will play a role in the coincidence PET imaging field purely from a sensitivity point of view. The perpendicular drift design could in theory mitigate part of the problem transaxially, but the requirement of putting the ASIC close to the detection material in both the sandwich or the connector (T) approaches depicted in Figure 3 will prevent efficient axial packing of the detectors.

F. Spatial Resolution

The spatial resolution of the CdTe family of detectors is limited by the range of the photoelectron involved in the event detection, which in turn is determined by the photon energy. The electron range defines the extent of the charge cloud from which the electrons and holes begin their drift toward their respective electrodes. Kalemci and Matteson (2002) calculate that a 100-keV photoelectron has a range of 47 microns, with the width of the charge cloud being approximately one-half this value. This cloud increases to 100 microns over a 2-mm drift in a 1000-V/cm electric field due to diffusion of the electrons. A 5-mm drift results in a 158-micron electron cloud, and a 10-mm drift gives a 224-micron cloud (Bellwied et al., 1999).

Although it appears that the extent of the charge cloud is less than 250 microns, other considerations must be made before designing the pixel layout. As has been previously stated, the smaller the pixel, the more impact the pixel–pixel boundaries have on the spectral tail. Also, the smaller the pixel, the greater the density of electronics and the more challenging the packaging and interconnection of the signals. In the nuclear medicine application, in which the extrinsic spatial resolution specification is dominated by the lead collimator, it makes sense to begin by using a relatively large pixel such as the 2.46-mm pixel of Imarad to make a modest improvement over NaI performance without making serious demands on packaging and electronics.

G. Collimation

All semiconductor imaging devices reported thus far have been based on pixels—resolution elements of known size and location, having a defined center position and boundary areas. The Anger camera concept is based on a multiple-PMT signal-weighting scheme for localization of the centroid of each event, and this position is not in a discrete locale but rather is an analog *xy* position. This difference in the resolution elements introduces the possibility for changes in collimator designs in the new semiconductor detectors. Mention

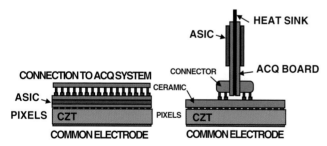

FIGURE 3 (a) The sandwich package (e.g., the Imarad module of Figure 2) and (b) the T package on. The sandwich package places a signal routing board between the CZT crystal (at the bottom of the drawing) and the ASIC circuitry. This scheme is more compact top to bottom, but presents a greater thermal management challenge compared to the T package. The T package uses a connector to place distance between the heat-generating ASIC from the CZT crystal. This allows upgrades of electronic circuitry to take place without changing the crystal. The sandwich package tends to be a permanent connection, through bonding, between the electronics and the crystal. The sandwich approach may better preserve the electrical integrity of the low-level signal through bonding and shorter connections, whereas the T package may eliminate the need for an active cooling subsystem.

has been made of the registered collimator—that is, a collimator with individual holes corresponding to individual pixels in the detector. Siemens has designed and tested three registered collimator designs for use with the Imarad module. The main advantage of the registered collimator for the semiconductor detector is that the collimator walls, or septa, cover the pixel–pixel boundaries and thereby eliminate some of the tail from the spectrum. The disadvantages include higher production cost than the hexagonal hole collimator and difficulty in alignment of the collimator with the pixel array. The idea of image processing through collimator beam modeling or resolution recovery remains to be exploited. The initial thought is that registered collimators enjoy some advantage for resolution recovery compared with hexagonal hole collimators.

Kalinosky et al. (1999) of Siemens investigated the effects of aliasing between a hexagonal-hole collimator and the square pixel array using a Monte Carlo simulation. Fourier analysis of the simulated response confirmed our intuition that a 15-degree rotation of the hexagonal collimator flat side from alignment with one of the square pixel directions produces minimal aliasing. Also confirmed was the fact that the aliasing artifact is observed only at unrealistically high count densities and will most likely be masked by errors in uniformity correction and other nonideal performance in the detection system.

The University of Arizona CZT system has very small, 380-micron pixels. Their team has designed a special registered collimator made by stacking tungsten foils to a depth of 7 mm with a bore size of 260 microns. This collimator was made by ThermoElectron/Tecomet (Woburn, MA) and is described by Kastis et al. (2000). It is likely that this high-resolution collimator will have septal penetration in addition to the high resolution. This allows for a new kind of nuclear imaging, described by B. M. W. Tsui (pers. comm., 2001), in which leaky or penetrable collimators are used to produce not a Gaussian point-spread response but rather a star artifact point-spread response. The operator then deconvolves the star artifact from the image to produce the high-resolution processed image. The concept is based on acquiring the additional needed counts to allow for a reliable deconvolution to be performed. Registered collimators of a completely new resolution scale such as those under investigation at the University of Arizona will allow new imaging techniques to be exploited. Nearly registered collimators can be designed to produce aliasing between the collimator and the pixel array—a concept of exploiting aliasing known as super-resolution (Lettington et al., 1996). Again, additional counts are needed to get the high-frequency information from the low-frequency pixel array.

The introduction of new detector technology in nuclear medicine, especially one with improved intrinsic spatial resolution, necessitates a reconsideration of general collimation schemes because the collimation will continue to dominate the ultimate system resolution. How can extrinsic resolution be improved? What about sensitivity? Single pinholes give high resolution, but also have the lowest sensitivity. Multiple pinholes (Yamamura et al. 2003; Schramm et al. 2001) have more sensitivity, but the information value of detected photons decreases as the complexity of the object being imaged increases. Therefore, for a few isolated small hot spots, multiple pinholes have advantages over single pinholes. Many new applications of nuclear medicine in the field of molecular imaging involve small isolated hot spots, so high-sensitivity, multiple pinholes should be developed for these applications. Coded apertures also involve many small holes and improve sensitivity relative to single pinholes. Meikle et al. (2001) and Accorsi and Lanza (2001) report on the use of coded apertures for small animal nuclear medicine imaging. Again, the object being imaged must remain simple, otherwise the information overlap is excessive and the value of the sensitivity increase is lowered. Jeanguillaume et al. (2001) describe efforts to design large-hole collimators specifically for pixellated detectors.

Another way of improving sensitivity while maintaining resolution performance is to use slat or one-dimensional collimation instead of the conventional hexagonal hole pattern (Gagnon et al., 2002). The increase in sensitivity from this design can be realized by applying a more sophisticated, fully three-dimensional plane-integral tomographic reconstruction technique to extract the information contained in the acquired projection data.

IV. NUCLEAR MEDICINE APPLICATIONS

Investigation into the medical applications of the CdTe family of detectors began in the field of X-ray imaging. This is primarily because detectors for X-ray imaging applications include film (in mammography and chest X-ray) and arrays of scintillators coupled to multiple-channel electronics in computerized tomography (CT). Film is cumbersome and nondigital, and the replacement of film is an obvious goal. The multiple-channel CT scanners also can be improved by using a compound semiconductor in the current mode. Digital mammography has the requirement of high-spatial-resolution performance down to 50-micron pixels (Yin et al., 2002). Yin et al. used a 0.2-mm thickness of CZT to measure a 65% detective quantum efficiency in the digital mammography application. Bertolucci et al. (1998) report on the use of CZT for digital radiography. Digital X-ray and CT imaging is done in the current mode, whereas nuclear medicine imaging is done in pulse-counting spectroscopic mode. Assuming radiographic X-ray fluences remain constant (or decrease to reduce patient dose) in the coming years, the trend toward higher-density electronics and smaller semiconductor pixels results in more processing per pixel as well as fewer photons per pixel. This presents the possibility of photon counting to replace the current mode in radiography (Fischer *et al.*, 2001). Investigations into the CT scanner application are presented in Section IVE.

Scheiber and colleagues have published reviews of the medical applications of CdTe and CZT (Scheiber,1996; Scheiber and Chambron 1992; Scheiber and Giakos, 2001). In 1992, Scheiber and Chambron note that progress was slowed by technical challenges of large arrays of electronics and that the crystal material at that time was good for miniaturized probes. In 1996, Scheiber states that it was too early to tell if CdTe:Cl or CZT will be the better performing crystal. In their 2001 review, Scheiber and Giakos report on CZT as compared with amorphous selenium (a-Se) and gas-filled detectors.

Other X-ray applications for these crystals include the detection of soft X-rays (e.g., 5.9-keV ^{55}Fe). This use requires low temperatures to get leakage current down into the picoampere level (Niemelae and Sipilae, 1994). CZT also has been used to do X-ray spectroscopy (Miyajima *et al.*, 2002). Long-term, continuous use of semiconductor detectors in an X-ray imaging device or in a nuclear medicine gamma camera requires the consideration of radiation damage to the crystal. It has been reported that radiation doses on the order of 30 kGy are necessary to degrade the spectral performance of CZT, and therefore these effects are not observed in the medical applications, in which the lifetime doses of the detectors are orders of magnitude lower than 30 kGy (Cavallini *et al.*, 2002).

A. Prototypes and Products

The first reports of a CdTe prototype reached the nuclear medicine community in the mid-1990s. This created great excitement because the promise of solid-state imaging cameras had existed for decades but the Anger camera technology of the late 1950s remains entrenched to this day. Perhaps the earliest mention of a CdTe-based gamma camera in the nuclear medicine field was given by Entine *et al.* (1979). This camera was based on a linear array of electrodes and rotated to give *xy* coordinates. Butler *et al.* (1993) describe CZT for nuclear medicine imaging. Butler *et al.* (p. 567, 1998) state that an image of a thyroid phantom in this article is "believed to be the first radionuclide image acquired with a room temperature semiconductor detector array and demonstrates that images of clinical quality can be achieved with such arrays." Digirad exhibited a CZT camera at the fall 1995 RSNA meeting in Chicago. The performance of Digirad's Model 2020tc CZT imager was reported by Butler *et al.* (1998).

At the 1996 Society of Nuclear Medicine meeting in Denver, General Electric Medical Systems (GEMS) presented a CdTe camera prototype in its exhibition booth. Verger *et al.* (1997) describe this prototype as a collaborative effort involving eV Products, GEMS, and Isorad (Yavne, Israel) that commenced in 1996. Eisen *et al.* (1996) of Isorad call this prototype version NUCAM. Eisen *et al.* compare the CdTe performance with an Anger-type camera in nuclear medicine applications of cardiac and thyroid imaging in patients. Scheiber and Giakos (2001) call these the first patient images acquired with this technology. The NUCAM was 16×16 cm^2 with 4-mm pixels. NUCAM operated more than 2 years in various hospitals in Israel and the United States with performance comparable to that of the NaI Anger camera (Eisen and Shor, 1998). Eisen *et al.* (1999) report on efforts to configure contacts to get good spectroscopic results and to do depth-of-interaction corrections. They also report on the advantages and disadvantages of CdTe and CZT in medical diagnostic systems. The NUCAM3 was announced by Eisen *et al.* (2002). It is CZT-based, 2.1-mm pixels of 5-mm thickness, and has a 18.5 cm $\times$ 20.1 cm^2 useful field of view. The NUCAM demonstrated usefulness in cardiac SPECT, breast, thyroid, and small organ evaluation.

Lund *et al.* (1997) produced a publicly available report on Sandia Laboratory's (Albuquerque, NM) miniature gamma-ray camera for tumor localization based on CZT. The University of Arizona has built a 64×64 CZT prototype with 380-micron pixels (Matherson *et al.*, 1998; Kastis *et al.*, 2000). The Bicron and Crismatec entities of Saint-Gobain (La Defense, France) and LETI are collaborating on a prototype called PEGASE (Mestais *et al.*, 2001). This prototype employs a rise-time compensation scheme devised by the CEA/LETI electronics group (Grenoble, France) to correct for depth-of-interaction effects and reduce tailing. The PEGASE prototype has 4-mm HPB CZT pixels. A brain phantom image demonstrates imaging performance comparable to that of NaI. The European BIOMED consortium (Strasbourg, France) has created a 15×15 cm^2 CdTe detector with 3-mm pixels (Scheiber *et al.*, 1999; Chambron *et al.*, 2000). It has been used for cardiac imaging (see Section IVB). Philips Medical Systems has presented a slat-collimated CZT imaging device called Solstice in collaboration with eV Products (Gagnon *et al.*, 2002; Griesmer *et al.*, 2002; De Geronimo *et al.*, 2002; Zeng *et al.*, 2002). This device has a 5-mm system-spatial resolution when reconstructed using the principle of plane integral tomography. The Solstice prototype has a 5% energy resolution at 140 keV and operates at room temperature without cooling. General Electric has worked with Imarad to create module-based prototypes for breast imaging (O'Connor *et al.*, 2002) and intraoperative applications (Blevis *et al.* 2002; Keidar *et al.*, 2002).

At Siemens, our collaboration with Imarad has produced the CZT cassette. A photograph of this prototype is shown in Figure 4. This device has a 12×20 cm^2 field of view. A comparison between this CZT prototype and NaI was performed (Wagenaar *et al.*, 2003) using the Hoffman brain phantom. Figure 5 shows the results of this comparison. Each image contains approximately 1 million counts in a 15% energy window. The phantom was in direct contact with the collimator, thereby emphasizing the intrinsic resolution ad-

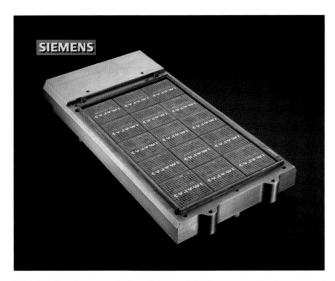

FIGURE 4 Siemens Medical Solutions' CZT cassette, based on Imarad modules. This device has a 12 × 20 cm² field of view. The pictured cassette has been shielded for low energy (< 200 keV).

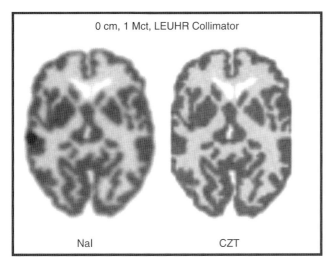

FIGURE 5 Hoffman brain phantom comparison between NaI (left) and CZT (right). Each image contains 1 million counts in a 15% energy window centered on 140 keV. Note the superior spatial resolution of the CZT image. The hot spot on the far left in the NaI image is from activity spilled on the fill hole; the edges of the CZT image are cropped by the 12-cm field of view of the prototype. (Reprinted from Nuclear Instruments and Methods in Physics Research Section A: Accelerators, Spectrometers, Detectors and Associated Equipment, Vol. 505, Wagenaar et al., Planar image quality comparison between a CdZnTe prototype and a standard NaI(Tl) gamma camera, pages 586–589, 2003, with permission from Elsevier.)

vantage of CZT (2.46 mm vs. ~ 3.6 mm). Lower count images also demonstrated the same performance advantage at low source–collimator distances.

CdTe and CZT detectors are more expensive per unit area than NaI, mainly because of the crystal growth and preparation. Therefore, these high-performance devices are being used in specialized, small field-of-view applications such as tumor, organ, or operating room imaging. It can be foreseen that medical applications of these devices will begin in this specialty niche as opposed to large-field-of-view, whole-body imaging.

B. Cardiac Imaging

Approximately one-half of all nuclear medicine studies performed clinically involve cardiac imaging. This large market must be considered first when investigating a new technology. The small volume of CdTe-type detectors is attractive for the new market of office-based cardiologists entering nuclear imaging. The BIOMED project has reported on two applications of cardiac imaging. The first is the continuous imaging of blood pool volume using a CdTe-based monitor strapped to the patient (Chambron et al., 1997). Scheiber et al. (1999) report on the use of the BIOMED camera for cardiac imaging. Siemens Medical Engineering (Erlangen, Germany) is part of the European BIOMED consortium.

C. Breast Imaging

Singh (1998) describes a CZT-based breast imaging device with 2- and 3-mm pixels. O'Connor et al. (2002) have reported on the Imarad-module-based GE breast prototype with a 20 × 20 cm² field of view. They observe reliable detection of 5- to 10-mm breast lesions in patient studies. Patt et al. (2002) of Gamma Medica (Los Angeles, CA) also uses an Imarad-based prototype of 16 × 20 cm² field of view for breast imaging. A compression device has been designed for this prototype, and the system is ready for clinical evaluation. For collimated nuclear medicine imaging, the influence of the detector's intrinsic spatial resolution is greatest at the lowest object-to-detector distance. Furthermore, the line-spread function (LSF) of the pixellated detector is very nearly a square-wave function with negligible spatial tail, whereas the LSF of the Anger camera is Gaussian-shaped with an extended spatial tail. The spatial resolution advantage demonstrated in Figure 5 should allow superior lesion detection in breast cancer compared to standard NaI gamma camera performance due to both the closer proximity allowed by the smaller camera geometry and the square-wave 2.46-mm intrinsic spatial resolution.

D. Handheld and Surgical Imagers

Surgical imagers are used to find radioactively labeled lymph nodes during surgery for excision and biopsy. Lund et al. (1997) describe a 1.5 × 1.5 cm² miniature gamma-ray camera design for tumor localization with an 8 × 8 CZT pixel array. Tsuchimochi et al. (2001) built a 32 × 32 CdTe pixel array with a 4.48 × 4.48 cm² FOV. Eurorad offers a 16 × 16 pixel CdTe product called the "Minicam" (http://www.eurorad.com/commun/00PDF/euro-minicam.pdf) for handheld imaging applications. Researchers at General Electric

(Blevis *et al.*, 2002; Keidar *et al.*, 2002) discuss the use of a handheld imager based on one Imarad module (4.0 × 4.0 cm^2) in the operating suite. This device allows a lesion at a depth of 6 cm to be visualized, and the 3D configuration of two or three hot lesions in close proximity can be discerned using only two different (orthogonal) views. Keidar *et al.* (2002) review the use of CdTe or CZT probes in surgery and for sentinel node imaging in breast cancer and melanoma. Anzai International (Tokyo, Japan) sells a product called eZ-SCOPE (Parnham *et al.*, 2002). Another intraoperative probe is described by Hai-Peng *et al.* (2002).

E. Dual-Modality Single-Photon Emission Computed Tomography and X-ray Computed Tomography

Dual modality imaging is currently very popular with the combination of PET and CT. Combined single-photon emission computerized tomography (SPECT) and CT devices are beginning to be used as well. Perhaps the best dual-modality application for CdTe is the combination of magnetic resonance imaging (MRI) and SPECT because the semiconductor detector contains no PMTs and therefore is not susceptible to the high magnetic field of the MRI (Iwata *et al.*, 1999). Because of the developmental history of these detectors, there has been much effort to create X-ray imagers and combined X-ray (current mode) and single-photon (pulse-counting mode) imaging systems.

Baldazzi *et al.* (1995) investigate the use of CdTe in high-energy, industrial CT applications. During this time, these detectors were investigated for use in next-generation medical CT scanners. They were rejected because polarization effects (Sato *et al.*, 1995) limited the speed of acquisition at a time when CT designers were pressing for higher acquisition rates (Scheiber, 1996; Matz and Weidner, 1998; Jahnke and Matz, 1999). Since that time, it has been realized that not all CT applications require high speed (Claesson *et al.*, 2002a, 2002b).

Another application for these detectors has been the dual-use, selectable configuration for SPECT and CT. These devices use the same detector and switch between current mode (CT) and pulse-counting mode (SPECT). Jakobson and Nemirovsky (1997) describe a low-noise, switched readout electronics scheme. A Sigma-Delta ADC circuit has been presented by Iwata *et al.* (1999, p. 743) for a combined SPECT/CT device. This ADC circuit allows a "trade between temporal resolution and bandwidth, i.e., a trade between 20 bit/ 2 kHz CT performance and 8 bit/500 kHz SPECT performance by varying digital postfiltering in the digital signal processor following the Sigma-Delta ADC."

F. Small Animal Imaging

The field of small animal imaging is the proving ground for the highest resolution devices. When SPECT imaging is performed on the 1- to 2-mm resolution scale, this is referred to as microSPECT. Because of CZT's superior intrinsic spatial resolution, it is particularly well suited to microSPECT. It is expected that CZT microSPECT devices will eventually find their way into the human imaging clinic, and therefore it is important to be aware of this field of research. Kastis *et al.* (2002) of the University of Arizona describe a small animal microSPECT device based on CZT detectors. This detector uses the 380-micron pixels described earlier to achieve SPECT resolution in the 1- to 2-mm range. Paulus *et al.* (1999) use CZT to do microCT imaging.

G. Compton Camera

Imaging of ^{18}F-fluorodeoxyglucose (^{18}F is a positron emitter) uptake in cancer patients using PET scanners has had a dramatic impact on the field of nuclear medicine. Because of the clinical importance of this contrast agent, detector developers have been investigating alternative ways to image the 511-keV-positron annihilation photons from the decay of ^{18}F and other positron emitters. Wagenaar *et al.* (1999) performed Monte Carlo calculations to determine the optimum pixel size for 140 keV and 511 keV in CZT pixellated detectors. The Compton camera is an especially attractive device for imaging 511 keV with CZT because Compton scatter is five times as likely to occur than photoelectric effect due to the effective atomic number of this material. Also, the relatively high energy of the gamma ray makes the Doppler broadening of the Compton scatter less severe than at 140 keV. Z. He *et al.* (1999) of the University of Michigan describe a 1-cm^3 volumetric spectroscopic CZT crystal for use in a Compton telescope. This device has 11 × 11 pixels and 20 depth-of-interaction resolution elements obtained from a cathode signal and a coplanar grid. Compton cameras and telescopes require the high energy and spatial resolution of CZT. Z. He *et al.* (2002) has reported point- and line-source imaging results with the three-dimensional CZT spectrometer. Du *et al.* (2001) carry out a Monte Carlo system performance prediction of a CZT Compton camera. Kroeger *et al.* (2001) conjecture that CZT may prove useful for a Compton telescope or a Compton camera, in which both the position and energy of detected events must be known with high resolution. Based on current work and performance capabilities, it appears that progress in Compton camera and telescope development will probably involve CdTe or CZT detector technology.

V. CONCLUSION

In the near term, the CZT crystal growers who aspire to enter the medical imaging field must achieve the best balance between low-cost production of high yields of contiguous operating crystals, on the one hand, and acceptable spectral performance in terms of energy resolution (which is linked ultimately to resistivity) and spectral stability over time, on the other. Furthermore, we can expect the spectral tail on the low-energy side of the photopeak to diminish over time as more research groups employ grids, novel contacts, improved surface preparations, and miniaturized and powerful electronic corrections—all to reduce the tail and improve the photofraction. The appeal of the CdTe family of semiconductor detectors boils down to the size of the charge cloud, which is less than 250 microns for realistic crystal sizes (cloud drift distances) and gamma-ray energies in the medically useful range. It does not take a great leap of credulity to imagine one day a CZT detector with nearly 100% active detector volume, less than 0.5 mm volume-of-interaction spatial resolution, a high photofraction (i.e., the tail has been eliminated), spectroscopic-grade energy resolution at room temperature, and constant reliable spectral performance over years of continuous operation. To complete the nuclear medicine imaging system, these perfected semiconductor detectors must be matched with collimation (or other image formation devices not yet invented) that is worthy of their fine detection performance. We are not yet near the physical limits of nuclear medicine image quality, and the CdTe and CZT family of semiconductors promise to help us make progress toward this important goal.

APPENDIX

Appendix A   Research Institutions Involved in CdTe or CZT Research

Institution	Division/Department	Location
Arizona, University of	Radiology; Optical Sciences	Tucson, AZ, USA
Bologna, University of	Physics; INFN	Bologna, Italy
California-Berkeley, University of	Bio Eng; Elect. Eng. and Comp. Sci.	Berkeley, CA, USA
California Institute of Technology	Space Rad. Lab; Jet Prop. Lab	Pasadena, CA, USA
California-Los Angeles, University of	Crump Inst. for Molecular Imaging; Mat. Sci. and Eng.	Los Angeles, CA, USA
California-Riverside, University of	Inst. Geophys. and Planetary Physics	Riverside, CA, USA
California-San Diego, University of	Med. Ctr. Astrophys. and Space Sci.	La Jolla, CA, USA
California-San Francisco, University of	Radiology	San Francisco, CA, USA
Carnegie Mellon University	Elect. and Comp. Eng. Dept.	Pittsburgh, PA, USA
Charles University	Institute of Physics	Prague, Czech Republic
Chicago, University of	Physics	Chicago, IL, USA
Coimbra, University of	Physics Dept., Elect. and Instrum.	Coimbra, Portugal
Columbia University	Astrophysics Lab	New York, NY, USA
Fachhochschule Munich		Munich, Germany
Faculty of Medicine	Institute of Physical Biology	Strasbourg, France
Federico University	Physics; INFN	Napoli, Italy
Fisk University	Physics	Nashville, TN, USA
Freiburg University	Kristallographisches Insitut	Freiburg, Germany
Harvard University	Center for Astrophysics	Cambridge, MA, USA
Hebrew University		Jerusalem, Israel
Hokkaido Institute of Technology		Sapporo, Japan
Ishikawa National College of Technology		Ishikawa, Japan
Johns Hopkins University	Applied Physics Laboratories	Laurel, MD, USA
Karlsruhe, University of	Crystals and Materials	Karlsruhe, Germany
Korea University	Department of Physics	Seoul, South Korea
Kyoto Institute of Technology		Kyoto, Japan
Louisiana State University	Physics and Astronomy	Baton Rouge, LA, USA
Maryland, University of		College Park, MD, USA
Michigan, University of	Nuclear Engineering	Ann Arbor, MI, USA
Minnesota, University of	Chemical Eng.; Materials Sci.	Minneapolis, MN, USA

(*Continued*)

Appendix A Research Institutions Involved in CdTe or CZT Research (Continued)

Institution	Division/Department	Location
Montreal, University of	Department of Physics	Montreal, Canada
New Hampshire, University of		Durham, NH, USA
Nippon Dental University	School of Dentistry	Niigata, Japan
Osaka University	Medical School	Osaka, Japan
Russian Academy of Science	Institute of Solid State Physics	Chernogolovka, Moscow, Russia
Academy Sinica	Institute of Nuclear Research	Shanghai, China
Shizuoka University	Graduate School of Elect. Sci. and Tech.	Hamamatsu, Japan
Sichuan University	Materials Science Dept.	Chengdu, China
Sinica Academy	Institute of Technical Physics	Shanghai, China
Solid State Physics Lab		Delhi, India
Stockholm Inst of Technology	Department of Physics	Stockholm, Sweden
Sunnybrook and Women's Health		Toronto, Ontario, Canada
Surrey, University of		Surrey, United Kingdom
Technion-Israel Institute of Technology	Electrical Engineering	Haifa, Israel
Tokyo Institute of Technology	Res. Ctr. for Quantum Effect Elect.	Tokyo, Japan
Waseda University	Adv. Research Inst. for Sci. and Eng.	Tokyo, Japan
Washington State University	Department of Physics	Pullman, WA, USA

Appendix B Government Agencies Involved in CdTe or CZT Research

Agency	Division/Department	Location
Air Force Research Laboratory		Kirtland AFB, NM, USA
Agency for Defense Development		Taejeon, South Korea
Argonne National Laboratory	Material Science Division	Argonne, IL, USA
Berkeley (EO Lawrence) National Lab		Berkeley, CA, USA
Brookhaven National Lab		Long Island, NY, USA
CEA Technologies Avancées	LETI	Grenoble, France
CERN	ISOLDE Collaboration	Geneva, Switzerland
European Commission	Joint Research Center	Ispra, Italy
European Space Agency (ESA)		Noordwijk, Netherlands
Japan Energy Corporation	Compound Semicond. Materials	Ibaraki, Japan
Lawrence Livermore National Lab		Livermore, CA, USA
Los Alamos National Lab	Safeguards Sci. and Tech. Group	Los Alamos, NM, USA
NASA	Goddard Space Flight Center	Greenbelt, MD, USA
NASA	Marshall Space Center	Huntsville, AL, USA
National Institute for Radiation Sciences	Anagawa, Inage-Ku	Chiba, Japan
National Institute of Standards and Technology (NIST)		Gaithersburg, MD, USA
Naval Research Laboratory		Alexandria, VA, USA
Orbital Sciences Corporation		Greenbelt, MD, USA
Sandia National Laboratories	Adv Materials Research Division; Reliability and Electrical Systems	Livermore, CA, USA
Sandia National Laboratories		Albuquerque, NM, USA
Soreq NRC		Yavne, Israel
U.S. Army	HPC Research Center	Minneapolis, MN, USA

Appendix C Companies Involved in CdTe or CZT Development

Company	Division/Department	Location
Acrorad		Okinawa, Japan
Amptek Inc.		Bedford, MA, USA
Anzai International		Tokyo, Japan
Baltic Scientific Instruments		Riga, Latvia
Bechtel		Las Vegas, NV, USA
Cardiac Mariners		Menlo Park, CA, USA
Cleveland Crystals, Inc		Cleveland, OH, USA
Crystal		Berlin, Germany
Digirad[a]		San Diego, CA, USA
EG&G Energy Measurements		Goleta, CA, USA
Eurorad		Strasbourg, France
General Electric Medical Systems	Elgems, Ltd.	Haifa, Israel
Grumman Aerospace Corp.		Bethpage, NY, USA
Hamamatsu Photonics KK	Electron Tube R&D Center	Iwata, Japan
Hughes Aircraft Company	Santa Barbara Research Center	Santa Barbara, CA, USA
Hughes Research Labs		Malibu, CA, USA
II-VI, Inc.	eV Products	Saxonburg, PA, USA
Imarad Imaging Systems, Ltd.		Rehovot, Israel
Integrated Detector Electronics AS		Oslo, Norway
Japan Energy Corporation	Acrotec	Toda, Saitama, Japan
Johnson Matthey Electronics		Spokane, WA, USA
Metorex International Oy		Espoo, Finland
Mitsubishi Heavy Industries		Yokohama, Japan
NEC Research Institute		Princeton, NJ, USA
Nova R&D		Riverside, CA, USA
Outokumpu Instruments Oy		Espoo, Finland
Philips Medical Systems, Inc.	Nuclear Medicine Division	Cleveland, OH and Milpitas, CA, USA
Philips Research	Imaging Systems	Aachen, Germany
Photon Imaging	Gamma Medica	Los Angeles, CA, USA
Radiation Monitoring Devices		Watertown, MA, USA
Raytheon Systems Company		Goleta, CA, USA
Ritec		Riga, Latvia
Shimadzu Corporation	Central Research Laboratory	Kyoto, Japan
Siemens Corporate Technology		Munich, Germany
Siemens Medical Solutions	Nuclear Medicine Group	Hoffman Estates, IL, USA
Spire Corporation		Bedford, MA, USA
Saint-Gobain	Crismatec/Bicron entities	La Defense, France and Solon, OH, USA
Sumitomo Metal Mining, Co., Ltd.		Tokyo, Japan
Texas Instruments		Dallas, Texas, USA

[a]Formerly Aurora Technologies Corporation.

References

Accorsi, R., Gasparini, F., and Lanza, R. C. (2001). A coded aperture for high-resolution nuclear medicine planar imaging with a conventional Anger camera:experimental results. *IEEE Trans. Nucl. Sci.* **48**(6): 2411–2417.

Akutagawa, W., and Zanio, K. (1969). Gamma response of semi-insulating material in the presence of trapping and detrapping. *J. Appl. Phys.* **40**(9): 3838–3854.

Arques, M., Baffert, N., Lattard, D., Martin, J. L., Masson, G., Mathy, F., Noca, A., Rostaing, J. P., Trystram, P., Villard, P., Cretolle, J., Lebrun F., Leray, J. P., and Limousin, O. (1999). A basic component for ISGRI, the CdTe gamma camera on board the INTEGRAL satellite. *IEEE Trans. Nucl. Sci.* **46**(3): 181–186.

Auricchio, N., Caroli, E., De Cesare, G., Dusi, W., Grassi, D., Hage-Ali, M., Perillo, E., Siffert, P., and Spadaccini, G. (1998). Spectroscopic response versus interelectrodic charge formation position in CdTe detectors. *Mat. Res. Soc. Symp. Proc.* **487**: 309–314.

Baldazzi, G., Rossi, M., Querzola, E., Guidi, G., Chirco, P., Scannavini, M. G., Zanarini, M., and Casali, F. (1995). A radiation detection system for high-energy computerized tomography using CdZnTe detectors. *IEEE Trans. Nucl. Sci.* **42**(4): 575–579.

Barrett, H. H., Eskin, J. D., and Barber, H. B. (1995). Charge transport in arrays of semiconductor gamma-ray detectors. *Phys. Rev. Lett.* **75**(1): 156–159.

Bellwied, R., Beuttenmuller, R., Brandon, N., Caines, H., Chen, W., DiMassimo, D., Dyke, H., Hoffmann, G. W., Humanic, T. J., Kotova, A. I., Kotov, I. V., Kraner, H. W., Li, Z., Lynn, D., Minor, B., Ott, G., Pandey, S. U., Pruneau, C., Rykov, V. L., Schambach, J., Sedlmeir, J., Sugarbaker, E., Takahashi, J., and Wilson, W. K. (1999). Two dimensional studies of dynamics of electron clouds in silicon drift detectors. *IEEE Trans. Nucl. Sci.* **46**(30): 176–180.

Bertolucci, E., Conti, M., Marcello, L., Russo, P., Chirco, P., and Rossi, M. (1998). Imaging performance of single-element CdZnTe detectors for digital radiography. *IEEE Trans. Nucl. Sci.* **45**(3): 406–412.

Blevis, I. M., Reznik, A., Wainer, N., Kopelman, D., Chaikov, A., Iosilevsky, G., Israel, O., and Hashmonai, M. (2002). Evaluation of a CZT intraoperative gamma camera (abstract). *J. Nucl. Med.* **43**(5): 232P.

Bolotnikov, A. E., Cook, W. R., Harrison, F. A., Wong, A.-S., Schindler, S. M., and Eichelberger, A. C. (1999). Charge loss between contacts of CdZnTe pixel detectors. *Nucl. Instrum. Methods A* **432**: 326–331.

Burger, A., Chen, H., Chattopadhyay, K., Shi, D., Morgan, S. H., Collins, W. E., and James, R. B. (1999). Characterization of metal contacts on [sic] and surfaces of cadmium zinc telluride. *Nucl. Instrum. Methods A* **428**(1): 8–13.

Burger, A., Chen, H., Tong, J., Shi, D., George, M. A., Chen, K.-T., Collins, W. E., James, R. B., Stahle, C. M., and Bartlett, L. M. (1997). Investigation of electrical contacts for $Cd_{1-x}Zn_xTe$ nuclear radiation detectors. *IEEE Trans. Nucl. Sci.* **44**(3): 934–938.

Butler, J. F., Friesenhahn, S. J., Lingren, C., Apotovsky, B., Simchon, R., Doty, F. P., Ashburn, W. L., and Dillon, W. (1993). CdZnTe detector arrays for nuclear medicine imaging. *1993 IEEE Nucl. Sci. Symp. Med. Imag. Conf. Rec.* 565–568.

Butler, J. F., Lingren, C. L., and Doty, F. P. (1992). $Cd_{1-x}Zn_x$ Te gamma-ray detectors. *IEEE Trans. Nucl. Sci.* **39**(4): 605–609.

Butler, J. F., Lingren, C. L., Friesenhahn, S. J., Doty, F. P., Ashburn, W. L., Conwell, R. L., Augustine, F. L., Apotovsky, B., Pi, B., Collins, T., Zhao, S., and Isaacson, C. (1998). CdZnTe solid-state gamma camera. *Nucl. Instrum. Methods* **45**(3): 359–363.

Cavallini, A., Fraboni, B., Dusi, W., Auricchio, N., Chirco, P., Zanarini, M., Siffert, P., and Fougeres, P. (2002). Radiation effects on II-VI compound-based detectors. *Nucl. Instrum. Methods A* **476**: 770–778.

Chambron, J., Arntz, Y., Echlancher, B., Scheiber, C., and Siffert, P. (2000). A pixellated gamma-camera based on CdTe detectors clinical interest and performances. *Nucl. Instrum. Methods A* **448**: 537–549.

Chambron, J., Dumitresco, B., Echlancer, B., Karman, M., and Nemeth, L. (1997). A portable cadmium telluride multidetector probe for cardiac function monitoring. *In* "Proceedings of Computers in Cardiology Conference." Available at www.cinc.org.

Chattopadhyay, K., Ma, X., Ndap, J.-O., Burger, A., Schlesinger, T. E., Greaves, M., Glass, H. L., Flint, J. P., and James, R. B. (2000). Thermal treatments of CdTe and CdZnTe detectors. *Proc. SPIE* **4141**: 303–308.

Chen, C. M. H., Boggs, S. E., Bolotnikov, A. E., Cook, W. R., Harrison, F. A., and Schindler, S. M. (2002). Numerical modeling of charge sharing in CdZnTe pixel detectors. *IEEE Trans. Nucl. Sci.* **49**(1): 270–276.

Claesson, T., Kerek, A., Molnar, J., and Novak, D. (2002a). A CT demonstrator based on a CZT solid state detector. *2001 IEEE Nucl. Sci. Symp. Med. Imag. Conf. Rec.* **3**: 1644–1646.

Claesson, T., Kerek, A., Molnar, J., and Novak, D. (2002b). An X-ray computed tomography demonstrator using a CZT solid-state detector. *Nucl. Instrum. Methods A* **487**(1–2): 202–208.

Cook, W. R., Boggs, S. E., Bolotnikov, A. E., Burnham, J. A., Fitzsimmons, M. J., Harrison, F. A., Kecman, B., Matthews, B., and Schindler, S. M. (2000). High resolution CdZnTe pixel detectors with VLSI readout. *IEEE Trans. Nucl. Sci.* **47**(4): 1454–1457.

Cook, W. R., Boggs, S. E., Bolotnikov, A. E., Burnham, J. A., Harrison F. A., Kecman, B., Matthews, B., and Schindler, S. M. (1999). First test results from a high resolution CdZnTe pixel detector with VLSI readout. *Proc. SPIE* **3769**: 92–96.

De Geronimo, G., O'Connor, P., Kandasamy, A., and Grosholz, J. (2002). Advanced readout ASICs for multielement CZT detectors. *Proc. SPIE* **4784**: 105–118.

Dharmadasa, M., Blomfield, C. J., Gregory, G. E., and Haigh, J. (1994). Aging effects and auger depth profiling studies of Sb/n-CdTe contacts. *Semicond. Sci. Technol.* **9**: 185–187.

Du, Y. F., He, Z., Knoll, G. F., and Wehe, W. L. (2001). Evaluation of a Compton scattering camera using 3D position sensitive CdZnTe detectors. *Nucl. Instrum. Methods A* **457**: 203–211.

Du, Y. F., He, Z., Li, W., Knoll, G. F., and Wehe, D. K. (1999). Monte Carlo investigation of the charge sharing effects in 3-D position sensitive CdZnTe gamma-ray spectrometers. *IEEE Trans. Nucl. Sci.* **46**(4): 634–637.

Eisen, Y., Mardor, I., Shor, A., Baum, Z., Bar, D., Feldman, G., Cohen, H., Issac, E., Haham-Zada, R., Blitz, S., Cohen, Y., Glick, B., Falk, R., Roudebush, S., and Blevis, I. (2002). NUCAM3-A gamma camera based on segmented monolithic CdZnTe detectors. *IEEE Trans. Nucl. Sci.* **49**(4): 1728–1732.

Eisen, Y., and Shor, A. (1998). CdTe and CdZnTe materials for room-temperature X-ray and gamma-ray detectors. *J. Crys. Growth* **184–185**: 1302–1312.

Eisen, Y., Shor, A., Gilath, C., Tsabarim, M., Chouraqui, P., Hellman, C., and Lubin, E. (1996). A gamma camera based on CdTe detectors. *Nucl. Instrum. Methods A* **380**: 474–478.

Eisen, Y., Shor, A., and Mardor, I. (1999). CdTe and CdZnTe gamma-ray detectors for medical and industrial imaging systems. *Nucl. Instrum. Methods A* **428**: 158–170.

Entine, G., Luthmann, R., Mauderli, W., Fitzgerald, L. T., and Williams, C. M. (1979). Cadmium telluride gamma camera. *IEEE Trans. Nucl. Sci.* **26**(1): 552–553.

Fischer, P., Kouda, M., Kruger, H., Lindner, M., Sata, G., Takahashi, T., Watanabe, S., and Wermes, N. (2001). A counting CdTe pixel detector for hard x-ray and gamma-ray imaging. *IEEE Trans. Nucl. Sci.* **48**(6): 2401–2404.

Franc, J., Hoeschl, P., Belas, E., Grill, R., Hlidek, P., Moravec, P., and Bok, J. (1999). CdTe and CdZnTe crystals for room temperature gamma-ray detectors. *Nucl. Instrum. Methods A* **434**: 146–151.

Gagnon, D., Zeng, G. L., Links, J. M., Griesmer, J. J., and Valentino, F. C. (2002). Design considerations for a new solid-state gamma-camera: Solstice. *2001 IEEE Nucl. Sci. Symp. Conf. Rec.* **2**: 1156–1160.

Griesmer, J. J., Kline, B., Grosholz, J., Parnham, K., and Gagnon, D. (2002). Performance evaluation of a new CZT detector for nuclear medicine: Solstice. *2001 IEEE Nucl. Sci. Symp. Conf. Rec.* **2**: 1050–1054.

Hai-Peng, W., Ying-Wu, L., Xi-Zheng, L., and Qiu, S. (2002). A small dimension intraoperative probe. *Nucl. Sci. Tech.* (China) **13**(3): 177–181.

Hamel, L. A., Macri, J. R., Stahle, C. M., Odom, J., Birsa, F., Shu, P., and Doty, F. P. (1996). Signal generation in CdZnTe strip detectors. *IEEE Trans. Nucl. Sci.* **43**(3): 1422–1426.

He, Z., Lehner, C., Feng, Z., Wehe, D. K., Knoll, G. F., Berry, J., and Du, Y. (2002). Hand-held gamma-ray imaging sensors using room-temperature 3-Dimensional position-sensitive semiconductor spectrometers. *AIP Conference Proceedings* **632**(1): 209–215.

He, Z. (2001). Review of the Shockley-Ramo theorem and its application in semiconductor gamma-ray detectors. *Nucl. Instrum. Methods A* **463**: 250–267.

He, Z., Knoll, G. F., Wehe, D. K., and Du, Y. F. (1998). Coplanar grid patterns and their effect on energy resolution of CdZnTe detectors. *Nucl. Instrum. Methods A* **411**: 107–112.

He, Z., Knoll, G. F., Wehe, D., and Miyamoto, J. (1997). Position sensitive single carrier CdZnTe detectors. *Nucl. Instrum. Methods A* **388**: 180–185.

He, Z., Li, W., Knoll, G. F., Wehe, D. K., Berry, J., and Stahle, C. M. (1999). 3-D position sensitive CdZnTe gamma-ray spectrometers. *Nucl. Instrum. Methods A* **422**: 173–176.

Iwata, K., Hasegawa, B., Heanue, J. A., Bennett, P. R., Shah, K. S., Boles, C. D., and Boser, B. E. (1999). CdZnTe detector for combined X-ray CT and SPECT. *Nucl. Instrum. Methods A* **422**: 740–744.

Jahnke, A., and Matz, R. (1999). Signal formation and decay in CdTe x-ray detectors under intense irradiation. *Med. Phys.* **26**(1): 38–48.

Jakobson, C. G., and Nemirovsky, Y. (1997). CMOS low-noise switched charge sensitive preamplifier for CdTe and CdZnTe X-ray detectors. *IEEE Trans. Nucl. Sci.* **44**(1): 20–25.

James, R. B., Brunett, B., Heffelfinger, J., van Scyoc, J., Lund, J., Doty, F. P., Lingren, C. L., Olsen, R., Cross, E., Hermon, H., Yoon, H., Hilton, N., Schieber, M., Lee, E. Y., Toney, J., Schlesinger, T. E., Goorsky, M., Yao, W., Chen, H., and Burger, A. (1998). Material properties of large-volume cadmium zinc telluride crystals and their relationship to nuclear detector performance. *J. Elect. Mat.* **27**(6): 788–799.

James, R. B., Schlesinger, T. E., Lund, J., and Schieber, M. (1995). Cd1-xZnxTe spectrometers for gamma and X-ray applications. *In* "Semiconductors and Semimetals" (T. E. Schlesinger and R. B. James, eds.), Vol. 43. Academic Press, San Diego, CA.

Jeanguillaume, C., Douiri, A., Quartuccio, M., Begot, S., Franck, D., Tence, M., and Ballongue, P. (2001). CACAO, a Collimation means well suited for pixellated gamma camera. *2001 IEEE Nucl. Sci. Symp. Med. Imag. Conf. Rec.* **4**: 2291–2294.

Kalemci, E., and Matteson. J. L. (2002). Investigation of charge sharing among electrode strips for a CdZnTe detector. *Nucl. Instrum. Methods A* **478**: 527–537.

Kalinosky, M. A., Wagenaar, D. J., Burckhardt, D., and Engdahl, J. C. (2000). Pixel-pixel interactions in a square pixel array CdZnTe detector (abstract). *Med. Phys.* **27**(6): 1380.

Kalinosky, M. A., Wagenaar, D. J., Pawlak, J., Rempel, T., and Engdahl, J. C. (1999). Aliasing in pixellated detectors. Proc of "Proceedings of Future Directions in Nuclear Medicine Physics and Engineering," p. 41. University of Chicago, Chicago, IL.

Kastis, G. A., Barber, H. B., Barrett, H. H., Balzer, S. J., Lu, D., Marks, D. G., Stevenson, G., Woolfenden, J. M., Appleby, M., and Tueller, J. (2000). Gamma-ray imaging using a CdZnTe pixel array and a high-resolution, parallel-hole collimator. *IEEE Trans. Nucl. Sci.* **47**(6): 1923–1927.

Kastis, G. A., Wu, M. C., Balzer, S. J., Wilson, D. W., Furenlid, L. R., Stevenson, G., Barber, H. B., Barrett, H. H., Woolfenden, J. M., Kelly, P., and Appleby, M. (2002). Tomographic small-animal imaging using a high-resolution semiconductor camera. *IEEE Trans. Nucl. Sci.* **49**(1): 172–175.

Keidar, Z., Frenkel, A., Iosilevsky, G., and Bar-Shalom, R. (2002). First clinical experience with a new hand held nuclear medicine imaging probe (abstract). *J. Nucl. Med.* **43**(5): 225P–226P.

Keyser, R. M. (2001). Characterization of room temperature detectors using the proposed IEEE standard. *2001 IEEE Nucl. Sci. Symp. Med. Imag. Conf. Rec.* **1**: 315–318.

Khusainov, A. K. (1992). Cadmium telluride detectors with thermoelectric cooling. *Nucl. Instrum. Methods A* **322**: 335–340.

Khusainov, A. K. (2001). Approaching cryogenic Ge performance with peltier cooled CdTe. *Proc. SPIE* **4507**: 50–56.

Knoll, G. F. (1999). "Radiation Detection and Measuremen," 3rd ed. John Wiley & Sons, New York.

Knoll, G. F., and McGregor, D. S. (1993). Fundamentals of semiconductor detectors for ionizing radiation. *Mat. Res. Soc. Symp. Proc.* **302**: 3–17.

Kroeger, R. A., Johnson, W. N., Kurfess, J. D., Philips, B. F., and Wulf, E. A. (2001). Three-compton telescope: theory, simulations, and performance. *2001 IEEE Nucl. Sci. Symp. Med. Imag. Conf. Rec.* **1**: 68–71.

Kurczynski, P., Krizmanic, J., Stahle, C., Parsons, A., and Palmer, D. (1997). CZT strip detectors for imaging and spectroscopy: Collimated beam and ASIC readout experiments. *IEEE Trans. Nucl. Sci.* **44**(3): 1011–1016.

Lee, E. Y., James, R. B., Olsen, R. W., and Hermon, H. (1999). Compensation and trapping in CdZnTe radiation detectors studied by thermoelectric emission spectroscopy, thermally stimulated conductivity, and current-voltage measurements. *J. Elect. Mat.* **28**(6): 766–773.

Lettington, A. H., Hong, Q. H., and Tzimopoulou-Frickle, S. (1996). Superresolution by spatial-frequency aliasing. *Proc. SPIE* **2744**: 583–590.

Levi, D., Albin, D., and King, D. (2000). Influence of surface composition on back-contact performance in CdTe/CdS photovoltaic devices. *Prog. Photovoltaics* **8**(6): 591–602.

Li, L., Lu, F., Shah, K., Squillante, M., Cirignano, L., Yao, W., Olson, R. W., Luke, P., Nemirovsky, Y., Burger, A., Wright, G., and James, R. B. (2001). A new method for growing detector-grade cadmium zinc telluride crystals. *2001 IEEE Nucl. Sci. Symp. Med. Imag. Conf. Rec.* **4**: 2396–2400.

Li, W., He, Z., Knoll, G. F., Wehe, D. K., and Berry, J. E. (2001). Experimental results from an Imarad 8×8 pixellated CZT detector. *Nucl. Instrum. Methods A* **458**(12): 518–526.

Llopart, X., Campbell, M., Dinapoli, R., San Segundo, D., and Pernigotti, E. (2002). Medipix2, a 64 k pixel readout chip with 55 micron square elements working in single photon counting mode. *IEEE Trans. Nucl. Sci.* **49**(5): 2279–2283. Also available at http://www.cern.ch/medipix.

Luke, P. N. (1995). Unipolar charge sensing with coplanar electrodes—application to semiconductor detectors. *IEEE Trans. Nucl. Sci.* **42**(4): 207–213.

Luke, P. N., and Eissler, E. E. (1996). Performance of CdZnTe coplanar-grid gamma-ray detectors. *IEEE Trans. Nucl. Sci.* **43**(3): 1481–1486.

Lund, J. C., Olsen, R. W., James, R. B., Cross, E., Hermon, H., McGregor, D. S., Hilton, N. R., McKisson, J. E., Van Scyoc, J. M., Yoon, H., Brunett, B. A., Moses, W. W., Beuville, E., Kelley, J. G., Doty, F. P., Patt, B. E., and Wolfe, D. (1997). Miniature gamma-ray camera for tumor localization. Sandia Report SAND97–8278, UC-408. National Technical Information Service, Springfield, VA.

Lund, J. C., Olsen, R., Van Scyoc, J. M., and James, R. B. (1996). The use of pulse processing techniques to improve the performance of CdZnTe gamma-ray spectrometers. *IEEE Trans. Nucl. Sci.* **43**(3): 1411–1416.

Macri, J. R., Apotovsky, B. A., Butler, J. F., Cherry, M. L., Dann, B. K., Drake, A., Doty, F. P., Guzik, T. G., Larson, K., Mayer, M., McConnell, M. L., and Ryan, J. M. (1996). Development of an orthogonal-strip CdZnTe gamma radiation imaging spectrometer. *IEEE Trans. Nucl. Sci.* **43**(3): 1458–1462.

Mardor, I., Shor, A., and Eisen, Y. (2001). Edge and corner effects on spectra of segmented CdZnTe detectors. *IEEE Trans. Nucl. Sci.* **48**(4): 1033–1040.

Matherson, K. J., Barber, H. B., Harrett, H. H., Eskin, J. D., Dereniak, E. L., Marks, D. G., Woolfenden, J. M., Young, E. T., and Augustine, F. L. (1998). Progress in the development of large-area modular 64×64 CdZnTe imaging arrays for nuclear medicine. *IEEE Trans. Nucl. Sci.* **45**(3): 354–358.

Matz, R., and Weidner, M. (1998). Charge collection efficiency and space charge formation in CdTe gamma and X-ray detectors. *Nucl. Instrum. Methods A* **406**: 287–298.

McGregor, D. S., He, Z., Seifert, H. A., Rojeski, R. A., and Wehe, D. K. (1998). CdZnTe semiconductor parallel Frisch grid radiation detectors. *IEEE Trans. Nucl. Sci.* **45**(3): 443–449.

McGregor, D. S., Rojeski, R. A., He, Z., Wehe, D. K., Driver, M., and Blakely, M. (1999). Geometrically weighted semicondcutor Frisch grid radiation spectrometers. *Nucl. Instrum. Methods A* **422**: 164–168.

Meikle, S. R., Fulton, R. R., Eberl, S., Dahlbom, M., Wong, K. P., and Fulham, M. I. (2001). An investigation of coded aperture imaging for small animal SPECT. *IEEE Trans. Nucl. Sci.* **48**(3): 816–821.

Mescher, M. J., Schlesinger, T. E., Toney, J. E., Brunett, B. A., and James, R. B. (1999). Development of dry processing techniques for CdZnTe surface passivation. *J. Elect. Mat.* **28**(6): 700–704.

Mestais, C., Baffert, N., Bonnefoy, J. P., Chapuis, A., Koenig, A., Monnet, O., Ouvrier, Buffet, P., Rostaing, J. P., Sauvage, F., and Verger, L. (2001). A new design for a high resolution, high efficiency CZT gamma camera detector. *Nucl. Instrum. Methods A* **458**: 62–67.

Miyajima, S., Imagawa, K., and Matsumoto, M. (2002). CdZnTe detector in diagnostic X-ray spectroscopy. *Med. Phys.* **29**(7): 1421–1429.

Narita, T., Bloser, P. F., Grindlay, J. E., and Jenkins, J. A. (2000). Development of gold contacted flip-chip detectors with IMARAD CZT. *Proc. SPIE* **4141**: 89–96.

Nemirovsky, Y., Asa, G., Gorelik, J., and Peyser, A. (2000). Spectroscopic evaluation of n-type CdZnTe gamma-ray spectrometers. *J. Elect. Mat.* **29**(6): 691–698.

Nemirovsky, Y., Ruzin, A., Asa, G., Gorelik, Y., and Li, L. (1997). Study of contacts to CdZnTe radiation detectors. *J. Elect. Mat.* **26**(6): 756–764.

Niemelae, A., and Sipilae, H. (1994). Evaluation of CdZnTe detectors for soft X-ray applications. *IEEE Trans. Nucl. Sci.* **41**(4): 1054–1057.

Noda, D., Nakamura, A., Aoki, T., and Hatanaka, Y. (2000). Fabrication of CdZnTe Based compound semiconductor diode for radiation detectors. *Bull. Res. Inst. Electronics* (Japan) **35**: 73–79.

O'Connor, M. K., Mueller, B. A., Blevis, I. M., Rhodes, D. J., Smith, R., and Collins, D. A. (2002). CZT detector for scintimammography: Evaluation in phantom and patient studies (abstract). *J. Nucl. Med.* **43**(5): 289P.

Okada, Y., Takahashi, T., Sato, G., Watanabe, S., Nakazawa, K., Mori, K., and Makishima, K. (2001). CdTe and CdZnTe detectors for timing measurement. *2001 IEEE Nucl. Sci. Symp. Med. Imag. Conf. Rec.* **4**: 2429–2433.

Okamoto, T., Yamada, A., and Konagai, M. (2000). Optical and electrical characterizations of highly efficient CdTe thin film solar cells prepared by close-spaced sublimation. *J. Crys. Growth* **214–215**: 1148–1151.

Parnham, K. B., Davies, R. K., Vydrin, S., Ferraro, F., Jimbo, M., and Ryou, H. (2002). Development, design, and performance of a CdZnTe-based nuclear medical imager (abstract). *J. Nucl. Med.* **43**(5): 229P.

Patt, B. E., Iwanczyk, J. S., and MacDonald, L. R. (2002). New 16×20 cm^2 CZT gamma camera for scintiammography (abstract). *J. Nucl. Med.* **43**(5): 231P.

Paulus, M. J., Sari-Sarraf, H., Gleason, S. S., Bobrek, M., Hicks, J. S., Johnson, D. K., Behel, J. K., Thompson, L. H., and Allen, W. C. (1999). A new X-ray computed tomography system for laboratory mouse imaging. *IEEE Trans. Nucl. Sci.* **46**: 558–564.

Prettyman, T. H., Ameduri, F. P., Burger, A., Gregory, J. C., Hoffbauer, M. A., Majerus, P. R., Reisenfeld, D. B., Soldner, S. A., and Szeles, C. S. (2001). Effect of surfaces on the performance of CdZnTe detectors. *Proc. SPIE* **4507**: 23–31.

Prettyman, T. H., Feldman, W. C., Fuller, K. R., Storms, S. A., Soldner, S. A., Szeles, C., Ameduri, F. P., Lawrence, D. J., Brown, M. C., and Moss, C. E. (2001). CdZnTe gamma-ray spectrometer for orbital planetary missions. *2001 IEEE Nucl. Sci. Symp. Med. Imag. Conf. Rec.* **1**: 63–67.

Qifeng, L., Shifu, Z., Beijun, Z., Li, C., Deyou, G., and Yingrong, J. (2002). Growth and properties of high resistivity CdZnTe single crystals. *Chin. J. Semiconductors* **23**(2): 157–160.

Raiskin, E., and Butler, J. F. (1988). CdTe low level gamma detectors based on a new crystal growth method. *IEEE Trans. Nucl. Sci.* **35**(1): 82–84.

Ramo, S. (1939). Currents induced by electron motion. *In* "Proceedings of the IRE," p. 584. Columbia University, New York.

Ramsey, B., Sharma, D. P., Austin, R., Gostilo, V., Ivanov, V., Loupilov, A., Sokolov, A., and Sipila, H. (2001). Preliminary performance of CdZnTe imaging detector prototypes. *Nucl. Instrum. Methods A* **458**: 55–61.

Rhoderick, E. H. (1978). "Metal-Semiconductor Contacts." Clarendon Press, Oxford.

Richards, J. D., Vigil, R., Grovatski, G., Baker, J., and Devito, R. (2001). Large area mercuric iodide photodetectors. *2001 IEEE Nucl. Sci. Symp. Med. Imag. Conf. Rec.* (Abstract #N22–86), 69.

Riley, G. A., and Bornstein, S. (2001). Room temperature flip chip assembly of CdZnTe pixel detector arrays. *Proc. SPIE* **4428**: 331–334.

Rothschild, R. E., Matteson, J. L., Heindl, W. A., and Pelling, M. R. (2002). Design and performance of the HEXIS CZT detector module. *Proc. SPIE* **4784**: 45–53.

Sato, T., Sato, K., Ishida, S., Kiri, M., Megumi, H., Yamada, M., and Kanamori, H. (1995). Local polarization phenomena in in-doped CdTe X-ray detector arrays. *IEEE Trans. Nucl. Sci.* **42**(5): 1513–1518.

Scheiber, C. (1996). New developments in clinical applications of CdTe and CdZnTe detectors. *Nucl. Instrum. Methods A* **380**: 385–391.

Scheiber, C., and Chambron, J., (1992). CdTe detectors in medicine: A review of current applications and future perspectives. *Nucl. Instrum. Methods A* **322**: 604–614.

Scheiber, C., Eclancher, B., Chambron, J., Prat, V., and Kazandjan, A. (1999). Heart imaging by cadmium telluride gamma camera european program "BIOMED" consortium. *Nucl. Instrum. Methods A* **428**: 138–149.

Scheiber, C., and Giakos, G. C. (2001). Medical applications of CdTe and CdZnTe detectors. *Nucl. Instrum. Methods A* **458**: 12–25.

Schlesinger, T. E., Brunett, B., Yao, H., van Scyoc, J. M., James, R. B., Egarievwe, S. U., Chattopadhyay, K., My, X- Y., Burger, A., Giles, N., El-Hanany, U., Shahar, A, and Tsigelman, A. (1999). Large volume imaging arrays for famma-ray spectroscopy. *J. Elect. Mat.* **28**(6): 864–868.

Schlesinger, T. E., Toney, J. E., Yoon, H., Lee, E. Y., Brunett, B. A., Franks, L., and James, R. B. (2001). Cadmium zinc telluride and its use as a nuclear radiation detector material. *Mat. Sci. Eng.* **32**: 103–189.

Schramm, N. U., Wirrwar, A. and Halling, H. (2001). Development of a multi-pinhole detector for high-sensitivity SPECT imaging. *2001 IEEE Nucl. Sci. Symp. Med. Imag. Conf. Rec.* **3**: 1585–1586.

Shao, Y., Barber, H. B., Balzer, S., and Cherry, S. R. (2000). Measurement of coincidence timing resolution with CdTe detectors. *Proc. SPIE* **4142**: 254–264.

Shockley, W. (1938). Currents to conductors induced by a moving point charge. *J. Appl. Phys.* **9**: 635.

Shor, A., Eisen, Y., and Mardor, I. (2001). Spectroscopy with pixelated CdZnTe gamma detectors—experiment versus theroy. *Nucl. Instrum. Methods A* **458**: 47–54.

Siffert, P., Cornet, A., Stuck, R., Triboulet, R., and Marfaing, Y. (1975). Cadmium telluride nuclear radiation detectors. *IEEE Trans. Nucl. Sci.* **22**(1): 211–225.

Singh, M. (1998). Design of a CZT based breast SPECT system. *IEEE Trans. Nucl. Sci.* **45**(3): 1158–1165.

Szeles, C., Cameron, S. E., Ndap, J.-O., and Chalmers, W. C. (2001). Advances in crystal growth of semi-insulating CdZnTe for radiation detector applications. *2001 IEEE Nucl. Sci. Symp. Med. Imag. Conf. Rec.* **4**: 2424–2428.

Toney, J. E., Schlesinger, T. E., Brunett, B. A., and James, R. B. (1998). Elementary analysis of line shapes and energy resolution in semiconductor radiation detectors. *Mat. Res. Soc. Symp. Proc.* **487**: 193–198.

Tsuchimochi, M., Sakahara, H., Hayama, K., Funaki, M., Shirahata, T., Orskaug, T., Maehlum, G., Yoshioka, K., and Nygard, E. (2001). Performance of a small CdTe gamma camera for radioguided surgery. *Proc. SPIE* **4508**: 74–87.

Verger, L., Bonnefoy, J. P., Glasser, F., and Ouvrier-Buffet, P. (1997). New developments in CdTe and CdZnTe detectors for X and γ-ray applications. *J. Elect. Mat.* **26**(6): 738–744.

Wagenaar, D. J., Brooks, D. J., Kalinosky, M. A., and Engdahl, J. C. (1999). Effects of detector scatter on photopeak fraction in pixellated solid state detectors. *Nucl. Instrum. Methods A* **422**: 463–468.

Wagenaar, D. J., Chowdhury, S., Engdahlt, J. C., and Burckhardt, D. D. (2003). Planar image quality comparison between a CdZnTe prototype and a standard NaI(Tl) gamma camera. *Nucl. Instrum. Methods A* **505**(1–2): 586–589.

Wagenaar, D. J., and Terwilliger, R. A. (1995). Effects of induced charge in the kinestatic charge detector. *Med. Phys.* **22**(5): 627–634.

Weigel, E., and Mueller-Vogt, G. (1996). Comparison of bridgman and THM methods regarding the effect of In doping and distribution of Zn in CdTe. *J. Crys. Growth* **161**: 40–44.

Wolf, H., Deicher, M., Osteheimer, V., Schachtrup, A. R., Stolwijk, N. A., and Wichert, T. (2001). The strange diffusivity of Ag atoms in CdTe. *Physica B* (Netherlands) **308–310**: 963–966.

Wright, G. W., James, R. B., Chinn, D., Brunett, B. A., Olsen, R. W., van Scyoc, J., Clift, M., Burger, A., Chattopadhyay, K., Shi, D., and Wingfield, R. (2000) Evaluation of NH4F/H2O2 effectiveness as a surface passivation agent for Cd1-xZnxTe crystals. *Proc. SPIE* **4141**: 324–335.

Yamamura, N., Uritani, A., Watanabe, K., Kawarabayashi,, J., and Iguchi, T. (2003). Development of three-dimensional gamma camera with imaging plates and multi-pinhole collimators. *Nucl. Instrum. Methods A* **505**(1–2): 577–581.

Yin, S., Tumer, T. O., Maeding, D., Mainprize, J., Mawdsley, G., Yaffe, M. J., Gordon, E. E., and Hamilton, W. J. (2002). Direct conversion CdZnTe and CdTe detectors for digital mammography. *IEEE Trans. Nucl. Sci.* **49**(1): 176–181.

Yoon, H., Goorsky, M. S., Brunett, B. A., van Scyoc, J. M., Lund, J. C., and James, R. B. (1999). Resistivity variation of semi-insulating Cd1-xZnxTe in relationship to alloy composition. *J. Elect. Mat.* **28**(6): 838–842.

Zanio, K., (1978). "Semiconductors and Semimetals, Vol. 13; Cadmium Telluride." Academic Press, New York.

Zeng, G. L., Gagnon, D., Matthews, C. G., Kolthammer, J. A., Radachy, J. D., and Hawkins, W. G. (2002). Image reconstruction algorithm for a rotating slat collimator. *Med. Phys.* **29**(7): 1406–1412.

CHAPTER 16

Application-Specific Small Field-of-View Nuclear Emission Imagers in Medicine

CRAIG LEVIN

Department of Radiology, Stanford University School of Medicine, Stanford, California

I. Overview of Application-Specific Small Field-of-View Imagers
II. Scintillation Detector Designs of Small Field-of-View Imagers
III. Semiconductor Detector Designs of Small Field-of-View Imagers
IV. Review of Designs and Applications for Small Field-of-View Imagers

I. OVERVIEW OF APPLICATION-SPECIFIC SMALL FIELD-OF-VIEW IMAGERS

A. Motivation for the Small Camera Concept

Traditional modes of radiology, such as computed tomography (CT), magnetic resonance imaging (MRI), ultrasound, and mammography, generate images that mainly exhibit the structural properties and parameters of the portion of the subject within the imaging field of view (FOV). These modalities usually have high detection sensitivity for the particular structural signature(s) of the disease of interest. However, the specificity of those structural signatures to the disease is often relatively low. The power of nuclear radioisotope imaging is its ability to image the biochemical parameters of the particular disease of interest.

Because of this functional information, the specificity for disease identification is often significantly higher in nuclear emission imaging than in other radiology modalities, while achieving similar sensitivity.

There is often interest in focusing functional nuclear medicine imaging studies on a particular organ. Unfortunately, because of the size (up to 50–70 cm across) and awkwardness of the conventional camera, it images a relatively large area of the body with a limited proximity to the organ of interest and often accepts a significant fraction of background activity from other organs. The result is a spatial-resolution and detection-sensitivity limitation for imaging that organ. Consider breast imaging with ^{99m}Tc-labeled sestamibi as an example. Because the standard scintillation camera cannot image the breast in close proximity, there is a limitation in the spatial resolution for breast tumor detection. The inadequate spatial resolution of standard scintillation cameras limits their sensitivity for the detection of tumors that are < 1 cm in diameter. In addition, the camera also sees a background haze from the hot myocardium that further reduces detection sensitivity. In fact, there are data (Maublant *et al.*, 1996; Khalkhali *et al.*, 1995) indicating that a portion of lesions missed in breast scintillation imaging with ^{99m}Tc-labeled sestamibi can be explained by either small size or lower isotope uptake of the particular lesions, tissue attenuation, and myocardial background. Such problems might be remedied in part by a dedicated, small, higher-resolution camera that can image the breast in close proximity for higher spatial resolution at multiple orientations that avoid unnecessary body background for higher detection sensitivity.

By "small FOV" we mean a FOV significantly smaller than that of the standard scintillation camera. Whereas a conventional camera is designed to image everything from thin bones to the entire trunk of the body, a small-FOV camera focuses only on a specific region or organ of the body. In addition to medicine, small-FOV imagers have been used in the fields of gamma-ray astronomy and radiation monitoring (Hailey *et al.*, 1989; Carter *et al.*, 1995; Rossi *et al.*, 1996; He *et al.*, 1993, 1997; Bird and Ramsden, 1990; Guru *et al.*, 1994).

With an imager designed specifically for breast imaging, for example, the source–collimator distance will be typically less than 5 cm and, depending on detector design, under such conditions the detector may play a larger role in determining the spatial resolution. We learn in Parts II and III that close proximity means higher spatial resolution for single- as well as coincidence-photon applications.

There are other advantages to using a small camera. Using the breast imaging example again, a small imager could be configured with a size and shape compatible with standard mammography fixtures for compressing and immobilizing breasts during tumor localization or core biopsy. For biopsy, the placement of the small camera over a lesion could be then guided by mammography. In this case, the X-ray and emission images could be overlaid for a registered functional–structural image to aid in diagnosis or staging of the disease. In addition, with breast compression, the spatial resolution and signal-to-noise ratio (SNR) should be superior to that attainable with conventional scintillation-camera studies simply due to less gamma-ray attenuation and closer proximity to the detector. This configuration is unavailable to the large standard scintillation camera. A small imager is also by definition very mobile and can be taken into other clinical environments, such as surgery or cardiology, where the bulkiness of the conventional scintillation camera prohibits its use.

Another advantage of the small camera concept is that, because of the small FOV, the camera can be designed to have significantly higher intrinsic spatial resolution than the conventional scintillation camera at a reasonable cost. These designs may involve custom-made collimators and alternative detector materials or photodetection devices. The ultimate result of these special designs is further detection sensitivity improvements.

Thus, the rationale and potential importance of the small, application-specific camera concept is clear. Because there have been many recent small radioisotope imager developments for breast imaging uses, (Scopinaro *et al.*, 1999; Pani, Pellegrini, *et al.*, 1997; Pani *et al.* 1994, 1996, 1998, 1999; Pani, Scopinaro *et al.*, 1997; "Dedicated" 1993; Weinberg *et al.*, 1996, 1997; Wojcik *et al.*, 1998; Patt *et al.*, 1996, 1998; Patt, Iwanczyk, *et al.*, 1997; Tornai, *et al.*, 1997; Levin *et al.*, 1996; Levin, Hoffman, Tornai, *et al.*, 1997; Levin, Hoffman, Silverman, *et al.*, 1997; Gruber *et al.*, 1998, 1999; Williams *et al.* 1997, 1999, 2000; Soares *et al.*, 1998, 1999, 2000; Fiorini *et al.*, 1997, 1999; Fiorini, Longoni, *et al.*, 1999; MacDonald *et al.*, 1999; Thompson *et al.*, 1994, 1995; Rober *et al.*, 1997; Bergman *et al.*, 1998; Raylman *et al.*, 1999; Zavarzin *et al.*, 1999; Hutchins and Simon, 1995; Moses *et al.*, 1995; Freifelder and Karp, 1997) we use this application example throughout this chapter. In the next sections, we discuss general design and performance issues for small FOV nuclear emission imager development. A majority of this chapter concentrates on describing various detector designs. In the last section of the chapter, we review several current designs and applications of small FOV imagers.

B. General Design Principles
1. Nuclear Emissions of Interest

The strength of radioisotope imaging is in its variety of radiopharmaceutical tracers that exist or can be developed to study specific processes. The agent should be easily extracted from the blood into the tissue region of interest (tumor, for example) with high affinity compared to surrounding regions and remained fixed for the duration of the imaging. Chapter 4 discusses radiopharmaceutical properties.

The most commonly used tracers in nuclear medicine imaging are gamma-ray emitters and many of the small emission imagers developed for medicine have been for imaging gamma rays (Scopinaro *et al.*, 1999; Pani, Pellegrini, *et al.*, 1997; Pani *et al.* 1994, 1996, 1998, 1999; Pani, Scopinaro *et al.*, 1997; "Dedicated" 1993; Weinberg *et al.*, 1996, 1997; Wojcik *et al.*, 1998; Patt *et al.*, 1995, 1996, 1998; Patt, Iwanczyk, *et al.*, 1997; Patt, Tornai, *et al.*, 1997; Tornai *et al.*, 1997; Levin *et al.*, 1996; Levin, Hoffman, Tornai, *et al.*, 1997; Levin, Hoffman, Silverman, *et al.*, 1997; Gruber *et al.*, 1998, 1999; Williams *et al.*, 1997, 1999, 2000; Soares *et al.*, 1998, 1999, 2000; Fiorini *et al.*, 1997, 1999; Fiorini, Longoni, *et al.*, 1999; MacDonald *et al.*, 1999; Kume *et al.*, 1986; Nagai and Hyodo, 1994; Aarsovold *et al.*, 1998, 1994; Mathews *et al.*, 1994; Antich *et al.*, 1992; Yasillo *et al.*, 1990; Milster *et al.*, 1984, 1990; Weisenberger *et al.*, 1998; de Notaristefani *et al.*, 1996; Daghighian *et al.*, 1994a, 1994b, Menard *et al.*, 1998; Tornai *et al.*, 1997; Butler *et al.*, 1993; Doty *et al.*, 1993; Barber *et al.*, 1984, 1994; Eskin *et al.*, 1999; Marks *et al.*, 1996, Matherson *et al.*, 1998; Hartsough *et al.*, 1995). For example, certain ^{99m}Tc-labeled compounds ($E_\gamma = 140$ keV, $t_{1/2} = 6$ hours) have been recently used for breast cancer imaging (Maublant *et al.*, 1996; Khalkhali *et al.*, 1993, 1994, 1995; Khalkhali, Cutrone, *et al.*, 1995; Baines *et al.*, 1986; Tabar *et al.*, 1982; Piccolo *et al.*, 1995;

Lastoria *et al.*, 1994; Waxman *et al.*, 1994; Taillefer *et al.*, 1995; Villanueva-Meyer *et al.*, 1996) and many small imager designs for breast imaging have focused on this isotope (Scopinaro *et al.*, 1999; Pani, Pellegrini, *et al.*, 1997; Pani *et al.* 1994, 1996, 1998, 1999; Pani, Scopinaro *et al.*, 1997; "Dedicated" 1993; Weinberg *et al.*, 1996, 1997; Wojcik *et al.*, 1998; Patt *et al.*, 1996, 1998; Patt, Iwanczyk, *et al.*, 1997; Tornai *et al.*, 1997; Levin *et al.*, 1996; Levin, Hoffman, Tornai, *et al.*, 1997; Levin, Hoffman, Silverman, *et al.*, 1997; Gruber *et al.*, 1998, 1999; Williams *et al.*, 1997, 1999, 2000; Soares *et al.*, 1998, 1999, 2000; Fiorini *et al.*, 1997, 1999; Fiorini, Longoni, *et al.*, 1999; MacDonald *et al.*, 1999). Small-animal imager designs have focused on other single-photon isotopes (Williams *et al.*, 1999; Weisenberger *et al.*, 1998). The desired properties of gamma-ray-emitting isotopes are basically the same whether one is using a small FOV imager or a conventional scintillation camera. The energy of the gamma ray emitted should be high in order to penetrate the body and create a robust signal in the detector. The half-life of the emitter should be relatively short to minimize the imaging time and the duration in which the patient is radioactive. The chemistry of the isotope should allow efficient labeling of multiple compounds.

Imaging the distribution of positron emitters in the body is also of interest in medicine. The technique of positron emission tomography (PET) discussed in Part III, is concerned with imaging the distribution of positron-emitting nuclei. In this method, the positrons are detected indirectly through the annihilation photon radiation created. The oppositely directed 511-keV photons resulting from the positron's annihilation are detected in coincidence, which gives indirect information about the location of the emitted positron. There has been much success in using ^{18}F-labeled ($t_{1/2}$ = 110 minutes) fluorodeoxyglucose (FDG) for breast imaging with PET (Hoh *et al.*, 1997; Tse *et al.*, 1992; Nieweg *et al.*, 1993; Wahl *et al.*, 1991; Avril *et al.*, 1996; 1999; Adler *et al.*, 1997). As a result, small FOV annihilation photon imagers have been and are currently being developed for positron annihilation coincidence detection in breast imaging (Thompson *et al.*, 1994, 1995; Robar *et al.*, 1997; Bergman *et al.*, 1998; Raylman *et al.*, 1999; Zavarzin *et al.*, 1999; Hutchins and Simon, 1995; Moses *et al.*, 1995; Freifelder and Karp, 1997). For breast and axillary lymph node imaging, typically two small FOV detector arrays are configured on either side of the breast. Planar projection or limited-angle tomographic images can be formed from the acquired data from these small FOV coincidence cameras. If the camera can be rotated, full tomographic imaging is possible. Small PET devices have also been developed for small-animal imaging with FDG and other PET tracers (Del Guerra *et al.*, 1998; Weber *et al.*, 199; Correia *et al.*, 1999; Jeavons *et al.*, 1999; Pichler *et al.*, 1998; Lecomte *et al.*, 1996; Bruyndonckx *et al.*, 1997; Watanabe *et al.*, 1997; Green *et al.*, 1994; Siegel *et al.*, 1999; Johnson *et al.*, 1997). We leave the discussion of small-animal PET to Chapter 10.

Small-FOV, direct positron imagers are being developed for surgical and autoradiographic applications where the charged particle is directly detected at or very near the site of emission (Daghighian *et al.*, 1995; Ljunggren and Strand, 1988, 1990, 1994; Ljunggren *et al.*, 1993, 1995; MacDonald *et al.*, 1995a, 1995b, Levin *et al.*, 1996, 1997; Tornai, MacDonald, Levin, Siegel, and Hoffman, 1996; Tornai, MacDonald, Levin, Siegel, Hoffman *et al.*, 1996; Tornai *et al.*, 1997, 1998; Hoffman *et al.*, 1997a, 1997b, 1998; Yamamoto *et al.*, 1997, 1999, Lees *et al.*, 1997, 1998, 1999; Overdick *et al.*, 1997). We refer here to both the positron and electron as *beta particles* or simply *betas*. Because of their small range in tissue and detector materials, betas are naturally collimated for *in situ* imaging within a surgical cavity or for *in vitro* autoradiographic imaging of a tissue sample. For these same reasons, it is not possible to image the distribution of positrons directly from outside the body with the exception of the possibility of studying subcutaneous processes with high-energy positron emitters (Yamamoto *et al.*, 1999). Because betas interact continuously, unlike photons, the detection efficiency is nearly perfect with close proximity of the detector to the tissue region of interest. If positron emitters are used, annihilation photon background will be present and some means for its rejection might need to be incorporated within the imager (Levin *et al.*, 1997; Hoffman *et al.*, 1998; Tornai *et al.*, 1998) So from an imaging perspective, it would be preferable in these designs to use an electron-emitting pharmaceutical. However, because of radiopharmaceutical development for PET, positron-emitting imaging agents are more widely available. For direct positron imagers, only planar projection imaging is available.

The production of gamma and positron emitters, their interactions in the imaged object and the detector materials, and the effect of these interactions on imaging are described in Chapter 4. Detector configurations for imaging of the various nuclear emissions are described in the next section.

2. Detector Configurations of Interest

In contrast to conventional gamma-ray cameras used in the clinic, small-FOV cameras are usually designed with a specific application and corresponding nuclear emission in mind. Because of the small size of these cameras, a variety of available detector technologies can be investigated and implemented at a reasonable cost. For gamma-ray emissions, a small-FOV imager consists of one or two detector heads that look essentially like a miniaturized Anger camera (Anger 1958). Many of the design principles of the Anger camera were covered in Part II of this book. In this chapter, we cover those design issues relevant to small FOV imagers. A schematic drawing of such a small-FOV

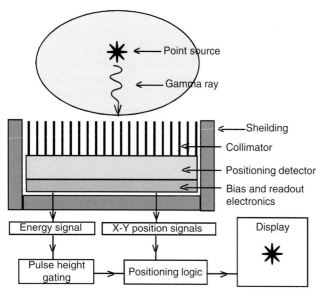

FIGURE 1 Schematic drawing of a small FOV gamma-ray imager. The image of a point source shown comprises a 2D histogram of *x-y* positions of pulse-height- (energy) selected detector events. Typical FOV dimensions might be less than or equal to 15×15 cm^2.

gamma-ray camera is shown in Figure 1. The camera comprises a collimator, a position-sensitive detector, bias and readout electronics, processing electronics, and a computer. Background shielding is important for a small imager to reduce undesired activity from outside the FOV. An example of this is the high myocardial background seen in ^{99m}Tc sestamibi breast imaging due to the close proximity and typically high tracer uptake of the heart.

The position-sensitive detector might be a single crystal, or an array of scintillation crystals coupled to an array of photomultipiler tubes (PMT), or a semiconductor photodetector array. The detector might also be a pixellated array of high-stopping-power semiconductor crystals for intrinsic detection of the radiation. Detector configurations involving gases are also possible, but they often suffer from low sensitivity for typical photon energies of interest as a result of lower detection efficiency and are not discussed here (see Bolozdynya *et al.*, Chapter 18 in this volume). Some of the most common designs for the position-sensitive detector for small FOV imagers will be presented in the next sections.

The energy and position signals from the detector are digitized and can be collected in *list mode* on an event by event basis. The *x-y* positions of the events to be included in the image-formation process are then selected in post-processing if their total energy is within a chosen desirable range (around the photopeak energy) for reduction of Compton scatter effects. This favors detector technologies that have good energy resolution. Pulse-height discrimination and gating can also be incorporated into the electronics (Figure. 1).

The dimensions of a small-FOV gamma-ray camera depend on the application. In breast imaging, the camera should be small enough to perform close-proximity imaging at various orientations to improve spatial resolution and reduce background from other organs. If the nuclear emission imaging will be performed in conjunction with mammography, the camera should be small enough to fit within the confines of the mammography unit. In surgery, the camera should be compact and light for ease of use during a procedure. Typical FOV dimensions, including shielding of a small camera, might be less than or equal to 15×15 cm^2. The thickness of the camera is also important in those cases where space is confined during imaging, as it is in a digital mammography unit, for example. In that case, the detector or photodetector technology must be chosen for compactness. Detector thickness, including shielding, of a few centimeters is possible with the careful choice of detector materials to facilitate compactness. In principle, if the detector head is rotatable, tomographic imaging may be performed. However, for most small-FOV imager applications, planar projection imaging is more typical because very close proximity to the body is a primary motivation.

A schematic drawing of a typical small-FOV positron-annihilation photon-coincidence camera is shown in Figure 2. The 511-keV annihilation photon is not well collimated for imaging using standard absorption collimation. Instead, the oppositely directed annihilation photons are collimated electronically using two opposing detectors set in electronic coincidence without physical collimators (although some sort of slat collimator might be inserted axially to reduce the scatter and randoms as discussed in Chapter 10). Because small-FOV coincidence imagers are typically designed to image single organs or small regions, the detector separation for imaging is usually as small as possible. The spatial resolution, sensitivity, and other parameters benefit from this close separation in ways similar to the case for small animal PET systems due to reduced annihilation photon noncolinearity and other geometric effects (see Chapter 4 and Part III).

The processing of the detector signals for small-FOV coincidence cameras is somewhat similar to that depicted in Figure 1, except there are now two sets of coincident detector signals to process. Event selection, or gating, by the total energy deposited in the two detectors is performed in a way similar to that in large-FOV coincidence cameras (Chapter 10). The energy gating can select events in the two detectors corresponding to two photopeak events, one photopeak and one scatter event, or even two scatter events. Random coincidence estimation and subtraction are typically performed using the delayed coincidence window method. The coincidence criterion favors detector technologies with fast signal generation for good coincidence time resolution, low random-coincidence rate, and high count rate performance. Because a high Compton scatter is expected for 511 keV photons, good energy resolution is required.

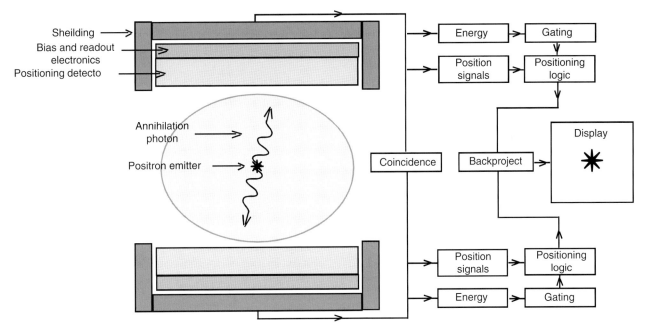

FIGURE 2 Schematic drawing of a small FOV positron annihilation photon coincidence imager. The resulting image comprises back-projected x-y positions of energy-selected detector events coincident in both detectors.

Unlike the commercial systems, the small-FOV coincidence cameras are typically not configured to rotate, and, thus, only planar projection or limited-angle tomographic imaging can be performed. As for the small-FOV gamma-ray camera, a variety of detector technologies are available for the small-FOV coincidence camera that may not be cost-effective for the larger commercial systems. However, because of the criteria of high detection efficiency, good spatial-, energy-, and coincidence- time resolutions; and relatively low cost, most small-FOV coincidence imager designs have used the scintillator–PMT combination for the position-sensitive detector. Because of the variation in detector crystal penetration depths for the 511-keV photons, some means for interaction-depth determination might be incorporated into the detector for optimal and more uniform spatial-resolution performance.

Small-FOV imagers that image positrons directly usually contain a single detector head, similar to that shown in Figure 1, with a few notable differences. Because of the beta's short range in tissue and detector material, no collimator is required, only a low-Z, low-density fraction of a millimeter-thick detector is required for very high-efficiency detection, and imaging with much higher spatial resolution is possible. Also, only beta activity very close to the tissue surface can be imaged and one must construct the imager with minimal material in front to minimize beta absorption prior to entering the detector. Because of the surgical or autoradiographic applications of beta imagers, the FOV size is usually smaller than its gamma-ray counterpart. If positrons are imaged, annihilation photon interactions in the detector can mimic beta

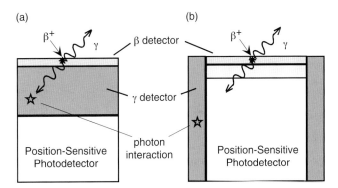

FIGURE 3 Examples of annihilation photon background rejection schemes for a positron scintillation imager. A β^+-γ coincidence in two separate crystals signifies a good positron interaction. In the phoswich design (a), the coincidence in the two distinct scintillation crystals is established by pulse shape discrimination. In the annulus design (b), two separate detector electronic channels are set in coincidence.

interactions causing significant loss in contrast. In that case some sort of annihilation-photon background rejection scheme should be configured into the system. Figure 3 shows examples of two such configurations for a positron scintillation imager (Levin, Tornai, MacDonald, et al., 1997; Tornai et al., 1998). In addition to the beta detector, a second, contiguous, high-Z, high-density detector is present for annihilation photon absorption. Because annihilation photons are essentially emitted in opposite directions, no matter what their emission orientation, that second detector has a high geometric detection efficiency for at least one of the annihilation photons. A true positron event would then be signified as a hit in the beta detector in coincidence with one in the second detector. This design results in a significant annihilation-photon background reduction, but a loss of sensitivity due to the coincidence requirement and the variation in detection efficiency of the second detector (Levin, Tornai, MacDonald, *et al.*, 1997). In principle, this annihilation-photon background rejection concept can use semiconductor instead of scintillation crystal detector technology for compactness. This concept is discussed again at the end of this chapter and in Hoffman *et al.* (Chapter 17 in this volume), where the topic of surgical probes is discussed. For surgical use, hand-held probe geometry is preferred.

For completeness, we also mention the concept of a Compton scatter camera for which a few small-FOV designs have been proposed for medical gamma-ray imaging (Singh *et al.*, 1995; Rohe and Valentine, 1995; LeBlanc *et al.*, 1999; Turner *et al.*, 1997; Bolozdynya *et al.*, 1997). Compton scatter cameras are discussed more in depth in Chapter 19 (in this volume), and so we do not discuss their design principles in detail here. The Compton scatter camera relies on two sets of detector arrays. The first, the scatter array, is composed of a pixellated array of relatively low-Z, low-density detectors that have a high probability of producing Compton scatter for the photon energy of interest. The second array of detectors, the absorption array, is composed of high-Z, high-density detectors that have a high probability for photoabsorption for the scattered photon energies of interest. The Compton scatter imager requires that there are at least two interactions detected per gamma-ray event: a scatter in the low-efficiency scatter detector array, followed by a photoabsorption in the high-efficiency absorption array. Due to the kinematics of Compton scatter, these two detected interactions constrain the original direction of the incoming photon to lie somewhere along a cone of ambiguity. The formation of a Compton scatter image consists of overlaying the cones of ambiguity for many such events. For a point source, the image appears in 2D as the intersection of many projected cones. The spatial resolution of a Compton scatter system depends on how precisely that cone of ambiguity can be defined. Because this cone was determined from the kinematics of Compton scatter, high spatial resolution imaging requires fine array pixel sizes and high energy resolution for both detector arrays. Both single-photon and coincidence systems can be formed with a Compton scatter detector configuration.

3. Performance Issues for Small Cameras

There are several important performance issues for small FOV imagers. These issues are related to the spectral, spatial, and temporal responses of the system and how they are affected by the presence of physical factors that degrade performance. Some of these factors are discussed in Parts II and III of this book. Here we discuss how these topics impact small camera imaging performance.

a. Detection Efficiency

High *detection sensitivity* for the emission energy of interest is important. Radioisotope imaging in medicine is count-limited and a nuclear emission camera must be capable of acquiring many events in a short period of time for high statistical significance and optimal image quality. This high detection sensitivity is achieved by ensuring that both the geometric and intrinsic detection efficiencies are high and that the discrimination electronics are optimized for the particular imaging situation. For gamma-ray imaging, because the sensitivity is limited by the collimator design, the collimator should be carefully chosen for high sensitivity. For high gamma-ray stopping power, the detector material should have high Z, high density, and be adequately thick for the energy range of interest. An advantage of a small-FOV camera is that it can be used in close proximity to the imaging subject and at favorable orientations. For breast imaging, a small gamma-ray camera can be placed directly against the breast and facing away from the heart for optimal geometric efficiency in detecting emissions from the breast and not from Compton scatter events from photons emitted outside the FOV. Because the lesions of interest in breast imaging may not be deep within the tissue, the collimator thickness may be reduced for improved detection sensitivity with acceptable spatial resolution loss for objects relatively shallow within tissue.

b. Spectral Response

The spectral or energy response of a positioning detector refers to its capabilities of detecting and identifying specific energies of interest from the radioisotope distribution to be imaged. It is important that the imaging system have good *energy resolution* to reduce energy-dependent-background physical effects that can degrade the imaging. For example, Compton-scatter photons in the object being imaged and in the materials making up the camera can result in mispositioned events and significant blurring in gamma-ray imaging. As with conventional systems, good energy resolution can help to discriminate between Compton-scatter and photoabsorption events by their total energy deposited in the detector. This discrimination is usually not perfect because there is often significant overlap between the photopeak and Compton-scatter energy regions in typical emission energy spectra. A small and compact camera designed to perform imaging at very close proximity to the body may have a high geometric efficiency for Compton-scatter event detection, depending on orientation. For breast imaging, Compton-scatter activity from the heart is often a dominating physical factor that limits the detectability of small tumors (Maublant *et al.*, 1996; Khalkhali, Cutrone, *et al.*, 1995; Scopinaro *et al.*, 1999; Pani *et al.*, 1996; Pani, Scopinaro, *et al.*, 1997; Levin, Hoffman, Silverman, *et al.*, 1997). Good energy resolution is especially important in this case. For high energy resolution in scintillation crystal designs, the number of photoelectrons

N_e created and detected should be high and the electronic noise component should be low.

An advantage of a small-FOV camera is that the detector design can be chosen specifically for high energy resolution within a reasonable budget. This allows one to use an exotic high-performance detector technology that may not be cost-effective or practical for a large-FOV commercial camera.

Another important aspect of spectral response of the imaging system is the *energy linearity*. The camera may be required to image low- and high-energy emissions such as those emitted from ^{99m}Tc or ^{131}I, respectively, or the camera may need to select certain energy regions of interest for Compton-scatter background rejection. These situations require a linear response between the emission and detected energies for best results. Good energy response *uniformity* assures that the different spectral-response parameters (pulse height, energy resolution, sensitivity, etc.) are similar for all portions of the position-sensitive detector's FOV. Very different spectral response over the FOV can produce spot artifacts and spatial irregularities that will degrade performance. Certain small-FOV camera designs suffer from significant variations in global energy response due to the nonuniform behavior of detector components used in the relatively small position-sensitive area. For example, if the small-camera design uses a large-area, cross-wire position-sensitive PMT (PSPMT), the variations in the photocathode quantum efficiency are often large, producing variations in spectral response for flood source irradiation that are difficult to correct.

c. Spatial Response

Spatial response refers to the camera's ability to position events accurately and precisely. This performance aspect is strongly correlated with the spectral-response parameters. *Spatial resolution* determines how precisely a camera can localize an emission event. Higher spatial resolution means finer structures can be resolved in the radioisotope distribution. Spatial resolution is affected by factors such as collimator design (if imaging gamma rays), detector material, degree of pixellation of the detector, electronic noise properties of the detector (Iwanczyk and Patt, 1995; Nicholson, 1974), and physical effects of the emission process. For single-photon imaging, the extrinsic spatial resolution is defined for the entire system, including collimator, whereas the intrinsic spatial resolution is the contribution due to the detector alone. A small-FOV gamma-ray imager can be designed to have less than a 2-mm full-width-at-half-maximum (FWHM) intrinsic spatial resolution for identifying tiny regions of high focal uptake at a reasonable cost. The extrinsic spatial resolution with the collimator blurring included might be 3 mm FWHM at 2-cm source–collimator distance. Thus, depending on the collimator design and the source–collimator distance, a small-FOV camera could have the spatial resolution determined by the detector pixel size rather than limited by the collimator properties. This is especially true for breast imaging where source–collimator distances are typically less than 5 cm. A conventional nuclear medicine camera has intrinsic spatial resolution of 3.5 mm FWHM or greater. Typically the corresponding extrinsic spatial resolution is collimator limited and typically greater than 7 mm FWHM or greater for a LEHR collimator at 5 cm source-detection distance.

Spatial linearity refers to the spatial relationship between events positioned by the camera and the true location of the emission. Good linearity is desirable in order to create a faithful representation of the radioisotope distribution without distortion of important structures. Standard nuclear medicine cameras typically have a linear spatial response out to the edges of the useful FOV. The crystal is usually masked off by a lead absorber to eliminate interactions in any outer nonlinear regions. Any small nonlinearities in the inner FOV that may occur can be easily corrected for in the software of the conventional camera. Nonlinearities in a small-FOV camera are more detrimental to the camera performance. A nonlinear region near the edges of a small-FOV camera may occupy a significant fraction of the entire camera's sensitive area. For example, some PSPMT small-camera designs have spatial nonlinearities near the edges of the sensitive detector. This effectively dead area could be detrimental to the camera's utility in an application of the small-FOV camera that requires sensitivity all the way to the detector edge. An example of such an application is imaging close to the chest wall in prone or lateral breast imaging. The sensitive or useful FOV should be as large as possible with little dead area for good image statistics in a short period of time. A related parameter is the *dynamic range* of the imaging system, which relates factors such as source energy and the positioning electronics to the spatial extent of objects seen in the formed image. A high dynamic range is desirable for a more useful representation of the activity distribution.

Good spatial response *uniformity* is important in order to avoid producing image artifacts. Spatial uniformity is measured by flood-irradiating the sensitive FOV for studying both local and global responses. In typical small-FOV imagers, there are significant variations in the spatial responses (pulse height, spatial resolution, sensitivity, etc.) due to variations in the detector responses within a relatively small FOV. For example, a PSPMT camera design suffers from relatively high spatial nonuniformity due to large variations in the photocathode response. There may also be nonuniformities near the detector edges due to the presence of nonlinear electronic responses. Typically uniformity correction is established in small-FOV designs by dividing out a raw acquired image using a high-statistics flood image.

d. Temporal Response

If very high data rates are anticipated, detector signals should be fast in order to avoid dead-time effects that can also degrade performance. In this case an imaging detector should be designed for high *count-rate capability*. For small scintillation cameras, this implies that the scintillation mechanism in the crystal and the photodetector response should be relatively fast both in rise and decay. For semiconductor detector designs (see Chapter 15 in this volume), the charge carrier drift velocity should be high and the drift time short. Slow signal formation times means limited count rate capability and pulse pile-up can occur. For coincidence cameras, these fast intrinsic signal formation times are also important for high *coincidence time resolution*. Good coincidence time resolution reduces the effects of random coincidences. In scintillation crystal designs of coincidence cameras, for good coincidence time resolution the scintillator decay time τ should be short, the detected number of photoelectrons N_e should be high, and the noise component should be low. Thus, there is some correlation between coincident time and energy resolution performances.

II. SCINTILLATION DETECTOR DESIGNS OF SMALL FIELD-OF-VIEW IMAGERS

Small scintillation cameras come in various shapes and forms. There have been numerous designs for medical imaging applications presented in the literature (Scopinaro *et al.*, 1999; Pani, Pellegrini, *et al.*, 1997; Pani, Scopinaro, *et al.*, 1997; Pani *et al.*, 1994, 1996; 1998, 1999; "Dedicated," 1993; Weinberg *et al.*, 1996, 1997; Wojcik *et al.*, 1998; Patt *et al.*, 1996, 1997, 1998; Tornai *et al.*, 1997; Levin, Hoffman, Tornai, *et al.*, 1997; Levin, Hoffman, Silverman, *et al.*, 1997; Levin *et al.*, 1996; Gruber *et al.* 1998, 1999; Williams *et al.*, 1997, 1991, 2000; Soares *et al.*, 1998, 1999, 2000; Fiorini *et al.*, 1997, 1999; Fiorini, Longoni, *et al.*, 1999; MacDonald *et al.*, 1995a, 1995b, 1999; Thompson *et al.*, 1994, 1995; Robar *et al.*, 1997; Bergman *et al.*, 1998; Raylman *et al.*, 1999; Zavarzin *et al.*, 1999; Hutchins and Simon, 1995; Moses *et al.*, 1995; Freifelder and Karp, 1997; Kume *et al.*, 1986; Nagai and Hyodo, 1994; Aarsovold *et al.*, 1998, 1994; Matthews *et al.*, 1994; Antich *et al.*, 1992; Yasillo *et al.*, 1990; Milster *et al.*, 1984, 1990; Weisenberger *et al.*, 1998; de Notaristefani *et al.*, 1996; Daghighian *et al.*, 1994a, 1994b, 1995; Menard *et al.*, 1998; Tornai *et al.*, 1997; Ljunggren and Strand, 1988, 1990, 1994; Ljunggren *et al.*, 1993, 1995; Levin *et al.*, 1996, 1997; Tornai, MacDonald, Levin, Siegel, and Hoffman 1996; Tornai, MacDonald, Levin, Siegel, and Hoffman *et al.*, 1996; Tornai *et al.*, 1997, 1998; Hoffman *et al.*, 1997a, 1997b, 1998; Yamamoto *et al.*, 1997, 1999). The most common single-photon imager design is essentially a miniaturization of the standard Anger camera design that appears in the nuclear medicine clinic. A schematic representation of such a design is shown in Figure 1.

A. Scintillation Crystal Design

1. Single Continuous Crystal

As with the standard nuclear medicine camera, a single scintillation crystal used in a small-FOV camera design must satisfy several criteria. The crystal must be adequately thick, with an appropriate Z and density for high absorption of the radiation of interest. The thinner the crystal, the better the light transmission into the photodetector and the better the energy and spatial resolutions. For imaging high-energy photons, there is also a smaller variation in depth of interaction of the photon in thin crystals, which improves the overall spatial resolution. The crystal must also have high intrinsic light yield. The mechanism for light yield for various scintillators is discussed in Wilkinson (Chapter 13). Typically, inorganic scintillation crystals are the brightest. Small-FOV scintillation camera designs for 140-keV ^{99m}Tc gamma rays commonly use 3- to 10-mm-thick NaI(Tl), CsI(Tl), or CsI(Na) (see Table 1).

TABLE 1 Properties of Common Scintillation Crystals Used in Small-FOV Imager Designs

Scintillator	Effective Z	Density (g/cc)	Radiation Length (mm)[a]	Relative Light Yield	Refractive Index	Decay Time (ns)	Peak Emission Wavelength (nm)	Hygroscopic?	Rugged?
NaI(Tl)	51	3.67	3.4	100	1.85	230	410	Yes	No
CsI(Tl)	54	4.51	2.2	135	1.79	1000	530	No	Yes
CsI(Na)	54	4.51	2.2	75	1.79	650	420	No	Yes
BGO	74.2	7.13	10.5	15	2.15	300	480	No	Yes
LSO(Ce)	65.5	7.4	11.6	75	1.82	40	420	No	Yes
CaF$_2$(Eu)[b]	16.9	3.17	N/A	50	1.43	940	435	No	Yes

[a] Radiation lengths for NaI(Tl), CsI(Tl) and CsI(Na) are for 140-keV photons; Values for BGO and LSO are at 511 keV.
[b] CaF$_2$(Eu) is used in beta imaging.

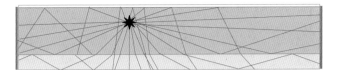

FIGURE 4 Representation of light ray propagation from a point source of light within a scintillation crystal slab with a white diffuse Lambertian reflector on top, absorbing sides, and a light diffuser on the bottom leading into the photodetector.

For imaging there are two requirements for the optics of a continuous crystal. The first is that the particular design promotes efficient light collection, for high SNR for detection, and ultimately better energy and spatial resolutions. The second requirement is that the distribution or cone of collected light be relatively narrow in spatial extent for better spatial resolution. There are a few design features of a single scintillation crystal that enhances efficient light collection. The use of reflectors, absorbers, special crystal surface finishes, and diffusers can facilitate light collection for imaging. These optical parameters are selected for their abilities to preferentially direct light rays into the photodetector while preserving the spatial localization of the light flash.

Figure 4 depicts the propagation of several light rays emanating from a point source of light from within a thin and wide scintillation crystal with the photodetector at the bottom. The use of a back reflector on the top face of the crystal significantly improves light-collection characteristics by reflecting any light refracted at the top back into the crystal. In a specular reflector, the angle of reflection is essentially equal to the angle of incidence; a standard mirror is an example of a specular reflector. Although the use of a mirror-like surface as the back reflector would be suitable for high light-collection purposes, it may not be ideal for use in an imaging crystal for a small FOV camera because it would effectively broaden the light cone subtended at the photodetector. A diffuse Lambertian reflector has the distribution of reflected light, $R(\theta_r)$ related to the angle of incidence, θ_i by:

$$R(\theta_r) \sim \cos \theta_i \qquad (1)$$

so that the reflected photons are preferentially forward directed. A diffuse reflector is desirable for a single-crystal design of a small imager because reflected light is preferentially focused toward the photodetector face of the crystal. This reduces light losses associated with multiple reflections off the relatively close crystal surfaces and limits the width of the light distribution for higher spatial resolution imaging. The reflector should be white for highest reflectivity. White epoxy, Teflon tape, and white paint are commonly used as diffuse reflectors. Figure 4 includes a depiction of a few preferentially forward-directed light rays reflected off the white, diffuse reflector on top.

The choice of crystal surface treatments is also critical. A ground, or diffuse, rather than polished top surface (contiguous with top reflector in Figure 4) may reduce the probability of total internal reflection off that crystal surface and the bottom crystal face as well. Assuming an air gap ($n_2 = 1$) between the crystal face and top reflector, Snell's law gives for the critical angle θ_c of total internal reflection:

$$\theta_c = \sin^{-1}(n_2/n_1) = \sin^{-1}(1/n) \qquad (2)$$

where $n_1 = n$ is the crystal index of refraction. Because the large crystal faces are close together, total internal reflection at the top face may trap the light and prevent it from entering the photodetector. If total internal reflection at the top surface were desired, that surface would be polished with an air gap.

Unlike standard commercial camera designs, in a single-crystal, small-FOV imager the crystal sides play more of a role in the resulting light distribution seen by the photodetector, especially for thicker crystals. Reflected light from the sides can contribute to the width and tail of the light spread distribution for most light source locations, especially near the crystal edges. Looking at Figure 4, it is easy to see that this side reflection problem becomes worse as the crystal thickness perpendicular to the detection plane increases and the cross-sectional detection area decreases. In this chapter, we call the ratio of the thickness to the cross-sectional area for such an optical light guide the *aspect ratio*. The aspect ratio is important for light-collection properties of small-FOV scintillation camera designs. For single-crystal scintillation camera designs, as the aspect ratio increases, at some point continuous positioning is no longer possible. For high-aspect-ratio crystals, it is more difficult to preserve the spatial location information of the origin of the scintillation light flash because of the relatively narrow solid angle subtended by the detector plane to the average light source location and significant role of side reflections. As the thickness increases, the point at which the single crystal is no longer useful for imaging occurs much more rapidly for typical small-FOV cameras. This is why most small modular PET detector designs that require the use of relatively thick crystals (for better absorption efficiency of high energy photons) must segment or pixellate the crystal to achieve good spatial resolution. We return to this issue in the next section. With a favorable aspect ratio, as depicted in Figure 4, to prevent the effects of side reflection, an optimal crystal side treatment might be a highly absorbing, dark coating to preferentially absorb light rays hitting the sides and ground diffuse surfaces to break up internal reflections.

The light distribution exiting the bottom of the crystal is sampled by the position-sensitive photodetector. Although one goal of the crystal design is to limit the extent of the light distribution, depending on the photodetector design and the thickness of the crystal, the light distribution shape may have to be manipulated before entering the photodetector to enhance positioning capabilities. For example, if a matrix of individual PMTs is used, the design might have to incorporate a light diffuser to broaden the light cone in order to involve more PMTs in the positioning of that event. For example, if an array of 5-cm-diameter PMTs were coupled directly to a crystal without any light diffuser present, there may be preferential positioning at certain spots in the resulting image. In addition to fewer PMTs being involved in the positioning, this preferential positioning effect is enhanced by the PMT photocathode's higher sensitivity in the center. These factors can cause the appearance of a tube pattern in raw flood images.

For more smooth continuous positioning, a light diffuser is used to share the light over more PMTs and more uniformly across the face of a given PMT. This is an example of positioning by light multiplexing. The cost of producing more-continuous imaging through enhanced light sharing is lower intrinsic spatial resolution due to a broader light cone. For best light transmission properties, the index of refraction of the light diffuser should be chosen somewhere near the geometric mean of that of the scintillation crystal and photodetector interface. A good figure of merit for the light diffuser thickness is on the order of one-half the photodetecting element size. For a multiple PMT camera, that thickness is roughly one-half a PMT diameter. For position-sensitive PMTs or semiconductor photodetector arrays, one-half the anode or pixel pitch will typically be most appropriate. It should also be noted that if a NaI(Tl) crystal is used, it is usually hermetically sealed with a glass transmission window, which may be an appropriate light guide depending on its thickness and the photodetector element spacing. If the position-sensitive light detector has finely spaced photodetecting elements for fine sampling of the light distribution, a light diffuser may not be necessary because the inherent light distribution width may be already comparable to the photodetector sampling distance.

The amount of light detected, the photodetector pitch, and the shape of the light cone subtended at the photodetector–crystal interface of a single scintillation crystal design all have a strong effect on the intrinsic spatial resolution of the system. The linear intrinsic spatial resolution R (FWHM) for a continuous scintillation crystal that has a light distribution shared over multiple photodetector elements with an Anger-type positioning scheme is governed by Poisson statistics (Tanaka et al., 1970; Barrett and Swindell, 1981):

$$R = 2.35 \left(\frac{(\sum_i s_i^2 n_i)^{1/2}}{\sum_i s_i (dn_i / dx)} \right) \quad (3)$$

where s_i is the weighting factor of the ith photodetector on the position signal, n_i is the number of photoelectrons created in that photodetector, and x is the linear coordinate. For the intrinsic spatial resolution limit, $R_{\lim}$, for a fixed scintillation point we minimize R in Eq. (3) with respect to the factors s_i and obtain (Tanaka et al., 1970; Barrett and Swindell, 1981):

$$R_{\lim} = 2.35 \left(\sum_i \frac{(dn_i / dx)^2}{n_i} \right)^{-1/2} \quad (4)$$

From this formula we see that the spatial resolution in single-crystal cameras improves essentially with brighter scintillation light bursts, sharper light distributions, and higher photodetector sampling rates.

The number of photoelectrons detected by each photodetector in Eq. (4) can be either estimated by the solid angle subtended by each detector or can be measured by moving a point source across the face of the scintillation crystal. A first approximation of the light distribution width can be estimated from Snell's law by looking at Figure 4. The shape of the light distribution seen by the detector will be roughly constrained by the critical angle, θ_c for total internal reflection at the crystal–diffuser interface (see Eq. 2). For a narrower light spread function, the critical angle should be as small as possible. Thus, the scintillation crystal should have a high index of refraction and the diffuser a lower value. Assuming the location of the point source of light is fixed, from basic trigonometry one can obtain a first approximation to the width of the light distribution at FWHM using Eq. 2. The thinner the crystal or the closer the point source of light to the photodetector, on average, the narrower the measured light distribution. For example, using NaI(Tl) ($n = 1.85$) coupled to a glass light diffuser ($n_1 = 1.5$), we obtain $\theta_c = 54°$. For interactions that occur at the top 6-mm-thick crystal, this translates to maximum light distribution width of roughly 8 mm FWHM. For interactions that occur 3 mm deep into the crystal, the width is roughly 4 mm. Without a light diffuser ($n_1 = 1$), this last value becomes approximately 2 mm FWHM.

A more accurate estimation of the shape of the light distribution or the light spread function can be obtained by measurements. By observing the number of photoelectrons measured by one photodetector as a function of position of that photodetector about the light source created from a point source of activity above the crystal, one can get a good estimate of the light spread function. This measurement includes the broadening due to the detector thickness, variation in interaction depth, and any contribution due to electronic noise. One may also estimate the light spread function through Monte Carlo simulations. Such computations determine the distribution of gamma-ray interactions within the crystal. Separate simulations model the optics of the system and track the history of all the resulting optical photons created in the crystal all the way into the photodetector (Knoll et al., 1988). Figure 5 shows the results of

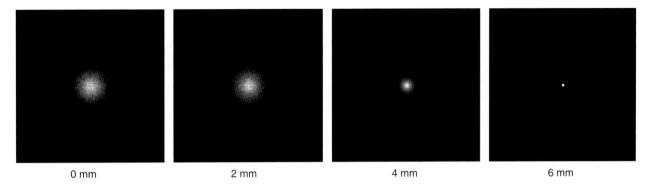

FIGURE 5 Light distribution seen at bottom of a $6 \times 6 \times 0.6$ cm^3 NaI(Tl) crystal for four depths from the top of the crystal of a point light source. The light spread widths are 9.2, 7.2, 3.5, and 0.7 mm FWHM, respectively.

using the DETECT code (Knoll *et al.*, 1988) to model the light spread function seen by the photodetecting surface as a function of depth of a point source of light (a.k.a. *depth of interaction*) in a $6 \times 6 \times 0.6$ cm^3 NaI(Tl) crystal with no light guide. The resulting light spread function profiles can typically be fit to a Lorentzian distribution:

$$F(x) = \frac{Aw}{(x-x_0)^2 + (w/2)^2} \quad (5)$$

where x_0 is the position of the profile peak, w is the FWHM, and A the height of the distribution. The light spread function is narrower for deeper interaction points that are closer to the photodetecting surface, as one might expect from geometry. Thus, assuming a high light distribution sampling rate, the intrinsic spatial resolution would be best if all interaction points were close to the photodetector surface as they would be in a thin crystal.

We note that the preferred single-crystal design parameters for a small FOV camera (ground top with white diffuse reflector, ground sides with black absorber, light diffuser, etc.) discussed here are examples that may not be optimal for every situation. One should test different combinations of potential parameters with experimental measurements and simulations to determine the optimal performance in terms of light collection, spatial resolution, uniformity, sensitivity, and so on.

2. Segmented Multiple Crystals

Another option for a small-FOV scintillation camera design is to segment the scintillation crystal into many small, optically isolated crystals. This segmented (a.k.a. *discrete* or *pixellated*) crystal scheme has four obvious advantages over the large continuous crystal design:

1. A gamma-ray photoabsorption within a given individual crystal in the array creates scintillation photons that are confined to that crystal and focused onto a small spot on the photodetector array (see Figure 6b). Because fewer photodetector elements are involved in positioning, this could improve the SNR for positioning compared to that for the scheme of light sharing among several photodetectors (Figure 6a).
2. Because the spatial resolution in this case is set by the crystal width or the degree of pixellation, using discrete crystals allows one to choose and fix the intrinsic spatial resolution of the system independent of the scintillator light yield (see Figure 6b).
3. Because the light collection properties are approximately identical for all crystals within the segmented array, the positioning of an event is set by the crystal locations and is linear up to the edge of the sensitive FOV, unlike in the continuous crystal scheme. Gain balancing will still be necessary due to inherent intercrystal light yield variations.
4. Detector scatter events that cause positioning errors and blur resolution can be rejected if the individual crystals are each coupled to individual photodetectors.

Disadvantages of the discrete crystal design compared to the continuous crystal design are:

1. It has a lower sensitivity per area due to the dead area between crystals.
2. Because the spatial resolution is determined by the size of the crystal, higher resolution requires smaller crystals and more of them to cover the same area.
3. With narrower crystals, because the aspect ratio is high, light transmission and collection into the photodetector degrades.
4. A matching square-hole collimator might be necessary for single-photon cameras.
5. The images created will be pixellated, and linear interpolation, smoothing, or detector subsampling may be necessary to deal with pixellation artifacts.
6. It is more expensive to manufacture.

It should be noted that one may configure the discrete crystal design shown in Figure 6b with a light diffuser between the crystals and the photodetector array, if desired, to read out more crystals with fewer photodetector array elements. Another way to achieve this is to only partially cut through the crystals for a pseudodiscrete array that allows light sharing near the bottom. These are other examples of

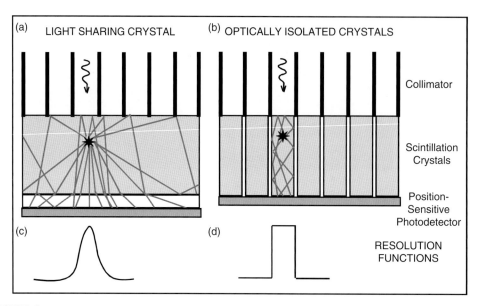

FIGURE 6 Schematic drawings of the light ray propagation for typical (a) single and (b) multiple scintillation crystal camera designs. For the pixellated design, scintillation light is confined to one crystal and focused on one spot of the photodetecting array. (c,d) Approximate shapes of the resulting theoretical intrinsic point spread functions for single-photon imaging.

positioning by light multiplexing. In such hybrid multiplexed designs, light is shared among several photodetector elements for positioning of the crystal interaction (see also Figure 12 later in the chapter). Because there is some light loss involved in such multiplexing schemes, each photodetector should have high sensitivity and low noise to maintain a reasonable SNR.

Discrete crystal design is typically chosen in single-photon imaging designs of a small-FOV camera that require a fixed spatial resolution and good spatial linearity. The single-photon breast imaging is an application in which high spatial resolution is needed to see small lesions and good spatial linearity is needed over the entire sensitive FOV so that prone or lateral breast imaging can be performed all the way to the chest walls, if desired, without distortion.

The use of pixellated crystals is also a popular choice for a coincidence imager. For high detection efficiency for 511-keV photons, in addition to a high Z and high density, the individual crystals should be relatively long. As discussed in the last section, light detection aspect ratio considerations do not favor a thick continuous crystal for good position sensitivity. To retain high spatial resolution in a 511-keV coincidence imager, either the crystal thickness must be reduced, reducing detection sensitivity, or the crystal should be segmented. For a segmented crystal design, there is a balance between spatial resolution and light collection efficiency (and energy resolution) due to poor aspect ratio. It should also be noted that for discrete crystal designs of any type the crystal pixel size should not be too small because of the corresponding drop in counts per intrinsic resolution element. As always, there is also a delicate balance between spatial resolution and counting efficiency for high image SNR. Because of the variation in interaction depths within a given crystal for 511-keV annihilation photons, there may be spatial resolution blurring in coincidence imagers caused by mispositioning photons that enter the crystals at oblique angles. Depth of interaction determination can reduce this blurring effect. Crystals such as LSO and BGO are commonly used in small-FOV coincident imagers (see Table 1).

A disadvantage of a very fine discrete crystal design is that incomplete energy deposition in a single crystal is more likely (for example, due to Compton scatter or, to a lesser degree, K-shell X-ray or electron escape (Levin, Tornai, Cherry, et al., 1997); see Chapter 4 in this volume), potentially causing event positioning errors. If the individual crystals are coupled to individual photodetectors, such events can be rejected by rejecting events with more than one element hit, but this could result in a significant loss in overall sensitivity. Because a small-FOV beta imager requires a very thin crystal due to the short range of betas, the light transmission aspect ratio is favorable for the use of a continuous rather than segmented crystal.

The scintillation light collection efficiency of a given discrete crystal can also be estimated from Snell's law and geometry. Figure 7 depicts the propagation of several scintillation light rays from a point interaction in a long and narrow crystal. If we assume that the length of the crystal is significantly greater than the width (high aspect ratio is required for high detection efficiency and spatial resolution), a perfect back reflector is used on the top of the crystal, an air gap surrounds all five nondetecting polished

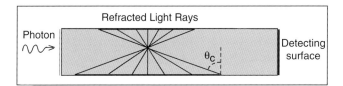

FIGURE 7 Depiction of the fraction of scintillation light lost from refraction through the sides of a long and narrow scintillation crystal for all incident angles less than the critical angle, θ_c. High-energy photons enter the crystal from the left and only a fraction of the scintillation light created is piped down to the detecting surface at the right.

sides, and the interactions occur not too close to the light sensor, then a majority of the light that is lost (not collected at the detector) is due to refraction out of the four long crystal side faces. This is similar in concept to the case for light entering an optical fiber, except that in typical fiber applications the light originates from outside the fiber. To obtain a rough estimation of the scintillation light lost through the sides and not collected at the crystal end, we first determine the solid angle Ω_L subtended to an interaction point by the cone of lost refracted rays (the cone of light rays impinging at angles less than the critical angle) centered on one crystal side face in Figure 7:

$$\Omega_L = \pi \tan^2 \theta_c \approx \pi \left(\frac{1}{n^2} \right) \quad (6)$$

where θ_c is the critical angle for total internal reflection in the crystal; for the second expression, we have used Eq. (2) to write this in terms of n, the crystal index of refraction, and expanded the result to first order in $1/n^2$. Note that this approximate expression for the solid angle subtended by the lost rays through one crystal side is independent of the light source location and crystal dimensions for a high aspect ratio crystal.

The fraction of all available light that is lost through all four sides is then estimated by multiplying Eq. (6) by 4 and dividing the result by 4π. The estimated fraction of available light f, which is collected at the photosensor in high aspect ratio crystals, is then:

$$f = 1 - \frac{1}{n^2} \quad (7)$$

Thus, for better light piping to the crystal ends, a higher refractive index crystal should be used. For example, using LSO ($n = 1.82$) crystal fingers, the light collection efficiency is roughly 70%, whereas for BGO ($n = 2.15$) it is nearly 80%. Equation (7) for the light collection efficiency in high-aspect-ratio crystals should be considered an upper limit. In reality, because the crystal surface conditions are never ideal, neither are the reflections and the fraction of the available light that is collected is typically lower than given in Eq. (7), especially for very high aspect ratio crystals. For low aspect ratio crystals, a majority of the light rays created may enter the photosensor without interactions with the crystal sides and the light collection efficiency may be higher than given by Eq. (7).

In general, the amount of light collected is also dependent on the depth at which the flash of light is created relative the photodetector. The larger the solid angle subtended by the photodetector at the interaction point in the crystal, the less one has to depend on reflections and piping of the light for light collection and the better the overall light signal. Because the most likely interaction depth for long crystals is near the photon entrance side of the crystal (see Figure 7), only a fraction of the available light will be collected. The probability of a photon of energy E interacting between depth x and $x + \Delta x$ in a crystal of density ρ and atomic number Z is a decreasing exponential function:

$$P(x; \Delta x) = e^{-\mu x}(1 - e^{-\mu \Delta x}) \quad (8)$$

where $\mu(E,\rho,Z)$ is the crystal attenuation coefficient. Thus, when the crystal aspect ratio is high, there is a significant fluctuation in the light collected from event to event, degrading the energy resolution.

The resulting energy spectrum and light collection efficiency can be accurately calculated first by using a Monte Carlo simulation of photon interactions in the crystal (GEANT, DERN, Geneva, Switzerland; MCNP, Los Alamos National Laboratory, NM; EGA, Stanford Linear Accelerator Center, Palo Alto, CA; see also Shao et al., 1996), that determines the point source origins of the scintillation light for each event. These light source positions are then used as input for the optical-photon-tracking Monte Carlo simulations (Knoll et al., 1998) to predict the system light collection efficiency and energy spectrum (Shao et al., 1996).

The individual crystal treatments for optimal light collection efficiency in the segmented crystal design are different than for the single continuous crystal design. The top of each individual long and thin crystal finger should be ground with a white diffuse reflector applied in order to break up internal reflections and preferentially direct reflected light toward the photodetector, as in the continuous crystal case. However, due to the high aspect ratio of these long and thin crystals, all spatial information of a point interaction occurring within a given crystal is lost. The photodetector positions that event only by the physical location of that crystal in the array. This point is discussed in a later section. As such, for optimal SNR for event positioning it is important to collect as much light as possible from the small bottom surface of the individual crystal. To enhance light collection efficiency, the long and thin sides of the crystal may be highly polished to increase the probability of internal reflections and enhance the piping of the light toward the photodetector end of the crystal. Wrapping the crystal sides in a high reflectivity reflector (white Teflon tape, for example) is a common strategy used to redirect a portion of refracted light lost from the sides back into the crystal.

B. Collimation for Scintillation Crystal Designs

Collimation considerations are crucial for small-FOV imager design. Physical collimation for gamma rays is discussed in Chapter 8. Here we mention collimation issues especially relevant to small-FOV imager design. Collimation design dictates how well a camera can determine the direction of incoming radiation and optimal design is crucial for good spatial resolution and high sensitivity. Collimation for positron annihilation cameras is achieved electronically by accepting only events generating a coincidence in both detector heads (see Figure 2). The extent of the annihilation photon collimation, or collimation aperture, is determined by the collinearity of the oppositely opposing photons, and the detector element size (see Chapters 4 and 11 in this volume). Thus, for high collimation of incoming annihilation photons, the detector heads should be close together and the detector elements should be fine.

Electronic collimation is also used in gamma-ray imaging using Compton scatter imagers (see Chapter 19 in this volume). A hit is required in the scatter array in coincidence with one in the absorption array. This collimation method constrains the incoming gamma ray to the surface of a cone. The pixel spatial resolutions and corresponding energy resolutions of the two detector arrays determine the angular resolution of the apex angle of this cone, or the collimation aperture in this case. The finer the detector pixellation and the higher quality the detectors used, the better the spatial resolution of the Compton camera. These aspects are discussed further in Chapter 19 in this volume.

Physical collimation for direct-positron imaging is unnecessary due to the short range of the betas in tissue and detector materials and typically close proximity to the beta emitter during imaging.

For a small-FOV gamma-ray imager, there are a few properties of the physical collimator that can be exploited in typical applications, such as breast imaging. We focus our discussion here on parallel-hole collimators because that is the most common configuration used in small-FOV single-photon imagers that have been developed. Because the application-specific, small-FOV imager is often designed with ultrahigh intrinsic spatial resolution in mind, investigating high spatial resolution collimators is necessary. Small-FOV cameras can use custom-made, high-spatial-resolution collimator designs that would not be practical or necessary for a typical large-FOV commercial camera. In Chapter 8, we learn that the collimator contribution to spatial resolution varies linearly with the collimator hole size and distance from the object of interest and varies inversely with the length of the collimator. The sensitivity, on the other hand, is roughly independent of the object distance, but strongly increases with hole diameter and decreases with hole length and septal thickness.

For a small-FOV imager, the collimator and scintillation crystal might be designed for an extrinsic spatial resolution below 4 mm FWHM. This cannot be accomplished practically for all applications with the standard clinical gamma-ray camera because a higher collimator spatial resolution means lower sensitivity. A small application-specific camera might accomplish this ultrahigh spatial resolution imaging by using a smaller-hole collimator and moving the detector very close to the object (e.g., directly against the breast). To compensate for the lost sensitivity, the collimator could be made thinner, which implies the images will only come into focus for relatively shallow objects. This limited focus might not be a major disadvantage in a situation such as breast imaging because most tumors will not typically be more than 3 cm deep into tissue from either side of a compressed breast. Figure 8 shows images of a ^{57}Co (122-keV gamma rays) point source taken with a 6-cm-FOV NaI(Tl)-PSPMT camera (Levin, Hoffman, Tornai, *et al.*, 1997) using varying thickness of a low-energy high-resolution (LEHR) parallel-hexagonal-hole collimator. Using a 2-cm-thick LEHR collimator with this small camera, structures less than 2.5 cm deep into tissue remain in focus with an overall spatial resolution of less than 4 mm FWHM. The standard clinical LEHR collimator is over 3 cm thick and the overall system spatial resolution is greater than 7 mm FWHM. Higher-resolution collimators further improve focusing capabilities for shallow lesions in tissue.

The chosen collimator design also depends on the scintillation crystal design. For a continuous crystal design, the standard hexagonal-hole collimator might suffice, depending on the desired performances in spatial resolution and sensitivity. For the discrete crystal design, depending on the crystal size, however, the collimator hole may need to match the square crystal shape; otherwise there may be fluctuations in sensitivity over the detector face (see Figure 6). For high aspect ratio, discrete crystals, depending on the object–collimator distance, the crystal rather than the collimator might actually limit the spatial resolution. Figure 9 shows a novel square-hole collimator developed for a small-FOV imager (Thermo Electron Tecomet, Woburn, MA).

C. Photodetector Design

The position-sensitive photodetector array reads out the collected scintillation light and turns it into a distribution of

FIGURE 8 Images of a ^{57}Co point source using a 6-cm-FOV NaI(Tl) PSPMT-based camera for different LEHR collimator thickness, varying from 0 to 2.0 cm. The relative collimator sensitivities are, from left to right, 1.00:0.113:0.016:0.011:0.009. The corresponding resolutions are 13.6, 5.7, 3.6, 3.3, and 3.3 mm FWHM, respectively. Relative scaling is used for display.

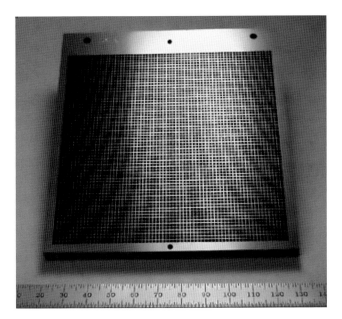

FIGURE 9 A high-resolution square-hole tungsten collimator fabricated by a stacked lamination technique in which multiple submillimeter tungsten foil sheets are stacked and bonded in a precise alignment to form the desired thickness. The square pixel pattern is formed into each layer by precision photolithography etching techniques. The collimator shown is 11 cm wide and 9 mm thick with 1.25-mm square holes and a 0.25-mm thick septa. (Courtesy of Patrick Kelly, Thermo Electron/Tecomet, Inc., Woburn, MA.)

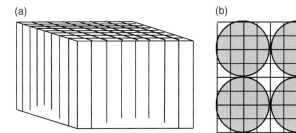

FIGURE 10 A depiction of (a) the pseudodiscrete crystal array (left) and (b) top view of the four-PMT readout scheme for a PET block detector (Casey and Nutt, 1986).

electrical pulses. These pulses determine x and y coordinates corresponding to the apparent position of the scintillation light flash created by an interaction. A small-FOV imager can use a variety of exotic photodetector technologies at a relatively low cost that would not be practical for a large FOV standard camera. For tumor-imaging applications, often the technology is chosen for high spatial resolution. Basic photodetector principles are discussed in Part IV of this book and will not be repeated here.

1. Multiple Photomultiplier Tubes

The simplest small-FOV scintillation camera design is essentially a miniaturization of the standard Anger camera. The scintillation light pulses can be read out by an array of individual PMTs. However, due to the small FOV, the standard 5-cm-diameter PMT is too large for adequate sampling of the light distribution in a small area and too bulky for a compact imager. Smaller PMTs are available and have been used for this purpose (Milster et al., 1985; Casey and Nutt, 1986). Scintillation light created from an interaction is shared over individual PMTs for positioning of that event (positioning by light multiplexing). An example of this is found in the PET block detector module design (Casey and Nutt, 1986), in which an array of pseudodiscrete scintillation crystals are read out economically by only four PMTs, as depicted in Figure 10. Because there are only four PMTs, positioning logic is simplified and the electronic multiplexing of the PMTs for reading out is unnecessary, unlike in the standard Anger camera (Anger, 1958; Barrett and Swindell, 1981). Single-photon cameras could also in principle use such a pseudodiscrete crystal, block detector design (Dahlbom et al., 1997). However, there is a limit to how many crystals can be decoded in this economical scheme and, hence, to the achievable intrinsic spatial resolution. To further improve on spatial resolution, the tube size can be reduced; however, the world's smallest PMT is about 1 cm in diameter. A PSPMT design is preferred because it has a tiny anode pitch for a high light distribution sampling rate and higher intrinsic spatial resolution results.

2. Position Sensitive PMTs

In this chapter, a PSPMT is defined as any single-tube PMT with positioning capabilities within its sensitive area. A small-FOV scintillation camera can be constructed from a single large-area PSPMT or multiple small PSPMTs. A small-area (< 12-cm FOV) gamma-ray camera, for example, can be constructed from a single PSPMT. A depiction of a large-area PSPMT and its signal formation process is shown in Figure 11. Early PSPMT designs that were used in small medical radionuclide imagers were the large-area varieties that use a grid or fine mesh dynode structure and cross-wire anode configuration. The grid or mesh dynode, cross-wire anode variety was the first PSPMT to be used in medical imaging. The anode plane typically comprises two planes of many finely spaced parallel wires, one for each direction for x-y positioning of the charge distribution created from a scintillation event (see Figure 11). The amplification of one photoelectron at the photocathode is shown in the figure, but typically several hundred to thousands of photoelectrons are created at the photocathode per scintillation event, depending on the photon energy, scintillator, and photocathode quantum efficiency. The resulting charge distribution created from a scintillation event is a concatenation of thousands of such charge avalanches. For example, a 140-keV photon absorption in NaI(Tl) creates approximately 5000 photons or 1000 photoelectrons, assuming a bialkali photocathode quantum efficiency of 20%. With a charge amplification

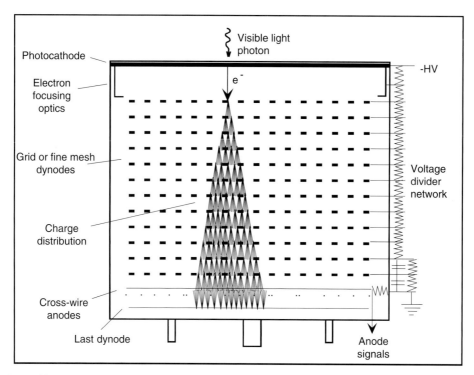

FIGURE 11 Depiction of the cross-wire anode PSPMT and its signal formation process. For simplicity, the charge avalanche created from only a single photoelectron is shown. A 10^5 photoelectron amplification factor is typical. Setting the cathode at ground and biasing the anode at +HV through a coupling capacitor is also possible.

factor of 10^5 that would lead to a charge signal comprising 10^8 electrons.

Note that in this design, a charge distribution rather than light distribution is shared over many finely spaced anode wires for positioning. This is an example of charge multiplexing. Because of the large number of electrons compared to light photons created, positioning by charge sharing is typically more precise than by light sharing. Typical anode wire pitch in these designs is 3–4 mm. The x-y anode wire signals are used to determine the mean x-y location of the resulting charge distribution for each scintillation event. This x-y position gives an estimate of the corresponding projected position of the interaction in the scintillation crystal. Because several anode wires are typically involved in the charge collection process, continuous positioning with spatial resolution significantly better than the wire pitch is achievable, depending on the crystal design and light output.

The spatial confinement of an event is somewhat limited in this PSPMT design because of significant *cross-talk*, or charge spreading, between successive dynode planes due to transverse electron diffusion during the multiplication process in the dynodes. Another disadvantage is the poor photocathode uniformity over the relatively large area, leading to nonuniformity in pulse height and sensitivity across the tube face. The relatively long tube length is also a problem for a compact design. For reduced cross-talk through more spatially constrained charge amplification, the microchannel plate (MCP) dynode mutianode PSPMT design has also been used for medical radioisotope imaging (Ljunggren and Strand, 1988; Lees *et al.*, 1997). A disadvantage of the MCP design is their unreliability at high data rates. Advances in micromachining technology have led to the development of the metal channel dynode configuration, which reduces the cross-talk between adjacent locations by constraining the charge multiplication process to within individual nearly isolated metal channels. The metal channel dynode design is similar in concept to the MCP concept, but the channels are shaped differently and the multiplication process is more stable at high data rates. Cost and mechanical considerations, however, have limited the area of commercially available metal channel dynode PMTs to roughly 2.5×2.5 cm^2. These PSPMTs are also quite thin, enhancing compactness compared to the bulky mesh dynode PSPMT. For a larger-area camera, however, many of these small units must be assembled in an array. A potential disadvantage of this configuration is the dead area present between each module.

3. Semiconductor Photodetector Arrays

There have been significant investigations into alternative photodetectors to the PMT in nuclear emission scintillation imaging. Most attempts thus far have been in relatively small-FOV imager designs. For a small compact imager, the PMT is not ideal due to its high cost (even for the more recent designs, many parts are still manufactured by hand);

relatively large size, bulkiness, and dead area (due to the many components and mechanical engineering required; see Figure 11); and relatively low photocathode quantum efficiency (~15–25%). Pixellated arrays of semiconductor photodetectors (SPD) have been investigated as an option (Patt et al., 1996, 1997, 1998; Tornai et al., 1997; Levin et al., 1996; Gruber et al., 1998, 1999; Fiorini et al., 1997, 1999; Fiorini, Longoni, et al., 1999; MacDonald et al., 1999; Casey et al., 1998; Shao et al., 1999). The use of SPDs allows high compactness for the imaging device. The three most common SPD technologies that have been investigated for use in small-FOV imagers in medicine are based on silicon: the silicon PIN photodiode (PD) (Patt et al., 1997, 1998; Levin, Hoffman, Tornai, et al., 1997; Gruber et al., 1998, 1999), the silicon avalanche photodiode (APD) (Casey et al., 1998; Shao et al., 1999), and more recently the silicon drift detector (SDD) (Fiorini et al., 1997, 1999; Fiorini, Longoni, et al., 1999; MacDonald et al., 1999), although other SPD technologies have been investigated for use in small-FOV nuclear emission imagers as well (Patt et al., 1996; Tornai et al., 1997; MacDonald et al., 1995a, 1995b; Suyama et al., 1998). The principles of operation of these devices are described in Chapter 15.

Unfortunately, the standard scintillation crystal used in nuclear medicine imaging, NaI(Tl) is not ideal for use with the SPDs under discussion because the light emission spectrum of NaI(Tl) does not match the absorption spectrum of silicon. The peak wavelength of emission for NaI(Tl) is 410 nm, well matched to that of the PMT bialkali photocathode absorption spectrum. The peak absorption wavelength of silicon is typically above 700 nm, although some quantum efficiency enhancement toward visible light wavelengths can be achieved with proper design (Suyama et al., 1998; Moszynski et al., 1997). CsI(Tl) (peak emission, 530 nm) has been proposed as an alternative scintillation crystal for use with silicon SPDs (Carter et al., 1995; Rossi et al., 1996; Patt et al., 1997; Levin et al., 1996; Gruber et al., 1998; Fiorini et al., 1997; MacDonald et al., 1999). CsI(Tl) has 35% higher intrinsic light yield at 530 nm than NaI(Tl) at 410 nm and a 65% higher stopping power at 140 keV. The higher stopping power allows the use of thinner crystals, improving light spread localization and transmission, with no loss in sensitivity. Table 1 gives some properties of CsI(Tl) and NaI(Tl). CsI(Tl) crystals are relatively nonhygroscopic, making them much easier to handle, and no hermetic seal or glass entrance window is required. We have also listed the crystal CsI(Na), a popular choice for use with PSPMTs, which has similar properties to NaI(Tl) except it is less hygroscopic (but is deliquescent) and does not need to be hermetically sealed, and so no entrance window is necessary.

Thus, an alternate camera design would replace the NaI(Tl) with CsI(Tl) and replace the PMT or PSPMT array with an array of SPDs. Appealing characteristics of the CsI(Tl)-SPD design are the SPD's inherent compactness, potentially smaller achievable spacing between elements for higher spatial resolution and low dead area, and higher (~80%) quantum efficiency compared to the PMT; and CsI(Tl)'s higher intrinsic light yield, higher stopping power, and better spectral match to silicon compared to NaI(Tl). Compared to a PSPMT, the SPD array is significantly more compact. An SPD array is typically less than 1 mm thick and a preamplifier array based on an application-specific integrated circuit (ASIC) would need less than 1 cm of additional space, reducing the space requirements for the CsI(Tl)-SPD imager to a few centimeters thick with only a few millimeters of dead area at the outer edges. In contrast, a typical NaI(Tl)-PSPMT device might be 15 cm thick with over 1 centimeter of surrounding dead area at the FOV edges, although certain multianode, metal channel dynode PSPMT designs are quite compact with relatively low dead area. Compactness is especially desirable for utility and versatility in applications such as breast imaging. Another impetus for the CsI(Tl)-PD approach is the seemingly inevitable reduction in the cost of silicon-based products compared to PMTs. Thus, the CsI(Tl)-PD concept will typically be more compact and, in principle, with an appropriate design and low noise readout could have a better SNR than a NaI(Tl)-PMT device, comparable sensitivity, and lower cost. The PMT, however, fundamentally has lower noise per readout area and a higher gain than the SPD.

For any SPD–preamplifier combination, for the optimum SNR at the preamplifier input the SPD terminal capacitance is kept to a minimum (Iwanczyk and Patt, 1995; Nicholson, 1974). In addition, the SPD leakage current generally increases with area. These facts, combined with the merit of using arrays of small crystals, as discussed in Section IIA2, imply that the best SNR for a CsI(Tl)-SPD imaging array is achieved when it is composed of tiny (≤ 2 mm) CsI(Tl) crystals coupled to correspondingly small SPD array elements. There are a few drawbacks to such a design. To cover a relatively large area with such a highly pixellated SPD array requires a customized and expensive readout. In addition, this concept requires a mosaic of several SPD arrays to cover a relatively large area, which might introduce dead regions between arrays and artifacts in the images. Also, typically in pixellated SPD array production, one expects perhaps 10% dead pixels.

The basic principles of the SPDs of interest are discussed in Chapter 15. Both the PD and SDD are essentially gain 1 devices. For the same sensitive area, the SDD has an advantage over the PD with regard to terminal capacitance. In the SDD design, charge created near the cathode from a light photon interaction drifts towards a much smaller, low capacitance readout anode for low noise current. Thus, the SDD offers the possibility of covering a relatively large imaging area with relatively large and hence fewer photosensor channels, each with very low noise current. PD elements have roughly the same size anode as cathode, and if low noise is desired, the PD elements must be very small and therefore numerous in order to cover an equivalent area. The APD is similar in concept to the PD except there is a separate, high electric field region in the device where drifting charge is amplified. APD gain typically is anywhere from 50 to 1000, so the signals are easier to process than for the PD, but, like the PD, due to noise considerations many small elements are required to cover a relatively large scintillation imaging area. Each channel of a SPD array typically requires a separate low-noise charge-sensitive preamplifier, which can make initial development of SPD arrays quite expensive.

If a continuous scintillation crystal design is used for positioning, light may be shared among several elements in the SPD array using a light diffuser, similar to that shown in Figure 6a. The PD, APD, or SDD can be used in such a design. The choice of the SPD pixel size depends on the interplay between SNR per pixel spatial resolution and the desired light distribution sampling rate. The discrete crystal design allows one to choose the desired intrinsic spatial resolution. In this case, either the crystals are matched one to one with the photodetectors (Figure 12a) using a PD or APD configuration or the light from the discrete crystals is spread over larger SPD pixels, as with the SDD configuration (Figure 12b). The decision of whether the scintillation crystal should be segmented depends on the aspect ratio and surface treatment of the individual crystals and noise level of the SPD pixels. Due to uncorrelated SPD element noise in an array, a discrete crystal design with one-to-one coupling to SPD pixels may have a better SNR even though the continuous crystal may have a better aspect ratio for light collection.

D. Electronic Readout of Position-Sensitive Photodetectors

1. PSPMT Readout

We focus our discussion of PSPMT electronic readout schemes on the cross-wire anode type design as an example, but these principles hold for pixellated anode designs as well. In the most commonly used readout schemes for cross-wire anode PSPMT designs of small-FOV imagers, the x-y position is calculated from the anode wire signals using two basic methods: the resistive charge division readout technique and the parallel readout technique.

a. Resistive Charge Division Readout

The resistive charge division technique (see Figure 13) incorporates resistors between each anode wire in the x and y wire planes, thus coupling or multiplexing many individual wire signals into four signals: x_+, x_- and y_+, y_-. By measuring the resulting currents generated on either end of each wire plane (Figure 13), these signals give a sensitive measure of the location of the charge distribution in each direction. This is another example of charge multiplexing for positioning. This method may not be suitable for imaging very low energy sources where the charge created per event might be significantly lower. Also, coupling signals from many wires limits the achievable spatial resolution. However, positioning involving sharing the charge (consisting of 10^8 electrons, for example) among several wires adds less of a blurring effect

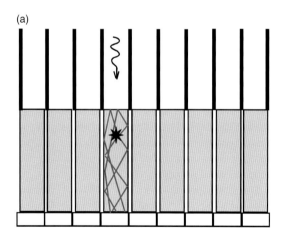
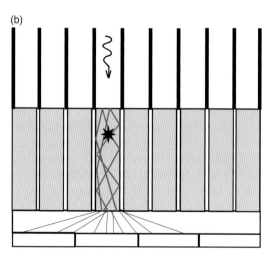

FIGURE 12 Schematic depiction of semiconductor photodetector readout schemes for discrete crystal designs. (a) Each crystal is coupled one to one to an individual photodiode or avalanche photodiode pixel in an array. (b) Light from a crystal is shared among several SDD channels for positioning. The SDD design requires fewer pixels, and hence electronic channels, to cover the same FOV, but because all the light is focused on one photodetector in (a), the signal-to-noise ratio per channel may be superior.

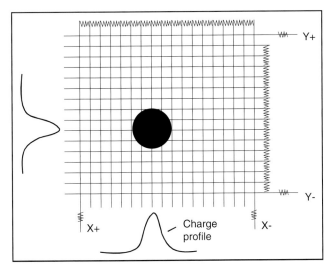

FIGURE 13 Example depiction of resistive charge division readout scheme for event positioning with a cross-wire anode PSPMT.

than that from light sharing (involving 5000 photons, for example) among several photodetector elements. For multianode PSPMT designs, there are other possible choices of resistor divider topology (Siegel *et al.*, 1996). Capacitive charge division is another option (Barrett and Swindell, 1981).

b. Parallel Readout

The parallel method of readout overcomes the drawbacks of resistive division. By reading out each anode individually, the full charge can be used for the mean *x-y* position calculation and the fine spatial sampling of the charge distribution is preserved for higher intrinsic spatial resolution. The disadvantages of this parallel scheme are that the signal per anode element may be relatively low, introducing noise into the position estimation, and more electronic readout channels are required. There are two basic solutions to the lower SNR per wire problem. In the first solution, small subgroups of multiple wires are summed together to improve the SNR for a wire subgroup (Wojcik *et al.*, 1998; Weisenberger *et al.*, 1998). However, this reduces the effective sampling pitch of the charge distribution. The second remedy only accepts those wire signals (or small wire subgroup signals) above a certain threshold and does not involve the low-signal wires in the position calculation. Thus, effectively the mean position would be calculated using only a local group of wires in the region of the charge distribution peak. For PSPMT designs with finely spaced anode wires, the parallel readout method has the advantage of a more linear response near the sensitive FOV edges than does the resistive charge division.

2. SPD Readout

Resistive charge division readout of an SPD array is difficult because the multiplexing means that uncorrelated noise from each photodetector element is added into the position estimate, thus potentially degrading intrinsic spatial resolution performance. Resistive charge division may also not be desirable for PDs and SDDs because they are gain 1 devices. Resistive charge division for an APD array is possible provided the SNR per pixel is high (Casey *et al.*, 1998; Shao *et al.*, 1999). Depending on the APD gain and noise level, one may need to first read out the signals by charge-sensitive preamplifiers before multiplexing by resistive charge division.

Typically parallel readout of individual SPD array pixels provides the best SNR because the noise contributions remains uncoupled, but this can involve a large number of electronic channels.

E. Electronic Processing and Data Acquisition for Imaging

Figure 14 shows very basic examples of potential signal processing and digitization data acquisition setups for imaging with either resistive charge division or parallel readout. The basic topology of the two schemes is similar: photodetector signals are split into two parts, one for digitization of the pulses (slow) and the other for the timing trigger or gate of the digitization process (fast). The resistive charge division network typically has four analog outputs; the parallel readout has multiple signals,

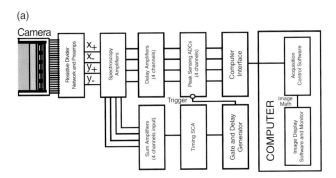
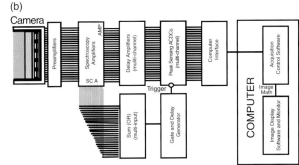

FIGURE 14 Basic examples of electronic processing topologies for image acquisition for the (a) resistive charge division readout and (b) parallel readout schemes.

depending on the total number of anode or pixel elements and whether any degree of multiplexing is used. Typically for the PSPMT and especially for the SPD array designs, each output channel is first taken through a preamplifier. For an SPD array, a low-noise charge-sensitive preamplifier is used for impedance matching (a.k.a. a *line driver*) and charge-to-voltage conversion of the pixel anode signal.

In both examples shown in Figure 14, the analog signals to be digitized are first sent through a spectroscopy amplifier to amplify and shape the signals. The resulting signals are sent into the analog-to-digital converters (ADCs). For the ADC trigger or gating portion of the signal processing in this example, the outputs are summed and sent into a single channel analyzer (SCA) for pulse height discrimination and then into a gate and delay generator. This latter component is necessary to align the fast and slow portions of the processing scheme in time. If desired, the range of pulse heights can be hardware selected (e.g., corresponding to the photopeak) by the SCA and only pulses of the correct total energy will be digitized, similar to the scheme depicted in Figure 1. Otherwise, the entire spectrum of energies can be accepted by simply maintaining an SCA discriminator level just above the noise level.

In the case where there is a significant noise contribution per channel such as in the readout of a pixellated SPD array, initial discrimination of individual channels could be accomplished by sending the detector signals through independent SCAs, providing digital outputs from each channel for signals exceeding a specified threshold set just above the noise in each channel. In this case, gating for the ADC is provided by a logic signal produced by performing a sum or logical OR of these digital outputs. Because of variations in photodetector sensitivity and electronics, multiple readout channels are typically gain balanced with the adjustable amplifiers prior to image acquisition using flood irradiation calibration procedures.

The setups shown in Figure 14 are for acquiring data in list mode, in which all the relevant detector signals from each scheme are digitized and these data are stored as a list of all the digitized signals for each event. The resulting data are then postprocessed with positioning arithmetic for image formation. This allows flexibility in signal postprocessing for image formation. For example, for parallel readout of an SPD array, list mode acquisition provides flexible pulse height discrimination on the individual pixel signals for determining which channels to include in the position estimate. List mode acquisition allows various energy windows to be studied without reacquiring the data. Once the optimal processing scheme (energy window, number of signals involved in the position calculation, etc.) has been chosen from processing list mode data, the selected processing logic can be incorporated into the hardware for more efficient acquisition in a clinical device.

For coincidence cameras, the positioning of the interactions in each head are performed in the same manner as shown in Figure 14. The difference is that the events are not accepted unless there is an electronic coincidence established between the two detector heads (see Chapter 10 in this volume). A second coincidence unit with a large delay between detector logic pulses may be used for an online subtraction of random coincidence events.

F. Event Positioning Schemes and Image Formation

1. Ratio Calculation

The four signals resulting from multiplexed designs such as resistive charge division are statistical in nature. Ignoring this fact for the moment, one estimate of the mean x coordinate of the charge distribution created in a PSPMT is obtained by an expression analogous to Anger logic (for origin at the center; see Figure 13):

$$X = \frac{x_+ - x_-}{x_+ + x_-} \quad (9)$$

with a corresponding formula for the y coordinate. A problem with this position estimate is that unlike the weighted outputs of individual PMT arrays in conventional gamma cameras, the outputs are not necessarily linear with respect to position and the resulting images may be distorted and the imaging dynamic range somewhat limited. Figure 15a is a graph showing the limited dynamic range and nonlinear positioning response near the edges of a standard cross-wire anode PSPMT readout of a 6×6 cm^2 NaI(Tl) crystal using resistive charge division and Eq. (9) for positioning (Levin, Hoffman, Tornai, *et al.* 1997). It may be possible to improve the dynamic range and linearity somewhat by modifications to the resistive divider chain shown in Figure 13 (Clancey *et al.*, 1997).

2. Maximum Likelihood

The maximum-likelihood (ML) position estimation scheme has been used to generate images from the four PSPMT outputs generated from resistive division with more linear behavior and better imaging dynamic range by treating the four output signals as statistical quantities. The ML position estimate determines the x-y location that is most likely to have created the four particular outputs measured. Because the four outputs are digitized and the images are a matrix of pixels, there are only a finite number of possible combinations of the four output signals (x_+, x_-, y_+, and y_-) and the x-y locations in the image. The ML positioning technique determines the appropriate mapping between these two spaces, by maximizing a corresponding likelihood function $P(x_+, x_-, y_+, y_- | x, y)$, commonly taken to be Poisson in nature (Milster *et al.*, 1985; Gray and Macovski, 1976;

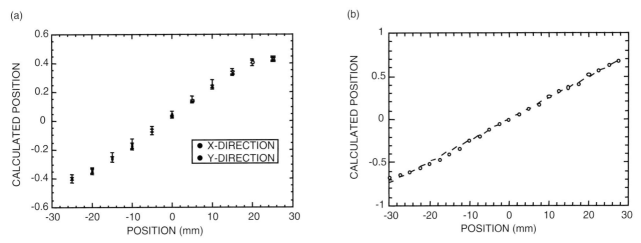

FIGURE 15 Plots of the calculated positions in the image of a ^{57}Co point source using Eq. (9) as a function of true source position for 6 × 6-cm^2 FOV continuous (a) and discrete (b) NaI(Tl) crystals coupled to a PSPMT with resistive charge division readout.

Clinthorne and Rogers, 1987). In practice, this is accomplished by generating a lookup table (LUT) that maps each of the possible combinations of the four output signals to the event location most likely to have produced them. The LUT may be generated with high count point source distribution measurements for $P(x_+, x_-, y_+, y_-|x, y)$ at all possible event locations. For example, for a 64 × 64 image, 4096 distribution measurements must be made. A search through this matrix of measured distributions is performed for the ML position estimate x,y for every possible set of the four camera output signals x_+, x_-, y_+, y_-. This LUT is loaded into memory and during image acquisition, the position estimate for each event is obtained by comparing the four signals to the LUT. If desired, the acquisition may be performed in list mode, storing the four signals for each event and the position estimate from the LUT for each event calculated in postprocessing. A disadvantage to this approach is a relatively lengthy calculation process for each event positioning and a tedious calibration process.

3. Positioning for Discrete Crystal Cameras with Light or Charge Multiplexed Readout

It is interesting to note that using discrete or segmented crystals instead of a continuous crystal can also improve the linearity and dynamic range of the images for multiplexed readout schemes such as resistive charge division. Figure 15b shows this improvement for an array of 2 × 2 × 6 mm^3 NaI(Tl) crystals coupled to a PSPMT. The reason for this improvement is that, assuming the crystals all produce roughly the same amount of scintillation light, this available scintillation light is focused on a spot on the position-sensitive photodetector face. The light distribution and collection efficiency from this optically isolated crystal is roughly the same for all positions, even near the edges, unlike in a continuous crystal.

If a discrete crystal array design is used in conjunction with a light or charge multiplexed readout, in which multiple channels are used to read out any given crystal, an additional step must be introduced before event positioning and image formation can be accomplished. As discussed in Section IIA2, for an interaction in a long and narrow discrete scintillation crystal (Figure 7) all information as to precisely where in a transverse plane through that crystal an interaction occurred is lost due to a poor aspect ratio for light transmission. One can only determine that the interaction occurred somewhere within that crystal width (The longitudinal interaction depth, on the other hand, can be determined by various methods; see Schmand et al., 1999; Rogers, 1995; Huber et al., 1998. For a stationary single-photon camera with discrete crystals, the intrinsic spatial resolution and the image pixel size are roughly the size of the crystals used. Many small crystals are required for high spatial resolution.

Determining which crystal is hit and positioning the event at the corresponding pixel position forms the image. This is accomplished with a LUT. A high statistics calibration flood image of the crystals in the array is first acquired (Figure 16a). This image is essentially a map of the crystal locations determined by the light confined to the individual crystals and is stored in a chosen matrix size. The spots represent a probability distribution of positioned scintillation locations using the multiplexed readout. As such, the FWHM of the spots in the flood field image may be less than the actual crystal size. The width of the spots is proportional to the width of the light spread distribution seen by the PMT and inversely proportional to the square root of the number of photoelectrons detected by the PMT. For very bright crystals, the spot size will be significantly smaller than the crystal size and there may be large enough spaces between spots to justify using even smaller crystals to further improve spatial resolution. For poor-light-yield

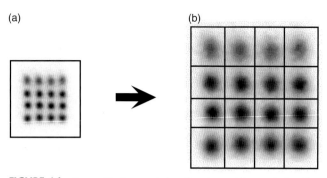

FIGURE 16 For positioning events in a discrete crystal array design (4 × 4 crystals shown), (a) the flood field calibration image of crystal array is used to draw (b) electronic borders for the positioning lookup table. The lookup table determines the image pixel to be incremented when a particular x-y position is read from the camera.

crystals, significant electronic noise, or both, depending on the crystal size, one may barely be able to separate the spots.

Digital region of interest borders are drawn electronically between crystals that appear in the image (Figure 16b). Those borders define the positions assigned to the individual crystals in the positioning LUT. The LUT may be formed, for example, by creating a matrix of identical numbers corresponding to positions inside borders that represent a particular crystal and associated pixel in the image. If the x-y coordinates of the positioned event falls inside the x-y coordinates of a particular defined crystal number in the LUT, that particular crystal number is read and the corresponding image pixel is incremented in intensity. We emphasize that this process is needed only when the number of array crystals is greater than the number of electronic readout channels.

For single-photon imagers using discrete crystals, the intrinsic resolution function will be on the order of the crystal size (see Figure 6). For coincidence imagers, the intrinsic spatial resolution function depends on other factors in addition to detector size (see Chapter 3 in this volume). For a point positron source at a distance significantly greater than the width of the individual detector crystals, the detector component of the resolution function will be triangular in shape with the FWHM equal to one-half the detector width. We note that the intrinsic spatial resolution may be improved by translating the array by a fraction of a crystal width for subsampling of the activity distribution and adding in these interleaved data sets. However, this involves extra imaging time and larger data sets.

4. Positioning with Parallel Readout

For positioning events that use parallel readout of individual or small subgroups of coupled anode or pixel signals (see Figure 14b), typically a hardware or software threshold determines which digitized signals to include in the event positioning calculation. The calculation of the position $<x>$ then follows the standard weighted mean (a.k.a. *centroid*) calculation:

$$<x> = \frac{\sum_{i=1}^{n} x_i S_i}{\sum_{i=1}^{n} S_i} \quad (10)$$

where x_i is the x coordinate of the ith anode element (wire, pixel, etc.), S_i its corresponding pulse height, and the sum is performed over the number n of anode elements or (small groups of elements) that have a pulse height above a particular predefined threshold. A similar expression holds for the y-coordinate calculation. Equation (10) is similar in structure to Eq. (9), but is used for positioning in the cross-wire anode PSPMT or SPD array designs where multiple anode elements are involved in the position calculation for each event.

In a discrete crystal–discrete SPD array design such as depicted in Figure 12a, where each crystal is coupled to only one SPD pixel, Eq. (10) is unnecessary because all the light from a given event is focused onto one photodetector element. In this case, the image is formed by incrementing the image pixel intensity corresponding to the physical location of the hit crystal for each event. If more than one crystal is hit, it is common to position the event at that crystal location with the largest signal provided the total energy deposited in the array is within the accepted range. An alternative for processing these multiple hit events in a individual crystal–photodetector design is to assume they are due to incomplete energy deposition [as a result of processes such as Compton scatter, K-shell X-ray escape, or multiple scatter (for betas)], and reject them. However, this could result in a significant sensitivity loss.

5. Event Positioning in Small-FOV Coincidence Imagers

The positioning schemes described thus far are for single-photon events that interact in a single-imager head. For imaging positron annihilation photons emanating from an object between two detector heads, events that hit the two heads in coincidence (see Figure 2) are first checked for their deposited energy. If the energy falls within the acceptable range, the two events are positioned within their respective detector head, depending on the type of crystal–photodetector array readout scheme. It is assumed that the origin of the annihilation is then somewhere on the line between the two detected interaction positions, or *line of response* (LOR). The simplest method for image formation is to organize the data into simple planar projections parallel to the detector faces through the object being imaged. The positioned data within a given projection plane make up the intersection of all possible back-projected LORs with the image pixels forming that plane. The coordinates of the positioned event within a plane depend on the depth of that plane between the two heads and the angle the LOR makes with the normal to that plane. A stack of such parallel back-projected planes can be formed, representing different planes of focus within the object being imaged. This is an example of a technique referred to as *focal plane tomography*. A more challenging technique that makes more comprehensive use of the data is to use a limited angle tomographic reconstruction algorithm. The success of such an approach depends on the extent to which the angular projection set is incomplete. For relatively large detector heads with close separation, there is only a relatively small subset of missing projection data. Some of the details of the focal plane tomography method for a small FOV coincidence imager, including attenuation and efficiency corrections to the LOR data, are presented in Section IVB of this chapter.

III. SEMICONDUCTOR DETECTOR DESIGNS OF SMALL FIELD-OF-VIEW IMAGERS

The topic of semiconductor absorption detectors in medical radioisotope imaging is covered in Chapter 15. For completeness, we include a brief discussion of these imaging devices here. Issues such as collimation, electronic processing and data acquisition for imaging, and event positioning schemes are similar in concept to that presented in previous sections and are not discussed in detail here. Currently, because of cost considerations of developing high-resolution semiconductor arrays and associated electronics for imaging and because of the limitation on the wafer size that can be manufactured, most semiconductor imaging arrays inherently have a small FOV. Larger-FOV images are constructed by a mosaic of the individual array modules or submodules.

A. Semiconductor versus Scintillation Crystals

Using semiconductors for the detection and imaging of radiation rather than the scintillation crystal–photodetector combination has several advantages in small imagers: (1) they are compact and have low dead area compared to the scintillation crystal–photodetector combination, especially if PMTs are used; (2) their potential ease of manufacturing might lead to lower costs; (3) a high intrinsic spatial resolution is achievable with the chosen array pixellation; and (4) a higher energy resolution is achievable with intrinsic detection of the electron–hole pairs created. To illustrate this last point, we use cadmium telluride (CdTe) and silicon (Si) semiconductor detectors as an example. The average energy, w, required to create an electron–hole pair in CdTe is 4.43 eV. Thus, a 140-keV gamma-ray photoabsorption in this crystal creates roughly 32,000 electron–hole pairs on average. The same gamma-ray interaction produces roughly 5300 photons on average in a NaI(Tl) scintillation crystal, from which only approximately 1100 photoelectrons emerge from a PMT photocathode. So, in theory the basic signal used for event position and energy determination is nearly 30 times larger for CdTe and hence more precise imaging results are expected. A 300-keV beta particle impinging on CaF_2(Eu) (a standard scintillation crystal used for beta detection) creates roughly 7200 light photons and perhaps only 1400 photoelectrons for positioning with a PMT. The same beta hitting Si, with a w value of 3.62 eV, creates roughly 83,000 electron–hole pairs, nearly a factor of 60 larger than the basic signal. Some properties of common semiconductors that have been used in small-FOV imagers for medicine are shown in Table 2.

Current challenges of semiconductor imaging array design in general include: (1) uniformity of spatial and energy response over the array area; (2) low yield of detector quality crystals (this especially applies to CdZnTe); (3) the low mobility of holes leads to hole trapping and lower energy tailing on the photopeak in energy spectra as a result of the incomplete charge collection; (4) the relatively low achievable electric field strengths limits available detector thickness and hence sensitivity for gamma rays; and (5) custom collimators may be required to match the electrode pattern pixellation.

TABLE 2 Properties of Common Semiconductors Used in Small-FOV Medical Imager Designs[a]

Semiconductor	Atomic Number Z	Density ρ (g/cm^3)	Band Gap E_g (eV)	Ave. e-h Creation Energy w (eV)	Resistivity (Ω-cm)	Electron Mobility μ_e (cm^2/V/s)	Hole Mobility μ_h (cm^2/V/s)	Electron Lifetime τ_e(s)	Hole Lifetime τ_h (s)
Si	14	2.33	1.12	3.62	2.3×10^5	1500	600	3×10^{-3}	3×10^{-3}
CdTe	48–52	6.06	1.47	4.43	3.0×10^9	1000	80	10^{-6}	10^{-6}
Cd$_8$Zn$_2$Te	30–52	5.81	1.5–2.2	5.0	2.5×10^{11}	1350	120	6×10^{-7}	2.5×10^{-8}
HgI$_2$	80–53	6.4	2.13	4.15	10^{13}	100	4	10^{-7}	10^{-8}

[a] All values are at 300 K. From Schlesinger and James (1995).

B. Semiconductor Imaging Array Configurations

Unlike scintillation crystal designs, a semiconductor imager is typically constructed out of a single monolithic crystal due to manufacturing constraints. Thin anode and cathode electrode patterns are deposited on either side of the crystal. Rather than scintillation light creation, transport, and collection, the concerns now are of charge carrier creation, transport, and collection. The positioning capabilities and spatial resolution are established by the particular electrode pixellation scheme chosen. Figure 17 shows common semiconductor detector electrode pattern configurations. The *cross-grid electrode design* (Figure 17a) uses two orthogonal sets of electrode strips on opposite faces of a monolithic crystal for charge carrier collection and *x-y* position readout. A high-voltage bias applied across the detector produces an electric field. When a gamma ray is stopped in the detector, the electrons and holes created move in opposite directions and are collected at the nearest row/column electrodes. The pixels are defined by the intersection of the row/column electrodes. Because the signals produced on the row and column electrodes are due to the motion of both the electron and hole components, the induced signals are identical in amplitude but opposite in polarity. Thus, in principle, all energy information could be extracted from either the row or column electrodes. Typically the high-voltage-side electrodes are each capacitively decoupled and connected to charge-sensitive preamplifiers. The low-voltage electrodes are read out by low-noise charge-sensitive preamplifiers (one for each strip) for high energy resolution and are thus used for energy information. An outer guard ring structure (not shown) is typically used to maintain a uniform electric field distribution at near the edges of the device. To optimize charge carrier collection, electric field simulations (Patt *et al.*, 1995) of the particular electrode configuration are usually performed.

The data can then processed in a way similar to the example shown in Figure 14b, except only the high-resolution, low-voltage electrodes are summed and used to generate the ADC gate. More elegant approaches (Barber *et al.*, 1994; Marks *et al.*, 1996; Matherson *et al.*, 1998) use multiplexed readouts, in which digitization is performed on only one or very few channels directly involved in the charge collection. For list mode acquisition, typically the SCA levels are set just above the noise level of the low-voltage electrodes. Signals from all row/column electrodes are digitized and the data may be stored in list mode for postprocessing. An event may be positioned in postprocessing, for example, by first determining which events have total row and total column electrode signals within the desired energy range. From these events, one can select, for each event, the row

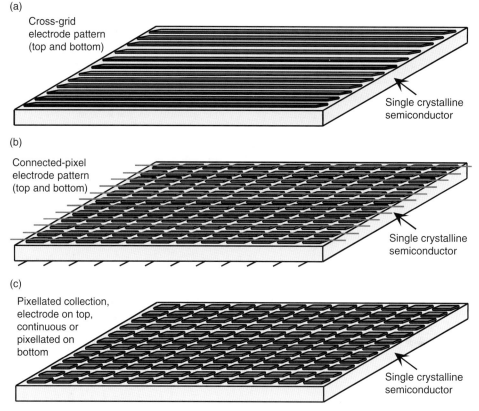

FIGURE 17 Common semiconductor imaging array designs include (a) cross-grid electrode, (b) connected-pixellated electrode, and (c) completely pixellated designs. For each, only the top collection electrode (anode) is shown.

and column electrode with the largest signals. The image pixel corre-sponding to that particular row/column intersection is then incremented. Due to variations in electric field, semiconductor crystal uniformity, and electronics, a gain balancing calibration using flood irradiation is typically performed using adjustable-gain amplifiers on each electronic channel.

The *discrete-pixel electrode design* (Figure 17b–c) uses a pixellated electrode pattern. The pattern can be on both sides (Patt *et al.*, 1995, 1997), or the collection side can be pixellated and the other side continuous (Matherson *et al.*, 1998). In the former case, for less electronic channels, the pixels in a particular row and column can be connected by fine wire bonds (Figure 17b middle), making it similar in concept to the cross-grid design. Advantages of connecting the pixels in this electrode configuration compared to the cross-grid design are improved electric field distribution for improved charge collection, reduced cross-talk, and lower leakage currents per resolution element (Patt *et al.*, 1995, 1997). Although square electrode elements are shown, other shapes have been used (Patt *et al.*, 1995). The processing and event positioning steps are similar to those described for the cross-grid design.

The completely pixellated discrete-electrode design (Figure 17c) has electrically isolated collection electrode pixels on one side and is either continuous or pixellated on the other. A drawback to this is an increased number of electronic channels, which increases complexity and cost of the readout. For very small pixel sizes, an ASIC preamplifier array is typically required. Short wire or bump bonds are needed between the crystal and preamplifier channels to limit the capacitive input to the preamplifier. The signals are processed in parallel similar to that depicted in Figure 14b; however, because a pixel hit is positioned by its *x-y* location in the array, only the electrode signals on the low voltage side of the array need to be processed. Due to the potentially large number of channels, a multiplexed digitizer significantly reduces the processing task (Barber *et al.*, 1994; Marks *et al.*, 1996; Matherson *et al.*, 1998). The event may be positioned by determining which hit pixel has the largest signal, similar in concept to positioning events in the discrete scintillator–discrete SPD design.

IV. REVIEW OF DESIGNS AND APPLICATIONS FOR SMALL FIELD-OF-VIEW IMAGERS

A. Small-FOV Gamma-Ray Imagers in Medicine

A majority of small imager development in the literature has been for imaging gamma rays (Scopinaro *et al.*, 1999; Pani, Pellegrini, *et al.*, 1997; Pani, Scopinaro, *et al.* 1997; Pani *et al.*, 1994, 1996, 1998, 1999; Weinberg *et al.*, 1996, 1997; Wojcik *et al.*, 1998; Patt, Iwanczyk, *et al.*, 1997; Patt, Tornai, *et al.*, 1997; Patt *et al.*, 1995, 1996, 1998; Tornai *et al.*, 1997; Levin, Hoffman, Tornai, *et al.*, 1997; Levin, Hoffman, Silverman, *et al.*, 1997; Levin *et al.*, 1996; Gruber *et al.*, 1998, 1999; Williams *et al.*, 1997, 1999, 2000; Soares *et al.*, 1998, 1999, 2000; Fiorini *et al.*, 1997, 1999; Fiorini, Longoni, *et al.*, 1999; MacDonald *et al.*, 1999; Kume *et al.*, 1986; Nagai and Hyodo, 1994; Aarsovold *et al.*, 1988, 1994; Matthews *et al.*, 1994; Antich *et al.*, 1992; Yasillo *et al.*, 1990; Milster *et al.*, 1984, 1990; Weisenberger *et al.*, 1998; de Notaristefani *et al.*, 1996; Daghighian *et al.*, 1994a, 1994b; Menard *et al.*, 1998; Tornai *et al.*, 1997; Butler *et al.*, 1993; Doty *et al.*, 1993; Barber *et al.*, 1994; Eskin *et al.*, 1999; Marks *et al.*, 1996; Barber *et al.*, 1984; Matherson *et al.*, 1998; Hartsough *et al.*, 1995). This is due to the prevalence of useful single-photon-emitting radiopharmaceuticals in nuclear medicine. A majority of this small camera work has been for breast cancer, the most common and second most deadly type of cancer found in women (California Cancer Researh Program, Sacramento, CA, 1998). There are two FDA-approved ^{99m}Tc-labeled tracers that have shown promise for breast cancer imaging (a.k.a. *scintimammography* or *mammoscintigraphy*) of single photons: ^{99m}Tc-sestamibi (MIBI) and methylene diphosphonate (MDP). Using these tracers, the breast tumor sensitivity values reported are comparable to and the specificity is roughly three to four times that for standard X-ray mammography (Maublant *et al.*, 1996; Khalkhali, Cutrone *et al.*, 1995; Khalkhali *et al.*, 1993, 1994, 1995; Baines *et al.*, 1986; Tabar *et al.*, 1982; Piccolo *et al.*, 1995; Lastoria *et al.*, 1994; Waxman *et al.*, 1994; Taillefer *et al.*, 1995; Villanueva-Meyer *et al.*, 1996). The false negative rates observed for breast cancer are only somewhat lower to that reported with biopsy/pathology (Hasselgren *et al.*, 1993; C. Kimme-Smith, UCLA Department of Radiology, pers. comm.).

1. Photomultiplier Tube Designs

A University of Arizona group (Milster *et al.*, 1984) has developed a small modular camera based on multiple PMTs. The module comprises a 10×10 cm^2 NaI(Tl) crystal, a light guide, four 5×5 cm^2 Hamamatsu square PMTs (see Figure 18a), and line driver–amplifier electronics. The main advantage of this design is for single-organ imaging applications where one may need to get closer to the body than possible with the standard Anger camera. Figure 18b shows that, due to this advantage, the modular camera produces higher-quality images than the standard Anger camera for the same imaging time.

Single-photon PSPMT camera designs for breast imaging have shown promising results (Scopinaro *et al.*, 1999; Pani, Pellegrini, *et al.*, 1997; Pani, Scopinaro, *et al.* 1997; Pani *et al.*, 1994, 1996, 1998, 1999; Weinberg *et al.*, 1996, 1997; Wojcik *et al.*, 1998; Levin, Hoffman, Tornai, *et al.*, 1997; Levin, Hoffman, Silverman, *et al.*, 1997; Williams

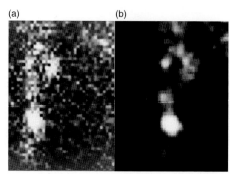

FIGURE 18 Left: Small "modular" camera consists of a 10 × 10 cm² NaI(Tl) crystal coupled through a light guide to 4 square PMTs. Right: Images of uptake of a ^{99m}Tc tracer in lymph glands in the upper thigh with a standard Anger camera (a) and the small modular camera (b). Due to improved positioning capabilities, the small module creates higher quality images than the standard Anger camera. (Courtesy of Brad Barber, University of Arizona, Tuscon, AZ).

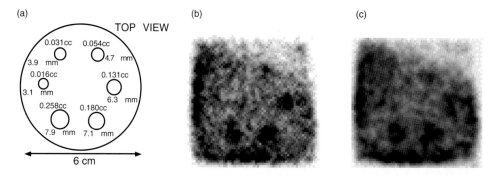

FIGURE 19 (a) Drawing of a tumor phantom with various sizes of spherical cavities that can be filled with ^{99m}TcO$_4$ and placed within a container filled with background activity. Ten-minute image acquisition with a 5:1 tumor-to-background concentration ratio with the filled spheres (b) 1 cm and (c) 2 cm deep within the warm background and scatter medium. The 4.7 mm (0.05 cc) lesion can just be resolved at both depths (Levin, Hoffman, Silverman, et al., 1997).

et al., 1997, 1999, 2000; Soares et al., 1998, 1999, 2000). In order to systematically study the limitations of tumor detectability, phantom imaging is commonly used. Figure 19 shows tumor phantom imaging results for a continuous NaI(Tl) crystal–single PSPMT design (Hamamatsu R3941, 7.6 × 7.6 cm²) that incorporated a 2-cm-thick LEHR collimator (Levin, Hoffman, Tornai, et al., 1997; Levin, Hoffman, Silverman, et al., 1997). The system had sufficient energy resolution (extrinsic, 11% FWHM at 140 keV), spatial resolution (extrinsic, 3.4 mm FWHM at 1 cm from face), and sensitivity (50 cts/s/μCi) to resolve lesions as small as 6 mm in diameter, or 0.1 cc in volume, for a 5:1 tumor to background uptake-to-concentration ratio, at a depth of up to 3–4 cm in a 10-minute acquisition time. For both shallower tumor depths and higher uptake ratios, the lesion resolvability improved. Tumor phantom study measurements (Levin, Hoffman, Silverman, et al., 1997) with this camera determined that the largest factors affecting tumor detectability using a small FOV imager are, in order of importance: (1) the tumor size, due to spatial resolution effects; (2) the depth into breast tissue, due to spatial resolution limitations (collimator blurring with source distance), photon attenuation in tissue, and contrast effects (breast tissue activity dominates measured activity); and (3) the energy window of acceptance, due to the presence of significant Compton scatter background. Secondary factors that affect detectability are uniformity correction, count time, tumor-to-tissue uptake ratio, and presence of background activity.

Figure 20 shows the design and results from a University of Rome group that has developed a single-photon camera for breast cancer imaging using ^{99m}Tc-MIBI (Scopinaro et al., 1999; Pani, Pellegrini, et al., 1997; Pani, Scopinaro, et al. 1997; Pani et al., 1994, 1996, 1998, 1999) and successfully implemented it in the clinic. Their design comprises a discrete array of 2 × 2 × 3 mm³ CsI(Tl) crystals with 0.25 mm spacing coupled to a 12.7-cm round cross-wire anode PSPMT (Hamamatsu R3292) with resistive charge division readout of the 28x and 28y cross-wire anodes. The general purpose collimator used has 1.5-mm-diameter and 2.5-cm-long hexagonal holes with 0.25-mm septa. They report an intrinsic spatial resolution between 1.6 and 1.8 mm FWHM, and 2.7 and 4.0 mm FWHM extrinsic spatial resolutions at 0 and 2–3 mm source to collimator distances, respectively, for a ^{99m}Tc point source. The energy resolution reported is < 19% FWHM at 140 keV, and the efficiency (excluding collimator) is 25% including the spaces between detectors (18%), the lower intrinsic detection efficiency

FIGURE 20 (a, b) Clinical system comprising the CsI(Tl)-PSPMT (12.7-cm-diameter FOV) camera detector with electronics, a support arm with counter balance and a data acquisition system. (c) The 11-cm-diameter CsI(Tl) array consists of 2300 crystal elements each $2 \times 2 \times 3$ mm^3 separated by 0.25 mm of white epoxy. (d) The camera is placed on top of a plastic paddle that allows mild breast compression. (e) Both the small camera and (f) a commercially available Anger camera can see a 13-mm-size lesion. A 7-mm lesion is not well resolved with the small camera (g) without compression but is (h) with mild compression. (i) Prone imaging with a standard scintillation camera cannot resolve the same lesion. (Courtesy of Roberto Pani, University of Rome, Italy.)

(60%) of 3-mm-thick CsI(Tl) for 140-keV photons, and a 27% asymmetric energy window. The collimated system sensitivity is 16 cpm/kBq low energy general purpose (LEGP). The camera is shielded by 1 cm lead and by 5 mm tungsten on the side facing the patient. The overall dead zone between the chest wall and camera FOV is 2 cm.

The group has observed that their small camera can adequately see malignant breast lesions that are <1 cm in diameter, which cannot typically be seen for prone imaging with a standard scintillation camera, due mainly to its inadequate positioning capability. With their system specifically designed for breast imaging (they prefer the terminology *single-photon emission mammography*, SPEM), the source–collimator distance is less than 5 cm and the scintillation detector (rather than the collimator) limits the spatial resolution. Furthermore, the group has shown the benefits of increased sensitivity by using mild breast compression during imaging. These results are demonstrated in Figure 20g–i. Their clinical studies report a sensitivity of 80% for ≤1-cm-diameter lesions with mild breast compression and attributed the improvement over the standard camera to the high spatial resolution and close camera–tumor distance.

A group at Thomas Jefferson National Laboratory has developed a small, high-resolution, single-photon scintillation camera for small animal research (Weisenberger et al., 1998). The camera is also based on a 12.7-cm-diameter PSPMT (Hamamatsu R3292-02) and uses an 11-cm-diameter array of discrete $1 \times 1 \times 3$ mm^3 CsI(Na) crystals separated by 0.2 mm white epoxy, and a custom high-resolution copper collimator (0.2-mm square holes, 0.05-mm septa, 5-mm thick; Weisenberger et al., 1998; Thermo Electron Tecomet, Woburn, MA). The PSPMT has 28×28 crossed anode wires. Connecting anode wires in small subgroups of two wires reduces the number of readout channels for amplification and digitization. No loss in spatial resolution was observed with this coarser sampling of the light distribution due to an improved SNR per wire subgroup. A mapping LUT extracted from a flood field image defines the crystal elements. The output of each crystal is treated individually to generate a gain map that corrects for crystal–crystal scintillation output variations as well as local PSPMT gain variations. For a 1.5-cm source-to-collimator distance, the measured spatial resolution is 1.5 mm FWHM; the geometric acceptance is approximately 0.008%, including dead space between crystals; and the point source sensitivity is 68/cpm/μCi.

Figure 21 shows imaging results of the *in vivo* uptake of the ^{125}I-labeled tracer methyl-3-4-iodophenyl-tropane-2-carboxic acid methyl ester (RTI-55) in mice. This compound is a cocaine analog used to dynamically study brain dopamine and seratonin transporters in mice. Compounds bind to these transporters, which are protein receptor molecules that facilitate the movement of these compounds across the cell membranes in the brain. At later imaging times, preferential accumulation in the brain was observed.

IntraMedical Imaging, LLC (IntraMedical Imaging, LLC, Santa Monica, CA) has developed a small battery-operated gamma camera (Figure 22a) for surgical use (Fig. 22b–c) in conjunction with ^{99m}Tc-, ^{123}I-, and ^{111}In-labeled compounds. The device comprises a continuous NaI(Tl) crystal coupled

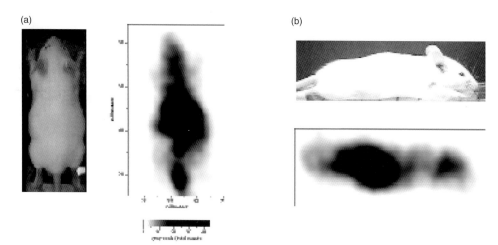

FIGURE 21 Images of the biodistribution of RTI-55 in a 25-gm male CD-1 mouse. (a) Coronal view taken with 40-minute acquisition time started 30 minutes after a tail vein injection of ~15 mCi of ^{125}I-labeled RTI-55. (b) Sagittal view with 45-minute acquisition time started 80 minutes after the same tail vein injection. (Courtesy of Andrew Weisenberger, Jefferson National Lab, Newport News, VA.)

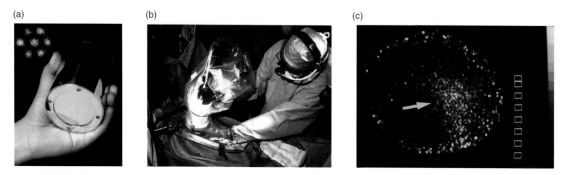

FIGURE 22 (a) IntraMedical small intraoperative gamma camera. The unit has a 9-cm diameter, and 10-cm length, weighs 1.5–2.0 kg depending on the collimator, and is battery operated. Inset in the picture is the image of a 2-mm spot pattern of ^{99m}Tc. (b) The camera being used inside a colorectal cancer patient's abdominal cavity. The tracer used is an ^{125}I-labeled monoclonal antibody (Mab)-A33, injected 1 week prior to surgery. (c) Portal lymph node imaged for 5 minutes showing high uptake in the colorectal cancer patient. (Courtesy of Farhad Daghighian, Intra-Medical Imaging LLC, Santa Monica, CA.)

to a 7.6-cm-FOV cross-wire anode PSPMT (Hamamatsu R2486) with approximately 6 mm of surrounding lead shielding and a 5.5-cm-diameter useful FOV. The spatial resolution quoted is 3.5 mm FWHM and the sensitivity is 1500 cpm/µCi for a collimated ^{99m}Tc point source.

The typical disadvantages of the single-tube PSPMT camera designs are poor photocathode uniformity, which causes large variations in gain and sensitivity over the FOV; significant charge spread in the dynode stages, which limits the intrinsic spatial resolution; relatively large dead area near the edges, which reduces camera sensitivity and usefulness; and relatively large thickness, which limits the compactness.

An improvement to the single-tube PSPMT small imager design is possible with the use of new-generation metal channel dynode PSPMTs. The metal channel dynode removes lateral charge diffusion in the charge amplification stages for higher spatial resolution performance with a cross-strip or multianode readout (Pani *et al.*, 1999; Hamamatsu Photonics, 1995). Because these devices are only 2.5 cm across, they are grouped together for position sensitivity over a relatively large area, similar to the conventional Anger Camera except that each tube is position sensitive. The tube packaging can be modified to further decrease the dead area between closely packed units. With this configuration, it is possible to read out the narrow light distributions from a fine array of discrete scintillation crystals for high spatial resolution and linearity. This design was incorporated into a small-FOV camera (Photon Imaging, Inc., Northridge, CA) that has become commercially available (shown in Figure 23).

This camera has intrinsic and extrinsic (2 cm from standard LEAP collimator) spatial resolutions of <2 mm and 3.4 mm FWHM, respectively, at 140 keV (both NaI(Tl) and CsI(Na) have been used). System sensitivities are 192, 266, and 488 cpm/µCi, respectively, for the LEHR, LEAP, and high-sensitivity collimators, respectively, comparable to corresponding values measured in large-FOV standard scintillation cameras. The best crystal energy resolution is 10% FWHM at 140 keV (median 15% for >3000 crystals). The camera is highly compact with less than 1 cm dead area at the FOV edge and is less than 6 cm thick, excluding collimator, allowing close placement of the camera and easy maneuverability for most imaging views. It can easily be mounted onto a clinical mammography unit. The device is currently undergoing clinical breast imaging trials at the UCLA Medical Center. The camera, of course, may be also used in applications where a small, high resolution imaging FOV is needed, such as in parathyroid or bone imaging, and in intraoperative imaging applications including pancreatic, gastrointestinal, lung, and lymph nodes. For the latter application, a 2.5-cm-FOV, hand-held version (shown in Figure 24), with intrinsic spatial resolution between 1.0 and 1.5 mm FWHM, is also available (Photon Imaging, Inc., Northridge, CA).

A University of Virginia and Jefferson National Laboratory collaboration (Williams *et al.*, 2000) have developed a small-FOV single-photon breast imager based on a discrete CsI(Na) crystal array coupled to an array of compact PSPMTs (Hamamatsu R7600-C8; similar to the

FIGURE 24 A 2.5-cm-FOV imager designed for imaging of sentinel lymph node involvement. The device is compatible with a variety of easily exchangeable collimators. (Courtesy of Photon Imaging, Inc., Northridge, CA.)

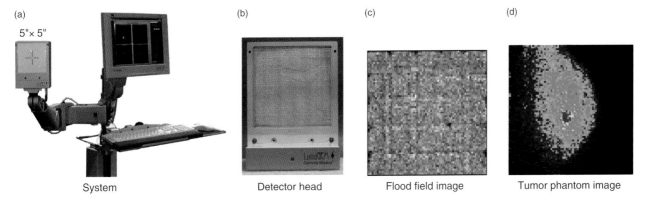

FIGURE 23 (a) A mobile scintillation camera shown mounted on an articulating arm with 5 degrees of freedom. (b) Camera head shown with the scintillation crystal array exposed. (c) Intrinsic flood field uniformity image (σ = 2.7%) revealing individual PSPMT pattern. (d) Tc-99m breast phantom image with 15-mm-diameter (2.5-cc) lesion 5 cm from the low energy all-purpose (LEAP) collimator with a lesion-to-background activity concentration ratio of 10:1. (Courtesy of Lawrence MacDonald, Photon Imaging, Inc., Northridge, CA.)

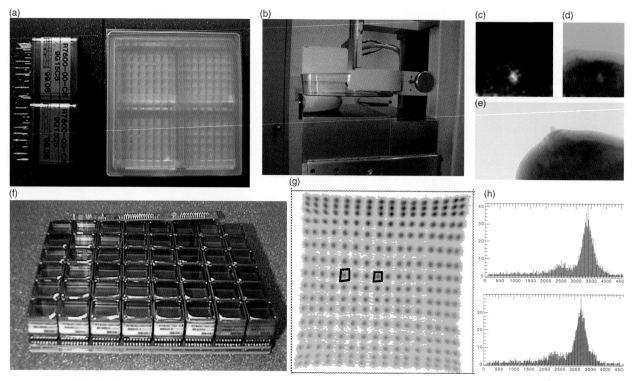

FIGURE 25 (a) Side view of compact metal channel dynode PSPMTs and top view of CsI(Na) pixellated crystal array module coupled to four PMTs. Each discrete crystal is $3 \times 3 \times 6$ mm^3 CsI(Na). The small camera comprises four such modules with a 10×10 cm FOV. (b) Incorporation of camera into digital mammography unit. (c) Mammoscintigraphy with Tc-99m sestamibi indicating a high focal uptake of a breast tumor. (d) Co-registered image of digital mammogram and mammoscintigraphy showing combined structural and metabolic activity images. (e) Corresponding digital mammogram image. (f) Picture of 8×6 array of compact PSPMTs for larger-FOV camera. (g) Flood field image of a detector module consisting of the 16×16 array submodule with $3.5 \times 3.5 \times 6$ mm^3 NaI(Tl) pixels coupled to 2×2 PSPMTs. (h) Energy spectra extracted from two crystal pixels in the flood image. (Courtesy of Stanley Majewski, Jefferson National Laboratory, Newport News, VA, and Mark Williams, University of Virginia, Charlottesville, VA.)

R5900 except with less housing material). The camera uses a LEHR collimator and surrounding shielding made of tungsten. The camera is incorporated into a digital X-ray mammography unit developed at Brandeis University to perform dual modality functional–structural imaging. Figure 25(a–e) shows the basic design of a detector module, the camera incorporated into the mammography unit, and an imaging result from their clinical trials being performed at the University of Virginia Medical Center. The system spatial resolution with breast compression is 5 mm FWHM using a high-efficiency collimator.

The Jefferson Lab group is also developing a second larger unit comprising a pixellated NaI(Tl) array (for more light) coupled to 48 compact PSPMTs. Figure 25f–h shows a picture of the PSMT array, a flood-field image of one of the detector submodules, and extracted individual pixel energy spectra. To reduce the number of electronic channels required, the PSPMT array is read out by interconnecting the anode wires from each of the 48 tubes and forming a set of $32x$ by $24y$ wires as inputs to the centroiding readout system for event positioning. The measured sensitivity was 2.8 counts/s/μCi with a 2.5-cm-thick LEHR hexagonal-hole collimator. A specially designed etched tungsten collimator has been developed for this system (see Figure 9). The system spatial resolution for a line source varied from 3.5 to 8.0 mm FWHM for source distances from 2.5 to 10 cm. The best energy resolution was 17% FWHM at 122 keV.

The University of Virginia–Jefferson Laboratory collaboration has adopted a similar approach for dual-modality animal imaging (Williams *et al.*, 1999). Figure 26 shows fused X-ray transmission and gamma ray emission images of a rat obtained during a study of bone growth resulting from gene therapy using a bone morphogenetic protein (BMP). The aim the study was to determine the fine structural characteristics of ectopic bone formation induced by gene therapy using a replication-defective BMP adenoviral vector. The vector containing the therapeutic gene sequence was injected directly into the muscle cells, resulting in the *in vivo* production of BMPs at the injection site. The figure show the same animal, 12 and 24 days postinjection. Rats were injected with an average of 310 μCi of ^{99m}Tc methylene diphosphonate (MDP) in the tail vein 3 hours prior to

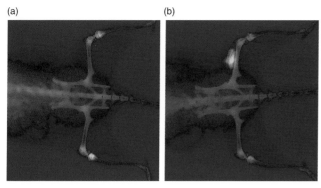

FIGURE 26 Dual modality MDP study (bone scan) overlaid on digital radiograph of a rat at (a) 12 days and (b) 24 days postinjection of BMP. (Courtesy of Mark Williams, University of Virginia.)

imaging. Following acquisition of the digital radiograph, gamma-ray images were acquired for 45 minutes, using a 50% window centered on the ^{99m}Tc 140-keV photopeak. Each image of the pelvis was taken anteroposterior to the gamma camera, using a 64 × 64 matrix, with a pixel size of 2.25 mm. The X-ray and gamma-ray images were scaled and superimposed based on a set of stored co-registration parameters. Intensity levels in the fused images are based on the products of the corresponding pixel values of each image, following histogram-based window and level adjustment. Thus appreciable pixel values in the fused image correspond to regions of both high X-ray attenuation and high MDP retention.

A group at University College London (Soares et al., 1998, 1999, 2000) has developed an unusual prototype gamma-ray imager for breast imaging that uses a continuous slab of scintillation crystal coupled to sets of wavelength shifting fibers (WSFs). Figure 27 shows their design and an imaging result. The camera consists of a single 12.5 × 12.5 × 0.3 cm^3 CsI(Na) crystal coupled with silicon optical compound to two orthogonal ribbons of 128, 1-mm-diameter, blue-to-green WSFs on opposite faces of the crystal to provide 2D position information. The position of gamma-ray interaction is determined by measuring the profiles of the distribution of light created by the WSFs on either crystal face as a result of absorption of the CsI(Na) light. The WSF light from the 16 largest fiber signals per ribbon is read out by multianode PMTs (Hamamatsu R5900-M16) and the position of the gamma-ray interaction in each direction is estimated by a centroid calculation. The other ends of the WSF ribbons were aluminized to improve light collection. The energy signal is obtained by a 2 × 2 array of square PMTs coupled to one face of the crystal through a fiber ribbon and a 5-mm-thick light diffuser, as seen in the figure. The group reported an energy resolution of 23% FWHM at 122 keV and an intrinsic spatial resolution of 4 mm FWHM. Improvements are expected with the use of NaI(Tl) (more light) and the use of avalanche photodetectors (higher quantum efficiency). Currently, the positioning is achieved with an average of less than one photoelectron detected in each WSF. The intrinsic detector efficiency is nearly 70%.

2. Semiconductor Detector Designs

A few groups have made progress toward the development of silicon-based semiconductor photodetectors for small scintillation imagers (Patt, Iwanczyk, et al., 1997; Patt et al., 1998; Gruber et al., 1998, 1999; Fiorini et al., 1997, 1999; Fiorini, Longoni, et al., 1999; MacDonald et al., 1999) in medicine. Two approaches have yielded encouraging results. The first is the use of silicon PIN PDs and the second uses the more sophisticated SDD design. Photon Imaging, Inc. (Northridge, CA), has developed a monolithic 8 × 8 array of PD pixels, each with active area 1.5 × 1.5 mm^2 (0.25 mm spacing) coupled to a matched array of discrete 1.5 × 1.5 × 6 mm CsI(Tl) crystals forming a completely pixellated detector system (Figure 28a). Short wire bonds couple a custom ASIC low-noise charge-sensitive preamplifier chip (Patt, Iwanczyk, et al., 1997; Patt et al., 1998) to the PD array. The group measured a pixel energy resolution of < 8% FWHM at 140 keV for all pixels with an electronic noise of 41 e$^-$ root mean square (rms) (3% contribution to the photopeak). A single-pixel

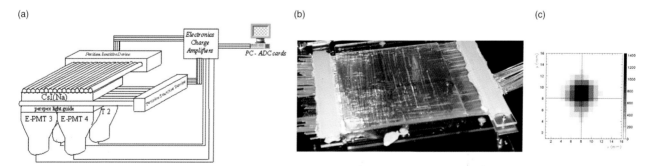

FIGURE 27 (a, b) Small prototype CsI(Na)-WSF imager consists of a 12.5 × 12.5 × 0.3 cm^3 continuous crystal, two sets of orthogonal WSF ribbons coupled to multianode PMTs for positioning, and four square PMTs shown coupled to one crystal face for energy measurement. (c) Image of a collimated ^{57}Co point source yields intrinsic spatial resolution of 4 mm FWHM (Courtesy of Antonio Soares, University College London.)

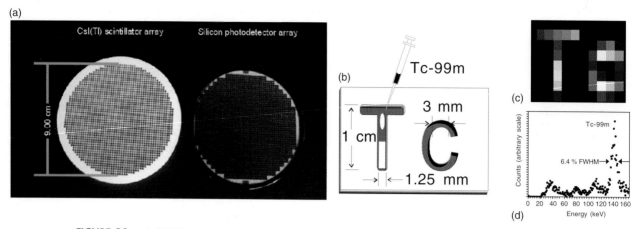

FIGURE 28 (a) CsI(Tl) array coupled to a monolithic array of silicon PIN photodiodes. (b,c) Transmission phantom imaging result using a portion of the array and (d) a single-pixel energy spectrum demonstrate high spatial and energy resolutions. (Courtesy of Larry MacDonald Photon Imaging Inc., Northridge, CA.)

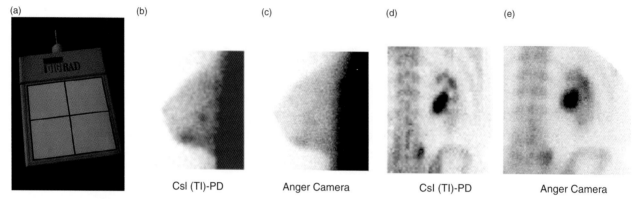

FIGURE 29 (a) Digirad Corporation's CsI(Tl)-PIN photodiode camera head measures $22 \times 22 \times 0.7$ cm^3 with only a 12-mm surrounding dead area (standard Anger camera typically has a ~75-mm surrounding dead space). (b,c) Breast and (d,e) lower back images taken with their CsI(Tl)-PD camera are of superior quality compared to the standard Anger camera. (Courtesy of Richard Conwell, Digirad Corporation, San Diego, CA.)

energy spectrum is shown in Figure 28d (6.4% FWHM at 140 keV). Pixel uniformity measured with a ^{99m}Tc flood source was 4.3% (standard deviation) for a 10% energy window. Initial imaging results (Figure 28c) using a transmission phantom (Figure 28b) demonstrate intrinsic spatial resolution of 1.53 mm FWHM as dictated by the crystal pixel size.

Digirad Corporation (San Diego, CA) developed a device based on a similar PD design. The 22×22 cm^2 FOV camera (see Figure 29a) comprises small submodules of $3 \times 3 \times 6$ mm^3 CsI(Tl) crystal arrays coupled to matching PD arrays. The better intrinsic spatial and energy resolutions and the closer proximity to the patient made possible with this compact camera allows superior quality imaging compared to the standard Anger scintillation camera (Figure 29b–e). The pixel size in this design must be kept small for a high SNR, which means a large number of small pixels and a relatively expensive and complex readout to cover a significant imaging FOV. A previous version of the camera incorporated submodules containing 3×3 mm^2 pixels of CdZnTe.

SDDs have a much better SNR for a given area than the PDs. In fact it is possible to manufacture a scintillation camera using a relatively sparse array of SDDs coupled to a continuous scintillation crystal in a light-sharing readout scheme similar to that used with PMTs in the standard scintillation camera. Figure 30 shows such a design investigated at Photon Imaging, Inc. (Northridge, CA).

A Politecnico di Milano group (C. Fiorini, et al., 1997, 1999, 1999) has assembled a prototype of another SDD design, comprising a monolithic array of seven hexagonal (2.4-mm-diameter), closely packed SDD pixels (total area 35 mm^2) (Figure 31a) coupled to a 7-mm-diameter, 1.3-mm-thick CsI(Tl) single crystal (Figure 31b). The output capacitance of each SDD pixel is on the order of 0.1 pF. Each SDD has on-chip, front-end JFET for low noise readout. The electronic noise of all seven units was measured to be approximately ~30 e$^-$ rms at 0°C. The measured absolute gain of the imager using a ^{57}Co source was roughly 13 e$^-$/keV. Gamma-ray interactions in the crystal are positioned with a centroid calculation using the seven SDD signals.

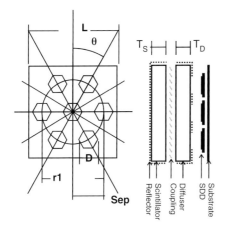 

FIGURE 30 Design of a silicon drift detector–based scintillation camera. The design comprises seven detectors, each ~1 cm in diameter, with a separation of twice the diameter for a 30-cm FOV. (Courtesy of Lawrence MacDonald, Photon Imaging, Inc., Northridge, CA.)

Intrinsic spatial resolution on the order of 0.6 mm FWHM and energy resolution of 16% was measured using a collimated beam of 122-keV photons moved across the border between two SDD pixels (Figure 31c). Imaging results (Figure 31e–f) using a transmission phantom (Figure 31d) indicate the small device has the capability of resolving fine structures with good linearity.

A University of Arizona group (Matherson *et al.*, 1998) has developed a prototype small-imaging system comprising four 64 × 64 element CdZnTe detector arrays, each mated to a custom multiplexer readout circuit for ultrahigh spatial resolution nuclear medicine imaging applications (Figure 32a; see also Chapter 15). The gamma-ray absorption occurs directly in the CdZnTe elements for high SNR. Each detector array module consists of a 2.5 × 2.5 cm² area, 0.15-cm-thick slab of CdZnTe patterned using photolithography into a 64 × 64 array of 380-μm-square-pixel gold electrodes on one side and a continuous gold electrode biased at –150 V on the other (Figure 32b). This pixel size was chosen for optimal charge carrier collection (Eskin *et al.*, 1999). Each pixel of the detector array is electrically connected to a gated integrator in the readout circuit by an indium-bump bond and multiplexed to a single output line, which is read out in a raster scan. The arrays are mounted on ceramic carriers and may be inserted into a heat exchanger for modest cooling down to roughly 7°C to reduce the leakage current and hence improve noise performance (see Chapter 15).

The four arrays that make up an imaging module can be operated in parallel and image data are acquired by a digital signal processing system. Each array has typically greater than 90% functioning pixels and detects individual gamma interactions with excellent energy resolution and a spatial resolution of 380 μm or better. An image of the 2D Hoffman brain phantom obtained with one 64 × 64 CdZnTe array coupled to a parallel-hole collimator with a 1-mm pitch is shown in Figure 32c. The phantom was filled with 110 mCi of ^{99m}Tc. Because of the small FOV of a single array, the array and collimator were mechanically stepped to 42 different positions to cover the phantom and imaged for 3 minutes per position. The entire image consists of 19 million counts. A custom high-resolution parallel-hole tungsten collimator that matched the pitch of the detector array was also fabricated (Matherson *et al.*, 1998; Thermo Electron Tecomet, Woburn, MA). The collimator is 7 mm thick, with a bore size of 260 μm, placed in a 64 × 64 square array on a 380-μm array pitch. A procedure was developed to carefully align the collimator on top of the detector array. Figure 32d also shows a prone position image taken of a 20-g mouse 3 hours after a tail-vein injection of 15 mCi of ^{99m}Tc-MDP. The image was formed with a single CdZnTe array coupled to the ultrahigh resolution collimator and seven 5-minute images were taken by longitudinally shifting the mouse. The final image shown on the right is a mosaic of all images. Each image in the mosaic has 30,000–50,000 counts.

B. Small-FOV Coincidence Imagers in Medicine

There are several groups developing small-FOV coincidence imagers for breast tumor and axillary lymph node imaging in conjunction with FDG (Thompson *et al.*, 1994, 1995; Robar *et al.*, 1997; Bergman *et al.*, 1998; Raylman *et al.*, 1999; Zavarzin *et al.*, 1999; Hutchins and Simon, 1995; Moses *et al.*, 1995; Freifelder and Karp, 1997). A group at the Montreal Neurological Institute (Thompson *et al.*, 1994, 1995; Robar *et al.*, 1997; Bergman *et al.*, 1998) was among the first to develop such a camera (positron-emission mammography, PEM) and implement it into the clinic. The imager comprises two detector heads in electronic coincidence, each consisting of four crystal modules coupled to a 76 × 76 mm² PSPMT (Hamamatsu R3941: mesh

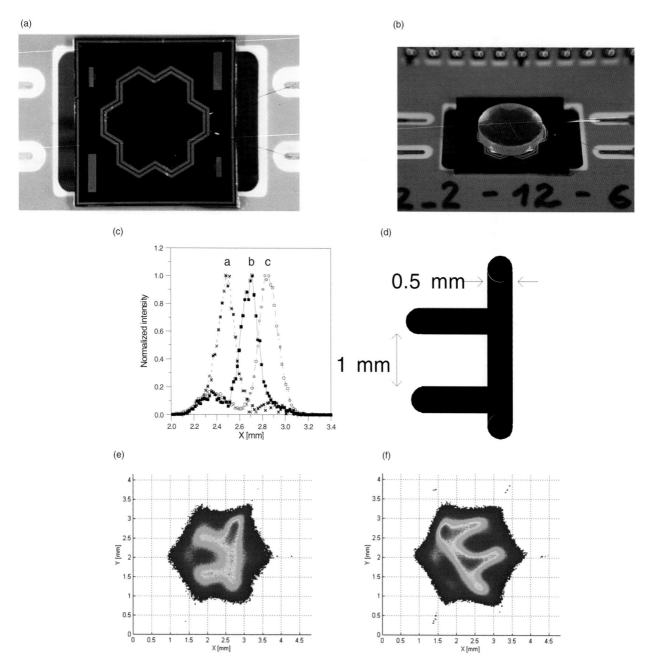

FIGURE 31 (a) Seven-element SDD array with 1.2-mm-diameter hexagonal pixels is coupled to (b) a 7-mm-diameter 1.3-mm-thick CsI(Tl) crystal for a prototype imager. (c) Intrinsic spatial resolution measurements yield 0.6 mm FWHM. (d) High-resolution transmission phantom measurements demonstrate (e,f) the imaging capabilities of the camera. (Courtesy of Carlo Fiorini and Antonio Longoni, Politecnico di Milano, Milan, Italy.)

dynode, cross-wire anode). Each crystal array comprises 2.0×2.0 mm^2 pseudodiscrete BGO crystals cut from a solid block with 0.25-mm-thick saw blades (see Figure 33a). The cutting pattern into the crystal is performed from both top and bottom faces with the two patterns offset by 1.0 mm to double the spatial sampling and improve the spatial resolution. The cuts are 7 mm deep on the top face (patient side) and 11 mm deep on the bottom face (PMT side), with a 2-mm gap between layers. The longer crystal elements further from the patient ensure equal probability for 511-keV photon detection in each crystal layer. Due to the offset between the top and bottom crystal patterns, scintillation light piped down a given crystal in the top layer is spread in the 2-mm gap between layers and piped down four (or more) crystals in the bottom layer before entering the PSPMT. This could lead to worse energy and spatial reso-

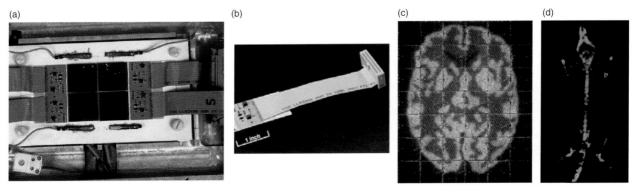

FIGURE 32 (a) Small CdZnTe imager comprises four 2.5×2.5 cm^2, 64×64 pixel arrays coupled to (b) a multiplexed readout. (c) High-resolution Hoffman brain phantom image and (d) ^{99m}Tc-MDP image of a 20-g mouse skeleton are formed using one array coupled to various high-resolution collimators. The array is translated through multiple positions to image the entire object. (Courtesy of Brad Barber, University of Arizona.)

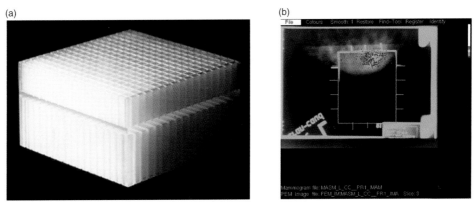

FIGURE 33 (a) A $36 \times 36 \times 20$ mm^3 BGO block cut into a 2×2 mm^2 crystal pattern on both faces for the PEM detector. The cuts are 7 mm deep into the top face (patient side) and 11 mm into the bottom face (PMT side) with a 2-mm gap. Four such modules mounted on a PSPMT make one detector head. (b) PEM image of a region of high focal uptake overlaid onto the mammogram. A rectangular wire grid facilitates the alignment of the two images. (Courtesy of Chris Thompson, Montreal Neurological Institute, Montreal, Canada.)

lutions for top layer (shallower) interactions. The intrinsic coincidence spatial resolution of the system is roughly 2.5 mm FWHM.

The two planar PEM detectors are incorporated into a conventional mammography system. The detectors are mounted on a sliding mechanism allowing them to pan over the breast surface. One detector is positioned above the compression paddle and the other directly below the compressed breast.

There are seven focal plane images formed by back-projecting measured LOR data onto seven parallel planes through the breast volume. The image plane thickness depends on the breast compression (the thickness of the breast between the plates). The lesion will appear most focused in the particular image plane corresponding to the true lesion depth within the tissue. Different efficiency LUTs are used to scale the image pixels based on the solid angle subtended by each pixel position in each image plane. An attenuation correction is calculated in real time using a μ value of 0.098 cm^{-1} and assuming that the breast fills the entire volume between detectors. Because there are two crystal layers per detector head and they are offset with respect to the PSPMT face, a depth of interaction correction can be performed that reduces the blurring associated with positioning oblique lines of response. A wire grid is used to center the small camera FOV over the suspected lesion seen in the mammogram and to scale the resulting PEM image appropriately so that the X-ray and emission images can be overlaid (Figure 33b). The point source sensitivity in air varies from 4.5% (1.7 kcts/s/μCi) to 1.0% (0.4 kcts/s/μCi) for detector separation (compression) of 4 cm to 14 cm for all energies. The spatial resolution is 2.5 and 2.0 mm FWHM when all and only normal lines of response, respectively, are accepted. The energy resolution per BGO crystal is on average 34% at 511 keV, with a range of 30–38%. Only a 75-MBq dose of FDG is required and imaging time is only 2–5 minutes.

Naviscan PET Systems, Inc. (Bethesda, MD), and the Jefferson National Laboratory group in collaboration with West Virginia University (WVU) (Raylman et al., 1999)

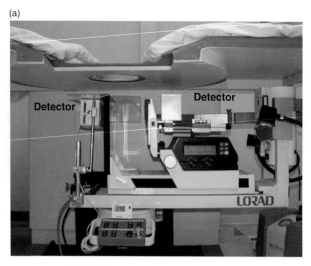
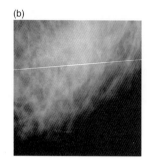
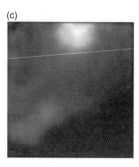

FIGURE 34 (a) PEM detector head incorporated into a LORAD stereotactic biopsy unit showing the importance of detector compactness. (Courtesy of Stanley Majewski, Jefferson National Laboratory, Newport News VA.) (b) Digital mammogram (included with the LORAD biopsy unit) showing a subtle breast lesion and (c) registered PEM image (4 minutes) indicating a region of high focal uptake. (Courtesy of Irving Weinberg, Naviscan, Bethesda, MD.)

have also independently produced similar coincidence imaging systems. Both groups incorporate their camera into a LORAD (Danbury, CT) stereotactic biopsy unit with digital mammography capabilities, as shown in Figure 34. Both groups use a discrete array of 3 × 3 mm GSO crystals coupled to a 4 × 4 array of compact metal channel dynode PSPMTs. Both groups organize the data in a way similar to the Montreal PEM camera, but their designs do not have depth of interaction determination. By limiting the LOR acceptance angles to small angles with respect to the normal to the detector plane, one can preserve spatial resolution uniformity within the FOV with a corresponding loss in sensitivity. In addition to using focal plane tomography for image formation, Naviscan has implemented an iterative limited angle tomographic reconstruction and has observed an improvement in lesion-to-background contrast with this algorithm compared to the focal plane method. Both groups quote a point source spatial resolution on the order of 3 mm at the FOV center. Naviscan quote a ^{18}F point source sensitivity of 1.0 kcts/s/μCi with a 300–600 keV energy window, and an energy resolution of < 20% FWHM at 511 keV for all crystals. The Jefferson group quotes similar sensitivity with an energy resolution of 16% FWHM at 511 keV.

A National Institutes of Health (NIH) group (Green *et al.*, 1994; Siegel *et al.*, 1999; Johnson *et al.*, 1997) has developed a small-FOV coincidence imager for animal research. They use an array of 26 × 22 discrete 2 × 2 × 10 mm^3 BGO crystals coupled to a single-tube cross-wire anode PSPMT (Hamamatsu R3941) with combined subgroups of two and three wires and use a resistive charge division readout (8X and 8Y). In addition to planar projection imaging, the camera has tomographic capabilities through the use of a

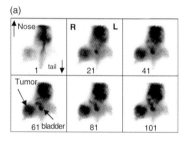
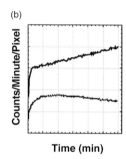

FIGURE 35 (a) Selected frames and (b) time activity curves (tumor and background) from a dynamic imaging study of a mouse taken with a small coincidence imager using ^{18}F-fluoromisonidazole. The high spatial resolution of the device allows one to observe characteristic uptake in small regions in the mouse. (Courtesy of Jurgen Seidel, National Institutes of Health, Bethesda MD.)

mechanically rotatable animal mount between the two detectors. Their quoted reconstructed spatial resolution (ramp filter) is 2.1 mm FWHM, the central point source sensitivity is 130 cts/s/μCi, and the coincidence time resolution is 6 ns FWHM. Figure 35 shows selected frames from dynamic planar projection images and an extracted time activity curve of ^{18}F-fluoromisonidazole, a hypoxic cell label, by an tumor xenograft implanted in the flank of an anesthetized 25-g mouse. The images demonstrate that the device is capable of detecting a heterogeneous distribution of tracer in the 1.0- to 1.5-cm-diameter tumors grown in this model.

C. Small-FOV Beta Imagers in Medicine

There are a few groups that have pursued small beta imager development (Daghighian *et al.*, 1995; Ljunggren and Strand, 1988, 1990, 1994; Ljunggren *et al.*, 1993,

1995; MacDonald et al., 1995a, 1995b; Levin et al., 1996, 1997; Tornai, MacDonald, Levin, Siegel, and Hoffman, 1996; Tornai, MacDonald, Levin, Siegel, Hoffman et al., 1996; Tornai et al., 1997, 1998; Hoffman et al., 1997a, 1997b, 1998; Yamamoto et al., 1997, 1999; Lees et al., 1997, 1998, 1999; Overdick et al., 1997;). As discussed in Section I, directly imaging betas has the advantages of high spatial resolution and sensitivity due to the fundamental confined nature of beta interactions in matter. However, many of the available beta-emitting tracers are positron emitters. Using these tracers, there may be annihilation photon background present.

IntraMedical Imaging, LLC (Santa Monica, CA) has developed a flexible beta imager for intraoperative imaging of tracers labeled with beta-emitting isotopes such as ^{131}I, ^{18}F, ^{11}C, ^{124}I, ^{90}Y, and ^{186}Re. The detector comprises a bundle of 1-mm plastic scintillating fibers coupled to a matching 1-meter-long bundle of clear flexible fibers coupled to a compact PSPMT at the other end (see Figure 36). The long flexible fiber bundle keeps the front end compact and isolates the patient from the PMT electronics, but also degrades the light signal measured from the scintillators. The device has a 1.5 × 2 cm^2 FOV, an intrinsic spatial resolution of less than 1 mm FWHM, and a sensitivity of 4 kcts/s/μCi. An image of a beta-emission spot phantom is shown in Figure 36b demonstrates the high-resolution imaging capabilities of the small beta camera.

A UCLA group also developed a prototype beta camera with fiber optic readout using low density and low effective Z materials such as CaF$_2$(Eu) as the scintillator (MacDonald

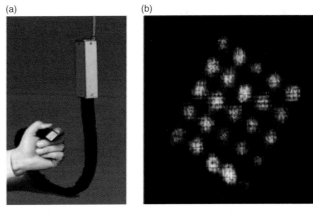

FIGURE 36 (a) Flexible intraoperative beta imager with 1.5 × 2 cm^2 FOV. (b) High-resolution image of ^{131}I beta-emission spot phantom. The Lucite emission phantom had 1-mm-diameter cavities spaced by 3 mm each filled with ^{131}I and covered with tape. (Courtesy of Farhad Daghighian, IntraMedical Imaging, Santa Monica, CA.)

et al., 1995a, 1995b; Levin et al., 1996, 1997; Tornai, MacDonald, Levin, Siegel, and Hoffman, 1996; Tornai, MacDonald, Levin, Siegel, and Hoffman et al., 1996; Tornai et al., 1997, 1998; Hoffman et al., 1997a, 1997b, 1998). The fiber optic bundle transfers the light to a multianode PSPMT (Philips XP1722). The camera is designed for imaging positron emissions from ^{18}F-labeled tracers and has incorporated two annihilation photon rejection schemes depicted in Figure 37 (Levin et al., 1997; Hoffman et al., 1998; Tornai et al., 1998; (see also Figure 3). In these schemes, a positron event is signified by a hit in the 0.5-mm-thick CaF$_2$(Eu) crystal, corresponding to a positron

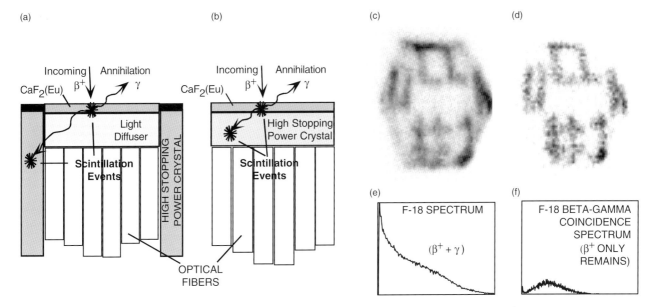

FIGURE 37 β^+–γ coincidence schemes for annihilation photon background rejection for a 1.2-cm^2 FOV positron imager involve a second, photon-absorbing crystal in either (a) a concentric annulus or (b) phoswich configuration. True positron events are signified by a hit in both the CaF$_2$(Eu) and an adjacent photon absorbing crystal. Images of a ^{18}F transmission phantom (c) before and (d) after photon background rejection activated. (e,f) The corresponding energy spectra demonstrate that the background-rejected image is composed of mainly β^+ interactions (Levin et al., 1997).

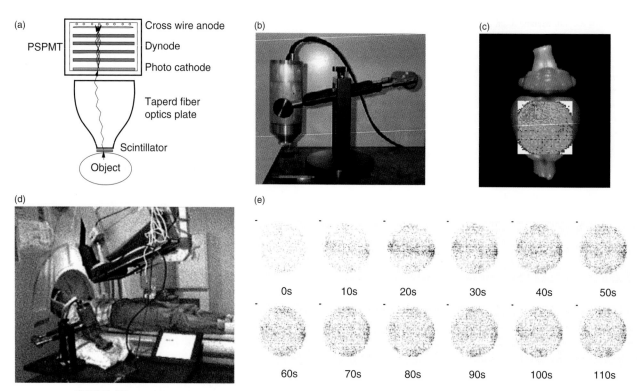

FIGURE 38 (a) Small-FOV beta camera uses a $CaF_2(Eu)$ crystal (10- to 20-mm in diameter), a tapered fiber optic bundle, and a PSPMT. (b) Counter-balanced arms for easy positioning support the unit. (c) The 2D distribution of ^{18}F-FDG on the brain surface of a rat. (d) Noninvasive ^{15}O water input function measurement through the skin in the arm of a patient during PET scan. (e) Dynamic arterial image frames (20-mm FOV; cumulative elapsed time shown underneath each image). Artery has horizontal orientation in images. (Courtesy of Seiichi Yamamoto, Kobe City College of Technology, Kobe, Japan.)

absorption, in coincidence with a hit in an adjacent high-Z, high-density crystal, corresponding to an absorption of an annihilation photon. The configuration of the second, photon-absorbing crystal is either a concentric annulus or a phoswich. In either case, no matter where the positron annihilation occurs in the primary detector, the adjacent second crystal has a high geometric detection efficiency for at least one of the back-to-back annihilation photons. The images in Figure 37(c–d) show the results of imaging a beta-transmission phantom, comprising a chosen pattern of letters and symbols, without and with the annihilation photon background-rejection coincidence circuit activated. The corresponding energy spectra are shown below each image (Figure 37e–f). A significant improvement in image contrast is seen by requiring the β^+–γ coincidence, corresponding to 99% annihilation photon rejection efficiency. However, the sensitivity is also significantly reduced as a result of the coincidence requirement.

A group at Kobe City College in collaboration with Osaka University (Yamamoto *et al.*, 1997, 1999) has also developed a small beta camera using $CaF_2(Eu)$, fiber optics and a PSPMT (Hamamatsu R2486) (Figure 38). The device uses a tapered fiber optic bundle to magnify the light distribution. The detector heads come in either 10- or 20-mm-diameter FOV $CaF_2(Eu)$ (0.5 mm thick). The tapered fiber bundle diameter matches that of the crystal on one side and is 50 mm on the PSPMT side, resulting in roughly a 30% light loss. The group has focused on using their camera for 2D imaging of the distribution of beta emitters over the surface of the brains of animals and for imaging an artery through the skin for noninvasive estimation of the input function for ^{15}O water in humans. Figure 38 shows examples of these applications.

References

Aarsvold, J. N., Barrett, H. H., *et al.* (1988). Modular scintillation cameras: A progress report. *Proc. SPIE* **914**: 319–325.

Aarsvold, J. N., Mintzer, R. A., Yasillo, N. J., *et al.* (1994). A miniature gamma camera in electrical injury; A multidisciplinary approach to therapy, prevention and rehabilitation. *NYAS Ann.* **720**: 192–205.

Adler, L., *et al.* (1997). Axillary lymph node metastases: Screening with F-18 2-deoxy-2-fluoro-D-glucose (FDG) PET. *Radiology* **203**: 323–327.

Anger, H. O. (1958). Scintillation camera. *Rev. Sci. Instr.* **29**(1): 27–33.

Antich, P. P., Kulkarni, P. V., Anderson, J., *et al.* (1992). High resolution I-125 imaging of rat brain blood flow with plastic scintillation detectors and position sensitive photomultipliers. *J. Nucl. Med.* **33**(5): 1003.

Avril, N., Schelling, M., Dose, J., Weber, W., and Schwaiger, M. (1999). Utility of PET in breast cancer. *Clin. Pos. Imaging* **2**(5): 261–271.

Avril, N., et al. (1996). Metabolic characterization of breast tumors with positron emission tomography using F-18 fluorodeoxyglucose. *J. Clin. Onc.* **14**: 1848–1857.

Baines, C. J., et al. (1986). Sensitivity and specificity of first screen mammography in the Canadian National Breast Screening Study: A preliminary report from five centers. *Radiology* **160**: 295–298.

Barber, H. B., Augustine, F. L., Barrett, H. H., et al. (1994). Semiconductor arrays with multiplexer readout for gamma-ray imaging: Results for a 48*48 Ge array. *Nucl. Instr. Meth. A* **353**: 361–365.

Barber, H. B., Barrett, H. H., Wild, W. J., et al. (1984). Development of small in vivo imaging probes for tumor detection. *IEEE Trans. Nucl. Sci.* **31**: 599–604.

Barrett, H. H., and Swindell, W. (1981). "Radiological Imaging," Vol. 1. Academic Press, New York.

Bergman, A., Thompson, C. J., Murthy, K., Robar, J. L., Clancy, R., English, M. J., Loutfi, A., Lisbona, R., and Gagnon, J. (1998). Technique to obtain positron emission mammography images in registration with X-ray mammograms. *Med. Phys.* **25**: 2119–2129.

Bird, A. J., and Ramsden, D. (1990). Images obtained with a compact gamma camera. *Nucl. Instr. Meth. Phys. Res. A* **A299**: 480–483.

Bolozdynya, A., Ordonez, C. E., and Chang, W. (1997). A concept of cylindrical Compton camera for SPECT. *1997 IEEE Nucl. Sci. Sym. Conf. Record* **2**: 1047–1051.

Bruyndonckx, P., Carnochan, P., Xuan, L., Yonggang, W., Visvikis, D., Duxbury, D., and Tavernier, S. (1997). Performance assessment and in-vivo imaging using the VUB-PET system. *1997 IEEE Nucl. Sci. Sym. Conf. Record* **2**: 1417–1422.

Butler, J. F., Doty, F. P., Apotovsky, B., et al. (1993). Progress in Cd(10x)Zn(x)Te (CZT) radiation detectors. *Mat. Res. Soc. Symp. Proc.* **302**:.

Carter, T. Bird, A. J., Dean, A. J., et al. (1995). An imager prototype for gamma-ray astronomy. *1994 IEEE Nucl. Sci. Sym. Conf. Record* **1**: 56–59.

Casey, M. E., Dautet, H., Waechter, D., Lecomte, R., Eriksson, L., and Schmand, M. (1998). An LSO BLOCK detector for PET using an avalanche photodiode array. *1998 IEEE Nucl. Sci. Sym. Med. Imaging Con.* **2**: 1105–1108.

Casey, M. E., and Nutt, R. (1986) A multicrystal two-dimensional BGO detector system for positron emission tomography. *IEEE Trans. Nucl. Sci.* **33**: 460–463.

Clancy, R. L., Thompson, C. J., Robar, J. L., and Bergman, A. M. (1997). A simple technique to increase the linearity and field-of-view in position sensitive photomultiplier tubes. *IEEE Trans. Nucl. Sci.* **44**: 494–498.

Clinthorne, N. H., and Rogers, W. L. (1987). A hybrid maximum likelihood position computer for scintillation cameras. *IEEE Trans. Nucl. Sci.* **34**(1): 97–101.

Correia, J. A., Burnham, C. A., Kaufman, D., and Fischman, A. J. (1999). Development of a small animal PET imaging device with resolution approaching 1 mm. *IEEE Trans. Nucl. Sci.* **46**: 631–635.

Daghighian, F., Eshaghian, B., and Shenderov, P. (1995). Intraoperative beta cameras. *J. Nucl. Med.* **(suppl.)**: 110P.

Daghighian, F., Miodownik, S., Shenederov, P., and Eshaghian, B. (1994a). Flexible intraoperative radiation imaging camera. U.S. Patent Number 5,325,855.

Daghighian, F., Miodownik, S., Shenederov, P., and Eshaghian, B. (1994b). Radiation imaging device having an enlarged uniform field of view. U.S. Patent Number 5,338,937.

Dahlbom, M., MacDonald, L. R., Eriksson, L., Paulus, M., Andreaco, M., Casey, M. E., and Moyers, C. (1997). Performance of a YSO/LSO phoswich detector for use in a PET/SPECT system. *IEEE Trans. Nucl. Sci.* **44**: 1114–1119.

de Notaristefani, F., Pani, R., Scopinaro, F., Barone, L. M., Blazek, K., De Vincentis, G., Malatesta, T., Maly, P., Pellegrini, R., Pergola, A., Soluri, A., and Vittori, F. (1996) First results from YAP:Ce gamma camera for small animal studies. *IEEE Trans. Nucl. Sci.* **43**(6): 3264–3271.

Dedicated Apparatus and Method for Emission Mammography, U.S. Patent 5252830, 1993.

Del Guerra, A., Di Domenico, G., Scandola, M., and Zavattini, G. (1998). YAP-PET: A small animal positron emission tomograph based on YAP:Ce finger crystals. *IEEE Trans. Nucl. Sci.* **45**(6): 3105.

Doty, F. P., Friesenhahn, S. J., and Butler, J. F. (1993). X-ray and gamma ray imaging with monolithic CdZnTe detector arrays. *Proc. SPIE-OE* **1945**: 145–151.

Eskin, J. D., Barrett, H. H., and Barber, H. B. (1999). Signals induced in semiconductor gamma-ray imaging detectors. *J. Appl. Phys.* **85**(2): 647–659.

Fiorini, C., Longoni, A., Perotti, F., Labanti, C., Rossi, E., Lechner, P., and Strüder, L. (1999). First prototype of a gamma-camera based on a single CsI(Tl) scintillator coupled to a silicon drift detector array. Paper presented at 1999 IEEE NSS/MIC Conference, Seattle, Washington.

Fiorini, C., et al. (1997). Gamma ray spectroscopy with CsI(Tl) scintillator coupled to silicon drift chamber. *IEEE Trans. Nucl. Sci.* **44**(6): 2553–2560.

Fiorini, C., et al. (1999). Position and energy resolution of a new gamma-ray detector based on a single CsI(Tl) scintillator coupled to a silicon drift chamber array. *IEEE Trans. Nucl. Sci.* **46**(4): 858–864.

Freifelder, R., and Karp, K. (1997). Dedicated PET scanners for breast imaging. *Phys. Med. Biol.* **42**: 2463–2480.

Gray, R. M., and Macovski, A. M. (1976). Maximum a posteriori estimation of position in scintillation cameras. *IEEE Trans. Nucl. Sci.* **23**: 849–852.

Green, M. V., Seidel, J., and Gandleer, W. R. (1994). A small animal system capable of PET, SPECT, and planar imaging. *J. Nucl. Med.* **35**(5): 61P.

Gruber, G. J., Moses, W. W., and Derenzo, S. E. (1999). Monte Carlo simulation of breast tumor imaging properties with compact, discrete gamma cameras. *IEEE Trans. Nucl. Sci.* **46**: 2119–2123.

Gruber, G. J., Moses, W. W., Derenzo, S. E., et al. (1998). A discrete scintillation camera module using silicon photodiode readout of CsI(Tl) crystals for breast cancer imaging. *IEEE Trans. Nucl. Sci.* **45**: 1063–1068.

Guru, S. V., He, Z., Ferreria, J. C., Wehe, D. K., and Knoll, G. F. (1994). A high energy gamma camera using a multiple hole collimator and PSPMT. *Nucl. Instr. Meth. Phys. Res.* **353**: 328–333.

Hailey, C. J., Harrison, F., Lupton, J. H., et al. (1989). An inexpensive, hard x-ray imaging spectrometer for use in x-ray astronomy and atomic physics. *Nucl. Instr. Meth. A* **276**: 340–346.

Hamamatsu Photonics (1995). Position sensitive photomultiplier tube R5900U-00-C8. Preliminary Data Sheet; catalog no. TPMH1139E01, Japan.

Hartsough, N. E., Barrett, H. H., Barber, H. B., and Woolfenden, J. M. (1995). Intraoperative tumor detection: Relative performance of single-element, dual-element, and imaging probes with various collimators. *IEEE Trans. Med. Imaging* **14**(2): 259–265.

Hasselgren, P., Hummel, R. P., Georgian-Smith, O., et al. (1993). Breast biopsy with needle localization: Accuracy of specimen x-ray and management of missed lesions. *Surgery* **114**: 836–842.

He, Z., Bird, A. J., Ramsden, D., and Meng, Y. (1993). A 5 inch diameter position-sensitive scintillation counter. *IEEE Trans. Nucl. Sci.* **40**: 447–451.

He, Z., Smith, L. E., Wehe, D. K., and Knoll, G. F. (1997). The CSPD-2 gamma-ray imaging system. *IEEE Trans. Nucl. Sci.* **44**: 911–915.

Hoffman, E. J., Tornai, M. P., Levin, C. S., MacDonald, L. R., and Holdsworth, C. H. (1998). A dual detector β-ray imaging probe with γ-ray background suppression for use in intra-operative detection of radiolabeled tumors. *Nucl. Instr. Meth. A* **409**: 511–516.

Hoffman, E. J., Tornai, M. P., Levin, C. S., MacDonald, L. R., and Siegel, S. (1997a). Design and performance of gamma and beta intra-operative imaging probes. *Phys. Medica* **13**(suppl. 1).

Hoffman, E. J., Tornai, M. P., Levin, C. S., MacDonald, L. R., and Siegel, S. (1997b). Gamma and beta intra-operative imaging probes. *Nucl. Instr. Meth. A* **392**: 324–329.

Hoh, C. K., et al. (1997). PET in oncology: Will it replace the other modalities? *Sem. Nucl. Med.* **27**: 94–106.

Huber, J. S., Moses, W. W., and Virador, P. R. G. (1998). Calibration of a PET detector module that measures depth of interaction. *IEEE Trans. Nucl. Sci.* **45**: 1268–1272.

Hutchins, G. D., and Simon, A. J. (1995). Evaluation of prototype geometries for breast imaging with PET radiopharmaceuticals. *J. Nucl. Med.* **36**(5): 69P.

Iwanczyk, J. S., and Patt, B. E. (1995). Electronics for X-ray and gamma ray spectrometers. In "Semiconductors for Room Temperature Nuclear Detector Applications" (T. E. Schlesinger and R. B. James, eds.), pp. 531–539. Academic Press, New York.

Jeavons, A. P., Chandler, R. A., and Dettmar, C. A. R. (1999). A 3D HIDAC-PET camera with sub-millimetre resolution for imaging small animals. *IEEE Trans. Nucl. Sci.* **46**: 468–473.

Johnson, C. A., Seidel, J., Carson, R. E., Gandler, W. R., Sofer, A., Green, M. V., and Daube-Witherspoon, M. E. (1997). Evaluation of 3D reconstruction algorithms for a small animal PET camera. *IEEE Trans. Nucl. Sci.* **44**(3): 1303–1308.

Khalkhali, I., Cutrone, J. A., Mena, I. G., et al. (1995). Scintimammography: The complementary role of ^{99m}Tc sestamibi prone breast imaging for the diagnosis of breast carcinoma. *Radiology* **196**: 421–426.

Khalkhali, I., et al. (1993). Tc-99m-SestaMIBI prone imaging in patients (PTS) with suspicion of breast cancer (Ca). *J. Nucl. Med.* **24**: 140P.

Khalkhali, I., et al. (1994). Clinical and pathologic follow up of 100 patients with breast lesions studied with scintimammography. *J. Nuc. Med.* **35**: 22P.

Khalkhali, I., et al. (1995). Tc-99m-sestamibi scintimammography of breast lesions: Clinical and pathological follow-up. *J. Nuc. Med.* **36**: 1731.

Knoll, G. F., Knoll, T. F., and Henderson, T. M. (1988). Light collection in scintillating detector composites for neutron detection. *IEEE Trans. Nucl. Sci.* **35**(1): 872–875.

Kume, H., Muramatsu, S., and Ida, M. (1986). Position-sensitive photomuliplier tubes for scintillation imaging. *IEEE Trans. Nucl. Sci.* **33**(1): 359–363.

Lastoria, S., Varrella, P., Mainolfi, C., Vergara, E., Maurea, S., et al. (1994). Tc-99m sestamibi scintigraphy in the diagnosis of primary breast cancer. *J. Nuc. Med.* **35**: 22P.

LeBlanc, J. W., Clinthorne, N. H., Hua, C. -H., Nygard, E., Rogers, W. L., Wehe, D. K., Weilhammer, P., and Wilderman, S. J. (1999). Experimental results from the C-SPRINT prototype Compton camera. *IEEE Trans. Nucl. Sci.* **46**(3): 201–204.

Lecomte, R., Cadorette, J., Rodrigue, S., Lapointe, D., Rouleau, D., Bentourkia, M., Yao, R., and Msaki, P. (1996). Initial results from the Sherbrooke avalanche photodiode positron tomograph. *IEEE Trans. Nucl. Sci.* **43**: 1952–1957.

Lees, J. E., Fraser, G. W., and Carthew, P. (1998). Microchannel plate detectors for C-14 autoradiography. *IEEE Trans. Nucl. Sci.* **45**: 1288–1292.

Lees, J. E., Fraser, G. W., and Dinsdale, D. (1997). Direct beta autoradiography using microchannel plate (MCP) detectors. *Nucl. Instr. Meth. A* **392**: 349–353.

Lees, J. E., Pearson, J. F., Fraser, G. W., et al. (1999). An MCP-based system for beta autoradiography. *IEEE Trans. Nucl. Sci.* **46**(3): 636–638.

Levin, C. S., Hoffman, E. J., Silverman, R. W., Meadors, A. K., and Tornai, M. P. (1997). Breast tumor detectability with small scintillation cameras. Paper presented at the 1997 IEEE Medical Imaging Conference, Albuquerque, New Mexico.

Levin, C. S., Hoffman, E. J., Tornai, M. P., and MacDonald L. R. (1996). Design of a small scintillation camera with photodiode readout for imaging malignant breast tumors. *J. Nucl. Med.* **37**: 52.

Levin, C. S., Hoffman, E. J., Tornai, M. P., and MacDonald L. R. (1997). PSPMT and photodiode designs of a small scintillation camera for imaging malignant breast tumors. *IEEE Trans. Nucl. Sci.* **44**: 1513–1520.

Levin, C. S., MacDonald, L. R., Tornai, M. P., Hoffman, E. J., and Park, J. (1996). Optimizing light collection from thin scintillators used in a beta-ray camera for surgical use. *IEEE Trans. Nucl. Sci.* **43**(3): 2053–2060.

Levin, C. S., Tornai, M. P., Cherry, S. R., MacDonald, L. R., and Hoffman, E. J. (1997). Compton scatter and X-ray crosstalk and the use of very thin intercrystal septa in high-resolution PET detectors. *IEEE Trans. Nucl. Sci.* **44**(2): 218–224.

Levin, C. S., Tornai, M. P., MacDonald, L. R., and Hoffman, E. J. (1997). Annihilation gamma-ray background characterization and rejection for a small surgical beta-ray camera imaging positron emitters. *IEEE Trans. Nucl. Sci.* **44**(4): 1120–1126.

Ljunggren, K., and Strand, S. E. (1988). Development of a digital imaging detector based on microchannel plates for biomedical samples emitting uncharged and charged particles. *Nucl. Instr. Meth. A* **273**: 784–786.

Ljunggren, K., and Strand, S. E. (1990). Beta camera for static and dynamic imaging of charged-particle emitting radionuclides in biologic samples. *J. Nucl. Med.* **31**: 2058–2063.

Ljunggren, K., and Strand, S. E. (1994). Reduction of noise in the beta camera for low activity applications. *IEEE Trans. Nucl. Sci.* **41**: 1666–1669.

Ljunggren, K., Strand, S. E., Ceberg, C. P., et al. (1993). Beta camera low activity tumor imaging. *Acta Oncol.* **32**: 869–872.

Ljunggren, K., Tagesson, M., Erlandsson, K., and Strand, S. E. (1995). Beta camera and pinhole SPECT imaging for dosimetry applications. *1995 IEEE Nucl. Sci. Sym. Med. Imaging Conf. Record* **3**: 1563–1566.

MacDonald, L. R., Patt, B. E., Iwanczyk, J., Shao, Y., and Hoffman, E. J. (1999). Silicon drift photodetector anger camera. Paper presented at the 1999 IEEE Medical Imaging Conference, Seattle, Washington.

MacDonald, L. R., Tornai, M. P., Levin, C. S., Park, J., Atac, M., Cline, D. B., and Hoffman, E. J. (1995a). Investigation of the physical aspects of beta imaging probes using scintillating fibers and visible light photon counters. *IEEE Trans. Nucl. Sci.* **42**(4): 1351–1357.

MacDonald, L. R., Tornai, M. P., Levin, C. S., Park, J., Atac, M., Cline, D. B., and Hoffman, E. J. (1995b). Small area, fiber coupled scintillation camera for imaging beta-ray distributions intra-operatively. *Proc. SPIE* **2551**: 92–101.

Marks, D. G., Barber, H. B., Barrett, H. H., Dereniak, E. L., Eskin, J. D., Matherson, K. J., Woolfenden, J. M., Young, E. T., Augustine, F. L., Hamilton, W. J., Venzon, J. E., Apotovsky, B. A., and Doty, F. P. (1996). A 48*48 CdZnTe array with multiplexer readout. *IEEE Trans. Nucl. Sci.* **43**: 1253–1259.

Matherson, K. J., Barber, H. B., Barrett, H. H., Eskin, J. D., Dereniak, E. L., Marks, D. G., Woolfenden, J. M., Young, E. T., and Augustine, F. L. (1998). Progress in the development of large-area modular 64*64 CdZnTe imaging arrays for nuclear medicine. *IEEE Trans. Nucl. Sci.* **45**(3): 354–358.

Matthews, K. L., Block, T. A., Aarsvold, J. N., et al. (1994). Assessment of tissue damage induced by electrical shock using a miniature gamma camera. *Med. Phys.* **21**: 932.

Maublant, J., de Latour, M., Mestas, M., et al. (1996). Technetium-99m-Sestamibi uptake in breast tumors and associated lymph nodes. *J. Nucl. Med.* **37**: 922–925.

Menard, L., Charon, Y., Solal, M., Laniece, P., Mastrippolito, R., Pinot, L., Ploux, L., Ricard, M., and Valentin, L. POCI: A compact high resolution gamma camera for intra operative surgical use. *IEEE Trans. Nucl. Sci.* **45**: 1293–1297.

Milster, T. D., Aarsvold, J. N., Barrett, H. H., et al. A full-field modular gamma camera. *J. Nucl. Med.* **31**(5): 632–639.

Milster, T. D., Selberg, L. A., Barrett, H. H., et al. (1984). A modular scintillation camera for use in nuclear medicine. *IEEE Trans. Nucl. Sci.* **31**(1): 578–580.

Milster, T. D., Selberg, L. A., Barrett, H. H., *et al.* (1985). Digital position estimation for the modular scintillation camera. *IEEE Trans. Nucl. Sci.* **32**: 748–752.

Moses, W., *et al.* (1995). PET camera designs for imaging breast cancer and axillary node involvement. *J. Nucl. Med.* **36**: 69P.

Moszynski, M., Kapusta, M., Wolski, D., Szawlowski, M., and Klamra, W. (1997). Blue enhanced large area avalanche photodiodes in scintillation detection with LSO, YAP and LuAP crystals. *IEEE Trans. Nucl. Sci.* **44**: 436–442.

Nagai, Y., and Hyodo, T. (1994). Position sensitive detector for gamma-rays using GSO crystal. *Nucl. Instr. Meth. A* **349**: 285–288.

Nicholson, P. W. (1947). "Nuclear Electronics." John Wiley & Sons, New York.

Nieweg, O. E., *et al.* (1993). Positron emission tomography of glucose metabolism in breast cancer: Potential for tumor detection, staging, and evaluation of chemotherapy. *Ann. N.Y.A. Sci.* **698**: 423–448.

Overdick, M., Czermak, A., Fischer, P., *et al.* (1996). "Bioscope" system using double-sided silicon strip detectors and self-triggering read-out chips. *Nucl. Instr. Meth. Phys. Res. A* **392**: 173–177.

Pani, R., Blazek, K., De Notaristefani, F., Maly, P., Pellegrini, R., Pergola, A., Scopinaro, F., and Soluri, A. (1994). Multicrystal YAP:Ce detector system for position sensitive measurements. *Nucl. Instr. Meth. A* **348**: 551.

Pani, R., De Vincentis, G., Scopinaro, F., Pellegrini, R., Soluri, A., Weinberg, I. N., and Pergola, A. (1998). Dedicated gamma camera for single photon emission mammography (SPEM). *IEEE Trans. Nucl. Sci.* **45**: 3127–3133.

Pani, R., Pellegrini, R., Scopinaro, F., De Vincentis, G., Pergola, A., Soluri, A., Corona, A., Filippi, S., Ballesio, P. L., and Grammatico, A. (1997). A scintillating array gamma camera for clinical use. *Nucl. Instr. Meth. A* **392**: 295–298.

Pani, R., Scopinaro, F., Pellegrini, R., Pergola, A., De Vincentis, G., Soluri, A., Weinberg, I. N., and Polli, N. S. A. (1996). The role of Compton scattering in scintimammography. *1996 IEEE Nucl. Sci. Sym. Conf. Record* **2**: 1382–1386.

Pani, R., Scopinaro, F., Pellegrini, R., Soluri, A., Weinberg, I. N., and DeVincentis G. (1997). The role of Compton background and breast compression on cancer detection in scintimammography. *Anticancer Res.* **17**: 1645–1649.

Pani, R., Soluri, A., Pergola, A., Pellegrini, R., Scafé, R., De Vincentis, G., and Scopinaro, F. (1999). Multi PSPMT scintillating camera. *IEEE Trans. Nucl. Sci.* **46**: 702–708.

Patt, B. E., Iwanczyk, J. S., Rossington Tull, C., *et al.* (1998). High resolution CsI(Tl)/Si-PIN detector development for breast imaging. *IEEE Trans. Nucl. Sci.* **45**: 2126–2131.

Patt, B. E., Iwanczyk, J. S., Tornai, M. P., Hoffman, E. J., and Rossington, C. (1997). Dedicated breast imaging system based on a novel solid state detector array. *J. Nucl. Med.* **38**(5): 535.

Patt, B. E., Iwanczyk, J. S., Tornai, M. P., Levin, C. S., and Hoffman, E. J. (1995). Development of a mercuric iodide detector array for medical imaging applications. *Nucl. Inst. Meth. A* **366**: 173–182.

Patt, B. E., Iwanczyk, J. S., Wang, Y. J., Tornai, M. P., Levin, C. S., and Hoffman, E. J. (1996). Mercuric iodide photodetector arrays for gamma-ray imaging. *Nucl. Inst. Meth. A* **380**: 295–300.

Patt, B. E., Tornai, M. P., Iwanczyk, J. S., Levin, C. S., and Hoffman, E. J. (1997). Development of an intraoperative gamma camera based on a 256-pixel mercuric iodide detector array. *IEEE Trans. Nucl. Sci.* **44**(3): 1242–1248.

Piccolo, S., *et al.* (1995). Technetium-99m-methylene diphosphonate scintimammography to image primary breast cancer. *J. Nuc. Med.* **36**: 718–724.

Pichler, B., Boning, C., Lorenz, E., Mirzoyan, R., Pimpl, W., Schwaiger, M., and Ziegler, S. I. (1998). Studies with a prototype high resolution PET scanner based on LSO-APD modules. *IEEE Trans. Nucl. Sci.* **45**: 1298–1302.

Raylman, R., Majewski, S., Wojcik, R., Weisenberger, A., Kross, B., Popov, V., and Bishop, H. A. (1999). Positron emission mammography-guided stereotactic breast biopsy. Paper presented at the Society of Nuclear Medicine 46th Annual Meeting, Los Angeles.

Robar, J. L., Thompson, C. J., Murthy, K., Clancy, R., and Bergman, A. M. (1997). Construction and calibration of detectors for high resolution metabolic breast cancer imaging. *Nucl. Instr. Meth. A* **392**: 402–406.

Rogers, J. (1995). A method for correcting the depth-of-interaction blurring in PET cameras. *IEEE Trans. Nucl. Sci.* **14**(1): 146–150.

Rohe, R., and Valentine, J. D. (1995). A novel Compton scatter camera design for in-vivo medical imaging of radiopharmaceuticals. 1995 IEEE *Nucl. Sci. Sym. Med. Imaging Conf. Record* **3**: 1579–1583.

Rossi, E., *et al.* (1996). A hexagonal multi-element CsI(Tl)-photodiode module for gamma-ray imaging. *1995 IEEE Nucl. Sci. Sym. Medical Imaging Conf. Record* **1**: 60.

Schlesinger, T. E., and James, R. B. (eds.) (1995). "Semiconductors for Room Temperature Nuclear Detector Applications, Vol. 43, Semiconductors and Semimetals." Academic Press, New York.

Schmand, M., Eriksson, L., Casey, M. E., Wienhard, K., Flugge, G., and Nutt, R. (1999). Advantages using pulse shape discrimination to assign the depth of interaction information (DOI) from a multi layer phoswich detector. *IEEE Trans. Nucl. Sci.* **46**: 985–990.

Scopinaro, F., Pani, R., De Vincentis, G., Soluri, A., Pellegrini, R., and Porfiri, L. M. (1999). High resolution scintimammography improves the accuracy of technetium-99m methoxyisobutylisonitrile scinti-mammography: Use of new dedicated gamma camera. *Eur. J. Nucl. Med.* **26**: 1279–1288.

Shao, Y., Cherry, S. R., Siegel, S., and Silverman, R. W. (1996). A study of inter-crystal scatter in small scintillator arrays designed for high resolution PET imaging. *IEEE Trans. Nucl. Sci.* **43**(3): 1938–1944.

Shao, Y., Silverman, R., Farrell, R., *et al.* (1999). Design studies of a high resolution PET detector. Paper presented at the 1999 IEEE MIC/NSS Conference, Seattle, Washington.

Siegel, S., Silverman, R. W., Yiping Shao, and Cherry, S. R. (1996). Simple charge division readouts for imaging scintillator arrays using a multi-channel PMT. *IEEE Trans. Nucl. Sci.* **43**: 1634–1641.

Siegel, S., Vaquero, J. J., Aloj, L., Seidel, J., Jagoda, E., Gandler, W. R., Eckelman, W. C., and Green, M. V. (1999). Initial results from a PET/planar small animal imaging system. *IEEE Trans. Nucl. Sci.* **46**(3): 571.

Singh, M., Doty, F. P., Friesenhahn, S. J., and Butler, J. F. (1995). Feasibility of using cadmium-zinc-telluride detectors in electronically collimated SPECT. *IEEE Trans. Nucl. Sci.* **42**: 1139–1146.

Soares, A. J., *et al.* (1998). Assessment of light distribution at the photocathode of 5″ and 1″ position sensitive photomultiplier tubes. *SPIE Proc. Med. Imaging Conf. 1998*, **3336**: 818–828.

Soares, A. J., *et al.* (1999). Development of a small gamma camera using wavelength-shifting fibres coupled to inorganic scintillation crystals for imaging 140 keV gamma rays. *IEEE Trans. Nucl. Sci.* **46**(3): 576–582.

Soares, A. J., *et al.* (2000). Development of a wavelength-shifting fibre gamma camera. *IEEE Trans. Nucl. Sci.* **47**(3): 1058–1064.

Suyama, M., Hirano, K., Kawai, Y., Nagai, T., Kibune, A., Saito, T., Negi, Y., Asakura, N., Muramatsu, S., and Morita, T. (1998). A hybrid photodetector (HPD) with a III-V photocathode. *IEEE Trans. Nucl. Sci.* **45**: 572–575.

Tabar, L., *et al.* (1982). Mammographic parenchymal patterns: risk indicator for breast cancer? *JAMA* **247**: 185–189.

Taillefer, R., *et al.* (1995). Technetium-99m-sestamibi prone scintimam-mography to detect primary breast cancer and axillary lymph node involvement. *J. Nuc. Med.* **36**: 1758.

Tanaka, E., Hiramoto, T., and Hohara, N. (1970). Scintillation cameras based on new position arithmetics. *J. Nucl. Med.* **11**(9): 542–547.

Thompson, C. J., Murthy, K., Picard, Y., Weinberg, I. N., and Mako, F. M. (1995). Positron emission mammography (PEM): A promising technique to detect breast cancer. *IEEE Trans. Nucl. Sci.* **42:** 1012–1017.

Thompson, C. J., Murthy, K., Weinberg, I. N., and Mako, F. (1994). Feasibility study for positron emission mammography. *Med. Phys.* **21**(4): 529–537.

Tornai, M. P., Hoffman, E. J., Levin, C. S., and MacDonald, L. R. (1997). Characterization of fluor concentration and geometry in organic scintillators for in situ beta imaging. *IEEE Trans. Nucl. Sci.* **43**(6): 3342–3347.

Tornai, M. P., Levin, C. S., MacDonald, L. R., Hoffman, E. J., and Park, J. (1997). Investigation of crystal geometries for fiber coupled gamma imaging intra-operative probes. *IEEE Trans. Nucl. Sci.* **44**(3): 1254–1261.

Tornai, M. P., Levin, C. S., MacDonald, L. R., Holdsworth, C. H., and Hoffman, E. J. (1998). A miniature phoswich detector for gamma-ray localization and beta imaging. *IEEE Trans. Nucl. Sci.* **45**(3): 1166–1173.

Tornai, M. P., MacDonald, L. R., Levin, C. S., Siegel, S., and Hoffman, E. J. (1996). Design considerations and performance of a 1.2 cm^2 beta imaging intra-operative probe. *IEEE Trans. Nucl. Sci.* **43**(4): 2326–2335.

Tornai, M. P., MacDonald, L. R., Levin, C. S., Siegel, S., Hoffman, E. J., Park, J., Atac, M., and Cline, D. B. (1996). Miniature nuclear emission imaging system for intra-operative applications. In "Proceedings from UCLA International Conference on Imaging Detectors in High Energy & Astroparticle Physics," pp. 133–147. World Scientific Publishing, River Edge, New Jersey.

Tornai, M. P., Patt, B. E., Iwanczyk, J. S., Levin, C. S., and Hoffman, E. J. (1997). Discrete scintillator coupled mercuric iodide photodetector arrays for breast imaging. *1996 IEEE Med. Imaging Conf. Record* **II**, 1034–1038; *IEEE Trans. Nucl. Sci.* **44–43:** 1127–1133.

Tse, N. Y., *et al.* (1992). The application of positron emission tomographic imaging with fluorodeoxyglucose to the evaluation of breast disease. *Ann. Surg.* **216:** 27–34.

Tumer, T. O., Shi Yin, and Kravis, S. (1997). A high sensitivity, electronically collimated gamma camera. *IEEE Trans. Nucl. Sci.* **44:** 899–904.

Villanueva-Meyer, J., Leonard, M. H., Jr., *et al.* (1996). Mammoscintigraphy with technetium-99m-sestamibi in suspected breast cancer. *J. Nucl. Med.* **37**(6): 926–930.

Wahl, R. L., *et al.* (1991). Primary and metastatic breast carcinoma: Initial clinical evaluation with PET with the radiolabeled glucose analogue 2-[F-18]-fluoro-2-deoxy-D-glucose. *Radiology* **179:** 765–770.

Watanabe, M., Okada, H., Shimizu, K., Omura, T., Yoshikawa, E., Kosugi, T., Mori, S., and Yamashita, T. (1997). A high resolution animal PET scanner using compact PS-PMT detectors. *IEEE Trans. Nucl. Sci.* **44:** 1277–1282.

Waxman, A., *et al.* (1994). Sensitivity and specificity of Tc-99m methoxyisobutal isonitrile (MIBI) in the evaluation of primary carcinoma of the breast: Comparison of palpable and non-palpable lesions with mammography. *J. Nuc. Med.* **35:** 22P.

Weber, S., Herzog, H., Cremer, M., Engels, R., Hamacher, K., Kehren, F., Mühlensiepen, H., Ploux, L., Reinartz, R., Reinhart, P., Rongen, F., Sonnenberg, F., Coenen, H. H., and Halling, H. (1999). Evaluation of the TierPET scanner. *IEEE Trans. Nucl. Sci.* **46**(4): 1177.

Weinberg, I., Majewski, S., Weisenberger, A., Markowitz, A., Aloj, L., Majewski, L., Danforth, D., Mulshine, J., Cowan, K., Zujewski, J., Chow, C., Jones, E., Chang, V., Berg, W., and Frank, J. (1996). Preliminary results for positron emission mammography: Real-time functional breast imaging in a conventional mammography gantry, *Eur. J. Nucl. Med.* **23**(7): 804–806.

Weinberg, I. N., Pani, R., Pellegrini, R., Scopinaro, F. DeVincentis, G., Pergola, A., and Soluri, A. (1997). Small lesion visualization in scintimammography. IEEE *Trans. Nucl. Sci.* **44**(3): 1398–1402.

Weisenberger, A. G., Kross, B., Majewski, S., Wojcik, R., Bradley, E., and Saha, M. (1998). Design features and performance of a CsI(Na) array based gamma camera for small animal gene research. *IEEE trans. nucl. Sci.* **45**(6): 3053–3058.

Williams, M. B., Galbis-Reig, V., Goode, A. R., Simoni, P. U., Majewski, S., Weisenberger, A. G., Wojcik, R., Phillips, W., and Stanton, M. (1999). Multimodality imaging of small animals. *RSNA Electron. J.* http://ej.rsna.org/ej3/0107-99.fin/dual99.htm.

Williams, M. B., Goode, A. R., Galbis-Reig, V., Majewski, S., Weisenberger, A. G., and Wojcik, R. J. (2000) Performance of a PSPMT based detector for scintimammography. *Phys. Med. Biol.* **45**(3): 781–800.

Williams, M. B., *et al.* (1997). Gamma-ray detectors for breast imaging. *Proc. Int. Soc. Opt. Eng.* **3115:** 226–234.

Wojcik, R., Majewski, S., Steinbach, D., and Weisenberger, A. G. (1998). High spatial resolution gamma imaging detector based on 5″ diameter R3292 Hamamatsu PSPMT. *IEEE Trans. Nucl. Sci.* **45:** 487–491.

Yamamoto, S., Matsuda, T., Hashikawa, K., and Nishimura, T. (1999). Imaging of an artery from skin surface using beta camera. *IEEE Trans. Nucl. Sci.* **46**(3): 583–586.

Yamamoto, S., Seki, C., Kashikura, K., Fujita, H., Matsuda, T., Ban, R., and Kanno, I. (1997). Development of a high resolution beta camera for a direct measurement of positron distribution on brain surface. *IEEE Trans. Nucl. Sci.* **44**(4): 1536–1542.

Yasillo, N. J., Beck, R. N., and Cooper, M. (1990). Design considerations for a single tube gamma camera. *IEEE Trans. Nucl. Sci.* **37**(2): 609–615.

Zavarzin, V. G., Kudrolli, H. A., Stepanov, P. Y., Weinberg, I. N., Adler, L. P., and Worstell, W. A. (1999). First results with PEM-X, a compact modular PET breast imager. *IEEE Med. Imaging Conf. Record.*

CHAPTER

17

Intraoperative Probes and Imaging Probes

EDWARD J. HOFFMAN*, MARTIN P. TORNAI[†], MARTIN JANECEK*, BRADLEY E. PATT[‡], and JAN S. IWANCZYK[‡]

*UCLA School of Medicine, Division Of Nuclear Medicine, Department of Pharmacology, Los Angeles, California
[†]Duke University Medical Center, Department of Radiology, Durham, North Carolina
[‡]Photon Imaging Inc., Northridge, California

I. Introduction
II. Early Intraoperative Probes
III. Clinical Applications
IV. The Future: Imaging Probes?
V. Discussion
VI. Conclusion

Intraoperative probes have been employed to assist in the detection and removal of tumors since the 1950s. Since the 1960s, essentially every detector type that could be miniturized has been tested or at least suggested for use as an intraoperative probe. These detectors have included basic Geiger-Müller (GM) tubes, scintillation detectors, and even state-of-the-art solid-state detectors. The radiopharmaceuticals have progressed from $^{32}PO_4^-$ injections for brain tumors to sophisticated monoclonal antibodies labeled with ^{125}I for colorectal cancers. The early work was mostly anecdotal, primarily interdisciplinary collaborations between surgeons and physical scientists. These collaborations produced a few publications, but never produced an ongoing clinical practice. In the mid 1980s, several companies offered basic gamma-detecting intraoperative probes as products. This led to the rapid development of radioimmunoguided surgery (RIGS) and sentinel node detection as regularly practiced procedures to assist in the diagnosis and therapy of cancer. More recently, intraoperative imaging probes have been developed. These devices add the ability to see the details of the detected activity, giving the potential of using the technique in a low-contrast environment. Intraoperative probes are now established as clinical devices. They have a commercial infrastructure to support their continued use, and there is ongoing research, both commercial and academic, that ensures continued progress and renewed interest in this slowly developing field.

I. INTRODUCTION

The concept of employing a hand-held radiation detector to assist a surgeon in locating and removing tumors was part of the fallout from the Manhattan project. In the years following World War II (1946–49), several investigators (Low-Beer, 1946; Moore, 1948; Selverstone and Solomon, 1948; Selverstone, Solomon, et al., 1949; Selverstone, Sweet, et al., 1949) realized that if one could label a tumor by making an intravenous injection with a compound labeled with a radioactive isotope it should be possible to locate the tumor by surveying the region under suspicion with a radiation detector. In this early work, the isotope was ^{32}P (14.3 days half-life, 1.71 MeV β^-) and the detector was a simple Geiger-Müller (GM) tube. This combination worked because the GM tube had a very high detection efficiency for the β^- from ^{32}P, and the relatively high energy of the β^- allowed penetration of up to 8 mm of tissue to reach the GM tube. The success of this approach was limited in part because the absorbed dose per unit activity for ^{32}P is 300–600 times that of the currently most frequently used isotope, ^{99m}Tc. For example, a 1.2–2.4 MBq (30–60 mCi) dose is equivalent to the 740 MBq (20 mCi) dose that is typical in diagnostic nuclear medicine. This meant that there were either very low count rates from the tumor or the physician needed to inject larger doses. In the early work with ^{32}P [3–5], 37–148 MBq (1–4 mCi) doses were injected for intraoperative probe–assisted surgeries. It should be pointed out that this work was performed before guidelines had been established for medical applications of radiation.

The usefulness of procedures using an intraoperative probe must be considered in terms of a measurement system. This system consists of the radiation detector, the radioisotope, and the compound that is labeled. First, the compound must have the properties that (1) it is extracted from the blood into the tumor, (2) its amount in the tumor must be significantly higher than the surrounding tissue, and (3) it must be extracted into the tumor and remain in the tumor in a time frame consistent with the requirements of the surgery and the half-life of the isotope. Second, the radioisotope must have (1) the chemical properties that allow it to form a stable labeled compound, (2) an appropriate half-life, and (3) emitted radiation that is compatible with the detector system and patient safety. Third, the detector system must have some combination of the following properties: (1) the efficient detection of the emitted radiation, (2) insensitivity to background radiation or the ability to suppress or measure and subtract background events, (3) good energy resolution to discriminate against background radiation, (4) collimation to shield against background radiation, and (5) good spatial resolution to resolve the tumor from the adjacent background.

II. EARLY INTRAOPERATIVE PROBES

A. Geiger-Müller Counters

The earliest intraoperative probes were GM counters, which are among the most basic kind of gas-filled radiation detectors, and ^{32}P was the isotope of choice (Low-Beer, 1946; Moore, 1948; Selverstone and Solomon, 1948; Selverstone, Solomon, *et al.*, 1949; Selverstone, Sweet, *et al.*, 1949; Nakayama, 1959; Thomas *et al.*, 1952). Fortuitously, the properties of the GM tube plus ^{32}P were a good match to the requirements of the good measurement system previously outlined. First, the compound, inorganic phosphate, was extracted into and remained in the tumor for a long period of time (many days). The tumor-to-background activity ratio was measured to be from 5.5 to 110 (Selverstone and Solomon, 1948; Selverstone, Solomon, *et al.*, 1949; Selverstone, Sweet, *et al.*, 1949) for various types of brain tumors. The isotope, ^{32}P, was used as a label in the PO_4^- ion, which is the most stable form of phosphorus. The emitted radiation is a beta particle. Beta radiation is not monoenergetic, as is the case with gamma rays, but is emitted in a continuous distribution of energies, in this case from 0 to 1.71 MeV. This maximum energy beta can penetrate up to 8 mm in tissue, although the average penetration is 2.8 mm. The properties of beta radiation reduces the technical requirements of the probe. The short range of the beta means that all detected radiation comes from a source that is almost touching the probe. Thus, there is no background from distal parts of the body, and the probe does not require collimation or other shielding. The simplest gas detectors (ionization chambers) simply collect the ions produced in the gas by the radiation. The GM tube is designed with a fine wire central electrode, which gives a very high electric field near the electrode. The high field can be produced by a simple 100-volt external supply (e.g., 11 9-volt batteries in series). The high field accelerates the electrons released by the absorbed radiation in the gas to a high enough velocity to cause secondary ionizations, releasing more electrons. The secondary electrons are also accelerated and cause more ionizations, with the process continuing until enough charge builds up around the electrode to cause a reduction in the electric field, stopping the ionization. The process can be initiated by a single ionization. Thus, the GM tube requires only that a small fraction of the energy of the beta particle be absorbed, causing ionization of gas atoms in its sensitive volume. A few initial ionizations in the counter gas leads to the same cascade of ionizations in the gas as a large number of initial ionizations and produces a signal just as strong. This means that essentially all betas striking the detector are counted, giving close to 100% efficiency or sensitivity.

The early probes were more invasive than the modern intraoperative probes, which tend to be used on exposed surfaces. Figure 1 is a photograph of two of the earliest

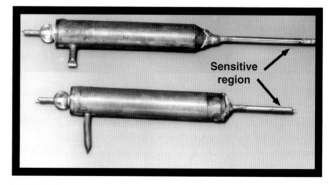

FIGURE 1 Photograph of the earliest GM tubes used as intraoperative probes. (From Robinson and Peterson, 1948; courtesy of Melvin Griem, M.D., assistant to C. Robinson in 1948.)

probes. The GM tube is about 3 mm in diameter, about the same diameter as the needle used in ventriculography, and only approximately 1–1.5 cm at the tip are sensitive to radiation. After exposing the brain, the probe is introduced into a convolution of the brain as far from the expected tumor location as possible to get a control reading. Then the probe is introduced into convolutions of the brain near the expected location of the tumor. A large increase in the count rate indicates the location of the tumor.

The GM tube is limited in that its sensitive volume is a gas, which is essentially transparent to gamma rays. Typically less than 1% of the gamma rays that pass through a GM counter are detected. A higher density detector consisting of higher atomic number materials is required for efficient detection of gamma rays.

A technical improvement in gas detector based probes was made by Robinson and Selverstone (1958), who created a probe based on the proportional counter. However, the use of the gas detector was only rarely reported in the literature (Nakayama, 1959; Thomas *et al.*, 1952; Ackerman *et al.*, 1960; Nelson *et al.*, 1964; Williamson *et al.*, 1986). Because one rarely reports on the failure of a technique, unless it has caused serious harm, the reasons for the phasing out of gas detectors are not well documented. It is presumed that the use of doses that were on the order of 100 times that normally accepted in nuclear medicine might have been a factor, as well as the fact that these detectors could not be used with gamma-emitting isotopes.

B. Scintillation Detectors

The only readily available solution for a gamma-ray sensitive probe in the 1950's was the scintillation detector (Harris *et al.*, 1956). The typical configuration is diagrammed in Figure 2 and is basically the configuration used by Harris et al. In this type of detector, the scintillator absorbs the radiation and emits visible light in proportion to the energy absorbed. The visible light must be measured with a photon detector, usually a photomultiplier tube (PMT). In the early versions, CsI(Tl) was used as the scintillator because it was easy to handle and was only slightly hygroscopic. The reliable PMTs of the day were on the order of 2.5 cm (1 inch) in diameter or larger. Therefore, a light pipe (Harris *et al.*, 1956; Morris *et al.*, 1971) or fiber optics (Swinth and Ewins, 1976; Colton and Hardy, 1983; Barber *et al.*, 1980; Harvey and Lancaster, 1981; Woolfenden *et al.*, 1984) were used to optically couple the small scintillators of the probe to these relatively large PMTs. Eventually, reliable PMTs became available in sizes as small as 10 mm in diameter, and modern probes have eliminated the light pipe (Knoll *et al.*, 1972; "Technical Specifications," 1999). The elimination of the light pipe or fiber optic allowed a significant improvement in light collection from the scintillator. This in turn led to improved energy resolution and, because a scattered gamma

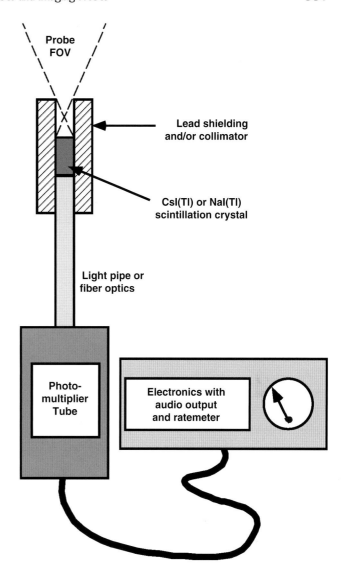

FIGURE 2 The basic configuration of scintillation detector–based intraoperative probe. The system consists of a simple shield/collimator; a scintillation detector, initially CsI(Tl) and eventually NaI(Tl); a light pipe or fiber optics; a photomultiplier; and electronics with a ratemeter and audio output.

ray loses energy, this also enabled the rejection of some of the scattered gamma rays.

The high penetration power of gamma rays means that background events could come from any part of the patient. These events could be reduced to some extent by absorption by the patient and to a greater extent by the low solid angle for detection of distant radiation sources, which is governed by an inverse square law relation with distance. However, because it is rare to have more than a few percent of the isotope in the tumor, the total background from the rest of the body can be very significant. The early scintillation detectors were shielded by varying amounts of platinum (Harris *et al.*, 1956; Morris *et al.*, 1971) or lead (Knoll *et al.*, 1972; Colton and Hardy, 1983; "Technical Specifications," 1999).

The collimators/shielding were little more than cylinders of absorber designed to shield the detector from most of the body and provide a modest amount of collimation for the detector. A typical collimator for a scintillation camera reduces the number of gamma rays that strike the detector by a factor of 10^3 to 10^4. This type of collimation would be too severe for the probe, which depends on its high sensitivity for its usefulness. The fact that a gamma probe has a large distal FOV is a factor that must be remembered in its employment in surgery. The final elements in the system are the electronics and readout. Because the scintillation detector provides a signal proportional to the energy of the gamma ray, it is possible to do some crude spectroscopy to set the sensitive energy range of probe to the select the desired gamma-ray energy and eliminate at least part of the scattered gamma rays. The count rate of the probe is then fed to a ratemeter, which in turn drives an audio output. The surgeon uses this output, either an increase in loudness or frequency, to locate the tumor. The ratemeter itself is used to check the level of increased uptake after the tumor has been located. These basic elements are still used in a modern system, as shown in Figure 3.

FIGURE 3 Scintillation detector–based intraoperative probe. (Top) Control unit with electronics, audio tone generator, ratemeter, and LCD readout of counts. (Bottom) ^{99m}Tc probe and collimator. (C) ^{111}In probe and collimator. (From "Technical Specifications," 1999; courtesy of Carewise Medical Products Corp., Morgan Hill, CA.)

More recently, plastic scintillators have been employed to directly detect the positrons, rather than detect the annhilation radiation from positron-emitting radiopharmaceuticals. One such device (Daghighian *et al.*, 1994), essentially the same as the design in Figure 2, was developed after the very high tumor-to-tissue ratio of [F-18]-5-fluorodeoxyuridine [23] was seen in positron emission tomography (PET) scans. A second system employs a long fiber optic cable (Raylman and Wahl, 1994; Raylman *et al.*, 1995) and has been used to detect ^{18}F-fluorodeoxyglucose (FDG) in abdominal tumors.

C. Solid-State Detectors

The reduction in the use of GM tubes was due in part to the introduction of solid-state detectors. Some advantages of solid state detectors are: (1) they have better energy resolution, (2) they can be manufactured in very small sizes, and (3) they can have very thin entry windows to allow the counting of very low-energy beta and gamma rays (Knoll, 1989). When radiation is absorbed by a solid-state detector, ionization occurs by promoting electrons from the bound state in the crystal to a free state called *a conduction band*, in which the electrons can flow almost as freely as in a metal. The positive charge is carried in the opposite direction by the electric field as the holes left by the excited electrons move to the opposite electrode. The motion of the holes in the crystal lattice is a key property of semiconductors (e.g., silicon and germanium). The free movement of the holes tends to occur only in very pure well-formed crystals because impurities or disruptions in the crystal structure tend to trap or slow down the migration of the holes. As a result, solid-state detectors tend to be very small and expensive.

If the material is not pure, it is sometimes possible to manipulate the semiconductor with the judicious addition of impurities. For instance silicon, which shares four valence electrons in the crystal, always has some level of impurities. If, as in Figure 4, 1 part per million boron is added to the silicon, the boron (with one less outer electron than silicon) will fit into the matrix, creating excess holes. The net effect will be to reduce the number of electrons that are free to conduct current. If one face has a thin layer of phosphorus diffused into the surface, the pentavalent phosphorus will provide local suppression of the holes and an excess of electrons (Figure 4a). A bias across the device that draws the electrons to the left and holes to the right sweeps an area in the middle free of both kinds of charge carriers (Figure 4b). After the initial application of bias sweeps the central region free of charge carriers, the bias is trying to extract electrons from a sea of holes and holes from a sea of electrons. The net effect is a very low current through the crystal. At this point, if radiation is absorbed in the central region, the number of electron–hole pairs produced is large compared to the leakage current, and the resulting pulse is collected and amplified for counting and energy measurement.

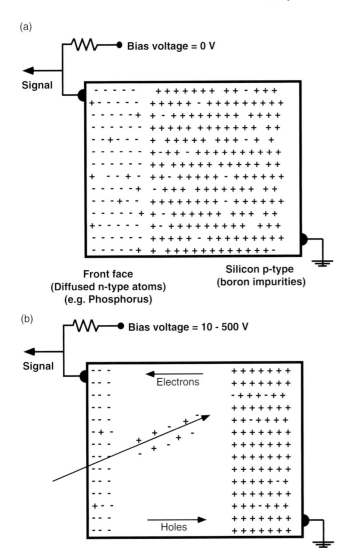

FIGURE 4 Diagram of one method used to create a detector from a semiconductor. One surface of silicon with 1 ppm boron is diffused with phosphorus. A bias voltage pulling the electrons to the phosphorus surface and the holes to the other end creates a carrier-free region. Radiation from gammas or betas will cause ionization in this region, and the electrons and holes are collected to provide the signal.

The earliest uses of solid-state probes were as direct replacements for GM tubes to detect ^{32}P in eye tumors (Pircher *et al.*, 1967; Larose *et al.*, 1971). The possibilities for miniaturization were demonstrated by Lauber (1972) by placing a detector inside a needle with an outside diameter of 1.1 mm and multiple detectors inside a 1.5-mm-diameter needle. These detectors were variations of the silicon device in Figure 4.

1. Avalanche Detectors

The solid-state detectors discussed so far are analogous to gas ionization counters in that the electron–hole pairs are created and simply collected by the electrodes. As in gas detectors, it is possible to design solid-state detectors that have regions of very high electric field that accelerate the electrons to high velocities. At medium fields the amplification is controlled and the output is proportional to the input energy, and at high field the output is large and independent of the input energy, much like a GM tube. This internal amplification is primarily amplification of the true signal and the device has potentially better signal-to-noise ratios than the standard silicon detector. These devices are referred to as *avalanche detectors*.

The initial application of avalanche detectors was for plutonium contamination in the lungs of nuclear power plant workers (Moldofsky and Swinth, 1972). The emissions were 13.5 to 20-keV L X-rays from the uranium daughter for plutonium decay. This application took advantage of the gain and good signal-to-noise ratio of the avalanche detector to detect these very-low energy emissions. Production of avalanche detectors has been fraught with difficulties, and only recently have these devices reemerged for use in some medical imaging applications.

2. Cadmium Telluride Detectors

Cadmium telluride (CdTe) detectors have a relatively high atomic number with respect to silicon, 48, 52, and 14, respectively. This translates into a factor of 100 improvement in the photoelectric stopping power for the 140-keV gamma ray from ^{99m}Tc. Thus, whereas the silicon-based probes were used in beta detection and low-energy gamma-ray detection, the CdTe detectors could be used with the relatively high-energy gamma rays from ^{99m}Tc. One difficulty that has been seen with CdTe, the similar CdZnTe and CdHgTe, and HgI_2 is the trapping of holes in the material. This has had the practical effect of limiting the thickness of the detectors. As long as the radiation was absorbed near the surface, the holes had only a short distance to travel. This resulted in complete charge collection and narrow, symmetric photopeaks. At higher energy, the interactions occur throughout the crystal and a significant fraction of the charge was uncollected. This resulted in photopeaks with low-energy tails that essentially nullify the potential high resolution of this material. Even with this limitation, CdTe is the detector of choice for one of the most widely used intraoperative probe systems. The key to that success is the use of ^{125}I-labeled radiopharmaceuticals—^{125}I has emissions that are ideally suited for CdTe's best energy range (27–37 keV). By limiting the system to ^{125}I, the detector could be relatively thin and low cost and the shielding/collimation requires only about 0.2 mm of lead for a factor of 1000 absorption of ^{125}I photons.

D. Detector Configurations

So far the discussion has been about the basic detectors. There have been a number of ways in which multiple

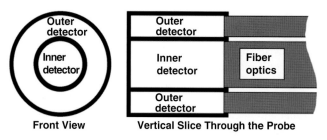

FIGURE 5 Dual-detector configuration with cylindrical detector inside annular detector. Fiber optic bundles transport light to PMTs.

detectors have been configured to improve sensitivity or to measure or suppress background.

1. Dual Detector Probes

The simplest configuration is the dual-detector probe (Hickernell *et al.*, 1988; Daghighian *et al.*, 1994), in which Hickernel and Daghighian use the second detector to provide a measure of the background. The basic configuration is shown in Figure 5. In both cases there was an inner cylindrical detector and an outer annular detector with optical fiber bundles providing the optical coupling to the PMTs.

Hickernel's system was for gamma-ray detection and had lead shielding and collimation. The operating principle was that, if the central detector had a higher count rate than the outer detector, the tumor was in the field of view (FOV).

Daghighian's system was designed to detect positrons and the background was primarily the annihilation radiation produced in other parts of the body. Thus, the outer detector was shielded from the positrons, and its count rate was assumed to be only the annihilation radiation. The system was calibrated with annihilation radiation before use to determine the ratio of the efficiencies of inner and outer detector for the 511-keV radiation. The count rate in the outer detector was multiplied by this ratio and subtracted from the inner detector online to give a net β^+ signal.

2. Coincident Dual-Detector Probes

If the radioisotope of interest emits two gamma rays simultaneously, the probability that both gamma rays will strike the same detector is the square of the probability that one will strike the detector. That is, if there is a 10% chance of one striking the detector, the second has essentially an equal 10% chance. Thus the probability of two gammas striking at the same time is 10% of 10%, or 1%. For a 1-cm-diameter probe the sensitivity would be about 10% at 2 mm from the detector, 3% at 4 mm, 0.6% at 8 mm, and 0.05% at 16 mm. The drop in sensitivity for the detection of both gammas is a factor of 200 in 14 mm. This provides a system that is very sensitive to activity at or just below the surface and essentially immune to background events more than 2 cm away.

The simplest implementation of this technique is to use the sum peak on a single detector (Matherson *et al.*, 1993; Hartsough *et al.*, 1995). For instance ^{111}In emits a 171-keV and a 245-keV gamma ray almost simultaneously. If both strike the detector, the event will register as a 416-keV sum peak. By placing the energy window on the sum peak, the system is effectively selecting those events that are in coincidence. One problem with this method is that if the second gamma ray is from another decay and falls within approximately 1 microsecond of the first, the two unrelated events will sum to the same energy and will give a false event. One way to minimize these events is to use two detectors and accept only those events that are electronically measured to be within 10–100 ns of one another. This reduces the error rate by a factor of 10 to 100. Saffer *et al.* (1992) describe and model an array of detectors using the coincidence concept, extending the idea to include any pair of detectors in an array of detectors. Of course, a critical limitation to this approach is the requirement of an isotope with coincident gamma rays. This is the antithesis of the ideal isotope for most nuclear medicine procedures. The ideal choice has been considered to be a single gamma ray in the range 100–200 keV. Multiple gamma rays tend to lower image quality and increase patient dose.

Watabe *et al.* (1993) use the dual coincidence probe concept in an unusual manner. The rate of accidental coincident events is proportional to the product of the rates on the two detectors. For a source that is moved to various distances from the detector pair, the accidental count rate will vary in the same manner as the coincidence count rate already described. Thus, if you have an isotope that only emits one gamma ray there can be no true coincidences only accidental coincidences. Monitoring the accidental rate would then have the same shallow depth of field as previously described. In addition, if one were to survey a region with tumors that were the same size, but with different amounts of isotope, the count rate over the hotter tumors would increase with the square of the activity. This would give an enhanced visibility to hotter tumors or, conversely, a diminished visibility to cooler tumors.

3. Stacked Silicon Detectors for Beta Detection

Raylman and Wahl (1996, 1998) have developed a solid-state version of the dual detector of Daghighian *et al.* (1994). This device consists of two ion-implanted silicon detectors with an 8-mm-diameter active area. The two detectors are stacked so that the second detector is shielded from all direct betas or positrons, ensuring that all events seen by the second detector are due only to the annihilation radiation. The two detectors are essentially identical in geometry and physical position relative to the flux of annihilation radiation from the body. Although there must be some calibration between the detectors, the count rate on the second detector is the background for the first detector. This means there is no variation due to a significantly different detector shape, which could distort the energy spectrum, as could be the case for the Daghighian et al. (1994) device.

E. Summary of Intraoperative Probe Instrumentation

In this section of this review, the basic principles of intraoperative probe detectors are introduced. Each type of detector is described as a general category. There are many variations of the gas detector, solid-state detector and scintillation detector, and the references at the end of the chapter provide a good starting point for those requiring more detailed knowledge of a particular device. Most of the early work with probes was essentially anecdotal. Usually, a surgeon and an engineer or physicist found they had a mutual interest, and they worked together to build a probe. We usually heard of this as part of a paper covering a few patients with some preliminary and usually promising results.

III. CLINICAL APPLICATIONS

The key to the continued use of a technology, such as an intraoperative probe, is a commercial interest in the product. Each surgeon cannot depend on developing a relationship with a nuclear engineer or physicist to allow him accesses to detector technology. An industrial backing has been developing for the use of scintillator and cadmium telluride probes. A relatively widespread use of intraoperative probes has been developed based on these technologies.

A fairly extensive literature has been accumulating on the use of intraoperative probes in two specific applications. The first, called radioimmunoguided surgery (RIGS), seems to be the most widely used application (Arnold, Schneebaum, Berens, Petty, et al., 1992; Arnold, Schneebaum, Berens, Mojzisik, et al., 1992; Arnold et al., 1995, 1996; Bell et al., 1990; Kim et al., 1993; Martin and Carey, 1991; Nieroda et al., 1989, 1990, 1995; O'Dwyer et al., 1986; Sardi et al., 1989; Waddington et al., 1991; Schneebaum et al., 1995; Nabi et al., 1993; Adams et al., 1998; Aftab et al., 1996; Badalament et al., 1993; Benevento et al., 1998; Burak et al., 1995; Cohen et al., 1991; Cote et al., 1996; de Nardi et al., 1997; Di Carlo et al., 1995; 1998; Greiner et al., 1991; Heij et al., 1997; Krag et al., 1992; Manayan et al., 1997; Martelli et al., 1998; Peltier et al., 1998; Roveda, 1990; Rutgers, 1995; Schneebaum et al., 1997; Stella et al., 1991; Stella et al., 1994; Williams et al., 1995; Barbera-Guillem et al., 1998; Martin et al., 1998). The second, a technique that allows the surgeon to identify the first lymph node downstream from a tumor (usually a melanoma or a breast tumor), has been rapidly gaining on RIGS as the most used procedure. The first node is the most likely place to find cancer cells and, if there are no cancer cells in this sentinel node, the probability that the cancer has spread is low (Alazraki et al., 1997; Albertini et al., 1996; Alex and Krag, 1993, 1996; Alex et al., 1993, 1998; Ames et al., 1998; Bartolomei et al., 1998; Berclaz et al., 1998; Bombardieri et al., 1998; Borgstein et al., 1998; Brobeil et al., 1997; Büchels et al., 1997; Crossin et al., 1998; de Hulla et al., 1998; Dresel et al., 1998; Glass et al., 1996; Greco et al., 1998; Gulec et al., 1997; 1998; Hader et al., 1998; Kapteijn, Nieweg, Liem, et al., 1997; Kapteijn, Nieweg, Muller, et al., 1997; Koch et al., 1998; Krag et al., 1995, 1998; Lieber et al., 1998; Meijer et al., 1996; Mudun et al., 1996; Nieweg et al., 1998; O'Brien et al., 1995; Offodile et al., 1998; Paganelli et al., 1998; Pijpers, Borgstein, et al., 1997; Pijpers, Meijer, et al., 1997; Pijpers et al., 1998; Reintgen et al., 1995, 1997; Rettenbacher et al., 1997; Schneebaum et al., 1998; Snider et al., 1998; Stewart and Lyster, 1997; Thompson et al., 1997; Tiourina et al., 1998; van der Veen et al., 1994; Veronesi et al., 1997; Vidal-Sicart et al., 1998; Cohen et al., 1998; Guliano et al., 1994).

A. Radioimmunoguided Surgery

RIGS was developed as a measurement system by a commercial vendor (Neoprobe Corp., Columbus, OH). The radiolabeled compounds were monoclonal antibodies (MAbs) (Kohler and Milstein, 1975) of colorectal cancers. Many attempts to develop MAbs as nuclear medicine imaging agents were only partially successful. Among the difficulties was the time required for tumor uptake and blood clearance of the isotope. The times required varied from several days to more than 3 weeks (Sardi et al., 1989). Iodine was the labeling agent of choice, but the best imaging agent of the iodine isotopes is ^{123}I, which has only a 13.2-hour half-life, much too short for these clearance times.

The RIGS system allows for this long biological time course by using a relatively long-lived isotope, ^{125}I, which has a half-life of 60 days. ^{125}I decays by electron capture to a 35.5-keV gamma-emitting state, most of which is internally converted. The emissions per decay are 6% 35.5-keV gammas, 115% 27-keV K_α X-rays and 25% 31-keV K_β X-rays. The low energy of the emissions is useful for localizing activity because of the relatively short attenuation length. A 30-keV X-ray is attenuated by a factor of 2 in only 1.9 cm of water. The low energy provides two other benefits in terms of the instrumentation. First, the shielding/collimation requires only the equivalent of 0.2–0.3 mm of lead (1–1.5 mm stainless steel) to reduce background by a factor of 1700 to 70,000. Second, RIGS uses CdZnTe (an improved version of CdTe), which, at these energies, is a very good detector with essentially no trapping to degrade resolution. Both of these factors reduce the weight requirements of the probe. The CdTe detector requires no PMT, just signal and bias lines and the thin shielding, which means that the probe can weigh less than 0.3 kg (Thurston et al., 1991).

The primary reason that the probe gives satisfactory results relative to imaging is the high sensitivity of the probe, which has open collimation similar to that shown in Figure 2. Thus, at close ranges the sensitivity per unit area is a factor of 1000 higher than a gamma camera.

RIGS has been used with some form of MAb labeled with ^{125}I since the mid-1980s (O'Dwyer et al., 1986). Essentially all the reports in the literature have been positive (Arnold, Schneebaum, Berens, Petty, et al., 1992; Arnold, Schneebaum, Berens, Mojzisik, et al., 1992; Arnold et al., 1995, 1996; Bell et al., 1990; Kim et al., 1993; Martin and Carey, 1991; Nieroda et al., 1990, 1995; O'Dwyer et al., 1986; Sardi et al., 1989; Waddington et al., 1991; Schneebaum et al., 1995; Adams et al., 1998; Badalament et al., 1993; Burak et al., 1995; Cohen et al., 1991; Cote et al., 1996; de Nardi et al., 1997; Di Carlo et al., 1995; Greiner et al., 1991; Heij et al., 1997; Manayan et al., 1997; Martelli et al., 1998; Schneebaum et al., 1997; Stella et al., 1991, 1994; Williams et al., 1995; Martin et al., 1998; Thurston et al., 1991). One critical article about radioimmunotargeting (RIT) (Rutgers, 1995), which includes RIGS as a special case, states that RIGS with ^{125}I was the most successful of all the related RIT techniques. One key criticism is that it had not been shown that there was any improvement in patient survival. Because the bulk of the work was performed in the 1990s, and Rutgers (1995) written in 1994, it was probably too early to make a strong argument about patient survival. The other major criticism was that there was a lack of a randomized clinical trial with patients treated by standard or RIGS-based procedures. Each group of patients has been chosen according to the criteria of the individual investigator to answer a different question. Most of the conclusions are positive but vague, with statements that RIGS seems very useful and that it should be investigated further.

The fact that the CdZnTe probe is available has led a number of investigators to look at other isotopes such as ^{99m}Tc, ^{111}In, ^{123}I, and ^{131}I (Heij et al., 1997; Roveda, 1990; Schneebaum et al., 1997). The term *RIGS* is actually a registered trademark of Neoprobe, Corp.; however, this term seems to have been adopted by the field as a general term for any MAb gamma-guided intraoperative procedure (e.g., Peltier et al., 1998). Neoprobe has adjusted to the use of other radionuclides by producing probes, electronics, and collimation appropriate for their use with the higher-energy isotopes (Figure 6 and Table 1).

B. Location of Sentinel Nodes

The sentinel node concept (Morton et al., 1992) is that in the spread of melanoma or breast cancer the metastases will pass down the lymph system and deposit some cells in

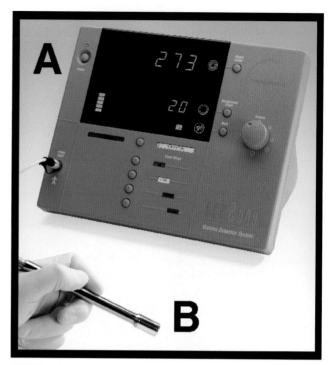

FIGURE 6 Neoprobe Model 2000 RIGS® system. (A) Neoprobe neo2000™ Control Unit with digital readout instead of ratemeter and automated energy windowing. (B) One of the two probes currently available. Custom shielding and collimation is available for each isotope. (Courtesy of Neoprobe Corp., Dublin, Ohio.)

the first, or sentinel, lymph node. If the location of this node can be identified and it can be biopsied, the spread (or the lack of spread) of the cancer can be identified and many times the lymphatic drainage can be spared (Guliano et al., 1994). The sentinel node can be identified by injecting a ^{99m}Tc-labeled colloid in or around the tumor and collecting a series of images to follow the motion of the activity through the lymphatic system (Alex and Krag, 1993; Alex et al., 1993; Krag et al., 1993). The dynamic part of the study allows the identification of the sentinel node in the cases in which there are multiple visible nodes. The static image can be used to locate the node. The location is marked on the skin for the surgeon, who then uses a gamma probe to locate the node surgically. Figure 7 is the static image from a lymphoscintigraphy study.

One of the key features of the image is the high contrast of the sentinel node. The activity is confined to the lymph system and the tumor, with little activity anywhere else in the body. The exceptions, of course, are the cases in which the anatomy of the patient places the sentinel node within or near to the injection site. With the nuclear medicine images as a guide, the surgeon can be confident in locating the node with a gamma probe. Again, the whole system must be considered in the procedure. In this case, the contrast is

TABLE 1 Manufacturers of Intraoperative Probes and Their Critical Properties

Manufacturer	Location	Model	Detector	Detector Size
Capintec in United States	6 Arrow Rd. Ramsey, NJ	GAMMED II B	CdTe CsI(Tl) (Si Photodiode)	$5 \times 5 \times 3$ mm 5 mm diam $\times$ 10 mm
Eurorad in Europe	23 rue du Loess, BP20, F-67037 Strasbourg Cedex 2, France	GAMMED II	CdTe CsI(Tl) (Si Photodiode)	$5 \times 5 \times 3$ mm 5 mm diam x 10 mm
Carewise	P.O. Box 1655, Morgan Hill, CA 95038–1655	C-Trak®	Indium Probe Technetium Probe NaI(Tl) Mini-ProbeEurorad	25 mm Outer diameter 19 mm of Probe Casing 15 mm
DAMRI/ORIS	Gif sur Yvette France	Modelo 2	NaI(Tl)	5 mm diam. $\times$ 15 mm
Diagnostic Technologies	150 Glover Ave. Norwalk, CT 06856	Navigator GPS	CdTe	10 mm diam.
Intra Medical Imaging, LLC	1444 Carmelina Ave. Suite 227, Los Angeles, CA 90025	Node Seeker™	LSO	8 mm diameter $\times$ 7 mm deep
Neoprobe (RIGS®)	425 Metro Place North, Suite 300 Dublin, OH 43017	Neoprobe 1500 neo2000™	CdZnTe	14 mm diam 19 mm diam
STRATEC Biomedical Systems AG	Gewerbestraβe 11 D-75217 Birkenfeld 2 Germany	Tec Probe 200	CsI(Na)	9.5 mm diam. $\times$ 15 mm

From Penkin's (1993); Tiourina *et al.*, (1998).

so high that the requirements of the instrumentation are reduced. A simple scintillation probe is more than adequate for this type of procedure.

The earliest and greatest quantity of work has been in the diagnosis and treatment of melanoma (Krag *et al.*, 1995; Mudun *et al.*, 1996; O'Brien *et al.*, 1995; Pijpers, Borgstein, *et al.*, 1997; Pijpers *et al.*, 1995, 1998; van der Veen *et al.*, 1994; Albertini *et al.*, 1996; Brobeil *et al.*, 1997; Büchels *et al.*, 1997; Glass *et al.*, 1996; Kapteijn, Nieweg, Liem, *et al.*, 1997; Kapteijn, Nieweg, Muller, *et al.*, 1997; Thompson *et al.*, 1997; Reintgen *et al.*, 1997; Rettenbacher *et al.*, 1997; Nieweg *et al.*, 1998; Lieber *et al.*, 1998; Alex *et al.*, 1998; Vidal-Sicart *et al.*, 1998; Cohen *et al.*, 1998; Bartolomei *et al.*, 1998; de Hulla *et al.*, 1998). The early work on sentinel nodes used a blue dye as the tracer, and the surgeon used visual inspection of the lymph nodes to locate the sentinel node (Morton *et al.*, 1992). This technique was tricky to use (Stewart and Lyster, 1997; Kapteijn, Nieweg, Liem, *et al.*, 1997) and, not only was the use of the gamma probe more accurate, it was easier to learn how to use it effectively (Pijpers, Borgstein, *et al.*, 1997; Gulec *et al.*, 1997). There has been a strong interest in applying the sentinel node technique to breast cancer (Meijer *et al.*, 1996; Gulec *et al.*, 1997, 1998; Veronesi *et al.*, 1997; Pijpers, Meijer, *et al.*, 1997; Crossin *et al.*, 1998; Borgstein *et al.*, 1998; Snider *et al.*, 1998; Berclaz *et al.*, 1998; Greco *et al.*, 1998; Offodile *et al.*, 1998; Paganelli *et al.*, 1998; Schneebaum *et al.*, 1998). A large multicenter study for the use of the sentinel node technique in breast cancer was completed in 1998 (Krag *et al.*, 1998). The conclusions of the study were positive but still somewhat skeptical, basically saying that the method worked but that the accuracy depended on the physician. The technique has also been applied to squamous cell carcinoma (de Hulla *et al.*, 1998; Koch *et al.*, 1998) and Merkel cell carcinoma (Ames *et al.*, 1998).

Table 1 lists the manufacturers and some of the properties of intraoperative probes. Although Neoprobe originally designed its system for ^{125}I, many sentinel node studies were done with Neoprobe systems. Neoprobe has added new models to its line of detectors. It should be noted that these companies have an evolving product line.

C. Summary of Clinical Applications

The manufacturers in Table 1 make various types of intraoperative probes readily available to the surgeon. They provide instruction in the use of the probes and service the product when there are problems. The industrial support has led to the availability of thousands of these devices at all types of hospitals, although most are in large medical centers. The use of the sentinel node technique requires only the support of a small nuclear medicine service to provide the ^{99m}Tc-labeled colloid and a set of images to help guide the surgeon. This relatively straightforward approach has led to widespread its use and has already spawned a number of editorials (Perkins, 1993; Alazraki, 1995) that advocate the concept and a number of review

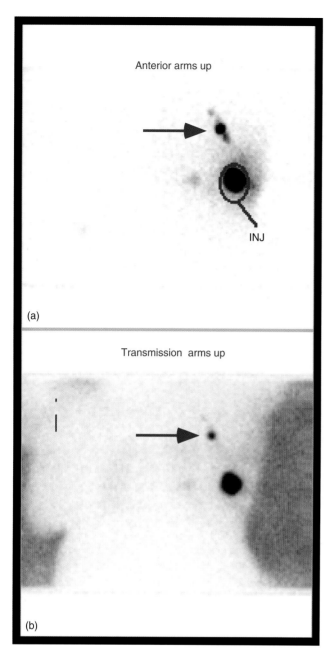

FIGURE 7 A lymphoscintigraphy study to locate the sentinel node. In this case, the activity is injected in a breast tumor and (a) the sentinel node is visualized (arrow), as well as the injection site and part of the activity still in the lymph system. (b) A transmission image of the patient, which gives the general layout of the anatomy. If the patient is scheduled for a biopsy of the sentinel node, she is marked on the skin, and the surgeon uses that mark and the image for the initiation of a gamma-probe-guided procedure.

articles (Alazraki et al., 1997; Stewart and Lyster, 1997; Gulec et al., 1998; Tiourina et al., 1998) that help teach and evaluate the concepts. In a number of cases, the technique has been accepted as being the preferred approach.

The RIGS® approach is a more difficult procedure. It requires the ability to work with the MAbs, and apparently the cost is relatively high (Roveda, 1990). Thus, although it is being used in thousands of procedures, the added difficulty of working with MAbs has slowed its general acceptance.

IV. THE FUTURE: IMAGING PROBES?

It is difficult to imagine what improvements in intraoperative probes could make a significant improvement in their performance. Even if the detector were perfect with 100% photopeak efficiency and perfect resolution, the procedures are basically limited by the amount of isotope that can be injected. The next improvement may be the development of imaging probes. An imaging probe can cover a larger area than a nonimaging probe and still pinpoint the location of the activity. In cases of relatively a poor signal-to-noise ratio, the image can show the distribution of the activity to pinpoint the hot area, whereas the nonimaging probe simply gives a tone or an activity reading representing the average activity in the region.

There are two types of devices that can be considered for the job of an imaging probe: (1) a small probe, 1–2 cm in diameter, to be used as a beta imager, designed for tissue contact readings and employing low-atomic-number detectors to minimize interactions with background gamma rays or X-rays, and (2) a gamma imager of the same dimensions as the beta camera, that uses higher-atomic-number detectors and a collimator.

A. Early Intraoperative Imaging Probes

The earliest imaging probes were a set of prototypes that were developed and evaluated by Barber et al. (1984). The devices were based on the concept of the coded aperture. Each probe was cylindrical with a stack of coaxial detectors and a cylindrical axial collimator with holes that form the coded aperture. They studied systems with aperture cylinders that rotated about the central detectors to give additional sampling and systems that were stationary to avoid the complication of the motion in system design. A possible configuration is shown in Figure 8. In this case the coded aperture rotates about a central stack of detectors, which are isolated by axial collimators. The rotating mechanism must provide feedback so that the system knows the location of each aperture when a gamma ray strikes one of the detectors. After the data are collected, the known response of the system allows the images to be reconstructed by a decoding procedure. The prototypes were capable of imaging high-contrast objects, but gave less satisfactory results when the contrast was low. Because one point of going to the added sophistication of imaging was to allow the user to have an instrument that would allow additional discriminating power in lesion detection, the device did not fulfill its purpose and the development has not continued.

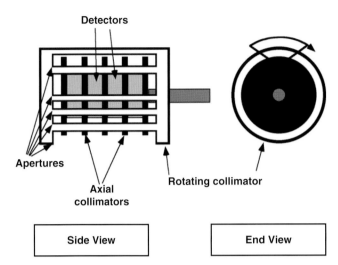

FIGURE 8 Configuration of a possible coded aperture imaging probe. The rotating collimator would be inside a protective sleeve, and the diameter of the device could be relatively narrow.

Woolfenden and Barber (1989) mention the first of a series of semiconductor imaging probes in their review article. The system had an array of 21 2-mm by 2-mm individual CdTe detectors. The problem with individual detectors is that each pixel requires its own set of electronics, which can be very expensive for even a small imaging devices. The obvious, but difficult to achieve, answer is to have a detector array with some sort of multiplexed readout. Barber et al. (1994) describe and evaluate such an array. The device was a 48 by 48 Ge array with 125 micron pixels that used a multiplexer for data readout. The authors have used the technique with CdZnTe, but plan to use these devices in a SPECT imager, not for imaging probes.

B. Beta Imaging Probes

One problem that brought back an interest in beta intraoperative probes was the difficulty of determining the margins of the tumor in brain surgery. It is critical in such surgeries that essentially all tumor tissue be removed, but because of the possibility of injury to the patient it is also important to minimize the loss of healthy tissue. The immediate impetus for a beta detector came from PET images of brain tumors with ^{18}F-flurodeoxyuridine (Fuknda et al., 1983) (Figure 9). The high contrast of ^{18}F-flurodeoxyuridine made it an ideal agent to label a tumor for an intraoperative probe procedure. However, the high energy (511 keV) of the annihilation radiation required a relatively large and unwieldy detector with heavy shielding. A more attractive approach was to detect the positrons directly, using a low Z detector, such as a plastic scintillator (Daghighian et al., 1994). The low Z of the plastic greatly reduced the background from the annihilation radiation, and the dual-detector technique (see Figure 5) allowed the measurement of and

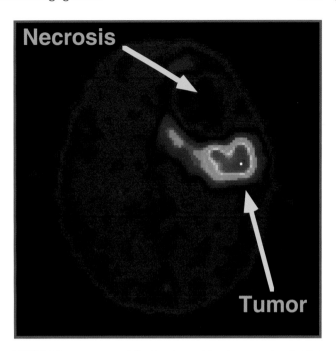

FIGURE 9 PET image of ^{18}F-flurodeoxyuridine demonstrating the high contrast of the tumor relative to healthy tissue in the brain.

correction for the residual 511-keV background. The short range of the positrons would theoretically allow localization on the order of 1 to 2 mm.

1. Scintillator-Based Beta Imaging Probes

It was anticipated that the bulk of the tumor be removed before the probe was employed to locate residual bits of the tumor. In order to have good localization, it was necessary for the detector to be small (~ 3-mm diameter), but this meant a large number of readings on the tissue surface. One way to reduce the problem of the large number of readings is to use a relatively large imaging detector (~ 12-mm diameter), which allows good localization while covering a much larger area (by approximately a factor of 16).

The initial concept of imaging probes was based on the use of scintillators coupled to photodetectors through fiber optics. A major problem with fiber optics was that only a small fraction of the randomly oriented scintillation light was actually transmitted down the fiber. Generally, fiber optics are noted for their ability to carry light long distances, but in such instances the light is from a source, such as a laser, directed down the center of the fiber. The situation for the probe is illustrated in Figure 10.

An optical fiber consists of a core, which transports the light, and a thin cladding, which reflects light back into the core. In order for light to travel down a fiber optic, it must be traveling in a direction such that it will strike the cladding beyond the critical angle for total internal reflection between the cladding and core ($n_{Core} > n_{Cladding}$). This limits the fraction of the light from the scintillator that will travel down the

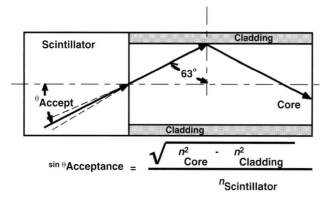

FIGURE 10 A scintillator attached to a single-claw optical fiber. For this case $n_{Core} = 1.59$; $n_{Cladding} = 1.42$; and $n_{Scintillator} = 1.43$ for CaF$_2$(Eu), 1.85 for NaI(Tl), and 1.59 for plastic scintillators. The solid arrow in scintillator indicates the acceptance angle for plastic scinitllator (26.7°) and dashed lines are for NaI(Tl) (22.7°) and CaF$_2$(Eu) (29.9°).

fiber optic. The fraction depends on the index of refraction of each component, and a cone of acceptance is defined by the angle given in the equation in Figure 10. Thus, if the angle of the light at the interface falls within the cone of acceptance, the light will be beyond the angle of total internal reflection in the fiber and it will be transmitted. The angles of acceptance for NaI(Tl), plastic, and CaF$_2$ are 22.7°, 26.7° and 29.9°, respectively, which translate into the respective geometric efficiencies of 3.9%, 5.2%, and 6.7%.

Plastic scintillators are the obvious low Z detector of choice for beta imaging, but the problem of getting a strong enough signal at the end of the fiber optic had to be solved. A plastic scintillator produces about 1000 scintillation photons for 100 keV of deposited energy. At 5% geometric efficiency, 50 photons exit the fiber optic and a PMT detects about 10 of these. This signal strength is too low for a reliable clinical device, and improvement of the signal was given top priority.

In the work of MacDonald *et al.* (1995; MacDonald, 1996), a very sensitive solid-state photon counter termed a visible light photon counter (VLPC) (Petroff and Staepelbroek, 1989) was used instead of a PMT. The VLPC has a 60–80% counting efficiency and is able to detect on the order of 30–40 photons per 100 keV. In addition, the VLPC has essentially no noise and is capable of resolving events that consisted of 1, 2, or 3 photons. Thus, the signal-to-noise ratio was excellent for the 100- to 200-keV beta energies that dominate the signal from ^{18}F. The major drawback of the VLPC was that it required cooling to just above liquid helium temperatures, and the problems associated with the cooling meant it was not a practical clinical device.

Fortunately, other aspects of the optimization process brought about an increase in the number of photons transported down the fiber optics. One of the primary improvements came from the realization that CaF$_2$ had almost ideal properties for this application. Its photon yield is 63% of NaI and 231% of a good plastic scintillator. Its low index of refraction means that more of the signal is transported down the fiber. The light yield at the PMT is 109% of a NaI(Tl) scintillator and 312% of a plastic scintillator or 30–60 photoelectrons in a PMT for 100- to 200-keV beta particles. In addition, the choice of reflector and detector thickness was optimized to increase the total photon flux down the fiber.

The design of an intraoperative beta imaging probe based on a scintillator and fiber optics is shown in Figure 11 (Tornai, MacDonald, Levin, Siegel, and Hoffman, 1996; Tornai, MacDonald, Levin, Siegel, Hoffman, *et al.*, 1996). In this system a thin disk of CaF$_2$ is the scintillator and a disk of transparent plastic diffuses the light among a number of fiber optics arranged in a hexagon in the manner of the PMTs on early scintillation cameras.

The diffuser thickness was chosen to provide a uniform flood field image as well as maintain good resolution. Figure 12 shows the flood field image as a function of diffuser. In this series of images, the fiber optics are clearly visible with no diffuser in place and they disappear as the diffuser thickness is increased. The uniformity is essentially constant for thickness greater than 1.7 mm. However, the resolution was also seen to gradually decrease with the thickness of the diffuser, although at a surprisingly slow rate.

The spatial resolution was found to be 0.63 ± 0.1 mm full width at half maximum (FWHM), when measured with a 0.1 mm slit transmission source. The system is designed to be placed in contact with the tissue of the patient with no collimator. The betas are emitted in all directions and ^{18}F was found to have an intrinsic spread on the order of 1.2 mm FWHM. The resolving power of this beta imaging system can be seen in the image in Figure 13.

A potential problem with a beta imaging probe is the annihilation radiation that is present when the beta happens to be a positron (Levin *et al.*, 1997). It is possible to get a measurement of this background by simply placing a plastic cap over the face of the imaging probe and taking a second image. In practice this would be cumbersome in a surgery and it would be difficult to reproduce the position of the probe for the second measurement. A number of methods were considered for suppression of this background (Levin *et al.*, 1997; Tornai *et al.*, 1998; Hoffman *et al.*, 1998). One method that was implemented and worked is shown in Figure 14.

This type of system, generally referred to as a *phoswich*, takes advantage of the difference in decay time of the scintillation light between the CaF$_2$(Eu) and the high Z scintillator (gadolinium orthosilicate, GSO; bismuth germanate, BGO; or lutetium orthosilicate LSO). When the positron is absorbed by the CaF$_2$(Eu), the scintillation light passes through the high Z scintillator and down through the optical fibers to give a positioning signal. The positron also annihilates in the CaF$_2$(Eu) and gives off two 511-keV photons at

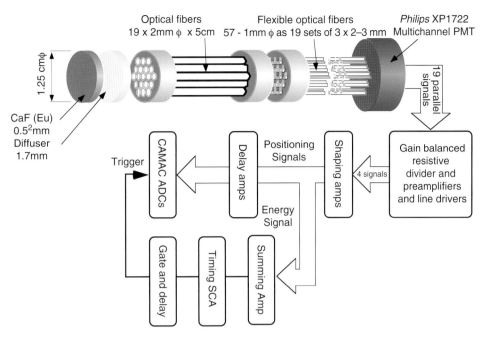

FIGURE 11 Diagram of beta imaging probe. The CaF$_2$(Eu) disk is optically coupled to a diffuser, which is then coupled to 19 2-mm-diameter optical fibers. This set of thick fibers with holders provides a rigid handle for the surgeon to hold. These are then coupled to 2–3 meters of flexible optical fibers that bring the signal to a multichannel PMT (MCPMT). The signals from the MCPMT are fed into a resistive divider network that provides X_+, X_-, Y_+, and Y_- positioning signals of the type used in scintillation cameras. These signals are then shaped, amplified, and fed to analog-to-digital converters (ADCs), while a sum of the signals is used to set an energy threshold and provide a trigger for the ADCs. This system is still in the prototype stage and all electronics and data collection are performed with modular research systems.

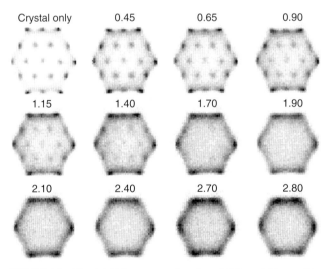

FIGURE 12 Flood field images as a function of diffuser thickness. The data are taken with a ^{204}Tl point source at a distance. ^{204}Tl is a pure beta emitter with a maximum beta energy of 763 keV. It is used because it has a half-life of 3.8 years and the energy is conveniently close to that of ^{18}F.

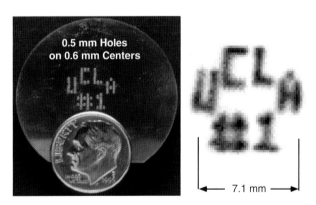

FIGURE 13 Transmission image of holes in a copper disk with a ^{204}Tl beta source. The holes have 0.5-mm diameters on 0.6-mm centers, which is essentially the same as the resolution of the probe.

180 degrees to one another. Because the high Z scintillators are coupled almost directly to the CaF$_2$(Eu), there is a very high probability that the high Z scintillator will detect one of these photons. The system requires that both types of events occur simultaneously before an event is considered valid. The manner in which this is achieved can be appreciated from the diagram in Figure 15, which is a digitized version of the signal from the PMT.

As shown in Figure 15, the fast GSO signal rises and falls in approximately 200 ns and because of this the amplitude is much greater than that of the CaF$_2$(Eu) signal. The problem with the GSO signal is that it bears very little

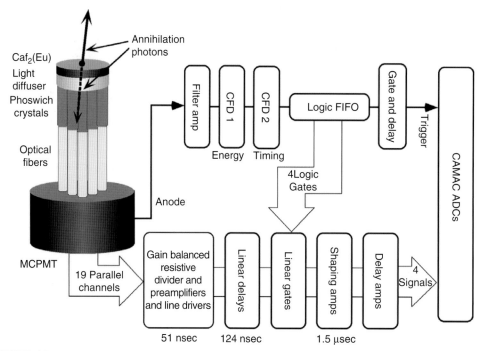

FIGURE 14 Diagram of a gamma-suppressing beta imaging probe. This system is based on the beta probe shown in Figure 11, except for two key differences. (1) The first 1–2 cm of the 2-mm-diameter fiber optic is replaced by a high Z scintillation detector. (2) The signals are taken from the MCPMT anode, which is the sum of all channels, and these signals are used for energy discrimination and a time pickoff to determine when a gamma interacted with the high Z scintillator.

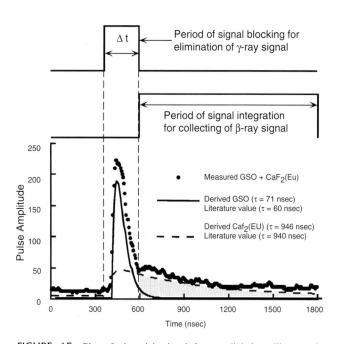

FIGURE 15 Plot of phoswich signal from a digital oscilloscope in which the beta (CaF$_2$(Eu)) and gamma (GSO) components are both present (dots). A least-squares fitting program was used to extract the two components from the signal with (τ) as a floating parameter. The results are given as solid line for GSO and a dashed line for CaF$_2$(Eu). In the diagram, it can be seen that a large fraction of the beta signal occurs after the GSO signal is gone.

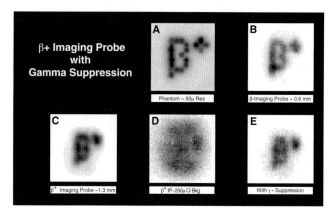

FIGURE 16 Images of β^+ phantom with various beta imaging systems using ^{18}F. (A) A high-resolution autoradiographic system. (B) Original beta imaging prototype (Figure 11). (C) Phoswich prototype with no out-of-field activity. (D) Phoswich prototype with 250 µCi out-of-field activity, but with no gamma suppression. (E) Phoswich prototype with 250 µCi out-of-field activity, but with gamma suppression.

relationship to the position of the beta signal, yet it must travel down the same set of optical fibers. In order to eliminate the GSO signal, the ADC is blocked for approximately 250 ns, as indicated in Figure 15, and then is opened for 1 µs to collect the CaF$_2$(Eu) signal.

The results from a prototype of this system are shown in Figure 16. Cylindrical high Z scintillators were not available and rectangular 2 mm × 2 mm × 10 mm crystals were employed in the optical path for proof of principle. The better geometric cylindrical crystals will undoubtedly provide higher resolution results. The phantom for Figure 16 was a series of shallow 0.5 mm holes on 0.6 mm centers in the form of a β^+. The variable intensity is due to the difficulty of accurately pipeting volumes on the order of 0.2 microliters. Image A was taken with an autoradiography system with 50-micron resolution. Image B was made with the system that produced Figure 13. Image C was taken with the phoswich system with only ^{18}F in the phantom. The loss of resolution due to the square high Z scintillators is obvious. Image D was taken with 250 µCi in a 400-ml beaker behind the phantom but without gamma suppression. Image E is the same as D, but with suppression. The image is obliterated without suppression (D), and the suppression brings back an image comparable in resolution to C but noisier (E).

2. Solid-State Beta Imaging Probes

The use of solid-state detectors for gamma probes has had problems due to hole trapping for high Z detectors and low sensitivity in the case of silicon. However, silicon has more than enough stopping power for betas. Significant progress in this area has been made in collaboration between our laboratory and Photon Imaging, Inc. (North Ridge, CA). The use of silicon as a direct detector for betas has been used for years. The application in this case is to create an imaging array from a small silicon wafer. The device is the same in principle as the one in Figure 4, except that the electrodes on the front and backsides are made of 16 isolated strips. The back side collects the electron signal from horizontal strips (Figure 17) to determine the y location of the event and the hole signal is collected on the front side from vertical strips to determine the x position of the event. Using strip detectors requires only 32 sets of electronics instead of the 256 channels required of a fully pixellated device.

Point sources of ^{18}F gave image resolutions of 1.5 ± 0.07 mm FWHM. The energy resolution was found to be 3.6 keV. Additional preliminary results are given in Figure 18. The spectrum of ^{18}F is shown directly and after a thin absorber blocked the positrons themselves (Figure 18a). The gamma signal is seen to be approximately 2.5% of the total. A simple imaging task was done with the stick figure transmission phantom shown on the strip detector. The holes on the arms are at 1-mm centers.

Although there is good suppression of the gamma background, it will probably be necessary to provide some method for suppression or subtraction of this background. One way to do this is to sacrifice the corners of the device with small beta shields. This would allow only gammas to be detected at the corners, and the average gamma rate could then be subtracted from the image. This assumes there is no structure in the gamma data. A more accurate method is similar to the phoswich method but without the use of the common signal path. PMTs of 10- to 13-mm diameters are available that could easily be incorporated into the handle of an imaging probe. Thus, a high Z scintillator could be coupled to a small PMT and have a strip detector mounted on its front end. Again, when a positron annihilates in the strip detector, the annihilation photons will strike the high Z scintillator, and the only valid events will have both signals. In this case, the events are recorded only if they occurred within 10–20 ns of one another. By comparison, the phoswich method required a wider time window of 1 microsecond, which meant a 50–100 times greater probability that an accidental event will be accepted in the background than a true coincidence technique.

3. Radiopharmaceuticals for Beta Probes

An important part of the measurement system for beta imaging probes is the radiopharmaceutical. Most radiopharmaceutical development has emphasized gamma emitters. The fact that positrons were also betas has been a problem rather than an advantage for PET. With beta probes, imaging or nonimaging, the fact that a PET image can be made with the tumor-seeking compound can be an important advantage in planning the surgery. Thus far we have seen that [F-18]-5-fluorodeoxyuridine (Abe et al., 1983) and 18F-Fluorodeoxyglucose (Raylman and Wahl, 1994, 1996, 1998; Raylman et al., 1995) have shown promise for beta probes.

^{99m}Tc has been the isotope of choice for most of nuclear medicine. The reason for that choice has been the fact that it has optimal gamma energy and it has no beta particle associated with its decay. A potential solution for this problem has become available in the form of a generator-based beta-emitting analog of ^{99m}Tc. The isotope in question is ^{188}Re, which is the 17-hour daughter of ^{188}W

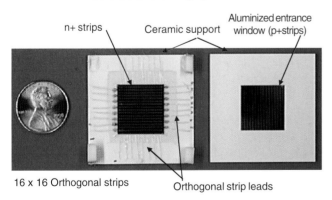

FIGURE 17 Photograph of silicon strip prototype for beta imaging probe (front and back). The device is mounted on a large support for convenience in testing. The front face is covered with a thin layer of aluminum to allow the betas to penetrate while blocking out light that would also produce a signal in the device. The strips have a pitch of 1 mm.

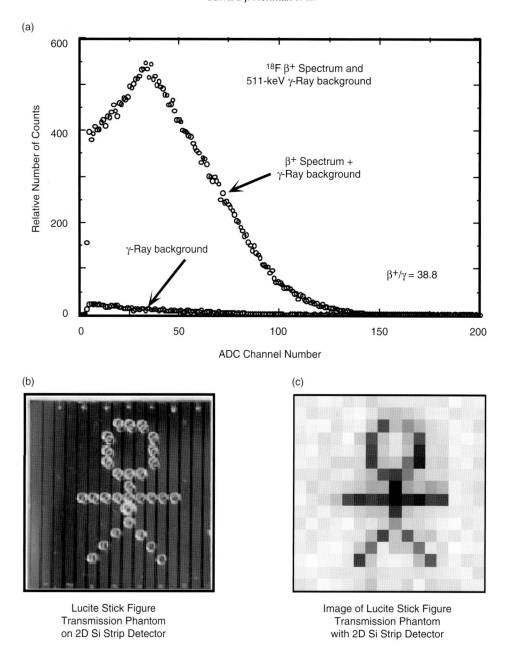

FIGURE 18 Preliminary results from silicon strip beta imaging probe. (a) The spectrum of betas and gammas from ^{18}F showing the suppression of gammas. (b) The stick phantom on the strip detector. (c) Image of phantom.

(half-life 69 days). The isotope has only a small gamma component, which can be useful in its quality control. ^{188}W is available from double neutron capture on ^{186}W in a high flux reactor (Knapp *et al.*, 1997; Ehrhardt *et al.*, 1998). A ^{188}W/^{188}Re generator is available (Isotope Products Laboratories, Burbank, CA). Re is in the same column in the periodic table as Tc and, because it is above the lanthanides and is subject to a phenomenon known as the lanthanide contraction, it has a size almost identical to Tc. Its chemistry is in many cases almost identical to Tc (e.g., the generator column and eluant are essentially the same as Tc). A number of groups are working with this generator system to produce tumor-seeking compounds that can be used for radiotherapy (Kairemo, 1996; Kairemo *et al.*, 1998; Prakash *et al.*, 1996; Rhodes *et al.*, 1996; Truitt *et al.*, 1997; Zamora, Bender, Gulhke, *et al.*, 1997; Zamora, Bender, Knapp, *et al.*, 1997). The tumor-to-tissue ratio requirement is much higher for therapy than diagnosis because of the very high doses used in therapy. Thus, a compound that might be considered inadequate for therapy could be almost perfect for use with a beta imaging probe. Its similarity to technetium means that preliminary localization can be done with ^{99m}Tc and a scintillation camera, as in the sentinel node work.

C. Gamma Imaging Probes

Barber and Woolfenden first introduced the concept of gamma imaging probes (Barber *et al.*, 1994; Woolfenden and Barber, 1990), but used the technology to pursue other avenues. Another approach that used a HgI$_2$ crystal set up as a 16 × 16 array was successful as a prototype, but was abandoned for more practical approaches based on silicon technology (Patt *et al.*, 1997). In our laboratory, the gamma imaging probe was approached in a manner similar to the beta probe. Instead of CaF$_2$(Eu), NaI(Tl) was the scintillator of choice. Initially a continuous detector was considered, but a 6- to 8-mm-thick by 12-mm diameter crystal did not allow for good image resolution and thinner crystals did not have the required sensitivity (Tornai *et al.*, 1997). Crystal arrays allowed for both good resolution and high efficiency, but required borrowing the principle of the PET block for implementation (Tornai *et al.*, 1997; Hoffman *et al.*, 1997).

The basic physical arrangement is diagrammed in Figure 19. The light coming from the crystal array is detected as an image of the crystals. The counts in each spot correspond to events in the particular crystal and a large number of small crystals are required for an image. In order to turn the spot pattern into an image the pattern is decoded in a lookup table (LUT). The data from the system are simple *x* and *y* values corresponding to the apparent position of the scintillation light. The image of the spot pattern from a flood source is stored in an appropriate size matrix. This image is shown in Figure 20. On this calibration image, a set of regions of interest are drawn, and a matrix of numbers with dimensions the same as the image size is created. The values in the matrix correspond to the detector number that is associated with that part of the image. Thus when an *x* and *y* combination from the ADCs provides an address location in the matrix, that location is

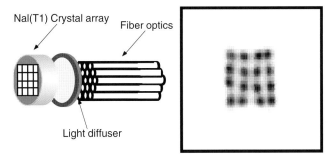

FIGURE 19 Diagram of gamma imaging probe. The diffuser and fibers produce an image of the crystal and all counts in each spot correspond to events absorbed in that crystal.

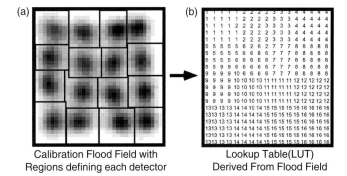

FIGURE 20 The technique for creating a detector identification LUT for a discrete crystal imaging system. (a) The calibration allows the drawing of regions of interest on the image matrix. A detector number is assigned to each region of interest, and the matrix elements corresponding to each region of interest are set equal to these numbers. (b) When image data are collected, the *x* and *y* data correspond to the LUT elements and are used fetch the detector number from the table.

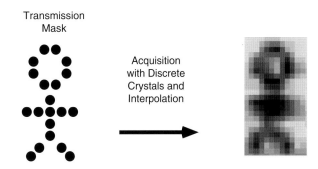

FIGURE 21 Image of stick figure similar to that in Figure 18, except the absorber is lead instead of plastic. Because the array was only 4 × 4, it was necessary to scan it over the phantom to get an image of the whole figure, and interpolation was done to get a less pixellated image.

read and contains the detector number. The location in the image being collected is then incremented by one.

A sample of an image taken with this 4 × 4 matrix is shown in Figure 21. In this case the crystal array was scanned to get the full image. The intrinsic resolution of a discrete crystal system is defined by the crystal size, and the system resolution is then defined by the collimator geometry and to some extent by the septal thickness.

A commercial imaging probe has become available (LumaGem, Northridge, CA). The device, shown in Figure 22, consists of a 25 mm by 25 mm position-sensitive photomultiplier coupled to a 16 × 16 array of CsI(Na) crystals. The crystals are 1 × 1 × 3 mm^3 and are separated from one another by an optically opaque material that is 0.25 mm thick. The scintillator array is covered with a diffusely reflective top and coupled directly to the PSPMT with an

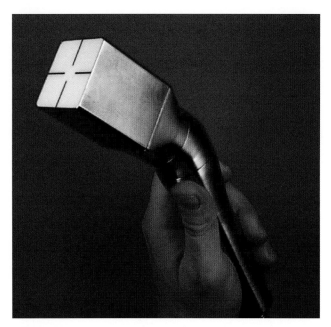

FIGURE 22 A commercial imaging probe (LumaGem, Northridge, CA). The device consists of a 25 mm by 25 mm position-sensitive photomultiplier coupled to a 16×16 array of CsI(Na) crystals. The crystals are $1 \times 1 \times 3$ mm^3, separated by optically opaque material 0.25 mm thick. The system, shown with a parallel-hole collimator, also has diverging and pinhole collimators.

optical coupling agent. The crystal of interaction is identified by the method illustrated in Figure 20. The system has a set of interchangeable collimators (parallel-hole, diverging, and pinhole collimators). The system is connected to a laptop computer. The device can operate in a simple detection mode identical to the standard probe, but as it is counting the pattern of the activity it analyzes for concentrated regions of activity. When one or two hot spots are detected, there is a distinctive tone added to the audio output of the system. Once the activity is located, the probe can then record an image to guide the surgeon.

D. Imaging Probe Summary

The technology for both beta and gamma intraoperative imaging probes is in place and has been tested on prototype units. Methods have been developed based on scintillation crystals coupled to PMTs with optical fibers and with direct coupling to small PSPMTs. The technology has been developed for suppression of the background due to the gamma rays and annihilation radiation that is associated with a number of isotopes. The resolutions achieved are compatible with ^{18}F, which is the lowest energy beta that is likely to be useful with beta imaging probes. If ^{188}Re is developed as a radiopharmaceutical for intraoperative beta probes, there should be a general improvement in the method. ^{188}Re has maximum beta energies of 1.98 and 2.13 MeV, which will cause some blurring of the image but also means that there is greater penetration of the tissue. However, the higher energy means the detector window can be stronger and the signal produced in the detector will be higher.

For the future, the silicon strip detector has superior properties for the beta imaging probe. The intrinsic resolution, which is already excellent, can be improved by simply making narrower strips. The energy resolution for the prototype was on the order of 3.6 keV, which is superior to any imaging detector being considered at this time. The excellent energy resolution allows for the possibility of using energy windowing to provide depth information (i.e., the lower energy betas tend to originate in deeper tissue, while high energy betas tend to come from the surface.).

Gamma imaging probes require a collimator for imaging, which causes a drop in sensitivity. Thus the pinhole collimator, which has high efficiency when directly over the hot area, is currently the collimator of choice. The strength of essentially all the intraoperative probes discussed in this review is their high sensitivity. The detectors are only collimated by a tube of absorber to eliminate radiation from the side. Even the beta imaging probe is wide open and, when placed against the tissue, it has almost 50% geometric efficiency.

The most likely nuclear medicine intraoperative imaging system is one of the small scintillation cameras being developed, for breast imaging (Levin, 1997; Pani *et al.*, 1998; Weinberg *et al.*, 1997; Patt *et al.*, 1998). The imaging probe described here is a derivative of this technology. It is one of the detector units of a 5 by 5 camera system. There are a number of other cameras reported in the literature that are based on PSPMTs (Levin, 1996; Pani *et al.*, 1998; Weinberg *et al.*, 1997), and these have fields of view ranging from 5 cm by 5 cm to a 10-cm diameter. These devices have more bulk than is desirable, but they are much smaller than the portable scintillation camera that is common in the nuclear medicine clinic. In the long run, PMT cameras should be replaced by cameras consisting of scintillators coupled to solid-state photodiode arrays (Patt *et al.*, 1998). In these systems the photodetector is an array of photodiodes on a single wafer of silicon. The scintillator can be a continuous crystal or a matrix of crystals designed to be coupled one to one with the elements on the wafer. Early testing shows energy resolutions averaging 8.7% FWHM for ^{99m}Tc for a set of 16 pixels and best resolution of the set of 6.4% FWHM. The crystal plus electronics need to only be 1–2 cm thick for such a system. With a shallow depth of field collimator, the system could be less than 5 cm thick by perhaps 12 cm in diameter. It would be small enough to bring into a surgery and large enough to image the whole area under suspicion. The high-efficiency shallow-field collimator and the fact that it might only require a single view of the surgical field might make it the choice relative to a small intraoperative gamma imaging probe.

V. DISCUSSION

Intraoperative probes have been used for over 50 years. Yet it is only in the 1990s that their use has had an impact. In the first 40 years (1946–1986), the field consisted primarily of physician–physical scientist collaborations in the fine tradition of interdisciplinary collaboration. Individuals met and found out that together they had the potential solution for a problem, in this case locating tumors in surgery. They built a probe system, employed it on a few cases, and published a paper. Then there was little follow-up. A wide variety of technologies were employed from the simplest of GM tubes to scintillation detectors to the most advanced solidstate detectors systems. Essentially every radiation detector system that could be miniaturized was tested or at least suggested as an intraoperative probe. Many of these systems were good solutions for the instrumentation aspects of the problem.

Since the mid-1980s, commercial support has developed for the concept. Neoprobe and Carewise seem to dominate the market, but a number of other companies (see Table 1) are also in the market. Because of this support, a surgeon can simply buy the probe, and the company will supply basic training and technical support; it is the surgeon, not the nuclear medicine physician, who is driving the research and development in this field (Alaraki, 1995). This can also be inferred from the journals dominating the reference list for this chapter.

There are two types of procedures dominating the literature and, presumably, clinical practice. Each of these has the properties of a good measurement system that ensure a good level of success. The first, the RIGS system, solves some of the problems of MAbs by using a long-lived label (^{125}I), which gives a much higher contrast at the time of surgery. The low energy of the emissions gives better localization because the absorption of the radiation in tissue causes a rapid drop in signal with distance. The good signal-to-noise ratio and energy resolution of the detector allows it to be used reliably at such low energy. Any lack of total acceptance of RIGS is due to the checkered history of MAbs as a reliable imaging agent:

The second procedure is the localization of the sentinel node. Again the nature of the localization is key. In many cases (e.g., Figure 7), the sentinel node is a lightbulb with very little background. With the imaging study as a guideline, the localization and biopsy of the sentinel node with an intraoperative probe is becoming a standard procedure. It is sometimes the preferred procedure. Part of the success of this procedure is that the process that is being observed is the well-understood drainage through the lymph system. Thus, the surgeon is confident of the interpretation of the images and the count rates that he sees with the probe.

It is not clear that improvement in the detectors for intraoperative probes will significantly improve their usefulness. Therefore, a completely different capability, that of imaging, will most likely provide the next major advance in intraoperative probes. Prototypes of beta imaging probes demonstrate the potential of this approach. The use of beta imaging is efficient because no collimator is employed and is highly localizing because of the short range of the beta. On the other hand, an equivalent gamma imaging probe may have some problems with sensitivity, particularly if a number of images are required over the area of the surgical field. The solution for gamma imaging during surgery is likely to be small scintillation cameras with low-resolution pinhole collimation or cameras that can image all or most of the surgical field in a single view while still being small enough to be relatively unobtrusive.

VI. CONCLUSION

The development of the various intraoperative probes systems is another example of a success of interdisciplinary collaborations. However, until there is commercial support for such a system, little more than anecdotal research is produced. Many of the early probe systems had the same capability as the systems that are now supplied by commercial vendors, but in the past there was no infrastructure to support the nonexpert user. The supplying of the probes and support to the user are equally important to the success of this technique as is the original development of the probes.

The RIGS and sentinel node procedures have established intraoperative probes as useful procedures. Whether imaging probes can have a similar impact has yet to be seen. If the past can be used as a predictor, the commercial support of imaging probes will be the key to their success.

Acknowledgements

This chapter is an updated version of the review article Hoffman EJ, Tornai MP, Janacek M, Patt BE, Iwanczyk JS, (1999), Intraoperative probes and imaging probes. *Eur. J. Nucl. Med.* **26**, 913–935 and is printed with the permission of the *European Journal of Nuclear Medicine*. This work was supported in part by DOE Contract DE-FC03-87ER60615.

References

Abe, Y., Fukuda, H., Ishiwata, K., Yoshioka, S., Yamada, K., Endo, S., Kubota, K., Sato, T., Matsuzawa, T., Takahashi, T., and Ido, T. (1983). I. Tumor uptakes of [F-18]-5-fluorouracil, [F-18]-5-fluorouridine, and [F-18]-5-fluorodeoxyuridine in animals. *Eur. J. Nucl. Med.* **8**: 258–261.

Ackerman, N. B., Shahon, D. B., and Marvin, J. F. (1960). The diagnosis of thyroid cancer with radioactive phosphorus. *Surgery* **47**: 615–622.

Adams, S., Baum, R. P., Hertel, A., Wenisch, H. J., Staib-Sebler, E., Herrmann, G., Encke, A., and Hör, G. (1998). Intraoperative gamma probe detection of neuroendocrine tumors. *J. Nucl. Med.* **39**: 1155–1160.

Aftab, F., Stoldt, H. S., Testori, A., Imperatori, A., Chinol, M., Paganelli, G., and Geraghty, J. (1996). Radioimmunoguided surgery and colorectal cancer. *Eur. J. Surg. Oncol.* **22:** 381–388.

Alazraki, N. (1995). Lymphoscintigraphy and the intraoperative gamma probe [editorial; comment]. *J. Nucl. Med.* **36:** 1780–1783.

Alazraki, N. P., Eshima, D., Eshima, L. A., Herda, S. C., Murray, D. R., Vansant, J. P., and Taylor, A. T. (1997). Lymphoscintigraphy, the sentinel node concept, and the intraoperative gamma probe in melanoma, breast cancer, and other potential cancers. *Sem. Nucl. Med.* **27:** 55–67.

Albertini, J. J., Cruse, C. W., Rapaport, D., Wells, K., Ross, M., DeConti, R., Berman, C. G., Jared, K., Messina, J., Lyman, G., Glass, F., Fenske, N., and Reintgen, D. S. (1996). Intraoperative radio-lymphoscintigraphy improves sentinel lymph node identification for patients with melanoma. *Ann. Surg.* **223:** 217–224.

Alex, J. C., and Krag, D. N. (1993). Gamma-probe guided localization of lymph nodes. *Surg. Oncol.* **2:** 137–143.

Alex, J. C., and Krag, D. N. (1996). The gamma-probe-guided resection of radiolabeled primary lymph nodes. *Surg. Oncol. Clin. N. Am.* **5:** 33–41.

Alex, J. C., Krag, D. N., Harlow, S. P., Meijer, S., Loggie, B. W., Kuhn, J., Gadd, M., and Weaver, D. L. (1998). Localization of regional lymph nodes in melanomas of the head and neck. *Arch. Otolaryngol.—Head Neck Surg.* **124:** 135–140.

Alex, J. C., Weaver, D. L., Fairbank, J. T., Rankin, B. S., and Krag, D. N. (1993). Gamma-probe-guided lymph node localization in malignant melanoma. *Surg. Oncol.* **2:** 303–308.

Ames, S. E., Krag, D. N., and Brady, M. S. (1998). Radio-localization of the sentinel lymph node in Merkel cell carcinoma: a clinical analysis of seven cases. *J. Surg. Oncol.* **67:** 251–254.

Arnold, M. W., Hitchcock, C. L., Young, D., Burak, W. E., Jr., Bertsch, D. J., and Martin, E. W., Jr. (1996). Intra-abdominal patterns of disease dissemination in colorectal cancer identified using radioimmunoguided surgery. *Dis. Colon Rectum* **39:** 509–513.

Arnold, M. W., Schneebaum, S., Berens, A., Mojzisik, C., Hinkle, G., and Martin, E. W., Jr. (1992). Radioimmunoguided surgery challenges traditional decision making in patients with primary colorectal cancer. *Surgery* **112:** 624–630.

Arnold, M. W., Schneebaum, S., Berens, A., Petty, L., Mojzisik, C., Hinkle, G., and Martin, E. W., Jr. (1992). Intraoperative detection of colorectal cancer with radioimmunoguided surgery and CC49, a second-generation monoclonal antibody. *Ann. Surg.* **216:** 627–632.

Arnold, M. W., Young, D., Hitchcock, C. L., Schneebaum, S., and Martin, E. W., Jr. (1995). Radioimmunoguided surgery in primary colorectal carcinoma: an intraoperative prognostic tool and adjuvant to traditional staging. *Am. J. Surg.* **170:** 315–318.

Badalament, R. A., Burgers, J. K., Petty, L. R., Mojzisik, C. M., Berens, A., Marsh, W., Hinkle, G. H., and Martin, E. W., Jr. (1993). Radioimmunoguided radical prostatectomy and lymphadenectomy. *Cancer* **71:** 2268–2275.

Barber, H. B., Augustine, F. L., Barrett, H. H., Dereniak, E. L., Matherson, K. L., Meyers, T. J., Perry, D. L., Venzon, J. E., Woolfenden, J. M., and Young, E. T. (1994). Semiconductor arrays with multiplexer readout for gamma-ray imaging: results for a 48*48 Ge array. *Nucl. Instru. Meth. Phys. Res. A* **353:** 361–365.

Barber, H. B., Barrett, H. H., Wild, W. J., and Woolfenden, J. M. (1984). Development of small *in vivo* imaging probes for tumor detection. *IEEE Trans. Nucl. Sci.* **31:** 599–604.

Barber, H. B., Woolfenden, J., Donahue, D. J., and Nevin, W. S. (1980). Small radiation detectors for bronchoscopic tumor localization. *IEEE Trans. Nucl. Sci.* **27:** 496–502.

Barbera-Guillem, E., Arnold, M. W., Nelson, M. B., and Martin, E. W., Jr. (1998). First results for resetting the antitumor immune response by immune corrective surgery in colon cancer. *Am. J. Surg.* **176:** 339–343.

Bartolomei, M., Testori, A., Chinol, M., Gennari, R., De Cicco, C., Leonardi, L., Zoboli, S., and Paganelli, G. (1998). Sentinel node localization in cutaneous melanoma: lymphoscintigraphy with colloids and antibody fragments versus blue dye mapping. *Eur. J. Nucl. Med.* **25:** 1489–1494.

Bell, J., Mojzisik, C., Hinkle, G., Derman, H., Schlom, J., and Martin, E. W. (1990). Intraoperative radioimmunodetection of ovarian cancer using monoclonal antibody B72.3 and a portable gamma-detecting probe. *Obstet. Gynecol.* **76:** 607–611.

Benevento, A., Dominioni, L., Carcano, G., and Dionigi. R. (1998). Intraoperative localization of gut endocrine tumors with radiolabeled somatostatin analogs and a gamma-detecting probe. *Sem. Surg. Oncol.* **15:** 239–244.

Berclaz, G., Crazzolara, A. O., Altermatt, H. J., Aebi, S., Fey, M. F., Haenggi, W., and Dreher, E. (1998). Sentinel lymphadenectomy: An alternative to axillary lymphadenectomy in breast cancer. *Schweizerische Medizinische Wochenschrift* **128:** 1730–1736.

Bombardieri, E., Crippa, F., Maffioli, L., Draisma, A., Chiti, A., Agresti, R., and Greco, M. (1998). Nuclear medicine approaches for detection of axillary lymph node metastases. *Q. J. Nucl. Med.* **42:** 54–65.

Borgstein, P. J., Pijpers, R., Comans, E. F., van Diest, P. J., Boom, R. P., and Meijer, S. (1998). Sentinel lymph node biopsy in breast cancer: Guidelines and pitfalls of lymphoscintigraphy and gamma probe detection. *J. Am. Coll. Surgeons* **186:** 275–283.

Brobeil, A., Kamath, D., Cruse, C. W., Rapaport, D. P., Wells, K. E., Shons, A. R., Messina, J. L., Glass, L. F., Berman, C. G., Puleo, C. A., and Reintgen, D. S. (1997). The clinical relevance of sentinel lymph nodes identified with radiolymphoscintigraphy. *J. Florida Med. Ass.* **84:** 157–160.

Büchels, H. K., Vogt, H., and Bachter, D. (1997). [Scintillation probe guided sentinel lymphadenectomy in malignant melanoma]. *Chirurg!* **68:** 45–50.

Burak, W. E., Jr., Schneebaum, S., Kim, J. A., Arnold, M. W., Hinkle, G., Berens, A., Mojzisik, C., and Martin, E. W., Jr. (1995). Pilot study evaluating the intraoperative localization of radiolabeled monoclonal antibody CC83 in patients with metastatic colorectal carcinoma [see comments]. *Surgery* **118:** 103–108.

Cohen, A. M., Martin, E. W., Jr., Lavery, I., Daly, J., Sardi, A., Aitken, D., Bland, K., Mojzisik, C., and Hinkle, G. (1991). Radioimmunoguided surgery using iodine 125 B72.3 in patients with colorectal cancer. *Arch. Surg.* **126:** 349–352.

Cohen, M., Gat, A., Haddad, R., Avital, S., Even-Sapir, E., Skornick, Y., Shafir, R., and Schneebaum, S. (1998). Single-injection gamma probe-guided sentinel lymph node detection in 40 melanomatous lymphadenectomies [In process citation]. *Ann. Plast. Surg.* **41:** 397–401.

Colton, C., and Hardy, J. G. (1983). Evaluation of a sterilizable radiation probe as an aid to the surgical treatment of osteoid-osteoma. *J. Bone Joint Surg.* **65A:** 1019–1022.

Cote, R. J., Houchens, D. P., Hitchcock, C. L., Saad, A. D., Nines, R. G., Greenson, J. K., Schneebaum, S., Arnold, M. W., and Martin, E. W., Jr. (1996). Intraoperative detection of occult colon cancer micrometastases using 125 I-radiolabled monoclonal antibody CC49. *Cancer* **77:** 613–620.

Crossin, J. A., Johnson, A. C., Stewart, P. B., and Turner, W. W., Jr. (1998). Gamma-probe-guided resection of the sentinel lymph node in breast cancer [with discussion]. *Am. Surg.* **64:** 666–669.

Daghighian, F., Mazziotta, J., Hoffman, E. J., Shenderov, P., Eshaghian, B., Siegel, S., and Phelps, M. E. (1994). Intraoperative beta probe: A device for detecting tissue labeled with positron or electron emitting isotopes during surgery. *Med. Phys.* **21:** 153–157.

de Hullu, J. A., Doting, E., Piers, D. A., Hollema, H., Aalders, J. G., Koops, H. S., Boonstra, H., and van der Zee, A. G. (1998). Sentinel lymph node identification with technetium-99m-labeled nanocolloid in squamous cell cancer of the vulva. *J. Nucl. Med.* **39:** 1381–1385.

de Nardi, P., Stella, M., Magnani, P., Paganelli, G., Mangili, F., Fazio, F., and di Carlo, V. (1997). Combination of monoclonal antibodies for radioimmunoguided surgery. *Internatl. J. Colorectal Dis.* **12**: 24–28.

Di Carlo, V., De Nardi, P., Stella, M., Magnani, P., and Fazio, F. (1998). Preoperative and intraoperative radioimmunodetection of cancer pretargeted by biotinylated monoclonal antibodies. *Sem. Surg. Oncol.* **15**: 235–238.

Di Carlo, V., Stella, M., De Nardi, P., and Fazio, F. (1995). Radioimmunoguided surgery: Clinical experience with different monoclonal antibodies and methods. *Tumori* **81**: 98–102.

Dresel, S., Weiss, M., Heckmann, M., Rossmüller, B., Konz, B., Tatsch, K., and Hahn, K. (1998). [Diagnosis of sentinel lymph node in malignant melanoma: preoperative lymphoscintigraphy and intraoperative gamma probe guidance]. *Nuklearmedizin* **37**: 177–182.

Ehrhardt, G. J., Ketring, A. R., and Ayers, L. M. (1998). Reactor-produced radionuclides at the University of Missouri Research Reactor. *Appl. Rad. Isotopes* **49**: 295–297.

Glass, L. F., Messina, J. L., Cruse, W., Wells, K., Rapaport, D., Miliotes, G., Berman, C., Reintgen, D., and Fenske, N. A. (1996). The use of intraoperative radiolymphoscintigraphy for sentinel node biopsy in patients with malignant melanoma. *Dermatol. Surg.* **22**: 715–720.

Greco, M., Agresti, R., and Giovanazzi, R. (1998). Impact of the diagnostic methods on the therapeutic strategies. *Q. J. Nucl. Med.* **42**: 66–80.

Greiner, J. W., Smalley, R. V., Borden, E. C., Martin, E. W., Guadagni, F., Roselli, M., and Schlom, J. (1991). Applications of monoclonal antibodies and recombinant cytokines for the treatment of human colorectal and other carcinomas. *J. Surg. Oncol.* (Suppl.) **2**: 9–13.

Gulec, S. A., Moffat, F. L., Carroll, R. G., and Krag, D. N. (1997). Gamma probe guided sentinel node biopsy in breast cancer. *Q. J. Nucl. Med.* **41**: 251–261.

Gulec, S. A., Moffat, F. L., Carroll, R. G., Serafini, A. N., Sfakianakis, G. N., Allen, L., Boggs, J., Escobedo, D., Pruett, C. S., Gupta, A., Livingstone, A. S., and Krag, D. N. (1998). Sentinel lymph node localization in early breast cancer. *J. Nucl. Med.* **39**: 1388–1393.

Guliano, A. E., Kirgan, D. M., Guenther, J. M., and Morton, D. L. (1994). Lymphatic mapping and sentinel lymphadenectomy for breast cancer. *Ann. Surg.* **220**: 391–401.

Hader, D., Moss, K., and Geier, N. (1998). Sentinel lymph node biopsy using lymphoscintigraphy. *Aorn J.* **68**: 572–576, 579–582, 585–588.

Harris, C. C., Bigelow, R. R., Francis, J. E., Kelley, G. G., and Bell, P. R. (1956). A CsI(Tl)-crystal surgical scintillation probe. *Nucleonics* **14**: 102–108.

Hartsough, N. E., Barrett, H. H., Barber, H. B., and Woolfenden, J. M. (1995). Intraoperative tumor detection: Relative performance of single-element, dual-element, and imaging probes with various collimators. *IEEE Trans. Med. Imag.* **14**: 259–265.

Harvey, W. C., and Lancaster, J. L. (1981). Technical and clinical characteristics of a surgical biopsy probe. *J. Nucl. Med.* **22**: 184–186.

Heij, H. A., Rutgers, E. J., de Kraker, J., and Vos, A. (1997). Intraoperative search for neuroblastoma by MIBG and radioguided surgery with the gamma detector. *Med. Pediatr. Oncol.* **28**: 171–174.

Hickernell, T. S., Barber, H. B., Barrett, H. H., and Woolfenden, J. M. (1988). Dual-detector probe for surgical tumor staging. *J. Nucl. Med.* **29**: 1101–1106.

Hoffman, E. J., Tornai, M. P., Janacek, M., Patt, B. E., and Iwanczyk, J. S. (1999). Intraoperative probes and imaging probes. *Eur. J. Nucl. Med.* **26**: 913–935.

Hoffman, E. J., Tornai, M. P., Levin, C. S., MacDonald, L. R., and Holdsworth, C. H. (1998). A dual detector beta-ray imaging probe with gamma-ray background suppression for use in intra-operative detection of radiolabeled tumors. *Nucl. Instru. Meth. Physi. Res. A* **409**: 511–516.

Hoffman, E. J., Tornai, M. P., Levin, C. S., MacDonald, L. R., and Siegel, S. (1997). Design and performance of gamma and beta intra-operative imaging probes. *Physica Medica* **13**: S243–247.

Kairemo, K. J. (1996). Radioimmunotherapy of solid cancers: A review. *Acta Oncologica* **35**: 343–355.

Kairemo, K. J., Strömberg, S., Nikula, T. K., and Karonen, S. L. (1998). Expression profile of vascular cell adhesion molecule-1 (CD106) in inflammatory foci using rhenium-188 labelled monoclonal antibody in mice. *Cell Adhesion Comm.* **5**: 325–333.

Kapteijn, B. A., Nieweg, O. E., Liem, I., Mooi, W. J., Balm, A. J., Muller, S. H., Peterse, J. L., Valdés Olmos, R. A., Hoefnagel, C. A., and Kroon, B. B. (1997). Localizing the sentinel node in cutaneous melanoma: gamma probe detection versus blue dye. *Ann. Surg. Oncol.* **4**: 156–160.

Kapteijn, B. A., Nieweg, O. E., Muller, S. H., Liem, I. H., Hoefnagel, C. A., Rutgers, E. J., and Kroon, B. B. (1997). Validation of gamma probe detection of the sentinel node in melanoma. *J. Nucl. Med.* **38**: 362–366.

Kim, J. A., Triozzi, P. L., and Martin, E. W., Jr. (1993). Radioimmunoguided surgery for colorectal cancer. *Oncology* **7**: 55–64.

Knapp, F. F., Jr., Beets, A. L., Guhlke, S., Zamora, P. O., Bender, H., Palmedo, H., and Biersack, H. J. (1997). Availability of rhenium-188 from the alumina-based tungsten-188/rhenium-188 generator for preparation of rhenium-188-labeled radiopharmaceuticals for cancer treatment. *Anticancer Res.* **17**: 1783–1795.

Knoll, G. (1989). "Radiation Detection and Measurement," 2nd ed. New York, John Wiley & Sons.

Knoll, G. F., Lieberman, L. M., Nishiyama, H., and Beierwaltes, W. H. (1972). A gamma ray probe for the detection of ocular melanomas. *IEEE Trans. Nucl. Sci.* **19**: 76–80.

Koch, W. M., Choti, M. A., Civelek, A. C., Eisele, D. W., and Saunders, J. R. (1998). Gamma probe-directed biopsy of the sentinel node in oral squamous cell carcinoma. *Arch. Otolaryngol.—Head Neck Surg.* **124**: 455–459.

Kohler, G., and Milstein, C. (1975). Continuous cultures of fused cell secreting antibody of predefined sensitivity. *Nature* **256**: 495–497.

Krag, D. N., Haseman, M. K., Ford, P., Smith, L., Taylor, M. H., Schneider, P., and Goodnight, J. E. (1992). Gamma probe location of 111indium-labeled B72.3: an extension of immunoscintigraphy. *J. Surg. Oncology* **51**: 226–230.

Krag, D. N., Meijer, S. J., Weaver, D. L., Loggie, B. W., Harlow, S. P., Tanabe, K. K., Laughlin, E. H., and Alex, J. C. (1995). Minimal-access surgery for staging of malignant melanoma [with discussion]. *Arch. Surg.* **130**: 654–660.

Krag, D. N., Weaver, D. L., Alex, J. C., and Fairbank, J. T. (1993). Surgical resection and radiolocalization of the sentinel lymph node in breast cancer using a gamma probe [with discussion]. *Surg. Oncol.* **2**: 335–340.

Krag, D., Weaver, D., Ashikaga, T., Moffat, F., Klimberg, V. S., Shriver, C., Feldman, S., Kusminsky, R., Gadd, M., Kuhn, J., Harlow, S., and Beitsch, P. (1998). The sentinel node in breast cancer—a multicenter validation study. *N. Engl. J. Med.* **339**: 941–946.

Larose, J. H., Jarrett, W. H., Hagler, W. S., Palms, J. M., and Wood, R. E. (1971). Medical problems in eye tumor identification. *IEEE Trans. Nucl. Sci.* **18**: 46–49.

Lauber, A. (1972). Development of miniaturized solid state detectors for the measurement of beta and gamma radiation in superficial and deep parts of living tissue. *Nucl. Instr. Meth.* **101**: 545–550.

Levin, C. S., Tornai, M. P., MacDonald, L. R., and Hoffman, E. J. (1997). Annihilation gamma-ray background characterization and rejection for a small beta ray camera imaging positron emitters. *IEEE Trans. Nucl. Sci.* **44**: 1120–1126.

Lieber, K. A., Standiford, S. B., Kuvshinoff, B. W., and Ota, D. M. (1998). Surgical management of aberrant sentinel lymph node drainage in cutaneous melanoma [with discussion]. *Surgery* **124**: 757–762.

Low-Beer, B. (1946). Surface measurements of radioactive phosphorus in breast tumors as a possible diagnostic method. *Science* **104**: 399.

MacDonald, L. R. (1996). Beta-Ray Imaging Surgical Scintillation Detector with Visible Light Photon Counter Readout. Ph.D. dissertation, University of California, Los Angeles.

MacDonald, L. R., Tornai, M. P., Levin, C. S., Park, J., Atac, M., Cline, D. B., and Hoffman, E. J. (1995). Investigation of the physical aspects of beta imaging probes using scintillating fibers and visible light photon counters. *IEEE Trans. Nucl. Sci.* **42**: 1351–1357.

Manayan, R. C., Hart, M. J., and Friend, W. G. (1997). Radioimmunoguided surgery for colorectal cancer. *Am. J. Surg.* **173**: 386–389.

Martelli, H., Ricard, M., Larroquet, M., Wioland, M., Paraf, F., Fabre, M., Josset, P., Helardot, P. G., Gauthier, F., Terrier-Lacombe, M. J., Michon, J., Hartmann, O., Tabone, M. D., Patte, C., Lumbroso, J., and Gruner, M. (1998). Intraoperative localization of neuroblastoma in children with 123I- or 125I-radiolabeled metaiodobenzylguanidine. *Surgery* **123**: 51–57.

Martin, E. W., Jr., and Carey, L. C. (1991). Second-look surgery for colorectal cancer. The second time around. *Ann. Surg.* **214**: 321–327.

Martin, E. W., Jr., Mojzisik, C. M., Hinkle, G. H., et al. (1988). Radioimmunoguided surgery using monoclonal antibody. *Am. J. Surg.* **156**: 386–392.

Matherson, K. J., Barber, H. B., Barrett, H. H., Hartsough, N. E., and Woolfenden, J. M. (1993). Intraoperative coincidence probe for tumor detection with 111-in-labeled monoclonal antibodies. *Conf. Record of the 1993 IEEE Nucl. Sci. Symp. & Med. Imag. Conf.* **2**: 1312–1316.

Meijer, S., Collet, G. J., Pijpers, H. J., van Hattum, L., and Hoekstra, O. S. (1996). [Less axillary dissection necessary due to sentinel node biopsy in patients with breast carcinoma]. *Nederlands Tijdschrift voor Geneeskunde* **140**: 2239–2243.

Moldofsky, P. J., and Swinth, K. L. (1972). Avalanche detector arrays for *in vivo* measurement of plutonium and other low activity, low energy emitters. *IEEE Trans. Nucl. Sci.* **19**: 55–63.

Moore, G. E. (1948). Use of radioactive diiodofluorescein in the diagnosis and localization of brain tumors. *Science* **107**: 569–571.

Morris, A. C., Jr., Barclay, T. R., Tanida, R., and Nemcek, J. V. (1971). A miniaturized probe for detecting radioactivity at thyroid surgery. *Phys. Med. Biol.* **16**: 397–404.

Morton, D. L., Wen, D. R., Wong, J. H., Economou, J. S., Cagle, L. A., Storm, F. K., Foshag, L. J., and Cochran, A. J. (1992). Technical details of intraoperative lymphatic mapping for early stage melanoma. *Arch. Surg.* **127**: 392–399.

Mudun, A., Murray, D. R., Herda, S. C., Eshima, D., Shattuck, L. A., Vansant, J. P., Taylor, A. T., and Alazraki, N. P. (1996). Early stage melanoma: Lymphoscintigraphy, reproducibility of sentinel node detection, and effectiveness of the intraoperative gamma probe. *Radiology* **199**: 171–175.

Nabi, H., Doerr, R. J., Balu, D., Rogan, L., Farrell, E. L., and Evans, N. H. (1993). Gamma probe assisted ex vivo detection of small lymph node metastases following the administration of indium-111-labeled monoclonal antibodies to colorectal cancers. *J. Nucl. Med.* **34**: 1818–1822.

Nakayama, K. (1959). Diagnostic significance of radioactive isotopes in early cancer of the alimentary tract, especially the esophagus and the cardia. *Surgery* **39**: 736–759.

Nelson, R. S., Dewey, W. C., and Rose, R. G. (1964). The use of radioactive phosphorus ^{32}P and a miniature Geiger tube to detect malignant neoplasia of the gastrointestinal tract. *Gastroenterology* **46**: 8–15.

Nieroda, C. A., Milenic, D. E., Carrasquillo, J. A., Scholm, J., and Greiner, J. W. (1995). Improved tumor radioimmunodetection using a single-chain Fv and gamma-interferon: potential clinical applications for radioimmunoguided surgery and gamma scanning. *Cancer Res.* **55**: 2858–2865.

Nieroda, C. A., Mojzisik, C., Sardi, A., Ferrara, P., Hinkle, G., Thurston, M. O., and Martin, E. W., Jr. (1989). The impact of radioimmunoguided surgery (RIGS) on surgical decision-making in colorectal cancer. *Dis. Colon Rectum* **32**: 927–932.

Nieroda, C., Mojzisik, C., Sardi, A., Ferrara, P., Hinkle, G., Thurston, M. O., and Martin, E. W., Jr. (1990). Radioimmunoguided surgery in primary colon cancer. *Cancer Detect. Prevent.* **14**: 651–656.

Nieweg, O. E., Jansen, L., and Kroon, B. B. (1998). Technique of lymphatic mapping and sentinel node biopsy for melanoma. *Eur. J. Surg. Oncol.* **24**: 520–524.

O'Brien, C. J., Uren, R. F., Thompson, J. F., Howman-Giles, R. B., Petersen-Schaefer, K., Shaw, H. M., Quinn, M. J., and McCarthy, W. H. (1995). Prediction of potential metastatic sites in cutaneous head and neck melanoma using lymphoscintigraphy. *Am. J. Surg.* **170**: 461–466.

O'Dwyer, P. J., Mojzisik, C. M., Hinkle, G. H., Rousseau, M., Olsen, J., Tuttle, S. E., Barth, R. F., Thurston, M. O., McCabe, D. P., Farrar, W. B., and Martin, E. W. (1986). Intraoperative probe-directed immunodetection using a monclonal antibody. *Arch. Surg.* **121**: 1391–1394.

Offodile, R., Hoh, C., Barsky, S. H., Nelson, S. D., Elashoff, R., Eilber, F. R., Economou, J. S., and Nguyen, M. (1998). Minimally invasive breast carcinoma staging using lymphatic mapping with radiolabeled dextran. *Cancer* **82**: 1704–1708.

Paganelli, G., De Cicco, C., Cremonesi, M., Prisco, G., Calza, P., Luini, A., Zucali, P., and Veronesi, U. (1998). Optimized sentinel node scintigraphy in breast cancer. *Q. J. Nucl. Med.* **42**: 49–53.

Pani, R., De Vincentis, G., Scopinaro, F., Pellegrini, R., Soluri, A., Weinberg, I. N., Pergola, A., Scafe, R., and Trotta, G. (1998). Dedicated gamma camera for single photon emission mammography (SPEM). *IEEE Trans. Nucl. Sci.* **45**: 3127–3133.

Patt, B. E., Iwanczyk, J. S., Rossington Tull, C., Wang, N. W., Tornai, M. P., and Hoffman, E. J. (1998). High resolution CsI(Tl)/Si-PIN detector development for breast imaging. *IEEE Trans. Nucl. Sci.* **45**: 2126–2131.

Patt, B. E., Tornai, M. P., Iwanczyk, J. S., Levin, C. S., and Hoffman, E. J. (1997). Development of an intraoperative gamma camera based on a 256-pixel mercuric iodide detector array. *IEEE Trans. Nucl. Sci.* **44**: 1242–1248.

Peltier, P., Curtet, C., Chatal, J., Le Doussal, J. M., Daniel, G., Aillet, G., Gruaz-Guyon, A., Barbet, J., and Delaage, M. (1998). Radioimmunodetection of medullary thyroid cancer using a bispecific anti-CEA/anti-indium-DTPA antibody and an indium-111-labeled DTPA dimer. *J. Nucl. Med.* **34**: 1267–1273.

Perkins, A. (1993). Peroperative nuclear medicine [editorial]. *Eur. J. Nucl. Med.* **20**: 573–575.

Petroff, M., and Staepelbroek, M. (1989). Photon counting solid-state photomultiplier. *IEEEE Trans. Nucl. Sci.* **36**: 158–162.

Pijpers, R., Borgstein, P. J., Meijer, S., Hoekstra, O. S., van Hattum, L. H., and Teule, G. J. (1997). Sentinel node biopsy in melanoma patients: Dynamic lymphoscintigraphy followed by intraoperative gamma probe and vital dye guidance [with discussion]. *World J. Surg.* **21**: 788–793.

Pijpers, R., Borgstein, P. J., Meijer, S., Krag, D. N., Hoekstra, O. S., Greuter, H. N. J. M., and Teule, G. J. J. (1998). Transport and retention of colloidal tracers in regional lymphoscintigraphy in melanoma: Influence on lymphatic mapping and sentinel node biopsy. *Melanoma Res.* **8**: 413–418.

Pijpers, R., Collet, G. J., Meijer, S., and Hoekstra, O. S. (1995). The impact of dynamic lymphoscintigraphy and gamma probe guidance on sentinel node biopsy in melanoma. *Eur. J. Nucl. Med.* **22**: 1238–1241.

Pijpers, R., Meijer, S., Hoekstra, O. S., Collet, G. J., Comans, E. F., Boom, R. P., van Diest, P. J., and Teule, G. J. (1997). Impact of lymphoscintigraphy on sentinel node identification with technetium-99m-colloidal albumin in breast cancer. *J. Nucl. Med.* **38**: 366–368.

Pircher, F. J., Anderson, B., Cavanaugh, P. J., and Sharp, K. W. (1967). Experiences with a solid state detector for surface counting of phosphorus-32. *J. Nucl. Med.* **8**: 444–450.

Prakash, S., Went, M. J., and Blower, P. J. (1996). Cyclic and acyclic polyamines as chelators of rhenium-186 and rhenium-188 for therapeutic use. *Nucl. Med. Biol.* **23:** 543–549.

Raylman, R. R., Fisher, S. J., Brown, R. S., Ethier, S. P., and Wahl, R. L. (1995). Fluorine-18-fluorodeoxyglucose-guided breast cancer surgery with a positron-sensitive probe: Validation in preclinical studies. *J. Nucl. Med.* **36:** 1869–1874.

Raylman, R. R., and Wahl, R. L. (1994). A fiber-optically coupled positron-sensitive surgical probe. *J. Nucl. Med.* **35:** 909–913.

Raylman, R. R., and Wahl, R. L. (1996). Evaluation of ion-implanted-silicon detectors for use in intraoperative positron-sensitive probes. *Med. Phys.* **23:** 1889–1895.

Raylman, R. R., and Wahl, R. L. (1998). Beta-sensitive intraoperative probes utilizing dual, stacked ion-implanted-silicone detectors: proof of principle. *IEEE Trans. Nucl. Sci.* **45:** 1730.

Reintgen, D., Rapaport, D., Tanabe, K. K., and Ross, M. (1997). Lymphatic mapping and sentinel node biopsy in patients with malignant melanoma. *J. Florida Med. Ass.* **84:** 188–193.

Rettenbacher, L., Koller, I., Gmeiner, D., Kässmann, H., and Galvan, G. (1997). [Selective regional lymphadenectomy in malignant melanoma using a gamma probe]. *Acta Medica Austriaca* **24:** 79–80.

Rhodes, B. A., Lambert, C. R., Marek, M. J., Knapp, F. F., Jr., and Harvey, E. B. (1996). Re-188 labelled antibodies. *Appl. Rad. Isotopes* **47:** 7–14.

Robinson, C., and Peterson, R. (1948). A study of small ether agon Geiger-Müller counters. *Rev. Sci. Inst.* **19:** 911–914.

Robinson, C. V., and Selverstone, B. (1958). Localization of brain tumors at operation with radioactive phosphorus. *J. Neurosurg.* **15:** 76–83.

Roveda, L. (1990). [Radioimmunoscintigraphy with mono-clonal antibodies in recurrences and metastases of colorectal tumors]. *Medicina* **10:** 160–161.

Rutgers, E. J. (1995). Radio-immunotargeting in colorectal carcinoma. *Eur. J. Cancer* **31A:** 1243–1247.

Saffer, J. R., Barrett, H. H., Barber, H. B., and Woolfenden, J. M. (1992). Surgical probe design for a coincidence imaging system without a collimator. *Imag. Vision Comp.* **10:** 333–341.

Sardi, A., Workman, M., Mojzisik, C., Hinkle, G., Nieroda, C., and Martin, E. W. (1989). Intra-abdominal recurrence of colorectal cancer detected by radioimmunoguided surgery (RIGS system). *Arch. Surg.* **124:** 55–59.

Schneebaum, S., Arnold, M. W., Houchens, D. P., Greenson, J. K., Cote, R. J., Hitchcock, C. L., Young, D. C., Mojzisik, C. M., and Martin, E. W., Jr. (1995). The significance of intraoperative periportal lymph node metastasis identification in patients with colorectal carcinoma. *Cancer* **75:** 2809–2817.

Schneebaum, S., Papo, J., Graif, M., Baratz, M., Baron, J., and Skornik, Y. (1997). Radioimmunoguided surgery benefits for recurrent colorectal cancer [see comments]. *Ann. Surg. Oncol.* **4:** 371–376.

Schneebaum, S., Stadler, J., Cohen, M., Yaniv, D., Baron, J., and Skornik, Y. (1998). Gamma probe-guided sentinel node biopsy—optimal timing for injection. *Eur. J. Surg. Oncol.* **24:** 515–519.

Selverstone, B., and Solomon, A. K. (1948). Radioactive isotopes in the study of intracranial tumors. *Trans. Am. Neurol. Ass.* **73:** 115–119.

Selverstone, B., Solomon, A. K., and Sweet, W. H. (1949). Location of brain tumors by means of radioactive phosphorus. *JAMA* **140:** 227–228.

Selverstone, B., Sweet, W. H., and Robinson, C. V. (1949). The clinical use of radioactive phosphorus in the surgery of brain tumors. *Ann. Surg.* **130:** 643–651.

Snider, H., Dowlatshahi, K., Fan, M., Bridger, W. M., Rayudu, G., and Oleske, D. (1998). Sentinel node biopsy in the staging of breast cancer. *Am. J. Surg.* **176:** 305–310.

Stella, M., De Nardi, P., Paganelli, G., Magnani, P., Mangili, F., Sassi, I., Baratti, D., Gini, P., Zito, F., Cristallo, M., et al. (1994). Avidin-biotin system in radioimmunoguided surgery for colorectal cancer: Advantages and limits. *Dis. Colon Rectum* **37:** 335–343.

Stella, M., De Nardi, P., Paganelli, G., Sassi, I., Zito, M., Magnani, P., Baratti, D., Mangili, F., Spagnolo, W., Siccardi, A. G., et al. (1991). Surgery for colorectal cancer guided by radiodetecting probe: Clinical evaluation using monoclonal antibody B72.3. *Eur. J. Surg.* **157:** 485–488.

Stewart, K. C., and Lyster, D. M. (1997). Interstitial lymphoscintigraphy for lymphatic mapping in surgical practice and research. *J. Investigative Surg.* **10:** 249–262.

Swinth, K. L., and Ewins, J. H. (1976). Biomedical probe using a fiber-optic coupled scintillator. *Med. Phys.* **3:** 109–112.

Technical Specifications of the C-Trak Surgical Guidance System, 1999.

Thomas, C. I., Krohmer, J. S., and Storaasli, J. P. (1952). Detection of intraocular tumors with radioactive phosphorous. *Arch. Opthalmol.* **47:** 276–286.

Thompson, J. F., Niewind, P., Uren, R. F., Bosch, C. M., Howman-Giles, R., and Vrouenraets, B. C. (1997). Single-dose isotope injection for both preoperative lymphoscintigraphy and intraoperative sentinel lymph node identification in melanoma patients. *Melanoma Res.* **7:** 500–506.

Thurston, M. O., Kaehr, J. W., Martin, E. W., III, and Martin, E. W., Jr. (1991). Radionuclide of choice for use with an intraoperative probe. *Antibody Immunoconjug. Radiopharmaceut.* **4:** 595–601.

Tiourina, T., Arends, B., Huysmans, D., Rutten, H., Lemaire, B., and Muller, S. (1998). Evaluation of surgical gamma probes for radioguided sentinel node localisation. *Eur. J. Nucl. Med.* **25:** 1224–1231.

Tornai, M. P., Levin, C. S., MacDonald, L. R., and Hoffman, E. J. (1997). Investigation of crystal geometries for fiber coupled gamma imaging intra-operative probes. *IEEE Trans. Nucl. Sci.* **44:** 1254–1261.

Tornai, M. P., Levin, C. S., MacDonald, L. R., Holdsworth, C. H., and Hoffman, E. J. (1998). A miniature phoswich detector for gamma-ray localization and beta imaging. *IEEE Trans. Nucl. Sci.* **45:** 1166–1173.

Tornai, M. P., MacDonald, L. R., Levin, C. S., Siegel, S., and Hoffman, E. J. (1996). Design considerations and initial performance of a 1.2 cm(2) beta imaging intra-operative probe. *IEEE Trans. Nucl. Sci.* **43:** 2326–2335.

Tornai, M. P., MacDonald, L. R., Levin, C. S., Siegel, S., Hoffman, E. J., Park, J., Atac, M., and Cline, D. B. (1996). In "Imaging Detectors in High Energy, Astroparticle and Medical Physics"; (J. Park, ed.), pp. 133–148. World Scientific, Singapore.

Truitt, K. E., Nagel, T., Suen, L. F., and Imboden, J. B. (1997). Structural requirements for CD28-mediated costimulation of IL-2 production in Jurkat T cells. *J. Immunol.* **156:** 4539–4541.

van der Veen, H., Hoekstra, O. S., Paul, M. A., Cuesta, M. A., and Meijer, S. (1994). Gamma probe-guided sentinel node biopsy to select patients with melanoma for lymphadenectomy. *Br. J. Surg.* **81:** 1769–1770.

Veronesi, U., Paganelli, G., Galimberti, V., Viale, G., Zurrida, S., Bedoni, M., Costa, A., de Cicco, C., Geraghty, J. G., Luini, A., Sacchini, V., and Veronesi, P. (1997). Sentinel-node biopsy to avoid axillary dissection in breast cancer with clinically negative lymph-nodes [see comments]. *Lancet* **349:** 1864–1867.

Vidal-Sicart, S., Piulachs, J., Pons, F., Castel, T., Palou, J., Herranz, R., and Setoain, J. (1998). Detection of sentinel lymph nodes by lymphatic gammagraphy and intraoperative gamma-ray probe in patients with malignant melanoma. Initial results. *Rev. Esp. Med. Nucl.* **17:** 15–20.

Waddington, W. A., Davidson, B. R., Todd-Pokropek, A., Boulos, P. B., and Short, M. D. (1991). Evaluation of a technique for the intraoperative detection of a radiolabelled monoclonal antibody against colorectal cancer. *Eur. J. Nucl. Med.* **18:** 964–972.

Watabe, H., Nakamura, T., Takahashi, H., Itoh, M., Matsumoto, M., and Yamadera, A. (1993). Development of a miniature gamma-ray endoscopic probe for tumor localization in nuclear medicine. *IEEE Trans. Nucl. Sci.* **40:** 88–94.

Weinberg, I. N., Pani, R., Pellegrini, R., Scopinaro, F., DeVincentis, G., Pergola, A., and Soluri, A. (1997). Small lesion visualization in scintimammography. *IEEE Trans. Nucl. Sci.* **44:** 1398–402.

Williams, H. T., Sorsdahl, O. A., Lawhead, R. A., Letton, A. H., and Sommerville, J. C. (1995). Colorectal and gynecological malignancies: comparison of B72.3 monoclonal antibody imaging, CT, and serum tumor antigens. *Am. Surgeon* **61:** 195–196.

Williamson, M., Boyd, C. M., McGuire, E. L., Angtuaco, T., Westbrook, K. C., Lang, N. P., Alston, J., Broadwater, J. R., Navab, F., and Bersey, M. L. (1986). Precise intraoperative location of gastrointestinal bleeding with a hand-held counter (Work in progress). *Radiology* **159:** 272–273.

Woolfenden, J. M., and Barber, H. B. (1989). Radiation detector probes for tumor localization using tumor-seeking radioactive tracers. *Am. J. Roentgenol.* **153:** 35–39.

Woolfenden, J. M., and Barber, H. B. (1990). Edign and use of radiation detector probes for intraoperative tumor detection using tumor-seeking radiotracers. In "Nuclear Medicine Annual" (L. M. Freeman, ed.), pp. 151–173. Raven Press, New York.

Woolfenden, J., Nevin, W., Barber, H. B., and Donahue, D. J. (1984). Lung cancer detection using a miniature sodium iodide detector and cobalt-57 bleomycin. *Chest* **85:** 84–88.

Zamora, P. O., Bender, H., Gulhke, S., Marek, M. J., Knapp, F. F., Jr., Rhodes, B. A., and Biersack, H. J. (1997). Pre-clinical experience with Re-188-RC-160, a radiolabeled somatostatin analog for use in peptide-targeted radiotherapy. *Anticancer Res.* **17:** 1803–1808.

Zamora, P. O., Bender, H., Knapp, F. F., Jr., Rhodes, B. A., and Biersack, H. J. (1997). Targeting peptides for pleural cavity tumor radiotherapy: Specificity and dosimetry of Re-188-RC-160. *Hybridoma* **16:** 85–91.

CHAPTER 18

Noble Gas Detectors

ALEXANDER BOLOZDYNYA

Constellation Technology Corporation, Largo, Florida

I. Why Noble Gas Detectors Are Interesting for Medical Gamma-Ray Imaging
II. Basic Processes of Energy Dissipation and Generation of Signals
III. Earlier Developments of Gas Detectors for Medical Applications
IV. Luminescence Detectors
V. Technical Features of Luminescence Detectors
VI. Applications for Single-Photon Emission Computed Tomography
VII. Concluding Remarks

I. WHY NOBLE GAS DETECTORS ARE INTERESTING FOR MEDICAL GAMMA-RAY IMAGING

At first glance, it seems strange that gas media with relatively low densities could be competitive with solid-state detectors used for gamma-ray imaging in nuclear medicine. Below are a few arguments showing why that impression is not accurate.

A. Brief History of Gas Detectors

The first device used for the detection of ionizing radiation was the gas (air) ionization chamber known from the nineteenth century as a gold-leaf electroscope. Since Henri Becquerel's discovery of radioactivity in 1896 the electroscope has been used to measure the integral flux of ionizing radiation. A pulse ionization chamber was developed for detection of single particles by E. Rutherford and H. Geiger in 1908. Soon after that, H. Geiger built his very sensitive gas-discharge particle counter, which was used in experiments that led to the identification of the alpha particle as the nucleus of the helium atom and to development of Rutherford's model of the atom (1912).

The first position-sensitive device for particle track visualization was the cloud gas chamber built by C. T. R. Wilson in 1912. This device served for decades in experimental particle physics. Digital imaging with gas detectors began with the development of the multiwire proportional chambers (MWPC) introduced in experimental high-energy physics by G. Charpak in the 1960s. Tremendous progress in MWPC has been achieved due to the development of digital signal processing, integrated circuits, and computers. Since that time practically every experimental installation in high-energy physics has involved MWPC, allowing the discovery of new particles such as the J/Ψ by S. Ting or W and Z by C. Rubbia, who won Nobel Prizes in 1976 and 1984, respectively. For the invention of these electronic detectors, G. Charpak was awarded a Nobel Prize for Physics in 1992.

Georges Charpak and his collaborators (F. Sauli, S. Majewski, A. J. P. L. Policarpo, T. Ypsilantis, and A. Breskin) have originated many new devices such as gas drift chambers, proportional scintillation chambers, and parallel-plate avalanche chambers and have pioneered the development of X-ray digital imagers for medicine, biology, and industry. The development of liquid xenon ionization chambers by L. Alvarez, H. Zaklad, S. Derenzo, and others in the 1960s and 1970s seemed be a way of imaging 140 to 511-keV gamma rays for nuclear medicine needs. However, at the beginning of the 1980s it was recognized that the energy resolution of noble liquid ionization detectors in the low energy range is much worse than predicted from ionization statistics and the attention of researchers turned to the development of high-pressure gas detectors. Two methods were developed for getting information from these detectors. The more traditional one is to measure the charge liberated by ionizing radiation. The other one is to measure light or photon emission generated by ionization electrons drifting in sufficiently high electric fields. This process is called electroluminescence (EL) or proportional scintillation. In this chapter, we focus on the advantages of EL noble gas detectors that allow both the high-energy resolution and precise position sensitivity, as well as acceptable detection efficiency for gamma rays in the energy range required for single-photon emission computed tomography (SPECT).

B. Intrinsic Energy Resolution

The statistic-limited energy resolution of ionization detector is defined as

$$\Delta E_i = 2.35 \, (FWE)^{1/2} \quad (1)$$

where E is the measured energy deposition, F is the Fano factor, and W is the average energy required for electron–ion pair production. One can see that the best achievable energy resolution is proportional to $(FW)^{1/2}$. For noble gases of moderate density (< 0.55 g/cm^3 for xenon; Bolotnikov and Ramsey, 1997), this factor has a relatively low value, comparable to that of the best semiconductors (Table 1). Using the data presented in Table 1, the theoretical limit for the energy resolution of xenon-filled detectors may be estimated to be 1.1% full width at half maximum (FWHM) at 140 keV. Note that in high-density xenon (> 0.55 g/cm^3) and liquid noble gas detectors the intrinsic energy resolution is degraded by large fluctuations of electron–ion yield because of very effective recombination in δ-electron tracks (Machulin et al., 1983; Bolotnikov and Ramsey, 1997). The lowest Fano factor could be achieved with mixed noble gases as Ar + 0.5% Xe, Ne + 0.5% Xe, or Xe + 5% He (Alkhazov et al., 1979). Could the theoretical limit for the intrinsic energy resolution shown in Eq. (1) be achieved in practice?

In practice, energy resolution of the ionization detector includes a few more terms:

$$\Delta E_{tot}^2 = \Delta E_i^2 + \Delta E_e^2 + \Delta E_t^2 \quad (2)$$

where ΔE_e is the noise of readout electronics, and ΔE_t is transmission fluctuations due to incomplete collection or charge (or light).

In the low energy range typical for SPECT, the energy resolution of ionization detectors is determined by the electronics noise, that is, the Johnson (thermal) noise in the input circuit of the amplifier. With optimized filter, the noise of the amplifier is determined as

$$\Delta E_e \approx 4(2/3)^{1/2}(kTC_D)^{1/2}(t_c/t_m)^{1/2} \quad (3)$$

where k is Boltzman constant, T is the temperature, C_D is the capacitance of the detector, t_c is carrier transit time through the amplifying device and t_m is time of the measurement (electron collection time) (Radeka, 1988). So, in the low energy range the capacitance of the detector (electrode system) is the most important factor influencing the energy resolution. The capacitance of the detector is directly dependent on its linear dimensions. The most popular method achieving a good energy resolution and the approximately 1-mm position resolution required in medicine is to use arrays of small detectors (or segmented electrodes) with approximately 1 pF capacitance per channel. However, because the linear size of the one-channel detector is close to the desirable position resolution Δx, the detector read-out system requires up to $V/(\Delta x)^3$ number of channels to support 3D position-sensitivity in V-sensitive volume of the detector. Numerically, it means that tens or hundreds of thousands of channels are needed to cover a customized field of view (FOV) of imaging devices such as an Anger camera.

Drift chambers and silicon strip detectors could be used to reduce the number of readout channels. But the capacitance of single channels of these detectors is approximately 10 pF and the energy resolution cannot be good enough for many important tasks in medical imaging. The energy resolution of EL detectors is not affected by the capacitance of the detector electrode structure at all, and small-capacity and low-noise photosensors could be used for the readout of light signals from large FOV detectors.

TABLE 1 Parameters of Some Gas and Solid-State Detector Media

	F	W (eV[a])	$(FW)^{1/2}$ (eV)$^{1/2}$
Si (77 K)	0.08–0.13[a]	3.8	0.5–0.7
Ge (77 K)	0.06–0.13[a]	3.0	0.4–0.6
CdZnTe (-40°C)	0.14[b]	5.0[b]	0.7
Ne + 0.5% Ar	0.050[c]	26.2 (Ar)	1.1
Xe	0.13[d]	21.5	1.7
NaI(Tl)	1[a]	26	5.1

[a]From Knoll (1989).
[b]From Niemela and Sipila (1994)
[c]From Alkhazov et al. (1979).
[d]From Anderson et al. (1979).

C. Position Resolution

The primary ionization electron created, for example, by photoabsorbing a gamma ray, propagates through the noble gas and generates secondary ionization electrons along its track. The largest part of the energy of the primary electron is deposited at the end of the path. Secondary ionization electrons are distributed along the trajectory of the primary ionization electron and from ionization clusters as simulated in Bolozdynya and Mergunov (1998) and shown in Figure 1. One can see that the center of gravity of the electron cluster is shifted from the point of gamma-ray interaction. This effect causes the major limitation for the position resolution of noble gas detectors. The size of the ionization cluster could be determined as the distance from the original point of interaction and the position of the center of gravity of the electron cloud (see a distance between two peaks in Figure 1). The average sizes of ionization clusters created by 20 to 50-keV electrons stopped in pressurized Ne and Ar and by 20 to 140-keV electrons stopped in pressurized Xe are presented in Figure 2 as a function of energy of the primary electron. In detector simulations, the smearing distribution of ionization electrons in the cluster due to diffusion should be acquired (Figure 1b).

D. Technical Features

An important advantage of gas detectors is that internal signal amplification can be used. Internally amplified signals are relatively insensitive to the thermal noise of the preamplifier and the detector could be significant in size; for example, it could be constructed as an array of long striplike one-channel detectors. Precision position resolution with pointlike interactions is achievable via the weighing of signals read from the array, for example, from the plane of the anode wires in MWPCs. Such large-FOV gas detectors could provide precise spatial resolution with a very moderate number of channels.

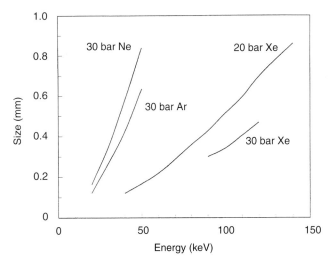

FIGURE 2 Average sizes of ionization clusters originated with electrons stopping in pressurized noble gases as a function of the energy of the electrons. (Reprinted from Bolozdynya and Mergunov, 1998, with permission from The Institute of Electrical and Electronics Engineers, Inc.)

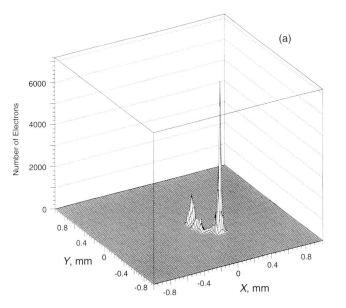
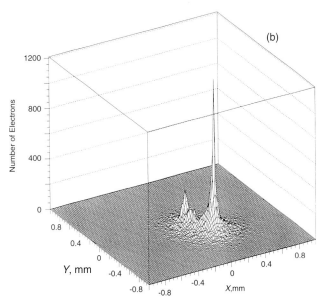

FIGURE 1 Distribution of secondary ionization electrons in an ionization cluster originated with 50-keV primary electron incoming perpendicularly to the xy-plane in the point of $\{x = 0, y = 0\}$ immediately (a) after stopping in 20 bar argon and (b) after 0.8 μs drift at 2 kV/cm/bar reduced electric field. (Reprinted from Bolozdynya and Mergunov, 1998, with permission from The Institute of Electrical and Electronics Engineers, Inc.)

TABLE 2 Energies for Production of Electron–Ion Pairs by Fast Electrons and Atom Ionization Potentials for Noble Gases at Normal Condition and Their Condensed Phases near the Triple Points

	W_{gas} (eV)	I_{gas} (eV)	W_{liquid} (eV)	I_{liquid} (eV)	W_{solid} (eV)	I_{solid} (eV)
He	41.3	24.59				
Ne	29.2	21.56				
Ar	26.4	15.76	23.6	13.4		14.3
Kr	24.2	14.00	18.4	11.55		11.67
Xe	22.0	12.13	9.76	9.22	12.4	9.27

Pure noble gases demonstrate a unique combination of detection properties: highly efficient scintillation, high mobility of excess electrons, and high-yield of EL in xenon-containing mixtures. Combinations of these different detection properties can lead to the development of novel gamma-ray imaging techniques.

Noble elements enable a wide range of gamma-ray interactions useful for imaging. For example, the cross section (per atom) of photoabsorption of 140-keV gamma rays in xenon is close to that of NaI(Tl) crystal; the cross section of Compton scatter in argon and neon is close to that of silicon. Doppler broadening at Compton scattering in light noble gases is comparable to that for carbon and silicon, which are considered as the best detector materials for scatter detectors of Compton cameras. All noble elements have demonstrated exceptional radiation hardness. The working media of gas detectors are inexpensive and allow the construction of devices in a variety of configurations such as cylindrical, spherical, and multilayer.

Gas detectors demonstrate high uniformity of detection properties across detector volume. Detectors equipped with multilayer electrode structure (see Section IVB) can support good detection efficiency and enhanced count rate.

II. BASIC PROCESSES OF ENERGY DISSIPATION AND GENERATION OF LIGHT SIGNALS

A. Ionization and Scintillation

The energy of ionizing radiation absorbed in the detector medium is balanced among three factors: the energy expended by ionization, the energy expended by excitation of atoms, and the kinetic energy of the subexitation electrons with energy lower than the excitation threshold. The average energy required to produce an electron–ion pair in the detector medium W is a measure of the ionization efficiency of the material. This value is defined as

$$W = K_i I_i + <I_{ex}><N_{ex}>/<N_i> + <E> \quad (4)$$

where I_i is the ionization potential, $<I_{ex}>$ is the average energy of atom excitation, $<N_{ex}>$ is the average number of excited atoms, $<N_i>$ is the average number of electron–ion pairs, K_i is a factor taking into account multiple acts of ionization, and $<E>$ is the average kinetic energy of subexitation electrons. The parameters of Eq. (4) for noble gases can be found in Platzman (1961) and for liquid state in (Doke et al., 1976; Aprile et al., 1993). The experimental values of W and ionization potentials for major noble gases in different phase states are presented in the Table 2. For gaseous xenon, the W value is practically constant at densities below 0.2 g/cm^3 and gradually decreases by approximately 15% when the density approaches to 1.7 g/cm^3. This is due to the formation and evolution of the electronic bonds in dense xenon (Bolotnikov and Ramsey, 1997).

Passing though the noble gas, the ionization radiation R generates electrons, ions A^+, and excited atoms A^* as follows.

$$R + A \rightarrow e + A^+ + R'$$
$$R + A \rightarrow A^* + R'$$
$$e + A^+ \rightarrow A^*$$

In dense noble gases ($n > 10^{19}$ cm^{-3}), it takes 10^{-11}–10^{-12} s to produce excited molecules in the reaction

$$A^* + 2A \rightarrow A_2^* + A$$

Photons emitted by the radiative decays of excited atoms are effectively scattered and absorbed in dense media. But dense noble gases and their liquid and solid phases are transparent for photons emitted upon decays of excited molecules

$$A_2^* \rightarrow 2A + h\nu$$

Such photons from the molecular continuum of noble gas scintillation spectra may be used for detection purposes. The average energy required for the production of one photon is 15 eV in liquid krypton (Akimov et al., 1993) and 14.2 eV in liquid xenon (Miyajima et al., 1995). The average energy for the production of scintillation photons in dense Kr and Xe gases is about twice these values.

TABLE 3 Scintillation Properties of Noble Gases

	Ar	Kr	Xe
λ_{max} (nm)	128	147	173
ξ	0.11	0.40	0.7
a_1/a_2	~10	0.2–0.4	2
$\tau_1(^1\Sigma_u^+)$ (ns)	5–6	2	3–4
$\tau_2(^3\Sigma_u^+)$ (ns)	1,000	90	30

The first two excited states of the noble gas molecules are the most important in scintillations. The pulse shape of the scintillation flash can be described as

$$S(t) = a_1 \exp(-t/\tau_1) + a_2 \exp(-t/\tau_2)$$

The decay times of the excited molecular states $\tau_1(^1\Sigma_u^+)$ and $\tau_2(^3\Sigma_u^+)$ and magnitudes of decay transitions for major noble gases in condensed phases are presented in Table 3. The table contains data on λ_{max}, the maximum position of the molecular continuum (which is approximately 20 nm wide), and ξ, the scintillation light yield measured relative to that of NaI(Tl) (Northrop et al., 1958). The conversion scintillation efficiency of noble gases is estimated to be from a few percent in gas phases up to 40–50% in condensed phases of krypton and xenon near their triple points.

B. Drift of Charge Carriers

Drift is a process of motion of charge carriers under the influence of an electric field. In the absence of external forces, electrons in a gas of temperature T move around with a Maxwellian energy distribution with most probable value kT (~ 0.04 eV at room temperature). Under the action of an electric field E, carriers acquire a net motion in the direction of the electric field with stationary drift velocity v_d averaged from instantaneous velocities, that is, $v_d = \langle v(t) \rangle$. At weak enough electric fields, when the carriers elastically collide with atoms and molecules of medium, the drift velocity is proportional to the electric field:

$$v_d = \mu E$$

The factor of proportionality is called mobility, μ. The mobility of electrons is a constant over the range of small electric field. The mobility of ions is a constant in practically all important cases. The probability of inelastic collisions of ions is significant at instantaneous velocities comparable with the electron velocity in the atom, $v(t) \sim 10^8$ cm/s; that is, the ion should have an energy of more than 10 keV to collide inelastically. This is practically impossible at realistic electric fields.

The distribution of carriers in the volume of the ionization detector is mostly nonuniform. The gradient of concentration leads to the spreading of carriers over volume (diffusion). Usually, the density of carriers is sufficiently small that one can neglect Coulomb interactions between them. Then, the current density of carriers can be described as

$$J = -D \nabla n + n v_d$$

At weak electric fields and small carrier concentrations the mobility can be expressed by the Einstein equation:

$$\mu = eD/kT$$

The condition of weak electric field means that carriers exist in thermodynamic equilibrium with the medium or, in other words, the carriers are thermalized. This condition practically always applies for ions. Note, that ion mobilities and diffusion coefficients consist of ~10^{-3} electron mobilities and diffusion coefficients. For example, Xe^+ ion mobility in xenon at 25 bar pressure is approximately 0.37 cm^2/V/s. If electrons are not thermalized, the mobility depends on the electric field and the drift cannot be described in our simple model. Then, the diffusion coefficient appears to be a tensor value: D_T in directions perpendicular to the electric field and D_L in the direction of the drift. At the case of only elastic collisions, there is a semiempirical Robson's relationship between D_T and D_L:

$$D_L / D_T = \partial(\ln v_d) / \partial(\ln E) = \partial(\ln v_d) / \partial(\ln E / N)$$

Note that the magnetic field introduces more anisotropy in the diffusion coefficient tensor. For many practical applications the distribution of carriers in an originally ($t = 0$) pointlike cluster containing n_0 carriers may be described as follows.

$$n(t) = \{n_o / [e(4\pi D_T t)(4\pi D_L t)^{1/2}]\} \exp[-(x^2 + y^2)/4D_T t]$$
$$\exp[-(z + v_d t)^2 / 4D_L t]$$

This suggests that the electric field be directed against the z-direction (Huxley and Crompton, 1974).

Electron drift velocities and diffusion coefficients at the reduced electric field $E/p = 0.1$ V/cm/torr for most noble gases are presented in Table 4. (For more detailed information about kinetic parameters of electrons in noble gases see Huxley and Crompton, 1974; Pierset and Sauli, 1984).

TABLE 4 Kinetic parameters of Electrons in Noble Gases at the Reduced Electric Field $E/p = 0.1$ V/cm/torr

	v_d (10^5 cm/s)	D_T/μ (V)	D_L/μ (V)
He	2.8	0.25	0.16
Ne	3.5	1	
Ar	2.3	3.0	0.3
Kr	1.6	2.0	0.22
Xe	1.05	2.7	0.27

C. Gas Gain and Electroluminescence

In the absence of electric fields, electrons injected into gas are quickly termalized. In the presence of an electric field, electrons can gain a sufficient amount of energy from the field between two collisions to cause the ionization or excitation of atoms. The secondary electrons can also gain energy for more ionization. This process can lead to the avalanche multiplication of electrons. A gas gain (amplification factor) of up to 10^6 is possible in noble gases doped with quenching impurities. The typical gas gain is usually in the range from 10^4 to 10^5. In pure noble gases, the gas gain of approximately 10 has been achieved. If a drifting electron can ionize α atoms per unit length, an increasing number of electrons is calculated from

$$dN = \alpha N_o \, dx$$

to be

$$N(x) = N_o \exp(\alpha x),$$

if α does not depend on N_0, the initial concentration of electrons. The coefficient α is called the first Townsend coefficient and depends on the field strength. The reduced coefficient α/p has a value of approximately 1 ion pair per cm drift per torr at reduced electric field strength $E/p = 100$ V/cm/torr for all pure noble gases.

If the energy of the drifting electrons is slightly below the ionization threshold, they effectively excite pure noble gases and generate electroluminescence (or proportional scintillation):

$$e + A \rightarrow e + A^*$$
$$A^* + 2A \rightarrow A_2^* + 2A \quad (at\ gas densities\ n > 10^{10}\ cm^{-3})$$
$$A_2^* \rightarrow 2A + h\nu$$

The intensity of electroluminescence (in ultraviolet photons/electron cm drift) at a uniform electric field in xenon is empirically defined as

$$dn_{ph}/dx = 70\,(E/p - 1.0)\,p,$$

i.e., the number of photons, n_{ph}, generated by one drifting electron is proportional to the drift path x (in cm), to the reduced electric field strength E/p (in kV/cm/bar), and to the gas pressure p (in bars).

The best measured value of the intensity of EL is 1700 photons/cm at $E/p = 3.4$ kV/cm/bar in 5 bar of xenon (Belogurov, 1995). Taking into account that the energy of a single photon is 8.4 eV, one may deduce that the conversion efficiency of the EL of xenon is > 100% just before breakdown. Assuming that at extremely high electric fields in pure Xe the charge multiplication factor is approximately 10, we may estimate the conversion efficiency of the EL to be up to 50%. There are data showing that some additional part of the acquired energy is emitted in the infrared (IR) range (Carugno 1998). So, we have to conclude that the EL of xenon is an effective mechanism for transforming the energy of the applied electric field into light energy. EL has been observed in all noble gases and their mixtures. In gas mixtures containing > 0.1% Xe, the light output and spectrum of EL is very similar to pure xenon.

The process of EL is linear, in contrast to the avalanche-like charge (electron) multiplication. This is because the energy of drifting electrons is mostly dissipated via the emission of photons, which do not participate further in the EL. Due to this circumstance, EL can provide lower fluctuations and better energy resolution than the gas gain amplification process.

D. Electron Emission from Condensed Phases

Excess electrons, liberated by ionizing radiation and chemically unbound to molecules of condensed dielectrics, have been observed to be effectively emitted from several nonpolar dielectrics with high electron mobility (Table 5). In these dielectrics, excess electrons can exist in a quasi-free state with 100% probability in the condensed Ar, Kr, and Xe and with lower probability in liquid saturated hydrocarbons with thermoactivated electron mobility. Behavior of quasi-free electrons near the interface of nonpolar dielectrics may be described in terms of a one-dimensional potential energy distribution (Figure 3a):

$$V_1(z) = V_0 - eE_1 z + A_1, \quad z < 0$$
$$V_2(z) = -eE_2 z + A_2, \quad z > 0$$
$$A_{1,2} = -e^2(\varepsilon_1 - \varepsilon_2)/\{4\varepsilon_{1,2}(z + \beta z/|z|)(\varepsilon_1 + \varepsilon_2)\}$$

where V_0 is the energy of the ground state of quasi-free electrons in the nonpolar dielectric (Table 5), ε_1 and ε_2 are the dielectric constants of the condensed nonpolar dielectric, and the value of a cutting parameter β is approximately the thickness of the liquid–vapor transition layer, which is usually a few times the condensed-phase interatom distance. In terms of the thermoelectron emission model, only those electrons that have a z projection (orthogonal to the interphase surface) of momentum p exceeding a threshold value $p_0 \approx (2m\,|V_0|)^{1/2}$ can be emitted.

When the emission time t_e is much more than the relaxation time of the momentum distribution function, the number of electrons under the interface can be defined as $N(t) = N_0 \cdot \exp(-t/t_e)$. The emission time is controlled by the height of the potential barrier $|V_0|$ and the kinetic energy of quasi-free electrons. If the lifetime of quasi-free electrons t_c is reduced by capture on electronegative impurities, the number of electrons under the surface is described by $N(t) = N_o\exp[-t(1/t_e + 1/t_c)]$. Then, the probability for electron emission is $K = 1/[1 + t_e/t_c]$. This model describes the process of excess electron emission from room-temperature, liquid, saturated hydrocarbons, such as isooctane and

TABLE 5 Electronic Properties of Working Media for Emission Detectors.

	T (K)	μo (cm²/V/s)	V_o (eV)	E_c (kV/cm)	t_e^a
Emitters of cold electrons					
⁴He, liquid	1–2	0.03	+1		100-10 s (10-100V/cm)
n-hexane, liquid	300	0.09	+0.09	100	
Isooctane, liquid	300	7	−0.18	90	20 μs (1000 V/cm)
TMP, liquid	297	24	−0.3	50	
Ar, liquid	84	475	−0.21	0.2	700 μs (100 V/cm)
Ne, solid	24	600	+1.1		
Emitters of hot electrons					
CH₄, liquid	100	400	−0.18	1.5	
CH₄, solid	77	~1000	0		< 0.1 μs (> 1000 V/cm)
Ar, liquid	84	475	-0.21	0.2	< 0.1 μs (> 300 V/cm)
Ar, solid	83	1000	+0.3 (6 K)		< 0.1 μs (>100 V/cm)
Kr, liquid	116	1800	−0.4	0.08	< 0.1 μs (>1600 V/cm)
Kr, solid	116	3700	−0.25 (20 K)		< 0.1 μs (>1000 V/cm)
Xe, liquid	161	2200	−0.61	0.05	< 0.1 μs (>1800 V/cm)
Xe, solid	161	4500	−0.46 (40 K)		< 0.1 μs (>1300 V/cm)

Reprinted from Bolozdynya (1999) with permission from Elsevier Science.
[a]Electric field in condensed phase is shown in brackets.

n-hexane, and liquid argon. In these media, the height of the potential barrier is comparable with the thermal energy of excess electrons, which are in thermodynamic equilibrium with the medium. Electrons in the high-momentum tail ($p_z > p_o$) of the momentum distribution (Figure 3a) have sufficient energy for emission. The time scale for emission (t_e) is quite long compared to the relaxation time of the distribution, and the thermal energy of the medium maintains the population of electrons in the tail. In a time t_e, approximately 70% of the electrons will be extracted through the high-momentum tail of the distribution function. In this case of thermo-electron emission, the emission time t_e in the range of 10^{-3} to 10^{-6} s depends on the applied electric field approximately as $1/E$ (Bolozdynya, 1999).

In condensed Kr and Xe, the potential barrier is so high ($|V_o| \gg kT$) that, at levels of purification $t_e/t_c \gg 1$, the thermoactivated electron emission from these liquids and solids has not been observed. On the other hand, in these media it is easy to achieve the electric field strength E_c in which drifting electrons can be heated (Table 5). Hot electrons with momentum $p_z > p_o$ escape from the condensed phase without delay. Effective and fast electron emission from these media has been observed at $E > E_c$. Note that electrons that are not emitted cannot continue their drift and thus cannot be heated by the applied electric field. They will be quickly cooled down through elastic interactions with atoms until they achieve a thermodynamic equilibrium with the medium. Then, these electrons may be emitted as thermal electrons.

In solid argon and neon, $V_0 > 0$ and only a thin potential barrier has to be penetrated by electrons emitting from the condensed phase (Figure 3b). Effective electron emission from these media should be observed at very low electric fields. Otherwise, in dielectrics with $V_0 > 0$ excess electrons can exist in low-mobile autolocalized states (in vacuum bubbles in liquid helium, hydrogen, and neon). Electrons can tunnel from these localized states through the thin potential barrier into the gas phase. The escape time is thermoactivated and depends on an electric field suppressing localized electrons to the interface. The escape time from liquid ⁴He has been observed to be 10–100 s at 10–100 V/cm electric field and 1–2 K temperature. At lower temperatures, the probability of electron tunneling is very low. However, localized electrons can be emitted from super-fluid helium if they are captured by quantum vortices (see references in Bolozdynya, 1999).

Thus, quasi-free electrons may escape from condensed noble gases as hot electrons on short time scales or as thermal electrons on a time scale t_e that depends on the applied electric field and the temperature of the dielectric. The total number of all emitted electrons is controlled only by the lifetime of excess electrons before their capture by electronegative impurities and other deep traps. The effect of electron emission can be used for the amplification of ionization signals originated in liquid or solid nonpolar dielectrics.

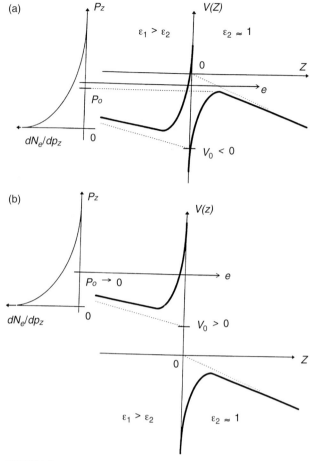

FIGURE 3 Potential energy distribution of excess electrons near the interface of condensed nonpolar dielectrics with (a) negative and (b) positive energy V_0 of ground state of excess electrons. (Reprinted from Bolozdynya, 1999, with permission from Elsevier Science.)

III. EARLIER DEVELOPMENTS OF GAS DETECTORS FOR MEDICAL APPLICATIONS

The development of new detection techniques is usually not directly applicable to medical imaging needs. However, it does often cultivate the ground for future revolutionary developments.

A. Ionization Chambers

Ionization chambers are the simplest of all gas-filled detectors. Their normal operation is based on the collection of the charge created by interaction of radiation in the gas. Ionization chambers operated in the current mode are often used in dosimetry (Attix and Roesch, 1966). In an ideal case, the amount of electric current generated in an ionization chamber is directly proportional to the intensity of the radiation field. Noble gas ionization chambers are simple, radiation-hard, and easily constructed in the 4π geometry used for accurate measurements of activity of gamma-ray sources (Suzuki et al., 1998).

Multichannel xenon ionization chambers pressurized up to 20 bar were developed in the 1970s and 1980s (Drost and Fenster, 1982, 1984) and successfully used in a number of clinical computed tomography (CT) scanners such as the 768-channel Philips model LX CT, the General Electric model CT 90000 Series II, and the Siemens model Somatom CR. The position resolution of multichannel ionization chambers has been measured at approximately 2 line pairs/mm, limited by the diffusion of electrons and the thickness of the absorption gap (parallax error). The position resolution is compromised with increased detection efficiency and sensitivity of the detectors.

B. Analog Imaging

Gas-filled (5- to 10-bar) chambers with a 1-cm gap have been used in the iconographic imaging system. The principle of operation relies on collecting charge carriers on an insulating Mylar foil stretched across the appropriate electrode. The charge pattern is developed off-line using a powder or liquid toner or is read out using a guarded electrometer probe to scan the potential image stored in the Mylar foil. The best position resolution of 10 line pairs/mm has been achieved with a xenon-filled chamber irradiated by 69-keV X-rays (Johns et al., 1974). The iconographic systems did not have wide use because of the bulky method of image processing.

A multiwire spark chamber with a TV-camera readout has been tested for gamma-ray imaging in the range 20–150 keV (Dubrovin et al., 1977). It has shown good position resolution (3–5 mm) but poor energy resolution and low count rate (< 5 kHz).

C. Digital Imaging with Multiwire Proportional Drift Chambers

MWPCs allow coordinates to be measured in digital form. The intrinsic position resolution can be approximately 1 mm and the time resolution can be approximately 10 ns. MWPC was one of the first techniques developed for digital X-ray imaging and autoradiography, and it has been tested for SPECT and positron emission tomography (PET). A review of MWPC developments for X-ray imaging can be found in Sauli (1999). Here we mention the most important of them for gamma-ray imaging (for additional information, see Ott et al. (1992)).

One of the first high-pressure MWPC gamma cameras for imaging low-energy photons was built at MIT and tested at Harvard Medical School (Zimmerman et al., 1981). The measured energy resolution was 10% FWHM for 81-keV ^{133}Xe. The count rate was found to be 70 kHz over a 10-cm FOV. The measured position resolution was 3.7 mm and the efficiency in the Anger camera mode was 70% at 81 keV. A larger camera has been built with a 25-cm FOV (Lacy et al., 1984). It was tested at 3-bar pressure of xenon–methane gas mixture. The detection efficiency is 50% at 40 keV and 10% at 100 keV, the intrinsic energy resolution is 33% at 22 keV, and the count rate capacity is 850 kHz for 50% dead time. The intrinsic spatial resolution is approximately 2.5 mm FWHM, but the system resolution with collimator is only 20–25 mm at 10 cm. System resolution degrades due to the parallax effect and the blurring of the image with fluorescent photons.

Although MWPCs have a relatively low efficiency at 511 keV, they can be built with sufficiently large FOV to obtain a sensitivity interesting for PET. The main advantage of this approach is the relatively low cost of the detector. Lead converters have been used to increase the detection efficiency of the MWPC from a fraction of 1% to approximately 10%. The most successful designs were developed by the British SERC Rutherford Appleton Laboratory (Bateman et al., 1984) and at CERN (Townsend et al., 1984). One of them used a stack of MWPC planes and thin lead-foil converters. Another used 5-mm-thick converters with a fine matrix of holes (0.8 mm in diameter and a 1 mm pitch). The system position resolution is achieved to be approximately 2 mm FWHM. However, because of their poor sensitivity only PET imaging for small organs such as the thyroid gland has been performed with these MWPC systems.

The first attempt to extend the technology of gas MWPC to more dense working media for radioisotope medical imaging was undertaken by L. Alvarez's group at the Lawrence Berkeley Laboratory (Zaklad et al., 1972). They investigated a liquid xenon multiwire chamber. The process of electron multiplication around thin wires in liquid xenon had been discovered, with an approximately 100× multiplication factor achieved. Large fluctuations of the amplification factor did not allow the achievement of the energy resolution acceptable for medical applications. Nevertheless, the group tested the first liquid xenon gamma-ray imaging camera with approximately 50-cm^2 FOV and 4 mm FWHM position resolution and published the first images acquired with noble gas detector (Zaklad et al., 1974). Later attempts to use a microstrip plate rather than a multiwire chamber in liquid xenon did not show a significant advantage (Policarpo, Chepel, Lopes, and Peskov, 1995).

The shortcomings of MWPCs used for gamma-ray imaging include low detection efficiency and poor energy resolution. Their advantages are high position resolution, high count rate capacity, and the possibility of construction of detectors with large FOVs.

IV. LUMINESCENCE DETECTORS

Luminescence detectors are photonic devices employing light emission for detection of ionizing radiation. Their primary advantage is the ability for fast data acquisition from large-FOV imaging systems.

A. Electroluminescence Drift Chambers

Since A. Policarpo's pioneering work on the development of the gas scintillating proportional counter (Policarpo, 1977), EL detectors have been considered a promising approach to the development of novel gamma-ray imaging instrumentation. EL allows for the transformation of ionization signals into amplified photon signals.

1. Principle and Review of Developments

One of the most successful developments of the EL imaging detector was the scintillation drift chamber (SDC), proposed by G. Charpak, H. N. Ngoc, and A. Policarpo (1981). Generally, the SDC consists of a pressure vessel that contains a low-field drift region, followed by a high-field light-generating gap, defined by two transparent wire electrodes (Figure 4). Sometimes, shaping rings are used to provide more-uniform electric field in the drift region. The light gap is viewed by a matrix of photodetectors. In many SDCs, photomultipliers have been placed outside the high-pressure vessel (Table 6). A matrix of seven small-size (0.5-in diameter) photomultiplier tubes (PMTs) have been successfully tested in a 50-mm FOV SDC filled with 20 bar of xenon (Belogurov et al., 1995). Other proven photosensors are wave-shifting fiber arrays coupled to PMTs (Parsons et al., 1990), an avalanche chamber with open CsI photocathodes (Bräuning et al., 1994), and microstrip chambers (Akimov et al., 1994).

One of the later, best performing SDCs (Bolozdynya, Egorov, et al., 1997a) contains a high-pressure chamber viewed through 19 2.5-cm-thick glass windows by a hexagonal array of 19 80-mm-diameter glass photomultipliers (Figure 5). A thin layer of wavelength shifter (0.5 mg/cm^2), p-terphenyl, is vacuum deposited on the internal surfaces of the glass windows to convert 170-nm ultraviolet (UV) photons to visible ones. The distance between the light gap and the

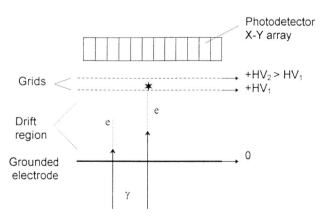

FIGURE 4 Schematic diagram of SDC design.

TABLE 6 Summary of the Characteristics of High-Pressure Xenon Imaging Scintillation Drift Chambers

Pressure [bar]	Photodetector array	Geometry		Energy resolution (% FWHM)		Position resolution at 60 keV (mm FWHM)		Source
		Diameter [cm]	Depth [cm]	60 KeV	122 KeV	xy	z	
5	5 PMT	10	10	3.2	4.9	~5	—	(Nguyen Ngoc, 1978; Nguyen Ngoc et al., 1988)
8	19 PMT	30	6	4.3		3.3	—	(Egorov, 1988)
15	Fibers	6	7		5.8	0.4	—	(Parsons et al., 1990)
4.5	Gas gain/CsI	10	6	4.1		2.6 (1.8)	—	(Bräuning et al., 1994)
9	19 PMT	30	3.8	3.6	2.6	3.7	0.6	(Bolozdynya, Egorov, et al., 1997a)

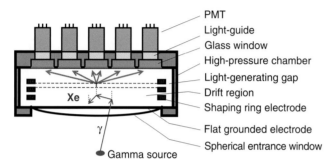

FIGURE 5 Schematic diagram of the gas scintillation drift chamber with 19-PMT-matrix readout, 30-cm-diameter FOV, and 9 bar xenon filling. (Reprinted from Bolozdynya, Egorov, et al., 1997a, with permission from Elsevier Science.)

windows is 4 cm. PMTs are coupled with silicon optical grease through light guides to the windows. The spherical front entrance window is made of 3-mm-thick aluminum and capable of operation up to 20 bar. An additional grounded 1-mm aluminum electrode, screening the window, is installed to keep a uniform electric field in the drift region. The active area of the detector (FOV) is 30 cm in diameter.

A readout system serves to select useful events and to acquire data about energy and the spatial distribution of pointlike ionization clusters (vertices) occurring in xenon filling while gamma radiation is absorbed in the drift gap. A useful event consists of a fast and a low-intensity scintillation signal followed by one, two, or three intensive electroluminescence flashes, generated by ionization clusters entering the light-generating gap from the drift region, where gamma radiation is absorbed and creates a few points of energy deposition. These may be points of photoabsorption of incoming gamma quanta, of photoabsorption of fluorescent photons, or of stopping of recoil electrons from Compton scattering. The block diagram of the electronics is shown in Figure 6. The electronics provide the following functions:

1. Produce a trigger by detecting prompt scintillation signals from the 19 PMTs in majority coincidence with a multiplicity of six with a gate of 50 ns.
2. Recognize EL signals and generate a gate signal for analog-to-digital convertors (ADCs), accepting electro-

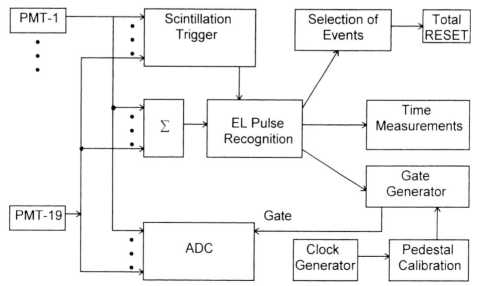

FIGURE 6 Block diagram of a readout electronics of the SDC shown in Figure 5. (ADC, analog-to-digital convertor) (Reprinted from Bolozdynya, Egorov, et al., 1997a, with permission from Elsevier Science.)

luminescent signals in the time interval of 40 μs after the scintillation occurred.

3. Select events with one, two, three, or more vertices.

The coordinates of pointlike ionization clusters projected on the plane of the PMT matrix entrance windows and the energy deposited, E, are determined by the following weighting procedure

$$E = \sum k_i A_i$$
$$x = \sum x_i A_i / \sum A_i$$
$$y = \sum y_i A_i / \sum A_i$$

where $i = 1, \ldots, 19$, A_i is the response signal from the PMT, centered at the point (x_i, y_i). The primary coefficients k_i were calculated assuming the amplification coefficients of 19 PMTs to be equal and using a specially measured response function for the single PMT of type FEU-139. Measurements with a multihole lead mask with regular holes of 4 mm in diameter and of 8 mm in pitch, placed just before the entrance window of the SDC and irradiated with a ^{57}Co point source located 1.5 m from the detector, showed up to 10% nonuniformity in the energy response and up to 3 mm coordinate nonlinearity at the edges of the FOV. After that, the k_i, x_i, and y_i coefficients were corrected using a number of linear iterations and less than 1% nonuniformity and less than 1 mm nonlinearity were achieved. Intrinsic spatial resolution of the SDC at 9 bar xenon filling was estimated by measuring the point spread function with collimated ^{57}Co and ^{241}Am gamma sources. It was found to be 4.6 mm FWHM for x-coordinate distribution of 122-keV (^{57}Co) photoabsorption vertices and 4.7 mm FWHM for x-coordinate distribution of 59.6-keV (^{241}Am) photoabsorption vertices. An image of a lead bar-phantom acquired with ^{241}Am gamma source is presented in Figure 7.

2. Three-Dimensional Position Sensitivity

In 3D mode, the SDC is triggered by the fast scintillation signals arising at the moment of absorption of the gamma rays in Xe gas. The scintillation flash gives a small number of primary photons. According to Parsons *et al.* (1990), the energy for creating one scintillation photon is approximately 76 eV in pressurized xenon. One 122-keV gamma photon absorbed in the drift region gives approximately 1630 scintillation photons. Light-collection efficiency of the SDC is estimated from the achieved energy resolution as $\varepsilon \approx 0.004$. To get a reliable trigger in a coincidence among 19 PMTs, a multiplication factor of 6 has been chosen. The efficiency of the scintillation triggering has been measured to be 60% with a ^{241}Am gamma source (60 keV) and 90% with a ^{57}Co gamma source (122 keV) at 9 bar xenon filling and is proportionally lower at lower pressures. The z position of the point of energy deposition is calculated as $z = \Delta t \cdot v_{dr}$, where Δt is a delay time between scintillation and the electroluminescent signal and v_{dr} is the

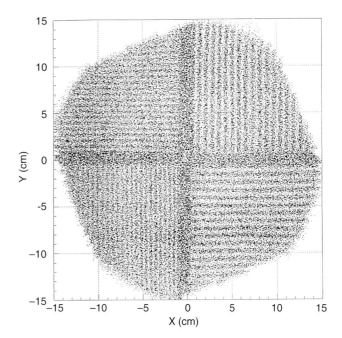

FIGURE 7 Image of lead bar-phantom with bars of 2.5 mm, 3.0 mm, 3.5 mm, and 4.0 mm width placed on the entrance window of the SDC (Figure 5) irradiated with a pointlike ^{241}Am gamma source placed 1.5 m distance from the window. (Reprinted from Bolozdynya, Egorov, *et al.*, 1997b, with permission from The Institute of Electrical and Electronics Engineers, Inc.)

drift velocity of the electrons in the drift region, which depends on electric field and gas pressure ($v_{dr} \sim 10^5$ cm/s in our case). Knowledge of the three coordinates and the energy of each vertex enables the recovery of the full picture of the interaction of the gamma rays with Xe in the drift region of the SDC. Intrinsic z-resolution of the SDC measures 0.6 mm FWHM at 9 bar xenon filling.

The ability of the SDC to measure x- and z- coordinate in 3D mode is illustrated by imaging the shadow of a lead mask distributed over the depth of the drift region (Figure 8b). The image of the lead mask in projection on the xy-plane is shown in Figure 8a. With knowledge of the z-coordinate it is possible to correct the effect of electron capture by electronegative impurities.

3. Energy Resolution

The experimentally measured energy resolution of the electroluminescence detector may be presented in the following form.

$$(\Delta E / E)_{\text{exper}}^2 = (\Delta E / E)_i^2 + R_c^2 + R_{\text{PD}}^2 \qquad (5)$$

where R_c accounts for the fluctuation of light collection and the conversion of light signal into electrical signal, and R_{PD} is the photodetector resolution. Modifying Policarpo's (1977) approximation for energy resolution of EL detector, we can describe the light-collection term as

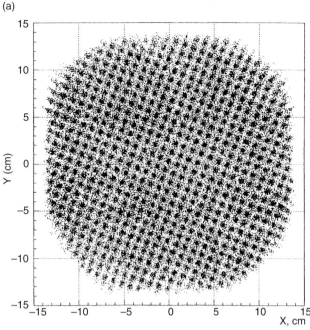

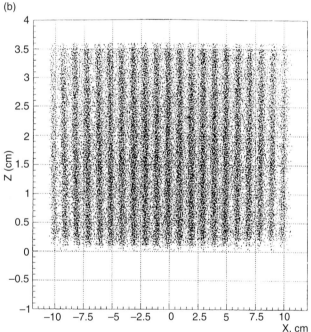

FIGURE 8 Image of a lead mask with regular holes of 4 mm in diameter and 8 mm in pitch placed just before the entrance window of the SDC (Figure 5) and exposed with a pointlike ^{57}Co gamma source located at a distance 1.5 m from the detector (a) in the xy-plane of the PMT matrix and (b) over the depth of the drift region in xz-projection. (Reprinted from Bolozdynya, Egorov, *et al.*, 1997a, with permission from Elsevier Science.)

$$R_c^2 = (l + f) / [\varepsilon Y(E_\gamma / W)] \quad (6)$$

where $f(K) \ll 1$ at gas gain $K < 10$ and $f(K) \sim 0.7$ at $K \sim 10^4$, ε is the efficiency of light collection and photon-electron conversion, and Y is the total light yield of EL measured in photons per drifting electron. Equation (6) shows that the energy resolution of an electroluminescence detector with a low gas gain will depend only on the fluctuation of the primary ionization and on the resolution of the photodetector if the following is achieved:

$$\varepsilon Y \geq 10 \quad (7)$$

In xenon-filled electroluminescence detectors, the total light yield of electroluminescence Y can reach >1000 photons/electron. Then, the theoretical limit on the energy resolution in the electroluminescence detector will be achievable using a low-noise photodetector with a total photon–photoelectron conversion efficiency of

$$\varepsilon \geq 1\% \quad (8)$$

For reference, the total conversion efficiency of the SDC with a fiber readout measures ~ 0.6% (Parsons *et al.*, 1990). The total conversion efficiency of the SDC with a PMT-matrix readout (Bolozdynya, Egorov, *et al.*, 1997a) is estimated at ~ 0.4%. It was concluded that the measured energy resolution of this camera was dominated by light-collection fluctuations. The energy spectrum, shown in Figure 9, has been measured with a noncollimated point source of ^{57}Co, which was located at the distance of 10 cm from the entrance window of the detector (Bolozdynya, Egorov, *et al.*, 1997a). The energy resolution of 2.7% measured within the 22-cm FOV indicates a good energy response correction. Further improvement of the energy resolution requires a

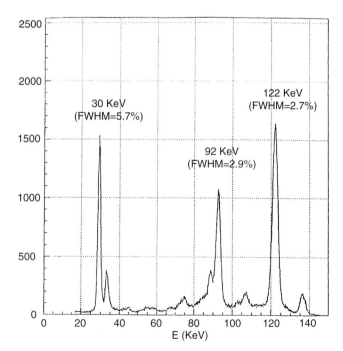

FIGURE 9 Energy spectrum measured with the SDC (Figure 5) irradiated with a noncollimated pointlike ^{57}Co gamma source located 10 cm distance from the entrance window. E/p = 1.2 kV/cm/bar. (Reprinted from Bolozdynya, Egorov, *et al.*, 1997a, with permission from Elsevier Science.)

more effective light-collection and photon–electron conversion system. More recently, thin (< 0.3 mm) backside-illuminated silicon photodiode arrays have demonstrated >70% quantum efficiency to light at our wavelengths of interest (> 400 nm) (Holland *et al.*, 1997). In an EL detector with electrodes covered by a mosaic of such photodiodes, the ε efficiency of light–electron conversion is estimated to be approximately 30%.

4. Count Rate and Efficiency

The count rate capability of the SDC depends on the operation mode. In the two-dimensional mode similar to the scintillation Anger gamma camera, the count rate is limited only by the duration of the electroluminescence signals (~1 μs) and may achieve some hundreds of kilohertz. In scintillation-triggered 3D mode, the rate performance is limited by the drift time of electrons drifting through the drift region. For a drift gap of 3.7 cm, a rate of 10–20 kHz can be achieved.

In general, the efficiency of the SDC is a function of thickness of the drift region and the gas pressure. For example, the SDC filled with 9 atm Xe has demonstrated 60% detection efficiency when triggered from prompt scintillation of 60 keV photons and 80% when triggered from the electroluminescence signal. With increasing pressure up to 20 atm, the efficiency of approximately 70% could be achieved at 140 keV. However, SDCs with higher efficiency has a reduced count rate capability. To achieve a high detection efficiency and a high count rate at the same time, a multilayer electroluminescence chamber is proposed.

B. Multilayer Electroluminescence Chamber

As we have seen, a classic SDC contains two active regions filled with a pressurized noble gas (see Figure 4). Measured radiation is absorbed in the drift region with a relatively weak electric field. The ionization electrons drift through this region into the region between two grided electrodes, where a high electric field is applied to generate the EL of the gas. So, EL is measured by the photodetector array placed on one side of the light-generating gap. Only because of this geometry, approximately 50% of the generated light is lost. In the SDC with fiber readout (Policarpo, 1977), another 25% is lost because the *x* and *y* fiber arrays shadow the light-production gap for one another. This is an inherent imperfection in the light-collection efficiency of the SDC design.

There is a proposed EL camera in which the light-generating gap and electron drift region are superposed (Bolozdynya and Morgunov, 1998). The working layer of the camera is formed by two electrodes with approximately 1 cm distance between them. The electrodes are made of a thin foil covered by thin photodetector arrays, for example, orthogonal arrays of wave-shifting fibers coupled to remote photomultipliers. As soon as ionization occurs between electrodes, the ionization electrons drift to the anode and excite EL of the gas. The 2D coordinates of the point of primary gamma interaction are determined by the centroid of electroluminescent signals distributed over the photodetector arrays. The *z*-coordinate of the point of energy deposition in the direction of the drift electric field is determined from the duration of the electroluminescent signal Δt and the well-known electron drift velocity v_d as follows:

$$z = v_d \cdot \Delta t \quad (9)$$

The total charge of photoelectrons measured from illuminated photodetectors is proportional to the *z*-coordinate and the energy deposited. The total charge of the electrons and ions collected on the readout fibers is negligible because there is no gas gain during the EL process. An important feature of electroluminescent signals in the fiber arrays is that the magnitude of the total electroluminescent signal, summing all *x*- and *y*-array signals, does not depend on the drift time. The magnitude of the total electroluminescent signal is proportional to the number of drifting electrons, that is, to E_γ/W. Thus, the deposited energy can be obtained from the magnitude of the total electroluminescent signal measured from all fibers of the same layer. The total time-integrated charge of photoelectrons measured from all fibers is proportional to E_{yz}.

One of the advantages of this design is high efficiency of electroluminescent light capture and conversion of light signals into electrical signals. The total efficiency of conversion of EL photons into photoelectrons in photodetectors coupled to fibers can be estimated as $\varepsilon > 1.5\%$ for photodetectors with quantum efficiency of $QE_{PD} > 60\%$. Thus the condition of Eq. (7) can be satisfied and the energy resolution of the camera may be limited only by fluctuations of primary ionization. Note that even using photomultipliers (QE_{PD} ~ 20–30%), this kind of EL camera will allow a much better energy resolution than that of any existing SDC design. Simulations have shown that the new detector can achieve position resolution approaching the primary ionization cluster size. Calculations have shown that N_z ~10 zones per square meter of FOV of such EL camera can be operating independently and simultaneously. To provide high detection efficiency, electrodes covered with scintillating fibers can be stacked into a multilayer structure (Figure 10) called a multilayer electroluminescent camera (MELC). The major gas (He, Ne, Ar, or Xe), gas pressure, number of layers, and total thickness of the detector can be chosen to achieve the required detection efficiency for different interactions. For light noble gases (He, Ne, and Ar) the Compton cross section is dominant for 140-keV gamma rays. Xenon provides the most effective photoabsorption of gamma rays and high light output of EL in noble gas mixtures. For example, a 20-cm-thick MELC consisting of 20 layers of 1 cm depth and filled with 20 bar Xe will have 85% photoabsorption efficiency for 140-keV gamma rays. This is comparable to the detection

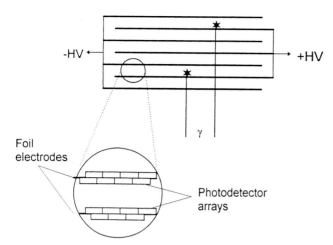

FIGURE 10 Schematic diagram of multilayer electroluminescence chamber.

efficiency of the 3/8-in-thick NaI(Tl) scintillators usually used in gamma cameras. The MELC filled with 20 bar Ar will be 8% effective for Compton scattering 140-keV gamma rays and practically transparent for photoabsorbing 140-keV gamma rays. The same efficiency for Compton scattering is achievable with a 3-mm-thick Si scatter detector.

In the range of electric fields that generate EL, the electron drift velocities in noble gases are approximately 10^6 cm/s. Then, a MELC with N_l layers can achieve a counting rate of $\sim N_l N_z$ MHz per square meter of FOV. Filled with 20 bar Xe, the MELC can operate as an Anger gamma camera with a collimator for 2D gamma imaging. An MELC filled with high-pressure Ar or Ne can be used as a fast and precise 140-keV gamma ray Compton scatter detector. An MELC filled with high-pressure Xe can be used as an absorption detector of scattered gamma rays. Together they may make up a complete Compton SPECT system (see Section VIC).

C. Liquid Xenon Detectors

The density of condensed xenon is 3.1 g/cm³ in the liquid phase and 3.6 g/cm³ in the solid phase at the temperature of its triple point (161 K). This makes xenon with high atomic number ($Z = 54$) a very effective absorber for gamma rays. A few attempts have been undertaken to use it for gamma-ray imaging.

1. Electroluminescence Emission Gamma Camera

As we have shown in Section IA, physical amplification of primary ionization signals improves the performance of ionization detectors. Earlier attempts to achieve effective charge or light multiplication in the vicinity of thin wires in condensed noble gases were not very successful (Derenzo *et al.*, 1974; Mijajima *et al.*, 1976). In some recent work, microstrip boards have also been investigated for this purpose (Akimov *et al.*, 1994; Policarpo, Chepel, Lopes, Peskov, *et al.*, 1995). But only a low amplification factor (of ~10–100) and a poor energy resolution have been achieved. In rarefied phases, however, electronic signals can be more easily amplified. Drifting electrons can be extracted from condensed dielectrics into the gas phase (see Section IID), where effective amplification processes can then be used. The emission detector works as follows. Measured radiation interacts with a condensed (liquid or solid) phase, excites and ionizes atoms, and generates a first signal via emission of photons or quasi-particles (scintillation in noble liquids and solids, phonons in crystals, and rotons in super-fluid helium). If an external electric field is applied, ionization electrons drift to the interface, escape into the rarefied (gas) phase, excite and ionize the atoms of the gas, and generate a second (amplified) signal. This signal is delayed relative to the first signal and can be used for two-dimensional coordinate and energy measurements. The third coordinate is defined by the delay time between the first and second signal. If the emission process is effective and the generated light is effectively measured, the energy resolution of the electroluminescence emission detector (EED) is not much differ from the resolution of liquid ionization detectors.

The first attempt to develop an EL emission detector for nuclear medicine imaging was in the middle of the 1970s (Lansiart *et al.*, 1976). It was experimentally shown that electron emission from liquid xenon followed by EL of the vapor allows significant gain in light output relative to the scintillation. Soon after this, a Moscow group developed an EL emission gamma camera with a 30-cm-diameter FOV. The camera was viewed with a matrix of 19 3-in-diameter photomultiplier tubes placed outside of the camera (Egorov *et al.*, 1983). Electrons extracted from 3.5-mm-thick solid xenon layer generate EL of the gas phase (~10 mm thick). The detector has shown 2.5 mm intrinsic 2D resolution measured with an alpha source (Bolozdynya *et al.*, 1985) and 15% FWHM energy resolution at 122 keV. In this experiment, the energy resolution was limited by fluctuations of electron collection due to the relatively low purity of xenon.

A small (~10 cm³ volume) EL emission drift chamber has been constructed for one-dimensional position registration of annihilation 511-keV quanta (Guschin *et al.*, 1982). A position resolution of 1-mm has been measured along the direction of electric field. Primarily the dimensions of the photoelectron tracks in liquid xenon determine the resolution. The detector has been used in experiments investigating angular distributions of annihilation quanta.

2. Liquid Xenon Scintillation Drift Chamber

The liquid xenon multiwire drift chamber triggered by scintillation has been proposed by V. Chepel *et al.* (1995) as a detector of 511-keV gamma rays for PET. In this detector, the scintillation is used as a trigger for the drift time measurement. The multiwire chamber collects the charge and determines the position of gamma quanta photoabsorption. One ionization cell is formed by two plane cathodes 10 mm apart and by 20 anode wires spaced by 2.5 mm, tied to pairs in 10 channels and read out independently. A cylindrical array of the multiwire ionization cells completes a tomograph ring. In the cell, the number of anode wires on which the ionization signal is induced gives the depth of interaction in the detector with accuracy of approximately 5 mm. The depth information allows the reduction of parallax error, decreasing the tomograph size and improving the solid angle resolution. A prototype of this detector is currently under development at University of Coimbra.

V. TECHNICAL FEATURES OF LUMINESCENCE DETECTORS

A. Gas Purification

Luminescence noble gas detectors require very pure working media. Electronegative impurities such as oxygen, water, carbon, and nitrogen monoxides effectively capture electrons. Any molecular impurities can change the energy distribution of drifting electrons, influence drift velocity and diffusion, and suppress the light yield of scintillation and electroluminescence. The allowed contamination lies between 10^{-6} and 10^{-9} relative concentration of impurities.

One of the main arguments against the development of commercial xenon detectors is the well-known necessity of providing the long-term purity of xenon against contamination by electronegative and organic molecules outgassed from inside materials. In fact, the problem is much like that of supporting a high vacuum inside sealed vacuum devices such as Röntgen tubes, kinescopes, or photomultipliers. Successful experience with the SAES Zr-Al getter, supporting the purity of the EL xenon detector during a one-half-year satellite experiment (Inoue et al., 1982) and the Ti-dust getter used for the same purpose in the ionization chamber Xenia for 9 years, operating on the Mir orbital station (Dmitrenko et al., 2002), has shown that the solution found long ago for vacuum devices works successfully for noble gas detectors. Active metal getters can support a purity of xenon in sealed detectors for years. Practically, more than 1 meter of electron drift length is achieved in liquid xenon and krypton and more than 20 m of drift length is achieved in liquid argon using different purification techniques. A review of purification methods could be found in (Barabash and Bolozdynya, 1993). In this section, we consider some most effective instruments and methods used for purification of noble gases.

Oxygen is the major electronegative impurity capturing free electrons. That is why for some simple cases, it is sufficient to purify the noble gas only of oxygen. The best instrument for removing oxygen from noble gases is the Oxysorb purifier by Messer Greisheim GmbH. The purifier contains chromium trioxide embedded on a silica gel matrix.

There are a number of evaporable and nonevaporable metal getters supporting high vacuum. Any of them could be used to support the purity of noble gases in sealed detectors. The most popular nonevaporable getter materials show optimum performance when operated at high temperatures (300–1000°C). Usually, they are used as chips or grains (Ca), have a porous structure (Ti), or are embedded (Zr) on porous structure such as graphite powder. The activity of hot metal getters depends on the value of the gas flux cooling them down.

Molecular sieves can be used for purification of organic (CH_4, DH_4, SiH_4) or molecular gases (N_2), as well as for the preliminary purification of large volumes of noble gases. Cobalt-exchanged, platinized zeolites 13X and 4A has been found capable of lowering the level of oxygen in noble gases to < 1 part per trillion in the temperature range 400–800°C (Sharma, 1995).

The most effective purification systems for high-pressure xenon include a titanium spark discharge purifier (Bolotnikov and Ramsey, 1996). Originally, this method was developed for liquid noble gases (Pokachalov et al., 1993). The principle of operation is as follows. A high-voltage spark discharge between titanium electrodes produces a large amount of titanium dust in the noble gas atmosphere. The titanium dust has a clean surface and is very chemically active. The dust absorbs most electronegative molecules capturing electrons (oxygen, water, carbon, nitrogen monoxide, etc.). In addition, the discharge breaks down big organic molecules and this enhances the purification process. The titanium dust covers the walls of the purifier and continues to absorb impurities for a long time without additional discharges. Detectors filled with xenon purified in spark purifiers usually contain a small amount of fine dust, and this residual dust continues to support high purity after the detector has been sealed for years (Dmitrenko et al., 2000).

One of the effective purification procedures is the circulation of gas in a closed loop between detector and purifiers. The procedure is important for washing inside surfaces from absorbed gases. Hot metal getters and Oxisorb have shown good performance in circulation gas systems. After some days of circulation with flow up to 100 liter/hour under 10-bar pressure, outgassing can be stabilized at a level where the energy resolution of the detector does not change during the experiments. Circulating purification procedure should be repeated each time after the chamber is open or has been out of use for a long time.

There are no absolutely universal purifiers. Obviously, a combination of different purifiers must be used to provide the needed level of purity of the gas. Even very effective purifiers, such as the titanium spark purifier, require preliminary purification for commercially available gases. So, a complete purification system should include Oxisorb (for example, Gross patron by Messer Giesheim) and a high-temperature metal getter (for example, Monotorr by SAES) as a first stage of purification and a spark purifier as a second stage. The purification system may include a full-metal circulation pump to provide multiple circulation of gas through active elements and a volume of the detector under a pressure up to 10 bar. Such a purification system has been successfully used for operation of high-pressure xenon scintillation drift chambers (Belogurov et al., 1995; Bolozdynya, Egorov, et al., 1997a).

B. Photosensors

The most common instrument for reading out the light from EL detectors is the PMT. In many SDC designs, the localization of EL and scintillation flash is provided by an array of PMTs connected to position-decoding electronics. It has been experimentally shown that ordinary glass PMTs may work in noble gases at 10-bar pressure (Belogurov et al., 1995) and even noble liquids (Akimov et al., 1995). At least two types of glass-tube PMTs have been tested inside high-pressure EL detectors (Table 7). Alternative methods such as photocathodes coupled to a microchannel plate (Charpak et al., 1980), wire chambers with triethylamine (TEA) or tetrakis (dimethylamine) ethylene (TMAE) gaseous photocathodes (Simons et al., 1985), and the solid CsI-photocathodes coupled to an avalanche chamber (Bräuning et al., 1994) have been proposed and tested.

Another prospective method of EL detection is to use wavelength-shifting fibers coupled to remote photodetectors. Early experiments with fibers showed that high-pressure noble gases can be absorbed in plastic fibers, affecting their luminescent properties; another danger is the contamination of the noble gas by outgassing plastic fiber (Parsons et al., 1990). However, authors of this first study used plastic fibers that were produced early in the history of fiber-optic development. Producers of modern plastic fibers affirm that their scintillating fibers are compatible with high vacuum, meaning they have a low vapor pressure and so may not contaminate the noble gases (Scintillating Optical Fibers, BICRON Corp.). More recent experiments with modern plastic scintillating fibers have not shown any negative effects and have confirmed good light-collection properties of plastic fiber arrays placed directly in pressurized xenon (Akimov et al., 1997) and argon (Carabello, et al., 1998).

C. UV Wavelength Shifting

The luminescence of the noble gases lies in the vacuum UV range of 120 to 170 nm (see Table 3). There are no known optical materials that are practical and sufficiently transparent for this optical range. The common way to work with UV luminescence of noble gases is to use wavelength shifters. For example, the UV luminescence of xenon (~170 nm) has been effectively measured with glass PMTs covered with the p-terphenyl ($C_{14}H_{18}$) wavelength shifter vacuum-deposited on the entrance window with a thickness of approximately 0.5 mg/cm^2 (Belogurov et al., 1995; Bolozdynya, Egorov, et al., 1997a; Akimov et al., 1995). The quantum efficiency of wave shifting 170 nm with p-terphenyl measures > 90% in pressurized xenon (Belogurov et al., 1995). The emission spectrum of p-terphenyl has two peaks, at 350 and 450 nm. An important property of this well-known scintillating dye is that it does not contaminate xenon. There are other effective wavelength shifters, such as buthyl-PBD, tetraphenyl butadiene (TPB), and diphenyl stilbene (DPS), with emission spectra peaks at 370, 440, and 409 nm, respectively, and a width of approximately 50 nm. They all can be deposited via vacuum evaporation with thicknesses from 0.01 to 1 mg/cm^2. Alternative deposition techniques include sprayed coatings and fluor-doped plastic films. The sprayed coating of 0.1–3 mg/cm^2 thickness may be prepared by airbrushing TPB solution in methyl alcohol or ethyl ether. Doped plastic films can be used for coating large surfaces. The techniques are described in (McKinsey et al., 1997). Plastic scintillating fibers can be used as a secondary wave shifters to adjust the quantum efficiency of the primary wavelength shifter to the photodetector (Bolozdynya and Morgunov, 1998; Parsons et al., 1990).

D. Construction

The energy and spatial resolutions of SDCs depend on how high an electric field can be applied to the light-production gap. Practically, a 2 kV/cm/bar reduced electric field is achieved in 20 bar xenon (Belogurov et al., 1995).

A few types of front entrance windows have been tested. A spherical window made of 3-mm-thick aluminum was

TABLE 7 Parameters of Glass-Tube Photomultipliers with SbCs Semitransparent Photocathodes (MELZ, Moscow) Tested in Pressurized Noble Gases

	FEU-85	FEU-60
Diameter of photocathode (mm)	25	10
Spectral sensitivity (nm)	300–600	300–600
Maximum sensitivity (nm)	340–440	380–490
Luminous sensitivity (mA/lm)	78	50
Number of stages	11	10
Tube diameter (nm)	30	15
Tube length (nm)	110	70
Mass (g)	50	12
Maximum outside pressure (bar)	8	45

From Belogurov et al., 1995; Akimov et al., 1995.

capable of operation up to 20 bar at a 30-cm diameter (Bolozdynya, Egorov, et al., 1997a). An additional grounded thin aluminum electrode, screening the window, was installed to keep a uniform electric field in the drift region. However, carbon-fiber epoxy could be considered as a preferable material for the construction of sophisticated-shape high-pressure chambers. For example, there is a positive report of the construction of a shell-like SDC of approximately 1-m FOV (Parsons et al., 1989).

Selection of proper materials for the construction of high-pressure noble gas detectors and gas systems is very important. The criteria of selection are high purity, low oxygen-absorbing ability, and chemical stability. In fact, any materials that are used in high-vacuum systems could be used to build noble gas detectors. The first candidates are stainless steel and ceramics. For example, the SDCs already mentioned (Belogurov et al., 1995; Bolozdynya, Egorov, et al., 1997a) and the gas systems were built from stainless steel 304. The inside insulators and HV feedthrough, soldered with metal flanges, are made of ceramics. Copper-indium gaskets are used for sealing the detector body, the entrance gamma-radiation windows, and the optical windows. The gas systems employ VCR gas connectors with silver-plated copper gaskets. High-strength alloys, such as the titanium–aluminum–vanadium alloy Ti-6Al-4V, may be used to build a relatively light high-pressure vessel (Mahler et al., 1998). Pure titanium is not acceptable for noble gas detectors containing hydrogen as accelerating doping because of its high reactivity with this gas.

VI. APPLICATIONS FOR SINGLE-PHOTON EMISSION COMPUTED TOMOGRAPHY

EL detectors have been tested in several gamma-ray imaging systems. These detectors possess inherently better energy resolution than the scintillation cameras that have been used in SPECT since the 1960s. The huge light output achievable with EL allows the development of compact imaging systems with enhanced spatial and contrast resolution. With fiber optics readout, high-efficiency multilayer gas detectors can be constructed in a variety of cylindrical configurations that would be difficult to achieve with crystal scintillators. In this section, we consider a few perspective applications of EL detectors for 140-keV gamma-ray imaging.

A. Small Gamma Camera

There are several areas in which small gamma cameras have important use. These include *in vitro* and *in vivo* imaging in nuclear medicine. Small *in vivo* imaging devices are of clinical interest in a number of body-surface imaging areas, including imaging of hard-to-get-at small organs. Examples include the thyroid for poorly functioning malignant nodules, the axillae for metastases from breast cancer, the testes for tumors versus inflammatory lesions, and eye-tumor localization studies. The development of new tracers often goes through animal testing prior to human studies. Animal studies are ideally suited to small detectors, especially when the FOV and the region of interest are comparable in size. The use of small portable imaging devices could play a beneficial role in the clean up of oil spills. The use of high-resolution energy-discriminating devices could be useful in assaying thin-layer chromatography strips as part of nuclear medicine quality control procedures.

A schematic drawing of a small high-pressure scintillation drift chamber with approximately 2-in-diameter FOV (Belogurov et al., 1995) is shown in Figure 11. The body of the chamber is workable up to 40 bar internal pressure. The electrode system of the chamber consists of two grid electrodes (50-μm-diameter stainless steel wires point-welded to a ring frame with pitch 0.5 mm) with a gap between them of

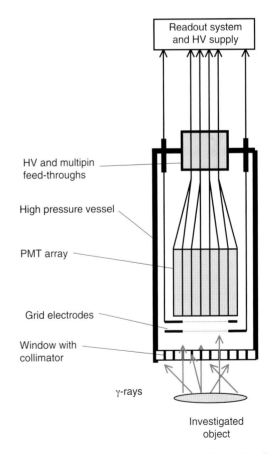

FIGURE 11 Schematic diagram of a small EL SDC with 2-in-diameter FOV.

4 mm and placed at a distance of 17 mm from the entrance window coupled to the collimator. The electrodes are separately installed on the top flange through Teflon isolators. PMTs are placed 17 mm above the light-production gap and screened with a grounded grid made of 70-μm stainless steel wires with a 2-mm pitch to assure that the electric field in the chamber does not influence the PMT photocathode. Alumna feedthroughs for 30 kV and two gas inputs are placed on the top flange. The PMT voltage dividers are placed outside of the chamber and connected to the PMT pins via a multipin feedthrough. Seven 0.5-in-diameter PMTs of type FEU- are placed inside of the chamber in a hexagonal arrangement at 2 mm distance from one another and are fed from the same voltage divider placed outside the chamber. The anodes of the PMTs are connected separately to individual Fera ADC inputs. The readout system is triggered by EL signals. Energy spectra of different X-ray and gamma-ray sources measured with this compact SDC are shown in Figure 12. One can see that a detection threshold of approximately 1 keV is achieved with this detector. The image of a pinhole collimator of 1 mm diameter was taken with 59.6-keV gamma-rays. The FWHM of the distribution was found to be 1.7 mm. Taking into account the diameter of the collimator (1 mm) and the geometry factor, one can estimate the intrinsic spatial resolution of 7 ×FEU-60 SDC as 1.2 mm FWHM. The cost of this system is estimated to be comparable with the cost of a small gamma camera with multichannel PMT readout, but the performance could be better than that of the classic Anger camera.

B. Cylindrical Gamma Camera

In SPECT imaging, a cylindrical detector geometry offers many advantages (Chang *et al.*, 1996). One of them is the possibility of avoiding the mechanical motion of the detector and providing very stable imaging performance. Another is the increased detector coverage leading to a high system sensitivity. There are a few state-of-the-art cylindrical scintillation gamma cameras developed for brain SPECT imaging (Chang, 1996). However, it is difficult and expensive to implement this configuration on the scale necessary for a whole-body SPECT system with NaI(Tl) scintillators. A more recently proposed multilayer noble gas EL camera with fiber optics readout (Bolozdynya and Morgunov, 1998) could be relatively easy to construct in the needed scale and configuration.

It is often stated that improved intrinsic detector resolution is of little use because the collimator, not the detector, limits the total system position resolution. However, H. Barrett and co-workers have shown that semiconductor detectors with fine position resolution coupled to a multipinhole collimator can provide both better system position resolution and higher sensitivity (Rogulsky *et al.*, 1993). Now this approach is developing for brain SPECT imaging.

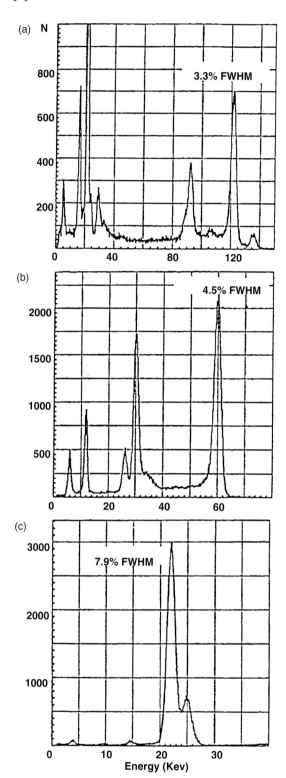

FIGURE 12 Energy spectra acquired with the small SDC filled with 20 bar xenon. (Reprinted from Belogurov *et al.*, 1995, with permission from The Institute of Electrical and Electronics Engineers, Inc.)

Another classical statement says that a detector cannot be thick because the collimator degrades with distance. How-

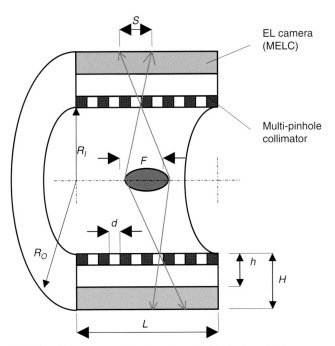

FIGURE 13 Conceptual design of a cylindrical electroluminescence camera for cardiac SPECT.

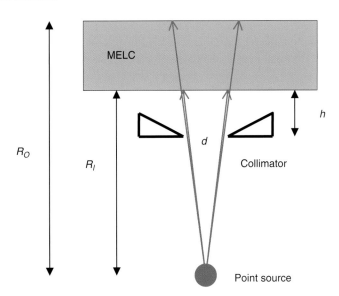

FIGURE 14 Determination of the average collimator resolution for the cylindrical MELC.

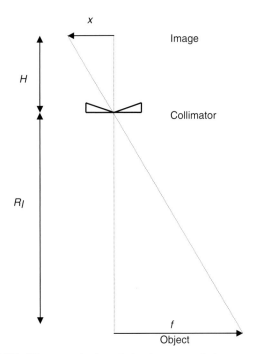

FIGURE 15 Determination of the detector resolution as an optical consideration.

ever, we can show that when large EL detectors are used instead of relatively small CdTe or CZT semiconductor arrays, the idea can be applied in one of the most popular nuclear medicine investigations—cardiac SPECT.

Let us consider a cylindrical MELC of thickness 10 cm, outside diameter $2R_o = 1$ m, and $L = 40$ cm long as shown in Figure 13. A multipinhole cylindrical collimator of internal radius $R_i = 30$ cm is enclosed inside the MELC. The collimator consists of $N_{p\text{-}h}$ pinholes of d diameter. The pinholes are focused to an 18-cm-diameter FOV located around the geometrical center of the camera. The FOV area $F = 254$ cm^2 is projected into the MELC to form an $S = FH/R_o$ image area on the back surface of the MELC. The number of pinholes needed can be estimated as $N_{p\text{-}h} = (2\pi R_o L)/S \approx 120$.

The system position resolution can be defined as $R_S^2 = R_C^2 + R_D^2$, where R_C is the collimator resolution, and R_D is the detector resolution. For a thick detector, the collimator resolution may be estimated by the average value between the dimensions of projections of a pinhole on front and back surfaces of the detector from the point source (as an ideal object) located at the center of the camera (Figure 14) as follows.

$$R_C \approx 0.5\,[R_o d/R_i + (R_i + h)d/R_i]$$
$$= 0.5\,d(R_o + R_i + h)/R_i = 1.5\,d$$

In the case of an ideal collimator, the uncertainty of the image due to the intrinsic detector resolution can be estimated as an optical consideration (Figure 15). Because the image dimension is defined via the object size in the ratio $f/x = R_i/H$, the accuracy of imaging with the detector can be estimated as an average value of the uncertainty in imaging object:

$$R_D = <\Delta f> = <[R_o/H]\Delta x> 2\,\Delta x$$

where Δx is the intrinsic MELC resolution. Then, the average system resolution can be calculated as

$$R_s^2 \approx (2\,\Delta x_i)^2 + (1.5\,d)^2$$

Assuming the MELC provides an approximately 1-mm three-dimensional position resolution (see Section IVD2), we calculate that the camera equipped with 4-mm-diameter pinholes provides a 6-mm position system resolution in the 18-cm-diameter central area of the FOV.

The total system detection efficiency can be defined as $\varepsilon_o = \varepsilon_d\, \varepsilon_{p\text{-}h}\, N_{p\text{-}h}$, where ε_d is the detection efficiency, and $\varepsilon_{p\text{-}h}$ is the pinhole geometrical efficiency. The detector efficiency of a 10-cm-thick camera filled with 20 bar Xe is calculated to be 62% at 140 keV (^{99m}Tc) and 99.8% at 77 keV (^{210}Tl). The pinhole geometrical efficiency can be estimated as $\varepsilon_{p\text{-}h} = (\pi\, d^2/4)/(4\,\pi\, R_i^2) = (\pi d R_i)^2$. Then, the system detection efficiency can be calculated as

$$\varepsilon_o = (\pi d^2 R_o L \varepsilon_d)/(8FHR_i)$$

Substituting the values in this formula, one can calculate that the detection efficiency (and the sensitivity of the system!) is respectively five or eight times better than that of the modern three-head SPECT system with a 30-cm radius of rotation. Note that the energy resolution of the MELC will be approximately three times better than that of an Anger camera in the range 77–140 keV. All these factors influence imaging quality; taken together they can lead to significantly enhanced cardiac imaging.

C. Compton Camera

Mechanical collimation is the primary factor that limits the imaging potential of SPECT. To overcome the limitations of mechanical collimation, Compton scatter imaging was proposed by Todd, Everett, and Nighhtingale in 1974 (for details see Chapter 19 in this volume). In the 1980s, Singh and co-workers developed and investigated the first Compton camera for imaging 140-keV single-photon emitters with electronic collimation. Since that pioneering work, almost all these experimental and theoretical studies have adopted a Compton camera design, with apertures limited by sizes and configuration similar to the design of a one-head Anger camera. However, for clinical SPECT imaging, it is desirable to sample emission projections from as many transaxial directions as possible. Detectors with limited apertures have been rotated around the source being imaged in order to meet the sampling requirement. Thus, the optimum configuration for a SPECT system should be cylindrical, which provides close to full solid-angle coverage of the gamma-ray flux from the source. Using these latest developments in 3D-sensitive EL detectors with fiber readout (Bolozdynya and Morgunov, 1998), a new concept for the cylindrical Compton camera has been proposed for SPECT (Bolozdynya, Ordonez, et al., 1997). The Compton camera consists of two concentric cylindrical detectors, as shown in Figure 14. The inner cylinder is the scatter detector. The outer cylinder is the absorption detector, which is in back of and to both sides of the scatter detector. This arrangement maximizes the coverage of photons scattered from the inner detector. The absorption detector needs to be properly shielded from the direct photon flux as indicated by the black bars in the drawing.

The events of interest are pairs of interactions—Compton scatter in the scatter detector and photoelectric absorption in the absorption detector—that are detected in coincidence. Both detectors should be capable of accurately providing information on the energy depositions and 3D positions of the photon interactions. The scatter interaction position defines the apex of a cone, which back-projects the corresponding Compton event into the imaging volume. The line that connects the scatter and absorption interaction positions defines the axis of the cone. Because of the azimuthal symmetry of Compton scattering, the source of the detected photon can be localized to lie on the surface of this cone. The accuracy of the determination of the cone axis is determined by the spatial-resolution performance of the two detectors. To enhance the scatter fraction of Compton events for the low energy photons used in SPECT, the scatter detector should be made of low-Z material. Potential scatter detectors are Si solid-state detectors or Ar-filled electroluminescence detectors.

The absorption detector is intended to totally absorb the energy of a scattered photon, hopefully in a single interaction. Thus, the absorption detector should be made of relatively high-Z material. Its thickness is determined by the stopping power needed for the energy of interest. Although Ge detectors meet the performance requirements for the absorption detector, they are impractical because of cryogenic requirements and cost considerations. As an alternative, a high-Z noble gas (xenon) detector, with adequate stopping power, has potential for use in the large cylindrical geometry. Two potential configurations for the cylindrical Compton camera are a MELC scatter detector filled with pressurized argon and a MELC absorption detector filled with pressurized xenon. The system position resolution of the Compton camera will be limited because of Doppler broadening in the both scatter and absorption detectors (Ordonez, et al., 1997). Still, the sensitivity of such a cylindrical Compton camera can be approximately 10 times higher than that of three-headed SPECT systems with 30 cm radius of rotation (ROR) at comparable or even better system position resolution (Bolozdynya, Ordonez, et al., 1997).

The cylindrical Compton camera can be built as a full-body SPECT system. The mature gas-detector technology developed in high-energy physics allows the building of cylindrical drift chambers that are several meters in dimension and that have many thousands of readout channels for time and amplitude measurement at tens of megahertz readout frequencies.

VII. CONCLUDING REMARKS

EL detectors noble gas are a unique detector technique, providing:

- Precise 3D position measurements for low-energy depositions from gamma-ray interactions
- A wide choice of working media for either Compton or photoabsorption interactions for ≤140-keV gamma rays
- An energy resolution comparable to room-temperature semiconductor detectors
- The detection threshold of ~1 keV, important for Compton scatter detectors and low-energy gamma-ray imaging
- A number of readout channels at least 100 times less than that of semiconductor arrays with comparable detection efficiency and FOV
- A flexible design for building imaging detectors with very large FOVs in cylindrical or spherical configurations
- Operation at high magnetic fields, for example inside magnetic resonance imaging (MRI) systems

EL detectors promise new benefits in energy resolution, three-dimensional position sensitivity, and reasonable cost for large-FOV systems in the development of new modalities in the twenty-first century.

Acknowledgments

I acknowledge A. B. Brill for many interesting conversations, moral support, and providing a great positive impact on the development of electroluminescence imaging detectors. I thank R. P. DeVito and V. V. Egorov for many years of cooperation in nuclear medicine detector research, W. Chang for the opportunity to study the technology of cylindrical scintillation gamma cameras, W. L. Rogers for many stimulating discussions about new imaging techniques, A. Bolotnikov for his expertise in the physics of high-pressure xenon ionization detectors, and D. Koltick for cooperation in the development of fiber optic readouts from high-pressure noble gas detectors.

References

Akimov, D., Belogurov, S., Bolozdynya, A., Churakov, D., Chumakov, M., Grishkin, Yu., Pozdnyakov, S., Solovov, V. (1994). Xenon scintillation drift chamber with microstrip readout. *In* "Proceedings of the International Workshop on Micro-Strip Gas Chambers" (Della Mea G. and Sauli F., eds.), pp.215–216. Edizioni Progetto, Padova.

Akimov, D., Belogurov, S., Burenkov, A., Churakov, D., Kuzichev, V., Morgunov, V., Smirnov, G., Solovov, V (1997). Scintillation proportional Xe counter with WLS fiber readout for low-energy X-rays. *Nucl. Instr. Meth.* **391A**: 468–470.

Akimov, D. Y., Bolozdynya, A. I., Churakov, D. L., Afonasiev, V. N., Belogurov, S. G., Brastilov, A. D., Burenkov, A. A., Dodohov, V. H., Gusev, L. G., Kuchenkov, A. V., Kuzichev, V. F., Lebedenko, V. N., Osipova, T. A., Rogovsky, I. A., Safronov, G. A., Simonychev, S. A., Solovov, V. N., Sopov, V. S., Smirnov, G. N., Tchernyshev, V. P. (1995). Scintillating LXe/LKr electromagnetic calorimeter. *IEEE Trans. Nucl. Sci.* **42**: 2244–2249.

Akimov, D., Bolozdynya, A., Churakov, D., Koutchenkov, A., Kuzichev, V., Lebedenko, V., Rogovsky, I., Chen, M., Chepel, V., Sushkov, V. (1993). Condensed krypton scintillators. *Nucl. Instr. Meth.* **327A**: 155–158

Alkhazov, G. D., Komar, A. P., and Vorob'ev, A. A. (1979). Ionization fluctuations and resolution of ionization chambers and semiconductor detectors. *Nucl. Instr. Meth.* **48**: 1–12.

Anderson, D. F., Hamilton, T. T., Ku, W. H.-M., and Novick, R. (1979). A large area, gas scintillation proportional counter. *Nucl. Instr. Meth.* **163**: 125–134.

Aprile, E., Bolotnikov, A., Chen, D., and Mukherjee, R. (1993). W value in liquid krypton. *Phys. Rev.* **48A**: 1313–1318.

Attix, F. H., and Roesch, W. C. (1966). Radiation Doisimetry. Academic Press, New York.

Barabash, A. S., and Bolozdynya, A. I. (in Rusian) (1993). "Ionization Liquid Detectors". Energoatomizdat, Moscow.

Bateman, J. E., Connolly, J. F., Stephenson, R., Tappern, G. J., and Fleshner, A. C. (1984). The Rutherford Appleton Laboratory's Mark I multiwire proportional chamber for positron camera. *Nucl. Instr. Meth.* **225**: 209–231.

Belogurov, S., Bolozdynya, A., Churakov, D., Koutchenkov, A., Morgunov, V., Solovov, V., Safronov, G., and Smirnov, G. (1995). High pressure gas scintillation drift chamber with photomultipliers inside of working medium. *1995 IEEE Nucl. Sci. Sym. Med. Imag. Conf. Rec.* **1**: 519–523. IEEE, San Francisco.

Bolotnikov, A. and Ramsey, B. (1996). Purification technique and purity and density measurements og high-pressure Xe. *Nucl. Instr. Meth.* **383A**: 619–623.

Bolotnikov, A. and Ramsey, B. (1997). The spectrometric properties of high pressure xenon. *Nucl. Instr. Meth.* **396A**: 360–370.

Bolozdynya, A. (1999). Two-phase emission detectors and their applications. *Nucl. Instr. Meth.* **422A**: 314–320.

Bolozdynya, A. I., Egorov, V. V., Kalashnikov, S. D., Krivoshein, V. L., Miroshnichenko, V. P., Rodionov, B. U. (in Russian) (1985). Emission electroluminescence camera with condensed xenon as a working medium. *Instr. Exper. Tech.* **4**: 43–45.

Bolozdynya, A., Egorov, V., Koutchenkov, A., Safronov, G., Smirnov, G., Medved, S., Morgunov, V. (1997a). High pressure xenon self-triggered scintillation drift chamber with 3D sensitivity in the range of 20-140 keV deposited energy. *Nucl. Instr. Meth.* **385A**: 225–238.

Bolozdynya, A. I., Egorov, V. V., Koutchenkov, A. V., Safronov, G. A.,

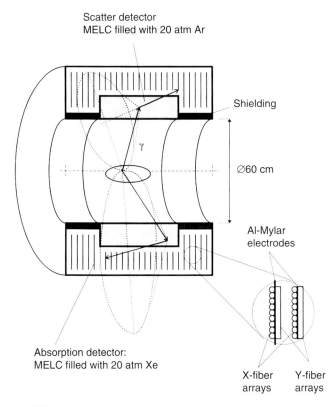

FIGURE 16 Schematic diagram of a cylindrical EL Compton camera.

Smirnov, G. N., Medved, S. A., and Morgunov, V. L. (1997b). High pressure xenon electronically collimated camera for low energy gamma ray imaging. *IEEE Trans. Nucl. Sci.* **44**: 2408–2413.

Bolozdynya, A. I. and Morgunov, V. L. (1998). Multilayer electroluminescence camera: concept and Monte Carlo study. *IEEE Trans. Nucl. Sci.* **45**: 1646–1655.

Bolozdynya, A., Ordonez, C. E., and Chang, W. (1997). A concept of cylindrical Compton camera for SPECT. *In* "1997 IEEE NSS & MIC Record", M06–08. IEEE, Albuquerque.

Bräuning, H., Breskin, A., Chechik, R., Dangendorf, V., Demian, A., Ulmann, K., Schmidt-Böcking., H. (1994). A large volume 3D imaging gas scintillation counter with CsI-based wire chamber readout. *Nucl. Instr. and Meth.* **348A**: 223–227.

Carabello, S., Koltick, D., Bolozdynya, A., Chang, W. (1998). Scintillating fiber performance in high pressure noble gases. *AIP Conf. Proic.* **450**: 70–79.

Carugno, G. (1998). Infrared emission in gaseous media induced by ionizing particles and by drifting electrons. *Nucl. Instr. Meth.* **419A**: 617–620.

Chang, W. (1996). Dedicated SPECT systems. *In* "Nuclear Medicine" (R. E. Henkin *et al.*, eds.) **1**: 247–253. Mosby, St.Louis.

Chang, W., Liu, J., Yu, D., and Loncaric, S. (1996). A cylindrical geometry for cardiac SPECT imaging. *IEEE Trans. Nucl. Sci.* **43**: 2219–2224.

Charpak, G., Ngoc, H. N., and Policarpo, A. (1981). Neutral radiation detection and localization. United States Patent 4,286,158, August 25.

Charpak, G., Policarpo, A., and Sauli, F. (1980). The photo-ionization proportional scintillation chamber. *IEEE Trans. Nucl. Sci.* **27**: 212–215.

Chepel, V.Yu., Lopes, M. I., Alves, M. A., Ferreira Marques, R., Policarpo, A. J. P. L. (1995). Liquid xenon multiwire chamber for positron tomography. *Nucl. Instr. Meth.* **367A**: 58–61.

Derenzo, S. E., Mast, T. S., Zaklad, H., and Muller, R. A. (1974). Electron avalanche in liquid xenon. *Phys. Rev.* **9A**: 2582–2591.

Dmitrenko, V. V., Grachev, V. M., Ulin, S. E., Uteshev, Z. M., Vlasik, K. F. (2000). High-pressure xenon detectors for gamma-ray spectrometry. *Appl. Rad. Isot.* **52**: 739–743.

Doke, T., Hitachi, A., Kubota, S., Nakamoto, A., Takahashi, T. (1976). Estimation Fano factor in liquid argon, krypton, xenon, and xenon-doped liquid krypton. *Nucl. Instr. Meth.* **134**: 353–357.

Dolgoshein, B. A., Lebedenko, V. N., and Rodionov, B. U. (1970). A new method of track registration of ionization particles in condensed matter. *JETP Lett.* **11**: 513–516

Drost, D. J. and Fenster, A. S. (1982). A xenon ionization detector for digital radiography. *Med. Phys.* **9**(2): 224–230.

Drost, D. J. and Fenster, A. S. (1984). A xenon ionization detector for scanned projection radiography: 95-channel prototype evaluation. *Med. Phys.* **11**(5): 602–609.

Dubrovin, S. A., Klyuch, V. E., Novikova, A. F., and Shishkanov, N. G. (1977). Spark ionization chamber for gamma ray imaging. *In* "10th All-Union Congress of Radiologists, Yerevan", pp 572–575. Moscow.

Egorov, V. (in Russian) (1988). Electroluminescence position-sensitive compressed xenon gamma-ray detector. *Instr. Exp. Techn.* **1**: 53–57.

Egorov, V. V., Miroshnichenko, V. P., Rodionov, B. U., Bolozdynya, A. I., Kalashnikov, S. D., Krivoshein, V. L. (1983). Electroluminescence emission gamma camera. *Nucl. Instr. Meth.* **205**: 373–374.

Guschin, E. M., Kruglov, A. A., Obodovsky, I. M., Pokachalov, S. G. (in Russian) (1982). Liquid xenon position-sensitive detector of gamma radiation. *Instr. Exper. Tech.* **3**: 49–52.

Holland, S. E., Wang, N. W., and Moses, W. W. (1997). Development of low noise, back-side illuminated silicone photodiode arrays. *IEEE Trans. Nucl. Sci.* **44**: 443–447.

Huxley, L. G. H., and Crompton, R. W. (1974). "The Diffusion and Drift of Electrons in Gases". John Wiley & Sons, New York.

Inoue, H., Koyama, K., Mae, T., Matsuoka, M., Ohashi, T., Tanaka, Y., and Waki, I. (1982). Gas scintillation proportional counters for Japanese astronomical satellites. *Nucl. Instr. Meth.* **196**: 69–72

Johns, H. E., Festner, A., Plewes, D., Boag, J. W., and Jeffery, P. N. (1974). Gas ionization methods of electrostatic image formation in radiology. *Br. J. Radiol.* **47**: 519–529.

Knoll, G. F.(1989). "Radiation Detection and Measurement", 2nd ed. John Wiley, New York.

Lacy, J. L., LeBlanc, A. D., Babich, J. W., Bungo, M. W., Latson, L. A., Lewis, R. M., Poliner, L. R., Jones R. H., and Johnson, P. C. (1984). A gamma camera for medical applications, using a multiwire counter. *J. Nucl. Med.* **25**: 1003–1021.

Lansiart, A., Seigneur, A., Moretti, J.-L., and Morucci, J.-P. (1976). Development research on a highly luminous condensed xenon scintillator. *Nucl. Instr. Meth.* **135**: 47–52.

Machulin, I. N., Miroshnichenko, V. P., Rodionov, B. U., Chepel, V.Yu. (1983). Feasibility of precision spectrometry of 1 MeV electrons in liquid xenon. *Sov. Tech. Phys. Lett.* **9**: 1128–1132.

Mahler, G. J., Yu, B., Smith, G. C., Kane, W. R., and Lemley, J. (1998). A portable gamma ray spectrometer using compressed xenon. *IEEE Trans. Nucl. Sci.* **45**: 1029–1033.

McKinsey, D. N., Brome, C. R., Butterworth, J. S., Golub, R., Habicht, K., Huffman, P. R., Lamoreaux, S. K., Mattoni, C. E. H., Doyle, J. M. (1997). Fluorescence efficiency of thin scintillating films in the extreme ultraviolet spectral region. *Nucl. Instr. Meth.* **132B**: 351–358.

Mijajima, M., Masuda, K., Hitachi, A., Doke, T., Takahashi, T., Konno, S., Hamada, T., Kubota, S., Nakamoto, A., and Shibamura, E. (1976). Proportional counter filled with highly purified liquid xenon. *Nucl. Instr. Meth.* **134**: 403–407

Miyajima M., Sasaki, S., Shibamura, E. (1995). Comments on "liquid xenon ionization and scintillation studies for a totally active-vector electromagnetic calorimeter". *Nucl. Instr. Meth.* **352A**: 548–551.

Nguyen Ngoc, H. (1978). Performance of an X-ray imaging gas scintillating proportional counter. *Nucl. Instr. Meth.* **154**: 597–601.

Ngueyn Ngoc, H., Jeanjean, J., Itoh, H., and Charpak, G. (1980). A xenon high-pressure proportional scintillation-camera for X- and gamma-ray imaging. *Nucl. Instr. Meth.* **172**: 603–608.

Niemela, A., and Sipila, H. (1994). Evaluation of CdZnTe detectors for soft X-ray applications. *IEEE Trans. Nucl. Sci.* **41**: 1054–1057

Northrop, J. A., Gursky, J. C. and Johnsrud, A. E. (1958). Further work with noble element scintillators. *IRE Trans. Nucl. Sci.* **5**: 81–87.

Ordonez, C. E., Bolozdynya, A., and Chang, W. (1997). Energy uncertainties in Compton cameras. *In* "1997 IEEE NSS & MIC Record", N26-05. IEEE, Albuquerque.

Ott, R. J., Flower, M. A., Babich, J. W., and Marsden, P. K. (1992). The physics of radioisotope imaging. *In* "The Physics of Medical Imaging" (S. Webb, ed.), pp. 142-318. The Institute of Physics, London,

Parsons, A., Edberg, T. K., Sadoulet, B., Weiss, S., Wilkerson, J., Hurley, K., Lin, R. P., and Smith, G. (1990). High pressure gas scintillation drift chambers with wave-shifter fiber readout. *IEEE Trans. Nucle. Sci.* **37**: 541–546.

Parsons, A., Sadoulet, B., Weiss, S., Edberg, T., Wilkerson, J., Smith, G., Lin, R. P., Hurley, K. (1989). High pressure gas scintillation drift chambers with wave-shifter fiber readout. *IEEE Trans. Nucl.Sci.* **36**: 931–935.

Pierset, A. and Sauli, F. (1984).Drift and diffusion of electrons in gases: a compilation: A compilation. Preprint CERN 84-08, Exp. Phys. Division, 13 July.

Platzman, R. L. (1961). Total ionization in gases by high-energy particles: An appraisal of our understanding. *Int. J. Appl. Rad. Isotopes* **10**: 116–127.

Pokachalov, S. G., Kirsanov, M. A., Kruglov, A. A., and Obodovski, I. M. (1993). *Nucl. Instr. Meth.* **327A**: 159–162.

Policarpo, A. J. P. L. (1977). The gas proportional scintillation counter. *Space Sci. Instr.* **3**: 77–107.

Policarpo, A. P. L., Chepel, V., Lopes, M. I., Peskov, V., Geltenbort, P., Ferreira Marques, R., Araujo, H., Fraga, F., Alves, M. A., Fonte, P.,

Lima, E. P., Fraga, M. M., Salete Leite, M., Silander, K., Onofre, A., Pinhao, J. M. (1995). Observation of electron multiplication in liquid xenon with a microstrip plate. *Nucl. Instr. Meth.* **365A:** 568–571.

Radeka, V. (1988). Noise in detectors. *Ann. Rev. Nucl. Part. Sci.* **38:** 217–277.

Rogulsky, M. M., Barber, H. B., Barret, H. H., Shoemaker, R. L. and Woolfendon, J. M. (1993). Ultra-high-resolution brain SPECT imaging: simulation results. *IEEE Trans. Nucl. Sci.* **40:** 1123–1129.

Sauli, F. (1999). Recent developments and applications of fast position-sensitive gas detectors. *Nucl. Instr. Meth.* **422A:** 257–262.

Sharma, P. K. (1995). Sorbents remove Oxygen at high temperatures. *NASA Tech Briefs* **April:** 52.

Simons, D. G., De Korte, P. A. J., Peacock, A., Smith, A., and Blecker, J. A. M. (1985). Performance of an imaging gas scintillation proportional counter with microchannel plate read-out. *IEEE Trans. Nucl. Sci.* **32:** 345–349.

Suzuki, H., Sibaike, K., Hashimoto, H., Kawada, Y. and Hino, Y. (1998). Analysis of 4-pi gamma ionization chamber response using EGS4 Monte Carlo code. *Appl. Radiat. Isot.* **49:** 1245–1249.

Townsend, D., Frey, P., Donath, A., Clark, R., Schorr, B., and Jeavons, A. (1984). Volume measurements *in vivo* using positron tomography. *Nucl. Instr. Meth.* **221:** 105–112.

Zaklad, H., Derenzo, S. E., Muller, R. A., Smadja, G., Smith, R. G., Alvarez, L. W. (1972). A liquid xenon radioisotope camera. *IEEE Trans. Nucl. Sci.* **19:** 429–431.

Zaklad, H., Derenzo, S. E., Muller, R. A., Smith, R. G. (1974). Initial images from a 24-wire liquid xenon gamma camera. *IEEE Trans. Nucl. Sci.* **20:** 362–367.

Zimmerman, R. E., Holman, B. L., Fahey, F. H., Lanza, R. C., Cheng, C., and Treves, S. (1981). Cardiac imaging with a high pressure low deadtime multiwire proportional counter. *IEEE Trans. Nucl. Sci.* **28:** 55–56.

C H A P T E R
19

Compton Cameras for Nuclear Medical Imaging

W. L. ROGERS,* N. H. CLINTHORNE,* and A. BOLOZDYNYA[†]

*Division of Nuclear Medicine, University of Michigan, Ann Arbor, Michigan
[†]Constellation Technology Corp., Largo, Florida

I. Introduction
II. Factors Governing System Performance
III. Analytical Prediction of System Performance
IV. Image Reconstruction
V. Hardware and Experimental Results
VI. Future Prospects for Compton Imaging
VII. Discussion and Summary

I. INTRODUCTION

A. Method and Motivation

At present, virtually all single-photon imaging in nuclear medicine relies on mechanical apertures of some sort to form projection images of the spatial distribution of gamma-ray-emitting radiolabeled compounds. The simplest aperture is a pinhole or pinhole array in a lead or tungsten sheet used to form pinhole images on a gamma camera. Gunter (Chapter 8) describes a variety of other apertures including parallel and converging-channel collimators. With the exception of coded aperture techniques, all these methods exhibit a limiting detection sensitivity that is inversely proportional to the spatial resolution. For single-photon emission computed tomography (SPECT) imaging, sensitivity is inversely proportional to the number of resolution elements in the object projection, that is, the reciprocal of the projection-space bandwidth (Rogers and Ackermann, 1992). Doubling either the object size or collimator resolution decreases sensitivity by a factor of four. This assumes that multiple projections of the object do not overlap, as they do for coded apertures. Performance of mechanical apertures also suffers with increasing gamma-ray energy. They perform well for ^{99m}Tc at 140 keV, but to image ^{131}I, septal thickness must be substantially increased to reduce gamma-ray penetration with a sacrifice in sensitivity and increase in weight. Despite the thicker septa, there remains considerable penetration background both from the principal 360-keV gamma rays and from the group of higher-energy emissions that extend to 723 keV and comprise about 8% of the emitted photons.

In contrast, the Compton camera provides information about the incoming photon direction electronically without restricting the detection solid angle in the way that mechanical apertures do. The method is illustrated in Figure 1.

The incoming gamma-ray Compton-scatters from an eletron in the position-sensitive first detector and is subsequently absorbed by the second position-sensitive detector. The recorded data consist of the interaction coordinates, X_1, Y_1, X_2, Y_2 and deposited energies for the two events that occur essentially simultaneously in the first and second detectors. The sum of the two energies provides an estimate of the incoming gamma-ray energy, and the energy deposited in the first detector, E_1, gives an estimate of the scattering angle, $\hat{\Phi}$. However, without knowledge of the recoil electron momentum vector, the azimuthal angle is undetermined and the actual point source location is only constrained to lie on

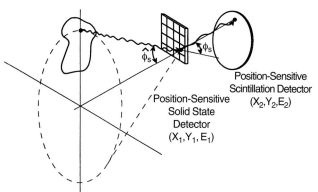

FIGURE 1 Principle of the Compton camera.

a conical shell with angular thickness $\Delta\hat{\Phi}$. This angular uncertainty is related to uncertainty in the recoil electron energy measurement and the electron's initial momentum. There is also uncertainty in the determination of the cone axis introduced by uncertainty in the estimates of the interaction points in the first and second detectors.

If the struck electron is assumed to be at rest, the relation between the incoming gamma-ray energy, $E_{\gamma 0}$, E_1, and $\hat{\Phi}$ is given by:

$$\cos\hat{\Phi} = 1 - \frac{E_1}{\alpha E_2} = 1 - \frac{E_1}{\alpha(E_{\gamma 0} - E_1)} \text{ where } \alpha = \frac{E_{\gamma 0}}{m_0 c^2} \quad (1)$$

Because the detection efficiency depends on the joint probability of an interaction in detector 1 followed by an interaction in detector 2, sensitivity depends on the size, type, and geometry of the two detectors. Angular resolution, on the other hand, depends on the noise characteristics of detector 1 and spatial resolution of the two detectors. We see that the direct relationship between spatial resolution and sensitivity that is the property of mechanical collimation does not exist for Compton cameras. Further, if the principal uncertainty in estimating Φ is the uncertainty in E_1 due to additive detector noise, we see from Eq. (2) that angular resolution improves with increasing gamma-ray energy, $E_{\gamma 0}$.

$$d\Phi = \frac{m_0 c^2}{\sin\Phi(E_{\gamma 0} - E_1)^2} dE_1 \quad (2)$$

These topics will be pursued more rigorously later, but from these two observations it is evident why Compton cameras have such an attraction for researchers in nuclear medicine and other fields in which gamma-ray imaging is a focus.

In this chapter we summarize the present status of Compton imaging in the nuclear medicine application. Following a brief history of the technique, we review the physical factors that influence system performance. In Section III, we discuss analytical predictions of Compton camera performance for quantification and detection tasks in comparison to collimated cameras and also show calculated values of detection sensitivity for two different system geometries.

Methods of image reconstruction are discussed in Section IV, and prototype hardware implementations are described in Section V. We conclude with a discussion of future possibilities for Compton imaging in nuclear medicine and outstanding questions.

B. Brief History

The Compton imaging technique was proposed independently by Pinkau (1966) and White (1968) for imaging solar neutrons. Pinkau employed spark chambers, and White described a method using scintillation detectors.

Schönfelder et al. (1973) described the application to gamma-ray astronomy, and over the next several years several groups flew Compton imagers on balloon flights. COMPTEL was described in 1984 (Schönfelder et al., 1984), and in 1992 the first satellite measurements from the Gamma-Ray Observatory made by COMPTEL were reported by Winkler et al. (1992). A number of groups are developing different versions of a second-generation Compton telescope. Most of these introduce tracking of the recoil electron as a means to reduce the conical ambiguity to a limited arc of the conical surface. MEGA is under development by the group at the Max Planck Institute for Extraterrestrial Physics in Munich (Schopper et al., 2000), TIGRE is under development at the University of California, Riverside (O'Neill et al., 1996; Battacharya et al., 1999), and similar work is under way at the Naval Research Laboratory (Kroeger et al., 1999).[1]

The possible application of the Compton imaging method to nuclear medicine was first described by Todd and Nightingale in 1974 (Todd, 1975; Todd et al., 1974; Everett et al., 1976). Singh and his colleagues published a number of seminal papers beginning in 1981 that described analytical and experimental results for a Compton camera composed of a pixelated germanium first detector and a standard Anger camera second detector (Singh, 1983; Singh and Doria, 1981, 1983, 1985; Singh et al., 1986, 1988; Doria and Singh, 1982; Hebert et al., 1987, 1988, 1990; Brechner and Singh, 1988, 1990). Their system geometry corresponded closely to that illustrated in Figure 1. The Compton scattering detector was designed and fabricated by R. H. Pehl and his colleagues at Lawrence Berkeley Laboratory and exhibited an energy resolution of 850 eV FWHM (Pehl et al., 1985). This detector was subsequently lent to the University of Michigan by Singh to investigate a ring geometry Compton camera and evaluate its potential for imaging radioactive spills, monitoring nuclear waste, and medical imaging. The results of this work have been reported by Martin and Gormley (Martin, 1994; Martin et al., 1993, 1994; Gormley, 1997; Gormley et al., 1997).

---

[1] For additional information, see http://www.gamma.mpe-garching.mpg.de/MEGA/mega.html, http://osse-www.nrl.navy.mil/, and http://tigre.ucr.edu/.

In the remainder of this chapter we summarize ongoing research at the University of Michigan and elsewhere relating to the application of Compton cameras to nuclear medical imaging.

II. FACTORS GOVERNING SYSTEM PERFORMANCE

The important attributes of an imaging system for nuclear medicine are spatial resolution and detection sensitivity for gamma rays in the energy range extending from 70 keV for ^{201}Tl to 511 keV, the positron annihilation energy. The factors governing resolution can be grouped into geometric effects, including the intrinsic resolution of the first and second detectors, electronic effects principally related to noise in the scattering detector and its readout electronics, and the physics of Compton scattering. This section examines the issue of spatial resolution. Sensitivity issues are discussed in later sections.

A. Geometric Effects

To rapidly evaluate various detector candidates and system geometries, it is necessary to have an approximate analytical model of system resolution. The propagation of the effects of first- and second-detector spatial resolution has been addressed by several authors, each with respect to their particular system design (Singh, 1983; Martin, 1994; Martin *et al.*, 1993; Gormley, 1997; LeBlanc *et al.*, 1998). More recently, Ordonez *et al.* (1999) have described a more generalized approach that can be used to predict geometric resolution for the planar geometry of Figure 1, the cylindrical geometry of C-SPRINT, and the conical geometry of the ring Compton camera. Results can be obtained for both isotropic and anisotropic detector response functions.

The independent effects of scatter-detector and absorption-detector resolution are illustrated in Figure 2. R_a represents the resolution of the absorption detector, which introduces uncertainty in the angular orientation of the backprojected cone. The first-detector resolution, R_s, causes uncertainty both in the position of the vertex and orientation of the cone. These effects are usually combined in quadrature under the assumption that they have Gaussian distributions.

The expression derived by Ordonez for cylindrical geometry was evaluated for an on-axis point source 10 cm from an on-axis scatter detector with isotropic 1-mm resolution and a long cylindrical second detector with 3-mm resolution on the surface of the cylinder (Hua, 2000). Results are shown in Figure 3 for two values of second-detector radius and two values of depth resolution. The effects of depth of interaction uncertainty in the second detector are very prominent in the neighborhood of 45° scattering for this geometry. Because the absorption detector is generally

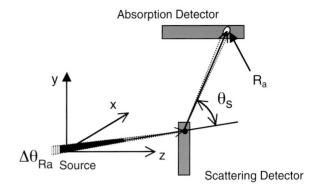

FIGURE 2 Illustration of individual scattering and absorption detector spatial resolution effects on the backprojected ray assuming no uncertainty in the scattering angle θ_s.

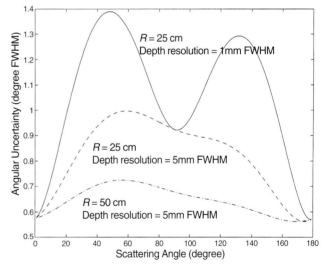

FIGURE 3 Angular uncertainty due to 3D detector resolution and system geometry for a small on-axis first detector with isotropic resolution of 1 mm and a long cylindrical second detector with 3-mm resolution on the cylindrical surface and depth resolution of 5 and 10 mm full width at half maximum (FWHM). Results are illustrated for cylindrical radii of 25 and 50 cm. (From Hua, 2000, © 2000 C-H Hua).

a thick detector, depth of interaction uncertainty will play an important role as gamma-ray energy increases and, as we see later, can dominate other effects. Increasing the distance between first and second detectors improves angular resolution in almost a linear fashion, as might be expected from Figure 2, but maintaining the same solid angle requires that the second-detector area increase as the square of the distance, and this has important cost implications.

B. Statistical and Electronic Effects

The ability to estimate the scattering angle from Eq. (1) depends strongly on the noise in the measurement of E_1 as shown in Eq. (2). The statistical noise component is related to the uncertainty in the number of electron–hole pairs produced by the recoil electron and subsequently collected. This imposes a physical limit on energy resolution. If the number of carriers, N, is proportional to the deposited energy and described by a Poisson process, then the energy resolution is given by

$$RES \equiv \frac{FWHM}{\bar{E}} = \frac{2.35k\sqrt{N}}{kN} = \frac{2.35}{\sqrt{N}} \quad (3)$$

Here, k is the proportionality constant and it is assumed that N is sufficiently large that the Poisson distribution is approximately Gaussian. In fact, the process is not Poisson, and if all the deposited energy were converted to charge carriers, the limiting energy resolution would approach zero. The departure from Poisson statistics depends in a complex fashion on the sequence of interactions by which initial recoil electron energy is split up between production of electron–hole pairs and other energy-loss mechanisms. The ratio of observed variance in N to that predicted by the Poisson model is termed the Fano factor. Measurements of the Fano factor for germanium detectors at 77 K ranges from 0.057 to 0.129 and for silicon detectors from 0.084 to 0.143 (Sansen and Chang, 1990), so solid-state detector performance is substantially better than one would predict from a Poisson model. For silicon, at 77 K, it requires 3.76 eV to create an electron–hole pair as opposed to 2.96 eV for germanium. A Poisson model predicts 14.4% and 12.8% full width at half maximum (FWHM) energy resolution for 1-keV gamma rays for silicon and germanium, respectively. Using the lowest measured Fano factors yields 1.2% and 0.73% FWHM energy resolution. At 140 keV, the limiting energy resolution for silicon is approximately 230 eV. As we see below, at the energies of interest statistical uncertainties are overwhelmed by electronic noise.

Electronic noise in sensor readout is usually expressed as the equivalent noise charge (ENC) at the input of a charge-sensitive preamplifier. The input stage in CMOS charge-sensitive devices is a field effect transistor (FET), and the noise is generally separated into series and parallel components. The series noise consists of a temperature-dependent channel noise and $1/f$ noise. The equivalent noise charge for the series component depends on the integration time constant and is proportional to the sum of the internal and external capacitance. As a specific example, consider the VA32C, a CMOS LSI preamplifier manufactured by Integrated Detector Electronics[2]:

$$\sqrt{\langle q^2_{series} \rangle} = 45e^- + 12e^-/pf_{ext} \quad (4)$$

The parallel equivalent noise charge is determined by the detector leakage current, the bias and feedback resistors, and the integration time constant. Again, for the VA32C with a 2 μs peaking time, this is approximately:

$$\langle q^2_{parallel} \rangle = \langle q^2_{leakage} \rangle + \langle q^2_{res} \rangle \cong 150^2 I_D + 1070^2 \frac{1}{R(M\Omega)} \quad (5)$$

Where I_D is the detector leakage current in nanoamps and R is the sum of the bias and feedback resistances in parallel. To see the relative importance of these various components, consider a pixelated silicon detector designed to have a 1.4-mm spatial resolution with a possible wafer thickness from .3 to 1.5 mm. The capacitance for 1 pixel including the traces and wire bond will be 3–4 pf. Most of the contribution comes from the traces and wire bond. Leakage current at room temperature can vary widely from manufacturer to manufacturer, and we have seen values ranging from 0.05 nA/pixel up to 0.4 nA/pixel for "good" devices. The feedback resistor in the VA32C is generally set in the neighborhood of 1000 MΩ, and a punch-through bias resistor for an AC-coupled detector has a similar value. Using these values in Eqs. (4) and (5) gives the following values for the series and parallel noise components:

$$\sqrt{\langle q^2_{series} \rangle} = 93e^- \quad (6)$$

$$\sqrt{\langle q^2_{leakage} \rangle} = 43 - 95e^- \quad (7)$$

$$\sqrt{\langle q^2_{res} \rangle} = 48e^- \quad (8)$$

Combining these in quadrature gives an overall noise equivalent charge of 113–140e^-. Multiplying by 3.62 eV per electron–hole pair in silicon at 300° K and by 2.35 to convert from σ to FWHM gives energy resolution of 961–1200 eV FWHM. It is clear that input capacitance and detector leakage currents are dominant effects. The reader is referred to Sansen and Chang (1990) and Knoll (2000) for further discussion of these effects.

C. Physics Effects

1. Doppler Broadening

The relationship between scattering angle and recoil energy of the struck electron described by Eq. (1) assumes that the initial momentum of the struck electron is zero. In

---

[2]Integrated Detector and Electronics (IDEAS ASA), Box 1, N-1330 Foruebu, Norway, Tel: +47 6782 7159, Specification sheets for the VA32C.2 and TA32CG.

fact, the incident gamma rays interact with bound electrons in the detector and these electrons have a momentum distribution that is specific to a given material and the physical state of the material. This precollision momentum distribution imposes an angular uncertainty of the scattered photon for a given measured recoil electron energy. In addition, the maximum Compton scattered energy is $E_{\gamma 0} - E_b$, the electron binding energy. The importance of the Doppler effect on limiting spatial resolution for the gamma-ray energies of interest in nuclear medicine was first pointed out by Ordonez et al. (1997), who showed that the FWHM of the Doppler-broadened energy spectrum corresponding to a 45° scattering of 140-keV photons from germanium is 1.39 keV and from atomic silicon is 0.93 keV. The value for germanium is 1.6 times worse than the intrinsic energy resolution of the detector used by Singh, whereas the value for silicon is close to the limit for silicon pad detectors at room temperature with currently available CMOS readout.

The electron binding energy can introduce a bias in estimating the mean energy lost by the scattered photon. However, the resultant low-energy X-rays and Auger electrons that follow the shell vacancy caused by the Compton event are very likely to be deposited in the same pixel and significant bias is not likely.

Clinthorne, Hua, et al. (1998) have examined Doppler broadening for a number of potential detector materials and have calculated the effects on image variance and spatial resolution. Figure 4 illustrates the angular broadening for 140-keV photons that deposit recoil electron energy corresponding to a 60° scatter from a free electron. Figure 4a shows the contributions of the various subshells in atomic silicon. The uppermost curve is the weighted sum of the subshell contributions. Compton profiles were taken from (Namito et al., 1994), and more recent results are available in (Namito et al., 2000). In Figure 4b the Doppler broadening for several possible detector materials is depicted normalized to the same area. Silicon is narrower than diamond at half maximum, but diamond has substantially reduced tails. Although one might expect that neon would exhibit reduced broadening because it has a large fraction of outer shell electrons, these electrons form a closed shell with relatively high binding energy. The curves for silicon and germanium are for the crystalline form. None of the curves is Gaussian, but can be fairly well represented by a superposition of three Gaussians.

2. Compton Scattering Cross Section

In the simple form of Compton camera illustrated in Figure 1, one wishes to maximize the probability of a single Compton interaction in the first detector followed by an escape and total residual energy deposition in detector 2. This requires that detector 2 subtend a large solid angle at the first detector and also be thick enough to stop the scattered gamma rays. The design of the second detector requires knowing the range of primary gamma-ray energies to be imaged and the range of scattering angles to be subtended by the second detector. The probability of a single scatter followed by an escape from the first detector depends on the Compton to total cross section for the first detector, the total amount of detector material, and its distribution about the source to be imaged. Figure 5 illustrates the Compton and Compton-to-total cross sections as a function of energy for silicon, germanium, and cadmium-zinc-telluride. Also shown are Compton angular distributions at three energies of interest.

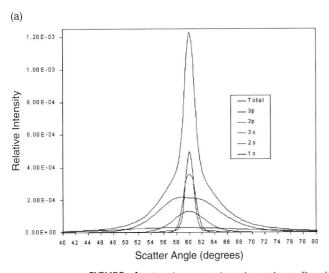

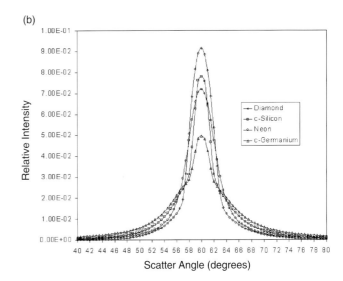

FIGURE 4 Angular uncertainty due only to Doppler broadening for scattering 140-keV photons about 60°. (a) Contributions broken down by electron subshell for atomic silicon. The large peak is the sum of the subshell contributions. (b) Doppler effect for four possible detector materials. The silicon and germanium curves are for the crystalline material. Note the increased broadening of the crystalline silicon compared to the atomic silicon at left. (Adapted from Clinthorne, Hua, et al., 1998.)

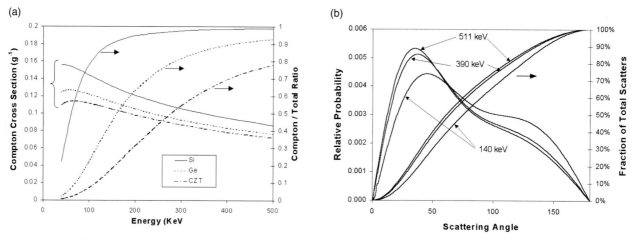

FIGURE 5 (a) Compton cross section and Compton/Total cross section ratio for Si, Ge, and CZT. The densities of Si, Ge, and CZT are 2.33, 5.32, and 5.86 g/cc, respectively. (b) Compton angular distributions, $d(\sigma)/d\theta$, for three photon energies normalized to Compton total cross section. Also shown are the integrals of the angular distributions as a function of scattering angle.

It is evident from Figure 5 that the probability for a scatter and escape at 140 keV will be substantially higher for silicon than for either germanium or CZT, even out to 511 keV. However, because of the lower density of silicon, approximately twice the volume of silicon compared to germanium is required to obtain the same probability of a Compton interaction. From the angular distribution and its integral, one can estimate the relative dependence of sensitivity on the gamma-ray energy and the angular range of scattered photons accepted by the Compton imaging system. If scatters between 20° and 90° are accepted, then 54%, 59%, and 60% of the scattered photons will be accepted for 140, 390, and 511 keV respectively. However, because the probability of Compton scattering is higher at 140 keV, the relative sensitivity at these three energies is 1, 0.79, and 0.73.

3. Polarization

Single gamma rays emitted from a nucleus are unpolarized. That is, all orientations of the electric vector are equally likely. When an unpolarized beam undergoes Compton scattering it becomes partially linearly polarized orthogonal to the scattering plane because the scattering cross section for the perpendicular and parallel components is not the same (Klein and Nishina, 1929). The degree of polarization is a function of scattering angle and initial gamma-ray energy as illustrated by Eq. (9) (McMaster, 1961) and Figure 6:

$$P = \frac{\sin^2 \theta}{1 + \cos^2 \theta + \left(\frac{E_0 - E_1}{m_0 c^2}\right)(1 - \cos\theta_1)} \quad (9)$$

Moreover, a subsequent Compton scattering of this partially polarized beam will produce an azimuthally asymmetric intensity in the scattered gamma rays that depends on the angle between the incident gamma-ray polarization, $\mathbf{e}_1$, and the scattered gamma-ray polarization, $\mathbf{e}_2$. Kamae et al.

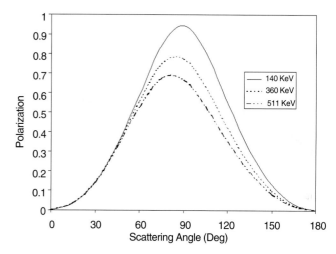

FIGURE 6 Gamma-ray polarization as a function of scattering angle at three primary gamma-ray energies.

(1987) have described an application of this method to measuring the energy, direction, and polarization of incident gamma rays.

Polarization influences two aspects of Compton camera performance. First, gamma rays that scatter in the patient will be partially polarized. Those scattered gamma rays that cannot be rejected by energy windowing will have an asymmetric azimuthal scattering distribution in the first detector and should therefore be back-projected using a weighted azimuthal distribution. Because the initial scattering plane is unknown, there is no frame of reference for the asymmetry, so these events must be modeled using a uniform azimuthal distribution. The effects of this mismodeling have not been evaluated. Second, polarization effects can be used to reduce the conical ambiguity for gamma rays that have not been scattered in the patient. For gamma rays that undergo two or more scatters in a Compton camera, Dogan (1993; Dogan

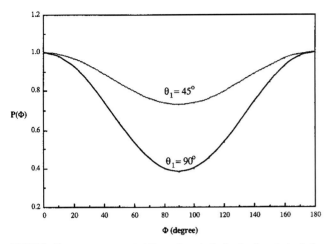

FIGURE 7 Azimuthal probability of conical distribution derived for double Compton scattering for 150-keV incident gamma rays. θ_1 is the first scattering angle, and Φ is the azimuthal angle around the cone measured with respect to the second scattering plane. (From Dogan et al., 1992, © 1992 IEEE.)

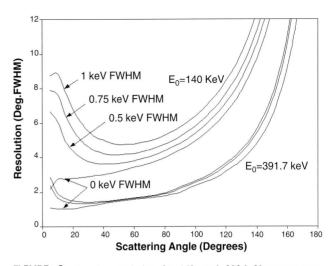

FIGURE 8 Angular resolution for 140- and 392-keV gamma rays scattered from crystalline silicon for different values of detector energy resolution plus Doppler broadening and excluding geometric effects. Doppler effects shown for angles less than 10° are unreliable. Note that an uncertainty of 2° corresponds to a spatial resolution of 3.5 mm FWHM at 10 cm. (From Hua et al., 1999, © 1999 IEEE.)

et al., 1992) has shown that it is possible to determine an azimuthal weighting function for the conical back-projection that reduces the ambiguity. One must first determine the sequence of interactions. Methods for sequencing have been described by Kamae and Hanada (1988), Dogan (1993; Dogan et al., 1990, 1992), and Durkee (Durkee, Antich, Tsyganov, Constantinescu, Fernando, et al., 1998; Durkee, Antich, Tsyganov, Constantinescu, Kulkarni, et al., 1998) and essentially consist of determining which set of energies and scattering angles calculated for each of the postulated sequences best fits the data. For n interactions, there are $n!$ sequences to test, so untangling more than three interactions could be very time consuming.

Figure 7 illustrates the probability of Compton double scattering as a function of azimuthal angle. Results are shown for 150-keV gamma rays and two different initial scattering angles (Dogan et al., 1992). Polarization effects have been included in a system design study by Chelikani et al. (2004), but the effect of this added information on improving image quality for Compton imaging has not been completely investigated to our knowledge. However, it appears from Figure 7 that one can substantially reduce the ambiguity in azimuth for low-energy gamma rays for the larger scattering angles.

D. Combined Effects

By examining the combined effects of detector noise, Doppler broadening and gamma-ray energy as a function of scattering angle, it is possible to gain substantial insight into the relative importance of these factors and how they influence the design of a Compton camera. Figure 8 illustrates the combined effect of detector noise and Doppler broadening over a range of scattering angles for 140-keV (Hua et al., 1999) and 392-keV gamma rays, corresponding to ^{99m}Tc and ^{113m}In. The curves for 0-keV detector noise show the effects attributable to Doppler broadening alone. For ^{99m}Tc this corresponds to approximately 5.5 mm at 10 cm, and for ^{113m}In approximately 2.6 mm for a 45° scattering.

Several important points are illustrated by Figure 8.
1. For higher-energy gamma rays, the effect of detector noise is almost negligible except for scattering angles below 20°.
2. Doppler broadening is an important effect at both high and low energies, but is substantially smaller for small scattering angles.
3. Because Doppler effects favor small angle scattering, there remains good reason to attempt to reduce detector noise even though the Doppler and detector-noise components are comparable in the neighborhood of 60° scattering for 140-keV gamma rays.

Based on these curves it appears that one should design a camera system such that the second detector will capture all scatters from the first detector out to 90 or 120°, depending on the energy range of interest.

Not only does detector noise affect angular uncertainty, it also determines the minimum scattering angle that can be detected because the detector threshold must be set above the noise level. From Eq. (1), the mean energy deposited for a 15° scatter of a 140-keV gamma ray is 1.29 keV. If the toe of the noise distribution (3σ) is 1.29 keV, then approximately one-half of the 15° scatters lie above this threshold. This corresponds to a detector energy resolution of approximately 1 keV FWHM.

Up to this point we have implicitly assumed that energy resolution in the second detector is only useful for accurately measuring total gamma-ray energy as a means for reducing the effects of scattering in the patient or imaging multiple tracers. However, it has been pointed out by Clinthorne (2001) that if the incident gamma-ray energy is known, the best estimate of E_1 is obtained from the measurement of both E_1 and E_2. If detectors of unequal energy resolution are used, the uncertainty in $\hat{E}_1$, $\sigma_{\hat{E}1}$ is

$$\sigma_{\hat{E}_1} = \sqrt{\frac{\sigma_1^2 \sigma_2^2}{\sigma_1^2 + \sigma_2^2}} \qquad (10)$$

If detectors 1 and 2 have equal energy resolution, the uncertainty in scattering angle can be reduced by a factor of $\sqrt{2}$.

III. ANALYTICAL PREDICTION OF SYSTEM PERFORMANCE

A. Noise Propagation and Lower Bound

Perhaps the most crucial question regarding the potential for Compton imaging in nuclear medicine is: How does Compton imaging compare to conventional imaging using Anger cameras with lead collimators? The comparison is complicated by the fact that projection images obtained from a collimated camera are a direct representation of the source distribution blurred by a Gaussian-like function characteristic of the system resolution. The raw image from a Compton camera is a superposition of conic sections that gives a point response function similar to a truncated $1/r$ function. As such, these images have the poor resolution and low contrast observed in SPECT images reconstructed by simple backprojection, and a reconstruction or decoding step is required to obtain usable images. Just as is the case for SPECT, the image reconstruction process amplifies image noise so that a simple comparison of raw detection efficiency between a collimated and Compton imaging system does not predict relative imaging performance. Furthermore, for Compton imaging systems we have shown that sensitivity and resolution depend on detector material, electronic noise, detector geometry, and gamma-ray energy in a complex manner. Thus, it is important to have an analytical method to compare the performance of different Compton camera system designs.

The question of noise propagation in Compton imaging compared to mechanical collimation was addressed by Singh and colleagues (1988) by a theoretical analysis using an annulus as an approximate response function for planar imaging. The annulus corresponds to a Compton camera response function for one scattering angle and an on-axis point source. They also describe simulations and experimental measurements using inverse filtering and algebraic reconstruction technique (ART) to reconstruct images. Their results showed a factor of 3 loss in signal-to-noise ratio (SNR) for a ^{137}Cs disk object as the diameter increased from 2 cm to 10 cm. For a 5-cm-diameter disk, the Compton camera image had a factor of 3 lower SNR than a pinhole image with the same number of counts. The authors conclude that for the 661-keV gamma rays of ^{137}Cs the effective sensitivity of a Compton camera is reduced by approximately a factor of 3 compared to mechanical collimation. Thus, to break even, the raw sensitivity of the Compton camera must be three times that of the collimated system. In the following sections, we present more recent results based on a more accurate system model and lower bound calculations that are independent of the particular reconstruction method.

1. Introduction to Lower Bound

The Cramer-Rao (CR) lower bound on estimator variance gives the lowest variance in a parameter of interest that can be achieved by any unbiased estimator. Thus, this measure of performance depends only on the object and the set of measurements and *not* on the estimator. The object must be represented by a finite set of parameters such as voxels or pixels, $\mathbf{o} \equiv [o_1, \ldots, o_m]$, and a statistical model must exist that relates the conditional probability of object parameters to the measurements:

$$f(\mathbf{y}|\mathbf{o}), \qquad (11)$$

where $\mathbf{y} \equiv [y_1, \ldots y_N]^T$ is the vector of N measurements. The bound is defined (Blahut, 1987) as the inverse of the Fisher information matrix, $\mathbf{F}$.

$$\mathbf{K}_{\hat{o}} \geq \mathbf{F}^{-1} \equiv E\left[\nabla_o^2 \log f(\mathbf{y}|\mathbf{o})\right]^{-1} \qquad (12)$$

For conditionally Poisson distributed measurements, as well as for Gaussian measurements in which the covariance is not a function of the parameters, the Fisher information matrix becomes:

$$\mathbf{F} = \left(\nabla \bar{\mathbf{y}}(\mathbf{o})\right)^T \mathbf{K}_y^{-1} \nabla \bar{\mathbf{y}}(\mathbf{o}) \qquad (13)$$

It is important to note that the CR bound can be asymptotically achieved by the maximum-likelihood estimator. Further, Clinthorne, Ng, *et al.* (1998) have shown that the CR bound is equivalent to the more familiar propagation-of-error formula applied to the maximum-likelihood estimator. The CR bound, however, has the advantage of being estimator-independent.

2. Uniform Cramer-Rao Bound

The difficulty with applying the CR bound to image reconstruction is that one can seldom use unbiased estimators because of the ill-conditioned nature of these problems. Instead one must use biased estimators that directly or indirectly impose smoothness constraints on the solution in order to reduce the mean-squared error in the images. Although one can define a CR bound for a particular biased estimator, Hero (1992; Hero *et al.*, 1996) has shown that it is possible to define a uniform CR (UCR) bound that is

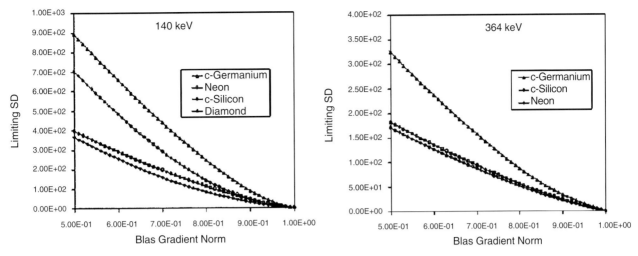

FIGURE 9 Uniform lower bounds for different detector materials, considering only the effects of Doppler broadening at 140 and 364 keV. Scattering angles from 45 to 90°. (From Clinthorne et al., 1998).

estimator-independent. The UCR bound depends on the norm of the bias gradient[3] rather than on the bias gradient specific to a given estimator.

The bias of the pth element of the object vector is $\bar{o}_p - o_p$ with gradient

$$\partial \bar{o}_p / \partial o_i - \partial o_p / \partial o_i, \quad i = 1...p...M \quad (14)$$

For the imaging problem, the second term will be 0, except when $i = p$, in which case, it is 1. The elements of the bias gradient are then:

$$\left[\partial \bar{o}_p / \partial o_1, ..., \partial \bar{o}_p / \partial o_p - 1, ..., \partial \bar{o}_p / \partial o_M \right] \quad (15)$$

The bound calculation yields the optimal bias gradient in the sense that no other bias gradient with the same norm will give a lower variance in the estimated parameter. If the system is spatially invariant,[4]

$$\partial \bar{o}_i / \partial o_p = \partial \bar{o}_p / \partial o_i \quad (16)$$

By adding 1 to the pth element, one obtains the mean gradient of the estimate that is also seen to be the impulse response function. We see that the norm of the bias gradient is closely related to the norm of the impulse response function.

---

[3] Also referred to as the bias gradient length.
[4] More precisely this is true if $(\mathbf{F} + \alpha\mathbf{R})^{-1}\mathbf{F} = \mathbf{F}(\mathbf{F} + \alpha\mathbf{R})^{-1}$ where $\mathbf{F}$ is the Fisher information matrix, $\mathbf{R}$ is the penalty function and α is the strength of the penalty. If $\mathbf{R}$ is the identity matrix (0th-order Tikhonov regularization), this is the case, and it will be approximately true for $\mathbf{R}$ equal to a roughness penalty based on squared differences between a pixel and its nearest neighbors.

3. Applications of Lower Bounds

a. Doppler Broadening for Different Materials

Uniform bounds for estimating the center pixel in a 7.5-cm disk source have been calculated to evaluate the effects of the Doppler broadening for the detector materials illustrated in Figure 4 at 140 and 360 keV (Clinthorne et al., 1998). The source was located 10 cm from a point detector and scattering angles from 45 to 90° were accepted. Electronic and statistical noise components were taken as zero, and perfect geometric resolution was assumed.

The results are illustrated in Figure 9. For lower-energy gamma rays, diamond slightly outperforms crystalline silicon, neon is substantially inferior to both diamond and silicon, and germanium is poorest. At 364 keV, the standard deviation for all detector materials decreases, but neon slightly outperforms crystalline silicon, and germanium clearly remains worst.

These curves demonstrate the importance of the tails of the line spread function relative to the width of the central peak as gamma-ray energy changes. Neon has a wider central peak compared to silicon because of the higher energy of the valence electrons in the closed shell. However, the fraction of neon valence electrons is higher than silicon. As gamma-ray energy is increased, the effect of long-range tails becomes the dominant factor limiting performance for silicon and germanium. The contribution of these tails is explicitly included in the bias gradient norm and accounts for the changes in the UCR bound with gamma-ray energy. The actual computation of these bound curves was performed on eight nodes of an IBM SP2 parallel processor, and each curve required approximately 1 week of computation.

b. Comparison of Compton Cameras to Collimated Systems

The UCR bound has also been used to quantify Compton camera planar imaging performance relative to conventional

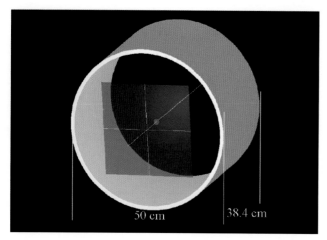

FIGURE 10 Ring Compton camera system used for UCR bound calculations. The source plane is 10 cm in front of the front edge of the cylindrical second detector. The point first detector lies on-axis 10 cm behind the object plane.

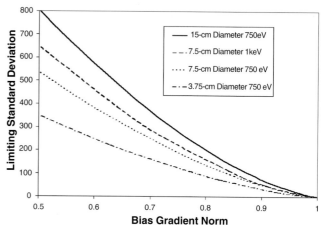

FIGURE 11 Comparison of lower-bound curves for a Compton camera as a function of object diameter normalized to the same number of detected photons per unit area. Minimum standard deviation in the estimate of the central pixel for various diameter disk sources is plotted as a function of bias gradient norm for 140-keV gamma rays. (From Hua et al. 2000, © 2000 C-H Hua)

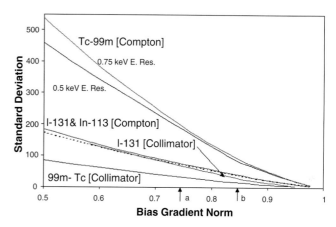

FIGURE 12 Comparison of lower-bound curves for a Compton camera and two conventional collimated camera systems normalized to the same number of detected photons. Minimum standard deviation in the estimate of the central pixel in a 7.5cm diameter disk source is plotted as a function of bias gradient norm. The bias gradient norm at point a corresponds to that for the measured camera point response function with a LEHR collimator at 10 cm. Point b is for a high-energy iodine collimator at 10 cm. (From Hua et al. 1999, © 1999 IEEE.)

imaging systems (Clinthorne et al., 1996; Hua, 2000; Hua et al., 1999). Bound calculations were made for a ring geometry Compton camera illustrated in Figure 10. This is an idealized version of an experimental ring camera to be described later. The point first detector is crystalline silicon, and the second detector is quantized into 512 pixels around the circumference and 128 pixels axially to give a pixel size of approximately 3 mm. Depth of interaction effects were not modeled. The 30 cm × 30 cm source plane is represented by an array of 64 × 64 pixels 4.7 mm square. Only scattering events between 45° and 90° were considered because they have the highest resolution (see Figure 8).

The system model included Doppler broadening for crystalline silicon, and bounds were calculated for first-detector electronic noise levels corresponding to energy resolution of 0.5 keV and 0.75 keV FWHM. The system parameters are optimistic and represent close to a best-case model. The collimator–camera response functions were based on measurements for pixel-sized sources. Resolution was 8.6 mm FWHM at 10 cm for the low-energy high-resolution (LEHR) collimator with ^{99m}Tc, and 11.3 mm FWHM for the high-energy collimator with ^{131}I. The bias gradient norms corresponding to these measured responses were 0.76 and 0.87, respectively.

Figure 11 illustrates the effect of object size on the limiting standard deviation in the estimate of the central pixel of a planar disk source. For reference, the effect of detector energy resolution for the 7.5-cm disk is also shown. The dependence on object size is a direct consequence of the fact that the signal from the central pixel is spatially multiplexed with the signals from all other pixels. As the size of the disk is increased, the variance of the added source volume is propagated to the pixel of interest. This is essentially the same sort of dependence of image noise on object diameter that one observes in tomographic imaging. At a bias gradient norm of 0.74, the standard deviation increases a factor of 1.47 from 133 to 196 when the disk diameter is doubled from 3.75 to 7.5 cm. This dependence on object size is similar to that obtained by Singh et al. (1988) for 5.33- and 10-cm diameter disks. Based on analytical SNR calculations for a ring aperture model of a Compton camera compared to an ideal pinhole, they obtain factor of 1.74 increase in standard deviation for the 10-cm disk compared to the 5.33-cm disk.

Figure 12 shows bound curves for the Compton camera at 140 keV(^{99m}Tc), 361 keV (^{131}I), and 391.7 keV (^{113m}In) (Hua et al., 1999). The first-detector energy resolution is 0.75 keV, and for Tc the effect of improving resolution to 0.5 keV is also illustrated. For the collimated camera, bounds for ^{99m}Tc and ^{131}I are shown. These curves are all normalized to an equal number of detected events, so they portray the relative information carried by a detected photon for each of the imaging systems. At 140 keV the collimator gives a substantially lower standard deviation than the Compton camera for the same number of detected events, whereas at 360 keV almost identical performance can be achieved.

Since the relative standard deviation for any of the systems will be inversely proportional to the square root of the number of detected events, one can estimate a noise-equivalent sensitivity for the Compton and collimated systems from these curves. We define *decoding penalty* as the variance of the Compton camera divided by the variance of the collimated system. The decoding penalty specifies how much greater sensitivity the Compton system must have in order to give relative standard deviation equal to that of the collimated system for planar imaging.

From Figure 13, the decoding penalty for imaging a 7.5-cm-diameter disk source with ^{99m}Tc is approximately 40. For a 15-cm Tc disk source the decoding penalty increases to 91 (Hua, 2000; Hua *et al.*, 1999). For a 7.5-cm ^{131}I disk the penalty varies from 1 to 1.5. This reflects the fact that Compton camera performance improves with increasing energy while collimator performance degrades. ^{113m}In, which emits a 391.7-keV gamma ray, has a decoding penalty of approximately 5 when compared to ^{99m}Tc imaged with a conventional camera system. This isotope of indium is of potential interest because it can be obtained from a ^{113}Sn generator and has a 100-minute half-life. The parent half-life is 119 days. Because of differences in radiation dosimetry, LeBlanc and co-workers (LeBlanc, Clinthorne, *et al.*, 1999) conclude that the effective sensitivity increase of a Compton camera must be increased to between 5 and 8, depending on tracer distribution in the patient, to obtain a comparable standard deviation for equal patient radiation dose.

The results shown in Figure 13 agree in spirit with the results of Singh *et al.* (1988) for the 662-keV gamma rays from ^{137}Cs. Based on the SNR computed for the ring aperture relative to the ideal pinhole collimator, they predict a decoding penalty of 4 for a 5.33-cm-diameter disk source. Experimental measurements with a 4.7-cm-diameter disk and image reconstruction using ART give a decoding penalty of approximately 2.3. These are slightly more pessimistic results than the previous predictions based on a lower-bound analysis for silicon detectors at 360 keV and a 7.5-cm-diameter disk source. Nevertheless, it is clear that a Compton camera must be substantially more sensitive than mechanical collimation to yield any gain in performance, especially for lower-energy gamma rays. Some estimates of sensitivity increases that might be obtained from Compton cameras are described in a later section.

B. Observer Performance

The lower-bound analysis of Compton cameras just given provides an assessment of relative performance of Compton imagers and collimated imagers for the task of quantification. It is of equal, or perhaps of greater, importance to examine how Compton cameras might perform relative to collimated systems for the lesion-detection task. The ideal Bayesian observer with signal known exactly and background known exactly (SKE and BKE) is one of the simplest, albeit overoptimistic, quantitative measures of lesion detectability. In this case the only noise source stems from the uncorrelated measurement noise, and the task is to determine the presence or absence of a lesion in a background distribution. Although actual human observer performance for a more complex signal and background representation is the ultimate measure of a medical imaging system performance, the ideal observer performance is readily calculated directly from the data and is appropriate for the early stages of instrumentation development when one wishes to evaluate the effects of a variety of system parameters.

1. Ideal Observer—Signal-to-Noise Ratio

The ideal observer makes the lesion-present–lesion-absent decision based on a function of the likelihood ratio of the lesion-present and lesion-absent raw data for a specified threshold (Fisher, 1925). The detectability index for an ideal observer is given by the SNR defined in Eq. (17).

$$\mathrm{SNR}^2 = \frac{\left[\bar{\lambda}_1 - \bar{\lambda}_0\right]^2}{\frac{1}{2}\left[\sigma_1^2 + \sigma_0^2\right]} \quad (17)$$

where $\bar{\lambda}_j$ and σ_j^2 represent the mean and variance of the decision variable under signal-absent and signal-present hypotheses ($j = 0,1$). Let **g** represent a column vector

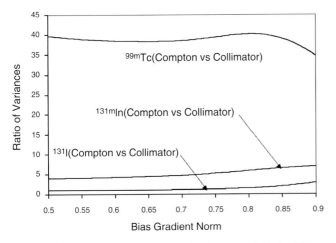

FIGURE 13 Plots of decoding penalty for the center pixel of a 7.5-cm-diameter disk source. (From Hua *et al.* 1999, © 1999 IEEE.)

of projection data from a given imaging system where $\mathbf{g} \approx$ Poisson $(\mathbf{Af})$. $\mathbf{A}$ represents the system matrix and $\mathbf{f}$ is a column vector representing the object. The likelihood ratio of the two hypotheses is (Fukunaga, 1972):

$$L(\mathbf{g}) = \frac{P(\mathbf{g}|H_1)}{P(\mathbf{g}|H_0)} \qquad (18)$$

One may also use the log of the likelihood ratio as the ideal decision variable; hence:

$$\overline{\lambda}_{ideal} = \log\left(P(\mathbf{g}|H_1) - P(\mathbf{g}|H_0)\right)$$

Following the methods of (Hua, 2000) for a large number of realizations,

$$\overline{\lambda}_0 = E\left(\lambda_{ideal}|H_0\right) = \left(\log \overline{\mathbf{g}}_1 - \log \overline{\mathbf{g}}_0\right)^T \overline{\mathbf{g}}_0 \qquad (19)$$

$$\overline{\lambda}_1 = E\left(\lambda_{ideal}|H_1\right) = \left(\log \overline{\mathbf{g}}_1 - \log \overline{\mathbf{g}}_0\right)^T \overline{\mathbf{g}}_1 \qquad (20)$$

$$\sigma_1^2 = \left(\log \overline{\mathbf{g}}_1 - \log \overline{\mathbf{g}}_0\right)^T \text{diag}\left(\overline{\mathbf{g}}_1\right)\left(\log \overline{\mathbf{g}}_1 - \log \overline{\mathbf{g}}_0\right) \qquad (21)$$

$$\sigma_0^2 = \left(\log \overline{\mathbf{g}}_1 - \log \overline{\mathbf{g}}_0\right)^T \text{diag}\left(\overline{\mathbf{g}}_0\right)\left(\log \overline{\mathbf{g}}_1 - \log \overline{\mathbf{g}}_0\right) \qquad (22)$$

and the SNR, Eq. (17), for an imaging system may be calculated for the two hypotheses.

2. Signal-to-Noise Ratios for Compton Cameras and Collimated Imaging Systems

Using the same system matrices for the Compton Camera and collimated camera described for the uniform-bound calculations, SNR^2 was calculated for the task of detecting Gaussian-shaped lesions centered on a larger-diameter Gaussian-shaped background (Hua, 2000). Effects of lesion and background diameter, lesion contrast, and gamma-ray energy were studied for both imaging systems under the constraint that both systems detect an equal number of photons. The SNR^2 depends linearly on the total number of counts and increases essentially linearly with lesion contrast over the range 0–0.5. Thus, the ratio of SNR^2 for the collimated camera to the Compton camera as a function of lesion contrast is almost constant and for the same total counts gives an estimate of how much more sensitive the Compton camera must be to give SNR^2 equal to the collimated camera for equal imaging time. These results are summarized in Table 1. These values have an interpretation for the detection task similar to the decoding penalty derived from the uniform bound for the quantification task, but for Gaussian-shaped lesions on a Gaussian-shaped background. The values of SNR^2 in brackets are for the same single pixel on a uniform disk used for the estimation task. It is of note that the relative performance of the Compton camera improves slightly for the single-pixel detection task. The same general trends are observed for the detection task and quantification task: (1) Relative Compton camera performance improves with higher gamma-ray energy and (2) relative Compton camera

TABLE 1   Ratio of SNR^2 for Collimators with Respect to Compton Camera for Equal Counts

Lesion Diameter (mm)	^{99m}Tc at 140 keV, Background Diameter (cm)			^{131}I at 364 keV, Background Diameter (cm)		
	3.75	7.5	15	3.75	7.5	15
5	4.2	10.1 [9.3]^a	30.0	1.9	4.0 [3.25]^a	11.5
10	2.8	6.6	19.5	1.8	3.7	10.4
15	2.0	4.6	13.5	1.5	3.2	9.0
20	1.6	3.5	10.5	1.3	2.7	7.5

^{a}From Hua (2000).
^{a}Numbers in brackets are SNR^2 for detecting a single 4.7-mm pixel on a 7.5-cm-diameter uniform disk. This is to enable comparison to the decoding penalty derived from uniform bound. All other values of SNR^2 are for Gaussian lesions on a Gaussian background.

performance decreases with object size. The decrease in relative performance between a 3.75-cm-diameter object and a 7.5-cm object is 2.4 using lower bounds and 2.2 using the ideal observer. However, the actual ratios derived for the detection task are more favorable to the Compton camera at 140 keV and, surprisingly, less favorable at 360 keV. Based on bound calculations, the decoding penalty is approximately 40 for ^{99m}Tc and 1.5–2.0 for ^{131}I for single-pixel estimation on a 7.5-cm disk. The detection task for the same object gives corresponding values of 9.3 and 3.25 for Tc and I, respectively. This is clearly a point requiring further investigation.

3. Ideal Observer—Area under Receiver Operating Curves

If the decision variable, λ, is Gaussian distributed, the area under the receiver operating curve (ROC) is given by (Barrett et al., 1998):

$$\text{Area} = \frac{1}{2} + \frac{1}{2}\text{erf}\left(\frac{SNR}{2}\right), \qquad (23)$$

where erf is the error function. This mapping of the SNR is illustrated in Figure 14. These curves quantify the lesion detectability for the two imaging systems. As expected, collimator performance is independent of object size and, in terms of lesion size, the collimator performance for a 5-mm lesion is almost the same as that of the Compton system for a 10-mm lesion for an equal number of detected photons.

C. Predicted Efficiency Gains for Various Geometries

It is abundantly evident from both the lower bound and observer comparisons of Compton cameras to collimated cameras that Compton imaging systems must have a raw gamma-ray detection sensitivity that exceeds that of the collimated system by a factor of 2 up to 100 or more

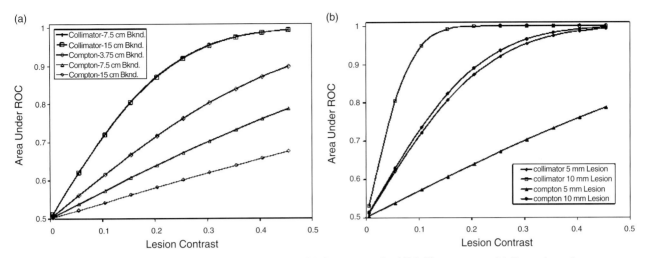

FIGURE 14 Area under the ROC as a function of lesion contrast for 140-keV gamma rays. (a) Comparison of Compton camera to collimated cameras for 5 mm FWHM Gaussian lesion and various diameter Gaussian backgrounds. (b) Relative camera performance for different lesion diameters and a 7.5-cm-diameter background. The Compton and collimated imaging systems each detected the same number of events. (From Hua, 2000, ©2000 C. H. Hua.)

depending on object size and gamma-ray energy (and the metric employed). In this section we examine the sensitivity gains that might be expected for two system geometries. The data are restricted to single scattering in the first detector followed by an escape and detection in the second detector.

1. Ring Compton Camera

The ring Compton camera, as portrayed in Figure 10, is hardly optimal, but it is a useful geometry for obtaining preliminary experimental data to verify the theoretical predictions. In particular, it is possible to shield the second detector from direct radiation by the source so that high activity sources may be used to obtain data from small first detector prototypes without saturating the second detector. It is also possible to vary the axial position of the first detector to investigate the dependence on scattering angles accepted and to restrict data to that with the highest resolution at a particular energy. To determine the sensitivity limits of the ring Compton camera geometry, Monte Carlo calculations using the SKEPTIC code (Wilderman, 1990) were carried out considering different size silicon detectors as a function of silicon thickness (Clinthorne *et al.*, 1996) for both 140 and 360 keV. Point sources were located on-axis 10 cm from the silicon detector. Sensitivities relative to low- and high-energy collimators were computed using a measured efficiency of 1.3×10^{-4} for the LEHR collimator and manufacturer-specified sensitivity of 1×10^{-4} for the high-energy collimator. For the Compton camera, only events that underwent a single scattering followed by detection in the ring detector were counted. The second detector is 1-cm-thick NaI, except for one case for ^{131}I in which the thickness was increased to 2.54 cm to illustrate the system sensitivity dependence on second detector thickness at higher energies. These results are plotted in Figure 15.

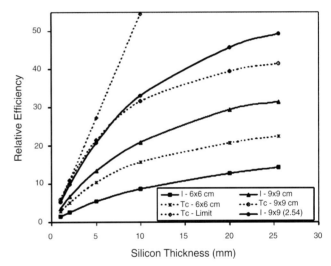

FIGURE 15 Relative efficiency of a Compton camera compared to a collimated camera for ^{99m}Tc and ^{131}I as a function of silicon thickness and area. The system geometry is the same as for Figure 10. The solid curves are for ^{131}I, and the increase in efficiency for the upper curve for the 9×9 cm silicon detector illustrates the effect of increasing the second-detector thickness from 1 cm to 2.54 cm of NaI. (From Clinthorne et al. 1996, © 1996 IEEE.)

At 140 keV, a sensitivity gain of approximately 40 can be realized for a 25-mm-thick stack of 9×9 cm arrays of silicon detectors, whereas at 365 keV a factor of 30 can be attained for a 1-cm-thick second detector. Increasing the second-detector thickness to 2.54 cm gives a sensitivity gain of almost 50 for iodine. The nonlinearity of the curves with thickness is due to the increased probability of absorption prior to the initial scattering or a second scattering or absorption following the initial scattering. Increasing first-detector size from 6×6 cm to 9×9 cm increases the area by a factor of 2.25 and increases the relative Tc sensitivity by a factor of 2.01 for 10-mm thickness, whereas increasing the 6 × 6 cm array from a thickness of 10 to 20 mm only increases the sensitivity by a factor of 1.3. This strongly suggests that one should consider distributing the same volume of first detector over a larger area with reduced thickness. The linear curve projects limiting gains that could be achieved by eliminating self-shielding and a second scattering or absorption following the initial scattering. In this case, one might achieve a sensitivity gain of 55 for ^{99m}Tc with a volume of 81 cc of silicon rather than a gain of 30.

2. Double-Ring Geometry

A double-ring geometry comprising a cylindrical first detector surrounded by a cylindrical or tire-shaped second detector has been described by several investigators (Ichihara, 1987; Rohe and Valentine, 1995, 1996; Valentine et al., 1996; Bolozdynya, Ordonez, et al., 1997). This geometry has several advantages. First, the scattering detector is distributed over a large area, and the problems of self-shielding and multiple scattering previously alluded to for large first-detector volume are greatly reduced and substantial sensitivity increases might be expected. Second, complete angular sampling of a source distribution is possible with no camera rotation. The major disadvantage is that the second detector directly views the source so that it must have very high count rate capability to avoid pile-up, dead time, and high accidental coincidence rates. The count rate will also be higher because the second detector views more small-angle scatters that do not carry high-resolution information and that will fall below the noise threshold of the first detector.

The system geometry illustrated in Figure 16 is configured such that up to 90° scatters from the mid-plane of the silicon ring will be intercepted. Again, the SKEPTIC Monte Carlo code was used for the simulation. Events were biased to intercept the silicon ring and were restricted to scattering events that deposited full energy in the system and whose scattering angles were between 20 and 90°. Calculations were performed for ^{99m}Tc and ^{131}I and were sorted into the following categories:
1. Single scattering in the silicon ring followed by absorption in the BGO ring.
2. Multiple scattering in the silicon ring followed by absorption in the BGO ring.
3. Single or multiple scattering in silicon followed by absorption in silicon.
4. The sum of 1 through 3.

Two points are evident from Figure 17. First, most of the sensitivity gain is reached for both Tc and I at a silicon thickness of approximately 16 mm if one restricts the events to single scattering in silicon. Second, at this thickness, one could achieve a further 50% sensitivity increase by including the second category of interaction, and appre-

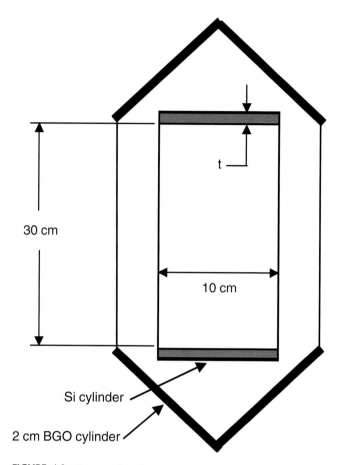

FIGURE 16 Cross section of a double-ring-geometry Compton camera used for Monte Carlo simulations of system sensitivity for a point source located on-axis.

ciable gains can then be achieved by further increasing silicon thickness.

In Figure 18, detection sensitivities for Tc and I are compared to those for the respective collimators. The data from Figure 17 were normalized to the fractional solid angle subtended by the silicon ring and divided by the appropriate collimator sensitivity. Considering single scattering in silicon followed by capture in the BGO ring we see that even for a 4-mm-thick Si ring, sensitivity gains of 153 and 133 for Tc and I, respectively, are predicted. Increasing the thickness to 16 mm increases these ratios to 298 and 420. These sensitivity gains are substantially greater than the decoding penalties calculated earlier, especially for the higher energy of ^{131}I. Although these relative sensitivities may be further increased by using multiple scattering events, this can only be realized at the expense of more complex data processing. Not only must one calculate the sequence of interactions by analyzing the kinematics, but one must also consider the effects of polarization, especially at lower energies. This is further complicated by the fact that accurate computation of the scattering angle between successive scattering events in

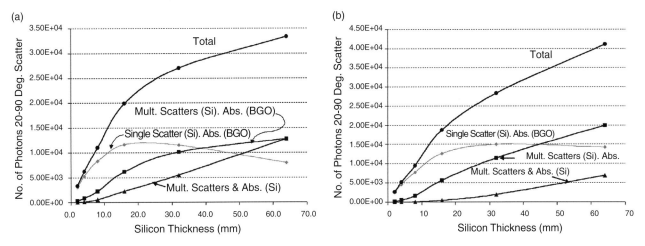

FIGURE 17 Number of photons per 100,000 incident on a silicon ring sorted into the three labeled interaction sequences. (a) ^{99m}Tc. (b) ^{131}I.

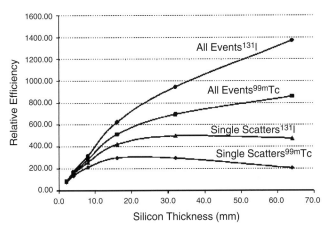

FIGURE 18 Relative sensitivities of the double ring Compton camera geometry and parallel hole collimators for ^{99m}Tc and ^{131}I.

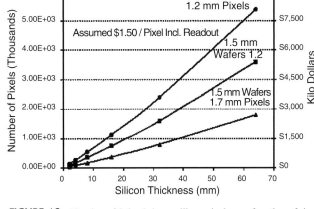

FIGURE 19 Number of 1.4 × 1.4 mm silicon pixels as a function of ring thickness for different silicon wafer thickness for pixel-voxel-type silicon scatter detectors. Estimated cost per pixel includes the readout electronics.

silicon depends both on the silicon voxel size and the distance between the successive events. Reducing the pixel voxel dimensions by a factor of 2 increases the system matrix a factor of 8, and for discrete pixel detectors one predicts a factor of 8 increase in price as depicted in Figure 19. On the plus side, using the polarization information is expected to reduce the conical ambiguity and thereby reduce the decoding penalty. A careful analysis will be required to quantify the trade-off.

The relation between cost and pixels shown in Figure 19 is based on the large-scale production of silicon detector arrays such as used in high-energy physics experiments. We have assumed detectors with discrete pixels rather than orthogonal strip detectors because they have lower noise due to their lower input capacitance and, as we have seen, this is of primary importance for imaging ^{99m}Tc. For applications to ^{131}I or higher-energy gamma rays, double-sided strip detectors might prove to be a good choice.

IV. IMAGE RECONSTRUCTION FOR COMPTON CAMERAS

A. Background

Ever since the Compton camera was proposed for medical applications by Todd, Nightingale, and Everett (1974), image reconstruction has been an issue. As in X-ray telescopes, the first image reconstruction methods proposed consisted simply of backprojecting event circles corresponding to the data uncertainty into a discretized image space with no additional filtering or processing. Such backprojection, while it may serve as a useful demonstration for point source reconstruction, is just as inappropriate for Compton camera reconstruction as it is for conventional SPECT or PET imaging. The tails of the each event circle, being positive, do not cancel and the result is a significantly blurred estimate of the radiotracer distribution in the object. More modern reconstruction methods have attempted to eliminate this blur using common techniques to solve the inverse problem more correctly. In this sense, the problem of reconstructing images from Compton camera data is no more difficult than for more conventional emission tomography—the main differences arise in the details of the implementation and in the sheer amount of computation. Rather than providing an exhaustive description of the various methods proposed for Compton camera reconstruction, this section embeds the reconstruction problem into the relevant mathematical framework in a non-rigorous manner. In this way, the relationship among the various reconstruction methods from analytic inverses to iterative statistical methods is evident. However, while we have attempted to keep things simple, it is hardly possible to draw the various reconstruction methods together without introducing some Hilbert space concepts. The second disclaimer is that many shortcuts have been taken in order to tie the widely different reconstruction techniques together. As an example, while a Compton camera measures energy as opposed to scattering angle, the reconstructions presented in section E assume that scattering angle is the measured parameter.

B. The Forward Problem

Assuming that $\mathbf{x} = [x, y, z]$ describes a 3D location in a continuous source distribution $f(\mathbf{x})$ and that $\mathbf{y} = [\mathbf{d}_1, \mathbf{d}_2, E_1, E_2]$ is a vector describing event detection locations and energies in the first and second detectors, the following formulation relates the expected value $g(\mathbf{y})$ of the projection measurements to the source distribution

$$g(\mathbf{y}) = (\mathcal{H}f)(\mathbf{y}) = \int_{\mathcal{X}} h(\mathbf{y},\mathbf{x}) f(\mathbf{x}) d\mathbf{x}, \quad (24)$$

where the kernel of the integral equation $h(\mathbf{y},\mathbf{x})$ represents the conditional probability of detecting an event at location $\mathbf{y}$ given that it was emitted at location $\mathbf{x}$ in the source. This conditional probability density models the effect of scattering-angle uncertainty due to electronic noise and Doppler-broadening as well as any spatial resolution limitations of the first or second detector. Non-linear aspects of the problem such as object-dependent γ-ray attenuation, are incorporated into the definition of $h(\mathbf{x},\mathbf{y})$ or, as in the case of system deadtime, often ignored. The source and projection functions f and g are assumed to be elements of the Hilbert spaces of square-integrable functions $L_2(\mathcal{X})$ and $L_2(\mathcal{Y})$ (i.e., $\int_{\mathcal{Y}} g^2(\mathbf{y}) d\mathbf{y} < \infty$) over finite regions of space (Keener, 1988).

Because of the finite regions of support $\mathcal{X}$ and $\mathcal{Y}$ and the fact that $h(\mathbf{y}, \mathbf{x})$ is bounded for all $\mathbf{x}$ and $\mathbf{y}$

$$\int_{\mathcal{X}}\int_{\mathcal{Y}} h^2(\mathbf{y},\mathbf{x}) d\mathbf{y} d\mathbf{x} < \infty. \quad (25)$$

Equivalently, $\mathcal{H}$ is a *Hilbert-Schmidt* operator and is therefore *compact*.[5]

Compact linear operators have many desirable properties. In particular, even though they are infinite-dimensional, they have an almost finite-dimensional quality and can be approximated to arbitrary accuracy by an appropriate sequence of degenerate linear operators or matrices. This justifies approximating the true, infinite-dimensional imaging problem by the matrix-vector approximation

$$\mathbf{g} = \mathbf{Hf} \quad (26)$$

where $\mathbf{H}$ is a $D \times B$ matrix approximating the operator $\mathcal{H}$, $\mathbf{f} = [f_1, ..., f_B]^T$ represents a discrete approximation (e.g., pixel values) to the source distribution f, and $\mathbf{g} = [g_1, ..., g_D]^T$ is the expected number of events in each of D discrete detector channels. In spite of the desirable properties of compact linear operators, there are undesirable side-effects, which will be become evident in the following sections.

In developing inversion methods to recover f from g, it is often useful to consider alternative representations. Note that equation (24) can also be represented by the following

$$g(\mathbf{y}) = \int_{\mathcal{Z}}\int_{\mathcal{X}} h_2(\mathbf{y},\mathbf{z}) h_1(\mathbf{z},\mathbf{x}) f(\mathbf{x}) d\mathbf{x} d\mathbf{z} \quad (27)$$

where the intermediate decomposition is often convenient from the viewpoint of solving the inverse problem (e.g., the decomposition may break the forward problem into two problems whose solutions are known). Along the same lines, a transformation can also be applied to the projection data in order to *adapt* the measurements into a system of equations that can be solved more easily

$$r(\mathbf{z}) = \int_{\mathcal{Y}}\int_{\mathcal{X}} h_3(\mathbf{z},\mathbf{y}) h_1(\mathbf{y},\mathbf{x}) f(\mathbf{x}) d\mathbf{x} d\mathbf{y}, \quad (28)$$

where $r(\mathbf{z})$ is the expected value of projection in the adapted geometry. Both techniques have been used extensively in developing Compton camera reconstructions.

Note that the above equations only represent the *expected value* of observing an event at a particular detector location. Quantum noise is significant in emission computed tomography and the primary reason for developing the Compton camera as an alternative to conventional SPECT where performance is limited by low efficiency. Unless the Compton advantage is overwhelming, one desires to use the measured data to estimate the source in a manner that is as statistically efficient as possible. In emission tomography, noise is dominated by counting statistics and it is common to embed the estimation problem within the framework of maximum-likelihood estimation using the conditionally Poisson likelihood function.

---

[5]More appropriately g and f should be restricted to the set of non-negative functions consistent with the fact that intensities must be non-negative; however, this would preclude the analytic solutions presented in Section E.

For continuous source and detector coordinates, the likelihood of observing a sequence of interactions with detector coordinates $\tilde{\mathbf{y}}_n$ (as above $\tilde{\mathbf{y}}$ includes the first and second detector locations and energies of the event) conditioned on the source distribution is given by

$$f(\{\tilde{\mathbf{y}}_n\}_{n=1}^{N_T}|f,T) = \prod_{n=1}^{N_T} g(\tilde{\mathbf{y}}_n) \exp\left[-\int_{\mathcal{Y}} g(\mathbf{y}) d\mathbf{y}\right], \quad (29)$$

where N_T is a Poisson distributed random variable denoting the total number of events observed in the time interval $[0, T]$ and $g(\mathbf{y})$, the expected intensity of the projection data during the observation interval is a function of the source distribution $f(\mathbf{x})$ as given by equation (1). Note that despite the fact that the source distribution may be a continuous function of location, there is never more than a finite number of measurements, N_T. Without restrictive conditions on the function space to which f belongs, f can never be recovered exactly from the measurements.

Somewhat more convenient to deal with from the viewpoint of image reconstruction is the likelihood function for the discrete formulation given by equation (26)

$$f(\tilde{\mathbf{g}}|\mathbf{f}) = \prod_{d=1}^{D} \frac{\left(\sum_{b=1}^{B} h_{db} f_b\right)^{\tilde{g}_d} \exp\left(-\sum_{b=1}^{B} h_{db} f_b\right)}{\tilde{g}_d!} \quad (30)$$

where $\tilde{\mathbf{g}} = [\tilde{g}_1, ..., \tilde{g}_D]^T$ represents the measured data $\{\tilde{\mathbf{y}}_n\}_{n=1}^{N_T}$ binned into D detector channels.

Note that because the adapted geometries specified by equation (28) in practice apply a transformation to the measured projection data, they also change the likelihood function—introducing correlations among the adapted measurements.

C. The Inverse Problem

Numerous techniques exist for developing inverses—or perhaps more appropriately, *pseudoinverses*—to the above integral equations. We first discuss a solution that has high relevance to both the continuous and discrete formulations of the problem.

Typically the expected value of the projections contain significant redundant information regarding the object—and this is certainly the case for the Compton camera. Without noise, the projection function g lies within the range of the system operator $\mathcal{H}$ and therefore has unique corresponding source distribution f. However, the presence of noise will almost surely result in the measured projections $\tilde{g}$ being *inconsistent* or not lying entirely within the range of $\mathcal{H}$. Assuming the measurement variance is the same at each point in $\tilde{g}$, for at least the finite-dimensional ($\mathcal{H} = \mathbf{H}$) reconstruction problem, the Gauss-Markov theorem (Rao and Toutenburg, 1999) states that the linear, unbiased estimator having the lowest variance is given by the least-squares solution, which can be written as (Keener, 1988)

$$\hat{f}_{LS} = (\mathcal{H}^*\mathcal{H})^{-1}\mathcal{H}^*\tilde{g}, \quad (31)$$

where

$$(\mathcal{H}^*\tilde{g})(\mathbf{x}) = \int_{\mathcal{Y}} h(\mathbf{y},\mathbf{x})\tilde{g}(\mathbf{y}) d\mathbf{y} \quad (32)$$

is the adjoint or "backprojection" operation and $(\mathcal{H}^*\mathcal{H})^{-1}$ is a self-adjoint operator that provides the "filtering" operation (no different than conventional filtered backprojection reconstruction in emission tomography).

Since the projection data are Poisson-distributed conditioned on the source distribution, the measurement noise in the measurement space is likely not equal at each point. For the situation in which this variance at each point in $\tilde{g}$ is known, the weighted least-squares solution is the best linear unbiased estimator for $\hat{f}$

$$\hat{f}_{WLS} = (\mathcal{H}^*\mathcal{W}\mathcal{H})^{-1}\mathcal{H}^*\mathcal{W}\tilde{g} \quad (33)$$

Here the appropriate weighting operator $\mathcal{W}$ is given by the inverse of the covariance operator at each point in the projection data. For conditionally Poisson data, the kernel of $\mathcal{W}$ would be

$$w(\mathbf{y},\mathbf{y}') = \begin{cases} g^{-1}(\mathbf{y})\delta(\mathbf{y}-\mathbf{y}') & g(\mathbf{y}) \neq 0 \\ 0 & otherwise \end{cases} \quad (34)$$

The weighted least-squares estimator essentially weights redundant data inversely with its uncertainty relative to the signal transfer characteristics $\mathcal{H}$ and then applies an appropriately modified filter $(\mathcal{H}^*\mathcal{W}\mathcal{H})^{-1}$ for deblurring.

Even though equations (31) and (33) are formal solutions of the forward problem, they have significant practical issues. As will be seen below, the stability of these solutions is poor, the finite set of measurements $\{\tilde{\mathbf{y}}_n\}_{n=1}^{N_T}$ cannot completely specify f, and the estimates $\hat{f}$ are not restricted to be non-negative. Nevertheless, these linear least-squares estimators have strong relationship to virtually all practical reconstruction methods as will be seen.

D. Overview of Compton Reconstruction Methods

With the above background, it is convenient at this point to provide a brief history of Compton image reconstruction for the near-field or 3D medical imaging application. In the sections that follow, recent developments are examined in more detail.

In the early 1980's Singh and Doria explored the idea of a two-stage reconstruction (the forward problem represented symbolically by equation (27) (Singh and Doria, 1983). In their approach, each first detector element was considered as a "virtual pinhole" aperture. In the absence of the Compton encoding (equivalently, at scattering angle zero), the projection of the object onto the second detector through each first detector element behaves as a pinhole projection of the object—albeit with different vignetting properties than a physical pinhole. For each first detector element their reconstruction recovered these pinhole projections using various iterative methods. The second stage reconstructed the 3D object distribution from the decoded pinhole projections.

Brechner and Singh (1990), and Hebert, et al. (1990) extended this work by noting that hemispherical second detectors centered on each first detector element could greatly simplify the reconstruction problem (the adapted geometry of equation [28]). The intersection of the cone ambiguity with a spherical second detector has a constant circular shape regardless of the cone direction. Equivalently, the "pinhole" projection of the object on this spherical surface is convolved with a rotationally-invariant function corresponding to the Compton ambiguity along with additional uncertainties in the scattering angle due to noise. This allowed use of the fast fourier transform (FFT) to improve the computational efficiency of part of the reconstruction through the use of its fast formula for circular convolutions. Hebert, et al. (1990), recognized the statistical inefficiency of the two-stage reconstruction and formulated the problem as one of maximum-likelihood estimation in which the 3D source distribution was estimated *directly* from the projection data.

The most significant issue in Compton camera reconstruction has proven to be computation, and more recently, several investigators have attempted to develop computationally efficient reconstructions. In particular, Rohe et al. (1997) has examined the point spread function (PSF) of the 3D backprojection of the coded data from a Compton camera consisting of concentric spherical first and second detectors. To the extent the backprojected PSF is spatially-invariant, the FFT and its corresponding fast convolution property for shift-invariant systems can be used to remove the blur in the backprojection. Sauve, et al. (1999) employed a multi-stage factorization of the system matrix *H*—again using the idea of hemispherical second detectors—and proposed several iterative reconstruction methods based on this factorization. Wilderman (1999, 2001) has developed a list-mode maximum-likelihood reconstruction method, which as shown below, results in a significant computational advantage over approaches using binned projection data.

Finally, considerable progress has been made in the effort to develop analytic reconstruction methods (Cree and Bones, 1994; Basko et al., 1998; Parra, 2000). While analytic methods generally have several disadvantages in comparison with their more general iterative counterparts, they are nevertheless important for the insight they bring to the Compton imaging problem and because they may provide practical reconstruction for specialized Compton imaging geometries.

E. Direct or "Analytic" Solutions

A common method for finding an inverse operator for the above integral equations is to expand the forward operator in a set of orthogonal functions that separates or "diagonalizes" it (analogous to the matrix case). For example, complex exponentials are a complete set of orthonormal functions that diagonalize shift-invariant operators in Cartesian coordinates to provide the familiar fast convolution formula. Once the operator has been diagonalized, solution is straightforward, although practical application often requires finding a computationally efficient method for computing the appropriate transforms.

Any compact linear operator—and in particular the emission tomography problem—can be represented by the following expansion (Naylor and Sell, 1982)

$$g(\mathbf{y}) = (\mathcal{H}f)(\mathbf{y}) = \int_{\mathcal{X}} \sum_{n=0}^{\infty} H_n u_n(\mathbf{y}) v_n^*(\mathbf{x}) f(\mathbf{x}) d\mathbf{x}$$
$$= \sum_{n=0}^{\infty} H_n F_n u_i(\mathbf{y}), \quad (35)$$

where

$$F_n = \int_{\mathcal{X}} v_n^*(\mathbf{x}) f(\mathbf{x}) d\mathbf{x}. \quad (36)$$

with $\{u_n(\mathbf{y})\}$ and $\{v_n(\mathbf{x})\}$ sets of orthonormal functions that span the range and domain of the operator $\mathcal{H}$, respectively (v^* being the complex conjugate of v). As noted in the previous section, compact linear operators are "nearly" finite-dimensional and the above decomposition is closely related to singular-value decomposition of the corresponding discrete approximation $\mathbf{H}$ to the imaging system

$$\mathbf{H} = \sum_{n=1}^{D} H_n \mathbf{u}_n \mathbf{v}_n^T \quad (37)$$

where $\{\mathbf{u}_n\}$ and $\{\mathbf{v}_n\}$ are orthonormal sets of vectors playing the same role as functions u and v. In both cases the set of coefficients $\{H_n\}$ is known as the *spectrum* of the operator.

Using the above expansion, and the orthonormality of $\{u_n(\mathbf{y})\}$ and $\{v_n(\mathbf{x})\}$, i.e.

$$\int_{\mathcal{X}} v_n^*(\mathbf{x}) v_m(\mathbf{x}) d\mathbf{x} = \begin{cases} 1 & n=m \\ 0 & otherwise \end{cases} \quad (38)$$

it is a simple matter to show that the inverse to the reconstruction problem can be approximated to arbitrary accuracy by the following series expansion

$$f(\mathbf{x}) \approx \sum_{n=0}^{N} H_n^{-1} G_n v_n(\mathbf{x}) \quad (39)$$

where

$$G_n = \int_{\mathcal{Y}} g(\mathbf{y}) u_n^*(\mathbf{y}) d\mathbf{y} \quad (40)$$

A most unfortunate property of compact linear operators—the one that essentially provides their close link with finite-dimensional matrices—is that as $N \to \infty$, the spectral coefficients $H_n \to 0$. It is immediately evident from equation (38) that small changes in g will induce arbitrarily large changes in the estimate of f for large enough N.[6] A common method of stabilizing the estimate, however, is to truncate the expansion at a value of N that provides an acceptable noise-fidelity tradeoff.

Several investigators have developed analytic reconstructions by finding decompositions of the form of equation (35) for specific adapted geometries. Three of these techniques are described in general terms below; refer to the corresponding articles to appreciate the details.

Cree and Bones (1994) develop a solution of the 3D source reconstruction problem from an adapted geometry consisting of parallel, planar first- and second-detectors with a parallel-hole collimator interposed between the two such that only scattered photons travelling orthogonal to detector plane are recorded.[7] Using the fact that the Fourier transform can be used to diagonalize a spatially-invariant convolution, in conjunction with the Hankel transform, they develop an inversion formula which shows that by collecting all scattering angles, the 3D source distribution can be recovered as the spatial extent of the detectors increases without limit. Practically, however, only limited-angle tomography is achievable with detectors of finite extent and the use of a parallel-hole collimator (either virtual or real) between the detectors results in an unacceptable loss of efficiency. Nevertheless, they suggest techniques for circumventing these limitations.

The methods of Parra and of Basko et al. are similar in two significant ways. First, as in the work of Singh and Doria, they adapt the geometry of the problem to consider each first detector element to be at the center of a spherical second detector. The second similarity, is that spherical harmonics are used as the sets of orthogonal functions to diagonalize the operator. Noting that when projected onto such a spherical second detector, the convolution kernel is rotationally-invariant depending only on the angle between the incident and scattered photon, the following expansion can be used

$$g(\theta, \Omega_2) = \sum_{n=0}^{\infty} H_n(\theta) \sum_{m=-n}^{n} Y_{nm}(\Omega_2) P_{nm},$$
$$P_{nm} = \int_{\Omega} p(\Omega_1) Y_{nm}^*(\Omega_1) d\Omega_1, \quad (41)$$

where $\Omega = (\omega, \phi)$ is a direction vector with $\omega \in [0, 2\pi)$ azimuth and $\phi \in [0, \pi)$ elevation, $p(\Omega)$ is the projection of the unscattered object distribution through the virtual pinhole, $Y_{nm}(\Omega)$ are spherical harmonics of degree n and order m, and the spectral coefficients $H_n(\theta)$ depend on the scattering angle θ. The above decomposition allow the expansion coefficients P_{nm} of the pinhole projections $p(\Omega)$ to be related to the expansion coefficients of the projection data at each scattering angle

$$P_{nm} H_n(\theta) = G_{nm}(\theta). \quad (42)$$

Using these ideas, Basko et al. propose the following reconstruction procedure:

1. Transform Compton camera data at each scattering angle into corresponding projections on a spherical detector concentric with each first detector element.
2. Calculate expansion coefficients $G_{nm}(\theta)$.
3. Compute the coefficients

$$G_{nm}(\frac{\pi}{2}) = \frac{H_n(\frac{\pi}{2})}{H_n(\theta)} G_{nm}(\theta) \quad (43)$$

and then

$$g(\frac{\pi}{2}, \Omega_1) \approx \sum_{n=0}^{N} \sum_{m=-n}^{n} G_{nm}(\frac{\pi}{2}) Y_{nm}^*(\Omega_1). \quad (44)$$

4. Noting that the $g(\pi/2, \Omega_1)$ form a set of (nearly) plane integrals through the 3D source distribution, the image can be reconstructed using the inverse 3D Radon transform (Natterer, 1986) with the assumption that the set of first detector locations or "cone-beam vertices" appropriately sample the Radon domain.[8]

Parra proposes several methods based loosely on the following procedure:

1. "Backproject" the data at each scattering angle onto a single spherical detector; i.e.

$$r(\Omega_1) = \int g(\theta, \Omega_2) h(\Omega_2, \Omega_1, \theta) d\Omega_2 d\theta. \quad (45)$$

Alternatively, the spherical harmonic expansion coefficients of the backprojected data can be expressed as

$$R_{nm} = \int_0^{\pi} H_n(\theta) G_{nm}(\theta) d\theta. \quad (46)$$

2. Apply the appropriate inverse filter to the backprojected data in order to recover the corresponding pinhole projections. This can be written in the following way in the transform domain

$$P_{nm} = R_{nm} / \int_0^{\pi} H_n^2(\theta) d\theta. \quad (47)$$

The operation can either be applied in the spherical harmonic domain or the transformed filter can be used for spherical deconvolution in the spatial domain.

3. From the P_{nm}, the pinhole projections $p(\Omega)$ can be recovered and used in a cone-beam image reconstruction technique in order to estimate the 3D source distribution.

---

[6] Consider the effect of the ramp filter for conventional emission tomography as the spatial frequency becomes large.

[7] For a Compton camera consisting of parallel first- and second-detectors, the kernel $h_3(z, y)$ in this case simply eliminates events in which the scattered photon does not travel perpendicularly to the detectors.

[8] Note that for the case of the real Compton camera, $g(\pi/2, \Omega)$ actually form weighted plane integrals through the object with the weight decreasing approximately as the square of the distance of the source location from the scattering detector.

Just as in filtered backprojection for reconstruction of conventional emission tomography images, there are numerous variations of the above approaches that differ mainly in the order and the domain in which the operations are performed.

Both Parra and Basko note the redundancy in the Compton projection data and Parra notes that different weighting of the angular information can lead to reconstructions having different invertibility or noise properties. Generally, a wide variety of non-negative weight functions $w(\theta)$ can be used

$$P_{nm} = \frac{\int_0^\pi w(\theta) G_{nm}(\theta) d\theta}{\int_0^\pi w(\theta) H_n(\theta) d\theta} \quad (48)$$

with the following choice being, perhaps especially significant for handling the fact that projection information at different scattering angles has different intrinsic uncertainties

$$P_{nm} = \frac{\int_0^\pi H_n(\theta) k(\theta) G_{nm}(\theta) d\theta}{\int_0^\pi H_n^2(\theta) k(\theta) d\theta} \quad (49)$$

where $k(\theta)$ is a function that weights the quality of the data at each scattering angle based on the inverse of its uncertainty and $H_n(\theta)$ quantifies how much "signal" is present in the projections. This has an obvious relationship with the weighted least-squares estimate of equation (10) but the class of admissible weighting functions is smaller and the estimator not as statistically efficient.

In spite of the aesthetics of analytic inversion methods, they have significant limitations:

1. It is difficult to handle γ–ray attenuation (especially non-uniform attenuation) because the operator $\mathcal{H}$, and therefore the orthogonal function expansion that diagonalizes it, becomes object-dependent.
2. Although some statistical weighting of redundant measurements can be done, it is typically less adequate than that provided by readily available alternatives such as maximum likelihood estimators.
3. Existence of an analytic inverse does not imply computational efficiency. Operations done in the spectral domain require the existence of a fast transform to be practical.
4. Analytic inverses typically depend on geometric adaptations of the imaging geometry for which the measurements often do not provide complete data (consider the case of a truncated second detector where measurements are not collected over 4π steradians as required for the above expansions). While iterative procedures can often be used to fill in this "missing" information, one of the reasons direct solutions are sought is to *avoid* iterative methods.
5. It is difficult to constrain the reconstruction to lie within a function space appropriate to source distributions (i.e., the set of *non-negative*, square-integrable functions on $\mathcal{X}$) other than by applying a trivial "clipping" operation (if $f(\mathbf{x}) < 0$ then set $f(\mathbf{x}) = 0$). Such estimates are in general not constrained solutions of the desired problem (e.g., a solution to the weighted least-squares problem subject to $\hat{f}$ being non-negative)

F. Statistically Motivated Solutions

By using an iterative reconstruction procedure virtually all of the aforementioned limitations can eliminated. Iterative methods allow simple incorporation of object attenuation, do not require restrictive system geometries, and can incorporate measurement uncertainty and constraints on the source distribution in a straightforward manner. In recent years, maximum likelihood (ML) estimation and the related *penalized* likelihood estimation have found widespread use in emission tomography for these reasons.

Recalling the discrete version of the likelihood function given by equation (30), the maximum likelihood (ML) estimator chooses the unknown source distribution $\hat{\mathbf{f}}$ that maximizes this function

$$\hat{\mathbf{f}} = \arg \max_{\mathbf{f}} f(\mathbf{g}|\mathbf{f}). \quad (50)$$

The ML estimate is typically found by differentiating the log-likelihood and solving the following system of equations subject to the condition $f_b \geq 0$ for $b = 1, ..., B$

$$\sum_{d=1}^{D} \frac{g_d h_{db}}{\sum_b h_{db} f_b} - \sum_{d=1}^{D} h_{db} = 0, \quad (51)$$

that must be done iteratively.

With certain restrictions (e.g., inverses must exist), the ML estimate $\hat{\mathbf{f}}$ satisfies

$$\hat{\mathbf{f}} = (\mathbf{H}^T \mathbf{W} \mathbf{H})^{-1} \mathbf{H}^T \mathbf{W} \mathbf{g} \quad (52)$$

subject to the constraint that all elements of the vector $\hat{\mathbf{f}}$ are non-negative. The weighting matrix $\mathbf{W}$ is given by the inverse of the *estimate* of the measurement covariance, i.e.,

$$\mathbf{W} = \text{diag}^{-1}(\mathbf{H}\hat{\mathbf{f}}). \quad (53)$$

While it proves to be difficult to calculate the ML estimate using these equations (due to non-negativity constraints and existence of the inverses) its relation with the weighted least-squares estimate given by equation (33) should be clear. The significant difference is that the estimated data covariance is used because the actual covariance is unknown.

It would be difficult to develop a direct reconstruction method based on expansion of either equation (33) or (52) because the orthogonal functions necessary for diagonalization depend on the object or the measurements. Moreover, for the ML estimator, the estimate $\hat{\mathbf{f}}$ appears on both sides of the equation. Fortunately, numerous iterative solution methods can be employed. The iterative expectation-maximization or EM-algorithm with the "usual" complete data space for emission tomography (Shepp and Vardi, 1982; Fessler et al., 1993) is convenient—if not particularly fast—for this problem

$$\hat{f}_b^{(k+1)} = \frac{\hat{f}_b^{(k)}}{\sum_d h_{db}} \sum_{d=1}^{D} \frac{g_d h_{db}}{\sum_b h_{db}\hat{f}_b^{(k)}} \qquad (54)$$

with the superscript k representing the iteration number.

Besides its relatively slow convergence rate, the above formulation of the algorithm suffers from an immediate problem: D, the number of possible detector elements in the vector **g** can be enormous—which, by the way, is also a significant problem for the analytic methods. A practical Compton camera might have 2^{16} first-detector elements, an equal number of second-detector elements, and 2^8 energy channels, resulting in approximately 10^{12} possible detector bins.

A typical dataset for this camera might likely have only 10^8 or so events. Obviously, most of the possible detector bins will contain zero, and for this case, an alternative formulation of the likelihood—the list-mode likelihood function—offers significant computational advantages (Barrett et al., 1997; Wilderman et al., 1999; Wilderman et al., 2001). While the EM algorithm can be rederived using equation (29) as a starting point, the computational advantage is immediately evident if, as above, we assume a very large number of possible detector channels D, most of them containing zero. Note that the above iteration can also be written as

$$\hat{f}_b^{(k+1)} = \frac{\hat{f}_b^{(k)}}{\sum_d h_{db}} \sum_{\{d: g_d \neq 0\}} \frac{g_d h_{db}}{\sum_b h_{db}\hat{f}_b^{(k)}} \qquad (55)$$

with the result that only the elements of the matrix **H** corresponding to nonzero g_d are needed and the sum ranges only over the number of events rather than all possible detector elements. The disadvantage of list-mode likelihood is that its complexity increases with the number of collected events. Nevertheless, it currently represents one of the more practical techniques for reconstructing Compton camera data and it incorporates the statistical uncertainties in the data in a way that will asymptotically achieve the performance predicted by Cramèr-Rao lower bound.

G. Regularization

Although, the scope of this section does not allow detailed treatment, the imaging problem is ill-posed. For the continuous formulation, small changes in the data can have an arbitrarily large effect on the reconstructions. Even in the finite-dimensional approximation, the matrix **H** is typically very poorly conditioned and the reconstruction problem *practically ill-posed*. In all cases, unbiased estimation is not possible and some bias needs to be incorporated into the reconstruction in order to reduce variance to acceptable levels (see Section III).

In conventional filtered backprojection, this biasing is accomplished by applying a window function to the ramp filter. Variance in reconstructed images is reduced with the undesirable side effect of increased image blur. Similar methods can be used in the linear least-squares estimators presented above as well as for the methods of Cree, Basko, and Parra. Other methods of reducing variance are to truncate the series expansion in equation (38) at a value of N that gives an acceptable resolution-noise tradeoff or by stopping an iterative reconstruction technique before convergence. The latter method is very often used to regularize maximum likelihood reconstructions.

The maximum likelihood estimator presented in the previous section has an additional advantage over the linear least-squares solutions of equations (31) and (33). The solutions are automatically constrained to be non-negative at each point in the reconstructed image. This constraint actually introduces a bias that significantly reduces variance in regions of the reconstructed image corresponding to low activity regions of the source.

Penalized likelihood estimation is becoming an increasingly popular method for regularization. A penalized maximum likelihood estimator is obtained by augmenting the likelihood with a penalty function

$$\hat{\mathbf{f}} = \arg \max_{\mathbf{f}} f(\tilde{\mathbf{g}}|\mathbf{f}) - \alpha R(\mathbf{f}), \qquad (56)$$

where R is a function penalizing undesirable characteristics of the reconstructed image such as large pixel-to-pixel intensity variations (or roughness) and the parameter α controls the degree to which the estimate corresponds to the measurements or to the penalty. Because its gradient is a linear function of **f**, a common choice for R in emission tomography problems is a quadratic roughness penalty

$$R(\mathbf{f}) = \mathbf{f}^T R \mathbf{f} \qquad (57)$$

where **R** is a matrix that that forms the difference of each pixel with its nearest neighbors. Although penalized likelihood estimation is becoming popular—and is a close parallel to Bayesian estimation using the maximum *a posteriori* (MAP) estimator—spatial resolution properties at each point in the reconstructed object can vary (Stayman and Fessler, 2002).

For more detailed treatments on regularization and the noise-resolution tradeoff, refer to (Demoment, 1989; Hero et al., 1996; Fessler and Rogers, 1996; and Qi and Huesman, 2001).

V. HARDWARE AND EXPERIMENTAL RESULTS

A. Silicon Pad Detectors

As described at the beginning of this chapter, the work of Singh and his colleagues was based on a pixelated germanium scattering detector. This selection was made primarily on the very high energy resolution that could be obtained (850 eV FWHM) combined with the fact that the sizable effects of Doppler broadening were not appreciated at that time. More recently, Du (Du *et al.*, 2001) has reported design of a Compton camera for high energies based on 3D CZT detectors, and the group at the Naval Research Laboratories (Kroeger *et al.*, 1999; Wulf *et al.*, 2002) has described germanium strip detectors for the Advanced Compton Telescope application. Because of the large uncertainties in scattering angle related to Doppler broadening for germanium, the cost and cooling requirements associated with germanium detectors, and the desire to maximize the probability of single scattering followed by escape from the first detector, current hardware development at the University of Michigan has focused on the development of low-noise silicon pad detectors (Weilhammer *et al.*, 1996). The most recent version of this detector and its readout electronics is shown in Figure 20 and fully described by Meier *et al.* (2000). The detector itself is 300-μm thick and consists of 256 1.4 × 1.4 mm pixels in an 8 × 32 array. The total sensitive area of the detector is 11.2 mm × 44.8 mm. The detector is AC coupled, and the pixel connections to wire bond pads on the long edges are accomplished by means of a fan-out structure in a second metal layer. The front-end readout is composed of eight 32-channel VA/TA chip sets manufactured by IDEAS[2].

The VA preamplifier is set to a 2-μs shaping time, whereas the TA has a short shaping time and generates a fast readout trigger pulse for events above a computer-set threshold. Each chip set reads out 32 channels for a valid trigger event. This detector and readout is identical to that described by LeBlanc (1999; LeBlanc *et al.*, 1998) and Hua (2000), except that detector and electronics are mounted on 10-μm-thick kapton foil and an improved version of the TA trigger generator is used in which the threshold values for all pads can be aligned to the same gamma-ray energy. The kapton foil printed circuit represents an attempt to mini-

FIGURE 20 Silicon pad sensor and VA/TA 32-channel readout chips mounted on 10-μm-thick kapton foil hybrid board (From Meier *et al.*, 2000, © 2000 IEEE.)

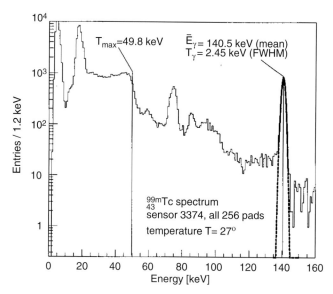

FIGURE 21 Energy spectrum for ^{99m}Tc summed over all pads. A 15-keV threshold was set just above the noise level. (From Meier *et al.*, 2000, ©2000 IEEE.)

mize nondetector mass in the first-detector assembly. Trigger times for all pads have a Gaussian distribution with a width of 17 ns FWHM without any compensation for differences in mean channel delay times or pulse height variation. A composite ^{99m}Tc spectrum for all pads is shown in Figure 21. A 15-keV trigger threshold could be set above the noise, and this permits triggering on 56° scattering events for ^{99m}Tc and 20° events for ^{131}I.

The energy resolution averaged over all pads was 2.45 keV for ^{99m}Tc. These detectors depleted at 55 V and exhibited a room-temperature leakage current averaging approximately 225 nA/256 pads at a bias of 60V. This resolution is less than desired; however, these detectors are certainly adequate for investigating Compton imaging in a prototype device. Measurements on a variation of this detector yielded noise values approaching 900 eV with the same readout chips.

B. C-Sprint

It is absolutely essential to confirm lower-bound and observer predictions of Compton camera performance experimentally. This is necessary to validate the system model used for performance calculation and for the image reconstruction. For this purpose, a silicon detector similar to that described has been combined with a single-photon ring tomograph (SPRINT) (Rogers *et al.*, 1982) to form a prototype Compton camera, referred to as C-SPRINT (LeBlanc, Bai, *et al.*, 1999).

SPRINT is composed of 11 modular NaI(Tl) scintillation cameras arranged in a 50-cm-diameter ring. Scintillator thickness is 13 mm, and the useful width of the detector ring is approximately 11 cm. The spatial resolution of the cameras is 3 mm FWHM. The silicon pad detector is located on-axis and is mounted on an optical bench to permit adjusting its axial position, as illustrated in Figure 22. The principal advantage of this ring geometry is that with appropriate shielding of the second detector ring one can use high activity sources with only a single silicon detector to obtain statistically significant data in reasonable collection times without

TABLE 2 Key system Parameters for C-SPRINT

Parameter	^{99m}Tc	^{131}I
Resolution for summed energy (FWHM)	17 keV (12%)	43 keV (12%)
Coincidence resolving time (ns, FWHM)	140	70
Coincidence window (ns)^a	90	90
Measured detection sensitivity^b	1.8×10^{-7}	1.2×10^{-6}

^{a}The maximum coincidence window is hardware-limited at present to 90 ns. From Le Blanc (1999).
^{b}Sensitivity is for an on-axis point source at 11 cm.

saturating the second detector. The key system parameters are summarized in Table 2, and details of system calibration and performance can be found in Le Blanc (1999).

The energy and timing resolution in Table 2 are primarily determined by the scintillation detector. Although detection sensitivity is substantially less than the expected sensitivity after correcting for known problems, including dead time, nonuniform silicon detector thresholds, and narrow coincidence window, it was sufficient to acquire image data as described next.

1. Planar Images

Images of isolated point sources of ^{99m}Tc and ^{131}I from C-SPRINT are shown in Figure 23. Images were reconstructed using 200 iterations of unregularized list-mode likelihood (Wilderman *et al.*, 2001). These two point source images clearly demonstrate the effect of gamma-ray energy on spatial resolution. The small side-lobe in the Tc image is unexplained. Because the list-mode likelihood algorithm enforces a positivity constraint, an isolated point source image is an overoptimistic measure of imaging performance. For this reason ^{99m}Tc and ^{131}I images for point sources superimposed on a 7.5-cm-diameter uniform disk source and

FIGURE 22 C-CPRINT. Si pad detector is on axis at the front edge of the NaI detector ring and disk phantom is 11 cm in front of the pad detector. (From LeBlanc, Bai *et al.*, 1999, ©1999 IEEE.)

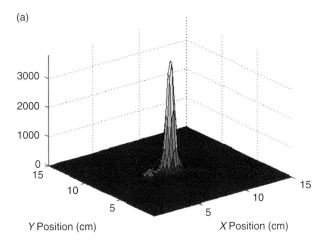
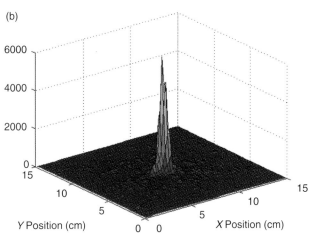

FIGURE 23 (a) ^{99m}Tc point source at 11 cm, 8.2 mm FWHM. (b) ^{131}I point source at 11 cm, 6.5 mm FWHM. 100 K counts each. (From LeBlanc, 1999, ©1999 JW LeBlanc.)

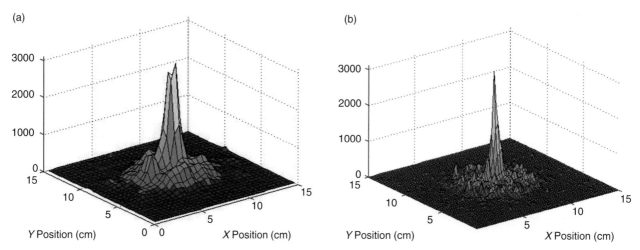

FIGURE 24 (a) ^{99m}Tc point source superimposed on a 7.5-cm-diameter uniform disk at 11 cm, FWHM = 11.7 mm. (b) ^{131}I point source superimposed on Monte Carlo–generated disk background at 11 cm, FWHM = 5.9 mm. (From LeBlanc, 1999, ©1999 J. W. LeBlanc.)

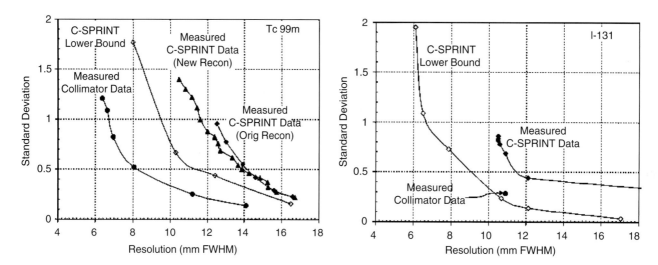

FIGURE 25 Standard deviation in center pixel normalized to the total number of detected photons versus FWHM spatial resolution for ^{99m}Tc and ^{131}I point sources on a uniform 7.5 cm diameter disk background. (Adapted from LeBlanc, 1999; LeBlanc, Bai, et al., 1999; ©1999 JW LeBlanc & IEEE.)

reconstructed with 200 iterations are shown in Figure 24. For ^{99m}Tc, 77,590 counts came from the 2.5-mm-diameter point source and 30,356 counts came from the disk. The data sets were acquired separately and then combined. In the case of ^{131}I, 50,000 counts were acquired from the point source and combined with 100,000 counts from a simulated disk source. The spatial resolution for all images was estimated from a 2D least squares fit to a Gaussian plus background.

2. Variance/Resolution Results

In order to compare the experimental results with predictions derived from the uniform bound calculations (see Section III A), variance in the central pixel computed from 10 realizations of experimental data is plotted in Figures 25 and 26 as a function of spatial resolution for C-SPRINT and a collimated Anger camera. This was done for both ^{99m}Tc and ^{131}I point sources on a 7.5-cm-diameter disk background spaced 11 cm from the first detector. FWHM and spectral resolution are both used as measures of resolution. Spectral resolution is defined in Eq. (58) (Lecomte et al., 1984) where $f(x)$ is the point response function. It is a useful additional measure of resolution because it is sensitive to the overall shape of the response function including the tails. Spectral resolution and FWHM for the uniform bound calculations are calculated from the mean gradient image (Section IIIA2).

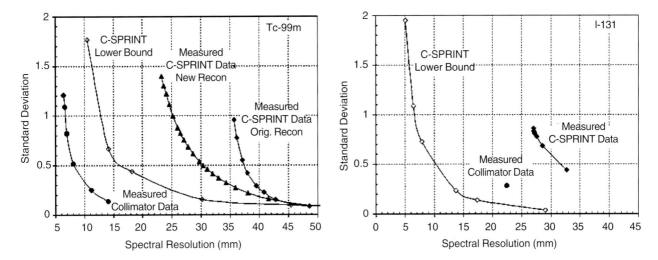

FIGURE 26 Same as Figure 25 except plotted as a function of spectral resolution to emphasize the effect of the tails of the point response function. (Adapted from LeBlanc, 1999; LeBlanc, Bai, et al., 1999; ©1999 JW LeBlanc & IEEE.)

$$R_s = \frac{\left[\int_x f(x)\,dx\right]^2}{\int_x [f(x)]^2\,dx} \qquad (58)$$

Recall that the uniform lower bound on variance is the variance in a given object parameter predicted for the maximum-likelihood estimator for a specified imaging system. Spatial resolution for the bound curve is implicitly governed by an identity matrix penalty function. Bound computation used measured system parameters for C-SPRINT, including actual system geometry, first-detector noise and Doppler broadening, and spatial resolution for first and second detectors excluding depth of interaction. For the Anger camera and collimator, measured point source responses were used.

The spatial resolution for the experimental data was varied by varying the number of iterations of the list-mode ML algorithm. For ^{99m}Tc, the variance in the C-SPRINT image is computed from 10 realizations consisting of 10,000 events from the point source and 7759 events from the disk source. The variance for the collimator is computed from 20 realizations, and spatial resolution was varied by varying the number of iterations of a maximum likelihood-expectation maximum (ML-EM) deconvolution algorithm. For the ^{131}I curves each realization consisted of 10,000 events from a measured point source and 10,000 events from a Monte Carlo–simulated disk source. Again, 10 realizations were used for the C-SPRINT data and 20 realizations for the iodine collimator. No attempt was made to deconvolve the iodine point source data because of the structure in the tails of the response function arising from septal penetration.

Two versions of the list-mode likelihood algorithm were used to reconstruct the ^{99m}Tc images. The first, (Original Recon) was that described by Wilderman et al. (2001) in which effects of finite resolution were approximated by back-projecting a central cone for the mean scattering angle and two neighboring cones with Gaussian-weighted amplitude at $\pm\Delta\theta_s$. The newer reconstruction weights the back-projected central cone with the sum of two Gaussians that approximates the true angular uncertainty. This function is truncated at approximately 5% of peak amplitude.

Figures 25 and 26 must be interpreted with caution because, in general, the shapes of the mean gradient image obtained from the bound calculation and the reconstructed point response functions will be different. Sensitivity to shape is particularly evident when comparing the new and original reconstructions of the Tc response function. When resolution is measured by FWHM, a relatively minor improvement is indicated. However, when spectral resolution is used as the metric, the new algorithm gives substantial improvement that is consistent with substantial reduction in the tails of the response function. Both the bound and experimental data for ^{131}I demonstrate the improved Compton camera performance at higher energy, both in the absolute sense and in comparison to the iodine collimator. Finally, it is quite evident that the experimental results differ substantially from the lower bound and that this difference is greater when spectral resolution, as opposed to FWHM, is used as a metric. This difference may arise from either the image reconstruction algorithm or from using an incorrect system model for the bound calculation. It is clearly important to resolve this question in order to be able to predict Compton camera performance and its dependence on system design parameters.

C. Scintillation Drift Chambers as Compton Cameras

Up to this point we have focused on the use of solid-state detectors, and in particular silicon pixel detectors, for the

Compton camera application. However, in one of the first pioneering efforts devoted to the subject, Todd, Nightingale, and Everett (1977) mention gas detectors as a possible alternative for medical Compton imaging. Later, Fujieda and Perez-Mendez (1987) conclude that noble gas electroluminescence drift chambers have an advantage over scintillators and semi-conductor detectors in Compton camera design due to their combination of fine position resolution, good energy resolution, and large field of view. Since these first publications, several noble gas detector designs for Compton imaging have been proposed, including a liquid-xenon drift chamber, two-phase emission camera, and high-pressure xenon electroluminescence detector (Chen and Bolozdynya, 1995). The idea to use liquid xenon (LXe) for SPECT Compton cameras has been rejected because of the relatively poor energy resolution provided by LXe at the lower energies of interest in nuclear medicine. Large fluctuations in the small number of δ-electrons created along tracks of ionization particles in LXe determine the intrinsic resolution of these detectors. For energies >1 MeV, the energy resolution of LXe ionization detectors is approximately 5% FWHM, and the idea of a LXe Compton telescope proposed by Alvarez (Alvarez *et al.*, 1973) can be realized for astrophysics gamma-ray imaging (Aprile *et al.*, 1998).

Compton-scattered 140-keV gamma rays generate small signals with a maximum energy of 50 keV and average energy of approximately 30 keV. For this reason low-threshold electroluminescence noble gas detectors attract attention. The product of approximately 20 years of research, a scintillation drift chamber (SDC) with 3D position sensitivity has been developed (Bolozdynya, Egorov, *et al.*, 1997b). The SDC is triggered by fast scintillation signals arising at the moment of gamma-ray interaction with pressurized xenon. The readout system of this detector is able to determine 3D coordinates and deposited energy of several point-like ionization clusters occurring in the sensitive volume. The design of the SDC is described in detail in Bolozdynya (Chapter 18 in this volume), where it is shown that electroluminescence (EL) detectors benefit from the intrinsic signal amplification at low energy deposition. Because of this gain, the energy resolution of EL detectors is not limited by the input capacitance at the preamplifier, as is the case for solid-state detectors (see Section IIB) or ionization chambers in which microphonic noise poses an additional-problem. For this reason, the energy resolution of EL detectors of approximately 1000 cm^2 in area is comparable to the energy resolution of room-temperature semiconductor detectors with an area of approximately 1 cm^2.

A schematic design of an SDC working as a Compton camera is shown in Figure 27. A single gamma quantum absorbed in noble gas may produce a few ionization clusters (vertices) resulting from photoabsorption of the gamma quantum, photoabsorption of the 29.7- and 33.8-keV fluorescent photons emitted by excited xenon atoms, Compton

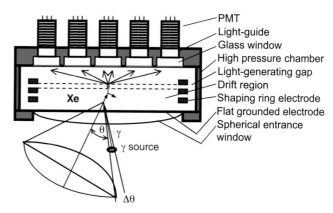

FIGURE 27 Schematic diagram of a scintillation drift chamber. (From Bolozdynya *et al.*, 1997b, © 1997 IEEE.)

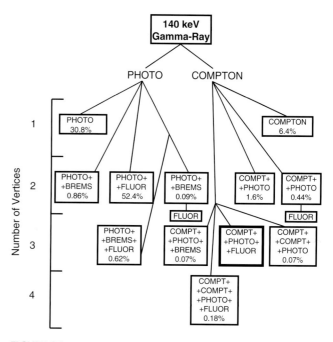

FIGURE 28 Probability of different interaction sequences for ^{99m}Tc in a 20-bar xenon chamber.

scattering of the gamma quantum, and photoabsorption of the Compton scattered or Bremsstrahlung photons.

Possible scenarios of 140-keV gamma-ray interactions with a 20-bar xenon SDC filling are illustrated in Figure 28. The probability of each process is shown as a percentage of the number of incident gamma rays entering the SDC along the axis of the device. One can see that three-vertex events contain the dominant portion of Compton interactions.

Figure 29a illustrates the energy spectrum of single-vertex events (pure photoabsorption) when the camera is exposed to a ^{99m}Tc pointlike source without collimator and triggered by the scintillation signals. The energy spectrum of the single-vertex events includes a 140-keV photoabsorption peak unaccompanied by a fluorescent emission, two approximately

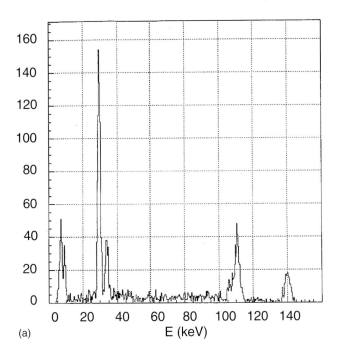
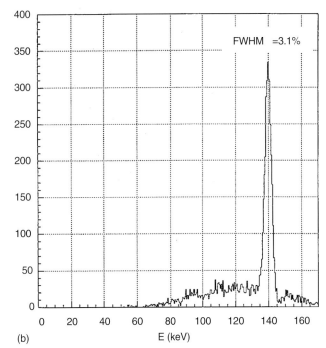

FIGURE 29 (a) Energy spectrum for single-vertex events for ^{99m}Tc. (b) Summed energies of three-vertex events for ^{99m}Tc. (From Bolozdynya et al., 1997b, © 1997 IEEE.)

110-keV overlapping escape peaks with energies of $E_0 - E_f$, and xenon fluorescent peaks in the neighborhood of $E_f = 30$ and 8 keV.

A significant fraction of two-vertex events contains a photoabsorption vertex of the primary photon and a photoabsorption vertex of the fluorescent photon emitted by the xenon atom. Because the energy of the Xe K-shell fluorescent photons is known, a fluorescent-gated energy spectrum can be acquired for two-vertex events using the well-determined energy of the fluorescent photon (Bolozdynya, Chapter 18).

Among three-vertex events, there are the Compton scatter events that can be recognized by the following selection criteria.

1. The total energy deposition for all three vertices should be equal to the energy of the incident gamma rays emitted by the known radioactive source.
2. One of the vertices should have energy deposition of approximately 30 keV (fluorescent photon signature).
3. The energy of the two other vertices should lie in the energy regions allowed by Compton kinematics:

$$\cos\theta_s = 1 - m_e c^2 \frac{E_1}{(E_0 E_2)} \quad (59)$$

where θ_s is the angle between the line connecting two conversion points and the direction of incoming gamma ray, m_e is the mass of the electron, E_1 is the energy deposition at the point of the Compton scattering, E_2 is the energy deposition at the point of the photoabsorption of the scattered gamma ray, and $E_0 = E_1 + E_2$ is the energy of the incident gamma ray.

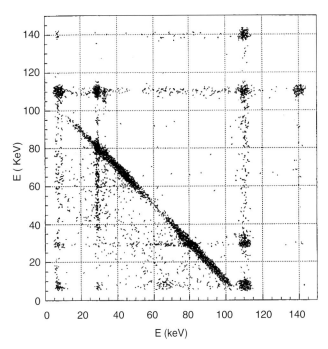

FIGURE 30 Two-dimensional plot of E_a versus E_b for three-vertex events, where a and b correspond to the two vertices that are not 30 keV. (From Bolozdynya et al., 1997b, © 1997 IEEE.)

A dual-energy plot of three-vertex events in which one of the vertices is an approximately 30-keV fluorescent event appears in Figure 30. The remaining two vertices, a and b, correspond to the recoil electron and photoabsorption of the scattered gamma ray. The photoabsorption is usually the source of the fluorescent photon. In this plot, a and b cannot be uniquely identified with these two events, but events containing a Compton scattering will be localized along the line of $E_a + E_b = E_0 - E_f$ and form a triangle with the E_a and E_0 coordinate axes. Three-vertex events, containing nonrelated vertices or combined one- and two-vertex events are localized in spots and along the lines with fixed deposited energy. To demonstrate the effectiveness of the selection criteria, compare the energy spectrum of random single-vertex events in Figure 29a with the spectrum of total energy deposited in the selected three-vertex Compton events Figure 29b. The three-vertex event spectrum has only one dominant peak corresponding to the incident photon energy. The tails correspond to misclassified three-vertex events such as those listed in Figure 28.

Using off-line processing, events from the Compton line in Figure 30 have been selected and Compton scatter angles have been calculated using Eq. (59). The angular distribution in Figure 31 shows an angular resolution of 15° FWHM. This unexpectedly poor angular resolution is largely due to Doppler broadening of the energy of gamma rays scattered from bound electrons in xenon (Ordonez et al., 1997; see also Section IIC1). Note that Doppler broadening has no influence on the total energy deposited in detectors, and this is why the peak of the spectrum in Figure 29b has a width close to the energy resolution of the SDC.

The following procedure was used to evaluate the spatial resolution of the SDC. For each three-vertex event that has been identified as a true Compton event, the 3D parameters of the conic surface determined by the direction of the incoming gamma ray have been calculated. The half-angle of the cone, θ in Figure 27, is calculated using Eq. (25), from the measured energies deposited in the Compton and scattered gamma-ray photoabsorption vertices. The apex of the cone is defined by the measured 3D position of the Compton vertex. The axis of the cone is determined by a line connecting the measured 3D positions of the Compton and photoabsorption vertices. This defines the 3D Compton cone for each selected three-vertex event. Knowing the coordinates of the source located 10.5 cm in front of the center of the grounded electrode, the shortest distance between the reconstructed 3D cone surface and the real position of the source is calculated. The statistical distribution of these distances is 25 mm FWHM, which is a measure of the 3D position accuracy of the SDC in the Compton camera mode (Bolozdynya, Egorov, et al., 1997a). This value, obtained with a 19-channel SDC, compares favorably to the 29-mm resolution calculated for a hypothetical Compton camera with a 1024-channel germanium scatter detector in a comparable configuration (Solomon and Ott, 1988). This clearly demonstrates the advantages of noble gas EL detectors.

This first experience with SDC clarified the high potential of EL noble gas detectors for the development of a practical Compton camera. It is now evident that it is impossible to provide low Doppler broadening of the scattered gamma-ray energy spectrum and high effective photoabsorption after scattering in the same medium. A Compton camera should consist of two separated detectors filled with different substances. Light noble gasses such as He, Ne, and Ar have a dominant Compton cross section for 140-keV gamma rays, exhibit low Doppler broadening for medium-energy gamma rays (see Figures 4 and 9 for Ne), and could be used as the scatter detector. Xenon provides the most effective photoabsorption of gamma rays and is an excellent candidate for the absorption detector.

Further progress in EL imaging cameras is expected with development of a multilayer camera with fiber optic readout and sub-millimeter 3D position resolution—the multilayer electroluminescent camera (MELC; Bolozdynya and Morgunov, 1998). The working gas, gas pressure, number of layers, and total thickness of the detector can be chosen to achieve the required detection efficiency for certain interactions. For example, a 20-cm-thick MELC consisting of 20 layers, each 1 cm deep, and filled with 20-bar Xe will have an 85% photoabsorption efficiency for 140-keV gamma rays. This is comparable to the detection efficiency of 3/8-in-thick NaI(Tl) scintillators usually used in gamma cameras. The MELC filled with 20-bar Ar will be 8% efficient for Compton scattering 140-keV gamma rays and practically transparent for photoabsorbing 140-keV gamma rays. The same efficiency for Compton scattering could be achieved with a 3-mm-thick Si detector. A high-sensitivity Compton camera can be built with a silicon or an argon-filled cylindrical scatter detector surrounded

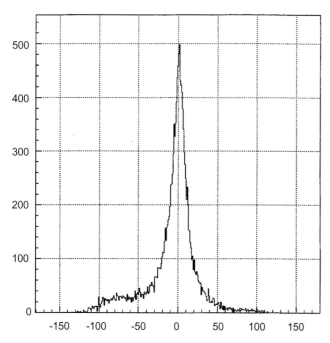

FIGURE 31 Angular resolution of the scintillation drift chamber in Compton camera mode. (© 1997 IEEE.)

by a xenon-filled cylindrical MELC (Bolozdynya, Ordonez, et al., 1997).

VI. FUTURE PROSPECTS FOR COMPTON IMAGING

We have seen that the physics underlying Compton scattering fixes the minimum angular uncertainty at 1–2° FWHM at 391 keV and 3–4° at 140 keV for silicon detectors. This corresponds to 1.75–3.5 mm and 5.25–7 mm at 10 cm, respectively, at these two energies. At 140 keV this resolution is further degraded by detector noise. A value of 1 keV FWHM adds an uncertainty between 5 and 5.5° over a scattering angle range of 30–85°. Raw sensitivity gains compared to the corresponding collimator for a single-scatter, double-ring Compton camera, as pictured in Figure 16, range from 300 at 140 keV to 400 at 360 keV for a 16-mm-thick silicon ring, and these figures could double if multiple scatter events were used along with thicker silicon. Using 1.5-mm-thick silicon pad detectors, the ring would contain about 750,000 channels with a current cost estimated to be in the vicinity of $1.25 million. The gains in raw sensitivity are reduced by a decoding penalty that is object-dependent and that is different depending on whether it is computed using the uniform CR bound or the ideal observer. At 140 keV, considering a 7.5-cm-diameter disk, the penalty is 10 for the ideal observer and 40 for the uniform bound. At 360 keV, the decoding penalty is approximately 1.5–2 according to the bound calculation and 3.25 for the ideal observer. Finally, preliminary noise measurements on a new silicon pad detector give noise values less than 1 keV FWHM. Given this state of affairs, what can we conjecture concerning the future of Compton imaging in nuclear medicine?

Certainly, before one would advocate the general-purpose application of Compton cameras for imaging ^{99m}Tc, it would be necessary to accomplish the following.

1. Performance predictions based on lower-bound or ideal observer analysis must be met experimentally, and the relationship of these two measures must be better understood.
2. Silicon imaging detectors of some type must be reliably fabricated with 500-eV energy resolution or better at a reasonable price.
3. Technical problems of detector packaging and heat removal for million pixel arrays must be solved.
4. Image reconstruction must be accomplished in acceptable times.
5. Three-dimensional images obtained using these detectors must be of high quality as verified by human observer studies and predicted sensitivity gains must be demonstrated.

It is evident that this is a long-term task requiring substantial research, and for the near term it seems reasonable to investigate other applications of Compton imaging that would help achieve some of these objectives and also offer more immediate gains with less uncertainty and reduced cost. In the next sections we look at special-purpose imaging tasks and techniques that appear particularly suited to Compton imaging and also discuss some ideas that can only be undertaken using the Compton technique.

A. Compton Probes

The underlying idea here is to consider imaging tasks in which the imaging instrument can be placed close to the object of interest so that a given angular uncertainty does not translate into an unacceptable spatial uncertainty. Furthermore, these are often situations in which the bulk and mass of a collimator proves a hindrance, and a large field of view can be an asset. A specific example is that of a transrectal prostate probe described by Zhang, Clinthorne et al. (2000) and Zhang, Wilderman et al. (2000). The probe, illustrated in Figure 32, serves as the first detector of a Compton camera, while the second detector is placed outside of the body, and could easily be a pair of coincidence cameras. Relative locations and orientations of the probe and coincidence cameras may be determined with six-degree-of-freedom motion trackers.

Monte Carlo simulations were run for a Compton imaging system composed of a $1 \times 1 \times 4$ cm^3 stack of silicon detector with 1-mm^3 voxels, and two 40-cm^2 second detectors placed 5 cm above and below a 40 cm $\times$ 40 cm $\times$ 20 cm thick model of a human body. A Y-shaped array of point sources with 5-mm spacing was placed at the location of the prostate 1 cm from the first detector. All effects of scattering, attenuation, detector penetration, and Doppler broadening were included in the simulation. Figure 33 illustrates planar images reconstructed by list mode likelihood for gamma-ray energies of 140, 364, and 511 keV. A 10% energy window on total deposited energy was used. Best performance was obtained at 364 keV for which the resolution was 2 mm FWHM and sensitivity was 1.2×10^{-3}. At 140 keV, resolution degraded to 2.5 mm.

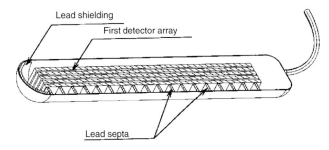

FIGURE 32 Section view of one possible prostate probe configuration. The first detector is composed of a stack of 1-cm-wide by 4-cm-long silicon pad detectors. Lead septa and shielding can be configured to limit sensitivity to body background.

FIGURE 33 Reconstructed images of point source arrays 1 cm from silicon detector: 140, 364, and 511 keV. Point source separation is 5 mm, and each point represents approximately 12,000 photons.

Based on these simulations, it appears that this probe could be used to image ^{111}In-labeled prostascint to monitor response to therapy or to detect recurrent prostatic disease. Similar small devices might find application in breast imaging (Zhang et al., 2004) or intraoperative tumor localization.

B. Combined PET-SPECT Imaging

Modern position emission tomography (PET) detectors have four important characteristics that make them almost ideal as second detectors for a Compton camera.

1. They accommodate high count rates, as required by the double ring geometry of Figure 16, and are capable of excellent coincidence timing.
2. Detectors have high stopping power for high-energy gamma rays.
3. Intrinsic spatial resolution can be 2–3 mm or better.
4. Methods have been developed for estimating depth of interaction in PET detectors (see Figure 3 for relevance to Compton imaging).

This suggests that a PET detector could be combined with a silicon ring detector for high-energy SPECT imaging. In addition, such a configuration could serve to augment the PET data by imaging a single annihilation photon when the second one has been absorbed or scattered outside the PET energy window. Furthermore, as described next, annihilation photons that interact with the silicon ring can add a very high resolution-data component to augment the standard PET data. At the moment, this is highly speculative but certainly deserving of future study.

C. Very High Resolution Animal PET

Another possible application related to Compton imaging and also to PET-SPECT imaging is that of high resolution positron imaging in small animals as described by Park et al. (2004). The physical resolution limit for small-diameter PET detector rings is imposed by the positron range in tissue. For ^{18}F, the range distribution is sharply peaked with a width of 200 μm FWHM. Current animal PET instruments, however, have resolution on the order of 1.5–2 mm FWHM. Although scintillation crystals can be reduced in size, spatial resolution will remain limited by the greater probability that the photon will scatter in the scintillator rather than be photoabsorbed. The maximum photoelectric efficiencies of BGO, LSO, and NaI at 511 keV are 41%, 33%, and 18%, respectively. The spatial resolution thus becomes limited by the scintillator volume over which the energy is deposited. Park et al. (2004) propose using a stack of silicon detectors with 0.5-mm^3 voxels or smaller to encourage Compton scattering followed by escape to a scintillation detector ring. In this manner, the initial interaction point of the photon can be fixed in three dimensions and resolution is limited only by recoil electron range (< 50μm in Si at the most likely scattering angle of ~35°; < 250μm at 90°). Figure 34 illustrates one possible configuration for a high resolution positron imager based on this principle.

In this arrangement, there are three basic initial interaction combinations for the two annihilation photons that carry useful information: (1) Si–Si coincidences, (2) Si–scintillator coincidences, and (3) scintillator–scintillator coincidences. The silicon interactions may be split further into photoabsorptions and single or multiple Compton interactions. The highest spatial resolution events will be Si–Si coincidences because both ends of the coincidence line are fixed within the silicon voxel size. In a 4-cm-diameter ring, position uncertainty arising from non-co-linearity of the annihilation photons is less than 100 μm, so this effect becomes important only toward the lower limit on voxel size as determined by recoil electron escape. For Si–scintillator events, resolution will degrade because the scintillator resolution will be 2–3 mm. However, resolution loss is small because the

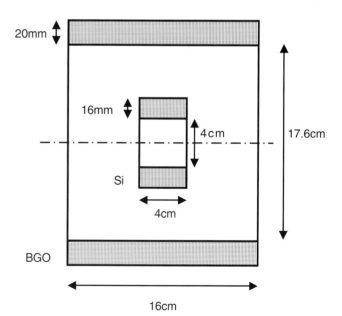

FIGURE 34 Section view of a very high resolution animal PET model used for Monte Carlo simulations. The silicon ring first detector has $0.3 \times 0.3 \times 1$ mm pixels, and the BGO ring is modeled as $3 \times 3 \times 20$ mm discrete elements.

high-resolution silicon interaction is closer to the field of view than the lower-resolution scintillator interaction. If the ratio of distances from the source point to the two detectors is 2:1, the position uncertainty due to scintillator resolution is reduced by a factor of 3. The scintillator–scintillator interactions correspond to the conventional interaction and have the lowest resolution.

Monte Carlo simulations of the system depicted in Figure 34 were used to determine the positron detection efficiency for point sources at different radial positions in the central plane and to generate data sets for a disc phantom consisting of six disks with diameters ranging from 1 mm to 10 mm (Park et al., 2001). Table 3 lists detection efficiencies for three subsets of the many possible combinations of interaction sequences. Single–single corresponds to a single interaction of each photon in the silicon detector. The interaction may be a photoabsorption or a Compton scatter followed by absorption in the BGO detector. In a Single–BGO, one photon has a single interaction in silicon while the other is absorbed by the BGO detector. BGO–BGO corresponds to the conventional PET event with no silicon interaction. The total efficiency is approximately 30%, and this increases significantly if multiple scatter events in silicon are included.

TABLE 3 Efficiency of Very High Resolution Animal PET System of Figure 3[a]

Radial Position (mm)	Detection Efficiency (%)		
	Single–single	Single–BGO	BGO–BGO
0	1.05	8.83	20.84
6	0.96	8.96	20.69
12	1.04	8.94	19.70
18	1.19	9.06	18.17

[a]Calculated for point source in center plane. Only single scattering or absorption interactions in the silicon detector are included. Back-scattered photons from BGO and events without full energy deposition are excluded.

The on-axis geometric resolution for a single–single event with 300-μm pixels is 150 μm FWHM without accounting for positron range, whereas the single–BGO events and BGO–BGO have a corresponding geometric resolution of 400 μm FWHM and 1500 μm FWHM, respectively, with 3-mm BGO elements if one knows which BGO element was struck first. Point source reconstructions by filtered back-projection gave 190 μm, 610 μm, and 1850 μm FWHM for cases when the BGO crystal with the first interaction was assumed to be the one with maximum deposited energy. Figure 35 shows three filtered back-projection images corresponding to the three interaction sequences described. It is important to note that each event defines an actual coincidence chord and there is no conical ambiguity. The image pixels are 156 μm, and a Hamming filter with cutoff of 0.5 pixels was used. The location of the BGO interaction is chosen as the crystal with maximum energy deposition and no correction is made for depth of interaction.

The 1-mm-diameter disk is clearly seen in only the single–single image. There is a hint of it in the Si–BGO image but it would be difficult to distinguish it from noise. Of the 100,000 events, only 450 counts come from this disk. In order to properly combine the data from the three types of interactions, a maximum likelihood algorithm will be required to appropriately weight the different data. It appears that this approach is capable of excellent spatial resolution with very satisfactory sensitivity. Actual implementation, however, will pose a number of challenges in hardware design and real-time data processing in order to identify the correct event sequences.

D. Coincidence SPECT

There are a number of radioisotopes that emit two or more cascade gamma rays per decay. These photons are, in effect, emitted simultaneously and depending on the level structure may be correlated in angle. For instance ^{111}In, with a 2.8-day half-life, emits two photons, 171 keV (90%) and 245 keV (94%); and ^{130}I, with a 12.4-hr. half-life emits three photons, 731.5 keV (90%), 668.5 keV (90%), and 536.1 keV (90%).

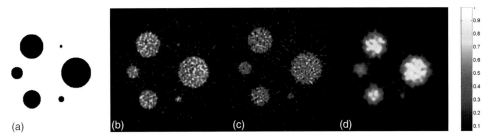

FIGURE 35 Filtered back-projection images of a phantom that consists of six disc sources: 1, 2, 4, 6, 8, and 10 mm in diameter. (a) object. (b) single–single interactions. (c) single–BGO interactions (d) BGO–BGO interactions. Each image contains 100,000 events. (From Park et al., 2001, © 2001 IEEE.)

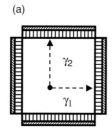 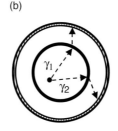 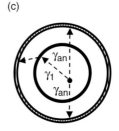

FIGURE 36 Three possibilities for imaging radionuclides that simultaneously emit multiple photons. (a) Collimated cameras imaging a pair of cascade gamma rays in coincidence. (b) A double ring Compton camera imaging a pair of cascade gamma rays. (c) A double ring Compton camera imaging a nuclide that emits a positron accompanied by one or more gamma rays in addition to the annihilation photons.

Several authors have pointed out that coincidence imaging of these photons could offer some advantages for nuclear medical imaging (Hart and Rudin, 1977; Boetticher et al., 1980; Liang and Jaszczak, 1990). In addition to cascade gamma rays, Liang and Jaszczak (1990) discuss the case for positron emitters that also emit one or more prompt gamma rays. They list nine such isotopes including ^{52}Fe, which has an 8.27-hour half-life. Figure 36 illustrates three possibilities for imaging these radionuclides.

In Figure 36a, if γ_1 and γ_2 are detected in coincidence, the source point can be located in three-dimensions without the need for image reconstruction. The problem is that the efficiency of these collimators is on the order of 10^{-4}, so the coincidence efficiency will be 10^{-8} if there is no angular correlation between the photons. This sensitivity is far too low, even though the image reconstruction step with its attendant noise amplification is eliminated.

In Figure 36b, the collimator is replaced by a double-ring-geometry Compton camera, such as the one illustrated in Figure 16. In this case, a fourfold coincidence restricts the source location to one or two lines in space as defined by the intersection of two cones and the known field of view. Because for single scatter events the Compton camera pictured in Figure 16 can have 300–400 times the sensitivity of a collimator, it might be expected to have a coincidence efficiency approaching 10^{-3}. It would now be necessary to compute the decoding penalty corresponding to the intersection of two cones and compare the image variance to that of an image reconstructed from standard projection data in order to determine how much real sensitivity gain is possible for this scheme.

Finally, Figure 36c illustrates the case for positron emission accompanied by a prompt gamma ray. In this example, the annihilation photons are detected in the second detector, but all the combinations and permutations described for the high-resolution animal PET application are possible. Nevertheless, in Figure 3c the source location is now constrained to lie on the intersection of a cone and line. This reduces to at most two points except for the unlikely case in which the line of the annihilation pair lies along the surface of the cone. This reduction of source location from a line to one or two points in space represents a large effective gain in SNR. As pointed out in Liang and Jaszczak (1990), it will substantially outperform time-of-flight PET. Although this illustration appears simple, in practice it will be necessary to consider all the possible interaction sequences and be able to uniquely identify which interactions correspond to the annihilation photons and which correspond to the prompt gamma ray. Again, it is necessary to compute the probabilities of the various interaction sequences and determine the magnitude of the effective sensitivity gain. Although some of these radionuclides might well offer improved imaging, a major drawback to their use is the need to develop new radiotracers. Indeed, the image improvements would need to be compelling to spark such a development effort.

E. Imaging of High Energy Radiotracers

More than 90% of clinical planar and SPECT imaging studies use ^{99m}Tc. It is inexpensive, is readily available from ^{99}Mo generators, is easily collimated, and may be imaged by Anger cameras. Furthermore, there is more than a quarter century of radioligand development aimed at addressing a number of specific clinical questions. On the other hand, we have seen that Compton imaging improves rapidly with energies above 140 keV. Furthermore, as energy increases from 140 keV to 360 keV and 511 keV, attenuation and scatter in the patient are reduced and the unscattered flux for 10 cm of water increases by 50% and 70%, respectively. It is thus reasonable to at least consider the applications for high-energy imaging and the possibility of developing new tracers based on high-energy gamma emitters.

Currently, the high-energy single-photon tracers are pretty much limited to the first three entries in Table 4. These isotopes are used both because they have unique chemical properties and also because their longer half-lives are necessary for labeling entities such as antibodies and liposomes that may require days to achieve high specific localization. ^{131}I is used for radiotherapy, and iodine has a number of properties that make it desirable as a label. It would be extremely valuable to be able to image it at high resolution and good sensitivity at diagnostic doses to evaluate tumor

TABLE 4 High Energy Radionuclides Used for Single Photon Imaging

Isotope	Half-Life	Percentage	Energy (keV)
^{131}I	8.04 days	81	364
		7	636
^{111}In	2.83 days	90	171
		94	245
^{67}Ga	3.26 days	35	93
		20	184
		16	300
^{113m}In	1.66 h	65	392

TABLE 5 Isotopes of Interest in the Study of Human Metabolism^a

Element	Half-Life	Gamma Energy (MeV)	Beta Energy (MeV)
^{22}Na	2.6 years	1.27 (99.9)	0.2 (90)
^{24}Na	15 h	2.75 (99.9)	0.55 (99.9)
^{42}K	12.36 h	1.5 (18)	1.56 (82)
^{28}Mg	20.9 h	1.34 (53)	0.16 (94)
^{59}Fe	44.5 days	1.29 (43)	0.15 (53)
^{58}Co	70.9 days	0.81 (99.5)	0.20 (15)
^{65}Zn	244 days	1.1 (51)	0.14 (1.4)
^{47}Ca	4.5 days	1.30 (74)	0.24 (81)

^{a}Values in parentheses are percentage abundance.

uptake and number and location of tumors, and after therapeutic treatment to evaluate tumor response. ^{113m}In was used for nuclear medical imaging before the widespread availability of ^{99m}Tc, and, as mentioned earlier, it is generator-produced from ^{113}Sn, which has a 119-day half-life. In parts of the world where ^{99m}Tc is not readily available, ^{113m}In is still used because the long parent half-life limits the need for generator replacement to approximately two per year. Because both of the indium isotopes can be well imaged with Compton cameras, the possibility exists for developing radiopharmaceuticals for applications requiring either long or short half-lives based on indium. Although the thought of developing a number of new radiolabled compounds using something other than technetium is certainly daunting, there is active research on using star-burst dendrimers as chelating agents that might enable the coupling of various radioactive metals to tumor-targeting agents (Balogh et al., 2000; Roberts et al., 1990; Kobayashi et al., 1999).

Other applications in which efficient imaging of high energy gamma-emitters would be highly desirable have been raised in discussions with A. B. Brill. He writes:

> The imaging of normal physiology remains an important goal, despite the current emphasis nuclear medicine places on the early detection of disease when abnormalities can best be treated successfully. Gamma emitting tracers such as ^{99m}Tc, ^{111}In and ^{123}I are the most important of the gamma emitters in current use. The imaging of beta emitters, including 511 keV positron emitters is limited by the higher dose they deposit in the body, and the use of electronic collimation and short-half life tracers makes PET imaging of ^{11}C, ^{15}O, ^{13}N, and ^{18}F feasible. Nature has not been kind to us in that many of the naturally occurring elements of interest in the body are long-lived beta/gamma emitters, examples of which are tabulated below [see Table 5]. Much of the effort in the past has been to determine body composition rather than biodistribution for many of these essential elements. However, there is good reason to know the distribution of functional bone marrow for example, and ^{59}Fe (or ^{52}Fe if it were available) would be very useful in determining regional distribution, rather than sub vertebral distribution. The same could be said for ^{28}Mg distribution in the heart. Changes in ^{24}Na distribution associated with anti hypertensive therapy would also be potentially useful. All of these questions could only be studied in man if there were a device without a collimator, such as the Compton Coincidence measurement system, which could image the small amounts of high energy gamma emitters that could be administered.[9]

VII. DISCUSSION AND SUMMARY

The history of new imaging developments in nuclear medicine demonstrates that there is a long road to general acceptance. SPECT, which has its roots in the early 1960s, required more than a quarter century to attain widespread acceptance, and the development of PET has followed a similar time course. Compton cameras make it possible to image a whole new range of radiotracers with higher resolution and higher sensitivity than collimated cameras. In addition they offer a method for improving the spatial resolution of PET close to the limits fixed by positron range. Further, the method permits investigating methods of direct 3D imaging of radionuclides without an image reconstruction step.

Compton cameras place a high demand on both detector technology and sophisticated methods of data acquisition that permit online sorting and processing of multiple coincidence events. Both of these areas have benefited enormously from

---

[9]Personal communication from A.B. Brill, M.D., Ph.D., Research Professor Radiology/Physics/BME, Vanderbilt University Medical School, Radiology Dept. MCN R-1302, December 2000.

high-energy physics research and gamma-ray astronomy, and this cross-fertilization will continue to assist the development of Compton imaging devices.

The emphasis in this chapter has been on PIN silicon pad detectors, but position-sensitive silicon drift detectors offer the potential for large-area, low-noise, position-sensitive detectors and have already been suggested for Compton imaging (Kuykens and Audet, 1988). Sizes up to 5 cm × 5 cm have been fabricated on a single wafer, and spatial resolution can be as good as 5 µm (Hall, 1988). At liquid-nitrogen temperatures, noise levels as low as 40 electrons have been demonstrated. Excellent energy resolution has also been obtained by Pullia and co-workers (1988) at Brookhaven for silicon pixel detectors. By bonding individual FET preamplifiers to each pad, they were able to reduce capacitance and obtain energy resolution of 380 eV at room temperature. These developments have the greatest importance for systems intended for use with technetium. Whether devices such as these can be commercially produced is not known, but they would certainly play an important role in establishing Compton camera feasibility. At higher gamma-ray energies, alternative selections such as neon gas detectors or CZT become worthy of consideration.

Projected cost is certainly an important factor, and it is difficult to predict. If one looks at PET, approximately 15 years ago the cost of a PET tomograph was approximately $4000 a channel. A couple of years ago one could purchase a block detector for approximately $6500, or approximately $100 a channel. Some educated guesses for large-scale production of silicon pad detectors place the cost at $1.50 per channel. Because of the low density of silicon and the limited thickness of undoped wafers, the total cost of a silicon ring, constructed of PIN silicon pad detectors would probably exceed $1 million. On the other hand, lithium-drifted silicon detectors, which can be much thicker, will reduce the number of channels and might offer a lower-cost alternative. At higher energies, double-sided silicon strip detectors or even neon gas detectors can be considered to reduce cost. It does not seem reasonable to consider cost as a primary stumbling block given the early stage of Compton camera investigations.

It seems fair to conclude that Compton imaging should be viewed as an enabling technology that will permit the investigation of a wide range of new tracers and imaging techniques for application to nuclear medicine. Cost does not appear prohibitive, and the various engineering challenges are stimulating but tractable.

Acknowledgments

The authors acknowledge the support by U.S. DHHS under NIH grants R01 CA32846 and CA54362 for much of the work described herein. In addition, computing services were provided by the University of Michigan Center for Parallel Computing, which is partially funded by NSF grant CDA-92-14296. We especially wish to credit Jim LeBlanc and Chia-ho Hua for our extensive use of their research results and Scott Wilderman for the numerous Monte Carlo simulations and image reconstructions. Finally we note that work at the University of Michigan related to Compton cameras involves a-long-standing collaboration among the Division of Nuclear Medicine (Les Rogers, Neal Clinthorne, and Dirk Meier), Nuclear Engineering (Glenn Knoll, David Wehe, and Scott Wilderman), EECS (Alfred Hero III and Jeffrey Fessler), CERN (Peter Weilhammer), LEPSI (Wojtec Dulinski), and Integrated Detector Electronics in Oslo, Norway (Einar Nygård), with support of the Norwegian Research Council.

References

Alvarez, L. W., Bauber, P. M., Smith, L. H., *et al.* (1973). The liquid xenon Compton telescope: A new technique for gamma-astronomy. Preprint Space Science Laboratory, Ser. 14, Issue 17, University of California, Berkeley.

Aprile, E., *et al.* (1998). XENA—a liquid xenon Compton telescope for gamma-ray astrophysics in the MeV regime. *Proc. SPIE* **3446**: 88.

Balogh, L., Tomalia, D. A., and Hagnauer, G. L. (2000). A revolution of nanoscale proportions. *Chem. Inn.* Vol. 30, No. 3. March 19–26.

Barrett, H. H., Abbey, C. K., and Clarkson, E. (1998). Objective assessment of image quality III: ROC metrics, ideal observers, and likelihood generating functions. *J. Opt. Soc. Am. A* **15**(6): 1520–1535.

Barrett, H. H., White, T., and Parra L. C. (1997). List-mode likelihood. *J. Opt. Soc. Am. A* **14**(11): 2914–2923.

Basko, R., Zeng, G. S. L., and Gullberg, G. T. (1998). Application of spherical harmonics to image reconstruction for the compton camera. *Phys. Med. Bio.* **43**(4): 887–894.

Bhattacharya, D., Dixon, D. D., Kong, V., O'Neill, T. J., White, R. S., Zych, A. D., Ryan, J., McConnell, M., Macri, J., Samimi, J., Akyüz, A., Mahoney, W., and Varnell, L. (1999). A Tracking and Imaging Gamma-Ray Telescope (TIGRE) for energies of 0.3–100 MeV. *In* "Proceedings of the 26th International Cosmic Ray Conference" (D. Kieda, M. Salamon, and B. Dingus, eds.), Vol. 5, pp. 72–75. American Institute of Physics, Melville, NY.

Blahut, R. (1987). "Principles and Practice of Information Theory." Addison-Wesley, Reading, MA.

Boetticher, H., Helmers, H., *et al.* (1980). Gamma-camera coincidence scintigraphy: Tomography without computerized image reconstruction. *In* "Medical Radionuclide Imaging," Vol. 1, pp. 147–148. IAEA, Vienna.

Bolozdynya, A., Egorov, V., Koutchenkov, A., Safronov, G., Smirnov, G., Medved, S., and Morgunov, V. (1997a). High pressure xenon electronically collimated camera for low energy gamma ray imaging. *IEEE Trans. Nucl. Sci.* **44**(6): 2408–2413.

Bolozdynya, A., Egorov, V., Koutchenkov, A., Safronov, G., Smirnov, G., Medved, S., and Morgunov, V. (1997b). High pressure xenon self-triggered scintillation drift chamber with 3-D sensitivity in the range of 20–140 keV deposited energy. *Nucl. Instru. Meth. Phys. Res. A* **385**: 225–238.

Bolozdynya, A. I., and Morgunov, V. L. (1988). Multilayer electroluminescence camera: Concept and Monte Carlo study. *IEEE Trans. Nucl. Sci.* **45**(3): 1646–1655.

Bolozdynya, A., Ordonez, C. E., and Chang, W. (1997). A concept of cylindrical Compton camera for SPECT. *1997 IEEE Nucl. Sci. Sym. Conf. Rec.* **2**: 1047–1051.

Brechner, R., and Singh, M. (1988). Comparison of an electronically collimated system and a mechanical cone-beam system for imaging single photons. *IEEE Trans. Nucl. Sci.* **35**(1): 649–653.

Brechner, R. R., and Singh, M. (1990). Iterative reconstruction of electronically collimated spect images. *IEEE Trans. Nucl. Sci.* **37**(3): 1328–1332.

Chelikani, S., Gore, J. C., Zubal, I. G. (2004). Optimizing Compton camera geometries. *Phys. Med. Biol.* **49**(8): 1387–1408.

Chen, M., and Bolozdynya, A. (1995). Radiation detection and tomography. U.S. Patent 5,665,971, August 8.

Clinthorne, N. H. (2001). A method for improving the spatial resolution of a Compton camera. U.S. Patent 6,323,492, Nov. 27.

Clinthorne, N. H., Hua, C.-H, LeBlanc, J. W., Wilderman, S. J., and Rogers, W. L. (1998). Choice of scattering detector for Compton scatter cameras. *J. Nucl. Med. Supp.* **5**: 51P.

Clinthorne, N. H., Meier, D., Hua, C., Weilhammer, P., Nygård, E., Wilderman, S. J., and Rogers, W. L. (2000). Very high resolution animal PET (abs). *J. Nucl. Med.* **41**(5) (Suppl S): 20P.

Clinthorne, N. H., Ng, C.-Y., Hua, C.-H., Gormley, J. G., LeBlanc, J. W., Wehe, D. K., Wilderman, S. J., and Rogers, W. L. (1996). Theoretical performance comparison of a Compton scatter aperture and a parallel-hole collimator. *1996 IEEE Nucl. Sci. Sym. and Med. Imag. Conf. Rec.* **2**: 788–792.

Clinthorne, N. H., Ng, C.-Y., Strobel, J., Hua, C.-H., LeBlanc, J. W., Wilderman, S. J., and Rogers, W. L. (1998). Determining detector requirements for medical imaging applications. *Nucl. Instru. Meth. Phys. Res. A* **409**: 501–507.

Clinthorne, N. H., Wilderman, S. J., Hua, C., LeBlanc, J. W., and Rogers, W. L. (1998). Choice of scattering detector for Compton scatter cameras. (abs). *J. Nucl. Med.* **39**: 51p.

Colombetti, L. G., Goodwin, D. A., and Hermanson, R. (1969). ^{113m}In labeled compound for liver and spleen studies. *J. Nucl. Med.* **10**(9): 597–602.

Cree M. J., and Bones, P. J. (1994). Towards direct reconstruction from a gamma camera based on compton scattering. *IEEE Trans. Med. Imag.* **13**(2): 398–407.

Demoment, G. (1989). Image reconstruction and restoration: overview of common estimation structures and problems. *IEEE Trans. Acous. Sp. Sig. Proc.* **37**(12): 2024–2036.

Dogan, N. (1993). Multiple Compton scatter camera for gamma-ray imaging. Ph.D. dissertation, University of Michigan.

Dogan, N., Wehe, D. K., and Akcasu, A. K. (1992). A source reconstruction method for multiple scatter Compton cameras. *IEEE Trans. Nuc. Sci.* **39**(5): 1427–1430.

Dogan, N., Wehe, D. K., and Knoll, G. F. (1990). Multiple Compton scattering gamma ray imaging camera. *Nucl. Instru. Meth. Phys. Res. A* **299**: 501–506.

Doria, D., and Singh, M. (1982). Comparison of reconstruction algorithms for an electronically collimated gamma camera. *IEEE Trans. Nucl. Sci.* **29**(1): 447–451.

Du, Y., He, Z., Knoll, G. F., Wehe, D. K., and Li, W. (2001). Evaluation of a Compton scattering camera using 3-D Position-sensitive CdZnTe detectors. *Nucl. Instru. Meth. Phys. Res. A.* **457**: 203–211.

Durkee, J. W., Antich, P. P., Tsyganov, E. N., Constantinescu, A., Fernando, J. L., Kulkarni, P. V., Smith, B. J., Arbique, G. M., Lewis, M. A., Nguyen, T., Raheja, A., Thambi, G., and Parkey, R. W. (1998). SPECT electronic collimation resolution enhancement using chi square minimization. *Phys. Med. Biol.* **43**: 2949–2974.

Durkee, J. W., Antich, P. P., Tsyganov, E. N., Constantinescu, A., Kulkarni, P. V., Smith, B., Arbique, G. M., Lewis, M. A., Nguyen, T., Raheja, A., Thambi, G., and Parkey, R. W. (1998). Analytic treatment of resolution precision in electronically collimated SPECT imaging involving multiple-interaction gammas. *Phys. Med. Biol.* **43**: 2975–2990.

Everett, D. B., Flemming, J. S., Todd, R. W., and Nightingale, J. M. (1976). A gamma-ray camera based on Compton scattering. In "Proceedings 7th L. H. Gray Conference," p. 89. John Wiley, New York.

Everett, D. B., Todd, R. W., and Nightingale, J. M. (1977). Gamma-radiation imaging system based on the Compton effect. *Proc. IEEE* **124**(11): 995–1000.

Fessler, J. A. Clinthorne, N. H., and Rogers, W. L. (1993). On complete data spaces for pet reconstruction algorithms. *IEEE Trans. Nuc. Sci.* **40**(4): 1055–1061.

Fessler, J. A. and Rogers, W. L. (1996). Spatial resolution properties of penalized likelihood image reconstruction methods: Space-invariant tomographs. *IEEE Trans. Imag. Proc.* **5**(9): 1346–1358.

Fisher, R. A. (1925). "Statistical Methods for Research Workers." Oliver & Boyd, Edinburgh.

Fujida, I., and Perez-Mendez, V. (1987). Theoretical considerations of a new electronically collimated gamma camera utilising gas scintillation. *Proc. SPIE* **767**: 84–89.

Fukunaga, K. (1972). "Introduction to Statistical Pattern Recognition." Academic Press, New York.

Gormley, J. E. (1997). Experimental comparison of mechanical and electronic gamma-ray collimation. Ph.D. dissertation, University of Michigan.

Gormley, J. E., Rogers, W. L., Clinthorne, N. H., Wehe, D. K., and Knoll, G. F. (1997). Experimental comparison of mechanical and electronic gamma-ray collimation. *Nucl. Instru. Meth. Phys. Res. A* **397**: 440–447.

Hall, G. (1988). Silicon drift chambers. *Nucl. Instru. Meth. Phys. Res. A* **273**: 559–564.

Hart, H., and Rudin, S. (1977). Three dimensional imaging of multi-millimeter sized cold lesions by focusing collimator coincidence scanning (FCCS). *IEEE Trans. Biomed. Eng.* **24**: 169–177.

Hebert, T., Leahy, R., and Singh, M. (1987). Maximum likelihood reconstruction for a prototype electronically collimated single photon emission system. *Proc. SPIE* **767**: 77–83.

Hebert, T., Leahy, R., and Singh, M. (1988). Fast MLE for SPECT using an intermediate polar representation and a stopping criterion. *IEEE Trans. Nucl. Sci.* **35**(1): 615–619.

Hebert, T., Leahy, R., and Singh, M. (1990). Three-dimensional maximum likelihood reconstruction for an electronically collimated single-photon-emission imaging system. *J. Opt. Soc. Am. A* **7**(7): 1305–1313.

Hero, A. O. (1992). A Cramèr-Rao type lower bound for essentially unbiased parameter estimation. Technical Report 890, M.I.T. Lincoln Laboratory, Lexington, MA, DYIC AD-A246666.

Hero, A. O., Fessler, J. A., and Usman, M. (1996). Exploring estimator bias-variance tradeoffs using the uniform cr bound. *IEEE Trans. Sig. Proc.* **44**(8): 2026–2041.

Hua, C.-H. (2000). Compton imaging system development and performance assessment. Ph.D. dissertation, University of Michigan.

Hua, C.-H., Clinthorne, N. H., Wilderman, S. J., LeBlanc, J. W., and Rogers, W. L. (1999). Quantitative evaluation of information loss for Compton cameras. *IEEE Trans. Nucl. Sci.* **46**(3): 587–593.

Ichihara, T. (1987). Ring type single photon emission CT imaging device. U.S. Patent 4,639,599, Jan. 27.

Kamae, T., Enomoto, R., and Hanada, N. (1987). A new method to measure energy, direction, and polarization of gamma-rays. *Nucl. Instru. Meth. Phys. Res. A* **260**: 254–257.

Kamae, T., and Hanada, N. (1988). Prototype design of multiple Compton camera. *IEEE Trans. Nucl. Sci.* **35**(1): 352–355.

Keener, J. P. (1988). "Principles of Applied Mathematics: Transformation and Approximation." Addison-Wesley, New York.

Klein, O., and Nishina, Y. (1929). Über de Streung von Strahlung durch freie Elektronen nach derneuen relativistischen Quantendynamik von Dirac. *Z. Physik.* **52**: 853.

Knoll, G. F. (2000). "Radiation Detection and Measurement," 3rd ed., chap. 11. John Wiley & Sons, New York.

Kobayashi, H., Wu, C., Kim, M.-K., Paik, C. H., Carrasquillo, J. A., and Brechbiel, M. W. (1999). Evaluation of the *in vivo* biodistribution of indium-111 and yttrium-88 labeled dendrimer-1B4M-DTPA and its conjugation with Anti-Tac monoclonal antibody. *Bioconjugate Chem.* **10**: 103–111.

Kroeger, R. A., Johnson, W. N., Kurfess, J. D., Phlips, B. F, Luke, P. N., Momayezi, M., and Warburton, W. K. (1999). Position Sensitive Germanium Detectors for the Advanced Compton Telescope. In "AIP Conference Proceedings of the 5th Compton Symposium," (McConnell, M. L., and Ryan, J. M., eds.), Vol. 510, pp. 794–798. American Institute of Physics, Melville, NY.

Kroeger, R. A., Johnson, W. N., Kurfess, J. D., Phlips, B. F., Wulf, E. A. (2000). Gamma ray energy measurement using the multiple Compton technique. *2000 IEEE Nucl. Sci. Sym. Med. Imag. Conf. Rec.* **1**: 8-7–8-11.

Kuykens, H. J. P., and Audet, S. A. (1988). A 3×3 silicon drift chamber array for application in an electronic collimator. *Nucl. Instru. Meth. Phys. Res. A* **273**: 570–574.

LeBlanc, J. W. (1999). A Compton camera for low energy gamma-ray imaging in nuclear medicine applications. Ph.D. dissertation, University of Michigan.

LeBlanc, J. W., Bai, X., Clinthorne, N. H., Hua, C., Meier, D., Rogers, W. L., Wehe, D. K., Weilhammer, P., and Wilderman, S. J. (1999). 99mTc imaging performance of the C-SPRINT Compton camera. *1999 IEEE Nucl. Sci. Sym. Med. Imag. Conf. Rec.* **1**: 545–552.

LeBlanc, J. W., Clinthorne, N. H., Hua, C.-H., Nygård, E., Rogers, W. L. (1998). C-SPRINT: A prototype Compton camera system for low energy gamma-ray imaging. *IEEE Trans. Nucl. Sci.* **45**(3): 943–949.

LeBlanc, J. W., Clinthorne, N. H., Hua, C., Rogers, W. L., Wehe, D. K., and Wilderman, S. J. (1999). A Compton camera for nuclear medicine applications using ^{113m}In. *Nucl. Instru. Meth. Phys. Res. A* **422**: 735–739.

Lecomte, R., Schmitt, D., and Lamoureax, G. (1984). Geometry study of a high resolution PET detection system using small detectors. *IEEE Trans. Nucl. Sci.* **31**(1): 556–561.

Liang, Z., and Jaszczak, R. (1990). Comparisons of multiple photon imaging techniques. *IEEE Trans. Nucl. Sci.* **37**(3): 1282–1292.

Martin, J. B. (1994). A compton scatter camera for SPECT imaging of 0.5 to 3.0 MeV gamma-rays. Ph.D. dissertation, University of Michigan.

Martin, J. B., Dogan, N., Gormley, J. E., Knoll, G. F., O'Donnell, M., and Wehe, D. K. (1994). Imaging multi-energy gamma-ray fields with a Compton scatter camera. *IEEE Trans. Nucl. Sci.* **41**(4): 1019–1025.

Martin, J. B., Knoll, G. F., Wehe, D. K., Dogan, N., Jordanov, V., Petrick, N., and Singh, M. (1993). A ring Compton scatter camera for imaging medium energy gamma-rays. *IEEE Trans. Nucl. Sci.* **40**(4): 972–978.

McMaster, W. H. (1961). Matrix representation of polarization. *Rev. Mod. Phys.* **33**(1): 8–28.

Meier D., Czermak, A., Jalocha, P., Sowicki, B., Kowal, M., Dulinski, W., Mæhlum, G., Nygård, E., Yoshioka, K., Fuster, J., Lacasta, C., Mikuz, M., Roe, S., Weilhammer, P., Hua, C.-H., Park, S.-J., Wilderman, S. J., Zhang, L., Clinthorne, N. H., and Rogers, W. L. (2000). Silicon detector for a Compton camera in nuclear medical imaging. *2000 IEEE Nucl. Sci. Sym. Med. Imag. Conf. Rec.*

Namito, Y, Ban, S., and Hirayama, H. (1994). Implementation of the Doppler Broadening of a Compton-scattered photon into the EGS4 code. *Nucl. Instru. Meth. Phys. Res. A* **349**: 489–494.

Namito, Y., Ban, S., and Hirayama, H. (1995). LSCAT: Low-energy photon scattering expansion for the EGS4 code. KEK International 95-10, August, Ibaraki-ken, Japan.

Namito, Y., Ban, S., and Hirayama, H. (2000). LSCAT: Low-energy photon-scattering expansion for the EGS4 code (inclusion of electron impact ionization). KEK International 2000-4, May, Ibaraki-ken, Japan.

Natterer, F. (1986). "The Mathematics of Computerized Tomography." Wiley, New York.

Naylor, A. W., and Sell, G. R. (1982). "Linear Operator Theory in Engineering and Science, volume 40 of Applied Mathematical Sciences." Springer-Verlag, New York.

O'Mara, R. E., Subramanian, McAfee, J. G., and Burgher, C. L. (1969). Comparison of ^{113m}In and other short-lived agents for cerebral scanning. *J. Nucl. Med.* **10**(1): 18–27.

O'Neill, T. J., Akyuz, A., Bhattacharya, D., Case, G. L., Dixon, D., Samimi, J., Tumer, O. T., White, R. S., and Zych, A. D. (1996). Tracking, imaging and polarimeter properties of the TIGRE instrument. *Astron. Astrophys. Supp. Series* **120**(4): C661–4.

Ordonez, C. E., Bolozdynya, A., and Chang, W. (1997). Doppler broadening of energy spectra in Compton scatter cameras. *1997 IEEE Nucl. Sci. Sym. Med. Imag. Conf. Rec.* **2**: 1361–1365.

Ordonez, C. E., Chang, W., and Bolozdynya, A. (1999). Angular uncertainties due to geometry and spatial resolution in Compton cameras. *IEEE Trans. Nucl. Sci.* **46**(4): 1142–1147.

Park, S., Han, L., Wilderman, S. J., Sukovic, P., Czermak, A., Jalocha, B., Kowal, M., Dulinski, W., Maehlum, G., Nygård, E., Yoshioka, K., Fuster, J., Lacasta, C., Mikuz, M., Roe, S., Weilhammer, P., Meier, D., Rogers, W. L., and Clinthorne, N. H. (2001). Experimental setup for a very high resolution animal PET based on solidstate detector. *2001 IEEE Nucl. Sci. Sym. Med. Imag. Conf. Rec.*

Park, S. J., Rogers, W. L., and Clinthorne, N. H. (2004). Design of a very high resolution small animal PET based on the Compton PET cencept. *IEEE Trans. Med. Imaging* (Submitted).

Park, S., Rogers, W. L., Wilderman, S. J., Sukovic, P., Han, L., Meier, D., Nygård, E., Weilhammer, P., and Clinthorne, N. H. (2001). Design of a very high resolution animal PET (abs). *J. Nucl. Med. Supp.* **42**: 102.

Parra, L. C. (2000). Reconstruction of cone-beam projections from compton scattered data. *IEEE Trans. Nucl. Sci.* **47**(4): 1543–1550.

Pehl, R. H., Madden, N. W., Landis, D. A., Malone, D. F., Cork, C. P. (1985). Cryostat and electronic development associated with multi-detector spectrometer system. *IEEE Trans. Nucl. Sci.* **32**(1): 22–28.

Pinkau, K. (1960). Die messung solarer und atmospharischer neutronen. *Z. Naturforschung A* **21a**(12): 2100–.

Pullia A., Kraner, H. W., Siddons, D. P., Furenlid, L. R., and Bertuccio, G. (1995). Silicon detector system for high rate EXAFS applications. *IEEE Trans. Nucl. Sci.* **42**: 585–589.

Qi, J., and Huesman, R. H. (2001). Theoretical study of lesion detectability of MAP reconstruction using computer observers. *IEEE Trans. Med. Imag.* **20**(8): 815–822.

Rao, C. R. and Toutenburg, H. (1999). "Linear Models: Least Squares and Alternatives" Second Edition. Springer Series in Statistics. Springer, New York.

Roberts, J. C., Adams, Y. E., Tomalia, D., Mercer-Smith, J. A., and Lavalee, D. K. (1990). Using starburst dendrimers as linker molecules to radiolabel antibodies. *Bioconjugate Chem.* **2**: 305–308.

Rogers, W. L., and Ackermann, R. J. (1992). SPECT instrumentation. *Am. J. Physiol. Imag.* **7**(3/4): 105–120.

Rogers W. L., Clinthorne N. H., Stamos J., Koral K. F., Mayans R., Keyes J. W., Jr Williams, J. J. (1982). SPRINT: A stationary detector single photon ring tomograph for brain imaging. *IEEE Trans. Med. Imag.* **MI-1**: 63–68.

Rohe, R. C., Sharfi, M. M., Kecevar, K. A., Valentine, J. D., and Bonnerave, C. (1997). Spatially-variant backprojection point kernel function of an energy-subtraction compton scatter camera for medical imaging. *IEEE Trans. Nucl. Sci.* **44**(6): 2477–2482.

Rohe, R., and Valentine, J. D. (1995). A novel Compton scatter camera design for *in-vivo* medical imaging of radiopharmaceuticals. *1995 IEEE Nucl. Sci. Sym. Med. Imag. Conf. Rec.* **3**: 1579–1583.

Rohe, R. C., and Valentine, J. D. (1996). Compton camera for *in vivo* medical imaging of radiopharmaceuticals. U.S. Patent 5,567,944, Oct. 22.

Sansen, W. M., and Chang, Z. Y. (1990). Limits of low electron noise performance of detector readout front ends in CMOS technology. *IEEE Trans. Circuits Sys.* **37**(11): 1375–1382.

Sauve, A. C., Hero, A. O., Rogers, W. L., Wilderman, S. J., and Clinthorne. N. H. (1999). 3d image reconstruction for a compton spect camera model. *IEEE Trans. Nucl. Sci.* **46**(6): 2075–2084.

Shepp, L. A. and Vardi, Y. (1982). Maximum likelihood reconstruction for emission tomography. *IEEE Trans. Med. Imag.* **1**(2): 113–122.

Schönfelder, V., Diehl, R., Lichti, G. G., Steinle, H., Swanenburg, B. N., Deerenberg, A. J. M., Aarts, H., Lockwood, J., Webber, W., Macri, J., Ryan, J., Simpson, G., Taylor, B. G., Bennett, K., and Snelling, M. (1984). The imaging COMPTON telescope, COMPTEL on the gamma ray observatory. *IEEE Trans. Nucl. Sci.* **31**(1): 766–770.

Schönfelder, V., Hirner, A., and Schneider, K. (1973). A telescope for soft gamma ray astronomy. *Nucl. Instru. Meth. Phys. Res.* **107**: 385.

Schopper, F., Andritschke, R., Kanbach, G., Kemmer, J., Lampert, M.-O., Lechner, P., Richter, R., Rohr, P., Schönfelder, V., Strüder, L., and

Zoglauer, A. (2000). Development of silicon strip detectors for a medium energy gamma-ray telescope. *IEEE Nucl. Sci. Symp. Med. Imag. Conf. Rec.* **1**: 3-122–3-126.

Schopper F., Andritschke R., Shaw H., Nefzger C., Zoglauer A., and Schonfelder V. (2000). CsI calorimeter with 3-D position resolution. *Nucl. Instru. Meth. Phys. Res. A* **442**: 394–399.

Singh, M., and Doria, D. (1983). An electronically collimated gamma camera for single photon emission computed tomography. part II: Image reconstruction and preliminary experimental measurements. *Medical Physics*, **10**(4): 428–35.

Singh, M. (1983). An electronically collimated gamma camera for single photon computed tomography. Part I: Theoretical considerations and design criteria. *Med. Phys.* **10**(4): 421–427.

Singh, M., Brechner, R. R., and Horne, C. (1986). Single photon imaging at high energies with electronic collimation. *Proc. SPIE* **671**: 184–188.

Singh, M., and Doria, D. (1981). Computer simulation of image reconstruction with a new electronically collimated gamma tomography system. *Proc. SPIE* **273**: 192–200.

Singh, M., and Doria, D. (1983). An electronically collimated gamma camera for single photon emission computed tomography. Part II: Image reconstruction and preliminary experimental measurements. *Med. Phys.* **10**(4): 428–435.

Singh, M., and Doria, D. (1985). Single photon imaging with electronic collimation. *IEEE Trans. Nucl. Sci.* **32**(1): 843–847.

Singh, M., Leahy, R., Brechner, R., and Hebert, T. (1988). Noise propagation in electronically collimated single photon imaging. *IEEE Trans. Nucl. Sci.* **35**(1): 772–777.

Solomon, C. J., and Ott, R. J. (1988). Gamma ray imaging with silicon detectors—a Compton camera for radionuclide imaging in medicine. *Nucl. Instru. Meth. Phys. Res. A* **273**: 787–792.

Stayman, J. W., and Fessler, J. A. (2002). Compensation for nonuniform resolution using penalized-likelihood reconstruction in space-variant imaging systems. *IEEE Trans. Med. Imag.* Submitted.

Todd, R. W. (1975). Methods and apparatus for determining the spatial distribution of a radioactive material. U.S. Patent 3,876,882, April.

Todd, R. W., Nightingale, J. M., and Everett, D. B. (1974). A proposed γ-camera. *Nature* **25**: 132.

Valentine, J. D., Bonnerave, C., and Rohe, R. C. (1996). Energy-subtraction Compton scatter camera design considerations: A Monte Carlo study of timing and energy resolution effects. *1996 IEEE Nucl. Sci. Sym. Conf. Rec.* **2**: 1039–1043.

Weilhammer, P., Nygård, E., Dulinski, W., Czermak, A., Djama, F., Gadomski, S., Roe, S., Rudge, A., Schopper, F., and Strobel, J. (1996). Si pad detectors. *Nucl. Instru. Meth. Phys. Res. A* **383**: 89–97.

White, R. S. (1968). An experiment to measure neutrons from the sun *Bull. Am. Phys. Soc. Ser. II* **13**(4): 714.

Wilderman, S. J. (1990). Vectorized algorithms for Monte Carlo simulation of kilovolt electron and photon transport (electron transport). Ph.D. dissertation, University of Michigan. Available at http://www.lib.uni.com/cr/umich/fullcit?p9116324.

Wilderman, S. J., Clinthorne, N. H., Fessler, J. A., and Rogers, W. L. (1999). List-mode maximum likelihood reconstruction of compton scatter camera images in nuclear medicine. *IEEE Nucl. Sci. Sym. Med. Imag. Conf.* **3**: 1716–1720.

Wilderman, S. J., Fessler, J. A., Clinthorne, N. H., and Rogers, W. L. (2001). Improved modeling of system response in list mode em reconstruction of compton scatter camera images. *IEEE Trans. Nucl. Sci.* **48**(1): 111–116.

Winkler, C., Bennett, K., Bloemen, H., Collmar, W., Connors, A., Diehl, R., van-Dordrecht, A., den-Herder, J. W., Hermsen, W., Kippen, M., Kuiper, L., Lichti, G., Lockwood, J., McConnell, M., Morris, D., Ryan, J., Schonfelder, V., Stacy, G., Steinle, H., Strong, A., Swanenburg, B., Taylor, B. G., Varendorff, M., and de-Vries, C. (1992). The gamma-ray burst of 3 May 1991 observed by COMPTEL on board GRO. *Astron. Astrophys.* **255**(1–2): L9–12.

Wulf, E. A., Johnson, W, N., Kroeger, R. A., Kurfess, J. D., and Philips, B. F. (2002). Germanium strip detector Compton telescope using three dimensional readout. *IEEE Nucl. Sci. Sym. Med. Imag. Conf. Rec.* **1**: 57–61.

Zhang, L, Clinthorne, N. H., Wilderman, S. J., Hua, C., Kragh, T. J., and Rogers, W. L. (2000). An innovative high efficiency and high resolution probe for prostate imaging (abs). *J. Nucl. Med.* **41**(5) (Suppl. S): 18P.

Zhang, L, Rogers, W. L., and Clinthorne, N. H. (2004). Potential of a Compton camera for high performance scintimammography. *Phys. Med. Biol.* **49**: 617–638.

Zhang, L., Wilderman, S. J., Clinthorne, N. H., and Rogers, W. L. (2000). An anthropomorphic phantom integrated EGS4 Monte Carlo code and its application in Compton probe. *Proc. 2000 IEEE Nucl. Sci. Sym. Med. Imag. Conf.* **3**: 119–122.

CHAPTER 20

Analytic Image Reconstruction Methods

PAUL E. KINAHAN*, MICHEL DEFRISE[†], and ROLF CLACKDOYLE[‡]
*Department of Radiology, University of Washington, Seattle, Washington
[†]Division of Nuclear Medicine, Vrije Universiteit Brussel, Brussels, Belgium
[‡]Department of Radiology, University of Utah, Salt Lake City, Utah

I. Introduction
II. Data Acquisition
III. The Central Section Theorem
IV. Two-Dimensional Image Reconstruction
V. Three-Dimensional Image Reconstruction from X-Ray Projections
VI. Summary

I. INTRODUCTION

The purpose of tomographic image reconstruction is to start with the data acquired by the scanner and arrive at a true cross-sectional image. (*Tomo* comes from the Greek for "slice.") In nuclear medicine with positron emission tomography (PET) and single-photon emission computed tomography (SPECT) we are typically forming images of a patient's *in vivo* radiolabeled tracer distribution to obtain functional information. The same principles of image reconstruction, however, apply to anatomic imaging, particularly X-ray computed tomographic imaging, and also to nonmedical imaging modalities as diverse as radioastronomy and electron microscopy.

This chapter has two purposes: to review the theory of standard two-dimensional and three-dimensional analytic image reconstruction methods and to summarize theoretical developments in three-dimensional reconstruction methods. We start by formulating a mathematical model of the acquisition process that allows us to propose solutions to the inverse problem. These solutions give some insight into tomography in general and also the key difference between two- and three-dimensional imaging—that of data redundancy.

There are several excellent general references on image reconstruction with a broader scope than nuclear medicine, both at the introductory (Herman, 1980; Kak and Slaney, 1988; Natterer and Wübbeling, 2001) and advanced levels (Varrett and Swindell, 1981; Deans, 1983; Natterer, 1986). For completeness we mention here some image reconstruction topics that are *not* covered in this chapter.

- Iterative methods are an important class of reconstruction algorithms, especially in cases in which more accurate modeling of the acquisition process is important. This topic is discussed in Lalush and Wernick (Chapter 21 in this volume) and the reader is also referred to the review articles by Ollinger and Fessler (1997) and Leahy and Qi (2000).
- Reconstruction methods for data collected from scanners that acquire complex divergent projections, such as cone-beam SPECT and multirow helical computed tomography (CT), is a rapidly evolving field. For more information on these approaches the reader is referred to the proceedings of the biannual International Meeting on Fully Three-Dimensional Image Reconstruction in Nuclear Medicine and Radiology ("1991 International Meeting," 1992; "1993 International Meeting," 1994; "1995 International Meeting," 1996; "1997 International Meeting," 1998; "1999 International Meeting," 2001; "2001 International Meeting," 2002) commonly known as the "3D Meetings."

- Efficient algorithms have been developed for various operators used in the reconstruction process, such as forward projection, backprojection, and filtering. Descriptions of these methods are scattered throughout the literature, with many in the following journals: *IEEE Transactions on Medical Imaging*, *IEEE Transactions on Nuclear Science*, *Physics in Medicine and Biology*, and *Medical Physics*.
- All functional imaging methods are necessarily dynamic at some level; however, in this chapter we assume a static tracer distribution to simplify our model. Four-dimensional (4D) reconstruction methods that incorporate the time-varying tracer distribution have also been developed (see Lalush and Wernick, Chapter 21 in this volume), as have retrospective kinetic modeling approaches to analyzing dynamic image sequences (see Morris, *et al.*, Chapter 23 in this volume).

The structure of this chapter is as follows. In Section II we describe a simple model of the data acquisition process. We then show in Section III how the Fourier transforms of the data and the original object are related through the central section theorem, also known as the central slice theorem. The central section theorem is used to explain two-dimensional image reconstruction in Section IV and three-dimensional image reconstruction in Section V. Section V also contains a presentation of the new theoretical results alluded to at the beginning of this introduction. Although the motivation of this work is the reconstruction of nuclear medicine images, the theoretical results can be applied to any modality that acquires line-integral data.

II. DATA ACQUISITION

The physics of photon generation and detection for the different imaging modalities are described elsewhere in this book. We use PET imaging here to motivate a line-integral model of the acquisition. The data acquisition processes for SPECT and X-ray CT can be related to the same model, as described, for example, by Shung *et al.* (1992).

We start by considering the parallelepiped joining any two detector elements as a volume of response (VOR) (Figure 1). In the absence of physical effects such as attenuation, scattered and accidental coincidences, and detector efficiency variations, the total number of coincidence events detected will be proportional to the total amount of tracer contained in the hypothetical tube or VOR, as indicated by the shaded area in Figure 1b. Because photon counting from radioactive decay is a random process, we have:

$$E\{\text{photons detected per second}\} = \iiint_{VOR} s(\mathbf{x}) f(\mathbf{x}) d\mathbf{x} \quad (1)$$

where $E\{\}$ is the expectation operator, $s(\mathbf{x})$ is the sensitivity within the tube at $\mathbf{x} = (x,y,z)$, and $f(\mathbf{x})$ is the three-dimensional distribution of radiotracer activity inside the patient. With Eq. (1) we can make a more accurate statement of our inverse problem: Given a set of noisy measurements for all volumes of response, what is the original $f(\mathbf{x})$? The process of estimating $f(\mathbf{x})$ from the measured data is called image reconstruction.

To develop a theoretical basis for analytic image reconstruction from line-integral data we require several assumptions about the scanning process, including continuous sampling and the absence of physical effects such as attenuation, scatter, radiotracer half-life, and collimator blurring. We assume the cross-sectional area of the parallelepiped tube becomes zero so that we are integrating along a line. In other words, the sensitivity $s(\mathbf{x})$ becomes a delta function, which is nonzero only along a specified line of response (LOR). We also assume that there is no statistical noise, which allows us to develop exact solutions. As discussed later, the presence of statistical noise means that no exact

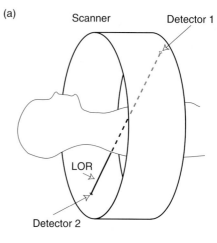
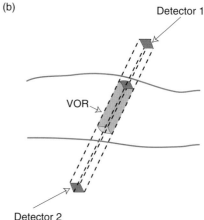

FIGURE 1 Tube, or volume of response, (VOR) corresponding to sensitive region scanned by two detector elements. (a) Overall scheme with the volume indicated as a line of response (LOR). (b) Detail showing VOR scanned by two of the rectangular detector elements (not to scale).

solution exists, so in practice some form of regularization is required. In addition, we only consider real-world functions $f(\mathbf{x})$ (compactly supported and square-integrable) that are nonpathological in any mathematical sense (Natterer, 1986).

All the simplifying assumptions are violated in practice. It is possible, however, to accurately model the physics of the acquisition, as described by Leahy and Qi (2000). These models typically lead to large sets of equations that can only be inverted with iterative techniques. Based on our line-integral model, however, we are able to derive theoretical insights that lead to image reconstruction algorithms that are useful in practice.

A. Two-Dimensional Imaging

For two-dimensional imaging we only consider LORs lying within a specified imaging plane. The acquired data are collected along a LOR through a two-dimensional object $f(x,y)$ as indicated in Figure 2, and we write Eq. (1) as a line integral. In two dimensions, we can express the line integral acquisition model by use of a rotated coordinate system, with a counter-clockwise rotation of ϕ (Figure 2). The rotated coordinates (subscript r) are related to the original coordinates by:

$$\begin{bmatrix} x \\ y \end{bmatrix} = \begin{bmatrix} \cos\phi & -\sin\phi \\ \sin\phi & \cos\phi \end{bmatrix} \begin{bmatrix} x_r \\ y_r \end{bmatrix} \quad (2)$$

With this notation we can compactly rewrite Eq. (1) as:

$$p(x_r, \phi) = \int_{-\infty}^{\infty} f(x,y) dy_r \quad (3)$$

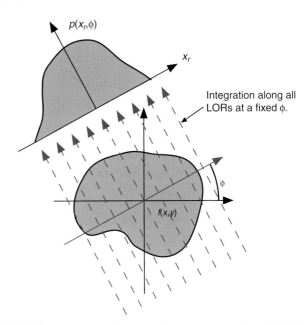

FIGURE 3 A projection formed from integration along all parallel LORs at an angle ϕ.

where the location of the LOR is described by x_r and ϕ, and $p(x_r, \phi)$ is the integral of $f(x,y)$ along the LOR. The image reconstruction problem can now be explicitly defined. Given $p(x_r, \phi)$ for all x_r and ϕ, what is the original function $f(x,y)$? In the presence of statistical noise, no exact solution is possible. Approaches constraining the solution by regularization methods are discussed after we have arrived at an exact solution for noiseless data.

For a fixed direction, ϕ, the set of line-integral data for all x_r forms a projection of $f(x,y)$ as shown in Figure 3. The collection of all projections for $0 \leq \phi < 2\pi$ as a two-dimensional function of x_r and ϕ has been termed a *sinogram* by the Swedish scientist Paul Edholm because the set of LORs passing through a fixed point (x_0, y_0) lie along a sinusoid described by $x_r = x_0 \cos\phi + y_0 \sin\phi$. This relationship is illustrated in Figure 4. A sinogram for a general object will be the linear superposition of all sinusoids corresponding to each point in the object.

At this point we make several comments about sinograms and projections:

1. The line-integral transform of $f(x,y) \to p(x_r, \phi)$ is called the *X-ray transform* (Natterer and Wübbeling, 2001), which in 2D is the same as the *Radon transform* as we describe later. The X-ray transform (or variants thereof) is a basis for a model of the data acquisition process of several modalities, including gamma cameras, SPECT, PET, and X-ray imaging systems, as well as several nonmedical imaging modalities.

2. Line-integral projections are periodic ($p(x_r, \phi + 2\pi) = p(x_r, \phi)$) and have an odd symmetry in the sense that $p(x_r, \phi) = p(-x_r, \phi + \pi)$. If this symmetry condition holds with measured data, then the angular range of sinograms

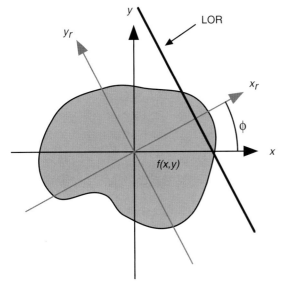

FIGURE 2 Two-dimensional LOR example. The value of $f(x,y)$ is integrated along the LOR to obtain $p(x_r, \phi)$.

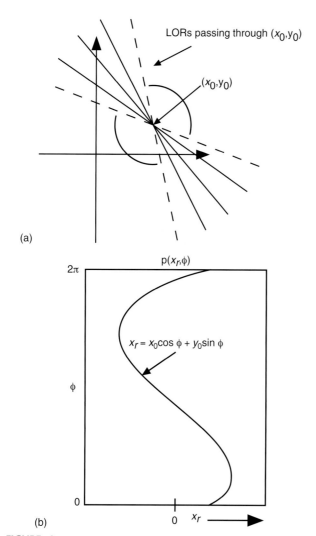

FIGURE 4 (a) Relation between LORs passing through a fixed point (x_0, y_0) and (b) the loci of integrated $p(x_r, \phi)$ values in the sinogram.

can, in principle, be constrained to $0 \leq \phi < \pi$. In many cases, such as SPECT imaging, in which the collimators introduce a depth-dependent response, we have $p(x_r, \phi) \neq p(-x_r, \phi + \pi)$, and so the full range $0 \leq \phi < 2\pi$ must be used.

3. The X-ray transform $f(x,y) \to p(x_r, \phi)$ is performed by the imaging process itself (that is, the scanner) and may at first glance appear to be a mapping from rectangular to polar coordinates. This is not the case, as is seen by considering that $p(0, \phi_1) \neq p(0, \phi_2)$ for $\phi_1 \neq \phi_2$. With the aid of the central section theorem, however, we will see that there is a mapping from rectangular to polar coordinates in the Fourier domain.

4. The integral of Eq. (3) is over the infinite real line, but as the tracer concentration in the patient is contained in the scanner field of view (FOV), with radius R_{FOV}, we have $f(x,y) \equiv 0$ for $x^2 + y^2 \geq R_{FOV}$, so the only contribution to the integral occurs for $-R_{FOV} < x_r < R_{FOV}$.

5. We have discussed data acquisition of a two-dimensional function, although any tracer distribution is three-dimensional. A more accurate statement would be to say that we have imaged one transverse plane at $z = z_0$, for which we can reexpress Eq. (3) as,

$$p(x_r, \phi, z = z_0) = \int_{-\infty}^{\infty} f(x, y, z = z_0) dy_r \quad (4)$$

We can thus image a volumetric object by repeating the acquisition for a desired range of z. If the sinogram for each value of z is reconstructed, we can stack the image planes together to form a three-dimensional image of $f(x,y,z)$. Although this can be considered a form of three-dimensional imaging, it is different from the fully three-dimensional acquisition model described in the next section.

B. Fully Three-Dimensional Imaging

In fully three-dimensional imaging we include both the two-dimensional line-integral data for all imaging planes perpendicular to the scanner or patient axis, called *direct planes*, as well as the line-integral data lying on oblique imaging planes that cross the direct planes, as shown in Figure 5. This type of data is typically collected by PET scanners to increase sensitivity and lower the statistical noise associated with photon counting, thus improving the signal-to-noise ratio (SNR) in the reconstructed image.

For the X-ray transform of a three-dimensional object, four parameters are needed to parameterize the LOR shown in Figure 5: two angles (ϕ, θ) to define the unit vector $\hat{z}_r(\phi, \theta) = (\cos\phi\cos\theta, \sin\phi\cos\theta, \sin\theta)$ parallel to the LOR,

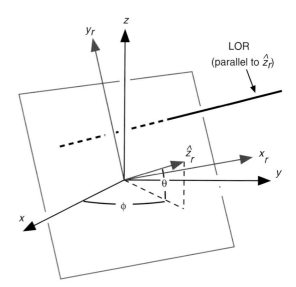

FIGURE 5 Parameterization of a three-dimensional LOR for the X-ray transform using a rotated coordinate frame. Note that the scanner axis, z, is vertical (for mathematical convention) rather than the implied horizontal direction in Figure 1.

and two coordinates (x_r, y_r) to locate the intersection of the LOR with the plane perpendicular to $\hat{z}_r(\phi, \theta)$.

By using the constraint $\hat{x}_r \cdot \hat{z} = 0$ and the definition of θ as the co-polar angle, the rotated coordinates are related to the original coordinates by:

$$\begin{bmatrix} x \\ y \\ z \end{bmatrix} = \begin{bmatrix} -\sin\phi & -\cos\phi\sin\theta & \cos\phi\cos\theta \\ \cos\phi & -\sin\phi\sin\theta & \sin\phi\cos\theta \\ 0 & \cos\theta & \sin\theta \end{bmatrix} \begin{bmatrix} x_r \\ y_r \\ z_r \end{bmatrix} \quad (5)$$

This choice of rotated coordinates and θ as the co-polar angle is customary because the equivalent two-dimensional data acquisition, or the direct planes, corresponds to $\theta = 0$. In addition, the (x_r, y_r) coordinates are in an appropriate orientation for viewing individual two-dimensional projections for a fixed $\hat{z}_r(\phi, \theta)$. Using the coordinate transform of Eq. (5), the line-integral projections along the LOR, located by (x_r, y_r, ϕ, θ), are easily expressed as:

$$p(x_r, y_r, \phi, \theta) = \int_{-\infty}^{\infty} f(x, y, z) dz_r \quad (6)$$

For a fixed direction, $\hat{z}_r(\phi, \theta)$, the set of line-integral data for all (x_r, y_r) forms a two-dimensional projection $p(x_r, y_r, \phi, \theta)$ of $f(x, y, z)$ as illustrated in Figure 6.

It is often convenient to regard $p(x_r, y_r, \phi, \theta)$ as a two-dimensional projection in the direction $\hat{z}_r(\phi, \theta)$. The full projection data set, however, is a four-dimensional function, so the X-ray transform of $f(x, y, z) \to p(x_r, y_r, \phi, \theta)$ has increased the number of dimensions by one, and we show later that this causes redundancies in the data.

The projection data set $p(x_r, y_r, \phi, \theta)$ is defined for $|x_r, y_r| < \infty$, $0 \le \phi < \pi$, and $|\theta| < \pi/2$ because there are similar symmetries as for the two-dimensional case:

$$\begin{aligned} p(x_r, y_r, \phi, \theta) &= p(x_r, y_r, \phi + 2\pi, \theta) \\ &= p(x_r, y_r, \phi, \theta + 2\pi) \\ &= p(-x_r, y_r, \phi + \pi, -\theta) \end{aligned} \quad (7)$$

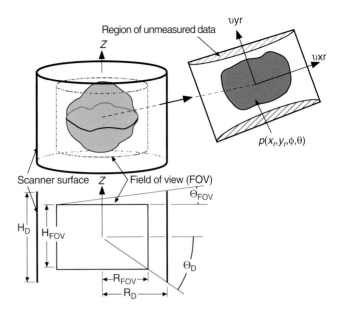

FIGURE 7 Range of measured data for $p(x_r, y_r, \phi, \theta)$ with the scanner detectors modeled as the curved wall of a cylindrical surface.

The range of nonzero values, however, is finite in x_r and y_r as indicated by Figure 7, with $|x_r| < R_{FOV}$ and $|y_r| < y_{r,\lim}$, where $y_{r,\lim} = y_{r,\lim}(x_r, \theta; R_{FOV}, H_{FOV})$. We also note that although the maximum value of θ for which LORs are detected is $\Theta_D = \arctan(H_D/2R_D)$, the projections will have regions of unmeasured data for $\Theta_{FOV} < |\theta| \le \Theta_D$, due to the finite axial extent of the scanner, where the limiting FOV angle, Θ_{FOV}, depends on the height and radius of the detector surface (H_D, R_D) and the FOV (H_{FOV}, R_{FOV}). The Fourier transform cannot be used with such truncated projections, which complicates the reconstruction process. In other words, the scanner response is no longer shift-invariant because the apparent intensity of a point source will depend on its location inside the field of view. To simplify our discussion, unless otherwise noted, we will assume that $|\theta| \le \Theta_{FOV}$.

Finally, we note that the special case of $p(x_r, y_r, \phi, \theta = 0)$ is equivalent to an axial stack (Eq. (4)) of two-dimensional sinograms with $p(x_r, \phi, z = y_r) = p(x_r, y_r, \phi, \theta = 0)$. In other words, the data for $p(x_r, y_r, \phi, \theta = 0)$ provide sufficient information to reconstruct a volumetric image of the tracer distribution. The purpose of including the data for $p(x_r, y_r, \phi, \theta \ne 0)$ is to increase the number of detected photons used in estimating the tracer distribution, which will lower the statistical noise and improve the image SNR.

C. The Relationship between the Radon and X-Ray Transforms

In fully three-dimensional imaging it is possible to collect another type of projection in which we integrate $f(x, y, z)$ across all two-dimensional planes. This is called the *Radon transform* and is illustrated in Figure 8, which also shows

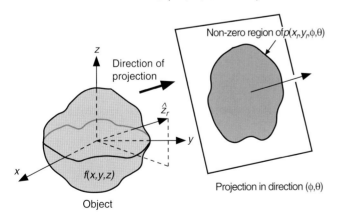

FIGURE 6 Illustration of a two-dimensional line-integral projection $p(x_r, y_r, \phi, \theta)$ of a three-dimensional object $f(x, y, z)$. The projection is formed from the set of all LORs parallel to $\hat{z}_r(\phi, \theta)$.

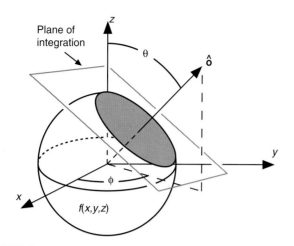

FIGURE 8 Illustration of a two-dimensional plane integral used for the Radon transform of a three-dimensional object. Note that in this case θ is used for the polar angle.

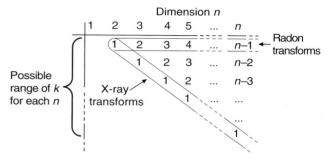

FIGURE 9 Schematic of the k-dimensional projection transforms for an n-dimensional object, where k ranges from $k = 1$ (the X-ray transform) to $k = n - 1$ (the Radon transform). For two-dimensional objects, the X-ray and Radon transforms are equivalent.

that nonzero values occur for only those planes that intersect the object.

For an n-dimensional object $f(\mathbf{x})$, $\mathbf{x} = (x_1, x_2, \ldots, x_n)$, the Radon transform of $f(\mathbf{x})$ is defined as the set of all integrals along the ($k = n - 1$)-dimensional hyperplanes intersecting the object. For our example, $n = 3$, so $k = 2$ and the Radon transform of a three-dimensional object is thus the set of integrals along all two-dimensional planes that intersect the object.

Using the same notation, the X-ray transform of an n-dimensional object is defined as the set of all ($k = 1$)-dimensional line-integrals through the object. For our three-dimensional example (Figure 5), the X-ray transform is the set of all line-integrals through the object.

A convenient representation for the location of the plane is based on the vector $t\hat{\mathbf{o}}(\phi,\theta)$, where $t \in R$ is the signed distance between the origin and the plane normal to $\hat{\mathbf{o}}(\phi,\theta) \in S^2$, where S^2 denotes the three-dimensional unit sphere. The formal expression for the Radon transform is compactly expressed by:

$$f(t,\hat{\mathbf{o}}) = f(t,\phi,\theta) = \iiint_{R^3} f(\mathbf{x})\delta(\mathbf{x}\cdot\hat{\mathbf{o}}(\phi,\theta) - t)d\mathbf{x} \quad (8)$$

where $\mathbf{x} = (x,y,z)$ and the Dirac delta function $\delta(x)$ has the sifting property $\int_a^c g(x)\delta(x - b)dx = g(b)$ if $a < b < c$. Here we use the nonitalicized $f(t,\phi,\theta)$ to represent the Radon transform of $f(x,y,z)$.

Because each planar integral is indexed by two angular variables (ϕ,θ) and one linear variable t, the Radon transform performs a mapping of a three-dimensional function $f(x,y,z)$ to a three-dimensional function $f(t,\phi,\theta)$ similar to the three-dimensional Fourier transform. In other words, each object is determined by its Radon transform and vice versa (in the absence of statistical noise)[1]. This relationship

---

[1] More precisely, the relationship is a unitary mapping between the derivative of the Radon transform and the object.

holds for any n-dimensional Radon transform, as discussed next.

For two-dimensional objects, the Radon and X-ray transform definitions coincide, as $k = n - 1 = 1$ (that is, line-integrals) for both, whereas for three- (or higher-) dimensional objects the X-ray transform remains a line-integral transform while the Radon transform becomes a plane integral for $n = 3$, and a hyperplane for $n > 3$. This family of k-dimensional projections for n-dimensional objects is illustrated in Figure 9, which also indicates the presence of other types of projections for $n \geq 4$. In nuclear medicine imaging the cases of most interest are, of course, $n = 2$ and $n = 3$, although the $n = 4$, $k = 2$ type of projection has found use in the *planogram* backprojection method (Brasse *et al.*, 2000).

For the $n = 3$ case, the $k = 1$ (X-ray transform) projections map $f(\mathbf{x})$, $\mathbf{x} \in R^3$ to $p(x_r, y_r, \phi, \theta)$, $(x_r, y_r) \in R^2$, and $(\phi,\theta) \in S^2$, where S^n is the unit sphere. For the $n = 3$ case, the $k = 2$ (Radon transform) projections map $f(\mathbf{x})$, $\mathbf{x} \in R^3$ to $f(t,\phi,\theta)$, $t \in R^1$, and $(\phi,\theta) \in S^2$. In other words, $f(\mathbf{x})$ is defined on R^3, $p(x_r, y_r, \phi, \theta)$ is defined on $R^2 \times S^2$, and $f(t,\phi,\theta)$ is defined on $R^1 \times S^2$. In general, the k-dimensional projection transforms for an n-dimensional object are defined on $R^{n-k} \times S^{n-1}$. We note that unlike the $n = 2$ case, the increase in dimensionality for the $n = 3$ X-ray transform implies redundancy in $p(x_r, y_r, \phi, \theta)$. The Radon transform $f(t,\phi,\theta)$, however, still has the same number of dimensions as $f(\mathbf{x})$. This is true for the Radon transform in any dimension, actually, because the n-dimensional Radon transform ($k = n - 1$) is defined on $R^1 \times S^{n-1}$. The general theory of k-dimensional projections is discussed by Keinert (1989).

III. THE CENTRAL SECTION THEOREM

The central section theorem (also known as the central slice or projection slice theorem) is the most important relationship in analytic image reconstruction. In this section we first derive the two-dimensional version, which is then easily extended to the three-dimensional X-ray transform. We then discuss the version for the three-dimensional Radon transform and generalizations for other transforms. For all the results presented here, we require that the imaging process be shift-invariant, which allows the use of Fourier transforms. By *shift-invariance* we mean that if we scan a shifted object, the projections are also shifted but are otherwise identical to the projections of the unshifted object. Shift-invariance is a natural property of two-dimensional imaging. For fully three-dimensional imaging situation is somewhat more complicated, but for now we assume the data are shift-invariant.

A. The Two-Dimensional Central Section Theorem

1. Derivation

We start by recalling the definition of the one-dimensional Fourier transform and the inverse transform:

$$F(v_x) = F_1\{f(x)\} = \int_{-\infty}^{\infty} f(x) e^{-i2\pi x v_x} dx$$

$$f(x) = F_1^{-1}\{F(v_x)\} = \int_{-\infty}^{\infty} F(v_x) e^{-i2\pi x v_x} dv_x \quad (9)$$

where we adopt the notation of capital letters for Fourier transformed functions, and, in general, the operator $F_n\{f(\mathbf{x})\}$ for the n-dimensional Fourier transform of $f(\mathbf{x})$, $F_n^{-1}\{F(\mathbf{u})\}$ for the inverse n-dimensional Fourier transform, and v_x as the Fourier space conjugate of x.

The one-dimensional Fourier transform of a projection is given by:

$$P(v_{xr}, \phi) = F_1\{p(x_r, \phi)\} = \int_{-\infty}^{\infty} p(x_r, \phi) e^{-i2\pi x_r v_{xr}} dx_r \quad (10)$$

An important step is the introduction of the definition of $p(x_r, \phi)$ from Eq. (3),

$$P(v_{xr}, \phi) = \int_{-\infty}^{\infty} p(x_r, \phi) e^{-i2\pi x_r v_{xr}} dx_r$$

$$= \int_{-\infty}^{\infty}\int_{-\infty}^{\infty} f(x, y) e^{-i2\pi x_r v_{xr}} dx_r dy_r \quad (11)$$

$$= \int_{-\infty}^{\infty}\int_{-\infty}^{\infty} f(x, y) e^{-i2\pi(x\cos\phi + y\sin\phi)v_{xr}} dxdy$$

$$= F(v_{xr}\cos\phi, v_{xr}\sin\phi)$$

where $F(v_x, v_y) = F_2\{f(x,y)\} = \int_{-\infty}^{\infty}\int_{-\infty}^{\infty} f(x,y) e^{-i2\pi(xv_x + yv_y)} dxdy$. Because the Fourier transform is invariant under rotation, we have from Eq. (2):

$$\begin{bmatrix} v_x \\ v_y \end{bmatrix} = \begin{bmatrix} \cos\phi & -\sin\phi \\ \sin\phi & \cos\phi \end{bmatrix} \begin{bmatrix} v_{xr} \\ v_{yr} \end{bmatrix} \quad (12)$$

We can use Eq. (12) to more concisely express Eq. (11) as:

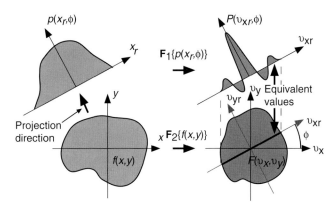

FIGURE 10 Two-dimensional central section theorem, showing the equivalency between the one-dimensional Fourier transform of a projection at angle ϕ and the central section at the same angle though the two-dimensional Fourier transform of the object.

$$P(v_{xr}, \phi) = F(v_x, v_y)\Big|_{v_{yr}=0} \quad (13)$$

Equation (13) is a key to understanding tomographic imaging. It shows that the Fourier transform of a one-dimensional projection is equivalent to a section, or profile, at the same angle through the center of the two-dimensional Fourier transform of the object. This is illustrated in Figure 10.

2. Data Sufficiency

Because any function is uniquely determined by its Fourier transform (that is, it can be computed via the inverse Fourier transform), the central section theorem indicates that if we know $P(v_{xr}, \phi)$ at all angles $0 \le \phi < \pi$, then we can somehow determine $F(v_x, v_y)$ and thus $f(x,y)$. This is the approach we take in the next section. We can also establish the conditions for necessary and sufficient data. To determine $F(v_x, v_y)$ and thus $f(x,y)$, we need $p(x_r, \phi)$ for $-R_{\text{FOV}} < x_r < R_{\text{FOV}}$ and $0 \le \phi < \pi$.

B. The Three-Dimensional Central Section Theorem for X-Ray Projections

1. Derivation

For fully three-dimensional imaging the X-ray and Radon transforms have different versions of the central section theorem. For the Radon transform version the reader is referred to Deans (1983, chap. 4) and Section IIIC. A version of the central section theorem valid for all n and k (Figure 9) is given by Keinert (1989).

To derive the central section theorem for the X-ray projection of a three-dimensional object we first compute the two-dimensional Fourier transform with respect to the first two (linear) variables:

$$P(v_{xr}, v_{yr}, \phi, \theta) = \int_{-\infty}^{\infty}\int_{-\infty}^{\infty} p(x_r, y_r, \phi, \theta) e^{-i2\pi(x_r v_{xr} + y_r v_{yr})} dx_r dy_r \quad (14)$$

If $F(\upsilon_x, \upsilon_y, \upsilon_z)$ is the three-dimensional Fourier transform of $f(x,y,z)$,

$$F(\upsilon_x, \upsilon_y, \upsilon_z) = \int_{-\infty}^{\infty}\int_{-\infty}^{\infty}\int_{-\infty}^{\infty} f(x,y,z) e^{-i2\pi(x\upsilon_x + y\upsilon_y + z\upsilon_z)} dx\, dy\, dz \quad (15)$$

then by using a derivation similar to Eq. (15), we arrive at the three-dimensional version of the central section theorem for X-ray projections:

$$P(\upsilon_{xr}, \upsilon_{yr}, \phi, \theta) = F(\upsilon_x, \upsilon_y, \upsilon_z)\big|_{\upsilon_{zr}=0} \quad (16)$$

where from Eq. (5) we have,

$$\begin{bmatrix} \upsilon_x \\ \upsilon_y \\ \upsilon_z \end{bmatrix} = \begin{bmatrix} -\sin\phi & -\cos\phi\sin\theta & \cos\phi\cos\theta \\ \cos\phi & -\sin\phi\sin\theta & \sin\phi\cos\theta \\ 0 & \cos\theta & \sin\theta \end{bmatrix} \begin{bmatrix} \upsilon_{xr} \\ \upsilon_{yr} \\ \upsilon_{zr} \end{bmatrix} \quad (17)$$

The meaning of the three-dimensional central section theorem for the X-ray projections is analogous to the two-dimensional case: A two-dimensional Fourier transform of a projection perpendicular to $\hat{z}_r(\phi,\theta)$ is equivalent to a section at the same orientation though the three-dimensional Fourier transform of the object. This is illustrated in Figure 11.

2. Data Sufficiency

As we did in the two-dimensional case, we can use the three-dimensional central section theorem to determine the conditions for necessary and sufficient data. It will be helpful to use the notation introduced by Orlov (1976a, 1976b). The orientation of a (nontruncated) projection can be identified by the location of the tip of the perpendicular unit vector $\hat{z}_r(\phi,\theta)$ on the unit sphere, and a set of such points associated with a set of projection orientations is designated Ω. Two sets of interest, $\Omega_0 = \{\hat{z}_r(\phi,\theta) | 0 \leq \phi < \pi, \theta = 0\}$ and $\Omega_\Theta = \{\hat{z}_r(\phi,\theta) | 0 \leq \phi < \pi, |\theta| \leq \Theta\}$, which correspond to multiplane two-dimensional and fully three-dimensional acquisitions, are shown in Figure 12.

Orlov's data sufficiency theorem states that a fully three-dimensional data set is sufficient if there is no great circle

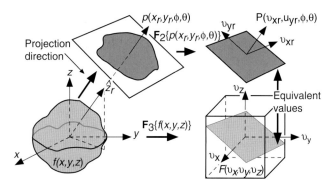

FIGURE 11 Three-dimensional central section theorem for X-ray transforms, showing the equivalency between the two-dimensional Fourier transform of a projection in direction $\hat{z}_r(\phi,\theta)$ and the central section at the same angle though the three-dimensional Fourier transform of the object.

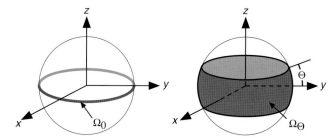

FIGURE 12 Sets of measured (and nontruncated) projections using Orlov's notation ($\Omega_0 = \{\hat{z}_r(\phi,\theta) | 0 \leq \phi < \pi, \theta = 0\}$ and $\Omega_\Theta = \{\hat{z}_r(\phi,\theta) | 0 \leq \phi < \pi, |\theta| \leq \Theta\}$) for two common cylindrical scanning geometries.

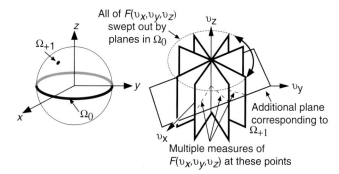

FIGURE 13 An example of data redundancy inherent in fully three-dimensional scanning even if only one additional projection Ω_{+1} is added to a two-dimensional data set (Ω_0).

on the unit sphere that does not intersect Ω (Orlov, 1976a). This is equivalent to saying that every part of the Fourier transform $F(\upsilon_x, \upsilon_y, \upsilon_z)$ is measured at least once.

The multiplane two-dimensional set Ω_0 is a subset of the fully three-dimensional set Ω_Θ, and because Ω_0 contains sufficient information to reconstruct $f(x,y,z)$ (because we know $P(\upsilon_{xr}, \phi)$ for each z), then Ω_Θ must contain redundant information. As a specific (albeit unusual) example, in Figure 13 the Fourier transforms of the set of projections corresponding to Ω_0 sweep out all of the volume of $F(\upsilon_x, \upsilon_y, \upsilon_z)$. A single additional projection, $\Omega_{+1} = \{\hat{z}_r(\phi_1, \theta_1) | \theta_1 \neq 0\}$, will contain redundant information along the corresponding oblique plane in Fourier space, and we can see that Ω_Θ contains a continuum of such redundant projections. As previously mentioned, we wish to use this redundant information to improve the image SNR.

3. Differences between Two-Dimensional and Fully Three-Dimensional Imaging

As a final comment, we note that we have seen two significant differences between two- and fully three-dimensional imaging: (1) the shift-variance of the scanner response when truncated projections are included, and (2) the redundancy of the fully three-dimensional data. Solutions for these issues are investigated after a review of two-dimensional image reconstruction.

C. Other Versions of the Central Section Theorem

So far we have only considered the $k = 1$ (X-ray transform) cases for $n = 2$ and $n = 3$. For the three-dimensional Radon transform, $f(t,\phi,\theta) = f(t,\hat{\mathbf{o}})$, we can compute the Fourier transform with respect to the linear variable $t \in R^1$:

$$F(v_t,\phi,\theta) = F_1\{f(t,\phi,\theta)\}$$

$$= \int_{\infty}^{\infty} \left[\iiint_{R^3} f(\mathbf{x})\delta(\mathbf{x}\cdot\hat{\mathbf{o}}(\phi,\theta)-t)d\mathbf{x}\right]e^{-i2\pi v_t t}dt \quad (18)$$

$$= |v_t| \int_{-\infty}^{\infty}\left[\iiint_{R^3} f(\mathbf{x})e^{-i2\pi v_t t}\delta(\mathbf{x}\cdot v_t\hat{\mathbf{o}}(\phi,\theta)-v_t t)d\mathbf{x}\right]dt$$

where we use the relationship $\delta(ax) = \delta(x)/|a|$. If we now define $v_t\hat{\mathbf{o}} = \mathbf{u} = (v_x,v_y,v_z)$ and $v_t t = \xi$, we have

$$F(v_t,\phi,\theta) = \iiint_{R^3}\left[\int_{-\infty}^{\infty} f(\mathbf{x})e^{-i2\pi\xi}\delta(\mathbf{x}\cdot\mathbf{u}-\xi)d\xi\right]d\mathbf{x}$$

$$= \iiint_{R^3} f(\mathbf{x})e^{-i2\pi\mathbf{x}\cdot\mathbf{u}}d\mathbf{x} \quad (19)$$

$$= F(\mathbf{u})$$

so along the line in frequency space defined by $v_t\hat{\mathbf{o}} = \mathbf{u}$, we have

$$\quad (20)$$

In words, Eq. (20) says that the one-dimensional Fourier transform along the line through the origin of the Radon transform, with direction $\hat{\mathbf{o}}$, is equivalent to values along the line $v_t\hat{\mathbf{o}}$ through the three-dimensional Fourier transform of the object. This is expressed in Figure 14 for the general n-dimensional Radon transform with added generalization of the direction vector to $\hat{\mathbf{o}} \in \mathbb{S}^{n-1}$, where $\mathbb{S}^{n-1}$ is the unit sphere in n-dimensions and in two-dimensions this reduces to Eq. (14). Finally, we note that appropriate versions of the central section theorem exist for all k-dimensional projections of an n-dimensional object (Figure 9) (Keinert, 1989).

IV. TWO-DIMENSIONAL IMAGE RECONSTRUCTION

A. Backprojection

An essential step in image reconstruction is backprojection, which is the adjoint to forward projection and is defined as:

$$b(x,y) = \int_0^{\pi} p(x_r,\phi)d\phi \quad (21)$$

where $x_r = x\cos\phi + y\sin\phi$ from the inverse of Eq. (2). Equation (21) implies that the appropriate value of $p(x_r,\phi)$ is summed over all ϕ for each image element (x,y) in turn. A more efficient procedure is illustrated in Figure 15, where for each value of ϕ all the image elements (x,y) are updated with the appropriate values of $p(x_r,\phi)$.

Conceptually this form of backprojection can be described as placing a value of $p(x_r,\phi)$ back into an image array along the appropriate LOR, but because the knowledge of where the values came from was lost in the integration with respect to y_r (see Eq. 3) the best we can do is place a constant value into elements along the LOR. This can be described as:

$$b^1(x,y;\phi) = p(x_r,\phi) \quad (22)$$

where $b^1(x,y;\phi)$ signifies the two-dimensional function formed by backprojecting the single projection indexed by

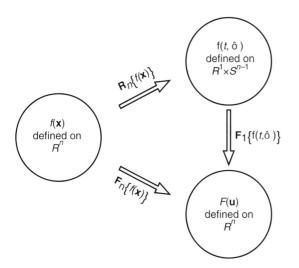

FIGURE 14 Outline of the central section theorem for the n-dimensional Radon transform. Note the correspondence to the central section theorem for X-ray transforms in Figures 10 and 11.

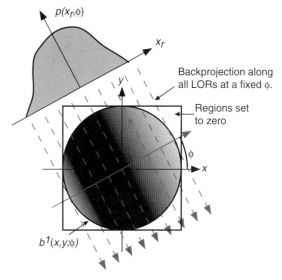

FIGURE 15 Backprojection into an image reconstruction array of all values of $p(x_r,\phi)$ for a fixed value of ϕ.

ϕ, with $x_r = x \cos\phi + y\sin\phi$. The full backprojection (see Eq. 21) is then formed by integration with respect to ϕ:

$$b(x,y) = \int_0^\pi b^1(x,y;\phi)d\phi \quad (23)$$

If we now consider the Fourier transform $B^1(v_x,v_y;\phi) = F_2\{b^1(x,y;\phi)\}$, and

$$\begin{aligned}B^1(v_x,v_y;\phi) &= \int_{-\infty}^\infty\int_{-\infty}^\infty b^1(x,y;\phi)e^{-2\pi i(xv_x+yv_y)}dx\,dy \\ &= \int_{-\infty}^\infty\int_{-\infty}^\infty p(x_r,\phi)e^{-2\pi i(x_rv_{xr}+y_rv_{yr})}dx_r\,dy_r \\ &= P(v_{xr},\phi)\delta(v_{yr})\end{aligned} \quad (24)$$

where the last result holds if we assume $b^1(x,y;\phi)$ is constant in the y_r direction (that is, not truncated to the FOV). In other words, the Fourier transform of a single backprojection is nonzero along a line through the origin. Along the line, the values are equal to the Fourier transform of the projection (and thus the object via the central section theorem of Eq. 13), multiplied by the delta function in the perpendicular direction. This result (illustrated in Figure 16) is sometimes known as the backprojection theorem and is a corollary to the central section theorem.

If we perform backprojections at two angles, say ϕ_1 and ϕ_2, and examine the Fourier transform of the result, we see that the contribution at the origin is doubled. In the limit of continuous sampling in ϕ we have:

$$\begin{aligned}B(v_x,v_y) &= \int_0^\pi B^1(v_x,v_y;\phi)d\phi \\ &= \int_0^\pi F(v_x,v_y)\delta(v_{yr})d\phi\end{aligned} \quad (25)$$

To interpret Eq. (25) we regard v_x and v_y as parameters and $\delta(v_{yr}) = \delta(v_{yr}(\phi))$ using the inverse of Eq. (12). Then by taking advantage of the identity $\delta(f(x)) = \Sigma_i(\delta(x - x_i)/|f'(x_i)|)$, where x_i are the zeros of $f(x)$, we come to the important result:

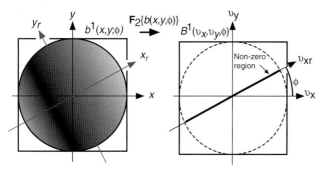

FIGURE 16 The Fourier transform of a backprojection at a single angle, $B^1(v_x,v_y;\phi)$, is equal to $P(v_{xr},\phi)\delta(v_{yr})$, the Fourier transform of the projection, multiplied by the delta function in the perpendicular direction. If $p(x_r,\phi)$ is the projection of $f(x,y)$, then from the central section theorem (Eq. 13) we have $B^1(v_x,v_y;\phi) = F(v_x,v_y)\delta(\theta_{yr})$, with $v_{yr} = -v_x\sin\phi + v_y\cos\phi$.

$$B(v_x,v_y) = \frac{F(v_x,v_y)}{v} \quad (26)$$

where $v = \sqrt{v_x^2 + v_y^2}$ is the radial frequency-space variable. Equation (26) says that the two-dimensional Fourier transform of the backprojection is the Fourier transform of the object weighted by the inverse distance from the origin. This inverse distance weighting, which comes from the polar coordinate oversampling of the Fourier transform, is a characteristic of backprojecting parallel X-ray transform data and is thus applicable to PET, SPECT, and X-ray CT imaging, as well as many nonmedical imaging modalities.

By computing the inverse Fourier transform of Eq. (26) we have:

$$b(x,y) = f(x,y) * h(x,y) \quad (27)$$

where $*$ represents the appropriate convolution, which is two-dimensional in this case. Using the two-dimensional Fourier transform of a rotationally symmetric function, which is also know as the Hankel transform (Bracewell, 1965), the linear system response function is given by

$$h(x,y) = F_2^{-1}\left\{\frac{1}{v}\right\} = \frac{1}{r} \quad (28)$$

where $r = \sqrt{x^2 + y^2}$. From Eq. (28) we see that the backprojection of X-ray transform data comprises a shift-invariant imaging system, as illustrated in Figure 17.

In imaging systems the linear system response function $h(x,y)$ is often called the point-response or point-spread function (PSF) because it describes the appearance of an ideal point-source object after the imaging process. Because the $1/v$ term in Eq. (26) amplifies low frequencies and attenuates high frequencies, the resulting output of the system in Figure 17 is a smoothed or blurred version of the input. Thus simple backprojection by itself is not sufficient for quantitatively accurate image reconstruction, and additional techniques must be to be employed to compensate for the $1/r$ blurring.

B. Reconstruction by Backprojection Filtering

Our goal is to compute $f(x,y)$ from $p(x_r,\phi)$, and an approach suggested by Eq. (26) is:

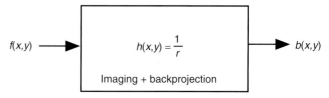

FIGURE 17 Illustration of combined imaging and backprojection as a linear shift-invariant system with $b(x,y) = f(x,y) * (1/r)$.

$$f(x,y) = F_2^{-1}\left\{\upsilon F_2\left\{b(x,y) = \int_0^\pi p(x_r,\phi)d\phi\right\}\right\} \quad (29)$$

This is known as the backprojection filtering (BPF) image reconstruction method, in which the projection data are first backprojected, filtered in Fourier space with the cone filter $\upsilon = \sqrt{\upsilon_x^2 + \upsilon_y^2}$, and then inverse Fourier transformed. Alternatively, the filtering can be performed in image space via the convolution of $b(x,y)$ with $F_2^{-1}\{\upsilon\}$. A disadvantage of this approach is that the function $b(x,y)$ has a larger support than $f(x,y)$ due to the convolution with the $(1/r)$ term, which results in gradually decaying values outside the support of $f(x,y)$. Thus any numerical procedure must first compute $b(x,y)$ using a significantly larger image matrix size than is needed for the final result. This disadvantage can be avoided by interchanging the filtering and backprojection steps shown next.

C. Reconstruction by Filtered Backprojection

If we interchange the order of the filtering and backprojection steps in Eq. (29), we obtain the useful filtered backprojection (FBP) image reconstruction method:

$$f(x,y) = \int_0^\pi p^F(x_r,\phi)d\phi \quad (30)$$

where the filtered projection, given by

$$P^F(x_r,\phi) = F_1^{-1}\{|\upsilon_{xr}| F_1\{p(x_r,\phi)\}\} \quad (31)$$

can be regarded as precorrected for the oversampling of the Fourier transform of $f(x,y)$. The one-dimensional ramp filter, $|\upsilon_{xr}|$, is a section through the rotationally symmetric two-dimensional cone filter $\upsilon = \sqrt{\upsilon_x^2 + \upsilon_y^2}$. The filtering step of Eq. (31) can also be performed by $p^F(x_r,\phi) = p(x_r,\phi) * h^{-1}(x_r)$, where the data-space version of the ramp filter, $h^{-1}(x_r)$, can be derived as (Kak and Slaney, 1988):

$$\begin{aligned}h^{-1}(x_r) &= F_1^{-1}\{|\upsilon_{xr}|\} \\ &= \lim_{\varepsilon \to 0} F_1^{-1}\{|\upsilon_{xr}|e^{-\varepsilon|\upsilon xr|}\} \\ &= \lim_{\varepsilon \to 0} \frac{\varepsilon^2 - (2\pi x_r)^2}{\left(\varepsilon^2 + (2\pi x_r)^2\right)^2}\end{aligned} \quad (32)$$

With either form of filtering, the prefiltered projection will have negative tails outside the support of $f(x,y)$ due to the $-(2\pi x_r)^{-2}$ term of Eq. (32) for large x_r. These tails remove the gradually decaying values outside the support of $f(x,y)$ that are otherwise present in a backprojection. An advantage of FBP is that the ramp filter is applied to each measured projection, which has a finite support in x_r, and we only need to backproject the filtered projections for $|x_r| \le R_{FOV}$. This means that with FBP the image can be efficiently calculated with a much smaller reconstruction matrix than can be used with BPF for the same level of accuracy. This is part of the reason for the popularity of the FBP algorithm.

A more direct, but perhaps less intuitive, derivation can be used to show that the ramp filter also arises from the Jacobian of the polar to rectangular sampling of the Fourier transform of the object. We start with the identity

$$f(x,y) = F_2^{-1}\{F(\upsilon_x,\upsilon_y)\} = \int_{-\infty}^{\infty}\int_{-\infty}^{\infty} F(\upsilon_x,\upsilon_y)e^{i2\pi(x\upsilon_x + y\upsilon_y)}dx\,dy \quad (33)$$

By transforming to standard polar coordinates for the two-dimensional plane $\{(\upsilon,\phi)|0 \le \upsilon \le \infty, 0 \le \phi < 2\pi\}$ we have,

$$f(x,y) = \int_0^{2\pi}\int_0^{\infty} F(\upsilon,\phi)e^{i2\pi\upsilon(x\cos\phi+y\sin\phi)}\upsilon\,d\upsilon\,d\phi \quad (34)$$

An alternate span of the two-dimensional plane is given by $\{(\upsilon,\phi)|-\infty \le \upsilon \le \infty, 0 \le \phi < \pi\}$, so we can change the limits of integration by using the symmetry relationship $F(\upsilon,\phi+\pi) = F(-\upsilon,\phi)$, but which requires $\upsilon d\upsilon \to |\upsilon|d\upsilon$, to obtain

$$f(x,y) = \int_0^\pi\int_{-\infty}^{\infty} F(\upsilon,\phi)e^{i2\pi\upsilon(x\cos\phi+y\sin\phi)}|\upsilon|\,d\upsilon\,d\phi \quad (35)$$

A key step is to recognize that $\upsilon = \upsilon_{xr}$, $x_r = x\cos\phi + y\sin\phi$, and $F(\upsilon,\phi) = P(\upsilon_{xr},\phi)$ (by the central section theorem), which allow us to write:

$$f(x,y) = \int_0^\pi\int_{-\infty}^{\infty} |\upsilon_{xr}| P(\upsilon_{xr},\phi)e^{i2\pi x_r \upsilon_{xr}}d\upsilon_{xr}\,d\phi \quad (36)$$

which is equivalent to Eq. (30).

We note that although Eq. (36) is correct as a continuous inversion formula, a direct discrete implementation of the ramp filter term as given by Eqs. (30) and (36) will, however, introduce a subtle problem with the removable singularity at $\upsilon_{xr} = 0$ because the filtered projection must necessarily have a zero average value. This, in turn, means that $\iint f(x,y)dx\,dy = 0$, which is clearly incorrect if, for example, $f(x,y)$ is a density function such that $f(x,y) \ge 0$ everywhere. This situation can be avoided in discrete implementations by calculating the filter function in image space (Kak and Slaney, 1988) or by careful calculation of residual offset terms to be applied to the ramp filter in frequency space (O'Sullivan, 1985).

1. Relation of Filtered Backprojection to the Inverse Radon Transform

We can explicitly relate the FBP solution of Eq. (36) to the inverse Radon transform, by use of the derivative theorem for Fourier transforms, that is $\mathcal{F}_1\{f'(x)\} = i2\pi\upsilon F(\upsilon)$, and the definition of $|x| = x\,\mathrm{sgn}(x)$, where $\mathrm{sgn}(x)$ (or the signum function) is defined as

$$\mathrm{sgn}(x) = \begin{cases} 1 & x > 0 \\ -1 & x < 0 \end{cases} \quad (37)$$

and has a Fourier transform of

$$F_1\{\text{sgn}(x)\} = \frac{i}{\pi v_{xr}} \quad (38)$$

We can then write Eq. (36) as

$$f(x,y) = \int_0^\pi \left[F_1^{-1}\{\text{sgn}(v_{xr})v_{xr}P(v_{xr},\phi)\} \right] d\phi$$

$$= \int_0^\pi \left[F_1^{-1}\{\text{sgn}(v_{xr})\} * F_1^{-1}\{v_{xr}P(v_{xr},\phi)\} \right] d\phi \quad (39)$$

and by computing the inverse Fourier transforms we have

$$f(x,y) = \frac{-1}{2\pi^2} \int_0^\pi \left[\frac{1}{x_r} * \frac{\partial p(x_r,\phi)}{\partial x_r} \right] d\phi \quad (40)$$

which is the inverse Radon transform, although a different approach was used by Radon (Deans, 1983). These results show that the solution to the inverse X-ray transform (in two dimensions) can be arrived at through several different routes. The use of the central section theorem, however, leads to the most insight about the nature of how the X-ray transform samples the Fourier transform of the object, as explicitly illustrated in Figure 18 for the discrete sampling that occurs in practice. For the discrete sampling shown in Figure 18, we can see that in the limit as $\Delta v_{xr} \to 0$ and $\Delta\phi \to 0$, the density of sampling of $F(v_x,v_y)$ varies as $1/v$, resulting in Eq. (26).

D. Reconstruction by Direct Fourier Methods

An alternate approach suggested by Figure 18 is directly interpolating from the polar sampling of the Fourier transform

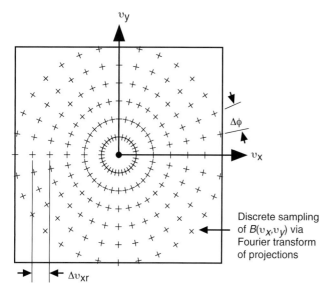

FIGURE 18 The discrete sampling pattern of $F(v_x,v_y)$ contained in $B(v_x,v_y)$, resulting from the use of discretely sampled projections.

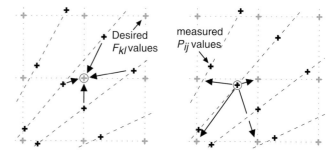

FIGURE 19 Image reconstruction by direct interpolation in the Fourier domain. For normal interpolation the nearest measured data values are pulled via interpolation to a desired grid point. In gridding, measured data values are pushed (also via interpolation) to the nearest rectangular grid points.

supplied by $P_{ij} = P(i\Delta v_{xr}, j\Delta\phi)$ to a rectangular sampling $F_{kl} = F(k\Delta v_x, l\Delta v_y)$ suitable for inverse discrete Fourier transforms, namely the fast Fourier transform (FFT). There are two possible approaches for this interpolation, which are illustrated in Figure 19. Using standard interpolation methods, the value at a desired rectangular grid point F_{kl} is estimated or pulled via interpolation from the nearest measured polar grid points P_{ij}. This approach makes poor use of the available data because, close to the origin, measured data points go unused.

In the gridding approach, each measured data point is pushed (via interpolation) to the nearest rectangular grid points, thus ensuring equal use of all measured data. To correct for the nonuniform sampling of $F(v_x,v_y)$ (described by Eq. 26), however, it is necessary to apply the cone filter $v = \sqrt{v_x^2 + v_y^2}$ to the gridded data. In addition, the accuracy of the reconstructed image depends strongly on the quality of the interpolation method, and unless the interpolation method is carefully chosen the resulting images will be less accurate or slower to compute than those produced by FBP (O'Sullivan, 1985). An interesting adjunct to the interpolation method is the use of alternative interpolating functions. With the standard contiguous pixel interpolating functions, the support of image space representation is finite, nonoverlapping, contiguous and easy to calculate and represent, but at the expense of an infinite and slowly decaying support in frequency-space. An alternative approach is to use basis functions that are constrained in extent and amplitude in both spaces. One such set of functions that have been shown to be very useful for interpolation methods are the modified Kaiser-Bessel (MKB) functions (Lewitt, 1992). If the direct Fourier method is implemented with gridding and carefully chosen MKB basis functions, then more accurate images can be produced in less time than with FBP algorithms. It should be noted, however, that FBP remains the most common method of image reconstruction due to its combination of accuracy (particularly with low-noise projection data), speed of computation, and simplicity of implementation.

E. Other Data Acquisition Formats

1. List Mode

The reconstruction methods considered so far are based on parallel X-ray transform projections in two dimensions. There are, however, other possible data acquisition formats, the most straightforward of which is list mode. In the list mode format, the LOR of each event is separately recorded in turn, possibly with other information, such as the time and photon energy. This is an efficient form of storage for otherwise sparse data histograms (that is, few photon events and many potential LORs), but it is not amenable to analytic reconstruction methods such as FBP without first histogramming the list-mode data into a standard sinogram or projection data format, which can be an inefficient procedure. Alternatively, each list-mode event can be backprojected into a reconstruction matrix, which is an efficient means of data storage but incurs the disadvantages of the BPF approach as previously discussed. For this reason, and because of the typically high levels of statistical noise, iterative methods are often used to reconstruct list-mode data.

2. Linograms

For FBP, the reconstruction of a point in the image is described by the integral along a sinusoid in the filtered sinogram (Figure 4b and Eq. 30); thus, two-dimensional backprojection involves a complex set of interpolations. An alternate approach was proposed by Edholm (1987), whose first insight for the linogram method was to choose "natural" parameters (u,v) to index a LOR for a pair of linear detectors, as shown in Figure 20(a), rather than the conventional parameters used for a circular tomograph. In the linogram acquisition format, all LORs passing through a point in the FOV lie on a straight line in the data space (Figure 20b), thus the term *linogram* for the data format.

The backprojection of all the LORs passing through a point in the (x_0,y_0) in the image will then correspond to integration along a straight line in the linogram, which is a simpler interpolation problem than the backprojection of sinograms. The next step is to note that integration along a set of parallel lines through the linogram will backproject the corresponding set of image points. Edholm's second insight was that to backproject such a set of points, the integration along the parallel lines through the linogram can be replaced by a section through the origin of the two-dimensional Fourier transform of the linogram. In other words we can take the unusual step of applying the two-dimensional central section theorem to the data space rather than to image space. The result is a fast and accurate method of two-dimensional backprojection using only Fourier transforms (Edholm and Herman, 1987; Edholm *et al.*, 1988; Magnusson, 1993). A disadvantage of the linogram method is that to adequately sample the object, the data must be collected from at least one more detector position, typically rotated by 90° so the linear detector arrays are parallel to the y-axis at $x = \pm 1$. This produces two linogram data sets, each of which must be filtered with the ramp filter, backprojected, and then summed.

3. Fan Beam

For many imaging modalities, such as X-ray CT or cardiac SPECT imaging, collecting parallel LOR data is relatively inefficient. In these situations using a fan-beam acquisition geometry (Figure 21) provides more efficient or higher resolution acquisitions.[2]

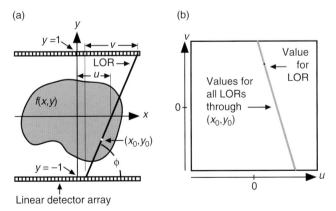

FIGURE 20 The linogram format. (a) Data acquisition with parallel linear detector arrays. (b) In the linogram data format, all LORs passing through a point (x_0,y_0) lie on a straight line in the data space.

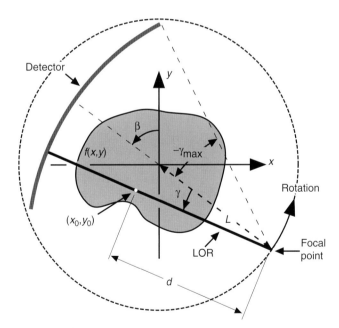

FIGURE 21 Fan-beam acquisition format in which each LOR is parameterized by two angles: the focal point angle $0 \leq \beta < 2\pi$ and the fan angle $-\gamma_{max} \leq \gamma \leq \gamma_{max}$. The focal point, which can correspond to the focal point of a SPECT collimator or the X-ray source for a CT scanner, rotates with the detector.

---

[2]For a circular PET tomograph, the native data format is actually fan beam, although it is almost always treated as parallel beam.

For $0 \leq \beta < 2\pi$ and $-\gamma_{max} \leq \gamma \leq \gamma_{max}$, individual LORs from a fan-beam projection $p(\gamma,\beta)$ correspond to LORs from a parallel-beam projection $p(x_r,\phi)$, where $x_r = -L \sin\gamma$ and $\phi = \beta + \gamma$, so there is clearly sufficient data to reconstruct $f(x,y)$. To reconstruct $f(x,y)$ from $p(\gamma,\beta)$, we use a change in variables for fan-beam projections in the convolution-backprojection version of Eqs. (30) and (31), with $0 \leq \phi < 2\pi$ and $-x_{r,max} \leq x_r \leq x_{r,max}$, where $x_{r,max} = -L \sin\gamma_{max}$:

$$f(x,y) = \frac{1}{2} \int_0^{2\pi} \int_{-x_{r,max}}^{x_{r,max}} h^{-1}(x\cos\phi + y\sin\phi - x_r')p(x_r',\phi)dx_r'd\phi \quad (41)$$

Using polar coordinates $x = r \cos\alpha$ and $y = r \sin\alpha$,

$$f(r,\alpha) = \frac{1}{2} \int_0^{2\pi} \int_{-x_{r,max}}^{x_{r,max}} h^{-1}(r\cos(\phi-\alpha) - x_r')p(x_r',\phi)dx_r'd\phi \quad (42)$$

Using $x_r' = -L \sin\gamma'$ and $\phi = \beta + \gamma'$, and after some simplification, we have:

$$f(r,\alpha) = \frac{1}{2} \int_0^{2\pi} \int_{-\gamma_{max}}^{\gamma_{max}} h^{-1}(r\cos(\beta+\gamma'-\alpha) + L\sin\gamma') \\ p(-L\sin\gamma', \beta+\gamma') L\cos\gamma' d\gamma' d\beta \quad (43)$$

For a fixed point (x_0,y_0) located at distance d along the LOR indexed by β and γ, we can use the relation $r\cos(\beta + \gamma' - \alpha) + L\sin\gamma' = d\sin(\gamma'-\gamma)$ and from Eq. (32) we can show that $h^{-1}(d \sin\gamma) = (\gamma/d \sin\gamma)^2 \, h^{-1}(\gamma)$, so if we recognize that $p(x_r = -L \sin\gamma', \phi = \beta + \gamma') = p(\gamma,\beta)$ for the specific LOR, we can write Eq. (43) as convolution followed by a distance-weighted back-projection:

$$f(r,\alpha) = \int_0^{2\pi} \frac{p^F(\gamma,\beta)}{d^2} d\beta \quad (44)$$

where $p^F(\gamma,\beta) = (p(\gamma,\beta) \cdot L\cos\gamma) * ((\gamma/\sqrt{2}\sin\gamma)^2 \cdot h^{-1}(\gamma))$.

With the advent of helical scanning protocols for fan-beam CT systems (during which there is both continuous motion of the patient through the scanner and rotation of the X-ray–detector assembly), there have been several advances made in helical/fan-beam reconstruction algorithms. Further information can be found in recent conference proceedings of the 3D Meetings ("1991 International Meeting," 1992; "1993 International Meeting," 1994; "1995 International Meeting," 1996; "1997 International Meeting," 1998; "1999 International Meeting," 2001; "2001 International Meeting," 2002) and the conference proceedings of the IEEE NSS/MIC meetings.

F. Regularization

The inverse problem of Eq. (4) is ill-posed, and its solution, (Eq. 36), is unstable in the sense that a small perturbation of the data, $p(x_r,\phi)$, can lead to an unpredictable change in the estimate of $f(x,y)$. Because photon detection is a stochastic process, some form of regularization is required to constrain the solution space to physically acceptable values.

The most common form of regularizing image reconstruction is via simple smoothing. With the FBP algorithm, Eq. (36), this can be written as

$$f(x,y) \approx \tilde{f}(x,y) = \int_0^\pi \int_{-\infty}^\infty W(\upsilon_{xr})|\upsilon_{xr}|P(\upsilon_{xr},\phi)e^{i2\pi x_r \upsilon_{xr}} d\upsilon_{xr}\, d\phi \quad (45)$$

where $\tilde{f}(x,y)$ is the reconstructed estimate of $f(x,y)$ and $W(\upsilon_{xr})$ is the apodizing (or smoothing) function. In the tomographic image reconstruction literature, the term *filter* is reserved for the unique ramp filter, $|\upsilon_{xr}|$, that arises from the sampling of the Fourier transform of the object (Figure 18). In other words, for two-dimensional image reconstruction there is only one possible filter (the ramp filter), whereas the apodizing function can take on any shape that is deemed most advantageous based on image SNR or other considerations. Note that unless $W(\upsilon_{xr}) = 1$, Eq. (45) yields a biased estimate of $f(x,y)$.

A simple model explains the principle of most apodizing functions. If we regard the measured projection data as the sum of the true projection data and uncorrelated random noise, that is $p^M(x_r,\phi) = p(x_r,\phi)+n(x_r,\phi)$, then the corresponding measured power spectrum is $P_{p^M}(\upsilon_{xr},\phi) = P_p(\upsilon_{xr},\phi) + P_n(\upsilon_{xr},\phi)$. With the finite frequency response of the detection system, the signal power, $P_p(\upsilon_{xr},\phi)$, will gradually roll off, whereas the noise power for $P_n(\upsilon_{xr},\phi)$ will remain essentially constant, as illustrated in Figure 22. The effect of the ramp filter, also shown in Figure 22, will be to amplify the high-frequency components of the power spectrum, which are dominated by noise at increasing frequencies. The use of a cosine apodizing window (also called a Von Hann, Hann, or Hanning window) is given by

$$W(\upsilon_{xr}) = \begin{cases} \frac{1}{2}\left(1 - \cos\left(\frac{\pi\upsilon_{xr}}{\upsilon_c}\right)\right), & |\upsilon_{xr}| < \upsilon_c \\ 0, & |\upsilon_{xr}| \geq \upsilon_c \end{cases} \quad (46)$$

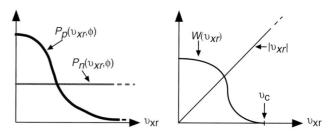

FIGURE 22 Illustration of the use of an apodized ramp filter $W(\upsilon_{xr})|\upsilon_{xr}|$ to suppress amplification of high-frequency noise power above the cutoff frequency υ_c, where $P_n(\upsilon_{xr},\phi) > P_p(\upsilon_{xr},\phi)$.

where v_c is a predetermined cut-off frequency. The specific shape of $W(v_{xr})$ will determine the noise/resolution trade-offs in the reconstructed image and also affect the noise covariance (noise texture) in the image. In practice it is difficult to rigorously justify a choice of $W(v_{xr})$. In principle, at least, $W(v_{xr})$ should be chosen to optimize a specific task, such as lesion detection.

V. THREE-DIMENSIONAL IMAGE RECONSTRUCTION FROM X-RAY PROJECTIONS

There are two important differences between two-dimensional and three-dimensional image reconstruction from X-ray transforms: the spatially varying scanner response and data redundancy. The data redundancy we have already noted in Section IIIB2 with the discussion of the set Ω_Θ (Figure 12), and we devote most of the remainder of this chapter to this subject, after discussing the spatial variance aspects. With our current understanding of two-dimensional image reconstruction, there are several different routes to understanding fully three-dimensional image reconstruction. We start with a straightforward extension of the two-dimensional results and then show how the data redundancy opens up many more possible methods of three-dimensional image reconstruction.

A. Spatial Variance and the Three-Dimensional Reprojection Algorithm

The spatially varying scanner response can be understood as the consequence of the finite axial extent of the scanner, leading to truncated projections (see Figure 7). In a two-dimensional ring scanner, the observed intensity of a point source, expressed as

$$I = \int_0^\pi \int_{-\infty}^{\infty} p(x_r, \phi) \, dx_r \, d\phi \qquad (47)$$

remains approximately constant, regardless of the position of the point source inside the scanner's FOV, as illustrated in Figure 23. For a three-dimensional cylindrical scanner, however, with truncated projections $p^T(x_r, y_r, \phi, \theta)$, the observed intensity of a point source

$$I = \int_{-\Theta}^{\Theta} \int_0^\pi \int_{-\infty}^{\infty} \int_{-\infty}^{\infty} p^T(x_r, y_r, \phi, \theta) \, dx_r \, dy_r \, d\phi \cos\theta \, d\theta \qquad (48)$$

varies depending on the position of the point source inside the scanner's FOV, particularly as the point source moves axially (Figure 23). For the impractical case of a nontruncated spherical scanner ($\Theta = \pi/2$), the scanner's response would be spatially invariant, although the issue of data redundancy would remain.

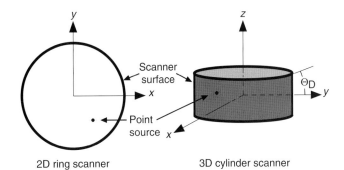

FIGURE 23 Illustration of the spatial invariance of two-dimensional imaging with X-ray transforms and the spatial variance of three-dimensional imaging with X-ray transforms. For the three-dimensional cylindrical scanner, the observed intensity of the point source varies depending on the position of the point source inside the scanner's FOV.

The spatially varying nature of the measured data complicates the image reconstruction process. For example, if regions of the support of $p(x_r, y_r, \phi, \theta)$ are unmeasured due to truncation (Figure 7), then accurately computing the two-dimensional Fourier transform of the projection, say for an FBP type of algorithm, is not possible by simply using FFTs.

A common method of restoring spatial invariance to measured three-dimensional X-ray projection data is the three-dimensional reprojection (3DRP) algorithm (Kinahan and Rogers, 1989). In this method, unmeasured regions of projections are estimated by numerically forward-projecting through an initial estimate of the image. The initial estimate is formed by reconstructing an image using the direct planes $p^T(x_r, y_r, \phi, \theta = 0)$ that are not truncated using two-dimensional FBP for each transverse plane. Further details on the implementation of this algorithm are given in Defrise and Kinahan (1998).

It is also possible to restrict the polar angle range to $|\theta| \leq \Theta_{FOV} < \Theta_D$ (Figure 7), that is, use only those projections that are not truncated. For medical imaging systems, however, this is usually not helpful because the patient extends the entire axial length of the scanner, ($H_{FOV} = H_D$ in Figure 7), so that $\Theta_{FOV} = 0$. However, by using either the 3DRP algorithm or restricting $|\theta| \leq \Theta_{FOV} < \Theta_D$, we assume that the projections are not truncated for the remainder of this chapter and that the set of complete projections correspond to the set Ω_Θ in Figure 12.

B. Three-Dimensional Backprojection.

Three-dimensional backprojection is analogous to the two-dimensional case (Eq. 21), and the backprojected image is given by,

$$b(x, y, z) = \int_{-\Theta}^{\Theta} \int_0^\pi p(x_r, y_r, \phi, \theta) \, d\phi \cos\theta \, d\theta \qquad (49)$$

Similar to the two-dimensional backprojection case, it is typically more efficient to backproject a two-dimensional view at a fixed direction (ϕ,θ), by placing the values of $p(x_r,y_r,\phi,\theta)$ back into a three-dimensional image array along the appropriate LORs. This can be expressed for a single view as:

$$b^1(x,y,z;\phi,\theta) = p(x_r,y_r,\phi,\theta) \qquad (50)$$

The full backprojection (Eq. 49) is then formed by integration with respect to ϕ and θ:

$$b(x,y,z) = \int_{-\Theta}^{\Theta}\int_0^{\pi} b^1(x,y,z;\phi,\theta)d\phi \cos\theta\, d\theta \qquad (51)$$

The three-dimensional backprojection is considerably more computationally intensive than the two-dimensional backprojection due to two factors: There is an extra level of interpolation in the axial direction implied by Eq. (50) (compared to Eq. 22) and there is the additional integration over the co-polar angle θ in Eq. (51) (compared to Eq. 23).

To develop some understanding of $b(x,y,z)$, analogous to two-dimensional backprojection, we start by deriving the three-dimensional version of the backprojection theorem by considering the Fourier transform of the backprojection of a single view

$$B^1(v_x,v_y,v_z;\phi,\theta) = \int_{-\infty}^{\infty}\int_{-\infty}^{\infty}\int_{-\infty}^{\infty} b^1(x,y,z;\phi,\theta)e^{-2\pi i(xv_x+yv_y+zv_z)}\,dx\,dy\,dz \qquad (52)$$

$$= \int_{-\infty}^{\infty}\int_{-\infty}^{\infty}\int_{-\infty}^{\infty} p(x_r,y_r,\phi,\theta)e^{-2\pi i(x_rv_{xr}+y_rv_{yr}+z_rv_{zr})}\,dx_r\,dy_r\,dz_r$$

$$= P(v_{xr},v_{yr},\phi,\theta)\delta(v_{zr})$$

where we assume $b^1(x,y,z;\phi,\theta)$ is constant in the z_r direction (that is, not truncated to the FOV). Thus the Fourier transform of a single backprojection is nonzero along a plane through the origin. Along the plane the values are equal to the Fourier transform of the projection (and thus the object via the central section theorem of Eq. 16), multiplied by the delta function in the perpendicular direction.

For the full backprojection $b(x,y,z)$ we can substitute Eq. (52) into Eq. (51) (similar to Eq. 25); this result was first derived by Colsher (1980). The result is more simply expressed in the frequency space spherical coordinates, $\mathbf{u}=(v,\varphi,\psi)$, shown in Figure 24, and is given by

$$B(v,\varphi,\psi) = F(v,\varphi,\psi)H(v,\varphi,\psi) \qquad (53)$$

where $F(v,\varphi,\psi)$ is the 3D Fourier transform of the object in spherical coordinates and $H(v,\varphi,\psi) = H(v,\psi)$ is the three-dimensional Fourier transform of the rotationally symmetric PSF, which has different behavior in the two regions indicated in Figure 24.

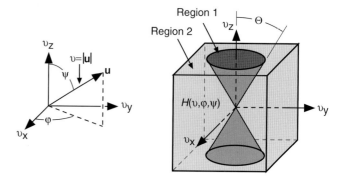

FIGURE 24 Frequency space spherical coordinates, $\mathbf{u} = (v,\varphi,\psi)$, and illustration of the two regions for the different behavior of the PSF, $H(v,\varphi,\psi)$, arising from the backprojection of three-dimensional X-ray projection data Ω_Θ in Figure 12.

$$H(v,\psi) = \begin{cases} \dfrac{2\pi}{v} & \text{Region }1:|\psi|\le\Theta \\[2ex] \dfrac{4\sin^{-1}\left(\dfrac{\sin\Theta}{|\sin\psi|}\right)}{v} & \text{Region }2:|\psi|>\Theta \end{cases} \qquad (54)$$

Similar to the two-dimensional case of Eq. (28), the three-dimensional backprojection is similar to a low pass filtering as a result of the $v^{-1}=|\mathbf{u}|^{-1}$ term in Eq. (54), and the backprojected image is a blurred version of the original object. The form of Eq. (54) can be understood by considering the three-dimensional version of the central section theorem for the X-ray projections (Eq. 16) in conjunction with the set of measured projections given by Ω_Θ (Figure 12). We derive Eq. (54) by using the filter equation, which gives more insight into the form of the blurring function. We can note however, that in the limit as $\Theta \to \pi/2$ Eq. (54) reduces to $H(v,\psi) = H(v) = 1/v$, corresponding to the (impractical) case of a spherical scanner system. As $\Theta \to 0$, which corresponds to Ω_0 (Figure 12) for a multiplane scanner acquiring only two-dimensional data with each imaging plane, Eq. (54) reduces to $H(v,\psi) = 1/(v|\sin\psi|) = 1/\sqrt{v_x^2+v_y^2}$, which corresponds to the two-dimensional case of Eq. (26). In other words, in Region 1 the oversampling of the backprojected image varies inversely as the distance from the origin, whereas in Region 2 the oversampling varies inversely as the distance from the v_z-axis. Thus the sampling of the Fourier transform of the object is determined by the data acquisition.

C. Three-Dimensional Reconstruction by Backprojection Filtering

The approach followed by Colsher (1980) was to develop a three-dimensional BPF image reconstruction algorithm by solving Eq. (53) for $F(\mathbf{u})$,

$$F(v,\varphi,\psi) = B(v,\varphi,\psi)G_c(v,\psi) \qquad (55)$$

where $G_c(\upsilon,\psi)$ is the Colsher reconstruction filter defined by

$$G_c(\upsilon,\psi) = \frac{1}{H(\upsilon,\psi)} \quad (56)$$

From Eq. (54) we see that $G_c(\upsilon,\psi)$ has ramp-type behavior similar to the two-dimensional reconstruction filter. The BPF algorithm implied by Eq. (55) is thus

$$f(\mathbf{x}) = F_3^{-1}\left\{G_c(\upsilon,\psi)F_3\left\{\int_{-\Theta}^{\Theta}\int_0^{\pi} p(x_r,y_r,\phi,\theta)\,d\phi\,d\theta\right\}\right\}\cos\theta \quad (57)$$

where the filtering by multiplication with $G_c(\upsilon,\psi)$ can be replaced by a convolution in image space with $F_3^{-1}\{G_c(\upsilon,\psi)\}$. With either approach, the filtering step can be seen as compensating for the sampling of the Fourier transform of $F(\mathbf{u})$ that is imposed by the collection of the X-ray transform data.

As in the two-dimensional case, the ramp filter term in Eq. (56) introduces a removable singularity at $\upsilon = 0$ because the filtered image has a zero average value. This situation can be avoided in discrete implementations by calculating the filter function in image space (Kinahan et al., 1988; Defrise et al., 1993), although this is considerably more complicated than for the two-dimensional case, or by oversampling the filter in frequency space (Stearns et al., 1994).

The implementation of Eq. (57) has the same difficulties as the two-dimensional BPF algorithm (Eq. 29); that is, the projection data must be backprojected into a much larger reconstruction volume than is needed for the final image volume. As in the two-dimensional case, this disadvantage can avoided by interchanging the filtering and backprojection steps.

D. Three-Dimensional Reconstruction by Filtered Backprojection

We can rewrite Eq. (57) as a FBP algorithm by interchanging the backprojection and filtering steps,

$$f(\mathbf{x}) = \iint_{\Omega_\Theta} F_2^{-1}\left\{P(\upsilon_{xr},\upsilon_{yr},\phi,\theta)G(\upsilon_{xr},\upsilon_{yr},\phi,\theta)\right\}d\phi\cos\theta\,d\theta \quad (58)$$

where $G(\upsilon_{xr},\upsilon_{yr},\phi,\theta)$ can be regarded as a two-dimensional filter $G(\upsilon_{xr},\upsilon_{yr})$ that varies with the projection and backprojection direction (ϕ,θ). For the Colsher filter, $G_c(\upsilon_{xr},\upsilon_{yr},\phi,\theta)$ can be derived using the central section theorem (Figure 15) by taking a two-dimensional section through the origin of $G_c(\upsilon,\psi)$ (Eq. 54) at the appropriate angle.

The filter used in three-dimensional image reconstruction can have forms other than the Colsher filter given in Eq. (56) and in addition, as we show next, the class of filters for FBP is more general than for BPF. In other words, for the three-dimensional X-ray transform the reconstruction filter is not unique, and FBP filters allow for the most general form (Clack, 1992). We show this through the use of the filter equation.

1. Filter Equation for the Three-Dimensional X-ray Transform

The filter equation is concisely derived using vector notation (Defrise et al., 1989) to parameterize the line-integrals first described by Eq. (6):

$$p(\mathbf{s},\hat{\mathbf{z}}_r) = \int_{-\infty}^{\infty} f(\mathbf{x}+\alpha\hat{\mathbf{z}}_r)d\alpha \quad (59)$$

where $\hat{\mathbf{z}}_r \in \Omega_\Theta$ (Figure 12), and $\mathbf{s} \in \hat{\mathbf{z}}_r^\perp$ such that $\mathbf{s} \in \mathbb{R}^3$ and $\mathbf{s}\cdot\hat{\mathbf{z}}_r = 0$. We note that $p(\mathbf{s},\hat{\mathbf{z}}_r) = p(\mathbf{s},-\hat{\mathbf{z}}_r)$. With this notation the central section theorem can be expressed as:

$$F(\mathbf{u}) = \iiint_{R^3} f(\mathbf{x})e^{-i2\pi(\mathbf{x}\cdot\mathbf{u})}d\mathbf{x} = P(\mathbf{u},\hat{\mathbf{z}}_r) \quad (60)$$

where $P(\mathbf{u},\hat{\mathbf{z}}_r)$ is the two-dimensional Fourier transform of the projection in direction $\hat{\mathbf{z}}_r$

$$P(\mathbf{u},\hat{\mathbf{z}}_r) = \iint_{\mathbf{s},\hat{\mathbf{z}}_r=0} p(\mathbf{s},\hat{\mathbf{z}}_r)e^{-i2\pi(\mathbf{s}\cdot\mathbf{u})}d\mathbf{s} \quad (61)$$

and where $\mathbf{u}$ is constrained to lie on the plane $\mathbf{u}\cdot\hat{\mathbf{z}}_r = 0$.

Starting from Eqs. (51) and (55) we have the reconstructed estimate of the Fourier transform of the object, $\tilde{F}(\mathbf{u})$,

$$F(\mathbf{u}) \approx \tilde{F}(\mathbf{u}) = B(\mathbf{u})G(\mathbf{u}) = \iint_{\Omega_\Theta} B^1(\mathbf{u};\hat{\mathbf{z}}_r)G(\mathbf{u})\,d\hat{\mathbf{z}}_r \quad (62)$$

and by using the three-dimensional backprojection theorem, Eq. (52),

$$\tilde{F}(\mathbf{u}) = \iint_{\Omega_\Theta} P(\mathbf{u},\hat{\mathbf{z}}_r)\delta(\upsilon_z)G(\mathbf{u})\,d\phi\cos\theta\,d\theta \quad (63)$$

Recognizing that $\upsilon_z = \mathbf{u}\cdot\hat{\mathbf{z}}_r$, we can write $\delta(\mathbf{u}\cdot\hat{\mathbf{z}}_r)G(\mathbf{u}) = \delta(\mathbf{u}\cdot\hat{\mathbf{z}}_r)G^1(\mathbf{u};\hat{\mathbf{z}}_r)$, where for the function $G^1(\mathbf{u};\hat{\mathbf{z}}_r)$ the vector $\mathbf{u}$ is restricted to $\hat{\mathbf{z}}_r^\perp = \{\mathbf{u}\in R^3|\mathbf{u}\cdot\hat{\mathbf{z}}_r = 0\}$. In other words $G^1(\mathbf{u};\hat{\mathbf{z}}_r)$ is zero except on the plane $\mathbf{u}\cdot\hat{\mathbf{z}}_r = 0$. With this change we have

$$\tilde{F}(\mathbf{u}) = \iint_{\Omega_\Theta} P(\mathbf{u},\hat{\mathbf{z}}_r)\delta(\mathbf{u}\cdot\hat{\mathbf{z}}_r)G^1(\mathbf{u};\hat{\mathbf{z}}_r)\,d\phi\cos\theta\,d\theta \quad (64)$$

As in Eq. (58), the function $G^1(\mathbf{u};\hat{\mathbf{z}}_r)$ used in FBP can be regarded as a two-dimensional filter that varies with the projection and backprojection direction $\hat{\mathbf{z}}_r(\phi,\theta)$. This is actually a more general form than $G(\mathbf{u})$ used in BPF (Clack, 1992). From Eq. (60) we have $P(\mathbf{u},\hat{\mathbf{z}}_r) = F(\mathbf{u})$ on the plane $\mathbf{u}\cdot\hat{\mathbf{z}}_r = 0$

$$\begin{aligned}\tilde{F}(\mathbf{u}) &= \iint_{\Omega_\Theta} F(\mathbf{u})\delta(\mathbf{u}\cdot\hat{\mathbf{z}}_r)G^1(\mathbf{u};\hat{\mathbf{z}}_r)\,d\phi\cos\theta\,d\theta \\ &= F(\mathbf{u})\iint_{\Omega_\Theta} \delta(\mathbf{u}\cdot\hat{\mathbf{z}}_r)G^1(\mathbf{u};\hat{\mathbf{z}}_r)\,d\phi\cos\theta\,d\theta \\ &= F(\mathbf{u})T_G(\mathbf{u})\end{aligned} \quad (65)$$

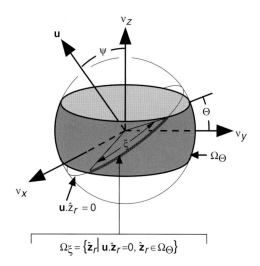

FIGURE 25 The Orlov sphere showing the set of directions of measured projections Ω_Θ and the set Ω_ξ of projections that contribute to the point **u**.

where $T_G(\mathbf{u})$ is the transfer function of the linear system of measured projections followed by FBP reconstruction using the filter $G^1(\mathbf{u};\hat{\mathbf{z}}_r)$. The reconstruction filter provides an accurate result if $\tilde{F}(\mathbf{u}) = F(\mathbf{u})$, which gives the filter equation (Defrise et al., 1989):

$$T_G(\mathbf{u}) = \iint_{\Omega_\Theta} \delta(\mathbf{u}\cdot\hat{\mathbf{z}}_r) G^1(\mathbf{u};\hat{\mathbf{z}}_r)\, d\hat{\mathbf{z}}_r = 1 \quad (66)$$

for all $\mathbf{u} \in R^3$.

The meaning of the filter equation is illustrated in Figure 25. For a given point in frequency space, $\mathbf{u}$, the delta function in Eq. (66) selects the set $\Omega_\xi = \{\hat{\mathbf{z}}_r | \mathbf{u}\cdot\hat{\mathbf{z}}_r = 0, \hat{\mathbf{z}}_r \in \Omega_\Theta\}$ of projection direction vectors within the measured set Ω_Θ that are perpendicular to $\mathbf{u}$. This represents the range of the measured projections that contribute to $\mathbf{u}$, or to any point along the line containing $\mathbf{u}$. This range is represented by ξ, the arc length of the great circle perpendicular to $\hat{\mathbf{z}}_r$ that intersects Ω_Θ. In other words, it represents the sampling density of the Fourier transform of the object at the point $\mathbf{u}$, and the filter $G^1(\mathbf{u};\hat{\mathbf{z}}_r)$ compensates for this, resulting in proper normalization of the sampling density of the Fourier transform of the object.

As a specific example of a valid filter, we assume that $G^1(\mathbf{u};\hat{\mathbf{z}}_r) = G(\mathbf{u})$, that is, the filter for a direction $\hat{\mathbf{z}}_r$ is a central section perpendicular to $\hat{\mathbf{z}}_r$ though $G(\mathbf{u})$ and that there is no additional dependence on direction. In that case, we have, where $(\iint_{\Omega_\Theta} \delta(\mathbf{u}\cdot\hat{\mathbf{z}}_r)\, d\phi \cos\theta\, d\theta) > 0$:

$$G(\mathbf{u}) = \frac{1}{\iint_{\Omega_\Theta} \delta(\mathbf{u}\cdot\hat{\mathbf{z}}_r)\, d\phi \cos\theta\, d\theta}$$
$$= \frac{|\mathbf{u}|}{\iint_{\Omega_\Theta} \delta(\hat{\mathbf{u}}\cdot\hat{\mathbf{z}}_r)\, d\phi \cos\theta\, d\theta} \quad (67)$$
$$= \upsilon / \xi$$

where we use the relations $\delta(\mathbf{u}\cdot\hat{\mathbf{z}}_r) = \delta(\hat{\mathbf{u}}\cdot\hat{\mathbf{z}}_r)/|\mathbf{u}|$ and $\upsilon = |\mathbf{u}|$ and recognize that the denominator is now simply the arc length ξ of Ω_ξ along the great circle. This arc length depends on the region of frequency space that $\mathbf{u}$ is in (Figures 24 and 25)

$$G(\mathbf{u}) = G_c(\mathbf{u}) = \begin{cases} \dfrac{\upsilon}{2\pi} & |\psi| \leq \Theta \\[2mm] \dfrac{\upsilon}{4\sin^{-1}\left(\dfrac{\sin\Theta}{|\sin\psi|}\right)} & |\psi| \leq \Theta \end{cases} \quad (68)$$

which is the Colsher filter of Eq. (56).

Equation (66) is a necessary and sufficient condition that must be met by any valid reconstruction filter. Although these filters will all produce accurate reconstructions with consistent projection data that are noiseless and continuous (Clack, 1992), the behavior of filters will differ in how they propagate inconsistent data introduced by noise or discretization, that is, data that do not fit the model of Eq. (59). There are two classes of filters that satisfy Eq. (66)—factorizable and nonfactorizable.

2. Factorizable Reconstruction Filters

The Colsher filter, Eq. (68), is an example of a *factorizable* filter, which is any FBP filter that can be expressed as the product of a central section through a three-dimensional filter function $G(\mathbf{u})$ and an even positive function $w(\hat{\mathbf{z}}_r)$ that depends only on the direction vector $\hat{\mathbf{z}}_r$, that is:

$$G^1(\mathbf{u};\hat{\mathbf{z}}_r) = G(\mathbf{u})w(\hat{\mathbf{z}}_r) \quad (69)$$

with $\mathbf{u}$ constrained to the plane $\mathbf{u}\cdot\hat{\mathbf{z}}_r = 0$. For the Colsher filter $w(\hat{\mathbf{z}}_r) = 1$, whereas for any other valid nonconstant $w(\hat{\mathbf{z}}_r)$ the filter is uniquely determined by Eqs. (66) and (69) (Schorr and Townsend, 1981):

$$G(\mathbf{u}) = \frac{1}{\iint_{\Omega_\Theta} \delta(\mathbf{u}\cdot\hat{\mathbf{z}}_r) w(\hat{\mathbf{z}}_r)\, d\phi \cos\theta\, d\theta} \quad (70)$$

as long as $\iint_{\Omega_\Theta} \delta(\mathbf{u}\cdot\hat{\mathbf{z}}_r) w(\hat{\mathbf{z}}_r) d\phi \cos\theta\, d\theta \neq 0$.

Among the important properties of factorizable FBP filters, such as the Colsher filter, is the ability to commute the order of filtering and backprojection. The FBP filter function on the plane $\mathbf{u}\cdot\hat{\mathbf{z}}_r = 0$ is a two-dimensional section through the origin (at the appropriate angle) of the three-dimensional filter used for BPF, as illustrated in Figure 26. This is similar to the equivalence of the two-dimensional FBP and BPF algorithms (Eqs. (36) and (29)).

An additional property of factorizable filters is the ramp-like behavior due to the relation $\delta(\mathbf{u}\cdot\hat{\mathbf{z}}_r) = \delta(\hat{\mathbf{u}}\cdot\hat{\mathbf{z}}_r)/|\mathbf{u}|$ used in Eq. (67). Finally, we note that factorizable filters, and the Colsher filter in particular, have optimally low noise-propagation properties due to the minimum norm properties of $G_c(\mathbf{u})$ (Defrise et al., 1990).

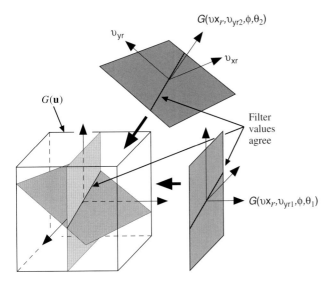

FIGURE 26 Factorizable FBP filter $G(v_{xr},v_{yr},\phi,\theta)$ at two different orientations $\hat{z}_r(\phi_1,\theta_1)$ and $\hat{z}_r(\phi_2,\theta_2)$. If $w(\hat{z}_r) = 1$ in Eq. (69), then the two filter sections agree along the line of intersection within a three-dimensional function $G(\mathbf{u})$. This property is not true for nonfactorizable FBP filters.

3. Nonfactorizable Reconstruction Filters

In contrast to the factorizable FBP filters, there is a more general class of FBP filters $G^1(\mathbf{u};\hat{z}_r)$ for which a factorization of the type in Eq. (69) (Figure 26) is not possible. In other words, using the notation $G(v_{xr},v_{yr},\phi,\theta) = G^1(\mathbf{u};\hat{z}_r)$, $G(v_{xr},v_{yr},\phi,\theta)$ is not equivalent to two-dimensional central sections of *any* function $G(\mathbf{u})$, even with a direction-weighting term $w(\hat{z}_r)$.

Although nonfactorizable filters have less favorable noise-propagation properties than factorizable filters (Defrise *et al.*, 1990), they do have the ability to allow for direct reconstruction of truncated projections (Defrise *et al.*, 1995), and two examples of this are the TTR filter (Ra *et al.*, 1992) and the FAVOR filter (Comtat *et al.*, 1993). These filters can be used with truncated X-ray transform data (Figure 7) without resorting first to the computationally intensive forward-projection step of the 3DRP algorithm (Section VA).

An interesting aspect of the relationship between factorizable and nonfactorizable filters is the existence of nontrivial null filters (Clack, 1992). If $G^1_{NF}(\mathbf{u};\hat{z}_r)$ is a nonfactorizable filter and $G^1_F(\mathbf{u};\hat{z}_r)$ is a factorizable filter, then the difference

$$N^1(\mathbf{u};\hat{z}_r) = G^1_{NF}(\mathbf{u};\hat{z}_r) - G^1_F(\mathbf{u};\hat{z}_r) \quad (71)$$

is a nonfactorizable filter that has the unusual property

$$\iint_{\Omega_\Theta} F_2^{-1}\{P(v_{xr},v_{yr},\phi,\theta)N(v_{xr},v_{yr},\phi,\theta)\}d\phi\cos\theta\, d\theta = 0 \quad (72)$$

for any consistent projection data that satisfy Eq. (6) or (59). Equation (72) can be regarded as a general class of consistency conditions for (nontruncated) three-dimensional X-ray transform data.

4. General Filter Functions and Properties

If there is an arbitrary function $Q^1(\mathbf{u};\hat{z}_r)$ such that $T_Q(\mathbf{u})$ exists for Eq. (66) but $T_Q(\mathbf{u}) \neq 1$, it is still possible to use $Q^1(\mathbf{u};\hat{z}_r)$ as the basis of a filter that satisfies the filter equation, Eq. (66) (Defrise *et al.*, 1993). The most straightforward approach is to use a multiplicative normalization in frequency space:

$$G^1(\mathbf{u};\hat{z}_r) = Q^1(\mathbf{u};\hat{z}_r) / T_Q(\mathbf{u}) \quad (73)$$

as long as $T_Q(\mathbf{u}) \neq 0$). An alternative approach that does not require $T_Q(\mathbf{u}) \neq 0$ is to use an additive normalization:

$$G^1(\mathbf{u};\hat{z}_r) = Q^1(\mathbf{u};\hat{z}_r) + \frac{w(\hat{z}_r)(1 - T_Q(\mathbf{u}))}{\iint_{\Omega_\Theta}\delta(\mathbf{u}\cdot\hat{z}_r)w(\hat{z}_r)d\hat{z}_r} \quad (74)$$

as long as $\iint_{\Omega_\Theta}\delta(\mathbf{u}\cdot\hat{z}_r)w(\hat{z}_r)d\hat{z}_r \neq 0$.

It is straightforward to verify that an FBP filter constructed using Eq. (73) or (74) is a solution to the filter equation and will thus correctly reconstruct objects from consistent projection data when used in a three-dimensional FBP algorithm. Such filters can be useful if, for example, the function $Q^1(\mathbf{u};\hat{z}_r)$ is nearly correct or has useful properties and only a small correction is needed. The approach was used in the derivation of the FAVOR filter, where the function $Q^1(\mathbf{u};\hat{z}_r)$ corresponded to simple (and fast) one-dimensional ramp-filtering of the projection data (Comtat *et al.*, 1993). Although this approach will work for any function $Q^1(\mathbf{u};\hat{z}_r)$, large correction terms will necessarily increase the propagation of statistical noise.

E. Other Three-Dimensional Reconstruction Methods

The alternative methods described in Sections IVC1, IVD, and IVE for two-dimensional image reconstruction are also applicable to three-dimensional image reconstruction, albeit with some significant differences and the addition of an entirely new class of methods using rebinning techniques.

1. Inverse Radon Transform

With a set of X-ray transform projection data (Eq. 59), the three-dimensional Radon transform (Eq. 8) can be calculated by summing all the line-integral data (Eq. 59) that lie on the plane defined by t and $\hat{o}$ (Figure 8) and are parallel to a fixed direction $\hat{z}_r$ that is perpendicular to $\hat{o}$ (Stazyk *et al.*, 1992):

$$f(t,\hat{o}) = \int_{-\infty}^{\infty} p(t\hat{o} + \alpha\hat{z}_r \times \hat{o}, \hat{z}_r)d\alpha \quad (75)$$

where $\hat{z}_r \times \hat{o}$ is the vector cross-product. We can improve our estimate of $f(t,\hat{o})$ by averaging Eq. (75) over all directions

$\hat{z}_r$ with $\hat{z}_r \cdot \hat{o} = 0$. This approach has the advantage that the averaged Radon transform incorporates the SNR improvements of the data redundancy of the three-dimensional X-ray transform but does not contain redundant information itself, and thus $f(\mathbf{x})$ can be computed with the inverse three-dimensional Radon transform (Deans, 1983). Effectively, the formation of the averaged Radon transform is a data reduction step because the dimensionality of the data goes from $\mathbb{R}^2 \times \mathbb{S}^2$ to $\mathbb{R}^1 \times \mathbb{S}^2$. The disadvantages of this procedure are similar to a direct Fourier interpolation method because complex interpolations are required. We note that there is a direct relationship between the filter equation and the averaged Radon transform (Defrise et al., 1993).

2. Direct Fourier Methods

As in two-dimensional image reconstruction, it is possible to directly interpolate from the sampling of the Fourier transform supplied by $P_{ijkl} = P(i\Delta v_{xr}, j\Delta v_{xy}, k\Delta\phi, l\Delta\theta)$ to a rectangular volumetric sampling $F_{i'j'k'} = F(i'\Delta v_x, j'\Delta v_y, k'\Delta v_z)$ suitable for inversion with FFTs. Such an approach has similar advantages and disadvantages compared to FBP algorithms as the two-dimensional case. One advantage of a three-dimensional direct Fourier method is that the sampling in the axial direction is similar to the linogram format, thus simplifying the interpolation procedure in the axial direction during the three-dimensional ray tracing of the backprojection process. Three-dimensional direct Fourier methods have been implemented via gridding methods using standard voxel basis functions (Stearns et al., 1990), and with MKB basis functions (Matej and Lewitt, 2001).

3. List-Mode Methods

As mentioned in Section IVD1, three-dimensional list-mode data can be reconstructed by first histogramming the data into projections. This can be followed by any of the three-dimensional reconstruction methods described in this chapter. For imaging systems with few events per scanner LOR, however, this can be a very inefficient mode of data processing and storage and can lead to high levels of statistical noise unless considerable regularization (smoothing) is applied. In these situations three-dimensional iterative methods are often used to reconstruct such list-mode data. Three-dimensional iterative methods also have the advantage of using the same functional form as two-dimensional methods, so they are often simpler to implement than three-dimensional analytic methods, albeit often at the expense of increased computation time.

4. Planogram Methods

The planogram data format is the generalization to three-dimensional imaging of the linogram format (Section IVD2), which is appropriate for flat panel detectors. As in the linogram format, backprojection can be efficiently carried out by using only FFT operations. The reconstruction of planogram data has been implemented with BPF algorithms, although FBP methods are also under development (Brasse et al., 2000).

5. Cone-Beam Methods

The cone-beam, or divergent three-dimensional X-ray, transform (Natterer, 1986) is not discussed here because the reconstruction procedures are considerably more complex than for the parallel X-ray transform data. There is no equivalent to the central section theorem for divergent projection data. There is, however, a direct relationship between cone-beam data and the derivative of the three-dimensional Radon transform of the object. Thus most cone-beam reconstruction methods are based on either iterative methods or analytic functions related to the three-dimensional Radon transform. There have been numerous recent developments in analytic cone-beam reconstruction methods, particularly in relation to the development of multirow X-ray CT scanners, and the interested reader is referred to the proceedings of the 3D Meetings ("1991 International Meeting," 1992; "1993 International Meeting," 1994; "1995 International Meeting," 1996; "1997 International Meeting," 1998; "1999 International Meeting," 2001; "2001 International Meeting," 2002).

6. Rebinning Methods

With three-dimensional X-ray transform data it is possible to use the measured data to estimate a stacked (volumetric) set of two-dimensional transverse sinograms (Eq. 4) by using some form of signal averaging, similar to the averaged three-dimensional Radon transform described in Section VE1. Such a procedure is called a rebinning algorithm. Rebinning algorithms have the advantage that each rebinned sinogram can be efficiently reconstructed with either analytic or iterative two-dimensional reconstruction methods. In addition, rebinning can significantly reduce the size of the data. The disadvantages of the rebinning methods are that they either pay a penalty in terms of a spatially varying distortion of the PSF or amplification of statistical noise or both. Although rebinning methods are not specifically a reconstruction procedure, they are an important addition to the array of techniques that bear on three-dimensional image reconstruction problems.

The simplest, but still often useful, rebinning method is the single-slice rebinning (SSRB) algorithm (Daube-Witherspoon and Muehllehner, 1987) in which the rebinned sinograms $p_R(x_r,\phi,z)$ are calculated as

$$P_R(x_r, \phi, z) = \frac{1}{2\theta_{\max}(z)} \int_{-\theta_{\max}(z)}^{\theta_{\max}(z)} p(x_r, y_r = z\cos\theta, \phi, \theta)\cos\theta\, d\theta \quad (76)$$

where $\theta_{\max}(z)$ is the maximum range of the co-polar angle for LORs crossing the transverse plane at distance z.

The summed LOR data are normalized by the term $1/[2\theta_{max}(z)]$ to (approximately) compensate for the nonuniform sampling of the rebinned sinograms.

Equation (76) ignores the co-polar angle, so that the LOR is placed in the transverse LOR that bisects the original LOR. This method is accurate for sources near the scanner axis (Kinahan and Karp, 1994), but there is increasing distortion of the PSF for sources further from the scanner axis (Sossi *et al.*, 1994).

A more accurate rebinning methods is the Fourier rebinning (FORE) algorithm (Defrise, 1995; Defrise *et al.*, 1997), which is based on a reasonably accurate equivalence between specific elements in the Fourier transformed oblique and transverse sinograms. In other words, the Fourier transformed oblique sinograms can be resorted into transverse sinograms and, after normalization for the sampling of the Fourier transform, inverse transformed to recover $p_R(x_r,\phi,z)$. The FORE method amplifies statistical noise slightly compared to SSRB, but yields a significantly less distorted PSF. The combination of FORE with appropriately weighted two-dimensional iterative reconstruction methods for each rebinned sinogram has successfully been employed for the reconstruction of three-dimensional whole-body PET data in clinically feasible times (Comtat *et al.*, 1998). One disadvantage of FORE relative to SSRB is that because the Fourier transform is computed, FORE cannot be applied to list-mode data, unlike SSRB.

VI. SUMMARY

The goals of this chapter are two. First, we have provided a theoretical basis for the core concepts of two- and three-dimensional analytic image reconstruction from X-ray projections. In particular, with the aid of the central section theorem in its various guises, we have shown how the image reconstruction problem is formed by Fourier transform sampling of the object. The use of the central section theorem also leads to the necessary and sufficient data conditions. Second, we have highlighted the literature that has evolved in the 1990s regarding more esoteric aspects of three-dimensional FBP reconstruction filters, which have a rich complexity not found in two-dimensional image reconstruction, and for which other uses may be found in the future.

References

1991 International Meeting on Fully Three-Dimensional Image Reconstruction in Nuclear Medicine and Radiology. (1992). *Phys. Med. Biol.* **37**.

1993 International Meeting on Fully Three-Dimensional Image Reconstruction in Radiology Nuclear Medicine. (1994). *Phys. Med. and Biol.* **39**.

1995 International Meeting on Fully Three-Dimensional Image Reconstruction in Radiology and Nuclear Medicine. (1996). *In* "Three-Dimensional Image Reconstruction in Radiology and Nuclear Medicine, Vol. 4, *Computational Imaging and Vision*" (P. Grangeat and J.-L. Amans, eds) Kluwer, Dordrect, The Netherlands.

1997 International Meeting on Fully Three-Dimensional Image Reconstruction in Radiology and Nuclear Medicine. (1998). *Phys. Med. Biol.* **43**.

1999 International Meeting on Fully Three-Dimensional Image Reconstruction in Radiology and Nuclear Medicine. (2001). *IEEE Trans. Med. Imag.* **45**.

2001 International Meeting on Fully Three-Dimensional Image Reconstruction in Radiology and Nuclear Medicine. (2002). *Phys. Med. Biol.* **47**.

Bracewell, R. (1965). "The Fourier Transform and Its Applications." McGraw-Hill, New York.

Brasse, D., Kinahan, P. E., Clackdoyle, R., Comtat, C., Defrise, M., and Townsend, D. W. (2004). Fast fully 3D image reconstruction in PET using planograms. *IEEE Trans. Med. Imag.* **23**: 413–425.

Clack, R. (1992). Towards a complete description of three-dimensional filtered backprojection. *Phys. Med. Biol.* **37**: 645–660.

Colsher, J. G. (1980). Fully three-dimensional positron emission tomography. *Phys. Med. Biol.* **25**: 103–115.

Comtat, C., Kinahan, P. E., Defrise, M., Michel, C., and Townsend, D. W. (1998). Fast reconstruction of 3D PET data with accurate statistical modeling. *IEEE Trans. Nucl. Sci.* **45**: 1083–1089.

Comtat, C., Morel, C., Defrise, M., and Townsend, D. W. (1993). The FAVOR algorithm for 3D PET data and its implementation using a network of transputers. *Phys. Med. Biol.* **38**: 929–944.

Crawford, C. R. (1991). CT filtration aliasing artifacts. *IEEE Trans. Med. Imag.* **10**: 99–102.

Daube-Witherspoon, M. E., and Muehllehner, G. (1987). Treatment of axial data in three-dimensional PET. *J. Nucl. Med.* **28**: 1717–1724.

Deans, S. R. (1983). "The Radon Transform and Some of Its Applications." John Wiley, New York.

Defrise, M. (1995). A factorization method for the 3D X-ray transform. *Inverse Probl.* **11**: 983–994.

Defrise, M., Clack, R., and Townsend, D. (1993). Solution to the three-dimensional image reconstruction problem from two-dimensional projections. *J. Opt. Soc. A* **10**: 869–877.

Defrise, M., Clack, R., and Townsend, D. W. (1995). Image reconstruction from truncated, two-dimensional parallel projections. *Inverse Probl.* **11**: 287–313.

Defrise, M., and Kinahan, P. (1998). Data acquisition and image reconstruction for 3D PET. *In* "The Theory and Practice of 3D PET, Vol. 32, Developments in Nuclear Medicine" (D. W. Townsend and B. Bendriem, eds), pp. 11–54. Kluwer Academic, Dordrecht.

Defrise, M., Kinahan, P. E., Townsend, D. W., Michel, C., Sibomana, M., and Newport, D. F. (1997). Exact and approximate rebinning algorithms for 3-D PET data. *IEEE Trans. Med. Imag.* **16**: 145–158.

Defrise, M., Townsend, D. W., and Clack, R. (1989). Three-dimensional image reconstruction from complete projections. *Phys. Med. Biol.* **34**: 573–587.

Defrise, M., Townsend, D. W., and Deconinck, F. (1990). Statistical noise in three-dimensional positron tomography. *Phys. Med. Biol.* **35**: 131–138.

Edholm, P. R., and Herman, G. T. (1987). Linograms in image reconstruction from projections. *IEEE Trans. Med. Imag.* **MI-6**: 301–307.

Edholm, P. R., Herman, G. T., and Roberts, D. A. (1988). Image reconstruction from linograms: Implementation and evaluation. *IEEE Trans. Med. Imag.* **7**: 239–246.

Herman, G. T. (1980). "Image Reconstruction from Projections." Academic Press, New York.

Kak, A. C., and Slaney, M. (1988). "Principles of Computerized Tomographic Imaging." IEEE Press, New York.

Keinert, F. (1989). Inversion of k-plane transforms and applications in computed tomography. *SIAM Rev.* **31**: 273–298.

Kinahan, P. E., and Karp, J. S. (1994). Figures of merit for comparing reconstruction algorithms with a volume-imaging PET scanner. *Phys. Med. Biol.* **39:** 631–638.

Kinahan, P. E., and Rogers, J. G. (1989). Analytic 3D image reconstruction using all detected events. *IEEE Trans. Nucl. Sci.* **36:** 964–968.

Kinahan, P. E., Rogers, J. G., Harrop, R., and Johnson, R. R. (1988). Three-dimensional image reconstruction in object space. *IEEE Trans. Nucl. Sci.* **35:** 635–640.

Leahy, R., and Qi, J. (2000). Statistical approaches in quantitative positron emission tomography. *Statist. Comp.* **10:** 147–165.

Lewitt, R. M. (1992). Alternatives to voxels for image representation in iterative reconstruction algorithms. *Phys. Med. Biol.* **37:** 705–716.

Magnusson, M. (1993). Linogram and other direct methods for tomographic reconstruction. Ph.D. thesis, Linköping University.

Matej, S., and Lewitt, R. M. (2001). 3D-FRP: Direct Fourier reconstruction with Fourier reprojection for fully 3D PET. *IEEE Trans. Nucl. Sci.* **48:** 1378–1385.

Natterer, F. (1986). "The Mathematics of Computerized Tomography." John Wiley, New York.

Natterer, F., and Wübbeling, F. (2001). "Mathematical Methods in Image Reconstruction." SIAM, Philadelphia.

Ollinger, J. M., and Fessler, J. A. (1997). Positron emission tomography. *IEEE Sig. Processing Mag.* **14:** 43–55.

Orlov, S. S. (1976a). Theory of three-dimensional image reconstruction I. Conditions for a complete set of projections. *Sov. Phys. Crystallog.* **20:** 429–433.

Orlov, S. S. (1976b). Theory of three-dimensional image reconstruction II. The recovery operator. *Sov. Phys. Crystallog.* **20:** 429–433.

O'Sullivan, J. D. (1985). A fast sinc function gridding algorithm for Fourier inversion in computer tomography. *IEEE Trans. Med. Imag.* **MI-4:** 200–207.

Ra, J. B., Lim, C. B., Cho, Z. H., Hilal, S. K., and Correll, J. (1992). A true 3D reconstruction algorithm for the spherical positron tomograph. *Phys. Med. Biol.* **27:** 37–50.

Schorr, B., and Townsend, D. (1981). Filters for three-dimensional limited-angle tomography. *Phys. Med. Biol.* **26:** 305–312.

Shung, K. K., Smith, M. B., and Tsui, B. M. W. (1992). "Principles of Medical Imaging." Academic Press, San Diego.

Sossi, V., Stazyk, M., Kinahan, P. E., and Ruth, T. (1994). The performance of the single-slice rebinning technique for imaging the human striatum as evaluated by phantom studies. *Phys. Med. Biol.* **39:** 369–375.

Stazyk, M., Rogers, J., and Harrop, R. (1992). Analytic image reconstruction in PVI using the 3D Radon transform. *Phys. Med. Biol.* **37:** 689–704.

Stearns, C. W., Chesler, D. A., and Brownell, G. L. (1990). Accelerated image reconstruction for a cylindrical positron tomograph using Fourier domain methods. *IEEE Trans. Nucl. Sci.* **37:** 773–777.

Stearns, C. W., Crawford, C. R., and Hu, H. (1994). Oversampled filters for quantitative volumetric PET reconstruction. *Phys. Med. Biol.* **39:** 381–388.

Barrett, H. H., and Swindell, W. (1981). "Radiological Imaging." Academic Press, New York.

CHAPTER 21

Iterative Image Reconstruction

DAVID S. LALUSH* and MILES N. WERNICK†

*Department of Biomedical Engineering, North Carolina State University, Raleigh, North Carolina and
University of North Carolina at Chapel Hill, North Carolina.
†Departments of Electrical and Computer Engineering and Biomedical Engineering, Illinois Institute of Technology, Chicago,
Illinois and Pridictek, LLC, Chicago, Illinois

I. Introduction
II. Tomography as a Linear Inverse Problem
III. Components of an Iterative Reconstruction Method
IV. Image Reconstruction Criteria
V. Iterative Reconstruction Algorithms
VI. Evaluation of Image Quality
VII. Summary
VIII. Appendices

I. INTRODUCTION

Traditionally, tomographic image reconstruction has been framed as a simple mathematical problem, namely that of inverting a discrete form of the Radon transform, as explained in Kinahan *et al.* (Chapter 20 in this volume). In this traditional approach, it is assumed that the data consist of line integrals of the object distribution, and no attempt is made to model explicitly the randomness of the gamma-ray counting process. This simplified version of the reconstruction problem is solved exactly by filtered backprojection (FBP), a method which allows emission tomography (ET) images to be computed very quickly. Unfortunately, because real ET data are not precisely described by the FBP model, the resulting images can exhibit significant inaccuracies. This is especially true in single-photon emission computed tomography (SPECT), where attenuation can cause severe artifacts if it is not suitably accounted for (see King *et al.*, Chapter 22 in this volume).

Avoiding the shortcomings of FBP requires that the reconstruction problem be framed in a way that more closely resembles reality. Rather than assuming a Radon model, modern reconstruction techniques use a more general linear model that can allow for a rich description of the blurring and attenuation mechanisms in the imaging process. Statistical reconstruction techniques in addition incorporate probabilistic models of the noise and, in the case of Bayesian methods, of the image itself.

The price of these enhancements is that the resulting mathematical problem is more difficult to solve than that of Radon transform inversion. Indeed, the solution generally cannot be written explicitly or, when it can be, the analytic form of the solution is impractical to compute. Thus, most reconstruction algorithms that attempt to incorporate an accurate imaging model are *iterative*, meaning that the estimated image is progressively refined in a repetitive calculation.

The principal trade-off between iterative techniques and FBP is one of accuracy versus efficiency. Iterative algorithms invariably require repeated calculations of projection and backprojection operations. Thus, they can require substantially greater computation time than FBP. Accurate modeling of physical effects in iterative algorithms can improve accuracy, but this added refinement can further compound the processing time. Initially, this disadvantage hampered the transition of iterative techniques from the research lab to the clinic. However, iterative techniques are now in widespread clinical use, owing to improvements in computer power and the development of efficient modeling techniques and fast reconstruction algorithms.

An additional distinction between iterative methods and FBP is in the appearance of the images they produce; in particular, the noise texture and image detail in an FBP image can look significantly different than that in an iterative reconstruction. Therefore, physicians must take these distinctions into account when interpreting the images.

There is not yet a consensus that iterative reconstructions are always superior to FBP images or, at least, that the benefits of iterative reconstructions always justify the increased computational costs; therefore, the two approaches will probably continue to coexist for some time.

This chapter presents the general principles of iterative image reconstruction and is intended to provide the reader with a starting point for further study. We loosely classify iterative methods into a few major types and provide details on some prominent examples of these major categories, but the reader should bear in mind that there are many variations on each theme in addition to the examples given. This chapter is intended as a tutorial on the basic concepts and not as a comprehensive survey of the literature. Regretfully, some algorithms deserving of more extensive discussion do not receive it, and some algorithms deserving of mention have no doubt been overlooked.

We begin with a formulation of the tomography problem as a linear inverse problem and define the statistical characteristics of the measured data. We then describe the two main components of any iterative method: (1) the criterion for selecting the best image solution and (2) the algorithm for finding that solution.

II. TOMOGRAPHY AS A LINEAR INVERSE PROBLEM

A. Linear Model of the Imaging Process

Any ET reconstruction problem can be formulated as the following estimation problem:

Find the object distribution **f**, given (1) a set of projection measurements **g**, (2) information (in the form of a matrix **H**) about the imaging system that produced the measurements, and, possibly, (3) a statistical description of the data and (4) a statistical description of the object (Fig. 1).

This problem statement applies equally to PET and SPECT and to all types of hardware configurations—for example, ring systems or dual rotating cameras in PET and

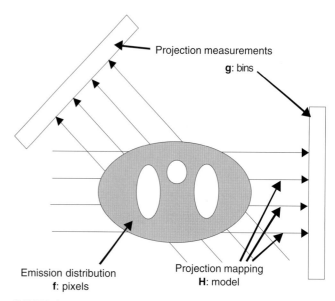

FIGURE 1 A general model of tomographic projection in which the measurements are given by weighted integrals of the emitting object distribution.

parallel, fan, cone, pinhole, or coded-aperture collimation in SPECT.

Indeed, if we assume that the imaging process is linear, the ET reconstruction problem is essentially similar to any linear inverse problem of the following form:

$$g_i = \int_{\mathbb{R}^D} d\mathbf{x} f(\mathbf{x}) h_i(\mathbf{x}), i = 1,...,P \quad (1)$$

where **x** is a vector denoting spatial coordinates in the image domain, g_i represents the ith measurement, and $h_i(\mathbf{x})$ is the response of the ith measurement to a source at **x**. In two-dimensional (2D) slice imaging, $D = 2$ and $\mathbf{x} = (x, y)$; in three-dimensional (3D) imaging, $D = 3$ and $\mathbf{x} = (x, y, z)$. The point-spread function (PSF) $h_i(\mathbf{x})$ can represent the effects of attenuation and all linear sources of blur. A good example of how various effects can be incorporated in the PSF can be found in Qi et al. (1996). For simplicity, we have suppressed additive contributions to the data, such as accidental coincidences (randoms) in PET.

What distinguishes tomography from other problems described by this linear model is that g_i are projections. Here, we use the term loosely to mean that the data are Radon-like, which implies a specific form for $h_i(\mathbf{r})$. In the ideal 2D Radon model on which FBP is based, $h_i(\mathbf{x}) = \delta(\mathbf{s}_i^T\mathbf{x} - t_i)$, in which case the measurements are simply line integrals of the form $g_i = \int_{L_i} f(\mathbf{x}) d\mathbf{x}$, where L_i is the line $\mathbf{s}_i^T\mathbf{x} = t_i$. Real ET data cannot be described by integrals over infinitely thin lines but instead must be thought of as integrals over finite strip- or cone-shaped regions of the object.

For computing purposes, we cannot represent the reconstructed image by a continuous-domain function; instead, we estimate a sampled version of the image, described in a discrete domain by column vector **f** (Fig. 2). Thus, each

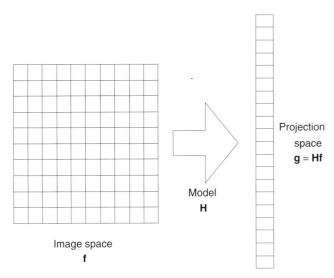

FIGURE 2 A discrete model of the projection process.

measurement in Eq. (1) can be approximated by the following system of linear equations:

$$g_i = \boldsymbol{h}_i^T \mathbf{f}, \quad i = 1,\ldots,P \tag{2}$$

which can be summarized by a single matrix equation as follows:

$$\mathbf{g} = \mathbf{H}\mathbf{f} \tag{3}$$

Here, $\mathbf{h}_i$ is the ith row of $\mathbf{H}$, and each element of $\mathbf{f}$, denoted by f_j, $j = 1, \ldots, N$, represents one pixel in the image space. For our purposes, the ordering of the pixels in $\mathbf{f}$ does not matter, but lexicographic ordering is usually assumed. In this general notation, $\mathbf{f}$ may represent either a 2D slice image or a 3D volume image, and complicated imaging systems can be readily represented within this approach by appropriate definition of $\mathbf{H}$.

We use the word *pixel* to refer to elements of the image domain, although it should be understood to encompass the term *voxel*, which is an element of a volume image. Each pixel is associated with a basis function $\phi_j(\mathbf{x})$, which transforms the continuous-domain function $f(\mathbf{x})$ into pixel values $f_j = \int_{\mathbb{R}^D} f(\mathbf{x})\phi_j(\mathbf{x})d\mathbf{x}$. The most commonly used pixel basis functions are ones that are constant within small, nonoverlapping rectangular (or cubic) regions arranged in a rectangular grid. In this case, the intensity of a pixel is intended to represent the expected number of emissions from within that pixel. However, there are benefits to using other types of basis functions, such as Gaussian basis functions (Lewitt, 1992) or finite-element models that adapt to the content of the image (Brankov *et al.*, 2004).

Note that the projection space is also discrete, with the projection data represented by the vector $\mathbf{g}$. The elements of $\mathbf{g}$ are referred to here as *projection bins* or simply *bins*, and every projection measurement is represented by one bin. Again, the ordering of the individual bins is not important.

Bins are generally sampled uniformly, although they need not be.

In Eq. (3), $\mathbf{H}$ is a $P \times N$ matrix called the *system matrix*, which describes the imaging process, and can represent attenuation and any linear blurring mechanisms. Each element of $\mathbf{H}$ (denoted by h_{ij}) represents the mean contribution of pixel j in the object to bin i in the projections. It is in the specification of $\mathbf{H}$ that the model of the projection process can become as simple or as complex as we require because the intensity of a projection bin is a weighted sum of intensities of the image pixels. To represent the Radon case, the matrix elements are defined so that a projection bin receives contributions only from pixels that are intersected by a given line and the contributions of pixels that do not intersect the line are set to zero. The linear model can also represent a more realistic case wherein a projection bin receives contributions from many pixels, each weighted according to the relative sensitivity of the projection bin to each pixel. These contributions are affected by physical factors such as attenuation, detector response, and scatter and can be estimated from knowledge of the system design and measurement of the patient attenuation distribution.

B. Statistical Model of Event Counts

So far, we have dealt only with the average behavior of the imaging system and have neglected the variability inherent in the photon-counting process used in ET. In consideration of the randomness in the projection data, Eq. (3) should truly be written as:

$$E[\mathbf{g}] = \mathbf{H}\mathbf{f} \tag{4}$$

where $E[\cdot]$ denotes expected value.

1. Poisson Model

Photon emissions are known to obey the Poisson distribution, and photon detections also obey the Poisson distribution, provided that detector dead time can be neglected and that no correction factors have been applied to the data. In this case, the numbers of events detected in the projection bins are independent of one another. Thus, the probability law for $\mathbf{g}$ is given by:

$$p(\mathbf{g}; \mathbf{f}) = \prod_{i=1}^{P} \frac{\bar{g}_i^{g_i} \exp(-\bar{g}_i)}{g_i!} \tag{5}$$

where $\bar{g}_i$ is the ith element of $E[\mathbf{g}] = \mathbf{H}\mathbf{f}$:

$$\bar{g}_i = \sum_{j=1}^{N} h_{ij} f_j \tag{6}$$

Although it is not exact for real imaging systems, the Poisson model is a good description of raw ET data and is the most commonly used model in the ET field. However, other probability models are often used. For example, when a PET imaging system internally corrects for random coincidences by subtracting an estimated randoms contribution, the statistics can be described by a shifted Poisson model (Yavus and Fessler, 1996). Gaussian models, which are described in the following section, are also often used because of their practical advantages.

2. Gaussian Model

In cases in which the mean number of events is reasonably high, the Poisson law in Eq. (5) can be approximated by the following Gaussian probability density function (PDF):

$$p(\mathbf{g};\mathbf{f}) = k\exp\left[-\frac{1}{2}\sum_{i=1}^{P}\frac{(g_i - \bar{g}_i)^2}{\bar{g}_i}\right] \quad (7)$$
$$= k\exp\left[-\frac{1}{2}(\mathbf{g}-\mathbf{Hf})^T\mathbf{C}^{-1}(\mathbf{g}-\mathbf{Hf})\right]$$

where k is a normalizing constant, and $\mathbf{C} = \text{diag}\{\bar{g}_1, \ldots, \bar{g}_P\}$ is the covariance matrix of $\mathbf{g}$. This approximation, which models the Poisson distribution to second order, is reasonably accurate when the mean numbers of events $\bar{g}_i$ are 20 or greater (Kalbfleisch, 1985); at low counts, the Poisson distribution becomes asymmetric about its peak, whereas the Gaussian distribution is always symmetric. Note that negative values of the elements of $\mathbf{g}$ have a probability of zero in the Poisson law, whereas the Gaussian approximation permits negative values. Thus, Poisson-based algorithms often have built-in constraints of nonnegativity, whereas Gaussian-based algorithms require additional constraints to achieve nonnegativity.

C. Spatiotemporal (4D) Imaging Model

In writing Eq. (1), we suppressed the fact that the activity distribution in the patient is actually a function of time. This fact is not considered in a static ET study, in which all the counts measured during the imaging session are summed together and used to produce a single static image of the patient. In this case, $f(\mathbf{x})$ in Eq. (1) should be interpreted as the time average of the spatiotemporal activity distribution $f(\mathbf{x},t)$.

Whereas a static study is concerned only with the spatial distribution of the tracer, gated studies and dynamic studies also measure temporal variations of the tracer concentration, as explained in Wernick and Aarsvold (Chapter 2 in this volume). In an ET study, there are two types of temporal variations of interest: (1) fluctuations caused by physiological interactions of the tracer with the body and (2) cardiac motion, which helps assess whether the heart is functioning normally. Other temporal variations, including respiratory motion, voluntary patient motion, and the steady decline of activity associated with radioactive decay, are uninformative effects to be corrected for, if possible.

For time-sequence imaging, as in the case of dynamic or gated studies, the imaging model may be expressed as follows:

$$g_{ik} = \frac{1}{\tau_k}\int_{I_k}dt\int_{\mathbb{R}^{D+1}}d\mathbf{x}\,f(\mathbf{x},t)h_i(\mathbf{x},t) \quad (8)$$
$$i = 1,\ldots,P, k = 1,\ldots,K$$

where I_k is the time interval of duration τ_k during which the kth frame of data is acquired. Equation (8), which describes the entire time sequence of data, can be written in discrete form as follows:

$$\tilde{\mathbf{g}} = \tilde{\mathbf{H}}\tilde{\mathbf{f}} \quad (9)$$

where the space-time system matrix $\tilde{\mathbf{H}}$ is given by:

$$\tilde{\mathbf{H}} = \begin{pmatrix} \mathbf{H}_1 & & \mathbf{0} \\ & \mathbf{H}_2 & \\ \mathbf{0} & & \ddots\, \mathbf{H}_K \end{pmatrix} \quad (10)$$

and the space-time data $\tilde{\mathbf{g}}$ and image $\tilde{\mathbf{f}}$ are concatenations of all the data and image frames, that is, $\tilde{\mathbf{g}} = (\mathbf{g}_1^T, \ldots, \mathbf{g}_K^T)^T$ and $\tilde{\mathbf{f}} = (\mathbf{f}_1^T, \ldots, \mathbf{f}_K^T)^T$. It can often be assumed that the system matrix is time-invariant, that is, $\mathbf{H}_1 = \mathbf{H}_2 = \ldots = \mathbf{H}_K$, in which case $\tilde{\mathbf{H}} = \mathbf{I} \otimes \mathbf{H}_1$, where $\otimes$ denotes the Kronecker product (Wernick *et al.*, 1999).

III. COMPONENTS OF AN ITERATIVE RECONSTRUCTION METHOD

It is important to recognize that any method for image reconstruction is composed of two related but distinct components. The first component, which we call the *criterion*, is the statistical basis or governing principle for determining which image is to be considered as the best estimate of the true image. The second component, which we call the *algorithm*, is the computational technique used to find the solution specified by the criterion. In short, the criterion is a strategy, and the algorithm is a set of practical steps to implement that strategy.

IV. IMAGE RECONSTRUCTION CRITERIA

A. Constraint Satisfaction

A simple approach to image reconstruction is to view the problem as one of finding an image that satisfies the constraints dictated by the measured data and prior knowledge (such as the nonnegativity of pixel intensity values). This is the basis for a variety of algorithms known in image reconstruction as algebraic reconstruction techniques (ART) (Herman, 1980); in engineering, these methods are known as *vector-space projection methods* (Stark and Yang, 1998).

To illustrate this idea, let us first suppose that the data are uncorrupted by noise. In this case, each projection measurement $g_i = \mathbf{h}_i^T\mathbf{f}$, $i = 1, \ldots, P$, defines a hyperplane in which the solution $\mathbf{f}$ must lie. In this sense, every hyperplane $g_i = \mathbf{h}_i^T\mathbf{f}$ is viewed as a set to which the solution is constrained to belong, and the solution must lie in the intersection of all these sets. If the number of measurements (equations) is greater than or equal to the number of pixels (unknowns), then the intersection of all the constraint sets may be a single point and the solution would be unique. A set of constraints that has a nonempty intersection is said to be *consistent*.

The essence of all constraint-satisfaction methods is that they aim to satisfy the known constraints. However, the problem is complicated when noise is present. In this case, each hyperplane (defined by a projection measurement) will be shifted by a random distance and the intersection of the hyperplanes (constraint sets) will be empty, unless the number of measurements is less than the number of pixels, in which case an infinite number of solutions may exist. The effects of noise can be alleviated by introducing constraints based on prior knowledge, usually representing image smoothness and non-negativity. Iterative algorithms for solving the constraint-satisfaction problem are described in Section VB.

The weakness of constraint-based methods is that they offer no mechanism for incorporating an explicit statistical model of the data. Thus, although these methods enjoyed considerable interest early in the development of ET, they have been largely supplanted by the maximum-likelihood and Bayesian methods described next.

B. Maximum-Likelihood Criterion

The maximum-likelihood (ML) criterion is a standard statistical-estimation criterion, proposed by R. A. Fisher (1921). In the ML criterion, it is presumed that the probability law $p(\mathbf{g};\mathbf{f})$ for the observation vector $\mathbf{g}$ is determined by some unknown deterministic parameter vector $\mathbf{f}$, which in our case is the object distribution we hope to reconstruct. In this context, $p(\mathbf{g};\mathbf{f})$ is called the likelihood function, which we denote by $L(\mathbf{f})$.

The ML criterion gives us a prescription for deciding which image, among all possible images, is the best estimate of the true object. The ML criterion can be stated simply as follows:

ML criterion: Choose the reconstructed image $\hat{\mathbf{f}}$ to be the object function $\mathbf{f}$ for which the measured data would have had the greatest likelihood $p(\mathbf{g};\mathbf{f})$.

In this sense, the ML criterion seeks a solution that has the greatest statistical consistency with the observations. Symbolically, the ML strategy can be written as follows:

$$\hat{\mathbf{f}} = \arg\max_{\mathbf{f}} p(\mathbf{g};\mathbf{f}) \quad (11)$$

that is, choose the value of $\mathbf{f}$ for which $p(\mathbf{g};\mathbf{f})$ is greatest.

ML estimators have some desirable properties that justify their use in many situations (Van Trees, 1968). First, ML estimators are *asymptotically unbiased*, meaning that, as the number of observations becomes large, the estimates become unbiased (i.e., $E[\hat{\mathbf{f}}] \to \mathbf{f}$). Second, ML estimators are *asymptotically efficient*, meaning that (again, for large data records) they yield the minimum variance among unbiased estimators. In other words, ML estimators are less susceptible to noise than other unbiased estimators.

Unfortunately, although ET images reconstructed using the ML criterion have the least variance (image noise) among unbiased estimators, the variance is still unacceptably high. Therefore, we invariably choose to permit a certain amount of bias in the reconstructed image in exchange for reduced variance. This is accomplished by introducing spatial smoothing in the images, which reduces image noise at the cost of reduced fidelity in the mean. As we see later, smoothing can achieved *explicitly* (by Bayesian methods or lowpass filtering) or *implicitly* (by prematurely stopping an iterative ML algorithm before it actually reaches the ML solution).

C. Least-Squares and Weighted-Least-Squares Criteria

In statistical estimation problems in which the likelihood function is unknown, one can instead use the least-squares (LS) principle to determine the best solution. In the context of image reconstruction, the LS criterion can be stated as follows:

LS criterion: Choose the value of $\mathbf{f}$ that, if observed through the system matrix $\mathbf{H}$, would yield projections $\mathbf{Hf}$ that are most similar to the observed projections $\mathbf{g}$ (in terms of Euclidiean distance).

Thus, the LS solution also aims to maximize the consistency between the observed data and the reconstructed image. The LS estimation method can be expressed symbolically as follows:

$$\hat{\mathbf{f}} = \arg\min_{\mathbf{f}} \|\mathbf{g} - \mathbf{Hf}\|^2 = \arg\min_{\mathbf{f}} \sum_{i=1}^{P}\left(g_i - \sum_{j=1}^{N} h_{ij} f_j\right)^2 \quad (12)$$

Equation (12) can be solved analytically to obtain the following closed-form solution:

$$\hat{\mathbf{f}} = \mathbf{H}^+\mathbf{g} \quad (13)$$

where $\mathbf{H}^+ = (\mathbf{H}^T\mathbf{H})^{-1}\mathbf{H}^T\mathbf{g}$ is the pseudoinverse of $\mathbf{H}$ (here we have assumed that $\mathbf{H}^T\mathbf{H}$ is invertible). This closed-form solution is not often used in ET because of the large dimension of **H**; therefore iterative procedures are normally employed.

When we have knowledge that some of the projection data g_i have greater variance than others, we can weight each of the error terms in the summation in Eq. (12) differently. This approach, called weighted-least-squares (WLS) estimation, can be written as follows:

$$\hat{\mathbf{f}} = \arg \min_{\mathbf{f}} (\mathbf{g} - \mathbf{Hf})^T \mathbf{D}(\mathbf{g} - \mathbf{Hf})$$
$$= \arg \min_{\mathbf{f}} \sum_{j=1}^{p} d_i \left(g_i - h_{ij} f_j \right)^2 \quad (14)$$

where **D** is a diagonal matrix, with elements d_i on the diagonal. The weights d_i are usually chosen to be $(\text{var}[g_i])^{-1}$. For ET data, which are Poisson-distributed, the variance equals the mean, so $d_i = \bar{g}_i^{-1}$. Like the LS solution, the WLS solution can be written in closed form as follows:

$$\hat{\mathbf{f}} = (\mathbf{H}^T \mathbf{D} \mathbf{H})^{-1} \mathbf{H}^T \mathbf{D}^{-1} \mathbf{g} \quad (15)$$

but, again, iterative methods are usually used instead because of the large dimension of **H**. Another WLS approach is the iteratively reweighted LS method, in which **D** is estimated from projections of the current image estimate.

It is important to recognize that, although the LS and WLS criteria do not explicitly refer to any probability model for the data, they are actually equivalent to ML estimation under a Gaussian model. In fact, WLS reconstruction was one of the earliest ML methods to be applied to ET reconstruction (Huesman *et al.* 1977). To see the connection between WLS and ML, compare the WLS function in Eq. (14) to the Gaussian model of ET data in Eq. (7). These functions are equivalent if we choose $\mathbf{D} = \mathbf{C}^{-1}$. In addition, the LS function is identical to ML estimation under an assumption that the observations g_i have equal variance, which is a poor assumption in ET.

D. Shortcoming of Maximum-Likelihood, Least-Squares, and Weighted-Least-Squares Methods

The shortcoming of the aforementioned statistical methods, when applied to ET image reconstruction, is that they tend to produce images that are exceedingly noisy. This problem arises because these classical criteria aim solely to enforce maximal consistency between the reconstructed image and the measured data. Unfortunately, because the data are noisy, the image that is most consistent with these data also tends to be noisy.

This is particularly serious in ET reconstruction because ET imaging systems are lowpass in nature (i.e., they produce data that are blurred), whereas the noise is broadband (i.e., it contains components at all frequencies).[1] To see intuitively why this presents a problem, let us consider the LS criterion. According to Eq. (13), the LS solution is obtained by applying the pseudoinverse of **H** to the data. If **H** is a lowpass operator, then its pseudoinverse is a highpass operator; thus it tends to amplify noise in the image. Specifically, if we view noise in the observed data as an additive zero-mean contribution **n** (so that $\mathbf{g} = \mathbf{Hf} + \mathbf{n}$) the LS solution is:

$$\hat{\mathbf{f}} = \mathbf{H}^+ \mathbf{g}$$
$$= \mathbf{H}^+ (\mathbf{Hf} + \mathbf{n})$$
$$= \mathbf{f} + \mathbf{H}^+ \mathbf{n}$$

The first term in the solution is the correct answer, but the second term consists of noise that has been subjected to a highpass operator that amplifies high-frequency components. The result, $\hat{\mathbf{f}}$, is generally an extremely noisy image.

E. Bayesian Methods

The ML, LS, and WLS methods are referred to as *classical* estimation criteria, which refers to their assumption that **f** is unknown but deterministic (not random). Classical methods are based on the notion that the data alone should determine the solution and that no prejudice from the experimenter should influence the estimate. In contrast, Bayesian methods assume that the unknown quantity **f** is itself random and can therefore be described by a PDF $p(\mathbf{f})$ that is known in advance of data collection. This permits the experimenter to inject his or her own prior expectations about **f** into the process. For example, if we image a patient's brain, we can imagine this person's brain to be a sample drawn randomly from some hypothetical population of brains, defined by $p(\mathbf{f})$. This PDF, called the *prior* (because it reflects information known in advance), permits us to modify the reconstructed image to conform to our expectations. We know that a positron emission tomography (PET) brain image should look something like a brain and not a car, a heart, or anything else that is not a brain. Thus, we can postulate in advance (*a priori*) that images that look like cars or hearts should have probability zero and those that look like brains should have some positive probability.

Philosophically speaking, introducing our prior beliefs may not be appropriate in a hypothesis-driven science experiment because it biases the outcome toward our preconceived expectations. However, the Bayesian approach is a very helpful practical tool for image reconstruction, provided that the prior is reasonable.

Ideally, the prior PDF might precisely define our prior knowledge about the true image, such as that the true image is

---

[1] Here we use the terms *lowpass* and *broadband* loosely because they apply properly only to space-invariant systems and stationary noise processes, which can be described by block-circulant systems and covariance matrices. However, analogous concepts can be discussed in the context of singular value decomposition.

a brain. Unfortunately, such a belief can be difficult to express mathematically. Therefore, Bayesian reconstruction methods usually do not make such ambitious statements about the true image **f**. Instead, these methods usually aim simply to encourage the reconstructed image to be smooth, so as to suppress the effect of noise. Specifically, a low probability is assigned to image solutions that have lots of fine detail on the assumption that these features are probably due to noise. This assumption is based on the knowledge that, because of its blurring effect, the imaging system **H** strongly suppresses image detail; therefore, any such detail that persists in the reconstruction is likely to have arisen from noise.

1. Theory of Bayesian Estimation

The main conceptual shortcoming of the ML criterion is that it fails to consider the consequences of our choosing one image solution over another. Bayes' theory, on the other hand, begins by stating these consequences explicitly in the form of a quantity called the loss function, denoted by $\lambda(\mathbf{f},\hat{\mathbf{f}})$. In the image reconstruction problem, this loss function measures the extent to which the reconstructed image $\hat{\mathbf{f}}$ deviates from the true image **f**. The loss function typically used in ET image reconstruction is called a *hit-or-miss* loss function:

$$\lambda(\mathbf{f},\hat{\mathbf{f}}) = \begin{cases} 0 & |\mathbf{f}-\hat{\mathbf{f}}| < \delta \\ 1 & \text{otherwise} \end{cases} \quad (16)$$

where δ is a positive constant, and $|\cdot|$ denotes L_1 norm, $|\mathbf{x}| = \Sigma_i |x_i|$. This loss function states that the reconstructed image $\hat{\mathbf{f}}$ is perfectly acceptable (zero loss) if it is sufficiently close to the true image **f** and that any less accurate result is equally unfavorable (unit loss).[2]

The aim of Bayesian methods is to find a criterion for choosing $\hat{\mathbf{f}}$ that will minimize the *average* loss that we experience when this criterion is applied. It is easy to show (Kay, 1993) that the minimization (on average) of Eq. (16) leads to the maximum *a posteriori* (MAP) criterion, which is stated as follows:

MAP criterion: Choose the value of **f** that maximizes the posterior PDF, $p(\mathbf{f}|\mathbf{g})$.

Symbolically, the MAP criterion can be expressed as:

$$\hat{\mathbf{f}} = \arg\max_{\mathbf{f}} p(\mathbf{f}\mid\mathbf{g}) \quad (17)$$

By using Bayes' law, given by:

$$p(\mathbf{f}\mid\mathbf{g}) = \frac{p(\mathbf{g}\mid\mathbf{f})\,p(\mathbf{f})}{p(\mathbf{g})} \quad (18)$$

the MAP criterion can be written in a more useful form as follows:

$$\hat{\mathbf{f}} = \arg\max_{\mathbf{f}} \frac{p(\mathbf{g}\mid\mathbf{f})\,p(\mathbf{f})}{p(\mathbf{g})} \quad (19)$$

By taking the logarithm of the quantity to be maximized, and omitting $p(\mathbf{g})$ (because it is not a function of **f**), the MAP criterion can be simplified to the following form:

$$\hat{\mathbf{f}} = \arg\max_{\mathbf{f}} \left[\ln p(\mathbf{g}\mid\mathbf{f}) + \ln p(\mathbf{f})\right] \quad (20)$$

From Eq. (20), we see that the MAP criterion is similar to the ML criterion; however, it uses the logarithm of the prior to penalize image solutions that do not have the expected properties. Specifically, the maximization in Eq. (20) attempts to balance consistency with the data **g** (as expressed by the likelihood term) with conformance to our prior expectations (expressed by the prior term). In other words, it aims to produce an image that is reasonably consistent with the data while not being too noisy. Note that the ML criterion can be viewed as a limiting case of the MAP criterion as the prior $p(\mathbf{f})$ tends toward a uniform PDF, which implies that *a priori* we do not prefer any solution over any other.

An additional benefit of the MAP criterion is that the function to be maximized can be sharper than the likelihood function and thus can make iterative reconstruction algorithms more efficient. Figure 3 illustrates the effect of the prior on a relatively flat ML objective function. The sharpness of the prior about its peak determines the sharpness of the MAP criterion. The resulting MAP solution is pushed away from the ML solution and toward the peak of the prior. Obviously, if the peaking of the prior is extreme, then the resulting solution will depend primarily on the prior and the measured data will be ignored. This undesirable result can be avoided by using a weak prior, so

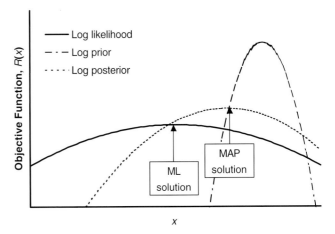

FIGURE 3 Comparison of objective functions in a simple 1D example. The MAP objective function (the log posterior) is sharper than the ML one (the likelihood), making the solution easier for an iterative algorithm to find.

---

[2]Other choices for the loss function lead to different reconstruction criteria. Notably, an L_2-error loss function implies a conditional mean, or minimum mean-square error, solution, whereas an L_1-error loss function leads to the median of the posterior as the solution.

that the solution captures the main properties of the ML solution while pushing the solution slightly in a direction that emphasizes smoothness.

2. Approaches to Defining the Prior

As we have seen, the MAP criterion is a simple one and not very different in form from the ML criterion. However, it can be difficult to express prior knowledge, by way of $p(\mathbf{f})$, in a mathematically meaningful and efficient way. One might be tempted, as we suggested earlier, to use the prior to express our knowledge that the true image should look like a brain, for example. However, this approach can be dangerous. For example, suppose that our prior model did not anticipate patients with a particular type of lesion. In this case, the reconstruction algorithm might view images containing this lesion as improbable and thus attempt to suppress the lesion in the reconstruction.

Owing to the difficulties associated with highly specific priors, the image reconstruction field has favored the use of priors that are simple and generic. The most prevalent approach in ET reconstruction is to use the prior simply to specify the belief that smooth images are more plausible than noisy ones. Such a prior makes a relatively bland statement about the true object distribution and can be applied generally because it does not aim to specify whether the object is a brain or a heart. The principal disadvantage of this approach is that legitimate image features can be obliterated along with the noise.

3. The Gibbs Distribution

A smooth image is one in which it is highly probable that neighboring pixels have similar intensity values. A prior that encourages this property attempts to suppress sharp transitions between pixels on the basis that such features are probably due to noise. By viewing the true pixel intensity values as random variables, a smooth image can be described as one in which intensity values of neighboring pixels are highly correlated while distant pixels are less so.

A simple mathematical model having this property is the Markov random field, which can be described by the Gibbs PDF (Geman and Geman, 1984):

$$p(\mathbf{f}) = \frac{1}{Z} \exp[-\beta U(\mathbf{f})] \quad (21)$$

where Z is a normalizing constant called the *partition function*; β is a scalar weighting parameter that determines the peaking of the distribution about its maxima; and $U(\mathbf{f})$ is the energy function. The energy function is a weighted sum of *potential functions*, which are functions of small sets of neighboring pixels, called *cliques*, denoted by S_c ($c = 1, \ldots, C$):

$$U(\mathbf{f}) = \sum_{c=1}^{C} \sum_{i,j,k\cdots \in S_c} w_{ijk\cdots} V_c(f_i, f_j, f_k, \cdots) \quad (22)$$

In Eq. (22), pixels indexed by $i,j,k,\ldots$, are elements of the same clique. This general model encompasses a number of priors proposed for ET reconstruction, including Gaussian (Levitan and Herman, 1987) and entropy (Liang et al., 1989) priors. Cliques may have any number of pixels. For example, the total intensity in the image can be constrained by using a clique that includes all the pixels in the image and an energy function based on the sum of the intensities of all pixels in the clique. It is permissible to mix clique sizes and energy functions in the Gibbs model, and a particular pixel may be a member of more than one clique.

In most ET reconstruction applications, there is a clique for each pixel i, and each clique consists of pixels that are nearest to pixel i. Methods using two-pixel cliques (e.g., Lee et al., 1995) generally use potential functions that are related to the difference in intensity between the two pixels, and the potential functions do not vary across the image. As a result, Eq. (22) is often simplified to:

$$U(\mathbf{f}) = \sum_{i=1}^{N} \sum_{j \in S_i, i<j} w_{ij} V_{ij}(f_i - f_j) \quad (23)$$

Although there are many possibilities for the clique structure and clique weights, usually cliques consist of local neighborhoods and the weights are usually determined by the inverse of the distance between the centers of the two pixels in the clique. The principal distinction between methods usually lies in the choice of potential function, which can strongly determine the smoothness properties of the MAP solution (Lalush and Tsui, 1992, 1993).

Clique potentials are generally bowl-shaped functions, as shown in Figure 4. The clique potential is zero at $r = f_i - f_j = 0$ and increases with increasing difference between the neighboring pixels. Thus, large intensity differences increase the energy function $U(\mathbf{f})$, which reduces the prior probability of the image. Clique energy functions fall roughly into two categories, based on how the behavior of the function for large r. Quadratic and higher-order functions (Fessler, 1994) apply increasing smoothing power as pixel differences increase. Linearly increasing functions (Green, 1990; Hebert and Leahy, 1992; Lalush and Tsui, 1993; Mumcuoglu et al., 1994) tend toward a linear function of r for large r. Priors that increase more slowly than a quadratic are sometimes called *edge-preserving* priors because they can have a selective effect on smoothing, removing small differences due to noise while retaining edge sharpness in the final result. Linearly increasing functions may be convex or nonconvex. Because exhaustive comparisons are difficult, it has not been established that any one type of clique potential is always preferred.

4. The Prior as a Penalty Term

Using the Gibbs prior of Eq. (21), the log posterior PDF becomes:

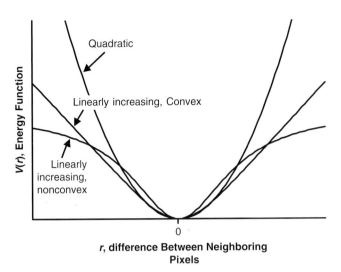

FIGURE 4 Potential functions used in Gibbs priors.

$$\ln p(\mathbf{f} \mid \mathbf{g}) = \ln L(\mathbf{f}) - \beta U(\mathbf{f}) \qquad (24)$$

where the likelihood function $L(\mathbf{f})$ may have either the Poisson or Gaussian form described earlier. Noisy images can produce large values of the likelihood function, but they will be penalized by the prior term $\beta U(\mathbf{f})$ and thus will tend not to be chosen when maximizing $\ln p(\mathbf{f}|\mathbf{g})$. In this way, a balance is struck between the requirements of the measured data, through the likelihood, and the requirements of the prior. This balance is governed by the weighting parameter β. If β is set to zero, the MAP solution is simply the ML solution; as β becomes large, the prior term dominates the maximization. Thus, β, which is called a *hyperparameter* of the optimization problem, is the most significant determinant of the degree of smoothing present in the MAP solution. Usually, β is set by the user based on experience and desired results, but there are several methods for automatically determining it (Higdon *et al.*, 1997; Johnson *et al.*, 1991; Zhou *et al.*, 1995).

The form of the prior determines whether Eq. (24) is unimodal or whether it may have disconnected local maxima. For priors with the form of Eq. (23), the clique potential function must be strictly convex in order to ensure that the objective function is unimodal (Lange, 1990). Thus, only convex priors (Fessler, 1994; Green, 1990; Mumcuoglu *et al.*, 1994) can guarantee convergence to a global maximum of the log posterior function $\ln p(\mathbf{f};\mathbf{g})$. Nonconvex priors (Hebert and Leahy, 1992; Lalush and Tsui, 1993) have been proposed and shown to have useful properties as well, despite the fact that they may converge only to local maxima.

The MAP method is also sometimes known as the *penalized ML method* because the prior term can be viewed as a penalty on solutions that have undesirable properties. By using this terminology one acknowledges that the chosen prior may not truly describe the PDF of the image, but may be only a practical device used to discourage unwanted image characteristics.

It should also be noted that the penalization approach can be extended to LS methods. For example, one can add a penalty term to the WLS function to obtain a method called penalized WLS (PWLS) (Fessler, 1994; Kay, 1993), which is equivalent to MAP reconstruction with a Gaussian likelihood function.

5. Anatomical Priors

Simple smoothing priors based on the Gibbs distribution are the predominant approach for MAP reconstruction. However, another approach involving what are called *anatomical priors* has also been widely studied (Chiao *et al.*, 1994; Gindi *et al.*, 1991; Leahy and Yan, 1991; Ouyang *et al.*, 1994; Brankov, Yang, Leahy, *et al.*, 2002). A major shortcoming of the basic Gibbs prior is that it penalizes all kinds of abrupt intensity variations in an effort to suppress noise. Unfortunately, in doing so it may also suppress legitimate image boundaries at the edges of anatomical features. Methods based on anatomical priors aim to identify valid image boundaries by using information gained from a second image of the patient obtained by an imaging modality such as magnetic resonance (MR) imaging. These methods have not gained wide acceptance, largely because of the difficulty of reconciling the different kinds of information that appear in the MR and ET images. For example, an anatomical boundary in an MR image may not have a corresponding functional boundary in the ET images, so imposing such a boundary would create a false image feature.

F. Criteria for Reconstruction of Image Sequences: 4D Reconstruction

Traditionally, image sequences have been reconstructed by applying the methods described thus far to each image frame, one by one. This approach misses an opportunity to reduce the effect of noise by encouraging smoothness between image frames. Between-frame smoothing can be used to great advantage in ET reconstruction because ET data usually vary slowly in time, whereas the noise is uncorrelated from frame to frame. In electrical engineering, image-recovery methods that exploit commonalities between images are called *multichannel methods*; a summary of these methods is given (Galatsanos *et al.*, 2000). The application of multichannel methods to ET has received considerable attention recently and is referred to as *spatiotemporal* or *4D reconstruction*.

1. Dynamic Imaging

Dynamic imaging, which may be achieved by PET or SPECT (see Wernick and Aarsvold, Chapter 2 in this volume), is used to identify temporal variations in the radiotracer concentration that reveal information about organ physiology. In dynamic imaging, the organ of interest is either reasonably stationary or is simply treated as such. In cardiac imaging, the result of this assumption is a motionless (and motion-blurred) image of the heart that does not depict cardiac motion but instead focuses on capturing important information about the kinetics of the tracer.

In dynamic imaging, between-frame smoothness can be achieved by enforcing smoothness in the reconstructed images along the time dimension of the image sequence. This is illustrated in Figure 5a, where, for simplicity, each image frame is depicted as a 2D slice. The connected pixels are those that are assumed to be correlated and may be grouped in the same clique in a Gibbs prior. The temporal variations in dynamic image sequences are usually very slowly varying, and the noise level can be exceedingly high; therefore, dynamic imaging can benefit greatly from temporal smoothing.

Many 4D reconstruction methods for dynamic imaging are based on representations of temporal variations in the image sequence using smooth time functions. For example, principal component (or Karhunen-Loève) basis functions were used in (Kao *et al.*, 1997; Wernick *et al.*, 1999). These basis functions, which are tailored to the specific data set, are designed to represent the data compactly (and thus yield a fast reconstruction) and to isolate and reject orthogonal noise components. Spline functions have also been used to model the variations in dynamic image sequences within the reconstruction algorithm (Nichols *et al.*, 1999; Reutter *et al.*, 2000). Many methods that estimate kinetic parameters from projection data assume exponential functions as the basis functions because these are solutions to standard kinetic models (See Morris *et al.*, Chapter 23 in this volume; Coxson *et al.*, 1990; Zeng *et al.*, 1995; Reutter *et al.*, 1998; Hebber *et al.*, 1997; Limber *et al.*, 1995). Several recent 4D reconstruction methods instead use various smoothing constraints on the temporal variations in the image sequence (Farncombe *et al.*, 1999, 2001), such as regional monotonicity of the time functions.

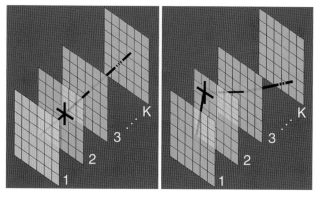

FIGURE 5 Smoothing used in spatiotemporal reconstruction (a) without motion compensation and (b) with motion compensation.

Dynamic SPECT has become a prominent research topic in recent years because traditional SPECT acquisitions have an important practical problem to overcome. For tracers with rapid kinetic properties, the tracer distribution in the body changes during the time required for the gamma camera to make a complete revolution about the patient. Thus, a complete set of projections is not available at any given instant of time. Therefore, images reconstructed by conventional means exhibit serious artifacts, and 4D reconstruction is essential.

2. Gated Imaging

In gated cardiac imaging, the dynamics of the tracer distribution are usually disregarded in favor of capturing cardiac motion. Whereas a dynamic study produces a long time sequence (like a movie), gated imaging produces a short, looping image sequence that represents a composite representation of a single heartbeat that summarizes data acquired over a large number of cardiac cycles.

In gated imaging, it is critical that cardiac motion be accounted for in the smoothing criterion. It is possible to accomplish this by viewing the heart's motion as a source of intensity fluctuations in each pixel (Narayanan *et al.*, 1999). However, a more appealing approach is to ensure that smoothing is performed between pixels in different frames that contain roughly equivalent tissue. The reasoning is that, if the myocardium (heart wall) were to occupy a given pixel in one frame but not in the next, the smoothing method would wrongly average (blur together) two distinct portions of the object. This will reduce the effect of noise, but will also distort the appearance of the object and its motion.

One approach uses the intensity-selective nature of certain Gibbs priors to enforce smoothing only among spatiotemporal neighboring pixels that share similar intensity (Lalush and Tsui, 1996, 1998a). This method does not require an explicit motion estimate and yet is able to resist the distortion.

Other 4D reconstruction methods for gated imaging bend the trajectories followed by the smoothing so as to follow the motion of the heart, as illustrated in Figure 5b. These techniques incorporate a motion estimation algorithm to determine the motion trajectories. Motion estimation has a long history of developments in the image-processing literature (Tekalp, 1995), but its application in gated ET imaging is relatively recent. Optical flow (Klein *et al.*, 1997) and deformable mesh modeling (Brankov, Yang, Wernick, *et al.*, 2002) have been studied, as well as simultaneous iterative methods of motion estimation and reconstruction (Mair *et al.*, 2002). Once motion is determined, smoothness can be enforced through the use of Gibbs priors, as demonstrated in (Lalush and Tsui, 1996, 1998a). Smoothness can also be enforced by using postreconstruction temporal filters (Brankov, Yang, Wernick, *et al.*, 2002), an idea that was first proposed (without motion estimation) by (King and Miller 1985).

The simultaneous reconstruction of gated and dynamic cardiac images has recently been proposed, allowing both wall motion and tracer dynamics to be assessed (Feng et al., 2003). This approach, which has been termed *5D reconstruction*, aims to gather more complete and accurate information about the heart from the available data.

V. ITERATIVE RECONSTRUCTION ALGORITHMS

Earlier we explained that an iterative image reconstruction method consists of a criterion for selecting the best image, combined with an algorithm for computing that image. Many different iterative approaches to solving the ET reconstruction problem have been studied, but they share some common traits. In the following section, we outline the common features of most iterative reconstruction algorithms and discuss some of their general properties.

A. General Structure of Iterative Algorithms

Most iterative reconstruction algorithms fit the general model shown in Figure 6. The process is begun with some initial estimate $\hat{\mathbf{f}}^{(0)}$ of the pixel intensity values in the image. A projection step is applied to the current image estimate $\hat{\mathbf{f}}^{(n)}$, which yields a set of projection values $\hat{\mathbf{g}}^{(n)}$ that would be expected if $\hat{\mathbf{f}}^{(n)}$ were the true image. The predicted projections $\hat{\mathbf{g}}^{(n)}$ are then compared with the actual measured data $\mathbf{g}$ to create a set of projection-space error values $\mathbf{e}_g$. These are mapped back to the image space through a backprojection operation to produce image-space error values $\mathbf{e}_f$ that are used to update the image estimate, which becomes the new estimate $\hat{\mathbf{f}}^{(n+1)}$. This process is repeated again and again until the iteration stops automatically or is terminated by the user. Each of these repetitions is called an *iteration*. At the conclusion of the process, the current image estimate is considered to be the final solution.

A critical but often-overlooked topic in image reconstruction is the practical computational choices involved in implementing the forward and backprojection steps involved in all iterative algorithms. In many cases, this issue can preclude the use of certain algorithm types because they are simply impractical to compute. Various approaches to defining the projection and backprojection operations are described in detail in Section VIIIA.

Note that we have not specified the details of the projection, comparison, backprojection, and update steps. It is principally in these steps that individual reconstruction algorithms differ. Note that direct reconstruction methods such as FBP use only the backprojection portion of the loop, so that there is no feedback about whether the image estimate, when projected, is consistent with the measured data. The power of iterative methods lies in the use of this feedback loop to refine the reconstructed image.

B. Constraint-Satisfaction Algorithms

Let us begin by reviewing the earliest iterative algorithms, which are based on the constraint satisfaction strategy introduced in Section IVA. The earliest algorithm of this kind, developed in 1937 by Kaczmarz (Rosenfeld and Kak, 1982; Kaczmarz, 1937), was designed for the solution of systems of linear equations. In the ET field, constraint-satisfaction methods are best known generally as ART (Gordon et al., 1970; Herman, 1980) and come in many varieties. Constraint-satisfaction methods have gained significant popularity in the electrical engineering community, where they are known as *vector-space projection methods* (Stark and Yang, 1998). Currently constraint-based methods are less prominent in the ET field than statistical methods, but they provide useful insights into the issues that underlie the reconstruction problem. These methods remain popular as solutions to other inverse problems, such as retrieval of Fourier phase information from magnitude-only data, and are worth understanding.

A brief summary of a general class of methods, called *projections onto convex sets* (POCS), is given in Section VIIIB. The following methods are all variations on this ap-proach, in which all the known information (data and prior knowledge) are based on constraints. The idea of POCS methods is to specify the solution as being a point in the solution space that satisfies all the constraints and thus lies in the intersection of all the sets that describe these constraints.

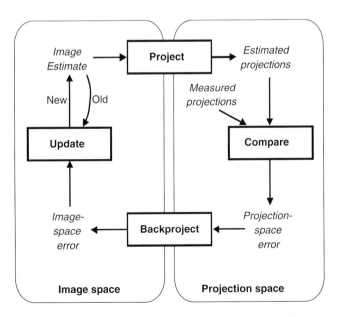

FIGURE 6 Flowchart of a generic iterative reconstruction algorithm.

1. The Kaczmarz Method/Algebraic Reconstruction Technique

Our imaging model, **g = Hf**, can be considered to be a set of simultaneous equations, one for each projection bin. Each linear equation in the imaging model, $g_i = \mathbf{h}_i^T \mathbf{f}$, defines a hyperplane in the vector space in which **f** is defined. Therefore, assuming that this set of simultaneous equations is consistent (which occurs when there is no noise), the solution is any point that lies in the intersection of all the hyperplanes. This point can be determined by a process in which, starting from an initial estimate $\hat{\mathbf{f}}^{(0)}$, the vector is repeatedly projected onto all the hyperplanes (here, the word *projection* is used in the linear algebra sense, not the tomography sense).

The operation of projecting a point onto a hyperplane is a simple one, given by:

$$\hat{f}_i^{(n+1)} = \hat{f}_i^{(n)} + h_{ji} \frac{\left(g_j - \sum_k h_{jk}\hat{f}_k^{(n)}\right)}{\sum_k h_{jk}^2} \quad (25)$$

Figure 7 shows a simple example of how the algorithm proceeds when there are only two pixels (two dimensions) and two measurements (two constraint sets).

A nonnegativity constraint set can also be introduced into the procedure. The operation of projecting onto this set is simple. In each iteration, any pixel having a negative value is set to zero:

$$\hat{f}_i^{(n+1)} = \begin{cases} \hat{f}_i^{(n)} & \text{if } \hat{f}_i^{(n)} \geq 0 \\ 0 & \text{if } \hat{f}_i^{(n)} < 0 \end{cases} \quad (26)$$

Similarly, pixels outside a particular image region S can be constrained to be zero by simply setting pixels outside S to zero in each iteration:

$$\hat{f}_i^{(n+1)} = \begin{cases} \hat{f}_i^{(n)} & \text{if } \hat{f}_i^{(n)} \in S \\ 0 & \text{if } \hat{f}_i^{(n)} \notin S \end{cases} \quad (27)$$

Various smoothing constraint sets may also be defined, as explained in Stark and Yang (1998).

The convergence rate of the Kaczmarz/ART method may be dependent on the orthogonality of the successive equations. Note in Figure 8a that when the equations are far from orthogonal, the process may require many iterations to reach the intersection. On the other hand, when the equations are more nearly orthogonal, the process requires relatively few iterations (Fig. 8b). It is important to note that when there is no unique solution the intersection of all the hyperplanes is empty, and so the iterative process oscillates among several solutions (Fig. 8c). This behavior can be avoided by underrelaxation, that is, by stopping short of each constraint set in each projection step. If done properly, this can lead to a LS solution to the problem (Censor *et al.*, 1983), which is a compromise solution lying in the middle of the triangular region in Figure 8c.

One disadvantage of ART is that it updates based on one measurement at a time. This requires a matrix-based ap-

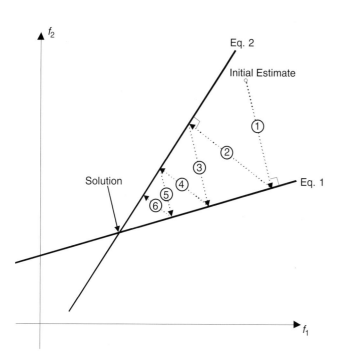

FIGURE 7 A simple example of the Kaczmarz procedure. The current estimate is projected successively onto each line by finding the point on each line that lies closest to the current estimate.

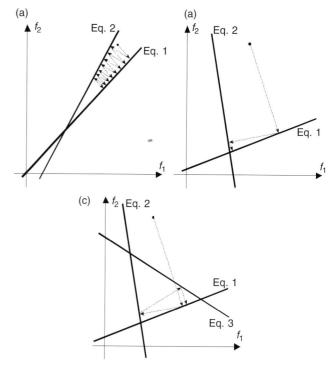

FIGURE 8 Illustrations of the Kaczmarz method in different situations, showing how the convergence of the method may be affected by the arrangement of the hyperplanes describing the linear equations.

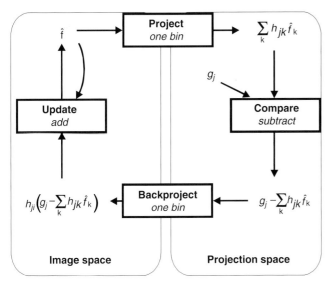

FIGURE 9 The algebraic reconstruction technique in the form of the general iterative model.

proach to the model so that individual elements of **H** can readily be accessed. This makes ART practical only for slice-by-slice reconstruction of parallel projection data.

2. Variations on ART

Several variations on ART have been proposed, and all fit the general iterative model of Figure 6, as shown in Figure 9. Multiplicative ART (MART) (Gordon *et al.*, 1970) uses a multiplicative error and update:

$$\hat{f}_i^{(n+1)} = \hat{f}_i^{(n)} \frac{g_j}{\sum_k h_{jk} \hat{f}_k^{(n)}} \text{ if } h_{ji} > 0 \qquad (28)$$

It has been shown that MART can be derived in a POCS framework (Mailloux *et al.*, 1993), but that its performance depends on the initial estimate. MART features automatic enforcement of nonnegativity: If the initial pixel intensities are nonnegative, then all iterated image estimates remain nonnegative. It is also simple to constrain pixels outside a region S to be zero by initializing them to zero.

Another variation of ART is the simultaneous iterative reconstruction technique (SIRT)[13], in which the update is performed for one pixel at a time using all equations that contribute to that pixel:

$$\hat{f}_i^{(n+1)} = \hat{f}_i^{(n)} + t \sum_j h_{ji} \frac{\left(g_j - \sum_k h_{jk} \hat{f}_k^{(n)}\right)}{\sum_k h_{jk}^2} \text{ if } > 0 \qquad (29)$$

The SIRT method is a parallel POCS algorithm, in which the projections are all performed simultaneously and then averaged. In SIRT, the relaxation parameter *t*, which may vary as the iterations progress, controls the convergence characteristics. An extension of SIRT, which has all pixels updated at the same time, allows the use of a projector-based model and removes the requirement for storing the matrix **H**.

In the following sections, we describe the most predominant iterative algorithms, WLS, ML, and MAP, which are based on statistical estimation criteria.

C. The Maximum-Likelihood Expectation-Maximization Algorithm

For several years, the leading iterative reconstruction algorithm for PET and SPECT has been the maximum-likelihood expectation-maximization (ML-EM) algorithm and its variations. The ML-EM algorithm was first proposed formally in 1977 as the solution to incomplete data problems in statistics (Dempster *et al.*, 1977) and has since found application in a wide range of statistical applications. Strictly speaking, the ML-EM algorithm presented in (Dempster *et al.*, 1977) is not an algorithm at all but rather a general prescription for developing algorithms that can be applied to a broad range of specific ML estimation problems (McLachlan and Krishnan, 1997).

Using this general prescription, an ML-EM algorithm for ET reconstruction was demonstrated in 1984 (Lange and Carson, 1984; Shepp and Vardi, 1982); however, the resulting iterative formula had been derived by a different approach in the 1970s (Lucy, 1974; Richardson, 1972) and was already known to the field of astronomy as the Richardson-Lucy algorithm. When applied to the ET reconstruction problem (or indeed to any linear inversion problem with Poisson noise), the ML-EM framework yields the following simple iterative equation, which is easy to implement and understand:

$$\hat{f}_j^{(n+1)} = \frac{\hat{f}_j^{(n)}}{\sum_{i'} h_{i'j}} \sum_i h_{ij} \frac{g_i}{\sum_k h_{ik} \hat{f}_k^{(n)}} \qquad (30)$$

Further explanation of the ML-EM algorithm, along with a derivation of the well-known iterative expression in Eq. (30), is provided in Section VIII C.

1. Properties of ML-EM

The ML-EM algorithm in Eq. (30) has a very simple form that fits our general model of an iterative algorithm (see Fig. 10). In fact, it is strikingly similar to MART (Eq. 28), the primary difference being that all pixels are updated simultaneously in the ML-EM algorithm. Its convergence behavior is consistent and predictable. Because the error and updates are multiplicative, ML-EM automatically imposes a nonnegativity constraint and allows for selected pixels to be preset to zero. The algorithm is simple to implement by computer using a projector-based model and eventually leads to a constrained ML solution.

The ML-EM algorithm for ET reconstruction has two main shortcomings. First, the convergence of the algorithm

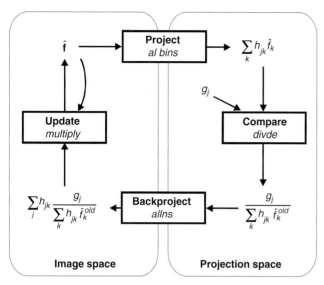

FIGURE 10 The maximum-likelihood expectation-maximization algorithm in the form of the general iterative model.

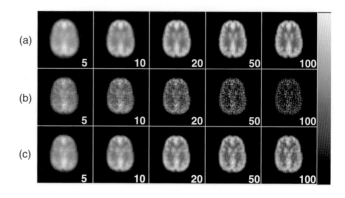

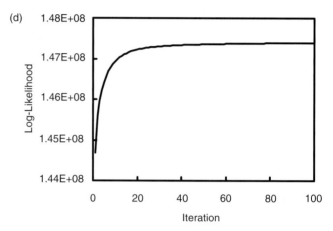

FIGURE 11 Convergence properties of the ML-EM algorithm. The images show the progression of iterated image estimates from simulated brain SPECT data. The algorithm modeled nonuniform attenuation in the matrix **H**. The cases shown are (a) noise-free, (b) noisy, and (c) the noisy reconstructions followed by a 3D Butterworth lowpass filter of order 4 and cut-off 0.2 cycles per pixel. The numbers in each cell indicate the iteration number. (d) The graph plots the log-likelihood function after each iteration up to 100 for the noisy data set.

is slow. A usable solution may require 30–50 iterations. Because each iteration requires one forward projection and one backprojection, the ML-EM algorithm can be expected to require one to two orders of magnitude more processing time than FBP. However, one should keep in mind that the ML-EM algorithm can perform much better than FBP because of its ability to solve the generic linear model and thus compensate for nonuniform attenuation, and so forth. Nevertheless, the required computation time initially hampered acceptance of the ML-EM method in clinical use.

The second shortcoming of the ML-EM algorithm is that the ML criterion on which it is based yields very noisy reconstructed images, as explained earlier. Thus, as the ML-EM iterations proceed, and the algorithm approaches the ML solution, the variance of the image estimate, which is manifested as noise, increases.

In practice, the ML-EM algorithm yields good results if the iterative procedure is stopped prematurely, and the results may, in addition, benefit from application of a postreconstruction lowpass filter, which is a common approach in clinical applications. Several approaches for deciding when to stop the iterations were proposed after the introduction of the ML-EM algorithm (e.g., Llacer and Veklerov, 1989). An alternative approach is the method of sieves (Snyder and Miller, 1985), which involves smoothing within each iteration to constrain the solution.

The behavior of the ML-EM algorithm with iterations is demonstrated in the example in Figure 11, which shows brain SPECT data realistically simulated using the Zubal brain phantom (Zubal *et al.*, 1994). ML-EM has been shown consistently to cause low spatial frequencies to appear first and then gradually to develop higher spatial frequencies as the iterations progress (Wilson *et al.*, 1994). Thus, as illustrated in the noise-free results of Figure 11a, early stoppage of ML-EM iterations amounts to an implicit form of smoothing of the reconstructed image. Figure 11b shows the image noise increasing as the estimate approaches the ML solution, but the postfiltered results (Fig. 11c) show little change after 50 iterations because the filter removes the frequency components that emerge in later iterations. The graph in Figure 11d shows the progression of the log-likelihood as a function of iterations, indicating that, although the log-likelihood appears to plateau early in the iterative process, the image estimates continue to change. The log-likelihood function is generally not a good measure of image quality for the same reason that the ML criterion is not an ideal reconstruction criterion, and it should be noted that images having nearly the same log-likelihood value can appear very different.

2. Variations on ML-EM

To address the problem of slow convergence, several methods for accelerating the ML-EM algorithm have been proposed. These include methods for extrapolating and increasing the magnitude of the change made at each iteration (Kaufman, 1987; Lange et al., 1987; Lewitt and Muehllehner, 1986; Rajeevan et al., 1992; Tanaka, 1987), using grids of different sizes to reduce processing time (Pan and Yagle, 1991; Ranganath et al., 1988), and increasing the number of updates by using only part of the data in each update (Browne and De Pierro, 1996; Byrne, 1996; Hudson and Larkin, 1994). Most of these methods improve the initial convergence of the algorithm; that is, they arrive at the log-likelihood plateau in fewer iterations. However, after reaching the plateau, the convergence rate is generally not increased greatly. An exception to this is the ordered-subsets EM (OS-EM) algorithm, which we discuss later.

The great success of the OS-EM algorithm in speeding up the reconstruction process, along with improvements in computer power, has lessened the practical need for additional acceleration strategies. However, there are other strategies worth noting. For example, the space-alternating generalized EM (SAGE) algorithm (Fessler and Hero, 1994) improves the convergence rate by updating each pixel individually and using a matrix-based projection model. SAGE is based on an alternate approach to the complete-data space, using instead a series of different complete-data spaces. The resulting algorithm converges to the ML estimate, obtains a usable result in fewer than 20 iterations, and can easily incorporate smoothing constraints. However, SAGE is inefficient for large 3D problems where projector-based models (Section VIIIA) must be used.

D. Least-Squares and Weighted-Least Squares Algorithms

The WLS criterion presented in Section IVC results in a quadratic function to be optimized. Optimization problems involving quadratic objective functions have been widely studied; therefore, there are many tools available for WLS reconstruction. The LS criterion is a special case of WLS, so the following discussion can be applied to LS reconstruction by choosing the weighting matrix to be the identity matrix.

1. General Structure of Methods

Many algorithms can be used to solve the WLS optimization problem in Eq. (14). All of them follow the same basic additive update formula:

$$\hat{\mathbf{f}}^{(n+1)} = \hat{\mathbf{f}}^{(n)} + t\,\Delta\mathbf{f}^{(n)} \tag{31}$$

where n is the iteration number; the scalar t is referred to as the *step size*; and the vector $\Delta\mathbf{f}^{(n)}$, which has the same dimensions as the image, is called the *step direction*. If we consider the multidimensional solution space, wherein each potential solution image is a point, then we can consider the iterative update as a movement through the solution space from $\hat{\mathbf{f}}^{(n)}$ to $\hat{\mathbf{f}}^{(n+1)}$.

2. Optimal Step Size

Algorithms for solving the WLS problem differ primarily in the manner in which the step direction is determined. For an arbitrary step direction, it is relatively simple to determine the step size that results in the greatest decrease in the weighted squared error. Let $J(\mathbf{f}) = (\mathbf{g} - \mathbf{H}\mathbf{f})^T \mathbf{D}(\mathbf{g} - \mathbf{H}\mathbf{f})$ denote the WLS objective function. Then we can optimize the step size by setting:

$$\left.\frac{\partial J(\mathbf{f})}{\partial t}\right|_{\mathbf{f}^{(n)}+t\Delta\mathbf{f}^{(n)}} = 0$$

and solving for t to obtain:

$$t = \frac{\Delta\mathbf{f}^{(n)T} \mathbf{H}^T \mathbf{D}\left(\mathbf{g} - \mathbf{H}\hat{\mathbf{f}}^{(n)}\right)}{\Delta\mathbf{f}^{(n)T} \mathbf{H}^T \mathbf{D}\mathbf{H}\Delta\mathbf{f}^{(n)}} \tag{32}$$

3. Steepest Descent

The steepest descent (SD) (or gradient descent) algorithm uses the gradient of the objective function as the step direction. In other words, the algorithm proceeds in each step in the direction in which the objective function decreases most quickly:

$$\Delta\mathbf{f} = -\left.\frac{\partial J(\mathbf{f})}{\partial \mathbf{f}}\right|_{\hat{\mathbf{f}}^{(n)}} = -\mathbf{H}^T \mathbf{D}\left(\mathbf{g} - \mathbf{H}\hat{\mathbf{f}}^{(n)}\right) \triangleq \mathbf{p} \tag{33}$$

To implement the algorithm efficiently requires a variation on the model of Figure 6, as shown in Figure 12. In this case, the projection space error $\mathbf{e} = (\mathbf{g} - \mathbf{H}\hat{\mathbf{f}}^{(n)})$ is updated incrementally along with the new image estimate and the next step direction. The initial gradient must be computed with an additional backprojection operation before the first iteration. Like the model in Figure 6, only a single backprojection and a single projection operation are required at each iteration. One iteration of SD takes about the same time as one iteration of ML-EM.

Although intuitive and relatively simple to implement, SD algorithms generally require more iterations than conjugate gradient algorithms to reach a usable solution. It is possible to enforce a nonnegativity constraint by setting negative pixels to zero after each iteration, but this requires recomputing the error vector $\mathbf{e}$ at the beginning of each iteration with an additional projection operation.

4. Conjugate Gradient

The conjugate gradient (CG) algorithm is more efficient than SD (Press et al., 1988) and has been applied effectively to reconstruction in ET (Huesman et al., 1977; Kaufman, 1993; Tsui et al., 1991). In CG, all step directions are chosen to be conjugate to one another:

$$\left(\mathbf{H}\Delta\mathbf{f}^{(i)}\right)^T \mathbf{D}\left(\mathbf{H}\Delta\mathbf{f}^{(j)}\right) = 0 \text{ if } i \neq j \tag{34}$$

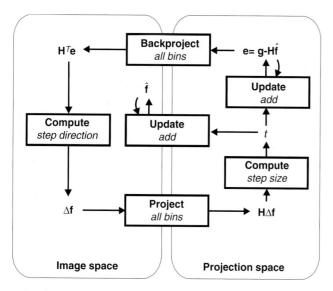

FIGURE 12 Weighted least-squares algorithms in a modified form of the general iterative model. The steepest descent, conjugate gradient, and coordinate descent forms differ only in how the step directions are computed.

where i and j are iteration numbers. Thus, minimization along a step direction does not interact with minimizations performed with respect to any previous step directions. In other words, each new step does not spoil the work of previous steps. For this reason, CG is guaranteed to find the minimum of a quadratic objective in N iterations, where N is the number of pixels in the image. However, a small 64×64 image will require 4096 iterations to reach the minimum! Fortunately, a usable image estimate is generally obtained in far fewer iterations.

The following relation, called the Fletcher-Reeves variant (Press *et al.*, 1988), can be used to compute conjugate gradient step directions:

$$\Delta \mathbf{f}^{(n+1)} = \mathbf{p}^{(n)} + \frac{\mathbf{p}^{(n)^T}\mathbf{p}^{(n)}}{\mathbf{p}^{(n-1)^T}\mathbf{p}^{(n-1)}} \Delta \mathbf{f}^{(n)} \qquad (35)$$

where the superscript n is the iteration number and the gradient $\mathbf{p}$ is computed as in Eq. (33). Here, the step direction depends on the previous step direction as well as the gradient directions at the current and previous iterations. This results in a slightly more complex computation of the step direction as compared to steepest descent, but a more efficient minimization of the weighted squared error. Enforcing a nonnegativity constraint is problematic, however, because manipulating negative pixels after one iteration effectively changes the step direction and ruins the delicate sequence of conjugate step directions generated. The implementation of CG can be done in the same form as Figure 12, substituting Eq. (35) for the step direction computation.

5. Coordinate Descent

A relatively simple approach to generating step directions is to update one pixel at a time. If a matrix-based model can be used (i.e., if the matrix $\mathbf{H}$ can be computed and stored), then a coordinate descent approach can be extremely efficient. If not, then coordinate descent is very inefficient. Coordinate descent has the added advantage of permitting the simple application of a nonnegativity constraint: If a pixel goes negative at one update, it is simply thresholded to zero with no real penalty to the algorithm. For coordinate descent, $\Delta \mathbf{f}$ is simply a vector with a 1 in the position of the pixel to be updated and zeros elsewhere. The optimal step size is again computed using Eq. (32), but it is only necessary to apply the elements of $\mathbf{H}$ that operate on the pixel in question, hence the need to be able to look up individual elements of $\mathbf{H}$. Coordinate descent has been applied successfully to ET reconstruction (Fessler, 1994) with the addition of a smoothing penalty, a method that is discussed later.

6. Properties of WLS Algorithms

The WLS-CG algorithm is not much faster than ML-EM, unless a transformation of the image space is first performed. This transformation matrix, called a *preconditioner* or *relaxation matrix* (Golub and Van Loan, 1989), seeks to equalize the curvature of the objective function along all its axes, making the algorithm more efficient in finding the minimum. The preconditioning process involves creating a transformed image space $\mathbf{v}$ as follows:

$$\mathbf{v} = \mathbf{Mf}, \quad \mathbf{M} = [\text{diag}(\mathbf{H}^T\mathbf{DH})]^{-1/2} \qquad (36)$$

The algorithm then solves for the gradients and step directions in the $\mathbf{v}$-space, converts the step direction back to the $\mathbf{f}$-space, and solves as usual for step size. The result (Lalush and Tsui, 1995) is that the computation of the current gradient $\mathbf{p}$ becomes:

$$\mathbf{p}^{(n)} = -\mathbf{M}^{-1}\mathbf{H}^T\mathbf{D}\left(\mathbf{g} - \mathbf{H}\hat{\mathbf{f}}^{(n)}\right) \qquad (37)$$

and the step direction in the $\mathbf{f}$-space is computed directly for preconditioned CG as:

$$\Delta \mathbf{f}^{(n+1)} = \mathbf{M}^{-1}\mathbf{p}^{(n)} + \frac{\mathbf{p}^{(n)^T}\mathbf{p}^{(n)}}{\mathbf{p}^{(n-1)^T}\mathbf{p}^{(n-1)}} \Delta \mathbf{f}^{(n)} \qquad (38)$$

The preconditioned WLS-CG algorithm has been claimed to converge up to 10 times faster than ML-EM (Tsui *et al.*, 1991). In general, from 8 to 15 iterations are required to reach a usable solution. Iterated image comparisons are shown in Figure 13. These show that, without preconditioning, convergence is still rather slow. With preconditioning, convergence is rapid, although rather significant changes occur from one iteration to the next at early iterations. Also, the spatial frequency recovery is not as gradual or predictable as in ML-EM. As in ML-EM, image noise increases with further iterations as the solution gets

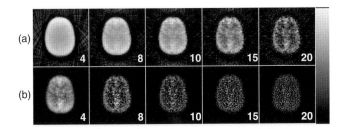

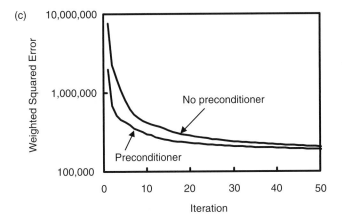

FIGURE 13 The convergence of the weighted least-squares conjugate gradient algorithm (a) without and (b) with a preconditioner. The images show the iterated image estimates from simulated brain SPECT data at the iteration numbers indicated. (c) The graph shows the weighted squared error as a function of iteration.

closer to the Gaussian ML solution. However, image noise with preconditioned WLS-CG tends to be somewhat higher than that obtained from ML-EM. This can be attributed to the lack of a nonnegativity constraint; when a pixel is affected by noise and tends to become negative, a nearby pixel may seek to cancel out this effect and increase in a positive direction. If pixel values are constrained from becoming negative, as in ML-EM, then large positive variations tend to be suppressed as well.

7. Other Poisson-Based ML Algorithms

Some of the optimization principles applied to the quadratic WLS objective have been applied to maximizing the Poisson likelihood function. One reason for this is to more accurately model the statistics in projection bins in which few events are acquired and the Gaussian approximation is less accurate. It is possible to derive a CG algorithm based on the nonquadratic Poisson likelihood (Mumcuoglu *et al.*, 1994). It is important to note that, with a nonquadratic objective, the conjugate gradient method no longer has the guarantee of convergence in *N* iterations, although this is of little practical use in the case of image reconstruction. Also, although the use of a Poisson objective constrains the projections of the image estimates to be nonnegative, individual pixel values may still become negative; thus, there is no inherent nonnegativity constraint on the pixels.

The biggest problem created by applying SD or CG to the Poisson-based objective is that there is not an explicit solution for the optimal step size as in Eq. (32). Thus, either a heuristic for step size must be used, or a 1D search method (Mumcuoglu *et al.*, 1994) must be employed to find the optimal step size. The former makes the algorithm difficult to use because the convergence will be data-dependent and dependent on the heuristic used. The 1D search is effective and does not significantly impact processing time, as shown in Mumcuoglu *et al.* (1994), but adds to the complexity of programming. Furthermore, preconditioning is required to achieve convergence rates comparable to those from WLS-CG.

E. Maximum A Posteriori Reconstruction Algorithms

As explained earlier, the ML solution is generally too noisy to be useful, so something must be done to control noise. In current clinical applications, the usual approach is to stop the iterations early and/or apply a linear filter. However, the MAP approach offers a more flexible and principled method of encouraging desirable properties in the reconstructed image.

1. MAP Algorithms

Owing to the similarities between the MAP and ML objective functions, most of the algorithms used for ML reconstruction have MAP counterparts. However, the prior term can often create complications that require approximations or additional steps. In other cases, the algorithm can be applied without difficulty, but only for certain forms of the prior.

The analog of the ML-EM algorithm is the following MAP-EM algorithm (Green, 1990):

$$\hat{f}_j^{(n+1)} = \frac{\hat{f}_j^{(n)}}{\sum_{i'} h_{i'j} + \beta \frac{\partial U(\mathbf{f})}{\partial \mathbf{f}}} \sum_i h_{ij} \frac{g_i}{\sum_k h_{ik} + \hat{f}_k^{(n)}} \quad (39)$$

This algorithm deviates from the ML-EM algorithm principally in that there is a prior term in the denominator. This term is problematic because it is supposed to be evaluated using the next estimate, $\hat{f}_j^{(n+1)}$, which is not yet available. MAP-EM algorithms differ in the approach they use to address this problem. The simplest approach is to evaluate the derivative term at the previous image estimate, a method called the one step late (OSL) procedure (Green, 1990). The MAP-EM OSL algorithm has been shown to converge to the MAP solution for only certain forms of the prior (Lange, 1990). Another type of approach, called a generalized EM (GEM) MAP algorithm (Hebert and Gopal, 1992), sequentially updates pixels and verifies that each update increases the posterior density, thus ensuring convergence to a maximum of the posterior density.

Several proposed algorithms have been based on the Gaussian likelihood function. The PWLS algorithm (Fessler 1994) employs a coordinate descent approach and uses a quadratic prior to solve explicitly for the optimal step size t. The algorithm is very efficient, usually requiring approximately 10–15 iterations, but requires a matrix-based approach and restricts priors to only quadratic forms. More general priors and projector-based models can be accommodated by a MAP conjugate gradient (MAP-CG) algorithm (Lalush and Tsui, 1995), which is analogous to the WLS-CG algorithm. This algorithm also requires approximately 10–15 iterations, but adds some programming complexity with the need to perform a local linear fit of the prior term in the step size calculation and has no nonnegativity constraint. It is also possible to use an EM algorithm with a Gaussian likelihood and a Gaussian prior (Levitan and Herman, 1987), but the resulting algorithm requires more iterations than PWLS or MAP-CG.

A Poisson-based CG algorithm, termed *preconditioned conjugate gradient* (PCG) (Mumcuoglu *et al.*, 1994), is the MAP analog of the Poisson-based CG algorithm. It requires line searching to optimize the step size, but its convergence rate is competitive with PWLS and MAP-CG and it does not have the convergence problems of the OSL method.

2. Properties of MAP Reconstructions

MAP methods alleviate the principal problems associated with ML algorithms. First, MAP reconstructions are smoother than their ML counterparts. Second, iterated MAP estimates tend to reach a point at which they change very little with further iterations, indicating approximate convergence (this behavior can be controlled by the weighting parameter β) (Lalush and Tsui, 1992). Figure 14 shows an example of the convergence behavior of MAP-EM OSL reconstructions with various values of β compared to ML-EM reconstruction. For moderate values of β, MAP reconstructions reach a point of effective convergence and are clearly smoother than their ML counterparts. As the value of β is decreased, the degree of smoothing is reduced because the criterion places less weight on the smoothing requirement of the prior. As β is increased, smoothing increases, but potentially important features in the image begin to degrade. Setting the weighting parameter is critical because excessive β values invariably result in a loss of contrast and detail and insufficient β values produce images that are too noisy.

Different types of priors produce different smoothing characteristics, as shown in Figure 15. Quadratic priors tend to result in smoothing that is qualitatively similar to that obtained by space-variant linear filtering (Fessler and Rogers, 1996). Linearly increasing priors (Green, 1990; Hebert and Leahy, 1992; Lalush and Tsui, 1992, 1993; Mumcuoglu *et al.*, 1994) usually have an adjustable *scaling*

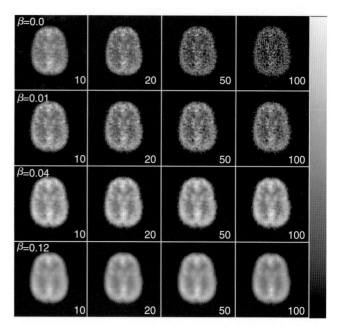

FIGURE 14 Iterated image estimates for the maximum *a posteriori* expectation-maximization one-step-late algorithm from noisy simulated brain SPECT data. Each row shows results for a different value of the weighting parameter, β, which determines the relative weight placed on the smoothing prior.

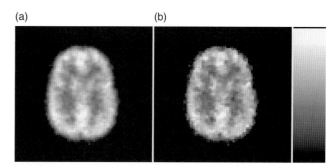

FIGURE 15 Results of using different types of potential functions with MAP-EM, zoomed to show detail. (a) Using a quadratic potential. (b) Using a linearly increasing, convex potential set to discourage small pixel differences.

parameter that sets the degree to which small pixel differences are discouraged. If strong smoothing of small pixel differences is applied, the result is a piecewise-continuous image with a number of regions of uniform intensity. This can result in spurious regions due to noise and false boundaries, which can be distracting to the viewer. Also, small and low-contrast image features may be lost. If the scaling parameter is set to allow small differences between neighboring pixels, the result can smooth noise while preserving a degree of sharpness in object boundaries.

Although MAP reconstruction successfully smoothes noise and improves convergence, it also has several disadvantages. First, success can be highly dependent on the choice of parameters. Unfortunately, it is not efficient to use a trial-and-error approach to parameter selection, as might be used with postreconstruction filters, because an entire iterative reconstruction must be performed to assess the result from one set of parameters. Second, excessive smoothing using Gibbs priors can result in a loss of image features or, in some cases, the creation of spurious features. This is a manifestation of the fact that the MAP estimator adds some bias to the ML problem in exchange for reduced noise variance. Finally, MAP algorithms apply smoothing properties that are somewhat different from the traditional Fourier-domain filters used in nuclear medicine, and so they may be difficult for physicians to interpret initially.

F. Subset-Based Reconstruction Algorithms

As noted earlier, the ML-EM algorithm has many desirable properties, but it suffers from slow convergence, a problem that has been addressed very successfully by subset methods (also called *block-iterative* or *row-action* methods). These techniques break up the full set of projection data into a series of mutually exclusive subsets and apply the reconstruction algorithm to each subset sequentially. Each pass through the data set is used to effect a greater number of iterative refinements, which results in significant acceleration compared with ML-EM, usually on the order of the number of subsets.

1. The OS-EM Algorithm

The OS-EM algorithm (Hudson and Larkin, 1994) is a simple modification of the ML-EM algorithm, given by:

$$\hat{f}_j^{new} = \frac{\hat{f}_j^{old}}{\sum_{i' \in S_n} h_{i'j}} \sum_{i \in S_n} h_{ij} \frac{g_i}{\sum_{l=1}^{N} h_{il} \hat{f}_l^{old}} \quad (40)$$

where the backprojections are performed only for the projection bins belonging to the subset S_n. At each update, a different subset of the projection data is used. Generally, one update in this algorithm is called a *subiteration* and one pass through all of the subsets is referred to as an *iteration*. Thus, the processing time for one iteration of OS-EM is comparable to that of one iteration of ML-EM.

The organization of the subsets is important to the performance of the algorithm. In addition, mathematical difficulties can result if any subset does not contain some contribution from every pixel in the field of view; in this case, the first summation in the denominator of Eq. (40) is zero. This is an important consideration in nonparallel projection geometries.

When reduced to a single subset encompassing all the projection data, the OS-EM algorithm reduces to the ML-EM algorithm. At the other extreme, one could define each subset to include a single projection bin, resulting in an algorithm similar to MART; however, this would preclude the use of projector-based models of **H**.

Usually, subsets are organized in groups of projection bins associated with one projection view or camera position, which is convenient for projector-based models. Figure 16 illustrates the typical procedure for a parallel projection model. First, a number of projection views are used to compute the update for the first subset. Then, more views are used to compute the next update, and so on. After all subsets have been used, the process begins again with the first subset.

Usually, the members of a subset are chosen to have maximum angular distance between them. For example, when creating 16 subsets from data with 128 projection views over a 360° arc, each subset will contain all the bins from eight views spaced at angular intervals of 45°. The sequence of updating subsets usually aims to maximize the angular spacing between successive subsets. In our example, the first subset may be chosen at cardinal directions, the second at 22.5° from the first, and the third subset offset one angular increment from the first (approximately 2.8°).

These are typical and intuitive approaches to organizing the subsets; however, evidence suggests that results are not

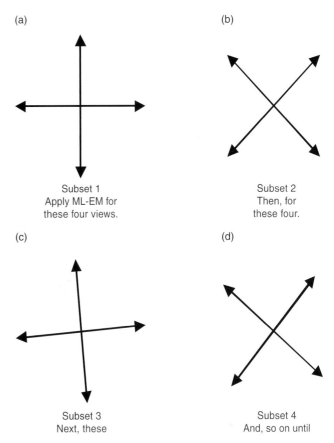

FIGURE 16 The process of using subset updates in a reconstruction algorithm. (a) Subset 1: Apply ML-EM for these four views. (b) Subset 2: Then apply the algorithm for these four. (c) Subset 3: Then apply the algorithm for these four. (d) Subset 4: Continue until all data are used.

very sensitive to the particular views chosen for a subset or to the ordering of the subsets, although suggestions for balancing the subsets are given in Hudson and Larkin (1994). On the other hand, the *number* of subsets, or alternately the number of views per subset, governs the degree of acceleration over ML-EM. A rule of thumb states that OS-EM at n iterations reaches roughly the same point of convergence as ML-EM at (Number of Subsets) × n iterations. The noise increases that much more quickly also, so the algorithm must be stopped and smoothing applied, just as in ML-EM.

2. Properties of OS-EM

Figure 17 shows examples of iterated image estimates using OS-EM for various numbers of subsets. Sustantial acceleration is achieved with little impact on the resolution and noise properties of the images. OS-EM appears to have nearly all the desirable properties of ML-EM, but with significant speed improvement. It generally requires fewer than seven iterations, compared to 8–12 for the fastest WLS-based algorithms. It has a built-in nonnegativity constraint and shares the property of ML-EM that low spatial frequencies converge first, with higher spatial frequencies improving with further iterations (Lalush and Tsui, 1998b). Its convergence properties are therefore very predictable and reproducible. The OS-EM algorithm is simple to implement, although there is the small added complexity of having to choose the number of subsets.

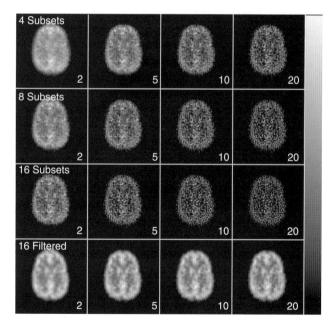

FIGURE 17 Iterated image estimates using the ordered-subsets expectation-maximization algorithm for simulated brain SPECT data. Iteration numbers are indicated in each cell. The original data had 128 views over a 360° arc and was broken up into 4, 8, and 16 subsets. The last row shows the results of filtering the 16-subset iterated estimates with a 3D Butterworth lowpass filter of order 4 and cut-off 0.2 cycles per pixel.

With all of its similarities to ML-EM, it is important to note that OS-EM is not really an EM algorithm and, at this writing, has no general proof of convergence. Experience with the algorithm has been generally good, and it has been shown, at least in some cases, to give results nearly identical to ML-EM in the mean (Lalush and Tsui, 2000). The principal cost of using the subset method is an increase in image noise for the same level of bias as compared to ML-EM (Lalush and Tsui, 2000). In other words, as more subsets are used in an effort to increase acceleration, image noise becomes worse. Thus, users should be wary of using a large number of subsets; modest acceleration of 8–10 times is possible with very little increase in noise.

3. Related Algorithms

Several other algorithms use the same subset concept as OS-EM to achieve usable results in a small number of iterations. The rescaled block-iterative (RBI) EM algorithm (Byrne, 1996) can be shown to converge to a consistent solution, provided one exists. This is a minor theoretical advantage, because most noisy data are bound to be inconsistent. In practical application, RBI-EM behaves almost identically to OS-EM, except that it tends to run approximately one-half as fast for the same number of subsets (Lalush and Tsui, 2000). The row-action maximum-likelihood algorithm (RAMLA) (Browne and De Pierro, 1996), developed independently, is almost identical to OS-EM, with the exception of a step-size adjustment factor. If this factor is decreased properly at successive iterations, then RAMLA will converge to the ML solution. Because the ML solution is not desirable due to its noise properties, this is again mostly a theoretical advantage. Several MAP analogs to the subset approach have also been developed (Hudson and Larkin, 1994; Lalush and Tsui, 1998a). For the most part, these cannot be proven to converge to the MAP estimate, but do appear to achieve the properties of MAP-EM, only faster.

G. Iterative Filtered Backprojection Algorithms

Many algorithms use an additive update that includes a filtered backprojection operation in the backprojection step. Although most of these iterative filtered backprojection (IFBP) methods were developed from intuitive arguments (i.e., without any specific governing criterion), they can be analyzed using some of the same principles that apply to WLS algorithms.

1. General IFBP Model

A general model for IFBP algorithms uses an additive update, as in Eq. (32), with the step direction determined by:

$$\Delta \mathbf{f}^{(n)} = \mathbf{B}\left(\mathbf{g} - \mathbf{H}\hat{\mathbf{f}}^{(n-1)}\right) \quad (41)$$

where the backprojection matrix $\mathbf{B}$ involves some use of a ramp filter or filtered backprojection. Contrast this with

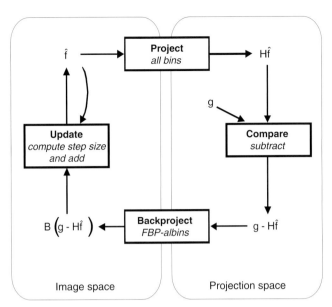

FIGURE 18 Iterative filtered backprojection algorithms in the form of the general iterative model.

the ML-EM and WLS algorithms, in which the backprojection step involves only the application of $\mathbf{H}^T$, which would be a conventional Radon backprojection in the case in which $\mathbf{H} = \mathbf{H}_{Radon}$. Different IFBP algorithms use different forms of the backprojection operation $\mathbf{B}$ and use different rules for determining step size. Figure 18 illustrates how IFBP algorithms follow the general iterative model.

2. Specific Algorithms

An early and widely used IFBP algorithm is the iterative Chang reconstruction method (Chang, 1979). This method was developed to facilitate reconstruction with correction for uniform or nonuniform attenuation in SPECT. The Chang algorithm uses $\mathbf{B} = \mathbf{W}\mathbf{B}_{FBP}$, where $\mathbf{W}$ is a diagonal weighting matrix having the same dimension as the image and composed of factors based on the average attenuation experienced by each pixel over the acquired arc, and $\mathbf{B}_{FBP}$ is the filtered backprojection operation. The step size t is always 1. This leads the Chang algorithm to be potentially unstable or oscillatory as iterations progress (Lalush and Tsui, 1994), a behavior that depends on the data. Still, if the attenuation distribution is relatively uniform, Chang's method can give reasonably accurate results in the first or second iteration.

The It-W iterative reconstruction algorithm (Wallis and Miller, 1993) performs the backprojection step by first applying a ramp filter to the projection-space error and then backprojecting with $\mathbf{H}^T$. The result is then weighted by the square of the Chang weighting matrix, $\mathbf{W}^2$. The step size in this case is applied to maintain a constraint on the total intensity in the reconstructed image, which should help prevent instability. The algorithm is fast (Wallis *et al.*, 1998), but it is complex to program and requires a number of adjustments to perform well.

Approximate SD and CG IFBP algorithms have also been developed (Lalush and Tsui, 1994). These methods are based on minimizing an unusual squared-error function whose gradient is approximately the step direction given in Eq. (41). Because the step size is optimized in this case, the algorithm is guaranteed to be stable. When the backprojection operator from the Chang algorithm is used in this context, the result is a stable version of the Chang algorithm. However, evidence does not indicate that IFBP-CG outperforms WLS-CG in any way.

Many other IFBP algorithms have been proposed (Liang, 1993, Maze *et al.*, 1993; Pan *et al.*, 1993; Walters *et al.*, 1981; Xu *et al.*, 1993). These use various methods for computing $\mathbf{B}$ and the step size. For the most part, they are either potentially unstable or are stable but less efficient than methods with optimal step size. As with the other iterative algorithms, all IFBP algorithms suffer from increasing image noise as iterations proceed, so they are generally stopped after a certain number of iterations and the results are filtered.

VI. EVALUATION OF IMAGE QUALITY

In this chapter we have reviewed a great many approaches to image reconstruction, which begs the question: Which method is best? Unfortunately, the answer is not a simple one; each method has its strengths and weaknesses, and algorithms may perform differently in different applications. Several of the mathematical tools of image evaluation, such as receiver operating characteristic (ROC) analysis, which are explained in the last portions of Kao *et al.* (Chapter 6 in this volume), should be studied in conjunction with the following discussion.

It is now generally accepted that the true test of an imaging system or image-reconstruction algorithm is its ability to produce images that help achieve a desired goal (Barrett, 1990; Barrett *et al.*, 1995; Metz, 1978, 1986, 1989). This approach to image evaluation is called *task-based assessment*. For example, if the images are to be used by physicians to diagnose disease, then an algorithm can be considered to perform well if it produces images that lead to accurate diagnoses. On the other hand, if the reconstruction algorithm is to be used for functional brain mapping, in which statistical methods are applied to detect regions of neuronal activation, then a "good" algorithm is one that causes the statistical detection methods to produce accurate results. Finally, if the task is quantitative, such as the measurement of tracer concentration in a given image region, then the reconstruction algorithm should be judged by its quantitative accuracy.

A sound test of image quality for performing a quantitative task can consist of the use of a physical phantom in which the true value of the quantity to be estimated (such as tracer concentration) is known and can be compared to the value reflected in the reconstructed image. However, it is sometimes difficult to make such a study truly representative of the clinical task, and so realistic simulation methods may be used instead. Ideal testing for a visual task, such as diagnosis by a physician, usually requires a human-observer study (Metz, 1978, 1986, 1989), in which trained observers view images for which the true state is known, permitting the diagnostic accuracy of the observers to be assessed. This generally requires a large number of realistic simulations or patient images with confirmed diagnosis.

The effort required to conduct these ideal performance tests can be substantial; therefore, it may not be practical in early stages of algorithm development to conduct exhaustive tests of this kind. Initially, simpler tests are often used to narrow the search for the "best" algorithm and imaging methodology. These tests include measures of statistical estimation performance (e.g., bias and variance), measures of effective spatial resolution, and numerical observers that act as surrogates for human observers.

A. Bias and Variance

Because ET data are random, an image reconstructed from these data is also random—every time an object is scanned, a slightly different image will be obtained. Borrowing from statistical-estimation theory, image-reconstruction algorithms are often evaluated in terms of the mean and variance of the average intensity within a region of interest. Ideally, we hope that the mean reconstructed image will be identical to the true image (or at least identical to a discrete pixel representation of the image). Our success in achieving this objective can be quantified by using a quantity called *bias*, which is the difference between the mean reconstructed image and the true image, $\mathbf{b} = E[\hat{\mathbf{f}}] - \mathbf{f}$. Furthermore, we want the variance of the reconstructed image to be identically zero, meaning that the reconstruction method produces exactly the same image from every noisy set of projections of the same object.

Of course, it is impossible to achieve both of these goals simultaneously. In fact, effort toward one goal tends to work against attainment of the other. This notion, which is a common theme throughout the statistical-estimation field, is commonly called the *bias-variance trade-off*. It is possible to achieve low variance by smoothing the reconstructed image to the point at which it has no visible detail. In this case, the cost of low variance is high bias; the image is inaccurate in the mean. Conversely, it is also possible to achieve low bias by using no smoothing at all (e.g., by running the ML-EM algorithm to convergence); however, the resulting image will exhibit high variance (and will appear very noisy).

In image reconstruction, the degree of enforced smoothness controls the bias-variance trade-off. Smoothness may be controlled implicitly (by the number of iterations of the ML-EM algorithm) or explicitly by adjusting the parameter β in MAP-EM. Either way, a curve in bias-variance space can be generated by sweeping through a range of values for the governing parameter. Figure 11a shows how the mean of the ML-EM estimate improves as the iterations progress. Figure 11b shows that, at the same time, the variance (noise level) of the image worsens. Therefore, we must always settle for the best compromise between these two competing factors.

Bias and variance can be computed in two ways. The obvious way is to generate many noise realizations of a reconstructed image and empirically calculate the mean and variance of these results (Wilson *et al.*, 1994). Owing to the computational burden of this approach, it is often preferred to compute these statistics directly when possible (Barrett *et al.*, 1994, Fessler, 1996, Qi and Leahy, 1999, Qi and Huesman, 2001, Xing and Gindy, 2002, and Jinyi, 2003).

B. Effective Spatial Resolution

Due to the principle of superposition, a linear imaging system (which includes a reconstruction algorithm in the case of ET) is characterized entirely by its PSF, that is, the image of a point object. Thus, the PSF provides a complete description of the spatial resolution produced by the system or algorithm in the absence of noise. However, when the reconstruction algorithm is nonlinear, the PSF does not have such a clear meaning because the image of a point source depends not only on the imaging system and reconstruction method, but also on the object itself. Nevertheless, the effective PSF is a helpful way to shed light on the spatial resolution properties of the reconstruction algorithm. Two methods that have been proposed for quantifying resolution are the effective local Gaussian (Liow and Strother, 1993) resolution and the local impulse response (Stamos *et al.*, 1988 and Fessler and Rogers, 1996).

The use of spatial resolution as a metric implies that the task requires the visibility of small details, which is not always the case. Further, it does not consider the local power spectrum of noise and how noise may interfere with the task to be performed. For this reason, some authors compute a local noise power spectrum and compare this to the system frequency response derived from the PSF (Wilson *et al.*, 1994; Lalush and Tsui, 1998b) to get a more complete picture of the noise-resolution trade-off.

C. Numerical Observers

If an image is to be assessed visually by a physician, it is widely accepted that the most conclusive test of a reconstruction algorithm is a human-observer study. When the image is to be used to determine the presence or absence of a lesion, a study can be designed to determine whether human observers (e.g., physicians) can make this decision reliably. However, human-observer studies are tedious and time-consuming; therefore, they are usually reserved for the final testing of algorithms. During the early development phases, it is helpful to have a computer algorithm, called a *numerical observer*, that can mimic human-observer performance and thus stand in as a surrogate for human observers. By far the most popular numerical observer is the channelized Hotelling observer (CHO) (Barrett *et al.*, 1993; Hotelling, 1931; Burgess *et al.*, 1997; Myers and Barrett, 1987; Myers *et al.*, 1990; Hutton and Strickland, 1997; Wollenweber *et al.*, 1999; Narayanan *et al.*, 2002).

The CHO uses the generalized likelihood-ratio test (GLRT) from statistical detection theory (Kay, 1993) as the lesion detector. The GLRT for the problem of an additive signal in Gaussian signal-independent noise is sometimes called the Hotelling detector. It so happens that the GLRT for this problem is a linear detector, which makes it particularly easy to implement. In the CHO, the features provided to the Hotelling detector are obtained by applying various bandpass filters to the image to coarsely simulate the receptive fields of the human visual system. The features so computed are called *channels*. Using this model of the human observer, it is possible to roughly predict how well humans will perform in detecting lesions from a class of images. It is possible to order reconstruction algorithms in terms of performance by comparing the lesion-detection accuracy predicted for each algorithm.

VII. SUMMARY

Iterative image reconstruction algorithms share a number of common traits. Most are general enough to address any tomographic problem that can be expressed by a linear relationship between the image pixels and the projection bins. Most fit reasonably well into a general model that involves repeating the processes of projecting an image estimate, comparing the estimated projections to the measured data to compute some form of error, backprojecting the error, and using the error to update the image estimate. The principle of feedback works in favor of the iterative algorithm, so that as iterations continue the image estimate gradually approaches some desired result. Different algorithms can be compared on the basis of a number of properties, but those that largely govern the acceptance of an algorithm are its stability, accuracy, speed, and ease of use or implementation.

Classical iterative algorithms, such as ART, are based on the Kaczmarz method of solving systems of linear equations. Although not widely used in nuclear medicine, they form an important basis for understanding the statistical algorithms that were developed later. Statistical algorithms have two basic parts: a statistical criterion (the basis for determining which image among the many possible is to be the solution) and a numerical algorithm (the method for finding the solution prescribed by the criterion). The most successful early statistical algorithm, ML-EM, is based on combining a Poisson statistical model with the EM algorithm. The positive traits of ML-EM spawned a number of variations, mostly seeking to improve on its slow convergence rate. Gaussian statistical models, related to WLS criteria, result in quadratic objective functions. These can efficiently use many established numerical algorithms, such as CG or coordinate descent, and result in faster reconstruction algorithms. Noise remains a problem, however.

IFBP algorithms, such as Chang's method, were developed intuitively, but are related to WLS algorithms. Because of the nontheoretical origins, some of these algorithms can be unstable. MAP criteria can be used to combat the noise sensitivity in ML solutions. MAP algorithms impose smoothness constraints, usually through establishing relationships between neighboring pixels and can therefore converge to usable solutions. The properties of smoothing are highly dependent on the choice of the smoothing distribution, and can lead to undesirable results if not carefully controlled. Ordered-subset approaches, such as OS-EM, offer all the positive features of ML-EM in far fewer iterations and have therefore come to dominate the field in recent years.

VIII. APPENDICES

A. Modeling the Projection of a Pixel

An often-overlooked topic in image reconstruction is the practical issue of computing forward projections of the current image estimate and backprojections of the predicted projections, which are integral steps in all the iterative algorithms we have discussed. Several different approaches for projecting a pixel were provided in the Donner algorithms library (Huesman *et al.*, 1977). These methods, which represent different trade-offs between efficiency, are illustrated in Figure 19. Among these are methods that model the activity in a pixel as being distributed in different ways, often referred to as *pixel-driven methods*. The simplest, most efficient, and least accurate is the point-projection model (Fig. 19a), wherein the intensity in a pixel is assumed to be contained in a point source at the center of the pixel. In projecting this model, it is only necessary to compute the bin location of the projection of the point and then to deposit all the intensity of that pixel in a single bin.

Slightly more complex is the convex-disk model in Figure 19b. In this case, the intensity in a pixel is assumed

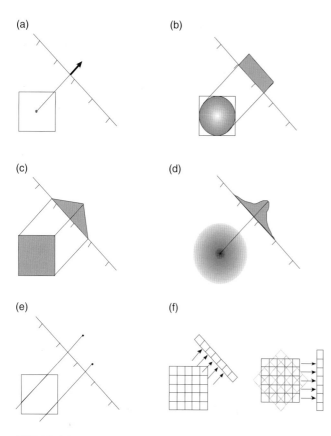

FIGURE 19 Approaches to modeling the projection of a pixel into a set of discrete bins. (a) Point-projection model. (b) Convex-disk model. (c) Area-weighted model. (d) Gaussian blobs. (e) Line-length approach. (f) Rotation-based projection model.

An alternative approach to modeling pixels is the use of Gaussian blobs (Fig. 19d). Because the parallel projection of a 2D Gaussian shape is a 1D Gaussian, it is a simple matter to compute the projection location of the center of the blob and then to distribute the projected Gaussian among the regions of the individual bins. The use of blobs also leads to a natural smoothing effect in the reconstructed image.

The pixel-driven methods are contrasted with ray-driven methods, which compute projections from the point of view of the projection bins. The line-length approach (Fig. 19e) was popular in early implementations because it computes projections with little storage requirement and also leads to a natural computation of attenuation factors. Here, a ray is cast out from the center of a given projection bin in the direction of projection. Every pixel that is intersected by the ray is assigned a weight proportional to the length of the ray in the region bounded by the pixel. To improve accuracy, it is possible to cast multiple rays out from each projection bin, but this makes the method much less computationally efficient.

Finally, an efficient approximation to the area-weighted model is the rotation-based projection model (Fig. 19f). In this case, we do not compute the projection of a given pixel onto the rotating bin array. Rather, the bin array remains fixed and the image is rotated by interpolation. Once in the rotated image frame of reference, the projection is always taken in the same direction. Backprojection can be accomplished by reversing the process. Several rotation-based approaches have been proposed (DiBella *et al.*, 1996; Wallis and Miller, 1997), the chief differences among them involving how the interpolation required for image rotation is accomplished.

For 3D imaging geometries, some of the methods described here have natural extensions. The area-weighted methods extend to volume-weighted methods. In converging-beam imaging, however, the computation of volume intersections is rather complex. It is possible to take advantage of rotation-based methods so that the volume intersections need only be computed in the rotated image frame of reference. A further extension results in a warping of the image space to convert converging projections into parallel projections (Zeng *et al.*, 1994).

Another important consideration in the modeling of projection is whether to explicitly compute and store the matrix **H** or to use a projection operator. In the first case, called a matrix-based model, the size of the matrix can result in enormous storage requirements if anything beyond a simple 1D projection of a 2D object is modeled. Also, the computation of the elements of the matrix can be time consuming except for simple cases. Furthermore, if a system has variable geometry, such as a SPECT system in which the radius of rotation varies from study to study, it becomes necessary to compute the matrix anew for each study per-

to be distributed in a disk fitting within the pixel. The disk intensity is not uniform but rather is lower in the center so that, when the disk is projected with a parallel-beam model, its projection becomes a rectangle, independent of the angle of projection. Because the projection is uniform, it is a simple matter to determine where the boundaries of the projection lie relative to the bin boundaries and to weight the pixel contributions to each bin accordingly. Note that whereas the point-projection method can model virtually any geometry, the convex-disk method is limited to parallel projection.

The most accurate example of pixel-driven methods is the area-weighted model (Fig. 19c). Here the pixel intensity is distributed uniformly over the space the pixel occupies. The projection of the pixel is angle-dependent, so the computation is more complex than in the previous cases. The contribution of a particular pixel to a particular bin is computed as the integral of the pixel projection that lies within the boundaries of the bin. In parallel projection, this is equal to the area of the intersection of a projection strip cast out from the projection bin with the space occupied by the pixel.

formed. An alternative is the more common projector-based approach. In this case, a subroutine is written to take an image and compute its projections without explicitly computing the individual elements of the matrix $\mathbf{H}$. Thus, the output of the subroutine is $\mathbf{Hf}$, even though the elements of $\mathbf{H}$ are never found. The projector-based approach is more flexible and has a smaller requirement for storage than the matrix-based approach, but it may require more computation time.

B. Projections onto Convex Sets

Let us suppose that the image $\mathbf{f}$ is unknown, but that we have knowledge about $\mathbf{f}$ in the form of constraint sets, C_i, $i = 1, \ldots, M$. In image reconstruction, most important constraints are those imposed by the data, but other prior information may also be represented as constraint sets. In the method of POCS, all the constraint sets are assumed to be convex, meaning that the line segment connecting any two points in the set also lies completely within the set.

In this formulation, the problem is to find an image vector $\mathbf{f}$ that satisfies all the constraints. In other words, the problem is to find a vector $\mathbf{f}$ that lies in the intersection of all the constraints, $\mathbf{f} \in \cap_{i=1}^{M} C_i$. Provided that the intersection is not empty and that all the sets are convex, the sequential POCS algorithm is guaranteed to converge to a point in the intersection. This is the basis for the standard ART algorithm.

The basic operation in the POCS algorithm is the projection (here, the term is used in the linear algebra sense, not in the tomography sense). Recall that the projection of a point $\mathbf{x}$ onto a set S is the point in S that is closest to $\mathbf{x}$.

The sequential POCS algorithm makes use of projections in the following way. Starting from any initial estimate for the image vector $\mathbf{f}$, the vector is projected onto each and every constraint set in sequence. The result of this sequence of steps becomes the new estimate, and the process is repeated until convergence. Each iteration can be expressed as $\hat{\mathbf{f}}^{(n+1)} = P_1 P_2 \ldots P_M \hat{\mathbf{f}}^{(n)}$, where P_i is the operation that projects any vector onto the set C_i.

A special case of the POCS algorithm is when all the constraint sets are hyperplanes, in which case the algorithm is known as the Kaczmarz method or ART.

C. Details of the Maximum-Likelihood Expectation-Maximization Algorithm

The ML-EM algorithm and its variants have dominated the field of iterative ET image reconstruction for many years; therefore, it is worth looking more closely at this widely used technique.

The EM algorithm, as proposed in Dempster *et al.* (1977), is a general prescription for developing iterative procedures for solving all sorts of ML estimation problems. The specific iterative formula in Lange and Carson (1984), the ML-EM algorithm for image reconstruction, is simply one application of this general problem-solving approach.

Let us begin by outlining the principles of the general EM methodology. There are many problems in statistics in which the ML solution is difficult to find, but would be easy to find if we had access to some additional data, which are sometimes called *missing data*. In some situations, these extra data are measurements that are literally missing (e.g., the experimenter failed to record them during data collection). In other situations, the notion of missing data is simply an abstraction that leads to a convenient solution to the problem. In these cases, the missing data are measurements that would be extremely helpful to have, but are not actually measurable. (ET reconstruction is an example of this latter use of the EM algorithm.) When applying the EM approach, the observed data that we actually have are called the *incomplete data*, denoted by $\mathbf{g}$, and the full complement of data we wish we had are called the *complete data*, denoted by $\mathbf{s}$.

In ET, the reason our photon count data are incomplete is that we do not know exactly where the photons collected in each projection bin came from. The process of tomographic projection mixes together the photons emitted from the object $\mathbf{f}$ into the projection bins g_j in a linear way described by the system matrix $\mathbf{H}$. The task of image reconstruction is, in a sense, to unmix the photon counts by inverting the effect of $\mathbf{H}$. Viewed in this way, our data would be complete if we knew not only how many counts were measured in projection bin j, but also how many of these came from each object pixel k. Thus, we can define an element of the complete data as a quantity s_{jk}, which denotes the number of photons emitted from pixel k that were detected in projection bin j. Of course, if we actually had access to data of this kind, the reconstruction problem would be trivial. The reason for postulating these unrealistic data is to allow us to use the EM mechanism for obtaining the ML solution.

1. Statement of the EM Algorithm

The EM algorithm consists of two alternating steps, which are repeated until convergence. These steps are called the *expectation step* (E-step) and the *maximization step* (M-step) and are defined as follows.

E-step Based on the current estimate $\hat{\mathbf{f}}^{(n)}$, compute:
$Q(\mathbf{f};\hat{\mathbf{f}}^{(n)}) = E[\ln p(\mathbf{s};\mathbf{f}) \mid \mathbf{g};\hat{\mathbf{f}}^{(n)}]$

M-step Choose the next estimate $\hat{\mathbf{f}}^{(n+1)}$ to maximize $Q(\mathbf{f};\hat{\mathbf{f}}^{(n)})$:

$$\hat{\mathbf{f}}^{(n+1)} = \arg \max_{\mathbf{f}} Q(\mathbf{f};\hat{\mathbf{f}}^{(n)})$$

The essence of the algorithm is as follows. If we had access to the complete data, the ML problem would be easy to solve; however, in the absence of the complete data, the log-likelihood function for the complete data cannot even be computed. Thus, in each iteration, the average of this log-likelihood function is computed in the E-step and maximized in the M-step. The following derivation shows that this procedure increases the likelihood $p(\mathbf{g};\mathbf{f})$ in each iteration (so long as that is possible).

2. EM Algorithm Produces Nondecreasing Likelihood

Recall that the aim of ML reconstruction is to find the image $\mathbf{f}$ that maximizes the likelihood function $p(\mathbf{g};\mathbf{f})$ or, equivalently, the logarithm of the likelihood (or log-likelihood) $l(\mathbf{f}) = \ln p(\mathbf{g};\mathbf{f})$.[3] In the context of the EM algorithm, we call this the incomplete-data log-likelihood.

To begin, let us relate the log-likelihood of the incomplete (observed) data $\mathbf{g}$ to that of the idealized complete data $\mathbf{s}$. From basic probability theory:

$$p(\mathbf{s};\mathbf{f}) = p(\mathbf{s}|\mathbf{g};\mathbf{f}) p(\mathbf{g};\mathbf{f}) \quad (42)$$

Taking the logarithm of both sides of Eq. (19) and rearranging, one obtains:

$$\begin{aligned}l(\mathbf{f}) &= \ln p(\mathbf{g};\mathbf{f}) \\ &= \ln p(\mathbf{s};\mathbf{f}) - \ln p(\mathbf{s}|\mathbf{g};\mathbf{f})\end{aligned} \quad (43)$$

Taking the expected value of both sides of Eq. (43) with respect to $p(\mathbf{s}|\mathbf{g};\hat{\mathbf{f}}^{(n)})$ yields:

$$\begin{aligned}l(\mathbf{f}) &= E[\ln p(\mathbf{s};\mathbf{f})|\mathbf{g};\hat{\mathbf{f}}^{(n)}] - E[\ln p(\mathbf{s}|\mathbf{g};\mathbf{f})|\mathbf{g};\hat{\mathbf{f}}^{(n)}] \\ &= Q(\mathbf{f};\hat{\mathbf{f}}^{(n)}) - R(\mathbf{f};\hat{\mathbf{f}}^{(n)})\end{aligned} \quad (44)$$

where $Q(\mathbf{f};\hat{\mathbf{f}}^{(n)})$ and $R(\mathbf{f};\hat{\mathbf{f}}^{(n)})$ are the two expectations in the first line of Eq. (44).

Now let us compare the value of the incomplete-data log-likelihood $l(\mathbf{f})$ for two successive iterations, n and $n + 1$:

$$l(\hat{\mathbf{f}}^{(n+1)}) - l(\hat{\mathbf{f}}^{(n)}) = \left[Q(\hat{\mathbf{f}}^{(n+1)};\hat{\mathbf{f}}^{(n)}) - Q(\hat{\mathbf{f}}^{(n)};\hat{\mathbf{f}}^{(n)})\right] + \left[R(\hat{\mathbf{f}}^{(n)};\hat{\mathbf{f}}^{(n)}) - R(\hat{\mathbf{f}}^{(n+1)};\hat{\mathbf{f}}^{(n)})\right] \quad (45)$$

We want to show that this difference is nonnegative, thus demonstrating that each iteration increases the likelihood when possible and, at worst, leaves it unchanged. The first term in square brackets in Eq. (45) is clearly nonnegative, because the M-step maximizes $Q(\mathbf{f};\hat{\mathbf{f}}^{(n)})$ and, thus, $Q(\hat{\mathbf{f}}^{(n+1)};\hat{\mathbf{f}}^{(n)}) \geq Q(\hat{\mathbf{f}}^{(n)};\hat{\mathbf{f}}^{(n)})$. The second term is also nonnegative, which can be seen from the following.

The following identity holds for any real number t: $\ln t \leq t - 1$. Thus, we can write:

$$E[\ln \pi(\mathbf{s})|\mathbf{g};\hat{\mathbf{f}}^{(n)}] \leq E[\pi(\mathbf{s})|\mathbf{g};\hat{\mathbf{f}}^{(n)}] - 1 \quad (46)$$

where

$$\pi(\mathbf{s}) = \frac{p(\mathbf{s}|\mathbf{g};\hat{\mathbf{f}}^{(n+1)})}{p(\mathbf{s}|\mathbf{g};\hat{\mathbf{f}}^{(n)})} \quad (47)$$

Evaluating the right-hand side of Eq. (46), we obtain:

$$\begin{aligned}E[\pi(\mathbf{s})|\mathbf{g};\hat{\mathbf{f}}^{(n)}] - 1 &= E\left[\frac{p(\mathbf{s}|\mathbf{g};\hat{\mathbf{f}}^{(n+1)})}{p(\mathbf{s}|\mathbf{g};\hat{\mathbf{f}}^{(n)})}\bigg|\mathbf{g};\hat{\mathbf{f}}^{(n)}\right] - 1 \\ &= \int \frac{p(\mathbf{s}|\mathbf{g};\hat{\mathbf{f}}^{(n+1)})}{p(\mathbf{s}|\mathbf{g};\hat{\mathbf{f}}^{(n)})} p(\mathbf{s}|\mathbf{g};\hat{\mathbf{f}}^{(n)}) d\mathbf{s} - 1 \\ &= 0\end{aligned} \quad (48)$$

From Eqs. (46)–(48), we obtain:

$$E\left[\ln \pi(\mathbf{s})|\mathbf{g};\hat{\mathbf{f}}^{(n)}\right] \leq 0$$
$$E\left[\ln p(\mathbf{s}|\mathbf{g};\hat{\mathbf{f}}^{(n+1)})|\mathbf{g};\hat{\mathbf{f}}^{(n)}\right] - E\left[\ln p(\mathbf{s}|\mathbf{g};\hat{\mathbf{f}}^{(n)})|\mathbf{g};\hat{\mathbf{f}}^{(n)}\right] \leq 0 \quad (49)$$
$$R(\hat{\mathbf{f}}^{(n+1)};\hat{\mathbf{f}}^{(n)}) - R(\hat{\mathbf{f}}^{(n)};\hat{\mathbf{f}}^{(n)}) \leq 0$$

Therefore, the second difference term in Eq. (45) is nonnegative, and we have established that:

$$l(\hat{\mathbf{f}}^{(n+1)}) \geq l(\hat{\mathbf{f}}^{(n)}) \quad (50)$$

In a variation of the EM algorithm, called the GEM *algorithm*, the M-step is replaced with a weaker step in which the expectation is not maximized. Instead, the GEM algorithm aims only to achieve $Q(\hat{\mathbf{f}}^{(n+1)};\hat{\mathbf{f}}^{(n)}) \geq Q(\hat{\mathbf{f}}^{(n)};\hat{\mathbf{f}}^{(n)})$.

3. EM Algorithm for ET Image Reconstruction

Now let us derive Eq. (30), which is the iterative formula for ML-EM image reconstruction in ET.[4] Recall that, in ET, a useful definition of the complete data is s_{im}, which is the (random) number of photons emitted from within pixel m and detected in projection bin i. Of course, this is an unmeasurable quantity, but it allows us conveniently to use the EM framework to solve the ML reconstruction problem. The complete data can be related to the observed projection data $\mathbf{g}$ and the image $\mathbf{f}$ as follows:

$$g_i = \sum_m s_{im} \quad (51)$$

$$E[s_{im}] = h_{im} f_k \quad (52)$$

The E-step of the EM algorithm requires an expression for the complete-data log-likelihood $\ln p(\mathbf{s};\mathbf{f})$. In ET, the counts s_{im} are independent Poisson-distributed random variables; therefore,

$$p(\mathbf{s};\mathbf{f}) = \prod_i \prod_m \frac{E[s_{im}]^{s_{im}} e^{-E[s_{im}]}}{s_{im}!} \quad (53)$$

and the log-likelihood is (using Eq. (52)):

$$\ln p(\mathbf{s};\mathbf{f}) = \sum_i \sum_m [s_{im} \ln(h_{im} f_m) - h_{im} f_m - \ln(s_{im}!)] \quad (54)$$

---

[3]This derivation is based partly on the derivation in McLachland and Krishnan (1997).

[4]This derivation is based on the one given in Lange and Carson (1984).

Now we are ready to compute the E-step of the EM algorithm.

E-step

Using Eq. (54), the E-step is computed as follows:

$$Q(\mathbf{f};\hat{\mathbf{f}}^{(n)}) = E[\ln p(\mathbf{s};\mathbf{f}) | \mathbf{g};\hat{\mathbf{f}}^{(n)}]$$
$$= \sum_i \sum_m \left\{ E[s_{im} | \mathbf{g};\hat{\mathbf{f}}^{(n)}] \ln(h_{im} f_m) - h_{im} f_m - E[\ln(s_{im}!)] \right\} \quad (55)$$

The conditional mean of s_{im} in Eq. (55) is given by:

$$E[s_{im} | \mathbf{g};\hat{\mathbf{f}}^{(n)}] = \frac{h_{im} \hat{f}_m^{(n)}}{\sum_k h_{ik} \hat{f}_k^{(n)}} g_i \triangleq p_{im} \quad (56)$$

which is simply the fraction of the detected counts in projection bin i that are expected to have emanated from pixel m, given that current image estimate $\hat{\mathbf{f}}^{(n)}$ is the source of these counts. Substituting Eq. (56) into Eq. (55), we obtain the final form of the E-step:

$$Q(\mathbf{f};\hat{\mathbf{f}}^{(n)}) = \sum_i \sum_m \left\{ p_{im} \ln(h_{im} f_m) - h_{im} f_m - E[\ln(s_{im}!)] \right\} \quad (57)$$

Next, we compute the M-step.

M-step

In the M-step, we find the next image estimate $\hat{\mathbf{f}}^{(n+1)}$ by maximizing $Q(\mathbf{f};\hat{\mathbf{f}}^{(n)})$ with respect to $\mathbf{f}$. We can accomplish by setting the derivative of $Q(\mathbf{f};\hat{\mathbf{f}}^{(n)})$ to zero and solving:

$$\frac{\partial Q(\mathbf{f};\hat{\mathbf{f}}^{(n)})}{\partial f_j} = 0 = \sum_i \left(\frac{p_{ij}}{\hat{f}_j^{(n+1)}} - h_{ij} \right) \quad (58)$$

Solving for $\hat{f}_j^{(n+1)}$ and using Eq. (56), we obtain the well-known ML-EM iteration for ET image reconstruction in Eq. (30):

$$\hat{f}_j^{(n+1)} = \frac{\hat{f}_j^{(n)}}{\sum_{i'} h_{i'j}} \sum_i h_{ij} \frac{g_i}{\sum_k h_{ik} \hat{f}_k^{(n)}}$$

Acknowledgments

Preparation of this chapter was supported in part by NIH/NHLBI grant HL65425.

References

Barrett, H. H. (1990). Objective assessment of image quality: effects of quantum noise and object variability. *J. Opt. Soc. Am. A* **7:** 1266–1278.

Barrett, H. H., Denny, J. L., Wagner, R. F., and Myers, K. J. (1995). Objective assessment of image quality, II. Fisher information, Fourier crosstalk, and figures of merit for task performance. *J. Opt. Soc. Am. A* **12:** 834–852.

Barrett, H. H., Wilson, D. W., and Tsui, B. M. W. (1994). Noise properties of the EM algorithm, I. Theory. *Phys. Med. Biol.* **39:** 833–846.

Barrett, H. H., Yao, J., Rolland, J. P., and Myers, K. J. (1993). Model observers for assessment of image quality. *Proc. Natl Acad. Sci. U.S.A.* **90:** 9758–9765.

Beekman, F., Kamphuis, C., and Viergever, M. (1996). Improved SPECT quantitation using fully three-dimensional iterative spatially variant scatter response compensation. *IEEE Trans. Med. Imaging* **15:** 491–499.

Brankov, J., Yang, Y., and Wernick, M. N. (2004). Tomographic image reconstruction based on a content-adaptive mesh model. *IEEE Trans. Med. Imaging.* **23:** 202–212.

Brankov, J. G., Yang, Y., and Wernick, M. N. (to appear). Tomographic image reconstruction based on a content-adaptive mesh model. *IEEE Trans. Med. Imaging.*

Brankov, J. G., Yang, Y., Wernick, M. N., and Narayanan, M. V. (2002). Motion compensated reconstruction of gated SPECT data. *2002 Conf. Rec. IEEE Nucl. Sci. Symp. Med. Imaging Conf.* **3:** 10–16.

Browne, J., and De Pierro, A. R. (1996). A row-action alternative to the EM algorithm for maximizing likelihoods in emission tomography. *IEEE Trans. Med. Imaging* **15:** 687–699.

Burgess, A. E., Li, X., and Abbey, C. K. (1997). Visual signal detectability with two noise components: Anomalous masking effects. *J. Opt. Soc. Am. A* **14:** 2420–2442.

Byrne, C. L. (1996). Block-iterative methods for image reconstruction from projections. *IEEE Trans. Imaging Processing* **5:** 792–794.

Censor, Y., Eggermont, P. P. B., and Gordon, D. (1983). Strong underrelaxation in Kaczmarz's method for inconsistent systems. *Numer. Math.* **41:** 83–92.

Chang, L. (1979). Attenuation and incomplete projection in SPECT. *IEEE Trans. Nucl. Sci.* **26:** 2780–2789.

Chiao, P. C., Rogers, W. L., Fessler, J. A., Clinthorne, N. H., and Hero, A. O. (1994). Model-based estimation with boundary side information or boundary regularization. *IEEE Trans. Med. Imaging* **13:** 227–234.

Coxson, P. G., Salmeron, E. M., and Huesman, R. H. (1990). A strategy for using closed form solutions for compartmental models of dynamic PET data. *1990 Conf. Rec. IEEE. Nucl. Sci. Symp.* 1577–1578.

Dempster, A. P., Laird, N. M., and Rubin, D. B. (1977). Maximum likelihood from incomplete data via the EM algorithm. *J. Royal Stat. Soc. B* **39:** 1–38.

DiBella, E., Barclay, A., Eisner, R., and Schaefer, R. (1996). A comparison of rotation-based methods for iterative reconstruction algorithms. *IEEE Trans. Nucl. Sci.* **43:** 3370–3376.

Farncombe, T., Celler, A., Noll, J., Maeght, D., and Harrop, R. (1999). Dynamic SPECT imaging using a single camera rotation. *IEEE Trans. Nucl. Sci.*

Farncombe, T., King, M. A., Celler, A., Blinder, S. (2001). A fully 4D expectation maximization algorithm using Gaussian diffusion based detector response for slow camera rotation dynamic SPECT. Paper presented at 2001 International Meeting on Fully 3D Image Reconstruction in Radiology and Nuclear Medicine, Pacific Grove, CA.

Feng, B., Pretorius, P. H., Farncombe, T. H., Dahlberg, S. T., Narayanan, M. V., Wernick, M. N., Celler, A. M., King, M.A., and Leppo, J.A. (2002). Imaging time-varying Tc-99m teboroxime localization and cardiac function simultaneously by five-dimensional (5D) gated-dynamic SPECT imaging and reconstruction. *Am. Soc. Nucl. Card.* **10:** S11–12.

Fessler, J. A. (1994). Penalized weighted least squares image reconstruction for positron emission tomography. *IEEE Trans. Med. Imaging* **13:** 290–300.

Fessler, J. (1996). Mean and variance of implicitly defined biased estimators (such as penalized maximum likelihood): Applications to tomography. *IEEE Trans. Image Processing* **5:** 493–506.

Fessler, J., and Hero, A. (1994). Space-alternating generalized expectation maximization algorithm. *IEEE Trans. Sig. Processing* **42:** 2664–2677.

Fessler, J. A., and Rogers, W. L. (1996). Spatial resolution properties of penalized-likelihood image reconstruction: Space-invariant tomographs. *IEEE Trans. Imaging Processing* **5:** 1346–1358.

Fisher, R. A. (1921). On the "probable error" of a coefficient of correlation deduced from a small sample. *Metron* **1:** 3–32.

Frey, E. C., Ju, Z. W., and Tsui, B. M. W. (1993). A fast projector-backprojector pair for modeling the asymmetric spatially-varying scatter response function for scatter compensation in SPECT imaging. *IEEE Trans. Nucl. Sci.* **40:** 1192–1197.

Galatsanos, N. P. Wernick, M. N., and Katsaggelos, A. K. (2000). Multichannel image recovery. *In* "Handbook of Image and Video Processing" (A. Bovik, ed.), pp. 155–168. Academic Press; San Diego, CA.

Geman, S., and Geman, D. (1984). Stochastic relaxation, Gibbs distributions, and the Bayesian restoration of images. *IEEE Trans. Patt Anal. Mach. Int.* **6:** 721–741.

Gilbert, P. (1972). Iterative methods for the reconstruction of three dimensional objects from their projections. *J. Theor. Biol.* **36:** 105–117.

Gindi, G., Lee, M., Rangarajan, A., and Zubal, I. G. (1991). Bayesian reconstruction of functional images using registered anatomical images as priors. *In* "Information Processing in Medical Imaging" (A. C. F. Colchester and D. J. Hawkes, eds.), pp. 121–131. Springer-Verlag, New York.

Golub, G. H., and Van Loan, C. F. (1989). "Matrix Computations." Johns Hopkins University Press, Baltimore.

Gordon, R., Bender, R., and Herman, G. T. (1970). Algebraic reconstruction techniques (ART) for three-dimensional electron microscopy and X-ray photography. *J. Theor. Biol.* **29:** 471–481.

Green, P. J. (1990). Bayesian reconstructions from emission tomography data using a modified EM algorithm. *IEEE Trans. Med. Imaging* **9:** 84–93.

Hebber, E., Oldenburg, D., Farncombe, T., and Celler, A. (1997). Direct estimation of dynamic parameters in SPECT tomography. *IEEE Trans. Nucl. Sci.* **44:** 2425–2430.

Hebert, T. J., and Gopal, S. S. (1992). The GEM MAP algorithm with 3D SPECT system response. *IEEE Trans. Med. Imaging* **11:** 81–90.

Hebert, T. J., and Leahy, R. (1992). Statistic-based MAP image reconstruction from Poisson data using Gibbs priors. *IEEE Trans. Sig. Processing* **40:** 2290–2303.

Herman, G. T. (1980). "Image Reconstruction from Projections: The Fundamentals of Computerized Tomography." Academic Press, New York.

Higdon, D. M., Bowsher, J. E., Johnson, V. E., Turkington, T. G., Gilland, D. R., and Jaczszak, R. J. (1997). Fully Bayesian estimation of Gibbs hyperparameters for emission computed tomography data. *IEEE Trans. Med. Imaging* **16:** 516–526.

Hotelling, H. (1931). The generalization of Student's ratio. *Ann. Math. Stati.* **2:** 360–378.

Hudson, H. M., and Larkin, R. S. (1994). Accelerated image reconstruction using ordered subsets of projection data. *IEEE Trans. Med. Imaging* **13:** 601–609.

Huesman, R. H., Gullberg, G. T., Greenberg, W. L., and Budinger, T. F. (1977). "User Manual, Donner Algorithms for Reconstruction Tomography." University of California, Lawrence Berkeley Laboratory.

Hutton, D. A., and Strickland, R. N. (1997). Channelized detection filters for detecting tumors in nuclear medical images. *Proc. SPIE* **3034:** 457–466.

Jinyi, Q. (2003). A unified noise analysis for iterative image estimation. *Phys. Med. Biol.* **48:** 3505–3519.

Johnson, V. E., Wong, W. H., Hu, X., and Chen, C. T. (1991). Image restoration using Gibbs priors: Boundary modeling, treatment of blurring, and selection of hyperparameters. *IEEE Trans. Patt. Anal. Mach. Int.* **13:** 413–425.

Kaczmarz, S. (1937). Angenaherte Auflosung von Systemen linearer Gleichungen. *Bull. Acad. Polon. Sci.* **A35:** 355–357.

Kalbfleisch, J. G. (1985). "Probability and Statistical Inference, Vol. 1. Probability." 2nd ed. Springer-Verlag, New York.

Kao, C.-M., Yap, J.-T., Mukherjee, J., and Wernick, M. N. (1997). Image reconstruction for dynamic PET based on low-order approximation and restoration of the sinogram. *IEEE Trans. Med. Imaging* **16:** 738–749.

Kaufman, L. (1987). Implementing and accelerating the EM algorithm for positron emission tomography. *IEEE Trans. Med. Imaging* **6:** 37–51.

Kaufman, L. (1993). Maximum likelihood, least squares, and penalized least squares for PET. *IEEE Trans. Med. Imaging* **12:** 200–214.

Kay, S. M. (1993). "Fundamentals of Statistical Signal Processing: Estimation Theory." Prentice Hall, New York.

King, M. A., and Miller, T. R. (1985). Use of a nonstationary temporal Wiener filter in nuclear medicine. *Eur. J. Nuce. Med.* **10:** 458–461.

Klein, G. J., Reutter, B. W., and Huesman, R. H. (1997). Non-rigid summing of gated PET via optical flow. *IEEE Trans. Nucl. Sci.* **44:** 1509–1512.

Lalush, D. S., and Tsui, B. M. W. (1992). Simulation evaluation of Gibbs prior distributions for use in maximum *a posteriori* SPECT reconstructions. *IEEE Trans. Med. Imaging* **11:** 267–275.

Lalush, D. S., and Tsui, B. M. W. (1993). A generalized Gibbs prior for maximum a posteriori reconstruction in SPECT. *Phys. Med. Biol.* **38:** 729–741.

Lalush, D. S., and Tsui, B. M. W. (1994). Improving the convergence of iterative filtered backprojection algorithms. *Med. Phys.* **21:** 1283–1286.

Lalush, D. S., and Tsui, B. M. W. (1995). A fast and stable maximum a posteriori conjugate gradient reconstruction algorithm. *Med. Phys.* **22:** 1273–1284.

Lalush, D. S., and Tsui, B. M. W. (1996). Space-time Gibbs priors applied to gated SPECT myocardial perfusion studies. *In* "Three-dimensional Image Reconstruction in Radiology and Nuclear Medicine" (P. Grangeat and J. L. Amans eds.), pp. 209–224. Kluwer Academic, Dordrecht.

Lalush, D. S., and Tsui, B. M. W. (1998a). Block-iterative techniques for fast 4D reconstruction using *a priori* motion models in gated cardiac SPECT. *Phys. Med. Biol.* **43:** 875–887.

Lalush, D. S., and Tsui, B. M. W. (1998b). Mean-variance analysis of block-iterative reconstruction algorithms modeling 3D detector response in SPECT. *IEEE Trans. Nucl. Sci.* **45:** 1280–1287.

Lalush, D. S., and Tsui, B. M. W. (2000). Performance of ordered-subset reconstruction algorithms under conditions of extreme attenuation and truncation in myocardial SPECT. *J. Nucl. Med.* **41:** 737–744.

Lange, K. (1990). Convergence of EM image reconstruction algorithms with Gibbs smoothing. *IEEE Trans. Med. Imaging* **9:** 439–446.

Lange, K., Bahn, M., and Little, R. (1987). A theoretical study of some maximum likelihood algorithms for emission and transmission tomography. *IEEE Trans. Med. Imaging* **6:** 106–114.

Lange, K., and Carson, R. (1984). EM reconstruction algorithms for emission and transmission tomography. *J. Comput. Assist. Tomogr.* **8:** 306–316.

Leahy, R., and Yan, X. (1991). Incorporation of anatomical MR data for improved functional imaging with PET. *In* "Information Processing in Medical Imaging" (A. C. F. Colchester and D. J. Hawkes, eds.), pp. 105–120. Springer-Verlag, New York.

Lee, S. J., Rangarajan, A., and Gindi, G. (1995). Bayesian image reconstruction in SPECT using higher order mechanical models as priors. *IEEE Trans. Med. Imaging* **14:** 669–680.

Levitan, E., and Herman, G. T. (1987). A maximum *a posteriori* probability expectation maximization algorithm for image reconstruction in emission tomography. *IEEE Trans. Med. Imaging* **6:** 185–192.

Lewitt, R. (1992). Alternatives to voxels for image representation in iterative reconstruction algorithms. *Phys. Med. Biol.* **37:** 705–716.

Lewitt, R. M., and Muehllehner, G. (1986). Accelerated iterative reconstruction for positron emission tomography based on the EM algorithm for maximum likelihood estimation. *IEEE Trans. Med. Imaging* **5:** 16–22.

Liang, Z. (1993). Compensation for attenuation, scatter, and detector response in SPECT reconstruction via iterative FBP methods. *Med. Phys.* **20:** 1097–1106.

Liang, Z., Jaszczak, R., and Greer, K. (1989). On Bayesian image reconstruction from projections: uniform and nonuniform *a priori* source information. *IEEE Trans. Med. Imaging* **8:** 227–235.

Limber, M. A., Limber, M. N., Celler, A., Barney, J. S., and Borwein, J. M. (1995). Direct reconstruction of functional parameters for dynamic SPECT. *IEEE Trans. Nucl. Sci.* **42:** 1249–1256.

Liow, J., and Strother, S. C. (1993). The convergence of object dependent resolution in maximum likelihood based tomographic image reconstruction. *Phys. Med. Biol.* **38**: 55–70.

Llacer, J., and Veklerov, E. (1989). Feasible images and practical stopping rules for iterative algorithms in transmission tomography. *IEEE Trans. Med. Imaging* **8**: 186–193.

Lucy, L. B. (1974). An iterative technique for the rectification of observed distribution. *Astrophys. J.* **79**: 745–754.

Mailloux, G., Noumeir, R., and Lemieux, R. (1993). Deriving the multiplicative algebraic reconstruction algorithm (MART) by the method of convex projections (POCS). *Proc. IEEE Int. Conf. Acoust., Speech Sig. Processing*, **5**: 457.

Mair, B. A., Gilland, D. R., and Cao, Z. (2002). Simultaneous motion estimation and image reconstruction from gated data. *Proc. IEEE Int. Symp. Biomed. Imaging* 661–664.

Malko, J. A., Van Heertum, R. L., Gullberg, G. T., and Kowalsky, W. P. (1986). SPECT liver imaging using an iterative attenuation correction algorithm and an external flood source. *J. Nucl. Med.* **27**: 701–705.

Maze, A., Le Cloirec, J., Collorec, R., Bizais, Y., Briandet, P., and Bourguet, P. (1993). Iterative reconstruction methods for nonuniform attenuation distribution in SPECT. *J. Nucl. Med.* **34**: 1204–1209.

McLachlan, G. J., and Krishnan, T. (1997). "The EM Algorithm and Extensions." John Wiley, New York.

Metz, C. E. (1978). Basic principles of ROC analysis. *Sem. Nucl. Medi.* **8**: 283–298.

Metz, C. E. (1986). ROC methodology in radiologic imaging. *Invest. Radiol.* **21**: 720–733.

Metz C. E. (1989). Some practical issues of experimental design and data analysis in radiological ROC studies. *Invest. Radiol.* **24**: 234–245.

Moore, S. C., Brunelle, J. A., and Kirsch, C. M. (1987). Quantitative multi-detector emission computerized tomography using iterative attenuation compensation. *J. Nucl. Med.* **23**: 706–714.

Mumcuoglu, E. U., Leahy, R., Cherry, S. R., and Zhou, Z. (1994). Fast gradient-based methods for Bayesian reconstruction of transmission and emission PET images. *IEEE Trans. Med. Imaging* **13**: 687–701.

Myers, K. J., and Barrett, H. H. (1987). Addition of a channel mechanism to the ideal-observer model. *J. Opt. Am. A* **4**: 2447–2457.

Myers, K. J., Rolland, J. P., and Barrett, H. H. (1990). Aperture optimization for emission imaging: Effect of a spatially varying background. *J. Opt. Soc. Am. A* **7**: 1279–1293.

Narayanan, M. V., Gifford, H. C., King, M. A., Pretorius, P. H., Farncombe, T. H., Bruyant, P., and Wernick, M. N. (2002). Optimization of iterative reconstructions of Tc-99m cardiac SPECT studies using numerical observers. *IEEE Trans. Nucl. Sci.* **49**: 2355–2360.

Narayanan, V. M., King, M. A., Soares, E., Byrne, C., Pretorius, H., and Wernick, M. N. (1999). Application of the Karhunen-Loeve transform to 4D reconstruction of gated cardiac SPECT images. *IEEE Trans. Nucl. Sci.* **46**: 1001–1008.

Nichols, T. E., Qi, J., and Leahy. R. M. (1999). Continuous time dynamic PET imaging using list mode data. *In* "Information Processing in Medical Imaging" (A. C. F. Colchester and D. J. Hawkes, eds.), pp. 98–111. Springer-Verlag, New York.

Ouyang, X., Wong, W. H., Johnson, V. E., Hu, X. P., and Chen, C. T. (1994). Incorporation of Correlated Structural Images in PET Image Reconstruction. *IEEE Trans. Med. Imaging* **13**: 627–640.

Pan, T.-S., and Yagle, A. E. (1991). Numerical study of multigrid implementations of some iterative image reconstruction algorithms. *IEEE Trans. Med. Imaging* **10**: 572–588.

Pan, X., Wong, W. H., Chen, C.-T., and Jun, L. (1993). Correction for photon attenuation in SPECT: Analytical framework, average attenuation factors, and a new hybrid approach. *Phys. Med. Biol.* **38**: 1219–1234.

Press, W. H., Flannery, B. P., Teukolsky, S. A., and Vetterling, W. T. (1988). "Numerical Recipes in C." Cambridge University Press, Cambridge, UK.

Qi, J. and Huesman, R. H. (2001). Theoretical study of lesion detectability of MAP reconstruction using computer observers. *IEEE Trans. Med. Imaging* **20**: 815–22.

Qi, J., Leahy, R. M., Hsu, C., Farquhar, T. H., and Cherry, S. R. (1996). Fully 3D Bayesian image reconstruction for the ECAT EXACT HR+. *IEEE Trans. Nucl. Sci.* **45**: 1096–1103.

Qi, J. and Leahy, R. M. (1999). Fast computation of the covariance of MAP reconstructions of PET images. *Proc. SPIE* **3661**: 344–355.

Rajeevan, N., Rajgopal, K., and Krishna, G. (1992). Vector-extrapolated fast maximum likelihood estimation algorithms for emission tomography. *IEEE Trans. Med. Imaging* **11**: 9–20.

Ranganath, M. V., Dhawan, A. P., and Mullani, N. (1988). A multigrid expectation maximization reconstruction algorithm for positron emission tomography. *IEEE Trans. Med. Imaging* **7**: 273–278.

Reutter, B. W., Gullberg, G. T., and Huesman, R. H. (1998). Kinetic parameter estimation from dynamic cardiac patient SPECT projection measurements. *1998 Conf. Rec. IEEE Nucl. Sci. Symp.* **3**: 1953–1958.

Reutter, B. W., Gullberg, G. T., Huesman, R. H. (2000). Direct least-squares estimation of spatiotemporal distributions from dynamic SPECT projections using a spatial segmentation and temporal B-splines. *IEEE Trans. Med. Imaging* **19**: 434–450.

Richardson, W. H. (1972). Bayesian-based iterative method of image restoration. *J. Opt. Soc. Am. A*, **62**: 55–59.

Rosenfeld, A., and Kak, A. C. (1982). "Digital Picture Processing." Academic Press, Orlando, FL.

Shepp, L. A., and Vardi, Y. (1982). Maximum likelihood estimation for emission tomography. *IEEE Trans. Med. Imaging* **1**: 113–121.

Smith, M. F., Floyd Jr., C. E., Jaszczak, R. J., and Coleman, R. E. (1992). Reconstruction of SPECT images using generalized matrix inverses. *IEEE Trans. Med. Imaging* **11**: 165–175.

Snyder, D. L., and Miller, M. I. (1985). The use of sieves to stabilize images produced with the EM algorithm for emission tomography. *IEEE Trans. Nucl. Sci.* **32**: 3864–3872.

Stamos, J. A., Rogers, W. L., Clinthorne, N. H., and Koral, K. F. (1988). Object-dependent performance comparison of two iterative reconstruction algorithms. *IEEE Trans. Nucl. Sci.* **35**: 611–614.

Stark, H., and Yang, Y. (1998). "Vector Space Projections: A Numerical Approach to Signal and Image Processing, Neural Nets, and Optics." John Wiley, New York.

Tanaka, E. (1987). A fast reconstruction algorithm for stationary positron emission tomography based on a modified EM algorithm. *IEEE Trans. Med. Imaging* **6**: 98–105.

Tekalp, M. A. (1995). "Digital Video Processing." Prentice-Hall, New York.

Tsui, B. M. W., Gullberg, G. T., Edgerton, E. R., Ballard, J. G., Perry, J. R., McCartney, W. H., and Berg, J. (1989). Correction of nonuniform attenuation in cardiac SPECT imaging. *J. Nucl. Med.* **30**: 497–507.

Tsui, B. M. W., Zhao, X.-D., Frey, E. C., and Gullberg, G. T. (1991). Comparison between EM and CG algorithms for SPECT image reconstruction. *IEEE Trans. Nucl. Sci.* **38**: 1766–1772.

Van Trees, H. L. (1968). "Detection, Estimation, and Modulation Theory, Part I." John Wiley & Sons, New York.

Wallis, J., and Miller, T. R. (1997). An optimal rotator for iterative reconstruction. *IEEE Trans. Med. Imaging* **16**: 118–123.

Wallis, J. W., and Miller, T. R. (1993). Rapidly converging iterative reconstruction algorithms in single-photon emission computed tomography. *J. Nucl. Med.* **34**: 1793–1800.

Wallis, J. W., Miller, T. R., and Dai, G. M. (1998). Comparison of the convergence properties of the It-W and OS-EM algorithms in SPECT. *IEEE Trans. Nucl. Sci.* **45**: 1317–1323.

Walters, E., Simon, W., Chesler, D. A., and Correia, J. A. (1981). Attenuation correction in gamma emission computed tomography. *J. Comput. Assist. Tomogr.* **5**: 89–94.

Wernick, M. N., Infusino, E. J., and Milosevic, M. (1999). Fast spatiotemporal image reconstruction for dynamic PET. *IEEE Trans. Med. Imaging* **18**: 185–195.

Wilson, D. W., Tsui, B. M. W., and Barrett, H. H. (1994). Noise properties of the EM algorithm, II. Monte Carlo simulations. *Phys. Med. Biol.* **39:** 847–871.

Wollenweber, S. D., Tsui, B. M. W., Lalush, D. S., Frey, E. C., LaCroix, K. J., and Gullberg, G. T. (1999). Comparison of Hotelling observer models and human observers in defect detection from myocardial SPECT imaging. *IEEE Trans. Nucl. Sci.* **46:** 2098–2103.

Xing, Y., and Gindi, G. (2002). Rapid calculation of detectability in Bayesian SPECT *Proc. IEEE Int. Symp. on Biomed. Imaging.* 78–81.

Xu, X.-L., Liow, J.-S., and Strother, S. C. (1993). Iterative algebraic reconstruction algorithms for emission computed tomography. *Med. Phys.* **20:** 1675–1684.

Yavus, M., and Fessler, J. A. (1996). Objective functions for tomographic reconstruction from randoms-precorrected PET scans. *1996 Conf. Rec. Nucl. Sci. Symp. Med. Imaging Conf.* **2:** 1067–1071.

Zeng, G. L., Gullberg, G. T., and Huesman, R. H. (1995). Using linear time-invariant system theory to estimate kinetic parameters directly from projection measurements. *IEEE Trans. Nucl. Sci.* **42:** 2339–2346.

Zeng, G. L., Gullberg, G. T., Tsui, B. M. W., and Terry, J. A. (1991). Three-dimensional iterative reconstruction algorithms with attenuation and geometric point response correction. *IEEE Trans. Nucl. Sci.* **38:** 693–702.

Zeng, G. L., Hsieh, Y., and Gullberg, G. (1994). A rotating and warping projector backprojector for fan-beam and cone-beam iterative algorithm. *IEEE Trans. Nucl. Sci.* **41:** 2807–2811.

Zhou, Z., Leahy, R. M., and Mumcuoglu, E. U. (1995). Maximum likelihood hyperparameter estimation for Gibbs priors with applications to PET. *In* "Information Processing in Medical Imaging" (Y. Bizais, C. Barillot, and R. DiPaola, eds.), pp. 39–52. Kluwer Academic, Dordrecht.

Zubal, I. G., Harrell, C. R., Smith, E. O., Rattner, Z., Gindi, G. R., and Hoffer, P. B. (1994). Computerized three-dimensional segmented human anatomy. *Med. Phys.* **21:** 299–302.

CHAPTER 22

Attenuation, Scatter, and Spatial Resolution Compensation in SPECT

MICHAEL A. KING,* STEPHEN J. GLICK,* P. HENDRIK PRETORIUS,* R. GLENN WELLS,* HOWARD C. GIFFORD,*
MANOJ V. NARAYANAN,* and TROY FARNCOMBE[†]

Division of Nuclear Medicine, Department of Radiology, University of Massachusetts Medical School, Worcester, Massachusetts
[†]*McMaster University, Hamilton, ON, Canada*

I. Review of the Sources of Degradation and Their Impact in SPECT Reconstruction
II. Nonuniform Attenuation Compensation
III. Scatter Compensation
IV. Spatial Resolution Compensation
V. Conclusion

I. REVIEW OF THE SOURCES OF DEGRADATION AND THEIR IMPACT IN SPECT RECONSTRUCTION

A. Ideal Imaging

Data acquisition for the case of ideal single-photon emission computed tomography (SPECT) imaging is illustrated in Figure 1, which portrays a single-headed gamma camera imaging a source distribution $f(x,y)$ at a rotation angle θ with respect to the x axis. The camera is equipped with a parallel-hole absorptive collimator, which for this ideal case only allows photons emitted from $f(x,y)$ in a direction parallel to the collimator septa to pass through to the NaI(Tl) crystal and be detected. The source distribution in Figure 1 is a cartoon drawing of a slice through the three-dimensional (3D) Mathematical Cardiac Torso (MCAT) phantom (Tsui *et al.*, 1994) with a Tc-99m sestamibi distribution. Let (t, s) be a coordinate system rotated by the angle θ counterclockwise with respect to the (x,y) coordinate system. Then t and s can be written in terms of x and y, and the rotation angle θ as (Boas, 1983):

$$t = x\cos\theta + y\sin\theta \\ s = -x\sin\theta + y\cos\theta \qquad (1)$$

The ray sum, $p(\theta,t')$, is the line integral over $f(t,s)$ with respect to s for $t = t'$, or

$$p(\theta,t') = \int_{-\infty}^{\infty} f_\theta(t',s)\,ds = \int_{-\infty}^{\infty}\int_{-\infty}^{\infty} f_\theta(t,s)\delta(t-t')\,dt\,ds$$

$$= \int_{-\infty}^{\infty}\int_{-\infty}^{\infty} f(x,y)\delta(x\cos\theta + y\sin\theta - t')\,dx\,dy \qquad (2)$$

where δ is the delta function, and Eq. (1) was used to express t in terms of $x,y,$ and θ. The function $p(\theta,t)$ for the case of ideal imaging is the Radon transform of $f(x,y)$, and is the parallel projection of $f(x,y)$ for a constant value of θ (Kak and Slaney, 1987). By rotating the camera head about the patient, a set of projections is acquired for different projection angles. This set of projections constitutes the data that will be used to estimate the source distribution from which they originated. Figure 2 shows the overlaid ideal projections of a point source in the liver of the MCAT phantom at projection angles of 0° (left lateral), 45° (left anterior oblique), 90° (anterior), and 135° (right anterior oblique) relative to the x axis in Figure 1. The location of the point source is indicated as the black circular point within the liver and on the x axis in Figure 1. Notice that in

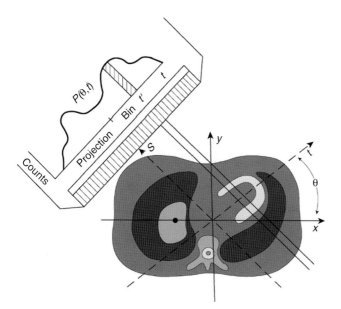

FIGURE 1 Geometry of ideal SPECT imaging.

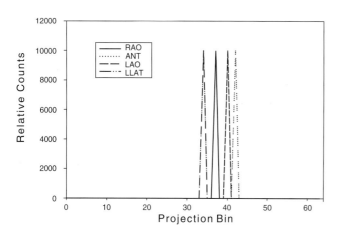

FIGURE 2 Right anterior oblique (RAO), anterior (ANT, left anterior oblique (LAO), and left lateral (LLAT) projections of a point source in the liver of the MCAT phantom as imaged with no sources of degradation. Note the consitency in size and shape of te projections.

the ideal case the projections are all of the same size and shape and vary with projection angle just in their positioning along the t axis. When the projections are stacked one on top of the other for viewing, they form a matrix called the sinogram. The matrix is so named because with 360° acquisitions each location traces out a sine function whose amplitude depends on the distance from the center of rotation (COR) and phase depends on its angular location.

B. Sources of Image Degradation

SPECT imaging is not ideal. Inherent in SPECT imaging are degradations that distort the projection data. This chapter focuses on three such degradations and the

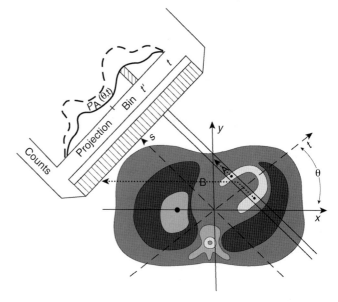

FIGURE 3 Impact of attenuation on SPECT imaging. Note that photon A is stopped photoelectrically and photon B is scattered so that it is not detected. The result is a decrease in the counts in the projection data as illustrated.

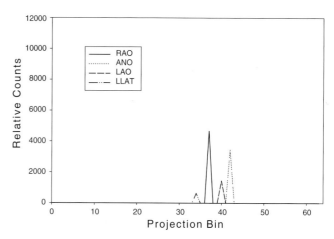

FIGURE 4 Right anterior oblique (RAO), anterior (ANT), left anterior oblique (LAO), and left lateral (LLAT) projections of a point source in the liver of the MCAT phantom as imaged with attenuation of the photons. Note the lack of consistency in size and shape of the projections and their decrease compared Figure 2.

compensation for them. The first is attenuation. In order for a photon to become part of a measured projection, it must escape the body. As illustrated in Figure 3, photons emitted that would otherwise be detected may be either photoelectrically absorbed (photon A) or scattered (photon B) so that they are lost from inclusion in the projections. Thus the attenuated projections, $p_A(\theta,t)$, will contain fewer events than the ideal projections. This is illustrated in Figure 4, which shows the attenuated projections of the point source of Figure 2. Note how the reduction in the number of detected events varies with the thickness and nature of material through which the photons must travel to be imaged. The extent of attenuation can be quantified mathematically by the transmitted fraction, $TF(t',s',\theta)$, which is the fraction of the photons from location (t',s') that will be transmitted through a potentially nonuniform attenuator at angle θ. The transmitted fraction is given by:

$$TF(t',s',\theta) = \exp\left(-\int_{s'}^{\infty} \mu(t',s)\,ds\right) \tag{3}$$

where $\mu(t,s)$ is the distribution of linear attenuation coefficients as a function of location. Equation (3) is accurate only for a monoenergetic photon beam and under the assumption that as soon as a photon undergoes any interaction it is no longer counted as a member of the beam. The latter is the good geometry condition (Attix, 1983; Soresenson and Phelps, 1987). The attenuated projections are obtained from the ideal projections by including TF within the integrals of Eq. (2). As a result of the differences in attenuation coefficient with type of tissue, the TF varies with the materials traversed even if the total patient thickness between the site of emission and the camera is the same. That is, it makes a difference if the photons are passing through muscle, lung, or bone. Similarly, a change in TF will occur if the amount of tissue that has to be traversed is altered. Thus, one needs to have patient-specific information on the spatial distribution of attenuation coefficients (an attenuation map or estimate of $\mu(t,s)$) in order to calculate the attenuation that occurs when photons are emitted from a given location in the patient and detected at a given angle.

The second source of degradation is the inclusion of scatter in the projections. This, as illustrated in Figure 5, leads to

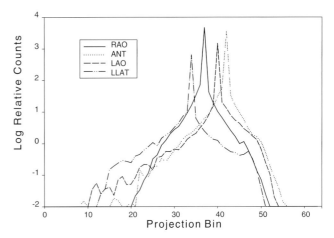

FIGURE 6 Right anterior oblique (RAO), anterior (ANT), left anterior oblique (LAO), and left lateral (LLAT) projections of a point source in the liver of the MCAT phantom as imaged with attenuation and scattering of the photons. Note the extended tails of the point-spread-functions and the use of logarithmic scaling to make these tails better visible.

the inclusion of photons in the projections that normally would not have been detected. Note that in Figure 6 a logarithmic scaling is used for the number of counts detected to better illustrate the contribution of scatter. To account for the presence of scatter, Eq. (3) can be modified by multiplying the exponential term by a buildup factor, B (Attix, 1983). The buildup factor is the ratio of the total number of counts detected within the energy window (primary plus scatter) to the number of primary counts detected within the window. In the good geometry case there is no scatter detected, so the buildup factor is 1.0. For the four point–source projections in Figure 6 the buildup factors were 1.40 for the right anterior oblique view, 1.52 for the anterior view, 1.84 for the left anterior oblique view, and 2.21 for the left lateral view. Photons undergoing classical scattering do not change energy during the interaction; thus, they can not be separated from transmitted photons on the basis of their energy. At the photon energies of interest in SPECT imaging, classical scattering only makes up a small percentage of the interactions in the human body. Compton scattering is the dominant mode of interaction under these conditions, and during Compton scattering the photon is reduced in energy as well as deflected from its original path. Thus one can use energy windowing to reduce the amount of scattered photons imaged, but not eliminate scatter, due to the presence of classically scattered photons and the finite energy resolution of current imaging systems. In fact, the ratio of scattered to primary photons in the photopeak energy window (scatter fraction) is typically 0.34 for Tc-99m (de Vries and King, 1994) and 0.95 for Tl-201 (Hademenous et al., 1993). When scatter is neither removed from the emission profiles prior to reconstruction nor incorporated into the reconstruction process, it can lead to overcorrection for attenuation because the detected scattered photons violate the good geometry assumption of Eq. (3).

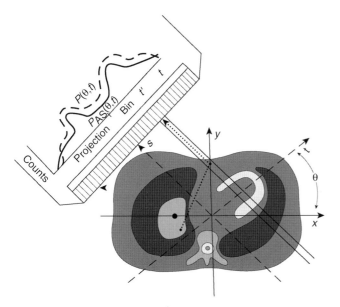

FIGURE 5 Impact of scatter on SPECT imaging. Notice the addition of the photon from the liver scattered such that it appears at the incorrect location within the projections.

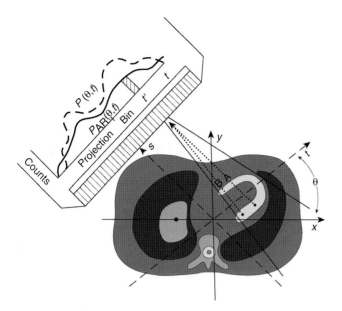

FIGURE 7 Impact of system spatial resolution on SPECT imaging. Notice the distance-dependent enlargement of the region from which primary photons can contribute to the projections without penetrating the collimator.

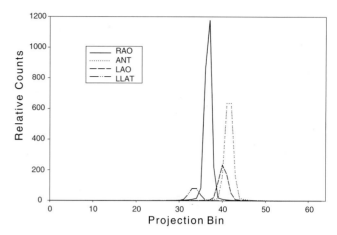

FIGURE 8 Right anterior oblique (RAO), anterior (ANT), left anterior oblique (LAO), and left lateral (LLAT) projections of a point source in the liver of the MCAT phantom as imaged with attenuation and scattering of the photons and nonstationary spatial resolution of the system. Note the variation in width and the use of different scaling compared to Figures 2 and 4. Note also that the low magnitude of the scatter tails with linear mapping of the relative counts.

The third source of degradation is the finite, distance-dependent spatial resolution of the imaging system. When imaging in air, the system spatial resolution consists of two independent sources of blurring (Sorenson, 1987). The first is the intrinsic resolution of the detector and electronics, which is well modeled as a Gaussian function. The second is the spatially varying geometrical acceptance of the photons through the holes of the collimator. This response is illustrated in Figure 7. Note in this figure that both photon A, which is emitted parallel to the s axis, and photon B, which is angled relative to the s axis, now make it through the collimator and are detected. Detailed theoretical analyses of the geometrical point-spread and transfer functions for multihole collimators have been published (Metz et al., 1980; Tsui and Gulberg, 1990; Frey et al., 1998; Formiconi, 1998). In the absence of septal penetration and scatter, the point–spread function (PSF) for parallel-hole collimators is typically approximated as a Gaussian function (Brookeman and Bauer 1973; Gilland et al., 1988) whose standard deviation (σ_C) is a linear function of distance given as:

$$\sigma_C(d) = \sigma_0 + \sigma_d d \qquad (4)$$

where d is distance from the face of the collimator, σ_0 is the standard deviation at the face of the collimator, and σ_d is the change in standard deviation with distance. Because the two sources of blurring are independent, they can be analyzed as subsystems in series (Metz and Doi, 1979). Therefore, the system PSF is the convolution of the collimator and intrinsic PSFs. The convolution of two Gaussian functions is a third Gaussian function. Thus, the system PSF can be approximated by a Gaussian function whose distance-dependent standard deviation (σ_S) is given as:

$$\sigma_S(d) = (\sigma_C^2(d) + \sigma_I^2)^{1/2} \qquad (5)$$

where σ_I is the standard deviation of the Gaussian function modeling the intrinsic spatial resolution. The system PSFs for the point source in the liver of the MCAT phantom is given in Figure 8 for the same four projection angles employed in the previous figures. Note the variation in the heights and widths of these responses due to the variation in distance between the point source and the face of the collimator.

Attenuation, scatter, and system spatial resolution are three important physical sources of image degradation in SPECT when imaging with Tc-99m or Tl-201, but they are not the only ones. Other sources include collimator scatter and X-ray production, septal penetration, patient and physiological motion, and nonstatic tracer distributions. The extent to which each of these is important varies with the energy of the photons emitted, the kinetics of the radiopharmaceutical, and the time required for imaging the patient. A full discussion of all these factors is beyond the scope of this chapter.

C. Impact of Degradations

As detailed in Chapter 20 of this book (Kinahan et al.), filtered backprojection (FBP) provides an exact analytical solution to reconstructing a slice from the projections in the sinogram. That is, FBP provides an inversion of the Radon transform (Kak and Slaney, 1987). However, as should now be clear, inverting the Radon transform is only part of the problem. Attenuation of the photons, the acquisition of scattered as well as transmitted photons, and the inherent

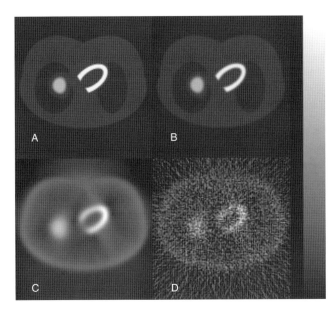

FIGURE 9 (A) Original source distribution from the MCAT phantom. (B) FBP reconstruction of ideal projections of source distribution. (C) FBP reconstructions of projections degraded by the presence of attenuation, scatter, and system resolution. (D) FBP reconstructions of projections degraded by the presence of attenuation, scatter, system resolution, and noise.

distance-dependent finite resolution of the imaging system have also distorted the acquired projections from the ideal projections. The result is that just inverting the Radon transform does not result in an accurate estimate of the original source distribution, even in the absence of noise. This is illustrated in Figure 9. Figure 9A shows the source distribution used to create the projections. Figure 9B shows the slice reconstructed using FBP from ideal projections; notice that an extremely good estimate of the original source distribution is obtained. Figure 9C shows the FBP reconstruction of the slice from projections that have been degraded by the presence of attenuation, scatter, and system spatial resolution. Notice the significant distortion in the intensities and shapes of the structures when the sources of degradation are included in creating the projections but are not included in the reconstruction process. For example, due to attenuation and the partial-volume effect (as detailed in Section IV), the heart wall is significantly reduced in apparent intensity compared to the soft tissues. Also, the soft tissue close to the body surface is now enhanced compared to that of the deeper tissues. Finally, note the significant distortion of the areas of low activity concentration around the heart and liver and the buildup of counts outside the body in the sternal notch region on the anterior surface. The goal of the compensation strategies discussed in this chapter is to return the slice estimate of Figure 9C to that of Figure 9A; however, the compensation strategies are limited in their ability to perform such a transformation by the presence of noise. The complication of the presence of noise is illustrated in Figure 9D, which shows the FBP reconstruction of the degraded projections to which has been added Poisson noise typical of that in perfusion imaging.

The cause of the distortions in Figure 9C is the inconsistency of the projection data with the model of imaging used in reconstruction. The model of the imaging system used with FBP reconstruction was that of an ideal imaging system (i.e., one with a PSF equal to a δ function whose integral is a constant independent of projection angle, as illustrated in Figure 2). The actual PSFs for SPECT imaging vary in shape and magnitude with location in the slice and projection angle, as illustrated in Figure 8. Without compensation for such variation prior to reconstruction or accounting for the variation as part of the reconstruction algorithm, the reconstructed PSF is anisotropic with long positive and negative tails (Manglos et al. 1988). The reason for this is that the use of the ramp filter in FBP to compensate for the blurring of backprojection results in negative values in the projections. These negative values cancel out the wrong guesses as to where the counts are located when backprojected. If the acquired PSFs are of different sizes and shapes, then the cancellation is not exact and a distorted reconstructed PSF results (King, Xia et al., 1996). The low-magnitude tails of the reconstructed PSFs from hot structures in images can add up, causing changes in the apparent count level of nearby structures. Clinically, this can result in decreased apparent localization in the inferior wall of the heart in perfusion images with significant hepatic activity (Genmano et al., 1994), the loss of the ability to visualize bone structures in bone scans at the level of significant bladder activity (O'Connor and Kelly, 1990; Gillen et al., 1991), and other artifacts. The distortion is not limited to FBP, but occurs with iterative reconstruction if the imaging model is inconsistent with the actual data collection (King, Xia, et al., 1996; Nuyts et al., 1995).

II. NONUNIFORM ATTENUATION COMPENSATION

A. Estimation of Patient-Specific Attenuation Maps

The determination of an accurate, patient-specific attenuation map is fundamental to performing attenuation compensation (AC). This fact has been appreciated for years (Jaszczak et al., 1979); however, it is the commercial interest in providing estimates of attenuation maps that will probably bring the benefits of AC into widespread clinical use. Three strategies have been employed for obtaining attenuation maps for use with AC: (1) importing and registering maps from another modality, (2) obtaining transmission data for estimating the attenuation maps with the gamma camera employed for emission imaging, and (3) estimating the attenuation map solely from the emission data. A previous review (King et al., 1995) of this topic can be consulted for more details.

High-resolution images from another modality can be imported and registered with the patient's SPECT data (Fleming, 1989; Meyer et al., 1996). Of course the task is made a lot simpler by acquiring the high-resolution slices while the patient is on the same imaging table (Baln Kespoor et al., 1996). The voxel values in the high-resolution images require scaling to the appropriate attenuation coefficients for the energy of the emission photon (LaCroix et al., 1994). In addition to their use for attenuation maps, these images can also provide anatomical contexts for the emission distributions and compensation for the partial-volume effect (Da Silva et al., 2001). Currently one imaging company is marketing a combined SPECT and computed tomography (CT) system that shares a single imaging table. The image noise in the attenuation maps from such a system is very low, and the in-plane resolution is very high compared to the SPECT slices. Such systems will most probably perform sequential emission and transmission imaging; however, the CT imaging time can be quite short if a high-end CT system is coupled with the SPECT system. One potential drawback to such a system is the cost, especially if it is desirable to use the CT system for diagnostic imaging as well as providing AC for the SPECT system.

The basic transmission source and camera collimator configurations are summarized in Figure 10 for a 50-cm field of view (FOV) single-headed SPECT system imaging a 50-cm diameter patient. The extent of truncation illustrated of course varies with the camera FOV and the patient diameter. With each of these configurations, the projections of the transmission source with the patient and table present in the FOV of the camera are combined with the transmission projections with the patient and table absent (blank scan) and input to reconstruction algorithms that solve Eq. (3) for the μs (attenuation map).

The first configuration, illustrated in Figure 10A, is a sheet transmission source opposite a parallel-hole collimator on the camera head. This configuration was investigated for a number of years for use with SPECT systems (Maeda et al., 1981; Malko et al., 1985; Bailey et al., 1987). Its major advantage is that it provides a transmission source that fully irradiates the camera head opposed to it and needs no motion of the source beyond that provided by the rotation of the gantry. The disadvantages of this configuration are that it is cumbersome to work with and requires source collimation in order for good geometry attenuation coefficients to be estimated (Cao and Tsui, 1992).

The second configuration, shown in Figure 10B, is the multiple-line-source array (Celler et al., 1998). With this configuration, the transmission flux comes from a series of collimated line sources aligned parallel to the axis of rotation of the camera. The spacing and activity of the line sources is tailored to provide a greater flux near the center of

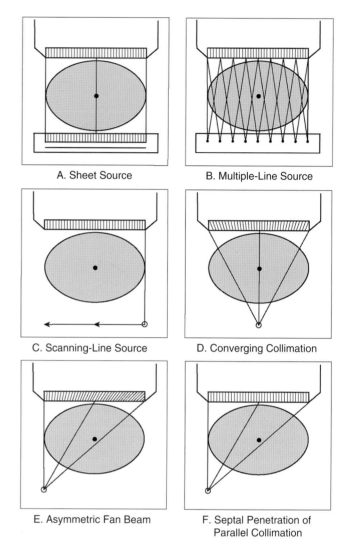

FIGURE 10 Six configurations for transmission imaging on a SPECT system.

the FOV, where the attenuation from the patient is greater. As the line sources decay, two new line sources are inserted into the center of the array. The rest of the lines are moved outward one position, and the weakest pair removed. The transmission profiles result from the overlapping irradiation of the individual lines, which varies with the distance of the source from the detector. The multiple-line-source array provides full irradiation across the FOV of the parallel-hole collimator employed for emission imaging without the need for translation of the source, as does use of a sheet source. Supporting, shielding, and collimating the source array is not quite as cumbersome as it is for the sheet source, but it is still more so than for a single-line or point source. A major advantage of the configuration is that no scanning motion of the sources is required. A major disadvantage of this system is the amount of cross-talk between the emission and transmission photons. Due to Compton scattering within the patient, scatter from whichever is higher in energy will contaminate the energy window for the other.

The sheet or multiple-line transmission source can be replaced by a scanning-line source (Tan *et al.*, 1993), shown in Figure 10C. The camera head opposed to the line source is electronically windowed to store only the events detected in a narrow region opposed to the line source in the transmission image. The electronic window moves in synchrony with the line thereby allowing transmission imaging across the entire FOV. The result is a significant reduction in the amount of scattered transmission radiation imaged in the transmission energy window. The transmission counts are concentrated in this moving spatial window thereby increasing their relative contribution compared to the emission events. Similarly using the portion of the FOV outside the transmission electronic window for emission imaging results in a significant reduction in the amount of cross-talk between transmission and emission imaging. Despite the significant reduction in cross-talk afforded by the scanning-line source methodology, compensation is still required for the residual cross-talk. The scanning-line source does have the disadvantage of requiring synchronized mechanical motion of the source with the electronic windowing employed to accept the transmission photons. The result can be an irradiation of the opposing head that changes with gantry angle or erratically with time. Also, communication and synchronization between the camera and moving the line source can be problematic. However, the advantages of this method compared to its disadvantages are such that it is currently the dominant configuration offered commercially.

The need for lateral motion of the transmission source can be avoided, without the need for a cumbersome sheet or multiple-line source, if convergent collimation is employed. Figure 10D depicts the use of a fan-beam collimator with a line transmission source at its focal distance (Tung *et al.*, 1992; Jaszczak *et al.*, 1993; Gullberg *et al.*, 1998). This configuration has three advantages: (1) near good geometry attenuation coefficients are measured because the collimator acts as an antiscatter grid, (2) a better spatial resolution sensitivity combination for small structures such as the heart is achieved than with parallel collimation for emission imaging, and (3) only line or point sources need to be handled, shielded, or mounted on the system. There are also potential disadvantages of this configuration: (1) an increased truncation of the FOV over that of parallel collimation on the same camera head, (2) the lack of electronic windowing to assist in correction of cross-contamination between emission and transmission imaging, (3) the need to keep the source at the convergence location of the collimation, and (4) the acquisition of the emission images by fan-beam collimators, which distort the projection images that physicians employ to look for breast shadows and lung uptake. Also, use of fan-beam collimators raises the possibility that both the transmission and emission projections of the heart may be truncated at some projection angles. One way around this on multi-headed SPECT systems is to do emission imaging with parallel-hole collimators on the heads that are not involved in transmission imaging (Narayanan *et al.*, 1998). The transmission source can be mounted on the system under motorized computer control to move the source radially with motion on the opposed camera head, thereby allowing body contouring acquisitions and overcoming this difficulty. With sequential transmission and emission imaging, there is no need to correct the emission images for the presence of the transmission source. Also, images acquired in the transmission window during emission imaging can be used to estimate the contamination of the transmission images with transmission source present by the emission photons. One disadvantage of sequential scanning is an increase in the potential for patient motion that can result in misalignment between the emission and transmission images introducing artifacts (Stone *et al.*, 1998). Methods have been developed to estimate the contamination of emission and transmission images by the other source of radiation when simultaneous emission and transmission imaging is employed (Tung *et al.*, 1992; Jaszczak *et al.*, 1993; Gullberg *et al.*, 1998; Barnden *et al.*, 1997; Narayanan, King, *et al.*, 2002). Thus, the influence of cross-contamination can be greatly reduced.

The major remaining potential problem with using convergent collimation for estimation of attenuation maps is truncation. Truncation of attenuation profiles results in the estimated attenuation maps exhibiting the cupping artifact. This artifact is the result of the pile up of information from the truncated region near its edge when the reconstruction is limited to reconstructing only the region within the FOV at every angle. This region is called the fully sampled region (FSR). If the reconstruction is not limited to the FSR, then the cupping artifact is not present, but the area outside the FSR is distorted and the area inside the FSR is slightly reduced in value. Significant improvement can be obtained by constraining the reconstruction to the actual area of the attenuator (Narayanan *et al.*, 1998). Because the goal of transmission imaging is to estimate attenuation maps for the correction of attenuation in the emission images, the important question is not whether the maps are distorted by truncation but rather whether the correction of attenuation is degraded by truncation. It has been reported that even though the attenuation map is significantly distorted outside the FSR, the attenuation sums (exponentials of the sums of the attenuation coefficients times path lengths) for locations in the FSR are estimated fairly accurately with iterative reconstruction (Tung *et al.*, 1992; Gullberg *et al.*, 1998). This is due to the iterative reconstruction algorithms employed in transmission reconstruction forcing the total attenuation across the maps to be consistent with the measured attenuation profiles and to all locations being viewed at least part of the time. Thus, for structures, such as the heart, that are within the FSR a reasonably accurate AC can be obtained. It has been shown that truncation may have an impact on the detection of cardiac lesions (Gregoriou *et al.*, 1998). Because both emission and transmission data were truncated in that work, it is unclear whether the decrease in the area under the receiver operating characteristic (ROC) curve was caused primarily by one, by the other, or by a combination of both types of truncation. There is no question, however, that truncation poses a serious problem for the AC of structures that are outside of the FOV at some angles.

Truncation can be eliminated, or at least dramatically reduced, by imaging with an asymmetric (as opposed to a symmetric) fan-beam collimator (Chang *et al.*, 1995; Gilland *et al.*, 1998; Hollinger *et al.*, 1998; LaCroix and Tsui, 1999). The use of asymmetric collimation, shown in Figure 10E, results in one side of the patient being truncated, instead of both. By rotating the collimator 360° around the patient, conjugate views will fill in the region truncated. If a point source with electronic collimator is employed instead of a line source, a significant improvement in cross-talk can be obtained (Beekman *et al.*, 1998). Other benefits are that point sources cost less than line sources and are easier to shield and collimate. The problem remains, however, that converging collimators acquire the emission profiles. This difficulty can be overcome by using photons from a medium-energy scanning-point source to create an asymmetric fan-beam transmission projection through a parallel-hole collimator by penetrating the septa of the collimator (Gagnon *et al.*, 2000), as illustrated in Figure 10F. With this strategy, transmission imaging is performed sequentially after emission imaging to avoid transmission photons from contaminating the emission data. This lengthens the period of time that the patient must remain motionless on the imaging table. Another problem with this method of transmission imaging is that it really is only useful for imaging with low-energy parallel-hole collimators. For imaging medium-energy and high-energy photon emitters, an asymmetric fan-beam collimator must be employed.

Alternatives to transmission imaging for the estimation of attenuation maps do exist. One method used with cardiac perfusion imaging is to inject Tc-99m macroaggregated albumin (MAA) after the delayed images and then reimage the patient. The lung region is obtained by segmentation of the MAA localization, and the body outline is obtained either by an external wrap soaked in Tc-99m (Madsen *et al.*, 1993) or by segmentation of scatter-window images (Wallis *et al.*, 1995). Assigning attenuation coefficients to the soft tissue and lung regions then forms the attenuation map. This method has the advantage of requiring no modification of the SPECT system to perform transmission imaging; however, a second pharmaceutical must be administered, and additional imaging must be performed. Another method of estimating attenuation maps without the use of transmission imaging is to segment the body and lung regions from scatter and photopeak energy window emission images of the imaging agent itself. This does not require any significant alteration or addition to current SPECT systems and imaging protocols beyond the simultaneous acquisition of a scatter window. The application of this method in patient studies has shown that approximate segmentation of the lung can be achieved interactively in many, but not all, patients (Pan, King, *et al.*, 1997). This lack of robustness has limited its clinical use. The emission data do contain information regarding photon attenuation, and efforts have been made at extracting information on the attenuation coefficients directly from the emission data. One approach for doing this is to solve iteratively for both the emission and attenuation maps from the emission data (Manglos and Young, 1994; Nuyts *et al.*, 1999). Another approach is to use the consistency conditions of the attenuated Radon transform to assist in defining an attenuation map that affects the emission projections in the same way as the true attenuation distribution (Welch *et al.*, 1997).

B. Compensation Methods for Correction of Nonuniform Attenuation

Not only has the ability to estimate attenuation maps improved greatly in the 1990s, but so has the ability to perform AC once the attenuation maps are known. In part this is due to the tremendous changes in computing power, making computations practical that could only be performed as research exercises 10 years ago. It is also due to an improvement in the algorithms used for correction and the efficiency of their implementations. An example is the development of ordered-subset or block-iterative algorithms for use with maximum-likelihood reconstruction (Hudson and Larkin, 1994; Byrne, 1998). Numerous AC algorithms have been proposed and investigated. For example, considerable effort has been directed towards the direct analytical solution for attenuation as part of reconstruction. Solutions to the exponential Radon transform (uniform attenuation within a convex attenuator) have been derived (Bellini *et al.*, 1979; Tretiak and Metz, 1980; Gullberg and Budinger, 1981; Hawkins *et al.*, 1988; Metz and Pan, 1995). These have been extended to correction of a convex region of uniform attenuation within a nonuniform attenuator (Markoe, 1984; Liang *et al.*, 1994; Glick *et al.*, 1995). The attenuated Radon transform, in the general case of nonuniform attenuation, has also been analytically solved (Natterer, 2001). Any comprehensive review of AC algorithms would require a chapter dedicated solely to this task. Therefore, we discuss at length here only the two most commonly used AC algorithms: the Chang algorithm (Chang, 1978) used with FBP and use of AC with the maximum-likelihood expectation-maximization reconstruction algorithm (MLEM) (Shepp and Vardi, 1982; Lange and Carson, 1984). The reader is referred to other reviews for more details and other algorithms (Parker, 1989; Tsui, Zhao, Frey, et al., 1994; King, Tsui et al., 1996).

To compare the algorithms we make use of the 3D MCAT phantom developed at the University of North Carolina at Chapel Hill (Tsui, Zhao, Gregoriou, et al., 1994). Source and attenuation maps from the MCAT phantom were input to the SIMIND Monte Carlo simulation of gamma camera imaging (Ljungberg, Chapter 20 in this book, and Ljungberg and Strand, 1989). SIMIND formed 128 × 128 pixel images for 120° angles about the source maps as imaged by a low-energy high-resolution parallel-hole collimator imaging Tc-99m. The primary and scatter components were recorded separately. This allowed the study of scatter-free images, that is, images on which "perfect" scatter compensation (SC) had been performed.

A simulation was made in which only the distance-dependent spatial resolution was simulated. This served as an example of "perfect" AC. Figure 11 shows the transverse slices and polar maps for 180° and 360° FBP reconstructions of this simulation. The slices were filtered prereconstruction with a 2D Butterworth filter of order 5.0 and cutoff frequency of 0.4 cycles/cm (0.125 of sampling frequency), as recommended for Tc-99m sestamibi rest images (Garica et al., 1990). Notice the absence of the artifacts outside the heart in the transverse slices. Also, notice the uniformity of counts in the polar map aside from the band of increased counts due to the joining of left ventricle

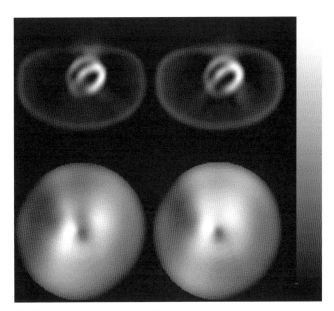

FIGURE 12 Comparison of FBP reconstructions without attenuation correction of MCAT phantom simulations that included influence of distance-dependent resolution and attenuation. Top row shows reconstructions of transverse slice through LV. Bottom row shows polar maps. Left column is for 180° reconstruction, and right row is for 360° reconstruction (Reproduced with permission from King, Tsui et al., 1996.)

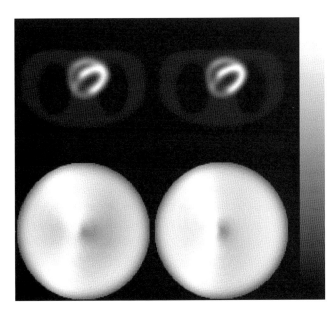

FIGURE 11 Comparison of FBP reconstructions without attenuation correction of MCAT phantom simulations that include solely the influence of distance-dependent resolution. Top row shows reconstructions of transverse slice illustrated in Figure 1. Bottom row shows polar maps. Left column is for 180° reconstructions, and right row is for 360° reconstruction (Reproduced with permission from King, Tsui et al., 1996.)

(LV) and right ventricle (RV), and the decrease in counts at the apex. These deviations from uniformity illustrate the impact of the partial-volume effect on the uniformity of cardiac wall counts. Where the wall is thicker, as with the joining of the RV and LV, the apparent counts will be higher. Where the wall is thinner, as with apical thinning (which was included in the simulated LV), the apparent counts will be lower. Thus, even with perfect AC and SC, one would not expect the wall of the LV to be uniform. Note also that there is slightly better uniformity of counts in the 360° reconstruction than the 180° reconstruction due to less variation in spatial resolution.

Figure 12 shows a transverse slice through the heart and bull's-eye polar map for 180° and 360° FBP reconstructions of the slices with no AC when attenuation was included in the simulation. Again, 2D Butterworth filtering of order 5.0 and a cutoff frequency of 0.4 cycles/cm were employed. Notice the artifacts outside the heart, due to the reconstruction of inconsistent projections, and the variation in intensity of the LV walls. The polar maps in this figure illustrate the fall off in counts from the apex to the base due to the increase in attenuation with depth into the body. A decrease in the anterior wall due to the breast attenuation for 180° reconstruction is observed. Notice the slightly better uniformity of the 360° reconstruction due to the reduction in the impact of projection inconsistencies by combining opposing views. The goal of AC is to return the reconstructed distributions of Figure 12 to those of Figure 11.

The best-known and most widely employed method that performs AC with FBP is due to Chang (1978). The multiplicative or zeroth-order Chang method is a postreconstruction correction method that compensates for attenuation by multiplying each point in the body by the reciprocal of the TF, averaged over acquisition angles, from the point to the edge of the body. That is, one calculates the correction factor $C(x', y')$ for each (x', y') location in the slice as:

$$C(x', y') = \left[\frac{1}{M}\sum_{i=1}^{M} \text{TF}(t', s', \theta_i)\right]^{-1} \quad (6)$$

where M is the number of projection angles θ_i, TF is calculated as per Eq. (3), and Eq. (1) is used to convert x' and y' into t' and s' for each θ_i. The correction is approximate because the TFs are not separable from the activity when summing over angle (Singh et al., 1988). An iterative correction can be obtained in the following way. The zeroth-order corrected slices are projected, mathematically simulating the process of imaging, and the resulting estimated profiles subtracted from the actual emission data. Error slices are reconstructed from the differences using FBP. After correction of the error slices for attenuation in the same manner as the zeroth-order correction, the error slices are added to the zeroth-order estimate of the slices to obtain the first-order estimate. Typically only the first-order correction is performed, but the process can be repeated any number of times to obtain higher-order estimates. One problem with doing this is that the method does not converge (i.e., does not reach a definite solution and then not change with further iteration). Instead, the higher-order estimates are characterized by elevated noise (Lalush and Tsui, 1994). Figure 13 shows MCAT transverse slices and polar maps for zeroth-, first-, and fifth-order Chang using

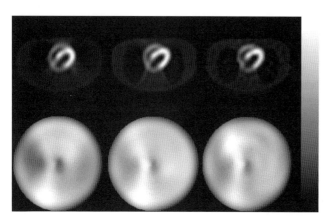

FIGURE 13 Comparison of FBP reconstructions with Chang attenuation correction of MCAT phantom simulations that include influence of distance-dependent resolution and attenuation. Top row shows reconstructions of transverse slice through LV. Bottom row shows polar maps. Left column is for zeroth-order or multiplicative Chang. Middle column is for one iteration of Chang. Right column is for five iterations of Chang. (Reproduced with permission from King, Tsui et al., 1996.)

the true nonuniform attenuation map to form the correction factors. Notice the considerable improvement of the poor correction of zeroth-order Chang after a single iteration. With five iterations, notice the increased noise evident in the slice and polar map.

As detailed in Chapter 21 (Lalush and Wernick), statistically based reconstruction algorithms start with a model for the noise in the emission data and then derive an estimate of the source distribution based on some statistical criterion. In MLEM, the noise is modeled as Poisson, and the criterion is to determine the source distribution that is most likely to have given the emission data using the expectation-maximization (EM) algorithm (Shepp and Vardi, 1982; Lange and Carson, 1984). Four advantages of MLEM that have led to its popularity are: (1) it has a good theoretical base, (2) it converges, (3) it readily lends itself to the incorporation of the physics of imaging such as attenuation, and (4) it compensates for nonuniform attenuation with a high degree of accuracy (Chornoboy et al., 1990).

$$f_j^{\text{new}} = \frac{f_j^{\text{old}}}{\sum_i h_{ij}} \sum_i \left(h_{ij} \frac{g_i}{\sum_k h_{ik} f_k^{\text{old}}} \right) \quad (7)$$

The MLEM algorithm is given in Eq. (7) and works in the following way. First, as represented in the denominator on the right side of Eq. (7), projections are made by summing the current (old) estimate of the voxel counts (f^{old}) for each slice voxel k that contributes to projection pixel i. The initial estimate is typically a uniform count in each voxel. The weights used when summing the voxel counts (h_{ik}) are the probabilities that a photon emitted in voxel k contributes to projection pixel i. These probabilities form a matrix (H) called the transition matrix. H has the number of voxels as the number of columns and the number of pixels in the projections as the number of rows. Thus for reconstruction of a $64 \times 64 \times 64$ source distribution being acquired into 64×64 projections at 64 angles, H is a $64^3 \times 64^3$ matrix. Each column in H is the unnormalized PSF for each acquisition angle stacked one on top of another for the associated voxel. If we form a column vector out of the voxel counts, then projection is given by the matrix multiplication of H times this vector, with the result being a column vector made up of the counts in each projection pixel. It is in this process of projecting (mathematically emulating imaging) that one includes the physics of imaging. For example, to include attenuation each h_{ik} would be formed as the product of the TF as calculated by Eq. (3) times the fractional contribution of voxel k to pixel i, based on the geometry used to model imaging. In reality, because of size one would not typically calculate and save the entire H matrix for 3D imaging. Instead, the needed terms are calculated on the fly by a procedure such as the following. Projecting along rows or columns is computationally inexpensive. Arbitrary projection angles can be placed in this orientation by appropriately rotating the estimated emission voxel slices and attenuation maps (Wallis and Miller, 1997). Given an aligned attenuation map for an estimate of an emission slice, one could include attenuation by starting with the voxel on the side opposite the projection being created and multiplying its value by the TF for passing through one-half the voxel distance of an attenuator of the given attenuation coefficient. The value of one-half the pixel dimension is usually used as an approximation for the self-attenuation of the activity in the voxel. One would then move to the next voxel along the direction of projection and add its value after correction for self-attenuation to the current projection sum attenuated by passing through the entire thickness of the voxel. One would then continue this process until having passed through all voxels along the path of projection. The result of the projection operation is an estimate of the counts in pixel i based on the current estimate of the voxel counts, and the model of imaging being used in the transition matrix. As shown by the division on the right side of Eq. (7), this estimate is then divided into the actual number of counts acquired in each pixel i. The ratio of the two indicates if the voxels along the given path of projection are too large (ratio less than 1), just right (ratio equals 1), or too small (ratio greater than 1). These ratios are then backprojected, as indicated by the summing over i in Eq. (7), to create an update for the estimate of voxel j. This update is the result of letting each ratio vote with a weight of h_{ij} on how the current estimate of voxel j should be altered. Note that the summing is now over the first subscript as opposed to the second as before. Thus, in terms of matrix algebra, we are multiplying by the transpose of H in the backprojection operation. The update is multiplied by the current voxel estimate to obtain the updated estimate after normalization by dividing by the backprojection of ones (division just to the right of the equal sign in Eq. 7).

The ordered-subset version of the MLEM (OSEM) (Hudson and Larkin, 1994) has accelerated reconstruction to the extent that clinically acceptable reconstruction times are now possible. With OSEM, one divides the projections into disjoint subsets. One then updates the estimate of the source distribution using just the projections within the given subset. A single iteration is complete after updating has occurred for all the subsets. The success of OSEM is such that it is now being used routinely for selected applications instead of FBP.

Figure 14 shows a comparison of the transverse slices and polar maps in the first column for 30 iterations of MLEM using the 180° of data from RAO to LPO, in the middle column for 30 iterations of MLEM using all 360° data, and at the right for 100 iterations using all 360° data. All the reconstructions were 3D postreconstruction filtered with a Butterworth filter with order of 5.0 and cut-off frequency of 0.64 cycles/cm. This cut-off is higher than the

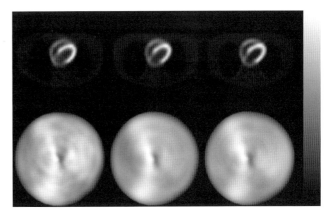

FIGURE 14 Comparison of MLEM reconstructions with attenuation correction of MCAT phantom simulations that include influence of distance-dependent resolution and attenuation. Top row shows reconstructions of transverse slice through LV. Bottom row shows polar maps. Left column is 30 iterations of 180° reconstruction. Middle column is for 30 iterations of 360° reconstruction. Right column is for 100 iterations of 360° reconstruction. (Reproduced with permission from King, Tsui et al., 1996.)

0.4 cycles/cm used with all the FBP reconstructions. Notice that all three yield excellent reconstructions; however, the 180° reconstruction is noisier (especially in the low-count areas behind the heart), and the uniformity of the polar map is slightly inferior compared to the 360° reconstructions. At 100 iterations, the noise has slightly increased, somewhat better resolution is apparent in the transverse slice, and the polar map is a little less uniform than at 30 iterations, probably due to the better resolution.

C. Impact of Nonuniform Attenuation Compensation on Image Quality

AC is required for accurate absolute quantitation of activity (Rosenthal et al., 1995). In addition to its altering quantitation, attenuation is a major cause of artifacts in SPECT slices, as discussed earlier. Thus, it is of interest to determine if AC can improve diagnostic accuracy in SPECT imaging. An area that has received a lot of attention as a candidate for the application of AC is cardiac perfusion imaging. Even though the need for AC from a physics point of view seems clear, the number of clinical studies reporting negative or equivocal results from the application of AC in perfusion imaging has lead to skepticism concerning its ultimate utility (Watson et al., 1999; Hendel et al., 1999). Two papers do show the potential for improvement in diagnostic accuracy with AC. LaCroix et al., (2000) conducted a ROC investigation using the MCAT phantom. They observed an improvement in defect detection when using AC with MLEM compared to FBP without AC, particularly for simulated patients with large breasts or a raised diaphragm. Ficaro et al., (1996) conducted a ROC investigation of AC in 60 patients who had undergone angiography. When coronary artery disease was defined as greater than 50% stenoses in the luminal diameter, the area under the ROC curve increased from 0.734 with no AC to 0.932 with AC. Thus, there is reason to believe that ultimately AC will be determined to make a significant contribution to improving the diagnostic accuracy of perfusion imaging. A position statement reviewing the literature and summarizing the current view of AC in cardiac imaging has been published (Hendel et al., 2002).

The question of the impact of attenuation and its compensation on tumor-detection accuracy is of significant current clinical interest (Whal, 1999). Both positive (Chan et al., 1997) and negative (Farquhar et al., 1999) results have been reported concerning the benefit of AC for F-18 labeled 2-fluoro-2-deoxy-D-glucose (FDG). Thus, there does not appear to be a definitive answer yet as to the role of AC in tumor detection for positron emission tomography (PET) FDG imaging.

Using simulated Ga-67 citrate SPECT imaging of the thorax for lymphoma, we (Wells et al., 2000) performed a localization receiver operating characteristics (LROC) comparison of (1) FBP reconstruction with no AC, (2) FBP reconstruction with multiplicative Chang AC, (3) FBP with one iteration of iterative Chang AC, (4) one iteration of OSEM with no AC, and (5) one iteration of OSEM with AC. The free parameters for each of the five strategies were optimized using preliminary LROC studies. To compare the strategies, 200 lesion sites were randomly selected from a mask of potential lymphoma locations. Of these, 100 were used in observer training and strategy optimization and 100 were used in data collection with five physician observers. In this study three different lesion contrasts were investigated; each image set contained an even distribution of lesion contrasts. We determined that AC does not significantly alter detection accuracy with FBP reconstruction. However, there was a significant improvement in detection accuracy when AC was included in OSEM (aggregate areas under the LROC curves A_L of 0.43, 0.39, 0.41, 0.41, and 0.58 for FBP with no AC, FBP with multiplicative Chang, FBP with iterative Chang, OSEM with no AC, and OSEM with AC, respectively). There were statistically significant differences between OSEM with AC and the other four reconstruction strategies for the aggregate and also for two lower-lesion contrasts. Numerically, OSEM with AC resulted in a larger A_L than did the other four strategies at the highest contrast considered, but the difference was no longer statistically significant. We believe this is an indication that once a lesion is of high enough contrast, OSEM with AC no longer provides a significant improvement in performance because there is little room for improvement—many of the lesions are obvious to begin with. We hypothesize that our positive finding for OSEM with AC is due in part to the use of the LROC methodology with no repetition of lesion locations in the comparison. This prevented the observers from memorizing the impact of attenuation artifacts on lesion appearance during training. Patient and lesion-site variability would also limit the physician's ability to do this clinically. Thus we believe that AC may improve tumor detection accuracy in SPECT imaging, especially when the tumors are in low-count regions in slices that also contain significant concentrations of the imaging agent. AC would then help clean up attenuation artifacts, which spread from the high-count regions and interfere with detection.

III. SCATTER COMPENSATION

The imaging of scattered photons degrades contrast and signal-to-noise ratio (SNR) and must be accounted for if AC is to be accurate (Jaszczak et al., 1981, 1982). The methods for estimating attenuation maps available commercially try to estimate good geometry attenuation coefficients. In the past, the use of an effective (reduced) attenuation coefficient approximately accounted for the presence of scattered as well as primary photons in the projections when performing AC. The coefficient was typically selected to undercorrect the primary content of the emission projections so that a uniform reconstruction of a tub with a uniform concentration of activity resulted when projections with both primary and scatter were reconstructed (Harris et al., 1984). The use of an effective attenuation coefficient to compensate for the presence of scatter is only, at best, a very approximate solution because the scatter distribution that is detected depends not only on the attenuator but also on the source distribution. Patient–activity distributions do not generally approximate a tub uniformly filled with activity. Thus instead of using an effective attenuation coefficient it is better to perform both SC and AC in conjunction because they are a team.

The best way to reduce the effects of scatter would be to improve the energy resolution of the imaging systems by using an alternative to the NaI(Tl) scintillator so that few scattered photons are acquired (Soresenson and Phelps, 1987). Figure 15 shows an energy spectrum obtained using the SIMIND Monte Carlo simulation package (Ljungberg and Strand, 1979) for an LAO view of the 3D MCAT with the source distribution that of a Tc-99m Sestamibi perfusion study, as illustrated in Figure 9. Notice the finite-width Gaussian response for the primary photons due to the finite energy resolution of the system, and the presence of scattered photons under this peak. The energy resolution (percent full width at half maximum, %FWHM) for the NaI(Tl) camera in this simulation was set at 10%. Figure 16 shows a plot of how the scatter fraction varies for %FWHMs of 1 to 10% when a symmetric window of width twice that of the %FWHM is employed in imaging. Note that both classical and Compton scattering were included in the simulation.

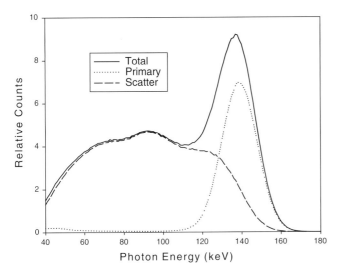

FIGURE 15 Total, primary, and scatter energy spectrums generated using the SIMIND Monte Carlo package for an LAO view of a simulated Tc-99 Sestamibi distribution in MCAT phantom.

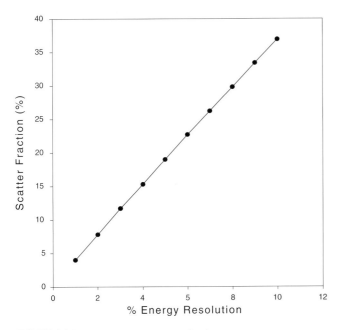

FIGURE 16 Plot of percentage scatter fraction versus percentage energy resolution of imaging system for LAO view of SIMIND simulated Tc-99m Sestamibi distribution in MCAT phantom.

From the plot it is evident that even with 1% energy resolution there is still a small amount of scatter imaged. Besides improving energy resolution to have less scatter within the imaging window, one can alter the placement of the energy window over the photopeak so that it covers less of the region below the peak itself (Atkins *et al.*, 1968). This reduces the amount of scatter collected, but also reduces the number of primary photons.

A number of SC algorithms have been proposed for SPECT systems that employ NaI(Tl) detectors. Generally, the methods of SC can be divided into two categories (Beekman, Kamphuis, *et al.*, 1997; Kadrmas, Frey, and Tsui, 1998). The first category, which we call scatter estimation, consists of those methods that estimate the scatter contributions to the projections based on the acquired emission data. The data used may be information from the energy spectrum or a combination of the photopeak data and an approximation of scatter PSFs. The scatter estimate can be used for SC before, during, or after reconstruction. The second category consists of those methods that model the scatter PSFs during the reconstruction process; we call this approach reconstruction-based scatter compensation (RBSC).

A. Scatter Estimation Methods

The difficulty in scatter estimation is illustrated in Figure 17, which shows that the acquired projection data are the result of the contributions from primary photons, scattered photons, and noise. In scatter estimation, one tries to obtain an accurate estimate of the scatter distribution avoiding any biases. However, even if this is obtained, its removal from the acquired projection still leaves the noise inherent in the detection of the scattered photons behind. Thus scatter estimation methods typically reduce the bias in the projections due to scatter, but enhance the noise level.

Energy-distribution-based methods seek to estimate the amount of scatter in a photopeak-energy-window pixel by using the variation of counts acquired, in the same pixel, in one or more energy windows. The pixel-based nature of this method allows for the generation of a scatter estimate for each pixel in the photopeak-window data. The Compton window subtraction method (Jaszczak *et al.*, 1984) is the classic example of this strategy. In this method, a second energy window placed below the photopeak window is used to record a projection consisting of scattered photons. This projection is multiplied by a scaling factor k and is then subtracted from the acquired projection to yield a scatter-corrected projection. This method assumes that the spatial distribution of the scatter within the Compton scatter window is the same as that within the photopeak window and that, once it is determined from a calibration study, a single scaling factor holds true for all applications on a given system. That the distribution of scatter in the two windows differs can be seen by noting that the average angle of scattering (and hence the degree of blurring) changes with energy. Also, the value of k varies depending on the radionuclide, energy-window definition, pharmaceutical distribution, region of the body, and other factors (Koral *et al.*, 1990).

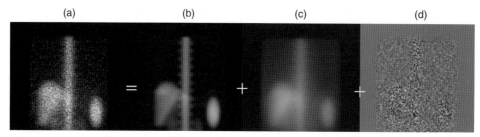

FIGURE 17 Monte Carlo simulated posterior views of a Ga-67 distribution within the MCAT phantom illustrating the problem with scatter estimation. (a) The noisy projection. (b) A noise-free primary-only projection. (c) A noise-free scatter-only projection. (d) The noise added to the middle two projections to produce the projection. Note each image is independently scaled, and the noise image in (d) contains both positives and negatives with the medium shade of gray around the outside representing zero.

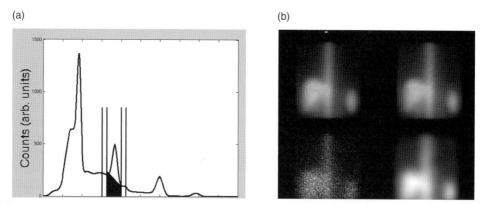

FIGURE 18 (a) A Ga-67 energy spectrum with TEW windows and scatter estimate indicated for its 185-keV photon. (b) The actual Monte Carlo–simulated scatter distribution in a posterior projection of a Ga-67 source distribution, the TEW estimate from noise-free projections, the TEW estimate for noisy projections, and the low-pass-filtered TEW estimate from the noisy projection data.

Making the scatter window smaller and placing it just below the photopeak window can minimize the difference in the distribution of scatter between it and the photopeak window. With this arrangement, one obtains the two-energy window variant of the triple-energy window (TEW) SC method (Ogawa et al., 1994). The two-energy window variant is useful when downscatter from a higher energy photon is not present, as in the case of imaging solely Tc-99m. In this method, the scatter in a pixel is estimated as the area under a triangle whose height is the average count per kiloelectronvolt (keV) in the window below the photopeak window and whose base is the width of the photopeak window in keV. When downscatter is present, a third small window is added above the photopeak as illustrated in Figure 18 (Ogawa et al., 1991). With the addition of this third window, scatter is estimated as the area under the trapezoid formed by the heights of the counts per keV in each of the two narrow windows on either side of the photopeak and a base with the width of the photopeak window. By making the windows smaller, one can estimate the scatter distribution from regions of the energy spectrum closer to the photopeak window, thereby improving the correspondence of the scattered-photon energies with those in the photopeak window. The price paid for this is that the estimation is based on fewer counts and therefore is noisier. Thus heavy low-pass filtering of the scatter estimate is required to reduce the noise in the estimate (King, deVries et al., 1997).

Energy-distribution-based methods that use a limited number of energy windows to estimate the primary counts do not completely compensate for scatter. This is in part due to their use of simple geometric shapes to estimate the scatter within the photopeak window. The main advantages of such methods are their speed and simplicity for clinical use. Performance can be improved through the use of more windows, which allows a greater degree of freedom when estimating the scatter within the photopeak window. A number of investigators have developed multiwindow methods (Koral et al., 1988; Gagnon et al., 1989; Haynor et al., 1995). The problem with methods that require more than a few energy windows is that the required number of windows is not available on many SPECT systems. Also, dividing the spectrum up into a large number of windows decreases the detected counts in each window, thus increasing the noise in the windows.

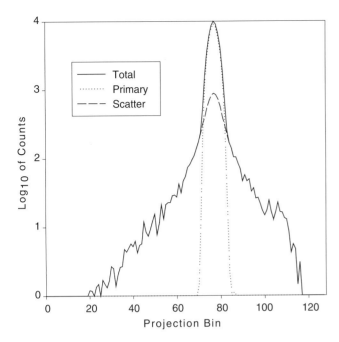

FIGURE 19 Plot of SIMIND simulated total, primary, and scatter PSFs for LAO view of a slice centered on axial location of a Tc-99m point source located in the left ventricular wall of the MCAT phantom. Note the asymmetry of the scatter response.

instead of a point. Convolution is performed one-dimensionally in the plane of the slice to be reconstructed. Scatter originating outside the plane is assumed to be included as a result of using the finite-length line.

Msaki et al. (1987) modeled the 2D scatter PSF and performed 2D convolution to account for across-plane scatter as well as in-plane scatter. This method was further refined by adapting the scatter PSF for the individual patient using photon transmission through an attenuation map (Meikle et al., 1994; Hutton et al., 1996). Convolution-subtraction methods offer a fast and reasonably accurate way of correcting for scatter. Their main disadvantages are (1) the subtraction of the scatter estimate elevates noise in the primary estimate, (2) the accuracy with which the scatter PSF is modeled is often poor, and (3) the estimation of scatter from sources not within the FOV of the camera poses a problem.

B. Reconstruction-Based Scatter Compensation Methods

RBSC starts with the estimated source and attenuation distributions and calculates the contribution of scatter to the projections by using the underlying principles of scattering interactions. With RBSC methods, compensation is achieved, in effect, by mapping scattered photons back to their point of origin instead of trying to determine a separate estimate of the scatter contribution to the projections (Kadrmas, Frey, and Tsui, 1998). All the photons are used in RBSC, and it has been argued that there should be less noise increase than with the other category of compensation (Beekman, Kamphuis, et al., 1997; Kadrmas, Frey, and Tsui, 1998). One disadvantage of this method, which RBSC shares with the convolution-based subtraction method discussed in the previous section, is that RBSC does not allow for the direct calculation of scatter from sources that are outside the reconstructed field of view.

The accuracy of RBSC, however, depends on the accuracy of modeling scatter, and accurate modeling of scatter is computation intensive. As with AC, the estimation of accurate patient-specific attenuation maps is essential because they are used to form the patient-specific scatter PSFs. It is also vital that interslice (3D) as well as intraslice (2D) SC be performed (Munley et al., 1991; Tsui, Frey, et al., 1994; Kamphuis et al., 1996). Thus, for reconstruction of a $64 \times 64 \times 64$ source distribution imaged at 64 angles one would need to form a transition matrix consisting of 64^4 PSFs, each consisting of 64×64 terms, or 64.7×10^9 terms in total. Because each patient presents a unique attenuation distribution, the transition matrix used in reconstruction ideally should be formed for the individual patient. Such a transition matrix could be obtained by manufacturing an attenuation distribution that matches the patient's and in which a point source can be positioned independently in each voxel to allow measurement of the PSFs. However, the point-source

Another subclass of scatter-estimation methods is the spatial-distribution-based methods. These methods seek to estimate the scatter contamination of the projections on the basis of the acquired photopeak window data, which serves as an estimate of the source distribution, and a model of the blurring of the source into the scatter distribution. The latter is typically an approximation to the scatter PSF; an illustration for one location in the 3D MCAT phantom is shown in Figure 19. Beck and colleagues (1968) conducted an early analysis of the contribution of scatter to the response observed for point and line sources.

An example of a spatial-distribution method for SC in SPECT is the convolution-subtraction method, which models the scatter response as decreasing exponentially from its center maximum value (Axelsson et al., 1984). The center value and slope of the exponential are obtained from measurements made with a finite-length line source. This function is convolved with the acquired projection data, and the result is used as an estimate of the scatter distribution in the projection. The estimated distribution is subtracted from the original data to yield scatter-corrected projection data. One problem with this method is that it assumes that the scatter model does not change with location. To overcome this problem, Monte Carlo methods are used to generate a set of scatter responses that are, in turn, used to interpolate the scatter response at a given location (Ljungberg and Strand, 1991). Both these approaches use scatter line-spread functions (LSFs) instead of PSFs. That is, the source modeled is a finite-length line imaging time is prohibitive for routine clinical use, and the storage in memory and on disk of such a matrix represents a significant problem even for current hardware standards. Thus, the terms in the matrix are normally calculated as needed without saving them and with various levels of approximation. Also, the contribution of scatter to the projections may be calculated directly without the actual formation of the PSFs, as in the case of Monte Carlo simulation.

For the case of a uniform convex attenuator, a parallel/serial model of the system PSF can be used to separate the distance-dependent camera response from the depth-dependent scatter response (Gilland *et al.*, 1988; Metz and Doi, 1979; Penny, King, *et al.*, 1990). In this approach, a distance-dependent Gaussian is used to model the system response to the primary photons. A second Gaussian is used to model the depth-dependent scatter response originating from the attenuator. This scatter response is convolved with the appropriate system response for the distance from the collimator because the two are in series and is summed after scaling with the response of the primary photons because primary and scatter are modeled as being imaged in parallel. The scaling is based on the scatter-to-primary ratio (scatter fraction) for the given depth in the attenuator. In this way PSFs can be quickly formed from the storage of just the regression models of the variation of the FWHM of the primary photons with distance, the scattered photons with depth, and the scatter fraction with depth.

The difficulty with the parallel/serial approach is that it accounts for only variations in the depth of the source and not variations in the attenuator to either side of the path of the primary photons to the camera. This approach has been refined by including parameterization of the scatter response as a function of the shape of the uniform attenuator relative to the position of the point source (Beekman *et al.*, 1993; Frey and Tsui, 1993). The result is system PSFs that match measured PSFs in uniform, convex attenuators exceedingly well but are not suitable for use with nonuniform attenuation. An attempt has been made to allow these slab-derived scatter responses to account for nonuniform attenuation, but the estimated responses have not achieved the same level of agreement as with uniform attenuation (Frey and Tsui, 1994; Beekmen, den Harder, *et al.*, 1997).

An alternative approach is to calculate the scatter response by integration of the Klein-Nishina formula for Compton scattering using the patient's attenuation maps for the attenuator distribution (Riauka and Gortel, 1994; Wells *et al.*, 1998). By calculating the PSFs analytically (as opposed to stochastically) these methods achieve noise-free estimates in a fixed amount of time, unlike Monte Carlo simulation methods in which noise in the estimate and calculation time are directly linked. Because the processing time increases dramatically with the order of the scatter interactions, only first-order (scattering once before detection) and second-order (scattering twice before detection) scatter are typically included in the calculation. The processing time can also be reduced by factoring the calculation so that a significant amount of it can be contained in precalculated camera-dependent lookup tables. These methods result in excellent agreement with experimental and Monte Carlo PSFs. For point sources, the computation times are significantly faster than with Monte Carlo simulation. This advantage over Monte Carlo, however, is dependent on the source configuration and is lost for large extended source distributions.

Still another approach is based on the calculation of an effective scatter source from which the contribution of scatter to the projections can be estimated using the same projector as for the primary photons (Kadrmas, Frey, and Tsui, 1998; Frey and Tsui, 1997; Zeng *et al.*, 1999; Bai *et al.*, 2000). With this approach, the estimated primary distribution is blurred into an effective scatter source distribution. The effective scatter source is formed by taking into account the probability that a photon emitted at a given location will reach the scattering site, that the photon there undergoes a scattering interaction that leads to it being detected, and finally that the scattered photon will interact in the crystal producing an event that is within the energy window used in imaging. The probability of it not being lost due to attenuation on its way from its last scattering site to the crystal is handled by the attenuated projector. By making some approximations in the calculation, excellent agreement with Monte Carlo simulation can be obtained in reconstruction times feasible for clinical use. For one method, the approximations made in order to attain this speed include assuming spatial invariance, so that the blurring kernels can be precalculated by Monte Carlo simulation, and truncating a Taylor series expansion of the exponential describing the probability of attenuation of the photon from the site of emission to the site of last scattering (Kadrmas, Frey, and Tsui, 1998; Frey and Tsui, 1997). With use of Monte Carlo precalculated kernels, the path of the photon from emission to last scattering interaction before detection can include scattering interactions up to any order desired. However, these intermediate scatterings are assumed to occur in a uniform medium. An alternative method formulates the effective scatter source distribution by incrementally blurring and attenuating each layer of the patient forward toward the detector (Zeng *et al.*, 1999; Bai *et al.*, 2000). The attenuation coefficients from the estimated attenuation map are used to do the layer-by-layer attenuation; thus, this method does not assume a uniform medium when correcting for the attenuation from the site of emission to scattering. The incremental blurring is, however, based on a Gaussian approximation to first-order Compton scattering as calculated from the Klein-Nishina equation. Thus only first-order scatter, making up between 80 and 90% of the scattering events in the Tc-99m photopeak window, are included. The effective scatter source image is created from the result of the incremental blurring and attenuation through multiplication by the voxel attenuation coefficient. This effective scatter source image is then incrementally projected, taking into account attenuation and system spatial resolution, to produce the estimate of scatter in the projection. In this final stage, the alteration of the attenuation coefficient resulting from the change in the energy of the photons upon scattering is not accounted for.

One final approach to including scatter in making the projections from the estimate of the source distribution during reconstruction is Monte Carlo simulation. Monte Carlo simulation was one of the first methods investigated for this purpose (Floyd et al., 1985; Bowsher and Floyd, 1991). However, it was too slow to consider for clinical use. This is no longer the case. Beekman and colleagues (Beekman et al., 2002) have developed a Monte Carlo approximation that can perform 10 iterations of OSEM using an unmatched projector/backprojector on a dual-processor PC in 10 minutes for a $64 \times 64 \times 64$ reconstruction of a Tc-99m distribution. The largest portion of the speed increase was obtained by combining stochastic photon transport of the interactions within the patient with an analytical model of the detection by the camera, a technique they call convolution-based forced detection (de Jong et al., 2001). In this method, the photon weight and the location of its last interaction in the patient are stored. On completion of the Monte Carlo simulation of each projection angle, the weights are grouped together according to their distance from the camera, convolved with the system PSF for that distance, and projected. This process accelerates Monte Carlo simulation by a factor of approximately 50. One can think of this as going back to the serial model of imaging in which Monte Carlo simulation of interactions within the patient are in series with an analytical model of the camera.

A number of changes in the iterative reconstruction algorithm have been combined with these approaches, with the result that the inclusion of 3D modeling of scatter during reconstruction is now possible in clinically feasible time frames. Such algorithmic improvements include the development of acceleration methods based on the updating of the reconstruction estimate by subsets of the data as previously discussed in Section II (Hudson and Larkin, 1994; Byrne, 1998; Kamphuis et al., 1998). The use of an unmatched projector/backprojector pair greatly speeds reconstruction, allowing one to model only scattering in the projection operation (Kamphuis et al., 1998; Zeng and Gullberg, 1992; Zeng and Gullberg, 2000). One can also use coarse-grid scatter modeling and either hold the scatter contribution fixed after a few iterations or update the scatter projection intermittently during reconstruction (Kadrmas, Frey, Karimi, et al., 1998).

C. Impact of Scatter Compensation on Image Quality

There is little question that SC is necessary for accurate activity quantitation (Rosenthal et al., 1995). However, the gains in contrast with SC are typically accompanied by altered noise characteristics so that the benefit of SC for tumor detection is uncertain. Using contrast-to-noise plots for simulated Tc-99m images, Beekman, Kamphuis, et al. (1997) compared ideal scatter rejection (the imaging of solely primary photons), ideal scatter estimation (provision of the actual scatter content in each projection bin), ideal scatter modeling (perfect knowledge of the scatter PSFs, which is the ideal for the RBSC methods), and no SC. They determined that ideal scatter rejection was the best, followed by ideal scatter modeling, and then ideal scatter estimation. Similar results were reported by Kadrmas, Frey, and Tsui (1998) for Tl-201 cardiac imaging. They noted that they had not performed a study of differences in noise correlations between the methods and that "such differences would be expected to affect the usefulness of reconstructed images for tasks such as lesion detection" (pp. 331, 332).

One method of accounting for noise correlation in a study of the impact of SC on image quality is to perform a human-observer ROC experiment. We have investigated the impact of scatter on cold- and hot-tumor detection for Tc-99m-labeled antibody fragments used for hepatic imaging (de Vries et al., 1997, 1999). Prior to performing this investigation, we hypothesized that scatter could: (1) degrade detection accuracy by decreasing contrast; (2) improve the accuracy of detection because it adds counts (information), some of which are in the correct location; or (3) have no impact because these two factors would cancel. The SIMIND Monte Carlo program (Ljungberg and Strand, 1989) was used to create high-count SPECT projections of the abdominal region as defined by the Zubal phantom (Zubal et al., 1994). The primary and scattered photons were stored in separate data files. Similarly, high-count projections of a 2.5-cm-diameter spherical tumor in each of three locations within the liver were also created via SIMIND. These projections were scaled and added to (hot tumors) or subtracted from (cold tumors) the background distribution. Projections were made with solely the primary photons present. These were used to assess the impact of perfect scatter rejection as might be approximated by imaging with a detector with extremely good energy resolution. Projections were also made with both the primary and scattered photons present (standard imaging of Tc-99m) and with 2.5 times the scattered Tc-99m photons present (an approximation to imaging with significantly more scattered photons, as would occur with Ga-67). Multiple noise realizations of the projections with no tumor present were created. Multiple realizations were also made with tumors at each of three locations. Energy-window-based SC was applied to the noisy projections that contained scatter. The projections were 2D low-pass filtered with a Butterworth filter whose parameters matched those used clinically in our department. The projections were then reconstructed using FBP and multiplicative Chang AC. The negative values were truncated from the slices, and upper-thresholding to place the average value of a voxel in the liver at the center of the gray scale was applied. A signal-known-exactly (SKE) ROC study was then conducted with five observers. The potential location of the tumor was indicated via removable crosshairs, and a continuous scale was used for the observers' confidence ratings. The LABROC1 program provided by C. E. Metz was used to estimate, from the ratings for each observer, a binormal ROC curve and the area under the curve (A_Z). Statistical analysis of the A_Z values indicated that only for the case of the hot tumors was there a statistically significant difference in detection accuracy between ideal scatter rejection (primary-only images) and the slices with the standard amount of scatter (uncorrected low-scatter case). In either case, the difference in A_Z values was small (0.88 versus 0.86 for cold tumors, and 0.84 versus 0.80 for hot tumors). For both tumor polarities, primary images gave a statistically significant increase in detection accuracy in comparison to the images with an artificially elevated amount of scatter. Here the difference in areas was larger (0.88 versus 0.81 for cold tumors, and 0.84 versus 0.74 for hot tumors). In no case did SC result in a statistically significant increase in detection accuracy over no compensation. However, with the artificially elevated amount of scatter the areas did increase with SC (0.81 with no compensation versus 0.83 with SC for the cold tumors, and 0.74 with no compensation versus 0.78 with SC for the hot tumors). These results indicate that scatter does decrease lesion detection accuracy, particularly when there is a significant amount of scatter present. They also suggest that scatter estimation methods may be able to improve lesion detection accuracy when there is a significant amount of scatter present. It awaits to be seen whether RBSC methods are superior to estimation methods for undoing the impact of scatter on lesion detection.

IV. SPATIAL RESOLUTION COMPENSATION

A. Restoration Filtering

For a shift-invariant linear system in the absence of noise, the image $g(x,y)$ is the convolution of the object $f(x,y)$ with the point spread function $h(x,y)$, or (Metz and Doi, 1979):

$$g(x,y) = \int\int_{-\infty}^{\infty} f(x',y') h(x-x', y-y') dx' dy' \quad (8)$$

By taking the 2D Fourier transform, Eq. (8) can be expressed in the frequency domain as:

$$G(u,v) = \text{OTF}(u,v) F(u,v) \quad (9)$$

where OTF is the optical transfer function that specifies the changes in both the magnitude and phase of the frequency components of the object by the imaging system, and u and v are the spatial frequencies.

We can solve Eq. (9) for F by dividing both sides by the OTF. Division by the OTF represents the inverse filter and is an example of unregularized restoration filtering (Gonzalez, 1976). In restoration filtering, one models the degradation (h or OTF) and applies the inverse process with the goal of recovering the original image. One problem with this approach, of course, is noise. For the case of spatially independent Poisson noise (Goodman and Belsher, 1976), Eq. (9) can be written as:

$$G(u,v) = \text{OTF}(u,v) F(u,v) + N(u,v) \quad (10)$$

where N is the contribution due to noise. For images degraded by Poisson noise, the average value of the noise power spectrum (P_N), which is proportional to the complex magnitude squared of N ($\|N\|^2$), is a constant equal to the total number of counts in the image (Goodman and Belsher, 1976; King et al., 1983). The object power spectrum (P_F) decreases rapidly with frequency, and the multiplication by the OTF further decreases its contribution to G. Thus, one reaches a point as the frequency increases where N will be the dominant contributor to G. Application of the inverse filter will result in a significant amplification of the noise at such frequencies. In restoration filtering, one usually follows the inverse filter at low frequencies, where the contributions of the signal dominate the image, and then switches to low-pass filtering to avoid excessive amplification of noise when the noise dominates. A number of filters have been proposed for regularizing restoration (Gonzalez, 1976).

One such filter is the Wiener filter, which uses the minimization of the mean squared error as the criterion for filter design (King et al., 1983). The Wiener filter can be written in the frequency domain as:

$$W(u,v) = \left(\frac{\text{OTF}^*(u,v)}{[\text{OTF}^2(u,v) + P_N(u,v)/P_F(u,v)]} \right)$$
$$= \left(\frac{1}{\text{OTF}(u,v)} \right) \left(\frac{\text{OTF}^2(u,v)}{[\text{OTF}^2(u,v) + P_N(u,v)/P_F(u,v)]} \right) \quad (11)$$

where * indicates the complex conjugate. The right-hand side of Eq. (11) explicitly shows the Wiener filter as the product of the inverse filter and a low-pass filter. To apply this filter, one needs estimates of P_N, P_F, and OTF. For 2D prereconstruction filtering of SPECT acquisition images, P_N can be estimated as the total image count, as stated previous. However, for 3D postreconstruction filtering, estimates of P_N must be modified to account for the impact of reconstruction (King et al., 1984). Once one has an estimate of P_N, an estimate of P_F can be obtained from the image power spectrum and knowledge of the OTF (Penny, King, et al., 1990; Penney, Glick, et al., 1990; King et al., 1983, 1984; Madsen, 1990). One is then left with estimation of the OTF. The problem here is that the OTF varies with distance from the face of the collimator. Thus there is no single OTF that correctly models the blurring for all the structures in the image.

One solution to the problem of a distance-dependent OTF is to approximate it as distance-independent, using a single OTF evaluated at a distance equal to the mean free path of the photons being imaged (King et al., 1983). A more accurate solution to the problem is offered by the frequency-distance relationship (FDR) (Edholm et al., 1986). The FDR states that in the 2D fast Fourier transform (FFT) of the sinogram, the amplitudes of the signal from any given distance relative to the COR are concentrated along a line that runs through the origin in the frequency domain. That is:

Angular frequency = − Distance · Spatial frequency   (12)

with the negative of the distance being the slope of the line, or:

Distance = − Angular frequency / Spatial frequency   (13)

where Distance is relative to the COR, so that the distance (d) from the camera face is COR + Distance. The FDR and the meaning of Eq. (13) can be visualized with the aid of Figure 20. Figure 20 shows two straight lines oriented at 45° passing through the origin of the 2D FFT of the sinogram. Each of these lines represents a constant value of the ratio of the angular to spatial frequency and, therefore, a specific distance from the COR. In the upper right quadrant of the 2D FFT, both the angular and spatial frequencies are positive. Any line that passes through the origin of the 2D FFT and is in this quadrant represents frequency locations at which the signal is concentrated in the source distribution closer to the camera than the COR. This can be seen by noting that in this quadrant, by Eq. (13), Distance is negative so that d is less than the COR. This quadrant is said to represent near-field locations. In the lower right quadrant, the angular frequencies are negative and the spatial frequencies are positive. This quadrant represents signal in the source emitted at distances farther away than the COR because now Distance is positive. This region represents the far-field locations. Thus FDR allows one to know the distance from the camera at which the signal was emitted, with the exception of the origin through which all lines or distances pass. Because all OTFs should be 1.0 at this location, this does not represent a problem when applying the FDR to a single slice. By modeling the detector point spread function as a Gaussian function whose width is dependent on d, as in Eq. (5), spatially variant restoration filters have been developed (Lewitt et al., 1989; Hawkins et al., 1991). A 3D implementation is achieved by also Fourier transforming in the between-slice direction (z axis of projections) (Glick et al., 1994). A problem is encountered in determining which distance to use to select the OTF for deblurring the z axis transform of the origin because all distances pass through this location. We have used the average source-to-detector distance for this purpose (Glick et al., 1994). Using the FDR, the Wiener filter can be expressed as:

$$W(u,v,w) = \left(\frac{1}{\text{OTF}(u,v,d)} \right) \left(\frac{\text{OTF}^2(u,v,d)}{[\text{OTF}^2(u,v,d) + P_N(u,v,w)/P_F(u,v,w)]} \right) \quad (14)$$

where w is the interslice or axial angular frequency in the 3D FFT of the sinograms, and d is the distance from the face of the collimator (COR − w/u). The FDR has been expanded to work with body contouring, as well as with noncircular camera orbits (Xia et al., 1995).

Application of the FDR to prereconstruction restoration filtering SPECT acquisitions has advantages in terms of computational load and being linear. It also has several disadvantages (Kohli, King, Glick, et al., 1998). It is limited as to how much resolution recovery can be obtained without excessive amplification of noise. Under certain conditions, after FDR prereconstruction filtering, transverse slices will contain circular noise correlations (Soares et al., 1996). Also, the FDR is an approximation, especially at low spatial frequencies. Finally, any form of prereconstruction filtering correlates the noise in the projections that will be input to reconstruction algorithms and MLEM assumes uncorrelated Poisson noise in the projection data.

B. Modeling Spatial Resolution in Iterative Reconstruction

Another method to correct for the distance-dependent camera response is the incorporation of a blurring model into iterative reconstruction (Kamphuis et al., 1996; Penny, King, et al., 1990; Beekman et al., 1993; Floyd et al., 1988; Tsui et al., 1988; Formiconi et al., 1989; Zeng et al., 1991; Liang, 1993). The problem with this method has been the immense increase in computational burden imposed when an iterative reconstruction algorithm includes such modeling in its transition matrix. The use of block-iterative recon-

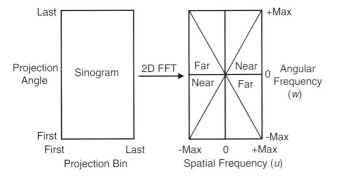

FIGURE 20 Frequency-distance relationship between sinogram and its two-dimensional fast Fourier transform.

struction algorithms has been shown to dramatically reduce the number of iterations required to reconstruct slices (Kamphuis et al., 1996; Pan, Luo, et al., 1997).

Blurring incrementally with distance using the method of Gaussian diffusion (McCarthy and Miller, 1991) can dramatically reduce the computational burden per iteration. With this method, system spatial resolution in the absence of scatter is modeled with a Gaussian function whose standard deviation, σ, varies with distance as per Eq. (5). A special property of Gaussians is that the convolution of two Gaussians produces a Gaussian whose σ equals the square root of the sum of the squares the σs of the original two Gaussians. Thus the system response at some distance d_{i+1} can be expressed as the convolution of the response at distance d_i with a Gaussian whose standard deviation (σ_{inc}) is given by:

$$\sigma_{inc} = \left(\sigma_S^2(d_{i+1}) - \sigma_S^2(d_i)\right)^{1/2} \qquad (15)$$

The advantage of the incremental approach is that σ_{inc} is much smaller than $\sigma_S(d_{i+1})$ and thus requires a smaller-dimension mask to approximate the Gaussian (i.e., less computation). In fact, one typically uses a mask made up of three terms to approximate the 1D incremental Gaussian. The first and third mask coefficients are the same due to the symmetry of Gaussians. Thus, 1D convolution can be implemented by summing the voxel values to either side of the center location of the mask, multiplying the result by the coefficient, and adding that result to the center voxel value multiplied by its coefficient. Making use of the separability of the Gaussian function, one can implement 2D convolution as a 1D horizontal convolution followed by a 1D vertical convolution. This requires only four multiplications and four additions per voxel. If a rotator is used to align the slices with the acquisition angles (Wallis and Miller, 1997; Di Bella et al., 1996), then the projection and backprojection steps can be implemented along columns (straight lines). The incremental steps in distance occur as one moves from one plane (distance from the camera) to the next. The projection can be implemented by convolving the farthest plane of the slices with the Gaussian for the increment in distance between that plane and the next plane closer to the detector. The result is then added to the next plane, and the sum is convolved with the Gaussian for the increment in distance to the next plane. The process is repeated until the plane nearest the detector is reached. The Gaussian for the distance between the detector and that plane is used in the final convolution. This will generally require a mask of larger than three terms to sample the Gaussian for this distance without significant truncation. The result is a projection in which the blurring varies with distance such that the farthest planes are blurred the most. With backprojection, one would start by first blurring the projection for the distance between the detector and the first plane of slices, and incrementally blurring as one proceeds along the slice columns thereafter. Using a step size of larger than one row can further reduce computational time (Bai et al., 1998).

Using a three-term mask to approximate a Gaussian can result in aliasing, distorting the desired response (King, Pan et al., 1997). For the Gaussian diffusion method (McCarthy and Miller, 1991), the mask can be determined from a linear difference equation approximation of the diffusion partial differential equation:

$$\frac{\partial q}{\partial z} = \alpha^2 \left(\frac{\partial^2 q}{\partial x^2} + \frac{\partial^2 q}{\partial y^2}\right) \qquad (16)$$

which describes the diffusion in the direction z (with a diffusion constant α^2) of the 3D activity distribution $q(x,y,z)$. The Gaussian diffusion mask that approximates the blurring from the plane at distance d_i to the plane at d_{i+1} is ($\sigma^2_{inc}/2$, $1 - \sigma^2_{inc}$, and $\sigma^2_{inc}/2$) (McCarthy and Miller, 1991). Use of these coefficients has been shown to minimize the impact of aliasing compared to alternative ways of defining the mask coefficients as a function of σ_{inc} (King, Pan et al., 1997).

Another way of decreasing the computational burden at each iteration is to model spatially variant resolution only in the projection step (Zeng and Gullberg, 1992). Zeng and Gullberg (2000) have reported on the conditions that are necessary in order to guarantee convergence of an iterative algorithm when the projector and backprojector are unmatched.

Even with all these computational enhancements, modeling the PSF in iterative reconstruction is still slower than restoration filtering. Furthermore, it results in reconstructions in which the extent to which a stationary and isotropic response results depends on the location of the source, the camera orbit, the number of iterations used, the reconstruction algorithm, and the source distribution (Pan, Luo, et al., 1997). However, it does avoid the approximation inherent in the FDR, does not alter the noise characteristics of the projections prior to reconstruction, and has yielded significant improvements in resolution without a significant increase in noise (Tsui, Frey, et al., 1994).

C. Impact of Resolution Compensation on Image Quality

One of the major impacts of spatial resolution is manifested in the partial-volume effect. Tomographic systems have a characteristic resolution volume that is determined by their 3D PSF (Soresenson and Phelps, 1987). Objects smaller than the extent of the PSF only partially occupy this volume. Therefore, the count determined at that location only partially reflects the object because it also contains a contribution from structures beyond this object. The result is that the apparent concentration for objects smaller than two to three times the FWHM of the PSF will depend on the size and shape of the object as well as the concentration of activity within the object (Hoffman et al., 1979; Kessler et al., 1984), and the object will be spread out into a region in the slices larger than its true size (King *et al.*, 1991). Objects larger in size will have their concentration distorted near their edges by the partial-volume effect and will be blurred out into the surrounding region. It has been observed that modeling of the PSF in iterative reconstruction can reduce the bias in quantitation better than can FDR-Wiener filtering (Pretorius *et al.*, 1998). However, there are limits to how far even modeling the system resolution in iterative reconstruction can go in correcting for the partial-volume effect. Thus, alternative strategies for partial-volume compensation are of great interest (Da Silva *et al.*, 2001; Hutton and Osiecki, 1998).

One question of significant clinical importance is whether restoration filtering or modeling the PSF in iterative reconstruction can improve the accuracy of lesion detection. We conducted an LROC investigation of this using the detection of tumors in simulated thoracic Ga-67 citrate SPECT slices as the task (Gifford *et al.*, 2000). When the reconstruction and filtering parameters were optimized for our detection task, we determined that OSEM with AC and detector resolution compensation (DRC) significantly improves detection accuracy over that with either OSEM with AC alone or OSEM with AC and FDR restoration filtering. In this investigation, five reconstruction strategies were compared: (1) FBP with multiplicative Chang AC (current clinical practice), (2) OSEM with AC, (3) OSEM with AC and FDR restoration filtering, (4) OSEM with AC and Gaussian-diffusion DRC (GD-DRC), and (5) OSEM with the AC of data simulated with a stationary detector response that just limited aliasing (ideal case). The last was investigated to serve as an upper limit for DRC. Figure 21 shows example reconstructions for several lesions for these five methods. Note that the visibility of tumors of the same contrast varies with location in the slices. This illustrates why we believe it is necessary to present lesions at numerous locations when assessing observer performance. A series of four observer studies, using five observers, was conducted. The first study compared the five methods for a single-tumor contrast. The second investigated the impact of number of iterations of OSEM with GD-DRC on detection accuracy. The third compared lesion detection for OSEM with and without GD-DRC for eight different lesion contrasts with observers reading images of a single contrast at a time. The fourth study compared the five reconstruction strategies for observers reading lesions of three different contrasts mixed together. In the fourth study the A_L for the three contrasts in aggregate were 0.55, 0.67, 0.67, 0.77, and 0.99, for the five methods, respectively. Note the tremendous difference between current clinical practice and the ideal case we investigated. Also notice that OSEM with AC and GD-DRC is able to improve detection to approximately halfway between the two. All differences between pairs of methods were statistically significant except OSEM with AC versus OSEM with AC and FDR restoration filtering. An

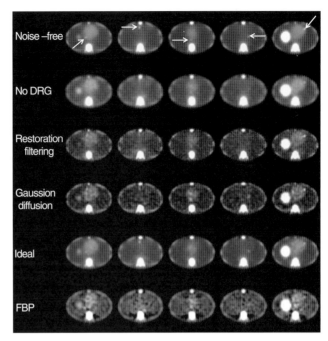

FIGURE 21 Examples of observer study images. Along top row arrows in the noise-free images reconstructed with the ideal DRC strategy indicate tumor locations. Noise realizations of the same slices are shown beneath for each of the DRC strategies. (Reproduced with permission from Gifford *et al.*, 2000.)

interesting observation in this study was that the number of iterations of OSEM for which the highest detection accuracy was achieved increased from 1 iteration with 8 subsets of 16 angles each for OSEM with AC to 8 iterations of 16 subsets of 8 angles each for OSEM with AC and GD-DRC. That is, more faithfully modeling the physics of imaging in the transition matrix employed in iterative reconstruction resulted in an increased number of iterations and improved detection. Eight iterations was the number determined to be optimal in the preliminary optimization studies. In the second observer study reported in the paper (Gifford *et al.*, 2000), six iterations resulted in a numerically, but not statistically, larger A_L. Eight (or even six) iterations of OSEM with AC and GD-DRC for 128 × 128 images is time consuming.

Our results for the improvement in detection accuracy with GD-DRC for simulated Ga-67 citrate images indicates that there may be a significant clinical impact from using modeling of the PSF with iterative reconstruction. It remains to be seen whether such an improvement is seen clinically and whether improvements in detection are observed for other clinical procedures. A significant improvement in image quality was reported for modeling the PSF in SPECT imaging of F-18 (Zeng *et al.*, 1998). Similarly, modeling the PSF was observed to improve quantitative measures of image quality for cardiac imaging (Kohli, King, Pan, *et al.*, 1998; Hutton and Lau, 1998) and brain imaging (Beekman *et al.*, 1996); however, it has yet to be determined to what extent such modeling improves the detection of cardiac or brain lesions.

V. CONCLUSION

From this review we believe it should be evident that solely compensating for attenuation is not enough to improve SPECT image quality to its fullest extent. Instead, attenuation, scatter, and resolution, as well as correction of patient and physiological motion and changes in localization during the course of acquisition, can impact image quality, and thus combined compensation is required. It is our opinion that as a result of improvements in algorithms and computer hardware, the best way to perform such compensations is to accurately model the degradations in iterative reconstruction. Evidence is emerging that these compensations do significantly improve the diagnostic utility of SPECT images. For example, in an ROC study of the accuracy of detection of coronary disease using cardiac-perfusion SPECT imaging with clinical images, we have determined that combined correction for attenuation, scatter, and spatial resolution is more accurate than FBP, solely AC, or AC and SC (Narayanan, Pretorius, *et al.*, 2002). However, it should be remembered that the corrected images look different than the uncorrected images physicians have grown accustomed to reading. Thus it will require a period of adjustment for the clinicians to develop a new mental picture of what "normal" looks like before the full potential of the compensations is realized clinically (Watson *et al.*, 1999; Hutton, 1997).

Acknowledgments

The authors thank the reviewer for suggesting additional material to be added to this chapter. This work was supported in part by U.S. Public Health grants HL50349 and CA42165 of the National Heart, Lung, and Blood Institute and the National Cancer Institute. Its contents are solely the responsibility of the authors and do not necessarily represent the official views of the National Heart, Lung, and Blood Institute or of the National Cancer Institute.

References

Atkins, F. B., Beck, R. N., Hoffer, P. N., and Palmer, D. (1968). Dependence of optimum baseline setting on scatter fraction and detector response function. *In* "Medical Radioisotope Scintigraphy," pp. 101–118. IAEA, Vienna.

Attix, F. H. (1983). "Introduction to Radiological Physics and Radiation Dosimetry." John Wiley & Sons, New York.

Axelsson, B., Msaki, P., and Israelsson. (1984). Subtraction of Compton-scattered photons in single-photon emission computed tomography. *JNM* **25:** 490–494.

Bai, C., Zeng, G. L., and Gullberg, G. T. (2000). A slice-by-slice blurring model and kernel evaluation using the Klein-Nishina formula for 3D scatter compensation in parallel and convergent beam SPECT. *Phys. Med. Biol.* **45:** 1275–1307.

Bai, C., Zeng, G. L., Gullberg, G. T., DiFilippo, F., and Miller, S. (1998). Slab-by-slab blurring model for geometric point response correction and attenuation correction using iterative reconstruction algorithms. *IEEE Trans. Nucl. Sci.* **45:** 2168–2173.

Bailey, D. L., Hutton, B. F., and Walker, P. J. (1987). Improved SPECT using simultaneous emission and transmission tomography. *JNM* **28:** 844–851.

Barnden, L. R., Ong, P. L., and Rowe, C. C. (1997). Simultaneous emission transmission tomography using technetium-99m for both emission and transmission. *EJNM* **24:** 1390–1397.

Beck, R. N., Schuh, M. W., Cohen, T. D., and Lembares, N. (1968). Effects of scattered radiation on scintillation detector response. *In* "Medical Radioisotope Scintigraphy," pp. 595–616. IAEA, Vienna.

Beekman, F. J., den Harder, J. M., Viergever, M. A., and van Rijk, P. P. (1997). SPECT scatter modelling in non-uniform attenuating objects. *Phys. Med. Biol.* **42:** 1133–1142.

Beekman, F. J., de Jong, H. W. A. M., and van Geloven S. (2002). Efficient fully 3D iterative SPECT reconstruction with Monte Carlo based scatter compensation. *IEEE Trans. Med. Imag.* **21:** 867–877.

Beekman, F. J., Eijkman, E. G. J., Viergever, M. A., Born, G. F., and Slijpen, E. T. P. (1993). Object shape dependent PSF model for SPECT imaging. *IEEE Trans. Nucl. Sci.* **40:** 31–39.

Beekman, F. J., Kamphuis, C., and Frey, E. C. (1997). Scatter compensation methods in 3D iterative SPECT reconstruction: A simulation study. *Phys. Med. Biol.* **42:** 1619–1632.

Beekman, F. J., Kamphuis, C., Hutton, B. F., and van Rijk, P. P. (1998). Half-fanbeam collimators combined with scanning point sources for simultaneous emission-transmission imaging. *JNM* **39:** 1996–2003.

Beekman, F. J., Kamphuis, C., and Viergever, M. A. (1996). Improved SPECT quantitation using fully three-dimensional iterative spatially variant scatter response compensation. *IEEE Trans. Med. Imag.* **15:** 491–499.

Bellini, S., Piacentini, M., Cafforio, C., and Rocca, F. (1979). Compensation of tissue absorption in emission tomography. *IEEE Trans. Acou. Sp. Sig. Proc.* **27:** 213–218.

Blankespoor, S. C., Wu, X., Kalki, K., Brown, J. K., Tang, H. R., Cann, C. E., and Hasegawa, B. H. (1996). Attenuation correction of SPECT using x-ray CT on an emission-transmission CT system: Myocardial perfusion assessment. *IEEE Trans. Nucl. Sci.* **43:** 2263–2274.

Boas, M. L. (1983). "Mathematical Methods in the Physical Sciences," 2nd ed. John Wiley & Sons, New York.

Bowsher, J. E., and Floyd, C. E. (1991). Treatment of Compton scattering in maximum-likehood, expectation-maximization reconstructions of SPECT images. *JNM* **32:** 1285–1291.

Brookeman, V. A., and Bauer, T. J. (1973). Collimator performance for scintillation camera systems. *JNM* **14:** 21–25.

Byrne, C. L. (1998). Accelerating the EMML algorithm and related iterative algorithms by rescaled block-iterative methods. *IEEE Trans. Imag. Proc.* **7:** 100–109.

Cao, Z. J., and Tsui, B. M. W. (1992). Performance characteristics of transmission imaging using a uniform sheet source with a parallel-hole collimator. *Med. Phys.* **19:** 1205–1212.

Celler, A., Sitek, A., Stoub, E., Hawman, P., Harrop, R., and Lyster, D. (1998). Multiple line source array for SPECT transmission scans: simulation, phantom, and patient studies. *JNM* **39:** 2183–2189.

Chan, M. T., Leahy, R. M., Mumcuoglu, E. U., Cherry, S. R., Czernin, J., and Chatziioannou, A. (1997). Comparing lesion detection performance for PET image reconstruction algorithms: A case study. *IEEE Trans. Nucl. Sci.* **44:** 1558–1563.

Chang, L. T. (1978). A method for attenuation correction in radionuclide computed tomography. *IEEE Trans. Nucl. Sci.* **25:** 638–643.

Chang, W., Loncaric, S., Huang, G., and Sanpitak, P. (1995). Asymmetric fan transmission CT on SPECT systems. *Phys. Med. Biol.* **40:** 913–928.

Chornoboy, E. S., Chen, C. J., Miller, M. I., Miller, T. R., and Snyder, D. L. (1990). An evaluation of maximum likelihood reconstruction for SPECT. *IEEE Trans. Med. Imag.* **9:** 99–110.

Da Silva, A. J., Tang, H. R., Wong, K. H., Wu, M. C., Dae, M. W., and

Hasegawa, B. H. (2001). Absolute quantification of regional myocardial uptake of Tc-99m- sestamibi with SPECT: Experimental validation in a porcine model. *J. Nucl. Med.* **42:** 772–779.

de Jong, H. W. A. M., Slijpen, E. T. P., and Beekman, F. J. (2001). Acceleration of Monte Carlo SPECT simulation using convolution-based forced detection. *IEEE Trans. Nucl. Sci.* **48:** 58–64.

de Vries, D. J., and King, M. A. (1994). Window selection for dual photopeak window scatter correction in Tc-99m imaging. *IEEE Trans. Nucl. Sci.* **41:** 2771–2778.

de Vries, D. J., King, M. A., Soares, E. J., Tsui, B. M. W., and Metz, C. E. (1999). Effects of scatter subtraction on detection and quantitation in hepatic SPECT. *JNM* **40:** 1011–1023.

de Vries, D. J., King, M. A., Tsui, B. M. W., and Metz, C. E. (1997). Evaluation of the effect of scatter correction on lesion detection in hepatic SPECT Imaging. *IEEE Trans. Nucl. Sci.* **44:** 1733–1740.

Di Bella, E. V. R., Barclay, A. B., and Schafer, R. W. (1996). A comparison of rotation-based methods for iterative reconstruction algorithms. *IEEE Trans. Nucl. Sci.* **43:** 3370–3376.

Edholm, P. R., Lewitt, R. M., and Lindholm, B. (1986). Novel properties of the Fourier decomposition of the sinogram. *Proc SPIE* **671:** 8–18.

Farquhar, T. H., Llacer, J., Hoh, C. K., et al., (1999). ROC and localization ROC analyses of lesion detection in whole-body FDG PET: Effects of acquisition mode, attenuation correction and reconstruction algorithm. *JNM* **40:** 2043–2052.

Ficaro, E. P., Fessler, J. A., Shreve, P. D., et al. (1996). Simultaneous transmission/emission myocardial perfusion tomography: Diagnostic accuracy of attenuation-corrected Tc-99m-sestamibi single-photon emission computed tomography. *Circulation* **93:** 463–473.

Fleming, J. S. (1989). A technique for using CT images in attenuation correction and quantification in SPECT. *Nucl. Med. Comm.* **10:** 83–97.

Floyd, C. E., Jaszczak, R. J., and Coleman, R. E. (1985). Inverse Monte Carlo: A unified reconstruction algorithm. *IEEE Trans. Nucl. Sci.* **32:** 799–785.

Floyd, C. E., Jaszczak, R. J., Manglos, S. H., and Coleman, R. E. (1988). Compensation for collimator divergence in SPECT using inverse Monte Carlo reconstruction. *IEEE Trans. Nucl. Sci.* **35:** 784–787.

Formiconi, A. R. (1998). Geometrical response of multihole collimators. *Phys. Med. Biol.* **43:** 3359–3379.

Formiconi, A. R., Pupi, A., and Passeri, A. (1989). Compensation of spatial response in SPECT with conjugate gradient reconstruction technique. *Phys. Med. Biol.* **34:** 69–84.

Frey, E. C., and Tsui, B. M. W. (1993). A practical method for incorporating scatter in a projector-backprojector for accurate scatter compensation in SPECT. *IEEE Trans. Nucl. Sci.* **40:** 1107–1116.

Frey, E. C., and Tsui, B. M. W. (1994). Modeling the scatter response function in inhomogeneous scattering media for SPECT. *IEEE Trans. Nucl. Sci.* **41:** 1585–1593.

Frey, E. C., and Tsui, B. M. W. (1997). A new method for modeling the spatially-variant, object-dependent scatter response function in SPECT. *Proc. 1966 MIC*, 1082–1086.

Frey, E. C., Tsui, B. M. W., and Gullberg, G. T. (1998). Improved estimation of the detector response function for converging beam collimators. *Phys. Med. Biol.* **43:** 941–950.

Gagnon, D., Todd-Pokropek, A., Arsenault, A., and Dapras, G. (1989). Introduction to holospectral imaging in nuclear medicine for scatter subtraction. *IEEE Trans. Med. Imag.* **8:** 245–250.

Gagnon, D., Tung, C. H., Zeng, L., and Hawkins, W. G. (2000). Design and early testing of a new medium-energy transmission device for attenuation correction in SPECT and PET. *Proc. 1999 Med. Imag. Conf. IEEE*, Paper M8–3.

Garica, E. V., Cooke, C. D., Van Train, K. F., Folks, R., Peifer, J., DePuey, E. G., Maddahi, J., Alazraki, N., Galt, J., Ezquerra, N., Ziffer, J., Areeda, J., and Berman, D. S. (1990). Technical aspects of myocardial SPECT imaging with technetium-99m sestamibi. *Amer. J. Card.* **66:** 29E–31E.

Germano, G., Chua, T., Kiat, H., Areeda, J. S., and Berman, D. S. (1994). A quantitative phantom analysis of artifacts due to hepatic activity in technetium-99m myocardial perfusion SPECT studies. *J. Nucl. Med.* **35:** 356–359.

Gifford, H. C., King, M. A., Wells, R. G., Hawkins, W. G., Narayanan, M. V., and Pretorius, P. H. (2000). LROC analysis of detector-response compensation in SPECT. *IEEE Trans. Med. Imag.* **19:** 463–473.

Gilland, D. R., Jaszczak, R. J., Greer, K. L., and Coleman, R. E. (1998). Transmission imaging for nonuniform attenuation correction using a three-headed SPECT camera. *JNM* **39:** 1005–1110.

Gilland, D. R., Tsui, B. M. W., McCarthy, W. H., Perry, J. R., and Berg, J. (1988). Determination of the optimum filter function for SPECT imaging. *JNM* **29:** 643–650.

Gillen, G. J., Gilmore, B., and Elliott, A. T. (1991). An investigation of the magnitude and causes of count loss artifacts in SPECT imaging. *JNM* **32:** 1771–1776.

Glick, S. J., King, M. A., Pan, T.-S., and Soares, E. J. (1995). An analytical approach for compensation of non-uniform attenuation in cardiac SPECT imaging. *Phys. Med. Biol.* **40:** 1677–1693.

Glick, S. J., Penney, B. C., King, M. A., and Byrne, C. L. (1994). Noniterative compensation for the distance-dependent detector response and photon attenuation in SPECT imaging. *IEEE Trans. Med. Imag.* **13:** 363–374.

Gonzalez, R. C., and Woods, R. E. (1992). "Digital Image Processing." Addison-Wesley, Reading, MA.

Goodman, J. W., and Belsher, J. F. (1976). Fundamental limitations in linear invariant restoration of atmospherically degraded images. *Proc. SPIE* **75:** 141–154.

Gregoriou, G. K., Tsui, B. M. W., and Gullberg, G. T. (1998). Effect of truncated projections on defect detection in attenuation-compensated fanbeam cardiac SPECT. *JNM* **39:** 166–175.

Gullberg, G. T., and Budinger, T. F. (1981). The use of filtering methods to compensate for constant attenuation in single-photon emission computed tomography. *IEEE Trans. Biomed. Eng.* **28:** 142–157.

Gullberg, G. T., Morgan, H. T., Zeng, G. L., Christian, P. E., Di Bella, E. V. R., Tung, C.-H., Maniawski, P. J., Hsieh, Y.-L., and Datz, F. L. (1998). The design and performance of a simultaneous transmission and emission tomography system. *IEEE Trans. Nucl. Sci.* **45:** 1676–1698.

Hademenous, G. J., King, M. A., Ljungberg, M., Zubal, G., and Harrell, C. R. (1993). A scatter correction method for Tl-201 images: A Monte Carlo investigation. *IEEE Trans. Nucl. Sci.* **30:** 1179–1186.

Harris, C. C., Greer, K. L., Jaszczak, R. J., Floyd, C. E., Fearnow, E. C., and Coleman, R. E. (1984). Tc-99m attenuation coefficients in water-filled phantoms determined with gamma cameras. *Med. Phys.* **11:** 681–685.

Hawkins, W. G., Leichner, P. K., and Yang, N. C. (1988). The circular harmonic transform for SPECT reconstruction and boundary conditions on the Fourier transform of the sinogram. *IEEE Trans. Med. Imag.* **7:** 135–148.

Hawkins, W. G., Yang, N.-C., and Leichner, P. K. (1991). Validation of the circular harmonic transform (CHT) algorithm for quantitative SPECT. *JNM* **32:** 141–150.

Haynor, D. R., Kaplan, M. S., Miyaoka, R. S., and Lewellen, T. K. (1995). Multiwindow scatter correction techniques in single-photon imaging. *Med. Phys.* **22:** 2015–2024.

Hendel, R. C., Berman, D. S., Cullom, S. J., et al. (1999). Multicenter clinical trial to evaluate the efficacy of correction for photon attenuation and scatter in SPECT myocardial perfusion imaging. *Circulation* **99:** 2742–2749.

Hendel, R. C., Corbett, J. R., Cullom, J., DePuey, E. G., Garcia, E. V., and Bateman, T. M. (2002). The value and pratice of attenuation correction for myocardial perfusion SPECT imaging: A joint position statemant from the American Society of Nuclear Cardiology and the Society of Nuclear Medicine. *J. Nucl. Card.* **9:** 135–143.

Hoffman, E. J., Huang, S.-C., and Phelps, M. E. (1979). Quantitation in positron emission computed tomography, 1. Effect of object size. *J. Comp. Assist. Tomogr.* **3:** 299–308.

Hollinger, E. F., Loncaric, S., Yu, D.-C., Ali, A., and Chang, W. (1998). Using fast sequential asymmetric fanbeam transmission CT for attenuation correction of cardiac SPECT imaging. *JNM* **39:** 1335–1344.

Hudson, H. M., and Larkin, R. S. (1994). Accelerated image reconstruction using ordered subsets of projection data. *IEEE Trans. Med. Imag.* **13:** 601–609.

Hutton, B. F. (1997). Cardiac single-photon emission tomography: Is attenuation correction enough? *EJNM* **24:** 713–715.

Hutton, B. F., and Lau, Y. H. (1998). Application of distance-dependent resolution compensation and post-reconstruction filtering for myocardial SPECT. *Phys. Med. Biol.* **43:** 1679–1693.

Hutton, B. F., and Osiecki, A. (1998). Correction of partial volume effects in myocardial SPECT. *J. Nucl. Cardiol.* **5:** 402–413.

Hutton, B. F., Osieck, A., and Meikle, S. R. (1996). Transmission-based scatter correction of 180° myocardial single-photon emission tomographic studies. *EJNM* **23:** 1300–1308.

Jaszczak, R. J., Chang, L. T., Stein, N. A., and Moore, F. E. (1979). Whole-body single-photon emission computed tomography using dual, large-field-of-view scintillation cameras. *Phys. Med. Biol.* **24:** 1123–1143.

Jaszczak, R. J., Coleman, R. E., and Whitehead, F. R. (1981). Physical factors affecting quantitative measurements using camera-based single photon emission computed tomography (SPECT). *IEEE Trans. Nucl. Sci.* **28:** 69–80.

Jaszczak, R. J., Gilland, D. R., Hanson, M. W., Jang, S., Greer, K. L., and Coleman, R. E. (1993). Fast transmission CT for determining attenuation maps using a collimated line source, rotatable air-copper-lead attenuators and fan-beam collimation. *JNM* **34:** 1577–1586.

Jaszczak, R. J., Greer, K. L., Floyd, C. E., Jr., Harris, C. C., and Coleman, R. E. (1984). Improved SPECT quantitation using compensation for scattered photons. *J. Nucl. Med.* **25:** 893–900.

Jaszczak, R. J., Whitehead, F. R., Lim, C. B., and Coleman, R. E. (1982). Lesion detection with single-photon emission computed tomography (SPECT) compared with conventional imaging. *J. Nucl. Med.* **23:** 97–102.

Kadrmas, D. J., Frey, E. C., Karimi, S. S., and Tsui, B. M. W. (1998). Fast implementations of reconstruction-based scatter compensation in fully 3D SPECT image reconstruction. *Phys. Med. Biol.* **43:** 857–873.

Kadrmas, D. J., Frey, E. C., and Tsui, B. M. W. (1998). Application of reconstruction-based scatter compensation to thallium-201 SPECT: Implementations for reduced reconstruction image noise. *IEEE Trans. Med. Imag.* **17:** 325–333.

Kak, A. C., and Slaney, M. (1987). "Principles of Computerized Tomographic Imaging." IEEE Press, New York.

Kamphuis, C., Beekman, F. J., van Rijk, P. V., and Viergever, M. A. (1998). Dual matrix ordered subsets reconstruction for accelerated 3D scatter compensation in single-photon emission tomography. *EJNM* **25:** 8–18.

Kamphuis, C., Beekman, F. J., and Viergever, M. A. (1996). Evaluation of OS-EM vs. ML-EM for 1D, 2D and fully 3D SPECT reconstruction. *IEEE Trans. Nucl. Sci.* **43:** 2018–2024.

Kessler, R. M., Ellis, J. R., and Eden, M. (1984). Analysis of emission tomographic scan data: limitations imposed by resolution and background. *J. Comp. Assist. Tomogr.* **8:** 514–522.

King, M. A., de Vries, D. J., Pan, T.-S., Pretorius, P. H., and Case, J. A. (1997). An investigation of the filtering of TEW scatter estimates used to compensate for scatter with ordered subset reconstructions. *IEEE Trans. Nucl. Sci.* **44:** 1140–1145.

King, M. A., Doherty, P. W., and Schwinger, R. B. (1983). A Wiener filter for nuclear medicine images. *Med. Phys.* **10:** 876–880.

King, M. A., Long, D. T., and Brill, A. B. (1991). SPECT volume quantitation: Influence of spatial resolution, source size and shape, and voxel size. *Med. Phys.* **18:** 1016–1024.

King, M. A., Schwinger, R. B., Doherty, P. W., and Penney, B. C. (1984). Two-dimensional filtering of SPECT images using the Metz and Wiener filters. *JNM* **25:** 1234–1240.

King, M. A., Tsui, B. M. W., and Pan, T.-S. (1995). Attenuation compensation for cardiac single-photon emission computed tomographic imaging, Part 1. Impact of attenuation and methods of estimating attenuation maps. *J. Nucl. Card.* **2:** 513–524.

King, M. A., Tsui, B. M. W., Pan, T.-S., Glick, S. J., and Soares, E. J. (1996). Attenuation compensation for cardiac single-photon emission computed tomographic imaging, Part 2. Attenuation compensation algorithms. *J. Nucl. Card.* **3:** 55–63.

King, M. A., Pan, T.-S., and Luo, D.-S. (1997). An investigation of aliasing with Gaussian-diffusion modeling of SPECT system spatial resolution. *IEEE Trans. Nucl. Sci.* **44:** 1375–1380.

King, M. A., Xia, W., de Vries, D. J., Pan, T.-S., Villegas, B. J., Dahlberg, S., Tsui, B. M. W., Ljungberg, M. H., and Morgan, H. T. (1996). A Monte Carlo investigation of artifacts caused by liver uptake in single-photon emission computed tomography perfusion imaging with Tc-99m labeled agents. *J. Nucl. Card.* **3:** 18–29.

Kohli, V., King, M. A., Glick, S. J., Pan, T.-S. (1998). Comparison of frequency-distance relationship and Gaussian-diffusion based methods of compensation for distance-dependent spatial resolution in SPECT imaging. *Phys. Med. Biol.* **43:** 1025–1037.

Kohli, V., King, M. A., Pan, T.-S., and Glick, S. J. (1998). Compensation for distance-dependent resolution in cardiac-perfusion SPECT: Impact on uniformity of wall counts and wall thickness. *IEEE Trans. Nucl. Sci.* **45:** 1104–1110.

Koral, K. F., Swailem, F. M., Buchbinder, S., Clinthorne, N. H., Rogers, W. L., and Tsui, B. M. W. (1990). SPECT dual-energy-window Compton correction: Scatter multiplier required for quantification. *J. Nucl. Med.* **31:** 90–98.

Koral, K. F., Wang, X. Q., Rogers, W. L., Clinthorne, N. H., and Wang, X. (1988). SPECT Compton-scattering correction by analysis of energy spectra. *JNM* **29:** 195–202.

LaCroix, K. J., and Tsui, B. M. W. (1999). Investigation of 90 dual-camera half-fanbeam collimation for myocardial SPECT imaging. *IEEE Trans. Nucl. Sci.* **46:** 2085–2092.

LaCroix, K. J., Tsui, B. M. W., Frey, E. C., and Jaszczak, R. J. (2000). Receiver operating characteristic evaluation of iterative reconstruction with attenuation correction in Tc-99m sestamibi myocardial SPECT images. *JNM* **41:** 502–513.

LaCroix, K. J., Tsui, B. M. W., Hasegawa, B. H., and Brown, J. K. (1994). Investigation of the use of x-ray CT images for attenuation compensation in SPECT. *IEEE Trans. Nucl. Sci.* **41:** 2793–2799.

Lalush, D. S., and Tsui, B. M. W. (1994). Improving the convergence of iterative filtered backprojection algorithms. *Med. Phys.* **21:** 1283–1285.

Lange, K., and Carson, R. (1984). EM reconstruction algorithms for emission and transmission tomography. *J. Comp. Assist. Tomogr.* **8:** 306–316.

Lewitt, R. M., Edholm, P. R., and Xia, W. (1989). Fourier method for correction of depth-dependent collimator blurring. *Proc SPIE* **1092:** 232–243.

Liang, Z. (1993). Compensation for attenuation, scatter, and detector response in SPECT reconstruction via iterative FBP methods. *Med. Phys.* **20:** 1097–1106.

Liang, Z., Ye, J., and Harrington, D. P. (1994). An analytical approach to quantitative reconstruction of non-uniform attenuated SPECT. *Phys. Med. Biol.* **39:** 2023–2041.

Ljungberg, M., and Strand, S.-E. (1989). A Monte Carlo program for the simulation of scintillation camera characteristics. *Comp. Meth. Prog. Biomed.* **29:** 257–272.

Ljungberg, M., and Strand, S.-E. (1991). Attenuation and scatter correction in SPECT for sources in nonhomogenous object: a Monte Carlo study. *JNM* **32:** 1278–1284.

Madsen, M. T. (1990). A method for obtaining an approximate Wiener filter. *Med. Phys.* **17:** 126–130.

Madsen, M. T., Kirchner, P. T., Edlin, J. P., Nathan, M. A., and Kahn, D.

(1993). An emission-based technique for obtaining attenuation correction data for myocardial SPECT studies. *Nucl. Med. Comm.* **14:** 689–695.

Maeda, H., Itoh, H., Ishii, Y., Makai, T., Todo, G., Fujita, T., and Torizuka, K. (1981). Determination of the pleural edge by gamma-ray transmission computed tomography. *JNM* **22:** 815–817.

Malko, J. A., Van Heertum, R. L., Gullberg, G. T., and Kowalsky, W. P. (1985). SPECT liver imaging using an iterative attenuation correction algorithm and an external flood source. *JNM* **26:** 701–705.

Manglos, S. H., Jaszczak, R. J., Floyd, C. E., Hahn, L. J., Greer, K. L., and Coleman, R. E. (1988). A quantitative comparison of attenuation-weighted backprojection with multiplicative and iterative postprocessing attenuation compensation in SPECT. *IEEE Trans. Med. Imag.* **7:** 128–134.

Manglos, S. H., and Young, T. M. (1994). Constrained IntraSPECT reconstruction from SPECT projections. *Proc. 1993 MIC*, 1605–1609.

Markoe, A. (1984). Fourier inversion of the attenuated x-ray transform. *SIAM J. Math. Anal.* **15:** 718–722.

McCarthy, A. W., and Miller, M. I. (1991). Maximum likelihood SPECT in clinical computation times using mesh-connected parallel computers. *IEEE Trans. Med. Imag.* **10:** 426–436.

Meikle, S. R., Hutton, B. F., and Bailey, D. L. (1994). A transmission-dependent method for scatter correction in SPECT. *JNM* **35:** 360–367.

Metz, C. E., Atkins, F. B., and Beck, R. N. (1980). The geometrical transfer function component for scintillation camera collimators with straight parallel holes. *Phys. Med. Biol.* **25:** 1059–1070.

Metz, C. E., and Doi, K. (1979). Transfer function analysis of radiographic imaging. *Phys. Med. Biol.* **24:** 1079–1106.

Metz, C. E., and Pan, X. (1995). A unified analysis of exact methods of inverting the 2-D exponential Radon transform, with implications for noise control in SPECT. *IEEE Trans. Med. Imag.* **14:** 643–657.

Meyer, C. R., Boes, J. L., Kim, B., Bland, P. H., Zasadny, K. R., Kison, P. V., Koral, K., Frey, K. A., and Wahl, R. L. (1996). Demonstration of accuracy and clinical versatility of mutual information for automatic multimodality image fusion using affine and thin-plate spline warped geometric deformations. *Med. Imag. Anal.* **1:** 195–206.

Msaki, P., Axelsson, B., Dahl, C. M., and Larsson, S. A. (1987). Generalized scatter correction method in SPECT using point scatter distribution functions. *JNM* **28:** 1861–1869.

Munley, M. T., Floyd, C. E., Jr., Tourassi, G. D., Bowsher, J. E., and Coleman, R. E. (1991). Out-of-plane photons in SPECT. *IEEE Trans. Nucl. Sci.* **38:** 776–779.

Narayanan, M. V., King, M. A., and Byrne, C. L. (2002). An iterative transmission reconstruction algorithm incorporating cross-talk correction for SPECT. *Med. Phys.* **29:** 694–700.

Narayanan, M. V., King, M. A., Pan, T.-S., and Dahlberg, S. T. (1998). Investigation of approaches to reduce truncation of attenuation maps with simultaneous transmission and emission SPECT imaging. *IEEE Trans. Nucl. Sci.* **45:** 1200–1206.

Narayanan, M. V., Pretorius, P. H., Dahlberg, S. T., Leppo, J., Spencer, F., and King, M. A. (2003). Human observer ROC evaluation of attenuation, scatter and resolution compensation strategies for Tc-99m myocardial perfusion imaging. *J. Nucl. Med.* **44:** 1725–1734.

Natterer, F. (2001). Inversion of the attenuated Radon transform. *Inver. Prob.* **17:** 113–119.

Nuyts, J., Dupont, P., Stroobans, S., Benninck, R., Mortelmans, L., and Suetens, P. (1999). Simultaneous maximum a posteriori reconstruction of attenuation and activity distributions from emission sinograms. *IEEE Trans. Med. Imag.* **18:** 393–403.

Nuyts, J., Dupont, P., Van den Maegdenbergh, V., Vleugels, S., Suetens, P., and Mortelmans, L. (1995). A study of liver-heart artifact in emission tomography. *J. Nucl. Med.* **36:** 133–139.

O'Connor, M. K. and Kelly, B. J. (1990). Evaluation of techniques for the elimination of "hot" bladder artifacts in SPECT of the pelvis. *J. Nucl. Med.* **31:** 1872–1875.

Ogawa, K., Harata, Y., Ichihara, T., Kubo, A., and Hasimoto, S. (1991). A practical method for position-dependent Compton-scatter correction in single photon emission CT. *IEEE Med. Imag.* **10:** 408–412.

Ogawa, K., Ichihara, T., and Kubo, A. (1994). Accurate scatter correction in single photon emission CT. *Ann. Nucl. Med. Sci.* **7:** 145–150.

Pan, T.-S., Luo, D.-S., Kohli, V., and King, M. A. (1997). Influence of OSEM, elliptical orbits, and background activity on SPECT 3D resolution recovery. *Phys. Med. Biol.* **42:** 2517–2529.

Pan, T.-S., King, M. A., Luo, D.-S., Dahlberg, S. T., and Villegas, B. J. (1997). Estimation of attenuation maps from scatter and photopeak window single photon emission computed tomographic images of technetium-99m labeled sestamibi. *J. Nucl. Card.* **4:** 42–51.

Parker, J. A. (1989). Quantitative SPECT: Basic theoretical considerations. *Sem. Nucl. Med.* **19:** 3–13.

Penney, B. C., Glick, S. J., and King, M. A. (1990). Relative importance of the error sources in Wiener restoration of scintigrams. *IEEE Trans. Med. Imag.* **9:** 60–70.

Penney, B. C., King, M. A., and Knesaurek, K. (1990). A projector, backprojector pair which accounts for the two-dimensional depth and distance dependent blurring in SPECT. *IEEE Trans. Nucl. Sci.* **37:** 681–686.

Pretorius, P. H., King, M. A., Pan, T.-S., de Vries, D. J., Glick, S. J., and Byrne, C. L. (1998). Reducing the influence of the partial volume effect on SPECT activity quantitation with 3D modelling of spatial resolution in iterative reconstruction. *Phys. Med. Biol.* **43:** 407–420.

Riauka, T. A., and Gortel, Z. W. (1994). Photon propagation and detection in single-photon emission computed tomography—an analytical approach. *Med. Phys.* **21:** 1311–1321.

Rosenthal, M. S., Cullom, J., Hawkins, W., Moore, S. C., Tsui, B. M. W., and Yester, M. (1995). Quantitative SPECT imaging: A review and recommendations by the focus committee of the society of nuclear medicine computer and instrumentation council. *JNM* **36:** 1489–1513.

Shepp, L. A., and Vardi, Y. (1982). Maximum likelihood reconstruction for emission tomography. *IEEE Trans. Med. Imag.* **1:** 113–122.

Singh, M., Horne, M., Maneval, D., Amartey, J., and Brechner, R. (1988). Non-uniform attenuation and scatter correction in SPECT. *IEEE Trans. Nucl. Sci.* **35:** 767–771.

Soares, E. J., Glick, S. J., and King, M. A. (1996). Noise characterization of combined Bellini-type attenuation correction and frequency-distance principle restoration filtering. *IEEE Trans. Nucl. Sci.* **43:** 3278–3290.

Sorenson, J. A., and Phelps, M. E. (1987). "Physics in Nuclear Medicine," 2nd ed. Grune & Stratton, Orlando, FL.

Stone, C. D., McCormick, J. W., Gilland, D. R., Greer, K. L., Coleman, R. E., and Jaszczak, R. J. (1998). Effect of registration errors between transmission and emission scans on a SPECT system using sequential scanning. *JNM* **39:** 365–373.

Tan, P., Bailey, D. L., Meikle, S. R., Eberl, S., Fulton, R. R., and Hutton, B. F. (1993). A scanning line source for simultaneous emission and transmission measurements in SPECT. *JNM* **34:** 1752–1760.

Tretiak, O., and Metz, C. (1980). The exponential Radon transform. *SIAM J. Appl. Math.* **39:** 341–354.

Tsui, B. M. W., Frey, E. C., Zhao, X. D., Lalush, D. S., Johnston, R. E., and McCartney, W. H. (1994). The importance and implementation of accurate 3D compensation methods for quantitative SPECT. *Phys. Med. Bio.* **39:** 509–530.

Tsui, B. M. W., and Gullberg, G. T. (1990). The geometric transfer function for cone and fan beam collimators. *Phys. Med. Biol.* **35:** 81–93.

Tsui, B. M. W., Hu, H.-B., Gilland, D. R., and Gullberg, G. T. (1988). Implementation of simultaneous attenuation and detector response correction in SPECT. *IEEE Trans. Nucl. Sci.* **35:** 778–783.

Tsui, B. M. W., Zhao, X. D., Frey, E. C., and McCartney, W. H. (1994). Quantitative single-photon emission computed tomography: Basics and clinical considerations. *Sem. Nucl. Med.* **24:** 38–65.

Tsui, B. M. W., Zhao, X. D., Gregoriou, Lalush, D. S., Frey, E. C., Johnston, R. E., and McCartney, W. H. (1994). Quantitative cardiac SPECT reconstruction with reduced image degradation due to patient anatomy. *IEEE Trans. Nucl. Sci.* **41:** 2838–2848.

Tung, C.-H., Gullberg, G. T., Zeng, G. L., Christian, P. E., Datz, F. L., and Morgan, H. T. (1992). Nonuniform attenuation correction using simultaneous transmission and emission converging tomography. *IEEE Trans. Nucl. Sci.* **39:** 1134–1143.

Wallis, J. W., and Miller, T. R. (1997). An optimal rotator for iterative reconstruction. *IEEE Trans. Med. Imag.* **16:** 118–123.

Wallis, J. W., Miller, T. R., and Koppel, P. (1995). Attenuation correction in cardiac SPECT without a transmission measurement. *JNM* **36:** 506–512.

Watson, D. D., Germano, G., DePuey, E. G., *et al.* (1999). Report summary of Wintergreen panel on instrumentation and quantification. *J. Nucl. Card.* **6:** 94–103.

Welch, A., Clack, R., Natterer, F., and Gullberg, G. T. (1997). Toward accurate attenuation correction in SPECT without transmission measurements. *IEEE Trans. Med. Imag.* **16:** 532–541.

Wells, R. G., Celler, A., and Harrop, R. (1998). Analytical calculation of photon distributions in SPECT projections. *IEEE Trans. Nucl. Sci.* **45:** 3202–3214.

Wells, R. G., King, M. A., Simkin, P. H., Judy, P. F., Brill, A. B., Gifford, H. C., Licho, R., Pretorius, P. H., Schneider, P., and Seldin, D. W. (2000). Comparing filtered backprojection and ordered-subsets expectation maximization for small lesion detection and localization in Ga-67 SPECT. *J. Nucl. Med.* **41:** 1391–1399.

Whal, R. L. (1999). To AC or not to AC: That is the question. *JNM* **40:** 2025–2028.

Xia, W., Lewitt, R. M., and Edholm, P. R. (1995). Fourier correction for spatially variant collimator blurring in SPECT. *IEEE Trans. Nucl. Sci.* **14:** 100–115.

Zeng, G. L., Bai, C., and Gullberg, G. T. (1999). A projector/backprojector with slice-to-slice blurring for efficient three-dimensional scatter modeling. *IEEE Trans. Med. Imag.* **18:** 722–732.

Zeng, G. L., and Gullberg, G. T. (1992). Frequency domain implementation of the three-dimensional geometric point response correction in SPECT imaging. *IEEE Trans. Nucl. Sci.* **39:** 444–453.

Zeng, L., and Gullberg, G. T. (2000). Unmatched projector/backprojector pairs in an alternative iterative reconstruction algorithm. *IEEE Trans. Med. Imag.* **19:** 548–555.

Zeng, G. L., Gullberg, G. T., Bai, C., Christian, P. E., Trisjono, F., Di Bella, E. V. R., Tanner, J. W., and Morgan, H. T. (1998). Iterative reconstruction of flourine-18 SPECT using geometric point response correction. *JNM* **39:** 124–130.

Zeng, G. L., Gullberg, G. T., Tsui, B. M. W., and Terry, J. A. (1991). Three-dimensional iterative reconstruction algorithms with attenuation and geometric point response correction. *IEEE Trans. Nucl. Sci.* **38:** 693–702.

Zubal, I. G., Harrell, C. R., Smith, E. O., Rattner, Z., Gindi, G., and Hoffer, P. B. (1994). Computerized three-dimensional segmented human anatomy. *Med. Phys.* **21:** 299–302.

CHAPTER 23

Kinetic Modeling in Positron Emission Tomography

EVAN D. MORRIS,* CHRISTOPHER J. ENDRES,[†] KATHLEEN C. SCHMIDT,[‡] BRADLEY T. CHRISTIAN,[§]
RAYMOND F. MUZIC JR.,[¶] RONALD E. FISHER[‖]

*Departments of Radiology and Biomedical Engineering, Indiana University School of Medicine, Indianapolis, Indiana
[†]Department of Radiology, Johns Hopkins University, Baltimore, Maryland
[‡]Laboratory of Cerebral Metabolism, NIMH, Bethesda, Maryland
[§]Wright State University, Kettering Medical Center, Dayton, Ohio
[¶]Departments of Radiology, Oncology and Biomedical Engineering, Case Western Reserve University, Nuclear Medicine,
University Hospitals Cleveland, Cleveland, Ohio
[‖]The Methodist Hospital and Department of Radiology and Neuroscience,
Baylor College of Medicine, Houston, Texas

I. Introduction
II. The One-Compartment Model: Blood Flow
III. Positron Emission Tomography Measurements of Regional Cerebral Glucose Use
IV. Receptor–Ligand Models
V. Model Simplifications
VI. Limitations to Absolute Quantification
VII. Functional Imaging of Neurochemistry—Future Uses
VIII. A Generalized Implementation of the Model Equations

I. INTRODUCTION

Positron emission tomography (PET) is a functional imaging technique that measures the local concentration of an exogenous tracer molecule in the target tissue. Thus, given a time sequence of PET images, one can quantify tracer kinetics *in vivo*. The power of PET lies in its molecular specificity. By using a particular radiotracer molecule, one can monitor the interaction of that molecule with the body's physiological processes. For example, blood flow can be measured by using radioactive water as a tracer; metabolism can be measured with radioactive glucose (FDG).

In this chapter we examine mathematical kinetic models used to analyze time sequences of PET images to gather quantitative information about the body. We include methods used in the two most ubiquitous applications of PET: imaging of blood flow and glucose metabolism. We also examine the use of PET to image specific receptor molecules, which capitalizes on the unique specificity of PET.

Supplied with single (static) PET images and no kinetic model to aid in their interpretation, one is confined to asking questions that can be answered only on the basis of spatial information such as: Where is glucose being used? Where are particular receptor molecules located within the brain or the heart?

With the acquisition of dynamic (time-sequence) imaging data and the application of kinetic models, one can pose many additional quantitative questions based on temporal information such as: How many receptors are there in a volume of tissue? What is the rate of trapping or the rate of influx of a tracer to the tissue? More recently, PET researchers have also begun asking questions about changes in particular parameters such as: When an endogenous ligand is perturbed, how does its level change? In this application, the radiotracer is used to infer information about an endogenous molecule that is not radioactive, exploiting the predictable nature of the competition between endogenous and exogenous ligands for specific binding sites.

Creative users of PET have also begun to ask questions about gene transcription. By linking a molecule that can be imaged to a therapeutic gene, one may be able to monitor the transcription of therapeutic genes with PET. The models for this application have not been worked out at this writing;

however we speculate in this chapter about the modeling issues that will arise if the application of PET imaging to gene therapy monitoring is to succeed as a quantitative tool.

All these applications can benefit from analysis based on modeling; but they require that the modeler understand how the data are generated and what the inherent inaccuracies in the data are.

Why do we need a model? In PET, the images are a composite of various superimposed signals, only one of which is of interest. The desired signal may describe, for example, the amount of tracer trapped at the site of metabolism or tracer bound to a particular receptor. In order to isolate the desired component of the signal (think of an image composed of layers), we must use a mathematical model relating the dynamics of the tracer molecule and all its possible states to the resultant PET image. Each of these states is known in the language of kinetic modeling as a *compartment*. For example, in a receptor-imaging study, the set of molecules that are bound to the target receptor can constitute one compartment. Each compartment is characterized by the concentration of the tracer within it as a function of time. These concentrations are related through sets of ordinary differential equations, which express the balance between the mass entering and exiting each compartment. By solving these simultaneous equations, we are able to determine the quantities of interest.

Kinetic models for PET typically derive from the one-, two-, or three-compartment model in which a directly measured blood curve (concentration of radiotracer in the blood as a function of time) serves as the model's input function.[1] The coefficients of the differential equations in the model are taken to be constants that are reflective of inherent kinetic properties of the particular tracer molecule in the system. By formally comparing the output of the model to the experimentally obtained PET data, one can estimate values for these kinetic parameters and thus extract information about binding, delivery, or any hypothesized process, as distinct from all other processes contributing to the PET signal.

In general, the information content of the PET data is inadequate to support models of great sophistication, and so adoption of a particular model is by necessity a great simplification of the truth. The interpretation of parameters comes with associated assumptions and conditions that must be satisfied and periodically revisited. Sometimes the overriding limitations are experimental. The most common such limitation is the inconvenience of, or outright prohibition against, obtaining direct measurement of the plasma input function (or input function). This limitation has of necessity given rise to a whole class of model simplifications, which are developed and examined in this chapter. Other experimental considerations are discussed with regard to possible future clinical and research uses of PET ligand studies. Finally, we conclude our treatment of PET kinetic modeling with a discussion of a generalized framework for the implementation and solution of the differential equations that constitute these models.

A. What's in a Compartment?

Compartmental modeling is used to describe systems that vary in time but not in space. Thus, one of the first assumptions of compartmental modeling is called the well-mixed assumption. We assume that there are no spatial concentration gradients (only gradients in time) within the area being sampled (e.g., a voxel in the case of a 3D image volume.) Within a voxel, we assume that whatever radioactive species contribute to the emanating radioactive signal are in uniform concentration and can be characterized as being in unique states. Each of these states is assigned a compartment, which in turn is described by a single ordinary differential equation.

Any system requires an input to drive it. We typically consider the input to the system to be the measured radioactivity concentration in the blood supply as a function of time. Although the blood supply is a compartment in the physical sense, it is not a compartment of the model in the mathematical sense because it is measured rather than solved for. However, the input concentration is often depicted as a box in graphical representations of kinetic models. We will adhere to this custom, but the reader should be aware that it is not strictly correct.

Compartmental models often arise in the course of describing the kinetics of a tracer injected into a physiological system. In tracer studies, we make certain assumptions about the behavior of the tracer molecule, often radioactive, but sometimes labeled with a flourochrome or other detectable marker. The key assumptions that must be satisfied are as follows.

1. The amount of tracer injected is a trace amount; that is, it causes no change in the physiology of the organism. For example, tracers are often drugs, but when injected as tracers they must cause no drug effects. We sometimes violate this assumption when giving low-specific-activity (near-saturation) doses of a radioligand in PET.
2. The tracer is in steady state with the endogenous molecule that the tracer seeks to emulate (which we call the *tracee*). That is, the tracer goes everywhere that the tracee goes. In PET, we often talk about freely diffusible tracers, which are tracers that easily cross into and out of the vasculature. Note that the tracee (e.g., glucose in the brain) can itself be in a non-steady-state condition with respect to the system if we perturb it, but the relationship between the tracer and tracee should be fixed.
3. There are no isotope effects. That is, the act of labeling the tracer molecule with a radionuclide does not alter its properties.

---

[1] There are modifications of the model (i.e., reference region models) that do not use an independently measured blood curve as input. These models are addressed later in this chapter.

There are a few additional assumptions to keep in mind. Usually, one assumes that the parameters of our model are time-invariant, at least over the duration of our study. One can consider the kinetic parameters to be a reflection of the steady state of the system and the innate properties of the ligand. The kinetic parameters that one estimates from a PET study provide a snapshot of the average state of the system and the ligand at a given time or over a short time period (i.e., the length of a study). To solve our differential equations one needs an input function, which drives the model, and initial conditions, which are usually zero because there is usually no radioactivity in the system before a tracer experiment begins. This might not be the case if one were to do multiple repeated studies, one right after the other. As a practical matter, in PET one typically assumes that all tissues of the body see the same input function, which is assumed to be the measured radioactivity in the blood plasma during the experiment. With zero initial conditions and without an input function, the concentration of tracer in the compartments of the model would remain at zero indefinitely.

B. Constructing a Compartmental Model

Compartmental models are used to describe the dynamics of PET measurements. The data are generally not of sufficient spatial resolution to support more-complicated mathematical descriptions that include spatial gradients (i.e., distributed models). Nevertheless, compartmental models have good predictive power so we continue to use them. How do we get from a complex physiological system to a highly simplified compartmental model?

Consider PET images of the brain (see Figure 1a), which are the primary source of data discussed in this chapter. Referring to a sample (in our case, a voxel), consider the primary states that a radioactive tracer molecule (blue diamonds on Figure 1b) can assume. First, the tracer in the brain might be in the blood. After all, that is how it is delivered to the brain. If it is freely diffusable, some tracer will eventually be in the extravascular state. If there are receptors in the brain specific for our tracer, some tracer may become bound to these receptors, at which time we consider it to be in the bound state. We use the term *state* because these

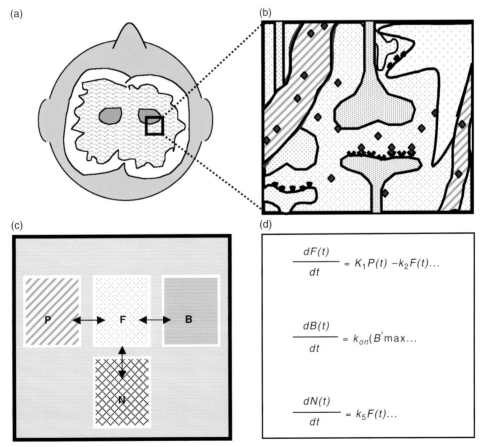

FIGURE 1 Schematic of the compartmental modeling process as applied to PET images. Steps: (a) identify a region of interest in an image, (b) conceptualize the contents of the region and the possible states of the tracer in the region, (c) further abstract the ligand states to compartments with a particular topology or connectedness, and (d) write the mass balances on each of the unknown compartments as ordinary differential equations.

compartments do not necessarily occupy distinct locations on the macroscopic scale. Figure 1b is our conception of the major possible states of the tracer molecule. In Figure 1c, we indicate a further abstraction of the schematic in Fig. 1b by identifying what we believe to be the primary states of the molecule and the connections between them.

Once the connections or paths that the tracer can take between compartments have been specified, the mass balances on each compartment can be written in the form of ordinary differential equations. One differential equation must be specified for each unknown quantity. If the concentration of tracer in the plasma is known, it is not a compartment but its (time-varying) concentration does factor into the mass balances of the compartments. The differential equations of the model (Figure 1d) are written to describe influx to, and efflux from, each compartment. They are functions of the rate constants for these processes and, in the case of saturable processes, the derivatives will also be a function of the number of available binding sites.

For more detailed treatments of compartmental modeling in physiology and in PET specifically, please consult E. Carson *et al.* (1983), Jacquez (1985), Berman *et al.* (1982), R. Carson (1996), and Huang and Phelps (1986).

We begin our discussion of specific models with an examination of the simplest model used in PET, the blood flow model, which is described by only one compartment.

II. THE ONE-COMPARTMENT MODEL: BLOOD FLOW

A. One-Tissue Compartmental Model

A simple compartmental model that frequently arises in PET applications is the one-tissue model (Figure 2), which describes the bidirectional flux of tracer between blood and tissue. Note that in kinetic modeling flow is generally not measured in terms of volume per time, as one might expect, but rather in terms of perfusion of tissue, which is described as volume per unit time per unit volume of tissue. Thus, the words *flow* and *perfusion* are used interchangeably in PET.

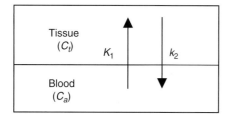

FIGURE 2 This one-tissue compartment model describes the bidirectional flux of tracer between blood (C_a) and tissue (C_t). The net tracer flux into tissue equals the flux entering the tissue ($K_1 C_a$) minus the flux leaving the tissue ($k_2 C_t$).

The model is characterized by the time-varying tracer concentration in tissue, $C_t(t)$, and arterial blood, $C_a(t)$ and two first-order kinetic rate constants (K_1, k_2). The tracer concentrations are measured in nanocuries per milliliter (nCi/ml). Throughout the remainder of this chapter, we suppress the explicit time dependence of the concentration functions for notational simplicity. For example, $C_a(t)$ and $C_t(t)$ will be written simply as C_a and C_t.

It is assumed that within each compartment (blood and tissue) the tracer is homogeneously distributed. The unidirectional tracer flux from blood to tissue is $K_1 C_a$, and the flux from tissue to blood is $k_2 C_t$; therefore, the net tracer flux into tissue is:

$$\frac{dC_t}{dt} = K_1 C_a - k_2 C_t \qquad (1)$$

Eq. (1) can be solved (see Appendix at the end of this chapter) for C_t to obtain:

$$C_t = K_1 C_a \otimes \exp(-k_2 t) \qquad (2)$$

where the symbol $\otimes$ denotes one-dimensional convolution.[2] If C_a and C_t are radioactivity concentrations, then k_2 (implicitly) includes a component for radioactive decay (see Section IVA). For a PET scan, C_t is the radioactivity concentration that is measured in a given tissue region. Blood samples may be drawn during a PET scan in order to measure C_a. If the tracer distribution in a tissue region is adequately described by the one-tissue model, then with serial measurements of C_t and C_a, Eq. (2) can be applied using standard nonlinear regression techniques to estimate the values of K_1 and k_2 for that region. We next develop an interpretation of these parameters in terms of the underlying physiology.

B. One-Compartment Model in Terms of Flow

The exchange of tracer between blood and tissue occurs either via diffusion or active transport across the capillary wall. Blood-flow tracers including $H_2^{15}O$ (also called ^{15}O-water) and ^{15}O-butanol pass via diffusion. In order to gain a better physiological understanding of the kinetic rate constants, it is instructive to first consider the passage of tracer within capillaries. For this purpose we can apply the model in Figure 2 on its smallest scale, such that the blood compartment represents a single capillary and the tissue compartment is the tissue in the immediate vicinity. Because some tracer is removed from the blood via extraction into tissue, the tracer blood concentration diminishes as it

---

[2] Elsewhere in this book, the symbol * is used to denote convolution. The symbols $\otimes$ and * have the same meaning. The symbol $\otimes$ is more customary in the kinetic modeling literature, whereas * is more commonly used in the signal-processing literature. Equation (2) can be written explicitly as $C_t(t) = K_1 \int_0^t C_a(t') \exp[-k_2(t - t')]\, dt'$. The limits of integration are confined to 0 to t because the system is causal, and C_a is 0 for $t < 0$.

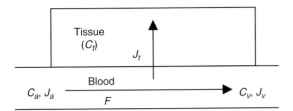

FIGURE 3 Tracer flow through a capillary. The values C_a and C_v are the tracer concentrations in arterial blood and venous blood, respectively, and C_t is the concentration in tissue. The tracer enters the capillary with flux J_a and leaves via tissue extraction (J_t) and the venous circulation (J_v). If tracer does not accumulate in the capillary, then the tracer flux into the capillary J_a equals the flux out of the capillary ($J_v + J_t$).

passes through the capillary. To include this effect in the one-tissue model, the blood compartment is extended as shown in Figure 3. The tracer enters the capillary from the arterial side with concentration C_a and exits the venous side with concentration C_v; blood flows through the capillary with flow rate F. To derive a relation between these quantities we apply the Fick principle that states that when a compartment is at steady state, the flux of material that enters the compartment equals the flux that exits the compartment. In steady state, the concentration of material in the compartment does not change. According to the Fick principle, a steady state exists in the capillary as long as the tracer does not accumulate in the capillary itself. This condition is usually well satisfied because a tracer molecule that enters the capillary will rapidly leave by tissue extraction or via the venous circulation. From the Fick principle we get $J_a = J_t + J_v$, where J_a and J_v are the arterial and venous tracer fluxes, respectively. The arterial tracer flux equals the product of blood flow and the arterial tracer concentration, that is, $J_a = FC_a$. Similarly, the venous tracer flux is $J_v = FC_v$. The net tracer flux into tissue is the difference between the arterial and venous fluxes:

$$J_t = \frac{dC_t}{dt} = J_a - J_v = F(C_a - C_v) \quad (3)$$

The net fraction of the incoming tracer that is extracted into tissue during one capillary pass is called the net extraction fraction (E_n), and is equal to:

$$E_n = \frac{C_a - C_v}{C_a} \quad (4)$$

The unidirectional extraction fraction (E_u) refers only to tracer that is extracted from blood to tissue. During the first pass of a tracer through a tissue site, the tracer flux from tissue to blood is effectively zero because $C_t = 0$. In this case E_n is equal to E_u, which is often referred to as the first-pass extraction fraction. By setting $C_t = 0$ in Eq. (1), Eqs. (1), (3), and (4) can be arranged to give the net flux into tissue during the first pass

$$\frac{dC_t}{dt} = (FE_u)C_a = K_1 C_a \quad (5)$$

Thus, the delivery rate constant K_1 equals the product of blood flow and the unidirectional (first-pass) extraction fraction. The interpretation of E_u has been further developed from the Renkin-Crone capillary model, which considers the capillary to be a rigid cylindrical tube (Renkin, 1959; Crone, 1964). Based on the Renkin-Crone model a theoretical expression for the unidirectional extraction fraction has been derived

$$E_u = 1 - \exp(-PS/F) \quad (6)$$

where P is the permeability of the tracer across the capillary membrane, and S is the capillary surface area per unit mass of tissue. The product of permeability and surface area (PS) has the same units as blood flow (e.g., ml/min/ml). PS for water has been estimated at 133 ml/(min·100 g) (Berridge et al., 1991) whereas the PS for butanol has been estimated to be 400 ml/(min·100 g) (Martin et al., 1999). From Eq. (6) we see that, when PS is large relative to F, E_u approaches a maximum value of 1 and $K_1 \cong F$. At the other extreme, when F is large relative to PS, then $E_u \cong PS/F$, and $K_1 \cong PS$. Although these formulas are derived from simplified arguments (Sorenson and Phelps, 1987), they provide useful rules of thumb that give physiological meaning to K_1. Figure 4a displays the relationship between extraction fraction as a function of flow for moderate and high PS products. Figure 4b plots extraction fraction times flow ($= K_1$) versus true flow.

To summarize, K_1 is closely related to blood flow when the extraction fraction is large, but is more related to permeability when the extraction fraction is small. Accordingly, the best tracers for studying blood flow have large extraction fractions (corresponding to $PS \gg F$). When blood flow is expected to be higher than normal, there is an added burden on the PS product to be high. Thus, Figure 5 shows a case of normal and high flow measured with two different tracers, ^{15}O-water and ^{15}O-butanol (Berridge et al., 1991). When flow is high (provoked by the breathing of carbon dioxide), the measured K_1 of the tracer with the higher capillary permeability is better able to reflect the true flow value. PS for water has been estimated at 133 ml/(min·100 g) (Berridge et al., 1991), whereas the PS for butanol has been estimated to be 400 ml/(min·100 g) (Martin et al., 1999). Hence, even at high flow induced by carbon dioxide, we would expect extraction E_u of butanol to be unity and the flow estimate under that condition to be faithful to its true value.

C. Volume of Distribution/Partition Coefficient

Consider an experiment in which the tracer blood concentration is maintained at a constant level over time. In the tissue, tracer will accumulate with a time course

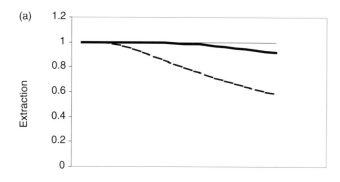

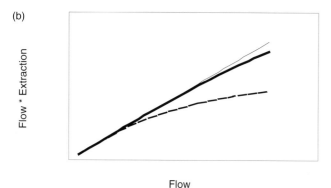

FIGURE 4 (a) Extraction fraction (E) versus flow (F). Solid curve is a highly extracted tracer. Dashed curve is a moderately extracted tracer. PS (permeability-surface area product) is higher for the former. Thin horizontal line is $E = 1$. (b) Flow times extraction fraction (i.e., apparent K_1 for blood flow model; see Eq. (5) versus true flow. Dashed curve is for moderately extraced ligand; thick solid curve is for a highly extracted ligand. Notice the increasing deviation at large flow values. Thin diagonal line is line of identity.

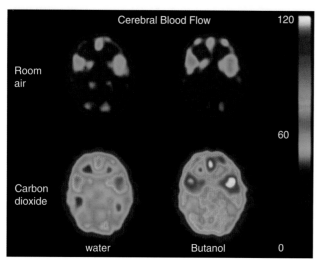

FIGURE 5 Blood flow scans using either ^{15}O-water or ^{15}O-butanol. At low flow rates (i.e., when subject breathes room air) the measured flow, K_1, is very similar across ligands, whereas at high flow rates (i.e., induced by subject breathing carbon monoxide) the measured flow using ^{15}O-water is lower than that using ^{15}O-butanol because of the lower permeability of the blood–brain barrier to water. See Figure 4 for an explanation. (Figure supplied courtesy of Dr. Marc Berridge, Case Western Reserve University.)

described by Eq. (2). After sufficient time the net tracer flux between compartments is zero, and the system is said to be in equilibrium. At this point the blood and tissue concentrations are constant and the ratio of these concentrations C_t/C_a is called the equilibrium volume of distribution V_D (alternatively called the distribution volume, or partition coefficient). Although V_D is a ratio of concentrations (and hence dimensionless), it is called a volume because it equals the volume of blood that contains the same activity as 1 ml of tissue (Carson, 1996). For example, $V_D = 2$ means that at equilibrium the tracer in tissue is twice as concentrated as that in blood. That is, 2 ml of blood has the same quantity of radioactivity as 1 ml of tissue. Because the net tissue flux (Eq. 1) at equilibrium is zero, V_D can be expressed as:

$$V_D = \frac{C_t}{C_a} = \frac{K_1}{k_2} \qquad (7)$$

For the one-tissue model, the volume of distribution is a macroparameter equal to K_1/k_2. In general, the precise formulation of V_D in terms of the kinetic rate constants ($K_1, k_2, \ldots$) depends on the particular compartmental model employed (see further discussions in Sections IV and V). From the physiological interpretations of K_1 (the product of blood flow and extraction) and V_D (the ratio of C_t/C_a at equilibrium), k_2 can be defined as the ratio of K_1 to V_D. In most experimental paradigms, it is easier to measure V_D than its constituent parameters, and this is often just as useful for interpreting results. A common experimental technique used to measure V_D is to deliver tracer via continuous infusion in order to maintain C_a at a constant level. Once tissue equilibrium is achieved, V_D is easily obtained from blood and tissue measurements using Eq. (7).

D. Blood Flow

An important application of the one-tissue model is the measurement of regional blood flow. Blood flow can be described as the volume of blood that passes through a blood vessel per unit time (e.g., ml/min). However, the regional measurement of blood flow is typically expressed as the blood flow per unit mass of tissue (e.g. ml/min/mg), which is also called *perfusion*. When referring to PET studies, perfusion or blood flow per unit mass is simply called *blood flow*. The tracer that is most commonly used to measure blood flow with PET is ^{15}O-water (Herscovitch et al., 1983; Raichle et al., 1983). Because ^{15}O-water is rapidly diffusible, the tissue concentration equilibrates with the venous outflow, such that these concentrations are related by $V_D = C_t/C_v$. Substitution of this expression into Eq. (3) gives:

$$\frac{dC_t}{dt} = F\left(C_a - \frac{C_t}{V_D}\right) \qquad (8)$$

Equation (8) can be solved for C_t as a function of F and V_D:

$$C_t(t) = FC_a(t) \otimes \exp\left(-\frac{F}{V_D}t\right) \quad (9)$$

For the practical application of Eq. (9), the distribution volume is generally fixed to some predetermined value (Iida et al., 1989), leaving F as the only unknown parameter. To estimate blood flow with Eq. (9), C_t is computed over a range of F values that encompass the expected range of blood-flow values. This procedure generates a lookup table that can be used to convert the tissue measurements to flow values. In the case of PET, which measures the integral of the tissue curve, Eq. (9) is integrated over the same time range as the PET acquisition to produce a lookup table that relates $\int C_t dt$ to F (Muzic et al., 1990). This technique is readily adaptable to parametric (pixel-by-pixel) calculations.

E. Dispersion, Delay Corrections

The measurement of blood flow with Eq. (9) is highly sensitive to the measurement accuracy of the arterial input function. To improve the accuracy, we need to account for subtle differences between the *measured* arterial input function (C_{am}), and the arterial concentration at the site of tissue measurement (C_a, considered the true input function). The discrepancies between the true and measured input functions are largely due to delay and dispersion. A time delay (Δt) occurs because the tracer generally requires different amounts of time to reach the sites of tissue measurement versus the sites of arterial sampling, due to differences in circulation distance. For example, ^{15}O-water usually reaches the brain before it is detected at the contralateral arterial sampling site. In addition, dispersion of the tracer will cause different degrees of smoothing of the input function, which results in C_{am} and C_a having different shapes. Dispersion (D) can be described as a monoexponential function of time:

$$D = \frac{1}{\tau}\exp\left(-\frac{t}{\tau}\right) \quad (10)$$

where τ is the dispersion time constant (Iida et al., 1986). Accounting for delay and dispersion, the true and measured input functions are related according to:

$$C_{am}(t) = C_a(t + \Delta t) \otimes \frac{1}{\tau}\exp\left(-\frac{t}{\tau}\right) \quad (11)$$

In order to determine the true input function C_a, two more parameters (τ and Δt) must be estimated. Several procedures have been developed to determine these values (Raichle et al., 1983; Meyer, 1989; Ohta et al., 1996; Muzic et al., 1993; Nelson et al., 1990). The delay and dispersion corrections must be applied before Eq. (9) is used to generate a lookup table. The discrepancy between C_a and C_{am} due to delay and dispersion is negligible after a few minutes. As a result, their effects are most evident in blood-flow studies, which are usually just a few minutes in duration. For most other applications (e.g., studies with receptor binding tracers), the study duration is 60 minutes or more and differences between C_a and C_{am} are generally ignored.

F. Noninvasive Methods

A method has been introduced to measure blood flow without an arterial input function. Instead, the method of Watabe et al. (1996) relies on measuring two independent tissue curves: one the whole brain curve and the other a gray-matter curve. The curves are assumed to have identical arterial inputs but different flows and volumes of distribution. Volume of distribution, V_D, in the grey matter is assumed to be known. Three remaining parameters are estimated: flow in whole brain, flow in grey matter, and V_D in whole brain. Unfortunately, if the two curves are not sufficiently independent, the method was found in practice to be highly sensitive to noise (H. Watabe, personal communication). Noninvasive methods (i.e., not requiring an arterial curve) have been used more successfully in receptor–ligand studies and are examined in detail in Section V.

G. Tissue Heterogeneity

As stated in Section IIA, the one-compartment model includes the assumption that tracer in the tissue compartment is homogeneously distributed. The spatial resolution of the current generation of PET scanners, however, is on the order of 4–6 mm full width at half maximum (FWHM) (see, e.g., Pajevic et al., 1998), whereas gray matter in human cortex forms a winding band only 2–3 mm wide. Most brain regions measured with PET, especially cortical regions, therefore contain mixtures of gray and white matter that have different concentrations of tracer due to their differing rates of blood flow. Analysis of data on a pixel-by-pixel basis, even though the pixels may be reconstructed to be ~2 mm on a side, does not eliminate the heterogeneity problem because the minimum object size for accurate quantification is 2 × FWHM (Hoffman et al., 1979). Because blood flow is higher in gray matter than in white matter, the partial-volume effect leads to underestimation of flow in gray matter when blood flow in the region as a whole is determined. Weighted average blood flow is underestimated as well and falls with increasing experimental duration if the kinetic model used in the analysis fails to account for the tissue heterogeneity (Schmidt et al., 1989). In simulations of PET H$_2$^{15}O studies, errors in weighted average flow due to heterogeneity have been estimated to range from < 5% in short-duration studies (Herscovitch et al., 1983; Koeppe et al., 1985) to approximately 20% in longer-duration studies (Koeppe et al., 1985).

Due to the unavoidable inclusion of white matter in each region of measurement, weighted average flow in a nominally gray matter region depends on the amount of white matter included. In order to determine gray matter flow, corrections for the partial-volume effect have been developed that are based on the use of a high-resolution magnetic resonance image of the subject coregistered to the PET image. More recently, methods to determine gray matter flow based on the PET data alone have been proposed (Iida et al., 2000; Law et al., 2000; Schmidt and Sokoloff, 2001).

III. POSITRON EMISSION TOMOGRAPHY MEASUREMENT OF REGIONAL CEREBRAL GLUCOSE USE

The [^{18}F]fluorodeoxyglucose ([^{18}F]FDG) method is used for the quantitative measurement of regional cerebral glucose use ($rCMR_{glc}$) in man and in animals. Because a close relationship has been established between functional activity and cerebral glucose use in discrete brain regions (Sokoloff, 1977), the method can be used to investigate regional changes in functional activity in various physiological and pathological states. The [^{18}F]FDG method was first developed for use with single-photon emission computerized tomography (SPECT) (Reivich et al., 1979) and shortly thereafter adapted for use with PET (Phelps et al., 1979). The [^{18}F]FDG method was derived from the autoradiographic [^{14}C]deoxyglucose ([^{14}C]DG) method for measuring local cerebral glucose use in animals (Sokoloff et al., 1977). [^{14}C]DG was specifically chosen as the tracer because its already well-known biochemical properties facilitated the design and analysis of a kinetic model that described the brain–plasma exchange of [^{14}C]DG and its phosphorylation by hexokinase in relation to those of glucose.

A. The Basic Compartmental Model

The model for the measurement of cerebral glucose use with [^{18}F]FDG in a homogeneous tissue in brain is based on that of [^{14}C]DG (Sokoloff et al., 1977) and includes two tissue compartments (Figure 6). [^{18}F]FDG and glucose in the plasma cross the blood–brain barrier by a saturable carrier-mediated transport process. Because [^{18}F]FDG and glucose compete for the same carrier, the rate of inward transport of [^{18}F]FDG can be described by the classical Michaelis-Menten equation modified to include the influence of the competitive substrate, that is,

$$v_i^*(t) = \frac{V_t^* C_p^*(t)}{K_t^* \left[1 + C_p(t)/K_t\right] + C_p^*(t)} \tag{12}$$

where v_i^* is the velocity of inward transport; C_p^* and C_p are the concentrations of [^{18}F]FDG and glucose,

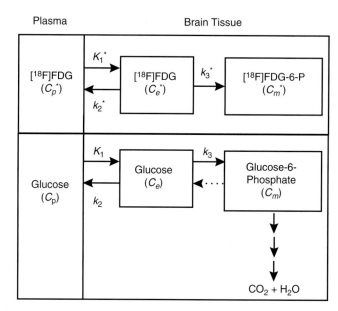

FIGURE 6 Model for measurement of cerebral glucose use with [^{18}F]fluorodeoxyglucose ([^{18}F]FDG) in a homogeneous tissue. C_p^* and C_p represent the concentrations of [^{18}F]FDG and glucose in the arterial plasma, respectively; C_e^* and C_e are the concentrations of [^{18}F]FDG and glucose in the exchangeable pool in the tissue; C_m^* is the concentration of metabolites of [^{18}F]FDG in the tissue, predominately [^{18}F]fluorodeoxyglucose-6-phosphate; and C_m is the concentration of metabolites of glucose in the tissue. The PET signal includes the sum of C_e^*, C_m^*, and the concentration of [^{18}F]FDG in the blood in the brain. The constants K_1^* and k_2^* represent the first-order rate constants for carrier-mediated transport of [^{18}F]FDG from plasma to tissue and back from tissue to plasma, respectively. k_3^* represents the first-order rate constant for phosphorylation of [^{18}F]FDG by hexokinase. The respective rate constants for glucose are K_1, k_2, and k_3. The model allows for the possibility that some of the phosphorylated glucose may return to the precursor pool (dashed arrow). (Adapted from Sokoloff et al., 1977.)

respectively, in the plasma; V_t^* is the maximal velocity of [^{18}F]FDG transport; K_t^* and K_t are the apparent Michaelis-Menten constants of the carrier for [^{18}F]FDG and glucose, respectively; and t is the variable time. Because [^{18}F]FDG is administered in tracer amounts, $C_p^*(t)$ can be considered negligible compared to $K_t^* [1 + C_p(t)/K_t]$. If the glucose concentration in the plasma is constant, then the velocity of inward transport can be expressed as $K_1^* C_p^*(t)$, where $K_1^* = V_t^*/[K_t^* (1 + C_p/K_t)]$ is a first-order rate constant. Once [^{18}F]FDG is in the tissue, it can either be transported back to the plasma or phosphorylated to [^{18}F]fluorodeoxyglucose-6-phosphate ([^{18}F]FDG-6-P), processes that follow Michaelis-Menten equations analogous to Eq. (12). When cerebral glucose use is in steady state, the concentration of glucose in brain (C_p) is constant and both outward transport and phosphorylation of [^{18}F]FDG can be considered first-order processes. Thus the rate of change of the [^{18}F]FDG concentration in the exchangeable pool in the tissue can be described by the equation:

$$\frac{dC_e^*}{dt} = K_1^* C_p^*(t) - \left(k_2^* + k_3^*\right) C_e^*(t) \tag{13}$$

where $C_p^*(t)$ (in nCi/ml plasma) represents the concentration of [^{18}F]FDG measured in the arterial plasma t minutes after its injection into the venous blood; $C_e^*(t)$ (in nCi/g brain tissue) is the concentration of unmetabolized tracer in the exchangeable pool in the tissue; K_1^* (in ml plasma/g brain/min) and k_2^* (in min^{-1}) represent the first-order rate constants for transport of [^{18}F]FDG from plasma to tissue and back from tissue to plasma, respectively; and k_3^* (in min^{-1}) represents the first-order rate constant for phosphorylation of [^{18}F]FDG by hexokinase. K_1^*, k_2^*, and k_3^* are constant with respect to time, but vary regionally and with different plasma and brain concentrations of glucose.

The advantage of using [^{18}F]FDG rather than labeled glucose as the tracer is that the primary and secondary products of its phosphorylation, unlike those of glucose, are trapped in the tissue and accumulate where they are formed for reasonably prolonged periods of time.[3] Therefore, the rate of change in the concentration of products of [^{18}F]FDG metabolism in the tissue, $C_m^*(t)$, is

$$dC_m^*/dt = k_3^* C_e^*(t) \quad (14)$$

and the total concentration of activity in the tissue, $C_i^*(t)$, is the sum of the concentrations in the precursor and product tissue pools:

$$C_i^*(t) = C_e^*(t) + C_m^*(t) \quad (15)$$

The PET scanner measures all activity in the field of view, both intra- and extravascular. Thus the total concentration of activity measured by the scanner, $C_t^*(t)$, is

$$C_t^*(t) = (1 - V_B) C_i^*(t) + V_B C_B^*(t), \quad (16)$$

where V_B is the fraction of the measured volume occupied by blood ($0 \leq V_B \leq 1$) and $C_B^*(t)$ is the concentration of label in whole blood. In the human, brain–blood volumes in gray and white matter are about 4 and 2%, respectively (Phelps et al., 1979).

The kinetic model for glucose uptake and metabolism in the tissue is equivalent to that of [^{18}F]FDG, except that glucose-6-phosphate is not trapped in the tissue but continues down the glycolytic pathway eventually to CO_2 and H_2O (Figure 6). The model allows for the possibility that some of the phosphorylated glucose may return to the precursor pool. In steady state, the constant fraction of the glucose that, once phosphorylated, continues down the glycolytic pathway is designated by Φ; in brain $\Phi \approx 1$ (Sokoloff et al., 1977).

---

[3] When the [^{18}F]FDG method was first adapted for use with PET, only relatively slow scanners were available. In some studies, tissue activity was measured as long as 14 hours after injection of [^{18}F]FDG when product loss was clearly observable, and a term for product loss was added to the kinetic model (Phelps et al., 1979; Huang et al., 1980). In studies of 2 h or less in duration there is no evidence for loss of metabolic products (Schmidt et al., 1992; Lucignani et al., 1993).

B. Protocol for Measurement of Regional Cerebral Glucose Use

Because the model describing the kinetics of [^{18}F]FDG and glucose *in vivo*, like all models representing biological events, is necessarily a simplified description that cannot fully account for all potentially relevant factors, the experimental procedure of the [^{18}F]FDG method, like that of the autoradiographic [^{14}C]DG method, was designed to minimize possible errors arising from limitations and/or imperfections of the kinetic model. The rationale for the procedure is made clear by the examination of the originally reported operational equation that applies to both methods (Sokoloff et al., 1977). Regional cerebral glucose use is calculated as:

$$rCMR_{glc} = \frac{C_t^*(t)\, T - K_1^* \int_0^T C_p^*(t) e^{-(k_2^* + k_3^*)(T-t)} dt}{\left(\dfrac{V_m K_m}{V_m^* K_m^*}\right)\left(\dfrac{\lambda}{\Phi}\right)\left(\int_0^T \dfrac{C_p^*(t)}{C_p} dt - \int_0^T \dfrac{C_p^*(t)}{C_p} e^{-(k_2^* + k_3^*)(T-t)} dt\right)} \quad (17)$$

where C_t^*, C_p^*, K_1^*, k_2^*, k_3^*, and C_p are as defined in Section IIIA. The first bracketed term in the denominator relates the relative preference of hexokinase for glucose over that for FDG; its components are V_m and K_m, the maximal velocity and Michaelis-Menten constant of hexokinase for glucose; and V_m^* and K_m^*, the equivalent kinetic constants of hexokinase for FDG. In the second term in the denominator λ represents the ratio of the distribution spaces of FDG and glucose, and Φ converts the rate of glucose phosphorylation into the net rate of glucose use. The first two terms together comprise the lumped constant of the method, so called because the six constants Φ, V_m, K_m, V_m^*, K_m^*, and λ have been lumped together. The lumped constant is discussed in Section IIIE.

In Eq. (17), the least reliable terms, those that depend on the kinetic model and rate constants, can be minimized by use of an optimally chosen experimental protocol. The numerator of Eq. (17) approximates the concentration of labeled products of [^{18}F]FDG phosphorylation formed in the tissue during the experimental period; this is determined as the difference between total radioactivity measured by PET (C_t^*) and a term that represents the estimated C_e^*, the free unmetabolized [^{18}F]FDG still remaining in the exchangeable pool in the tissue at the end of the experimental period. In order to minimize the free [^{18}F]FDG in the tissue, a long period of tracer circulation after a pulse of [^{18}F]FDG is used; C_t^* then represents mainly labeled products of [^{18}F]FDG phosphorylation. C_t^* should also be corrected for blood volume and residual blood activity, but these correction factors are negligible and usually omitted. In the denominator of the operational equation, a term that corrects for the lag in equilibration of the tissue precursor pool behind the plasma is subtracted from the integrated specific activity measured

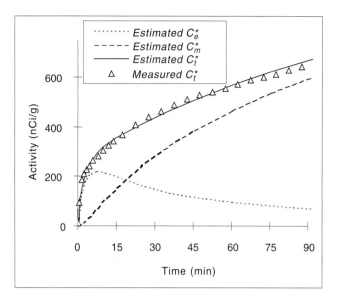

FIGURE 7 Time course of measured activity in visual cortex of a normal subject (open triangles), and model-estimated concentrations of total ^{18}F (C_t^*, solid line), and [^{18}F]FDG (C_e^*, dotted line), and [^{18}F]FDG-6-P (C_m^*, dashed line). Activity in the blood pool is not shown. Note that application of the homogeneous tissue [^{18}F]FDG model to the heterogeneous region results in systematic errors in the model-estimated total activity. All data are decay-corrected. (Data from Lucignani et al., 1993.)

in the plasma. The magnitude and, therefore, impact of this correction term is also minimized by the use of the long period of tracer circulation after the pulse. Because the total ^{18}F concentration in the tissue and the integrated plasma specific activity are increasing while the correction terms that are subtracted from them are decreasing with time after a pulse, the effects of errors in the estimates of these correction terms diminish with time. The experimental period cannot be extended indefinitely, however, because the loss of products of [^{18}F]FDG phosphorylation eventually becomes significant.

Estimated time courses of the activity in the various tissue components are shown in Figure 7.

C. Estimation of Rate Constants

When only single-ring (relatively inefficient and slow) PET scanners were available, several hours were required to scan the entire brain, and the strategy for determination of the rate constants was the same as that for the [^{14}C]DG method; they were estimated in a separate group of subjects by nonlinear least squares fitting of the model equation to the measured time courses of total ^{18}F activity in various brain regions. The rate constants estimated from several individuals were averaged, and the population average rate constants were subsequently used with each new subject's measured final ^{18}F concentration in brain regions of interest and the time course of arterial plasma specific activity to calculate $rCMR_{glc}$ by means of Eq. (17). More recent generations of scanners contain multiple rings that allow the entire brain to be scanned simultaneously; their increased efficiency also allows more rapid acquisition of multiple time frames of data. This has made it technically possible to obtain statistically reliable time courses of total radioactivity and to estimate rate constants in each region for each individual subject. It was expected that obtaining rate constants for each individual would lead to improvements in accuracy in the rate-constant-dependent terms of Eq. (17), which, in turn, would allow studies to be completed in shorter total scanning times. Problems arising from tissue heterogeneity, however, have limited the extent to which studies can be shortened, as discussed in Section IIIE.

D. The Lumped Constant

The lumped constant is a unitless physical constant used to convert rates of [^{18}F]FDG uptake and phosphorylation into rates of glucose use. The directly measured value of the lumped constant for [^{18}F]FDG in whole brain in humans under normal physiological conditions is 0.52 (Reivich et al., 1985). Although it is difficult to measure the lumped constant under nonphysiological conditions in humans, it is expected that the lumped constant for [^{18}F]FDG in humans should have similar properties to the lumped constant for deoxyglucose that has been extensively studied in animals. The lumped constant for deoxyglucose has been shown to be relatively stable over a wide range of plasma glucose concentrations in normal brain (Sokoloff et al., 1977). It is influenced mainly by plasma and tissue glucose levels, falling progressively but very gradually with increasing plasma glucose levels all the way to severe hyperglycemia (Schuier et al., 1990). When glucose supply is limiting, such as in hypoxemia, ischemia, or extreme hypermetabolism, the lumped constant may increase considerably (Suda et al., 1990). In such cases it is important to estimate the value of the lumped constant locally, possibly from measurements of the brain uptake of methylglucose (Gjedde et al., 1985; Dienel, Cruz, et al., 1991). Less than extreme changes in glucose metabolism have only small effects on the lumped constant (Dienel, Nakanishi, et al., 1991).

E. Tissue Heterogeneity

The kinetic model of Figure 6 was specifically designed for a homogeneous tissue within which rates of blood flow, glucose metabolism, transport of glucose and [^{18}F]FDG between tissue and blood, and so on are uniform. As described in Section IIG, the limited spatial resolution of the PET scanner assures that most brain regions, especially cortical regions, contain mixtures of gray and white matter. Figure 8 shows total activity in one image plane following a pulse of [^{18}F]FDG and a magnetic resonance image (MRI) of the same subject. A small region of interest 2 × FWHM

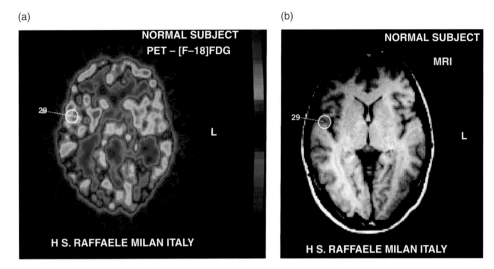

FIGURE 8 Total activity in one image plane obtained 115–120 min following (a) a pulse of [^{18}F]FDG and (b) a magnetic resonance image (MRI) of the same normal subject. PET image was obtained with a Siemens/CPS whole-body positron emission tomograph (model 931/04-12) with spatial resolution in the image plane of 8 mm FWHM. A small region of interest 2 FWHM in diameter in the superior temporal gyrus is shown; both gray and white matter are included. (Reproduced from Schmidt et al., 1992.)

in diameter, the minimum object size for accurate quantification (Hoffman et al., 1979), is shown in the superior temporal gyrus; it includes both gray and white matter. Because gray and white matter are kinetically heterogeneous, their unavoidable admixture in PET data must be taken into account in the kinetic model (Figure 9).

Ideally, one would like to determine the rate of glucose use separately in the gray and white matter in each region of interest, and various methods to correct for partial volume effects[4] have been proposed. These methods usually require a co-registered structural image (e.g., MRI) and may require some additional assumptions about the composition of the volume. An alternative is to determine the average value of the rate of glucose use or the average values of the kinetic parameters, properly weighted for the relative masses of the constituents in the mixed tissue. Use of the kinetic model designed for homogeneous tissues to analyze PET data results in errors in estimates of the rate constants and $rCMR_{glc}$; the magnitude of the errors depends on the total time over which the data are collected. Estimates of k_2^* and k_3^* initially decline with time and become constant and equal to their true mass-weighted average values only after the experimental periods have been extended long enough for the tissue pools of [^{18}F]FDG to equilibrate with the arterial plasma (Kuwabara et al., 1990; Schmidt et al., 1991, 1992). Overestimation of the rate constants k_2^* and k_3^*, because it leads to underestimation in the amount of free [^{18}F]FDG in

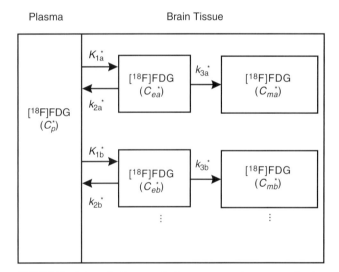

FIGURE 9 Schematic diagram of a heterogeneous tissue as a collection of multiple homogeneous subregions; the kinetic model of each homogeneous subregion is as presented in Figure 6. C_p^*, represents the concentration of [^{18}F]FDG in the arterial plasma; C_{ea}^*, C_{eb}^*, . . . represent the concentrations of [^{18}F]FDG in the exchangeable pool of each homogeneous subregion of the heterogeneous brain tissue; and C_{ma}^*, C_{mb}^*, . . . represent the concentrations of metabolites of [^{18}F]FDG in the corresponding homogeneous subregions. The PET signal includes the sum of the activities in all the subregions, each weighted by its relative mass, plus the concentration of [^{18}F]FDG in the blood in the brain. Rate constants for each homogeneous subregion are as defined in Figure 6. (Adapted from Schmidt et al., 1991.)

the exchangeable pool in the tissue (second term in the numerator of Eq. 17), results in overestimation of the rate of glucose use. In addition, if the homogeneous kinetic model is modified to include the possibility of loss of [^{18}F]FDG-6-P and the resultant model is applied to

---

[4]The partial-volume effect is called this because only part of the total volume is occupied by the tissue of interest (e.g., gray matter); bias is introduced into its measurement by the presence of other tissues within the volume.

heterogeneous tissues, estimates of the rate constant for product loss are artifactually high due to the tissue heterogeneity, even in the absence of any real product loss, at least up to 120 min following a pulse of [^{18}F]FDG (Schmidt et al., 1991, 1992; Lucignani et al., 1993). Overcorrection for product loss results in further overestimation of $rCMR_{glc}$. The optimal period for obtaining the best estimates of the mass-weighted average kinetic model rate constants is the period beginning at the time of tracer injection and continuing for a minimum of 60 min, but not longer than 120 min (Lucignani et al., 1993). Such prolonged dynamic scanning periods, however, are impractical on a routine basis, and population average rate constants must be used with Eq. (17) to determine $rCMR_{glc}$. In this case the scanning time to determine the final brain activity, $C_t^*(T)$, should be delayed as long as possible following the pulse of [^{18}F]FDG (but not longer than 120 min) to minimize the effects of errors in the rate constants. Alternatively, tissue heterogeneity can be modeled explicitly, or methods that do not depend on tissue homogeneity can be employed.

1. Compartmental Model of Tissue Heterogeneity

A compartmental model that explicitly takes tissue heterogeneity into account has been developed (Schmidt et al., 1991, 1992). In the first 120 min following the pulse of [^{18}F]FDG, the model can be used to describe accurately the time course of radioactivity in a heterogeneous tissue without any assumption of loss of metabolic product. Determination of the parameters of the model, however, requires dynamic scanning from the time of tracer injection until after all tissue precursor pools have equilibrated with the arterial plasma, a scanning period of not less than 60 min (Schmidt et al., 1992). For this reason, the model is of less practical use for determination of $rCMR_{glc}$ than the methods described next.

2. Multiple-Time Graphical Analysis Technique

An alternative to the use of specific compartmental models is the multiple-time graphical analysis technique, the Patlak plot (Patlak et al., 1983; Patlak and Blasberg, 1985), that applies equally well to heterogeneous as to homogeneous tissues. Application of the Patlak plot is restricted to an interval of time in which (1) the free [^{18}F]FDG in all tissue-exchangeable pools in the region of interest (e.g., in gray and white matter) has equilibrated with the [^{18}F]FDG in the plasma, and (2) there is no product loss. During this interval, the apparent distribution space for total ^{18}F in the tissue, $C_t^*(T)/C_p^*(T)$, when plotted against the normalized integrated tracer concentration in arterial plasma, $\int_0^T C_p^*(t)dt/C_p^*(T)$, increases linearly with slope K directly proportional to $rCMR_{glc}$ (Figure 10). The rate of glucose use can then be estimated as:

$$rCMR_{glc} = KC_p / LC \quad (18)$$

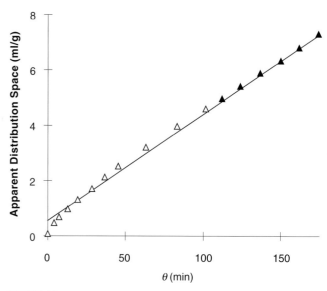

FIGURE 10 Patlak plot of data shown in Figure 7. Activity was measured in the visual cortex of a normal human subect. The normalized integrated tracer concentration in arterial plasma, $\theta(T) = \int_0^T C_p^*(t)dt/C_p^*(T)$, is plotted on the abscissa and the apparent distribution space for total ^{18}F in the region, $C_t^*(T)/C_p^*(T)$, is plotted on the ordinate. Following equilibration of all precursor pools in the tissue with the arterial plasma, the plot becomes linear as long as there is no product loss; the slope of the linear portion of the graph is directly proportional to $rCMR_{glc}$. The unbroken line is the best fit to the data measured 60–90 min following a pulse of [^{18}F]FDG (solid triangles). Open triangles represent measurements not included in the fit. (Data from Lucignani et al., 1993.)

where LC is the lumped constant of the method and C_p is the arterial plasma glucose concentration. Because either lack of equilibration with the plasma of any precursor pool in the region of interest or loss of metabolic products may cause nonlinearity of the graph, application of this technique requires identification of an appropriate linear segment of the curve. There is evidence that there are slowly equilibrating compartments until approximately 45 min after a pulse of [^{18}F]FDG, but in most tissue regions the estimated slope K, and hence the estimated value of $rCMR_{glc}$, is relatively stable in the interval between ~45 and ~105–120 min following the pulse of [^{18}F]FDG (Lucignani et al., 1993).

3. Spectral Analysis

A second alternative to the use of a fixed kinetic model is the spectral analysis technique that is based on a more general linear compartmental system (Cunningham and Jones, 1993; Cunningham et al., 1993; Turkheimer et al., 1994, 1998). The technique provides an estimate of the minimum number of compartments necessary to describe the kinetics of the system and hence does not require the number of compartments to be fixed a priori. It also provides estimates of the rate constant of trapping of tracer in the tissue, and the amplitudes and decay constants of the reversible components. This information can be used for subsequent specification of a kinetic model, or it can be used to estimate selected parameters of the system that do not depend on the specific model configuration, for example, the rate of trapping of tracer, the unidirectional clearance of tracer from blood to tissue, the total volume of distribution of the tracer, or mean transit times (Cunningham *et al.*, 1993). Spectral analysis does not require that the system be in steady state.

The spectral analysis technique assumes that the total concentration of tracer in the field of view can be described by the equation:

$$C_t^*(T) = \alpha_o \int_0^T C_p^*(t)dt + \sum_{j=1}^n \alpha_j \int_0^T C_p^*(t) e^{B_j(T-t)} dt + V_B C_B^*(T) \quad (19)$$

where C_t^*, C_p^*, C_B^*, and V_B are as defined in Section IIIA; n is the number of detectable compartments in the tissue; and α_j and β_j are scalar parameters that depend on exchange of tracer among the compartments. For radiotracers other than [^{18}F]FDG C_p^* in Eq. (19) may represent arterial blood concentration, arterial plasma concentration, or metabolite-corrected, nonprotein-bound plasma concentration of activity depending on which fraction of radioactivity in the blood is exchangeable with the tissue. C_B^*, on the other hand, always relates to the total radioactivity in the blood pool. The exponents β_j are assumed to be nonpositive real numbers and the coefficients α_j to be nonnegative real numbers. When the spectral analysis technique is applied to the analysis of [^{18}F]FDG data, the rate of glucose use is directly proportional to the coefficient α_0:

$$rCMR_{glc} = \alpha_0 C_p / LC \quad (20)$$

where C_p and LC are as previously defined. It is often necessary to apply a noise filter in order to obtain an accurate estimate of α_0 (Turkheimer *et al.*, 1994).

Spectral analysis does not apply to all linear compartmental systems, but a class of systems to which it does apply has been identified (Schmidt, 1999). These include noncyclic[5] strongly connected[6] systems that have an exchange of material with the environment confined to a single compartment, as well as certain noncyclic systems with traps. Furthermore, a noninterconnected collection of compartmental systems meets the conditions for application of spectral analysis when the individual subsystems making up the collection individually meet the spectral analytic conditions. This property renders the spectral analysis technique particularly useful for tracer kinetics studies in brain with PET where the limited spatial resolution leads to inclusion of kinetically dissimilar gray and white matter tissues in the field of view (See Figure 11). Spectral analysis is applicable to the one-compartment kinetic model used for the measurement of

---

[5] A cyclic compartmental system is one in which it is possible for material to pass from a given compartment through two or more other compartments and back to the starting compartment.

[6] In strongly connected systems it is possible for material to reach every compartment from every other compartment.

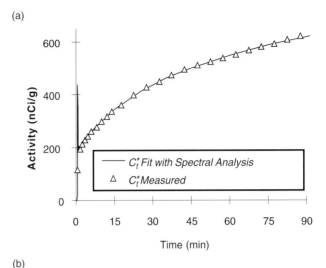

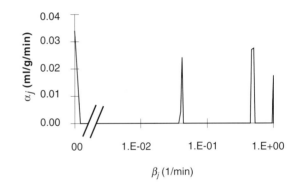

FIGURE 11 Spectral analysis of data shown in Figure 7. Activity was measured 0–90 min following a pulse of [^{18}F]FDG in the visual cortex of a normal human subject. (a) Total activity in the region, measured (open triangles) and estimated (line). Early spike is due to radioactivity in the blood pool. (b) Components detected by the analysis. The exponent β_j is plotted on the abscissa and the coefficient α_j on the ordinate. In addition to a blood component (not shown), a component corresponding to trapped tracer ($\beta_j = 0$) and three equilibrating components were found. The equilibrating components have been associated with gray matter ($\beta_j = 0.5$/min), white matter ($\beta_j = 0.04$/min), and noise in the input function ($\beta_j = 1.0$/min) (Data from Turkheimer *et al.*, 1994.)

cerebral blood flow with H$_2$^{15}O and PET (Huang *et al.*, 1983; Herscovitch *et al.*, 1983), the two-compartment model for measurement of glucose metabolism with [^{18}F]FDG, and some three-compartment models used in receptor–ligand binding studies, such as those designed to assess muscarinic cholinergic receptor binding by use of [^{11}C]scopolamine (Frey *et al.*, 1985) or to assess benzodiazepine receptor density by use of [^{11}C]flumazenil (Koeppe *et al.*, 1991). The heterogeneous counterparts of each of these models can also be analyzed with spectral analysis due to the additivity of the components in Eq. (19) and the observation that the activity in a heterogeneous mixture of tissues is simply the weighted sum of the activity in the constituent tissues making up the mixture.

IV. RECEPTOR–LIGAND MODELS

A. Three-Compartmental Model

The simplest of compartmental models applied to receptor–ligand studies postulates two tissue compartments. These two tissue compartments along with a plasma compartment[7] are arranged in series. The tracer is delivered—typically by intravenous (iv) injection—into the blood, and it traverses the free compartment on its way to interacting with the receptor. If these three states of the radioligand are inadequate to describe the data, sometimes a third tissue compartment is introduced (Figure 12), which is termed the nonspecifically bound compartment. The bound and nonspecific compartments are distinguished as follows. Specifically, a bound ligand (unlike a nonspecifically bound one) is both saturable and displaceable by nonradioactive molecules of the same tracer. The rate of change in radioactivity concentration in each tissue compartment in Figure 12 is given by the ordinary differential equations that describe the flux of ligand into (and out of) the respective compartments:

$$\frac{dF}{dt} = K_1 P - (k_2 + k_3)F + k_4 B - \lambda F - k_5 F + k_6 N \quad (21)$$

$$\frac{dB}{dt} = k_3 F - k_4 B - \lambda B \quad (22)$$

$$\frac{dN}{dt} = -\lambda N + k_5 F - k_6 N \quad (23)$$

where F, B, and N are the time-varying concentrations of tracer in the free, bound, and nonspecific tissue compartments, respectively. That is, $F = F(t)$, $B = B(t)$, and $N = N(t)$. $K_1, k_2, \ldots, k_6$ are the apparent first-order rate constants relating the transfer of ligand between the various compartments. These ks are the model parameters to be estimated from experimental data. The parameters, k_3 and k_4, are apparent first-order rate constants that characterize the binding and dissociation of the ligand to and from the receptor, respectively. The rate constant, λ, for radioactive decay, refers to the rate constant of conversion of the isotope labeling the ligand. As the isotope decays (with half-life = $\ln(2)/\lambda$) to a nonradioactive species, the labeled molecule is effectively lost from the system. P is the (measured) concentration of radioactive ligand in the plasma. P refers only to the concentration of native ligand molecules and not to any radioactive metabolites. The preceding model, in a

---

[7]Strictly speaking, the plasma is not a compartment of the model. The concentration of tracer in the plasma is measured independently and thus applied to the tissue model as a known input function.

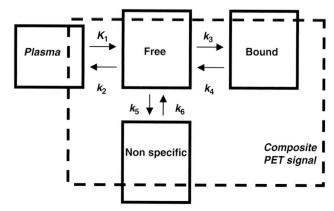

FIGURE 12 General three-compartment model. Some fraction of the radioactivity in each of the plasma, free, bound (and nonspecific) compartments sum to give the radioactivity in a PET image. Hence, the composite PET signal in a region or pixel includes a small fraction of the plasma signal and a large fraction of the tissue compartments, respectively. The model parameters $K_1, \ldots, k_6$ reflect exchange rates for a labeled ligand moving between the various compartments.

slightly different form, was introduced into the PET literature by Mintun *et al.* (1984).

One of the innovations of the original Mintun formulation was to collapse the nonspecific compartment into the free compartment. Mathematically, this can be justified if the rate constants for nonspecific binding, k_5 and k_6, are fast relative to the other rate constants of the system. In this case, ligand in the nonspecific compartment is defined as being in rapid equilibrium with the free compartment. If this condition is satisfied, then at any time a fixed fraction of the free plus nonspecific compartment is the effective free pool available for receptor binding. This fraction, $f_2 (= k_6/[k_5 + k_6])$, may not be strictly identifiable (see later discussion) in a receptor-rich region alone, although it may be estimatable from data in a receptor-free region where only two other parameters (K_1 and k_2, in addition to k_5 and k_6) are needed to describe the data.

1. Output Equation

Because it takes a finite frame time to acquire enough scintillation events to be able to reconstruct a PET image, the measured PET data comprise average numbers of counts over discrete intervals. These intervals need not be (and generally are not) constant. Counts are converted to concentration by way of a calibration. Once the model has been solved for the instantaneous activity in each compartment (F, B, and N), these concentration curves are summed and integrated to generate a theoretical output that is compatible with the measured PET activity over each acquisition time frame, $[t_i, t_{i+1}]$, as follows:

$$\text{PET}[t_i, t_{i+1}] = \frac{1}{\Delta t_i} \int_{t_i}^{t_{i+1}} (\varepsilon_v V + \varepsilon_f F + \varepsilon_b B + \varepsilon_n N) dt \quad (24)$$

The weights, ε_i, that premultiply each compartment concentration in Eq. (24) are the respective volume fractions. Usually the vascular volume fraction, ε_v is small, and the tissue volume fractions are set at unity or $(1-\varepsilon_v)$. This is indicated schematically in Figure 12: The composite PET signal (enclosed by a dashed-line rectangle) includes a small fraction of the radioactivity in the plasma compartment but a large fraction of the activities in the three tissue compartments. A further subtlety of Eq. (24) is that V refers to the vascular (i.e., whole-blood) contribution to the PET activity. Whereas the plasma concentration is the concentration that must be considered, and modeled as $P(t)$, when considering the driving force for a ligand crossing the blood–brain barrier, the whole-blood activity (a subset of P) is what determines the vascular contribution to the PET signal. We can use numerical methods to solve Eqs. (21)–(23) over time.

A group of simulations of the dynamics in the free, bound, and nonspecific compartments following bolus injection of a generic [^{11}C]-labeled ligand, is shown in the panels of Figure 13. As indicated by the icons at the top of each column, the simulations represent, respectively: two tissue compartments (free and bound) with reversible binding, three tissue compartments (free, bound, nonspecific), and two tissue compartments (free and bound) with irreversible binding. The middle row of simulated data is not decay corrected. Therefore, these represent the type of signal that is actually measured by the PET scanner. The bottom row of simulations certains the decay-corrected concentrations in the compartments for the corresponding uncorrected simulations. The half-life of [^{11}C] is 20.2 minutes, so by 60 minutes almost three half-lives have elapsed. All three simulations are generated from the same plasma input function. Despite the differences in the simulations and in the underlying compartment concentrations, the total radioactivity (not corrected for decay) directly related to what is observed by the PET scanner in each case is nearly the same. In the more realistic case where noise is present, the curves will be even less distinct. The similarity of the uncorrected data for all three model topologies is meant to indicate the potential

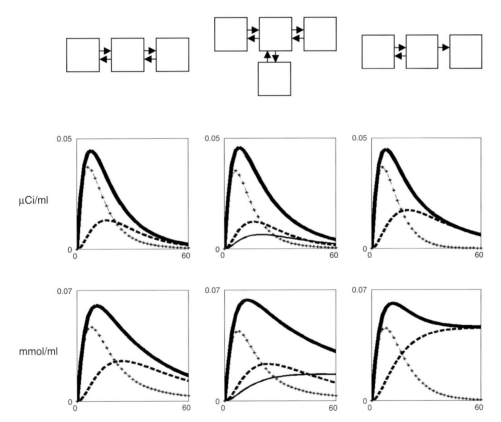

FIGURE 13 Simulations of two- and three-tissue compartment models commonly in use in PET. All simulations were generated with the same plasma input function, $P(t)$, not shown. Top row indicates model topology and number of parameters (indicated by arrows). Middle row shows simulations with radioactive decay (analogous to the signal measured by PET scanner). Bottom row shows respective simulations after correction for radioactive decay. Thin line with symbols (-+-+-+) is the free concentration; the heavy dashed line (- - -) is the bound concentration; thin solid line is the nonspecific binding; and heavy solid line is the weighted sum of all radioactivity (i.e., the continuous PET curve). Notice the similarity of the PET curves prior to decay correction despite differences in the simulations. Curves produced with COMKAT (Muzic and Cornelius, 2001).

difficulty of determining *a priori* the correct number of compartments or identifying the model parameters. In fact, the problem of parameter identifiability is seen to be even more difficult than suggested in Figure 13 when one considers that the part of the curve that indicates irreversibility of ligand binding is at the late end of the curve. Unfortunately, because the error in the PET data obeys Poisson statistics (the variance of the data is equal to the mean), these same late-time data with few counts are the least reliable. The issue of parameter identifiability is addressed further later.

2. Measuring the Blood Curve

The input or driving function P represents the amount of tracer ligand that is presented over time to the tissue of interest (i.e., the brain). It is often measured via the radioactivity counted in blood plasma samples drawn from the radial artery. The model given in Eqs. (21)–(24) and diagrammed in Figure 12 deals only with the behavior of a unique tracer molecule, such as [^{11}C]raclopride. The rate constants pertain, specifically, to this molecule. Therefore, if metabolism of the radioligand in the periphery leads to labeled metabolites in the plasma, we must invoke at least one correction and one assumption. First, the nonnative tracer species must be removed either physically from the blood samples or mathematically from the resulting radioactivity counts. Second, it must be assumed that the labeled metabolite(s) does(do) not cross the blood–brain barrier (BBB). Thus, one consideration for choosing among candidate tracers may be the generation (or not) of labeled metabolites. In some cases, the production of labeled metabolites may be inevitable. In the case of [^{18}F]fluorodopa, complicated models have been introduced to account for reentrance of blood-borne metabolite species to the tissue of interest (e.g., the brain) and, in doing so, their contributions to the PET signal (Huang *et al.*, 1991).

B. Modeling Saturability

To this point, we have not introduced any term into the model that reflects the saturable nature of the bound-ligand compartment. That is, we have imposed no explicit upper bound on B. And yet, a unique characteristic of receptor–ligand interactions is that, like enzyme–substrate reactions, they are saturable. The inclusion of an explicit receptor density term, B'_{max}, requires an elaboration of Eqs. (21) and (22). In Eqs. (25)–(27), we expand k_3 as $k_{on}(B'_{max} - B/SA)$:

$$\frac{dF}{dt} = K_1 P - (k_2 + k_5)F - k_{on}(B'_{max} - B/SA)F + k_{off}B + k_6 N - \lambda F \quad (25)$$

$$\frac{dB}{dt} = k_{on}(B'_{max} - B/SA)F - k_{off}B - \lambda B \quad (26)$$

$$\frac{dN}{dt} = k_5 F - k_6 N - \lambda N \quad (27)$$

where k_{off} is equivalent to our previous k_4; SA is the specific activity, which declines in time according to the radioactive decay rate, λ; and k_{on} is the rate constant for a bimolecular association of free ligand and available receptors. Thus, the binding rate is equal to the product of the concentrations of free ligand, F, available receptor sites, $(B'_{max} - B/SA)$, and rate constant, k_{on}. At the time of injection ($t = 0$), the bound ligand B is zero, therefore B'_{max} represents the concentration of available receptors in the absence of exogenous ligand. As additional receptors become bound to ligand molecules, the availability of free receptors drops and saturation is approached asymptotically. Because availability can be affected by either labeled or unlabeled ligand, normalization by SA converts B to the concentration of total ligand (i.e., Radioactivity concentration/Radioactivity per moles = Molar concentration). This form of the receptor–ligand model, which was first proposed explicitly by Huang and colleagues (Bahn *et al.*, 1989; Huang and Phelps, 1986; Huang *et al.*, 1989) is analogous to the bimolecular interaction between substrate and available binding sites on an enzyme.

C. Parameter Identifiability: Binding Potential or Other Compound Parameters

Despite the inclusion of an explicit B'_{max} term in the model representing the baseline level of available receptor sites, this value may not be discernible from the acquired data. In other words, it may not be estimated reliably. Consider that for a single bolus injection of sufficiently high SA ligand, the bimolecular binding term in Eqs. (25) and (26), $k_{on}(B'_{max} - B/SA)F$, reduces to $k_{on}(B'_{max})F$. That is, $B'_{max} \gg B/SA$. Because k_{on} and B'_{max} always appear as a product, never separately, they are said not to be identifiable and the model defaults to Eqs. (21)–(23). Attempting to estimate both k_{on} and B'_{max} in this case would be akin to trying to estimate m and n from data with the overparameterized model of a line $y = mnx + b$. Mintun and co-authors make this point in their 1984 paper, saying that k_{on} and B'_{max} are not separable from a single bolus experiment under tracer (i.e., high SA) conditions. For that reason, they choose to estimate the compound term, B'_{max}/K_D, which they refer to as *binding potential*; Koeppe *et al.* (1991) recommend further model simplifications when it is not possible even to identify B'_{max}/K_D, explicitly. Sometimes the data from a receptor–ligand study can be fit to a two-compartment (one-tissue) model (i.e., plasma and free) and inclusion of an explicit bound compartment is not justified. This will be true if the binding and dissociation rate constants for the ligand, k_3 and k_4, are much faster than the influx and efflux constants, K_1 and k_2. When this is true, the estimated k_2 will represent an apparent efflux rate constant, [$k_2'' = k_2/(1 + k_5/k_6 + B'_{max}/K_D)$]. This parameter, may still reflect the binding potential, as indicated by the B'_{max}/K_D term, and may be the only reliably estimatable index of binding. Koeppe and colleagues (1991, 1994) have done much work to characterize the kinetics of flumazenil and other muscarinic cholinergic ligands. They point out that quantification is difficult (as with these ligands) if the dissociation rate constant k_{off} is slow and cannot be reliably distinguished from zero. Conversely, quantification of these ligands' binding rate constants is difficult and highly sensitive to changes in blood flow when binding is very rapid relative to the rates of transport across the BBB (i.e., K_1 and k_2) (Zubieta et al., 1998).

There are many instances in research when binding potential (BP) is a satisfactory indicator of (available) receptor level. In these cases, BP, reported in table form by brain region or as a parametric map, is a reasonable end point for a functional imaging study. There have been a number of studies that monitor change in apparent BP as an indicator of change in the level of an endogenous competitor to the injected ligand (see discussion of neurotransmitter activation studies in Sections IV D and VII). A number of easy and robust methods for calculating this parameter are discussed here. However, implicit in the adoption of BP as a surrogate B'_{max} is the assumption that the affinity, K_D, of the ligand for the receptor is constant across subjects or treatments. For those cases in which changes in affinity cannot be ruled out or where some other mechanistic question is at stake, more than a single bolus injection is indicated. Delforge et al. (1989, 1990) proposed an experimental protocol involving three sequential bolus injections of varying specific activities. At least one of those injections also contains a nontrace amount of cold (i.e., unlabeled) ligand. In such a protocol, the number of available receptors is significantly perturbed, and it is this process of modulating availability over a broad range of values that serves to decorrelate k_{on} from B'_{max}. Delforge and co-authors show via experiments and simulations that it was possible to estimate unambiguously seven model parameters (including k_{on}, B'_{max}, and k_{off}) using a three-injection protocol. Thus, the individual parameters are said to be identifiable. An example of the data from a multiple-injection protocol, in which [^{11}C]-CFT (a dopamine transporter ligand) was administered to a rhesus monkey, is given in Figure 14. The authors of the study (Morris, Babich, et al., 1996) were able to estimate five parameters simultaneously (nonspecific binding parameters k_5 and k_6 had to be fixed based on cerebellum data). In designing experiments, one must beware that protocols that rely on nontrace doses of ligand may not be feasible in psychiatric patients—particularly children—because the large mass doses of ligand are likely to have pharmacological effects.

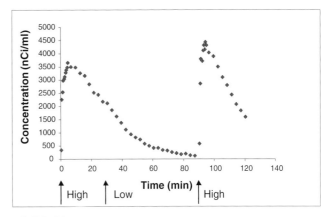

FIGURE 14 Dynamic PET activity from region drawn on striatum of rhesus monkey. Three-injection protocol used is described in Morris, Babich, et al. (1996a). Injections of [11C]-CFT were made at 0, 30, and 90 minutes (indicated by arrows). Second injection was low specific activity causing net displacement of labeled ligand from receptor by unlabeled ligand. Notice the difference in responses after the first and second high-specific-activity injections because of the presence of residual cold ligand on board. Data are not decay-corrected.

1. Modeling Multiple-Injection Data—Two Parallel Models

To analyze data from their multiple-injection experiments, Delforge and colleagues (1989, 1990) proposed a further modification of the three-compartment model. Because these protocols depend on at least one nontrace[8] injection of ligand, the researchers propose parallel models for the labeled and unlabeled species. The model can be described as follows.

$$\frac{dF^j}{dt} = K_1 P^j - (k_2 + k_5)F^j - (k_{on}/V_R)(B'_{max} - B^h - B^c)F^j + k_{off}B^j + k_6 N^j - \lambda^j F^j \quad (28)$$

$$\frac{dB^j}{dt} = \qquad\qquad (k_{on}/V_R)(B'_{max} - B^h - B^c)F^j - k_{off}B^j \qquad - \lambda^j B^j \quad (29)$$

$$\frac{dN^j}{dt} = \qquad k_5 F^j \qquad\qquad\qquad\qquad - k_6 N^j - \lambda^j N^j \quad (30)$$

where the superscript j refers to either hot ($j = h$) or cold ($j = c$) ligand, and λ^j is the radioactive decay constant for the isotope when j refers to the hot ligand and 0 when j refers to cold (i.e., there is no radiodecay of cold ligand.)

The Delforge nomenclature employs the compound term k_{on}/V_R as the apparent association rate constant. V_R is the volume of reaction of the receptor but k_{on}/V_R is treated as one indivisible parameter. If the microenvironment surrounding the receptor is unusually inaccessible to the ligand, the apparent binding rate will differ from the true binding rate (see Delforge et al., 1996 for a thorough discussion of the reaction-volume concept). In practice, the number of estimable

---

[8] *Nontrace* here means an amount great enough to have visible biological or pharmacological effects, that is, not insignificant, biologically.

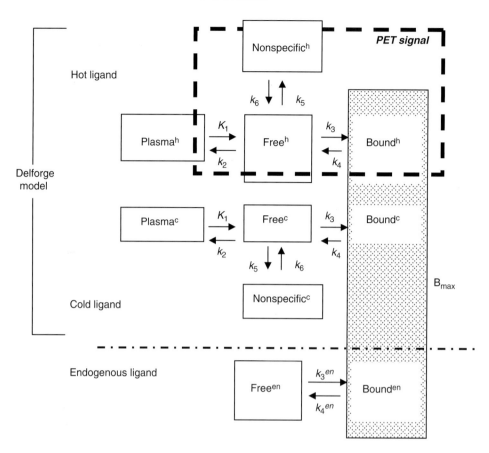

FIGURE 15 Schematic corresponding to model Eqs. (28)–(31). Notice that the fixed number B_{max}, the total population of possible ligand binding sites available to hot, cold, or endogenous ligand, is the vehicle by which the three parallel models are coupled. That is, the term B_{max}, more specifically $(B_{max}-B-B^{en}-B^c)$, appears in the mass balances for each of the Free and Bound compartments.

parameters in the model is seven. (From a parameter estimation standpoint, k_{on}/V_R is no different from the k_{on} term introduced in earlier equations.) No new parameters have been added to the model given earlier (Eqs. 24–26) because the hot and cold ligands are identical, biologically. In essence, the Delforge model is two models joined at the hip. One model is for the cold species and one is for the hot (i.e., labeled). The two models are coupled by the bimolecular binding term that occurs in both. The model of Eqs. (28–30) is diagrammed as the top two-thirds of Figure 15 (i.e., everything associated with hot and cold ligand.)

There is an additional experimental burden in performing multiple-injection experiments. Two plasma input functions must be measured. The first, P^h, is the standard curve of radioactivity in the plasma, P, that has been mentioned previously. The other, P^c, is the concentration (in pmol/ml) of cold ligand in the plasma. Because this is not readily measurable during the experiment, it must be reconstructed from the available plasma radioactivity measurements and knowledge of the specific activities of each injection. (See Morris, Babich, *et al.*, 1996, App.). Despite its greater complexity, the Delforge approach maximizes the number of model parameters that can be properly identified (Morris, Alpert, *et al*, 1996). The identifiability of parameters is closely tied to the concept of sensitivity coefficients. For a parameter from a given model to be identifiable from experimental data, the derivatives of the objective function (which is being minimized when fitting data) with respect to the different parameters must be linearly independent, that is, distinct. In other words, each parameter must contribute something distinct to the model curve. If a change in one parameter can be confused as a change in another parameter (as could be the case in the simulations of Figure 13) then it will not be possible to uniquely identify all the parameters. The strength of the Delforge approach, on a theoretical level, is that it perturbs the system so that we observe it at multiple set points. Figure 16 is a simulation of the Delforge model shown schematically as the top two-thirds Figure 15. Notice that the (hot) bound compartment concentration reaches a peak after the first high-specific-activity injection at time 0, but cannot reach the same peak after the second injection of low *SA* material has caused a significant number of receptor sites to be bound to cold exogenous ligand.

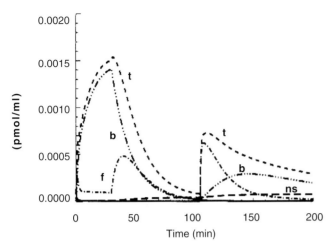

FIGURE 16 Simulation using the model shown in Figure 15 of the compartmental (f = free, b = bound, ns = nonspecific, and t = total tissue) concentrations of radioligand during a three-injection experiment. such simulations are able to fit the type of data shown in Figure 15 with an identifiable set of model parameters. In this simulation, the injections are at 0, 30, and 100 minutes. the response to high-specific-activity injections at 0 and 100 min are different because of the presence of residual cold ligand from the injection of low-specific-activity ligand at 30 minutes. See Morris, Alpert, *et al* (1996).

In fitting this data, which is more information-rich than those from a single-injection experiment, the different influences of each parameter on the model become apparent. For a discussion of the multiple-injection approach, sensitivity coefficients and why the multiple-injection approach works in some cases (See Morris, Bonab *et al.*, 1999). Unfortunately, the multiple-injection approach is experimentally and logistically much more complicated than a single-bolus study. Nevertheless, this complexity might be justified in the preliminary development and exploration of a new ligand when precise estimates of all parameters are needed. The parameter estimates resulting from such a multiple-injection study would then be used to develop and analyze a simpler clinical protocol.

D. Neurotransmitter Changes—Nonconstant Coefficients

Each of the foregoing models, regardless of complexity, is founded on some common assumptions. One is that the model parameters are constant over the duration of the experiment. This is a reasonable assumption if the system itself is at steady state. Whereas [^{11}C]-raclopride levels may rise and fall as it is taken up and eliminated from the tissue following a bolus injection, dopamine levels remain unchanged over the time scale of interest. The rate constants that are determined by the overall state of the system remain unchanged. Anything that violates these assumptions may invalidate the use of the constant-coefficient models given earlier and must be investigated. For example, the consequences of unwanted changes in blood flow and their impact on the estimation of raclopride binding have been investigated using both simulations (Frost *et al.*, 1989; Logan *et al.*, 1994; Laruelle *et al.*, 1994) and an experimental perturbation—hyperventilation (Logan *et al.*, 1994). In the hyperventilation study, Logan *et al.* (1994) find that the compound parameter, distribution volume ratio (see Section VC), was insensitive to absolute flow level and thus was perhaps a more robust parameter than individual rate constants (e.g., k_3, k_4, or k_{on}). In 1991, Logan and colleagues explored the consequences of changes in dopamine on the measurement of binding parameters of a D2 receptor ligand, methylspiroperidol.

There has been growing interest in detecting and quantifying transient changes in neurotransmitter concentrations that may be useful for understanding the etiology of neuropsychiatric diseases (see Laruelle *et al.*, 1996; Breier *et al.*, 1997 for evidence that schizophrenia is associated with abnormal amphetamine-induced dopamine release). The potential for such a measurement exists because the endogenous ligand (e.g., dopamine) competes with the exogenous ligand for binding to available receptor sites. Any alteration in the steady-state level of this ever-present competition could alter the PET signal. To model a system in which the neurotransmitter concentration is not constant, we must again extend the models developed here. Just as the Delforge model (Eqs. 28–30) extended its antecedents by adding terms for labeled and unlabeled exogenous ligand species, the changing neurotransmitter model must add terms for the endogenous ligand species. Morris and co-workers (1995) introduce a differential equation into the receptor–ligand model to describe the binding and release of endogenous species from the receptor:

$$\frac{dB^{en}}{dt} = \left(k_{on}^{en}/V_R^{en}\right)\left(B_{max} - B^h - B^c - B^{en}\right)F^{en} - k_{off}^{en}B^{en} \quad (31)$$

where the superscript en refers to the endogenous species. Note that B_{max} is the concentration of *all* the receptor sites (available and not) and the bimolecular binding terms for the other species (hot and cold tracer ligand) must be modified accordingly. F^{en} is a time-varying concentration of endogenous chemical available for binding to the receptor. k_{on}^{en}/V_R^{en} is the apparent binding rate constant for the endogenous ligand. Equation (31) can be solved simultaneously with the multiple-species system of Eqs. (28)–(30). A similar approach is taken by Endres and colleagues (1997), who retain the format of Eqs. (21)–(23) but introduce a time-varying binding rate parameter:

$$k_3 = k_{on}B_{free}(t)$$

where $B_{free}(t)$ is $B_{max} - B^h - B^c - B^{en}$. B_{max} is still constant, but B^h, B^c, and B^{en} can all vary with time. This enhanced model (as described by Eqs. 28–31) is diagrammed by the entire network of compartments in Figure 15.

There have been two experimental approaches to the problem of quantifying neurotransmitter changes.[9] The first is a natural extension of the studies already mentioned. It is based on the comparison of PET signals from two bolus injections of radiotracer. One bolus study is performed while neurotransmitter is at steady state; a second bolus study is carried out while endogenous neurotransmitter is undergoing an induced change. The alternative experimental approach also involves a perturbation of the neurotransmitter but it is done only after the radiotracer has been brought to a steady state in the tissue by a combination of bolus plus infusion administrations. Let us look briefly at this steady-state method.

1. Steady-State Technique

The bolus-plus-infusion method was developed for receptor–ligand characterization studies by Carson and colleagues at the National Institutes of Health (Kawai *et al.*, 1991; Carson *et al.*, 1993). The beauty of this method is that by properly combining a bolus with a constant infusion of radiotracer, the radiotracer level in the brain can be brought to a constant level in a minimum of time. With ligand levels in the tissue no longer varying in time, the derivatives on the left-hand sides of Eqs. (21)–(23) disappear and the models reduce to a set of algebraic equations. In fact, no modeling is required. For many ligands this technique is ideal for reliably estimating distribution volume (which is a linear function of binding potential). The steady-state method has been used quite extensively in SPECT studies (Laruelle *et al.*, 1994), which, because of long-lived isotopes, can be applied to data acquired the day following a bolus injection (no infusion needed). The bolus-plus-infusion method is easily extended to neurotransmitter change experiments by perturbing the endogenous neurotransmitter pharmacologically once the steady state level of radiotracer has been achieved. The measured parameter that has been used to indicate neurotransmitter alteration is the change in binding potential, ΔBP. Figure 17 shows a change in steady-state level of raclopride activity in the striatum of a monkey before and after amphetamine perturbation. The activity in the cerebellum (devoid of D2 receptors) is unaffected (Endres *et al.*, 1997).

2. Experimental Considerations

Whether one chooses the two-bolus or the constant infusion approach, there are two important questions to be answered: (1) Can the change in PET signal be reliably detected given the inherent variability in the method? (2) Can

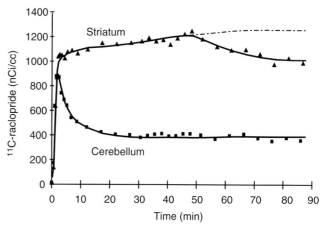

FIGURE 17 Bolus-plus-infusion method for detecting changes in neurotransmitter level. Amphetamine-induced perturbation of raclopride binding at 45 min. Top curve shows depression in [11C]-raclopride signal in striatum following heightened competition between raclopride and dopamine for binding at D2 receptors. Perturbation was 0.4 mg/kg injection of amphetamine given to rhesus monkey. Bottom curve shows no effect of dopamine modulation on cerebellum. Dotted continuation of top curve shows model-predicted continuation of curve absent amphetamine injection. Data are decay-corrected. (Data from Endres *et al.*, 1997.)

the apparent change in some parameter of interest be unambiguously linked to changes in neurotransmitter, given the inherent variability and covariability of the parameters? With regard to question 1, Endres *et al.* (1997) use simulations to show that in the case of the bolus-plus-infusion technique, ΔBP is proportional to the integral of the endogenous neurotransmitter curve (i.e., the additional amount of neurotransmitter, released transiently, above the steady-state level). Morris and Fisher and colleagues at Massachusetts General Hospital (Morris *et al.*, 1995; Fisher *et al.*, 1995) show that with the two-bolus method the sum of squares difference (X^2) between the two dynamic PET curves (with and without perturbation) increases with duration of perturbation. More recently, Aston *et al.* (2000) adopted a hybrid approach of modeling and statistical tests to analyze the two-bolus data to arrive at pixel-by-pixel maps of BP for each of the two dynamic studies. They then used paired *t*-tests to make a neurotransmitter activation map of those pixels whose BP was statistically different in rest and activated conditions. More important, the effect of transient neurotransmitter release is distinct from a transient increase in blood flow, and the precise timing of the perturbation relative to the tracer bolus can greatly affect the magnitude of the signal change. Endres *et al.* (1998) provide a mathematical formalism for understanding the importance of perturbation timing and its relation to the kinetics of the labeled ligand. The bolus-plus-infusion method depends on measuring BP at steady state in two conditions. If steady state is not reached in the pre- or postperturbation phase, the reliability of the measurements will be affected. The striatal (top) curve in Figure 17 may not

---

[9]An approach intermediate between those discussed here has been proposed by Friston *et al.* (1997). A single-bolus study is performed and any statistically significant deviation in time from the predicted dynamic curve is a measure of activation, in effect detecting a change in k_2. A pharmacological challenge was used as a test case.

have quite come to equilibrium when the amphetamine was administered. The time required to reach steady state can be minimized by shifting the fraction of tracer delivered in the bolus versus that delivered via infusion (Carson *et al.*, 1993). Along the same lines, one can optimize the precision of the measured parameter, ΔBP, by appropriately choosing the best range of times for calculating BP_{pre} and BP_{post} with the least amounts of noise. When a uniform optimal protocol used for measuring ΔBP was applied retrospectively to data acquired by Breier *et al.* (1997), the significance of the difference between the dopamine responses of schizophrenics and controls to amphetamine was improved ($p < 0.038$ to $p < 0.012$) (see Watabe *et al.*, 2000). The implications of the various techniques for imaging neurotransmitter change and some applications are examined in Section VII of this chapter.

E. Neurotransmitter Levels

Delforge *et al.* (2001) have proposed a possible scheme for measuring the steady-state level of endogenous neurotransmitter. They first observe that in many analyses of receptor–ligand data there is an unwanted correlation between receptor density B'_{max} and affinity K_D/V_R (in their terminology). The researchers interpret this correlation (or bias in K_D/V_R) as a deficiency in the compartmental model. The deficiency, they claim, is an ignored effect of endogenous neurotransmitter. They then model the bias in estimated K_D/V_R as first order in endogenous neurotransmitter concentration (K_D/V_R measured = K_D/V_r true + [Endogenous ligand concentration]*Correlation). This putative linear relationship involving the concentration of endogenous ligand can then be expressed in terms of its effect on occupied receptor sites through the *a priori* knowledge (i.e., a guess) of the K_D of endogenous ligand and the assumption that the endogenous ligand is in fast equilibrium between the free and receptor-bound states. The upshot is an equation for the steady-state occupancy of receptors by endogenous ligand:

$$\frac{B^{en}}{B_{max}} = \frac{\text{const} \times B'_{max}}{1 + \text{const} \times B'_{max}} \quad (32)$$

The paper is commended to the reader for further study with the following caveat: Equation (32) predicts that in the limit as the density of receptors goes to a very large number, the fraction occupied at steady state will be 1.

V. MODEL SIMPLIFICATIONS

A. Reference Region Methods

The arterial plasma time activity curve serves as the input function to the compartmental model that is used for parameter estimation (see Eq. 21). Often represented as a compartment in itself, $P(t)$ is multiplied by the rate constant K_1, and represents the delivery of the radiotracer from the arterial plasma to the tissue space. The radioconcentration of the plasma is typically sampled from the radial artery and assumed to be representative of the plasma concentration presented to the brain tissue.

Although $P(t)$ is often needed to drive a particular kinetic model, there are many burdens that accompany the acquisition of an arterial plasma input function. These include discomfort to the patient, increased risk of transferring a blood-borne disease (particularly with animal studies), and the need for additional personnel and equipment in withdrawing and assaying the plasma samples. For these reasons, there is considerable motivation for developing an analysis scheme that does not require sampling arterial plasma. This can frequently be accomplished through the use of radioactivity concentration measurements in a region of tissue that has negligible specific binding, called a reference region. Reference region models rely on two key devices: (1) the model of a receptor-rich region can be expressed in terms of the model of a reference region, and (2) the values for distribution volume (DV) in different regions can be constrained.

As shown in the standard compartmental model configuration (Eqs. 21–24), the movement of the radiotracer to and from the tissue is described by the rate constants K_1 and k_2. The ratio of these two parameters, K_1/k_2, represents the distribution volume, DV. By assuming DV to be constant across the regions of the brain, several simplifications can be made to the functional equations introduced in Section IV. The $P(t)$ term can be replaced in the model equations by using the radiotracer kinetics from an alternate region. This section presents several reference region models that have been validated against the standard compartmental analysis (i.e., including an arterial input function).

B. Compartmental Model Simplifications

Cunningham and co-workers (1991) proposed solving for the arterial input function in terms of the derivatives of a reference region and substituting this value into the equations for the regions with specific binding. They start with the state equation for a reference region (Eq. 21 for a region with no explicit B or N term) and solve for the plasma function:

$$P(t) = \frac{1}{K_1^R}\left(\frac{dF_R}{dt} + k_2^R F_R\right) \quad (33)$$

Assume that $P(t)$ is the same in reference region as in the receptor-rich regions and substitute:

$$\frac{dF}{dt} = \frac{K_1}{K_1^R}\left(\frac{dF_R}{dt} + k_2^R F_R\right) - (k_2 + k_3)F + k_4 B \quad (34)$$

$$\frac{dB}{dt} = k_3 F - k_4 B \qquad (35)$$

where F_R, F, and B represent the radioligand concentrations in the reference, free, and specifically bound regions, respectively. Although Eqs. (34) and (35) may look formidable, they are easily solved. Recall that once measurements have been taken in the designated reference region over time, F_R is merely a measured array of values and dF_R/dt can be calculated directly. A constraint is placed on this system that requires the free space distribution of the radioligand be equal in the reference and receptor-rich regions, that is, $K_1^R/k_2^R = K_1/k_2$. This model still allows for blood flow differences between regions by constraining only the ratio of K_1 to k_2 and not the individual parameters. With these substitutions, four parameters, $R_I(= K_1/K_1^R)$, k_2, k_3, and k_4, can be estimated iteratively. This model was originally applied to [^{3}H]-diprenorphine (Cunningham et al., 1991) and later modified for use with PET and [^{11}C]-raclopride in rats (Hume et al., 1992) and humans (Lammertsma et al., 1996).

Not all reference regions are ideal; that is, they may contain some receptor sites of interest. In such cases modifications can also be made to the reference region model to account for specific binding in the reference region (Cunningham et al., 1991). By assuming that the specific binding rate constants k_{on} and k_{off} are uniform across regions and by having prior knowledge of B'_{max} in the reference region, it is possible to calculate the binding constants for a region of interest. A summary of assumptions and other requirements inherent in each reference region model are summarized in Table 1.

The work of Gunn et al. (1997) extends the Cunningham model in an effort to further simplify the experimental analysis and generate voxel-by-voxel parametric images of several select rate constants. Gunn and colleagues define a set of basis functions that are linear in the parameter estimates in order to calculate a linear least squares solution. With this method, the time-dependent tissue concentration is described by:

$$C_t(t) = \theta_1 F_R(t) + \theta_2 B_i(t),$$

where $B_i(t) = F_R(t) \otimes e^{-\theta_i(BP)t}$ and θ_i are linear combinations of the parameter estimates. For this case, the number of estimated parameters is reduced to three: R_I, k_2, and BP ($= k_3/k_4$).

C. Logan Analysis

The method of Logan (1990) was initially developed to provide a technique for rapid data analysis. It implements an analytic solution that can be solved noniteratively to generate a parametric image of binding potentials. The original formulation was based on the multiple-time graphical method of Patlak et al. (1983) Like the Patlak technique (discussed in Section III), it does not assume a particular kinetic model in the calculation of the distribution volume. Using matrix notation, the kinetics of radiotracers in a reversible compartment can be formulated as (Hearon 1963):

$$\frac{d\mathbf{A}}{dt} = \mathbf{KA} + \mathbf{Q}P(t), \qquad (36)$$

where $\mathbf{A}$ is a vector of the amount of radioactivity of radioligand in each reversible compartment, matrix $\mathbf{K}$

TABLE 1 Assumptions for Using a Reference Region[a]

Method/Author	Model Dependence	DV_{F+NS}	A priori Information
Cunningham et al. (1991)	Assumes a two-tissue compartment model for the target region and either a one- or two-compartment model for the reference region	Assumes K_1/k_2' is constant across regions	
Logan et al. (1990, 1996) DV and DVR	Not dependent on a particular model structure	DVR in itself does not require DV_{F+NS} to be constant across regions; this assumption is enforced to translate it into BP	Requires a priori knowledge of a population average k_2 term for one of the methods and time of linearity for both.
Ichise et al. (1996)	Not dependent on particular model structure	Does not require constant ratio of target to reference activities to relate DVR to BP	Does not require mean k_2.
Lammertsma et al. (1996); Gunn et al. (1997)	Assumes that the target region can be adequately fit to a single-tissue compartment (or assumes the exchange between free and specifically bound regions is sufficiently rapid, which is not really the case for raclopride)	Assumes K_1/k_2' is constant across regions	

[a] All models assume a region exists that has negligible specific binding.

contains the intercompartment rate constants, and **Q** is a vector of the rate constants designating transport from the plasma to the exchangeable compartments. In applying this equation to PET data, it is assumed that the PET activity measured in a reconstructed volume consists of a sum of the three compartment states ($F + B + N$). Note that the contribution of the plasma radioactivity P is neglected. For this case, Eq. (36) can be expressed as:

$$\begin{bmatrix} \frac{dF}{dt} \\ \frac{dB}{dt} \\ \frac{dN}{dt} \end{bmatrix} = \begin{bmatrix} -(k_2+k_3+k_5) & k_4 & k_6 \\ k_3 & -k_4 & 0 \\ k_5 & 0 & -k_6 \end{bmatrix} \begin{bmatrix} F \\ B \\ N \end{bmatrix} + \begin{bmatrix} K_1 \\ 0 \\ 0 \end{bmatrix} P(t) \quad (37)$$

After integrating, premultiplying by the inverse of the **K** matrix, and premultiplying by the summing vector [1 1 1], the integral of the PET activity is given by:

$$\int_0^T PET(t)dt = \int_0^T (F+B+N)dt = -[1,1,1]\mathbf{K}^{-1}\mathbf{Q}\int_0^T P(t)dt + [1,1,1]\mathbf{K}^{-1}\mathbf{A} \quad (38)$$

Additional algebraic manipulation yields a linear equation with a slope and an intercept:

$$\frac{\int_0^T PET(t)dt}{PET(T)} = \text{Slope}\frac{\int_0^T P(t)dt}{PET(T)} + \text{Const} \quad (39)$$

where Slope $= -[1\ 1\ 1]\mathbf{K}^{-1}\mathbf{Q} = DV$, and DV is the distribution volume, $(K_1/k_2')(1 + k_3'/k_4)$, for the case where nonspecific binding is not modeled explicitly.

Typically, the nonspecific binding (N) is assumed to be rapid with respect to the transport and specific binding, and its effect can be incorporated into apparent model parameters of a reduced model. Thus, the rate constants are now represented as composite parameters, $k_2' = k_2(1 + k_5/k_6)^{-1}$ and $k_3' = k_3(1 + k_5/k_6)^{-1}$ (Koeppe et al., 1994). As a practical matter, the relative rapidity of nonspecific binding often renders k_5 and k_6 unidentifiable, and as a result k_2' is the only accessible parameter. Unfortunately, it is often improperly referred to simply as k_2 in the literature.

1. Logan Analysis with Reference Regions

It is also possible to manipulate Eq. 36 to eliminate the plasma-dependent term and replace it with the kinetics of a reference region (Logan et al., 1996). This formulation leads nicely to images of DVR (= BP + 1) at each pixel, as can be seen in Figure 18. The functional equation then takes the form:

$$\frac{\int_0^T C_t(t)dt}{C_t(T)} = DVR\frac{\int_0^T F_R(t)dt + [F_R(T)/\bar{k}_2]}{C_t(T)} + b' \quad (40)$$

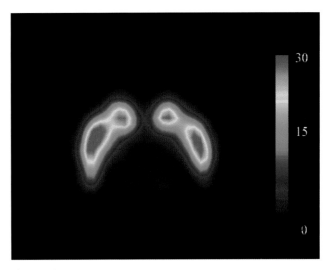

FIGURE 18 A DVR parametric image of [F-18]-fallypride binding in the brain of rhesus monkey. Each voxel in this image represents a calculation of DVR as given in Eq. 40. [F-18]-fallypride is a highly selective dopamine D2 antagonist that- binds to the D2 receptor-rich regions of the brain, as seen by the high uptake in the striatal regions (caudate and putamen) and lower uptake in the thalamus (below the striatum). (For details see Christian et al., 2000.)

where $C_t(t)$ is the PET-measured concentration of the radioligand in the tissue, $F_R(t)$ has been defined as the PET concentration in the reference tissue, DVR is defined as the distribution volume ratio, $\bar{k}_2$ is the regional average of k_2, and b' is a constant of integration. For reversible ligands, a time will be reached when the function becomes linear. This time is dependent on the individual rate constants of a radioligand. The slope of the resultant line represents the ratio of the distribution volumes of the target to reference region DVR. In the case of the three-compartment model (Eq. 37), DVR is equal to:

$$\frac{\frac{K_1}{k_2'}\left(1 + \frac{k_3'}{k_4}\right)}{\frac{K_1^R}{k_2'^R}}$$

If the free plus nonspecific distribution volumes ($DV_{F+NS} = K_1/k_2'$) are constant across regions, then the distribution volume *ratio* simplifies to:

$$DVR = 1 + \frac{k_3'}{k_4} = 1 + f_{NS}\frac{B'_{max}}{K_D}$$

where $f_{NS} = 1/(1 + k_5/k_6)$ (Logan et al., 1994). This simplification is noteworthy because it leads directly to the nearly ubiquitous use of Logan plots as a noniterative means of calculating some rather robust parameters that are proportional to B'_{max}. So we have what we wanted all along—a method of analysis that is easy to apply and apparently yields an estimate of something akin to B'_{max}. But are there drawbacks to this seemingly simple and popular method?

D. Limitations and Biases of the Reference Region Methods

The first step in applying a reference region method of analysis is to validate it against a gold standard. In general, the gold standard for PET radioligand studies is a two-tissue compartment kinetic model with measured arterial plasma as the input function. Knowledge of the individual parameter estimates can then be used to test the biases introduced by the simplifying assumptions of the reference region methods. For example, one of the assumptions of the simplified reference region method (Gunn *et al.*, 1997) is that the equilibration between the free and bound space be rapid, as is the case with many PET radioligands. However, for those radioligands with moderately nonstandard kinetics, a significant bias in the calculation of BP is seen if the rapid equilibration assumption is violated (Alpert *et al.*, 2000). For an example, see Figure 19.

Similarly, preliminary validation is needed when using the Logan DVR method (Eq. 40) to determine the mean k_2 parameter and to provide an estimate of when linearity is achieved. It is impractical to have an *a priori* estimate of the population average value for k_2, so the second method introduced by Logan *et al.* (1996) is more common and is valid as long as the regrouped term on the right-hand side of Eq. (41) is a constant:

$$\frac{\int_0^T C_t(t)dt}{C_t(T)} = \mathrm{DVR}\frac{\int_0^T F_R(t)dt}{C_t(T)} + \left[\frac{F_R(T)/\bar{k_2}}{C_t(T)} + b'\right] \quad (41)$$

If $F_R(T)/C_t(T)$ is a constant, then Eq. (41) will be a line after some time, as previously discussed. In this case, an average k_2 is not required. Ichise *et al.* (1996, 2001) recognized

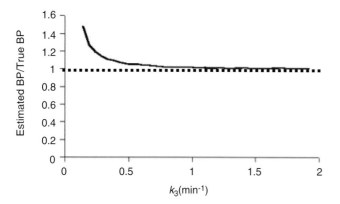

FIGURE 19 Bias in binding potential as estimated by simplified reference region technique. Estimates are based on simulated data. Ratio of estimated to true BP is plotted on y-axis versus true k_3. As k_3 (1/min) becomes smaller, rapid equilibration between free and bound compartments—a necessary assumption of the method—is violated. (Figure supplied courtesy of Dr. N. M. Alpert. See Alpert *et al.*, 2000).

that instead of requiring the term $F_R(T)/C_t(T)$ to become a constant, one could merely treat it as another independent variable in a multiple linear regression scheme. Doing so eliminates the need for an average k_2 and the requirement that $F_R(T)/C_t(T)$ be constant. This approach may eliminate some contributors to bias in the estimates of DVR that are introduced if the condition of constant ratio of tissue activities is never truly achieved.

An illustration of DVR for three D2 receptor antagonists with varying affinities is shown in Figure 20. It has also been demonstrated that the measurement of DVR is biased in the presence of statistical noise (Hsu *et al.*, 1997), which can be quite significant in PET imaging studies. This bias is in part a result of nonindependence of the observations used in the Logan plots. That is, integral quantities of the radioactivity in the target region are plotted on the y axis and integral quantities of the activity in the blood or reference tissue are plotted on the x axis. Each of the y values (and similarly, each of the x values) are thus correlated with each other. This correlation has a biased effect on the estimate of slope (DV or DVR) (see Slifstein and Laruelle, 2000). However, it has been shown that the bias in DVR can be minimized by using smoothing techniques (Logan 2000; Logan *et al.*, 2001).

E. Practical Use

The use of a reference region method of analysis is an intermediate step in the process of shepherding a newly developed radioligand from validation to application stage. When using PET for *in vivo* hypothesis testing, the absolutely simplest form of analysis is to compare radioactivity in different regions on a single static PET image. Obviously such a method eliminates the need for the invasive arterial input measurement and the time-consuming dynamic PET scan. Unfortunately, simple ratio methods can be fraught with bias and variability (Hoekstra *et al.*, 1999). Although the use of a reference region falls short of eliminating the need for dynamic data, it is often the optimal choice in the compromise between simplicity and accuracy.

VI. LIMITATIONS TO ABSOLUTE QUANTIFICATION

Model-based data analyses typically assume that tissue radioactivity is accurately measured in absolute units (i.e., µCi/ml). In reality, it is not so simple. Attenuation of radiation by tissue, scatter, randoms, dead time, and spatial resolution effects all degrade the accuracy of PET. For the most part, image reconstruction software provided by the PET scanner manufacturer provides the means to correct for all

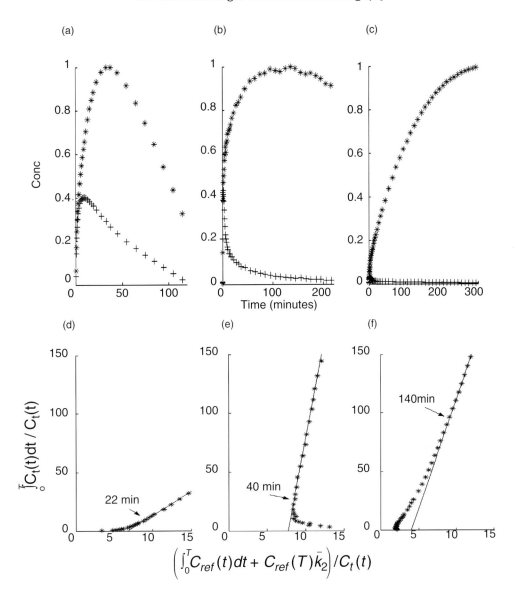

FIGURE 20 Simulated time activity curves and Logan DVR plots of three different D2 receptor antagonists with varying receptor affinities and kinetics in the striatum (*) and the cerebellum (+). (a) [^{11}C]-Raclopride, moderate D2 affinity (for simulation: k_3 = 0.33 min^{-1}, k_4 = 0.11 min^{-1}). (b)[^{18}F]-Fallypride, high affinity (for simulation: k_3 = 1.16 min^{-1}, k_4 = 0.04 min^{-1}). (c) [^{76}Br]-FLB 457, high affinity (for simulation: k_3 = 2.91 min^{-1}, k_4 = 0.02 min^{-1}) for the D2 site. (d,e,f) Logan DVR plots for the corresponding radioligands; [^{11}C]-Raclopride with a rapid linearization occurring at approximately 20 minutes, [^{18}F]-allypride with a moderate linearization time of 40 minutes, and [^{76}Br-FLB 457] taking a long time for linearization of approximately 140 minutes.

but the spatial resolution effects. Consequently, we discuss only spatial resolution effects here.

Spatial resolution effects are sometimes also called partial-volume effects. In this chapter, we use term *spatial resolution effects* because it is more correct and descriptive. Indeed, partial-volume effect is often quantified as the ratio of the measured-to-true radioactivity concentration. The term *partial voluming* is more correctly applied to the problem of tissue heterogeneity within a pixel or region of interest (see discussion in Section III of this chapter).

A. Spatial Resolution Effects

PET images exhibit relatively poor spatial resolution. In fact, the PET image can be thought of as a blurred version of the true radioactivity distribution. Often one approximates the blurring as a low-pass filtering via convolution with a Gaussian kernel (the point-spread function) whose width is specified according to its FWHM. Thus, FWHM is a measure of spatial resolution. The consequence of this convolution is that radioactivity is blurred from one area to adjacent areas. Noting that convolution can be considered to be a special case of matrix-vector multiplication, we can quantitatively express this relation as:

$$\mathbf{M} = \mathbf{CT} \qquad (42)$$

where $\mathbf{M}$ is a vector with values M_i that are the *measured* radioactivity concentrations in region or pixel i, and $\mathbf{T}$ is a vector with values T_j that are the *true* radioactivities in region or pixel j. (Henceforth, we refer only to the case where i and j denote regions and think of the pixel-by-pixel case as a special case of a region containing a single pixel.) $\mathbf{C}$ is the matrix:

$$\mathbf{C} = \begin{bmatrix} HSRC_1 & CSRC_{21} & \cdots & CSRC_{n1} \\ CSRC_{12} & HSRC_2 & & \vdots \\ \vdots & & \ddots & \\ CSRC_{1n} & \cdots & & HSRC_n \end{bmatrix} \qquad (43)$$

where $HSRC_i$ is the hot-spot recovery coefficient and reflects the fraction of the true radioactivity in region i that is actually *measured* in region i. On a pixel-wise basis, $CSRC_{ij}$ is the cold-spot recovery coefficient and reflects the fraction of the true radioactivity in region i that is measured region j. In an ideal system, all HSRC values are unity and all CSRC values are zero. In practice, this is not the case. Values of elements of $\mathbf{C}$ are dependent on size and relative location of pixels or regions.

Components of spatial resolution effects are sometimes described as spill-out and spill-in (Strul and Bendriem, 1999). Spill-out occurs when not all of the radioactivity from the region of interest is measured within its region. Some of the activity is said to spill out into adjacent regions. The amount of spill-out present is dependent on HSRC. Conversely, spill-in is the amount of activity measured in the structure of interest that is, in fact, radioactivity that originates in adjacent structures. The amount of spill-in is dependent on the CSRCs.

Figure 21 provides an illustration of these concepts. Figure 21a represents the true radioactivity concentration distribution of two small hot objects on a warm field. Figure 21b shows the image that would be obtained by PET. Note that the measurable radioactivity distribution appears blurred compared with the true. Intensity profiles are shown in the line plot in Figure 21e and correspond to the true and measured radioactivity concentrations, respectively, along a horizontal line (indicated by hash marks) passing through the centers of images in Figures 21a and 21b. The angular curve represents the Intensity profile in the true activity image, whereas the smooth curve corresponds to the simulated PET image in Figure 21b. In this case, the PET image will underestimate activity in the center of a hot ellipse while overestimating the activity in a background area just outside the hot spot. The activity from the hot ellipses is said to have spilled out from the ellipses and to have spilled in to adjacent areas. The remaining panels of Figure 21 indicate the placement of ROIs on simulated (Figure 21c) and on real PET images of rhesus monkey brains. Figure 21f shows a surface plot describing the complicated relationship between true activity, object size, and measurable activity for the specific ROI placements shown here. The simulation-generated plot in Figure 21f (the lookup table) was used to correct the PET activities observed in the study of age-related decline of D2 receptor function in monkeys by Morris, Chefer et al. (1999).[10]

B. Correcting for the Spatial Resolution Effect

In concept, the method for correcting for spatial resolution effects is straightforward. Given $\mathbf{M}$ and $\mathbf{C}$, solve Eq. (42) for $\mathbf{T}$. The difficulty arises in determining $\mathbf{C}$. Different approaches have been used depending on the availability of anatomic data (usually in the form of computerized tomography, CT, or MRI images) and on any *a priori* assumptions that might be appropriate.

1. Corrections Requiring Anatomic Information

In the case that anatomic information is available, the values of C may be calculated as follows.
1. Use regions of interest to mark structures of presumed uniform radioactivity concentration in the tissues of interest, as well as to mark adjacent background areas. For each such region j, create a synthetic image setting pixel values to unity in the given region and zero elsewhere.
2. Blur the image by convolution with a Gaussian kernel with a FWHM that emulates the spatial resolution of the particular PET scanner.
3. Interpret these two pieces of synthetic data in the context of Eq. (42).

Specifically, Steps 1–3 have simulated the case in which T_j is unity and all other elements of $\mathbf{T}$ are zero. Thus, the result of matrix-vector multiplication on the right-hand side of Eq. (42) yields a vector corresponding to column j of matrix $\mathbf{C}$. Equating this vector to the vector of measured radioactivity concentrations, one defines the HSRC and CSRC values on column j of $\mathbf{C}$.

4. Repeat this process for all regions to define values for all columns of $\mathbf{C}$.

With $\mathbf{C}$ determined, the true distribution $\mathbf{T}$ can be estimated for a given measured radioactivity distribution $\mathbf{M}$ by solving $\mathbf{M} = \mathbf{CT}$ for $\mathbf{T}$. The precision of the result is dependent on the determinant of $\mathbf{C}$ as well as the precision of $\mathbf{M}$. For details of the implementation, we refer the reader to Rousset *et al.* (1998).

In the case that the true radioactivity concentrations in all but one region are assumed to be uniform and known, an

---

[10]The primary finding of the Morris *et al.* (1999) study is that if spatial resolution effects are *not* corrected, the age-related decline of dopamine (D2) binding potential in monkeys will be overestimated by 50% of the true decline.

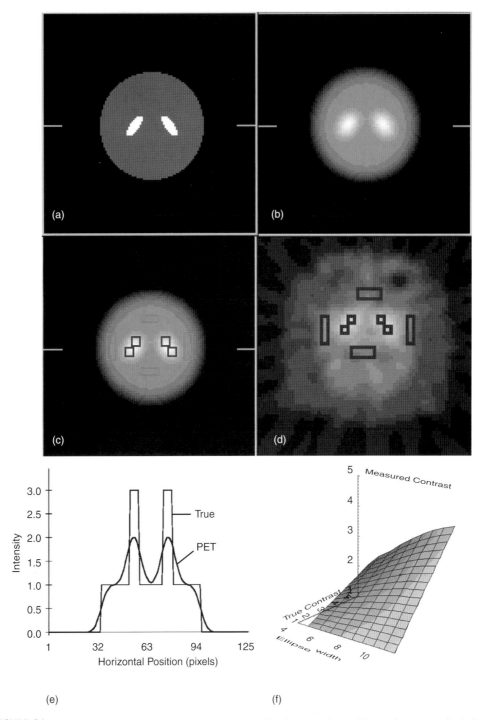

FIGURE 21 (a) Phantom image containing simulated straita. (b) Simulated PET image. Phantom image convolved with impulse response function of PC4096 PET scanner. (c) Placement of regions of interest (ROIs) on simulated PET image used to determine HSRC and CSRC. (d) Comparable placement ROIs on PET image of monkey brain containing striata. (e) Line profiles comparing intensity distributions across true image and simulated PET image. (f) Lookup table (presented as surface plot) relating measured to true activity for hot objects (such as those in image a) of different sizes. (See Morris, Chefer *et al.*, 1999 for details.)

alternative approach may be applied to correct for spatial resolution effects. Termed anatomically guided pixel-by-pixel algorithms (APAs) by Strul and Bendriem (1999), the techniques (Videen 1988; Meltzer *et al.*, 1990, 1996; Muller-Gartner 1992) differ in a few details, but the general principles are the same. In the anatomically guided methods, we confine ourselves to n number of different tissue types. n will be a number much less than the number of pixels in the image. Nevertheless, we treat each tissue type as we treated individual pixels. That is, the tissue types are distinct (masks) and do not overlap. We consider the area being imaged as being composed of a small number of different tissue types—for example, gray matter (GM), white matter (WM), and cerebral spinal fluid (CSF)—with all but one of the types being uniform with respect to radioactivity concentration. Next, the image is decomposed (i.e., segmented) into its constituent tissue types, and a mathematical model of the scanner resolution effects is applied to each in order to account for them so that the radioactivity concentration in the nonuniform tissue may be estimated.

Formally, let index $j = 0$ denote a tissue type with unknown concentration, whereas indices $j > 0$ denote tissue types whose concentrations are assumed to be known and uniform. For each tissue type j, use CT or MRI images to create its binary mask image $\mathbf{X}_j$ containing ones in pixels corresponding to the selected tissue and zeros elsewhere. Thus, the true or ideal radioactivity distribution image $\mathbf{I}_{true}$ can be described as a summation of its constituent parts:

$$\mathbf{I}_{true} = A_0 \mathbf{X}_0 + \sum_{j=1}^{N-1} \overline{A}_j \mathbf{X}_j \quad (44)$$

where A_0 is the radioactivity image of the tissue of nonuniform concentration, $\mathbf{X}_0$ is its mask image, $\overline{A}_j$ is a scalar denoting the activity concentration in tissue type j, and $\mathbf{X}_j$ is its mask image. The observed image $\mathbf{I}_{obs}$ is related to the true image using convolution with a Gaussian point-spread function G to model spatial resolution effects:

$$\mathbf{I}_{obs} = \mathbf{I}_{true} \otimes G \quad (45)$$

Inserting $\mathbf{I}_{true}$ from Eq. (44):

$$\mathbf{I}_{obs} = \left(A_0 \mathbf{X}_0 + \sum_{j=1}^{N-1} \overline{A}_j \mathbf{X}_j \right) \otimes G \quad (46)$$

Rearranging gives:

$$(A_0 \mathbf{X}_0) \otimes G = \mathbf{I}_{obs} - \left(\sum_{j=1}^{N-1} \overline{A}_j \mathbf{X}_j \right) \otimes G \quad (47)$$

In the case that region 0 is uniform, then A_0 is effectively a scalar, so that:

$$(A_0 \mathbf{X}_0) \otimes G = A_0 (\mathbf{X}_0 \otimes G) \quad (48)$$

in which case, we can solve the equation to determine the true radioactivity concentration in region 0:

$$A_0 = \frac{\mathbf{I}_{obs} - \left(\sum_{j=1}^{N-1} \overline{A}_j \mathbf{X}_j \right) \otimes G}{\mathbf{X}_0 \otimes G} \quad (49)$$

To the extent that region 0 is uniform, this relation can be used to approximate the concentration in region 0.

The two methods presented here for correction of spatial resolution effects, the APA methods (Eqs. 44–49) and the solution of a linear system of equations method ($\mathbf{M} = \mathbf{CT}$), are similar in a number of respects. First, consider the image $\mathbf{I}_{obs}$ and $\mathbf{I}_{true}$ as vectors whose elements correspond to pixel values. In this case the convolutions in Eq. (47) can be written as a matrix-vector multiplication:

$$\mathbf{C}(A_0 \mathbf{X}_0) = \mathbf{I}_{obs} - \mathbf{C} \left(\sum_{j=1}^{N-1} \overline{A}_j \mathbf{X}_j \right) \quad (50)$$

To the extent that tissue 0 is uniform, its activity distribution A_0 is a scalar and can be brought outside the matrix multiplication:

$$\mathbf{C}(A_0 \mathbf{X}_0) = A_0 (\mathbf{C} \mathbf{X}_0) \quad (51)$$

in which case:

$$A_0 (\mathbf{C} \mathbf{X}_0) = \left[\mathbf{I}_{obs} - \mathbf{C} \left(\sum_{j=1}^{N-1} \overline{A}_j \mathbf{X}_j \right) \right] \quad (52)$$

The (approximate) solution technique (used to arrive at Eq. 49) is equivalent to a pixel-by-pixel division of the image on the right-hand side of Eq. (52) by the blurred mask image, $\mathbf{CX}_0$, in order to estimate A_0. Clearly, the alternative approach of solving Eq. (50) as a linear system of equations to obtain A_0 is a more rigorous approach than Eq. (49). However, the number of computations needed for the rigorous approach is much greater than for the approximate solution.

A major limitation of the correction methods described thus far is that they require anatomic information. Often CT or MRI is not available to provide such information. Moreover, the methods require that the anatomic and functional images be spatially co-registered. Inaccuracies in registration affect the accuracy of subsequent corrections (Strul *et al.*, 1999).

2. Corrections Not Requiring Anatomical Information

Corrections that do not require anatomical data are loosely classified into those that are based on a pharmacokinetic model and those that are based on a model of the tissue boundary. Ones based on a pharmacokinetic model include parameters such as HSRC or CSRC in addition to the usual physiological parameters to be estimated. This approach is most prevalent in cardiac data analysis in which radioactivity in the ventricular cavity spills out into the surrounding myocardium (Herrero *et al.*, 1989). The difficulty with this approach is that estimates of the physiological parameters of interest are often unreliable due to a strong correlation between them and the correction parameters (Feng *et al.*, 1996; Muzic *et al.*, 1998).

Correction methods based on a model for tissue boundaries try to estimate the size and location of tissue boundaries simultaneously with the radioactivity concentration. For example, a cancerous lesion might be modeled as a sphere of unknown radioactivity concentration superimposed on a region that is locally uniform in radioactivity (Chen *et al.*, 1999). An optimization is then performed to find the combination of sphere diameter and radioactivity concentration that best matches the measured data. One pitfall of this approach is that when the diameter of the lesion is much smaller than the scanner resolution FWHM, the lesion's apparent size in the image is largely unrelated to its true size and it cannot be estimated reliably. Because lesion size and radioactivity concentration are highly correlated, neither size nor concentration can be estimated reliably (Chen *et al.*, 1999).

3. Comparison of Methods

Methods that use anatomical data have both advantages and disadvantages compared to those that do not require such data. Methods that use anatomical data are better able to estimate tissue radioactivity of very small lesions because the lesion size is specified. Thus, the correlation between lesion size and radioactivity is not an issue. The disadvantage of the anatomically based methods is this same requirement for anatomical data. Such data may not always be available or may be expensive to obtain. Moreover, the analyses presume perfect registration of the anatomical images to the PET, as well as proper segmentation of the anatomic images. These requirements add significant burdens to the data processor, and the final results are heavily dependent on how well each of these steps is performed.

VII. FUNCTIONAL IMAGING OF NEUROCHEMISTRY—FUTURE USES

Neuroimagers must resist the temptation to view the brain as composed of functional units approximately 1 cubic centimeter in size, that is, on the order of the resolution of their imaging device. On one crude measure, the brain does work that way—general functions of the brain often are organized on that sort of spatial dimension. However, many brain functions are carried out by diffuse neurochemical pathways—systems of neurons whose axons and dendrites spread through much of the brain but compose only a small portion of each brain region (Cooper *et al.*, 1996; Squire and Kandel, 1999). Some of these systems, those that use norepinephrine, acetylcholine, dopamine, or serotonin, are known to play critical roles in such varied functions as learning, memory, attention, pleasure, reward, and mood. These pathways are frequently activated independently of the surrounding brain tissue, and their activity therefore cannot be determined by conventional activation studies that rely on detecting gross changes in blood flow. Can brain imaging detect the activity of functional subpopulations of neuronal elements?

PET offers a possibility. It requires the combination of quantitative neurotransmitter receptor imaging, as described earlier, with techniques of conventional brain activation imaging. As has been mentioned, the premise is that competition occurs between an injected radioligand and a neurotransmitter for binding to a limited population of receptors. In practical terms, a patient is injected with a neurotransmitter receptor radioligand, such as a dopamine receptor ligand, while performing a brain activation task. If a task stimulates a particular neurotransmitter system under study (e.g., dopamine neurons), then we would expect more of the transmitter (dopamine) to be released into the synapse. Released intrasynaptic dopamine competes for receptor sites with an appropriately chosen radioligand. Hence, there should be a decrease in the amount of specific radioligand binding observed (Figure 22). By comparing the specific binding at rest with that during activation, we can demonstrate the activation of a subset of neurochemically distinct neurons—in this example, a dopaminergic pathway.

In principle, and given an appropriate ligand, our induced competition technique could be applied to any neurochemical pathway. Radioligands are currently being developed or are already available for imaging dopamine, serotonin, acetylcholine, GABA, histamine, norepinephrine, and other synapses. Thus, functional neurochemical maps of the brain could be created that complement the general activation maps being produced with conventional functional brain imaging (^{15}O-water studies in PET or fMRI.).

The site of the neurochemical competition in question is the synapse. Synapses are specialized structures that allow one neuron to communicate with another chemically. The transfer of chemical information takes anywhere from 2 milliseconds for fast (e.g., glutamate) synapses to tens or hundreds of milliseconds (or occasionally many seconds) for slow (e.g., serotonin or muscarinic acetylcholine) synapses. The steps of information transfer are schematized in more detail in Figure 23.

A. Can Neurotransmitter Activation Imaging Really Work?

Although the concept of competition between endogenous neurotransmitter and injected radioligand makes sense

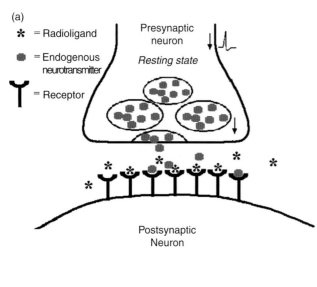

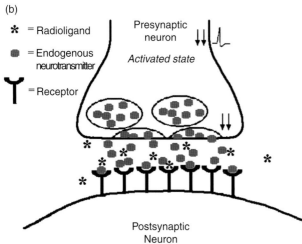

FIGURE 22 (a) Schematic of dopamine neuron at rest. Number of red circles sitting in receptor cups indcates the basal level of occupation of receptors by endogenous ligands. (b) Schematic of dopamine neuron during activation. Note increase in number of postsynaptic receptors occupied by endogenous ligands.

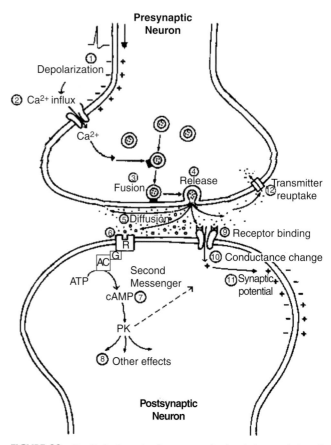

FIGURE 23 Detailed schematic of trans-synaptic chemical transmission of signal. (1) When an electrical signal, usually in the form of an action potential, reaches the presynaptic terminal, it depolarizes the membrane of the terminal, which normally rests at approximately -70 mV (inside compared to outside the cell). (2) This depolarization causes voltage-sensitive Ca^{2+} channels in the presynaptic membrane to open, allowing a brief influx of Ca^{2+} ions. (3) The Ca^{2+} ions interact with calcium-sensitive proteins to initiate a chain of events that causes the vesicles, which contain neurotransmitter molecules, to fuse with the presynaptic membrane. (4) This results in the release of the transmitter into the synaptic cleft. (5) The transmitter diffuses throughout the cleft. (Note that the schematic drawing does not properly indicate the extreme narrowness of the cleft—it is only 20 to 30 nm wide.) (6) Transmitter molecules bind to receptors on the postsynaptic membrane. In the left side of the figure, receptors coupled to intracellular second messenger systems are shown. These receptors are found in slow synapses, such as dopamine, norepinephrine, and muscarinic acetylcholine synapses. (7) In a slow synapse, binding of the transmitter to the receptor activates an enzyme, usually through a G-protein, which activates an enzyme that causes a change in concetration of a second messenger. The example shown here is adenylyl cyclase (AC) causing an increase in cAMP. (8) Many second-messenger systems work by activating a protein kinase (PK). The protein kinase then phosphorylates any of a large number of proteins. Transmitter receptors and membrane ionic channels are among the many possible targets of protein kinases. (9) At fast synapses, transmitter molecules bind to a different type of receptor. (10) This receptor itself is an ionic channel and when it binds transmitter it opens, allowing ions to flow in or out of the cell. (11) At fast excitatory synapses, as shown here, positively charged Na^+ ions, and sometimes Ca^{2+} ions too, flow in, depolarizing the postsynaptic cell. At fast inhibitory synapses, the pore is usually a Cl^- channel. (12) The action of the neurotransmitter is terminated by diffusion out of the synaptic cleft and by active reuptake into the presynpatic terminal. The membrane molecule responsible for this reuptake is called a transporter. (Adapted from Shepherd, 1998.)

intuitively, rigorous scientific proof of the validity of this activation-imaging technique is required. The strongest evidence that PET can be used to image neurotransmitter release comes from experiments in nonhuman primates in which a microdialysis system is surgically implanted in the brain to measure extracellular concentrations of dopamine (DA) (Breier *et al.*, 1997; Endres *et al.*, 1997; see analogous experiments using SPECT in Laruelle *et al.*, 1997). While in a PET scanner, monkeys were injected with amphetamine, which causes increased release of DA into the synapse. The DA receptor binding is assessed with PET while the extracellular concentration of DA is simultaneously measured by microdialysis. Microdialysis experiments detected increased extracellular DA while PET detected decreased binding of [^{11}C]-raclopride. The degree of the increase in DA was shown to correlate with the degree of reduction of receptor–ligand binding. This is strong evidence that the DA is affecting the PET signal by competing with the radioligand for binding sites.

B. Activation of the Dopamine System by a Task Can Be Imaged

Initially, this approach to synaptic imaging was performed using drugs that produce enormous increases in neurotransmitter concentrations. However, it has also been suggested via simulations that mental or physical tasks might be sufficient to produce a detectable change in the PET signal (Fisher *et al.*, 1995; Morris *et al.*, 1995). In 1998 there was an experimental breakthrough. Koepp and colleagues (1998) performed PET imaging of dopamine D2 receptors using [^{11}C]-raclopride. They first imaged subjects at rest and then again while subjects played a video game beginning just prior to injection and continuing throughout the PET imaging procedure. The video game required subjects to maneuver a tank through a battlefield by manipulating a joystick. To motivate the subjects, they were promised monetary rewards commensurate with their success at the game.

From their data, Koepp *et al.* (1998) estimated the BP of [^{11}C]-raclopride in the striatum at rest and compared it to the BP during the video game. The technique they used to calculating BP was developed by Gunn *et al.* (1997) and is discussed in Section VB. The researchers found that playing the game resulted in a decreased binding potential, which they interpreted as an increase in DA release. The task used by Koepp *et al.* included both motor and reward components, both of which have been implicated strongly in the function of the dopaminergic system (Schultz, 1997; Chase *et al.*, 1998). Unfortunately, it is not possible to say which aspect(s) of the task caused the DA release. In all likelihood, the motor and reward components were both involved, but this is a subject for future study.

More recently, a similar experiment has been reported using SPECT (Larisch *et al.*, 1999; Schommartz *et al.*, 2000). Subjects were injected with [^{123}I]-iodobenzamide, a D2 receptor ligand. Starting just before injection and continuing for 30 minutes afterward, the subjects performed a constant task of writing with their right hand. Imaging was performed after completion of the task (90 minutes postinjection). Binding potential was approximated by calculating the ratio of the activity in an ROI (striatum) to that in a reference region (frontal cortex). A note to modelers: Because binding potential equals the ratio of ligand in the bound to free compartments at steady state, this direct measurement of binding potential may be justified under certain conditions. If not, a model-based estimation of binding potential or some other relevant parameter is preferable.[11] Nonetheless, the analysis of Larisch and colleagues appeared to show a decrease in available D2 receptors in the striatum compared to control subjects who did not perform the handwriting task. If the data are valid, they suggest that the handwriting task activated dopaminergic neurons, and the consequent increase in DA release in the striatum interfered with radioligand binding. These results, while interesting, are less convincing than those of Koepp *et al.* (1998), first, perhaps because of their inferior analysis methods and, second, because the decreased binding that they did find was statistically significant only in the right striatum. This is a bit curious because we expect the effect to be bilateral. The authors of the study conjecture that postural movements that accompany the handwriting task might explain the right-sided striatal dopamine release. If this SPECT study or others like it can withstand more rigorous analysis, then we may begin to see neurotransmitter activation studies being performed with SPECT as well as PET.

Two intriguing developments in the area of neurotransmitter activation imaging were published too late for discussion in this chapter, each of which offers methods for producing images of dopamine activation from single dynamic PET scans (see Alpert *et al.*, 2002; Pappata *et al.*, 2002). Both methods are related, in principle, to the method of Friston *et al.* (1997), which uses the generalized linear model to look for statistically significant excursions from the dynamic [^{11}C]-raclopride PET curve that one would expect were no DA activation present.

C. Clinical Uses of Neurotransmitter Activation Studies

The most significant clinical use of neurotransmitter imaging so far has been in the field of schizophrenia. Schizophrenia is a relatively common psychiatric disease, striking approximately 1% of the population, and is usually devastating to the victim. Patients commonly experience hallucinations (including hearing voices) and delusions (false beliefs). Most schizophrenics eventually withdraw from society and suffer frequent hospitalizations (Kaplan

---

[11]Really, we must know that (a) a steady-state ratio of ligand in bound and free compartments has been reached. In addition, (b) the frontal cortex must be a valid reference region for this ligand, (c) there must not be any significant nonspecific binding of the ligand in either region, and (d) there must not be any significant plasma radioactivity at the time of scanning. If these four conditions are satisfied, then the activity in the reference region can be used as an approximation of the concentration of free ligand in the ROI. Further, the activity in the ROI *minus* the activity in the reference could be used as an approximation of the activity in the bound compartment alone in the ROI. Then the ratio of these two would be a fair approximation of the binding potential. For a good comparison of single-scan to more sophisticated modeling techniques see Koeppe *et al.* (1994).

and Sadock, 1998). The cause of schizophrenia remains unknown. One of the first and most widely known hypotheses, called the dopamine hypothesis, postulates that the underlying defect is an abnormality in the dopaminergic system (Carlsson and Lindqvist, 1963; Soares and Innis, 1999). It was formulated decades ago and is based primarily on two observations. (1) The most effective treatments for schizophrenia are drugs that block dopamine receptors; the ability of these drugs to relieve the psychotic symptoms of schizophrenia correlates with the affinity of the drugs for the D2 receptor. (2) Certain drugs (e.g., amphetamine) that increase DA release can cause psychotic symptoms that resemble symptoms of schizophrenia.

Sadly, there has been no definitive identification of the abnormality in the dopamine system of schizophrenics. Powerful support for the hypothesis appeared when Wong et al. (1986) published data demonstrating an increased number of dopamine receptors in schizophrenics compared to controls. However, this was followed by a report from Farde et al. (1990) demonstrating no difference in receptor number between schizophrenics and controls. The resolution of these conflicting results is still subject to debate. Perhaps the strongest evidence for the dopamine hypothesis comes from recent pharmacological neurotransmitter activation studies. Using amphetamine to stimulate the release of DA and using either SPECT (Laruelle et al., 1996) or PET (Breier et al., 1997), researchers have shown that the stimulated release of DA is greater in schizophrenics than it is in controls. If it can be shown that schizophrenics similarly overrespond to ordinary sensory stimuli or to stress or social interactions, then the dopamine hypothesis will be strengthened. In a sort of deactivation study with SPECT, subjects were given alpha-methyl-para-tyrosine (α-MPT)—a drug that temporarily depletes dopamine stores from the presynaptic terminals. The result was an increase in the available D2 receptors, presumably secondary to reduced levels of endogenous dopamine (Abi-Dargham et al., 2000). In this experiment, schizophrenics showed a significantly greater increase in B_{max} (19%) under α-MPT than did normal volunteers (9%), suggesting that schizophrenics have more baseline dopamine in their synapses than normals. The aforementioned experiments demonstrate how neurotransmitter activation studies can be used to test the dopamine hypothesis of schizophrenia.

D. Caution in the Interpretation of Neurotransmitter Activation Studies

The simplest interpretation of neurotransmitter activations studies, namely, that a task stimulates release of the transmitter that then competes with the radioligand for limited binding sites, has been challenged (Inoue et al., 1999; Tsukada et al., 1999; Laruelle, 2000). This competition-based hypothesis fails to account for three observations.

1. The degree of inhibition of radioligand binding does not always correspond to the degree of elevation of extracellular DA stimulated by certain pharmacological agents. In other words, there are examples of drugs producing greater DA release but apparently less inhibition.
2. The inhibition of radioligand binding often persists well beyond the time that extracellular DA is increased. The persistence of inhibition appears to be at variance with predictions of simulations using well-established kinetic parameters of raclopride.
3. The binding of some DA radioligands is not inhibited by DA release and might even increase as DA increases!

It may yet be possible to reconcile recent observations with the competition hypothesis, but other factors probably also contribute to the data. For example, it is likely that pharmacological stimulation often results in changes in the local microenvironment near the postsynaptic receptors (Inoue et al., 1999). Such changes would include the charges on the external surface of cell membranes, affecting the ability of polar radioligands or neurotransmitters to diffuse to postsynaptic receptors. (Consider that DA, serotonin, and many other neurotransmitters are charged molecules at neutral pH.)

A second concern is that the number of functional postsynaptic receptors can change during pharmacological or synaptic activity (Nicoll and Malenka, 1999; Zhou et al., 2001). This happens when available receptors can become internalized by the cell or, alternatively, new receptors can become inserted into the membrane. The functional properties of most receptors, including their binding affinities, can be altered by phosphorylation or other chemical modification, which can be a consequence of pharmacological or synaptic stimulation (Nicoll and Malenka, 1999). The postsynaptic membrane, in other words, is not a static structure but is always changing. Receptors are being added and taken away, and the properties of the receptors are changing as they become phosphorylated or dephosphorylated. The time course and magnitude of these fluctuations are not accurately known, but, if they are rapid relative to the time course of a PET study and large relative to the local receptor density, they could complicate the analysis of transmitter activation studies.

1. Cautions for Modelers of Neurotransmitter Activation Studies

Let us review the preceding discussion from the standpoint of a modeler who wants to explain a neurotransmitter activation experiment in three or fewer compartments. In the world of simplistic compartmental models, changes in the microenvironment of the synapse might mean that V_R, the volume of reaction (Delforge, 1996), is not a constant for the duration of the experiment. This is equivalent to saying that k_{on}/V_R (i.e., the effective association rate constant) varies during the experiment. If the number of postsynaptic receptors (or their efficiency) changes while we are conducting our experiment, then B_{max} is not a constant. If receptors are being phosphorylated while the image is being acquired, then their affinity for the ligand might be changing. Because affinity is $K_D = k_{off}/k_{on}$, then either k_{off} or k_{on}, or both, might be changing. Each of these possibilities contradicts one of our key model assumptions—that the model contains only time-invariant parameters. If any of our presumed time-invariant parameters changes during the image acquisition, then our model becomes decidedly more complicated—and our parameters become much more difficult to estimate—especially because some of our formerly constant parameters are now functions.

E. PET Imaging and Kinetic Modeling Issues in Gene Therapy

For several decades, gene therapy has been widely and eagerly anticipated as a potentially great advance in medicine. Gene therapy can be defined broadly as the introduction of functioning genes into patients in an effort to treat disease. It can be achieved in one of two general ways. One is by placement of a normal gene into the tissue of patients who, because of a mutation, lack a particular gene or possess a faulty one. An example of such a mutation is cystic fibrosis, which is caused by a defective gene for a protein that normally serves as both a chloride channel and a modulator of other ionic channels. The second strategy depends on introducing a deleterious gene directly into tumors. The expression of the gene by tumor cells leads directly to self-destruction or at least increased susceptibility to destruction by chemotherapeutic drugs or the immune system. In both cases, the therapeutic gene may be introduced into the tissue of interest by packaging the gene in a virus or liposome and then infecting the patient. Many clinical gene therapy trials are underway, although currently only a very moderate clinical success has been achieved. The introduction of the normal gene into the airways of patients with cystic fibrosis has yielded some therapeutic value, although the effect lasts only a few weeks (Albelda, et al., 2000).

One of the main problems hindering gene therapy research and development is a difficulty in monitoring the actual expression of the therapeutic gene. Levels of gene expression (measured as amount of protein being produced by transcription and translation of the therapeutic gene) and the distribution throughout the body must be quantified *in vivo*, both in animals and, ultimately, in human subjects. PET offers exciting potential in this area. As we have seen, it is well-suited to *in vivo* quantification and it is usually capable of scanning the whole body (MacLaren et al., 2000; Hustinx et al., 2000).

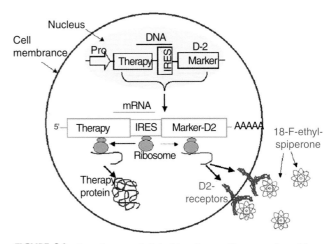

FIGURE 24 A marker gene is linked to a therapeutic gene and provides a marker of gene therapy success. A single piece of DNA contains both the therapeutic gene and a maker gene, separated by an internal ribosomal entry site (IRES). A single promoter (Pro) controls transcription of both genes. (Without the IRES, only the first gene would be reliably transcribed.) The marker gene in this case codes for a dopamine (D2) receptor, which becomes inserted into the cell membrane. Following injection of 18-F-ethylspiperone, a PET scan will quantify the number of dopamine receptors, which is proportional to the amount of therapy protein. (Adapted from Yu *et al.*, 2000.)

1. Monitoring Gene Expression—Two Methods

The first step in using PET to monitor gene therapy is to construct a piece of DNA that contains both the therapeutic gene and a second gene, called the marker gene (see Figure 24; note the term *reporter* is sometimes used instead of *marker*). The marker gene does not contribute to the therapy but instead encodes a protein that is capable of being measured *in vivo* by PET. The piece of DNA must be constructed in such a way that whenever the therapeutic gene is transcribed and translated into a protein molecule, so is the marker gene, so that the amounts of the therapy protein and the marker protein are the same or at least consistently proportional to one another. A PET scan would thus measure the amount of marker protein being expressed in the subject, from which it can be inferred how much of the therapy protein has been expressed. Constructing the requisite piece of DNA is not easy, but there has been considerable preliminary success (Tjuvajev *et al.*, 1998; MacLaren *et al.*, 1999). One of the DNA cassettes and its action is depicted in Figure 24.

The marker gene protein can function in one of two ways. It can be a cell surface receptor, such as a dopamine receptor (MacLaren *et al.*, 1999). This method is schematized in Figure 24. Cells that normally do not normally express dopamine receptors will do so if they have successfully taken up the DNA containing therapy and marker genes. The number of the resulting dopamine receptors can be quantified, as we have discussed at length earlier in this

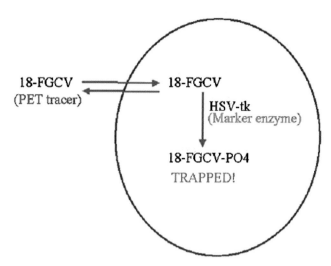

FIGURE 25 Marker gene codes for an enzyme. In this example, the marker gene codes for the viral enzyme thymidine kinase, abbreviated HSV-tk (herpes simplex virus thymidine kinase). When the PET radiotracer 18-F-GCV (fluoroganciclovir) is administered to the subject or tumor, it is taken up by a wide variety of cells. In normal cells, it quickly washes out of the cell. However, in cells that have received gene therapy and are producing viral thymidine kinase, GCV is phosphorylated and cannot leave the cell. A subsequent PET scan, therefore, will demonstrate the tissue location and magnitude of therapeutic gene expression.

chapter. A second type of marker protein is an intracellular enzyme that modifies (e.g., phosphorylates) a radioactive PET probe, rendering the probe impermeant to cell membranes (Tjuvajev *et al.*, 1998; Hustinx *et al.*, 1999; Bankiewicz *et al.*, 2000). (see Figure 25). The result is that cells that express the marker gene trap the PET probe, much the way that tumor cells and other cells trap 18-FDG (see Section III of this chapter). In the example shown, the marker protein is the viral enzyme HSV-tk, and the PET probe is 18-F-ganciclovir (18-F-GCV). It is worth noting that the ganciclovir–HSV-tk system is particularly attractive because the HSV-tk gene is both the therapeutic gene and the marker gene! Ganciclovir is a drug that after being phosphorylated by HSV-thymidine kinase becomes highly toxic. Therefore, ganciclovir can be administered intravenously to the subject and any tumor cells that have received and expressed the therapeutic gene will be killed. Prior to administering ganciclovir treatment, those same cells can be imaged with 18-F-GCV. This eliminates the potentially troublesome problem of deducing therapeutic gene expression from measurements of marker gene expression.

2. Quantification Issues

Many efforts are currently underway to monitor gene therapy with PET. To be useful, however, PET must not only visualize tissue that expresses the marker protein, but must also quantify the amount of protein present. Quantification will require kinetic modeling similar to the methods described earlier. There may also be some unique problems that must be addressed by the modeler.

1. When a cell that normally does not express dopamine receptors is stimulated to do so, what proportion of the produced receptors end up on the cell membrane? Does this vary from tissue to tissue?
2. Does the number of receptors implanted into the cell membrane saturate? In other words, is the number of receptors in the membrane (to be measured by PET) linearly related to the number of receptors (and hence the therapeutic protein) produced by the cell?
3. How does the number of available receptors in the tissue (B_{max}) vary over time? Suppose, for example, that PET imaging at 24 hours after introduction of the gene yields a good estimate of dopamine receptor production, but by 72 hours most of the receptors have been degraded or internalized. Imaging shortly after transfection would be required. On the other hand, early imaging after transfection may not allow enough time for all the expressed dopamine receptors to become positioned in the membrane where they will be accessible to a radioligand. In modeling terms, we have to assume B_{max} is constant even though, in gene therapy systems, it is actually a function of time. One must find a time after gene transfer at which B_{max} is relatively constant and is a reliable indicator of both marker and therapeutic gene expression. This issue is also relevant for the case of a marker gene that is an enzyme leading to trapping of the PET tracer.
4. Consider the case of the marker gene as enzyme. Does the transport of the PET tracer into the cells occur at the same rate *in vivo* as it does *in vitro*, and is the transport in tumor cells the same as in normal cells? In other words, how can we estimate a value for k_1^*? This is similar to the problem of calculating the lumped constant in FDG quantification.

We must keep in mind that the goal of this imaging exercise is to estimate the amount of expression of the therapeutic gene. As such, it will not be useful if one cannot establish a consistent relationship between marker gene expression and therapeutic gene expression. Expression of the marker gene must be detectable with PET. Establishing that these relationships are true and detectable is currently being explored in animal studies (see, for example, Yu *et al.*, 2000).

VIII. A GENERALIZED IMPLEMENTATION OF THE MODEL EQUATIONS

A. State Equation

The models and equations in the preceding sections may be generalized to consider the case of an arbitrary number

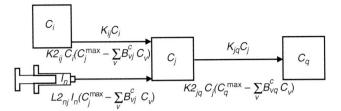

FIGURE 26 Building blocks used to define compartment models. Compartments are shown as rectangles. Arrows indicate unimolecular (labeled above arrow) and saturable bimolecular (labeled below arrows) fluxes. (Note: The graphical representation of a building block might appear on first glance to be equivalent to a compartmental model, but it is; a building block is the graphical depiction of a mass balance on a single compartment. Thus, the appropriate analogy is between a building block and an ordinary differential equation.) (Reprinted by permission of the Society of Nuclear Medicine from: Muzic R.F., et al. COMKAT: Compartment Model Kinetic Analysis Tool. *J Nucl Med* 2001; 42:636–645.)

of inputs and compartments. Consider a model as being constructed from a number of building blocks (Muzic and Cornelius, 2001). Specifically, in building block j, compartment j is the recipient of material from one or more inputs and one or more compartments, and compartment j also serves as the source for material for one or more compartments. This is depicted in Figure 26. Thus, a model with N compartments may be described using building blocks $j = 1, 2, \ldots, N$.

The state equation describing fluxes into and out of compartment j may be written as follows. First, consider the flux from as many as n inputs. If an exchange is unimolecular, the flux is proportional to the concentration in the source inputs and its contribution to the state equation for compartment j is:

$$+\sum_n L_{nj} I_n \quad (53)$$

where L_{nj} denotes the apparent unimolecular rate constant from input n to compartment j, I_n is the molar concentration of input n, and the summation is over all inputs n. If the exchange is saturable as a result of a limited number of binding sites in compartment j, then the contribution is usually a bimolecular term and is given by:

$$+\left(C_j^{\max} - C_j\right)\sum_n L2_{nj} I_n \quad (54)$$

where $L2_{nj}$ denotes the apparent bimolecular rate constant from input n to compartment j, $C_j^{\max}$ is the concentration of binding sites in compartment j, and C_j is the concentration in compartment j.[12] Normally compartment j

---

[12] An important point is that in this formulation the compartment concentrations are maintained as molar and not as radioactivity concentrations. This is because biochemical processes are dependent on molar concentration and not on radioactivity. Radioactivity is used only because it permits the measurements based on the external measurement of gamma rays.

may be considered to be nonsaturable or saturable, but not both. When it is nonsaturable all values $L2_{nj}$ are zero, whereas when it is saturable all values L_{nj} are zero.

Material may enter compartment j from other compartments. The terms describing these contributions are analogous to those shown earlier in which inputs served as the source of material:

$$+\sum_i K_{ij} C_i \quad (55)$$

when compartment j is nonsaturable, and:

$$+\left(C_j^{\max} - C_j\right)\sum_i K2_{ij} C_i \quad (56)$$

when it is saturable. Here K_{ij} and $K2_{ij}$ denote, respectively, the apparent unimolecular and bimolecular rate constants from compartment i to compartment j, and $(C_j^{\max} - C_j)$ denotes the concentration of sites available for binding.

Compartment j may serve as the source of material to other compartments q. Because this corresponds to a loss of material from compartment j, these terms are written with negative signs. The terms are:

$$-C_j \sum_q K_{jq} \quad (57)$$

for nonsaturable recipient compartments q, and:

$$-C_j \sum_q K2_{jq} \left(C_q^{\max} - C_q\right) \quad (58)$$

for saturable recipient compartments q. Compartment q may be considered as either nonsaturable or saturable, but not both. When it is nonsaturable all values $K2_{jq}$ are zero, whereas when it is saturable all values K_{jq} are zero.

The state equation for compartment j is thus written as the sum of all contributions and losses already described:

$$\frac{dC_j}{dt} = \sum_n L_{nj} I_n + \left(C_j^{\max} - C_j\right)\sum_n L2_{nj} I_n + \sum_i K_{ij} C_i$$
$$+ \left(C_j^{\max} - C_j\right)\sum_i K2_{ij} C_i - C_j \sum_q K_{jq} - C_j \sum_q K2_{jq}\left(C_q^{\max} - C_q\right) \quad (59)$$

A set of such equations, one equation for each compartmental building block, defines the exchanges in a multicompartment, multiinput model.

Using the preceding formulation, the model equations have been implemented in MATLAB (The MathWorks, Inc., Natick, MA) in a program called COMKAT: Compartment Model Kinetic Analysis Tool (Muzic and Cornelius, 2001), which is available for download at www.nuclear.uhrad.com/comkat. In this implementation it is convenient to maintain the compartment and input concentrations in vectors **C** and **I**, respectively, and the rate constants in matrices **L**, **L2**, **K**, and **K2** with sizes given in Table 2, in which *Num. Inputs* is the number of inputs and

TABLE 2 Matrix Dimensions in General Model Implementation

	Rows	Columns
I	Num. inputs	1
C	Num. Comptm'ts	1
K	Num. Comptm'ts	Num. Comptm'ts
K2	Num. Comptm'ts	Num. Comptm'ts
L	Num. inputs	Num. Comptm'ts
L2	Num. inputs	Num. Comptm'ts

Num. Comptm'ts is the number of compartments. The state equation for the set of compartments is then written as

$$\dot{C} = F(C) = L'I + (C^{max} - C).*(L2'I) + K'C + (C^{max} - C).*(K2'C) - C.*sum(K,2) - C.*(K2(C^{max} - C)) \quad (60)$$

where $\dot{C}$ denotes the time derivative of C, F is a function that takes C as input and returns $\dot{C}$, and $'$ denotes matrix transpose. Moreover, following MATLAB notation, the .* operator indicates element-by-element multiplication and sum(K,2) indicates a column vector obtained by summing each row in K.

In some cases it is desirable for a single saturable pool of binding sites to be represented by multiple compartments. Consider a study involving a particular receptor and more than one injection of ligand (as discussed previously). For each injection, the specific activity is different (Delforge *et al.*, 1990). For ligand that is bound to the receptor, one may assign separate compartments corresponding to each injection in order to differentiate between ligands from each injection. In this case, there is only one pool of binding sites, but the sites may be occupied by material from any injection. To accommodate this, our formulation is extended by introducing a coupling matrix B^c (a square matrix with dimensions: *Num. Comptm'ts*) with elements B_{ij}^c that equal unity when compartments i and j correspond to a common pool of receptors and that equal zero otherwise. Thus, the number of available binding sites is calculated as ($C^{max} - B^c C$) and the state equation is revised to:

$$\dot{C} = F(C) = L'I + (C^{max} - B^c C).*(L2'I) + K'C + (C^{max} - B^c C).* (K2'C) - C.*sum(K,2) - C.*(K2(C^{max} - B^c C)) \quad (61)$$

B. Output Equation

The output equation relates compartment and input concentrations to the model output. Most typically, model output is defined to have a correspondence to measured data and is used in data fitting. In the case of nuclear medicine applications, the measured data are the externally detected radioactivity. Thus, the output equation must relate radioactivity concentrations to the compartment concentrations that are maintained in terms of molar concentration. This is done by scaling the compartmental concentrations by the specific activity that expresses the radioactivity strength in terms of picomoles per microcurie (pmol/µCi). Specific activity from material u at time t is calculated as:

$$A_u = A_u^0 \exp(-d_u t) \quad (62)$$

where A_u^0 is the specific activity at $t = 0$, and d_u (= ln(2)/half-life) is the radioactivity decay constant. A_u^0 is measured in an aliquot of material that is not injected. Thus, the contribution of the compartments to output l is calculated as:

$$\sum_u W_{ul} A_u C_u \quad (63)$$

where W_{ul} is the weight that compartment u contributes to output l. For example, W_{ul} could be set to zero if there is no contribution. Alternatively, it could be set to a value indicating the tissue volume fraction that is represented by compartment u.

The contribution of the inputs to output l is calculated in an analogous fashion

$$\sum_u X_{ul} I_u \quad (64)$$

where X_{ul} is the weight that input u contributes to output l. For example, values of X_{ul} might be set to equal the tissue vascular space fraction so that the model output includes vascular radioactivity. Note, however, that there is no inclusion of the specific activity A_u in this expression. Separate inputs should be defined to account for input to compartments and for the vascular radioactivity. The values of X_{ul} corresponding to inputs to compartments are set to zero, whereas values of X_{ul} corresponding to vascular radioactivity are set to equal the vascular space fraction. This is appropriate because, most often, the inputs that serve as source for compartments are the plasma concentration of a substance, whereas for the intravascular radioactivity contribution blood concentration is required. Consequently, separate input functions are used to specify radioactivity concentration, and these should be expressed in terms of radioactivity concentration so that no multiplication by specific activity is needed.

The output equation is obtained by adding terms for the compartment and input contributions and integrating them over the scan interval, or frame, to compute the time-averaged value over the scan:

$$M_l(t_b, t_e) = \frac{1}{t_e - t_b} \int_{t_b}^{t_e} \left(\sum_u W_{ul} A_u C_u + \sum_u X_{ul} I_u \right) dt \quad (65)$$

where t_b and t_e are the beginning and ending time of the scan and M_l denotes output l.

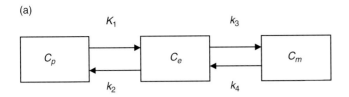

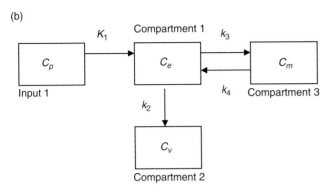

FIGURE 27 (a) Typical FDG model representation, reminiscent of Figure 6. (b) Mathematically, more precise presentation of FDG model in which the depiction is truer to the equations, which do not include a means of incrementing $C_p(t)$ based on efflux of tracer from compartment C_e.

C. Implementing an Arbitrary Model

To implement an arbitrary model within this framework is as easy as drawing compartments - as many as needed - and connecting them with arrows.

Consider for example, the fluorodeoxyglucose (FDG) metabolism model (Huang *et al.*, 1980; Phelps *et al.*, 1979) shown in its usual form in Figure 27a. As shown in this figure, normally the arrow labeled k_2 points from C_e back to the input C_p. Because an increased flux represented by this arrow does not cause C_p to increase in the model (recall that C_p is measured data), it is not correct to draw the arrow to point to C_p. Rather, this flux represents a loss of material to venous circulation, so the arrow should be drawn as a loss to a venous compartment. Thus, the model is more properly described as shown in Figure 27b. To implement this model in the framework described in the previous sections, number the compartments 1, 2, and 3, assigning the numbers as most convenient. For example, let $C_e = 1$, $C_v = 2$, and $C_m = 3$. Because the k_2 arrow is drawn from compartment 1 to compartment 2, the (1,2) element of **K** should be set equal to the numerical value for k_2. (The first index represents the source compartment, and the second represents the destination compartment.) Similarly, the k_3 arrow is drawn from compartment 1 to 3 so it corresponds to the (1,3) element of **K**, whereas k_4 corresponds to the (3,1) element of **K**. K_1 is a little different. The source is C_p, which is an input. These fluxes are represented in the **L** matrix. Because there is only one input, it is considered input number 1 and the flux from input 1 to compartment 1 corresponds to the (1,1) element of **L**. For models that include saturable compartments, the bimolecular rate constants are represented in **K2** and **L2** in an analogous fashion to how the unimolecular rate constants are represented in **K** and **L**.

Continuing with the glucose metabolism model, the model typically has a single output, and it is calculated as $C_e + C_m$. To implement this while also accounting for temporal averaging over the scan duration, simply assign values of elements of **W** to account for compartments 1 and 3 (C_e and C_m) contributing to output 1. Following the (source compartment, destination output) convention with **W** as well, simply set the (1,1) and (3,1) elements to unity. Alternatively, we could calculate the output as $x\, C_e + y\, C_m$ by setting $W_{1,1} = x$ and $W_{3,1} = y$. In the case in which vascular radioactivity is to be included in the output as well, define another input, number it as input 2, corresponding to the blood (as opposed to plasma) radioactivity, and then set the (2,1) value of **X** to equal the volume fraction attributed to vascular space.

Complex multiple-injection, receptor-saturation models (e.g., Delforge, *et al.*, 1990) may also be defined within this formulation. To do this entails defining separate inputs for each injection. For a discussion of how to describe multiple inputs mathematically for use with multiple-injection studies see Morris, Albert, *et al.*, (1996).

Acknowledgments

The authors acknowledge the gracious contributions of figures by Nat Alpert and Marc Berridge and the helpful correspondence of Hiroshi Watabe. The authors also thank Marc Normand for his proofreading assistance.

Appendix

To understand how we arrive at a convolution as a solution to Eq. (1), let us apply the method of Laplace transforms to solve it.

First, we take the transform of both sides:

$$L\left\{\frac{dC_t}{dt}\right\} = L\{K_1 C_A - k_2 C_t\}$$

In the transform domain, we have:

$$s\overline{C}_t - C_t(0) = K_1 \overline{C}_A - k_2 \overline{C}_t$$

where $\overline{C}_t$ is the transform of C_t and $C_t(0)$ is the initial condition in the time domain.

Because the initial condition is zero, we can rearrange to solve for $\overline{C}_t$ as follows:

$$\overline{C}_t = \frac{K_1 C_A}{s + k_2}$$
$$= K_1 C_A \frac{1}{s + k_2}$$

Recall that a product in the transform domain corresponds to a convolution in the time domain. Hence,

$$C_t(t) = L^{-1}\{K_1 \overline{C}_A\} \otimes L^{-1}\{1/(s+k_2)\}$$

where L^{-1} designates the inverse transform operation. And we arrive at a solution of the form:

$$C_t(t) = K_1 C_A \otimes e^{-k_2 t}$$
$$= K_1 \int_0^t C_A(\tau) e^{-k_2(\tau-t)} d\tau$$

References

Abi-Dargham, A., Rodenhiser, J., Printz, D., Zea-Ponce, Y., Gil, R., Kegeles, L. S., Weiss, R., Cooper, T. B., Mann, J. J., Van Heertum, R. L., Gorman, J. M., and Laruelle, M. (2000). From the cover: Increased baseline occupancy of D2 receptors by dopamine in schizophrenia. *Proc. Natl. Acad. Sci. U.S.A.* **97**(14): 8104–8109.

Albelda, S. M., Wiewrodt, R., and Zuckerman, J. B. (2000). Gene therapy for lung disease: Hype or hope? *Ann. Intern. Med.* **132**(8): 649–660.

Alpert, N. M., Badgaiyan, R., and Fischman, A. J. (2002). Detection of neuromodulatory changes in specific systems: Experimental design and strategy. *NeuroImage* **16**(3): S53.

Alpert, N. M., Bonab, A., and Fischman, A. J. (2000). The simplified reference region model for parametric imaging: The domain of validity. *J. Nucl. Med.* **41**(5 suppl.): 54P.

Aston, J. A., Gunn, R. N., Worsley, K. J., Ma Y., Evans, A. C., and Dagher, A. (2000). A statistical method for the analysis of positron emission tomography neuroreceptor ligand data. *Neuroimage* **12**(3): 245–256.

Bahn, M. M., Huang, S. C., Hawkins, R. A., Satyamurthy, N., Hoffman, J. M., Barrio, J. R., Mazziotta, J. C., and Phelps, M. E. (1989). Models for in vivo kinetic interactions of dopamine D2-neuroreceptors and 3-(2′-[18F]fluoroethyl)spiperone examined with positron emission tomography. *J. Cereb. Blood Flow Metab.* **9**: 840–849.

Bankiewicz, K. S., Eberling, J. L., Kohutnicka, M., Jagust, W., Pivirotto, P., Bringas, J., Cunningham, J., Budinger, T. F., and Harvey-White J. (2000). Convection-enhanced delivery of AAV vector in parkinsonian monkeys: In vivo detection of gene expression and restoration of dopaminergic function using pro-drug approach. *Exp. Neurol.* **164**(1): 2–14.

Berman, M., Grundy, S. M., and Howard, B. V. (1982). "Lipoprotein Kinetics and Modeling." Academic Press, New York.

Berridge, M. S., Adler, L. P., Nelson, A. D., Cassidy, E. H., Muzic, R. F., Bednarczyk, E. M., and Miraldi, F. (1991). Measurement of human cerebral blood flow with [15O]butanol and positron emission tomography. *J. Cereb. Blood Flow Metab.* **11**: 707–715.

Breier, A., Su, T-P., Saunders, R., Carson, R. E., Kolachana, B. S., de Bartolomeis, A., Weinberger, D. R., Weisenfeld, N., Malhotra, A. K., Eckelman, W. C., and Pickar, D. (1997). Schizophrenia is associated with elevated amphetamine-induced synaptic dopamine concentrations: Evidence from a novel positron emission tomography method. *Proc. Natl. Acad. Sci. U.S.A.* **94**: 2569–2574.

Carlsson, A., and Lindqvist, M. (1963): Effect of chlorpromazine or haloperidol on formation of 3-methoxytyramine and normetanephrine in mouse brain. *Acta Pharmacol. Toxicol.* **20**: 140–144.

Carson, E. R., Cobelli, C., and Finkelstein, L. (1983). "The Mathematical Modeling of Metabolic and Endocrine Systems". John Wiley & Sons, New York.

Carson, R. (1996). Mathematical modeling and compartmental analysis. In "Nuclear Medicine Diagnosis and Therapy" (J. C. Harbert, W. C. Eckelman, R. D. Neumann, eds.) Thieme Medical Publishers, New York.

Carson, R. E., Breier, A., de Bartolomeis, A., Saunders, R. C., Su T. P., Schmall, B., Der, M. G., Pickar, D., and Eckelman, W. C. (1997). Quantification of amphetamine-induced changes in [11-C]-raclopride binding with continuous infusion. *J. Cereb. Blood Flow Metab.* **17**: 437–447.

Carson, R. E., Channing, M. A., Blasberg, R. G., Dunn, B. B., Cohen, R. M., Rice, K. C., and Herscovitch, P. (1993). Comparison of bolus and infusion methods for receptor quantitation: application to [18-F]-cyclofoxy and positron emission tomography. *J. Cereb. Blood Flow Metab.* **13**: 24–42.

Chase, T. N., Oh, J. D., and Blanchet, P. J. (1998). Neostriatal mechanisms in Parkinson's disease. *Neurology* **51**(suppl 2): S30–S35.

Chen, C. H., Muzic, R. F., Jr., Nelson, A. D., and Adler, L. P. (1999). Simultaneous recovery of size and radioactivity concentration of small spheroids with PET data. *J. Nucl. Med.* **40**: 118–130.

Christian, B. T., Narayanan, T. K., Shi, B., and Mukherjee, J. (2000). Quantitation of striatal and extrastriatal D-2 dopamine receptors using PET imaging of [^{18}F]fallypride in nonhuman primates. *Synapse* **38**: 71–79.

Cooper, J. R., Roth, R., and Bloom, F. E. (1996). "The Biochemical Basis of Neuropharmacology," 7th ed. Oxford University Press, New York.

Crone, C. (1964). Permeability of capillaries in various organs as determined by use of the indicator diffusion method. *Acta Physiol. Scan.* **58**: 292–305.

Cunningham, V. J., Ashburner, J., Byrne, H., and Jones, T. (1993). Use of spectral analysis to obtain parametric images from dynamic PET studies. In "Quantification of Brain Function: Tracer Kinetics and Image Analysis in Brain PET" (K. Uemura, N. A. Lassen, T. Jones, and I. Kanno, eds.), pp. 101–108. Elsevier Science Publishers, Amsterdam.

Cunningham, V. J., Hume, S. P., Price, G. R., Ahier, R. G., Cremer, J. E., and Jones, A. K. P. (1991). Compartmental analysis of diprenorphine binding to opiate receptors in the rat in vivo and its comparison with equilibrium data in vitro. *J. Cereb. Blood Flow Metab.* **11**: 1–9.

Cunningham, V. J., and Jones, T. (1993). Spectral analysis of dynamic PET studies. *J. Cereb. Blood Flow Metab.* **13**: 15–23.

Delforge, J., Bottlaender, M., Pappata, S., Loch, C., and Syrota, A. (2001). Absolute quantification by positron emission tomography of the endogenous ligand. *J. Cereb. Blood Flow Metab.* **21**(5): 613–630.

Delforge, J., Syrota, A., and Bendriem, B. (1996). Concept of reaction volume in the in vivo ligand-receptor model. *J. Nucl. Med.* **37**: 118–125.

Delforge, J., Syrota, A., and Mazoyer, B. M. (1989). Experimental design optimisation: Theory and application to estimation of receptor model parameters using dynamic positron emission tomography. *Phys. Med. Biol.* **34**: 419–435.

Delforge, J., Syrota, A., and Mazoyer, B. M. (1990). Identifiability analysis and parameter identification of an in vivo ligand-receptor model form PET data. *IEEE Trans. Biomed. Eng.* **37**: 653–661.

Dienel, G. A., Cruz, N. F., Mori, K., Holden, J., and Sokoloff, L. (1991). Direct measurement of the λ of the lumped constant of the deoxyglucose method in rat brain: Determination of λ and lumped constant from tissue glucose concentration or equilibrium tissue:plasma distribution ratio for methylglucose. *J. Cereb. Blood Flow Metab.* **11**: 25–34.

Dienel, G. A., Nakanishi, H., Cruz, N. F., and Sokoloff, L. (1991). Determination of local brain glucose concentration from the brain:plasma methylglucose distribution ratio: Influence of glucose utilization rate. *J. Cereb. Blood Flow Metab.* **11**(suppl 2): S581.

Endres, C. J., and Carson, R. E. (1998). Assessment of dynamic neurotransmitter changes with bolus or infusion delivery of neuroreceptor ligands. *J. Cereb. Blood Flow Metab.* **18**: 1196–1210.

Endres, C. J., Kolachana, B. S., Saunders, R. C., Su, T., Weinberger, D., Breier, A., Eckelman, W. C., and Carson, R. E. (1997). Kinetic modeling fo [11C]raclopride: Combined PET-microdialysis studies. *J. Cereb. Blood Flow Metab.* **17**: 932–942.

Farde, L., Eriksson, L., Blomquist, G., and Halldin, C. (1989). Kinetic analysis of central [11C]raclopride binding to D2-dopamine receptors studied by PET: A comparison to the equilibrium analysis. *J. Cereb. Blood Flow Metab.* **9:** 696–708.

Farde, L., Wiesel, F. A., Stone-Elander, S., Halldin, C., Nordstrom, A. L., Hall, H., and Sedvall, G. (1990). D2 dopamine receptors in neuroleptic-naive schizophrenic patients: A positron emission tomography study with [11C]raclopride. *Arch. Gen. Psychiatry* **47**(3): 213–219.

Feng, D., Li, X., and Huang, S. C. (1996). A new double modeling approach for dynamic cardiac PET studies using noise and spillover contaminated LV measurements. *IEEE Trans. Biomed. Eng.* **43:** 319–327.

Fisher, R. E., Morris, E. D., Alpert, N. M., and Fischman, A. J. (1995). In vivo imaging of neuromodulatory synaptic transmission using PET: A review of the relevant neurophysiology. *Hum. Br. Map.* **3:** 24–34.

Frey, K. A., Hichwa, R. D., Ehrenkaufer, R. L. E., and Agranoff B. W. (1985). Quantitative *in vivo* receptor binding, III. Tracer kinetic modeling of muscarinic cholinergic receptor binding. *Proc. Natl. Acad. Sci. U.S.A.* **82:** 6711–6715.

Friston, K. J., Malizia, A. L., Wilson, S., Cunningham, V. J., Jones, T., and Nutt, D. J. (1997). Analysis of dynamic radioligand displacement or "activation" studies. *J. Cereb. Blood Flow Metab.* **17:** 80–93.

Frost, J. J., Douglass, K. H., Mayber, H. S., Dannals, R. F., Links, J. M., Wilson, A. A., Ravert, H. T., Crozier, W. C., and Wagner, H. N., Jr. (1989). Multicompartmental analysis of [11C]-carfentanil binding to opiate receptors in humans measured by positron emission tomography. *J. Cereb. Blood Flow Metab.* **9:** 398–409.

Gjedde, A., Wienhard, K., Heiss, W.-D., Kloster, G., Diemer, N. H., Herholz, K., and Pawlik, G. (1985). Comparative regional analysis of 2-fluorodeoxyglucose and methylglucose uptake in brain of four stroke patients. With special reference to the regional estimation of the lumped constant. *J. Cereb. Blood Flow Metab.* **5:** 163–178.

Gunn, R. N., Lammertsma, A. A., Hume, S. P., and Cunningham, V. J. (1997). Parametric imaging of ligand-receptor binding in PET using a simplified reference region model. *Neuroimage* **6:** 279–287.

Hearon, J. Z. (1963). Theorems on linear systems. *Ann. N.Y. Acad. Sci.* **108:** 36–68.

Herrero, P., Markham, J., and Bergmann, S. R. (1989). Quantitation of myocardial blood flow with H2 15O and positron emission tomography: Assessment and error analysis of a mathematical approach. *J. Comput. Assist. Tomogr.* **13:** 862–873.

Herscovitch, P., Markham, J., and Raichle, M. E. (1983). Brain blood flow measured with intravenous $H_2^{15}O$, I. Theory and error analysis. *J. Nucl. Med.* **24:** 782–789.

Hoekstra, C. J., Paglianiti, I., Hoekstra, O. S., Smit, E. F., Postmus, P. E., Teule, G. J., and Lammertsma, A. A. (1999). Monitoring response to therapy in cancer using [^{18}F]-2-fluoro-2-deoxyglucose and positron emission tomography: An overview of different analytical methods. *Eur. J. Nucl. Med.* **27**(6): 731–743.

Hoffman, E. J., Huang, S. C., and Phelps, M. E. (1979). Quantitation in poistron emission computed tomography, 1. Effect of object size. *J. Comput. Assist. Tomogr.* **3:** 299–308.

Hsu, H., Alpert, N. M., Christian, B. T., Bonab, A. A., Morris, E., and Fischman, A. J. (1997). Noise properties of a graphical assay of receptor binding. *J. Nucl. Med.* **38:** 204P.

Huang, S. C., Bahn, M. M., Barrio, J. R., Hoffman, J. M., Satyamurthy, N., Hawkins, R. A., Mazziotta, J. C., and Phelps, M. E. (1989). A double-injection technique for the in vivo measurement of dopamine D2-receptor density in monkeys with 3-(2'-[18F]fluoroethyl)spiperone and dynamic positron emission tomography. *J. Cereb. Blood Flow Metab.* **9:** 850–858.

Huang, S. C., Carson, R. E., Hoffman, E. J., Carson, J., MacDonald, N., Barrio, J. R., and Phelps, M. E. (1983). Quantitative measurement of local cerebral blood flow in humans by positron computed tomography and ^{15}O-water. *J. Cereb. Blood Flow Metab.* **3:** 141–153.

Huang, S. C., Dan-chu, Y., Barrio, J. R., Grafton, S., Melega, W. P., Hoffman, J. M., Satyamurthy, N., Mazziotta, J. C., and Phelps, M. E. (1991). Kinetcs and modeling of L-6-[18F]Fluoro-DOPA in human positron emission tomographic studies. *J. Cereb. Blood Flow Metab.* **11:** 898–913.

Huang, S. C., Mahoney, D. K., and Phelps, M. E. (1986). Quantitation in positron emission tomography, 8. Effects of nonlinear parameter estimation on functional images. *J. Cereb. Blood Flow Metab.* **6:** 515–521.

Huang, S. C., and Phelps, M. E. (1986). Principles of tracer kinetic modeling in positron emission tomography and autoradiography. *In* "Positron Emission Tomography and Autoradiography: Principles and Applications for the Brain and Heart". (M.E. Phelps, J. Mazziotta, and H. Schelbert, eds.) pp. 287–346. Raven Press, New York.

Huang, S. C., Phelps, M. E., Hoffman, E. J., Sideris, K., Selin, C. J., and Kuhl, D. E. (1980). Noninvasive determination of local cerebral metabolic rate of glucose in man. *Am. J. Physiol.* **238:** E69–E82.

Hume, S. P., Myers, R., Bloomfield, P. M., Opacka-Juffry, J., Cremer J. E., Ahier, R. G., Luthra, S. K., Brooks, D. J., and Lammertsma, A. A. (1992). Quantitation of carbon-11 labeled raclopride in rat striatum using positron emission tomography. *Synapse* **12:** 47–54.

Hustinx, R., Eck, S. L., and Alavi, A. J. (1999). Potential applications of PET imaging in developing novel cancer therapies. *J. Nucl. Med.* **40:** 995–1002.

Hustinx, R., Eck, S. L., and Alavi, A. J. (1999). Potential applications of PET imaging in developing novel cancer therapies. *J. Nucl. Med.* **40**(6): 995–1002.

Ichise, M., Ballinger, J. R., Golan, H., Vines, D., Luong, A., Tsai, S., and Kung, H. F. (1996). Noninvasive quantification of dopamine D2 receptors with iodine-123-IBF SPECT. *J. Nucl. Med.* **37**(3): 513–20.

Ichise, M., Meyer, J. H., and Yonekura, Y. (2001). An introduction to PET and SPECT neuroreceptor quantification models. *J. Nucl. Med.* **42**(5): 755–763.

Iida, H., Kanno, I., Miura, S., Murakami, M., Takahashi, K., and Uemura, K. (1986). Error analysis of a quantitative cerebral blood flow measurement using $H_2^{15}O$ autoradiography and positron emission tomography. *J. Cereb. Blood Flow Metab.* **6:** 536–545.

Iida, H., Kanno, I., Miura, S., Murakami, M., Takahashi, K., and Uemura, K. (1989). A determination of the regional brain/blood partition coefficient of water using dynamic positron emission tomography. *J. Cereb. Blood Flow Metab.* **9:** 874–885.

Iida, H., Law, I., Pakkenberg, B., Krarup-Hansen, A., Eberl, S., Holm, S., Hansen, A. K., Gundersen, H. J. G., Thomsen, C., Svarer, C., Ring, P., Friberg, L., and Paulson, O. B. (2000). Quantitation of regional cerebral blood flow corrected for partial volume effect using O-15 water and PET, I. Theory, error analysis, and stereological comparison. *J. Cereb. Blood Flow Metab.* **20:** 1237–1251.

Inoue, O., Kobayashi, K., and Gee, A. (1999). Changes in apparent rates of receptor binding in the intact brain in relation to the heterogeneity of reaction environments. *Crit. Rev. Neurobiol.* **13**(2): 199–225.

Jacquez, J. A. (1985). "Compartmental Analysis in Biology and Medicine," 2nd ed. University of Michigan Press, Ann Arbor.

Kaplan, H. I., and Sadock, B. J. (1998). "Synopsis of Psychiatry," 8th ed. Lippincott Williams and Wilkins, Baltimore.

Kawai, R., Carson, R. E., Dunn, B., Newman, A. H., Rice, K. C., and Blasberg, R. G. (1991). Regional brain measurement of Bmax and KD with the opiate antagonist cyclofoxy: Equilibrium studies in the conscious rat. *J. Cereb. Blood Flow Metab.* **11**(4): 529–544.

Koepp, M. J., Gunn, R. N., Lawrence, A. D., Cunningham, V. J., Dagher, A., Jones, T., Brooks, D. J., Bench, C. J., and Grasby, P. M. (1998). Evidence for striatal dopamine release during a video game. *Nature* **393:** 266–268.

Koeppe, R. A., Frey, K. A., Mulholland, G. K., Kilbourn, M. R., Buck, A., Lee, K. S., and Kuhl, D. E. (1994). [11C]tropanyl benzilate-binding to muscarinic cholinergic receptors: Methodology and kinetic modeling alternatives. *J. Cereb. Blood Flow Metab.* **14:** 85–99.

Koeppe, R. A., Holden, J. E., and Ip, W. R. (1985). Performance comparison of parameter estimation techniques for the quantitation of local cerebral blood flow by dynamic positron computed tomography. *J. Cereb. Blood Flow Metab.* **5:** 224–234.

Koeppe, R. A., Holthoff, V. A., Frey, K. A., Kilbourn, M. R., and Kuhl, D. E. (1991). Compartmental analysis of [^{11}C]flumazenil kinetics for the estimation of ligand transport rate and receptor distribution using positron emission tomography. *J. Cereb. Blood. Flow Metab.* **11:** 735–744.

Kuwabara, H., Evans, A. C., and Gjedde, A. (1990). Michaelis-Menten constraints improved cerebral glucose metabolism and regional lumped constant measurements with [^{18}F]fluorodeoxyglucose. *J. Cereb. Blood Flow Metab.* **10:** 180–189.

Lammertsma, A. A., Bench, C. J., Hume, S. P., Osman, S., Gunn, K., Brooks, D. J., and Frackowiak, R. J. (1996). Comparison of methods for analysis of clinical [^{11}C]raclopride studies. *J. Cereb. Blood Flow Metab.* **16:** 42–52.

Larisch, R., Schommartz, B., Vosberg, H., and Muller-Gartner, H. W. (1999). Influence of motor activity on striatal dopamine release: A study using iodobenzamide and SPECT. *Neuroimage* **10:** 261–268.

Laruelle, M. (2000). Imaging synaptic neurotransmission with in vivo binding competition techniques: A critical review. *J. Cereb. Blood Flow Metab.* **20**(3): 423–451.

Laruelle, M., Abi-Dargham, A., van Dyck, C. H., Gil, R., D'Souza, C. D., Erdos, J., McCance, E., Rosenblatt, W., Fingado, C., Zoghbi, S. S., Baldwin, R. M., Seibyl, J. P., Krystal, J. H., Charney, D. S., and Innis, R. B. (1996). Single photon emission computerized tomography imaing of amphetamine-induced dopamine release in drug-free schizophrenic subjects. *Proc. Natl. Acad. Sci. U.S.A.* **93**(17): 9235–9240.

Laruelle, M., Iyer, R. N., Al-Tikriti, M. S., Zea-Ponce, Y., Malison, R., Zoghbi, S. S., Baldwin, R. M., Kung, H. F., Charney, D. S., Hoffer, P. B., Innis, R. B., and Bradberry, C. W. (1997). Microdialysis and SPECT measurements of amphetamine-induced release in nonhuman primates. *Synapse* **25:** 1–14.

Laruelle, M., Wallace, E., Seibl, P., Baldwin, R. M., Zea-Ponce, Y., Zoghbi, S. S., Neumeyer, J. L., Charney, D. S., Hoffer, P. B., and Innis, R. B. (1994). Graphical, kinetic, and equilibrium analyses of in vivo [123I]β-CIT binding to dopamine transporters in healthy human subjects. *J. Cereb. Blood Flow Metab.* **14:** 982–994.

Law, I., Iida, H., Holm, S., Nour, S., Rostrup, E., Svarer, C., and Paulson, O. B. (2000). Quantitation of regional cerebral blood flow corrected for partial volume effect using O-15 water and PET, II. Normal values and gray matter blood flow response to visual activation. *J. Cereb. Blood Flow Metab.* **20:** 1252–1263.

Logan, J. (2000). Graphical analysis of PET data applied to reversible and irreversible tracers. *Nucl. Med. Biol.* **27:** 661–670.

Logan, J., Dewey, S. L., Wolf, A. P., Fowler, J. S., Brodie, J. D., Angrist, B., Volkow, N. D., and Gatley, S. J. (1991). Effects of endogenous dopamine on measures of [18F]-methylspiroperidol binding in the basal ganglia: Comparison of simulations and experimental results from PET studies in baboons. *Synapse* **9:** 195–207.

Logan, J., Fowler, J. S., Volkow, N. D., Ding, Y. S., Wang, G. J., and Alexoff, D. L. (2001). A strategy for removing the bias in the graphical analysis method. *J. Cereb. Blood Flow Metab.* **21**(3): 307–320.

Logan, J., Fowler, J. S., Volkow, N. D., Wang, G. J., Dewey, S. L., MacGregor, R. R., Schlyer, D. J., Gatley, S. J., Pappas, N., King, P., Hitzemann, R., and Vitkun, S. (1994). Effects of blood flow on [^{11}C]-raclopride binding in the brain: model simulations and kinetic analysis of PET data. *J. Cereb. Blood Flow Metab.* **14:** 995–1010.

Logan, J., Fowler, J. S., Volkow, N. D., Wang, G. J., Ding, Y. S., and Alexoff, D. L. (1996). Distribution volume ratios without blood sampling from graphical analysis of PET data. *J. Cereb. Blood Flow Metab.* **16:** 834–840.

Logan, J., Fowler, J. S., Volkow, N. D., Wolf, A. P., Dewey, S. L., Schlyer, D. J., MacGregor, R. R., Hitzemann, R., Bendriem, B., Gatley, S. J., and Christman, D. R. (1990). Graphical analysis of reversible radioligand binding from time-activity measurements applied to [N-^{11}C-methyl]-(-)-cocaine PET studies in human subjects. *J. Cereb. Blood Flow Metab.* **10:** 740–747.

Lucignani, G., Schmidt, K., Moresco, R. M., Stirano, G., Colombo F., Sokoloff, L., and Fazio, F. (1993). Measurement of regional cerebral glucose utilization with [^{18}F]FDG and PET in heterogeneous tissues: theoretical considerations and practical procedure. *J. Nucl. Med.* **34:** 360–369.

MacLaren, D. C., Gambhir, S. S., Satyamurthy, N., Barrio, J. R., Sharfstein, S., Toyokuni, T., Wu, L., Berk, A. J., Cherry, S. R., Phelps, M. E., and Herschman, H. R. (1999). Repetitive, non-invasive imaging of the dopamine D2 receptor as a reporter gene in living animals. *Gene Ther.* **6**(5): 785–791.

MacLaren, D. C., Toyokuni, T., Cherry, S. R., Barrio, J. R., Phelps, M. E., Herschman, H. R., and Gambhir, S. S. (2000). PET imaging of transgene expression. *Biol. Psych.* **48:** 337–348.

Martin, C. C., Jerabek, P. A., Nickerson, L. D. H., and Fox, P. T. (1999). Effect of partition coefficient, permeability surface product, and radioisotope on the signal-to-noise ratio in PET functional brain mapping: A computer simulation. *Hum. Br. Map.* **7**(3): 151–160.

Meltzer, C. C., Leal, J. P., Mayberg, H. S., Wagner, H. N., and Frost, J. J. (1990). Correction of PET data for partial volume effects in human cerebral cortex by MR imaging. *J. Comput. Assist. Tomogr.* **14:** 561–570.

Meltzer, C. C., Zubieta, J. K., Links, J. M., Brakeman, P., Stumpf, M. J., and Frost, J. J. (1996). MR-based correction of brain PET measurements for heterogeneous gray matter radioactivity distribution. *J. Cereb. Blood Flow Metab.* **16:** 650–658.

Meyer, E. (1989). Simultaneous correction for tracer arrival delay and dispersion in CBF measurements by the H2O15 autoradiographic method and dynamic PET. *J. Nucl. Med.* **30:** 1069–1078.

Mintun, M. A., Raichle, M. E., Kilbourn, M. R., Wooten, G. F., and Welch, M. J. (1984). A quantitative model for the *in vivo* assessment of drug binding sites with positron emission tomography. *Ann. Neurol.* **15:** 217–227.

Morris, E. D., Alpert, N. M., and Fischman, A. J. (1996). Comparison of two compartmental models for describing receptor ligand kinetics and receptor availability in multiple injection PET studies. *J. Cereb. Blood Flow Metab.* **16:** 841–853.

Morris, E. D., Babich, J. W., Alpert, N. M., Bonab, A. A., Livni, E., Weise, S., Hsu, H., Christian, B. T., Madras, B. K., and Fischman, A. J. (1996). Quantification of dopamine transporter density in monkeys by dynamic PET imaging of multiple injections of ^{11}C-CFT. *Synapse* **24:** 262–272.

Morris, E. D., Bonab, A. A., Alpert, N. M., Fischman, A. J., Madras, B. K., and Christian, B. T. (1999). Concentration of dopamine transporters: To Bmax or not to Bmax? *Synapse* **32**(2): 136–140.

Morris, E. D., Chefer, S. I., Lane, M. A., Muzic, R. F., Jr., Wong, D. F., Dannals, R. F., Matochik, J. A., Bonab, A. A., Villemagne, V. L., Ingram, D. K., Roth, G. S., and London, E. D. (1999). Decline of D2 receptor binding with age, in rhesus monkeys: Importance of correction for simultaneous decline in striatal size. *J. Cereb. Blood Flow Metab.* **19:** 218–229.

Morris, E. D., Fisher, R. E., Alpert, N. M., Rauch, S. L., and Fischman, A. J. (1995). In vivo imaging of neuromodulation using positron emission tomography: Optimal ligand characteristics and task length for detection of activation. *Hum. Br. Map.* **3:** 35–55.

Muller-Gartner, H. W., Links, J. M., Prince, J. L., Bryan, R. N., McVeigh, E., Leal, J. P., Davatzikos, C., and Frost, J. J. (1992). Measurement of radiotracer concentration in brain gray matter using positron emission tomography: MRI-based correction for partial volume effects. *J. Cereb. Blood Flow Metab.* **12:** 571–583.

Muzic, R. F., Jr., Chen, C. H., and Nelson, A. D. (1998). Method to account for scatter, spillover and partial volume effects in region of interest analysis in PET. *IEEE Trans. Med. Imag.* **17:** 202–213.

Muzic, R. F., Jr., and Cornelius, S. (2001). COMKAT: Compartment model kinetic analysis tool. *J. Nucl. Med.* **42:** 636–645.

Muzic, R. F., Jr., Nelson, A. D., and Miraldi, F. (1990). Mathematical simplification of a PET blood flow model. *IEEE Trans. Med. Imag.* **9:** 172–176.

Muzic, R. F., Jr., Nelson, A. D., and Miraldi, F. (1993). Temporal alignment of tissue and arterial data and selection of integration start times for the H$_2$^{15}O autoradiographic CBF model in PET. *IEEE Trans. Med. Imag.* **12:** 393–398.

Nelson, A. D., Muzic, R. F., Jr., Miraldi, F., Muswick, G. J., Leisure, G. P., and Voelker, W. (1990). Continuous arterial positron monitor for quantitation in PET imaging. *Am. J. Physiol. Imag.* **5:** 84–88.

Nicoll, R. A., and Malenka, R. C. (1999). Expression mechanisms underlying NMDA receptor-dependent long-term potentiation. *Ann. N.Y. Acad. Sci.* **30**(868): 515–525.

Ohta, S., Meyer, E., Fujita, H., Reutens, D. C., Evans, A., and Gjedde, A. (1996). Cerebral [15O]water clearance in humans determined by PET, I. Theory and normal values. *J. Cereb. Blood Flow Metab.* **16:** 765–780.

Pajevic, S., Daube-Witherspoon, M. E., Bacharach, S. L., and Carson, R. E. (1998). Noise characteristics of 3-D and 2-D PET images. *IEEE Trans. Med. Imag.* **17:** 9–23.

Pappata, J., Dehaene, S., Poline, J. B., Gregoire, M. C., Jobert, A., Delforge, J., Frouin, V., Bottlaender, M., Dolle, F., Di Giamberardino, L., and Syrota, A. (2002). In vivo detection of striatal dopamine release during reward: A PET study with [11C]Raclopride and a single dynamic scan approach. *NeuroImage* **16:** 1015–1027.

Patlak, C., and Blasberg, R. (1985). Graphical evaluation of blood-to-brain transfer constants from multiple-time uptake data: Generalizations. *J. Cereb. Blood Flow Metab.* **5:** 584–590.

Patlak, C., Blasberg, R., and Fenstermacher, J. (1983). Graphical evaluation of blood-to-brain transfer constants from multiple-time uptake data. *J. Cereb. Blood Flow Metab.* **3:** 1–7.

Phelps, M. E., Huang, S. C., Hoffman, E. J., Selin, C. J., Sokoloff, L., and Kuhl, D. E. (1979). Tomographic measurement of local cerebral glucose metabolic rate in humans with [^{18}F]2-fluoro-2-deoxy-D-glucose: Validation of method. *Ann. Neurol.* **6:** 371–388.

Raichle, M. E., Martin, W. R. W., Herscovitch, P., Mintun, M. A., and Markham, J. (1983). Brain blood flow measured with intravenous H2O15, II. Implementation and validation. *J. Nucl. Med.* **24:** 790–798.

Reivich, M., Alavi, A., Wolf, A., Fowler, J., Russell, J., Arnett, C., MacGregor, R. R., Shiue, C. Y., Atkins, H., Anand, A., Dann, R., and Greenberg, J. H. (1985). Glucose metabolic rate kinetic model parameter determination in humans: The lumped constants and rate constants for [^{18}F]fluorodeoxyglucose and [^{11}C]deoxyglucose. *J. Cereb. Blood Flow Metab.* **5:** 179–192.

Reivich, M., Kuhl, D., Wolf, A., Greenberg, J., Phelps, M., Ido, T., Casella, V., Fowler, J., Hoffman, E., Alavi, A., Som, P., and Sokoloff, L. (1979). The [^{18}F]fluorodeoxyglucose method for the measurement of local cerebral glucose utilization in man. *Circ. Res.* **44:** 127–137.

Renkin, E. M. (1959). Transport of potassium-42 from blood to tissue in isolated mammalian skeletal muscles. *Am. J. Physiol.* **197:** 1205–1210.

Rousset, O. G., Ma, Y., and Evans, A. C. (1998). Correction for partial volume effects in PET: Principle and validation. *J. Nucl. Med.* **39:** 904–911.

Schmidt, K. (1999). Which linear compartmental systems can be analyzed by spectral analysis of PET output data summed over all compartments? *J. Cereb. Blood Flow Metab.* **19:** 560–569.

Schmidt, K., Lucignani, G., Moresco, R. M., Rizzo, G., Gilardi, M. C., Messa, C., Colombo, F., Fazio, F., and Sokoloff, L. (1992). Errors introduced by tissue heterogeneity in estimation of local cerebral glucose utilization with current kinetic models of the [^{18}F]fluorodeoxyglucose method. *J. Cereb. Blood Flow Metab.* **12:** 823–834.

Schmidt, K., Mies, G., and Sokoloff, L. (1991). Model of kinetic behavior of deoxyglucose in heterogeneous tissues in brain: a reinterpretation of the significance of parameters fitted to homogeneous tissue models. *J. Cereb. Blood Flow Metab.* **11:** 10–24.

Schmidt, K., and Sokoloff, L. (2001). A computationally efficient algorithm for determining regional cerebral blood flow in heterogeneous tissues by positron emission tomography, *IEEE Trans. Med. Imag.* **20**(7): 618–632.

Schmidt, K., Sokoloff, L., and Kety, S. S. (1989). Effects of tissue heterogeneity on estimates of regional cerebral blood flow. *J. Cereb. Blood Flow Metab.* **9**(suppl. 1): S242.

Schommartz, B., Larisch, R., Vosberg, H., and Muller-Gartner, H. W. (2000). Striatal dopamine release in reading and writing measured with [123I]iodobenzamide and single photon emission computed tomography in right handed human subjects. *Neurosci. Lett.* **292**(1): 37–40.

Schuier, F., Orzi, F., Suda, S., Lucignani, G., Kennedy, C., and Sokoloff, L. (1990). Influence of plasma glucose concentration on lumped constant of the deoxyglucose method: Effects of hyperglycemia in the rat. *J. Cereb. Blood Flow Metab.* **10:** 765–773.

Schultz, W. (1997). Dopamine neurons and their role in reward mechanisms. *Curr. Opin. Neurobiol.* **7:** 191–197.

Shepherd, G. M. (1998). "The Synaptic Organization of the Brain," 4th ed. Oxford University Press, New York.

Slifstein, M., and Laruelle, M. (2000). Effects of statistical noise on graphic analysis of PET neuroreceptor studies. *J. Nucl. Med.* **41**(12): 2083–2088.

Soares, J. C., and Innis, R. B. (1999). Neurochemical brain imaging investigations of schizophrenia. *Biol. Psych.* **46:** 600–615.

Sokoloff, L. (1977). Relation between physiological function and energy metabolism in the central nervous system. *J. Neurochem.* **29:** 13–26.

Sokoloff, L., Reivich, M., Kennedy, C., DesRosier, M. H., Patlak, C. S., Pettigrew, K. D., Sakurada, O., and Shinohara, M. (1977). The [^{14}C]deoxyglucose method for the measurement of local cerebral glucose utilization: theory, procedure, and normal values in the conscious and anesthetized albino rat. *J. Neurochem.* **28:** 897–916.

Sorenson, J. A., and Phelps, M. E. (1987). "Physics in Nuclear Medicine," 2nd ed. Grune and Stratton, Orlando, FL.

Squire, L. R., and Kandel, E. R. (1999). "Memory: From Mind to Molecules." Scientific American Library, New York.

Strul, D., and Bendriem, B. (1999). Robustness of anatomically guided pixel-by-pixel algorithms for partial volume effect correction in positron emission tomography. *J. Cereb. Blood Flow Metab.* **19:** 547–559.

Suda, S., Shinohara, M., Miyaoka, M., Lucignani, G., Kennedy, C., and Sokoloff, L. (1990). The lumped constant of the deoxyglucose method in hypoglycemia: Effects of moderate hypoglycemia on local cerebral glucose utilization in the rat. *J. Cereb. Blood Flow Metab.* **10:** 499–509.

Tjuvajev, J. G., Avril, N., Oku, T., Sasajima, T., Miyagawa, T., Joshi, R., Safer, M., Beattie, B., DiResta, G., Daghighian, F., Augensen, F., Koutcher, J., Zweit, J., Humm, J., Larson, S. M., Finn, R., and Blasberg, R. (1998). Imaging herpes virus thymidine kinase gene transfer and expression by positron emission tomography. *Cancer Res.* **58:** 4333–4341.

Tsukada, H., Nishiyama, S., Kakiuchi, T., Ohba, H., Sato, K., and Harada, N. (1999). Is synaptic dopamine concentration the exclusive factor which alters the in vivo binding of [11C]raclopride? PET studies combined with microdialysis in conscious monkeys. *Br. Res.* **841**(1–2): 160–169.

Turkheimer, F., Moresco, R. M., Lucignani, G., Sokoloff, L., Fazio, F., and Schmidt, K. (1994). The use of spectral analysis to determine regional cerebral glucose utilization with positron emission tomography and [^{18}F]Fluorodeoxyglucose: Theory, implementation and optimization procedures. *J. Cereb. Blood Flow Metab.* **14:** 406–422.

Turkheimer, F., Sokoloff, L., Bertoldo, A., Lucignani, G., Reivich, M., Jaggi, J. L., and Schmidt, K. (1998). Estimation of component and parameter distributions in spectral analysis. *J. Cereb. Blood Flow Metab.* **18:** 1211–1222.

Videen, T. O., Perlmutter, J. S., Mintun, M. A., and Raichle, M. E. (1988). Regional correction of positron emission tomography data for the effects of cerebral atrophy. *J. Cereb. Blood Flow Metab.* **8:** 662–670.

Watabe, H., Endres, C. J., Breier, A., Schmall, B., Eckelman, W. C., and Carson, R. E. (2000). Measurement of dopamine release with continuous infusion of [11C]raclopride: optimization and signal-to-noise consideration. *J. Nucl. Med.* **41(3):** 522–530.

Watabe, H., Itoh, M., Cunningham, V., Lammertsma, A. A., Bloomfield, P., Mejia, M., Fujiwara, T., Jones, A. K., Jones, T., and Nakamura, T. (1996). Noninvasive quantification of rCBF using positron emission tomography. *J. Cereb. Blood Flow Metab.* **16(2):** 311–319.

Wong, D. F., Wagner, H. N. J., Tune, L. E., Dannals, R. F., Pearlson, G. D., Links, J. M., Tamminga, C. A., Broussolle, E. P., Ravert, H. T., Wilson, A. A., *et al.* (1986). Positron emission tomography reveals elevated D2 dopamine receptors in drug-naïve schizophrenics. *Science* **234:** 1558–1563.

Yu, Y., Annala, A. J., Barrio, J. R., Toyokuni, T., Satyamurthy, N., Namavari, M., Cherry, S. R., Phelps, M. E., Herschman, H. R., and Gambhir, S. S. (2000). Quantification of target gene expression by imaging reporter gene expression in living animals. *Nature Med.* **6:** 933–937.

Zhou, Q., Xiao, M. Y., and Nicoll, R. A. (2001). Contribution of cytoskeleton to the internalization of AMPA receptors. *Proc. Natl. Acad. Sci. U.S.A.* **98(3):** 1261–1266.

Zubieta, J.-K., Koeppe, R. A., Mulholland, G. K., Kuhl, D. E., and Frey, K. A. (1998). Quantification of muscarinic cholinergic receptors with [11C]NMPB and positron emission tomography: Method development and differentiation of tracer delivery from receptor binding. *J. Cereb. Blood Flow Metab.* **18:** 619–631.

CHAPTER

24

Computer Analysis of Nuclear Cardiology Procedures

ERNEST V. GARCIA, TRACY L. FABER, C. DAVID COOKE, and RUSSELL D. FOLKS

Emory University School of Medicine, Atlanta, Georgia

I. Introduction
II. Advances in Single-Photon Emission Computed Tomography Instrumentation
III. Advances in Computer Methods
IV. Conclusion

I. INTRODUCTION

State-of-the-art computer analysis methods in myocardial perfusion single-photon emission computed tomography (SPECT) imaging are now available to clinicians worldwide. Advancements in such computer methods have evolved in key areas such as total automation, data-based quantification of myocardial perfusion, measurement of parameters of left-ventricular global and regional function, three-dimensional displays of myocardial perfusion and function, image fusion, and the use of artificial intelligence and data mining techniques in decision support systems. These advancements, when applied to higher-quality images from instrumentation advancements, have resulted in tools that allow clinicians to make a more comprehensive and accurate assessment of their patients' myocardial status. The advancements continue to strengthen the already strong role of myocardial perfusion SPECT imaging in the diagnosis and prognosis of patients with coronary artery disease.

II. ADVANCES IN SINGLE-PHOTON EMISSION COMPUTED TOMOGRAPHY INSTRUMENTATION

SPECT system gantries are equipped with multiple detectors. The most popular systems are those that offer a selectable variable angle between two detectors. These detectors are positioned at 90° from one another for the 180° orbits used in cardiac imaging and 180° from one another for the 360° orbits used in most other applications (Garcia, 1994). The increased counting sensitivity offered by these multiple detector cameras has been applied to myocardial perfusion SPECT cardiac studies to reduce acquisition time and to allow for the additional assessment of myocardial function.

III. ADVANCES IN COMPUTER METHODS

Commercial nuclear medicine computer systems now have the capability of acquiring and reconstructing electrocardiogram (ECG)-gated myocardial perfusion studies. This feature has promoted the use of SPECT for the total automatic assessment of myocardial perfusion; global and regional function, including the assessment of left ventricular (LV) ejection fraction; LV end-diastolic and systolic volumes; and LV myocardial wall motion and thickening. These assessments are being primarily investigated using gated SPECT imaging of the Tc-99m myocardial perfusion tracers and more recently with thallium-201.

A. Automatic Oblique Reorientation and Reslicing

The principal axes of the left ventricle (LV) are not the same as those of the patient's body. Therefore, whereas tomographic image slices are generally aligned with body coordinates, cardiac images are usually viewed as slices that coincide with the axes of the LV. Images oriented parallel to the principal axes of the LV are called short-axis, horizontal long-axis, and vertical long-axis slices (examples for a normal patient are shown in Figure 1).

Automatic techniques for reorienting LV images from transaxial slices (perpendicular to the body's long axis) to these oblique slices are now used routinely (Germano, Kavanagh, et al., 1995; Mullick and Ezquerra, 1995). Two such approaches start by identifying the LV region in the transaxial images by using a threshold-based approach that makes use of knowledge about the expected position, size,

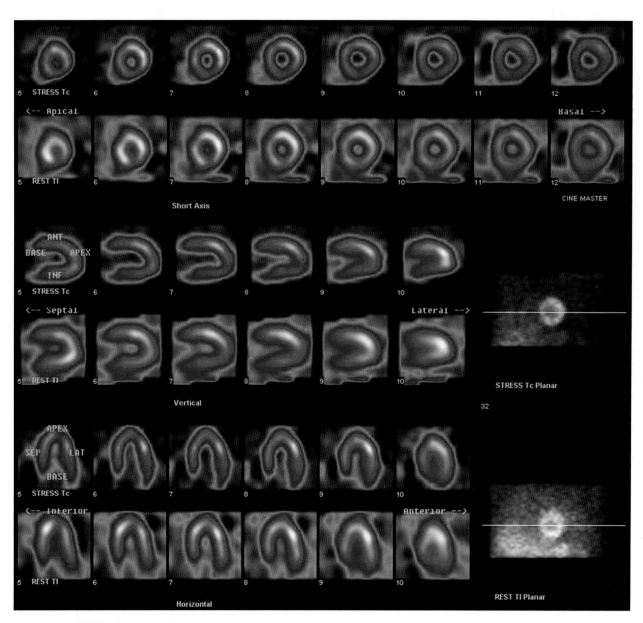

FIGURE 1 Tomographic myocardial perfusion images in various orientations. The top two rows show the stress and rest short-axis slices, the middle two rows show the stress and rest vertical long-axis slices, and the lower two rows show the stress and rest horizontal long-axis slices. The right-most column shows two representative stress and rest planar projections used in the reconstruction of the tomographic slices.

and shape of the LV. Once the LV has been isolated, one can use the original data to refine the estimate of the myocardial surface (Germano, Kavanagh, et al., 1995). In this approach, normals to an ellipsoid fit to the LV region are used to resample the myocardium, and a Gaussian function is fit to the profiles obtained at each sample. The best-fit Gaussian is used to estimate the myocardial center for each profile. Next, after further refinement based on image intensities and myocardial smoothness, the resulting mid-myocardial points are fit to an ellipsoid whose long axis is used as the final LV long axis. Another approach (Mullick and Ezquerra, 1995) involves a heuristic technique to determine the optimal LV threshold and isolate it from other structures. After this is accomplished, the segmented data are used directly to determine the long axis. The binary image is tessellated into triangular plates, and the normal of each plate on the endocardial surface is used to point to the LV long axis. The intersection of these normals (or near-intersections) are collected and fit to a three-dimensional (3D) line, which is then returned as the LV long axis.

B. Stress–Rest Studies

To provide context for the following discussion, let us briefly review the principle of stress–rest perfusion imaging, a technique for diagnosis and prognosis of coronary artery disease. In a stress–rest protocol, the patient is imaged twice: once while the patient is at rest, and once after the heart is stressed either by exercise or pharmacologically. These images may be acquired in either order, depending on the particulars of the protocol. The stress image is used to identify regions of the heart that exhibit diminished perfusion, which is indicative of disease. If the regions show perfusion deficit in the stress image, but appear normal in the rest image, then their function may be restored by surgical or other intervention. Regions that appear abnormal even during rest are in a more-severe state of disease.

C. Automated Perfusion Quantification

Data-based methods for identifying a patient's myocardial perfusion abnormalities from ^{201}Tl and ^{99m}Tc SPECT studies have been previously developed and are widely available (Garcia et al., 1985, 1990; DePasquale et al., 1988). Data-based methods use a statistically defined database of normal patients that serve as a basis against which to compare the 3D perfusion distribution from prospective patients.

One method (Garcia et al., 1990) uses several image identification techniques (e.g., image clustering, filtered thresholding, and specified threshold constraints) for isolation of the left myocardium from the remainder of an image. Once the left ventricular myocardium has been identified, the apical and basal image slices, the Cartesian coordinates of the central axis of the ventricular chamber, and a limiting radius for the maximum count circumferential profile search are determined automatically. In the majority of cases, operator interaction is required only for verification of the automatically determined parameters. If at any time the program fails to locate any of the features, it will branch to an interactive mode and require the operator to select the parameters manually.

This technique has been developed to generate count profiles from a two-part sampling scheme based on stacked short-axis slices, in which the apical region of the myocardium is sampled using spherical coordinates, and the rest of the myocardium is sampled using cylindrical coordinates. This approach, which promotes radial sampling that is mostly perpendicular to the myocardial wall for all points, helps produce an accurate representation of the perfusion distribution with minimal sampling artifacts. Following operator verification of the automatically derived features, the 3D maximum-count myocardial distribution is extracted from all stacked short axis tomograms. Maximum-count circumferential profiles are automatically generated from the short-axis slices using this two-part sampling scheme. These profiles are generated for the stress and rest myocardial perfusion distributions. A normalized percentage change between stress and rest conditions is also calculated as a reversibility circumferential profile (stress–rest protocols are introduced in section III B). The most normal region of the stress distribution is used to normalize the rest distribution to the stress distribution. These distributions are then compared to a gender-matched database generated from patients without coronary artery disease.

Other similar approaches have been implemented (Germano et al., 2000; Slonka et al., 1995; Ficaro, Kritzman, et al., 1999). These four approaches have in common that they are automated and are part of packages for quantifying both myocardial perfusion and function.

D. Image Display

Once perfusion has been quantified, the quantitative data must be displayed. Polar maps were developed as a way to display the quantified perfusion data of the entire left ventricle in a single picture. More recently, 3D displays have been adopted as a more natural way to present the information.

1. Polar Maps

Polar maps, or bull's-eye displays, are the standard for viewing circumferential profiles. An example display is shown in Figure 2. These displays allow a quick and comprehensive overview of the circumferential samples from all slices by combining them into a color-coded

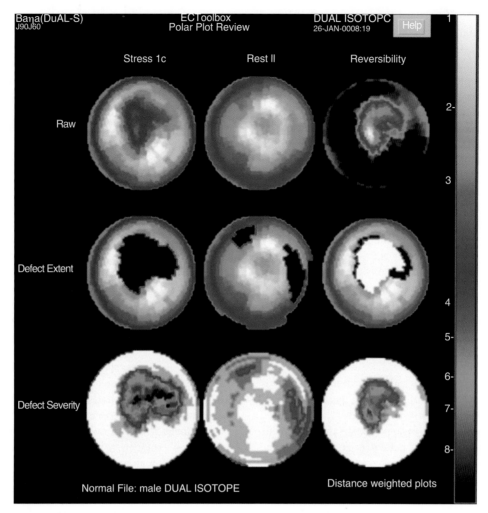

FIGURE 2 Polar maps displaying the results of perfusion quantification. Top row: The quantitated stress (left) and rest (center) perfusion polar maps, and reversibility polar map (right). Middle row: Stress (left) and rest (center) blackout maps, and whiteout reversibility map (right). Bottom row: The number of standard deviations below normal of each myocardial sample of the stress (left) and rest (center) perfusion. This patient had a reversible apical, anterior, and lateral perfusion defect.

image. The points of each circumferential profile are assigned a color based on normalized count values, and the colored profiles are shaped into concentric rings. The most apical slice processed with circumferential profiles forms the center of the polar map, and each successive profile from each successive short-axis slice is displayed as a new ring surrounding the previous. The most basal slice of the LV makes up the outermost ring of the polar map.

The use of color can help the clinician to identify abnormal areas at a glance. Abnormal regions from the stress study can be assigned a black color, thus creating a called blackout map. Blacked-out areas that normalize (recover) at rest are color-coded white, thus creating a whiteout reversibility map. This can be seen in Figure 2. Additional maps can aid in the evaluation of the study by indicating the severity of any abnormality. For example, a standard deviation map depicts, in terms of numbers of standard deviations, the degree to which each point in the circumferential profiles is below the normal value.

Polar maps, while offering a comprehensive view of the quantitative results, distort the size and shape of the myocardium and any defects. Three-dimensional displays have been introduced that are more realistic representations than polar maps.

2. Three-Dimensional Displays

Three-dimensional graphics techniques can be used to overlay results of perfusion quantification onto a representation of a specific patient's left ventricle. This representation is generated using endocardial or epicardial surface points extracted from the perfusion data.

For example, in one approach, the detection of the myocardial surface starts with the coordinates of the maxi-

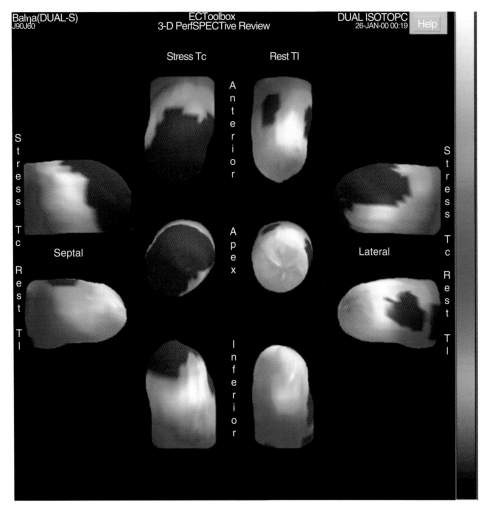

FIGURE 3 Three-dimensional display of the myocardial perfusion distribution. This is a 3D display of the same patient's myocardial perfusion distribution seen in Figure 2. The 3D left ventricle is depicted from different orientations after the distribution has been compared to a dual isotope normal database (^{99m}Tc sestamibi stress/ ^{201}Tl rest). The colors are indicative of the amount of tracer uptake that is related to myocardial perfusion, with the lighter colors representing higher myocardial flow. The blackout results highlight in black the regions of the stress study that were statistically abnormal. The reverse result highlights in white the regions of the rest study that statistically improved at rest and were found at stress to be abnormal. This patient exhibits extensive anterior, lateral, and apical stress perfusion defects with almost complete reversibility at rest.

mal myocardial count samples created during perfusion quantification (Garcia *et al.*, 1990). The coordinates of each sampled point are filtered to remove noise. Estimates of the endocardial (inner) surface and epicardial (outer) surfaces can be determined by subtracting and adding a fixed distance from an estimate of the center of the mycardium.

Once extracted, boundary points can be connected into triangles or quadrilaterals, which are in turn connected into a polygonal surface representing the endocardial or epicardial surface. Once a surface is generated, perfusion is displayed by assigning to each triangle or quadrilateral a color corresponding to the counts extracted from the myocardium at that sample during a quantification process. Colors can be interpolated between triangles to produce a continuum of colors across the myocardium, or the triangles can be left as patches of distinct color. Figure 3 displays the information seen in the polar maps of Figure 2, but using the 3D representation described in Garcia *et al.* (1990).

Surface models can generally be displayed using standard computer graphics packages, allowing rotations and translations of the model, positioning of one or more light sources, and adjustment of the model's opacity (the amount of light shining through the model) or reflectance (the amount of light reflecting off of the model). The speed of a 3D surface model display depends greatly on the type and number of individual elements from which the model is composed. In general, however, surface model displays are quite fast and often can be rotated and translated interactively. The accuracy of such displays depends primarily on the surface detection and the mapping of function onto the surface. In general, they have the advantage of showing the actual size and shape of the left ventricle, and the extent and location of any defect, in a very realistic manner.

E. Automatic Quantification of Global and Regional Function

The quantification of global and regional parameters requires knowledge of the endocardial and (depending on the parameter being determined) the epicardial surface throughout the cardiac cycle. Surface points may be manually assigned or automatically detected. Manual methods are subjective and time-consuming. Automated methods use either boundary (edge) detection or geometric modeling.

In one approach (Germano, Kiat, *et al.*, 1995), automatic border definition is done using thresholds of a Gaussian function precalibrated to a phantom. Although the spatial resolution of SPECT is too low to actually measure the edges of the endocardial surface, the precalibration of a system to a phantom and the high tolerance of the ejection fraction (EF) calculation to errors in these borders results in clinically useful results. This popular approach is the basis of the first commercial software package to offer totally automatic processing, from beginning to end (Germano, Kiat, *et al.*, 1995), for both ^{99m}Tc-sestamibi and ^{201}Tl imaging.

In another approach (Faber, Cooke, *et al.*, 1999), geometric modeling is used. The model uses the maximal count circumferential profile points obtained from the perfusion quantification of the LV as the center of the myocardium. The program assumes that the myocardial thickness at end diastole (ED) is 1 cm throughout and calculates myocardial thickening throughout the cardiac cycle by using Fourier analysis. This analysis assumes that the change in count is proportional to the change in thickness, a by-product of the partial-volume effect. By determining the center and absolute thickness of the LV myocardium at ED, the center of the myocardium throughout the cardiac cycle, and the percent thickening from frame to frame, 3D geometric models of the endocardial and epicardial surfaces are created for each time interval in the cardiac cycle. Once the orientation of the LV has been determined either manually or using commercially available automatic reorientation algorithms, the generation of the endocardial and epicardial surfaces is totally automatic. The generated surfaces are used for all the necessary measurements of global and regional function such as ED and end-systolic (ES) volumes, LV myocardial mass, and LV ejection fraction. Figures 4 and 5 show the functional results of the patient example illustrated in Figures 2 and 3. Combinations of the various approaches have also been implemented (Ficaro, Quaife, *et al.*, 1999).

F. Artificial Intelligence Techniques Applied to SPECT

Assistance in the interpretation of medical images by decision-support systems has made significant progress, mostly due to the implementation of artificial intelligence (AI) techniques. Expert systems and neural networks are two AI techniques that are currently being applied to nuclear medicine.

1. Expert-System Analysis of Perfusion Tomograms

Expert systems are becoming commercially popular because they are designed to circumvent the problem of having few experts in areas where many are needed. Thus, expert systems attempt to capture the knowledge or expertise of the human domain expert and provide it to a large number of nonexperts. An example of the power of expert systems is found in an expert system called PERFEX (Garcia *et al.*, 1995), which was developed to assist in diagnosing coronary artery disease from thallium-201 3D myocardial distributions. This type of approach has potential for standardizing the image interpretation process. After reviewing many studies from patients with angiographically documented coronary artery disease, heuristic rules were derived that best correlate the presence and location of perfusion defects on thallium-201 SPECT studies with coronary lesions. These rules operate on data that are input to the expert system from the SPECT quantification process, which identifies defects as portions of the myocardium in which normalized perfusion falls below a predetermined number of standard deviations when compared to a gender-matched normal file. The rules also identify reversibility as defects at stress that improve at rest. An automatic feature extraction program then describes the location and severity of each defect or reversibility. The location is expressed in the form of 32 possible descriptors, which are defined as coordinates of both depth (basal, medial, distal-apical, and proximal-apical) and angular location (eight subsets of the septal, inferior, lateral, and anterior myocardial walls). The severity is expressed in terms of certainty factors that range from –1 to +1 (–1 means there is definitely no disease, +1 means there is definitely disease, and the range from –0.2 to +0.2 means equivocal or indeterminable). Using these features, the expert system automatically determines the location, size, and shape of each defect/reversibility. This information is used to fire or execute over 200 heuristic rules in order to produce new facts or draw new inferences. For each input parameter and for each rule, a certainty factor is assigned, which is used to determine the certainty of the identification and location of a coronary lesion.

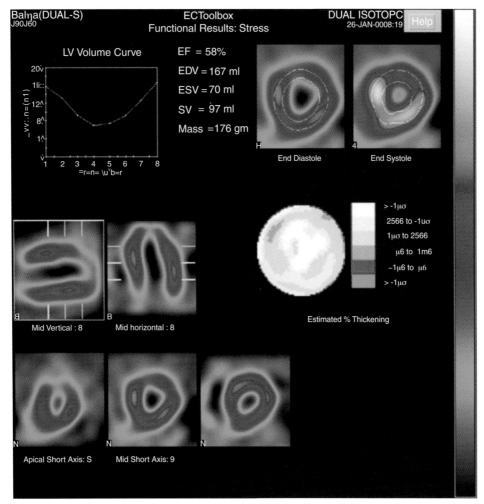

FIGURE 4 Display of myocardial function parameters for the patient shown in Figures 2 and 3. This comprehensive report includes the patient's LV volume–time curve, ejection fraction (EF), end-diastolic volume (EDV), end-systolic volume (ESV), stroke volume (SV), and total LV mass. The report also includes in the middle right panel a polar map of myocardial thickening as measured using amplitude and phase analysis. The map is color coded to correspond to the percent-thickening color bar on its right. The lower left panel may be used for a cinematic display of the LV to visually assess wall motion and wall thickening.

2. Neural Networks

Neural networks have been developed as an attempt to simulate the highly connected biological system found in the brain through the use of computer hardware and software. In the brain, a neuron receives input from many different sources. It integrates all these inputs and fires (sending a pulse down the nerve to other connected neurons) if the result is greater than a set threshold. In the same way, a neural network has nodes (the equivalent of a neuron) that are interconnected and receive input from other nodes. Each node sums or integrates its inputs and then uses a linear or nonlinear transfer function to determine whether the node should fire. A neural network can be arranged in many different ways; for example, it can have one or more layers of nodes, it can be fully connected (where every node is connected to every other node), or it can be partially connected. Also, it can have feed-forward processing (in which processing only travels one direction) or it can have feedback processing (in which processing travels both ways).

Another important aspect of neural networks is their ability to learn based on input patterns. A neural network can be trained in either a supervised or unsupervised mode. In the supervised mode, the network is trained by presenting it with input and the desired output. The error between the output of the network and the desired output is used to adjust the weights of the inputs of all the nodes in such a way that the desired output is achieved. This is repeated for many input sets of training data and for multiple cycles per training set. Once the algorithm has

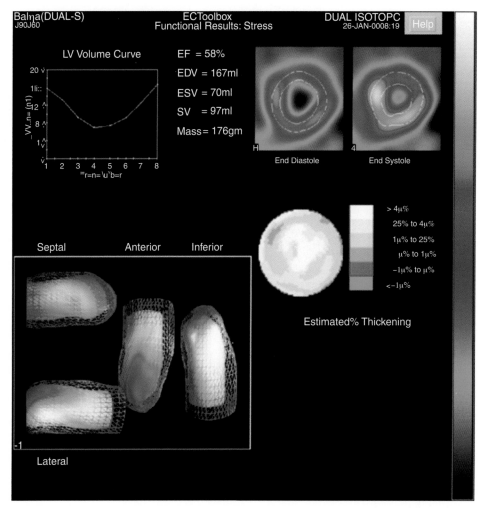

FIGURE 5 Three-dimensional display of myocardial function. This is a 3D display of the patient's myocardial endocardial surface at end diastole (mesh) and at end systole (solid surface) depicted from different orientations. This display may be viewed in a dynamic fashion to visually assess wall motion as the excursion of the wall and wall thickening as the change in color. The color is a function of both myocardial perfusion and myocardial thickness. A myocardial segment that increases in color brightness indicates thickening and thus viable myocardium. This patient exhibits abnormal resting wall motion and thickening in the septum.

converged (i.e., once the weights change very little in response to training), the network can be tested with prospective data. Unsupervised training is similar to supervised training except that the output is not provided during learning. Unsupervised training is useful when the structure of the data is unknown. The main advantage of a neural network is its ability to learn directly from training data rather than from rules provided by an expert. However, if the training data are not complete, the network may not give reliable answers.

Neural networks have been used by three different groups in nuclear cardiology to identify which coronary arteries are expected to have stenotic lesions for a specific hypoperfused distribution (Hamilton *et al.*, 1995). These methods differ in the number of input and output nodes used. The number of nodes that may be used (i.e., the complexity of the model) depends on the amount of available training data The output of the neural network encodes the information it is to report and can be as simple as a single node signifying that there is a lesion present in the myocardium.

G. Software Registration of Multimodality Imagery: Image Fusion

Often in clinical cardiology, more than one type of imaging procedure is used to evaluate patients for cardiac disease. Coronary angiography is the gold standard for diagnosis of coronary artery stenosis, magnetic resonance imaging is performed to determine gross anatomy, and SPECT is used to evaluate myocardial perfusion. However, image registration of the heart or its structures is very difficult because the heart is constantly in motion and because its orientation within the chest cavity may change with patient positioning. Thus, most of the work in cardiac registration is still in the research phase.

For example, accurate assessment of the extent and severity of coronary artery disease requires the integration of physiological information derived from SPECT perfusion images and anatomical information derived from coronary angiography. This integration can be performed by registering a 3D LV model representing myocardial perfusion with the patient's own 3D coronary artery tree and presenting both in a single unified display. The patient-specific 3D coronary arterial tree is obtained from a geometric reconstruction performed on simultaneously acquired, digital, biplane angiographic projections or from two single-plane projections acquired at different angles. The 3D myocardial surface can be generated using boundary detection or modeling techniques on the SPECT image.

In one approach, the 3D reconstructed arterial tree is registered with the myocardial perfusion surface model by automatically minimizing a cost function that describes the relationships of the coronary artery tree with the interventricular and atrioventricular groove and the surface of the myocardium (Faber *et al.*, 1996; Faber, Folks, *et al.*, 1999). Figure 6 illustrates this unified display. Another approach performs the alignment between the 3D left ventricle and coronary arteries by using acquisition parameters for SPECT and angiography to determine patient coordinates of the two models. Once these coordinates are known, the models can be easily aligned and a simple translation between the two models is applied if necessary to refine the match (Schindler *et al.*, 1999).

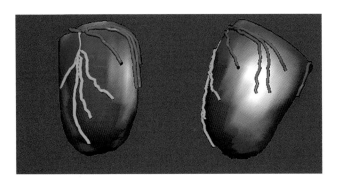

FIGURE 6 Lateral and anterior views of a subject's coronary artery trees reconstructed from biplane angiograms, overlaid on the epicardial surface obtained from SPECT perfusion images. In these images, left-ventricular perfusion is color coded onto the epicardial surface such that the warmer colors indicate higher perfusion and the blacked-out region indicates statistically significant perfusion decreases. The coronary arteries are displayed as green distal to the stenosis and red in normal portions of the tree. Note the overlap between the stenosed coronary artery and the anterior perfusion defect.

IV. CONCLUSION

State-of-the-art computer analysis methods in myocardial perfusion SPECT imaging are available to clinicians worldwide. Advancements in computer methods have evolved in key areas such as total automation, data-based quantification of myocardial perfusion, measurement of parameters of left-ventricular global and regional function, three-dimensional displays of myocardial perfusion and function, image fusion and the use of artificial intelligence, and data mining techniques in decision support systems. These advancements in computer methods applied to higher-quality images from instrumentation advancements have resulted in tools that allow clinicians to make a more comprehensive and accurate assessment of their patients' myocardial status. The advancements have continued to strengthen the already strong role of myocardial perfusion SPECT imaging in the diagnosis and prognosis of patients with coronary artery disease.

Acknowledgments

This work was supported in part by the National Institutes of Health, NHLBI grant number R01 HL42052. The authors received royalties from the sale of clinical application software related to some of the research described in this chapter. The terms of this arrangement have been reviewed and approved by Emory University in accordance with its conflict-of-interest policy.

References

DePasquale, E., Nody, A., DePuey, G., Garcia, E., Pilcher, G., Bredleau, C., Roubin, G., Gober, A., Gruentzig, A., D'Amato, P., and Berger, H. (1988). Quantitative rotational thallium-201 tomography for identifying and localizing coronary artery disease. *Circulation* **77**(2): 316–327.

Faber, T. L., Cooke, C. D., Folks, R. D., Vansant, J. P., Nichols, K. J., DePuey, E. G., Pettigrew, R. I., and Gacia, E. V. (1999). Left ventricular function and perfusion from gated SPECT perfusion images: An integrated method. *J. Nucl. Med.* **40**: 650–659.

Faber, T., Folks, R. D., Klein, J. L., Santana, C. A., Candell-Riera, J., *et al.* (1999). Myocardial area-at-risk validates accuracy of automatic unification of coronary artery trees with SPECT. *J. Nucl. Med.* **40**: 78P.

Faber, T., Klein, J. L., Folks, R. D., Hoff, J. G., Peifer, J. W., *et al.* (1996). Automatic unification of three-dimensional cardiac perfusion with three-dimensional coronary artery anatomy. In *Computers in Cardiology* pp. 333–336. IEEE Press, Indianapolis.

Ficaro, E. P., Kritzman, J. N., and Corbett, J. R. (1999). Development and clinical validation of normal Tc-99m sestamibi database: Comparison of 3D-MSPECT to CEqual. *J. Nucl. Med.* **40**: 125P.

Ficaro, E. P., Quaife, R. A., Kritzman, J. N., and Corbett, J. R. (1999). Accuracy and reproducibility of 3D-MSPECT for estimating left ventricular ejection fraction in patients with severe perfusion abnormalities. *Circulation* **100**: I-26.

Garcia, E. V. (1994). Quantitative myocardial perfusion single-photon emission computed tomographic imaging: Quo vadis? (Where do we go from here?), *J. Nucl. Cardiol.* **1**: 83–93.

Garcia, E. V., Cooke, C. D., Krawczynska, E., Folks, R., Vansant, J. P., deBraal, L., Mullick, R., and Ezquerra, N. F. (1995). Expert system interpretation of technetium-99m sestamibi myocardial perfusion tomograms: Enhacements and validation. *Circulation* **92**(8): 1–10.

Garcia, E. V., Cooke. C. D., Van Train, K. F., Folks, R. D., Peifer, J. W., DePuey, E. G., Maddahi, J., Alazraki, N., Galt, J. R., Ezquerra, N. F., Ziffer, J., and Berman D. (1990). Technical aspects of myocardial perfusion SPECT imaging with Tc-99m sestamibi, *Am. J. Cardiol.* **66:** 23E–31E.

Garcia, E. V., Van Train, K., Maddahi, J., Prent, F., Areeda, J., Waxman, A., and Berman, D.S. (1985). Quantification of rotational thallium-201 myocardial tomography. *J. Nucl. Med.* **26:** 17–26.

Germano, G., Kavanagh, P. B., Su, H. T., Mazzanti, M., Kiat, H., Hachamovitch, R., Van Train, K. F., Areeda, J. S., and Berman, D. S, (1995). Automatic reorientation of 3-dimensional transaxial myocardial perfusion SPECT images. *J. Nucl. Med.* **36:** 1107–1114.

Germano, G., Kavanagh, P. B., Waechter, P., Areeda, J., Van Kriekinge, S., Sharir, T., Lewin, H. C., and Berman, D. S. (2000). A new algorithm for the quantitation of myocardial perfusion SPECT. I: Technical principles and reproducibility. *J. Nucl. Med.* **41:** 712–719.

Germano, G., Kiat, H., Kavanagh, P. B., Moriel, M., Mazzanti, M., Su, H. T., Van Train, K. F., and Berman, D. S. (1995). Automatic quantification of ejection fraction from gated myocardial perfusion SPECT. *J. Nucl. Med.* **36:** 2138–2147.

Hamilton, D., Riley, P. J., Miola, U. J., and Amro, A. A. (1995). A feed forward neural network for classification of bull's-eye myocardial perfusion images. *Eur. J. Nucl. Med.* **22:** 108–115.

Mullick, R., and Ezquerra, N. F. (1995). Automatic determination of left ventricular orientation from SPECT data. *IEEE Trans. Med. Imag.* **14:** 88–99.

Schindler, T., Magosaki, N., Jeserick, *et al.* (1999). Fusion imaging: Combined visualization of 3D reconstructed coronary artery tree and 3D myocardial scintigraphic image in coronary artery disease. *Intl. J. Cardiac Imag.* **15:** 357–368.

Slomka, P. J., Hurwitz, G. A., Stephenson, J., and Cradduck, T. (1995). Automated registration and sizing of myocardial stress and rest scans to three-dimensional normal templates using an image registration algorithm. *J. Nucl. Med.* **36:** 1115–1122.

CHAPTER 25

Simulation Techniques and Phantoms

MICHAEL LJUNGBERG

Department of Medical Radiation Physics, The Jubileum Institute, Lund University, Sweden

I. Introduction
II. Sampling Techniques
III. Mathematical Phantoms
IV. Photon and Electron Simulation
V. Detector Simulation
VI. Variance Reduction Methods
VII. Examples of Monte Carlo Programs for Photon and Electrons
VIII. Examples of Monte Carlo Applications in Nuclear Medicine Imaging
IX. Conclusion

I. INTRODUCTION

The Monte Carlo (MC) method is a numerical technique useful for solving statistical problems and involves the use of random numbers to simulate, for example, measurements in a specific geometry. The name comes from Monte Carlo in Monaco where gambling is popular and refers to the random nature of such activities. The method is not used solely for nuclear medicine imaging—on the contrary, it can be found in numerous scientific applications. However, because nuclear medicine imaging deals with radioactive decay, the emission of radiation energy through photons and particles, and the detection of these quanta and particles in various materials, the MC method is an important simulation tool in both nuclear medicine research and clinical practice. This chapter describes the basics of MC simulations. Review papers and textbooks on the subject are listed in the References (Andreo, 1991; Ljungberg *et al.*, 1998; Raeside, 1976; Turner *et al.*, 1985; Zaidi, 1999).

A. Why Simulation? Limitations and Benefits

What is it that makes computer simulation so powerful compared with real measurements? First, there is a considerable advantage for the development of new equipment in that, if an accurate computer model of the system is available, it is very easy to study the effects of changes in individual parameters such as crystal thickness, collimator hole dimensions, and phantom size. This allows the optimization of complex systems without the need for expensive manufacturing and testing of equipment. One of the limitations, however, is that because the results are based on a computer model, the accuracy of the results will depend on the accuracy of the model. The question we should always ask is whether the model includes all the factors we would expect to affect the measurements in real situations. For example, if the aim is to simulate an energy spectrum measured with a NaI(Tl) scintillation detector in a laboratory geometry, it may be necessary to include in the model all the equipment and the walls around the detector to obtain as accurate an energy spectrum as possible. However, if we are not interested in the lower part of the energy distribution, then neglecting to simulate scattering from the walls may not affect the results in our primary region of interest (which is often the photopeak region).

Another major advantage of simulations in nuclear medicine imaging is that they allow studies of parameters that are not measurable in practice. The fraction of photons that are scattered in a phantom and their contribution to the image are such parameters, they can only be measured indirectly for a very limited number of geometries. In addition, in a computer model it is possible to turn off certain effects, such as photon attenuation and scattering in the phantom, which means that ideal images that include camera-specific parameters can be created and used as reference images. In combination with patient-like phantoms, the Monte Carlo method can be used to produce simulated images very close to those acquired from real measurements, and the method is therefore a very useful tool.

B. Probability Density Functions

It is of fundamental importance to obtain information about the physical process to be simulated. We need to know how various particles interact with the material in question and which types of interaction will occur under which conditions. This information is usually expressed as a probability density function (*pdf*). From this information, a stochastic variable can be sampled and used in the simulation procedure. A *pdf* is defined over a range [*a*, *b*] and it should, ideally, be possible to integrate the function so that it can be normalized by integration over its entire range.

C. The Random Number Generator

Another fundamental requirement in MC simulations is access to random numbers. In theory, a random number cannot be predicted or calculated. Despite this, the most common approach in computer simulations is to use a computer program to obtain randomly distributed numbers. These pseudorandom numbers are calculated from previous numbers, which implies that there must be an original number to initiate the random number sequence. Thus, starting two simulations with the same initial value will in the end produce the same results, even through random numbers are involved. Furthermore, because simulations are performed on digital machines that use numbers limited to a known range (2 bytes, 4 bytes, 8 bytes, etc.), there is always a chance that a random number sampled during the simulation will be equal to the initial value. The result will then be a repetition of the random number sequence, which may have serious consequences for the accuracy of the results.

II. SAMPLING TECHNIQUES

In all MC calculations, stochastic variables are sampled from the *pdf* using uniformly distributed random numbers. Two distinct methods, the *distribution function method* and the *rejection method*, are frequently used (Ljungberg, 1998a).

A. The Distribution Function Method

A cumulative probability distribution function *cpdf*(*x*) is obtained by integrating the *pdf*(*x*) over the interval [*a*,*x*], using:

$$cpdf(x) = \int_a^x pdf(x')\, dx' \quad (1)$$

A random variable x can then be sampled by replacing the value of the function *cpdf*(*x*) with a uniform random number R, distributed within the range [0,1] and then solve for x. The distribution function method is very efficient because it only requires a single random number for each sample.

B. The Rejection Method

Sometimes, it is virtually impossible to obtain an inverse of the analytical function. In such cases, the rejection method can be applied. Basically, the method is an iterative procedure in which an initial value from a normalized function $pdf^*(x) = pdf(x)/\max[pdf(x)]$ is sampled using a linearly distributed value of x within the range [*a*, *b*] from the relation $x = a + R_1(b - a)$. A second random number, R_2, will then decide if this guess is acceptable as a properly sampled value. The test is done by calculating the value of the function $pdf^*(x)$ using the linearly sampled value of x and then checking whether R_2 is lower than the value of $pdf^*(x)$. If it is not, then a new value of x is linearly generated. If the condition is fulfilled, x is accepted as a properly distributed stochastic value. The efficiency of the rejection method may vary because many samples of x may be required before a value is accepted.

III. MATHEMATICAL PHANTOMS

In a nuclear medicine MC simulation, a photon history often starts by defining the point of decay in an activity volume. The distribution of the simulated activity may be very simple, for example a point source, or it may be a more complicated distribution that mimics a distribution in organs of a patient. The technique for sampling a radionuclide distribution thus depends on the complexity of the distribution, and two distinct methods, based on different principles, are used. In the first, decays are generated from an analytical description of a source; in the second, the decays are generated from a discrete pixel-based image of the source.

A. Analytical Phantoms

The rejection method is used to determine whether a uniformly sampled decay is inside the shape to be simulated. Combining simple geometries, such as cylinders, spheres, and ellipsoids allows the simulation of more complicated volumes. Examples of such phantoms are the MIRD phantom (Snyder *et al.*, 1969), which forms the basis of MIRD S-value dosimetry, and the recently published brain phantom (Bouchet *et al.*, 1996). These types of phantoms have been widely used in diagnostic radiation dose estimation, but have not been used as frequently for MC simulation in nuclear medicine imaging.

B. Voxel-Based Phantoms

The simulation of patient-like activity and density distributions using analytical functions may be difficult, because it is not easy to model accurately the shapes of organs and structures in the body based on simple geometries. An alternative approach is, thus, to use digital images of the phantom. The relative location of each pixel determines its position in the *x*, *y*, and *z* directions, and the pixel value determines the activity, that is, the number of particles emitted from that location. Directly measured data from, for example, computed tomography (CT) scanners can be used to allow the simulation of patients. Zubal and Harrell (1993, 1994) have developed one of the most widely used voxel-based phantoms based on a detailed CT study of a male. To construct the phantom, important organs and structures in the images were segmented and each of these regions is assigned a number so that each voxel can be identified as being part of a specific organ. This byte-coded phantom can be used to define density and activity distributions in three dimensions. In a similar manner, Zubal and Harrell have also developed a voxel-based phantom of the brain, using data from a high-resolution magnetic-resonance imaging (MRI) study. Segmented arms and legs obtained from the NIH Visual Human Project (National Library of Medicine) have been added by Dawson *et al.* (1997) to the Zubal voxel man phantom. These data were derived from anatomically detailed, 3D representations of a normal male human cadaver. Our group has further developed this modified Zubal phantom by straightening out the arms to make the phantom more useful for whole-body simulations. Figure 1 shows rendered images of the original Zubal torso phantom (A), the modified version with the original Visual Human Project arms and legs (B), and our version with the arms straightened out (C). This modified phantom (Figure 1C) has been used in the simulation examples discussed in this chapter. It is difficult to change the size and shape of individual organs because

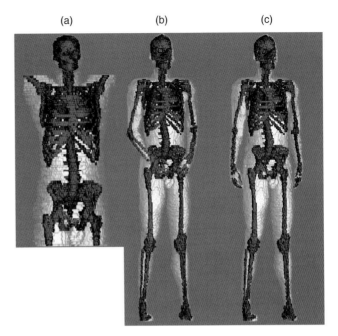

FIGURE 1 Rendered images showing (a) the original voxel-based torso phantom developed by Zubal and Harrell (1993, 1994) (b) the modified version with arms and legs from the Visual Human project added to the torso, and (c) our version with the arms straightened out

the phantom is essentially a series of images. However, by changing the pixel size it is possible to scale the overall phantom.

A popular four-dimensional (4D) phantom for medical imaging research is the Mathematical CArdiac Torso (MCAT) phantom, originally based on the MIRD phantom (Snyder *et al.*, 1969). The MCAT phantom has been shown to be very useful for the evaluation of image reconstruction methods (Pretorius *et al.*, 1999; Tsui *et al.*, 1993) and the study of the effects of anatomical variations on myocardial single-photon emission computed tomography (SPECT) images. With the addition of cardiac and respiratory motion, it has been used in the study of gated SPECT images (Lalush and Tsui, 1998) and the effects of respiratory motion on myocardial SPECT images (Tsui *et al.*, 2000). As shown in Figure 2A, the 4D MCAT phantom is based on simple geometric primitives but uses cut planes, intersections, and overlap to form complex biological shapes. Because it is defined by continuous surfaces, the MCAT phantom can be defined at any spatial resolution. Using the MCAT phantom, it is also possible to model anatomical variations by varying the parameters that define the geometric primitives and cut planes describing the internal organs (heart, liver, breast, etc). Because the anat-omy of the phantom can be modified in this way, the MCAT program can be used to simulate a patient population in patient studies. However, the geometric primitives limit the ability of the MCAT phantom to model the anatomy realistically.

A more realistic version of the 4D MCAT phantom is currently being developed (Segars, 2001; Segars *et al.*, 2001;

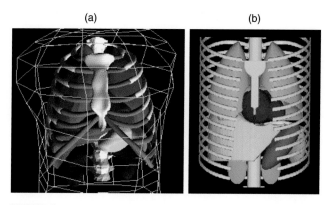

FIGURE 2 Surface renderings are shown of (a) the current geometry-based 4D MCAT phantom and (b) the NURBS-based 4D MCAT phantom. The NURBS-based MCAT phantom is more realistic than the geometry-based phantom while maintaining its flexibility for modeling anatomical variations and patient motion.

Tsui *et al.*, 2000). The organ shapes in this new MCAT phantom are modeled with nonuniform rational B-splines surfaces (NURBS) using actual human data as the basis for the construction of the continuous surfaces. NURBS are widely used in computer graphics to mathematically define complex curves or surfaces. The majority of organ shapes in the NURBS-based MCAT phantom are based on the Visible Human CT dataset. Figure 2B shows the anterior view of the new phantom. This phantom also includes the beating heart and respiratory motion. The beating heart model is based on a gated MRI cardiac scan of a normal patient, whereas the respiratory model is derived from known respiratory physiology. Because it is based on human data, the NURBS-based 4D MCAT phantom can be used to model organ shapes, anatomical variations, and patient motions more realistically than the geometry-based 4D MCAT phantom without sacrificing its flexibility. Fewer organs are defined than in the Zubal and Harrell phantom, but it is more flexible in terms of changing organ locations and sizes.

C. Other Types of Phantoms

The use of voxel-based phantoms, especially at high resolution, can be slow because the general principle is to step through the phantom collecting information along the way. When the voxel number becomes very large, the procedure becomes time-consuming. Suggestions for speeding up the calculations by introducing special structures, called octrees, have been made by Ogawa *et al.* (1995). Wang and colleagues (Wang, Jaszczak, and Coleman, 1993; Wang, Jaszczak, Gilland, *et al.*, 1993) have also developed a composed object for MC simulation to speed up the otherwise calculation-intense simulation when using large voxel-based phantoms. Some programs, such as the MCNP4B program, described later, use combinatorial geometry packages to construct complex structures from simple geometries.

IV. PHOTON AND ELECTRON SIMULATION

A. Cross-sectional Data for Photons

Good accuracy in the cross-sectional data for the various photon processes is essential because these data form the basis of the simulation. Several different sources of differential attenuation coefficients can be found in the literature. The data, published by Storm and Israel (1970), were incorporated into early software, and the work by Hubbell and Berger (Hubbell, 1969; Hubbell *et al.*, 1975) on this topic has also been used extensively. The latter group has developed a PC-based program called XCOM (Berger and Hubbell, 1987) that is very useful. The XCOM cross-sectional generator can also be found on the Internet (http://physics.nist.gov/PhysRefData/Xcom/Text/XCOM.html). Well-established sources of cross-sectional data and other links of interest can be found on the home page of the National Nuclear Data Center at Brookhaven National Laboratory (http://www.nndc.bnl.gov). When comparing results from different simulation studies, it is important to consider the source of the cross-sectional data because slight differences in these may affect the results.

B. Path-Length Simulation

An important stochastic parameter is the path length (Snyder *et al.*, 1969). It is dependent on the photon energy, the mass-attenuation coefficient and the density of the material. The probability function is given by

$$pdf(x) = \mu \exp(-\mu x) \quad (2)$$

The probability of a photon traveling a distance less than or equal to d is given by

$$cpdf(d) = \int_0^d \mu \exp(-\mu x)dx = [-\exp(-\mu x)]_0^d = 1 - \exp(-\mu d) \quad (3)$$

Using the distribution function method already described, a path length can be sampled by replacing $cpdf(d)$ by a uniformly distributed random number R, where

$$R = cpdf(d) = [1 - \exp(-\mu d)] \quad (4)$$

The stochastic variable d can be calculated from

$$d = -\frac{1}{\mu}\ln(1-R) = -\frac{1}{\mu}\ln(R) \quad (5)$$

In Eq. (5), $(1-R)$ is equal to R because $(1-R)$ has the same uniform distribution as R.

C. Sampling Interaction Types

The type of interaction at the end of the track needs to be sampled if it is determined that the photon will not escape the volume of interest. The probability of each type of photon interaction as a function of energy and material is given by the differential attenuation coefficients normalized to the total linear attenuation coefficient. A distribution function can then be created from these normalized coefficients and a uniform random number, R, will finally determine the actual type of interaction.

D. Photoabsorption

This interaction results in the absorption of a photon with an energy deposition of $h\nu$ in the material. Depending on the desired accuracy in the simulation model, it may be necessary to further simulate the effect of the absorption by including the emission of characteristic X-ray photons, for example the 30-keV K_α X-ray from iodine in the NaI(Tl) crystal, because these photons can escape the volume of interest, decreasing the amount of absorbed energy.

E. The Compton Process

In a Compton process, the photon energy lost by the interacting electron is only a fraction of its initial energy. The photon is also scattered in a direction different from its initial path. There is a direct relation between the initial photon energy, the scattered photon energy, $h\nu'$, and the scattering angle, θ:

$$h\nu' = \frac{h\nu}{1 + \frac{h\nu}{m_0 c^2}(1-\cos\theta)} \quad (6)$$

The differential cross section for photon scattering is given by the Klein-Nishina equation:

$$d\sigma_{\gamma,\mathcal{H}}^e = \frac{r_e^2}{2}\left(\frac{h\nu'}{h\nu}\right)^2\left(\frac{h\nu}{h\nu'} + \frac{h\nu'}{h\nu} - \sin^2\theta\right)d\Omega \quad (7)$$

A scattering angle can be sampled using the method developed by Kahn (1956). This method is a mixture of both the distribution function method and the rejection method. A description of the method can be found in Ljungberg et al. (1998a).

F. Bound Electrons

Equation (7) is based on the assumption that the interacting electron is free and at rest. For photon interactions with tightly bound atomic electrons in high-Z material, this assumption is not justified when the photon energy is in the same order as the binding energy of the electron. Correction for bound electrons must then be made in the sampling of the scattering angle (see, for example, Persliden, 1983; Ljungberg and Strand, 1989).

G. Coherent Scattering

In this process, the energy transfer to the electron is very small, but the photon path is changed (Evans, 1955). Scattering in the forward direction is most probable. Coherent scattering is often neglected. The reason is that, because the coherent scattering is mostly forward-peaked, it can be partly taken into account by allowing the photon to travel on average a little longer between interactions. In some cases, it is necessary to simulate coherent scattering explicitly. A sampling technique based on form-factor data for the material has been described by Persliden (1983).

H. Electron Simulation

In most simulations in the field of nuclear medicine imaging, the path length of the secondary electrons is so small that the assumption of local energy absorption is justified. It is, however, of interest to compare the difference in the methods of simulating electrons and photons. Like photons, electrons (and positrons) are subject to violent interactions. The transport theory of electrons is more complicated than that for photons. The major difference is the enormous number of interactions undergone by the electron before it is completely stopped in the material. These interactions can be classified into catastrophic interactions (large direction and energy changes) and noncatastrophic interactions (small direction and energy changes). The following interactions are categorized as catastrophic.

- Scattering with large energy loss (Möller scattering, $e^-e^- \rightarrow e^-e^-$)
- Scattering with large energy loss (Bhaba scattering, $e^+e^- \rightarrow e^+e^-$)
- Hard bremstrahlung emission (eN $\rightarrow$ eN + bremstrahlung)
- Positron annihilation at rest ($e^+e^- \rightarrow$ 2 annihilation photons)

These interactions can be sampled discretely in a reasonable amount of computing time. In the following types of interactions, the result of the interaction is not catastrophic (i.e., the direction and energy of the electron are almost the same after the interaction).

- Low-energy Möller scattering ($e^-e^- \rightarrow e^-e^-$)
- Atomic excitation
- Soft bremstrahlung emission (eN $\rightarrow$ eN + bremstrahlung)
- Elastic electron multiple scattering from atoms (eN $\rightarrow$ eN)

It is not realistic to sample these interactions discretely in a reasonable amount of computing time. Instead, statistical theories have been developed that describe these weak interactions by accounting for them in a cumulative manner by including the effects of many such interactions at the same time. These are called statistically grouped interactions. Multiple scattering can be handled in two ways. In Class I theory, the CSDA (continuous slowing down approximation) energy loss includes the energy loss in both discrete and catastrophic events. The change in angle of the primary particle is not explicitly modeled for discrete events, because it is included in the treatment of multiple scattering. In the Class II method, the path length to the point of a discrete event is determined in a way similar to that in the simulation of photon transport. At the point of interaction, the angular direction of the primary electron is determined in a way that depends on the direction of the scattered electron. This requires a modification of the LET (linear energy transfer) so that all events below a given cut-off energy are regarded as continuous and all those above are regarded as discrete.

V. DETECTOR SIMULATION

The simulation examples in this chapter have been made by the use of the SIMIND program (described in Section VII).

A. Simulation of Energy Resolution

In many scintillation camera simulation programs, the explicit simulation of an imaging detector stops after the determination of the energy deposition in the crystal. However, the actual detection of the scintillation light from the energy depositions involve statistical errors that depend on:
1. The amount of light emitted
2. The collection and guidance of the light to the photomultiplier tubes (PMTs)
3. The efficiency of the photo cathodes in the PMTs in converting the light into photoelectrons
4. The collection of the electrons on the PMT anode
5. The signal processing in the electronics

All these steps contribute to an overall uncertainty in the measurement of the energy, which is commonly referred to as the energy resolution. Most of these steps can be explicitly simulated using MC methods (Knoll *et al.*, 2000). To maintain flexibility in the program, however, the explicit simulation of the components is often disregarded and it is assumed that the energy deposition can be convolved with an energy-dependent Gaussian function. The Box-Muller transformation (Box and Muller, 1958) can be used to convolve an initial value of the imparted energy, E, online to produce a Gaussian distributed estimate of the energy signal, P, assuming a standard deviation, SD.

1. Obtain two uniformly distributed random numbers R_1 and R_2 within the range $[-1,1]$.
2. If $R_3 = R_1^2 + R_2^2$ is greater than unity, repeat Step 1.
3. Obtain P from the equation $P = E + \sqrt{2\dfrac{\ln\left(\dfrac{1}{R_3}\right)}{R_3}} R_1 \, \mathrm{SD}$

Note that the full width at half maximum (FWHM) = 2.35 SD and that the energy resolution is generally assumed to vary as the reciprocal of the energy squared. Figure 3 shows graphs from a simulation of photons from a ^{131}I distribution within the Zubal phantom that impinge on a scintillation camera. Graph (A) shows the energy imparted in the crystal and graph (B) shows the spectrum when a Gaussian function is applied to convolve the energy in order to take into account the energy resolution. Note that the events from

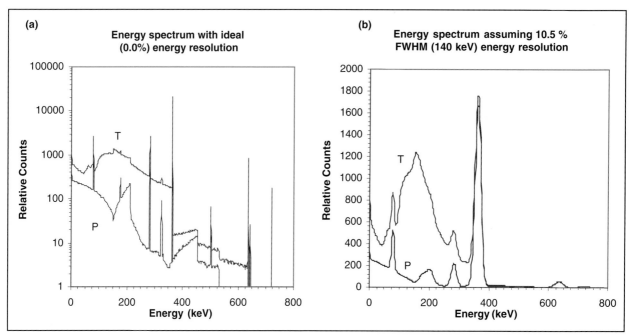

FIGURE 3 Simulated ^{131}I energy spectrum with (a) an ideal energy resolution and (b) an energy resolution of 10.5% FWHM at 140 keV. The events from the primary photons (P) and the total number of events (T) are shown separately.

primary photons, which have not been interacted in the phantom, have been plotted separately.

B. Simulation of Temporal Resolution

When imaging an activity distribution resulting in high count rates in the camera, a phenomenon called pile-up may occur, leading to limitations in both the energy and the spatial measurements. This is because of the decay time of the scintillation light. A limited number of MC investigations have been performed to study the effect of pile-up and mispositioning of events in nuclear medicine imaging. Strand and Lamm (1980) showed in an early study that a MC simulation could predict the degree of mispositioning when imaging four high-activity sources in each corner of a quadrant. By including light from earlier events, events appeared in the images in a distinct pattern between the sources. In these calculations, it was necessary to sample a time interval randomly between two consecutive events. The authors assumed an instantaneous rise time and an exponential decay time and showed that a Poisson-distributed sample of the time interval could be obtained from:

$$e^{(-T_n/TK)} = R^{1/(TK \cdot IA)} \tag{8}$$

where TK is the decay constant and IA is the expected count rate in the whole energy spectrum. The contribution to the signal from the previous scintillation event P_{n-1} can then be described by

$$Z_n = W_n + Z_{(n-1)} R^{1/(TK \cdot IA)} \tag{9}$$

where R is a uniformly distributed random number in the range [0,1] and W_n is the contribution from the current interaction.

Ljungberg *et al.* (1995) calculated the effect of count rate on scattering parameters, such as the scatter fraction, and concluded that scatter correction methods obtained from low-count-rate measurement cannot always be applied to high-count-rate situations because of the distortion in the energy pulse-height distribution. It was demonstrated that primary events appeared on the high-energy side of a fixed photopeak window and events in the Compton region piled up into the energy thus changing the fraction of primary and scattered events in the image. One of the difficulties when comparing these simulations with real measurements is that it can be very difficult to find out how pile-up rejection techniques work in the real camera and to create an accurate computer model of the process.

C. Backscattering of Photons behind the NaI(Tl) Crystal

Modeling the backscatter of photons behind the crystal is a difficult task because of the complex geometry of light guides and PMTs. Nevertheless, for some applications, especially for radionuclides that emit photons of energies higher than the principal energy, scattering of these photons back into the crystal can cause degradation of the image quality. To include this in the model, at least partly, an equivalent layer of a backscattering material can be defined that allows scattering back into the crystal. The difficult task is then to determine the thickness of the layer so that the same scattering conditioning as in the light-guide and PMTs can be achieved (De Vries *et al.*, 1990). Figure 4A shows a graph of the ^{131}I simulation when backscattering has been included.

D. Collimator Penetration and Scattering

It may be difficult to simulate collimators due to the shapes of the holes or the large number of holes. In many situations, it is sufficient to consider only geometrical collimation and assume that penetration and scattering may be neglected. The collimator response can be simulated by a distance-dependent Gaussian function. Explicit simulations of interactions in the collimator have been described by De Vries (1990), Moore (1995), and Yanch (1993). Figure 4B shows a graph of the ^{131}I simulation when both backscattering behind the crystal and photon interactions in the collimator have been included. Note the increase of the amplitudes for the 637-keV and 723-keV photons. This is a result of the penetration through the collimator septa.

VI. VARIANCE REDUCTION METHODS

A. The Idea behind the Weight

In real measurements, a detector responds to a photon interaction and an event occurs because of this interaction. In imaging devices, such as the scintillation camera, not only is the energy imparted to the detector measured but also the location of the event (by using the center of gravity of the emitted scintillation light). When an event occurs, the channel in the energy pulse-height analyzer corresponding to the energy imparted is updated by one count. This can be interpreted as the distribution being updated by a probability equal to unity because the event has actually occurred. If we call this value of unity a weight, then for physical measurements the weight can be either unity or zero (either an event is detected or it is not). The probability of detection depends on physical parameters, such as attenuation, collimation, and crystal thickness. In the world of computers, probability functions can be changed. For example, photon emission can be simulated in only one hemisphere, but the results from such simulations make sense only if we correct for the change in photon emission probability. Thus, because the probability of

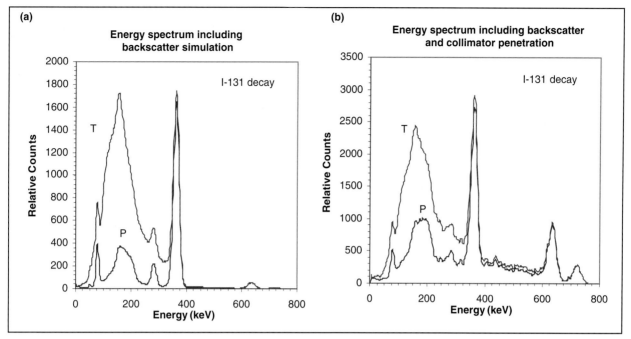

FIGURE 4 Energy spectrum from the ^{131}I simulation with (a) backscattering behind the crystal and (b) in addition to this photon interactions in the collimator (c). The contribution into the 364-keV energy window downscattered from the 637-keV and 723-keV photons can also be seen. These events together with events from photons that penetrate the hole septum will degrade the image quality.

emission into one hemisphere is 0.5, the value (or the weight) used to update the energy pulse-height distribution or the image in case of detection is equal to 0.5. If the computer model includes no scattering media in the other hemisphere, then the efficiency of the program will have been increased by a factor of 2 because we are not simulating photons traveling in the direction of the other hemisphere. It is not physically meaningful because there are no materials that can make these photons change direction toward the detector.

B. Forced Detection (Angular and Spatial)

One of the most efficient variance reduction methods in scintillation camera simulations is the forced detection method, in which the angular (or spatial) probability distribution function is altered in order to make as many of the photons impinge on the detector as possible. Because the scintillation camera has a collimator attached to it, the probability of a photon passing through the collimator holes per emission is on the order of 10^{-4}. If an MC simulation is applied without employing any variance reduction, then the majority of the simulated photons will be rejected by the collimator and will not contribute to the energy signal. Forced detection by allowing emission only into a limited solid angle along a certain direction is therefore a very efficient way of speeding up the calculation process in nuclear medicine imaging. Suitable distributed photon emission within a maximum polar angle θ_{max}, can be obtained from the relation:

$$\cos(\theta) = 1 - R\,[1 - \cos(\theta_{max})] \quad (10)$$

assuming a uniformly distributed azimuthal angle according to:

$$\varphi = 2\,R\,\pi \quad (11)$$

Because the probability distribution for these angles has been changed, it is necessary to correct for this forcing technique by modifying the photon weight:

$$W_n = W_{n-1}(1 - e^{-\mu d_{max}}) \quad (12)$$

Note, that when applying several variance reduction methods, the photon weight is the product of the weights from each method. In a similar way, the photon can be forced to interact in the crystal by changing the probability distribution function used to sample the path length. Normally, a path is sampled using Eq. (5). However, it is possible to force the photon to have a path length less than or equal to a maximum distance, d_{max}, by changing Eq. (5) to

$$d = -\frac{1}{\mu}\ln[1 - R\,(1 - e^{-\mu d_{max}})] \quad (13)$$

The weight, W, must also be modified here by multiplying it by the probability of detection:

$$W_n = W_{n-1}[1 - e^{-\mu d_{max}}] \quad (14)$$

An example of an increase in efficiency using this technique is the simulation of 511-keV photons impinging on a standard scintillation camera with a crystal thickness of 0.95 cm. Direct MC simulation will give approximately 110 interacting photons in the energy window per 1000 impinging photons on the crystal, but when Eq. (13) is applied all the photons will be recorded. In addition to the methods described here, other variance reduction methods exist such as photon splitting, Russian roulette, photon stratification, and important sampling. An excellent overview of the theory behind the variance reduction has been given by Haynor (1998).

VII. EXAMPLES OF MONTE CARLO PROGRAMS FOR PHOTON AND ELECTRONS

In Ljungberg et al. (1998), we describe some of the MC software that has been widely used in nuclear medicine simulations. In this section, we briefly outline the essential features of some of these programs.

A. The SIMIND Program

This program was originally designed for the calibration of whole-body counters, but soon evolved to simulate scintillation cameras (Ljungberg, 1998b; Ljungberg and Strand, 1989). It is now available in Fortran-90 and can be run on major computer platforms including PCs. The SIMIND program actually consists of two programs, CHANGE, which defines the parameters, and SIMIND, which performs the actual simulation. The program can simulate nonuniform attenuation from voxel-based phantoms and includes several types of variance reduction techniques. Transmission SPECT images can also be simulated. For particular projects, the user can write scoring routines that are linked to the main code. Documented COMMON blocks are used to access data. This allows considerable flexibility because the user does not need to modify the main code. The program has been modified to allow it to be run on parallel computers using the MPE command language. The use of SIMIND is well documented (http://www.radfys.lu.se/simind/).

B. The SimSET Program

The SimSET program (http://depts.washington.edu/~simset/html/simset_main.html) models the physical processes and instrumentation used in emission tomography imaging. It is written in a modular format with the Photon History Generator (PHG) as the core module. The PHG includes photon creation and transport and heterogeneous attenuators for both SPECT and positron emission tomography (PET). The Collimator module receives the photons from the PHG and tracks them through the collimator. The Detector module receives the photons from both the PHG module and the Collimator module and tracks them through the specified detector, recording the interactions within the detector for each photon. Each module can create a Photon History File in which information is recorded on the photons it has tracked. The Binning module is used to process photons and detection records and can be used to sort the photon or detection records into histograms, online, or from preexisting history files. The PHG, Detector, and Collimator modules are configured using parameter files and data tables. Digital phantoms for the PHG (Activity and Attenuation Objects) can be created using a function called the text-based utility in the Object Editor.

C. The EGS4 Package

The EGS4 computer code (http://www.slac.stanford.edu/egs) is a general-purpose package for Monte Carlo simulation of coupled transport of electrons and photons in an arbitrary geometry for particles with energies from above a few kilo electron volts up to several giga electron volts. The EGS4 code is actually a package of subroutines with a flexible user interface. It is necessary to write a user code in a preprocessor language, called Mortran-77, which then produces Fortran-77 code. The Mortran language makes use of macro definitions that allow the user to modify internal parameters in the EGS4 code. The simulation is based on Class II electron transport. Use of the EGS4 package requires the user to specify the following.

1. The physical characteristics of the incident particles (type, energy, location and direction)
2. The target regions in which particle transport is to be simulated (type of medium, equation describing the region boundaries)
3. The energy cut-off values ECUT and PCUT, below which the transport of electrons and photons, respectively, is terminated
4. The maximum energy loss per transport step in percent (ESTEPE) due to the continuous slowing down of electrons between hard collisions or due to irradiative energy-loss events

In the EGS4 system, each geometric region can be defined as a medium with different density. After sampling the physical characteristics of the incident particle, the SHOWER subroutine is called to go into EGS4, and EGS4 then controls the particle transport. EGS4 calls a subroutine, HOWFAR, which is programmed by the user to define the geometry (the region boundaries and the phantom geometry are defined here). When an event such as a Compton scattering event or hard collision with energy depositions takes place, EGS4 then calls a subroutine, AUSGAB, which is also programmed by the user to record the information required.

D. The MCNP4 Program

MCNP4 (http://laws.lanl.gov/x5/MCNP/index.html) is a general-purpose, continuous-energy, generalized-geometry, time-dependent Monte Carlo transport code (Briesmeister, 1997). It can be used for several particles, including neutrons, photons, and electrons, or for coupled combinations of these particles. MCNP4 allows the arbitrary configuration of material using a generalized geometry package. For example, collimator simulations can be performed relatively easily with the MCNP4 code because the combinatorial geometry package allows for the simulation of hexagonal holes in a symmetrical pattern. The simulation employs a Class I method. The program is written in Fortran-77.

VIII. EXAMPLES OF MONTE CARLO APPLICATIONS IN NUCLEAR MEDICINE IMAGING

A. General Detector Parameters and Energy Spectrum Analysis

One of the earliest studies on detector simulation is Anger and Davis (1964). In this investigation, the efficiency of the NaI(Tl) scintillator is evaluated for different photon energies. In addition, the intrinsic resolution resulting from multiple scattering is studied. Work in this direction has also been carried out by Beattie and Byrne (1972).

Because of the limited energy resolution of NaI(Tl) crystals, a significant number of scattered photons will be detected in the photopeak energy window. It is important to understand the distribution of these photons when developing correction methods. Work in this field has been published by Floyd et al. (1984) and Manglos et al. (1987).

B. Evaluation of Scatter and Attenuation Correction Methods

The Monte Carlo method has been particularly useful in the development of scatter and attenuation correction methods, mainly due to its ability to differentiate between scattered and primary events in the image. Comparing simulated projection data without scattering and attenuation effects with scatter- and attenuation-corrected data can give the accuracy of the activity recovery. Investigations of both the shape of the scatter response functions and the fraction of scatter in the images can be made using MC calculations, providing valuable information for the development of scatter correction methods. Studies have been carried out mainly for ^{99m}Tc (e.g., Beck et al., 1982, Ljungberg and Strand, 1990a, 1990b, 1991) and ^{201}Tl (Frey and Tsui, 1994, Kadrmas et al., 1998, 1999). Recently, studies have also been reported on ^{131}I (Dewaraja et al., 2000) and ^{67}Ga (De Vries et al., 1999). Comparative studies have been performed by several investigators (e.g., Buvat et al., 1995, Ljungberg, King, et al., 1994). Work has also been carried out to implement the MC method in the reconstruction method itself (Floyd et al., 1985).

C. Collimator Simulation

Studies of the interaction of photons in scintillation camera collimators have been performed for ^{111}In using specially designed programs (De Vries et al., 1990). More recently, a study has been published on the penetration of high-energy photons from ^{131}I decay (Dewaraja et al., 2000). In addition to the principal 364-keV photon energy, ^{131}I also emits two photons with energies 637 keV and 723 keV and with abundances of 7.2% and 1.8%, respectively. Despite these small abundances, it has been shown that the contribution of these photons to the main energy window can be significant due to their low absorption cross section in the collimator. Figure 5 shows two whole-body simulations with an HE collimator where images (A and B) do not include simulation of penetration and scattering in the collimator and image (C) includes both penetration and scattering using the algorithms described in De Vries et al. (1990). One can see the degradation in spatial resolution, and it is important to bear in mind that penetration also affects the accuracy in the attenuation and scatter correction.

^{67}Ga has proven to be useful in tumor imaging but potential difficulties exist because ^{67}Ga emits photons with a range

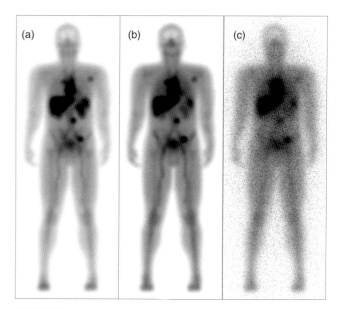

FIGURE 5 Simulated images of the Zubal phantom in which (a) no photon interactions in the phantom have been simulated, (b) photon interaction in the phantom and backscattering behind the crystal are included, and phantom interaction, backscattering, and (c) photon interaction in the collimator have been included. The degradation in the image quality caused by the septal penetration can clearly be seen.

of energies. De Vries et al. (1999) have studied the effects of collimator penetration, scatter, and Pb X-rays, which degrade the contrast, image resolution, and quantitation of the images. They developed and tested a MC program that simulated characteristic Pb X-rays. To validate the simulation of collimator penetration and scattering (and backscatter from components behind the NaI crystal) measurements were made using ^{51}Cr (320 keV). The simulation provided good estimates of both spectral and spatial distributions.

Moore et al. (1995) have used the Monte Carlo method for human observer studies in detecting lesions of unknown size in a uniform noisy background. A MC program was first run to estimate the contribution to the radial point-spread function for collimators of different spatial resolution containing tungsten of different contents by imaging a ^{99m}Tc source at the center of a 15-cm-diameter, water-filled phantom. They found that the optimum collimator design for detecting lesions of unknown size in the range 2.5–7.5 mm required a system resolution of approximately 8.5 mm FWHM.

D. Transmission SPECT Simulation

Transmission SPECT is a method used to determine the distribution of attenuation within the patient by measuring the fraction of photons transmitted from a source mounted on the one side of the patient to a camera mounted on the other side. By comparing projections with and without the patient, attenuation projections can be calculated and reconstructed in analogy with X-ray CT. The difference in the simulation technique lies in the way in which the emitted photons are transported to the phantom and to the detector. The MC method has proven to be useful for predicting the downscatter of events from emission photons into the transmission window (or vice versa) (Li et al., 1994; Ljungberg, Strand, et al., 1994; Welch et al., 1995).

E. MC Calculations and SPECT in Dose Planning of Radionuclide Therapy

An important area in nuclear medicine is oncological applications in which radionuclides are used to treat cancer patients. Accurate calculations of the absorbed dose to the tumor and to other critical organs are necessary. In external radiation treatment, the radiation beam is well known in terms of both dimension and characteristics. The distribution of an internally administered amount of activity is initially unknown. A measure of this distribution is thus required in order to calculate the absorbed dose distribution. The Monte Carlo method can be used to evaluate the accuracy in the scatter and attenuation correction methods for isotopes used for treatment (such as ^{131}I and ^{90}Y) and camera limitations, such as, the spatial resolution. The Monte Carlo method can also be used to explicitly calculate the absorbed dose in three dimensions using, for example, the EGS4 package

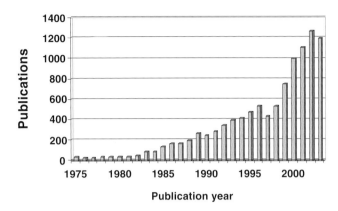

FIGURE 6   The number of publications in which the word *Monte Carlo* has been defined as a keyword is shown as a function of the publication year ranging from 1975 to 2001. (Source: PubMed.)

(Ljungberg et al., 2002, Ljungberg et al., 2003). Some programs apply methods to convolve the SPECT images to obtain a dose distribution using a point–dose kernel that has, in many cases, been MC calculated (Sgouros et al., 1990). Most tables found in the literature for conversion between cumulative activity and absorbed dose for diagnostic radiopharmaceuticals are based on MC calculations using a standard geometrical phantom.

IX. CONCLUSION

The use of the Monte Carlo method has increased significantly. Figure 6 shows the results of a PubMed search with the keyword *Monte Carlo* for each year ranging from 1975 to 2001. Note that this only covers the contents in the Medline database. The MC method is clearly widely used, and this is mainly due to the dramatic increase in the power, availability, and capacity of computers. Because the development of computers is still very rapid, a bright future can be foreseen for advanced MC modeling of nuclear medicine imaging. This will allow us to learn more about the limitations and potential of SPECT and planar imaging and to develop better correction methods to overcome some of these limitations.

Acknowledgments

The author acknowledges William Segars, University of North Carolina at Chapel Hill; Tommy Knöös, Lund University Hospital, Sweden; and Tom Harrison, University of Washington Imaging Research Laboratory, for valuable comments on the chapter and Katarina Sjögreen, Lund University Hospital, Sweden, for preparing some of the figures. This work has been supported by grants from the Swedish Cancer Foundation, The Gunnar Nilsson Foundation and the Bertha Kamprad Foundation.

References

Andreo, P. (1991). Monte Carlo techniques in medical radiation physics. *Phys. Med. Biol.* **36**: 861–920.

Anger, H. O., and Davis, D. H. (1964). Gamma-ray detection efficiency and image resolution in sodium iodine. *Rew. Sci. Inst.* **35**: 693–697.

Beattie, R. J. D., and Byrne, J. (1972). A Monte Carlo program for evaluating the response of a scintillation counter to monoenergetic gamma rays. *Nucl. Inst. Meth.* **104**: 163–168.

Beck, J., Jaszczak, R. J., Coleman, R. E., Stermer, C. F., and Nolte, L. W. (1982). Analysis of SPECT including scatter and attenuation using sophisticated Monte Carlo modeling methods. *IEEE. Trans. Nucl. Sci.* **29**: 506–511.

Berger, M. J., and Hubbell, J. R. (1987). XCOM: Photon Cross-sections on a Personal Computer. NBSIR 87-3597, National Bureau of Standards, Washington, DC.

Bouchet, L. G., Bolch, W. E., Weber, D. A., Atkins, H. L., and Poston, J. W., Sr. (1996). A revised dosimetric model of the adult head and brain. *J. Nucl. Med.* **37**: 1226–1236.

Box, G. E. P., and Muller, S. H. (1958). A note on the generation of random normal deviates. *Annals Math. Stat.* **29**: 610–611.

Briesmeister, J. F. (1997) MCNP—A general Monte Carlo n-particle transport code. LA-12625-M, Los Alamos National Laboratory, Los Alamos.

Buvat, I., Rodriguezvillafuerte, M., Todd-Pokropek, A., Benali, H., and Dipaola, R. (1995). Comparative assessment of nine scatter correction methods based on spectral analysis using Monte Carlo simulations. *J. Nucl. Med.* **36**: 1476–1488.

Dawson, T. W., Caputa, K., and Stuchly, M. A. (1997). A comparison of 60 Hz uniform magnetic and electric induction in the human body. *Phys. Med. Biol.* **42**: 2319–2329.

De Vries, D. J., King, M. A., and Moore, S. C. (1998). Characterization of spectral and spatial distributions of penetration, scatter and lead X-rays in Ga-67 SPECT. *IEEE Med. Imag. Conf. Rec.*, **M9-6**: 1701–1710.

De Vries, D. J., Moore, S. C., Zimmerman, R. E., Friedland, B., and Lanza, R. C. (1990). Development and validation of a Monte Carlo simulation of photon transport in an Anger camera. *IEEE. Trans. Med. Imag.* **4**: 430–438.

Dewaraja, Y. K., Ljungberg, M., and Koral, K. F. (2000). Characterization of scatter and penetration using Monte Carlo simulation in 131-I imaging. *J. Nucl. Med.* **41**: 123–130.

Evans, R. D. (1955). "The Atomic Nucleus." McGraw-Hill, New York.

Floyd, C. E., Jaszczak, R. J., and Coleman, M. (1985). Inverse Monte Carlo: A unified reconstruction algorithm for SPECT. *IEEE. Trans. Nucl. Sci.* **32**: 779–785.

Floyd, C. E., Jaszczak, R. J., Harris, C. C., and Coleman, R. E. (1984). Energy and spatial distribution of multiple order Compton scatter in SPECT: A Monte Carlo investigation. *Phys. Med. Biol.* **29**: 1217–1230.

Frey, E. C., and Tsui, B. M. W. (1994). Use of the information content of scattered photons in SPECT by simultaneously reconstruction from multiple energy windows. *J. Nucl. Med.* **35(5)**: 17P.

Haynor, D. R. (1998). Variance reduction techniques. In "Monte Carlo Calculation in Nuclear Medicine: Applications in Diagnostic Imaging" (M. Ljungberg, S.-E. Strand, and M. A. King, eds.), pp. 13–24. IOP Publishing, Bristol and Philadelphia.

Hubbell, J. H. (1969) Photon cross sections, attenuation coefficients and energy absorption coefficients from 10 keV to 100 GeV. NSRDS-NBS 29, National Bureau of Standards, Washington, DC.

Hubbell, J. H., Veigle, J. W., Briggs, E. A., Brown, R. T., Cramer, D. T., and Howerton, R. J. (1975). Atomic form factors, incoherent scattering functions and photon scattering cross sections. *J. Phys. Chem. Ref. Data.* **4**: 471–616.

Kadrmas, D. J., Frey, E. C., and Tsui, B. M. (1998). Application of reconstruction-based scatter compensation to thallium-201 SPECT: Implementations for reduced reconstructed image noise. *IEEE Trans. Med. Imag.* **17**: 325–333.

Kadrmas, D. J., Frey, E. C., and Tsui, B. M. (1999). Simultaneous technetium-99m/thallium-201 SPECT imaging with model-based compensation for cross-contaminating effects. *Phys. Med. Biol.* **44**: 1843–1860.

Kahn, H. (1956) Application of Monte Carlo. RM-1237-AEC, Rand Corp., Santa Monica, CA.

Knoll, G. F., Knoll, T. F., and Henderson, T. M. (2000). Light collection in scintillation detector composites for neutron detection. *IEEE Trans. Nucl. Sci.* **35**: 872–875.

Lalush, D. S., and Tsui, B. M. (1998). Block-iterative techniques for fast 4D reconstruction using a priori motion models in gated cardiac SPECT. *Phys. Med. Biol.* **43**: 875–886.

Li, J. Y., Tsui, B. M. W., Welch, A., and Gullberg, G. T. (1994). A Monte Carlo study of scatter in simultaneous transmission-emission data acquisition using a triple-camera system for cardiac SPECT. *J. Nucl. Med.* **94**: 81P.

Ljungberg, M. (1998a). Introduction to the Monte Carlo method. In "Monte Carlo Calculation in Nuclear Medicine: Applications in Diagnostic Imaging" (M. Ljungberg, S.-E. Strand, and M. A. King, eds.), pp. 1–12. IOP Publishing, Bristol and Philadelphia.

Ljungberg, M. (1998b). The SIMIND Monte Carlo program. In "Monte Carlo Calculation in Nuclear Medicine: Applications in Diagnostic Imaging" (M. Ljungberg, S.-E. Strand, and M. A., King, eds.), pp. 145–163. IOP Publishing, Bristol and Philadelphia.

Ljungberg, M., King, M. A., Hademenos, G. J., and Strand, S.-E. (1994). Comparison of four scatter correction methods using Monte Carlo simulated source distributions. *J. Nucl. Med.* **35**: 143–151.

Ljungberg, M., Frey, E.C., Sjögreen, K., Liu, X., Dewarja, Y., and Strand, S.-E. (2003). 3D absorbed dose calculations based on SPECT: evaluation for 111-In/90-Y therapy using Monte Carlo simulations. *Cancer Biotherapy and Radiopharmaceuticals.* **18**: 99–108.

Ljungberg, M., Sjögreen, K., Liu, X., Frey, E., Dewarja, Y., and Strand, S.-E. (2002). A 3-dimensional absorbed dose calculation method based on quantitative SPECT for radionuclide therapy: evaluation for 131-I using Monte Carlo simulation. *J. Nucl. Med.* **43**: 1101–1109.

Ljungberg, M., and Strand, S.-E. (1989). A Monte Carlo program simulating scintillation camera imaging. *Comp. Meth. Progr. Biomed.* **29**: 257–272.

Ljungberg, M., and Strand, S.-E. (1990a). Attenuation correction in SPECT based on transmission studies and Monte Carlo simulations of build-up functions. *J. Nucl. Med.* **31**: 493–500.

Ljungberg, M., and Strand, S.-E. (1990b). Scatter and attenuation correction in SPECT using density maps and Monte Carlo simulated scatter functions. *J. Nucl. Med.* **31**: 1559–1567.

Ljungberg, M. and Strand, S.-E. (1991). Attenuation and scatter correction in SPECT for sources in a nonhomogeneous object: A Monte Carlo study. *J. Nucl. Med.* **32**: 1278–1284.

Ljungberg, M., Strand, S.-E., and King, M. A. (1998). "Monte Carlo Calculation in Nuclear Medicine: Applications in Diagnostic Imaging." IOP Publishing, Bristol and Philadelphia.

Ljungberg, M., Strand, S.-E., Rajeevan, N., and King, M. A. (1994). Monte Carlo simulation of transmission studies using a planar sources with a parallel collimator and a line source with a fan-beam collimators. *IEEE. Trans. Nucl. Sci.* **41**: 1577–1584.

Ljungberg, M., Strand, S.-E., Rotzen, H., Englund, J. E., and Tagesson, M. (1994). Monte Carlo simulation of high count rate scintillation camera imaging. *IEEE Med. Imag. Conf. Rec.*, **4**: 1682–1686.

Manglos, S. H., Floyd, C. E., Jaszczak, R. J., Greer, K. L., Harris, C. C., and Coleman, R. E. (1987). Experimentally measured scatter fractions and energy spectra as a test of Monte Carlo simulations. *Phys. Med. Biol.* **22**: 335–343.

Moore, S. C., De Vries, D. J., Nandram, B., Kijewski, M. F., and Mueller, S. P. (1995). Collimator optimization for lesion detection incorporating prior information about lesion size. *Med. Phys.* **22:** 703–713.

Ogawa, K. and Maeda, S. (1995). A Monte Carlo method using octree structure in photon and electron transport. *IEEE. Trans. Nucl. Sci.* **42:** 2322–2326.

Persliden, J. (1983). A Monte Carlo program for photon transport using analogue sampling of scattering angle in coherent and incoherent scattering processes. *Comput. Progr. Biomed.* **17:** 115–128.

Pretorius, P. H., King, M. A., Tsui, B. M., Lacroix, K. J., and Xia, W. (1999). A mathematical model of motion of the heart for use in generating source and attenuation maps for simulating emission imaging. *Med. Phys.* **26:** 2323–2332.

Raeside, D. E. (1976). Monte Carlo principles and applications. *Phys. Med. Biol.* **21:** 181–197.

Segars, W. P. (2001). Development of a new dynamic NURBS-based cardiac-torso (NCAT) phantom. PhD diss., University of North Carolina at Chapel Hill.

Segars, W. P., Lalush, D. S., and Tsui, B. M. W. (2001). Modeling respiratory mechanics in the MCAT and spline-based MCAT phantoms. *IEEE Trans. Nucl. Sci.* **48:** 89–97.

Sgouros, G., Barest, G., Thekkumthala, J., Chui, C., Mohan, R., Bigler, R. E., and Zanzonico, P. B. (1990). Treatment planning for internal radionuclide therapy: Three-dimensional dosimetry for nonuniform distributed radionuclides. *J. Nucl. Med.* **31:** 1884–1891.

Snyder, W. S., Ford, M. R., Warner, G. G., and Fisher H. L., Jr. (1969). Estimates of absorbed fractions for monoenergetic photon sources uniformely distributed in various organs of a heterogeneous phantom: MIRD pamphlet no. 5. *J. Nucl. Med.* **10:** 5–52.

Storm, E., and Israel, H. I. (1970). Photon cross sections from 1 keV to 100 MeV for elements Z=1 to Z=100. *Nucl. Data Tables* **A7:** 565–681.

Strand, S.-E., and Lamm, I.-L. (1980). Theoretical studies of image artifacts and counting losses for different photon fluence rates and pulse-height distributions in single-crystal NaI(Tl) scintillation cameras. *J. Nucl. Med.* **21:** 264–275.

Tsui, B. M. W., Segars, W. P., and Lalush, D. S. (2000). Effects of upward creep and respiratory motion in myocardial SPECT. *IEEE Trans. Nucl. Sci.* **47:** 1192–1195.

Tsui, B. M. W., Terry, J. A., and Gullberg, G. T. (1993). Evaluation of cardiac cone-beam single photon emission computed tomography using observer performance experiments and receiver operating characteristic analysis. *Invest. Radiol.* **28:** 1101–1112.

Turner, J. E., Wright, H. A., and Hamm, R. N. (1985). A Monte Carlo primer for health physicists. *Health. Phys.* **48:** 717–733.

Wang, H. L., Jaszczak, R. J., and Coleman, R. E. (1993). A new compostite model of objects for Monte Carlo simulation of radiological imaging. *Phys. Med. Biol.* **38:** 1235–1262.

Wang, H. L., Jaszczak, R. J., Gilland, D. R., Greer, K. L., and Coleman, M. (1993). Solid geometry based modelling of non-uniform attenuation and Compton scattering in objects for SPECT imaging systems. *IEEE Trans. Nucl. Sci.* **40:** 1305–1312.

Welch, A., Gullberg, G. T., Christian, P. E., Datz, F. L., and Morgan, H. T. (1995). A transmission-map-based scatter correction technique for SPECT in inhomogeneous media. *Med. Phys.* **22:** 1627–1635.

Yanch, J. C., and Dobrzeniecki, A. B. (1993). Monte Carlo simulation in SPECT: Complete 3D modeling of source, collimator and tomographic data acquisition. *IEEE Trans. Nucl. Sci.* **40:** 198–203.

Zaidi, H. (1999). Relevance of accurate Monte Carlo modeling in nuclear medical imaging. *Med. Phys.* **26:** 574–608.

Zubal, I. G., and Harrell, C. R. (1993). Voxel based Monte Carlo calculations of nuclear medicine images and applied variance reduction techniques. *Imag. Vis. Comput.* **10:** 342–348.

Zubal, I. G., and Harrell, C. R. (1994). Computerized three-dimensional segmented human anatomy. *Med. Phys.* **21:** 299–302.

Index

A

Acyclovir, 99
ADCs. See Analog-to-digital converters
Adrenergic system, 95
AEC. See Atomic Energy Commission
Algebraic reconstruction technique, 446, 454–455
Algorithms. See Iterative algorithms
Aliasing errors, 103
Aluminum gamma counting, 241
Ammonia, ^{13}Na, 205
Ammonium molybdate, 62
Amplifier, linear, 246–247
Analog imaging, 366
Analog-to-digital converters, 312
Analytical phantoms, 553
Analytic image reconstruction
   central section theorem, 426–429
   Compton cameras, 400–402
   data acquisition, 422–423
   definition, 19
   fan beam, 433–434
   Fourier methods, 432, 440
   general references, 421–422
   linograms, 433
   list mode, 433
   radon/X-ray transforms, 425–426
   regularization, 434–435
   3D imaging, 424–425, 435–441
   2D imaging, 429–435
Anesthesia, 216–217
Anger cameras. See Gamma cameras
Angiogenesis, 47
Annihilation, 58
Antibody targeting, 46
Antineutrino, 59–60
Antisense RNA, 99
APDs. See Avalanche photodiodes
Application-specific integrated circuits, 273, 280
Application-specific small FOV imagers
   beta imagers, 328–330
   coincidence imagers, 325–328
   data acquisition, 311–312
   electronic processing, 311–312
   event positioning
      coincidence imagers, 314–315
      light multiplexed readout cameras, 313–314
      maximum likelihood, 312–313
      parallel readout, 314
      ratio calculation, 312
   gamma ray imagers, 317–323, 325
   general design principles
      detector configurations, 295–298
      nuclear emissions, 294–295
      performance, 298–300
      spatial response, 299
      temporal response, 299–300
   image formation, 312–313
   motivation for, 293–294
   PMT design
      multiple, 307
      position sensitive, 307–308
      readout, 310–311
   scintillation crystal design
      collimation for, 306
      segmented multiple, 302–304
      semiconductor designs vs., 315
      single continuous, 300–302
   semiconductor designs, 315–317
Approximators, 114
ART. See Algebraic reconstruction technique
Artifacts, 166–168
Artificial intelligence techniques, 546–547
ASICs. See Application-specific integrated circuits
Assessment of image quality
   detective quantum efficiency, 125
   hoteling observer, 125–126
   ideal observer, 124–125
   likelihood ratio, 124–125
   model observer, 125–126
   noise equivalent quanta, 125
   overview, 123–124
   ROC analysis, 124
Asymptotically efficient, 447
ATLAS tomography, 220
Atomic Energy Commission, 26
Attenuation
   corrections
      CT, 206–209
      methods, 560
      PET, 188–189
      PET/CT systems, 202–205
   gamma rays, 16
   high-energy photons, 74–75
   nonuniform compensation
      correction methods, 480–483
      impact, 483–484
      patient-specific maps, 477–480
   SPECT, 474–475
Auger electron, 55, 230
Avalanche detectors, 339
Avalanche photodiodes
   guard structures, 264
   multiplication theory, 262–263
   noise structure, 263–264
   -PMT comparison, 264–265
   properties, 262
   structure, 262

B

Backproject, 401–402
Backprojection. See also Filtered backprojection
   description, 18
   noise in, 19
   3D image, 435–436
   2D image, 429–430
Bayesian methods
   application, 448–449
   Gibbs distribution, 450–451
   MAP criterion, 449–450
   parameter estimation, 122
   priors
      anatomical, 451
      defining, 449–450
      as penalty, 451
   theory, 449–450

Beta detectors
  intraoperative
    radiopharmaceuticals, 349–351
    scintillator-based, 345–349
    solid-state, 349
  small FOV imagers, 328–330
  stacked silicon, 340–341
Beta emissions, 57–59, 69–70
Beta particle, dE/dx, 66–67
BGO. See Bismuth germanate
Bialkali materials, 256
Bias estimation, 123
Bias-variance trade off, 464
Binding energy, 54
Bismuth germanate
  advantages, 243
  characterization, 30, 181–182
  PET/Compton camera imaging, 413
  systems using, 185, 186
Blood flow
  curve measurement, 514
  dispersion delay, 505
  distribution volume, 503–504
  measurement, 504–505
  one-compartment model, 502–503
  partition coefficient, 503–504
  tissue heterogeneity, 505–506
  tumors, 47
BMP. See Bone morphogenetic protein
Bolus-plus infusion model, 518
Bone
  imaging, 45
  marrow, imaging, 45–46
  SPECT study, 145–146
  whole-body studies, 129–130
Bone morphogenetic protein, 322
Bound electrons, 555
Box-Muller transformations, 556
Bragg curve, 67
Brain
  adrenergic system, 95
  biochemical processes, 90–91
  -blood barrier, permeability, 91
  cholinergic system, 93–94
  development, 91
  diagnostics
    D-2 ligand levels, 96–97
    drug mapping, 96
    early techniques, 39–40
    enzymes, 96
    etiological studies, 97–92
    neurotransmitter systems, 97
    posttreatment monitoring, 97
    presynapse integrity, 96
    receptor mapping, 96
    receptor occupancy, 96
    substance abuse, 97
    transporter mapping, 96
  dopaminergic system
    characterization, 92–93
    endogenous ligand levels, 96–97
    mapping, 95
    rat, 222
  GABAergic system, 95
  glucose measurement, 221

  glucose use
    estimate rate, 508
    lumped constant, 508
    measurement model, 506–507
    protocol, 507–508
    tissue heterogeneity, 508–512
  glutamate system, 95
  opiod system, 95
  serotonergic system, 94–95
  SPECT study, 146
Breast imaging
  cadmium telluride, 283
  CTZ, 283
  false-positive rate, 46
  PET, 295
  single-photon emission, 318
Bremsstrahlung X-rays, 65, 70–71
Bull's-eye displays, 543–544

C

Cadmium telluride
  advantages, 267, 272–273
  breast imaging, 283
  cardiac imaging, 283
  collimation, 280–281
  Compton camera, 284
  contacts, 270–272
  cooling, 279–280
  crystal growth, 270–272
  depth-of-interaction, 276–277
  dual modality, 284
  electronics processing, 278–279
  energy spectrum, 274–277
  energy windowing, 279
  handheld imagers, 283–284
  intraoperative probes, 239
  photoelectric absorption, 266
  pixel effects, 275–276, 277
  products, 282–283
  properties, 280
  prototypes, 282–283
  resolution, 275
  scientific exchange, 273–274
  single processing, 278–279
  small animal imaging, 284
  surgical imagers, 283–284
  timing resolution, 280
  uniformity, 279
Cadmium zinc telluride
  advantages, 267, 272–273
  breast imaging, 283
  cardiac imaging, 283
  collimation, 280–281
  Compton camera, 284
  contacts, 270–272
  cooling, 279–280
  crystal growth, 270–272
  depth-of-interaction, 276–277
  dual modality, 284
  electronics processing, 278–279
  energy spectrum, 274–277
  energy windowing, 279
  handheld imagers, 283–284
  photoelectric absorption, 266

  pixel effects, 275–276, 277
  products, 282–283
  prototypes, 282–283
  resolution, 275
  scientific exchange, 273–274
  single processing, 278–279
  small animal imaging, 284
  surgical imagers, 283–284
  timing resolution, 280
  uniformity, 279
Calcium, sup47/supCa, 45
Cancer
  brain, F-FDG studies, 98
  imaging applications, 46–48
  lung, imaging, 46
Carbon, sup11/supC
  early development, 26
  pancreas imaging, 46
  tumor imaging, 46
Carbon, sup14/supC, 26
Carbon monoxide, sup11/supCO, 44
Carcinoembryonic antigen, 47
CdTe. See Cadmium telluride
CdZnTe. See Cadmium zinc telluride
Center of gravity method, 79
Central section theorem
  two-dimensional, 427, 428–429
  three-dimensional, 427–429
  X-ray projections, 427–428
CeraSPECT
  description, 169
  design, 170–171
  principles, 170–171
Cerenkov radiation, 65
Chain reaction, 60
Channel photomultiplier, 259–260
Charge division method, 79
Cholinergic system, 93–94
Clique potentials, 450
Coherent scattering, 555
Coincidence imagers, 314–315
Coincident annihilation photons, 85–86
Coincident dual-detector probes, 340
Collimation
  CdTe/CZT detectors, 280–281
  principles, 132
  small FOV imagers, 306
  types, 132–133
Collimators
  annular sensograde, 171–172
  -Compton cameras, comparisons, 391–393
  depth-dependent blur, 16–17
  design advances, 35–36
  design principles, 153–154
  geometry, 154–156
  imaging properties, 156–160
  imaging system, 154–156
  multisegment slant-hole, 149
  optimization, 27
  parallel-hole
    description, 15
    design, 155–156
    off-set-fan beam, 142–143
    optimal design, 162–164

    PSRF, 158
    resolution, 158–159
  penetration, 557
  resolution, 136
  rotating-slit, 149
  scattering, 557
  secondary constraints
    hole-pattern visibility, 166–168
    septal thickness, 165–166
    weight, 164–165
  septal penetration, 160–162
  simulation, 560–561
  SNR, 394
  SPECT, 15
Color coding, 42–43
Colsher filter, 438
Compartment modeling
  application, 500–501
  blood flow
    dispersion delay, 505
    distribution volume, 503–504
    one-tissue model, 502–503
    partition coefficient, 503–504
    tissue heterogeneity, 505–506
  cerebral glucose use
    basic model, 506–507
    estimate rate, 508
    lumped constant, 508
    protocol, 507–508
    tissue heterogeneity, 508–512
  characterization, 500
  construction, 501–502
  Logan analysis, 520–522
  practical use, 522
  reference region methods, 519, 522
  simplifications, 519–520
  tissue heterogeneity, 510
Composite scintillators, 30
Compton cameras
  advantages, 381–382
  CdTe/CZT detectors, 284
  -collimated systems, comparison, 391–393
  C-SPRINT, 404–406
  efficiency gains, 394–397
  EL detection, 378–379
  electronic effects, 385–386
  general purpose imaging, 411
  geometric effects, 385
  geometries, 394–397
  hardware, 150
  high energy radiotracers, 414–415
  history, 384–385
  image reconstruction
    background, 397–398
    backproject, 401–402
    computation, 400
    direct solutions, 400–402
    forward problem, 398–399
    inverse problem, 399
    regulation, 403
    statistical solutions, 402–403
    two-stage, 399–400
  lower bounds, 390–391
  motivation, 150

  noise propagation, 390–393
  performance
    factors, 385–390
    observer, 393–394
    prediction, 390–397
  -PET, 412–413
  -PET/SPECT imaging, 412
  physics effects
    combined, 389–390
    Doppler broadening, 386–387
    polarization, 388–389
    scattering cross section, 387–388
  probes, 411–412
  receiver operating curves, 394
  ring, 395–396
  SDC alternative, 407–411
  silicon pad detectors, 403–404
  SNR, 393–394
  -SPECT, 413–414
  -SPECT/ PET, 412
  statistical effects, 385–386
Compton edge, 72
Compton effect, 71–73, 231–232
Compton probes, 411–412
Compton process, 555
Compton scatter
  cross section, 387–388
  definition, 71
  FOV designs, 298
  photon, 135
  window, 485
Compton shoulder, 232
Compton suppression counting, 250–251
Conduction band, 338
Cone-beam methods, 440
Conjugate gradient algorithm, 457–458
Conversion efficiency, 234
Convex set projections, 467
Convolution interpolation, 111–112
Convolution theorem, 105
Coordinate descent algorithm, 458
Coulomb barrier, 63
Coulomb elastic scattering, 68–69
CPM. See Channel photomultiplier
Cramer-Rao bound, 390–391
Cross-grid electrode design, 316
Cross-sectional images, 21
Crystals
  growth, CdTe/CZT, 270–272
  hard oxide, 245
  housing, 244
  inorganic, 81–83
  material, 244
  NaI(Tl) scintillators, 309
  segmented multiple, 302–304
  single continuous, 300–302
  surface, 245–246
CsI(Tl) scintillators, 235–236
CT. See X-ray computed tomography
Cubic B-spline interpolation, 113
Cubic splines, 121
Cyclotrons, 30, 63–64
Cylindrical gamma camera, 376–378
CZT. See Cadmium zinc telluride

D

Daghighian's system, 340
Data acquisition. See also Image reconstruction
  formats, 433–444
  gantry motions, 141
  image reconstruction, 422–423
  small FOV imagers, 311–312
  timing, 14
Decay. See Radioactive decay
Degradation. See Image Degradation
Delay line technique, 79
dE/dx, 66–67
Delta. See Knock-on electron
Deoxyglucose, 221
Depth-dependent blur, 16–17
Depth of interaction
  approximations, 174
  definition, 169
  determination factors, 175
  effects, 276–277
  neurotome, 173, 174
  system function, 192
Detective quantum efficiency, 125
Detectors
  assembly, 246
  components, 244–245
  configurations, 249–251
  counting efficiency, 248–249
  fabrication, 244–245
  general parameters, 560
  performance factors, 255–256
  resolution compensation, 493
  simulation, 556–557
  variance reduction, 557–559
DFT-zero padding interpolation, 113–114
DICOM standard, 202
Diethylenetriaminepentaacetic acid, 129
Differentiation, 105
Digital imaging, 366–367
Diphenyl stilbene, 374
Direct planes, 424
Discrete Fourier transform, 110–111
Discrete-pixel electrode design, 317
Discrete sampling
  aliasing, 108–109
  -FT and DFT, relationship, 111
  theorem, 108–109
DOI. See Depth of interaction
Dopamine
  activation, 529
  measurement, 517
  receptors, 92–93
  schizophrenics, 530
  storage vesicles, 92
  system
    characterization, 92–93
    endogenous ligand levels, 96–97
    mapping, 95
    rat, 222
Doppler broadening
  characterization, 386–387
  energy effects, 389
  UCR, 391

Dosimetry
 MC calculations, 561
 PET/CT systems, 209–210
 small animal (PET), 215–216
DPS. *See* Diphenyl stilbene
Drift chambers. *See* Scintillation drift chambers
Drift diodes, 265–266
Drift of change carriers, 363
Drift speed, 276
Drug screening, 225
Dual detector probes, 340
Dynamic imaging, 452
Dynamic study, 20
Dynode structures, 257

E

ECG. *See* Electrocardiogram
Edge-preserving priors, 450–451
EGS4 package, 559
EL. *See* Electroluminescence detectors
Electrocardiogram, 20–21
Electroluminescence detectors
 emission gamma camera, 372
 multilayer chamber, 371–372
 SDC, 367–374, 410
 SPECT
  Compton camera, 378–379
  cylindrical gamma camera, 376–378
  small gamma camera, 375–376
Electromagnetic wave theory, 2
Electron-hole pairs, 79–80
Electronic processing, 311–312
Electrons
 auger, 230
 bound, 555
 coherent scattering, 555
 delta, 67–68
 emissions, 364–366
 free, capture, 373
 MC programs, 559–560
 multiplication, 257
 -positron interactions
  beta trajectories, 69–70
  energy loss effects, 65–66
  hard collisions, 67
  Moliere's theory, 68–69
 simulation, 555–556
Elementary particle theory, 59–60
Emission tomography
 bone imaging, 45
 categorization, 11
 characterization, 11, 139–140
 cross-sectional images, 21
 data acquisition, 17, 20–21
 definition, 1
 developments, 22–23
 early applications
  bone, 45
  brain, 39–40
  cardiac, 42–43
  parathyroid, 41
  pulmonary, 43–44
  renal, 43–44
  thyroid, 40–41

 evolution, 31–38
 molecular biology, 48–49
 pancreas imaging, 46
 process, 13–15
 reticuloendothelial imaging, 45–46
 tracer principle, 11–12
 tumor imaging, 46–48
 visualization methods, 12
End-point energy, 59
Energy
 band structure, 79–80
 binding, 54, 55
 end-point, 59
 inelastic collision, 66–67
 levels, 54
 loss effects, 65–66
 low tail, CdTe/CZT detectors, 275
 radioisotope levels, 54
 resolution
  detector, 247–248
  determination, 136
  EL, 369–371
  good, 298
  intrinsic, 360
  measurement, 76
  scintillator, 239–240
 spectra analysis, 560
 straggling, 70
 transition, 55
 windowing, 279
ET. *See* Emission tomography
Event positioning, 135
Excitation effects, 66–67
Expert-system analysis, 546–547
Extrinsic scintillators, 235

F

Fan beam, 433–434
FBP. *See* Filtered backprojection
F-FDG. *See* F-fluorodeoxyglucose
F-fluorodeoxyglucose
 application, 12
 bone imaging, 45
 brain imaging
  early studies, 40
  glucose uptake, 506–512
  processes studies, 90–91
  rat, 221
  research, 33
 cardiac imaging, 42–43
 characterization, 12
 PET/CT systems, 204–205
 small animals (PET), 217
 tumor imaging, 46, 47
Field of View
 data acquisition, 311–312
 single-head SPECT system, 478–479
 small imagers, application-specific
  beta imagers, 328–330
  coincidence imagers, 325–328
  electronic processing, 311–312
  event positioning, 312–315
  gamma ray imagers, 317–325
  general design principles, 295–300

 image formation, 312–313
 PMT design, 307–311
 scintillation crystal design, 300–304, 306, 315
Filtered backprojection
 inverse radon transform, 431–432
 iterative algorithms, 462–463
 iterative reconstruction *vs.*, 443–444
 PET, 18
 SPECT distortions, 477
 3D image
  algorithm, 437
  factorizable filters, 438
  functions/properties, 439
  X-ray transform, 437–438
 2D image, 429–430
Filtering. *See also* Smoothing
 pseudoinverse, 117
 Tikhonov regularization, 117–118
 Wiener, 118–119
Fission fragments, 60, 62
Fission reactor, 60
Flat panel photomultiplier tubes, 260
Fluoride, sup18/supFr, 45
Forced detection, 558–559
Fourier rebinning algorithm, 441
Fourier transform
 direct, reconstruction by, 432, 440
 discrete, 110–111
 properties, 110–111
 spectrum and, 105–106
FOV. *See* Field of view
FSR. *See* Fully sampled region
Full width at half maximum
 CeraSpect, 176–177
 estimating, 159–160
 gamma cameras, 154
 parallel-hole collimators, 162–164
 Sherbrooke system, 219
Fully sampled region, 474
Functional light output, 234
FWHM. *See* Full width at half maximum

G

GaAsP photocathodes, 256
GABAergic system, 95
Gadolinium oxyorthosilicate
 PET/CT systems, 199
 signals, problems with, 347–348
 systems using, 185, 186
Gain stabilization techniques, 251
Gain structures, 257
Gamma cameras
 anger, 130–131
 collimation, 132–133
 cylindrical, 376–378
 description, 15, 130–131
 development, 31
 early devices, 28–30
 EL emission, 372, 375–376
 event positioning, 135
 medical use, basic, 252
 performance, 136
 photon interactions, 135

pixel effects, 275–276
scintillation detection, 133–135
Gamma rays
  attenuation, 16
  cascade, 58–59
  detection
    direct, 266
    intraoperative probes, 351–352
    SPECT, 15–16
  imagers, small FOV, 317–323
  imagers, spatial resolution, 84–85
  isomeric transitions, 57
  laws governing, 17
  photons, 54
  scanning devices, 27
  -scintillator interaction
    Compton effect, 231–232
    decay processes, 229–230
    photoelectric processes, 230–231
Gantry, 200
Gantry motions, 141
Gas
  detectors, 76, 78–79
  gain, 364
  proportional scintillation counters, 79
Gated imaging, 452
Gated study, 20
Gaussian diffusion, 492, 493
Gaussian model, 446
Gaussian noise, 115–116
Geiger-Müller detectors
  application, 26
  brain imaging, 39
  description, 336–337
  limitations, 337
  properties, 336
Gene expression, 223, 531–532
Gene therapy, 531–532
Gibbs distribution, 450–451
Glucose
  cardiac level, 221
  cerebral, compartment modeling
    basic model, 506–507
    estimate rate, 508
    lumped constant, 508
    protocol, 507–508
    tissue heterogeneity, 508–512
  metabolism, 221–222
Glutamate system, 95
GM detectors. See Geiger-Müller detectors
GSO. See Gadolinium oxyorthosilicate

H

Half-life, 22, 57
Hamamatsu scanners, 217–218
Hammersmith system, 218
Hard collisions, 67
Heart
  angiography, 548
  CdTe/CZT imaging, 283
  gated imaging, 452
  glucose measurement, 221
  imaging, 42–43
  SPECT perfusion studies

global function quantification, 546
image display, 543–546
multimodality imagery, 546, 548–549
oblique reorientation, 542–543
protocol, 146–148
quantification, 543
regional function quantification, 546
reslicing, 542–543
stress-rest, 543
Hermetic seals, 244
Hickernel's system, 340
HIDAC detector, 219–220
High-pressure Bridgman method, 270–271
High-throughput phenotyping, 225
High-voltage input, 246
Hole-pattern visibility, 166–168
Hoteling observer, 125–126
HPB. See High-pressure Bridgman method

I

Ideal bandlimitied interpolation, 112
Ideal observer, 124–125
IEEE. See Institute of Electrical and Electronics Engineers
IFBP. See Iterative filtered backprojection algorithms
Image degradation, SPECT
  impact, 476–477
  nonuniform attenuation compensation, 477–484
  sources, 474–476
Image reconstruction. See also Data acquisition
  analytic
    central section theorem, 426–429
    Compton cameras, 400–402
    data acquisition, 422–423
    definition, 19
    fan beam, 433–434
    Fourier methods, 432, 440
    general references, 421–422
    linograms, 433
    list mode, 433
    radon/X-ray transforms, 425–426
    regularization, 434–435
    3D imaging, 424–425, 435–441
    2D imaging, 429–435
  bias, 464
  Compton cameras
    analytic solutions, 400–402
    background, 397–398
    inverse problem, 399
    statistically solutions, 402–403
  convex set projections, 467
  description, 14
  5D, 453
  iterative
    algorithms, 453–463
    Bayesian methods, 448–451
    constraint satisfaction, 446–447
    description, 19
    dynamic imaging, 452
    FBP vs., 443–444
    4D image model, 446, 451–453
    gated imaging, 452–453

least-squares, 447–448
linear model, 444–445
maximum-likelihood criterion, 447
sequences criteria, 451–452
spatial resolution, 491–492
statistical model, 445–446
weighted-least-squares, 447–448
MLEM details, 467–468
numerical observers, 464–465
PET, 18–19
pixel projection, 466–467
quality test, 463–464
spatial resolution, 464
SPECT, 16–17
variance, 464
Imaging science
  cross-sectional, 21
  -data acquisition, 5–6
  definition, 7
  digital advances, 3–4
  display, 7
  distribution, 7
  early devices
    gamma cameras, 28–30
    PET, 30–31
    scanning imagers, 26–27
  epistemology, 8
  evaluation, 7–8
  formation, 5
  fundamentals, 3, 5–8
  future issues, 8–9
  history, 1–2
  human vision, 7
  imperfections, 4–5
  interpretation, 7
  methodology, 8
  numerical, 34
  observation, 7
  optimization, 7–8
  optimum-weighted multichannel, 38
  PET, 19–20
  pioneers, 2–3
  quality, 34
  recording, 7
  recovery, 6–7
  SPECT, 14–15
  visualization, 7
Impurities, N type, 80
Inantimyosin, 43
Indium, ^{113m}In, 415
Inelastic collision energy, 66–67
Institute of Electrical and Electronics Engineers, 273
Integration, 105
Interface materials, 244
Internal conversion, 54
International Microelectronics and Packaging, 273
Interpolation
  approximators, 114
  convolution, 111–112
  cubic B-spline, 113
  cubic convolution, 113
  definition, 111
  DFT-zero padding, 113–114

Interpolation—(Continued)
  ideal bandlimited, 112
  linear, 112–113
  nearest-neighbor, 112
  nonuniform sampling, 114
Intraoperative probes
  beta imaging
    radiopharmaceuticals, 349–351
    scintillator-based, 345–349
    solid-state, 349
  clinical applications
    RIGS, 341–342
    sentinel node location, 342–343
  coincident dual detector, 340
  concept, 335–336
  detector configurations, 339–341
  development, 335
  dual detector, 340
  early designs, 336
  early prototypes, 344–345
  gamma imaging, 351–352
  instrumentation, 341
  scintillation detectors, 337–338
  solid-state detectors, 338–339
  stacked silicon detectors, 340
  summary, 352
  usefulness, 336
Inverse radon transform, 439–440
In vivo measurements, 253
Iodine
  ^{123}I, 415
  ^{128}I, 26
  ^{131}I
    Compton imaging, 414–415
    C-SPRINT image, 405–406
    renal imaging, 44
Ionization
  chambers, 77–79, 366
  effects, 66–67
  in gases, 77–79
  minimum, 67
  noble gas detectors, 362–363
  production statistics, 76
Iron, ^{50}Fe, 415
Iron, ^{52}Fe, 42, 415
Isobaric transitions, 57–58
Isomeric transitions, 57
Iterative algorithms
  constraint-satisfaction, 453–455
  general structure, 453
  IFBP, 462–463
  least-squares, 457–459
  MAP, 459–460
  OS-EM, 461–462
  subset-based, 461–462
  weighted-least-squares, 457–459
Iterative filtered backprojection algorithms, 462–463
Iterative image reconstruction
  algorithms, 453–463
  Bayesian methods, 448–451
  constraint satisfaction, 446–447
  description, 19
  dynamic imaging, 452
  FBP *vs.*, 443–444

  4D image model, 446, 451–453
    gated imaging, 452–453
    least-squares, 447–448
    linear model, 444–445
    maximum-likelihood criterion, 447
    sequences criteria, 451–452
    spatial resolution, 491–492
    statistical model, 445–446
    weighted-least-squares, 447–448

J

JFET. *See* Junction field effect transistor
Junction field effect transistor, 265

K

Kaczmarz method, 454–455
Kaiser-Bessel functions, 432–433
Kidney imaging, 44–45
Klein-Nishina formula, 488
Knock-on electron, 66–67
Krypton, ^{81m}Kr, 42, 43

L

Labeling isotopes, 21–22
scapL/scapDOPA, 92
Least-squares, 447–448, 457–459
LFT. *See* Light transfer function
Light
  absorption, 241, 260
  distribution, 301–302
  nature of, 2
  optics modification, 172–173
  pipe, 244
  scintillation
    detection, 81–83
    output, 234–239
    pulse, 237
    reflection, 241
    transmission, 241
  seals, 244
Light transfer function, 172–173
Likelihood ratio, 124–125
Linear amplifier, 246
Linear interpolation, 112–113
Linearity equation, 105
Linear shift-invariant systems, 104
Line-integral transform. *See* X-ray transforms
Line of response
  attenuation correction, 188–189
  coincidence imagers, 314–315
  defining, 188
  description, 181
  image reconstruction, 433–434
  scatter correction, 190
Line of stability, 55–56
Liquid drop model, 56
List mode, 433
List-mode methods, 445
Liver imaging, 45–46
LLED. *See* Lower-level energy discriminator
Logan analysis, 520–522
Longitudinal tomography, 29

LOR. *See* Line of Response
Low background counting, 250–251
Lower bounds, 390–393
Lower-level energy discriminator, 186
LSO. *See* Lutetium oxyorthosilicate
Luminescence. *See also* Electroluminescence
  centers, 235
  definition, 235
  nobel gas detectors, 364, 373–374
  photo, 235
  scintillator, 234–235
Lumped constant, 508
Lungs
  cancer, imaging, 46
  imaging technologies, 43–44
  physiology studies, 27
  respiratory gating, 203–204
Lutetium oxyorthosilicate
  advantages, 243
  characterization, 181–182, 184
  LSO(Ce) scintillators, 235–236
  semiconductor diode, 261
  systems using, 185, 186
Lymph node imaging, 45–46

M

Mab. *See* Monoclonal antibodies
Macroaggregated albumin, 129
Magic numbers, 56
Magnetic fields, 259
Magnetic resonance imaging
  -CeTe/CZT hybrids, 284
  PET/CT systems and, 210
  voxel-based phantoms, 553
Mammography
  CdTe/CZT imaging, 283
  false-positive rate, 46
  single-photon emission, 318
Mammospect, 175–176
MAP. *See* Maximum a posteriori probability
Materials Research Society, 273
Mathematical cardiac torso, 473–474, 476, 553–554
Maximum a posteriori probability
  algorithms, 459–460
  analogues, 462
  criterion, 449–450
  description, 122
  reconstruction, 460–461
Maximum likelihood estimation
  criterion, 447
  event positioning, 312–313
  related algorithms, 462
Maximum likelihood expectation-maximization
  iterative formula, 468–469
  nondecreasing likelihood, 468
  ordered subset, 483–484, 493
  properties, 455–457
  statement, 467–468
  subset method, 461–462
  variations, 457
Maxwell's theory, 2
MCA. *See* Multichannel analyzer
MCAT. *See* Mathematical cardiac torso

MCNP4 program, 560
MDP. *See* Methylene diphosphonate
Median filters, 120–121
MELC. *See* Multilayer electroluminescent camera
Mercury
    chlormerodrin, 31
    ^{197}Hg, 44
    ^{203}Hg, 44
Messenger RNA, 98–99
Metastabel states, 54
Methionine, ^{11}C, 46
Methylene diphosphonate, 129, 322
Methyl-3-4-iodophenyl-tropane-2-carboxic acid methyl ester, 319–320
Alpha-Methyl-para-tyrosine, 530
Michaelis-Menten constants, 506
Microchannel plate, 308
MicroPET, 219
Minimum mean-square error, 122
MKB. *See* Kaiser-Bessel functions
MMSE. *See* Minimum mean-square error
Model observer, 125–126
Molecular biology, 48–49
Moliere's theory, 68–69
Molybdenum, ^{90}Mo, 62–63, 414, 415
Monoclonal antibodies, 341
Monte Carlo simulations
    applications, 551, 560–561
    benefits, 551–552
    definition, 240
    drawbacks, 551–552
    electron calibration, 559–560
    mathematical phantoms, 552–554
    photon calibration, 559–560
    probability distribution, 552
    random number generator, 552
    sampling techniques, 552
    scatter compensations, 488–489
    SIMIND, 489
    temporal resolution, 557
    third-order moments, 174
MRI. *See* Magnetic resonance imaging
MRS. *See* Materials Research Society
Multialkali photocathodes, 256
Multichannel analyzer, 247
Multichannel method, 451
Multilayer electroluminescent camera, 371–372
Multi-pinhole coded aperture, 149
Multiple-injection data, 515–517
Multiple-time graphical analysis, 510
Multiplication, electron, 257–258
Multiplication, ADP, 262–263
Multisegment slant-hole collimators, 149
Multiwire proportional chambers
    description, 78–79
    digital imaging with, 366–367
    introduction, 359
    noble gas detectors, 219–220
Muscarinic receptors, 93
Myocardial ischemia, 42

N

NaI(Tl) scintillators
    advantages, 184–185
    crystals, 309
    drawbacks, 309
    electron energy, 239
    PET, 183
    photon backscattering, 557
    sub needed, 26–27, 134
    visible spectrum, 235–236
    whole-body counting, 251
Nearest-neighbor interpolation, 112
NEMA NU-1-2001 standard, 144
Neural networks, 547–548
NeurOtome, 173–175
Neurotransmitters
    activation studies
        clinical uses, 529–530
        effectiveness, 527–529
        interpretation, 530–531
    dopamine
        activation, 529
        measurement, 517
        schizophrenics, 530
        system, 92–93, 95–97
    PET imaging, 517–519
    serotonergic system, 94–95
    storage vesicles, 92
Neutrinos, 59–60
Neutrons
    activation, 60–61
    deficient, 56
    thermal, 60–61
Nicotinic receptors, 93
Noble gas detectors
    analog imaging, 366
    condensed phases, 364–366
    development, 359–360
    drift of change carriers, 363
    EL chambers, 367–371, 374–375
    electroluminescence, 364
    electron emission, 364–366
    gas gain, 364
    intrinsic energy resolution, 360
    ionization, 362–363
    ionization chambers, 366
    liquid xenon, 371–372
    luminescence
        photosensors, 374
        purification, 373–374
        UV wavelength shifting, 374
    MPDC digital imaging, 366–367
    position resolution, 361
    scintillation, 362–363
    technical features, 361–362
Noise
    APDs, 263–264
    backprojection, 19
    definition, 114
    electronic, PMTs, 257–259
    equivalent count rate, 191–192
    equivalent quanta, 125
    Gaussian, 115–116
    models, 115–116
    poisson, 115, 136
    propagation, Compton cameras, 390–393
    random variables, 114–115
    SPECT, 17
    stochastic processes, 115
    white, 117
Nonuniform sampling, 114
Nuclear de-excitation, 54–55
Nuclear emissions, 53–54
Nuclear generators, 62–63
Nuclear medicine
    CdTe/CTZ applications, 281–284
    history, 25–26
    oncology role, 46–47
    scintillation detectors, 251–253
    SPECT, 145–148, 542–543, 546–547
    theoretical developments, 33–34
Nuclear reactors
    design, 60–61
    fission fragments, 62
    fission in, 60
    neutron activation, 61
Nuclear stability, 55–56
Nuclear transmutation, 56
Numerical observers, 464–465

O

Offset-fan beam transmission, 142–143
Oncology, 222–223
One step late procedure, 459–460
Opiod system, 95
Ordered-subsets expectation-maximization algorithm, 461–462
OSL. *See* One step late procedure
Oxygen, 373

P

Pancreas imaging, 46
Parallax error, 180–181
Parallel-hole collimators
    description, 15
    geometry, 155–156
    off-set-fan beam, 142–143
    optimal design, 162–164
    PSRF, 158
    resolution, 158–159
Parallel readouts, 310–311, 314
Parathyroid imaging, 41
Parkinson's disease, 96
Parseval's theorem, 105–106
Partial voluming, 523
Partition functions, 450
Path-length simulation, 554
Patient handling system, 200–201
Patlak plot, 510
PD. *See* Photodiodes
PDT. *See* Photynamic therapy
Peak-to-valley ratios, 239–240
Penalized maximum likelihood method, 451
Pencil beam equation, 240
PET. *See* Positron emission tomography
PHA. *See* Pulse-height analyzer
Phantoms
    analytical, 553
    use of, 554
    voxel-based, 553–554
    Zubal, 556–557

Phenotyping, 225
Phoswich measurements, 251
Photodetectors. *See also* Photomultiplier tubes
   gamma ray detection, 266
   Pin, 261–262
   process, 255
   semiconductor diode, 260–261
Photodiodes
   avalanche
      description, 218–219
         guard structures, 264
         multiplication theory, 262–263
         noise structure, 263–264
         properties, 262
         structure, 262
      description, 81
      drift, 265–266
Photoelectric effect, 73–74
Photoelectric process, 230–231
Photoluminescence, 235
Photomultipliers
   description, 170
   LTF, 172–174
   optical overlapping, 175
Photomultiplier tubes
   absorption spectra, 81
   -ADP comparison, 264–265
   anger camera, 131
   block design, 184
   channel photomultiplier, 259–260
   dynode structures, 257
   EL detectors, 374
   electron conversion, 239
   electronic properties, 257–259
   electron multiplication, 257
   event positioning, 134
   flat panel, 260
   future trends, 259–260
   gain structures, 257
   hybrid, 259
   interface, 245
   light diffusion, 302
   magnetic fields, 259
   measurement function, 246
   multicathode, 192
   optical interface, 241
   performance variances, 183–184
   PET/CT systems, 199
   photocathode, 256
   position-sensitive, 259, 307–308
   quantum efficiency, 257
   replacement of, 272
   scintillation detectors, 134
   SDC, 363–364
   selection, 244–245
   small FOV imagers
      designs, 317–325
      multiple array, 307
      position sensitive, 307–308
      readout, 310–311
      semiconductor arrays, 308–310, 323–324
   SNR issues, 249
Photons
   absorption, 555

backscattering, 557
coincident annihilation, 85–86
Compton process, 555
conversion efficiency, 234
cross-sectional data, 554
high-energy, in matter
   attenuation, 74–75
   Compton effect, 71–73
   photoelectric effect, 73–74
   types, 70–71
interactions, 135
MC programs, 559–560
path-length simulation, 554
sampling interaction, 554–55
UV, 133–134
weight, 557–558
Photopeak-energy window, 485–486
Photynamic therapy, 222
PHS. *See* Patient handling system
Pin diodes, 261–262
Pixel effects
   CdTe/CZT detectors, 275–276
   projection, 466–467
   small, CdTe/CZT detectors, 277
Planar imaging, 103
Planogram methods, 440
PMs. *See* Photomultipliers
PMTs. *See* Photomultiplier tubes
Point source response function
   characteristics, 159
   description, 156–157
   individual rays, 161
   properties, 157
Point spread function
   application, 430, 444
   FWHM, 180
   linear imaging, 104–105, 464
   parallel/serial model, 488
   PET, 180
   SPECT distortions, 477
Poisson-based ML algorithms, 459
Poisson model, 445–446
Poisson noise, 115, 136
Polarity, 258–259
Polarization, 388–389
Polar maps, 543–544
Positron emission tomography
   advantage, 33
   attenuation correction, 188–189
   basic principles, 179–182
   basis for, 252–253
   bone imaging, 45
   breast imaging, 295
   cardiac imaging, 42–43
   -CdTe/CZT hybrids, 284
   characterization, 17
   compartment modeling
      application, 499–501
      blood flow, 502–506
      cerebral glucose use, 506–512
      characterization, 500
      Logan analysis, 520–522
      practical use, 522
      reference region methods, 519, 522
      simplifications, 519–520

-CT systems
   attenuation correction, 206–209
   benefits, 195–196
   computers, 202
   design considerations, 198–199
   development, 197–198
   dosimetry, 209–210
   drawbacks, 196–197
   fusion, 202–205
   future issues, 210
   gantry, 200
   image registration, 206
   MRI combinations, 210
   protocols, 202–205
   software, 202
   technical advantages, 197
detector designs, 182–184
development, 19–20
DOI system, 192
early devices, 30–31
electronics, 187–188
FBP, 18
function, 17
FWHM, 180
gene therapy, 531–532
geometry, 184–185
image reconstruction
   Bayesian methods, 448
   factors, 17–19
   linear inverse problem, 444–445
kinetic model equations
   arbitrary, 535
   output, 534
   state, 531–534
molecular biological studies, 48–49
mulilayer systems, 192
multimodality scanners, 193
NEC, 190–191
neurotransmitter activation studies, 527–531
oncology, 47–48
parallax error, 180–181
PMT system, 192–193
PSF, 180
pulmonary imaging, 43–44
receptor-ligand model
   modeling saturability, 514
   neurotransmitter changes, 517–519
   nonconstant coefficients, 517–519
   parameter identifiability, 514–517
   three component model, 512–514
resolution factors, 179–180
scatter correction, 190–191
scintillators, 186–187
small animal research
   ADP arrays, 221
   applications, 221–223
   ATLAS, 220
   benefits, 213–214
   Compton cameras, 412–413
   cost issues, 217
   drug screening, 225
   early development, 217–218
   future challenges, 223–224
   Hamamatsu system, 217–218
   Hammersmith system, 218

HIDAC, 219–220
injected dose, 215–216
injected mass, 215–216
input function measurement, 216
MADPET, 221
microPET, 219
multimodality, 225
opportunities, 214–215
resolution limits, 223–224
sensitivity, 215
Sherbrooke system, 218–219
spatial resolution, 215
TierPET, 220
tracer specificity, 216
UCLA system, 217–218
YAP, 220
spatial resolution effect
    components, 524
    correcting, 524–527
    quality, 523–524
-SPECT
    bridge between, 173–175
    comparison, 17–18
    Compton cameras, 412
    hybrids, 169–170
Positronium, 179
Positrons
-electron interactions
    beta trajectories, 69–70
    energy loss effects, 65–66
    hard collisions, 67
    Moliere's theory, 68–69
emitters, 21
-sensitive detectors
    electron-hole pair, 79–80
    gas, 78–79
    ionization in gases, 77–79
    photon distribution, 76–77
    scintillation, 82–83
Positron-sensitive photomultiplier tubes, 307–308
Positron volume imaging, 185
Postron-sensitive photomultiplier tubes, 259
Potassium, ^{42}P, 42
Potassium, ^{43}P, 42
Potential functions, 450
Power spectrum, 116–117
Preamp, 246
Preconditioned conjugate gradient, 460
Projection bins, 445
Projections onto convex sets, 453–454
Proportional chambers, 77–78
Prostate imaging, 46
Protein products, 99
Proton rich, 56
PSD. See Pulse-shape discrimination
Pseudoinverse filtering, 117
PSF. See Point spread function
PSPMT. See Positron-sensitive photomultiplier tubes
PSRF. See Point source response function
Pulse decay times, 238
Pulse-height analyzer, 187
Pulse-shape discrimination, 192
PVI. See Positron volume imaging

Q

Quantum efficiency, 257, 260

R

Radiation interactions
    detection
        basics, 76–79
        scintillators, 81–83
        semiconductors, 79–81
    electron/positron in matter
        beta trajectories, 69–70
        energy loss effects, 65–66
        excitation, 66–67
        hard collisions, 67
        inelastic collision energy, 66–67
        ionization, 66–67
        Moliere's theory, 68–69
    high-energy photons in matter
        attenuation, 74–75
        Compton effect, 71–73
        photoelectric effect, 73–74
        types, 70–71
    measurement, 76
Radioactive decay
    beta rays, 57–59
    definition, 54
    gamma rays, 57
    gamma ray/scintillators, 229–230
    half-life, 57
    isobaric transitions, 57–58
    isomeric transitions, 57
    nomenclature, 56
    probability, 56–57
Radioactive gases, 32–33
Radioimmuoguided surgery, 341–342
Radioisotopes. See also specific elements
    binding energy, 54
    decay probability, 56–57
    de-excitation, 54–55
    definition, 53
    energy levels, 54
    forces, 54
    fundamental spatial resolution
        coincident annihilation photons, 85–86
        importance, 83–84
        single-photon emissions, 84–85
    half-life, 22
    labeling, 21–22
    labeling methods, 26
    physical factors, 83
    stability, 55–56
    therapeutical use, 25–26
    transmutation, 56
Radionuclide production
    cyclotron, 63–64
    generators, 62–63
    reactors, 60–61
Radiopharmaceuticals
    administration, 14
    applications, 21–22
    beta probes, 349–351
    brain diagnostics
        applications, 96–98
        development, 91

genomic, 98–99
oncology, 98
in vivo criteria, 91–92
definition, 12
early development, 31–33
high energy, Compton imaging, 414–415
production, 14
small animal (PET), 216
Radiotracers. See Radiopharmaceuticals
Radon transforms
    definition, 423
    inverse, 439–440
    SPECT distortions, 477
    -X-ray transform, 425–426
Random variables, 114–115
Range straggling, 70
Rat
    brain, 221
    dopaminergic system, 222
    heart, 221
Ratio calculations, 312
RAYTRC, 161
RBI. See Rescaled block-iterative
RBSC. See Reconstruction-based scatter compensation
Readouts, 311, 313–314
Rebinning methods, 440–441
Receiver operating curves
    analysis, 124
    characteristic, 463
    ideal observer, 394
Receptor-ligand model
    modeling saturability, 514
    neurotransmitter changes, 517–519
    nonconstant coefficients, 517–519
    parameter identifiability, 514–517
    three component, 512–514
Reconstruction-based scatter compensation, 485, 487
Reflectors, ideal, 244
Region of Interest, 123
Reporter genes, 223
Reprojection algorithm, 435
Rescaled block-iterative, 462
Resistive charge division readout, 310–311
Resolution
    CdTe/CZT detectors, 275, 280
    collimator, 136
    C-SPRINT, 406–407
    energy
        definition, 247–248
        determination, 136
        EL, 369–371
        good, 298
        intrinsic, 360
        measurement, 76
        scintillator, 239–240
    FWHM
        CeraSpect, 176–177
        estimating, 159–160
        gamma cameras, 154
        parallel-hole collimators, 162–164
        penetration effect, 160–161
        Sherbrooke system, 219
    gamma camera, 136

Resolution—(*Continued*)
   noble gas detectors, 360–361
   parallel-hole collimator, 158–159
   small animal (PET), 223–224
   spatial (*See* Spatial resolution)
   timing, 251
Respiratory gating, 203–204
Rhenium, ^{188}Re, 349–350
RIGS. *See* Radioimmuoguided surgery
Ring Compton cameras, 395–396
ROC. *See* Receiver operating curves
ROI. *See* Region of Interest
Rotating-slit collimators, 149
Row-action maximum likelihood algorithm, 462
Rubidium, ^{81}Rb, 42
Ruggedized high temperature, 251

S

Sampling
   discrete
      aliasing, 108–109
      -FTand DFT, relationship, 111
      theorem, 108–109
   nonuniform, 114
   photon interaction, 554–55
   techniques, MC simulations, 552
   Whittaker-Shannon theorem, 108–109
Scaling parameter, 460
Scanning imagers, 26–27
Scatter
   coherent, 73
   collimator, 557
   Compton, 71–72, 135, 190, 298
   correction, 190, 560
   definition, 16
   detector transfer function, 37–38
   incoherent, 73
   multiple Coulomb elastic, 68–69
   Thomson, 72
Scatter compensation, SPECT
   estimation methods, 485–487
   impact, 484–485
   impact quality, 489–490
   reconstruction methods, 487–489
Schizophrenia, 529–530
Scintillation detectors
   applications, 81
   beta imaging, 345–349
   components, 244–245
   composite, 30
   conversion efficiency, 236–237
   description, 16, 133–135
   design, 242–243
   development, 229
   energy resolution, 239–240
   extrinsic, 235
   fabrication, 244–245
   -gamma-ray interaction
      Compton effect, 231–232
      decay processes, 229–230
      photoelectric processes, 230–231
   ideal, 234
   inorganic crystals, 81–83
   intraoperative probes, 337–338
   light
      emission, 255
      optics modification, 172–173
      output, 234–239
      pulse, 237
   luminescence, 234–235
   material density, 240–241
   measurement, 246–249
   mechanical properties, 241–242
   medical applications, 251–253
   NaI(TI), 26–27
   noble gas, 362–363
   optical properties, 241
   PET, 186–187
   segmented, 150
   small FOV imagers
      collimation for, 306
      segmented multiple, 302–304
      semiconductor *vs*., 315–316
      single continuous crystal, 300–302
   spatial resolution, 36–37
Scintillation drift chambers
   construction, 374–375
   count rate, 371
   description, 267
   development, 267
   energy resolution, 369–371
   liquid xenon, 372–373
   principles, 367–369
   3D position sensitivity, 369
SDCs. *See* Scintillation drift chambers
Segmented-scintillator detectors, 150
Selenium ^{75}Se, 46
Semiconductor detectors
   basic principles, 309–310
   CdTe/CZT
      advantages, 272–273
      breast imaging, 283
      cardiac imaging, 283
      collimation, 280–281
      Compton camera, 284
      contacts, 270–272
      cooling, 279–280
      crystal growth, 270–272
      depth-of-interaction, 276–277
      dual modality, 284
      electronics processing, 278–279
      energy spectrum, 274–277
      energy windowing, 279
      handheld imagers, 283–284
      pixel effects, 275–276, 277
      products, 282–283
      properties, 270
      prototypes, 282–283
      resolution, 275
      scientific exchange, 273–274
      single processing, 278–279
      small animal imaging, 284
      surgical imagers, 283–284
      timing resolution, 280
      uniformity, 279
   common technologies, 309
   description, 76
   diode, 260–261
   electron-hole pairs, 79–80
   position-sensitive, 80–81
   small FOV imagers, 315–317, 323–325
Sensograde collimators, 171–172
Sentinel node location, 342–343
Septal thickness, 165–166
Serotonin transport, 94–95
Shell model, 54
Sherbrooke system, 218–219
Signals
   charge carrier processing, 278–277
   processing, CdTe/CZT detectors, 278–279
   random variables, 114–115
   shape, PMT, 258
   stochastic processes, 115
Silicon pad detectors, 403–404
SIMIND, 489, 559
Simple gross counters, 250
SimSET program, 559
Simultaneous iterative reconstruction technique, 455
Single-photon emission computed tomography
   annular single-crystal systems
      CeraSpect, 169–171
      mammospect, 175–176
      neurotomes, 173–175
      overview, 169–170
      scintillation camera, 172–173
      sensograde collimators, 171–172
      small animal, 176–177
   basis for, 252
   bone imaging, 45
   -CeTe/CZT hybrids, 184
   characterization, 127
   clinical application
      bone protocol, 145–146
      brain protocol, 146
      myocardial perfusion, 146–148
   collimators, 15
   -CT systems, 143–144, 197
   data noise, 17
   development, 14–15
   dynamic, 452
   in early brain imaging, 40
   EL detectors
      Compton camera, 378–379
      cylindrical gamma camera, 376–378
      small gamma camera, 375–376
   gamma ray detection, 15–16
   ideal imaging, 473–474
   image degradation
      impact, 476–477
      nonuniform attenuation compensation, 477–484
      scatter compensation, 484–490
      sources, 474–476
      spatial resolution compensation, 490–494
   image reconstruction, 16–17, 444–445
   instrumentation advances, 541
   MC calculations, 561
   molecular biological studies, 48–49
   myocardial perfusion studies
      AI techniques, 546–547
      global function quantification, 546
      image display, 543–546
      oblique reorientation, 542–543

quantification, 543
regional function quantification, 546
reslicing, 542–543
stress-rest, 543
oncology, 47–48
-PET
bridge between, 173–175
comparison, 17–18
Compton cameras, 412
hybrids, 169–170
planar
characterization, 127–128
limitation, 130
perfusion studies, 129
thyroid studies, 128
ventilation studies, 129
whole-body bone studies, 129–130
pulmonary imaging, 43–44
simulation transmission, 561
systems
configurations, 141
early, 140–141
gantry motions, 141
performance, 144–145
tomography in, 136–137
transmission-source, 141–143
Single-photon emission mammography, 318
Single-photon emissions, 21–22, 84–85
Single-photon ring tomography
C-SPRINT
characterization, 404–405
resolution, 406–407
variance, 406–407
Single-slice rebinning, 440–441
Sinogram, 423
SIRT. See Simultaneous iterative reconstruction technique
Small animals
CdTe/CZT imaging, 284
CeraSpect imaging, 176–177
as human disease models, 214
PET imaging
ADP arrays, 221
anesthesia, 216–217
applications, 221–223
ATLAS, 220
benefits, 213–214
Compton cameras, 412–413
cost issues, 217
drug screening, 225
early development, 217–218
future challenges, 223–224
Hamamatsu system, 217–218
Hammersmith system, 218
HIDAC, 219–220
input function measurement, 216
MADPET, 221
microPET, 219
multimodality, 225
opportunities, 214–215
resolution limits, 223–224
sensitivity, 215
spatial resolution, 215
TierPET, 220
tracer specificity, 216

UCLA system, 217–218
YAP, 220
Smoothing. See also Filtering
definition, 119
filters, 119–120
median filters, 120–121
splines, 121
Solid-state detectors
beta imaging, 349
function, 150
intraoperative probes, 338–339
Space charge buildup, 78
Spatial detection, 558–559
Spatial frequency domain, 37
Spatial penetration, 37–38, 160–162
Spatial resolution effect
CdTe/CZT detectors, 280
coincident annihilation photons, 85–86
compensation
image quality, 492–494
iterative reconstruction, 491–492
restoration filtering, 490–491
definition, 523
FOV imagers, 299
image reconstruction, 464
PET
components, 524
correcting, 524–527
limitations, 179
quality, 523–524
scintillation detectors, 36–37
single-photon emissions, 84–85
small animal (PET), 215
3D X-ray image, 435
Spatial response, 299
Spatial uniformity, 258
SPECT. See Single-photon emission computed tomography
Spectral analysis, 510–511
Spectral response curve, 247
Spectrum power, 116–117
SPEM. See Single-photon emission mammography
Splines, cubic, 121
SPRINT. See Single-photon ring tomography
SSRB. See Single-slice rebinning
Stacked silicon detectors, 340–341
Static study, 20
Statistic-limited energy resolution, 360
Steepest descent algorithm, 457
Stochastic processes, 115
Stroke victims, 40
Strontium, ^{85}Sr, 45
Substance abuse, 97
System matrix, 445
System transfer functions, 106–107

T

Task-based assessment, 463
TEA. See Trimethylamine
Technetium
^{99m}Tc
beta probes, 349–350
bone imaging, 45

brain imaging, 39–40
C-SPRINT image, 405–406
parathyroid imaging, 41
pulmonary imaging, 43
radiotracers, 32
renal imaging, 44
thyroid imaging, 41
^{99m}Tc-MAG3, 44–45
$^{99m}Tc^{90}Mo$, 62–63
photon, 31
Temperature
CdTe/CZT detectors, 279–280
PMT, 258
ruggedized high, 251
Temporal response, 299–300
Tetrakis (dimethylamine) ethylene, 374
Tetraphenyl butadiene, 374
The Minerals, Metals, & Materials Society, 273
Thermal neutrons, 60–61
Thin windows, 250
THM. See Traveling heater method
Thomson scatter, 72
Thymidine kinase, 99
Thyroid
imaging
SPECT, 128
studies, 26
technologies, 40–41
therapy, 40–41
TierPET tomography, 222
Tikhonov regularization, 117–118
Time response, 258
Timing, 251, 280
Tin, ^{113}Sn, 415
Tissue heterogeneity
blood flow, 505–506
cerebral glucose use, 508–512
compartmental model, 510
TMAE. See Tetrakis (dimethylamine) ethylene
TMS. See The Minerals, Metals, & Materials Society
Tomographic imaging, 103
Tomography
emission imaging, 138–140
in SPECT, 136–137
systems, 3
transmission imaging, 137, 139
types, 136–137
Total scintillation efficiency, 236–237
TPB. See Tetraphenyl butadiene
Tracer principle, 11–12
Transfer efficiency, 235
Transient equilibrium, 62
Transition energy, 55
Transmission imaging, 137, 139
Transmission tomography, 139
Transmitted fraction, 475
Transmutation, 56
Traveling heater method, 270–271
Trimethylamine, 374
Tumors. See also specific organs
blood flow, 47
delectability, 34–35
imaging, 46–48
Tungsten, ^{188}W, 349–350

U

UCR. *See* Uniform Cramer-Rao bound
Uniform Cramer-Rao bound, 390–391
UV light, 133–134, 374

V

Variance estimation, 123
Variance reduction methods, 557–559
Vector-space projection methods, 446–447, 453
Ventilation studies, 129
Vision, 2, 7
Voltage divider, 245, 258–259
Voxel-based phantoms, 553–554

W

Wavelength shifting fibers, 323
Weighted-least-squares, 447–448
Well counters, 250
Whittaker-Shannon sampling theorem, 108–109
Whole body
  scans, 27, 28
Whole-body
  bone studies, 129–130
  counting, 251
Wiener filtering, 118–119
WSFs. *See* Wavelength shifting fibers

X

Xenon
  detector purification, 373–374
  liquid, detectors, 372–373
  photo absorption by, 371–372
  ^{133}Xe, 43
X-ray computed tomography
  fan beams, 434
  oncology, 47–48
  -PET systems
    attenuation correction, 206–209
    benefits, 195–196
    computers, 202
    design considerations, 198–199
    development, 197–198
    dosimetry, 209–210
    drawbacks, 196–197
    fusion, 206
    future issues, 210
    gantry, 200
    image registration, 206
    MRI combinations, 210
    protocols, 202–205
    software, 202
    technical advantages, 197
  -SPECT systems, 143–144, 197, 478
X-rays
  characteristic, 55
  discovery, 3
  discrete energies, 70–71
  emission characteristics, 234
  photons, 555
X-ray transforms
  central section theorem, 427–428
  definition, 423
  performance, 424
  -Radon transform, 425–426
  3D image
    backprojection, 437–439
    cone-beam method, 445
    direct Fourier method, 439–440
    inverse transform, 439–440
    list-mode method, 445
    planogram method, 445
    rebinning method, 440–441
    reprojection algorithm, 435
    spatial variance, 435

Y

YAP. *See* Yttrium aluminum perovskit
Yttrium aluminum perovskit, 222

Z

Z atom, 55
Z materials, 55, 241
Zubal phantom, 556–557